Card and English on Police Law

T0134193

Card and English on Police Law

Sixteenth Edition

Richard Card LLB LLM HON LLD FRSA

Emeritus Professor of Law
De Montfort University, Leicester

OXFORD
UNIVERSITY PRESS

Great Clarendon Street, Oxford, OX2 6DP,
United Kingdom

Oxford University Press is a department of the University of Oxford.
It furthers the University's objective of excellence in research, scholarship,
and education by publishing worldwide. Oxford is a registered trade mark of
Oxford University Press in the UK and in certain other countries

© Oxford University Press 2022

The moral rights of the author have been asserted

First Edition published in 1985
Fifteenth Edition published in 2017
Sixteenth Edition published in 2022

Impression: 1

Public sector information reproduced under Open Government Licence v3.0
(http://www.nationalarchives.gov.uk/doc/open-government-licence/open-government-licence.htm)

Published in the United States of America by Oxford University Press
198 Madison Avenue, New York, NY 10016, United States of America

British Library Cataloguing in Publication Data
Data available

Library of Congress Control Number
Data available

ISBN 978-0-19-286616-5

Printed in the UK by
Ashford Colour Press Ltd, Gosport, Hampshire

Preface

When Jack English and I wrote the first edition of this book 35 years ago, our aims were to cover comprehensively, and in terms easily understood by those who had received no legal training, those areas of the law and legal procedure with which all police officers are concerned. The material included in *Police Law* and its companion website has been prepared in a way which recognises the ordinary needs of police officers, of those in legal advice agencies, and of members of the public who wish to refer to a legal text. Now in its 16th edition, the work remains true to its original aims.

This is the fifth edition of this book which I have prepared on my own following Jack English's untimely death in January 2010. I had hoped to be joined by Fraser Sampson in preparing this edition but his appointment by the Home Secretary as the joint Biometrics and Surveillance Commissioner meant that he did not have the time to be involved. Before his appointment Fraser was involved in planning the new edition and I am most grateful for his input at that stage.

The previous edition, published in 2017, has been extensively amended in consequence of the many subsequent legislative changes.

I wish to thank all those who have assisted me in various ways in the production of this edition. In particular, I would like to thank my wife, Rachel, for her considerable assistance with the manuscript, Alex Johnson, Priya Kumaravel, Vijayalakshmi Kumar and Ashirvad Moses for their help in the production of the book, and Kim Harris and Benjamin Johnson for compiling the index and the tables of cases and statutes.

As a result of the large amount of relevant legislation which continues to be made, there was a risk that *Police Law* would become so large as to be physically unwieldy. The solution to this has been to continue the practice whereby some matters are dealt with in the companion website. Generally, these are matters of considerable detail not suited to the text to which they refer, and matters of minor importance. Reference to them, and to their inclusion on the companion website, is made at the appropriate points in the text. The companion website also contains regular updates to the chapters contained in the book.

The companion website can be found at <http://www.oup.com/police16>.

The law is stated as it was on 1 July 2021, except for the omission of any provision which it is known will no longer be in force at the planned publication date of this book.

1 July 2021   Richard Card

Contents

Table of Statutes

The European Convention on Human Rights (ECHR) is tabled under Sch 1 to the Human Rights Act 1998 (HRA 1998)

Table of Statutory Instruments

Table of EU Regulations and Directives

Abbreviations

AA 1967	Abortion Act 1967
ABH	actual bodily harm
ACPO	Association of Chief Police Officers
ADBA	American Dog Breeders' Association
AETR	European agreement concerning the work of crews of vehicles engaged in international transport
AEVA 2018	Automated and Electric Vehicles Act 2018
AHA 1981	Animal Health Act 1981
ASB	anti-social behaviour
A-sBA 2003	Anti-social Behaviour Act 2003
A-sB,C&PA 2014	Anti-social Behaviour, Crime and Policing Act 2014
AWA 2006	Animal Welfare Act 2006
BA 1976	Bail Act 1976
BA 2010	Bribery Act 2010
BNA 1981	British Nationality Act 1981
BTP	British Transport Police
CA 2003	Communications Act 2003
CAA 1981	Criminal Attempts Act 1981
CAA 1984	Child Abduction Act 1984
CA(YP)A 1997	Confiscation of Alcohol (Young Persons) Act 1997
CBO	criminal behaviour order
CCA 2013	Crime and Courts Act 2013
CDA 1971	Criminal Damage Act 1971
CDA 1998	Crime and Disorder Act 1998
C(IC)A 2003	Crime (International Co-operation) Act 2003
CJA 1967	Criminal Justice Act 1967
CJA 1988	Criminal Justice Act 1988
CJA 2003	Criminal Justice Act 2003
CJCA 2015	Criminal Justice and Courts Act 2015
C&JA 2009	Coroners and Justice Act 2009
CJIA 2008	Criminal Justice and Immigration Act 2008
CJPA 2001	Criminal Justice and Police Act 2001
CJPOA 1994	Criminal Justice and Public Order Act 1994
CLA 1967	Criminal Law Act 1967
CLA 1977	Criminal Law Act 1977
CMA 1990	Computer Misuse Act 1990
CMCHA 2007	Corporate Manslaughter and Corporate Homicide Act 2007
CPC	Certificate of Professional Competence
CPIA 1996	Criminal Procedure and Investigations Act 1996

CPM	Commissioner of Police of the Metropolis
CPN	community protection notice
CPS	Crown Prosecution Service
CSO	community support officer
C-TA 2008	Counter-Terrorism Act 2008
CYPA 1933	Children and Young Persons Act 1933
DA 1991	Deer Act 1991
DA 2005	Drugs Act 2005
DDA 1991	Dangerous Dogs Act 1991
DDES(E&W)O 2015	The Dangerous Dogs Exemption Schemes (England and Wales) Order 2015
DDTRO	Drug Dealing Telecommunication Restriction Order
DPA 2018	Data Protection Act 2018
DPP	Director of Public Prosecutions
DSCC	Defence Solicitor Call Centre
DVLA	Driver and Vehicle Licensing Agency
DVPN	domestic violence protection notice
DVPO	domestic violence protection order
DVSA	Driver and Vehicle Standards Agency
EA	Environment Agency
EA 1875	Explosives Act 1875
EC	European Community
ECE	UN Economic Commission for Europe
ECHR	European Convention on Human Rights
ELPB	Elected Local Policing Body
EPA 1990	Environmental Protection Act 1990
ESA 1883	Explosive Substances Act 1883
EU	European Union
FA 2006	Fraud Act 2006
FCA 1981	Forgery and Counterfeiting Act 1981
FiA 1968	Firearms Act 1968
Fi(A)A 1997	Firearms (Amendment) Act 1997
FLA 1996	Family Law Act 1996
F(O)A 1991	Football (Offences) Act 1991
FR 2004	Fireworks Regulations 2004
FSA 1989	Football Spectators Act 1989
GBH	grievous bodily harm
GCHQ	Government Communications Headquarters
GRC	gender recognition certificate
GV(LO)A 1995	Goods Vehicles (Licensing of Operators) Act 1995
HA 1980	Highways Act 1980
HA 2004	Hunting Act 2004

HPTRA 2018	Haulage Permits and Trailer Registration Act 2018
HRA 1998	Human Rights Act 1998
IDA 2010	Identity Documents Act 2010
ID(C)A 1981	Indecent Displays (Control) Act 1981
IPA 2016	Investigatory Powers Act 2016
LA 1872	Licensing Act 1872
LA 2003	Licensing Act 2003
LGV	large goods vehicle
LM(V)A 2018	Laser Misuse (Vehicles) Act 2018
MCA 1980	Magistrates' Courts Act 1980
MDA 1861	Malicious Damage Act 1861
MDA 1971	Misuse of Drugs Act 1971
MHA 1983	Mental Health Act 1983
MPV	mechanically propelled vehicle
MSA 2015	Modern Slavery Act 2015
MV(DL)R 1999	Motor Vehicles (Driving Licences) Regulations 1999
MV(IR)R 2011	Motor Vehicles (Insurance Requirements) Regulations 2011
MV(IR)(I, R and D)R 2011	Motor Vehicles (Insurance Requirements) (Immobilisation, Removal and Disposal) Regulations 2011
MV(WSBCFS)R 1993	Motor Vehicles (Wearing of Seat Belts by Children in Front Seats) Regulations 1993
MV(WSB)R 1993	Motor Vehicles (Wearing of Seat Belts) Regulations 1993
NCA	National Crime Agency
NHS	National Health Service
NSPCC	National Society for Prevention of Cruelty to Children
OAPA 1861	Offences against the Person Act 1861
OPA 1959	Obscene Publications Act 1959
PA 1911	Perjury Act 1911
PA 1996	Police Act 1996
PACE	Police and Criminal Evidence Act 1984
PBA 1992	Protection of Badgers Act 1992
PCA 1953	Prevention of Crime Act 1953
PCA 1978	Protection of Children Act 1978
PCA 2002	Proceeds of Crime Act 2002
P&CA 2009	Policing and Crime Act 2009
PCC	Police and Crime Commissioner
PCV	passenger-carrying vehicle

PEA 1977	Protection from Eviction Act 1977
PHA 1997	Protection from Harassment Act 1997
PII	public interest immunity
PNC	Police National Computer
PND	penalty notice for disorder
POA 1936	Public Order Act 1936
POA 1986	Public Order Act 1986
PPVA 1981	Public Passenger Vehicles Act 1981
PRA 2002	Police Reform Act 2002
PSA 2016	The Psychoactive Substances Act 2016
RD(A)A 1978	Refuse Disposal (Amenity) Act 1978
RDVR 1986	Removal and Disposal of Vehicles Regulations 1986
RIPA 2000	Regulation of Investigatory Powers Act 2000
RNLI	Royal National Lifeboat Institution
RSA 2006	Road Safety Act 2006
RSPB	Royal Society for the Protection of Birds
RTA 1988	Road Traffic Act 1988
RTOA 1988	Road Traffic Offenders Act 1988
RTRA 1984	Road Traffic Regulation Act 1984
RV(C&U)R 1986	Road Vehicles (Construction and Use) Regulations 1986
RV(DRM)R 2001	Road Vehicles (Display of Registration Marks) Regulations 2001
RV(R&L)R 2002	Road Vehicles (Registration and Licensing) Regulations 2002
SCA 2015	Serious Crime Act 2015
SE(CA)A 1985	Sporting Events (Control of Alcohol, etc) Act 1985 SI statutory instrument
SHPO	sexual harm prevention order
SMD	scrap metal dealer
SMDA 2013	Scrap Metal Dealers Act 2013
SOA 1956	Sexual Offences Act 1956
SOA 2003	Sexual Offences Act 2003
SOCPA 2005	Serious Organised Crime and Police Act 2005
SOPO	sexual offences prevention order
SPA 2019	Stalking Protection Act 2019
SRA	Solicitors Regulatory Authority
StOA 1959	Street Offences Act 1959
SVGA 2006	Safeguarding of Vulnerable Groups Act 2006
TA 1968	Theft Act 1968
TA 1978	Theft Act 1978
TA 1985	Transport Act 1985
TA 2000	Terrorism Act 2000
TA 2006	Terrorism Act 2006
TEN	temporary events notice
TfL	Transport for London

TPIM	terrorism prevention and investigation measures
TPIMA 2011	Terrorism Prevention and Investigation Measures Act 2011
TSR&GD 2016	Traffic Signs Regulations and General Directions 2016 Act 1992
VA 1824	Vagrancy Act 1824
V(C)A 2001	Vehicles (Crime) Act 2001
VCRA 2006	Violent Crime Reduction Act 2006
VERA 1994	Vehicle Excise and Registration Act 1994
VOO	violent offender order
WAO	witness anonymity order
WM(P)A 1996	Wild Mammals (Protection) Act 1996
YCC	youth conditional caution
YJCEA 1999	Youth Justice and Criminal Evidence Act 1999

POLICE LAW, POWERS, AND PROCEDURE

PART I

POLICE LAW, POWERS, AND PROCEDURE

CHAPTER 1

Introductory Matters

This book deals with those branches of the law of particular relevance to police officers **1.1** in England and Wales.

SOURCES

There are two principal sources of English law: legislation and common law. **1.2**

Legislation

In modern times legislation has played a preponderant part in the branches of the law **1.3** dealt with in this book.

Statute

'Statute' has traditionally meant an Act of the UK Parliament. Except for the very few **1.4** common law offences, all serious offences are defined by statute, as are many minor offences. By and large the law relating to police powers, criminal procedure and evidence is derived from statute.

Since 2011, what is now known as 'Senedd Cymru or the Welsh Parliament' (hereafter referred to simply as Senedd Cymru) has had power to enact its own Acts, which only apply in relation to Wales, on any matter within its legislative competence. Under the Wales Act 2017, the Senedd Cymru is not able to legislate on reserved matters or matters within a list of restrictions on its competence. Most of the matters in this book are matters of those types. The impact of the Acts of the Senedd on Police Law is, indeed, marginal.

Subordinate legislation

A statute may give power to some body, such as the Queen in Council, a minister of the **1.5** UK Government, the Welsh Ministers (the ministers of the Welsh Government), or a local authority or other public body, to make Orders in Council, regulations, or byelaw (respectively), and prescribe for their breach. Subordinate legislation made at central government level is normally required to be made by way of 'statutory instrument'.

Retained EU law

Following the UK's withdrawal from the EU, the effect in the UK of EU legislation **1.6** (which, in the context of Police Law, means EU Regulations and EU Directives) is as follows. So far as operative immediately before 31 December 2020 at 11.00 pm:

(a) EU derived legislation (ie EU legislation implemented by legislation in England and Wales, Scotland, and Northern Ireland), and
(b) EU legislation which had not been implemented or properly implemented by legislation in England and Wales, Scotland, and Northern Ireland but nevertheless had direct effect in the law of those countries if certain conditions were satisfied,

continues to have effect in the law of England and Wales, Scotland, and Northern Ireland on and after that point of time. Such continuing legislation is known as 're-tained EU law' (or, more simply, as 'retained law').

Common law

1.7   Common law is that part of the law which is not the result of legislation, ie it is the law which has been developed by the decisions and rulings of the judges. Very few offences are now governed and defined by the common law. Moreover, the judges do not now have the capacity to create new offences or widen existing ones. Examples of common law offences are murder, manslaughter and conspiracy to defraud.

Although almost all offences are now governed by legislation, the general prin-ciples of criminal liability dealt with later in this chapter are derived from the common law.

The common law has been built up on the basis of the doctrine of precedent, under which the reported decisions of certain courts are more than just authoritative legal statements whose effect is persuasive, since they can be binding (ie must be applied) in subsequent cases. In the context of courts relevant to criminal cases, the following have binding effect: decisions of the Supreme Court (and of the House of Lords, which was the supreme court of appeal until 2009), the Court of Appeal (Criminal Division), a divisional court of the Queen's Bench Division of the High Court, and a judge sitting in the Administrative Court of that Division. Whether or not one of their decisions is binding in a particular case depends on the relative standing of the court which made the decision and the court in which that decision is subsequently cited. The reason is that the doctrine of precedent depends on the principle that the courts form a hier-archy which, in the case of courts with criminal jurisdiction, is in the following des-cending order: *Supreme Court (House of Lords), Court of Appeal (Criminal Division), divisional court, judge in the Administrative Court,* Crown Court and magistrates' courts. The basic rule is that a decision by one of the courts italicised is binding on those courts below that in which it was given and, save in exceptional circumstances, will be followed by a court of equal status. A decision which is not binding under the above rules is nevertheless of persuasive authority, and may be cited to a court in a subsequent case.

European Convention on Human Rights and the Human Rights Act 1998

1.8   The European Convention on Human Rights (ECHR), taken together with the Human Rights Act 1998 (HRA 1998), is of major importance.

Impact of HRA 1998

1.9   There has for a considerable time been a right of individual petition to the European Court of Human Rights in Strasbourg against a breach by the UK of a right under the ECHR. Under HRA 1998, remedies are available in the courts of England and Wales in respect of breach of a Convention right set out in Sch 1 (the Convention rights). The Convention does not automatically take priority over English law. Our courts have not been given a power to disapply an inconsistent Act of Parliament (primary legislation), but the effect of the main provisions in the Act comes close to permitting disapplica-tion. On the other hand, subordinate legislation incompatible with a Convention right can sometimes be quashed.

The Convention rights   The Convention rights specified in HRA 1998, Sch 1 include rights:

(a) to life (ECHR, art 2);
(b) not to be subjected to torture or inhuman or degrading treatment or punishment (art 3);
(c) not to be deprived of liberty save in specified cases, eg after conviction or lawful arrest, and in accordance with a procedure prescribed by law (art 5);
(d) to a fair trial (including the presumption of innocence) (art 6);
(e) to respect for private and family life (art 8);
(f) to freedom of thought, conscience and religion (art 9);
(g) to freedom of expression (art 10); and
(h) to freedom of assembly and association (art 11).

The exercise of the last four rights may be restricted by the law on specified grounds if this is necessary in a democratic society (for example in the interests of national security, for the prevention of disorder or crime, for the protection of health or morals or for the protection of the rights and freedoms of others).

By ECHR, art 14 the enjoyment of the Convention rights must be secured without discrimination on any ground such as sex, race, opinion, national or social origin or other status.

In determining a question which has arisen in connection with a Convention right, our courts must take into account the case law of the European Court of Human Rights.

Statutory interpretation   Acts of Parliament and other legislation must, so far as possible, be read and given effect to in a way which is compatible with the Convention rights. The courts, where necessary, will prefer a strained but possible interpretation which is consistent with Convention rights to one more consistent with the statutory words themselves.

Declaration of incompatibility   If a court is unable to interpret a statutory provision compatibly with a Convention right, it will have to proceed as normal. The issue of incompatibility can be raised on appeal. As far as criminal cases are concerned, if satisfied that a provision of primary legislation (Act of Parliament) is incompatible with a Convention right, a judge of the Administrative Court, a divisional court, the Court of Appeal or the Supreme Court may then make a declaration of incompatibility. The court may also make such a declaration in respect of a provision of other legislation which is so incompatible if satisfied that (disregarding the possibility of revocation) the primary legislation prevents removal of that incompatibility.

If a declaration of incompatibility is made, the government and Parliament are not required to take remedial action.

Unlawful acts   It is unlawful for a 'public authority', a term which includes a police officer, to act in a way incompatible with a Convention right, unless:

(a) as a result of one or more provisions of primary legislation, the authority could not have acted differently; or
(b) in the case of one or more provisions of, or made under, primary legislation which cannot be read or given effect in a way which is compatible with the Convention rights, the authority was acting so as to give effect to or enforce those provisions.

This provides some protection for police officers who act in accordance with the law as it stands at the moment in question.

A person who is a victim of such an unlawful act may bring civil legal proceedings against a public authority in respect of any act by that authority. A court may grant such relief or remedy (including compensatory damages) or make an order as it considers appropriate within the terms of HRA 1998.

GENERAL PRINCIPLES OF CRIMINAL LIABILITY

Criminal liability

1.10 There are two elements of criminal liability:

(a) the outward conduct which must always be proved against the defendant (hereafter 'D') (the actus reus, 'the legally blameworthy act'); and

(b) the state of mind which, apart from in exceptional offences, it must be proved that D had at the time of the relevant conduct (the mens rea, the 'legally blameworthy state of mind').

Of course, the definition of these two things varies from offence to offence, but if the relevant conduct and state of mind can be proved by the prosecution D is guilty unless (s)he can rely successfully on a defence.

Actus reus

1.11 It would be wrong to think that proof of the relevant conduct required for an offence is limited to proof of an act on the part of D. There are two reasons.

First, some offences can be committed (and, indeed, some can only be committed) by a failure to do a particular act on the part of a person who was under a legal duty to do that act; an example of such an offence is that of failing to provide a specimen of breath for analysis, without reasonable excuse, when required to do so under the Road Traffic Act 1988, s 7. It should also be noted that there are some offences whose definitions require neither an act nor an omission but simply the existence of a state of affairs on the part of D; an example is the offence of being found drunk in a public place, contrary to the Licensing Act 1872, s 12.

The second reason why more than an act on D's part must be proved is that rarely, if ever, is a mere act (or omission or state of affairs) sufficient for criminal liability for a substantive offence. The definitions of offences often specify surrounding circumstances which are essential to render the act, etc criminal. Sometimes the definition requires a consequence to result from the act or omission, such as the consequence in murder that another human being dies. The need to prove that a requisite consequence resulted from D's act or omission is known as the requirement of causation. The legal rules about causation are dealt with at **10.7** in relation to homicide offences, but the principles described there are also generally applicable to the other offences with a causation requirement. The specified circumstances and/or consequences of an offence are part of its actus reus.

Mens rea

1.12 Mens rea is not concerned with whether D had an evil mind or knowledge of the wrongfulness of the act. D's ignorance of the criminal law is no defence, nor generally is the fact that D did not regard his/her conduct as immoral or know that it was so regarded by the bulk of society.

The expression mens rea refers to the state of mind expressly or impliedly required by the definition of the offence charged. This varies from offence to offence, but typical instances are intention, recklessness, knowledge and belief.

Intention

A number of offences require D to have acted with a particular intention: in murder **1.13** D must have acted with intent unlawfully to kill, or cause grievous bodily harm to, another human being; in theft D must have dishonestly appropriated another's property with intent permanently to deprive him/her of it. Generally speaking, 'intention' refers to a state of mind in relation to a potential consequence of one's act. In some crimes, that consequence must actually result in order for there to be criminal liability; in others, such as theft, it is not necessary that the intended consequence should occur.

In law, a person 'intends' a consequence of his/her conduct if it is his/her aim or purpose to achieve it. A person who does not have the aim or purpose of bringing about a particular consequence may nevertheless be found to have intended that consequence if it is proved that it was virtually certain to result from his/her conduct and (s)he foresaw that it was virtually certain to result.

Recklessness

In some offences 'recklessness', either as to the consequence required for the actus reus, **1.14** or as to a requisite circumstance of it, or as to some other risk, suffices for criminal liability as an alternative to some other mental state such as intention or knowledge.

A person acts recklessly with respect to:

(a) a circumstance when (s)he is aware of a risk that it exists, or will exist;
(b) a consequence when (s)he is aware of a risk that it will occur,

and it is, in the circumstances known to him/her, unreasonable to take the risk.

Transferred malice

Provided that D intends the specified consequence to occur, or (where recklessness suf- **1.15** fices) is reckless as to that consequence occurring, it is irrelevant that the actual object of that consequence is not the person or property whom D intends to harm, or is reckless as to harming. Thus, if D fires a gun intending to kill X, it is irrelevant that the person killed is Y. D will have intended to kill another human being, and will therefore have acted with the mens rea for murder. This is called the doctrine of transferred malice.

Knowledge

Many offences require 'knowledge' as to the material circumstances (ie the circum- **1.16** stances specified for the actus reus). A person knows that a circumstance exists if (s)he is certain that it exists.

Belief

In some offences 'belief' as to the material circumstances is required. A person believes **1.17** that a circumstance exists if (s)he is virtually certain that it does.

Statutory expressions

In some statutes the mens rea requirement, in whole or part, is denoted by adverbs such **1.18** as 'wilfully', 'maliciously' or 'dishonestly'. These terms have acquired special meanings in law; see further **13.3, 24.88, 9.16,** and **14.58.**

Effect of inadvertence or mistake

1.19 If D lacks the requisite intention, recklessness or other state of mind because (s)he did not think about the matter in issue or made a mistake which prevented him/her having that state of mind, D is not guilty of the offence, however unreasonable his/her inadvertence or mistake. Of course, the more unreasonable an allegation of inadvertence or mistake is, the less likely it is that the allegation will be accepted by the jury or magistrates.

Proof of a state of mind

1.20 In proving whether D had a requisite state of mind, regard must be had:

(a) to the statements of D; and

(b) to the conduct and circumstances of D, and the presence of any motive, since these *may* give rise to the inference that D had the necessary state of mind.

In addition, the Criminal Justice Act 1967, s 8 provides that a court or jury in determining whether a person has committed an offence:

(a) is *not bound* in law to infer that (s)he intended or foresaw a result of his/her actions by reason only of its being a natural and probable consequence of those actions, but

(b) must decide whether (s)he did intend or foresee that result by reference to all the evidence, drawing such inferences from the evidence *as appear proper* in the circumstances.

Knowledge or belief as to a circumstance *may be* inferred from proof that D was wilfully blind as to it (ie D realised the risk that it might exist but deliberately refrained from making inquiries).

Strict liability

1.21 In the case of many, generally minor, offences, the courts have held that a person can be convicted despite having no type of mens rea as to a particular element of the actus reus (or sometimes even though (s)he had no type of mens rea as to any element). These offences are known as offences of 'strict liability'.

All but three offences of strict liability are statutory offences. Clearly, an offence is not one of strict liability if the statutory definition expressly uses a word such as 'intentionally', 'recklessly', or 'knowingly'. The fact that a statutory definition does not use any word importing the concept of mens rea does not necessarily mean that the offence is one of strict liability; it depends on whether or not the courts are prepared to imply a requirement of mens rea into the definition. The basic rule is that, where a statute is silent on the point, it is presumed that Parliament intended a requirement to prove mens rea (so that the offence is not one of strict liability), unless this is rebutted by clear evidence that Parliament intended the contrary.

Justifications

1.22 An offence is not committed if D has a legal justification such as the use of reasonable force in the prevention of crime, effecting a lawful arrest, self-defence or defence of property (see Chapter 9) which renders the conduct lawful. Police officers are often empowered to do something which would otherwise be an offence. Such powers are another example of a legal justification.

Infancy

A child under 10 is irrebuttably presumed not to be guilty of an offence.   **1.23**

Insanity

Everyone is presumed sane until the contrary is proved. The mere fact that a person is   **1.24**
medically insane is no defence, but a person proved to be legally insane has a defence.
The test of insanity for the purposes of legal responsibility is provided by the *M'Naghten
Rules*, which comprise three elements which must be proved by D if (s)he raises the
defence of insanity. At the time of the conduct in question:

(a) *D must have been suffering from a 'disease of the mind'*, ie an impairment of the
    mental faculties of reason, memory and understanding due to a disease as opposed
    to some external factor like a blow on the head;
(b) *D must have been suffering a 'defect of reason' due to disease of the mind*, ie a depriv-
    ation of reasoning power (as opposed to momentary confusion or absent-minded-
    ness); and
(c) *as a result, D must not have known the physical nature and quality of his/her act or,
    if D did know this, not have known (s)he was doing 'a legal wrong'.* The mere fact
    that, because of a defect of reason due to a disease of the mind, D acted under an
    irresistible impulse is not enough.

Defendants found not guilty by reason of insanity in the Crown Court must be made
subject to one of three orders, one of which is indefinite detention in hospital until this
is no longer necessary for the protection of the public.

Automatism

Generally, it is a defence that D was in a state of automatism at the time of the act   **1.25**
in question. An act is done in a state of automatism if it is done by the muscles
without any control of the mind (eg a reflex action, or a spasmodic or convulsive
act) or if it is done during a state involving lost consciousness (eg concussion,
or a hypoglycaemic coma resulting from insulin taken by a diabetic). Where the
automatism directly results from a disease the relevant defence is insanity, not
automatism.

D does not have to prove the defence of (non-insane) automatism. Instead D simply
has the burden of adducing sufficient evidence to raise the issue; D's evidence will
very rarely be sufficient unless it is supported by medical evidence. If D adduces sufficient evidence, D must be acquitted unless the prosecution disproves the alleged
automatism.

Sometimes, non-insane automatism is self-induced in that it results from something
done or not done by D (as where a diabetic becomes an automaton as a result of taking
insulin, or, having taken insulin, failing to eat sufficiently thereafter). In such a case,
D cannot be convicted of an 'offence of specific intent' (which is defined below). Nor
can D be convicted of an 'offence of basic intent' (below), *unless* either the automatism
was caused by his/her voluntary intoxication *or*, before becoming an automaton, D was
aware that something done or not done was likely to make him/her aggressive, unpredictable or uncontrollable (as opposed simply to becoming unconscious), and (s)he
deliberately disregarded the risk.

Intoxication

1.26   The rules relating to the effect of intoxication in criminal liability are the same whether the intoxication was caused by drink or drugs. Intoxication is not in itself a defence. It is no excuse that because of intoxication D's power to judge between right and wrong, or to exercise self-control, was impaired.

Voluntary intoxication

1.27   In most cases a person's intoxication is regarded as 'voluntary'. Intoxication is voluntary if it results from knowingly taking alcohol or some other drug. There are two exceptions: intoxication is not voluntary where it is caused by something taken under and in accordance with medical advice, or where it is caused by a non-dangerous drug (ie a sedative or soporific drug), provided that D is not aware of the risk of becoming unpredictable or aggressive when taking it. These cases of involuntary intoxication are dealt with below.

Where D was voluntarily intoxicated, D may rely on the intoxication as evidence that (s)he lacked mens rea if, but only if, the offence requires proof of a specific intent. This rule applies not only where D's alleged lack of mens rea was a direct result of being intoxicated at the time of the offence but also where it was a direct and proximate result of immediate prior intoxication (as where D triggered an immediate psychotic illness as a consequence of proximate ingestion of drugs which did not remain in D's system at the time of the offence). Even if an offence is one of specific intent, an intoxicated, mistaken belief that (s)he is acting in self-defence or the like will not excuse D.

Where an offence is one of basic intent, D may be convicted of it if (s)he was voluntarily intoxicated at the time of his/her conduct, even though because of his/her intoxication (s)he did not have the mens rea required for that offence, and even though (s)he was then in a state of automatism, provided that (s)he would have been aware of the risk in question had (s)he been sober. However, the fact that an offence is one of basic intent does not prevent D relying on evidence of his/her voluntary intoxication as evidence that his/her act (as opposed to its consequences) was an 'accident', such as where D stumbles into someone in an intoxicated state.

The following have been held to be offences of specific intent: murder; wounding or causing grievous bodily harm with intent contrary to the Offences Against the Person Act 1861 (OAPA 1861), s 18; theft; robbery; and attempt to commit an offence.

Conversely, the following are examples of basic intent: manslaughter; maliciously wounding or inflicting grievous bodily harm contrary to OAPA 1861, s 20; assault occasioning actual bodily harm; assault on a constable in the execution of his/her duty; sexual assault; taking a conveyance without lawful authority; and criminal damage contrary to the Criminal Damage Act 1971, s 1(1) or (2) (unless only an intentional offence is alleged).

Involuntary intoxication

1.28   The situation is different where a person is involuntarily intoxicated. Intoxication is involuntary in the two exceptional cases outlined in the first paragraph of **1.27**, and also where it is not self-induced (as where a person has been secretly drugged).

Where D was involuntarily intoxicated, D can use evidence of this intoxication as evidence that (s)he lacked the mens rea for the offence in question (whether or not it is an offence of specific intent). However, if it can be proved that D had the necessary mens rea when the offence was committed, it is no defence that involuntary intoxication led D to commit an offence which (s)he would not have committed when sober.

Duress

The defences of duress by threats and duress of circumstances are restricted to cases where D was impelled to act as (s)he did because, on the facts as (s)he reasonably believed them to be, (s)he had good cause to believe that (s)he or someone for whom (s)he felt responsible was subject to a threat of imminent death or serious physical injury unless (s)he acted as (s)he did to avoid the threatened harm. The defence of duress by threats deals with the case where the threat comes from another person and, expressly or impliedly, is in the form of an order to do a particular, nominated act or suffer the harm. The threat need not be conveyed to D directly; instead it can be conveyed indirectly by a third party. The defence of duress of circumstances deals with the case where the threat of death or serious physical injury is of any other type (as where the threat comes from the surrounding circumstances).

The threat must be such that an ordinary, sober person of reasonable firmness sharing D's characteristics would have responded as D did. Because this 'ordinary person' is someone of reasonable firmness (s)he is not invested with a characteristic of D which did not make D less able to resist the threat than an ordinary person of reasonable firmness. In addition, the ordinary person of reasonable firmness is not invested with a characteristic of D, such as pliancy, vulnerability to pressure or timidity, since it would be a contradiction in terms to invest an ordinary person with these. On the other hand, if D is in a category of persons who might be less able to resist pressure than people outside that category, the characteristic which puts him/her in that category may be a relevant one. Obvious examples are age (a young person may not be as robust as a mature person); pregnancy (added fear for the unborn child); serious physical disability (may inhibit self-protection); and a recognised mental illness or psychiatric condition (may make the person more susceptible to pressure).

The defences of duress by threats and duress of circumstances are not applicable in respect of conduct committed after a threat has ceased to be operative or if D could have neutralised the threat by seeking police protection. With a few exceptions, the defences are available in respect of any offence. The principal exceptions are murder and attempted murder.

The defence of duress by threats is not available to a person who foresaw or ought to have foreseen that voluntary association with other people involved a risk of being subjected to any compulsion by threat of violence. Thus, the defence is not available to a person who joins a criminal gang, knowing that other members might bring pressure to bear on him/her to rob someone, and who is subsequently put under such pressure.

PARTIES TO A CRIME

Perpetrators

A perpetrator is otherwise known as the principal. The perpetrator is normally the person who, with the relevant mens rea, commits the actus reus of the offence, eg by firing the fatal shot in murder, or having intercourse in rape. Of course, there can be more than one perpetrator, as where a group of men enter a building as trespassers in order to steal therein; in such a case there are said to be joint perpetrators of the offence.

A person who makes use of an innocent agent in order to procure the commission of an offence is the perpetrator of the offence, even though (s)he does nothing with his/her own hands. Thus, a person who kills V by posting a time bomb to V which is delivered by an innocent postman, or who employs a child under ten (the age of criminal responsibility) to remove goods from a shop, may be convicted of murder or theft, as the case may be, if (s)he acts with the appropriate mens rea.

Accomplices

1.31   A person who aids, abets, counsels or procures the commission of an offence (an accomplice) is liable to be tried and punished for that offence as a principal offender.

Aiding, abetting, counselling and procuring

1.32   The terms 'aiding' and 'abetting' are often used together, but they refer to different things; 'aid' describes the activity of a person who assists the perpetrator to commit the principal offence, and 'abet' describes the activity of a person who encourages the perpetrator to commit it, whether or not in either case present at the time of commission. Although liability as an accomplice does not require an agreement between an alleged accomplice and the perpetrator, an agreement between the two is by its nature a form of encouragement by the former. 'Counsel', which means 'encourage', does not add anything strictly but is used to describe encouragement before the commission of the principal offence. A person 'procures' the commission of the principal offence where (s)he sets out to see that it is committed and takes appropriate steps to produce its commission.

It can be seen from the above that basically there must be some assistance or encouragement by a person in the commission of an offence by another before (s)he can be convicted as an accomplice to it. That assistance or encouragement must be given before, or at the time of, the commission of the offence. Someone who assists the perpetrator after the commission of an offence is not liable as a party to it, but one who assists the perpetrator to escape detection or arrest may commit the offence of assisting offenders (**22.1**).

Mens rea

1.33   The mens rea required of an accomplice is not the same as that required of the perpetrator of the principal offence. Instead, what is required to be proved is an intent to assist or encourage the commission by the perpetrator of the act which constitutes or results in the principal offence.

This requires the following to be proved against an alleged accomplice:

(a)   an intention to assist or encourage conduct by the perpetrator which constitutes or results in the commission of the principal offence;

(b)   knowledge of the facts necessary to constitute the principal offence;

(c)   an intention to assist or encourage the perpetrator to commit the principal offence with the mens rea required for it. The alleged accomplice is not required to know the details of that offence. Where a party to a joint criminal venture to commit an offence intentionally perpetrates a further offence in doing so, another party to the venture is liable for the further offence only if (s)he intends to assist or encourage the perpetrator in committing the further offence in accordance with the above rules. When the further offence is murder, but the person is not guilty of it as an accomplice under the above rules, (s)he may be convicted of constructive manslaughter if (s)he has intentionally assisted in or encouraged the unlawful act from which the victim died and any sober and reasonable person would realise that it carried the risk of some harm resulting.

Miscellaneous points

A person who cannot in law perpetrate a particular offence may nevertheless be con-   **1.34**
victed as an accomplice to it. For example, a woman can be convicted of rape as an accomplice where she has encouraged or assisted a man to rape another woman.

Where D is alleged to be an accomplice to an offence, the charge may allege that D
aided, abetted, counselled or procured it, and D will be convicted if (s)he is proved to
have participated in one or more of these four ways.

Vicarious liability

Vicarious liability means liability for the acts of another person which D has not   **1.35**
authorised and of which D was ignorant.

Vicarious *criminal* liability is exceptional; it is imposed only on employers and, in
some cases, certain other people with a similar status. It arises in three ways:

(1) Where a statute expressly states that a specified person is vicariously liable for the
    prohibited act of another. This is rare.
(2) Where a statutory strict liability offence uses a word like 'use' or 'sell' which con-
    notes an activity which can be performed by an employee on behalf of his/her
    employer, an employer can be vicariously liable for a prohibited 'using', etc by the
    employee in the course of his/her employment.
(3) Certain statutes impose duties on someone who is a licensee and make it an
    offence for someone with that status knowingly to contravene them. If a licensee
    completely delegates his/her statutory responsibilities to someone else such as
    an employee, and the delegate knowingly contravenes one of these duties, the
    conduct and the state of mind of the delegate are imputed to the person with
    the specified status, who is consequently vicariously liable for the offence in
    question.

CORPORATE LIABILITY

A company or other corporate body may be criminally liable: **1.36**

(a) on the basis of vicarious liability (above); or
(b) for breach of a statutory duty, such as that imposed on it as an 'occupier' of prem-
    ises or as an 'employer', etc,

in the same way as a natural person. The offences to which these principles apply are
essentially ones of strict liability.

A company or other corporate body may also be liable for most offences (murder
is a notable exception) requiring proof of mens rea where the acts and state of mind
of an individual who physically committed the offence can be attributed to it. In the
case of a *common law offence*, such as conspiracy to defraud, such attribution is only
possible in respect of a person who is a 'controlling officer', ie someone who represents
the 'directing mind and will' of the corporation (such as a director). Whether or not
the conduct of a particular employee of a corporation can be attributed to the corporation for the purposes of criminal liability for a *statutory offence* depends on the proper
construction of the particular rule of law concerned; a particular statute may be interpreted so that the act and state of mind of someone lower down the company or other
corporate body's hierarchy can be attributed to it.

ENCOURAGING OR ASSISTING CRIME

1.37   The Serious Crime Act 2007 (SCA 2007), Part 2 (ss 44 to 67) provides three offences of encouraging or assisting crime.

Intentionally encouraging or assisting an offence

1.38   A person commits an offence against SCA 2007, s 44 if (s)he:

(a) does an act capable of encouraging or assisting the commission of an offence other than encouraging or assisting suicide; and

(b) intends to encourage or assist its commission.

In s 44 (and ss 45 and 46, below), the reference to the doing of an act includes a reference to a failure to act, the continuation of an act already begun and an attempt to do an act (except an act amounting to the commission of the offence (**1.55**) of attempting to commit *another* offence).

Encouraging or assisting an offence believing it will be committed

1.39   A person commits an offence against SCA 2007, s 45 if (s)he:

(a) does an act (see above) capable of encouraging or assisting the commission of an offence; and

(b) believes that the offence will be committed and that his/her act will encourage or assist its commission.

For this offence, it need not be proved that D intended to encourage or assist the commission of the offence which his/her act is capable of assisting or encouraging. Instead, it is sufficient to prove that D believed that an act would be done which would amount to the commission of that offence and that his/her act would encourage or assist the doing of that act.

In reckoning whether (for the purposes of s 45) an act is capable of encouraging or assisting the commission of an offence, specified offences are to be disregarded. As a result, a person cannot be guilty under s 45 of encouraging or assisting a specified offence. The offences specified are:

(a) an offence under SCA 2007, s 44, 45, or 46, and

(b) an offence listed in SCA 2007, Sch 3, Parts 1, 2, or 3 (mainly various statutory offences of incitement referred to later in this book). They also include the following offences:

    (i)   solicitation of murder (Offences Against the Person Act 1861, s 4);

    (ii)   encouraging or assisting suicide (Suicide Act 1961, s 2(1));

    (iii)   assisting an offender (Criminal Law Act 1967, s 4(1));

    (iv)   concealing an offence for reward (Criminal Law Act 1967, s 5(1));

    (v)   participating in an organised crime group (Serious Crime Act 2015, s 45);

    (vi)   statutory conspiracy (Criminal Law Act 1977, s 1(1));

    (vii)   common law conspiracy; and

    (viii) criminal attempt (Criminal Attempts Act 1981, s 1(1)).

It is sufficient if D believes that the offence will be committed once certain conditions are met. Thus, if D encourages E to beg some money from V and tells E that, if V refuses, E should take some money from V by force, D can be convicted under s 45 of encouraging or assisting robbery because D believes that the offence (of robbery) will be committed if certain conditions are met (ie V refuses to hand over money).

If it is alleged under s 45 that D believed that an offence would be committed and that his/her act would encourage or assist its commission, it is sufficient to prove that D believed:

(a) that an act would be done which would amount to the commission of that offence; and

(b) that his/her act would encourage or assist the doing of that act.

Encouraging or assisting an offence believing one or more will be committed

By SCA 2007, s 46, a person commits an offence if (s)he: **1.40**

(a) does an act (**1.38**) capable of encouraging or assisting the commission *of one or more of a number of offences*; and

(b) believes:
   (i) that one or more of those offences will be committed (but has no belief as to which), and
   (ii) that his/her act will encourage or assist the commission of one or more of them.

It is immaterial for the purposes of (b)(ii) whether the person has any belief as to which offence will be encouraged or assisted. The Court of Appeal has held that s 46 should only be used when it may be that D, at the time of doing the act, believes that one or more of, for example, *either* offence X, *or* offence Y, *or* offence Z will be committed, but has no belief as to which one or ones of the three will be committed.

An offence under s 44, 45, or 46, or an offence under Sch 3, Parts 1, 2, or 3 (**1.39**), is to be disregarded in reckoning whether an act is capable of encouraging or assisting the commission of one or more of a number of offences.

If it is alleged under (b) that D believed that one or more of a number of offences would be committed and that his/her act would encourage or assist the commission of one or more of them, it is sufficient to prove that D believed that one or more of a number of acts would be done which would amount to the commission of one or more of those offences; and that his/her act would encourage or assist the doing of one or more of those acts.

Points relating to ss 44 to 46 in general

Act capable of encouraging or assisting

SCA 2007, s 65 provides that the reference to a person's doing an act that is capable of   **1.41** encouraging the commission of an offence includes a reference to his/her doing so by:

(a) threatening another person or otherwise putting pressure on another person to commit the offence;

(b) taking steps to reduce the possibility of criminal proceedings being brought in respect of that offence; or

(c) failing to take reasonable steps to discharge a duty.

However, s 65 also provides that a person is not to be regarded as doing an act that is capable of encouraging or assisting the commission of an offence merely because (s)he fails to respond to a constable's request for assistance in preventing a breach of the peace.

Further provisions about mens rea

1.42   SCA 2007, s 47(5) sets out the mens rea that must be proved under s 44, 45, or 46 if an offence that it is alleged D intended or believed would be committed requires proof of mens rea or of circumstances or consequences. It provides that, in proving whether an act or omission is one which, if done, would amount to the commission of an offence:

(a) if the offence is one requiring proof of fault, it must be proved that:
   (i)   D believed that, were the act to be done, it would be done with that fault,
   (ii) D was reckless as to whether or not it would be done with that fault, or
   (iii) D's state of mind was such that, were D to do it, it would be done with that fault; and
(b) if the offence is one requiring proof of particular circumstances or consequences, it must be proved that:
   (i)   D believed that, were the act to be done, it would be done in those circumstances or with those consequences, or
   (ii) D was reckless as to whether or not it would be done in those circumstances or with those consequences.

In the case of an offence under s 44, (b)(i) is to be read as if the reference to 'D believed' were a reference to 'D intended or believed'.

It follows from (a) that, if an offence that D is alleged to have intended or believed would be encouraged or assisted requires mens rea, D must believe or be reckless to the possible consequence that, if the act encouraged or assisted is done, it would be done with the necessary mens rea by the person encouraged or assisted, or (by (a)(iii)) D must have the necessary mens rea for that offence. The latter point is important where D thinks that the other person will lack the mens rea. Thus, the offences under s 44, 45, or 46 apply to the encouragement or assistance of an innocent agent. Section 47(6) provides that, for the purposes of (a)(iii), D is assumed to be able to do the act in question. Thus, D cannot escape liability in respect of (a)(iii) simply because it is impossible for him/her to commit the offence encouraged or assisted.

It follows from (b) that a requirement to prove mens rea on D's part applies in respect of any circumstances or consequences of the actus reus of the offence that D is alleged to have intended or believed would be encouraged or assisted, and that this is so even though that offence is one of strict liability as to that particular element. For example, the offence of driving a motor vehicle on a road or other public place with excess alcohol, contrary to the Road Traffic Act 1988, s 5, is one of strict liability, for which no mens rea is required. If D asks E to drive him/her to the station, in ignorance that E has been drinking and is 'over the limit', D cannot be convicted of an offence of encouraging or assisting crime (the only relevant one would be that under s 44) because (s)he neither believes that E is over the limit, nor is (s)he reckless of the consequence that if E complies with his/her request E would prove to be over the limit, despite the fact that if E did drive D as requested E would commit the substantive offence.

In the case of an offence under s 46 it is sufficient to prove the matters referred to in s 47(5) by reference to one offence only.

Commission of offence encouraged or assisted not required

SCA 2007, s 49 provides that D may be guilty of offences under s 44, 45, or 46 whether **1.43** or not any offence capable of being encouraged or assisted is in fact committed. Thus, an offence under s 44, 45, or 46 is complete on the giving of the encouragement or assistance.

Where, however, the offence encouraged or assisted by D is *committed*, an offence under s 44, 45, or 46 exists alongside D's liability for that offence as a party to it.

Defence of acting reasonably

SCA 2007, s 50 provides that D is not guilty of an offence under s 44, 45, or 46 if (s) **1.44** he proves that (a) (s)he knew certain circumstances existed, and that it was reasonable for him/her to act as (s)he did in those circumstances; or (b) (s)he reasonably believed certain circumstances to exist, and it was reasonable for him/her to act as (s)he did in the circumstances as (s)he believed them to be.

The factors which might be considered, in determining whether it was reasonable for a person to act as (s)he did, include: the seriousness of the anticipated offence; any purpose for which (s)he claims to have been acting; and any authority by which (s) he claims to be acting. However, there might be other factors which would be equally relevant.

Protective offences: victims not liable

In the case of 'protective offences' D does not commit an offence under SCA 2007, s **1.45** 44, 45, or 46 if D falls within the category of a 'protected person' and D is the person in respect of whom the offence is committed or would have been committed. A 'protective offence' is one that exists (wholly or partly) for the protection of a particular category of persons, eg the child sex offences included in the Sexual Offences Act 2003. Thus, where a child encourages a man to commit a child sex offence against her, that child is not guilty of an offence of encouraging or assisting that offence, as it was fashioned to protect her. Nor could she be convicted as an accomplice to the sex offence itself if it was committed.

Trial and punishment

An offence under s 44 or 45 is triable in the same way as the anticipated offence. An **1.46** offence under s 46 is only triable on indictment.

Where D is convicted of an offence under s 44 or 45, or of an offence under s 46 by reference to *only one* offence ('the reference offence'), D is liable to any penalty for which he would be liable for the anticipated or reference offence.

Where D is convicted under s 46 by reference to *more than one* reference offence, and one or more of the reference offences is punishable with imprisonment, the maximum sentence is the maximum term of imprisonment for any of the offences (taking the longer or longest maximum as the limit for this purpose if the terms differ); alternatively, an unlimited fine may be imposed (except where one of the reference offences is murder). In any other case a fine may be imposed.

OFFENCE OF PARTICIPATING IN ACTIVITIES OF ORGANISED CRIME GROUP

The Serious Crime Act 2015, s 45(1) provides that a person (D) who participates in **1.47** the criminal activities of an organised crime group commits an offence triable only on indictment and punishable with up to five years' imprisonment. For this purpose, D participates in the criminal activities of an organised crime group if (s)he takes part in any activities that (s)he knows or reasonably suspects (a) are criminal activities of an organised crime group, or (b) will help an organised crime group to carry on criminal activities.

'Criminal activities' are activities carried on with a view to obtaining (directly or indirectly) any gain or benefit, which are activities:

(a) carried on in England and Wales, and which constitute an offence in England and Wales punishable on conviction on indictment with imprisonment for seven years or more; or

(b) carried on outside England and Wales, and which constitute an offence under the law of the country where they are carried on, and would constitute an offence in England and Wales of the kind mentioned in (a) if the activities were carried on in England and Wales.

'Organised crime group' means a group that has as its purpose (or as one of its purposes) the carrying on of criminal activities, and consists of three or more persons who act (or agree to act) together to further that purpose.

For D to be guilty of an offence under s 45(1) it is not necessary:

(a) for D to know any of the persons who are members of the organised crime group;

(b) for all of the acts or omissions comprising participation in the group's criminal activities to take place in England and Wales (so long as at least one of them does); or

(c) for the gain or benefit referred to above to be financial in nature.

By s 45(8), it is a defence for D to prove that his/her participation was necessary for a purpose related to the prevention or detection of crime.

CONSPIRACY: INTRODUCTION

1.48   There are two offences of conspiracy:

(1) It is a *statutory* offence to agree with any other person or persons to commit an offence. Statutory conspiracy is governed by the Criminal Law Act 1977 (CLA 1977).

(2) It is a *common law* offence to agree to defraud.

Both types of conspiracy are only triable on indictment. The maximum punishment for statutory conspiracy is the same as the maximum term of imprisonment for the offence conspired to (or the longer or longest if there is more than one such offence and their maxima differ). If no such offence is imprisonable a person convicted of conspiracy cannot be sentenced to imprisonment, although a fine can be imposed. The maximum imprisonment for common law conspiracy to defraud is 10 years.

Both types of conspiracy require a concluded agreement between two or more people. It is not sufficient for one person to have spoken in the presence of another of his/her intention to commit a crime—that other person must agree with him/her that the crime be committed. Immediately such an agreement has been reached, the offence of conspiracy is complete; no steps need be taken in furtherance of it, although it will usually be most unlikely that the conspiracy will be discovered or capable of proof if no further steps are taken. A party cannot escape liability for conspiracy by withdrawing from the agreement.

STATUTORY CONSPIRACY

CLA 1977, s 1(1) provides that if a person agrees with any other person or persons that   **1.49**
a course of conduct shall be pursued which, if the agreement is carried out in accordance with their intentions, either:

(a) will necessarily amount to or involve the commission of any offence or offences by
    one or more of the parties to the agreement (s 1(1)(a)); or
(b) would do so but for the existence of facts which render the commission of the
    offence or any of the offences impossible (s 1(1)(b)),

he is guilty of conspiracy to commit the offence or offences in question.

The offence which the agreed course of conduct will necessarily (or would) amount
to or involve may be of any type, including (with one exception) a summary offence.

The exception is provided by the Trade Union and Labour Relations (Consolidation)
Act 1992, s 242. It is that where, in pursuance of any agreement, the acts constituting
the offence are to be done in contemplation or furtherance of a trade dispute, that
offence is not an 'offence' for the purposes of conspiracy if it is triable only summarily
and *not* punishable with imprisonment. This exemption is provided to remove the
potential danger of conspiracy charges in relation to the many minor offences which
might be committed by union members in the course of a dispute, for example, simple
obstructions of highways.

By CLA 1977, s 1A, s 1(1) applies in certain cases to an agreement to pursue a course
of conduct outside England and Wales which would amount to an offence under the
law of the country concerned.

Section 1(1)(a)

The key issue under s 1(1)(a) is whether, when the agreement is made, the course of   **1.50**
conduct agreed on by the parties will necessarily amount to or involve the commission
of an offence by one or more of them if it is carried out in accordance with their intentions. Thus, if D1 and D2 agree to have intercourse with a woman without her consent,
there is a statutory conspiracy to rape because the course of conduct agreed on—penile
penetration—will necessarily amount to the commission of rape if it was carried out
in accordance with their intentions, namely to have penile penetration with a woman
without her consent.

The reference in s 1(1) to 'the commission of any offence or offences by one or more
parties to the agreement' is not limited to the physical commission by one or more of
them but also includes its commission by an innocent agent of one or more of the parties, as where two individuals agree to send an innocent agent to deliver a parcel bomb.
The principle of innocent agency is also applicable on the context of common law conspiracy to defraud.

Section 1(1)(b)

CLA 1977, s 1(1)(b) deals with agreements whose criminal objectives are impossible of   **1.51**
fulfilment. It provides that an agreement on a course of conduct which, if the agreement
is carried out in accordance with the parties' intentions, would necessarily amount to
or involve the commission of an offence or any offences by one or more of the parties
but for the existence of facts which render the commission of the offence or any of the

offences impossible is a statutory conspiracy. This is a sensible provision; the essence of the offence is hatching a plot to commit a crime and this is not made less blameworthy because the actual commission of that crime is rendered impossible because of some fact.

As a result of s 1(1)(b), it is clear, for example, that D1 and D2 can be convicted of conspiracy to murder even though their intended victim was already dead when they agreed to kill him/her.

Mens rea

1.52   At least two parties to the agreement must share the mens rea set out below.

If the actus reus of the substantive offence (ie the agreed offence) requires the existence of a fact or circumstance, a defendant and at least one other party to the agreement (X) must intend or know that it shall or will exist.

There cannot be a statutory conspiracy unless there are at least two parties to an agreement who intend that the agreement be carried out and that the offence they are alleged to have conspired to commit be committed. Surprisingly, because it does violence to the wording of CLA 1977, s 1(1), the House of Lords has held that an individual party to an agreement can be convicted of conspiracy, even though (s)he did not intend that it be carried out and the offence intended by the other parties be committed. In practice this ruling has been ignored and the courts require an intention that the agreement be carried out and that the offence be committed to be proved against an individual defendant.

A defendant and at least one other party (X) must have any additional mens rea (eg dishonesty and an intent permanently to deprive on a charge of conspiracy to steal) required for the substantive offence.

Exemptions from liability for statutory conspiracy

1.53   CLA 1977, s 2 provides the following exemptions:

(a) a person cannot be guilty of a conspiracy to commit any offence if (s)he is the intended victim of that offence;

(b) a person is not guilty of a conspiracy to commit any offence if the *only* other person or persons with whom (s)he agrees (both initially and at all times during the currency of the agreement) is or are:
   (i)   his/her spouse or civil partner;
   (ii) under the age of criminal responsibility (ie under 10); or
   (iii) an intended victim or victims of the substantive offence.

The Court of Appeal has held that, for the purposes of (i), 'spouse' includes a party to a polygamous marriage which is valid under the law of the place of celebration and is not regarded as void under English law.

COMMON LAW CONSPIRACY TO DEFRAUD

1.54   There can be a conspiracy to defraud without any element of deception since it is sufficient to prove an agreement by dishonesty to deprive a person of something which is his/hers (or to which (s)he is or would be or might be entitled) or an agreement by dishonesty to injure some proprietary right of his/hers. Where the intended victim of an agreement is a public official performing public duties, as distinct from a private individual, there can also be a conspiracy to defraud if the agreement is dishonestly to deceive such a person into acting contrary to his/her duty (eg in granting a licence or giving information). The causing of economic loss or prejudice, or (as the case may be) the deceiving of a public official into acting contrary to his/her duty, need not be the purpose of the parties to the agreement, since it suffices if they have dishonestly agreed to bring about a state of affairs which they realised would or might have such a result. 'Dishonesty' bears the same meaning as in offences against property (**14.58**).

To be a conspiracy to defraud an agreement must incorporate some unlawfulness in its object or means.

Clearly, the definition of conspiracy to defraud is wide enough to cover cases where the object or means of the agreement would itself be an offence as well as those where the object or means would not be criminal but would otherwise be unlawful (eg as a tort). Agreements to steal, or to forge, and so on, may, of course, be prosecuted as statutory conspiracies to steal, etc but, by virtue of the Criminal Justice Act 1987, s 12, they may instead be charged as common law conspiracies to defraud. However, they should ordinarily be charged as a statutory conspiracy unless there is good reason otherwise. Spouses (but not civil partners) are not guilty of common law conspiracy if they are the only parties to the agreement.

ATTEMPT

The Criminal Attempts Act 1981 (CAA 1981), s 1(1) provides that if, with intent to commit an offence to which CAA 1981, s 1 applies, a person does an act which is more than merely preparatory to the commission of the offence, he is guilty of attempting to commit the offence. The offence under s 1(1) is triable in the same way as the offence attempted. The maximum sentence is the same as for the offence attempted.   **1.55**

The offences to which CAA 1981, s 1 applies are described by s 1(4) which provides that s 1 applies to any offence which, if completed, would be triable in England and Wales as an indictable offence, except:

(a) conspiracy (whether common law or statutory);
(b) aiding, abetting, counselling, procuring, or suborning the commission of an offence;
(c) offences under the Criminal Law Act 1967, s 4(1) (assisting offenders) or s 5(1) (concealing offences); and
(d) encouraging or assisting suicide under the Suicide Act 1961, s 2(1).

It follows that all attempts to commit indictable offences, other than those specified above, are offences contrary to s 1 but attempts to commit summary offences are not. It should be noted that a number of statutes providing summary offences also provide specific offences of attempt in relation to them.

Mens rea

The requirement of mens rea plays a particularly important role in the crime of attempt because whether or not a particular act amounts to an attempt might well hinge on the intent with which it is done. For example, to strike a match near a haystack might or might not be attempted arson of a haystack, depending on whether there is an intent to set fire to the haystack or to light a cigarette: the intent colours the act.   **1.56**

The mens rea specified by CAA 1981, s 1(1) is an 'intent to commit an offence to which this section [ie s 1] applies' and which the defendant (D) is alleged to have attempted. This apparently straightforward statement needs further explanation since 'an intent to commit the offence attempted' may involve a number of mental states.

D must, of course, intend to commit an act or to continue with a series of acts which, when successfully completed, will amount to or lead to an offence. In addition, if the crime attempted requires some consequence to result from his/her conduct, D must *intend* to cause that consequence. This is so even though some other type of mens rea (eg recklessness) is required or suffices for the full offence. This requirement of mens rea in attempt can be illustrated as follows:

(1) On a charge of attempted murder, it must be proved that D intended the unlawful death of another human being. By way of comparison, if D had actually killed someone D could be convicted of murder merely because D intended his/her act unlawfully to cause grievous bodily harm to another person.

(2) On a charge of attempted criminal damage, it must be proved that D intended the destruction or damaging of property belonging to another, even though if D had actually destroyed or damaged that property D could have been convicted of criminal damage if (s)he was reckless as to the risk that his/her act might possibly have this effect.

Where the actus reus of the crime attempted includes some circumstance, such as the fact that the goods handled are stolen in handling stolen goods, D will have sufficient mens rea as to that circumstance on a charge of attempt if (s)he intends that it will exist, or knows or believes that it does. Moreover, although there are conflicting Court of Appeal decisions, there is authority that, where some lesser mental state as to a circumstance suffices for the full offence or no mental state as to it is required at all, recklessness as to it suffices on a charge of attempt.

The requirement that D must intend to commit the offence attempted means that D must have any other mental element, additional to mens rea as to the elements of the actus reus of the crime attempted, required for that crime. Thus, to be convicted of attempted theft D must not only have intended to appropriate the property belonging to another but have acted dishonestly and with intent permanently to deprive the 'owner' of it.

Actus reus

1.57   CAA 1981, s 1(1) requires 'an act that is *more than merely preparatory* to the commission of the offence', ie the full offence which D intends to commit. Appellate courts have held that this means that D must have got as far as having embarked on the commission of the full offence (ie taken the steps immediately preceding the final act by D necessary to effect D's plan and bring about the commission of the full offence). If D has so acted it is irrelevant that something more remains to be done by some other, innocent person(s), as where D posts a parcel bomb to his/her intended murder-victim.

Impossibility

1.58   CAA 1981, s 1(2) provides that a person may be guilty of attempting to commit an offence to which s 1 applies, even though the facts are such that the commission of the offence is impossible. Therefore, a person who attempts to murder by an inadequate dosage of poison is guilty of attempted murder.

Section 1(3) purports to reinforce the rule in s 1(2), by declaring that where:

(a) apart from s 1(3) D's intention would not be regarded as having amounted to an intent to commit an offence; but

(b) if the facts of the case had been as D believed them to be, D's intention would have been so regarded,

D is to be regarded as having an intent to commit that offence.

By virtue of s 1(2) and (3), a person who intends to kill another by shooting him/her in a bed, but who discovers after (s)he has fired the shot that the assumed and intended victim was a pillow, can be convicted of attempted murder. Likewise, a person who handles some goods mistakenly believing that they are stolen can be convicted of attempting to handle stolen goods.

Other statutory offences of attempt

CAA 1981, s 3 applies the provisions of the Act to other existing statutory attempts to commit offences, by stating that the same provisions apply to those attempts. Section 3 therefore ensures a common approach to all offences of attempted crime, even where the specific offence of attempt is created by a particular statute and even where that specific offence relates to a summary offence.   **1.59**

CHAPTER 2

Police Officers, Staff, and Volunteers

JURISDICTION OF CONSTABLES

2.1 In law every member of a police force holds the office of constable, whatever his/her rank. Although (s)he is not 'a member of a police force', so does a special constable.

The Police Act 1996 (PA 1996), s 30 provides that every member of a police force or a special constable has all the powers and privileges of a constable throughout England and Wales and the adjacent UK waters. 'UK waters' means the sea and other waters within the seaboard limits of the territorial seas.

IMPERSONATION, ETC AND CAUSING DISAFFECTION AMONGST MEMBERS OF A POLICE FORCE

Impersonation

2.2 A person who, with intent to deceive, impersonates a member of a police force or a special constable, or makes any statement or does any act calculated falsely to suggest that (s)he is such a member or constable, commits an offence against PA 1996, s 90(1). It is also an offence against s 90(2) for a person who, not being a member of a police force or a special constable, wears any article of police uniform in circumstances where it gives him/her an appearance so nearly resembling that of a member of a police force as to be calculated to deceive. Section 90(3) makes it an offence to possess an article of police uniform, unless it was obtained lawfully and is possessed for a lawful purpose. Offences under s 90 are summary offences punishable as follows:

- up to six months' imprisonment and/or an unlimited fine (s 90(1));
- a fine not exceeding level 3 on the standard scale (s 90(2));
- a fine not exceeding level 1 on that scale (s 90(3))

Causing disaffection

2.3 A person who causes, or attempts to cause, or does any act calculated to cause, disaffection amongst members of any police force, or induces, or attempts to induce, or does any act calculated to induce, any member of a police force to withhold his/her services commits an indictable (either-way) offence against PA 1996, s 91(1). The section also applies to special constables as it does to members of a police force. An offence against s 91(1) is punishable with up to two years' imprisonment and/or an unlimited fine.

EXERCISE OF POLICE POWERS, ETC BY CIVILIAN STAFF AND VOLUNTEERS

Police powers for civilian staff and volunteers

The Police Reform Act 2002 (PRA 2002), s 38(1) authorises a chief officer of po-   **2.4**
lice of any police force to designate a relevant employee as either or both of the
following:

(a) a community support officer (CSO);
(b) a policing support officer (PSO).

For these purposes, 'relevant employee' means:

(a) in the case of:
    (i) a police force maintained for an area listed in PA 1996, Sch 1, or
    (ii) the police force maintained for the Metropolitan Police District,
    a member of the civilian staff of that force;
(b) in the case of any other police force, a person who is:
    (i) employed by the police authority maintaining that force, and
    (ii) under the direction and control of the chief officer making the designation.

Section 38(1A) authorises a chief officer of police of any police force to designate a
police volunteer as either or both of the following:

(a) a community support volunteer (CSV);
(b) a policing support volunteer (PSV).

'Police volunteer' means a person who is under the direction and control of the
chief officer making such designation otherwise than because the person is a constable, special constable or a relevant employee. A person is treated as a relevant
employee only in relation to times when (s)he is acting in the course of his/her
employment.
    A person designated under s 38 has the powers and duties conferred on him/her by
the designation.

Powers and duties of persons designated under s 38

The powers and duties that may be conferred or imposed on a person designated under   **2.5**
s 38 are:

(a) any power or duty of a constable, *other than a power or duty specified in Sch 3B,
    Part 1, namely, any power or duty of a constable*:
    (i)   to make an arrest;
    (ii)   to stop and search someone or a vehicle or other thing;
    (iii) to perform a custody officer's functions at a designated police station if a
          custody officer is not readily available;
    (iv) to conduct an intimate search when authorised by an inspector under
          PACE, s 55;
    (v)   to perform a function only exercisable by a constable of a specified rank;
    (vi) under terrorism legislation (as defined at **21.1**);
    (vii) under the Official Secrets Acts 1911 and 1989; or
    (viii) to make an application for a targeted interception warrant or mutual assist-
          ance warrant on behalf of the Metropolitan Police Commissioner;

(b) where the person is designated as a *community support officer (CSO) or a community support volunteer (CSV)*, any additional power or duty described in Sch 3C as a power or duty of a community support officer or community support volunteer, namely:

(i) issue of fixed penalty notices under powers given to an 'authorised officer' by certain legislation relating to local authority matters (eg penalty notices for litter, graffiti or fly-posting);

(ii) requirement of name and address if the CSO or CSV has reason to believe that the person has committed a 'relevant offence' (eg a fixed penalty offence in respect of which (s)he may give a penalty notice, an offence under the Vagrancy Act 1824, s 3 or 4, an offence which appears to have caused injury, alarm, or distress to another, or any loss of or damage to another's property, or certain other offences) or a 'relevant licensing offence'. It is a summary offence for a person to fail to comply with such a requirement (PRA 2002, s 38(6B)(b) and Sch 3C, para 3(2));

(iii) search for and seize alcohol or tobacco from a young person. It is a summary offence to fail without reasonable excuse to consent to such a search (s 38(6B)(b) and Sch 3C, para 4(4));

(iv) seize and retain a controlled drug or a psychoactive substance found in someone's possession and that possession is reasonably believed to be unlawful. If a controlled drug is so found and that possession is reasonably believed to be unlawful, the CSO or CSV may require that person's name and address. It is a summary offence to fail to comply with such a requirement (s 38(6B)(b) and Sch 3C, paras 5(5) (CD) and 6(6) (PS));

(v) where the CSO or CSV has required a person (P) to give his/her name and address under a power to do so and there is non-compliance or the CSO or CSV has reasonable grounds for suspecting that the person has given a false or inaccurate name or address, the CSO or CSV may require P to remain with him/her for a period, not exceeding 30 minutes, for the arrival of a constable. (S)he may also make such a requirement where a person fails to comply with a requirement to stop doing something constituting an offence under the Vagrancy Act 1824, s 3 or 4. When making such a requirement, the officer may enforce it by a gesture. The alleged offender may elect to accompany the CSO to a police station as an alternative to waiting. Such a person, on arrival at a police station, is under a duty to remain there until his/her custody is transferred to a constable. Failure to wait, or making off en route to a police station, is a summary offence (s 38(6B)(b) and Sch 3C, para 7(7));

(vi) search a person asked to wait under (b)(v) if the CSO or CSV reasonably believes that the person may endanger him/herself or has anything concealed which might assist his/her escape;

(vii) use reasonable force to prevent an individual subject to a requirement to wait under (b)(v) making off, or to conduct a search under (b)(vi);

(viii) exercise of powers under the Removal and Disposal of Vehicles Regulations 1986 (abandoned vehicles).

The maximum punishment for the offences at (ii), (iii), (iv), and (v) is a level 3 fine.

Police powers for civilian employees, etc under collaboration agreements

By s 38B, the chief officer of police of a police force (the 'assisted force') may designate **2.6**
a person (C) who is:

(a) a civilian employee of, or a police volunteer with, another police force (the 'assisting force'),

(b) designated under s 38 by the chief officer of police of the assisting police force (the 's 38 designation'), and

(c) permitted, under relevant police collaboration provision, to discharge powers and duties specified in that provision for the purposes of the assisted force.

Such a designation (the 'collaboration designation') must designate C as a CSO or PSO or both (if C is a civilian employee of the assisting force) or as a CSV or PSV or both (if C is a police volunteer with the assisting force). The collaboration designation may designate C a CSO or PSO, or as a PSO or PSV, only if the s 38 designation designates C as an officer of that description.

C will have the powers and duties conferred or imposed on C by the collaboration designation. However, a power or duty may be conferred or imposed on C by the collaboration designation only if C is permitted, under the relevant police collaboration provision, to discharge that power or duty for the purposes of the assisted force.

The collaboration designation must specify the restrictions and conditions to which C is subject in the discharge of the powers and duties conferred or imposed by the collaboration designation. Those restrictions and conditions must include the restrictions and conditions specified in the relevant police collaboration provision.

POLICE POWERS FOR CONTRACTED-OUT STAFF

PRA 2002, s 39 deals with the situation where a local policing body (police and crime **2.7**
commissioner or the like) has contracted with a person ('the contractor') for the provision of services relating to the detention or escort of persons who have been arrested or are otherwise in custody.

Under s 39, the relevant chief officer of police may designate any employee of the contractor as a detention officer, or as an escort officer, or as both. A person so designated has the powers under Sch 4, Parts 3 and 4 respectively conferred or imposed on him/her by the designation.

A chief officer of police must not so designate a person unless satisfied that the contractor is a fit and proper person to supervise the carrying out of the functions for the purposes of which that person is designated.

Powers of persons designated under s 39

Detention officer Where a designation applies a power or duty of a constable listed in **2.8**
Sch 4, Part 3 below to any person, it may authorise a detention officer: to require a person to attend a police station for the taking of a sample or fingerprints (and to take those fingerprints without consent); to photograph a detained person; to carry out intimate or non-intimate searches; to search a person answering live link bail; to carry out a search and examination to establish identity (including the taking of a photograph of an identifying mark); to take impressions of footwear; and to carry out a search of someone attending a police station to answer live link bail, and to seize articles found.

A designation may authorise a detention officer to keep under control detainees at a police station (or to assist others in doing so) and to secure that persons in detention do not escape; reasonable force may be used for these purposes. A designation may authorise a detention officer to carry out specified duties to give information in relation to intimate searches, X-rays, and ultrasound scans carried out in respect of Class A drug searches.

Escort officer   Where a designation applies a power or duty of a constable listed in Sch 4, Part 4 below to a person, it may authorise him/her to take an arrested person to a police station or to escort a person in police detention. An escort officer designated to escort a person in police detention may also use reasonable force to keep the detainee under control. A designated escort officer remains in control until the detainee is handed over to the custody officer or other responsible person, and may use reasonable force to prevent that person escaping and to keep him/her under his/her control. A designated escort officer may carry out a non-intimate search of the detainee.

General   A detention officer or escort officer is not authorised to engage in designated conduct otherwise than in the course of his/her employment.

COMMUNITY SAFETY ACCREDITATION SCHEMES

2.9   PRA 2002, s 40 authorises a chief officer to establish and maintain a 'community safety accreditation scheme' for the exercise in his/her area, by persons accredited by him/her under s 41, of powers related to community safety and security and (in co-operation with the local police) combating crime and disorder, public nuisance, and other forms of anti-social behaviour. Any such scheme must contain provisions for arrangements to be made with employers who are carrying on a business in the police area for those employers to supervise the carrying out of such functions for the purposes of which accreditation powers are provided. Chief officers must ensure that the employer has made satisfactory arrangements for the handling of complaints relating to the exercise of those powers.

The powers which may be conferred on accredited persons are set out in PRA 2002, Sch 5, namely powers to give fixed penalty notices:

(a)   under the Criminal Justice and Police Act (CJPA) 2001 (**3.40**), except in respect of offences of drunkenness, theft, criminal damage, or leaving litter (contrary to EPA 1990, s 87);

(b)   in respect of offences relating to cycling on a footpath, failing to secure attendance of a pupil at school, being an excluded pupil in a public place, graffiti or fly-posting, and litter under EPA 1990, s 88.

Accredited persons may also be given powers to deal with alcohol consumption in a designated public place, confiscation of liquor or tobacco from those under 18, abandoned vehicles, stopping vehicles for testing, and controlling traffic for the purpose of escorting a load of exceptional dimensions.

Accredited persons may be authorised to require names and addresses in relation to certain traffic offences, to control traffic, and to photograph persons given fixed penalty notices.

ACCREDITATION OF WEIGHTS AND MEASURES INSPECTORS

2.10   PRA 2002, s 41A provides that weights and measures inspectors (trading standards officers) may be accredited ('an accredited officer') by a chief officer of police. An accredited officer may be accredited to issue fixed penalty notices under CJPA 2001, to require the name and address of someone reasonably believed to have committed a fixed penalty offence for the purposes of CJPA 2001 and/or to photograph someone to whom (s)he has given such a notice. The powers may only be exercised by such an accredited officer in the exercise of his/her duties as a trading standards officer.

GENERAL ABOUT DESIGNATED OR ACCREDITED PERSONS

2.11 A designated person or person accredited under PRA 2002, s 41 or s 41A must produce evidence of his/her designation or accreditation if requested to do so when exercising one of his/her powers. In addition, such a person is restricted in the exercise of powers to times when (s)he is wearing approved uniform or badge. However, a police inspector (or above) may direct a particular policing support officer or policing support volunteer not to wear uniform for the purposes of a particular operation.

Section 46(1) provides that it is an offence to assault a designated person, an accredited person, an accredited inspector, or a person assisting a designated or accredited person or an accredited inspector, in each case in the execution of his/her duty. By s 46(2), it is an offence to resist or wilfully obstruct such a person.

Under s 46(3) a person commits an offence if, with intent to deceive, he—

(a) impersonates a designated person, an accredited person or an accredited inspector;
(b) makes any statement or does any act calculated falsely to suggest that (s)he is such a person; or
(c) makes any statement or does any act calculated falsely to suggest that (s)he has powers of such a person that exceed his/her actual powers.

The offences under s 46 are summary offences punishable with up to six months' imprisonment and/or an unlimited fine (s 46(1) and (3)), or up to one month's imprisonment and/or a fine nor exceeding level 3 (s 46(2)).

These auxiliary personnel must have regard, in the discharge of their duties, to the provisions of the PACE Codes of Practice. A person lawfully in the custody of such a person must be treated as in police detention.

Where any power exercised by one of the above types of person includes power to use reasonable force to enter premises, that power may only be exercised in the company, and under the supervision, of a constable, or to save life or limb or to prevent serious damage.

NATIONAL CRIME AGENCY OFFICERS

2.12 The National Crime Agency (NCA), established by the Courts and Crime Act 2013 (CCA 2013) is not a police force but its officers are referred to for convenience. The NCA has:

(a) the 'crime-reduction function' of securing that efficient and effective activities to combat organised crime and serious crime are carried out (whether by the NCA, other law enforcement agencies, or other persons);
(b) the 'criminal intelligence function' of gathering, storing, processing, analysing, and disseminating information that is relevant to activities to combat organised crime or serious crime, to activities to combat any other kind of crime, or to exploitation proceeds investigations, exploitation proceeds orders (and applications for such orders);

(c) the functions relating to the recovery of assets conferred by the Proceeds of Crime Act 2002; and

(d) the other functions conferred by CCA 2013 (eg assistance to a foreign government) and by other enactments.

The crime-reduction function does not include the function of the NCA itself prosecuting offences.

Apart from its Director General, the NCA's officers are persons appointed by the Director General as NCA officers, persons seconded to the NCA to serve as NCA officers, and NCA specials (ie persons appointed by the Director General as NCA officers on a part-time basis).

If a person with operational powers, eg a constable (other than a special constable), Revenue and Customs officer, or immigration officer, becomes an NCA officer, his/her existing powers are suspended during the time when (s)he is seconded as an NCA officer. This does not apply to a member of the British Transport Police, Civil Nuclear Constabulary, or Ministry of Defence Police who is an NCA officer by virtue of a secondment.

The Director General may designate any other NCA officer as a person having one or more of the following:

(a) the powers and privileges of a constable;
(b) the powers of an officer of Revenue and Customs;
(c) the powers of a general customs official;
(d) the powers of an immigration officer.

The Director General may not so designate an NCA officer as having particular operational powers unless satisfied that (s)he is capable of, and adequately trained for, exercising those powers, and is otherwise suitable.

The following summary offences are provided by CCA 2013, s 10 and Sch 5. It is an offence to resist or wilfully obstruct a designated officer acting in the exercise of an operational power, or a person assisting him/her in the exercise of such a power (CCA 2013, Sch 5, para 21), or to assault such a designated officer or such a person (Sch 5, para 22).

A person commits an offence if, with intent to deceive, (s)he:

(a) impersonates a designated officer;
(b) makes any statement or does any act calculated falsely to suggest that (s)he is a designated officer; or
(c) makes any statement or does any act calculated falsely to suggest that (s)he has powers as a designated officer that exceed the powers (s)he actually has (Sch 5, para 23).

These offences are punishable with up to six months' imprisonment and/or a fine as follows: a level 3 fine (Sch 5, para 21) or an unlimited fine (Sch 5 para 22 or 23).

HANDLING OF COMPLAINTS AND CONDUCT MATTERS, ETC RELATING TO PERSONS SERVING WITH THE POLICE, AND POLICE DISCIPLINARY OR PERFORMANCE PROCEEDINGS

2.13   These matters are dealt with in Chapter 2 on the companion website.

CHAPTER 3

Elements of Criminal Procedure

INSTITUTION OF CRIMINAL PROCEEDINGS

Responsibility for prosecutions

Prosecutions (criminal proceedings) are normally instituted on behalf of: **3.1**

(a) a police force;
(b) the Director of Public Prosecutions (DPP); or
(c) a body with statutory powers to do so in specific matters, such as the Serious Fraud Office, a Secretary of State, or a local authority.

The DPP is the head of the Crown Prosecution Service (CPS). The DPP is under the general supervision of the Attorney General. The DPP's functions are discharged on his/her behalf and under his/her direction by Crown Prosecutors working in the CPS.

Where prosecutions are instituted on behalf of a police force (whether by a police officer or other person) the DPP is obliged to take over their conduct.

By way of exception, and with the qualifications below, the DPP is not obliged to take over a prosecution instituted on behalf of the police for a range of minor offences such as the following:

fixed penalty road traffic offences (eg speeding); failing to comply with a traffic direction; failing to stop, report an accident, or give information or documents; using or keeping vehicle without a vehicle licence; careless or inconsiderate driving; criminal damage, not including arson and only if the value involved is no more than £5,000; low-value shop-lifting (goods under £200 in value); being drunk in a highway, other public place, or licensed premises; throwing fireworks in a thoroughfare; trespassing or throwing stones on the railway; disorderly behaviour while drunk in a public place; behaviour likely to cause harassment, alarm or distress; depositing and leaving litter; contravention of a prohibition or failure to comply with a requirement imposed by or under fireworks regulations or making false statements; knowingly giving a false alarm of fire; failing to give a sample for the purposes of testing for the presence of Class A drugs; and failing to attend an assessment following testing for the presence of Class A drugs. With one exception, all these offences are summary offences. The exception is criminal damage, which is an indictable (either-way) offence.

The rule that the DPP is not obliged to take over a prosecution instituted on behalf of the police for any of the above offences does not apply if the proceedings are commenced by the defendant (D) being charged by a custody officer at a police station, or if D is under 16 when the proceedings are commenced. In addition, the rule ceases to apply if:

(a) a case summary or copy of the witness statement(s) upon which the prosecution intends to rely in the event that the court proceeds under the 'pleading guilty by post' procedure (**3.27**: which is not available in respect of criminal damage) is not served with the requisition or summons; or

(b) a case summary or copy of the witness statement(s) upon which the prosecution intends to rely if D pleads guilty is not served with a requisition or summons in respect of criminal damage; or

(c) at any time a magistrates' court begins to receive evidence, other than evidence in proceedings held in D's absence or under the 'pleading guilty by post' procedure, or evidence adduced in an attempt to avoid a 'totting up' disqualification; or

(d) a magistrates' court indicates that it is considering imposing a custodial sentence.

The DPP is not under a duty to take over a prosecution dealt with in accordance with the single justice procedure described at **3.28**.

There is generally nothing to prevent a private individual or an organisation which is not governmental or public from instituting and conducting criminal proceedings, and such 'private' prosecutions are occasionally instituted. The DPP may take over a private prosecution even if his/her purpose is to offer no evidence against D in the public interest and thereby to abort those proceedings.

A qualification to the general rule that the police, a private individual or an official of a public or private body may institute criminal proceedings is that there are offences where there is a requirement that proceedings for an offence may only be instituted by or with the consent of the Attorney General or of the DPP. In the latter case, the DPP's functions can, of course, be exercised by a Crown Prosecutor on the DPP's behalf. Such a requirement does not prevent D's arrest or remand in custody or on bail but if the requisite consent has not been given before the hearing at which D indicates his/her plea or intended plea the subsequent proceedings will be null and void.

Alternatives to prosecutions

3.2   Traditionally, the criminal process has involved prosecution, trial (if D pleaded not guilty) and sentence. Other methods are increasingly being used to deal quickly and proportionately with low-level, often first-time, offending.

Where the offender admits the offence the following out-of-court disposals are available in an appropriate case:

(a) community resolution, which involves a minor offence being resolved, in lieu of more formal action, by an agreement between the parties instigated by the police; it may, for example, involve the offender offering an apology, making reparation (such as fixing damage) or paying compensation;

(b) a cannabis or khat warning in respect of simple possession;

(c) simple cautions (**6.83**);

(d) youth cautions (**6.84**);

(e) conditional cautions (**6.86**); and

(f) youth conditional cautions (**6.87**).

In addition, whether or not the offender admits his/her guilt, many road traffic offences can be dealt with by the issue (and satisfaction) of a fixed penalty notice, and so may over a score or so of offences involving 'disorderly behaviour' (a penalty notice for disorder: PND). The former are dealt with in Chapter 25, and the latter at the end of this chapter (**3.39–3.44**). (There are also other offences in respect of which penalty notices may be issued under other legislation and by officials not associated with the police.)

Ways of instituting criminal proceedings

Criminal proceedings may be instituted in one of four ways: **3.3**

(1) a 'relevant prosecutor' may institute criminal proceedings against a person (D) by issuing a document (a written charge) which charges D with an offence. Where a relevant prosecutor so institutes proceedings, it must at the same time issue:
   (a) a requisition securing D's presence before a magistrates' court (if necessary by an arrest under warrant); or
   (b) a single justice procedure notice;
(2) a person may institute criminal proceedings by laying an oral or written information alleging that D has, or is suspected of having, committed an offence. That person must then secure D's presence before a magistrates' court by a summons or by an arrest under a warrant. This method of instituting criminal proceedings by laying an information for the purpose of obtaining a summons is no longer available to a relevant prosecutor authorised to issue a requisition;
(3) by an arrest under a warrant, followed by a charge; or
(4) by an arrest without a warrant, followed by a charge.

Relevant prosecutors   **3.4**

A 'relevant prosecutor' is:

(a) a police force;
(b) the DPP;
(c) the Attorney General;
(d) the Director of the Serious Fraud Office;
(e) the Commissioners for Revenue and Customs, or of Inland Revenue;
(f) the Director General of the National Crime Agency;
(g) a Secretary of State;
(h) any person specified in an order made by the Secretary of State, eg a local authority (other than a parish council or its equivalent), the Environment Agency, the National Resources Body for Wales, and Transport for London;

or, in each case above, a person authorised by such a body or person to institute criminal proceedings.
   With the exception of Transport for London, those specified by the Secretary of State are only authorised to issue written charges and single justice procedure notices. If they wish to proceed otherwise than by such a notice, they must proceed by going down the laying of information route referred to at **3.3(2)**.

Written charge or information   **3.5**

A written charge or information must contain:

(a) a statement of the offence which describes the offence 'in ordinary language' and (if the offence is created by statute) identifies the legislation that creates it; and
(b) sufficient particulars of the conduct constituting the commission of the offence to make clear what the prosecutor alleges against D.

Where a number of incidents, taken together, amount to a course of conduct (having regard to the time, place, or purpose of commission), those incidents may be included in the allegation. Moreover, a single document may contain more than one charge.

By the Magistrates' Courts Act 1980 (MCA 1980), s 127, a written charge or information relating to an offence triable only summarily cannot generally be tried unless it has been issued or laid, respectively, within six months from the time when the offence was committed. A written charge can be regarded as issued when a document comprising it is completed, with all relevant details and in the form needed for service. Provided this is done within the time limit, the written charge will have been issued in time, even though served later.

An example of an exception to the general rule under s 127 is provided by the Vehicle Excise and Registration Act 1994, which permits a written charge to be issued, or an information to be laid, in respect of offences of using or keeping a vehicle without a vehicle licence up to three years after the commission of the offence.

Requisition, single justice procedure notice, or summons

3.6   A requisition requires the person on whom it is served to appear before a magistrates' court to answer the written charge.

A single justice procedure notice requires the person on whom it is served to serve on the designated officer for a magistrates' court specified in the notice a written notification stating:

(a) whether (s)he desires to plead guilty or not guilty; and
(b) if (s)he desires to plead guilty, whether or not (s)he desires to be tried in accordance with the 'trial by single justice on the papers' provisions below.

Where a relevant prosecutor issues a written charge and a requisition, both documents must be served on the person concerned, and a copy of both must be served on the appropriate court officer. A corresponding rule applies where a relevant prosecutor issues a written charge and a single justice procedure notice. If a single justice procedure notice is served on a person, the relevant prosecutor must:

(a) at the same time serve on him/her such documents as prescribed by Criminal Procedure Rules; and
(b) serve copies of those documents on the designated officer specified in the notice.

A summons is a written order issued by a justice of the peace or by authorised court staff on behalf of a justice. It is directed to the person named in an information and requires him/her to appear before a magistrates' court to answer the charge against him/her.

A requisition or summons must contain a notice setting out when and where D must attend the court, and must specify each offence in respect of which it has been issued. Additionally, a summons must identify the issuing court, and a requisition must identify the person under whose authority it is issued.

Other types of summonses

3.7   These are mentioned here for convenience.

Summonses for a breach of the peace

3.8   A summons may also be issued as a result of a complaint. A complaint is a written or verbal allegation made before a justice in respect of anything which is within the civil jurisdiction of a magistrates' court to make an order. Of particular importance to a police officer is the fact that a complaint may be made to the effect that a person has committed a breach of the peace. Breach of the peace is not in itself a criminal offence. If the complaint is proved, that person may be bound over to keep the peace. A magistrates' court has jurisdiction to hear any complaint.

Summonses in respect of a witness

Where a justice of the peace or an authorised member of court staff is satisfied that **3.9** any person in England or Wales, or outside the British Isles, is likely to be able to give material evidence, or to produce any document or thing likely to be material evidence, at the summary trial of a written charge or information or the hearing of a complaint by a magistrates' court, and it is in the interests of justice to issue a summons to secure the attendance of that person to give evidence or produce the document or thing, the justice may issue a summons requiring that person to attend before the court to give evidence or to produce the thing or document. Unlike other summonses, such a summons may not be served by post. A justice may refuse to issue a witness summons if (s)he is not satisfied that the application was made as soon as reasonably practicable after D pleaded not guilty. Crown Courts are also empowered to issue witness summonses in prescribed circumstances.

Service of requisitions, single justice procedure notices, or summonses

A requisition, single justice procedure notice or summons may be served on an individual by: **3.10**

(a) handing it to him/her (except that if (s)he is under 18, it must be handed to a parent, guardian, or other appropriate adult if such a person is readily available);
(b) leaving it at the appropriate address for service (ie an address in England and Wales where it is reasonably believed that (s)he will receive it); or
(c) sending it to that address by first class post or by the equivalent of first class post;

and may be served on a corporation by:

(i) handing it to a person holding a senior position in that corporation;
(ii) leaving it at the appropriate address for service (ie the corporation's principal office in England and Wales, and if there is no readily identifiable principal office then any place in England and Wales where it carries on its activities or business); or
(iii) sending it to that address by first class post or the equivalent of first class post.

In addition, service of a requisition, single justice procedure notice, or summons can be effected by the use of document exchange (DX) or by electronic means.

Service by document exchange

Where the person to be served has given a DX box number, and has not refused to accept service by DX, a requisition, single justice procedure notice, or summons may be served by: **3.11**

(a) addressing it to that person at that DX box number
(b) leaving it at the DX at which the addressee has that DX box number, or a DX at which the person serving it has a DX box number.

Service by electronic means

3.12   Where the person to be served has given an electronic address and has not refused to accept service at that address, or is given access to an electronic address at which a document may be deposited and has not refused to accept service by the deposit of a document at that address, a requisition, single justice procedure notice, or summons may be served by:

(a) sending it by electronic means to that electronic address; or

(b) depositing it at an electronic address to which the recipient has been given access and:

  (i)   in every case, making it possible for the recipient to read the document, or view or listen to its content, as the case may be;

  (ii) unless the court otherwise directs, making it possible for the recipient to make and keep an electronic copy of the document; and

  (iii) notifying the recipient of the deposit of the document (which notice may be given by electronic means).

Where a document is so served, the person serving it need not provide a paper copy as well.

Time of service

3.13   A document served by handing it over is served on the day it is handed over. Unless something different is shown, a document served on a person by any other method is served:

(a) in the case of a document left at an address, on the next business day after the day on which it was left;

(b) in the case of a document sent by first class post or by the equivalent of first class post, on the second business day after the day on which it was posted or dispatched;

(c) in the case of a document served by DX, on the second business day after the day on which it was left at a DX as permitted under the above rule;

(d) in the case of a document served by electronic means:

  (i)   on the day on which it is sent by sending it by electronic means to the address which the recipient has given, if that day is a business day and if it is sent by no later than 14.30 on that day;

  (ii) on the day on which notice of its deposit is given, if that day is a business day and if that notice is given by no later than 14.30 that day; or

  (iii) otherwise, on the next business day after it was sent or such notice was given; and

(e) in any case, on the day on which the addressee responds to it, if that is earlier.

Unless something different is shown, a document produced by a computer system for dispatch by post is to be taken as having been sent by first class post, or by the equivalent of first class post, to the addressee on the business day after the day on which it was produced.

General

3.14   Proof of service is usually by means of a certificate of service signed by the person who effected the service, explaining how and when service was made.

A requisition, single justice procedure notice, or summons may be served, in accordance with rules of court, in Scotland or Northern Ireland. The same is true vice versa in respect of a Northern Irish summons. A Scottish citation may be served in England and Wales in the same way as may be done in Scotland.

Special provision is made as to service on a person outside the UK by the Crime (International Co-operation) Act 2003 and the Criminal Procedure Rules 2020, and as to service on an overseas company by the Companies Act 2006, s 1139.

Warrants of arrest

3.15 A justice of the peace may issue a warrant to arrest when a written information is laid before him/her to the effect that a person has, or is suspected of having, committed an offence.

No such warrant may be issued for the arrest of an adult unless:

(a) the offence to which the warrant relates is an indictable offence or is punishable with imprisonment, or
(b) his address is not sufficiently established for a summons, or a written charge and requisition, to be served on him.

Where the offence charged is an indictable offence, a warrant may be issued at any time notwithstanding that a written charge and requisition have, or a summons has, previously been issued.

An arrest warrant directs its addressee: (a) to arrest the accused; and (b) bring him/her before the magistrates' court named in it. However, by way of exception in respect of (b), an arrest warrant may be endorsed with a direction that on arrest the person to be arrested shall be released on bail, with or without sureties. This is known as 'backing for bail'.

An arrest warrant remains in force until executed or withdrawn. If the original is lost a justice may issue a replacement. As to charging a person arrested under a warrant, see Chapter 6.

Other warrants

Warrant to arrest a witness

3.16 Such a warrant may be issued by a justice, who is satisfied that a person who could give material evidence, etc is unlikely to attend court voluntarily. Such a warrant will only be issued where a summons would be ineffective.

Warrants of control or commitment

3.17 Under MCA 1980, s 76, a magistrates' court may issue a warrant of control or of commitment where a person is in default of payment of a sum adjudged to be paid by a conviction or order of a magistrates' court.

A warrant of control empowers an authorised civil enforcement officer or approved enforcement agency to take control of goods (other than 'exempt goods') and sell them to recover that sum. 'Exempt goods' include the defaulter's clothing, bedding, furniture, household equipment and provisions, and (up to an aggregate value of £1,350) items necessary for use by the defaulter for his/her trade, employment or education. Under MCA 1980, s 78(4), it is a summary offence, punishable with a fine not exceeding level 1 on the standard scale, for any person to remove the goods, or the marks placed on them to signify seizure.

A warrant of commitment orders the defaulter to be arrested and committed to prison. It may be issued *either* where it appears on the return to a warrant of control that the defaulter's assets are insufficient to pay the amount outstanding *or* instead of a warrant of control. If the magistrates' court considers this expedient, it may fix a term of imprisonment and suspend the issue of the warrant until such time and on such conditions as it thinks fit. A receipt must be obtained for the defaulter when (s)he is handed over to a prison pursuant to a warrant of commitment.

Search warrant

3.18   This type of warrant is described in Chapter 4.

Execution of warrants

3.19   A warrant of arrest, or of control, or of commitment, or a search warrant, may be executed anywhere in England and Wales by any person to whom it is directed, or by any constable acting in his/her police area, or by a civilian enforcement officer or approved enforcement agency.

In addition, by the Criminal Justice and Public Order Act 1994, s 136 a warrant issued in England, Wales, or Northern Ireland for the arrest or commitment of a person (or a like warrant issued in Scotland) may, without endorsement, be executed in one of the other home countries by a constable of a police force of the country of issue or of the country of execution or by a British Transport Police constable, as well as by any other persons within the directions of the warrant. Where a constable executes a warrant issued in Scotland in any other home country, any rule of law which concerns:

(a) the powers and duties of that constable;
(b) the rights of the person arrested; and
(c) the procedures to be followed after the arrest,

applies in relation to the arrest (subject to certain modifications set out in s 137ZA) as though the warrant had been executed in Scotland and, if the constable who executed it is not a constable of the Police Service of Scotland, as though the constable were.

A search warrant may not be executed by a person entitled to execute it who does not have it in his/her possession at the time.

The following types of warrant not issued by a justice may be executed by a constable, although (s)he is not in possession of the warrant:

(a) warrant to arrest a person in connection with an offence;
(b) warrant under the armed forces legislation (desertion, etc);
(c) warrant relating to the non-appearance of a defendant, warrants of control, warrants of commitment, and warrants issued to arrest a potentially unwilling witness;
(d) warrant of arrest under the Family Law Act 1996, s 47(8) (failure to comply with occupation order or non-molestation order); and
(e) warrant of arrest under the Sentencing Code, Sch 4, para 3 (offender referred to court by youth offender panel).

However, a warrant referred to in (a)–(e) must, on the demand of the person concerned, be shown to him/her as soon as practicable.

Arrest warrants issued by judicial authorities in EU countries or Norway or Iceland may be executed in the UK.

Entry to execute warrants

The Police and Criminal Evidence Act 1984 (PACE), s 17 (**4.63**) permits a constable to   **3.20**
enter and search premises, including any vehicle, vessel, aircraft or tent, in specified circumstances for the purpose of executing a warrant of arrest issued in connection with
or arising out of criminal proceedings, or a warrant of commitment.

Arrest without a warrant, followed by charge

This is dealt with later in Chapters 4 and 6.   **3.21**

CLASSIFICATION OF OFFENCES BY METHOD OF TRIAL

Offences can be classified as indictable offences or summary offences, according to   **3.22**
their mode of trial.

'Indictable offence' means an offence which, if committed by an adult, is triable in
the Crown Court, whether it is exclusively so triable (eg murder or robbery) or triable
either way, ie in the Crown Court or a magistrates' court (eg unlawful wounding or
fraud).

'Summary offence' means an offence which, if committed by an adult, is triable only
in a magistrates' court (eg most driving offences).

The vast majority of offences are summary ones.

COURTS OF CRIMINAL JURISDICTION

There are two methods of trying persons accused of criminal offences. One is by judge   **3.23**
and (almost invariably) jury in the Crown Court; the other is summarily by a magistrates' court without a jury. With a few exceptions, all criminal proceedings in the
Crown Court begin in a magistrates' court since a defendant (D) tried in the Crown
Court must normally have been sent for trial there by a magistrates' court.

MAGISTRATES' COURTS

A magistrates' court is normally composed of two or three lay justices of the peace.   **3.24**
Some functions can be discharged by a single justice; a notable example is the trial by a
single justice on the papers procedure (**3.28**).

A District Judge (Magistrates' Courts), a legally qualified, salaried magistrate, may sit
alone and has all the powers of two lay justices.

Some matters authorised to be done by, to or before a single justice may be done by
authorised court staff. A magistrates' court has jurisdiction in relation to a range of civil
matters, some of which are described later in this book.

In criminal matters a magistrates' court has jurisdiction in relation to the following
matters.

Offences triable summarily only

A magistrates' court has jurisdiction to try any summary offence alleged to have been   **3.25**
committed by a person who appears or is brought before the court.

A person appearing before a magistrates' court does so either in answer to a requisition or summons, or under arrest.

Guilty plea

3.26 A person (D) who appears before the court will have the charge read over to him/her and will be asked if (s)he pleads guilty or not guilty. If D pleads guilty the court must be satisfied that it is a clear and unequivocal plea. On a plea of guilty being entered, the court may convict without hearing evidence. In practice, however, the facts of the case are outlined by the prosecution, and D or his/her legal representative is at liberty to dispute those facts should (s)he wish to do so and may also put before the court any mitigating facts which (s)he feels that the court should consider before passing sentence. Usually the court will also hear any evidence by the prosecutor of any previous recorded convictions of D and of his/her general character in order to enable it to decide the appropriate penalty. D may also ask the court to take into consideration, when passing sentence, other offences which D has committed.

Pleading guilty by post

3.27 The Magistrates' Courts Act 1980 (MCA 1980), s 12 permits a plea of guilty of a summary offence to be entered at a magistrates' court (or where D is 16 or 17 when the summons or requisition is issued, at a youth court) without the necessity for D to attend the proceedings, or for any witnesses to be called. There are certain conditions which must be met in respect of the offence:

(a) the proceedings must be by way of requisition or summons;

(b) D, when served with the requisition or summons for the offence, must also have been served with:

    (i) a notice explaining the prescribed procedure;

    (ii) *either* a concise statement of such facts relating to the charge as will be placed before the court if D pleads guilty without appearing before the court, *or* a copy of such signed written statement(s) as to truth complying with the Criminal Justice Act 1967, s 9 (**8.43**: proof by written statement) as will be so placed in those circumstances; and

    (iii) if any information relating to D (ie information relevant to sentence) will, or may in the circumstances, be placed before the court by or on behalf of the prosecutor, a notice containing or describing the information; and

(c) D must have notified the designated officer that (s)he wishes to plead guilty.

Where D does not appear and proof of service of the documents referred to in (b) is given, the court may hear and dispose of the case in the absence of D, the prosecutor or both. The documents are read to a court before a conviction is registered.

Should D appear even though the notification referred to in (c) has been received, the court may, with his/her consent, proceed as just described but D may make oral submissions with a view to mitigation of sentence, in place of any written submission which (s)he may have sent.

If the court thinks it inappropriate to accept D's plea, it must adjourn the case and it proceeds as if no notification of a guilty plea had ever been given.

Trial by single justice on the papers

3.28 Under MCA 1980, s 16A, a magistrates' court which may consist of a single justice may try a written charge on the papers if the offence charged is a summary offence *not punishable with imprisonment*, and the defendant (D) was at least 18 when charged. In such a case, the prosecution may serve on D a written charge and requisition and a single justice procedure notice, a summary of the prosecution's evidence, a copy of any written statements, and a notice containing or describing any information relevant to sentence. If that is done, the single justice procedure will apply unless D serves a written notice indicating a desire either to plead not guilty or not to be dealt with under the single justice procedure notice. If D fails to respond to the documents served, the single justice procedure can still be used even though D has not admitted the offence.

The court may try the charge in the absence of the parties. It may not hear any oral evidence and proceeds on the basis of the documents specified above and any written submission that D makes with a view to mitigation of sentence.

If a magistrates' court decides that the s 16A procedure is inappropriate in D's case, or it proposes to order a disqualification for driving and D (having been informed of the opportunity to make representations) expresses a wish to do so, the court must adjourn the case and issue a summons.

Not guilty plea

If D pleads not guilty the court must hear the evidence and, at the request of either **3.29** party, the court will order all witnesses out of court so that evidence may be presented independently. The proceedings will be opened by the prosecution outlining the facts of the case and calling its witnesses one by one to give evidence to the court. Each witness will either take the oath or affirm, and will give evidence during examination-in-chief. After the prosecution has asked questions to introduce and identify the witness, it may not ask leading questions in relation to the facts in issue. Leading questions are those which may usually be answered 'yes' or 'no'. For example, a prosecutor may not ask, 'Did you see the defendant standing outside the premises at 2 East Street?' but instead should ask, 'Where was the defendant when you saw him?'.

When a witness has given evidence for the prosecution (s)he may be *cross-examined* by the defence upon any aspect of the evidence which (s)he has given. Leading questions may be asked. Questions put in cross-examination (a) must be calculated to elicit answers which are directly relevant to an issue in the case, or (b) must be calculated to test the credibility of the evidence or to diminish its value. In respect of (a), cross-examination is not limited to matters arising during the examination-in-chief; it may relate to anything relevant to an issue in the case. Questions may be asked in respect of new facts unexpectedly introduced in the witness's testimony which do not form part of the cross-examining side's case and could not reasonably have been anticipated.

At the conclusion of the cross-examination the prosecution has a right to *re-examine* the witness upon *any new facts* which have come to light, or to clear up ambiguities which might have arisen, during cross-examination. Leading questions may not be put, nor may new evidence be introduced.

D or his/her legal representative may then address the court, whether or not (s)he calls witnesses. PACE 1984, s 79, which also applies to trials in the Crown Court, requires that, if the defence intends to call two or more witnesses to the facts of the case and those witnesses include D, D must be called before the other witness or witnesses unless the court in its discretion otherwise directs. Such a direction is likely where, for instance, a witness is to speak of an occurrence before the matters about which D is to give evidence.

When a witness for the defence, including D, has given his/her evidence, (s)he may be cross-examined by the prosecution and may then be re-examined by the defence on any new facts or ambiguities which have arisen during the cross-examination.

When the prosecution and defence have finished calling their evidence, each side has a right to address the court (except that the prosecution does not have the right if D has not called witnesses and is not legally represented), the prosecution's closing speech coming before that of the defence.

The court may adjourn to consider its verdict and may seek the advice of authorised court staff by specific request. If it decides to convict, its pronouncement of this verdict will usually be followed by evidence by the prosecutor of any previous recorded convictions of D and of D's general character in order to enable the court to decide on the appropriate penalty. A plea in mitigation may be made by D or his/her legal representative. D may also take the opportunity to ask the court to take into consideration, when passing sentence, other offences which D has committed.

Before beginning to try an offence, or at any time during such proceedings, a magistrates' court may adjourn the proceedings, and may remand D, either in custody or on bail.

Offences triable either way: where to be tried

3.30   An offence which is triable either way may be tried either on indictment in the Crown Court or summarily in a magistrates' court. Where it is tried depends on the following rules.

Indication of intended plea   Unless a notice has been given under the Crime and Disorder Act 1998 (CDA 1998), s 51B or s 51C (below), where someone aged 18 or over (D) appears or is brought before a magistrates' court charged with an either-way offence, the court must explain to D that D may indicate whether (if the offence were to proceed to trial) (s)he would plead guilty or not guilty; and explain that if (s)he indicates that (s)he would plead guilty:

(a) the court must proceed as if the proceedings constituted from the beginning the summary trial of the offence, and the court had asked whether (s)he pleaded guilty or not guilty; and

(b) (s)he may be committed for sentence to the Crown Court if the court is of the opinion that a Crown Court sentence should be available.

The court must then ask D whether (if the offence were to proceed to trial) (s)he would plead guilty or not guilty. If D indicates that (s)he would plead guilty the court must proceed as if the proceedings constituted from the beginning the summary trial of the offence and D had pleaded guilty. If the court is of the opinion that one of the specified grounds exists the court may commit D in custody or on bail to the Crown Court for sentence. If D indicates that (s)he would plead not guilty, or fails to indicate how (s)he would plead, the court must go on to the next stage.

Allocation of trial proceedings   The next stage is that the court, which may for this purpose, consist of a single justice, must decide in 'allocation of trial proceedings' whether the offence appears to be more suitable for summary trial or for trial on indictment. Before making a decision the court must give the prosecutor an opportunity to disclose any previous convictions and hear any representations from both parties about the mode of trial. The court must consider whether the sentencing powers of a magistrates' court would be adequate for the offence(s).

Where a summary trial appears to be more suitable the court must explain to D that summary trial appears to be more suitable and that D can either consent to be so tried or, if (s)he wishes, be tried on indictment and that, if tried summarily and convicted, (s)he may be committed to a Crown Court for sentence (3.33). At this stage, D may request an indication of whether a custodial or a non-custodial sentence would be more likely to be imposed if (s)he were to be tried summarily and plead guilty. The court may or may not give such an indication. If it does, it must ask D whether, on the basis of the indication, (s)he wishes to change his/her plea. If D indicates that (s)he would plead guilty the court must proceed to summary trial. Where the court does not give an indication as to sentence, or D does not indicate that (s)he would plead guilty, the court must ask D whether (s)he consents to summary trial or wishes to be tried on indictment. If D consents, the court must proceed to summary trial.

If D does not consent to summary trial, or the court decides that the offence appears to be more suitable for trial on indictment, or the prosecution requests that the offence be tried on indictment, the court must proceed to send D to the Crown Court for trial.

Children and young persons   The above provisions about indication of intended plea and allocation of proceedings do not apply in the case of a child or young person (ie someone under 18), but similar provisions apply in respect of an indication of intended plea. As to sending a juvenile to the Crown Court, see **3.31**.

Notice under CDA 1998, s 51B or 51C   If a notice is served on a magistrates' court under CDA 1998, s 51B or s 51C, the effect is that the court must forthwith send D, regardless of his/her age, to the Crown Court for trial; it cannot allocate the case for summary trial if the offence is an either-way one. By s 51B, in serious or complex fraud cases a notice may be given by a designated authority (the DPP, the Director of the Serious Fraud Office, or the Secretary of State) in respect of an indictable offence where the authority is satisfied that there is sufficient evidence for the person charged to be put on trial and that a case of fraud of such seriousness or complexity is revealed making it appropriate that the management of the case should, without delay, be taken over by the Crown Court. By s 51C, in sexual offences and some other offences involving child witnesses, the DPP may give such a notice if (s)he is of the opinion that the evidence is sufficient for the person to be put on trial for the offence; that a child would be called as a witness at the trial; and that, for the purpose of avoiding prejudice to the welfare of the child, the case should be taken over by the Crown Court and proceed without delay. For these purposes a child is a person who is under the age of 17, or a person of whom a video recording was made when (s)he was under the age of 17 with a view to its admission as evidence-in-chief in the trial.

Sending cases to Crown Court for trial

Where an *adult* (D) appears or is brought before a magistrates' court charged with an **3.31** offence which is triable on indictment only, or where (under the above provisions) an offence triable either way is not to be tried by a magistrates' court, or where a notice under CDA 1998, s 51B or s 51C has been given, the court must send D forthwith to the Crown Court for trial. It may at the same time send D for trial for any related either-way offence, and for a related summary offence if it satisfies the 'requisite conditions', namely is punishable with imprisonment or involves obligatory or discretionary disqualification from driving. Where another adult (A) is jointly charged with D with an either-way offence, and D is sent for trial, provision is made about sending A for trial for that offence and any related either-way or summary offence. Provision is also made for the corresponding case where A is a child or young person (ie under 18), substituting 'indictable' for 'either-way' in the previous sentence.

Generally, *children and young persons* (ie persons under 18) have to be tried summarily in a youth court (or sometimes a magistrates' court other than a youth court);

see **3.32**. However, in the following cases someone under 18 may be sent to the Crown Court for trial. The first was that mentioned in the previous paragraph. The second is where a child or young person appears or is brought before a magistrates' court charged with an offence for which any of certain conditions is satisfied:

(a) the offence is one of homicide;
(b) the offence is one of a number of particular types of offence of possessing a prohibited weapon or using someone to mind a weapon, and the accused young person was aged 16 or 17 when (s)he committed it;
(c) the offence is one in respect of which a long period of detention may be ordered and the court considers that such a sentence ought to be possible;
(d) a notice has been served under CDA 1998, s 51B or s 51C in respect of the child or young person; or
(e) the offence is a 'specified offence', namely one of a range of violent or sexual offences, and it appears to the magistrates' court that if the defendant child or young person is found guilty of the offence the criteria for an extended sentence would be met.

If one of conditions (a)–(e) is satisfied, the magistrates' court must send the child or young person for trial. Provision is made about sending him/her for trial for any related indictable or summary offence.

Youth courts

3.32   A youth court is a summary court (ie a magistrates' court) and is composed of justices who are specially appointed because of their qualifications. It sits for the purpose of hearing any charge against a child or young person or to exercise any other jurisdiction conferred on youth courts by the Courts Act 2003 or any other enactment.

The general public do not have a right of access to proceedings in a youth court.

A magistrates' court before which person under 18 (ie 'juvenile') appears charged with an offence which, in the case of an adult, is triable only on indictment or triable either way must deal with it *summarily* unless the provisions at **3.31** relating to a juvenile apply.

With certain exceptions, no charge against a juvenile may be heard summarily by a magistrates' court other than a youth court. The exceptions, which allow trial by an 'adult' magistrates' court, are where:

(a) the juvenile is charged jointly with an adult;
(b) an adult is charged with aiding, abetting, counselling, procuring, allowing, or permitting an offence with which a juvenile is charged;
(c) a juvenile is charged with aiding, etc an offence committed by an adult;
(d) the fact that the person is a juvenile is discovered in the course of proceedings in a magistrates' court; and
(e) the charge against a juvenile arises out of circumstances which are the same as, or are connected with, those which give rise to an offence by an adult (eg a theft by a youth and a handling by an adult).

The law recognises the specialist nature of youth courts by requiring a Crown Court which finds a juvenile guilty of an offence other than homicide to send the case to a youth court for sentencing as if that juvenile had been found guilty by that court, unless satisfied that it would be undesirable to do so. With specified exceptions, an adult magistrates' court which finds a juvenile guilty must likewise send him/her to a youth court for sentencing.

The sentencing powers of courts generally are limited in relation to children and young persons.

CROWN COURT

The Crown Court has jurisdiction over all offences which are triable only on indictment and also over offences triable either way in respect of which D has been sent for trial on indictment. Trials on indictment are heard with a jury who determine the issue of guilt after a direction on the relevant law by the judge. A trial can be without a jury where there is a risk of jury tampering.   **3.33**

A person who has been convicted by a magistrates' court may appeal to the Crown Court against sentence if (s)he pleaded guilty, or against conviction or sentence if (s)he pleaded not guilty. Appeals are heard by a court consisting of a judge and no more than four justices of the peace.

The Crown Court also deals with people committed to it by a magistrates' court for sentence for an offence (offence X). A magistrates' court can so commit a person (D) if (a) it has convicted D of offence X and considers that a sentence beyond its powers is merited, or (b) D, having indicated a guilty plea to offence X in allocation of trial proceedings, has been sent for trial for related offences. These cases are determined by a judge sitting alone.

The Central Criminal Court in London is a Crown Court.

Live links at preliminary hearings

The Crime and Disorder Act 1998 provides for participation in preliminary hearings in a magistrates' court or the Crown Court via a live-link. The relevant rules and their temporary substitution (until the end of 24 March 2022 unless that period is extended) by much broader provisions by the Coronavirus Act 2020 are set out in Chapter 3 of the companion website.   **3.34**

APPELLATE COURTS

Queen's Bench Division of the High Court

One function of the High Court is to hear appeals by either party on points of law from magistrates' courts or from the Crown Court in respect of appeals from magistrates' courts to the Crown Court. Such an appeal is known as appeal by case stated, since the magistrates' court or Crown Court is asked to 'state a case', that is set out its reasons for its finding on the basis of its interpretation of the law in relation to the facts found by it. The High Court is not concerned with any form of retrial.   **3.35**

Cases of the above type are dealt with by the Administrative Court, which is part of the Queen's Bench Division of the High Court. They are heard by a divisional court consisting of two (or sometimes three) judges.

Judicial review   Judges in the Administrative Court also deal with the *judicial review* of the decisions of a magistrates' court, the Crown Court in its appellate capacity, or a range of other bodies. Cases are heard by a single judge or a divisional court. In such cases the court will examine whether there was authority or power by which the decision made could have been properly reached; whether the procedure followed the rules of natural justice; or whether the exercise of discretion was lawful. Thus the court is not examining the correctness of the decision on its merits but whether it was lawfully and reasonably reached.

Court of Appeal (Criminal Division)

3.36   Appeals to this division are normally heard by a court consisting of three judges.

If convicted of an offence on indictment before the Crown Court, a person may appeal to this court against *conviction* with the leave of the Court of Appeal or the trial judge. On an appeal against conviction, the Court of Appeal can dismiss the appeal, allow it and quash the conviction, or substitute a conviction for another offence if it appears that D should have been convicted of that offence, rather than the one of which D was actually convicted. If fresh evidence has come to light, the court may order a new trial. It may also do so in any other case where the interests of justice so require.

A person convicted on indictment before the Crown Court may appeal against *sentence* with the leave of the Court of Appeal, and so may a person who has been sentenced by the Crown Court on committal for sentence. On *appeal* against sentence, the Court of Appeal may vary a sentence but cannot increase it. However, the Attorney General may (in certain cases) refer a sentence to the Court of Appeal with its leave if it appears to him/her to be unduly lenient. On such a reference, the court may impose a more severe sentence.

CJA 2003, Part 9 permits a prosecutor to appeal to the Court of Appeal in relation to any ruling made by the judge within a trial on indictment, other than a ruling that a jury be discharged. Following notice, proceedings may be adjourned and the judge's ruling has no effect within the period of adjournment.

Supreme Court

3.37   The Supreme Court took over the House of Lords' judicial functions in 2009.

The Supreme Court is the highest court in the land and hears criminal appeals by either party from the Court of Appeal or the Queen's Bench Division, but only where:

(a) the court has certified that a question of law of general public importance is involved; and

(b) the court or the Supreme Court is satisfied that the point of law is one which ought to be considered by the Supreme Court; and

(c) leave to appeal to the Supreme Court has been given by the court or the Supreme Court itself.

IMPRISONMENT AND FINES: MAXIMUM SENTENCES

3.38   Where an offence triable in the Crown Court, whether because it is indictable only or triable either way, is referred to in this book, reference is made to the maximum term of imprisonment which can be imposed by the Crown Court on conviction before it. That term also applies to someone committed for sentence by a magistrates' court. In some, relatively few, offences, special provision is made for a minimum prison sentence in certain circumstances or for specified circumstances aggravating the prison term imposed. Except in the case of murder and certain other cases, the Crown Court may fine the offender instead of, or in addition to, imprisoning him/her or imposing some other sentence. There is no limit on the fines which the Crown Court may impose.

Where an offence triable summarily only is referred to in this book, the maximum sentence is set out. Many summary offences are not imprisonable. Where they are, the maximum term varies from offence to offence but the maximum for any one offence is currently six months' imprisonment. The maximum which can be imposed on summary conviction for an either-way offence is normally six months' imprisonment.

There are also limits on the fines which magistrates' courts may impose in respect of many summary offences. The maximum fine for a summary offence is normally governed by the level on the standard scale of fines hereafter simply referred to as 'level' assigned to the offence in question. The standard scale is as follows:

Level	Amount of fine
1	£200
2	£500
3	£1,000
4	£2,500
5	No maximum; referred to in the text as an unlimited fine

Special maxima apply in the case of children under 18 or under 14.

There is normally no limit on the fine which a magistrates' court can impose in respect of an indictable (either-way) offence; the maximum imprisonment which a magistrates' court impose in respect of such an offence is six months unless a statute expressly overrides this rule.

PENALTY NOTICES FOR 'DISORDERLY BEHAVIOUR'

The Criminal Justice and Police Act 2001 (CJPA 2001), s 2 provides that a constable **3.39** who has reason to believe that someone aged 18 or over has committed a 'penalty offence' *may* issue a 'penalty notice for disorder' (PND).

If the penalty offence is a 'relevant penalty offence', ie a penalty offence in relation to which there is an approved educational course, the constable may issue a penalty notice with an education option. A PND with an education option is a penalty notice that also offers recipients the opportunity to discharge their liability to be convicted of the penalty offence by paying for and completing an approved educational course related to the offence for which notice was given. An approved educational course means a course run as part of an educational course scheme established by the chief constable.

Penalty offences

'Penalty offences' under CJPA 2001 fall into two groups: **3.40**

Offence creating provision	Description of offence
First group:	
Explosives Act 1875, s 80	Throwing fireworks in a thoroughfare
Criminal Law Act 1967, s 5(2)	Wasting police time or giving false report
Criminal Justice Act 1967, s 91	Disorderly behaviour while drunk in a public place

Offence creating provision	Description of offence
Theft Act 1968, s 1	Theft (limited by *guidance* to retail thefts of goods not exceeding £100 inc VAT in value)
Criminal Damage Act 1971, s 1(1)	Destroying or damaging property (limited by *guidance* to damage not exceeding £300)
Misuse of Drugs Act 1971, s 5(2)	Possession of various forms of cannabis and cannabis derivatives, etc
Public Order Act 1986, s 5	Behaviour likely to cause harassment, alarm or distress
Licensing Act 2003, s 141	Sale of alcohol to person who is drunk
Licensing Act 2003, s 146(1), (3)	Sale of alcohol to person under 18
Licensing Act 2003, s 149(3), (4)	Purchase of alcohol on behalf of person under 18, etc
Licensing Act 2003, s 151	Delivery of alcohol to person under 18 or allowing such delivery
Communications Act 2003, s 127(2)	Using a public electronic communications network in order to cause annoyance, inconvenience, or needless anxiety
Fireworks Act 2003, s 11	Contravention of a prohibition or failure to comply with a requirement imposed by or under fireworks regulations or making false statements
Fire and Rescue Services Act 2004, s 49	Knowingly giving a false alarm to person acting on behalf of a fire and rescue authority
Second group:	
Licensing Act 1872, s 12	Being drunk in a highway, other public place, or licensed premises
Parks Regulation (Amendment) Act 1926, s 2(1) so far as it creates an offence against Parks Regulation Act 1872 relating to Royal Parks and Other Open Spaces Regulations 1997:	Failing to comply with, or contravening, those Regulations by:
(a) reg 3(3)	Littering
(b) reg 3(4)	Cycling except on park road or designated area
(c) reg 3(6)	Dog fouling
British Transport Commission Act 1949, s 55	Trespassing on a railway
British Transport Commission Act 1949, s 56	Throwing stones, etc at trains or other things on railways
Misuse of Drugs Act 1971, s 5(2)	Possession of khat and any preparation or other product containing khat
Environmental Protection Act 1990, s 87	Depositing and leaving litter
Licensing Act 2003, s 149(1)	Purchase of alcohol by person under 18 on relevant premises

Offence creating provision	Description of offence
Licensing Act 2003, s 150	Consumption of alcohol by a person under 18 on relevant premises or allowing such consumption

Amount of penalty

The Penalties for Disorderly Behaviour (Amount of Penalty) Order 2002 prescribes **3.41** penalties of £90 for the first group of offences, and £60 for the second group.

Effect of penalty notice

CJPA 2001, s 4 provides that, if a PND is given to a person (A) under s 2, and A asks **3.42** to be tried for the alleged offence, proceedings may be brought against him. Such a request must be in the specified manner before the end of the suspended enforcement period (21 days beginning with the date the notice was given). If, in the case of a PND that is not one with an education option, by the end of the suspended enforcement period the penalty has not been paid, and A has not made a request to be tried, a sum equal to one and a half times the amount of the penalty may be registered under s 8 for enforcement against A as a fine. In the case of a PND with an education option, a sum equal to one and a half times the amount of the penalty may be registered under s 8 for enforcement against A as a fine if, by the end of the suspended enforcement period:

- (a) (i) A does not ask to attend an approved educational course relating to the offence to which the notice relates,
  - (ii) A does not pay the penalty, or
  - (iii) A does not request to be tried; or
- (b) (i) A has asked to attend an approved education course, and
  - (ii) A does not, in accordance with regulations, pay the course fee, start such a course, or complete such a course.

General restrictions on proceedings

The relevant provisions are in CJPA 2001, s 5. **3.43**

Proceedings for the offence to which a PND relates may not be brought during the suspended enforcement period. If the penalty is paid before the end of the suspended enforcement period, no proceedings may be brought for the offence.

Proceedings for an offence to which a PND with an education option relates may not be brought against a person who has, by the end of the suspended enforcement period, asked to attend an approved educational course relating to the offence, unless (b) above applies. If the person to whom a PND with an education option is given:

- (a) completes, in accordance with regulations, an approved educational course relating to the offence to which the notice relates, and
- (b) pays the course fee in accordance with those regulations,

no proceedings may be brought for the offence.

Miscellaneous points

3.44   The provisions in respect of registration of penalties and sums payable in default, and enforcement, are the same as those provided in relation to the more widely available fixed penalties for road traffic offences: see Chapter 25.

It is not intended that a PND should always be issued in respect of the specified cases; the constable has discretion. PNDs are designed for minor and straightforward cases. More serious cases should be dealt with by using the traditional criminal process.

Community support officers or community support volunteers may issue PNDs under CJPA 2001. So may persons accredited under a community safety accreditation scheme, except that they may not issue a notice in respect of theft, depositing and leaving litter, criminal damage, or the two offences involving drunkenness in a public place.

The issue of a PND for an offence and the payment of the penalty under it is only a bar to proceedings for the offence specified in the notice. Payment of the penalty is not an admission of guilt for that offence or any proof that it has been committed by A. It does not prevent proceedings against A for another offence. This is of obvious importance where it subsequently becomes apparent that that offence was committed in the course of the same incident.

A victim should be consulted about the possibility of a PND being given, but a witness cannot insist that the case is disposed of in a particular way. The Ministry of Justice has issued guidance on PNDs.

JURISDICTION OF ARMED SERVICE AUTHORITIES TO DEAL WITH CRIMINAL OFFENCES

3.45   The Acts relating to armed service personnel make provisions for concurrent jurisdiction in respect of civil and service courts. As a result, arrangements have been made between civilian and military prosecuting and investigating authorities governing how they will work together, decide issues of jurisdiction, and, where necessary, decide how an allegation will be investigated.

CHAPTER 4

Police Powers

The description of police powers in this book focuses on the powers of police offi- **4.1** cers, ie 'constables' (members of police forces hold the status of constable, whatever their rank, and special constables). This is done for convenience because it must not be forgotten that some powers possessed by constables are exercisable under the Police Reform Act 2002 by police civilians and accredited persons as explained at **2.4–2.11**.

Conduct in breach or in excess of a relevant statutory or common law power may constitute a breach of a 'Convention right' under the Human Rights Act 1998 and may cause the police officer concerned to be liable in civil law or in criminal law. It may also result in disciplinary proceedings and render inadmissible evidence obtained as a result.

Police powers must be used fairly, responsibly, and with respect for the people to whom they apply. It is unlawful to discriminate, harass, or victimize someone because (s)he has or is perceived to have a 'protected characteristic'. Under the Equality Act 2010, s 149, when police officers are carrying out their functions, they must also have due regard to the need to eliminate unlawful discrimination, harassment, and victimisation; to advance equality of opportunity between people who share a relevant protected characteristic and people who do not; and to take steps to foster good relations between those persons. The 'protected characteristics' are age, disability, gender re-assignment, race, religion or belief, sex, sexual orientation, marriage and civil partnership, and pregnancy and maternity.

PACE

Throughout this book, but particularly in this and the next four chapters, references are **4.2** made to the Police and Criminal Evidence Act 1984, hereafter PACE. Its provisions are law and therefore must be followed. Although non-compliance by a police officer is not in itself an offence or civil wrong, it may render unlawful the exercise of a police power or other conduct. It may also result in disciplinary proceedings or render inadmissible evidence obtained by it.

PACE CODES OF PRACTICE

PACE, s 66 requires the Secretary of State to issue Codes of Practice, approved by both **4.3** Houses of Parliament, in connection with the exercise of a wide range of powers by police officers. The Secretary of State may at any time revise a Code of Practice then in force. Such a revision does not require the approval of the Houses of Parliament.

Eight PACE Codes of Practice have been issued and revised. They apply throughout England and Wales. They are: Code A (stop and search); Code B (search and seizure); Code C (detention, treatment and questioning of persons); Code D (identification); Code E (audio recording of interviews); Code F (visual recording of interviews); Code

G (arrest); and Code H (detention, treatment, and questioning of persons detained under the Terrorism Act 2000, s 41 and Sch 8).

Annexes to a Code are provisions of the Code but guidance notes printed with the Code are technically not. They are, however, frequently referred to by the courts when considering whether there has been a breach of a Code.

The powers covered by the Codes must be used fairly, responsibly, with respect for people suspected of committing offences, and without unlawful discrimination.

Apart from the PACE Codes of Practice, there are a number of codes under other statutes which are referred to at the appropriate points.

Compliance with PACE Codes

4.4   The provisions of the PACE Codes are not law but they should be followed. By s 67(11), the Codes are admissible in evidence and may be taken into account by a court where relevant to any question in the proceedings. Conduct by a police officer which involves a breach of a Code may result in disciplinary proceedings.

Any person with police powers conferred by a designation or accreditation under the Police Reform Act 2002 must have regard to relevant provisions of a Code in the exercise of powers and duties. Breach of a Code's provisions does not of itself give rise to criminal or civil proceedings (s 67(10)). Evidence obtained in a way involving a breach of a Code may be ruled inadmissible.

Wherever the Codes require the prior authority or agreement of an officer of at least inspector or superintendent rank, a sergeant is treated as holding the rank of inspector, and a chief inspector as holding the rank of superintendent, if (s)he has been authorised by a superintendent to perform the functions of the higher rank (PACE, s 107).

Gender-related issues

4.5   All searches, procedures, and requirements must be carried out with courtesy, consideration, and respect for the person concerned.

Certain provisions of the PACE Codes explicitly state that searches and other procedures may only be carried out by, or in the presence of, persons of the same sex as the person subject to the search or other procedure, or require action to be taken or information to be given which depends on whether the detainee is treated as being male or female.

Annexes to Codes A, C, and H contain general provisions dealing with the implications of this in respect of transgender individuals (including transsexuals) and transvestites. Police officers should show particular sensitivity when dealing with such persons. 'Transgender' is generally used as an umbrella term to describe people whose gender identity (self-identification as being a woman, man, neither or both) differs from the sex they were registered as at birth. The term includes, but is not limited to, transsexual people. 'Transsexual' means a person who is proposing to undergo, is undergoing, or has undergone a process (or part of a process) for the purpose of gender re-assignment by changing physiological or other attributes of that person's sex. This includes aspects of gender such as dress and title. 'Transvestite' means a person of one gender who dresses in the clothes of a person of the opposite gender; a transvestite does not live permanently in the gender opposite to that person's birth sex.

Legally, the *gender* (and, accordingly, the sex) of an individual (P) is P's gender as registered at birth unless P possesses a gender recognition certificate (GRC), in which case P's gender is the acquired gender.

When establishing whether someone should be treated as being male or female the following approach, designed to maintain the person's dignity, minimise embarrassment and secure the person's co-operation, should be followed:

(1) The person (P) must not be asked whether P has a GRC.
(2) If there is no doubt as to whether P should be treated as being male or female, P should be dealt with as being of that sex.
(3) If at any time (including during the search, procedure or requirement) there is doubt as to whether P should be treated or continue to be treated as being male or female:
   (a) P should be asked what gender P considers P to be. If P expresses a preference to be dealt with as a particular gender, P should be asked to confirm that preference by signing the custody record or, if a custody record has not been opened, the search record or the officer's notebook. Subject to (b) below, P should be treated according to that preference except with regard to the requirements to provide P with information concerning menstrual products and P's personal needs relating to health, hygiene, and welfare described at **6.9**. In these cases, if P's confirmed preference is to be dealt with as being male, *P should be asked in private whether P wishes to speak in private with a member of the custody staff of a gender of P's choosing about the provision of menstrual products and P's personal needs*, notwithstanding P's confirmed preference;
   (b) if there are grounds to doubt that the preference in (a) accurately reflects P's predominant lifestyle, eg if P asks to be treated as a woman but documents and other information make it clear that P lives predominantly as a man, P should be treated according to what appears to be P's predominant lifestyle;
   (c) if P is unwilling to express a preference, efforts should be made to determine P's predominant lifestyle and P should be treated as such. For example, if P appears to live predominantly as a woman, P should be treated as being female except with regard to the requirements to provide P with information concerning menstrual products and P's personal needs relating to health, hygiene, and welfare. In these cases, if P's predominant lifestyle has been determined to be male, *the italicised words in (a) apply*, notwithstanding P's determined predominant lifestyle;
   (d) if none of the above apply, P should be dealt with according to what reasonably appears to have been P's sex as registered at birth.

Once a decision has been made about which gender P is to be treated as having, each officer responsible for the search or procedure should where possible be advised before the search or procedure of any doubt as to P's gender and P informed the doubts have been disclosed.

Where the gender of P is established, the decision should be recorded either on the search record, in the officer's notebook, or, if applicable, in P's custody record. Where P elects which gender P considers P to be but is not treated in accordance with that preference, the reason must be recorded.

Information about an application for a GRC or about a successful applicant's gender before it became that person's acquired gender is 'protected information' and must not be disclosed by a police officer or police staff in contravention of the Gender Recognition Act 2004. Disclosure includes making a record which is read by others.

POLICE POWERS DESCRIBED IN THIS CHAPTER

4.6   Police powers are dealt with in the following order in this chapter:

(a) stop and search;

(b) road check;

(c) arrest;

(d) entry and search in relation to an arrest;

(e) search under a search warrant;

(f) investigatory powers under the Investigatory Powers Act 2016 (interception of communications, etc),

(g) surveillance and the conduct and use of 'covert human intelligence sources';

(h) authorisation of action in respect of property; and

(i) removal to a place of safety of mentally disordered persons.

A judge in the Administrative Court has held that, where a police officer effects a stop, search, or arrest under a power to do so, which the officer is justified in using, there is no requirement at common law for him/her to be aware of the legal origin of that power in order for its exercise to be lawful.

An act is not unlawful because a police officer does not ask him/herself or forgets which power (s)he had, provided that (s)he had the power to do what (s)he did with knowledge and belief which (s)he had.

POWERS TO STOP AND SEARCH

4.7   Various statutes (eg PACE, s 1) give *every* constable (**2.1**) the power on reasonable suspicion to stop and search for particular articles, for example, stolen or prohibited articles, controlled drugs and firearms.

Stop and search by a constable may be authorised under the Criminal Justice and Public Order Act 1994, s 60 (**4.19**) if the necessary reasonable belief exists.

Code A: the Code of Practice for the exercise by Police Officers of Statutory Powers of Stop and Search, and by Police Officers and Police Staff of Requirements to Record Public Encounters applies to the above powers of stop and search and also to:

(a) powers to search a person who has not been arrested in the exercise of a power to search premises; and

(b) specified powers in the Terrorism Prevention and Investigation Measures Act 2011, Sch 5, paras 6, 8 and 10 to search an individual who has not been arrested.

On the other hand, Code A does not apply to the powers of stop and search under PACE, s 6 (powers of constables employed by statutory undertakers (bodies authorised by statute to carry out any railway, road transport, inland navigation, or harbour undertaking) on premises of statutory undertaker), or under the Aviation Security Act 1982, s 24B (hijacking), or to searches carried out for the purposes of examination under the Terrorism Act 2000 (TA 2000), Sch 7 (port and border controls) and to which the Code of Practice for Examining Officers issued under that Act applies, or to the powers of stop and search to which the Code of Practice issued under TA 2000, s 47AB (**21.21**) applies.

Nothing in the provisions about stop and search affects the ability of an officer to speak to or question a person in the ordinary course of his/her duties without detaining the person or exercising any element of compulsion. A police officer trying to discover whether, or by whom, an offence has been committed may question anyone from whom useful information might be obtained, subject to the restrictions imposed by Code C (Chapter 5). A person's unwillingness to reply does not alter this entitlement, but in the absence of a power to arrest or to detain in order to search, the person is free to leave at will and cannot be compelled to remain with the officer.

Power under PACE, s 1 to stop and search

4.8

PACE, s 1(2) states that a constable:

(a) may search any person or vehicle ('vehicle' includes vessels, aircraft, and hover-craft), *and* anything which is in or on a vehicle, for stolen or prohibited articles or an article to which PACE, s 1(8A) applies or any firework to which s 1(8B) applies; and

(b) may detain persons or vehicles for the purpose of such a search.

By s 1(3), a constable only has this power to search if (s)he has *reasonable grounds for suspecting* (see further **4.11**) that (s)he will find stolen or prohibited articles or an article to which s 1(8A) applies or any firework to which s 1(8B) applies; (s)he must of course have such grounds *before* carrying out a search. Reasonable force may be used in the exercise of these powers, but every effort should be made to persuade a person to co-operate and force should only be used as a last resort. A compulsory search may only be made if it is established that the person is unwilling to co-operate.

The courts have held that the test of lawfulness of a search under s 1 is:

(a) whether the searching constable suspected that (s)he would find stolen or prohibited articles or anything to which s 1(8A) or (8B) applies;

(b) assuming that that constable had the necessary suspicion, whether there was reasonable cause for that suspicion; and

(c) if the answer to (a) and (b) is 'yes', whether that constable exercised his/her discretion to make a search in accordance with the general principles for the exercise of executive discretion, ie the decision to search must have been made taking into account relevant matters and excluding irrelevant ones, it must have been made for a proper purpose, and it must not have been one that no reasonable constable (taking account of the relevant matters) could have made.

It is not necessarily essential that the searching constable reasonably suspects each and every individual member of a group to be carrying a prohibited article etc before (s)he can search the individual members of the group. A divisional court has held that, if (a) and (b) above have been satisfied in respect of the group, the issue in relation to searching the individual members goes to (c), namely whether the constable's decision to search the individuals, given his/her suspicion, reasonably held, went beyond the bounds of decision open to him/her under (c) in the circumstances in question.

Where a police officer has reasonable grounds to suspect that a person is in innocent possession of a specified article, the power to stop and search exists notwithstanding that there would be no power of arrest. However, every effort should be made to secure the voluntary production of the article before the power is resorted to.

A '*prohibited article*' is defined by s 1(8) as:

(a) an offensive weapon, defined by s 1(9) as any article made or adapted for use for causing injury to persons, or intended by the person having it with him/her for such use by him/her or by some other person; or

(b) an article made or adapted for use in the course of or in connection with an offence of burglary, theft, taking a conveyance without authority, fraud, or destroying or damaging property, or intended by the person having it with him/her for such use by him/her or by some other person.

An *article to which PACE, s 1(8A) applies* is any article with a blade or point in relation to which a person has committed, or is committing or is going to commit, an offence under the Criminal Justice Act 1988, s 139 or s 139AA (**19.52–19.53**).

An *article to which s 1(8B) applies* is any firework which a person possesses in contravention of a prohibition imposed by the Fireworks Regulations 2004 (**19.65–19.67**). '*Firework*' means a device which is a firework for the purposes of the relevant British Standard specification or which would be if intended as a form of entertainment.

Places in which powers to search can be exercised

4.9   PACE, s 1(1) states that a constable may exercise any powers under s 1:

(a) in any place to which at the time when (s)he proposes to exercise the power the public or any section of the public has access, on payment or otherwise, as of right or by virtue of express or implied permission; or

(b) in any other place to which people have ready access at the time when (s)he proposes to exercise the power but which is not a dwelling.

Persons have a right to use streets and highways; they have express permission to access cinemas, theatres, or football grounds subject to paying an entry fee. There is an implied permission for persons to enter shops and other buildings to carry out business transactions.

'Place to which people have ready access' in (b) is wide in meaning. It extends the power of search to any place (other than a dwelling) to which the public have access in fact, whether lawfully or not; for example, a private field or grounds into which people regularly gain access as trespassers.

Section 1(4) states that, if a person is in a garden or yard occupied with, and used for, the purposes of a dwelling or on other land so occupied and used, a constable may not search him/her under s 1 unless the constable has reasonable grounds for believing that:

(a) (s)/he does not reside in the dwelling; and

(b) (s)/he is not in the place in question with the express or implied permission of a person who resides in the dwelling.

Similar restrictions apply to the search of a vehicle (or anything in or on it) in a garden, yard, or land occupied with and used for the purpose of a dwelling. By s 1(5), a constable may not search it unless (s)he has reasonable grounds for believing that:

(a) the person in charge of the vehicle does not reside in the dwelling; and

(b) the vehicle is not in the place in question with the express or implied permission of a person who resides in the dwelling.

In effect, the only places in which a constable cannot exercise s 1 powers are dwelling houses, the curtilage of dwelling houses if the person or vehicle is there lawfully, or any other place which is secure and does not permit ready access.

Seizure of articles

4.10   PACE, s 1(6) provides that suspected stolen or prohibited articles, or an article suspected to be one to which s 1(8A) or s 1(8B) applies, found in a search may be seized.

Reasonable suspicion

Code A gives the following guidance on what may be 'reasonable suspicion' for the purposes of statutory powers of stop and search which require a reasonable suspicion. 'Reasonable suspicion' requires (a) that the officer suspects that the relevant matter is satisfied, and (b) that there are reasonable grounds for that suspicion.

'Suspicion' exists where a person thinks that there is a possibility, which is more than fanciful, that the relevant matter is satisfied. It can be contrasted with 'belief' which requires more than mere suspicion, since belief refers to having no substantial doubt about the matter in question.

General

The test of reasonable suspicion must be applied to the particular circumstances in each case and is in two parts:

(a) the officer must have formed a *genuine* suspicion in his/her own mind that (s)he will find the object for which the search power being exercised allows him/her to search; and
(b) the suspicion that the object will be found must be *reasonable*. This means that there must be an objective basis for that suspicion based on facts, information and/or intelligence which are relevant to the likelihood that the object in question will be found, so that a reasonable person would be entitled to reach the same conclusion based on the same facts and information and/or intelligence.

Officers must therefore be able to explain the basis for their suspicion by reference to intelligence or information about, or some specific behaviour by, the person concerned.

The exercise of the 'reasonable suspicion stop and search powers' depends on the likelihood that the person searched is in possession of an item for which (s)he may be searched; it does not depend on the person concerned being suspected of committing an offence in relation to the object of the search. A police officer who has reasonable grounds to suspect that a person is in *innocent possession* of a stolen or prohibited article, controlled drug or other item for which the officer is empowered to search, may stop and search the person even though there would be no power of arrest. This provision would apply when a child under the age of criminal responsibility (10 years) is suspected of carrying any such item; (s)he would be in innocent possession of it, even if (s)he knew (s)he had it. It is not uncommon for children under 10 to be used by older children and adults to carry stolen property, drugs, and weapons and, in some cases, firearms, for the criminal benefit of others, either in the hope that police may not suspect they are being used for carrying the items, or knowing that, if they are suspected of being couriers and are stopped and searched, they cannot be arrested or prosecuted for any criminal offence. The availability of stop and search powers in such a case therefore allow the police to intervene effectively to break up criminal gangs and groups that use young children to further their criminal activities.

A guidance note to Code A states that whenever a child under 10 is suspected of carrying unlawful items for someone else, or is found in circumstances which suggest that his/her welfare and safety may be at risk, the facts should be reported and actioned in accordance with established force safeguarding procedures. This will be in addition to treating the child as a potentially vulnerable or intimidated witness in respect of his/her status as a witness to the serious criminal offence(s) committed by those using him/her as courier. Safeguarding considerations will also apply to other persons aged under 18 who are stopped and searched under any of the powers to which Code A applies.

Personal factors can never support reasonable grounds for suspicion

4.13   Reasonable suspicion can never be supported on the basis of personal factors. This means that unless the police have information or intelligence which *provides a description* of a person suspected of carrying an article for which there is a power to stop and search, the following *cannot be used*, alone or in combination with each other, or in combination with any other factor, as the reason for stopping and searching any individual, including any vehicle which (s)he is driving or in which (s)he is being carried:

(a) a person's physical appearance with regard, for example, to any of the 'relevant protected characteristics' set out in the Equality Act 2010, s 149 (**4.1**), or the fact that the person is known to have a previous conviction; and

(b) generalisations or stereotypical images that certain groups or categories of people are more likely to be involved in criminal activity.

Reasonable grounds for suspicion based on information and/or intelligence

4.14   Reasonable grounds for suspicion should normally be linked to accurate and current intelligence or information relating to articles for which there is a power to stop and search being carried by individuals or being in vehicles in any locality. This would include reports from members of the public or other officers describing:

(a) a person who has been seen carrying such an article or a vehicle in which such an article has been seen;

(b) crimes committed in relation to which such an article would constitute relevant evidence, for example, property stolen in a theft or burglary, an offensive weapon, or bladed or sharply pointed article used to assault or threaten someone or an article used to cause criminal damage to property.

Searches based on accurate and current intelligence or information are more likely to be effective. Targeting searches in a particular area at specified crime problems not only increases their effectiveness but also minimises inconvenience to law-abiding members of the public. It also helps in justifying the use of searches both to those who are searched and to the public.

Reasonable grounds for suspicion and searching groups

4.15   Where there is reliable information or intelligence that members of a group or gang habitually carry knives unlawfully or weapons or controlled drugs, and wear a distinctive item of clothing or other means of identification (eg jewellery, insignias, tattoos, or other features known to identify members of the particular gang or group) in order to identify themselves as members of that group or gang, that distinctive item of clothing or other means of identification may provide reasonable grounds to stop and search any person believed to be a member of that group or gang.

   A similar approach would apply to particular organised protest groups where there is reliable information or intelligence that:

(a) the group in question arranges meetings and marches to which one or more members bring articles intended to be used to cause criminal damage and/or injury to others in support of the group's aims;

(b) at one or more previous meetings or marches arranged by that group, such articles have been used and resulted in damage and/or injury; and

(c) on the subsequent occasion in question, one or more members of the group have brought with them such articles with similar intentions.

These circumstances may provide reasonable grounds to stop and search any members of the group to find such articles. A guidance note to Code A states that a decision to search individuals believed to be members of a particular group or gang must be judged on a case-by-case basis according to the circumstances applicable at the time of the proposed searches and, in particular, having regard to: (i) the number of items suspected of being carried; (ii) the nature of those items and the risk they pose; and (iii) the number of individuals to be searched. A group search will only be justified if it is a necessary and proportionate approach based on the facts and having regard to the nature of the suspicion in these cases. The extent and thoroughness of the searches must not be excessive. The size of the group and the number of individuals it is proposed to search will be a key factor and steps should be taken to identify those who are to be searched to avoid unnecessary inconvenience to unconnected members of the public who are also present. The onus is on the police to be satisfied and to demonstrate that their approach to the decision to search is in pursuit of a legitimate aim, necessary and proportionate.

Reasonable grounds for suspicion based on behaviour, time, and location

Reasonable suspicion may exist without specific information or intelligence and on **4.16** the basis of the behaviour of a person. For example, if an officer encounters someone on the street at night who is obviously trying to hide something, the officer may (depending on the other surrounding circumstances) base such suspicion on the fact that this kind of behaviour is often linked to stolen or prohibited articles being carried. An officer who forms the opinion that a person is acting suspiciously or that (s)he appears to be nervous must be able to explain, with reference to specific aspects of the person's behaviour or conduct which (s)he observed, why (s)he formed that opinion. A hunch or instinct which cannot be explained or justified to an objective observer can never amount to reasonable suspicion.

Securing public confidence and promoting community relations

Searches are more likely to be effective, legitimate, and secure public confidence when **4.17** their reasonable grounds for suspicion are based on a range of objective factors. The overall use of these powers is more likely to be effective when up-to-date and accurate intelligence or information is communicated to officers and they are well informed about local crime patterns. Local senior officers have a duty to ensure that those under their command who exercise stop and search powers have access to such information, and the officers exercising the powers have a duty to acquaint themselves with that information.

Questioning to decide whether to carry out a search

An officer who has reasonable grounds for suspicion may detain the person concerned **4.18** in order to carry out a search. Before carrying out the search the officer may ask questions about the person's behaviour or presence in circumstances which gave rise to the suspicion. As a result of questioning the detained person, the reasonable grounds for suspicion necessary to detain that person may be confirmed or, because of a satisfactory explanation, be dispelled. Questioning may also reveal reasonable grounds to suspect the possession of a different kind of unlawful article from that originally suspected. However, reasonable grounds for suspicion cannot be provided retrospectively by such questioning during a person's detention or by refusal to answer any questions asked.

If, as a result of questioning before a search, or other circumstances which come to the attention of the officer, there cease to be reasonable grounds for the necessary suspicion, no search may take place. In the absence of any other lawful power to detain, the person is free to leave at will and must be so informed.

There is no power to stop or detain a person in order to find grounds for a search. Police officers have many encounters with members of the public which do not involve detaining people against their will and do not require any statutory power for an officer to speak to a person. However, if reasonable grounds for suspicion emerge during such an encounter, the officer may detain the person to search him/her, even though no grounds existed when the encounter began. As soon as detention begins, and before searching, the officer must inform the person that (s)he is being detained for the purpose of a search and take action in accordance with the procedure at **4.25**.

Authorisation of stop and search in anticipation of violence or after violence

4.19 The Criminal Justice and Public Order Act 1994 (CJPOA 1994), s 60 provides further powers to stop and search. Section 60(1) provides that where a police officer of the rank of inspector (or above) reasonably believes that:

(a) incidents involving serious violence may take place in any locality in his police area, and that it is expedient to give an authorisation under s 60 to prevent their occurrence; or

(b) (i) an incident involving serious violence has taken place in England and Wales in his police area;

    (ii) a dangerous instrument or offensive weapon used in the incident is being carried in any locality in his police area by a person; and

    (iii) it is expedient to give an authorisation under s 60 to find the instrument or weapon; or

(c) persons are carrying dangerous instruments or offensive weapons in any locality in his police area without good reason,

he may give an authorisation that the powers conferred by s 60 are to be exercised at any place within that locality for a specified period not exceeding 24 hours.

'Offensive weapon' in (b) is extended in the case of an incident of the kind mentioned in (b) to include 'any article used in the incident to cause or threaten injury to any person or otherwise to intimidate'.

The extent of a 'locality' within which these powers may be exercised will differ in relation to the nature of the incident. An anticipated serious disorder at a pub would be in a small 'locality' while one within a housing estate would be in an extensive 'locality'. Provided it is proved that thought has been given to defining the locality, a court is unlikely to rule that the authorisation was invalid. The authorising officer should not set a geographical area which is wider than that (s)he believes necessary for the purpose of the authorisation's objective. The locality must be carefully specified and officers must be aware of the geographical area enclosed. If the powers are to be used in response to a threat or incident which straddles police force areas, an officer from each of the forces affected must give an authorisation.

Although the word 'necessary' does not appear in s 60(1), the effect of the European Convention on Human Rights, art 8 is that an authorisation made under s 60 must be made only if the officer believes that it is necessary, as well as expedient, to prevent serious violence, or to find dangerous instruments or offensive weapons after an incident involving serious violence, or to apprehend persons carrying weapons.

The period during which these powers may be exercised must be the minimum considered necessary to deal with the risk of violence or the carrying of knives or offensive weapons.

Where an inspector gives an authorisation (s)he must, as soon as it is reasonably practicable to do so, cause a superintendent (or above) to be informed.

The initial authorisation may be extended once only for a further 24 hours on the authority of a superintendent (or above), where expedient, having regard to offences committed, or reasonably suspected to have been committed, in connection with an activity falling within the authorisation. Thereafter further use of the powers requires a new authorisation.

Police powers

Stop and search Where an authorisation is in force under CJPOA 1994, s 60 a constable in uniform is empowered to stop: **4.20**

(a) any pedestrian and search him/her and anything carried by him/her, for offensive weapons or dangerous instruments; and

(b) any vehicle and search the vehicle, its driver, and any passenger for offensive weapons or dangerous instruments.

A driver of a vehicle which has been stopped is entitled to obtain a written statement to that effect, if (s)he applies within 12 months. The same rights apply to pedestrians or a person in a vehicle.

These stops and searches may be carried out *whether or not* the *constable in question* has any grounds for suspecting that the person or vehicle is carrying weapons or articles of the specified kind. A constable may seize any dangerous instrument or any article which (s)he has reasonable grounds for suspecting to be an offensive weapon. A 'dangerous instrument' is one which has a blade or is sharply pointed; a vehicle includes a 'caravan'; and an 'offensive weapon' is as defined by PACE, s 1(9) (**4.8**). A person carries a dangerous instrument or an offensive weapon if (s)he has it in his/her possession.

The selection of persons and vehicles under s 60 to be stopped and, if appropriate, searched should reflect an objective assessment of the nature of the incident or weapon in question and the individuals and vehicles thought likely to be associated with that incident or those weapons. The powers under s 60 must not be used to stop and search persons and vehicles for reasons unconnected with the purpose of the authorisation. When selecting persons and vehicles to be stopped in response to a specific threat or incident, officers must take care not to discriminate unlawfully against anyone on the grounds of any of the protected characteristics (set out at **4.1**).

Require removal of masks etc   Where an authorisation under CJPOA 1994, s 60 is in force in relation to any locality, by s 60AA(2) a constable in uniform is given the power in that locality:

(a) to require any person to remove any item (eg a face mask) which the constable reasonably believes that person is wearing wholly or mainly for the purpose of concealing his/her identity; and

(b) to seize any item which the constable reasonably believes any person intends to wear wholly or mainly for that purpose.

Where an authorisation under s 60 is not in force in relation to a locality, an authorisation may be given by an inspector (or above) for the exercise of the powers under s 60AA(2) if (s)he reasonably believes that activities may take place in any locality in his/her police area that are likely (if they take place) to involve the commission of offences, and that it is expedient, in order to prevent or control the activities to give the following authorisation. In such a case that officer may authorise under s 60AA(3) that the powers under s 60AA(2) shall be exercisable at any place within that locality for a specified period not exceeding 24 hours. Where it then appears to a superintendent (or above) that, in view of offences which have been, or are reasonably suspected to have been, committed in connection with the activities to which the authorisation relates, (s)he may direct that the authorisation continues in force for a further 24 hours. Where an inspector gives the initial authorisation under s 60AA(3), as soon as practicable (s)he must inform a superintendent (or above).

Further points

4.21   In the case of the British Transport Police references to a 'locality' or to 'a locality in his/her police area' are references to any locality in or in the vicinity of any policed premises, or to the whole or any part of such premises.

An authorisation under CJPOA 1994, s 60 or 60AA must be in writing and signed and must specify the grounds for it, the locality and the period during which the powers are exercisable. By way of exception, an authorisation under (b) in **4.19**(b) or a s 60AA authorisation need not be given in writing where it is not practicable to do so but any oral authorisation must state the matters which otherwise would have to be specified in respect of an authorisation under s 60 or s 60AA and must be recorded in writing as soon as practicable.

An extension of the initial period must be signed by its maker, or where that is not practicable, recorded in writing as soon as practicable.

As indicated above, an officer exercising the power to require the removal of a face covering must reasonably believe that someone is wearing the item wholly or mainly for the purpose of concealing his/her identity. There is no power to stop and search for a face covering. However, a face covering may be seized if it is discovered when searching for something else, or is seen to be carried in circumstances in which an officer reasonably believes it is intended to be used for the purpose of concealing identity.

A person who fails to stop or (as the case may be) to stop his/her vehicle when required to do so by a constable exercising these powers commits an offence (s 60(8)). Failure to remove an item when required is an offence (s 60AA(7)). Both offences are summary and punishable with up to one month's imprisonment and/or a fine not exceeding level 3.

Authorisation of stop and search for the prevention of terrorism

4.22   This is dealt with in Chapter 21.

Searches where no power to search

4.23   An officer must not search a person, even with his/her consent, where no power to search is applicable. Code A specifies, as a sole exception, that an officer does not require a specific power to search persons with their consent given as a condition of entry to sports grounds or other premises.

Where it transpires that search is not required or is impracticable

PACE, s 2(1) provides that, if a constable detains a person or vehicle in the exercise of the power under s 1 or any similar power to stop and search, (s)he need not subsequently carry out that search if it appears to him/her that no search is required or that a search is impracticable. These circumstances will frequently arise: a person will often be detained on valid grounds for the purpose of a search and then satisfy the constable of his/her bona fides by answering his/her questions, or because of other circumstances which come to the attention of the officer. In such a case it is unnecessary to search and the detention will not be unlawful merely because the search was not carried out.   **4.24**

Procedure before carrying out a search

The following procedure is provided by Code A. Before any search of a detained person or attended vehicle takes place the officer must take reasonable steps, if not in uniform, to show his/her warrant card to the person to be searched or in charge of the vehicle to be searched, and whether or not in uniform, to give that person the following information: **4.25**

(a) that (s)he is being detained for the purposes of a search;

(b) the officer's name (except in the case of enquiries linked to the investigation of terrorism, or otherwise where the officer reasonably believes that giving his/her name might put him/her in danger, in which case a warrant or other identification number must be given and the name of the police station to which the officer is attached);

(c) the legal search power which is being exercised; and

(d) a clear explanation of the object of the search in terms of the article or articles for which there is a power to search; and in the case of:

   (i) the power under the CJPOA 1994, s 60, the nature of the power, the authorisation and the fact that it has been given;

   (ii) the powers under the Terrorism Prevention and Investigation Measures Act 2011, (TPIMA 2011) Sch 5, the fact that a terrorism prevention and investigation measures (TPIM) notice is in force or is being served (as the case may be), and the nature of the power being exercised. (For a search for compliance purposes, the search warrant must be produced and the person provided with a copy.);

   (iii) all other powers requiring reasonable suspicion, the grounds for that suspicion;

(e) that (s)he is entitled to a copy of the record of the search if one is made if (s)he asks within three months from the date of the search and:

   (i) if (s)he is not arrested and taken to a police station as a result of the search and it is practicable to make the record on the spot, that immediately after the search is completed (s)he will be given, if (s)he requests, either a copy of the record, or a receipt which explains how (s)he can obtain a copy of the full record or access to an electronic copy of the record, or

   (ii) if (s)he is arrested and taken to a police station as a result of the search, that the record will be made at the station as part of his/her custody record and (s)he will be given, if (s)he requests, a copy of his/her custody record which includes a record of the search as soon as practicable whilst (s)he is at the station.

A divisional court has held that a failure to take the above reasonable steps renders a search unlawful.

The above requirements apply even if the police officer reasonably believes that the suspect is about to dispose of the item in question. The Court of Appeal has suggested that they can be satisfied by simply saying, eg 'Jones, Central; drug search; spit it out'.

The person should also be given information about police powers to stop and search and the individual's rights in these circumstances.

If the person to be searched, or in charge of a vehicle to be searched, does not understand what is being said, or there is any doubt about his/her ability to understand English, the officer must take reasonable steps to bring to that person's attention his/her rights and any relevant provisions of Code A. If that person is deaf or cannot understand English and has someone with him/her, the officer must establish whether that person can interpret or otherwise help him/her to give the required information.

A constable asking a person to remove a mask pursuant to the CJPOA, s 60AA is not performing a search, so the above requirements do not apply.

Conduct of search

4.26   Every reasonable effort must be made to minimise the embarrassment that a person being searched may experience. By PACE, s 2(8) a person may be detained for a search for such time as is reasonably required to permit a search to be carried out either at the place where the person or vehicle was first detained or nearby.

The co-operation of the person to be searched must always be sought, even if (s)he initially objects to being searched. A forcible search may be made only if it has been established that the person is unwilling to co-operate (eg by opening a bag) or resists. A judge in the Administrative Court has held that a pat-down search does not constitute a forcible search, and that neither does placing a hand on a person's body without applying pressure. Reasonable force may be used as a last resort, but only if this is necessary to detain the person or to search him/her. The length of time for which a person or vehicle may be detained will depend on the circumstances, but it must be reasonable and not extend beyond the time taken for the search.

Code A states that, where the exercise of the power requires reasonable suspicion, the extent of the search will be related to the nature of the article sought and the circumstances. If a person is seen to put an offensive weapon into a particular pocket, then, unless there are grounds for suspecting that it has been moved elsewhere, the search must be confined to that pocket; whereas, if the article sought may easily be concealed anywhere on the person, the search may have to be more thorough.

An officer who is not in uniform may not stop a vehicle for the purpose of a search.

Removal of clothing

4.27   Code A restricts searches in public to a 'superficial examination of outer clothing'. A constable is not authorised under PACE, s 1, or under any other power of stop and search, to *require* a person to remove any of his/her clothing in public other than an outer coat, jacket, or gloves, or a mask, etc under CJPOA 1994, s 60AA (**4.20**). Where there might be religious sensitivities about asking someone to remove a face covering, as in the case of a Muslim woman, the police officer should permit the item to be removed out of public view (for example, in a police van or police station if there is one nearby). Where practicable, the item should be removed in the presence of an officer of the same sex and out of sight of anyone of the opposite sex. A search in public of clothing which has not been removed must be restricted to a superficial examination of outer garments. This does not prevent an officer placing his/her hands inside the pockets of outer clothing, or feeling around the inside of collars, socks, and shoes if this is reasonably necessary. Subject to the restrictions upon removal of headgear (below), a person's hair may be searched in public.

If, on reasonable grounds, a more extensive search than a superficial examination of outer clothing is considered necessary (eg by requiring a person to take off a T-shirt or headgear) it must be done out of view of the public, for example in a police van or a police station if there is one nearby. Any search involving more than the removal of an outer coat, jacket or gloves, headgear or footwear, or any other item concealing identity, may only be made by an officer of the same sex as the person searched and may not be made in the presence of anyone of the opposite sex (see **4.5**) unless the person being searched specifically requests it.

Searches which involve exposure of intimate parts of the body must not be conducted as a routine extension of a less thorough search simply because nothing is found in the course of the initial search. Searches involving such exposure may be carried out only at a nearby police station or other nearby location which is out of the public view.

A search in a street itself should be regarded as being in public, even though the street is empty at the time the search begins.

Recording requirements

Searches which do not result in an arrest

4.28   A record of the search must be made, electronically or on paper, unless there are exceptional circumstances making this wholly impracticable (eg in situations involving public disorder or when the recording officer's presence is urgently required elsewhere). If a record is to be made, the officer carrying out the search must make the record on the spot unless this is not practicable, in which case, the officer must make the record as soon as practicable after the search is completed.

If the record is made at the time, the person searched or in charge of the vehicle searched must be asked if (s)he wants a copy; if (s)he does, (s)he must be given immediately (a) a copy of the record, or (b) a receipt which explains how (s)he can obtain a copy of a full record or access to an electronic copy of the record.

An officer is not required to provide a copy of the full record or a receipt at the time if (s)he is called to an incident of higher priority. In such a situation, the officer should consider giving the person details of the station which (s)he may attend for a copy of the record. A receipt may take the form of a simple business card which includes sufficient information to locate the record should the person ask for a copy, for example, the date and place of the search, a reference number, or the name of the officer who carried out the search (unless the exception at **4.30**(5) in the case of terrorism investigations etc applies).

Searches which result in an arrest

4.29   If a search results in a person being arrested and taken to a police station, the officer carrying out the search must ensure that a record of the search is made as part of the custody record. The custody officer must then ensure that the person is asked if (s)he wants a copy of the record, and if (s)he does, that (s)he is given a copy as soon as practicable.

The requirement to make the record of the search as part of the person's custody record does not apply if the person is granted 'street bail' after arrest (**4.57**) to attend a police station and is not taken in custody to the police station. An arrested person's entitlement to a copy of the search record which is made as part of his/her custody record does not affect his/her entitlement to a copy of his/her custody record or any other provisions of PACE Code C: the Code of Practice for the Detention, Treatment, and Questioning of Persons by Police Officers about custody records.

Record of search

4.30   The record of a search must always include the following information:

(1) *A note of the self-defined ethnicity, and, if different, the ethnicity as perceived by the officer making the search, of the person searched or of the person in charge of the vehicle searched (as the case may be).*

Officers should record the self-defined ethnicity of every person stopped according to the categories used in the 2001 census question listed in Annex B to Code A. The person should be asked to select one of the five main categories representing broad ethnic groups and then a more specific cultural background from within this group. The ethnic classification should be coded for recording purposes using the coding system in Annex B. An additional 'Not stated' box is available but should not be offered to respondents explicitly. Officers should be aware and explain to members of the public, especially where concerns are raised, that this information is required to obtain a true picture of stop and search activity and to help improve ethnic monitoring, tackle discriminatory practice, and promote effective use of the powers. If the person gives what appears to the officer to be an 'incorrect' answer (eg a person who appears to be white states that (s)he is black), the officer should record the response that has been given and then record his/her own perception of the person's ethnic background by using the Police National Computer (PNC) classification system. If the 'Not stated' category is used the reason for this must be recorded on the form.

(2) *The date, time and place the person or vehicle was searched.*

(3) *The object of the search in terms of the article or articles for which there is a power to search.*

(4) *In the case of:*

(a) *the power under the CJPOA 1994, s 60, the nature of the power, the authorisation, and the fact that it has been given;*

(b) *the powers under TPIMA 2011, Sch 5, the fact that a TPIM notice is in force or is being served (as the case may be), and the nature of the particular power being exercised. In addition, for a search for compliance purposes, the date of the search warrant, and the fact that it was produced and a copy provided must be recorded, and the warrant must also be endorsed by the constable executing it to state whether anything was found and whether anything was seized;*

(c) *all other powers requiring reasonable suspicion, the grounds for that suspicion.*

(5) *Subject to one exception, the identity of the officer carrying out the search.*

Where a stop and search is carried out by more than one person the identity of all the officers must be recorded in the record. The exception referred to is that the names of police officers are not required to be shown on the search record or any other record required to be made under Code A in the case of enquiries linked to the investigation of terrorism or otherwise where an officer reasonably believes that recording names might endanger the officers. In such cases the record must show the officers' warrant or other identification number and duty station.

For the purposes of completing the search record, there is no requirement to record the name, address, and date of birth of the person searched or the person in charge of a vehicle which is searched and the person is under no obligation to provide this information. In addition, the person should not be asked to provide the information for the purpose of completing the record.

A record is required for each person and each vehicle searched. However, only one record is required where a person who is in a vehicle is searched in addition to the vehicle, and the object and grounds of the search are the same. Where only a vehicle is searched, the self-defined ethnic background of the person in charge of the vehicle must be recorded, unless the vehicle is unattended.

The record of the grounds for making a search must, briefly but informatively, explain the reason for suspecting the person concerned, by reference to information and/or intelligence about, or some specific behaviour by, the person concerned.

Where officers detain an individual with a view to performing a search, but the need to search is eliminated as a result of questioning the person detained, a search should not be carried out (see **4.18**) and a record is not required.

Unattended vehicles

If an unattended vehicle, or anything in or on it, is searched the constable is required   **4.31**
to leave a notice:

(a) stating that (s)he has searched it;
(b) giving the name of the police station to which (s)he is attached;
(c) stating that any application for compensation for any damage caused by the search may be made to that police station; and
(d) stating that the person in charge of the vehicle may obtain a copy of the record of the search (if there is one) if (s)he asks for it within three months.

The notice should explain how (if applicable) an electronic copy of the record may be accessed. The notice must be left inside the vehicle (or on it, if things have been searched without opening it). A vehicle which has been searched must, if practicable, be left secure.

Monitoring and supervising the use of stop and search powers

General

Supervising officers must monitor the use of stop and search powers and should consider   **4.32**
whether there is evidence of such powers being exercised on the basis of stereotyped images or inappropriate generalisations. They must satisfy themselves that officers are acting in accordance with Code A. They must also examine whether the records reveal any trends or patterns which give cause for concern, and, if they recognise such trends, take appropriate action. Senior officers with area or force-wide responsibilities must also monitor such matters on a broader basis. Disproportionate use in relation to particular sections of the community should be investigated. Arrangements must be made, in consultation with local policing bodies, for records to be scrutinised by community representatives. Such scrutiny must take account of an individual's right to confidentiality.

Suspected misuse of power by individual officers

Police supervisors must monitor the use of stop and search powers by individual of-   **4.33**
ficers to ensure that they are being applied appropriately and lawfully. Examples of monitoring are direct supervision of the exercise of the powers, examining stop and search records and asking the officer to account for the way in which (s)he conducted

and recorded particular searches, or through complaints about a stop and search that the officer has carried out. Where a supervisor identifies issues with the way that an officer has used a stop and search power, the facts of the case will determine whether the standards of professional behaviour as set out in the Code of Ethics have been breached and which formal action is pursued.

POWERS TO CONDUCT A ROAD CHECK

Statutory power

4.34   PACE, s 4 governs the conduct of road checks by police officers for the purpose of ascertaining whether a vehicle is carrying a person:

(a) who has committed an offence other than a road traffic offence or a vehicle excise offence;

(b) who is a witness to such an offence;

(c) intending to commit such an offence; or

(d) unlawfully at large.

A 'road check' consists of the exercise in a locality of the power conferred by RTA 1988, s 163 (power of a constable in uniform to stop a mechanically propelled vehicle or pedal cycle on a road) in such a way as to stop all vehicles or vehicles selected by any criterion (as, for example, the person sought is known to be in a black Volkswagen Golf and therefore only vehicles similar to that description will be selected).

Section 4 does not in any way affect an officer's powers to deal with road traffic matters and (s)he may stop as many vehicles as (s)he thinks necessary for that purpose. An officer who wishes to stop a single vehicle, which (s)he reasonably suspects may be carrying a person who has committed or intends to commit an offence, etc, may do so under the powers given by RTA 1988, s 163, which are unaffected in such circumstances.

Authorisation of road checks

4.35   A road check must be authorised in writing. Normally, the authorisation must be given by an officer of the rank of superintendent (or above). PACE, s 4 limits the instances in which authorisation may be given. An officer may only authorise a road check under s 4 for the purpose specified:

(a) in **4.34**(a) or (c), if (s)he has reasonable grounds:–
   (i)   for believing that the offence is an indictable offence; and
   (ii) for suspecting that the person is, or is about to be, in that locality in which vehicles would be stopped if the road check were authorised;

(b) in **4.34**(b), if (s)he has reasonable grounds for believing that the offence is an indictable offence;

(c) in **4.34**(d), if (s)he has reasonable grounds for suspecting that the person is, or is about to be, in that locality.

If it appears to an officer below the rank of superintendent that a road check is required as a matter of urgency for one of the above purposes, (s)he may authorise it, but that officer must, as soon as practicable:

(a) make a written record of the time of authorisation; and

(b) cause a superintendent (or above) to be informed.

The superintendent (or above) may then authorise, in writing, the road check to continue. If (s)he decides it should not continue, (s)he must record the fact that it took place, and its purpose (including any relevant indictable offence).

Time limits, records, and searches

The maximum period for which a road check may be authorised is seven days, but **4.36** written authorisation may be given for a further period not exceeding seven days if a superintendent (or above) believes that a road check ought to continue.

The authorisation may be for a road check to be carried out throughout the 24 hours of each day, or may be limited to specified times.

The record (ie written authorisation) which must be kept of a road check must show:

(a) the period during which the road check is authorised to continue;
(b) the name of the officer who authorised it;
(c) the purpose of the road check (including any relevant indictable offence); and
(d) the locality in which vehicles are to be stopped.

The person in charge of a vehicle stopped in a road check is entitled to obtain a written statement *of the purpose* of the road check if (s)he applies for it within 12 months.

PACE does not give direct powers to a constable to search a vehicle stopped in a road check. However, (s)he is empowered to do so:

(a) if (s)he has reasonable grounds for suspecting that it contains stolen or prohibited articles (PACE, s 1: **4.8**); or
(b) for the purpose of arresting someone for an offence or certain other offences if (s)he has reasonable grounds to suspect that the person is there (PACE, s 17: **4.63**); or
(c) under powers granted by any other statute, eg the Firearms Act 1968.

Common law powers

The power under PACE, s 4 does not affect or replace the existing common law power to **4.37** set up road checks where an imminent breach of the peace is reasonably apprehended.

Having stopped the vehicle under this power, the police may also search it thereunder. Anyone who insists on continuing his/her journey is liable to arrest under the common law power to arrest for an apprehended breach of the peace (**18.1**).

Clearly, where the imminent breach of the peace would involve the commission of an indictable offence a road check can be operated either under PACE, s 4 or under the common law power. The procedural safeguards laid down for s 4 road checks do not, of course, apply to those under the common law power.

Prevention of terrorism: police cordons and prohibitions or restrictions on parking

The Terrorism Act 2000 (TA 2000), ss 33 to 36 and 48 to 51 provide powers to impose **4.38** a police cordon and to prohibit or restrict parking for the purposes of a 'terrorist investigation' (as defined by TA 2000, s 32). These provisions are dealt with in Chapter 21.

POWERS TO ARREST

An arrest (**4.50**) may be authorised by a warrant or be lawfully made without war- **4.39** rant. The law relating to the issue and execution of warrants of arrest issued by a justice of the peace is explained at **3.15** and **3.19**. The law relating to arrest without warrant is as follows.

ARREST WITHOUT WARRANT: CONSTABLES

4.40   PACE, s 24, supplemented by PACE Code G: the Code of Practice for the Statutory Power to Arrest by Police Officers deals with this.

A lawful arrest under s 24 requires two elements:

(a) a person's involvement, or attempted involvement, or suspected involvement, in the commission of a criminal offence; and

(b) reasonable grounds for believing that the person's arrest is necessary for one of a number of specified reasons.

Involvement or suspected involvement in the commission of an offence

4.41   By PACE, s 24(1) to (3):

(1) A constable may arrest without warrant:
   (a) anyone who is *about to* commit an offence;
   (b) anyone who *is in* the act of committing an offence;
   (c) anyone whom *(s)he has reasonable grounds for suspecting to be about* to commit an offence;
   (d) anyone whom *(s)he has reasonable grounds for suspecting to be* committing an offence.

(2) If a constable has *reasonable grounds for suspecting* that an offence *has been* committed (s)he may arrest without warrant anyone whom *(s)he has reasonable grounds* to suspect of being guilty of it.

(3) If an offence *has been committed*, a constable may arrest without warrant:
   (a) anyone who *is* guilty of the offence;
   (b) anyone whom *(s)he has reasonable grounds for suspecting to be* guilty of it.

The following may be noted in relation to 1(c) and (d), (2) and (3).

'Suspicion' exists where a person thinks that there is a possibility, which must be more than fanciful, that the relevant matter is satisfied.

There must be some reasonable, objective grounds for the suspicion, based on known facts or information which are relevant to the likelihood that the offence has been or will be committed and to the fact that the person to be questioned is committing, committed or will commit it. There may be grounds for arresting a suspect on the sole basis of the word of an informant but any police officer should treat that information with considerable reserve.

As in the case of most other police powers where phrases like 'reasonable grounds for suspecting' are used, it is not enough that such reasonable grounds for suspicion exist; the constable must actually suspect the matter in question. The Court of Appeal has emphasised that whether a police officer has reasonable grounds for suspicion depends on the information which (s)he has, but that information need not be based on first-hand knowledge. So long as it comes from a source which it is reasonable to rely on, it can form the basis of a reasonable suspicion; the officer is not required to check it. The Court of Appeal has held that, when an arresting officer's suspicion is formed on the basis of a PNC entry, that entry is likely to provide him/her with a reasonable suspicion. An officer instructed to arrest a person but without being given reasonable grounds to suspect that person's guilt cannot assume that his/her superior who instructed him/her must have other information providing reasonable grounds which has not been disclosed to him/her.

The necessity criteria

By PACE, s 24(4), the power of arrest without warrant conferred by s 24(1) to (3) is only exercisable if the constable has *reasonable grounds for believing* that it is *necessary* for specified reasons (the 'necessity criteria' set out in s 24(5)) to arrest the person. The test in, s 24(4) has to be applied on the basis of what was known at the time by the constable and the situation faced by him/her without benefit of hindsight. There must not only be reasonable grounds for believing the relevant matter but also the constable must actually believe that matter. 'Belief' is more than 'suspicion'; it refers to having no substantial doubt about the relevant matter. As a divisional court has pointed out, the concept of necessity requires more than that an arrest is merely desirable or convenient for one of the specified reasons. The context is important to the application of the necessity criteria. A judge in the Administrative Court has explained the law as follows. The test in s 24(4) must be applied to the particular circumstances of the arrestee. Context is also why the question of alternatives inevitably arises. The test does not mean that there should be no feasible or viable alternative, or that arrest must in every case be a matter of last resort. Rather, it must be 'the practical and sensible option'. Thus, for example, while it may not require interrogating a suspect as to whether (s)he will attend a police station voluntarily, it is at least relevant to see whether that is a practical alternative. A belief not based on some evaluation of the options is unlikely to be found to be based on reason.

4.42

The *specified reasons* are set out in s 24(5) as follows:

(a) to enable the name of the person in question to be ascertained (in the case where the constable does not know, and cannot readily ascertain, the person's name, or has reasonable grounds for doubting whether a name given by the person as his/her name is his/her real name);

(b) correspondingly as regards the person's address;

(c) to prevent the person in question:
   (i) causing physical injury to him/herself or any other person;
   (ii) suffering physical injury;
   (iii) causing loss of or damage to property;
   (iv) committing an offence against public decency (but only where members of the public going about their normal business cannot reasonably be expected to avoid the person in question); or
   (v) causing an unlawful obstruction on the highway;

(d) to protect a child or other vulnerable person from the person in question;

(e) to prevent any prosecution for the offence from being hindered by the disappearance of the person in question, as where there are reasonable grounds for believing that if the person is not arrested (s)he will not attend court, or that street bail after arrest would be insufficient to deter the suspect from trying to evade prosecution;

(f) to allow the prompt *and* effective investigation of the offence or of the conduct of the person in question. A divisional court has held that 'prompt' must be considered in context, taking account of all the relevant circumstances, and that 'effective' is not the same as efficient or cost-effective but means tending to achieve its purpose.

Code G: the Code of Practice for the Statutory Power of Arrest by Police Officers provides examples of cases which head (f) may cover, such as:

(i)   where there are reasonable grounds to believe that the person concerned:
    (a) has made false statements and/or presented false evidence;
    (b) may steal or destroy evidence;
    (c) may make contact with co-suspects or conspirators;
    (d) may intimidate or threaten or make contact with witnesses; or
    (e) may have information which may be obtained by questioning; or
(ii)   when, in connection with an indictable offence there are reasonable grounds to believe that, there is a need to:
    (a) enter and search any premises occupied or controlled by a person;
    (b) search the person; or
    (c) prevent contact with others; or
(iii) when in connection with a recordable offence, it is reasonably believed that it is necessary to take fingerprints, footwear impressions, samples, or photographs of the suspect.

Additional Points

4.43   Code G adds the following points set out in the next three paragraphs.

An arrest interferes with the right to liberty recognised by the European Convention on Human Rights, art 5. The power must be used fairly, responsibly, with respect for the suspect, and without unlawful discrimination. It remains an operational decision at the discretion of the officer as to:

(a) which one or more of the necessity criteria (if any) applies to the individual; and
(b) *if any of the criteria do apply*, whether to arrest, grant street bail after arrest, report for written charge etc, issue a penalty notice, or take any other action that is open to the officer.

In applying the criteria, the arresting officer has to be satisfied that at least one of the reasons supporting the need for arrest is satisfied.

Applying the necessity criteria requires the constable to examine and justify the reason or reasons why a person needs to be arrested or (as the case may be) further arrested.

The circumstances that may satisfy those criteria remain a matter for the operational discretion of individual officers. In considering the individual circumstances, the constable must take into account the situation of the victim, the nature of the offence, the circumstances of the suspect, and the needs of the investigative process.

A police officer must consider carefully the necessity to arrest when deciding whether or not to do so. For example, not every refusal to provide a name and address at the officer's first request necessitates an arrest. Such a refusal should be followed by an explanation of the consequences of a continued refusal to provide an identity and an address which the officer believes to be true, and only if there is a continued refusal to provide such an identity and address should the arrest be made. However, a divisional court has accepted that where a person, whose name was unknown to a constable, was reasonably suspected by the constable of having committed an offence, these requirements were satisfied by the questions, 'What is your name?' and 'What is your address?' followed by a refusal to answer. In the event of refusal to answer, the offender should be told that (s)he is being arrested in relation to the particular offence *and* for refusing to give his/her name and address.

Action to be taken in relation to an arrest

See **4.51-4.57**. **4.44**

ARREST WITHOUT WARRANT: OTHER PERSONS

PACE, s 24A provides that: **4.45**

(1) A person other than a constable may arrest without warrant:
   (a) anyone who is in the act of committing an indictable offence;
   (b) anyone whom (s)he has reasonable grounds for suspecting to be committing an indictable offence.
(2) Where an indictable offence has been committed, a person other than a constable may arrest without warrant:
   (a) anyone who is guilty of that offence;
   (b) anyone whom (s)he has reasonable grounds for suspecting to be guilty of it.
(3) The power of arrest conferred by (1) and (2) is only exercisable if:
   (a) the person making the arrest has reasonable grounds for believing that for any of the reasons mentioned in (4) it is necessary to arrest the person in question; and
   (b) it appears to the person making the arrest that it is not reasonably practicable for a constable to make it instead.
(4) The reasons are to prevent the person in question:
   (a) causing physical injury to him/herself or any other person;
   (b) suffering physical injury;
   (c) causing loss of or damage to property; or
   (d) making off before a constable can assume responsibility for him/her.

The above provisions do not apply in relation to an offence under the Public Order Act 1986, Parts III or IIIA (racial etc hatred offences).

As can be seen, the conditions under which an arrest can be made by some 'other person' are more limited than in the case of a constable.

OTHER POWERS OF ARREST WITHOUT WARRANT

Specific statutory power of arrest without warrant

PACE preserves a small number of specific powers of arrest, such as those under the **4.46** Bail Act 1976, s 7 (**6.80**).

Cross-border powers of arrest

CJPOA 1994, s 137 provides that a constable of an English or Welsh police force, who **4.47** has reasonable grounds for suspecting that an offence has been committed or attempted in England and Wales and that the suspected person is in Scotland or Northern Ireland, may arrest without warrant the suspected person wherever (s)he is in Scotland or Northern Ireland, provided the conditions under the law of England and Wales for a lawful arrest are satisfied. Where a person is arrested in Scotland or Northern Ireland under s 137, (s)he must be taken to the nearest convenient designated police station in England or Wales or to a designated police station in a police area where the offence is being investigated. This must be done as soon as reasonably practicable. Constables of the police service in Scotland or Northern Ireland have a similar power of arrest within England and Wales in respect of offences committed or attempted in Scotland or Northern Ireland respectively.

By *s 137A*, similar powers are granted to a constable of a police force in England and Wales to arrest a person without warrant in England and Wales if the constable:

(a) has reasonable grounds for suspecting that P has *committed a specified offence in Scotland or in Northern Ireland*; and

(b) also has reasonable grounds for believing that it is necessary to arrest P to allow the prompt and effective investigation of the offence, or to prevent any prosecution for the offence from being hindered by the disappearance of P.

An offence committed in Scotland or Northern Ireland is a specified offence if it is an offence:

(a) that is punishable by statute with imprisonment or detention for ten years or more;

(b) specified in Sch 7A, Part 2 (Scotland) or 3 (Northern Ireland);

(c) of attempting or conspiring to commit, or of inciting the commission of, an offence mentioned in (a) or (b); or

(d) committed in Northern Ireland under the Serious Crime Act 2007, Part 2 (encouraging or assisting crime) in relation to an offence under (a) or (b) (s 137B).

By s 137C, P (a person arrested under s 137A) in respect of a specified offence may be detained only for the purpose of enabling an arrest warrant under s 136 (**3.19**) to be obtained and executed, or enabling P to be re-arrested under s 137. P may be detained for this purpose for:

(1) an initial period of three hours beginning with the time of the arrest;

(2) a second period of no more than 21 hours beginning with the end of the initial period, but only if detention for that period is authorised by an inspector (or above) in the arresting force and in the investigating force;

(3) a third period of no more than 12 hours beginning with the end of the second period, but only if detention for that period is authorised by an officer above the rank of inspector in the arresting force and in the investigating force.

An officer of the arresting force may give an authorisation for the purpose of (2) or (3) only if satisfied that it is in the interests of justice to do so. An officer of the investigating force may give an authorisation for the purpose of (2) only if satisfied that:

(i) there are reasonable grounds to suspect that P has committed the specified offence;

(ii) a constable intends that P be arrested as soon as is reasonably practicable (whether under a s 136 warrant or under s 137) and is acting expeditiously for that purpose; and

(iii) it is in the interests of justice to give the authorisation.

An officer of the investigating force may give an authorisation for the purpose of (3) only if satisfied that there continue to be reasonable grounds to suspect that P has committed the specified offence, and satisfied in respect of (ii) and (iii).

If, at any time while P is detained, an 'appropriate officer' in the investigating force is satisfied that it is no longer in the interests of justice for P to be detained, the officer must notify the arresting force, and P must be released immediately. 'Appropriate officer' means in relation to the person's detention:

(a) for the initial period, any constable;
(b) for the second period, an inspector (or above); or
(c) for the third period, an officer above the rank of inspector.

By s 137D, a person (P) arrested under s 137A must be informed as soon as practicable of the purpose for which (s)he may be detained under s 137C, and of the periods for which (s)he may be detained. Provisions under Scottish or Northern Irish Law (as the case may be) concerning matters such as the information to be given on arrest and the rights of arrested persons apply (with specified modifications) in relation to persons arrested under s 137A in respect of a specified offence committed in Scotland or Northern Ireland.

Constables who are members of the police services in Scotland or Northern Ireland have a similar power of arrest under s 137A within one of the other home countries in respect of specified offences (under their law) committed or attempted in England or Northern Ireland (Police Service of Scotland) or England or Scotland (Police Service of Northern Ireland) respectively, in which cases safeguards (under the law of country of commission) of the type just referred to apply as modified.

Reasonable force may be used to effect an arrest under *s 137* or *137A*. Generally, the powers of a constable ancillary to arrest under *s 137* or *137A* are those which apply in the country in which the suspected offence was committed. Powers of search are available in respect of an arrest under s 137 or 137A.

Section 140 provides that, where a constable in England and Wales would have power to arrest without warrant under PACE, s 24 (**4.41–4.42**), a constable of the police service in Scotland or Northern Ireland who is in England and Wales has the same power. Likewise, a power of arrest without warrant in Scotland or Northern Ireland applies in favour of a constable from an English force where a constable of the police service in Scotland or Northern Ireland (as the case may be) would have a power of arrest without warrant.

The powers under *ss 137, 137A*, and *140* may also be exercised in England and Wales and Scotland (but not Northern Ireland) by a British Transport Police officer.

Arrest of service personnel—absentees without leave and deserters

Under the Armed Forces Act 2006, s 314, a constable who is a member of a police force **4.48** or of the Ministry of Defence police has power to arrest without warrant any person whom (s)he has reasonable cause to suspect of being a person subject to service law who has deserted or is absent without leave. Warrants authorising such arrests may also be issued by justices of the peace. Whether or not a person so arrested admits to being an absentee or deserter (s)he *must* be brought before a court as soon as practicable.

Immediately such an arrest has been effected, the appropriate service authority must be informed and provided with all particulars of the case, stating whether the identity of the person arrested is disputed and whether his/her behaviour is non-compliant.

Power to arrest at common law

4.49   At common law a police officer or anyone else has power to arrest without warrant in certain circumstances, and this power is not affected by PACE. The common law power exists where:

(a) a breach of the peace is committed by the person arrested in the presence of the person making the arrest; or

(b) no breach of the peace has occurred in the presence of the person making the arrest, but (s)he reasonably believes that such a breach by the person arrested is actually imminent.

'Breach of the peace' is not an offence in itself. It is defined at **18.1**.

These common law powers of arrest are important to police officers. They are in no way restricted by the nature of the place in which the breach of the peace occurs or is anticipated. The arrest within the above terms is a preventive measure and should be effected wherever it is necessary to preserve the peace, even on private premises and even if no member of the public is present. If two men are fighting, that is a breach of the peace regardless of all other considerations and an arrest by anyone who sees it occur is justified. It might be that the stage at which a fight actually occurs has not been reached but a constable reasonably believes that it will occur in the immediate future; in such a case (s)he may arrest those in dispute. Likewise, where an actual fight might have been discontinued but a constable reasonably believes that the argument is not at an end and that the fight is likely to be resumed, and a breach of the peace to occur, an arrest is justified.

Once a constable reasonably foresees a breach of the peace, (s)he is entitled to remain on premises in which, until that time, (s)he has been a trespasser.

Since a breach of the peace does not necessarily require anyone to be acting, or threatening to act, unlawfully, the power to arrest for breach of the peace is not limited to cases where unlawful conduct occurs, or is suspected or anticipated. By way of a limit on the arrest of those who are not acting unlawfully, the Court of Appeal has set out four factors to be considered in such a case.

(1) There must be a real and present threat to the peace justifying depriving a citizen, not at the time acting unlawfully, of his/her liberty.

(2) The threat must come from the person arrested.

(3) The conduct must clearly interfere with the rights of others and its natural consequence must be 'not wholly unreasonable violence' from a third party.

(4) The conduct of the person to be arrested must be unreasonable.

OTHER POINTS ABOUT ARREST

The nature of an arrest

4.50   An arrest involves the intentional deprivation of a person's liberty to go where (s)he pleases as an initial stage of the criminal process. A police officer who (a) physically restrains someone or (b) restricts his/her movements but does not at that time intend or purport to arrest him/her, eg because (s)he simply wants to detain him/her temporarily to establish his/her identity, is criminally and civilly liable for false imprisonment (and in the first case, for assault), even if (s)he would have been justified in arresting that person; there is no power to detain someone short of an arrest.

An arrest is normally effected by the seizing or the touching of a person's body with a view to his/her detention. It is possible, however, to effect an arrest merely by words if they bring to the person's notice that (s)he is under restraint and will be compelled to remain, and (s)he submits to that compulsion. Reasonable force may be used to effect an arrest (**9.24**). Where unreasonable force is used the arrest is unlawful.

If the arrest occurs outside, it is probable that some form of physical restraint will be applied as the constable will be anxious to ensure that his/her prisoner does not escape. However, if the person arrested is a quiet, elderly person who is not physically capable of such escape, physical restraint would be unnecessary and undesirable. In the same way, persons already in custody who are arrested for other offences do not require a show of restraint, nor would it serve any purpose.

Information to be given on arrest

PACE, s 28 provides that where a person is arrested, otherwise than by being informed **4.51** that (s)he is under arrest, the arrest is unlawful unless, as soon as practicable after his/ her arrest, the person arrested is informed that (s)he is under arrest. This is so whether or not the fact of his/her arrest is obvious. In addition, no arrest is lawful unless the person arrested is informed of the ground for arrest at the time of arrest, or as soon as practicable thereafter. This is a separate requirement from that just mentioned. It applies regardless of whether the ground for arrest is obvious. An arrested person must be given sufficient information to enable him/her to understand that (s)he has been deprived of his/her liberty and the reason for his/her arrest, eg when a person is arrested on suspicion of committing an offence, (s)he must be informed of the suspected offence's nature and when and where it was committed. (S)he must also be informed of the reason or reasons why arrest is considered necessary. Vague or technical language should be avoided. The adequacy of the information must be judged objectively having regard to the information reasonably available to the arresting officer.

While it is preferable that the person arrested is informed of the precise offence for which (s)he is being arrested, for example: 'I am arresting you for an offence of burglary which I have seen you commit', this is not essential (although the ground given must be a valid one). It suffices if, by the use of commonplace language, (s)he is informed of the type of offence for which (s)he is being arrested, so that (s)he has the opportunity to give information which would avoid the arrest. Thus, where a constable making an arrest reasonably suspected that the offence was theft, handling, or taking a conveyance (a car), her statement that she was arresting the suspect for unlawful possession of a car was held to be sufficient. The Court of Appeal has held that the requirements of s 28 are satisfied where a suspect is arrested by one officer, but informed by another of the reason for his/her arrest.

Section 28 exempts an officer from the necessity to give either item of information if it was not reasonably practicable to do so by reason of the person's escape before the information could be given.

Caution

Code G requires that a person must be cautioned upon arrest for an offence unless: **4.52**

(a) it is impracticable to do so by reason of his/her condition or behaviour at the time; or

(b) he has already been cautioned immediately prior to arrest and before questions, or further questions, were put to him/her under Code C (**5.27**) as a person suspected of an offence.

The caution must be in the following terms:

'You do not have to say anything. But it may harm your defence if you do not mention when questioned something which you later rely on in court. Anything you do say may be given in evidence.'

Records of arrest

4.53   A police officer who makes an arrest for an offence must record in his/her pocket book or by any other method used for recording information:

(a) the nature and circumstances of the offence leading to arrest;
(b) the reason or reasons why arrest was necessary;
(c) the giving of the caution; and
(d) anything said by the person at the time of arrest.

The reference to 'any other method' includes any official report book or electronic recording device issued to a constable that enables the record to be made. The record should be made at the time of the arrest unless impracticable to do so. If not made at the time, it must be completed as soon as possible thereafter.

On arrival at the police station, or after being arrested at a police station, the arrested person must be brought before the custody officer as soon as practicable. The custody officer must open a custody record and the information given by the arresting officer on the circumstances and reason or reasons for arrest must be recorded as part of the custody record. As an alternative, a copy of the entry made by the officer (as required above) may be attached to the custody record. The custody record stands as a record of the arrest.

Young people

4.54   A guidance note in Code C states that someone under 18 should not be arrested at his/her place of education unless this is unavoidable. In the event of this happening, the principal or his/her nominee must be informed.

Arrest elsewhere than at a police station

4.55   PACE, s 30(1) and (1A) requires that, where a person is, at any place other than a police station, arrested by a constable for an offence or is taken into custody by a constable having been arrested for an offence by someone other than a constable, (s)he must be taken to a police station as soon as practicable. Section 30(1) and (1A) are subject to s 30A (**4.57**)) and s 30(7) and (7A) (**4.56**).

The policy of s 30 is to bring the arrested person within the protective rules of Code C as soon as practicable. A deliberate breach of s 30(1) and (1A) may lead to evidence obtained during the delay being excluded on the ground that the fairness of the proceedings has been affected. The inclusion of 'as soon as practicable' allows for the circumstances in which it would be unrealistic to take such a person directly to a police station. This is expressly recognised by s 30(10) and (10A), which provides that a constable may delay taking an arrested person to a police station (or releasing him/her under s 30A) if that person's presence elsewhere is necessary in order to carry out such investigations as it is reasonable to carry out immediately; in such a case any questions put to the arrested person should be confined to those investigations. Thus, where a person is arrested for a theft which (s)he was seen to commit but the property was not in his/her possession when arrested after a chase, a constable carrying out one of the duties of his/her office, to protect property, could quite properly retrace the route of the chase with his/her prisoner in order to recover the stolen property as soon as possible.

If there is any delay in taking an arrested person to a police station, the reason must be recorded on first arrival there or, as the case may be, when (s)he is released on bail. With certain exceptions, the police station to which the person arrested must be taken under s 30 must be a *designated police station* (s 30(2)), ie a police station approved for the purpose of detention of arrested persons (**6.2**). In exceptional circumstances, a person arrested may be taken to *any police station*. These circumstances are where:

(a) the constable is working in a locality covered by a police station which is not a designated police station; or

(b) (s)he is a constable belonging to a police force maintained by an authority other than a local policing body (eg the British Transport Police) (s 30(4)).

Neither exception applies if it appears to the constable that it may be necessary to keep the arrested person in police detention for more than six hours. In that case the person arrested must be taken to a *designated* police station (s 30(3)).

In addition, by s 30(5), any constable may take an arrested person to *any* police station if the constable has:

(a) arrested him/her without the assistance of any other constable and no other constable is available to assist him/her; or

(b) taken him/her into custody from a person other than a constable without the assistance of any other constable and no other constable is available to assist him/her,

and (in either case) it appears to the constable that (s)he will be unable to take the arrested person to a designated police station *without the arrested person injuring him/herself, the constable or some other person*.

When the person is arrested and taken to a non-designated police station, and there is no officer present at that police station to act as a custody officer, the inspector at the designated police station to which that person would have been taken must be informed. If the first police station to which an arrested person is a non-designated station, (s)he must be taken to a designated station within six hours unless (s)he is previously released (s 30(6)).

A separate custody record must be opened as soon as practicable for each person who is brought to a police station under arrest. Custody records are discussed in the next chapter.

Release without bail

PACE, s 30(7) and (7A) requires that a person arrested by a constable at a place other than a police station must be released without bail if a constable is satisfied, before the person arrested reaches a police station, that there are no grounds for keeping him/her under arrest or releasing him/her on bail. The constable must record the fact of the release as soon as practicable (s 30(8) and (9)).   **4.56**

Release of a person arrested elsewhere than at a police station: street bail

4.57   PACE, s 30A(1) provides that, where a person (P) has been arrested for an offence by a constable, or taken into custody after arrest by someone other than a constable, a constable may release P at any time before P arrives at a police station:

(a) without bail unless s 30A(1A) applies; or
(b) on bail ('street bail') if s 30A(1A) applies.

Section 30A(1A) applies if:

(i) the constable is satisfied that releasing P on bail is necessary and proportionate in all the circumstances (having regard, in particular, to any conditions of bail which would be imposed); and
(ii) an inspector (or above) authorises the release on bail (having considered any representations made by P).

A person may be released on street bail at any time before (s)he arrives a police station.

A person released on street bail must be required to attend at a police station. Where a constable releases such a person (P) on bail:

(a) no recognisance for P's surrender to custody may be taken from him/her;
(b) no security for P's surrender to custody may be taken from him/her or from anyone else on his/her behalf;
(c) P may not be required to provide a surety or sureties for his/her surrender to custody; and
(d) no requirement to reside in a bail hostel may be imposed as a condition of bail.

Subject to (a)–(d), a constable may impose, as conditions of bail, requirements which (s)he considers to be necessary:

(i) to secure that P surrenders to custody;
(ii) to secure that P does not commit an offence while on bail;
(iii) to secure that P does not interfere with witnesses or otherwise obstruct the course of justice, whether in relation to him/herself or any other person;
(iv) for P's own protection or, if P is under the age of 18, for P's own welfare or in P's own interests.

No other requirements may be imposed as a condition of bail.

Where conditions are applied to street bail, the person released on bail (P) has the right to apply for variation of the conditions to a custody officer or to a magistrates' court (ss 30CA, 30CB).

By s 30B, a notice in writing must be given to P by the constable releasing him/her under s 30A. The notice must state the offence for which P was arrested, the ground on which (s)he was arrested and whether P is being released without bail or on bail. A notice given where release is on street bail must inform P that (s)he is required to attend a police station. It must also specify:

(a) the police station which P is required to attend; and
(b) the time on the bail end date when P is required to attend the police station.

Where street bail has been granted subject to conditions, the above notice must specify the requirements imposed by those conditions, and must explain the provisions relating to variation of conditions. P may be required to attend at a different police station from that specified in the notice or to attend at a different time or an additional time. P may not be required to attend at a time which is after the 'bail end date' in relation to him/her, ie the last day of the 28-day period beginning with the day after the day on which (s)he was arrested for the offence in relation to which street bail is granted.

Section 30C provides that the requirement to attend a police station may be cancelled by written notice and that, if P is required to attend a non-designated police station, (s)he must be released or taken to a designated police station not more than six hours after his/her arrival. Nothing in the Bail Act 1976 applies to street bail, nor do the provisions of PACE prevent re-arrest without warrant of a person released on street bail if new evidence justifying a further arrest has come to light since his/her release.

By s 30D, a constable may arrest without warrant a person who has been released on street bail but:

(a) fails to attend the specified police station at the appointed time; or
(b) is reasonably suspected by the constable to have broken a condition of bail.

A person so arrested must be taken to any police station as soon as practicable after the arrest. An arrest under s 30D is treated as an arrest for the purposes of s 30 (above) and s 31 (below).

Arrest for a further offence

Where a person is under arrest at a police station for an offence and it appears that, if released, (s)he would be liable to arrest for some other offence, PACE, s 31 requires that (s)he be arrested for that other offence. The usual procedures to be followed after an arrest must be carried out; the person must be told that (s)he is being arrested for that other offence and what the reasons are for that arrest, and (s)he must be cautioned. This must be done whether or not it is intended to release him/her at that time.   **4.58**

POWERS OF ENTRY AND SEARCH IN RELATION TO AN ARREST

Search and seizure on arrest for an offence

Search

PACE, s 32(1) provides that, where a person has been *arrested elsewhere than at a police station*, a constable may search him/her if the constable has reasonable grounds for believing that the arrested person may present a danger to him/herself or others.   **4.59**

Section 32(2) goes further and also empowers a constable in such a case, whether or not (s)he has such grounds:

(a) to *search* the *arrested person* for anything which (s)he might use to assist him/her to escape from lawful custody or which might be evidence relating to an offence (s 32(2)(a)); *and*
(b) if the offence for which (s)he has been arrested is an indictable offence, to *enter and search* any *premises* in which (s)he was when arrested or immediately before (s)he was arrested for evidence relating to the offence (s 32(2)(b)).

This power to search only extends to a search which is reasonably required for the purpose of discovering any such thing or any such evidence. A constable may not search a person under (a) unless (s)he has reasonable grounds for believing that the person may have concealed on him/her anything for which a search is permitted under (a) (s 32(5)). Similarly, a constable may not search premises under (b) unless (s)he has reasonable grounds for believing that there is evidence for which a search is permitted under (b) on the premises (s 32(6)).

(a) gives statutory authority to 'on the spot' searches at the time of an arrest. The power to search for evidence is particularly important since, if such property is not recovered at the time of such arrest, it is likely to be disposed of by the arrested person if an opportunity presents itself.

'*Premises*' in (b) and elsewhere in the PACE provisions in this chapter includes any place and, in particular, includes any vehicle, vessel, aircraft, or hovercraft, any offshore installation, any renewable energy installation and any tent or movable structure.

PACE, s 32(7) deals with the problem, in relation to (b), of communal occupation (including bedsit premises) by stating that where there are two or more separate dwellings, the search must be limited to the dwelling in which the arrest took place or in which the arrested person was immediately before his/her arrest *and* any part of the premises which the occupier of the dwelling uses in common with the other occupiers of other dwellings comprised in the premises. In the case of a bedsit, therefore, the constable would be able to search the room in which the arrested person was found and the common kitchen and lounge area used by all of the residents. A divisional court has stated that there is no decisive test to determine when a place of occupation is to be treated as divided into separate dwellings and that each case must be decided on its own facts. In the case in question, the bedrooms which homeless people had a licence to occupy for a day in a local authority hostel, each of which had a lock, were separate dwellings, because each licensee had a sufficient degree of exclusive possession.

The powers under s 32 do not authorise the removal of clothing in a public place other than an outer coat, jacket, or gloves, but they do authorise a search of a person's mouth (s 32(4)).

Section 32 does not provide any general powers of search either of persons or premises when an arrest is made. The particular circumstances described in s 32 must exist. If someone is arrested for a sexual assault and is known to be of a non-violent disposition, there are no grounds for an 'on the spot' search; nor would there be, in most circumstances, reason to search premises unless articles had been used in the commission of the offence.

Seizure from a person

4.60   A constable searching under PACE, s 32(1) may seize and retain anything which (s)he has reasonable grounds for believing that the person searched might use to cause physical injury to him/herself or another (s 32(8)).

A constable searching a person under PACE, s 32(2)(a) may seize and retain anything, other than an item subject to legal privilege (**4.68**), if (s)he has reasonable grounds for believing that the person might use it to escape from lawful custody or that it is evidence of an offence or has been obtained in consequence of the commission of an offence (s 32(9)).

The Criminal Justice and Police Act 2001 (CJPA 2001), s 51 provides additional powers of seizure from the person which apply to the power of seizure under PACE, s 32.

CJPA 2001, s 51(1) deals with the case where a person (P) carrying out a lawful search of another person finds something which (s)he reasonably believes may be, or may contain, something for which (s)he is authorised to search and to seize. It provides that if, in the circumstances, *it is not reasonably practicable for it to be determined whether what P has found (or its contents) is (or are) something which (s)he is entitled to seize*, P's powers of seizure include seizing so much of what (s)he has found as is necessary to remove from the place of the search to enable that to be determined.

Section 51(2) provides a power of seizure where a person carrying out a lawful search of someone finds anything ('the seizable property') which (s)he would be entitled to seize but for its being comprised in something else that (s)he has (apart from s 51(2)) no power to seize. Section s 51(2) provides that where:

(a) the power under which that person would have power to seize the seizable property is a power to which s 51 applies; and
(b) in all the circumstances *it is not reasonably practicable for the seizable property to be separated, at the time and place of the search, from that in which it is comprised,*

that person's power of seizure includes power to seize both the seizable property and that from which it is not reasonably practicable to separate it.

The factors to be taken into account when considering whether it is reasonably practicable to achieve one of the things italicised in s 51(1) or (2) are the same as in respect of s 50 (**4.61**), with the modifications necessary to recognise that s 51 applies to a search of a person.

Where a person exercises this power of seizure under s 51, (s)he must give written notice to the person from whom such property is seized:

(a) specifying what has been seized;
(b) specifying the grounds under which the power was exercised;
(c) setting out the effect of ss 59–61 (remedies and safeguards);
(d) specifying the name and address of the person to whom any application for the return of the seized property must be made; and
(e) specifying the name and address of the person to whom an application may be made to be allowed to attend the initial examination of the seized property which must be held to determine how much of it is property for which the person seizing had a power to search (s 52(4)).

A divisional court has held that non-compliance with s 52 does not necessarily render unlawful a seizure under s 50 (**4.61**) or 51; it is a factor to be considered by the court on an application under s 59 (**4.89**), or in judicial review proceedings, or under PACE, s 78 (exclusion of unfair evidence) in any later trial.

Section 51 also applies to a number of other powers of seizure from the person listed in CJPA 2001, Sch 1, Part 2, some of which are referred to later in this book.

'Seize' (and cognate expressions) in ss 50 (below) and 51 include 'take a copy of'.

Seizure from premises

The powers of seizure under PACE, s 19 (**4.86**) apply to a search of premises under   **4.61**
PACE, s 32(2)(b).

In addition, the CJPA 2001, s 50(1) deals with situations where a person who is lawfully on any premises finds objects on those premises which (s)he reasonably believes to contain something for which (s)he is authorised to search and in respect of which there would be a power of seizure under s 32 or under one of the more than 70 other statutory provisions listed in CJPA 2001, Sch 1, Part 1 to which s 50 applies. Some of these are referred to later in this book. If, in the circumstances, *it is not reasonably practicable for it to be determined whether what the person has found (or its contents) is (or are) something which (s)he is entitled to seize,* his/her powers of seizure under s 50(1) include seizing so much of what (s)he has found as is necessary to remove from the premises to enable that to be determined.

Section 50(2) provides a power of seizure where a person lawfully on any premises finds anything on those premises ('the seizable property') which (s)he would be entitled to seize but for its being comprised in something else that (s)he has (apart from s 50(2)) no power to seize. Section 50(2) provides that where:

(a) the power under which that person would have power to seize the seizable property is a power to which s 50 applies; and
(b) in all the circumstances it is *not reasonably practicable for the seizable property to be separated in those premises from that in which it is comprised,*

that person's power of seizure includes power to seize both the seizable property and that from which it is not reasonably practicable to separate it.

Section 50(3) states that the factors to be taken into account in considering whether it is *reasonably practicable to achieve one of the things italicised in s 50(1) or (2)* are confined to:

(a) how long it would take to carry out the determination or separation on those premises;
(b) the number of persons who would be required to carry out that determination or separation on those premises within a reasonable period;
(c) whether the determination or separation would (or would if carried out on those premises) involve damage to property;
(d) the apparatus or equipment that it would be necessary or appropriate to use for the carrying out of the determination or separation; and
(e) in the case of separation, whether the separation:
    (i) would be likely, or
    (ii) if carried out by the only means that are reasonably practicable on those premises, would be likely,
    to prejudice the use of some or all of the separated seizable property for the purpose for which something seized under the power in question is capable of being used.

A divisional court has held that the *reasonable practicability of separation* test referred to above is a broad practical test and not a narrower test of technical or physical separation. Physical or technical considerations might play a part in the analysis, but they were by no means the only criteria for assessment. In the case before the court separation would have been hugely time consuming, expensive, and disruptive, and for no discernible good reason; the judge's decision that separation was not reasonably practicable was upheld.

Where a power of seizure under s 50 has been exercised, written notice containing the same information as under s 52(4) (**4.60**) must be given to the occupier, or some other person on the premises who is in charge (s 52(1) and (2)). Where there is no person on the premises a notice must be attached, in a prominent place, to the premises (s 52(3)). As to the effect of a breach of s 52, see **4.60**.

Examination and return of property seized under s 50 or 51

4.62   CJPA 2001, s 53(2) requires that at a subsequent initial examination, which must take place as soon as reasonably practicable, property seized under s 50 or 51 which does *not* fall within s 53(3) must be separated from the rest of the seized property and returned as soon as reasonably practicable. By s 53(3), property does not have to be returned if:

(a) it is property for which there was a power of search and it is not an item subject to legal privilege (**4.68**);

(b) there are reasonable grounds for believing that it is property obtained in consequence of an offence, or evidence in relation to any offence, and that it is necessary for it to be retained in order to prevent its concealment, loss, alteration, or destruction; or

(c) it is property which cannot be reasonably practically separated from that in (a) or (b).

By s 54(1) or 55(1), (i) property which is or contains an item subject to legal privilege, and (ii) property which is or contains excluded or special procedure material (**4.68**), respectively must be returned unless (by s 54(2) or 55(2) respectively) its retention is authorised by the grounds set out in (b) or, where comprised in something else, it cannot reasonably practicably be separated from other property whose return is not required or which satisfies (b).

A divisional court has held that the obligation in s 53 to return seized property extends to data copied from seized electronic devices which are, following copying, restored to their owners, and that such data can be returned through deletion or destruction.

Entry to premises to effect an arrest

Apart from the common law power to enter premises to deal with a breach of the peace or to prevent it (**18.1**), all powers of entry to effect an arrest are statutory.   **4.63**

PACE, s 17 provides that, without prejudice to any other enactment, a constable may enter and search any premises (as defined at **4.59**) for the purpose of:

(a) executing a warrant of arrest issued in connection with or arising out of criminal proceedings, or a warrant of commitment issued under the Magistrates' Courts Act 1980, s 76;

(b) arresting someone for an indictable offence;

(c) arresting someone for an offence under the Public Order Act 1936, s 1 (prohibited uniforms), the Public Order Act 1986, s 4 (fear or provocation of violence), RTA 1988, s 4 (driving, etc under the influence of drink or drugs) or s 163 (failure to stop when required to do so by a police constable in uniform), the Transport and Works Act 1992, s 27 (drink or drugs on public guided transport systems), or the Animal Welfare Act 2006, s 4, 5, 6(1) or (2), 7, and 8(1) or (2) (unnecessary suffering, mutilation, docking of tails, administering poisons, and animal fighting);

(d) arresting someone for an offence under CJPOA 1994, s 76 (failure to comply with an interim possession order), the Criminal Law Act 1977, ss 6 to 8, or 10 (offences of entering and remaining on property), or the Legal Aid, Sentencing and Punishment of Offenders Act 2012, s 144 (squatting in a residential building)—in these cases the arresting constable must be in uniform;

(e) arresting, in pursuance of the Children and Young Persons Act 1969, s 32(1A), any child or young person who has been remanded to local authority accommodation or youth detention accommodation;

(f) arresting someone for an offence to which the Animal Health Act 1981, s 61 applies (powers of entry and search in relation to rabies offences);

(g) arresting a person under any of the following provisions:

(i) PACE, s 30D (**4.57**);

(ii) PACE, s 46A (**6.79**);

(iii) the Bail Act 1976, s 5B(7) (arrest where a person fails to surrender to custody in accordance with a court order);

(iv) the Bail Act 1976, s 7 (**6.80**);

(v) the Legal Aid, Sentencing and Punishment of Offenders Act 2012, s 97 (arrest where a child is suspected of breaking conditions of remand);

(h) recapturing someone who is, or is deemed for any purpose to be, unlawfully at large while liable to be detained:

(i) in a prison, young offender institution, secure training centre or secure college; or

(ii) in pursuance of the Powers of Criminal Courts (Sentencing) Act 2000, s 92 or the Sentencing Code, s 260 (detention of children and young persons guilty of grave crimes), in any other place;

(i) recapturing someone unlawfully at large whom (s)he is pursuing (ie chasing); or

(j) saving life or limb or preventing serious damage to property. A divisional court has held that the words 'life or limb' are wide enough to cover saving someone from him/herself as well as saving someone from a third party.

Except in the case of (j), the powers of entry and search under PACE, s 17: (a) are only exercisable where a constable has reasonable grounds for believing that the person whom (s)he is seeking is on the premises, and (b) are limited in respect of premises consisting of two or more dwellings in the same way as with the powers to search under s 32 (**4.59**).

A search under s 17 may only be made to the extent reasonably required for the purpose for which the power of entry is required. The courts have ruled that, when exercising his/her right of entry by force under s 17, a police officer must give any occupant present the reason why (s)he is seeking entry even if that reason is apparent in the circumstances. Otherwise the forcible entry will be unlawful, unless circumstances make giving reasons impossible, impracticable or undesirable. If the real reason is to arrest a person inside for an offence, for instance, it is insufficient to tell the occupants that the officer wishes to 'speak to' that person about the offence.

Entry, search, and seizure after an arrest

Entry and search

4.64   PACE, s 18(1) empowers a constable to enter and search any premises (**4.59**) which *are* occupied or controlled by a person under arrest for an indictable offence before that person is taken to a police station or released under s 30A (**4.57**), if (s)he has reasonable grounds for suspecting that there is on the premises evidence, other than items subject to legal privilege (**4.68**), which relates to that offence or to another indictable offence connected with or similar to that offence (eg, if the person has been arrested for burglary, the proceeds of other burglaries or large-scale theft, but not drugs). A search under s 18 is only permitted to the extent reasonably required for the purpose of discovering such evidence (s 18(3)). Because the power under s 18 only applies to premises *occupied or controlled* by the arrested person, a search of premises not so occupied or controlled is unlawful—even if they are reasonably believed to be so occupied or controlled (eg because the arrested person has given a false name and address).

Unless the following exception applies, a search under s 18 may not be made without the written authorisation of an officer of the rank of inspector (or above). The exception is that such a search may be carried out by a constable without an authorisation before a person is taken to a police station or released under s 30A if the presence of that person elsewhere is necessary for the effective investigation of the offence (s 18(4)–(5A)). In such a case, the constable who conducts the search must notify an inspector (or above) of it as soon as practicable.

PACE Code B: the Code of Practice for Searches of Premises by Police Officers and the Seizure of Property Found by Police Officers on Persons or Premises provides that if possible the authorising officer should record the authority on the Notice of Powers and Rights (which should be given or left as explained at **4.83**). The record of the grounds for the search and the nature of the evidence sought should be made in the custody record if there is one. Otherwise it should be made in the officer's pocket book or any official report book or electronic recording device issued to a constable that enables the record to be made, or the search record. If the person in occupation or control of the premises is in police custody at the time that the search is carried out, the record must be made in the custody record.

An officer who proposes to enter and search premises under a written authorisation under s 18 must explain to the occupier, in so far as it is practicable to do so, the reason why (s)he intends so to act. Failure to do so will render his/her subsequent conduct unlawful. A divisional court has held that simply offering to show the written author-isation at a window to an occupier who refused to come to the window was insufficient; the reason must be stated.

A search of any person who has not been arrested, which is carried out during a search of premises, must be carried out in accordance with Code A.

A divisional court has held that there is no priority or preference between investiga-tors obtaining a search warrant and their using post-arrest powers of search under s 18 or 32 (**4.60**). It stated that the powers were distinct powers with distinct criteria; where the criteria for both processes could be fulfilled, investigators had a choice.

Seizure

A constable may seize and retain anything on the premises for which (s)he may search **4.65** under the above power (PACE, s 18(2)). (Although a vehicle, vessel, or tent is 'prem-ises' for the purpose of PACE, this does not prevent the seizure of the vehicle, vessel or tent itself.) The additional powers of seizure under CJPA 2001, s 50 (**4.61**) apply to this power of seizure.

POWERS TO SEARCH UNDER A SEARCH WARRANT

Power to issue warrant

Many statutes provide *particular* powers to enter and search premises under the au- **4.66** thority of a warrant. An example is provided by the Terrorism Act 2000, Sch 5 (**4.72**). In addition, PACE, s 8 provides justices with a *general* power to issue warrants to enter and search premises (as defined at **4.59**).

Applications for a search warrant are served on the court officer (or the court, if the court office is closed), and not on a person who would be affected by the warrant. Applications are heard in private, unless the court otherwise directs, in the absence of any person who would be affected by the warrant.

Search warrants under PACE, s 8

Requirement for issue of search warrant under s 8

PACE, s 8(1) provides that if, on an application made by a constable, a justice is satisfied **4.67** that there are reasonable grounds for believing that:

(a) an indictable offence (or a 'relevant offence' under the Immigration Act 1971) has been committed; and
(b) there is material on premises mentioned in PACE, s 8(1A) (below) which is likely to be of substantial value (whether by itself or with other material) to the investigation of the offence; and
(c) the material is likely to be relevant (ie admissible) evidence; and
(d) it does not consist of or include items subject to legal privilege, excluded material or special procedure material (**4.68**); and
(e) *any* of the conditions specified in s 8(3) below applies, in relation to each set of premises specified in the application;

he may issue a warrant authorising a constable to enter and search the premises.

A divisional court has pointed out that that the words 'reasonable grounds for believing that an indictable offence has been committed' mean what they say; they do not mean that an offence has been committed.

A divisional court has held that a computer and its hard disc are 'material' for the purposes of s 8(1).

Section 8(1A) provides that the 'premises' mentioned in (b) above are:

(a) one or more sets of premises specified in the application (in which case the application is for a 'specific premises warrant'); or
(b) any premises occupied or controlled by a person specified in the application, including such sets of premises as are so specified (in which case the application is for an 'all-premises warrant').

Section 8(1B) provides that, if the application is for an all-premises warrant, the justice of the peace must also be satisfied:

(a) that, because of the particulars of the offence referred to in s 8(1)(a) above, there are reasonable grounds for believing that it is necessary to search premises occupied or controlled by the person in question which are not specified in the application in order to find the material referred to in s 8(1)(b) above; and
(b) that it is not reasonably practicable to specify in the application all the premises which (s)he occupies or controls and which might need to be searched.

These additional provisions assist in the investigation of serious organised crime. Crime syndicates may occupy or control many sets of premises and the search of one might lead to the discovery of other premises upon which the material being sought is situated. Provided that the search continues within buildings which are within the syndicate's control, it will be authorised by an 'all-premises warrant'.

It is vital to remember that the justice must be satisfied that all the requirements in s 8(1)(a) to (d) above are satisfied. If (s)he is, it is necessary for him/her also to be satisfied of the existence of any one of the conditions set out in s 8(3) referred to in s 8(1)(e), namely that:

(a) it is not practicable to communicate with any person entitled to grant entry to the premises;
(b) it is practicable to communicate with a person entitled to grant entry to the premises but it is not practicable to communicate with any person entitled to grant access to the evidence;
(c) entry to the premises will not be granted unless a warrant is produced; or
(d) the purpose of the search may be frustrated or seriously prejudiced unless a constable arriving at the premises can secure immediate entry to them.

If the application for the warrant does not identify which of these conditions is being relied on, the issue of the warrant will be unlawful.

The obtaining of a search warrant is never to be treated as a formality. All the material necessary to justify the grant of the warrant should be contained in the information on the application form. The Supreme Court has held that the material relied on as giving rise to reasonable grounds for believing the matters set out in s 8(1) need not take any particular nature or form, nor itself be admissible in evidence at any trial that might be envisaged. In addition, it held, the justice can have regard to material which cannot on public interest grounds subsequently be disclosed to a person affected by the warrant who wishes to challenge it.

A divisional court has held that as a matter of good practice, the applicant for a search warrant under s 8 should, save in exceptional circumstances, be an officer directly involved in the investigation.

Items for which a search warrant under PACE, s 8 cannot be issued

As noted above, a search warrant cannot be issued under s 8 where the material in **4.68** question is reasonably believed to consist of or include items prima facie subject to legal privilege, excluded material or special procedure material. In such a case application should be made to a judge under the provisions of PACE, Sch 1 described below.

Various terms included in PACE, s 8(1) are defined by ss 10 to 13 as follows.

Items subject to legal privilege   These are defined as:
(a) communications between a professional legal adviser and client (or representative) made in connection with the giving of legal advice;
(b) communications between a professional legal adviser and client (or representative), or between any such adviser or his/her client (or representative) and any other person, made in contemplation of legal proceedings and for the purpose of them; and
(c) items enclosed with or referred to in such communications.

The same definition also applies for the purpose of the other legislation referred to in this chapter.

The exemption relating to items subject to legal privilege only applies if the material is in the possession of a person entitled to possess it. Items held with the intention of furthering a criminal purpose are specifically excluded from being subject to legal privilege, even though that intention does not exist on the part of the holder (eg a solicitor) but does on the part of another (eg a solicitor's client).

Excluded material   This means:
(a) personal records (ie documentary and other records relating to the health of an individual or counselling given to him/her) acquired or created by a person in the course of a trade, business, etc or for the purpose of any paid or unpaid office and held in confidence by him/her;
(b) human tissue or tissue fluid taken for diagnosis or medical treatment and held in confidence; or
(c) journalistic material (ie material acquired or created for the purpose of journalism and held by the person acquiring or creating it for that purpose) consisting of documents or other records and held in confidence.

Special procedure material   This means:
(a) journalistic material, other than excluded material, and

(b) material, other than items subject to legal privilege and excluded material, held in confidence and acquired or created in the course of any trade, business, etc.

Production orders and search warrants under PACE, Sch 1

4.69   Although a justice may not issue a search warrant under PACE, s 8 in respect of material of any of the three types set out above, s 9 recognises that there may be occasions in which it is necessary to obtain access to excluded material or special procedure material by providing that a constable may obtain access to such material by making an application under Sch 1. Section 9 does not provide for access to items subject to legal privilege. However, a divisional court has held that a mobile phone can properly be the subject of a search warrant under s 9 and Sch 1 even where material subject to legal privilege may be found on it, provided that the wording of the warrant clearly excludes any such material from that which can be sought or seized.

Under PACE. Sch 1, a Circuit judge may make a 'production order' or, in certain cases, issue a search warrant in respect of excluded material or special procedure material.

These provisions are intended to protect the confidence of the maker or holder of the record, etc, and not that of the suspect. Consequently, it is open to the maker or holder voluntarily to disclose the material. It is only where there is no consent on his/her part to do so that the special provisions of s 9 and Sch 1 come into play.

Production order

4.70   An application for a production order must be made by a constable. The person who has custody of the material—but not the suspected person unless (s)he is in custody of the material—must be notified of the application (and of the material sought) and allowed to attend the hearing of the application and make representations. The Supreme Court has held that this means that each party should know what material the other party is asking the judge to take into account in making his/her decision and should have a fair opportunity to respond to it.

A production order requires the person apparently in possession of the material in question *either* to produce it to the constable for him/her to take away *or* to give the constable access to it not later (*in either case*) than the end of a seven-day period from the date of the order or the end of such longer period as the order may specify. Where the material consists of information stored in electronic form, an order to produce has effect as an order to produce the material in a form in which it can be taken away and in which it is visible and legible, and an order to give access has effect as an order to give a constable access to the material in a form in which it is visible and legible.

The making of a production order depends on one or other of two sets of 'access conditions' being fulfilled. If the judge is satisfied of this, (s)he may (not must) make a production order.

The two sets of access conditions   The first set of access conditions is fulfilled if:
(a) there are reasonable grounds for believing that:
    (i) an indictable offence has been committed;
    (ii) there is material which consists of special procedure material or includes special procedure material and does not also include excluded material on premises specified in the application, or on premises occupied or controlled by a person specified in the application (including all such premises on which

there are reasonable grounds for believing that there is such material as it is reasonably practicable so to specify);

(iii) the material is likely to be of substantial value (whether by itself or together with other material) to the investigation in connection with which the application is made; and

(iv) the material is likely to be relevant evidence (held by a divisional court to mean that the material is likely to be such that, if produced, it would be immediately admissible in evidence at a trial without more);

(b) other methods of obtaining the material:

(i) have been tried without success; or

(ii) have not been tried because it appeared that they were bound to fail; and

(c) it is in the public interest having regard:

(i) to the benefit likely to accrue to the investigation if the material is obtained; and

(ii) to the circumstances under which the person in possession of the material holds it,

that the material should be produced or that access to it should be given.

The second set of access conditions is met if:

(i) there are reasonable grounds for believing that there is material which consists of or includes special procedure material or excluded material on premises specified in the application or on premises occupied or controlled by a person specified in the application (including all such premises on which there are reasonable grounds for believing that there is such material as it is reasonably practicable so to specify);

(ii) prior to the enactment of PACE, s 9, a search of such premises for that material could have been authorised by the issue of a warrant to a constable under one of the enactments replaced by s 9; and

(iii) the issue of such a warrant would have been appropriate.

It will be noted that the second set of access conditions is not limited to cases where it is reasonably believed that an indictable offence has been committed.

A production order can be made even if it infringes or might infringe the privilege against self-incrimination of the person ordered to make production.

A divisional court has held that, before granting the application, the judge must be satisfied that it is substantially the last resort, ie that other practicable methods of obtaining the material have been exhausted without success.

Search warrant

4.71 The requirement of notice, together with the seven (or more) days' grace referred to above, gives a person time to dispose of incriminating material.

PACE, Sch 1 provides that a judge, on application by a constable, may issue a specific premises or all-premises search warrant in two types of case.

First, a judge may do so if satisfied that:

(a) either set of access conditions is met; and

(b) any of the four further conditions is also met in relation to each set of premises specified in the application:

(i) that it is not practicable to communicate with any person entitled to grant entry to the premises;

(ii) that it is practicable to communicate with a person entitled to grant entry to the premises but it is not practicable to communicate with any person entitled to grant access to the material;

(iii) that the material contains information which is subject to a statutory restriction on disclosure or obligation of secrecy and is likely to be disclosed in breach of it if a warrant is not issued;

(iv) that service of notice of an application for a production order may seriously prejudice the investigation.

A search warrant (as opposed to a production order) will not be granted in practice in respect of a solicitor's office, unless proceeding by way of production order might seriously prejudice the investigation.

Second, a judge may also issue a search warrant under Sch 1 if satisfied that:

(a) the second set of access conditions is met; and
(b) a production order relating to the material has not been complied with.

The judge may not issue an all-premises warrant under Sch 1 unless (s)he is satisfied:

(a) that there are reasonable grounds for believing that it is necessary to search premises occupied or controlled by the person in question which are not specified in the application, as well as those which are, in order to find the material in question; and
(b) that it is not reasonably practicable to specify all of the premises which (s)he occupies or controls which might need to be searched.

Code B requires that an inspector (or above) be present and in charge of a search under a warrant issued under PACE, Sch 1 or TA 2000, Sch 5 (below). The Code also contains further provisions concerning such a search.

Search warrants under Terrorism Act 2000, Sch 5

4.72   By TA 2000, Sch 5, a constable may apply to a justice of the peace for the issue of a warrant for the purposes of a terrorist investigation authorising a constable to enter specified premises, to search those premises and persons found there, and to seize and retain relevant material. The 'premises' may be one or more sets of premises specified in the application (in which case the application is for a specific premises warrant); or any premises occupied or controlled by a person specified in the application, including such sets of premises as are so specified (in which case the application is for an 'all-premises' warrant). A search warrant under the present provision does not authorise the seizure and retention of items subject to legal privilege.

Code A points out that, although the power to search persons found on premises to which a search warrant issued under TA 2000, Sch 5 applies does not require prior specific grounds to suspect that the person to be searched is in possession of an item for which there is an existing power to search, it is still necessary to ensure that the selection and treatment of those searched under these powers is based upon objective factors connected with the search of the premises, and not upon personal prejudice.

Schedule 5 also provides for production orders and search warrants in respect of excluded or special procedure material for the purposes of a terrorist investigation.

Procedure before application is made for a search warrant or a production order

The relevant provisions are contained in Code B. Where information is received appearing to justify an application for any search warrant or for a production order, the officer concerned must take reasonable steps to check that the information is accurate, recent, and not provided maliciously or irresponsibly. An application is not permitted on the basis of information from an anonymous source where corroboration has not been sought. The nature of the articles and their location must be established as specifically as possible.

4.73

The officer must also make reasonable inquiries to establish what, if anything, is known about the likely occupier of the premises and the nature of the premises themselves, and whether they have previously been searched (and, if so, how recently); (s)he must also obtain any other information relevant to the application.

An application to a justice of the peace for a search warrant or to a judge for a search warrant or production order may not be made without the signed written authority of an inspector (or above) or, in a case of urgency where no officer of this rank is readily available, the senior officer on duty. Where the application is made to a judge under TA 2000, Sch 5 it must be supported by a signed written authority from a superintendent (or above). In addition, other than in a case of urgency, the community relations officer must be consulted before a search takes place which might have an adverse effect on police/community relations. In urgent cases, the local police community liaison officer should be informed of the search as soon as possible after it has been made.

Making an application

Between them, PACE, s 15 and Code B provide the following rules for *all* search warrants. An application for a search warrant must be supported by information in writing, specifying:

4.74

(a) the enactment under which the application is made;
(b) (i) whether the warrant is to authorise entry and search of:
    (a) one set of premises, or
    (b) if the application is under PACE s 8, or Sch 1, more than one set of specified premises or all premises occupied or controlled by a specified person;
  (ii) the premises to be searched;
(c) the object of the search;
(d) the grounds on which the application is made (including, when the purpose of the search is to find evidence of an alleged offence, an indication of how the evidence relates to the investigation);
(e) where the application is under s 8 or Sch 1, for a single warrant to enter and search:
  (i) more than one set of specified premises, the officer must specify each set of premises which it is desired to enter and search;
  (ii) all premises occupied or controlled by a specified person, the officer must specify:
    (a) as many sets of premises which it is desired to enter and search as it is reasonably practicable to specify,
    (b) the person who is in occupation or control of those premises and any others which it is desired to search,

(c) why it is necessary to search more premises than those which can be specified, and

(d) why it is not reasonably practicable to specify all the premises which it is desired to enter and search;

(f) whether an application under s 8 is for a warrant authorising entry and search on more than one occasion; if so, the officer must state the grounds for this and whether the desired number of entries authorised is unlimited or a specified maximum;

(g) that there are no reasonable grounds for believing that material to be sought consists of or includes:

(i) items subject to legal privilege (when applying to a justice of the peace or a judge); or

(ii) excluded material or special procedure material (when applying to a justice of the peace);

however, this does not affect the additional powers of seizure in the Criminal Justice and Police Act 2001, ss 50 and 51: (**4.60–4.61**);

(h) if applicable, a request for the warrant to authorise a person or persons to accompany the officer who executes the warrant.

Although the identity of an informant need not be disclosed, the officer must be prepared to deal with questions about the accuracy of previous information provided by that source or other related matters. The Court of Appeal has held that an applicant for a search warrant under PACE, s 8 must make a full and frank disclosure of material facts within his/her knowledge; in effect, the applicant for the warrant must put on a 'defence hat' and ask him/herself what, if (s)he was representing the person who would be affected by the warrant, (s)he would say to the justice, and then state it.

A divisional court has held that police officers would be well advised to seek legal advice before applying for a search warrant under Sch 1, especially if the material sought is more than likely to contain legally privileged material. It added that the application must:

(a) set out each of the statutory requirements which has to be satisfied in the particular case before the warrant in question can be granted, although only the set of access conditions which has been selected in a particular case should be dealt with;

(b) explain how each of the relevant requirements is satisfied by setting out all the relevant facts relied on, including all facts and matters which are said to show that a particular 'reasonable belief' is justified and (in the case of the first set of access conditions) why it is in the public interest for the material to be produced (having regard to **4.70**(c));

(c) state whether there might be a claim for legal privilege in respect of any communication sought and, if so, how and why that would arise together with precise details of the arrangements which are to be taken to ensure that there will be an independent supervising lawyer present at the time of the search;

(d) make full and frank disclosure (as in the previous paragraph).

An application under TA 2000, Sch 5 must indicate why a production order would be inappropriate.

The constable must answer on oath any question asked by the justice or judge. If an application is refused, no further application may be made for a warrant to search those premises unless supported by additional grounds.

The search warrant

PACE, s 8(1C) provides that a warrant may authorise entry to and search of premises   **4.75**
on more than one occasion if, on the application, the justice of the peace is satisfied that
it is necessary to authorise multiple entries in order to achieve the purpose for which
(s)he issues the warrant. PACE, s 8(1D) provides that, if multiple entries are so authorised, the number of entries authorised may be unlimited, or limited to a maximum.

Section 8(2) authorises a constable to seize and retain anything for which a search
has been authorised.

Although PACE distinguishes between a 'specific premises warrant' and an 'all-
premises warrant', a single warrant may include both types provided that the relevant
information is given in the application.

By s 15(6), a warrant (which term includes a schedule to the warrant):

(a) must specify:
  (i)   the name of the person who applies for it (it is not enough to state, for example,
         that that applicant is 'the paedophile intelligence unit'),
  (ii) the date on which it is issued,
  (iii) the enactment under which it is issued, and
  (iv) each set of premises to be searched, or (in the case of an all-premises warrant)
         the person who is in occupation or control of premises to be searched, together
         with any premises under his/her occupation or control which can be specified
         and which are to be searched; and
(b) must identify, so far as is practicable, the articles or persons to be sought. (The pur-
      pose of (b) is to enable anyone interested in the execution of the warrant to know
      the limits of the powers granted, and to enable him/her to challenge the lawfulness
      of the seizure of a particular item. A divisional court has held that specifying the
      mobile phone referred to in **4.69** as the item sought, rather than the material found
      on it, was capable of satisfying (b).)

In relation to (a)(iv), if the police want to keep information about the search of other
sets of premises from the occupier, separate warrants should be obtained; alternatively,
the addresses of the other sets of premises can be redacted. A divisional court has held
that whether or not a warrant satisfies the requirements of s 15(6) must be judged exclusively by reference to the terms of the warrant and not by reference to any other
source such as the information supporting the grant of the warrant.

A search based on a warrant which does not satisfy the requirements of s 15(6) is
unlawful.

Section 15(7) provides that there must be two copies of a specific premises warrant
which specifies only one set of premises and does not authorise multiple entries. In
the case of any other warrant, as many copies as are reasonably required may be made.
Where a warrant authorises multiple entries it must specify whether the number of entries authorised is unlimited, or limited to a specified maximum.

A divisional court has held that 'a copy' of a search warrant means a full copy, and
that therefore a copy which omitted the premises to be searched and left that information to be filled in by hand in respect of particular premises by the officer executing the
warrant on those premises was not a valid copy of the warrant.

Execution of a search warrant

PACE, s 15(1) provides that an entry on or search of premises under a search warrant is   **4.76**
unlawful and there is no right to retain anything seized, unless ss 15 (above) and 16 are

complied with. Section 16 sets out the following provisions which apply to all search warrants.

A warrant may be executed by any constable. A warrant may authorise a person, other than a constable or designated person, to accompany the constable who executes the warrant (s 16(2)); for example, an expert in a particular field to help accurately identify the material sought. Such a person has the same powers as a constable in relation to the execution of the warrant and the seizure of anything to which it relates, but may exercise them only in the company, and under the supervision, of a constable (s 16(2A), (2B)). Entry and search must be within three calendar months of issue (s 16(3)) (or within the period specified in the enactment under which the warrant is issued, if shorter) and must be at a reasonable hour unless this would frustrate the purpose of the search (s 16(4)). However, this does not apply to warrants issued under TA 2000, Sch 5 which are exercisable only within 24 hours of issue. Section 16(3A) covers the execution of all-premises warrants. It provides that in the case of an all-premises warrant, no premises which are not specified in the warrant may be entered or searched unless an inspector (or above) has in writing authorised them to be entered. Section 16(3B) provides that no premises may be entered or searched for the second or any subsequent time under a warrant which authorises multiple entries unless an inspector (or above) has authorised that entry in writing. Code B provides that the inspector (or above) referred to in s 16(3A) and (3B) must not be involved in the investigation.

A divisional court has held that, where a search warrant authorises entry 'on one occasion' to search for specified goods, the word 'occasion' does not equal one day, or require the search to be completed in one calendar day. Thus, it is not a breach of such a warrant for the police to occupy the premises for a number of days while the search is carried out.

Where the occupier of the premises is present, the constable must, when seeking to execute a search warrant:

(a) identify him/herself (by warrant or other identification number in the case of terrorism inquiries) and, if not in uniform, show his/her warrant card (but in so doing in the case of terrorism inquiries, (s)he need not reveal his/her name);

(b) produce the warrant to the occupier; and

(c) supply him/her with a copy of it.

The constable must act likewise in relation to a person apparently in charge of the premises in the absence of the occupier.

A divisional court has held that requirements (a) to (c) existed so that the occupier of premises might know that those who executed a search warrant had lawful authority to do so and could see before the search began what property might be seized under the search warrant. It held that a failure by a constable executing a search warrant to hand over a copy of it until after the search had been completed was unlawful as being in breach of the requirements. Likewise, another divisional court has held that a search was unlawful where a warrant stated that the premises to be searched were described in an attached schedule but the copy given to the occupier did not include the schedule. This would be the case even if an uncertified photocopy of the schedule was attached to the warrant.

In any other case, the constable must leave a copy of the warrant in a prominent place on the premises.

The premises may only be searched to the extent required for the purpose for which the warrant was issued. A divisional court has held that a breach of this provision does not necessarily render the entirety of an entry, search and seizure unlawful.

A divisional court has held that an independent supervising lawyer should be present at a search where there is a potential issue as to legal professional privilege.

The Court of Appeal has held that the Equality Act 2010 requires the police to make reasonable adjustments to the practice of conducting searches in spoken English so that it does not have a detrimental effect on deaf persons, eg by adopting a policy of a sign language interpreter on standby every time there is a search of premises of a person known or believed to be deaf, or taking other reasonable steps, such as having available officers skilled in lip reading and sign language.

Save in exceptional circumstances the media should not be invited to be present when a search warrant (or any other investigative procedure) is being executed, nor should they be invited to police briefings prior to the execution of a warrant, because reports emanating from such involvement are liable to prejudice a fair trial.

SEIZURE

A constable who is searching any person or premises under any statutory power or **4.77** with the consent of the occupier may seize anything, other than a legally privileged item, which:

(a) is covered by a warrant;
(b) the officer has reasonable grounds for believing is evidence of an offence or has been obtained in consequence of the commission of an offence but only if seizure is necessary to prevent the items being concealed, lost, disposed of, or tampered with; or
(c) is covered by the additional powers of seizure in CJPA 2001, s 50 and 51 (**4.60-4.61**) allowing an officer to seize property and retain it for sifting or examination elsewhere.

GENERAL PROVISIONS ON ENTRY, SEARCH, SEIZURE, ACCESS, AND RETENTION

In addition to the specific rules already described, Code B lays down a number of rules **4.78** which apply to searches by police of premises (as defined at **4.59**), whether those of the suspect or anyone else including the victim:

(a) undertaken for the purposes of an investigation into an alleged offence, with the occupier's consent, other than searches made in the following circumstances:
    (i) routine scenes-of-crime searches;
    (ii) calls to a fire or burglary made by or on behalf of an occupier or searches following the activation of fire or burglar alarms;
    (iii) searches where it is unnecessary to seek consent because in the circumstances this would cause disproportionate inconvenience to the person concerned;
    (iv) bomb threat calls;
(b) under powers conferred by PACE, ss 17 (entry to arrest/search), 18 (entry and search after arrest) and 32 (search on arrest);
(c) undertaken in pursuance of any search warrant issued in accordance with PACE, s 15 or 16 (**4.74–4.76**); or
(d) under any other police power to enter premises with or without a search warrant for any purpose connected with the investigation into an alleged or suspected offence or with the enforcement of a TPIM notice.

The exception for routine scenes-of-crime searches ceases to apply if the search develops into more than just a routine scenes-of-crime search. At that point, at the latest, Code B applies.

Entry other than with consent

4.79   The officer in charge must first attempt to communicate with the occupier or any other person entitled to grant access to the premises by explaining the authority under which (s)he seeks entry to the premises and ask the occupier to allow him/her to do so, unless:

(a) the premises are unoccupied;
(b) the occupier and any other person entitled to grant access are absent; or
(c) there are reasonable grounds to believe that to alert the occupier or any other person entitled to grant access by attempting to communicate with him/her would frustrate the object of the search or endanger the officers concerned or other persons.

The circumstances in (c) might exist where there are known to be a number of persons on the premises all of whom are suspected of being involved in the offence, any of whom could dispose of the evidence whilst these procedures are being followed; or, in the case of entry to search for an armed criminal, where danger might arise if (s)he was warned of imminent arrest. Although, in the circumstances in (c), an officer need not comply with the requirements as to identification, production of his/her warrant card (if not in uniform) and search warrant (if any) before effecting entry, (s)he must do so before conducting the search (except that if the search is under a search warrant it is enough if (s)he gives a copy of the search warrant at the first reasonable opportunity).

Unless (c) applies, where the premises are occupied, the officer in charge of the search must before the search begins:

(i)   identify him/herself (by warrant or other identification number in a terrorist investigation) and, if not in uniform, show his/her warrant card (although (s)he may do so without revealing his/her name in a terrorist investigation);
(ii) state the purpose of the search and the grounds for undertaking it; and
(iii) identify and introduce any person accompanying the officer on the search (such persons should carry identification for production on request) and briefly describe that person's role in the process.

Searches with consent

4.80   Code B provides that, if it is proposed to search premises with the consent of a person (X) entitled to grant entry, the consent must, if practicable, be in writing. Before seeking such consent the officer in charge must state the purpose of the search and its extent. This information must be as specific as possible, particularly regarding the articles or persons sought and the parts of the premises to be searched. X must be clearly informed that (s)he is not obliged to consent, that any consent may be withdrawn at any time, including before the search starts or while it is underway, and that anything seized may be produced in evidence. If, at the time, X is not suspected of an offence the officer must tell him/her so when stating the purpose of the search. An officer cannot enter and search premises or continue to search premises if consent has been given under duress or is withdrawn before the search is completed.

In the case of a lodging house or other similar accommodation, a search should not be made on the basis solely of the landlord's consent.

It is unnecessary to seek consent where this would cause disproportionate inconvenience to the person concerned; for example, where it is reasonable to assume that innocent occupiers would agree to, and expect that, police would take the proposed action. Examples are where a suspect has fled from the scene of the crime and it is necessary quickly to check surrounding gardens and the like to see whether (s)he is hiding; or where police have arrested someone in the night after a pursuit and it is necessary to make a brief check of gardens along the route of the pursuit to see whether stolen or incriminating articles have been found.

Use of force

Where the police are acting under a warrant or under one of the above statutory **4.81** powers, reasonable and proportionate force may be used if necessary in the following cases if the officer in charge is satisfied that the premises are those specified in the warrant where:

(a) the occupier or another person entitled to grant access has refused to allow entry;
(b) it is impossible to communicate with such a person;
(c) the premises are known to be unoccupied or the occupier, etc is known to be absent; or
(d) there are reasonable grounds to believe that to alert the occupier, etc would frustrate the object of the search or endanger someone.

It is permissible to restrict the movement of occupants to one room while another room is being searched.

Designated persons

If a power conferred on a designated person under the Police Reform Act 2002, Part 4: **4.82**

(a) allows reasonable force to be used when exercised by a police officer, a designated person exercising that power has the same entitlement to use force; and
(b) includes power to use force to enter premises, that power is not exercisable by that designated person except:
   (i) in the company and under the supervision of a police officer; or
   (ii) for the purpose of:
      (a) saving life or limb; or
      (b) preventing serious damage to property.

Notice of Powers and Rights

Under Code B, an officer conducting a search of premises under a search warrant issued under PACE or TA 2000 or under a power given by PACE (see (a) below), or with the occupier's consent, must, unless it is impracticable to do so, provide the occupier with a copy of a notice in a standard format: **4.83**

(a) specifying whether the search is made under warrant, or with consent, or in the exercise of powers under PACE, s 17, 18, or 32 (**4.63**, **4.64**, and **4.59**);
(b) summarising the extent of the powers of search and seizure conferred in PACE and other relevant legislation as appropriate;
(c) explaining the rights of the occupier and of the owner of property seized;

(d) explaining that compensation may be payable in appropriate cases for damage caused in entering and searching premises, and giving the address to which an application should be directed; and

(e) stating that a copy of Code B is available for consultation at any police station.

If the occupier is present, copies of the notice, and of the warrant (if the search is made under warrant), should if practicable be given to him/her before the search begins, unless the officer in charge of the search reasonably believes that to do so would frustrate the object of the search or endanger the officers concerned or other persons. If the occupier is not present, copies of the notice, and of the warrant where appropriate, should be left in a prominent place on the premises or appropriate part of the premises and should be endorsed with the names of the officer in charge of the search and of his/her police station and the date and time of the search. The officer's warrant number, not his/her name, should be given in the case of a terrorism investigation. The warrant itself should be endorsed to show that this has been done.

Conduct of searches

4.84   Code B provides as follows.

Premises may be searched only to the extent necessary to achieve the object of the search, having regard to the size and nature of what is sought. This means that a search cannot be made in places where the articles specified in a warrant could not possibly be found or for longer than necessary to find those articles. For example, a warrant to search for television sets would not authorise the examination of the contents of small drawers.

A search may not continue under the authority of a warrant once everything specified in it has been found, or the officer in charge is satisfied that things specified are not on the premises.

The search must be conducted with due consideration for the property and privacy of the occupier, and with no more disturbance than necessary. Reasonable force may only be used to conduct the search where it is necessary because the co-operation of the occupier cannot be obtained or is insufficient for the purpose. An occupier must not be discouraged or prevented from securing the services of a friend, neighbour, or other person to witness the search, unless the officer in charge has reasonable grounds to believe that this would seriously hinder the investigation or endanger the officers concerned or other people. A search need not be unreasonably delayed for this purpose. If the premises have been entered by force the officer in charge must satisfy him/herself, before leaving, that they are secured either by arranging for the occupier or his/her agent to be present or by any other appropriate means.

A person need not be cautioned before being asked questions solely concerned with the proper and effective conduct of a search; for example, to locate the key to a locked drawer or otherwise to seek co-operation during the search. If questioning goes beyond this point it may amount to an interview and would require the associated safeguards.

In determining when to make a search, the officer in charge must always give regard to the time of day at which the occupier is likely to be present, and should not search at a time when the occupier or any other person on the premises is likely to be asleep unless this is unavoidable. If the wrong premises are mistakenly searched, everything possible should be done to allay any sense of grievance. In appropriate cases assistance should be given to obtain compensation.

The local police/community consultative group or its equivalent should be informed as soon as practicable after a search where there is reason to believe that it might have had an adverse effect on relations between the police and the community.

Records of searches

Code B requires that, if premises have been searched, a record of the search must be **4.85** made. That record must be made by or on behalf of the officer in charge of the search, on his/her return to the police station. The record must include:

(a) the address of the premises;

(b) the date, time, and duration of the search;

(c) the authority for it (including a copy of the warrant and the written authority to apply for it, or the written consent where the search was made thereunder);

(d) the names of officer(s) in charge of, and of the other officers and other authorised persons conducting, the search (except in the case of inquiries linked to the investigation of terrorism, or where it is reasonably believed that disclosing their names might endanger them; here the record should state the warrant or other identification number and duty station of each officer and identification number of any police staff);

(e) the names of any persons on the premises (if known);

(f) any grounds for refusing the occupier's request to have someone present during the search;

(g) a list of articles seized (or a note of its location) and, if not covered by a warrant, the reason for seizure;

(h) whether force was used and, if so, the reason;

(i) a list of any damage caused, and the circumstances in which it was caused;

(j) if applicable, the reason it was not practicable to give the occupier a copy of the 'Notice of Powers and Rights'; and

(k) when the occupier was not present, the place where copies of the Notice of Powers and Rights and search warrant were left on the premises.

Search registers must be maintained at each sub-divisional or equivalent police station and the above record must be made, copied, or referred to in the register.

On each occasion when premises are searched under a warrant, the warrant must be endorsed to show:

(i) whether any articles specified in the warrant were found and the address where found;

(ii) whether any other articles were seized;

(iii) the date and time at which it was executed and, if present, the name and address of the occupier, or if (s)he was not present, the name of the person in charge of the premises;

(iv) the names of the officers who executed it and any authorised persons who accompanied them (with the exception in (d) above); and

(v) whether a copy, together with a copy of the Notice of Powers and Rights, was handed to the occupier; or whether it was endorsed as required (**4.83**).

A warrant must be returned to the appropriate person within three months from issue or sooner on completion of the search(es) authorised by it. The appropriate person is:

(a) if the warrant was issued by a justice of the peace, the designated officer for the local justice area; or

(b) if it was issued by a judge, the appropriate officer of the court from which (s)he issued it.

General powers of seizure

4.86   PACE, s 19 gives the police wide powers of seizure in addition to those otherwise provided by the Act or elsewhere. It provides that a constable lawfully on any premises (defined at **4.59**) (eg with the occupier's consent) may seize anything on the premises if (s)he has reasonable grounds for believing that:

(a) it has been obtained in consequence of the commission of an offence; or

(b) it is evidence in relation to an offence which (s)he is investigating or any other offence; *and* (in either case)

(c) it is necessary to seize it in order to prevent it from being concealed, lost, altered or destroyed.

Although a vehicle, vessel, or tent is premises for the purpose of PACE, the reference to anything 'on the premises' does not prevent the seizure of a vehicle, vessel, or tent as a whole.

In the same circumstances (ie (a) to (c)), a constable may require any information stored in any electronic form and accessible from the premises to be produced in a form in which it can be taken away and in which it is legible or from which it can readily be produced in such a form.

Section 19 also provides that no relevant statutory power of seizure authorises the seizure of an item which the constable has reasonable grounds for believing to be subject to legal privilege (**4.68**). Generally, a constable will have to examine an item to test a claim to legal privilege and, if no claim is made, the constable will not have reasonable grounds for believing it to be privileged until (s)he has examined it. If the constable later obtains reasonable grounds, the item must be returned forthwith, but the seizure does not become unlawful.

A divisional court has held that, where property has been obtained by an unlawful search and is stored at a police station, s 19 cannot be relied on to enable a constable to seize it there and thereby to convert unlawful possession into lawful possession.

Additional powers of seizure

4.87   The additional powers of seizure under CJPA 2001, s 50 and related provisions (**4.61**) apply where there is a right of seizure under a search warrant or under PACE, ss 18 (**4.65**), 19 (above), and 32 (**4.60**).

Record of seizure and access

4.88   PACE, s 21 provides that a constable who seizes anything *under any statutory power* must, on request, provide within a reasonable time the occupier of premises where it was seized or a person who had custody or control of the thing immediately before seizure with a record of what (s)he seized. A divisional court has held that the duty to provide a record of what was seized does not extend to the provision of a composite item by item breakdown of the contents of a computer disc which has been seized.

Section 21 also provides for access to a thing which has been seized, and for the photographing or copying of it, by a person who had custody or control of it immediately before seizure.

Retention

Unless a court exercises a statutory power of forfeiture of property on conviction, the situation is as follows.   **4.89**

PACE, s 22 provides that anything seized or taken away under s 19 may be retained so long as is necessary in all the circumstances. 'So long as necessary' in s 22 means so long as necessary for carrying out the purposes for which the powers of seizure under s 19 are conferred. In particular, anything seized under s 19 for the purposes of a criminal investigation may be retained (unless a photograph or copy would suffice) for use as evidence at a trial for an offence, or for forensic examination or investigation in connection with an offence. Moreover, anything may be retained to establish its owner, where there are reasonable grounds for believing that it has been obtained in consequence of the commission of an offence.

Section 22 has been construed by the Court of Appeal as meaning that, if the police consider that, for 'criminal law enforcement purposes' (ie the four matters set out in (a) to (d) at **7.70**), it is necessary to retain a DNA profile derived from a swab taken under s 19 from a scene of crime, they cannot use it for any other unconnected purpose. Thus, for example, the police cannot disclose a profile for use to resolve a paternity issue in civil proceedings.

Nothing seized on the grounds that it may be used to:

(a) cause physical injury;
(b) damage property;
(c) interfere with evidence; or
(d) assist in escape from police detention or lawful custody,

may be retained after the person from whom it was seized has been freed from detention or custody or has been bailed.

There is no power to retain seized property for any purpose other than that for which it was originally seized.

Any person who had custody or control of property immediately prior to its seizure must, if it is retained, be provided on request within a reasonable time with a list or description of the property, and (s)he or his/her representative must be allowed supervised access to examine the property or have it photographed or copied (or must be provided with a photograph or copy) within a reasonable time and at his/her own expense, unless the officer in charge of the investigation has reasonable grounds for believing that this would prejudice the investigation of the offence or criminal proceedings, or lead to the commission of an offence by providing access to unlawful material such as pornography. A record of such grounds must be made if access is denied.

The Court of Appeal has ruled that the police may retain seized property:

(a) for a short period, after detention under s 22 is no longer necessary, while they consider their position; or
(b) so long as necessary in all the circumstances after the CPS has decided not to prosecute but a private prosecution is being contemplated or taking place.

According to a divisional court, s 22 does not entitle the police, once an investigation is over, to retain CCTV footage on the basis that its owner might use it to identify someone in order to carry out a revenge attack.

A person claiming property seized by the police may apply for its possession to a magistrates' court under the Police (Property) Act 1897 and should, where appropriate, be advised of this procedure. The court has a discretion under the 1897 Act.

In addition to the provisions in PACE 1984, s 22, retention of seized property is also governed by CJPA 2001, s 56. Section 56 applies to property seized by any constable who is lawfully on premises (or by a person authorised by a search warrant to accompany a constable), and to property seized by a constable carrying out a lawful search of any person. Retention of such property is authorised by s 56 if there are reasonable grounds for believing that it is property obtained in consequence of the commission of an offence and that its retention is necessary to prevent it being concealed, lost, damaged, altered or destroyed. The power under s 56 also extends to property that is believed to be evidence in relation to any offence where its retention is required for the same purposes.

By CJPA 2001, s 59, any person with a relevant interest in property seized under CJPA 2001 s 50 or 51 (**4.60-4.61**) or under any other statutory powers of seizure conferred on a constable may apply to a judge of the Crown Court for its return on one or more of the grounds that:

(a) there was no power of seizure;

(b) the seized material is or contains matter subject to legal privilege not comprised in property falling within s 54(2) (**4.62**);

(c) (with qualifications) the seized material is or contains excluded or special procedure material; or

(d) the seized property is or contains something seized under s 50 or 51 which does not fall within s 53(3) (**4.62**).

The term 'any person with a relevant interest in property' refers to the person from whom the property was seized, any person with an interest in it or any other person if (s)he had custody or control of it immediately before the seizure.

INVESTIGATORY POWERS UNDER INVESTIGATORY POWERS ACT 2016

4.90   The Investigatory Powers Act 2016 (IPA 2016), which consists of 272 sections and 10 schedules, sets out the extent to which certain investigatory powers may be used to interfere with privacy. Because of the size of the Act it is not possible in this book to do more than give an overview; a fuller account appears in Chapter 4 of the companion website.

The text below is written on the basis of the provisions of IPA 2016 as they apply to the police and the Secretary of State. Although other specified public officers and bodies have powers under the Act, a discussion of them falls outside the scope of Police Law.

The investigatory powers

4.91   IPA 2016 provides the following powers:

(a) Part 2 sets out circumstances in which the interception of communications is lawful in pursuance of a targeted interception warrant or mutual assistance warrant;

(b) Part 3 sets out circumstances in which the obtaining of communications data is lawful in pursuance of such an authorisation or under a warrant and makes further provision about the obtaining and treatment of such data; and

(c) Part 5 deals with targeted equipment interference warrants.

Prohibitions

Unlawful interception

IPA 2016, s 3(1) provides that a person commits an indictable (either-way) offence **4.92** (**6.48**), punishable with up to two years' imprisonment, if (s)he intentionally intercepts in the UK, without lawful authority, a communication in the course of its transmission by means of a public telecommunication system, a private telecommunication system, or a public postal service.

'Transmission' includes the period when a transmissions system stores a communication in such a manner as to enable the intended recipient to access it (eg where it is stored on voicemail); this is so even if it has been accessed by him/her. The recording by a covert listening device in a motor car of one person's voice speaking on a telephone is not an interception. Nor is the audio recording by an undercover police officer of a telephone conversation with a suspect, although it does amount to surveillance (dealt with later in this chapter).

By s 3(2), it is not an offence under s 3(1) for a person to intercept a communication in the course of its transmission by means of a private telecommunication system if (s)he is someone with a right to control the operation or use of the system, or has the express or implied consent of such a person to carry out the interception.

Lawful authority   A person has lawful authority to carry out an interception if, and only if, the interception:

(a) is carried out in accordance with a targeted interception warrant or mutual assistance warrant under IPA 2016, Part 2, Chapter 1 (**4.94**);

(b) is authorised by Part 2, Chapter 2 (eg interception with the consent of sender, or by a provider of postal or telecommunication service);

(c) in the case of a communication stored in or by a telecommunication system, the interception-

    (i) is carried out in accordance with a targeted equipment interference warrant under Part 5 (**4.97**);

    (ii) is in the exercise of any statutory power that is exercised for the purpose of obtaining information or taking possession of any document or other property; or

    (iii) is carried out in accordance with a court order made for that purpose.

Conduct which has lawful authority for the purposes of IPA 2016 by virtue of (a) or (b) is treated as lawful for all other purposes.

Any other conduct which is carried out in accordance with a warrant under Part 2, Chapter 1, or is authorised by Part 3, Chapter 2 is treated as lawful for all purposes.

Prohibition against unlawful obtaining of communications data

IPA 2016, s 11(1) provides that a relevant person who, without lawful authority, know- **4.93** ingly or recklessly obtains communications data from a telecommunications operator or a postal operator is guilty of an indictable (either-way) offence punishable with up to two years' imprisonment. A 'relevant person' is someone who holds an office, rank, or position with a relevant public authority. Section 11(1) does not apply to a relevant person who shows that (s)he acted in the reasonable belief that the person had lawful authority to obtain the communications data.

Targeted interception warrants and mutual assistance warrants

4.94   IPA 2016, Part 2, Chapter 1 deals with targeted interception warrants and mutual assistance warrants.

A *targeted interception warrant* is a warrant which authorises or requires its addressee to secure, by conduct described in it, any one or more of the following:

(a) the interception, in the course of their transmission by a postal service or telecommunication system, of communications described in the warrant;

(b) the obtaining of secondary data (postal systems data, or identifying data which can be separated from a telecommunication without revealing the meaning of the communication's content, associated with or attached to the communication being attached) from communications transmitted by means of a postal service or telecommunication system and described in the warrant;

(c) the disclosure, in any manner described in the warrant, of anything obtained under the warrant to the person to whom the warrant is addressed or to any person acting on that person's behalf.

A *mutual assistance warrant* is a warrant authorising or requiring its addressee to secure, by any conduct described in the warrant, any one or more of the following:

(a) the making of a request, in accordance with an international mutual assistance agreement, for the provision of any assistance in connection with, or in the form of, an interception of communications;

(b) the provision to the competent authorities of a country or territory outside the UK, in accordance with such agreement, of any assistance in connection with, or in the form of, an interception of communications;

(c) the disclosure of anything obtained under the warrant to its addressee or to any person acting on his/her behalf.

A targeted interception warrant or mutual assistance warrant also authorises (in addition to the conduct described in the warrant) any necessary ancillary conduct by its addressee or a third party.

Issue

4.95   A targeted interception warrant or mutual assistance warrant is issued by the Secretary of State with—except in urgent cases—the prior approval of a Judicial Commissioner on the application of an 'intercepting authority', for example, of the Commissioner of the Metropolitan Police or the Director General of the NCA (through whom applications by other chief officers of police must be channelled). In urgent cases, a Judicial Commissioner must approve the decision to issue the warrant within three working days; otherwise it lapses.

Authorisations for obtaining communications data

4.96   IPA 2016, Part 3 provides a regulatory framework for targeted authorisations for obtaining 'communications data' (ie data about the use made of a communication service—the 'who', 'when', 'where' and 'how' of a communication (eg an internet connection record) —as opposed to the content of a communication), the obtaining of which is dealt with by the provisions relating to the warrants referred to above.

The Investigatory Powers Commissioner and designated senior officers of a police force have powers to grant authorisations under Part 3.

An authorisation may not authorise any conduct consisting in the interception of communications in the course of their transmission by means of a telecommunication system.

Targeted equipment interference

IPA 2016, Part 5 provides for the issuing by a 'law enforcement chief' (a term including a chief officer of police, Metropolitan Police Assistant Commissioner and the Director General of the NCA) or appropriate delegate of targeted equipment interference warrants.

4.97

A targeted equipment interference warrant authorises or requires its addressee to interfere with any equipment for the purpose of obtaining communications, equipment data or other information. It must also authorise or require its addressee to secure such obtaining of the communications, equipment data, or other information to which the warrant relates, and may also authorise that person to secure the disclosure, in any manner described in the warrant, of anything so obtained. For these purposes, the obtaining of communications or other information through a targeted equipment interference warrant includes doing so by (a) monitoring, observing, or listening to communications or activities, or (b) recording anything which is monitored, observed or listened to. As a result of (a), it will not be necessary for such activity to be authorised separately under RIPA 2000, Part II referred to at **4.98**.

Directed surveillance, covert human intelligence sources, and intrusive surveillance

The Regulation of Investigatory Powers Act 2000 (RIPA 2000), Part II (ss 26–48) creates a regulatory framework for three types of activity: directed surveillance; the conduct and use of covert human intelligence sources; and intrusive surveillance. A fundamental feature of each of these is that it is 'covert'.

4.98

Each of the three types of activity involves a different authorisation procedure. RIPA 2000, s 27 provides that conduct falling within one of these three types is lawful for all purposes if it is authorised and carried out in accordance with the authorisation. No civil liability will be incurred in respect of conduct 'incidental' to such lawful conduct, which is not in itself conduct in respect of which an authorisation or warrant is capable of being granted under a relevant enactment and might reasonably have been expected to be sought. A 'relevant enactment' means RIPA 2000; the Intelligence Services Act 1994, s 5 (warrants for the intelligence services); or an enactment contained in the Police Act 1997 (PA 1997), Part III (powers of covert entry and interference with property by the police (**4.110**)).

Directed surveillance

Definition

Surveillance is directed if it is covert, but not intrusive, and is undertaken:

4.99

(a) for the purpose of a specific investigation or a specific operation;

(b) in such a manner as is likely to result in the obtaining of private information about a person (whether or not one specifically identified for the purposes of the investigation or operation); and

(c) otherwise than by way of an immediate response to events or circumstances the nature of which is such that it would not be reasonably practicable for an authorisation under RIPA 2000, Part II to be sought for the carrying out of the surveillance.

'Private information' includes any information relating to a person's private or family life.

Surveillance is *covert* if it is carried out in a manner that is calculated to ensure that the persons who are subject to the surveillance are unaware that it is or may be taking place.

Surveillance is *intrusive* if, and only if, it is covert surveillance that:

(a) is carried out in relation to anything taking place on any residential premises (a police cell is such) or in any private vehicle (a police vehicle is not a private vehicle); and

(b) involves the presence of an individual on the premises or in the vehicle or is carried out by means of a surveillance device.

However, there are qualifications to this. Surveillance is *not intrusive* if it involves no more than the placing of a vehicle location device, or it is surveillance involving the interception of a communication which is sent by, or intended for, a person who has consented to the interception. In addition, it is *not intrusive* if it is carried out by a device which is not present upon the residential premises or private vehicle concerned, unless the device is such that it consistently provides information of the same quality and detail as might be expected to be obtained from a device which was actually present on the premises or in the vehicle. Lastly, surveillance by devices designed to catch TV licence dodgers is neither directed nor intrusive.

Authorisation

4.100 By RIPA 2000, s 28, a 'designated person' may grant an authorisation for the carrying out of directed surveillance if (s)he believes it is necessary on specified grounds, including national security, the prevention or detection of crime, and the prevention of disorder. The designated person must believe that the authorised surveillance is proportionate to what is sought to be achieved by it.

Among the persons designated for the purposes of s 28 (and s 29, below, with the exceptions at **4.104**) are police officers of the rank of superintendent (inspector in urgent cases). For the purposes of the grant of an authorisation that combines an authorisation under s 28 or 29 and an authorisation under s 30 for carrying out intrusive surveillance (**4.107**), the Secretary of State is *the* person designated.

A written authorisation under s 28 has effect for three months. Urgent oral authorisations or written authorisations granted by a person who is entitled to act only in urgent cases only have effect for 72 hours.

Code of practice

4.101 The Investigatory Powers (Codes of Practice and Miscellaneous Amendments) Order 2018 brought into force in August 2018 a revised code of practice relating to covert surveillance which applies to directed surveillance.

Use of a covert human intelligence source

Definition

A person is a covert human intelligence source if (s)he: **4.102**

(a) establishes or maintains a personal or other relationship with a person, for the covert purpose of facilitating:
   (i) the covert obtaining of information or the provision of access to information to another person; or
   (ii) the covert disclosure of information obtained by the use of such a relationship or in consequence of such a relationship;
(b) covertly uses such a relationship to obtain information or to provide such access; or
(c) covertly discloses information obtained by the use of such a relationship, or as a consequence of the existence of such a relationship.

The 'use of a covert human intelligence source' refers to inducing, asking or assisting a person to engage in the conduct of such a source, or to obtain information by means of such a source.

A purpose is covert in relation to the establishment or maintenance of a personal or other relationship only if the relationship is conducted in a manner that is calculated to ensure that one of the parties to the relationship is unaware of the purpose. A relationship is used covertly, and information obtained is disclosed covertly, only if it is used or disclosed in a manner that is calculated to ensure that one of the parties to the relationship is unaware of the use or disclosure in question.

Authorisation

RIPA 2000, s 29 governs the authorisation of a covert human intelligence source. An **4.103** authorisation is generally valid for 12 months; for exceptions see **4.104** and **4.105**. It may be granted by a designated person (see **4.100**).

The criteria which apply to the grant of an authorisation for directed surveillance (**4.100**) also apply to covert surveillance. However, there are additional requirements concerning satisfactory arrangements for the supervision of the source depending on whether or not a source is a source of a relevant collaborative unit, and there are further requirements where matters subject to legal privilege are involved, and also about records relating to the use of the source.

An authorisation under s 29 may not have the effect of authorising a covert human intelligence source who is a person designated under the Police Reform Act 2002, s 38 (**2.4**) to establish contact in person with another person.

Notification and prior approval regime

The Regulation of Investigatory Powers (Covert Human Intelligence Sources: Relevant **4.104** Sources) Order 2013 introduced a notification and prior approval regime in respect of an authorisation of a covert human intelligence source (CHIS) under RIPA 2000, s 29 (s 29 authorisation) where that source is a *relevant source*.

'Relevant source' means a CHIS holding an office, rank, or position with a listed law-enforcement body, including a police force maintained under the Police Act 1996, the City of London Police Force, the Metropolitan Police Force, the Ministry of Defence Police, and the British Transport Police.

Notification   Except in the case of a long-term authorisation (below), the 2013 Order requires a person granting a s 29 authorisation for the conduct or use of a relevant source to give notice of that grant to a Judicial Commissioner, appointed by the Prime Minister, save where the authorisation is for the grant or renewal of a 'long-term authorisation' (defined below). The notice must be given or transmitted electronically in writing. It must specify the grounds for the authorisation and the conduct authorised.

Prior approval   must be sought from a Judicial Commissioner in respect of the grant or renewal of a long-term authorisation. Specified criteria apply, both in respect of any request for approval and the Commissioner's decision. There is an appeal mechanism where a request for prior approval is refused.

A s 29 authorisation for the conduct or use of a relevant source is a long-term authorisation if the periods mentioned in (a) and (b) below, when taken together, exceed 12 months:

(a) *the period for which the relevant source will be authorised under the authorisation,* having regard to a provision that, where the period(s) in respect of which a relevant source has previously been authorised is less than 12 months in total, any further authorisation will cease to have effect, unless renewed, at the end of a period of 12 months less the total period for which the source has previously been authorised, but disregarding any period of authorisation which ceased to have effect more than three years prior to the intended commencement date of the authorisation; and

(b) *any period or periods for which the relevant source has previously been authorised as a source in relation to the same investigation or operation,* other than
   (i) a period or periods where the authorisation was granted orally or by a person whose entitlement to act is confined to urgent cases, or
   (ii) any period of authorisation which ceased to have effect more than three years prior to the intended commencement date of the authorisation referred to in (a).

The 12-month period is reduced to three months in respect of authorisations involving access to legally privileged (**4.68**) material.

Who can authorise?   An authorisation of a relevant source who holds an office, rank, or position in a police force may be made by an assistant chief constable, or by a commander in the case of the City of London or Metropolitan Police Forces. In urgent cases, an authorisation of a relevant source may be made by a superintendent instead of such a senior officer.

A long-term authorisation of a CHIS who holds an office, rank, or position in a police force may only be made by a chief constable, the Commissioner of the City of London Police or an assistant commissioner of the Metropolitan Police Service.

Matters subject to legal privilege: additional requirements

4.105   Additional requirements where matters subject to legal privilege are involved are as follows in relation to police operations.

The Regulation of Investigatory Powers (Covert Human Intelligence Sources: Matters Subject to Legal Privilege) Order 2010 (the 2010 Order) applies where any conduct that is, or is to be, authorised in an authorisation under s 29 consists in any activities involving conduct or a source, or the use of a source to:

(a) obtain matters,
(b) provide access to another person to any matters, or
(c) disclose matters,

subject to legal privilege (**4.68**).

An authorisation for such conduct must not be granted unless it satisfies the following additional requirements under the 2010 Order.

Before a person grants or renews an authorisation for specified conduct, (s)he must in accordance with arrangements made by a Judicial Commissioner give or transmit electronically written notice to the Commissioner. The notice must state that the requisite approval is sought and include the specified matters.

An authorisation for conduct to which the 2010 Order applies must not be granted or renewed until it has been approved by the Commissioner, and written notice of the Commissioner's decision to approve the grant or renewal of the authorisation has been given or transmitted electronically.

The Commissioner must give his/her approval to the grant or renewal of the authorisation if, and only if, satisfied that there are reasonable grounds for believing that the authorisation is necessary and the requirements of proportionality and satisfactory arrangements which apply to authorisations are met.

Where an authorisation authorises conduct to which the 2010 Order applies, it has effect for three months. The authorisation will specify the conduct which is authorised and require that it is carried out in accordance with the authorisation. In the case of informants the authorisation will be specific to the individual and the particular investigation.

Where a single authorisation under s 29 authorises conduct to which the 2010 Order applies and other conduct falling within RIPA 2000, Part II, the additional requirements only apply in relation to those parts of the combined authorisation which authorise obtaining, giving access to, or disclosing, matters subject to legal privilege.

Code of Practice

The Investigatory Powers (Codes of Practice and Miscellaneous Amendments) Order 2018 brought into force in 2018 a revised code of practice relating to the use or conduct of covert human intelligence sources and the handling of any information obtained by the use of CHIS.   **4.106**

Intrusive surveillance

As already stated, surveillance is *intrusive* if, but only if, it is *covert* surveillance (see above) carried out in relation to anything taking place on any residential premises (a police cell is such) or in any private vehicle (a police vehicle is not a private vehicle), and involves the presence of an individual on the premises or in the vehicle or is carried out by means of a surveillance device.   **4.107**

Directed surveillance that is carried out in relation to anything taking place on so much of a police station, court building, lawyer's business premises, or a prison or similar place as is, at any time during the surveillance, used for the purpose of legal consultations is also treated as intrusive surveillance.

Authorisation

RIPA 2000, s 32 states that intrusive surveillance authorisations may be granted by the Secretary of State (valid for six months) and senior authorising officers (chief constables and equivalents, plus assistant commissioners of the Metropolitan Police) (three months). Such authorisations will only be granted where it is believed that this is necessary in the interests of national security or of the economic well-being of the UK,   **4.108** or for the purpose of preventing or detecting serious crime, *and* that the authorised surveillance is proportionate to what is sought to be achieved.

Authorisations valid for 72 hours for intrusive surveillance may be granted under s 34, by the appropriate deputy chief constable or an assistant chief constable who is designated to act, or their equivalents (specified as a commander in the Metropolitan Police) in urgent cases where it is not reasonably practicable for the senior authorising officer to do so.

When a police authorisation for intrusive surveillance is granted or cancelled, notice must be given to a Judicial Commissioner.

Except in cases of urgency, the approval of a Judicial Commissioner is needed before a police authorisation for intrusive surveillance may take effect.

Code of Practice

4.109 The Investigatory Powers (Codes of Practice and Miscellaneous Amendments) Order 2018 brought into force in August 2018 a revised code of practice relating to covert surveillance which applies to intrusive surveillance.

AUTHORISATION OF ACTION IN RESPECT OF PROPERTY

4.110 The Police Act 1997 (PA 1997), Part III introduced a system whereby authorisations may be given in respect of entry on or interference with property or with wireless telegraphy. No act carried out in accordance with an authorisation is unlawful. Although authorisations may be given by 'authorising officers' (which term includes, among others, a chief constable, the Commissioner or an Assistant Commissioner of Police of the Metropolis, and the Commissioner of the City of London Police) they are subject in some cases to approval by a Judicial Commissioner.

The law is as follows where the authorising officer is a chief constable, or the Commissioner or an Assistant Commissioner of the Metropolitan Police, or the Commissioner of the City of London Police. Where an authorising officer believes that:

(a) it is necessary for the action specified to be taken to prevent or detect serious crime; and

(b) that action is proportionate to what the action seeks to achieve,

(s)he may authorise the taking of such action:

(i) in respect of such property in his/her police area (or area specified in a collaboration agreement with another force), as (s)he may specify;

(ii) in respect of property outside that area for the purpose of maintaining or retrieving any equipment, apparatus or device the placing of which in the relevant area has been authorised under PA 1997, Part III, RIPA 2000, Part II or IPA 2016 (above); or

(iii) in that area as (s)he may specify, in respect of wireless telegraphy.

Provision is made for authorisations to be given by specified senior officers (assistant chief constables or commander in the case of police forces) where it is not reasonably practicable for an authorising officer or his/her designated deputy to consider an application for an authorisation.

Authorisations must be in writing, although in urgent cases they may be given orally. Unless renewed, they cease to have effect after 72 hours if given orally, or by someone other than the actual authorising officer or his/her designated deputy. In any other case an authorisation may last for three months.

As soon as is reasonably practicable, the giving or cancellation of an authorisation must be notified to a Judicial Commissioner.

Where the person who gives an authorisation believes that any of the property specified in it is used wholly or mainly as a dwelling or as a bedroom in a hotel, or constitutes office premises, or that the action authorised is likely to result in someone acquiring knowledge of legally privileged matters (**4.68**), confidential personal information, or confidential journalistic material, the authorisation is ineffective until approved by a Judicial Commissioner and the person who gave the authorisation has been notified of that approval. However, this does not apply in a case of urgency.

Confidential personal information is (a) personal information which a person has acquired or created in the course of any trade, business, profession or other occupation, or for the purpose of any paid or unpaid office, and which (s)he holds in confidence, and (b) communications as a result of which personal information is so acquired or created and is held in confidence.

Confidential journalistic material is (a) material acquired or created for the purposes of journalism which is in the possession of persons who acquired or created it for those purposes, is held in confidence and has been continuously held (by one or more persons) subject to such an undertaking, restriction or obligation since it was first acquired or created for those purposes; and (b) communications as a result of which information is acquired for those purposes and so held.

A person may not, for the purpose of obtaining 'communications', 'private information', or 'equipment data', make an application under the Police Act 1997, Part III for authorisation to engage in conduct which could be authorised by a targeted equipment interference warrant under IPA 2016, Part 5 (**4.97**) if the applicant considers that the conduct would (unless done under lawful authority) constitute one or more offences under the Computer Misuse Act 1990, ss 1–3A (computer misuse (**15.2–15.8**).

Code of Practice

The Investigatory Powers (Codes of Practice and Miscellaneous Amendments) Order **4.111** 2018 provides guidance on entry on or interference with property or with wireless telegraphy by public authorities under PA 1997, Part III.

COMBINATION OF WARRANTS AND AUTHORISATIONS

By IPA 2016, s 248 and Sch 8, the Secretary of State may, on an application made by or **4.112** on behalf of 'a relevant intercepting authority' (**4.95**), issue a warrant that combines a targeted interception warrant with one or more of the following:

(a) a targeted equipment interference warrant;
(b) an authorisation to interfere with property under the Police Act 1997, Part III;
(c) an authorisation of directed surveillance under RIPA 2000, s 28;
(d) an authorisation of intrusive surveillance under RIPA 2000, s 32.

IPA 2016, s 248 and Sch 8 also provide that a 'law enforcement chief' (**4.97**) may, on an application made by a person who is an 'appropriate law enforcement officer' (a term including a chief officer of police or Assistant Commissioner of the Metropolitan Police) in relation to the chief, issue a warrant that combines a targeted equipment interference warrant with one or more of the following:

(a) an authorisation under the Police Act 1997, Part III;

(b) an authorisation under RIPA 2000, s 28;

(c) an authorisation under RIPA 2000, s 32.

SURVEILLANCE CAMERAS

4.113 The Regulation of Investigatory Powers Act 2000 does not apply to surveillance by the police which is not covert, such as the overt taking of photographs by police of protesters. Thus, it does not apply to the overt use by police of live automated facial recognition (AFR) technology to create biometric data based on digitised measurements of images (facial features) of members of the public captured by surveillance cameras. From the images, the software identifies faces and extracts unique facial features to create a unique biometric template which is then compared with those features contained in a database or 'watchlist'.

There is, however, a legal framework for AFR. It includes the Data Protection Act 2018 (DPA 2018), the Code of Practice for Surveillance Camera Systems, and local policies promulgated by the police. The Code of Practice provides guidance on the appropriate use of surveillance camera systems by the police. Operators of surveillance camera systems must have regard to the Code. A court may take into account a failure to do so when determining an issue before it.

The use of AFR must not be incompatible with the European Convention on Human Rights, art 8 (right to respect for private life) and therefore the legal framework in question must be sufficiently certain to satisfy art 8's legality requirement. To satisfy that requirement, the framework must be sufficiently precise for an individual to predict with certainty who, for example, was added to the watchlist or where the cameras would be.

The Court of Appeal has stated in relation to the 'who question' (ie who was added to the watchlist) that DPA 2018 was insufficient on its own, and, although it appeared possible that the Code of Practice could have dealt with who should be on a police force watchlist, it did not. Consequently, the deficiency would need to be made up by local policies. The Court thought that it might be prudent for there to be at least consistency in the content of local policies and that might be the appropriate subject of an amendment to the Code. In the case before the Court, policy governing inclusion on the watchlist specified not only those wanted on suspicion of an offence or on a warrant, and vulnerable people, but also 'other persons where intelligence is required'. The Court held that, while the first three categories were objective, the final was not, and left too broad a discretion in individual officers. Turning to the 'where question' (ie where the cameras were stationed), the Court held that the local policies provided no clear criteria as to where deployment could take place, and the range was very broad and without apparent limits; too much discretion was left to individual officers. In the context of the legality requirement, the court emphasised that it was a crucial feature that images which did not produce a match were automatically and almost instantaneously deleted without human observation.

There is a public sector equality duty (PSED) under the Equality Act 2010, s 149, whose requirements are important. In the above case, there was evidence that some available AFR systems were capable of discriminating against black and minority ethnic people and women, as a result of inadequacies in the data sets used to 'train' the systems. The Court held that the police force in question had breached the PSED. They had not done all that they could reasonably have done to fulfil the PSED by ensuring that the system in use did not discriminate.

The Surveillance Camera Commissioner, appointed by the Home Secretary, has the functions of encouraging compliance with the Code of Practice, reviewing the operation of the code, and providing advice about the code (including changes to it or breaches of it).

REMOVAL TO A PLACE OF SAFETY OF MENTALLY DISORDERED PERSONS

Warrant to search for and remove patients

The Mental Health Act 1983 (MHA 1983), s 135 provides as follows. **4.114**

By s 135(1), if it appears to a justice of the peace, on information on oath laid by an approved mental health professional, that there is reasonable cause to suspect that a person (P) believed to be suffering from mental disorder:

(a) has been, or is being, ill-treated, neglected, or kept otherwise than under proper control, in any place within the jurisdiction of the justice, or
(b) being unable to care for him/herself, is living alone in any such place,

the justice may issue a warrant authorising any constable to enter, if need be by force, any premises specified in the warrant in which that person is believed to be, and, if thought fit, to remove P to a place of safety (**4.116**) with a view to the making of an application in respect of P of compulsory admission to hospital or guardianship, or of other arrangements for P's treatment or care. If the premises specified in the warrant are a place of safety, the constable may, instead of so removing P, keep P at those premises for the purpose mentioned in s 135(1) (s 135(1A)). A patient may be detained at the place of safety to which (s)he is removed or kept for a period not exceeding 'the permitted period', ie:

(i) the period of 24 hours beginning with (where P has been removed to a place of safety) the time when P arrives at that place, or (where P has been kept at such a place) the time when the constable first entered the premises to execute the warrant; or
(ii) by s 136B, where an authorisation in order to permit the assessment of P to be completed is given by the responsible registered medical practitioner, that period of 24 hours and such further period (up to 12 hours) as specified in the authorisation.

If P is detained at a police station, and the assessment would be carried out or completed at the station, the registered medical practitioner may not give such an authorisation unless a superintendent (or above) approves it.

In the execution of the warrant a constable must be accompanied by an approved mental health professional and by a registered medical practitioner.

By s 135(2), if it appears to a justice, on information on oath laid by any constable or other person who is authorised to take a patient to any place, or to take into custody or retake a patient who is liable under MHA 1983 (or the equivalent Scottish provision) to be so taken or retaken:

(a) that there is reasonable cause to believe that P is to be found on premises within the justice's jurisdiction; and
(b) that admission to the premises has been refused or that a refusal of such admission is apprehended,

the justice may issue a warrant authorising any constable to enter the premises, if need be by force, and remove the patient.

In the execution of such a warrant a constable may be accompanied by a registered medical practitioner and/or by any person authorised by or under MHA 1983 (or its Scottish equivalent) to take or retake the patient.

Removal etc of mentally disordered persons without a warrant

4.115 By MHA 1983, s 136, if a person (P) appears to a constable to be suffering from mental disorder and to be in immediate need of care or control, the constable may, if (s)he thinks it necessary to do so in the interests of P or for the protection of other persons, remove P to a place of safety (**4.116**), or, if P is already at a place of safety, keep P there or remove P to another such place. This power may be exercised where P is at any place, other than:

(a) any house, flat or room where P, or any other person, is living, or
(b) any yard, garden, garage or outhouse used in connection with the house, flat or room, other than one used in connection with one or more other houses, flats, or rooms.

For these purposes of exercising this power, a constable may enter any place where the power may be exercised, if need be by force.

Before deciding to remove P to, or to keep P, at, a place of safety, the constable must, if it is practicable to do so, consult a registered medical practitioner, a registered nurse, an approved mental health professional, or a registered occupational therapist or paramedic.

By s 136(2), P may be detained at the place of safety for a period not exceeding the permitted period of detention for the purpose of enabling him/her to be examined by a registered medical practitioner and to be interviewed by an approved mental health professional and of making any necessary arrangements for his/her treatment or care. The permitted period of detention is:

(a) the period of 24 hours beginning with—
  (i) in a case where P is removed to a place of safety, the time when P arrives there;
  (ii) in a case where P is kept at a place of safety, the time when the constable decides to keep the person there; or
(b) where an authorisation is given under s 136B (**4.114**), that period of 24 hours and such further period as is specified in the authorisation.

A constable, an approved mental health professional or a person authorised by either of them for these purposes may, before the end of the permitted period of detention, take P to one or more other places of safety. If P is so taken, P may be detained there no later than the end of the permitted period of detention (s 136(4)).

Place of safety

4.116 For the above purposes, 'place of safety' means residential accommodation provided by a local social services authority, a hospital to which MHA 1983, Part II applies, a police station, an independent hospital or care home for mentally disordered persons or any other suitable place. For these purposes:

(a) a house, flat or room where the person believed to be suffering from a mental disorder (P) is living is not a suitable place unless—

(i) if P is the sole occupier of the place, P agrees to the use of the place as a place of safety;

(ii) if P is an occupier of the place but not the sole occupier, both P and one of the other occupiers agree to the use of the place as a place of safety;

(iii) if P is not an occupier of the place, both P and the occupier (or, if more than one, one of the occupiers) agree to the use of the place as a place of safety;

(b) a place other than one mentioned in (a) may not be regarded as a suitable place unless a person who appears to the constable to be responsible for the management of the place agrees to its use as a place of safety.

By s 136A, a child (under 18) may not be removed to, kept at or taken to a place of safety that is a police station.

Under the MHA 1983 (Places of Safety) Regulations 2017, an adult (A) may only be removed to, kept at, or taken to, a place of safety that is a police station where—

(a) the decision-maker is satisfied that—

(i) A's behaviour poses an imminent risk of serious injury or death to A or another person;

(ii) because of that risk, no place of safety other than a police station in the relevant police area can reasonably be expected to detain A; and

(iii) A's welfare is checked by a healthcare professional at least every 30 minutes, and appropriate care and treatment-action taken, and, so far as practicable, a healthcare professional is available throughout A's detention at the station; and

(b) the decision-maker (below) is not an officer of the rank of inspector or above, an officer of that rank or above authorises that A may be removed to, kept at, or taken to a place of safety that is a police station.

If either part of (a)(iii) is not fulfilled the custody officer must arrange for D to be taken to another place of safety.

Before determining that the circumstances in (a)(i)–(iii) exist, a decision-maker (the constable exercising a power under s 135 or 136 or the constable or approved mental health professional who authorises a person to exercise that power) must, if it is reasonably practicable to do so, consult a registered medical practitioner, a registered nurse, an approved mental health professional, a registered occupational therapist, or a paramedic.

Protective searches

Provision is made for these by s 136C.   **4.117**

Where a warrant is issued under s 135, a constable may search the person (P) to whom the warrant relates (P), if (s)he has reasonable grounds for believing *that P may present a danger to him/herself or to others, and is concealing on his/her person an item that could be used to cause physical injury to him/herself or to others.* The power to search conferred may be exercised:

(a) where a warrant is issued under s 135(1), at any time during the period beginning with the time when a constable enters the premises specified in the warrant and ending when P ceases to be detained under s 135;

(b) in a case where a warrant is issued under s 135(2), at any time while P is being removed under the authority of the warrant.

Where a person (P) is detained under s 136(2) or (4), a constable may search P, at any time while P is so detained, if the constable has reasonable grounds for believing the matters italicised above.

These powers do not authorise a constable to require P to remove any of his/her clothing other than an outer coat, jacket, or gloves, but do authorise a search of P's mouth. A constable may seize and retain anything found, if (s)he has reasonable grounds for believing that P might use it to cause physical injury to him/herself or to others.

CHAPTER 5

Police Questioning and the Rights of Suspects

INTRODUCTION

Code C

In this and the following chapter, frequent reference is made to PACE Code C: the Code **5.1** of Practice for the Detention, Treatment and Questioning of Persons by Police Officers. This Code applies to persons in custody at police stations, whether or not they have been arrested, and (except for its provisions as to reviews and extensions of detention) to those who have been removed to a police station as a place of safety under the Mental Health Act 1983, s 135 or 136) (**4.114–4.117**); see further **5.10**. Persons who are voluntarily at a police station or other location must be treated with no less consideration.

Matters to which Code C does not apply

Code C does not apply to the following persons in custody: **5.2**

(a) persons arrested on warrants issued in Scotland by police officers under the Criminal Justice and Public Order Act 1994 (CJPOA 1994), s 136 (**3.19**), or arrested or detained without warrant by officers from a police force in Scotland under s 137. In these cases, police powers and duties and the person's rights and entitlements whilst at a police station in England and Wales are the same as those in Scotland;

(b) persons arrested for the purpose of fingerprinting under the Immigration and Asylum Act 1999, s 142;

(c) persons whose detention is authorised by an immigration officer under the Immigration Act 1971, Sch 2 or 3 or the Immigration and Asylum Act 2002, s 62;

(d) persons who are convicted or remanded prisoners held in police cells on behalf of the prison service; and

(e) persons detained for searches under stop and search powers except as required by PACE Code A (stop and search).

No part of Code C applies to a detained person:

(a) to whom PACE Code H (terrorism) (**4.3**) applies because:
  (i) he is detained following arrest under the Terrorism Act 2000 (TA 2000), s 41 and not charged, or
  (ii) an authorisation has been given for post-charge questioning of him/her as a terrorist suspect—matters dealt with in Chapter 21;

(b) to whom the Code of Practice for Examining Officers and Review Officers applies because (s)he is detained for examination under TA 2000, Sch 7.

Where Code C requires the prior authority or agreement of an officer of at least inspector or superintendent rank, that authority may be given by a sergeant or chief inspector, respectively, authorised to perform the functions of the higher rank by a superintendent (or above).

General

5.3 *Nothing in Code C requires the identity of officers or other police staff to be recorded or disclosed if the persons concerned reasonably believe that recording or disclosing their names might put them in danger. In such cases they must use their warrant or other iden-tification number and the name of their police station. The same anonymity rule applies to Codes E and F (dealt with later), substituting 'interviewer' for 'persons concerned'.*

Evidence obtained in breach of the procedures set out below is liable to be excluded in any subsequent court proceedings. In addition, a failure to comply with Code C may lead to a disciplinary offence.

Code C provides that references in Code C and any other PACE Code to written records, forms and signatures include electronic records and forms and electronic con-firmation that identifies the person making the record or completing the form.

Vulnerable persons

5.4 In Code C, 'vulnerable' applies to any person who, because of a mental health condition or mental disorder:

(i) may have difficulty understanding or communicating effectively about the full im-plications for him/her of any procedures and processes connected with—
   -   his/her arrest and detention; or (as the case may be)
   -   his/her voluntary attendance at a police station or his/her presence elsewhere, for the purpose of a voluntary interview; and
   -   the exercise of his/her rights and entitlements;

(ii) does not appear to understand the significance of what (s)he is told, of questions (s)he is asked or of his/her replies:

(iii) appears to be particularly prone to:
   -   becoming confused and unclear about his/her position;
   -   providing unreliable, misleading, or incriminating information without knowing or wishing to do so;
   -   accepting or acting on suggestions from others without consciously knowing or wishing to do so; or
   -   readily agreeing to suggestions or proposals without any protest or question.

Code C provides that if at any time an officer has any reason to suspect that a person of any age may be vulnerable, in the absence of clear evidence to dispel that suspicion, that person must be treated as such for the purposes of Code C. To establish whether any such reason may exist in relation to a person suspected of committing an offence, the custody officer in the case of a detained person, or the officer investigating the offence in the case of a person who has not been arrested or detained, must take, or cause to be taken, the following action:

(a) reasonable enquiries must be made to ascertain what information is available that is relevant to any of the factors described below as indicating that the person may be vulnerable might apply. Examples of relevant information that may be available include:

- the behaviour of the adult or juvenile;
- the mental health and capacity of the adult or juvenile;
- what the adult or juvenile says about him/herself;
- information from relatives and friends of the adult or juvenile;
- information from police officers and staff and from police records;
- information from health and social care (including liaison and diversion services) and other professionals who know, or have had previous contact with, the individual and may be able to contribute to assessing his/her need for help and support from an appropriate adult. This includes contacts and assessments arranged by the police or at the request of the individual or (as applicable) his/her appropriate adult or solicitor;

(b) a record must be made describing whether any of the factors in (a) appear to apply and provide any reason to suspect that the person may be vulnerable or (as the case may be) may not be vulnerable; and

(c) the record mentioned in (b) must be made available to be taken into account by police officers, police staff and any others who, in accordance with Code C or any other Code, are required or entitled to communicate with the person in question. This would include any solicitor, appropriate adult and health care professional and is particularly relevant to communication by telephone or by means of a live link.

When a person is under the influence of drink and/or drugs, it is not intended that (s)he is to be treated as vulnerable and requiring an appropriate adult for the above purpose unless other information indicates that any of the factors just described may apply to him/her. When the person has recovered from the effects of drink and/or drugs, (s)he should be re-assessed in accordance with the provisions above.

A person may be vulnerable as a result of a having a mental health condition or mental disorder. Similarly, simply because an individual does not have, or is not known to have, any such condition or disorder, does not mean that (s)he is not vulnerable for the purposes of Code C. It is therefore important that the custody officer in the case of a detained person or the officer investigating the offence in the case of a person who has not been arrested or detained, as appropriate, considers on a case by case basis, whether any of the above factors might apply to the person in question. In doing so, the officer must take into account the particular circumstances of the individual and how the nature of the investigation might affect him/her and bear in mind that juveniles, by virtue of their age, will always require an appropriate adult.

The Mental Health Act 1983 Code of Practice describes the range of clinically recognised conditions which can fall within the meaning of mental disorder for the above purposes. The Code is published here: https://www.gov.uk/government/publications/code-of-practice-mental-health-act-1983.

Juvenile

A 'juvenile' is someone under 18. If a person appears to be under 18, (s)he must, in the **5.5** absence of clear evidence that (s)he is older, be treated as a juvenile for the purposes of Code C and any other PACE Code.

Custody officer

5.6   Code C makes frequent reference to 'the custody officer'; such references include any police officer performing the functions of a custody officer. References in the Code to a police officer include a 'designated person' (ie a person designated under the Police Reform Act 2002, s 38 or 39 (**2.4, 2.7**) acting in the exercise or performance of the powers and duties conferred or imposed by their designation. Nothing in Code C prevents a custody officer, or other police officer or designated person given custody of the detainee by the custody officer, from allowing another person (see (a) and (b) below) to carry out individual procedures or tasks at the police station if the law allows. However, the officer or designated person remains responsible for making sure the procedures and tasks are carried out correctly in accordance with the PACE Codes. The other person who is allowed to carry out the procedures or tasks must be someone who *at that time* is:

(a) under the direction and control of the chief officer of the force responsible for the police station in question; or

(b) providing services under contractual arrangements (but without being employed by the chief officer of the police force), to assist a police force in relation to the discharge of its chief officer's functions.

Designated persons and others mentioned in (a) and (b) must have regard to any relevant provisions of the PACE Codes.

A custody officer must perform the functions in Code C as soon as practicable. A custody officer is not in breach of the Code if delay is justifiable (as where a large number of suspects are brought to a police station simultaneously to be placed in custody, or interview rooms are all in use, or there are difficulties in contacting the appropriate adult, solicitor, or interpreter) and reasonable steps are taken to prevent unnecessary delay. The custody record must show when a delay has occurred and the reason.

Appropriate adult

5.7   Code C (and the other PACE Codes referred to in this chapter) refer to 'the appropriate adult'.

The role of the appropriate adult is to safeguard the rights, entitlements and welfare of juveniles and vulnerable persons to whom the provisions of Code C and any other PACE Code apply. For this reason, the appropriate adult is expected, amongst other things, to:

(a) support, advise and assist them when, in accordance with Code C or any other PACE Code, they are given or asked to provide information or participate in any procedure;

(b) observe whether the police are acting properly and fairly to respect their rights and entitlements, and inform an officer of the rank of inspector or above if (s)he considers that they are not;

(c) assist them to communicate with the police whilst respecting their right to say nothing unless they want to as set out in the terms of the caution (see **5.27**);

(d) help them to understand their rights and ensure that those rights are protected and respected

In the case of a juvenile, 'the appropriate adult' means:

(a) his/her parent or guardian (or, if (s)he is in the care of a local authority or voluntary organisation, a person representing that authority or organisation);

(b) a local authority social worker; or

(c) failing either of the above, another responsible adult aged 18 or over who is not

  (i) *a police officer;*

  (ii) *employed by the police;*

  (iii) *under the direction or control of the chief officer of a police force; or*

  (iv) *a person who provides services under contractual arrangements (but without being employed by the chief officer of a police force), to assist that force in relation to the discharge of its chief officer's functions,*

  whether or not that person is on duty at the time.

A person, including a parent or guardian, should not be the appropriate adult if (s)he is suspected of involvement in the offence in question, is the victim, is a witness, is involved in the investigation, or has received admissions. If the parent or guardian of a juvenile is estranged from him/her, (s)he should not be asked to be the appropriate adult if the juvenile expressly and specifically objects to his/her presence. The fact that a parent participates in questioning a juvenile during the interview does not disqualify the parent from being the appropriate adult. If a juvenile admits an offence to a social worker or member of a youth offending team other than when that person is acting as the appropriate adult, another appropriate adult should be appointed.

Although 'the appropriate adult' is referred to in the singular, cases can arise where it is appropriate for more than one adult to be present as 'appropriate adults' during the interview of a juvenile etc, as where both parents are present or where a parent with language difficulties is present and a second adult is present to assist in questions of language.

In the case of a person who is vulnerable, 'the appropriate adult' means:

(a) a relative, guardian, or other person responsible for his/her care or custody;

(b) someone who has experience of dealing with vulnerable persons but who is not a person italicised in (c) above, whether or not (s)he is on duty at the time; or

(c) failing these, some other responsible adult aged 18 or over.

In the case of people who are mentally disordered or otherwise mentally vulnerable it may in some cases be more satisfactory if the appropriate adult is someone who has experience or training in their case rather than a relative lacking such qualifications. However, if the person him/herself prefers a relative to a better-qualified stranger, his/her wishes should if practicable be respected.

A solicitor or independent custody visitor who is present at the police station in a professional capacity may not act as the appropriate adult.

A person should always be given an opportunity, when the appropriate adult is called to a police station, to consult privately with a solicitor in the absence of the appropriate adult if (s)he wishes to do so. The appropriate adult is not subject to legal privilege.

DOCUMENTATION

Code C provides as follows.

5.8

When a person is brought to a police station under arrest, or is arrested at a police station having attended there voluntarily, or attends a police station in answer to bail, (s)he must be brought before the custody officer as soon as practicable after arrival, or following arrest at the police station. This equally applies to designated and non-designated police stations. Such a person is 'at a police station' if (s)he is anywhere upon the premises or enclosed yards forming part of the premises.

A separate custody record must be opened as soon as practicable for each person who is brought to a police station under arrest or who is arrested at the police station, having attended there voluntarily.

Where the arresting officer is not physically present when a detainee is brought to the police station, the arresting officer's account must be made available to the custody officer remotely or by a third party on the arresting officer's behalf. All information which is required to be recorded under Code C must be recorded as soon as practicable in the custody record unless otherwise specified. Any audio or visual recording made in the custody area is not part of the custody record. It is a matter for the custody officer to determine whether a record should be made of the property a detained person has with him/her or had taken from him/her on arrest. Any record made is not required to be kept as part of the custody record but the custody record should be noted as to where such a record exists. Whenever a record is made the detainee must be allowed to check and sign it as correct. Any refusal to sign must be recorded.

Where a person is answering street bail, the custody officer should link any documentation held in relation to the arrest with the custody record and any further action must be recorded in accordance with the Code.

In the case of any action requiring the authority of an officer of a specified rank, his/her name and rank must be recorded in the custody record, *except where the person is detained under TA 2000, or where there are reasonable grounds to believe that naming the officer would endanger him/her (in which case the record must state the officer's warrant or other identification number and duty station).*

All entries in the custody record must be timed and signed by the maker. In the case of a record held on a computer, this should be timed and contain the operator's identification. Warrant or other identification numbers and the name of the duty station should be used rather than names in the cases italicised above where the special rule applies.

If someone is arrested and taken to a police station as a result of a search in the exercise of any stop and search power to which Code A or the 'Terrorism Search Powers Code' (**21.21**) issued under TA 2000 applies, the officer carrying out the search is responsible for ensuring that the record of the stop and search is made as part of the custody record. The custody officer must then ensure that the person is asked if (s)he wants a copy of the search record and, if (s)he does, that (s)he is given a copy as soon as practicable.

The custody officer is responsible for the accuracy and completeness of the custody record and for ensuring that the record (or a copy) accompanies a detained person if (s)he is transferred to another police station. The record must show the time of, and reason for, a transfer and the time a person is released from detention. The detainee's solicitor and appropriate adult must be permitted to inspect the custody record as soon as practicable after arrival at the police station and at any other time during his/her detention. When someone leaves police detention or is taken to court, (s)he or his/her legal representative or appropriate adult must be given, on request, a copy of the custody record as soon as practicable. This entitlement lasts for twelve months after release.

The fact and time of any refusal by any person to sign a custody record when asked to do so in accordance with Code C must be recorded.

INITIAL ACTION

Detained persons: normal procedure

When a person is brought to a police station under arrest or arrested at the station **5.9** having gone there voluntarily, the custody officer must ensure that (s)he is told clearly about the right to be informed about the offence and (as the case may be) any further offences for which (s)he is arrested while in custody and why (s)he has been arrested and detained, and about the following *continuing rights* which may be exercised at any stage during the period in custody:

(a) the right to have someone informed of his/her arrest;
(b) the right to consult privately with a solicitor and that free independent legal advice is available;
(c) the right to consult the PACE codes;
(d) if applicable, the right to interpretation and translation and the right to communicate with his/her High Commission, Embassy or Consulate.

The detainee must also be given a *written notice*:

(a) setting out:
    (i) the above rights;
    (ii) the arrangements for obtaining legal advice;
    (iii) the right to a copy of the custody record;
    (iv) the right to remain silent as set out in the caution in the terms set out at **5.27**);
    (v) the right to have access to materials and documents which are essential to effectively challenging the lawfulness of his/her arrest and detention for any offence and (as the case may be) any further offences for which (s)he is arrested whilst in custody;
    (vi) the maximum period for which the detainee may be kept in police detention without being charged, when detention must be reviewed and when release is required;
    (vii) the right to medical assistance in accordance with the requirements at **6.9** and **6.10**;
    (viii) the right, if prosecuted, to have access to the evidence in the case before his/her trial; and
(b) briefly setting out his/her entitlements while in custody.

The detainee must be given an opportunity to read the notice and be asked to sign the custody record to acknowledge receipt of these notices. An 'easy read' illustrated version should also be provided if available.

The custody officer must:

(a) record the offence(s) that the detainee has been arrested for and the reason(s) for the arrest;
(b) note on the custody record any comment the detainee makes in relation to the arresting officer's account but must not invite comment. If the arresting officer is not physically present when the detainee is brought to a police station, the arresting officer's account must be made available to the custody officer remotely or by a third party on the arresting officer's behalf. If the custody officer authorises a person's detention, (s)he must record the grounds for detention in the detainee's presence and, at the same time, inform the detainee of them unless the detainee

is incapable of understanding, or is (or may become) violent, or in urgent need of medical attention. In such a case, the information must be given as soon as practicable. The detainee must be informed of the grounds for his/her detention before (s)he is questioned about any offence;

(c) note any comment the detainee makes in respect of the decision to detain him/her but must not invite comment;

(d) not put specific questions to the detainee regarding his/her involvement in any offence, nor in respect of any comments (s)he may make in response to the arresting officer's account or the decision to place him/her in detention. Such an exchange is likely to constitute an interview and require the associated safeguards set out at **5.35**.

These provisions also apply to any further offences and grounds for detention coming to light while the person is detained.

Documents and materials essential to effectively challenging the lawfulness of the detainee's arrest and detention must be made available to the detainee or his/her solicitor. (Documents and materials will be 'essential' for this purpose if they are capable of undermining the reasons and grounds which make the detainee's arrest and detention *necessary*.) The decision about whether particular documents or materials must be made available for this purpose rests with the custody officer who determines whether detention is necessary, in consultation with the investigating officer who has the knowledge of the documents and materials in a particular case necessary to inform that decision. A note should be made in the detainee's custody record of the fact that documents or materials have been made available under this provision and when. The investigating officer should make a separate note of what is made available and how it is made available in a particular case. This provision also applies (with modifications) for the purposes of the provisions about reviews and extensions of detention, and charging detained persons.

The custody officer or other custody staff as directed by the custody officer must:

(a) ask the detainee whether at this time, (s)he:
 (i) would like legal advice;
 (ii) wants someone informed of his/her detention;

(b) ask the detainee to sign the custody record to confirm his/her decisions in respect of (a);

(c) determine whether the detainee:
 (i) is, or might be, in need of medical treatment or attention;
 (ii) is a juvenile and/or vulnerable and therefore requires an appropriate adult;
 (iii) wishes to speak in private with a member of the custody staff who may be of the same sex about any matter concerning their personal needs relating to health, hygiene and welfare (see **6.9**);
 (iv) requires help to check documentation, or an interpreter;

(d) if the detainee is a female aged 18 or over, ask if she requires or is likely to require any menstrual products whilst in custody; and

(e) record the decision and actions in respect of (c) and (d).

If the appropriate adult is:

(a) already at the police station, the above provisions must be complied with in the appropriate adult's presence;

(b) not at the station when these provisions are complied with, they must be complied with again in the presence of the appropriate adult when (s)he arrives,

and a copy of the notice given to the detainee must also be given to the appropriate adult.

The custody officer must ensure that at the time the copy of the notice is given to the appropriate adult, or as soon as practicable thereafter, the appropriate adult is advised of the duties of the appropriate adult.

When the needs mentioned in (c) above are being determined, the custody officer is responsible for initiating an assessment to consider whether the detainee is likely to present specific risks to custody staff, any individual who may have contact with detainee (eg legal advisers, medical staff) or him/herself. This risk assessment must include the taking of reasonable steps to establish the detainee's identity and to obtain information about the detainee that is relevant to the detainee's safe custody, security, and welfare, and risks to others. It should always include a Police National Computer (PNC) check. It may be necessary for the custody officer to consult and involve others, eg the arresting officer or an appropriate healthcare professional. Other records held by or on behalf of the police and other UK law enforcement authorities that might provide information relevant to the detainee's safe custody, security, and welfare, and risk to others, and to confirming his/her identity, should also be checked. It may be necessary for the custody officer to consult and involve others, eg the arresting officer or an appropriate healthcare professional. Risk assessments must follow a structured process which clearly defines the categories of risk to be considered and the results must be incorporated in the detainee's custody record. The *Detention and Custody Authorised Professional Practice* produced by the College of Policing provides detailed guidance on risk assessments and identifies key risk areas which should always be considered. The custody officer must ensure those responsible for the detainee's custody are appropriately briefed about the risks. The content of any risk assessment and any resulting analysis need not be shown to the detainee or any person acting on his/her behalf. However, information should not be withheld from anyone acting on the detainee's behalf, if to do so might put him/her at risk. The custody officer is responsible for implementing the response to any specific risk assessment.

The grounds for a person's detention must be recorded in his/her presence if practicable.

Detained persons—special groups

If a person appears to be blind or seriously visually impaired or unable to read, deaf, or unable to speak or has difficulty orally because of a speech impediment, (s)he should be treated as such for the purpose of Code C in the absence of clear evidence to the contrary.

5.10

If a person is blind, seriously visually impaired, or unable to read, the custody officer must ensure that his/her solicitor, relative, the appropriate adult, or some other person likely to take an interest in him/her (and not involved in the investigation) is available to help in checking any documentation. Where Code C requires consent or signing, then the person who is assisting may be asked to sign instead, if the detained person so wishes. However, there is no requirement that the appropriate adult be called solely to assist in checking and signing documentation for a person who is not a juvenile or vulnerable.

If the detainee appears not to speak or understand English or to have a hearing or speech impediment, the custody officer must ensure that:

(a) without delay, an interpreter is called for assistance in the initial action referred to at **5.9**. If the detainee appears to have a hearing or speech impediment, the reference to 'interpreter' includes appropriate assistance necessary to comply with the required initial action;

(b) in addition to the continuing rights set out at **5.9**, the detainee is told clearly about his/her right to interpretation and translation;

(c) the written notice given to the detainee referred to at **5.9** is in a language the detainee understands and includes the right to interpretation and translation together with information about the provisions referred to at **5.80-5.88**, which explain how the right applies; and

(d) if the translation of the notice is not available, the information in the notice is given through an interpreter and a written translation provided without undue delay.

If the detainee is a citizen of an independent Commonwealth country or a national of a foreign country, including the Republic of Ireland, the custody officer must ensure that in addition to the continuing rights set out above, the detainee is informed as soon as practicable about his/her rights of communication with his/her High Commission, Embassy, or Consulate set out at **5.26**. This right must be included in the written notice given to the detainee referred to above.

If a juvenile is known to be subject to a court order under which a person or organisation is given any degree of statutory responsibility to supervise or otherwise monitor him/her, reasonable steps must also be taken to notify that person or organisation (the 'responsible officer'). The responsible officer is normally a member of a youth offending team, except in the case of a curfew order which involves electronic monitoring, when the monitoring contractor is normally the responsible officer.

If the detainee is a juvenile, the custody officer must ascertain the identity of a person responsible for the detainee's welfare, eg his/her parent or guardian, local authority with care, or someone who has temporarily assumed responsibility for his/her welfare, and must, as soon as practicable, inform that person of the arrest, the reason and the place of detention.

If the detainee is a juvenile or a vulnerable person, the custody officer must, as soon as practicable, ensure that:

(a) the detainee is informed of the decision that an appropriate adult is required and the reason for that decision;

(b) the detainee is advised of the duties of the appropriate adult as described above, and that (s)he can consult privately with the appropriate adult at any time;

(c) the appropriate adult, who in the case of a juvenile may or may not be a person responsible for his/her welfare, as explained above, is informed of the grounds for his/her detention and his/her whereabouts; and

(d) the attendance of the appropriate adult at the police station to see the detainee is secured.

A person detained under the Mental Health Act 1983, s 135 or 136 must be assessed as soon as possible within the permitted period of detention specified in that Act (**4.114** and **4.115**). A police station may only be used as a place of safety for an adult, and then only in accordance with the Mental Health Act 1983 (Places of Safety) Regulations 2017 (**4.116**). If that assessment is allowed to take place at the police station, and does, an approved mental health professional and a registered medical practitioner must be called to the station as soon as possible to carry it out. The appropriate adult has no role in the assessment process and his/her presence is not required. Once the detainee has been assessed and suitable arrangements made for his/her treatment or care, (s)he can no longer be detained under s 135 or 136. A detainee must be immediately discharged from detention if a registered medical practitioner, having examined him/her, concludes that (s)he is not mentally disordered within the meaning of the Act.

The Children and Young Persons Act 1933, s 31 requires that arrangements must be made for ensuring that a girl under 18, while detained in a police station, is under the care of a woman. Code C states that the custody officer must ensure that the woman under whose care the girl is, makes the enquiries and provides the information concerning personal needs relating to the girl's health, hygiene, welfare, and menstrual products referred to at **6.9**. Section 31 also requires that arrangements must be made for preventing any person under 18 (P), while detained in a police station, from associating with an adult charged with any offence, unless that adult is a relative or is jointly charged with the same offence as P. Guidance for police officers and police staff on the operational application of s 31 has been published by the College of Policing.

Actions taken under the provisions relating to detained persons—special groups must be recorded.

The above requirements to inform apply even if Annex B to Code C (**5.16**) applies.

Persons attending a police station or elsewhere voluntarily

PACE, s 29 provides that where, for the purpose of assisting with an investigation, a person attends voluntarily at a police station or at any other place where a constable is present, or accompanies a constable to a police station or any other place without having been arrested, (s)he is entitled to leave at will unless (s)he is placed under arrest. If the interview is at 'any other place' where the interviewer requires the informed consent of the person and/or occupier (if different) to remain (eg that person's home) 'entitled to leave' means 'require the interviewer to leave'. **5.11**

If, during a person's voluntary attendance at a police station or other location, it is decided that it is necessary to arrest him/her, (s)he must be informed at once that (s)he is under arrest and the grounds and reasons (**4.51**), and be brought before the custody officer at that police station or one to which (s)he is then taken. The same rules apply thereafter as in the case of other detainees. The interviewer's responsibilities reflect that of the custody officer with regard to detained suspects including determining whether the suspect requires an appropriate adult, help to check documentation, an interpreter and the provision of interpretation and translation services.

Information to be given when arranging a voluntary interview

If the suspect's arrest is not necessary but (s)he is cautioned as required (**5.27**), the person who, after describing the nature and circumstances of the suspected offence, gives the caution must at the same time, inform him/her that (s)he is not under arrest and that (s)he is not obliged to remain at the station or other location. **5.12**

The suspect must not be asked to give his/her informed consent to be interviewed until after (s)he has been informed of the rights, entitlements, and safeguards (set out below) that apply to voluntary interviews. The interviewer is responsible for ensuring that the suspect is so informed and for explaining these rights, entitlements, and safeguards.

The interviewer must inform the suspect that the purpose of the voluntary interview is to question him/her to obtain evidence about his/her involvement or suspected involvement in the offence(s) described when (s)he was cautioned and told that (s)he was not under arrest. The interviewer must then inform the suspect that the following matters will apply if (s)he agrees to the voluntary interview proceeding:

(a) His/her right to information about the offence(s) in question by providing sufficient information to enable him/her to understand the nature of any such offence(s) and why (s)he is suspected of committing it. This applies whether or not (s)he asks for legal advice and includes any further offences that come to light and are pointed out during the voluntary interview and for which (s)he is cautioned.

(b) His/her right to free legal advice by:

  (i) explaining that (s)he may obtain free and independent legal advice if (s)he wants it, and that this includes the right to speak with a solicitor on the telephone and to have the solicitor present during the interview;

  (ii) asking if (s)he wants legal advice and recording his/her reply; and

  (iii) if the person requests advice, securing its provision before the interview by contacting the Defence Solicitor Call Centre and explaining that the time and place of the interview will be arranged to enable him/her to obtain advice and that the interview will be delayed until (s)he has received the advice unless, in accordance with (c) and (d) at **5.25**, (nominated solicitor not available and duty solicitor declined, and change of mind) an inspector (or above) agrees to the interview proceeding; or

  (iv) if the person declines to exercise the right, asking him/her why and recording any reasons given.

  Note: When explaining the right to legal advice and the arrangements, the interviewer must take care not to indicate, except to answer a direct question, that the time taken to arrange and complete the voluntary interview might be reduced if the suspect does not ask for legal advice or does not want a solicitor present when (s)he is interviewed, or if the suspect asks for legal advice or (as the case may be) asks for a solicitor to be present when (s)he is interviewed, but changes his/her mind and agrees to be interviewed without waiting for a solicitor.

(c) His/her right, if the interviewer determines:

  (i) that (s)he is a juvenile or is vulnerable; or

  (ii) that (s)he needs help to check documentation,

  to have the appropriate adult present or (as the case may be) to have the necessary help to check documentation; and that the interview will be delayed until the presence of the appropriate adult or the necessary help, is secured.

(d) If (s)he is a juvenile or vulnerable and does not want legal advice, his/her appropriate adult has the right to ask for a solicitor to attend if this would be in his/her best interests and the appropriate adult must be so informed. In this case, action to secure the provision of advice if so requested by the appropriate adult will be taken without delay in the same way as if requested by the person (see (b)(iii) above). However, (s)he cannot be forced to see the solicitor if (s)he is adamant that (s)he does not wish to do so.

(e) His/her right to an interpreter, if the interviewer determines that (s)he requires an interpreter and that if an interpreter is required, making the necessary arrangements and that the interview will be delayed to make the arrangements.

(f) That interview will be arranged for a time and location that enables:

  (i) the suspect's rights described above to be fully respected; and

  (ii) the whole of the interview to be recorded using an authorised recording device in accordance with Code E (Code of Practice on Audio Recording Interviews with Suspects: **5.51**) or (as the case may be) Code F (Code of Practice on Visual Recording with Sound of Interviews with Suspects: **5.71**).

(g) That his/her agreement to take part in the interview also signifies agreement for that interview to be audio-recorded or (as the case may be) visually recorded with sound.

The provision by the interviewer of factual information just described above and, if asked by the suspect, further such information, does not constitute an interview for the purpose of Code C and when that information is provided:

(a) the interviewer must remind the suspect about the caution as required (see **5.31**) but must not invite comment about the offence or put specific questions to the suspect regarding his/her involvement in any offence, nor in respect of any comments (s)he may make when given the information. Such an exchange is itself likely to constitute an interview (see **5.33**) and require the associated interview safeguards described at **5.35-5.44**.

(b) Any comment the suspect makes which are outside the context of an interview but which might be relevant to the offence, must be recorded and dealt with in accordance with **5.42**.

(c) The suspect must be given a notice summarising the matters described in the previous paragraph and which includes the arrangements for obtaining legal advice. If a specific notice is not available, the notice given to detained suspects with references to detention-specific requirements and information redacted, may be used.

(d) For juvenile and vulnerable suspects:
    (i) the information must be provided or (as the case may be) provided again, together with the notice, in the presence of the appropriate adult;
    (ii) if cautioned in the absence of the appropriate adult, the caution must be repeated in the appropriate adult's presence;
    (iii) the suspect must be informed of the decision that an appropriate adult is required and the reason;
    (iv) the suspect and the appropriate adult must be advised that the duties of the appropriate adult include giving advice and assistance in accordance with that adult's role, and that (s)he can consult privately at any time;
    (v) his/her informed agreement to be interviewed voluntarily must be sought and given in the presence of the appropriate adult and for a juvenile, the agreement of a parent or guardian of the juvenile is also required.

Action taken under the provisions described above under the present heading must be recorded. The record must include the date, time and place the action was taken, who was present and anything said to or by the suspect and to or by those present.

Before asking the suspect any questions about his/her involvement in the offence (s) he is suspected of committing, the interviewing officer must ask him/her to confirm that (s)he agrees to the interview proceeding. This confirmation must be recorded in the written interview record or in accordance with Code E (audio recording) or Code F (audio visual recording).

Persons answering street bail

When someone is answering street bail, the custody officer should link any documentation held in relation to arrest with the custody record. Any further action must be recorded on the custody record.   **5.13**

Alleged failure to give information about rights to suspect

5.14   If a complaint is made by or on behalf of a suspect who has been cautioned that the requisite information and (as the case may be) access to records and documents has not been provided, the matter must be reported to an inspector to deal with as a complaint for the purposes of the provisions at **5.78** and **6.11** if the challenge is made during an interview.

RIGHT NOT TO BE HELD INCOMMUNICADO

Notification of detention

5.15   A detainee has a right under PACE, s 56 to have someone informed at public expense and as soon as practicable of his/her detention. However, this extends only to one friend, relative, or other person who is known to him/her or is likely to take an interest in his/her welfare. If the person cannot be contacted, the detainee may choose up to two alternatives. If they cannot be contacted the custody officer may allow further attempts. If a person is moved from one police station to another, the right to have someone informed of his/her whereabouts arises again.

Delay

5.16   Annex B to Code C (which largely mirrors PACE, s 56 in this respect), provides that a delay in informing someone may be authorised by an inspector (or above) where the person is in police detention for an indictable offence and has not been charged with it. Such authorisation may only be given in two types of case. The first is where the inspector has reasonable grounds for believing that telling the person named of the arrest will:

(a)   lead to interference with, or harm to, evidence connected with an indictable offence or interference with or physical harm to other persons; or

(b)   lead to the alerting of other persons suspected of having committed such an offence but not yet arrested for it; or

(c)   hinder the recovery of any property obtained as a result of such an offence.

The second type of case is that an inspector may authorise delay where (s)he has reasonable grounds for believing that the person detained for an indictable offence has benefited from his/her criminal conduct (decided in accordance with the Proceeds of Crime Act 2002, Part 2) and that the recovery of the value of the property constituting that benefit will be hindered by telling the named person of the arrest.

If a delay is authorised, the detainee must be informed, as soon as practicable, of the reason for it and the reason must be noted on his/her custody record. If given orally, the authorisation must be confirmed in writing as soon as practicable. When the reasons for delay have been removed, eg by the arrest of other persons, the detainee must be asked if (s)he wishes to exercise the right to communicate (and the custody record must be noted accordingly) and any request to communicate must be granted. The right to communicate cannot be delayed for more than 36 hours.

Any delay authorised under the above provisions should be proportionate and last no longer than necessary.

Visits, letters and phone calls

5.17   If a detainee agrees, (s)he may, at the discretion of the custody officer, receive visits from friends, family, or others likely to take an interest in his/her welfare, or in whose welfare the detainee has an interest. Where inquiries by interested persons are received concerning a detainee's whereabouts, the information must be given if the detainee agrees and an inspector (or above) has not authorised delay in the release of such information. The custody officer must exercise his/her discretion as to visits in the light of the availability of sufficient manpower to supervise a visit and any possible hindrance to the investigation.

The detainee must also be supplied on request with writing material. Letters and messages must be sent (at his/her expense) as soon as practicable, but all letters, other than those to his/her solicitor, may be read. The detainee may speak for a reasonable time to one person on the telephone. Whether or not the call can be made at police expense is a matter for the custody officer's discretion. Unless the call is to a solicitor, a police officer may listen to the call and may terminate it if it is being abused. The detainee must be cautioned that what (s)he says in a letter, call, or message (other than to his/her solicitor) may be read or listened to and may be given in evidence. An interpreter may make a call on behalf of a detainee.

Where an inspector (or above) considers that the sending of a letter or the making of a telephone call might result in any of the consequences referred to in the list referred to at **5.16**, and the person is detained in connection with an indictable offence, that officer can deny or delay the exercise of either or both these privileges.

Any delay or denial of these rights should be proportionate and should last no longer than necessary.

Prisoners

Where a prisoner has been transferred to police custody for specific purposes and periods under the Crime (Sentences) Act 1997, Sch 1, the exercise of the above rights is subject to any additional conditions specified in the transfer direction for the purpose of regulating his/her contact and communication with others whilst in police custody. **5.18**

Documentation

A record must be kept of: **5.19**

(a) any request made in relation to matters of communication and of the action taken in consequence of that request;
(b) letters or messages sent, calls made, or visits received; and
(c) any refusal by the detainee to have information about him/herself or his/her whereabouts given to an outside inquirer. The detainee must be asked to countersign the record accordingly.

LEGAL ADVICE

By PACE, s 58 a person arrested and held in police custody is entitled, if (s)he so requests, to consult a solicitor privately at any time. This is a fundamental right. Unless Annex B (delay) applies, all detainees must be informed of this right. Code C provides that the consultation may be in person, in writing, or by telephone, and that free independent legal advice is available. **5.20**

A detainee has a right to free legal advice and to be represented by a solicitor. What follows explains the arrangements which enable detainees to obtain legal advice. The arrangements also apply, with appropriate modifications, to persons attending a police station or other location voluntarily who are cautioned prior to being interviewed.

When a detainee asks for free legal advice, the Defence Solicitor Call Centre (DSCC) must be informed of the request.

Free legal advice will be limited to telephone advice provided by CDS Direct if a detainee is:

(a) detained for a non-imprisonable offence;
(b) arrested on a bench warrant for failing to appear and being held for production at court (except where the solicitor has clear documentary evidence available that would result in the client being released from custody);
(c) arrested for drink-driving (driving/in charge with excess alcohol or while unfit through drink, or failing to produce a specimen); or
(d) detained in relation to breach of police or court bail conditions,

unless one or more exceptions apply, in which case the DSCC should arrange for advice to be given by a solicitor at the police station.

Examples of exceptions are:

(a) the police want to interview the detainee or carry out an identification parade;
(b) the detainee needs an appropriate adult, is unable to communicate over the telephone, or alleges serious maltreatment by the police;
(c) the investigation includes another offence not included in the list;
(d) the solicitor to be assigned is already at the police station.

When free advice is not limited to telephone advice, a detainee can ask for free advice from a solicitor (s)he knows or, if (s)he does not know a solicitor or the solicitor (s)he knows cannot be contacted, from the duty solicitor.

To arrange free legal advice, the police should telephone the DSCC. The call centre will decide whether legal advice should be limited to telephone advice from CDS Direct, or whether a solicitor known to the detainee or the duty solicitor should speak to the detainee.

A guidance note states that when a detainee wants to pay for legal advice him/herself:

(a) the DSCC will contact a solicitor of the detainee's choice on his/her behalf;
(b) the detainee may, when free advice is only available by telephone from CDS Direct, still speak to a solicitor of his/her choice on the telephone for advice but the solicitor would not be paid by legal aid and may ask the person to pay for the advice;
(c) the detainee should be given an opportunity to consult a specific solicitor or another solicitor from that solicitor's firm. If this solicitor is not available, (s)he may choose up to two alternatives. If these alternatives are not available, the custody officer has discretion to allow further attempts until a solicitor has been contacted and agreed to provide advice;
(d) the detainee is entitled to a private consultation with his/her chosen solicitor on the telephone or the solicitor may decide to come to the police station;
(e) if the detainee's chosen solicitor cannot be contacted, the DSCC may still be called to arrange free legal advice.

Apart from carrying out duties necessary to implement these arrangements, an officer must not advise the suspect about any particular firm of solicitors.

In the case of a person who is a juvenile or is vulnerable, an appropriate adult should consider whether legal advice from a solicitor is required. If such a detainee wants to exercise the right to legal advice, the appropriate action should be taken and should not be delayed until the appropriate adult arrives. If the detainee indicates that (s)he does not want legal advice, the appropriate adult has the right to ask for a solicitor to attend if this would be in the best interests of the detainee and must be so informed. In this case, action to secure the provision of advice if so requested by the appropriate adult must be taken without delay in the same way as when requested by the detainee. However, the detainee cannot be forced to see the solicitor if (s)he is adamant that (s) he does not wish to do so.

Whenever legal advice is requested (and unless delay is permitted under Annex B (**5.21**)), the custody officer must act without delay to secure the provision of legal advice. If the detainee declines to exercise his/her right to speak to a solicitor in person, the officer should point out that the right to legal advice includes the right to speak with a solicitor on the telephone. If the detainee continues to waive his/her right, or a detainee whose right to free legal advice is limited to telephone advice from CDS Direct declines to exercise that right, the officer should ask him/her why. Any reasons must be recorded on the custody record or the interview record as appropriate. Once it is clear that a person neither wishes to speak to a solicitor in person, nor by telephone, (s)he should cease to be asked his/her reasons.

Except as allowed in the case of persons arrested under TA 2000, s 41 (**21.46**) if the requirement for privacy of consultation is compromised because what is said or written by the detainee or the solicitor for the purpose of the giving or receiving of legal advice is overheard, listened to, or read by others without the informed consent of the detainee, the right will effectively have been denied. Where a detainee speaks to a solicitor on the telephone, (s)he should be allowed to do so in private unless this is impractical because of the design and layout of the custody area or the location of the telephones.

Delay

Access to legal advice may be delayed if the person is in police detention for an indictable offence, has not yet been charged, and a superintendent (or above) authorises the delay. An oral authorisation must be confirmed in writing as soon as practicable. Delay may only be authorised in two types of case specified in PACE, s 58 and Annex B to Code C.   **5.21**

The first is where the superintendent (or above) has reasonable grounds for believing that access to a solicitor at a time when the detainee wishes to have access will:

(a) lead to interference with or harm to evidence connected with an indictable offence or interference with or physical injury to other persons; or

(b) lead to the alerting of other persons suspected of having committed such an offence but not yet arrested for it; or

(c) hinder the recovery of any property obtained as a result of such an offence.

The Court of Appeal has held that an authorisation on such a ground can only be justified by reference to specific circumstances, including evidence as to the person detained or the *actual solicitor* sought to be consulted. Thus, if it is believed that a particular solicitor may bring about one of the consequences at (a) to (c) above, that belief would only be relevant to that particular solicitor and would not apply to others selected by the detainee.

The second type of case is that a superintendent (or above) may authorise delay where (s)he has reasonable grounds to believe that the person detained for an indictable offence has benefited from his/her criminal conduct (decided in accordance with the Proceeds of Crime Act 2002, Part 2) and that the recovery of the value of the property constituting that benefit will be hindered by the exercise of the right to consult a solicitor privately.

Annex B adds that access to a solicitor may not be delayed on the ground that (s)he might advise the detainee not to answer any questions or that (s)he was initially asked to attend by someone other than the detainee, provided the detainee wishes to see him/her. In the latter case the detainee must be told that the solicitor has come to the police station at another person's request, and must be asked to sign the custody record to signify whether or not (s)he wishes to see the solicitor.

If delay is authorised the detainee must be told the reason and the reason must be noted in his/her custody record as soon as practicable. Once the reason for delay ceases, no further delay in permitting access to a solicitor is permissible and the detainee must be asked if (s)he wishes to exercise his/her right to legal advice (and the custody record must be noted accordingly). *In any case* the detainee must be permitted to consult a solicitor within 36 hours from the 'relevant time' (a term defined at **6.40**).

Where a delay has been authorised under Annex B and an interview takes place during it, a court or jury may not draw adverse inferences from a suspect's silence.

Arrival of solicitor at police station

5.22   Unless Annex B (above) applies, when a solicitor arrives at a police station to see a particular person, that person must be informed of the solicitor's arrival whether or not (s)he is being interviewed, and asked if (s)he wishes to see the solicitor. This applies even if the detainee has declined legal advice or, having requested it, subsequently agreed to be interviewed without receiving advice. The attendance and the detainee's decision must be recorded. Where a consultation is permitted, a solicitor is entitled to be present whilst the detainee is interviewed if the detainee so wishes.

Accredited or probationary representatives

5.23   For the purposes of Code C, a 'solicitor' is a person who holds a current practising certificate, or an accredited or probationary representative included on the register of representatives maintained by the Legal Aid Agency.

An accredited or probationary representative sent to provide advice on an actual solicitor's behalf must be admitted to the police station for this purpose, unless an inspector (or above) considers that such a visit will hinder the investigation of crime and directs otherwise. (Hindering the investigation of crime does not include giving the detainee proper legal advice.) In determining this issue, an inspector must take into account in particular whether the identity or status of the accredited or probationary representative has been satisfactorily established; whether (s)he is of suitable character to provide legal advice (a person with a criminal record is unlikely to be suitable unless the conviction was for a minor offence and is not of recent date); and any other matters in any written letters of authorisation provided. The Court of Appeal has held that a chief constable is not entitled to make a blanket order banning a particular solicitor's probationary representative from all police stations in his/her area. (S)he may advise his/her officers that a particular representative is likely to hinder an investigation but the officer dealing with the case must decide whether, in the particular circumstances, the representative should be excluded.

If an inspector refuses access to an accredited or probationary representative or a decision is taken that such a person must not remain at an interview, (s)he must forthwith notify the solicitor on whose behalf that person was to have acted, or was acting, and give him/her an opportunity of making other arrangements. The detainee must also be informed and the custody record noted. If an inspector considers that a particular firm of solicitors is persistently sending probationary representatives who are unsuited to provide legal advice, (s)he should inform a superintendent (or above), who may wish to take the matter up with the Solicitors Regulation Authority (SRA).

Removal of 'solicitor'

A 'solicitor' may only be required to leave an interview if his/her conduct is such that **5.24** the investigating officer is unable properly to put questions to the suspect.

The solicitor's only role in the police station is to protect and advance the legal rights of his/her client. On occasions this may require the solicitor to give advice which has the effect of avoiding the client giving evidence which strengthens a prosecution case. The solicitor may intervene to seek clarification or to challenge an improper question to his/her client or the manner in which it is put, or to advise his/her client not to reply to particular questions, or if (s)he wishes to give his/her client further legal advice. (S)he may only be required to leave if his/her approach or conduct prevents or unreasonably obstructs proper questions being put to the detainee or the detainee's response being recorded, as where a solicitor answers on his/her client's behalf or provides written replies for his/her client to quote. If the investigating officer considers that a solicitor is acting in such a way, (s)he will stop the interview and consult a superintendent (or above), if one is readily available, and otherwise an officer not below the rank of inspector who is not concerned with the investigation. After speaking to the solicitor the officer who has been consulted will decide whether or not the interview should continue in the presence of the solicitor. If (s)he decides that it should not, the detainee will be given an opportunity to consult another solicitor before the interview continues and that solicitor will be given an opportunity to be present.

Code C points out that the removal of a solicitor from an interview is a serious step and, if it occurs, the superintendent (or above) who took the decision will consider whether the incident should be reported to the Solicitors Regulatory Authority (and to the Legal Aid Agency in the case of a duty solicitor). If the decision was taken by an officer below the rank of superintendent, a superintendent (or above) must consider whether to make such a report. A guidance note in the Code points out that where an officer takes the decision to exclude a solicitor, (s)he must be in a position to satisfy the court that the decision was properly made. In order to do this, (s)he may need to witness what is happening him/herself.

Other points

In the absence of compelling reasons, delaying access to a solicitor is likely to amount to **5.25** a breach of the right to a fair trial under the European Convention on Human Rights, art 6.

Any request for legal advice, and the action taken on it, must be recorded in the custody record. If a person has asked for legal advice and an interview has commenced in the absence of his/her solicitor (or the solicitor is required to leave) a record must be made in the interview record.

A detainee who wants legal advice *may not be interviewed or continue to be interviewed* until (s)he has received it unless:

(a) a delay has been authorised in accordance with Annex B to Code C (**5.21**), in which case the restriction on drawing adverse inferences from silence (**5.28**) will apply because the detainee is not allowed an opportunity to consult a solicitor; or

(b) a superintendent (or above) has reasonable grounds to believe that:

    (i) the consequent delay might lead to interference with, or harm to, evidence or other people, or to serious loss of, or damage to, property, or to alerting other people suspected of having committed an offence but not yet arrested; or might hinder the recovery of property obtained in consequence of the offence; or

    (ii) when a solicitor, including a duty solicitor, has been contacted and has agreed to attend, awaiting his/her arrival would cause unreasonable delay to the process of investigation;

in which instances the restriction upon drawing adverse inferences from silence will apply; or

(c) the solicitor nominated or selected from a list by the detainee cannot be contacted, has previously indicated that (s)he does not wish to be contacted, or having been contacted has declined to attend, and the detainee has been advised of the Duty Solicitor Scheme but has declined to ask for the duty solicitor; in which instances the restriction upon drawing inferences from silence will not apply because the detainee has been allowed an opportunity to consult a solicitor; or

(d) the detainee changes his/her mind about wanting legal advice or (as the case may be) about wanting a solicitor present at the interview and states that (s)he no longer wishes to speak to a solicitor. In this case, the interview may be started or continued without delay provided that:

    (i) an inspector (or above) asks the detainee why (s)he has changed his/her mind, and makes reasonable efforts to ascertain the solicitor's expected time of arrival and to inform the solicitor that the suspect has stated that (s)he wishes to change his/her mind and the reason (if given);

    (ii) the detainee's reason for his/her change of mind (if given) and the outcome of the action in (i) are recorded in the custody record;

    (iii) the detainee after being informed of the outcome of the action in (i) confirms in writing that (s)he wants the interview to proceed without speaking or further speaking to a solicitor or (as the case may be) without a solicitor being present and does not wish to wait for a solicitor by signing an entry to this effect in the custody record;

    (iv) an inspector (or above) is satisfied that it is proper for the interview to proceed in these circumstances, and

        (*a*) gives authority in writing for the interview to proceed; if the authority is not recorded in the custody record, the officer must ensure that the custody record shows the date and time of the authority and where it is recorded, and

        (*b*) takes or directs the taking of reasonable steps to inform the solicitor that the authority has been given and the time when the interview is expected to commence, and records or causes to be recorded the outcome of this action in the custody record;

    (v) when the interview starts and the interviewer reminds the suspect of his/her right to legal advice, the interviewer must ensure that the following is recorded in the written interview record or the interview record made in accordance with PACE Code E (audio recording) or F (visual recording with sound):

- confirmation that the detainee has changed his/her mind about wanting legal advice or (as the case may be) about wanting a solicitor present and the reasons for it if given;
- the fact that authority for the interview to proceed has been given and, with the usual exception, the name of the authorising officer;

- that, if the solicitor arrives at the station before the interview is completed, the detainee will be so informed without delay and a break will be taken to allow him/her to speak to the solicitor if (s)he wishes, unless (b) above applies;
- that, at any time during the interview, the detainee may again ask for legal advice and that, if (s)he does, a break will be taken to allow him/her to speak to the solicitor, unless (a), (b), or (c) applies.

In these circumstances there will be no restriction on drawing adverse inferences from silence because the detainee is allowed an opportunity to consult a solicitor if (s) he wishes.

If (a) applies, where the reason for authorising the delay ceases to apply, there may be no further delay in permitting the exercise of the right in the absence of a further authorisation unless (b), (c), or (d) applies. Where (b)(i) applies, once sufficient information has been obtained to avert the risk, questioning must cease until the detainee has received legal advice unless (a), (b)(ii), (c), or (d) applies. Where (c) applies, the interview may be started or continued without further delay provided that an inspector (or above) has agreed. Where (d) applies, the interview may be started or continued without further delay provided that the detainee has given an agreement in writing, or on recording media, to being interviewed without receiving legal advice and that an inspector (or above), having inquired into the detainee's reasons for his/her change of mind, has given authority for the interview to proceed. Confirmation of the detainee's agreement, his/her change of mind, his/her reasons where given, and the name of the authorising officer, except where it might endanger the officer to do so, must be recorded in writing or in the interview record made in accordance with Code E or F, at the beginning or recommencement of the interview. Such authorisation can be given over the telephone, if the authorising officer is able to satisfy him/herself as to the reason for the detainee's change of mind and is satisfied that it is proper to continue the interview in those circumstances.

In considering whether awaiting the arrival of a solicitor would cause unreasonable delay (see (b) above), the superintendent should, where practicable, ask the solicitor for an estimate of the time (s)he is likely to take in coming to the station, and relate this to the time for which detention is permitted, to whether a required period of rest is imminent, and to the requirements of other investigations in progress. If the solicitor says that (s)he is on his/her way or that (s)he will set off immediately, it will not normally be appropriate to begin an interview before (s)he arrives. If it appears that it will be necessary to begin an interview before the solicitor's arrival (s)he should be given an indication of how long the police could wait so that (s)he has the opportunity to make arrangements for someone else to provide legal advice.

A breach of the requirement that detainees should generally be allowed access to a solicitor does not necessarily justify exclusion of evidence obtained at the interview. Where a challenge is made, the court must establish whether a request was made and whether it had been refused. If this is so it must then consider whether a delay in compliance with such a request was permissible under the above rules. If it was not, the court must then consider whether the evidence obtained at the interview should be excluded under s 76 (confessions) or s 78 (unfairly obtained evidence (**8.52** and **8.74**) or by any rule of common law.

The Supreme Court has decided that the effect of art 6 of the European Convention on Human Rights (right to a fair trial) is that legal advice must be available not only to a person in custody but also to anyone else who has been 'charged' for the purposes of art 6, which occurs where the person's situation has been substantially affected.

This situation is likely to occur where questions are addressed to a person which the police have reason to think may well elicit an incriminating response, ie when (s)he has ceased to be a potential witness and become a suspect. However, even if someone not in police custody has been 'charged' in this sense, this does not necessarily mean that (s)he has been deprived of a fair trial contrary to art 6; it would depend on the circumstances.

ALIENS, ETC

5.26   Code C provides that a detainee who is a citizen of an independent Commonwealth country or a foreign national is entitled on request to communicate at any time with his/her High Commission, Embassy, or Consulate. Such a person must be informed as soon as practicable of this right and asked if (s)he wants his/her High Commission etc told of his/her whereabouts and the grounds for his/her detention. Such a request should be acted on as soon as practicable.

A detainee who is a national of a country with which a consular treaty is in force requiring notification of arrest (for the list of countries, see <https://gov.uk/governm ent/publications/table-of-consular-conventions-and-mandatory-notification-obligati ons>) must also be informed that, with the following qualification, notification of his/ her arrest will be sent to his/her High Commission etc as soon as practicable, whether or not (s)he requests it. The qualification is that, if the detainee claims that (s)he is a refugee or has applied or intends to apply for asylum, the custody officer must ensure that UK Visas and Immigration (UKVI) is informed as soon as practicable of the claim. UKVI will then determine whether compliance with relevant international obligations requires notification of the arrest to be sent and will inform the custody officer as to what action police need to take.

Consular officers may visit to advise one of their nationals. Those visits must take place out of the hearing of a police officer.

A record must be made:

(a) when a detainee is informed of the above rights and of any requirement to notify under a consular treaty;

(b) of any communications with a High Commission, etc; and

(c) of any communications with UKVI about a detainee's claim to be a refugee or to be seeking asylum and the resulting action taken by police.

CAUTIONS

When a caution must be given

5.27   Code C provides as follows.

A person whom there are grounds to suspect of an offence must be cautioned before any questions about an offence, or further questions if the answers provide the grounds for suspicion, are put to him/her if either his/her answer or silence (ie failure or refusal to answer or answer satisfactorily) may be given in evidence to a court in a prosecution. The grounds for suspicion must be reasonable grounds, based on known facts or relevant information. The Court of Appeal has held that there was no obligation to caution when at the time of the interview the police officers were unaware of the existence of a byelaw which could have given rise to a reasonable suspicion of an offence, and were simply conducting the interview to find out about what had happened.

Cautions may be given in the Welsh language where use of the Welsh language is appropriate.

A person need not be cautioned if questions are put to him/her for other purposes, for example:

(a) solely to establish his/her identity or his/her ownership of any vehicle; or
(b) to obtain information in accordance with any relevant statutory requirement;
(c) in furtherance of the proper and effective conduct of a search (for example, to determine the need to search in the exercise of powers to stop and search or to seek co-operation while carrying out a search); or
(d) to seek verification of a written record.

Whenever a person not under arrest is initially cautioned, or is reminded that (s)he is still under caution after a break, (s)he must at the same time be told that (s)he is not under arrest, be informed about how (s)he may obtain legal advice, and be told about the other rights and entitlements (5.12) that apply to a voluntary interview. The officer must point out that the right to legal advice includes the right to speak with a solicitor over the telephone. The officer must ask the person whether (s)he wishes to do so.

A person who is arrested, or further arrested, must be informed that (s)he is under arrest and given the grounds for that arrest. (S)he must also be cautioned unless it is impracticable because of his/her condition or behaviour at the time, or unless (s)he has already been cautioned immediately before arrest.

Since Code C obliges police officers to administer a caution if the answer to a question may be offered in evidence in criminal proceedings, it is necessary to caution motorists when pointing out to them that they have committed an offence as their reply is almost certainly going to be relevant to the proceedings. Indeed, although there is no direct requirement to caution when informing a person that (s)he will be reported for an offence, the admissibility of his/her reply in evidence may depend upon whether or not (s)he has been cautioned at some stage. Practice may still require that such a reply is included in an officer's report and, if this is so, the offender should be cautioned at some stage before (s)he makes it.

Nothing in Code C requires a caution to be given or repeated when informing a person not under arrest that (s)he may be prosecuted for an offence. However, a court will not be able to draw any inferences under the Criminal Justice and Public Order Act 1994, s 34 (8.57), if the person was not cautioned.

The caution prescribed by Code C is:

> You do not have to say anything. But it may harm your defence if you do not mention when questioned something which you later rely on in court. Anything you do say may be given in evidence.

Minor deviations do not constitute a breach of this requirement provided that the sense of the caution is preserved.

Other forms of caution are prescribed for use where the restriction on drawing adverse inferences from silence applies (5.29).

After any break in questioning under caution, the person being questioned must be made aware (s)he remains under caution. If there is any doubt the relevant caution should be given again in full.

When, despite being cautioned, a person fails to co-operate or to answer particular questions which may affect his/her immediate treatment, (s)he should be informed of any relevant consequences and that those consequences are not affected by the caution. Examples are when a person's refusal to provide his/her name and address when charged may make him/her liable to detention, or when his/her refusal to provide particulars in accordance with a statutory requirement, such as under the Road Traffic Act 1988, may amount to an offence or may make him/her liable to a further arrest.

Special warnings under CJPOA 1994, ss 36 and 37

5.28   When a suspect who is interviewed at a police station or authorised place of detention after arrest fails or refuses to answer certain questions, or to answer them satisfactorily, *after due warning*, a court or jury may draw such inferences as appear proper under CJPOA 1994, ss 36 and 37 (**8.59–8.60**). Such inferences may only be drawn when:

(a) the restriction on drawing adverse inferences from silence (below) does not apply; and

(b) the suspect is arrested by a constable and fails or refuses to account for any objects, marks, or substances, or marks on such objects, found on his/her person, or in or on his/her clothing or footwear, or otherwise in his/her possession, or in the place where (s)he is arrested, and the person fails or refuses to account for the objects, marks, or substances found; or

(c) the arrested suspect was found by a constable at a place at or about the time the offence for which (s)he was arrested is alleged to have been committed, and (s)he fails or refuses to account for his/her presence at that place.

When the restriction on drawing adverse inferences from silence applies, the suspect may still be asked to account for any of the matters in (b) or (c) but the special warning described in the next paragraph will not apply and must not be given.

For an inference to be drawn when a suspect fails or refuses to answer a question about one of these matters or to answer it satisfactorily, the suspect must first be given the *special warning*. (S)he must be told in ordinary language the following *specified information*:

(a) what offence is being investigated;

(b) what fact (s)he is being asked to account for;

(c) this fact may be due to him/her taking part in the commission of the offence;

(d) a court may draw a proper inference if (s)he fails or refuses to account for this fact; and

(e) a record is being made of the interview and it may be given in evidence if (s)he is brought to trial.

Cautions when the restriction on drawing adverse inferences from silence applies

5.29   CJPOA 1994, ss 34, 36, and 37 (**8.57, 8.59,** and **8.60**) describe the conditions under which adverse inferences can be drawn from a person's failure or refusal to say anything about his/her involvement in the offence when interviewed, or on being charged or informed that (s)he may be prosecuted. These provisions are subject to an overriding *restriction* on the ability of a court or jury to draw inferences from a person's silence. This restriction applies:

(a) to any detainee at a police station who, before being interviewed, or charged, or informed that (s)he may be prosecuted, has:

(i) asked for legal advice;

(ii) not been allowed an opportunity to consult a solicitor, as in Code C; and

(iii) not changed his/her mind about wanting legal advice;

(Condition (ii) will apply when a detainee who has asked for legal advice is interviewed before speaking to a solicitor, but will not apply if the detained person declines to ask for the duty solicitor.);

(b) to anyone charged with, or informed that (s)he may be prosecuted for, an offence who:

(i) has had brought to his/her notice a written statement made by another person or the content of an interview with another person which relates to that offence;

(ii) is interviewed about that offence; or

(iii) makes a written statement about that offence.

Where a requirement to caution arises at a time when the restriction on drawing adverse inferences from silence applies, the caution must be:

You do not have to say anything, but anything you do say may be given in evidence.

Whenever the restriction begins or ceases to apply after a caution has been given, the person must be re-cautioned in the appropriate terms. The changed position in relation to inferences and the fact that the previous caution no longer applies must be explained to the detainee in ordinary language.

Code C suggests, in a guidance note, that where the restriction on drawing adverse inferences begins to apply it should be explained that the caution previously given no longer applies because after caution the detainee asked to speak to a solicitor but has not yet had an opportunity to do so (restriction (a)); or because the detainee has been charged with or informed that (s)he may be prosecuted for the offence (restriction (b)). (S)he should be informed that this means that, from now on, adverse inferences cannot be drawn at court and his/her defence will not be harmed just because (s)he chooses to say nothing. (S)he should be asked to note that the new caution does not say anything about his/her defence being harmed.

Where restriction (a) ceases to apply before or at the time the person is charged or informed that (s)he may be prosecuted, the detainee should be told that the caution previously given no longer applies, and that this is because after that caution (s)he has been allowed an opportunity to speak to a solicitor. (S)he should be told to listen carefully as the caution now being given explains how his/her defence at court may be affected by his/her choosing to say nothing.

Special groups

The *specified information* referred to at **5.28** must not be given to a suspect who is a juvenile or a vulnerable person unless the appropriate adult is present.   **5.30**

If a juvenile or a vulnerable person is cautioned in the absence of the appropriate adult, the caution must be repeated in the appropriate adult's presence.

General

It is important that the appropriate caution is given; if an inappropriate one is chosen the court is liable to exclude evidence subsequently obtained.   **5.31**

Where there is a break in questioning under caution the officer in charge must ensure that the person being questioned is aware that (s)he remains under caution; if there is any doubt the caution should be given again in full when the interview resumes. This is important because the officer may have to satisfy a court that the person understood that (s)he was still under caution when the interview resumed.

If it appears that a person does not understand what the caution means, the officer who has given it should go on to explain it in the officer's own words.

If a person is not under arrest when an initial caution (or reminder) is given, the officer must tell him/her that (s)he is not under arrest and is not obliged to remain with the officer. The officer must also tell him/her that (s)he is free to leave if (s)he wishes and remind him/her that (s)he may obtain free legal advice if (s)he wishes. See further **5.11**.

Documentation

5.32   A record must be made when a caution is given, either in the officer's official report book or in the interview record as appropriate. As already indicated, references in Code C and any other PACE Code to written records, forms, and signatures include electronic records and forms and electronic confirmation that identifies the person making the record or completing the form.

INTERVIEWS

General provisions concerning interviews

What is an 'interview'?

5.33   Code C defines an interview as 'the questioning of a person regarding his involvement or suspected involvement in a criminal offence or offences which, by virtue of the Code, is required to be carried out under caution'. For example, where, after arresting a man for possessing an offensive weapon, police officers asked him in the police car why he had the knife and he told them there had been some trouble and that he had it for his own protection, it was held that this amounted to an interview for the purposes of Code C, so that a record should have been made of it.

Procedures undertaken under the Road Traffic Act 1988, s 7 (**26.66**) or the corresponding provisions under the Transport and Works Act 1992 (railways, tramways, and guided transport systems) and the Railways and Transport Safety Act 2003 (aviation activities or professional maritime activities) do not constitute interviews for the purposes of the Code. An informal conversation at or near the scene of a crime is not an 'interview', but it will be if it descends into detailed questioning.

Witnesses

5.34   The provisions of Code C and Codes E and F (below) which govern the conduct and recording of interviews do not apply to interviews with, or taking statements from, witnesses.

Action

5.35   Before a suspect is interviewed, (s)he and, if (s)he is represented, his/her solicitor must be given sufficient information to enable them to understand the nature of the offence, and why (s)he is suspected of committing it, in order to allow for the effective exercise of the rights of the defence. However, whilst the information must always be sufficient for the person to understand the nature of any offence, this does not require the disclosure of details at a time which might prejudice the criminal investigation. The decision about what needs to be disclosed for the purpose of this requirement therefore rests with the investigating officer who has sufficient knowledge of the case to make that decision. What is sufficient will depend on the circumstances of the case, but it should normally include, as a minimum, a description of the facts relating to the suspected offence that are known to the officer, including the time and place in question. The officer who discloses the information must make a record of what was disclosed, and when it was disclosed. This record may be made in the interview record, his/her pocket book or other form provided for this purpose.

Following a decision to arrest a suspect (s)he must not be interviewed about the relevant offence except at a police station (or other authorised place of detention) unless the consequent delay would be likely to:

(a) lead to interference with or harm to evidence connected with an offence, or interference with or physical harm to other persons, or serious loss of, or damage to, property; or

(b) lead to the alerting of other persons suspected of having committed an offence but not yet arrested for it; or

(c) hinder the recovery of property obtained in consequence of the commission of an offence.

Interviewing in any of these exceptions should cease once the relevant risk has been averted or the necessary questions have been put in order to attempt to avert that risk.

Immediately prior to the commencement or recommencement of any interview at a police station (or other authorised place of detention), the interviewing officer must remind the suspect of his/her entitlement to free legal advice and that the interview can be delayed for him/her to obtain legal advice (unless one of the exceptions where delay is permitted applies (**5.21**). By way of a modification where live-link interpretation (**5.89**) is to be used, before the reminder is given, the operation of live-link interpretation must be explained and demonstrated to the suspect, his/her solicitor and appropriate adult, unless it has been previously explained and demonstrated.

At the beginning of an interview carried out in a police station, the interviewing officer, after cautioning the suspect, must put to him/her any significant statement or silence which occurred in the presence and hearing of a police officer or civilian interviewer before the start of the interview and which has not been put to the suspect in the course of a previous interview, and must ask him/her whether (s)he confirms or denies that earlier statement or silence and whether (s)he wishes to add anything. This does not prevent an interviewer from putting significant statements and silences to a suspect again at a later stage of a further interview. A '*significant statement*' is one which appears capable of being used in evidence against the suspect, in particular a direct admission of guilt. A '*significant silence*' is a failure or refusal to answer a question or answer satisfactorily when under caution which might, allowing for the restriction on drawing adverse inferences from silence, give rise to an inference under CJPOA 1994. The interviewing officer must ensure that all such reminders are noted in the record of the interview.

No interviewer may try to obtain answers to questions or to elicit a statement by the use of oppression. Except as provided by Code C, no police officer may indicate, except in answer to a direct question, what action will be taken on the part of the police if the interviewee answers questions, makes a statement, or refuses to do either. If the interviewee asks the officer directly what action will be taken in the event of his/her answering questions, making a statement or refusing to do either, the officer may inform him/her what action the police propose to take in that event provided that the action is itself proper and warranted.

An interview concerning an offence with which the interviewee has not been charged or for which (s)he has not been informed that (s)he may be prosecuted must cease when:

(a) the officer in charge of the investigation is satisfied that all relevant questions to obtain accurate and reliable information about the offence have been put to the interviewee, including giving an opportunity for an innocent explanation to be put forward and asking any questions to test the accuracy or reliability of such explanation (eg to clear up ambiguities);

(b) the officer in charge of the investigation has taken account of any other available evidence; and

(c) the officer in charge of the investigation, or (in the case of a detained suspect) the custody officer, reasonably believes that there is sufficient evidence to provide a realistic prospect of conviction for that offence if the interviewee was prosecuted.

This does not, however, prevent officers in revenue cases or acting under the money laundering confiscation provisions from inviting suspects to complete a formal question and answer record after the interview is concluded.

In addition, according to the Court of Appeal, provided the officer has an open mind and is prepared not to charge (or not to refer to another officer responsible for charging) if the suspect produces a convincing account, the interviewer may proceed in order to see whether the suspect can produce an explanation.

Interview records

5.36 An accurate record must be made of each interview with a suspect, whether or not the interview takes place at a police station. The record must state the place of the interview, the times of its start and end, any breaks in it, and the names of those present. The requirement to record names of those present is subject to the special rule at **5.3**.

An interview record:

(a) enables the prosecutor to make informed decisions;

(b) is capable of being exhibited to an officer's witness statement and used pursuant to the Criminal Justice Act 1967, s 9 (**8.43**);

(c) enables the prosecutor to comply with advance disclosure rules; and

(d) where the record is accepted by the defence, facilitates the conduct of the case by the prosecution, the defence and the court.

The record must be made in the written interview record or in accordance with PACE Code E: the Code of Practice on Audio Recording Interviews with Suspects or PACE Code F: the Code of Practice on Visual Recording with Sound of Interviews with Suspects.

Codes E and F do not apply to the conduct and recording in England and Wales of interviews of persons detained under the Terrorism Act 2000, s 41 (or Sch 7 (port and border controls)), or post-charge questioning of persons authorised under the Counter-Terrorism Act 2008, s 22, dealt with in Chapter 21, both of which must be video recorded with sound in accordance with the provisions of a separate Code of Practice. If, during the course of an interview or questioning under Code E or F, it becomes apparent that the interview or questioning should be conducted under that separate Code, the interview should only continue in accordance with that Code.

Codes E and F provide safeguards (a) for suspects against inaccurate recording of the words used in questioning them and of their demeanour during the interview, and (b) for police interviewers against unfounded allegations made by, or on behalf of, suspects about the conduct of the interview and what took place during the interview which might otherwise appear credible. Recording of interviews must therefore be carried out openly to instil confidence in its reliability as an impartial and accurate record of the interview.

In what follows authorised recording device means a removable recording media device (see below) or a secure digital recording network device (see below) that the chief officer of police has authorised interviewers under his/her direction and control to use to record the interview in question at the place in question, provided that the interviewer in question has been trained to set up and operate the device, in compliance with the manufacturer's instructions and subject to the operating procedures required by the chief officer.

Requirement to use authorised audio-recording device when available

By para 2.1 of Code E, if an authorised recording device (*audio* or visual with *sound*) is **5.37** in working order and an interview room or other location suitable for that device to be used is available, that device must be used to record the following matters:

(a) any interview with a person cautioned in accordance with Code C, in respect of any summary offence, any indictable offence, or any indictable (either-way) offence, when:
   (i) that person (the suspect) is questioned about his/her involvement or suspected involvement in that offence and (s)he has not been charged or informed (s)he may be prosecuted for that offence; and
   (ii) exceptionally, further questions are put to a person about any offence after (s) he has been charged with, or told that (s)he may be prosecuted for, that offence (see **6.51**);
(b) when a person who has been charged with, or informed that (s)he may be prosecuted for, any offence, is told about any written statement or interview with another person and (s)he is handed a true copy of the written statement or the content of the interview record is brought to his/her attention in accordance with Code C (**6.50**).

The whole of each of the matters just mentioned must be audio-recorded, including the taking and reading back of any statement as applicable.

(Note: Nothing in Code E is intended to preclude audio-recording at police discretion at police stations or elsewhere when persons are charged with, or told they may be prosecuted for, an offence or they respond after being so charged or informed.)

Written record A written record of the matters described in (a) and (b) above must be made in accordance with the general provisions about interviews in Code C (see 5.33–5.36 (first three paragraphs) and 5.40) only if:

(i) an authorised recording device in working order is not available; or
(ii) such a device is available but a location suitable for using that device to make the audio recording of the matter in question is not available; and
(iii) the 'relevant officer' (described below) considers on reasonable grounds, that the proposed interview or (as the case may be) continuation of the interview or other action, should not be delayed until an authorised recording device in working

order and a suitable interview room or other location become available and decides that a written record must be made;

(iv) in accordance with provisions referred to below, the suspect or the appropriate adult on his/her behalf objects to the interview being audibly recorded and the 'relevant officer', after having regard to the nature and circumstances of the objections, decides that a written record must be made;

(v) in the case of a detainee who refuses to go into or remain in a suitable interview room and in accordance with the provisions of Code C about interviews in police stations (**5.45**), the custody officer directs that interview be conducted in a cell and considers that an authorised recording device cannot be safely used in the cell.

Note: When the suspect appears to have a hearing impediment, these provisions do not affect the separate requirement (below) for the interviewer to make a written note of the interview at the same time as the audio recording. Also note: A decision not to audio-record an interview for any reason may be the subject of comment in court.

In (iii) above, 'relevant officer' means:

(a) if the person to be interviewed is arrested elsewhere than at a police station for an offence and before (s)he arrives at a police station, an urgent interview in accordance with Code C (**5.35**) is necessary to avert one or more of the three risks specified there, the 'relevant officer' means the interviewer, who may or may not be the arresting officer, who must have regard to the time, place and urgency of the proposed interview;

(b) if the person in question has been taken to a police station after being arrested elsewhere for an offence or is arrested for an offence whilst at a police station after attending voluntarily and is detained at that police station or elsewhere in the charge of a constable, the 'relevant officer' means the custody officer at the station where the person's detention was last authorised. The custody officer must have regard to the nature of the investigation and ensure that the detainee is dealt with expeditiously, and released as soon as the need for his/her detention no longer applies;

(c) In the case of a voluntary interview (**5.11**) which takes place:

   (i) at a police station and the offence in question is an indictable offence, the 'relevant officer' means a sergeant (or above), in consultation with the investigating officer;

   (ii) at a police station and the offence in question is a summary offence, the 'relevant officer' means the interviewer in consultation with the investigating officer if different,

   (iii) elsewhere than at a police station and the offence is one of the four indictable offence types which satisfy the conditions in the Annex to Code E, the 'relevant officer' means the interviewer in consultation with the investigating officer, if different;

   (iv) elsewhere than at a police station and the offence in question is an indictable offence which is not one of the four indictable offence types which satisfy the conditions in the Annex to Code E, the 'relevant officer' means a sergeant (or above), in consultation with the investigating officer;

   (v) elsewhere than at a police station and the offence in question is a summary only offence, the 'relevant officer' means the interviewer in consultation with the investigating officer, if different.

The Annex to Code E provides as follows:

The four specified indictable offence types—two conditions   The first condition is that the indictable offence in respect of which the person has been cautioned is one of the following: (i) possession of herbal cannabis or cannabis resin, contrary to the Misuse of Drugs Act 1971 (MDA 1971), s 5(2); (ii) possession of khat, contrary to MDA 1971, s 5(2); (iii) low-value shoplifting, contrary to the Theft Act 1968, s 1; (iv) criminal damage to property, contrary to the Criminal Damage Act 1971, s 1(1). The reference to each of these offences includes an attempt to commit that offence.

The *second* condition is that: where the person has been cautioned in respect of an offence described in (i) or (ii), the requirements of *(a)* and *(b)* below are satisfied; or where the person has been cautioned in respect of an offence described in (iii), the requirements of *(a)* and *(c)* are satisfied; or where the person has been cautioned in respect of an offence described in (iv), the requirements of *(a)* and *(d)* are satisfied.

The requirements of *(a)* that apply to all four offences described above are that:

(i)   with regard to the person suspected of committing the offence: (s)he appears to be 18 or over; there is no reason to suspect that (s)he is a vulnerable person for whom an appropriate adult is required; (s)he does not appear to be unable to understand what is happening because of the effects of drink, drugs or illness, ailment or condition; (s)he does not require an interpreter in accordance with Code C (**5.82**) and—in accordance with PACE, s 24 and Code G (Arrest) (**4.43**) which refers to s 24—his/her arrest is not necessary in order to investigate the offence;

(ii) it appears that the commission of the offence: has not resulted in any injury to any person; has not involved any realistic threat or risk of injury to any person; and has not caused any substantial financial or material loss to the private property of any individual; and

(iii) the person is not being interviewed about any other offence.

These requirements will cease to apply if, as a result of what the suspect says or other information which comes to the interviewing officer's notice: it appears that the suspect is under 18, does require an appropriate adult, is unable to appreciate the significance of questions and his/her answers, is unable to understand what is happening because of the effects of drink, drugs, or illness, ailment or condition, or requires an interpreter; or the police officer decides that the suspect's arrest is now necessary.

The requirements of *(b)* that apply to the offences described in (i) and (ii) are that a police officer who is experienced in the recognition of the physical appearance, texture and smell of herbal cannabis, cannabis resin, or (as the case may be) khat, is able to say that the substance which has been found in the suspect's possession by that officer or, as the case may be, by any other officer not so experienced and trained is either herbal cannabis, cannabis resin, or khat, and to say that the quantity of the substance found is consistent with personal use by the suspect and does not provide any grounds to suspect an intention to supply others.

The requirements of *(c)* that apply to the offence described in (iii) are that it appears to the officer: that the value of the property stolen does not exceed £100 inclusive of VAT, that the stolen property has been recovered and remains fit for sale unless the items stolen comprised drink or food and have been consumed, and that the person suspected of stealing the property is not employed (whether paid or not) by the person, company or organisation to which the property belongs.

The requirements of *(d)* that apply to the offence described in (iv) are that it appears to the officer that the value of the criminal damage does not exceed £300, and that the person suspected of damaging the property is not employed (whether paid or not) by the person, company, or organisation to which the property belongs. (Note that whilst a court must proceed as if damage of this value is a summary offence (**14.148**), the offence is technically an indictable one.)

If it appears to the interviewing officer that before the conclusion of an interview, any of the requirements in *(a)* to *(d)* that apply to the offence in question have ceased to apply, the Annex ceases to apply. The person being interviewed must be so informed and a break in the interview must be taken. The reason must be recorded in the written interview record and the continuation of the interview must be audio- recorded.

A decision in relation to a particular indictable offence that the conditions and requirements in the Annex to Code E are satisfied is an operational matter for the interviewing officer according to all the particular circumstances of the case. These circumstances include the outcome of the officer's investigation at that time and any other matters that are relevant to the officer's consideration as to how to deal with the matter.

The purpose of allowing the interviewer to decide that a written record of interview is to be made is to support the policy which gives police in England and Wales options for dealing with low-level offences quickly and non-bureaucratically in a proportionate manner.

Duties of the 'relevant officer' and the interviewer   When, in accordance with the above provisions, a written record is made:

(a) the relevant officer must record the reasons for not making an audio recording and the date and time the relevant officer's decision in a non-availability or objection case was made; and ensure that the suspect is informed that a written record will be made;

(b) the interviewer must ensure that the written record includes the date and time of such decision, who made it and where the decision is recorded, and also includes the fact that the suspect is informed; and

(c) the written record must be made in accordance with the general provisions about interviews in Code C (**5.33-5.36** (first three paragraphs) and **5.48**).

Remote monitoring of interviews   If the interview room or other location where the interview takes place is equipped with facilities that enable audio-recorded interviews to be remotely monitored as they take place, the interviewer must ensure that suspects, their legal representatives and any appropriate adults are fully aware of what this means and that there is no possibility of privileged conversations being listened to.

With this in mind, the following safeguards should be applied:

(a) the remote monitoring system should only be able to operate when the audio recording device has been turned on;

(b) the equipment should incorporate a light, clearly visible to all in the interview room, which is automatically illuminated as soon as remote monitoring is activated;

(c) interview rooms and other locations fitted with remote monitoring equipment must contain a notice, prominently displayed, referring to the capacity for remote monitoring and to the fact that the warning light will illuminate whenever monitoring is taking place;

(d) at the beginning of the interview, the interviewer must explain the contents of the notice to the suspect and if present, to the solicitor and appropriate adult and that explanation should itself be audio-recorded;

(e) the fact that an interview, or part of an interview, was remotely monitored should be recorded in the suspect's custody record or, if the suspect is not in detention, the interviewer's pocket book, a term which includes any official report book or electronic recording device issued to the police. That record should include the names of the officers doing the monitoring and the purpose of the monitoring (e.g. for training, to assist with the investigation, etc).

Use of live link—Interviewer not present at the same station as the detainee   Code C sets out the conditions which, if satisfied allow a suspect in police detention to be interviewed using a live link by a police officer who is not present at the police station where the detainee is held. These provisions also set out the duties and responsibilities of the custody officer, the officer having physical custody of the suspect and the interviewer and the modifications that apply to ensure that any such interview is conducted and audio- recorded in accordance with Code E or (as the case may be) visually recorded in accordance with Code F.

When interviews and matters to which Code F applies may be visually recorded with sound

There is no statutory requirement to make a *visual recording*. However, the provisions of Code F must be followed on any occasion that the relevant officer (above) considers that a visual recording of any matters mentioned in para 2.1 of Code E (**5.37**) should be made. Having regard to the safeguards for suspects and police interviewers described above, examples of occasions when the relevant officer is likely to consider that a visual recording should be made include when:

5.38

(a) the suspect (whether or not detained) requires an appropriate adult;
(b) the suspect or his/her solicitor or appropriate adult requests that the interview be recorded visually;
(c) the suspect or other person whose presence is necessary is deaf or deaf/blind or speech impaired and uses sign language to communicate;
(d) the interviewer anticipates that when asking the suspect about his/her involvement in the offence concerned, (s)he will invite the suspect to demonstrate his/her actions or behaviour at the time or to examine a particular item or object which is handed to him/her;
(e) the officer in charge of the investigation believes that a visual recording with sound will assist in the conduct of the investigation, for example, when briefing other officers about the suspect or matters coming to light during the course of the interview; and
(f) the authorised recording device that would be used in accordance with para 2.1 of Code E (above) incorporates a camera and creates a combined audio and visual recording and does not allow the visual recording function to operate independently of the audio recording function.

Other points about audio or video recordings

If a detainee refuses to go into or remain in a suitable interview room, and the custody officer considers, on reasonable grounds, that the interview should not be delayed the interview may, at the custody officer's discretion, be conducted in a cell using portable recording equipment or, if none is available, recorded in writing as in Code C. The reasons for this must be recorded.

5.39

In the application of Code E to the conduct and recording of an interview of a suspect who has not been arrested, note that:

(a) references to the 'custody officer' include references to an officer of the rank of sergeant (or above) not directly involved in the investigation of the offence(s);

(b) if the interview takes place elsewhere than at a police station, references to 'interview room' include any place or location which the interviewer is satisfied will enable the interview to be conducted and recorded in accordance with Code E and where the suspect is present voluntarily; and

(c) provisions in addition to those which expressly apply to such interviews must be followed insofar as they are relevant and can be applied in practice.

An interviewer who is not sure, or has any doubt, about the suitability of a place or location of an interview to be carried out elsewhere than at a police station should consult an officer of the rank of sergeant (or above) for advice.

The whole of an audibly or visually recorded interview must be so recorded, including the taking and reading back of any statement. If, during the course of any audibly or visually recorded interview, it becomes apparent that the interview should be conducted under the terrorism code of practice for video recording with sound (**21.50**), the interview should continue in accordance with that code.

The recording of interviews must be carried out openly. It must be made clear to the suspect that there is no opportunity to interfere with the equipment or recording media. Where the recording is not by secure digital network, the 'master recording' will be sealed before it leaves the presence of the suspect. A master recording may be either one of two recording media used in a twin-deck/drive machine or the only medium used in a single-deck/drive machine. A second medium will be used as a working copy and it may be either the other medium in the case of a twin-deck/drive machine, or a copy of the master used in a single-deck/drive machine. Such a copy must be made in the presence of the suspect and without the master recording leaving his/her sight.

As to the procedure for audio or video recording, see **5.51–5.79**.

Written record

5.40   Written interview records must be signed and timed by the maker.

Any written record must be made during the interview, unless in the investigating officer's view this would not be practicable or would interfere with the conduct of the interview, and must constitute either a verbatim record or, failing this, an account of the interview which adequately and accurately summarises it. If the record is not made during the interview, it must be made as soon as practicable thereafter and the reason must be recorded in the interview record.

Checking the interview record

5.41   Unless it is impracticable, the person interviewed must be given the opportunity to read the interview record and to sign it as correct or to indicate the respects in which (s)he considers it inaccurate. When a suspect agrees to read records of interviews and of other comments and to sign them as correct, (s)he should be asked to endorse the record with words such as 'I agree that this is a correct record of what was said' and add his/her signature. Where the suspect does not agree with the record, the officer should record the details of any disagreement and then ask the suspect to read these details and then sign them to the effect that they accurately reflect his/her disagreement.

If the person concerned cannot read, or refuses to read the record or to sign it, the senior police officer present must read it over to him/her and ask him/her whether (s)he would like to sign it as correct or to indicate the respects in which (s)he considers it inaccurate. The police officer must then certify on the interview record itself what has occurred.

If the interview is audibly or visually recorded the arrangements set out in Code E or Code F apply (**5.60**, **5.68** and **5.72**).

Records of comments outside interview

A written record must be made of any comments made by a suspected person, including unsolicited comments, which are outside the context of a formal interview but which might be relevant to the offence. A judge in the Administrative Court has held that this requirement does not apply only to incriminating comments or only where a caution has been given. Any such record must be timed and signed by the maker. Where practicable the person must be given the opportunity to read that record and to sign it as correct or to indicate the respects in which (s)he considers it inaccurate. Any refusal to sign must be recorded.   **5.42**

Juvenile and vulnerable persons

A juvenile or vulnerable person must not be interviewed regarding his/her involvement or suspected involvement in a criminal offence or offences, or be asked to provide or sign a written statement under caution or record of an interview, in the absence of the appropriate adult, subject to the special provisions below about urgent interviews authorised by a superintendent.   **5.43**

A juvenile or vulnerable person may be particularly open to suggestion. Consequently, special care must always be exercised in questioning such a person, and it is important to obtain corroboration of any facts admitted wherever possible.

A juvenile may only be interviewed at his/her place of education in exceptional circumstances and then only when the principal or his/her nominee agrees and is present. Every effort should be made to contact both the parents, or other person responsible for his/her welfare, and the appropriate adult (if a different person). A reasonable time should be allowed to enable the appropriate adult to attend. Where this would cause undue delay, and unless the offence is against the educational establishment, the principal or his/her nominee can act as the appropriate adult for the purposes of the interview.

An interview is an indivisible process. A failure to involve the appropriate adult in some part of it is a breach of Code C and may render the whole of the interview inadmissible.

If an appropriate adult is present at an interview, (s)he must be told that (s)he is not expected to act simply as an observer. (S)he must also be informed that the purposes of his/her presence are:

(a) to advise the person being interviewed and to observe whether or not the interview is conducted fairly and properly; and

(b) to facilitate communication with the person being interviewed.

The appropriate adult may be required to leave the interview if his/her conduct is such that the interviewer is unable properly to put questions to the suspect. This will include situations where the appropriate adult's approach or conduct prevents or unreasonably obstructs proper questions being put to the suspect or the suspect's responses being recorded. A guidance note points out that the appropriate adult may intervene if (s)he considers it is necessary to help the suspect understand any question asked and to help the suspect to answer any question, and that the appropriate adult may only be required to leave if his/her approach or conduct prevents or unreasonably obstructs proper questions being put to the suspect or the suspect's response being recorded. Examples of unacceptable conduct include answering questions on a suspect's behalf or providing written replies for the suspect to quote. An officer who takes the decision to exclude an appropriate adult must be in a position to satisfy the court that the decision was properly made. In order to do this (s)he may need (a) to witness what is happening, and (b) if the suspect has a solicitor, and (s)he witnessed what happened, to give that solicitor an opportunity to comment.

If the interviewer considers an appropriate adult is acting in such a way, (s)he will stop the interview and consult an officer not below superintendent rank, if one is readily available, and otherwise an officer not below inspector rank not connected with the investigation. After speaking to the appropriate adult, the officer consulted must remind the adult that his/her role does not allow him/her to obstruct proper questioning and give the adult an opportunity to respond. The officer consulted will then decide whether the interview should continue without the attendance of that appropriate adult. If the officer decides it should, another appropriate adult must be obtained before the interview continues, unless the provisions below about urgent interviews at police stations apply.

If a person is blind, seriously visually impaired, or unable to read, the custody officer must ensure that his/her solicitor, relative, the appropriate adult or some other person likely to take an interest in him/her (and not involved in the investigation) is available to help in checking any documentation. Where Code C requires consent or signing, then the person who is assisting may be asked to sign instead, if the detained person so wishes. However, there is no requirement that the appropriate adult be called solely to assist in checking and signing documentation in such circumstances.

If the appropriate adult or the person's solicitor is present during the interview, (s)he should also be given an opportunity to read and sign the interview record or any written statement taken down during the interview. Any refusal to sign when asked must itself be recorded on the interview record. If the interview is audibly or visually recorded the arrangements set out in Code E or Code F apply.

Vulnerable suspects: urgent interviews at police stations

5.44   The following interviews may take place only if a superintendent considers that delaying the interview will lead to the consequence (potential interference, harm, loss, etc) set out at **5.25**(b)(i), and is satisfied that the interview would not significantly harm the person's physical or mental state:

(a) an interview of a detained juvenile or vulnerable person without the appropriate adult being present;

(b) an interview of any detained person other than in (a) who appears unable to:
   (i)   appreciate the significance of questions and his/her answers; or
   (ii) understand what is happening because of the effects of drink, drugs or any illness, ailment or condition;

(c) an interview, without an interpreter being present, of a detained person whom the custody officer has determined requires an interpreter (**5.82**) which is carried out by an interviewer speaking the suspect's own language or (as the case may be) otherwise establishing effective communication which is sufficient to enable the necessary questions to be asked and answered in order to avert the consequences referred to above.

These interviews may not continue once sufficient information has been obtained to avert the consequences referred to above. A record must be made of the grounds for any decision urgently to interview a person.

Interviews in police stations

When interviewer and suspect are present at same police station

Code C provides that if a police officer wishes to interview a detained person, or to conduct inquiries which require the presence of a detained person, it is the custody officer who must decide whether to deliver his/her prisoner into that officer's custody. An investigating officer who is given custody of a detainee takes over responsibility for the detainee's care and safe custody until (s)he returns the detainee to the custody officer when (s)he must report the manner in which (s)he complied with the Code whilst having custody of the detainee.   **5.45**

In any period of 24 hours a detained person must be allowed a continuous period of at least eight hours' rest, free from questioning, travel or any interruption by police officers in connection with the investigation concerned. This period should normally be at night or other appropriate time which takes account of when the detainee last slept or rested. If a detainee is arrested at a police station after going there voluntarily, the 24-hour period runs from the time of arrival at the police station. The period of rest may not be interrupted or delayed except:

(a) when there are reasonable grounds for believing that not delaying or interrupting that period would:
   (i)   involve a risk of harm to people or serious loss of, or damage to, property;
   (ii) delay unnecessarily the detainee's release from custody;
   (iii) otherwise prejudice the outcome of the investigation;
(b) at the request of the detainee, his/her appropriate adult, or his/her legal representative;
(c) when a delay or interruption is necessary in order to comply with the legal obligations and duties under Code C relating to reviews and extensions of detention or as to the care and treatment of detained persons who are intoxicated or in need of clinical treatment and attention.

If such a period of rest is interrupted in accordance with (a), a fresh period must be allowed. Interruptions under (b) and (c) do not require that a fresh period be allowed.

As far as practicable, an interview must take place in a properly heated, lit, and ventilated interview room. The detained person must not be required to stand.

A suspect whose detention without charge has been authorised under PACE, because his/her detention is necessary so that (s)he may be interviewed with a view to obtaining evidence of the offence for which (s)he was arrested, may choose not to answer questions. However, the police do not require a suspect's consent to interview him/her for this purpose. If a suspect takes steps to prevent an interview, such as refusing to leave his/her cell, or trying to leave the interview room, (s)he must be cautioned and be told that refusal to co-operate could lead to the interview taking place in the cell and that such failure to co-operate may be given in evidence. A further invitation should then be given to co-operate by going to the interview room. If the suspect refuses and the custody officer considers, on reasonable grounds, that the interview should not be delayed, the custody officer has discretion to direct that the interview be conducted in a cell.

With the italicised exception at **5.3**, before the start of an interview, each interviewing officer must identify him/herself and any other officers present to the interviewee.

In all interviews, there must be breaks at the recognised meal times and there must also be short breaks for refreshment at intervals of approximately two hours, subject to the interviewing officer's discretion to delay a break if there are reasonable grounds for believing that it would:

(a) involve a risk of harm to persons or serious loss of, or damage to, property; or
(b) delay unnecessarily the person's release from custody; or
(c) otherwise prejudice the investigation.

A guidance note states that meal breaks should normally last at least 45 minutes and shorter breaks after two hours should last at least 15 minutes. If a break is delayed as permitted by Code C and prolongs the interview, a longer break should then be provided. If there is a short interview, and a subsequent short interview is contemplated, the length of the break may be reduced if there are reasonable grounds to believe that this is necessary to avoid any of the consequences set out at (a) to (c) above.

Interviewer not present at same station as the detainee—use of live link

5.46   PACE, s 39 allows a person in police detention to be interviewed using a live link by a police officer who is not at the police station where the detainee is held. For these purposes, 'live link' means an arrangement by means of which the interviewing officer who is not present at the police station where the detainee is held, is able to see and hear, and to be seen and heard by, the detainee concerned, the detainee's solicitor, appropriate adult, and interpreter (as applicable) and the officer who has custody of that detainee.

Subject to (a) to (f) below, the custody officer is responsible for deciding on a case by case basis whether a detainee is fit to be interviewed and should be delivered into the physical custody of an officer who is not involved in the investigation, for the purpose of enabling another officer who is investigating the offence for which the person is detained and who is not at the police station where the person is detained, to interview the detainee by means of a live link.

(a) The custody officer must be satisfied that the live link to be used provides for accurate and secure communication with the suspect. In respect of communications between the interviewing officer, the suspect and anyone else whose presence at the interview, or, (as the case may be) whose access to any communications between the suspect and the interviewer, has been authorised by the custody officer or the interviewing officer, chief officers must be satisfied that live-link interpretation used in their force area for the purposes of audio and visual communication and audio and visual or audio without visual communication provides for accurate and secure communication with the suspect. This includes ensuring that at any time during which live link interpretation is being used: a person cannot see, hear or otherwise obtain access to any communications between the suspect and interpreter or communicate with the suspect or interpreter unless so authorised or allowed by the custody officer or, in the case of an interview, the interviewer and that as applicable, the confidentiality of any private consultation between a suspect and his/her solicitor and appropriate adult is maintained.
(b) Each decision must take account of the age, gender and vulnerability of the suspect, the nature and circumstances of the offence, and the investigation and the impact on the suspect of carrying out the interview by means of a live link. For this reason, the custody officer must consider whether the ability of the particular suspect, to

communicate confidently and effectively for the purpose of the interview is likely to be adversely affected or otherwise undermined or limited if the interviewing officer is not physically present and a live-link is used. In considering whether the use of the live link is appropriate in a particular case, the custody officer, in consultation with the interviewer, should make an assessment of the detainee's ability to understand and take part in the interviewing process and make a record of the outcome. If the suspect has asked for legal advice, his/her solicitor should be involved in the assessment and in the case of a juvenile or vulnerable person, the appropriate adult should be involved. Although a suspect for whom an appropriate adult is required may be more likely to be adversely affected as described, it is important to note that a person who does not require an appropriate adult may also be adversely impacted if interviewed by means of a live link.

(c) If the custody officer is satisfied that interviewing the detainee by means of a live link would not adversely affect or otherwise undermine or limit the suspect's ability to communicate confidently and effectively for the purpose of the interview, the officer must so inform the suspect, his/her solicitor and (if applicable) the appropriate adult. At the same time, the operation of the live-link must be explained and demonstrated to them, they must be advised of the chief officer's obligations (above) concerning the security of live-link communications and they must be asked if they wish to make representations that the live-link should not be used or if they require more information about the operation of the arrangements. They must also be told that at any time live-link is in use, they may make representations to the custody officer or the interviewer that its operation should cease and that the physical presence of the interviewer should be arranged.

(d) If:
   (i) representations are made that a live-link should not be used to carry out the interview, or that at any time it is in use, its operation should cease and the physical presence of the interviewer arranged; and
   (ii) the custody officer in consultation with the interviewer is unable to allay the concerns raised;
   then live-link may not be used, or (as the case may be) continue to be used, unless authorised in writing by an officer of the rank of inspector or above in accordance with (e).

(e) Authority may be given if the officer is satisfied that interviewing the detainee by means of a live link is necessary and justified. In making this decision, the officer must have regard to:
   (i) the circumstances of the suspect;
   (ii) the nature and seriousness of the offence;
   (iii) the requirements of the investigation, including its likely impact on both the suspect and any victim(s);
   (iv) the representations made by the suspect, his/her solicitor and (if applicable) the appropriate adult that a live-link should not be used (see (b));
   (v) the impact on the investigation of making arrangements for the physical presence of the interviewer; and
   (vi) the risk if the interviewer is not physically present, evidence obtained using link interpretation might be excluded in subsequent criminal proceedings; and
   (vii) the likely impact on the suspect and the investigation of any consequential delay to arrange for the interviewer to be physically present with the suspect.

(f) The officer given custody of the detainee and the interviewer take over responsibility for the detainee's care, treatment and safe custody for the purposes of Code C until the detainee is returned to the custody officer. On that return, both must report the manner in which they complied with the Code during period in question.

For the purpose of (e)(v), factors affecting the arrangements for the interviewer to be physically present will include the location of the police station where the interview would take place and the availability of an interviewer with sufficient knowledge of the investigation who can attend that station and carry out the interview.

When a suspect detained at a police station is interviewed using a live link in accordance with the above provisions, the officer given custody of the detainee at the police station and the interviewer who is not present at the police station, take over responsibility for ensuring compliance with the provisions in Code C concerning interviews in general and interviews in police stations, or Code E (audio recording) or Code F (audio visual recording) that govern the conduct and recording of that interview. In these circumstances:

(a) *the interviewer who is not at the police station where the detainee is held* must direct the officer having physical custody of the suspect at the police station, to take the action required by those provisions and which the interviewer would be required to take if (s)he was present at the police station;

(b) *the officer having physical custody of the suspect at the police station* must take the action required by those provisions and which would otherwise be required to be taken by the interviewer if (s)he was present at the police station. This applies whether or not the officer has been so directed by the interviewer but in such a case, the officer must inform the interviewer of the action taken;

(c) during the course of the interview, the officers in (a) and (b) may consult each other as necessary to clarify any action to be taken and to avoid any misunderstanding. Such consultations must, if in the hearing of the suspect and any other person present with the suspect (for example, a solicitor, appropriate adult or interpreter) be recorded in the interview record.

General: documentation

5.47 A record must be made of times during which a detainee was not in the custody of the custody officer, and why; of any reason for refusal to deliver the detainee out of that custody; of any reason why it was not practicable to use an interview room; of any action taken in consequence of a refusal to be interviewed, and of the actions, decisions, authorisations, representations, and outcomes arising from the live link requirements above.

Written statements under caution

5.48 All written statements made at police stations under caution must be written on the forms provided for the purpose. All written statements under caution must be taken in accordance with the rules in Annex D to Code C ((1)–(12) below). Before a person makes a written statement under caution at a police station (s)he must be reminded about the right to legal advice. It is not normally necessary to ask for a written statement if the interview was recorded in writing and the record signed or audibly or visually recorded in accordance with Code E or Code F (**5.51–5.77**) respectively.

Written by a person under caution

(1) A person must always be invited to write down him/herself what (s)he wants to say.

(2) A person who has not been charged with, or informed that (s)he may be prosecuted for, any offence to which the statement relates, must:

    (a) unless the statement is made at a time when the restriction on drawing adverse inferences from silence applies, be asked to write out and sign the following before writing what (s)he wants to say:

> **'I make this statement of my own free will. I understand that I do not have to say anything but that it may harm my defence if I do not mention when questioned something which I later rely on in court. This statement may be given in evidence;'**

    (b) if the statement is made at a time when the restriction on drawing adverse inferences from silence applies (**5.29**), (s)he must be asked to write out and sign the following before writing what (s)he wants to say:

> *'I make this statement of my own free will. I understand that I do not have to say anything. This statement may be given in evidence.'*

(3) When a person, on the occasion of being charged with, or informed that (s)he may be prosecuted for, any offence, asks to make a statement which relates to any such offence and wants to write it (s)he must, unless the restriction on drawing adverse inferences from silence applied when (s)he was so charged or informed that (s)he might be prosecuted, be asked to write out the words in bold above and sign them before writing what (s)he wants to say. If the restriction on drawing adverse inferences from silence applied when the person was so charged or informed that (s)he would be prosecuted, (s)he must be asked to write out the words italicised above and sign them before writing what (s)he wants to say.

(4) Where a person, who has already been charged with or informed that (s)he may be prosecuted for any offence, asks to make a statement which relates to any such offence and wants to write it, (s)he must be asked to write out and sign the words italicised above before writing what (s)he wants to say.

(5) Any person writing his/her own statement must be allowed to do so without any prompting except that a police officer or civilian interviewer may indicate to him/her which matters are material or question any ambiguity in the statement.

Written by a police officer or other police staff

(1) If a person says that (s)he would like someone to write it for him/her, a police officer or other member of police staff must write the statement.

(2) If the person has not been charged with, or informed that (s)he may be prosecuted for, any offence to which the statement (s)he wants to make relates, (s)he must, before starting, be asked to sign, or make his/her mark, to the appropriate one of the following endorsements.

    (a) Unless the statement is made at a time when the restriction on drawing adverse inferences from silence applies, the endorsement is:

> *'I, …, wish to make a statement. I want someone to write down what I say. I understand that I do not have to say anything but that it may harm my defence if I do not mention when questioned something which I later rely on in court. This statement may be given in evidence.'*

(b) If the statement is made at a time when the restriction on drawing adverse inferences from silence applies, the endorsement is:

'I, ..., wish to make a statement. I want someone to write down what I say. I understand that I do not have to say anything. This statement may be given in evidence.'

(3) If, on the occasion of being charged with or informed that (s)he may be prosecuted for any offence, the person asks to make a statement which relates to any such offence, (s)he must before starting be asked to sign, or make his/her mark to, the following as appropriate. Unless the restriction on drawing adverse inferences from silence applied when (s)he was so charged or informed that (s)he may be prosecuted, the words are the same as in (2)(a). If the restriction on drawing adverse inferences from silence applied when (s)he was so charged or informed that (s)he may be prosecuted, the words are the same as in (2)(b).

(4) If, having already been charged with or informed that (s)he may be prosecuted for any offence, a person asks to make a statement which relates to such an offence, (s)he must, before starting, be asked to sign or make his/her mark to the form of words in (2)(b).

(5) The person writing the statement must take down the exact words spoken by the person making it and (s)he must not edit or paraphrase it. Any questions that are necessary (eg to make it more intelligible) and the answers given must be recorded contemporaneously on the statement form.

(6) When the writing of a statement is finished the person making it must be asked to read it and to make any corrections, alterations, or additions (s)he wishes. When (s)he has finished reading it (s)he must be asked to write and sign or make his/her mark on the following certificate at the end of the statement:

'I have read the above statement, and I have been able to correct, alter, or add anything I wish. This statement is true. I have made it of my own free will.'

(7) If the person making the statement cannot read, or refuses to read it, or to write the above-mentioned certificate at the end of it or to sign it, the person taking the statement must read it over to him/her and ask him/her whether (s)he would like to correct, alter, or add anything and put his/her signature or make his/her mark at the end. The person taking the statement must then certify on the statement itself what has occurred.

Audio recorded interviews with suspects

5.51   Code E provides for two methods for the audio recording of interviews with suspects:

- interview recording using removable recording media device; and
- interview recording using secure digital recording network device.

In addition to Code E's provisions at **5.36-5.37** about both methods, the separate rules for each method are set out below.

Interview recording using removable recording media device

Removable recording media device

5.52   'Removable recording media device' means a recording device which, when set up and operated in accordance with the manufacturer's instructions and the operating procedures required by the chief officers, uses removable, physical recording media (such as magnetic tape, optical disc, or solid state memory card) for the purpose of making a clear and accurate, audio recording or (as the case may be) audio-visual recording, of the interview in question which can then be played back and copied using that device or any other device. A sign or indicator on the device which is visible to the suspect must show when the device is recording.

Recording and sealing master recordings—general

5.53 When using an authorised removable recording media device, one recording, the master recording, will be sealed in the suspect's presence. A second recording will be used as a working copy. The master recording is any of the recordings made by a multi-deck/drive machine or the only recording made by a single deck/drive machine. The working copy is one of the other recordings made by a multi-deck/drive machine or a copy of the master recording made by a single deck/drive machine. The purpose of sealing the master recording before it leaves the suspect's presence is to establish his/her confidence that the integrity of the recording is preserved. If a single deck/drive machine is used the working copy of the master recording must be made in the suspect's presence and without the master recording leaving his/her sight. The working copy must be used for making further copies if needed.

Commencement of interviews

5.54 When the suspect is brought into the interview room or arrives at the location where the interview is to take place, the interviewer must, without delay but in the suspect's sight, unwrap or open the new recording media, load the recording device with new recording media and set it to record. A modification of Code E by Code C provides that, for the purposes of live-link interpretation, before the interview commences, the operation of live-link interpretation (**5.89**) must be explained and demonstrated to the suspect, his/her solicitor and appropriate adult, unless it has been previously explained and demonstrated.

The interviewer must point out the sign or indicator which shows that the recording equipment is activated and is recording, and must then:

(a) tell the suspect that the interview is being audibly recorded using an authorised removable recording media device and outline the recording process. When doing so, the interviewer should refer to the definition of removable recording media device and briefly describe how it is operated and how recordings are made;

(b) subject to the 'anonymity provision' (**5.3**), give his/her name and rank and that of any other interviewer present;

(c) ask the suspect and any other party present, eg the appropriate adult, a solicitor, or interpreter, to identify themselves for the purpose of voice identification;

(d) state the date, time of commencement, and place of the interview;

(e) tell the suspect that (s)he will be given a copy of the recording of the interview in the event that (s)he is charged or informed that (s)he will be prosecuted but if (s)he is not charged or informed that (s)he will be prosecuted (s)he will only be given a copy as agreed with the police or on the order of a court; and that (s)he will be given a written notice at the end of the interview setting out his/her right to a copy of the recording and what will happen to the recording; and

(f) if equipment for remote monitoring of interviews is installed, explain the contents of the notice to the suspect, solicitor, and appropriate adult as required above and point out the light that illuminates automatically as soon as remote monitoring is activated.

Any person entering the interview room after the interview has commenced must be invited by the interviewer to identify him/herself for the purpose of the audio recording and state the reason why (s)he has entered the interview room.

The interviewer must:

(a) caution the suspect (**5.27–5.32**); and
(b) if the suspect is detained, remind him/her of his/her entitlement to free legal advice (**5.20**); or
(c) if the suspect is not detained under arrest, explain this and his/her entitlement to free legal advice (**5.12**), and ask the suspect to confirm that (s)he agrees to the voluntary interview proceeding.

The interviewer must then put to the suspect any 'significant statement or silence' (as to which see **5.35**).

It must be remembered that a special warning (**5.28**) must be given before inferences can be drawn under the Criminal Justice and Public Order Act 1994, s 36 or 37 and of the fact that adverse inferences cannot be drawn where legal advice has been requested but has not yet been received.

Objections by the suspect

5.55   If the suspect or an appropriate adult on his/her behalf objects to the interview being audibly recorded either at the outset, during the interview or during a break, the interviewer must explain that the interview is being audibly recorded and that Code E requires the objections to be recorded on the audio recording. When any objections have been audibly recorded or the suspect or appropriate adult has refused to have his/her objections recorded, the relevant officer must decide, after having regard to the nature and circumstances of the objections, whether a written record of the interview or its continuation, is to be made and that audio recording should be turned off. Following a decision that a written record is to be made, the interviewer must say that (s)he is turning off the recorder and must then make a written record of the interview as in Code C (**5.36** and **5.40**). Following a decision that a written record is not to be made, the interviewer may proceed to question the suspect with the audio recording still on.

The above provisions are meant to apply to objections based on the suspect's genuine and honestly held beliefs and to allow officers to exercise their discretion to decide that a written interview record is to be made according to the circumstances surrounding the suspect and the investigation. Objections that appear to be frivolous with the intentions of frustrating or delaying the investigation would not be relevant.

The relevant officer should be aware that a decision to continue recording against the wishes of the suspect may be the subject of comment in court.

If the suspect indicates (s)he wants to tell the interviewer about matters not directly connected with the offence of which (s)he is suspected and (s)he is unwilling for these matters to be audio recorded, the suspect should be given the opportunity to tell the interviewer about these matters after the conclusion of the formal interview

Changing recording media

5.56   When the recorder shows the recording media only has a short time left to run, the interviewer must so inform the person being interviewed and round off that part of the interview. If the interviewer leaves the room for a second set of recording media, the suspect must not be left unattended. The interviewer will remove the recording media from the recorder and insert the new recording media which must be unwrapped or opened in the suspect's presence. The recorder should be set to record on the new media.

To avoid confusion between the recording media, the interviewer must mark the media with an identification number immediately after it is removed from the recorder.

Taking a break during interview

When a break is taken, the fact that a break is to be taken, the reason for it and the time   **5.57**
must be recorded on the audio recording. When the break is taken and the interview room vacated by the suspect, the recording media must be removed from the recorder and the procedures for the conclusion of an interview (**5.60**) followed.

When a break is a short one and both the suspect and an interviewer remain in the interview room, the recording may be stopped. There is no need to remove the recording media and when the interview recommences the recording should continue on the same recording media. The time the interview recommences must be recorded on the audio recording.

After any break in the interview the interviewer must, before resuming the interview, remind the person being questioned of his/her right to legal advice if (s)he has not exercised it and that (s)he remains under caution or, if there is any doubt, give the caution in full again. The interviewer should bear in mind that (s)he may have to satisfy a court that the person realised that the caution still applied. (S)he may also have to show that nothing occurred during a break in an interview or between interviews which influenced the suspect's recorded evidence. In view of this, (s)he should consider, at recommencement or at a subsequent interview, summarising on recording media the reason for the break and confirming this with the suspect. As to lengths of breaks, see **5.45**.

Equipment failure

If there is an equipment failure which can be rectified quickly, eg by inserting new   **5.58**
recording media, the interviewer must follow the procedures about the changing of recording media. When the recording is resumed the interviewer must explain what happened and record the time the interview recommences. However, if it is not possible to continue recording using the same recording device or by using a replacement device, the interview should be audio-recorded using a secure digital recording network device, if the necessary equipment is available. If it is not available, the interview may continue and be recorded in writing in accordance with the general provisions about interviews in Code C as directed by the relevant officer.

Removing recording media from the recorder

Recording media which is removed from the recorder during the interview must be   **5.59**
retained and the procedures which apply to them at the conclusion of the interview (below) followed.

Conclusion of interview

At the conclusion of the interview, the suspect must be offered the opportunity to   **5.60**
clarify anything (s)he has said and asked if there is anything (s)he wants to add.

At the conclusion of the interview, including the taking and reading back of any written statement, the time must be recorded and the recording stopped.

The interviewer must seal the master recording with a master recording label and treat it as an exhibit. The interviewer must sign the label and ask the suspect and any third party present during the interview to sign it. (A modification of Code E made by Code C provides that, if live-link interpretation (**5.89**) has been used, the interviewer should ask the interpreter to observe the removal and sealing of the master recording and to confirm in writing that (s)he has seen it sealed and signed by the interviewer. A clear legible copy of the confirmation signed by the interpreter must be sent via the live link to the interviewer. The interviewer is responsible for ensuring that the original confirmation and the copy are retained with the case papers for use in evidence if required and must advise the interpreter of his/her obligation to keep the original confirmation securely for that purpose.)

If the suspect or third party refuse to sign the label, an inspector (or above), or if not available the custody officer, or if the suspect has not been arrested, a sergeant, must be called into the interview room and asked, subject to the 'anonymity provision' (**5.3**), to sign it.

The suspect must be handed a notice explaining how the audio recording will be used, the arrangements for access to it, and that if (s)he is charged or informed (s)he will be prosecuted, a copy of the audio recording will be supplied as soon as practicable or as otherwise agreed between the suspect and the police or on the order of a court.

After the interview

5.61   The interviewer must make a note in his/her pocket book that the interview has taken place and that it was audibly recorded, the time it commenced, its duration and date and identification number of the master recording. If no proceedings follow in respect of the person whose interview was recorded, the recording media must be kept securely as in accordance with the provisions below about master recording security. Where proceedings follow, the interviewer must prepare a written record of the interview and sign it. Any such record should be made in accordance with the national guidelines for police officers, police staff, and CPS prosecutors concerned with the preparation, processing, and submission of prosecution files. Before preparing it the interviewer may refresh his/her memory by listening to the working copy of the recording media to check its accuracy. The interview record must be exhibited to any written statement prepared by the interviewer. If the interviewer's evidence of the interview is accepted by the defence, the evidence must refer to the fact that the interview was audio recorded and may be presented to the court in the form of the interview record. Where the interviewer's evidence is not accepted by the defence, the interviewer must refer to the fact that the interview was audibly recorded and produce the master recording medium of the whole interview, as an exhibit, informing the court of any transcription which has been made of which (s)he is aware.

Master recording security

5.62   The officer in charge of each police station at which interviews with suspects are recorded or as the case may be, where recordings of interviews carried out elsewhere than at a police station are held, must make arrangements for master recordings to be kept securely and their movements accounted for on the same basis as material which may be used for evidential purposes, in accordance with force standing orders.

A police officer must not break the seal on a master recording which is required for criminal trial or appeal proceedings. If it is necessary to gain access to the master recording, the police officer must arrange for its seal to be broken in the presence of a representative of the CPS. In such a case, the defendant or his/her legal adviser should be informed and given a reasonable opportunity to be present. If the defendant or his/her legal representative is present (s)he must be invited to re-seal and sign the master recording. If either refuses or neither is present this should be done by the representative of the CPS. If the recording has been delivered to the Crown Court following sending for trial, the Crown Prosecutor will apply to the Chief Clerk of the Crown Court for the release of the media for unsealing by the Crown Prosecutor.

The chief officer of police is responsible for establishing arrangements for breaking the seal of the master copy where no criminal proceedings result, or the criminal proceedings to which the interview relates have been concluded, and it becomes necessary to break the seal. These arrangements should be those which the chief officer considers are reasonably necessary to demonstrate to the person interviewed and any other party who may wish to use or refer to the interview record that the master copy has not been tampered with and that the interview record remains accurate.

With the exception italicised below, a representative of each party must be given a reasonable opportunity to be present when the seal is broken and the master recording copied and re-sealed. If one or more of the parties is not present when the master copy seal is broken because (s)he cannot be contacted or refuses to attend, or in the case italicised below, arrangements should be made for an independent person such as a custody visitor, to be present. Alternatively, or as an additional safeguard, arrangements should be made to visually record the procedure.

A person need not be given an opportunity to be present when;

(a) *it is necessary to break the master copy seal for the proper and effective further investigation of the original offence or the investigation of some other offence; and*

(b) *the officer in charge of the investigation has reasonable grounds to suspect that allowing an opportunity might prejudice such an investigation or criminal proceedings which may be brought as a result or endanger any person.*

The italicised words could apply, for example, when one or more of the outcomes or likely outcomes of the investigation might be: (i) the prosecution of one or more of the original suspects; (ii) the prosecution of someone previously not suspected, including someone who was originally a witness; and (iii) any original suspect being treated as a prosecution witness and when premature disclosure of any police action, particularly through contact with any parties involved, could lead to a real risk of compromising the investigation and endangering witnesses.

When the master recording seal is broken, a record must be made of the procedure followed, including the date, time, place, and persons present.

Interview recording using secure digital recording network device

Secure digital recording network device

'Secure digital recording network device' means a recording device which, when set **5.63** up and operated in accordance with the manufacturer's instructions and the operating procedures required by the chief officers, enables a clear and accurate original audio recording or (as the case may be) audio-visual recording, of the interview in question, to be made and stored using non-removable storage, as a digital file or a series of such files that can be securely transferred by a wired or wireless connection to a remote secure network file server system (which may have cloud based storage) which ensures that access to interview recordings for all purposes is strictly controlled and is restricted to those whose access, either generally or in specific cases, is necessary. Examples of access include playing back the whole or part of any original recording and making one or more copies of, the whole or part of that original recording. A sign or indicator on the device which is visible to the suspect must show when the device is recording.

An authorised secure digital recording network device does not use removable media. As will be seen, some of the provisions relating to removable recording media device apply to the use of secure digital recording network devices.

Commencement of interviews

5.64　When the suspect is brought into the interview room or arrives at the location where the interview is to take place, the interviewer must, without delay and in the sight of the suspect, switch on the recording equipment and in accordance with the manufacturer's instructions start recording. A modification of Code E made by Code C provides that, for the purposes of live-link interpretation (**5.89**), before the interview commences, the operation of live-link interpretation must be explained and demonstrated to the suspect, his/her solicitor and appropriate adult, unless it has been previously explained and demonstrated.

The interviewer must point out the sign or indicator which shows that the recording equipment is activated and is recording and must then:

(a) tell the suspect that the interview is being audibly recorded using an authorised secure digital recording network device and outline the recording process in the same way as required in the case of a removable recording media device;

(b) subject to the 'anonymity provision' (**5.3**), give his/her name and rank and that of any other interviewer present;

(c) ask the suspect and any other party present, eg the appropriate adult, a solicitor or interpreter, to identify themselves for the purpose of voice identification;

(d) state the date, time of commencement, and place of the interview; and

(e) inform the person that (s)he will be given access to the recording of the interview in the event that (s)he is charged or informed that (s)he will be prosecuted but if (s)he is not charged or informed that (s)he will be prosecuted (s)he will only be given access as agreed with the police or on the order of a court, and that (s)he will be given a written notice at the end of the interview setting out his/her rights to access the recording and what will happen to the recording.

(f) If equipment for remote monitoring of interviews as described above is installed, explain the contents of the notice to the suspect, solicitor, and appropriate adult as required above and point out the light that illuminates automatically as soon as remote monitoring is activated.

The provisions in the last four paragraphs of **5.54** apply.

Objections by the suspect

5.65　The same provisions apply as those in **5.55**.

Taking a break during interview

5.66　When a break is taken, the fact that a break is to be taken, the reason for it and the time must be recorded on the audio recording. The recording must be stopped and the procedures for the conclusion of an interview using a removable recording media device (**5.60**) followed. When the interview recommences the procedures which apply for commencing an interview using a removable recording media device interview must be followed to create a new file to record the continuation of the interview. The time the interview recommences must be recorded on the audio recording. After any break in the interview the interviewer must, before resuming the interview, remind the person being questioned of their right to legal advice if (s)he has not exercised it and that (s)he remains under caution or, if there is any doubt, give the caution in full again.

The interviewer should bear in mind that (s)he may have to satisfy a court that the person realised that the caution still applied. (S)he may also have to show that nothing occurred during a break in an interview or between interviews which influenced the suspect's recorded evidence. In view of this, (s)he should consider, at recommencement or at a subsequent interview, summarising on recording media the reason for the break and confirming this with the suspect. As to lengths of breaks, see **5.45**.

Failure of recording equipment

If there is an equipment failure which can be rectified quickly, eg by commencing a **5.67** new secure digital network recording using the same device or a replacement device, the interviewer must follow the appropriate procedures as in the case of taking a break during interview. When the recording is resumed, the interviewer must explain what happened and record the time the interview recommences. If, however, it is not possible to continue recording on the same device or by using a replacement device, the interview should be audio-recorded on removable media, commencing as required for an interview by such means, if the necessary equipment is available. If it is not available, the interview may continue and be recorded in writing as directed by the relevant officer

Conclusion of interview

At the conclusion of the interview, the suspect must be offered the opportunity to **5.68** clarify anything (s)he has said and asked if there is anything (s)he wants to add.

At the conclusion of the interview, including the taking and reading back of any written statement, the time must be orally recorded, and the suspect must be handed a notice which explains how the audio recording will be used and the arrangements for access to it, and explains that if (s)he is charged or informed that (s)he will be prosecuted (s)he will be given access to the recording of the interview either electronically or by being given a copy on removable recording media. The notice must explain that if the suspect is not charged or informed that (s)he will be prosecuted, (s)he will only be given access as agreed with the police or on the order of a court. The notice should provide a brief explanation of the secure digital network and how access to the recording is strictly limited. The notice should also explain the access rights of the suspect, their legal representative, the police, and the prosecutor to the recording of the interview. Space should be provided on the form to insert the date, the identification number, filename, or other reference for the interview recording. The suspect must be asked to confirm that (s)he has received a copy of the notice. If the suspect fails to accept or to acknowledge receipt of the notice, the interviewer will state for the recording that a copy of the notice has been provided to the suspect and that (s)he or she has refused to take a copy of the notice or has refused to acknowledge receipt. The time must be recorded and the interviewer must ensure that the interview record is saved to the device in the presence of the suspect and any third party present during the interview and notify them accordingly. The interviewer must then explain that the record will be transferred securely to the remote secure network file server (see below). If the equipment is available to enable the record to be transferred there and then in the suspect's presence, then it should be so transferred. If it is transferred at a later time, the time and place of the transfer must be recorded. The suspect should then be informed that the interview is terminated.

After the interview

The interviewer must make a note in his/her pocket book that the interview has taken **5.69** place and that it was audibly recorded, the time it commenced, its duration and date and the identification number, filename, or other reference for the recording. If no proceedings follow in respect of the person whose interview was recorded, the recordings must be kept securely (see below). Where proceedings follow, the requirements as to recording are the same as apply after an interview by removable recording media device.

Security of secure digital network interview records

5.70   The recordings are first saved locally on the device before being transferred to the remote network file server system. The recording remains on the local device until the transfer is complete. If for any reason the network connection fails, the recording will be transferred when the network connection is restored. The interview record files are stored in read only form on non-removable storage devices, for example, hard disk drives, to ensure their integrity.

Access to interview recordings, including copying to removable media, must be strictly controlled and monitored to ensure that access is restricted to those who have been given specific permission to access for specified purposes when this is necessary. For example, police officers and CPS lawyers involved in the preparation of any prosecution case, persons interviewed if (s)he has been charged or informed (s)he may be prosecuted and their legal representatives.

Visual recording with sound of interviews with suspects

Visual recording with sound

5.71   For the purpose of Code F, a visual recording with sound means an audio recording of an interview or other matter made in accordance with the requirement in para 2.1 of Code E (**5.37**) during which a simultaneous visual recording is made which shows the suspect, the interviewer and those in whose presence and hearing the audio recording was made.

Provisions that apply to visual recording with sound

5.72   Code F provides that, for the purpose of making such a visual recording, the provisions of Code E and the relevant notes for guidance apply equally to visual recordings with sound as they do to audio-only recording, subject to the following additional provisions which apply exclusively to visual recordings.

One of those additional provisions states that, following a decision made by the relevant officer that an interview or other matter mentioned in para 2.1 of Code E (**5.37**) should be visually recorded, the relevant officer may decide that the interview is not to be visually recorded if it no longer appears that a visual recording should be made or because of a fault in the recording device. However, a decision not to make a visual recording does not detract in any way from the requirement for the interview to be audio-recorded in accordance with para 2.1 of Code E. Such a decision may be the subject of comment in court. The 'relevant officer' responsible should therefore be prepared to justify that decision.

Video provisions

5.73   The device used to make the visual recording at the same time as the audio recording must ensure coverage of as much of the room or location where the interview takes place as it is practically possible to achieve whilst the interview takes place. Interviewers will wish to arrange that, as far as possible, visual recording arrangements are unobtrusive.

It must be clear to the suspect, however, that there is no opportunity to interfere with the recording equipment or the recording media.

Where the 'anonymity provision' (**5.3**) applies, the officers and staff may have their backs to the visual recording device, and when, in accordance with Code E, as it applies to Code F, arrangements are made for the suspect to have access to the visual recording, the investigating officer may arrange for anything in the recording that might allow the officers or police staff to be identified to be concealed.

Before commencement

Before visual recording commences, the interviewer must inform the suspect that, in accordance with the provisions (**5.38**) about when to make a visual recording, a visual recording is being made and explain the visual and audio recording arrangements. If the suspect is a juvenile or a vulnerable person, the information and explanation must be provided or (as the case may be) provided again, in the presence of the appropriate adult.   **5.74**

A modification of Code F made by Code C provides that, for the purposes of live-link interpretation (**5.89**), before the interview commences, the operation of live-link interpretation must be explained and demonstrated to the suspect, his/her solicitor and appropriate adult, unless it has been previously explained and demonstrated.

Objections by the suspect

If the suspect or an appropriate adult on his/her behalf objects to the interview being visually recorded either at the outset or during the interview or during a break in the interview, the interviewer must explain that the visual recording is being made in accordance with the provisions about when to make a visual recording and that Code F requires the objections to be recorded on the visual recording. When any objections have been recorded or the suspect or the appropriate adult has refused to have his/her objections recorded visually, the relevant officer must decide, in accordance with the previous paragraph and having regard to the nature and circumstances of the objections, whether visual recording should be turned off. These provisions are meant to apply to objections based on the suspect's genuine and honestly held beliefs and to allow officers to exercise their discretion to decide whether a visual recording is to be made according to the circumstances surrounding the suspect and the investigation. Objections that appear to be frivolous with the intentions of frustrating or delaying the investigation would not be relevant. Following a decision that visual recording should be turned off, the interviewer must say that (s)he is turning off the visual recording. The audio recording required to be maintained in accordance with Code E must continue and the interviewer must ask the person to record his/her objections to the interview being visually recorded on the audio recording. If the relevant officer considers that visual recording should not be turned off, the interviewer may proceed to question the suspect with the visual recording still on. If the suspect also objects to the interview being audio-recorded, the relevant officer must decide, after having regard to the nature and circumstances of the objections, whether a written record of the interview or its continuation, is to be made and that audio recording should be turned off. Following a decision that a written record is to be made, the interviewer must say that (s)he is turning off the recorder and must then make a written record of the interview as in Code C (**5.36** and **5.40**). Following a decision that a written record is not to be made, the interviewer may proceed to question the suspect with the audio recording still on. The relevant officer should be aware that a decision to continue visual recording against the wishes of the suspect may be the subject of comment in court.   **5.75**

Talk of other matters

5.76   If the suspect indicates that (s)he wishes to tell the interviewer about matters not directly connected with the offence of which (s)he is suspected and that (s)he is unwilling for these matters to be visually recorded, (s)he should be given the opportunity to tell the interviewer about these matters after the conclusion of the formal interview.

Equipment failure

5.77   If there is a failure of equipment and it is not possible to continue visual recording using the same type of recording device (ie a removable recording media device or a secure digital network device) or by using a replacement device of either type, the relevant officer may decide that the interview is to continue without being visually recorded. In these circumstances, the continuation of the interview must be conducted and recorded in accordance with the provisions of Code E (**5.52-5.70**).

A guidance note states the following. Where the interview is being visually recorded and the media or the recording device fails, the interviewer should stop the interview immediately. Where part of the interview is unaffected by the error and is still accessible on the media or on the network device, that part must be copied and sealed in the suspect's presence as a master copy or saved as a new digital network recording as appropriate. The interview should then be recommenced using a functioning recording device and new recording media as appropriate. Where the media content of the interview has been lost in its entirety, the media should be sealed in the suspect's presence and the interview begun again. If the visual recording equipment cannot be fixed and a replacement device is not immediately available, the interview should be audio-recorded in accordance with Code E.

Complaints

5.78   If, during an interview at a police station, the detained person makes a complaint concerning a matter covered by Code C, Code E or Code F, it must be recorded in the interview record and the interviewing officer must inform the custody officer who must deal with it in accordance with Code C (**6.11**). (These obligations also apply if it comes to the interviewer's notice that the interviewee may have been treated improperly.) Where the interview is being audibly recorded or visually recorded with sound, the recording should be left running until the custody officer has entered the room and has spoken to the interviewee. Continuation or termination of the interview should be at the discretion of the interviewing officer pending action by an inspector.

Where, during the course of an interview which is being recorded, a complaint is made about a matter not connected with Code C, Code E, or Code F, the decision to continue the interview or terminate it is at the discretion of the interviewing officer. If continued, the officer must inform the complainant that the complaint will be brought to the attention of the custody officer at the conclusion of the interview, which must be done as soon as practicable thereafter.

Other points

5.79   The custody record must show the times during which the prisoner was not in the custody of the custody officer and the reason why (s)he was removed from that custody; if a request for the delivery of a prisoner out of that custody was refused, the reason for refusal must be recorded. Decisions to defer breaks must be recorded, with grounds, in the interview record. Assuming (s)he is still at the police station when it is made, the person interviewed must be allowed to read the interview record and to sign it as being correct or to indicate what (s)he considers to be inaccurate, but no person may be kept in custody for this sole purpose.

In conclusion, it should be noted that the Codes of Practice do not prevent a police officer from asking questions at the scene of a crime to elicit an explanation which could provide the arrested person with the opportunity to show that (s)he was innocent. The reason is obvious: since the questions are not asked with a view to establishing admissions on which proceedings can be founded, there is technically no interview for the purposes of the Codes. If, in the course of asking such questions, a suspect makes a confession, it is prima facie admissible (even if the suspect is a juvenile and no appropriate adult is present).

In addition, there is no universal rule that, whenever there is a breach of a Code of Practice in a police interview, all subsequent interviews must be tainted and evidence of them excluded. Such a rule would fetter a judge's discretion under PACE, s 78 (**8.74**), which depends upon the facts of the particular case.

PACE, s 60B provides that, where:

(a) a person (S) suspected of the commission of a criminal offence is interviewed by a police officer but is not arrested for the offence; and
(b) the police officer in charge of investigating the offence determines that—
    (i) there is not sufficient evidence to charge S with an offence; or
    (ii) there is sufficient evidence to charge S with an offence but S should not be charged with an offence or given a simple, conditional, youth or conditional youth caution in respect of an offence,
    a police officer must give S notice in writing that S is not to be prosecuted.

This provision does not prevent the prosecution of S for an offence if new evidence comes to light after the notice was given.

INTERPRETERS

Code C contains provisions about interpreters which are set out at **5.81-5.91**.   **5.80**

General

Chief officers of police are responsible for making arrangements to provide appropri-   **5.81** ately qualified independent persons to act as interpreters and to provide translations of essential documents for:

(a) detained suspects who the custody officer has determined require an interpreter, and
(b) suspects who are not under arrest but are cautioned and who the interviewer has determined require an interpreter. In these cases, the responsibilities of the custody officer are, if appropriate, assigned to the interviewer. An interviewer who has any doubts about whether and what arrangements for an interpreter must be made or about how the interpretation provisions of Code should be applied to a suspect who is not under arrest should seek advice from a sergeant (or above).

If the suspect has a hearing or speech impediment, references to 'interpreter' and 'interpretation' include appropriate assistance necessary to establish effective communication with that person. With regard to persons in Wales, nothing in Code C or any other Code affects the application of the Welsh Language Schemes produced by police and crime commissioners in Wales in accordance with the Welsh Language Act 1993.

Chief officers have discretion when determining the individuals or organisations they use to provide interpretation and translation services for their forces provided that these services are compatible with the requirements of the Directive. One example which chief officers may wish to consider is the Ministry of Justice Framework Agreement for interpretation and translation services.

All reasonable attempts should be made to make the suspect understand that interpretation and translation will be provided at public expense.

References in Code C and any other PACE Code to making arrangements for an interpreter to assist a suspect mean making arrangements for the interpreter to be physically present in the same location as the suspect unless the live-link interpretation provisions (**5.89**) allow live link interpretation to be used.

Interviewing suspects: foreign languages

5.82   Unless a superintendent (or above) considers that the provisions about urgent interviews at **5.35** and **5.44** apply, a suspect who requires an interpreter because (s)he does not appear to speak or understand English must not be interviewed unless arrangements are made for a person capable of interpreting to assist the suspect to understand and communicate.

If a person who is a juvenile or a vulnerable person is interviewed and the person acting as the appropriate adult does not appear to speak or understand English, arrangements must be made for an interpreter to assist communication between the person, the appropriate adult and the interviewer, unless the interview is urgent and a superintendent (or above) considers that delay will lead to the consequences set out at **5.35** and **5.44**.

When a written record of the interview is made, the interviewer must make sure that the interpreter makes a note of the interview at the time in the person's language for use in the event of the interpreter being called to give evidence, and certifies its accuracy. The interviewer should allow sufficient time for the interpreter to note each question and answer after each is put, given, and interpreted. The person should be allowed to read the record or have it read to him/her and sign it as correct or indicate the respects in which (s)he considers it inaccurate. If an audio or visual record of the interview is made, the arrangements in Code E or Code F apply (**5.51–5.77**).

In the case of a person making a statement under caution to a police officer or other police staff other than in English:

(a) the interpreter must record the statement in the language it is made;
(b) the person must be invited to sign it;
(c) an official English translation must be made in due course.

The application of this rule to live-link interpretation is dealt with at **5.90**.

Interviewing suspects who have a hearing or speech impediment

5.83   Unless a superintendent (or above) considers that the provisions about urgent interviews at **5.35** and **5.44** apply, a suspect who requires an interpreter or other appropriate assistance to enable effective communication with him/her because (s)he appears to have a hearing or speech impediment must not be interviewed without arrangements having been made to provide an independent person capable of interpreting or of providing other appropriate assistance.

An interpreter should also be arranged if a person who is a juvenile or a vulnerable person is interviewed and the person present as the appropriate adult appears to have a hearing or speech impediment, unless the interview is urgent and a superintendent (or above) considers that the provisions about urgent interviews apply.

If a written record of the interview is made, the interviewer must ensure that the interpreter is allowed to read the interview record and certify its accuracy in the event of the interpreter being called to give evidence. If an audio or visual record of the interview is made, the arrangements in Code E or Code F apply (**5.51-5.77**).

Additional rules for detained persons

If the detainee cannot communicate with a solicitor under his/her entitlement to legal **5.84** advice because of language, hearing, or speech difficulties, an interpreter must be called. A police officer or any other police staff may not be used for this purpose.

After the custody officer has determined that a detainee requires an interpreter and following the initial action referred to at **5.9**, arrangements must also be made for an interpreter to:

(a) explain (i) the grounds and reasons for any authorisation for the detainee's continued detention, before or after charge, and (ii) any information about the authorisation given to the detainee by the authorising officer and which is recorded in the custody record;

(b) be present at the magistrates' court for the hearing of an application for a warrant of further detention or any extension or further extension of such warrant to explain (i) any grounds and reasons for the application and (ii) any information about the authorisation of the detainee's further detention given to him/her by the court; and

(c) explain any offence with which the detainee is charged or for which (s)he is informed that (s)he may be prosecuted and any other information about the offence given to him/her by or on behalf of the custody officer.

Translations of essential documents

Written translations, oral translations, and oral summaries of essential documents in a **5.85** language the detainee understands must be provided in accordance with the following provisions:

> *Essential documents comprise records required to be made in accordance with Code C which are relevant to decisions to deprive a person of his/her liberty, to any charge, and to any record considered necessary to enable a detainee to defend him/herself in criminal proceedings and to safeguard the fairness of the proceedings.* Passages of essential documents which are not relevant need not be translated. It is not necessary to disclose information in any translation which is capable of undermining or otherwise adversely affecting any investigative processes, for example, by enabling the suspect to fabricate an innocent explanation or to conceal lies from the interviewer.

The Table of Essential Documents below lists the documents considered essential for the purposes of Code C and when (subject to the provisions below) written translations must be created and provided. See the live-link interpretation material below for the Table's application to live-link interpretation.

Essential Documents for the Purposes of this Code	When Translation to be Created	When Translation to be Provided
(i) The grounds for each of the following authorisations to keep the person in custody as they are described and referred to in the custody record: (a) authorisation for detention before and after charge given by the custody officer and by the review officer; (b) authorisation to extend detention without charge beyond 24 hours given by a superintendent; (c) a warrant of further detention issued by a magistrates' court and any extension(s) of the warrant; (d) an authority to detain in accordance with the directions in a warrant of arrest issued in connection with criminal proceedings including the court issuing the warrant.	As soon as practicable after each authorisation has been recorded in the custody record.	As soon as practicable after the translation has been created, whilst the person is detained or after (s)he has been released. *There is no power under PACE to detain a person or to delay his/her release solely to create and provide a written translation of any essential document.*
(ii) Written notice on charge showing particulars of the offence charged or the offence for which the suspect has been told (s)he may be prosecuted.	As soon as practicable after the person has been charged or reported.	As soon as practicable after the person has been charged or reported.
(iii) Written interview records. Written statement under caution.	To be created contemporaneously by the interpreter for the person to check and sign.	As soon as practicable after the person has been charged or told (s)he may be prosecuted.

The custody officer may authorise an oral translation or oral summary of documents (i) to (ii) in the table (but not (iii)) to be provided (through an interpreter) instead of a written translation. Such an oral translation or summary may only be provided if it would not prejudice the fairness of the proceedings by in any way adversely affecting or otherwise undermining or limiting the ability of the suspect in question to understand his/her position and to communicate effectively with police officers, interviewers, solicitors, and appropriate adults with regard to his/her detention and the investigation of the offence in question and to defend him/herself in the event of criminal proceedings. The quantity and complexity of the information in the document should always be considered, and specific additional consideration given if the suspect is a vulnerable person or a juvenile. The reason for the decision must be recorded.

Subject to the next three paragraphs, a suspect may waive his/her right to a written translation of the essential documents described in the table but only if (s)he does so voluntarily after receiving legal advice or having full knowledge of the consequences and gives his/her unconditional and fully informed consent in writing.

The suspect may be asked if (s)he wishes to waive his/her right to a written translation, and before giving his/her consent (s)he must be reminded of his/her right to legal advice and asked whether (s)he wishes to speak to a solicitor.

No police officer or police staff should do or say anything with the intention of persuading a suspect who is entitled to a written translation of an essential document to waive that right. No police officer or police staff shall indicate to any suspect, except to answer a direct question, whether the period for which (s)he is liable to be detained, or, if not detained, the time taken to complete the interview, might be reduced: (a) if (s)he does not ask for legal advice before deciding whether (s)he wishes to waive his/her right to a written translation of an essential document; or (b) if (s)he decides to waive his/her right to a written translation of an essential document. There is no power under PACE to detain a person or to delay his/her release solely to create and provide a written translation of any essential document.

For the purpose of the waiver:

(a) The consent of a person who is vulnerable is only valid if the information about the circumstances under which (s)he can waive the right, the reminder about his/her right to legal advice, and his/her consent are given in the presence of the appropriate adult.

(b) The consent of a juvenile is only valid if his/her parent's or guardian's consent is also obtained unless the juvenile is under 14, when his/her parent's or guardian's consent is sufficient in its own right, and the requisite information and reminder and the juvenile's consent are also given in the presence of the appropriate adult (who may or may not be a parent or guardian).

The detainee, his/her solicitor or appropriate adult may make representations to the custody officer that a document which is not included in the table is essential and that a translation should be provided. The request may be refused if the officer is satisfied that the translation requested is not essential for the purposes described in italics above.

If the custody officer has any doubts about

(a) providing an oral translation or summary of an essential document instead of a written translation;

(b) whether the suspect fully understands the consequences of waiving his/her right to a written translation of an essential document; or

(c) refusing to provide a translation of a requested document,

the officer should seek advice from an inspector (or above).

Action taken in accordance with the above provisions must be recorded in the detainee's custody record or interview record as appropriate.

Complaints

5.86   If a detainee complains that (s)he is not satisfied with the quality of interpretation, or of a translation, the custody officer, or (as the case may be) the interviewer, is responsible for deciding whether a different interpreter should be called, or a different translation provided, in accordance with the procedures set out in the arrangements made by the chief officer.

Decisions not to provide interpretation and translation

5.87   If a suspect challenges a decision:

(a) made by the custody officer or (as the case may be) by the interviewer, that (s)he does not require an interpreter; or

(b) not to provide a different interpreter or another translation, or not to translate a requested document,

the matter must be reported to an inspector to deal with as a complaint (see **5.78** and **5.86**) if the challenge is made during an interview.

Documentation

5.88   The following must be recorded in the custody record or, as applicable, the interview record:

(a) action taken to arrange for an interpreter, including the live-link requirements in Annex N to Code C (**5.91**) as applicable;

(b) action taken when a detainee is not satisfied about the standard of interpretation or translation provided;

(c) when an urgent interview is carried out in the absence of an interpreter;

(d) when a detainee has been assisted by an interpreter for the purpose of providing or being given information or being interviewed;

(e) action taken in accordance with the prescribed procedure when:

(i)   a written translation of an essential document is provided;

(ii) an oral translation or oral summary of an essential document is provided instead of a written translation and the authorising officer's reason(s) why this would not prejudice the fairness of the proceedings;

(iii) a suspect waives his/her right to a translation of an essential document;

(iv) representations that a document which is not included in the table above is essential and that a translation should be provided are refused and the reason for the refusal.

Live-link interpretation

Definition

5.89   For the purposes of the following text, 'live-link interpretation' means an arrangement to enable communication between the suspect and an interpreter who is not physically present with the suspect. The arrangement must ensure that anything said by any person in the suspect's presence and hearing can be interpreted in the same way as if the interpreter was physically present at that time. The communication must be by audio and visual means for the purpose of an interview, and for all other purposes it may be (a) by audio and visual means, or (b) by audio means only, as follows:

Audio and visual communication This applies for the purposes of an interview conducted and recorded in accordance with Code E (audio recording) or Code F (visual recording) (**5.51–5.77**) and, during that interview, live link interpretation must:

(a) enable the suspect, the interviewer, solicitor, appropriate adult and any other person physically present with the suspect at any time during the interview, and an interpreter who is not physically present, to see and hear each other; and

(b) enable the interview to be conducted and recorded in accordance with the provisions of Codes C, E and F, subject to the modifications in Part 2 of Annex N to Code C below.

Audio and visual or audio without visual communication This applies to communication for the purposes of any provision of Code C or any other PACE Code, except as described in respect of *Audio and visual communication* above, which requires or permits information to be given to, sought from, or provided by a suspect, whether orally or in writing, which would include communication between the suspect and his/her solicitor and/or appropriate adult. For these cases, live link interpretation must:

(i) enable the suspect, the person giving or seeking that information, any other person physically present with the suspect at that time and an interpreter who is not so present, to either see and hear each other, or to hear without seeing each other (eg by using a telephone); and

(ii) enable that information to be given to, sought from, or provided by, the suspect in accordance with the provisions of Code C or any other PACE Code that apply to that information, as modified for the purposes of the live-link, by Part 2 of Annex N to Code C.

General

The requirement in (b) and (ii) above, that live-link interpretation must enable compliance with the relevant provisions of the Codes C, E and F, means that the arrangements must provide for any written or electronic record of what the suspect says in his/her own language which is made by the interpreter, to be securely transmitted without delay so that the suspect can be invited to read, check, and, if appropriate, sign or otherwise confirm that the record is correct or make corrections to the record.   **5.90**

Chief officers must be satisfied that live-link interpretation used in their force area for the purposes of audio and visual communication and audio and visual or audio without visual communication provides for accurate and secure communication with the suspect. This includes ensuring that at any time during which live link interpretation is being used: a person cannot see, hear or otherwise obtain access to any communications between the suspect and interpreter or communicate with the suspect or interpreter unless so authorised or allowed by the custody officer or, in the case of an interview, the interviewer and that as applicable, the confidentiality of any private consultation between a suspect and his/her solicitor and appropriate adult is maintained.

Annex N

Part 1 of Annex N to Code C determines whether the use of a live-link is appropriate in any particular case. Under Part 1, decisions in accordance with Annex N that the physical presence of the interpreter is not required and to permit live-link interpretation must   **5.91** be made on a case by case basis. Each decision must take account of the age, gender, and vulnerability of the suspect, the nature and circumstances of the offence and the investigation, and the impact on the suspect according to the particular purpose(s) for which the suspect requires the assistance of an interpreter and the time(s) when that assistance is required. Thus, the custody officer in the case of a detained suspect, or—in the case of a suspect who has not been arrested—the interviewer (subject to the provisions at **5.81**(b)), must consider whether the ability of the particular suspect, to communicate confidently and effectively for the purpose in question (see the next paragraph) is likely to be adversely affected or otherwise undermined or limited if the interpreter is not physically present and live-link interpretation is used. Although a suspect for whom an appropriate adult is required may be more likely to be adversely affected as described, it is important to note that a person who does not require an appropriate adult may also be adversely impacted by the use of live-link interpretation.

Examples of purposes referred to in the previous paragraph include:

(a) understanding and appreciating the suspect's position having regard to any information the suspect has given to, or sought from, him/her, in accordance with Code C or any other PACE Code including
  (i)   the caution,
  (ii) the special warning,
  (iii) information about the offence,
  (iv) the grounds and reasons for detention,
  (v)   the translation of essential documents,
  (vi) his/her rights and entitlements,
  (vii) intimate and non-intimate searches of detained persons at police stations, and
  (viii)provisions and procedures to which Code D (Identification) applies concerning, for example, eye-witness identification, taking fingerprints, samples, and photographs;
(b) understanding and seeking clarification from the interviewer of questions asked during an interview conducted and recorded in accordance with Code E or Code F and of anything else that is said by the interviewer and answering the questions;
(c) consulting privately with his/her solicitor and (if applicable) the appropriate adult
  (i)   to help decide whether to answer questions put during interview, and
  (ii) about any other matter concerning his/her detention and treatment whilst in custody; and
(d) communicating with practitioners and others who have some formal responsibility for, or an interest in, his/her health and welfare.

If the custody officer or the interviewer (subject to **5.81**(b)) is satisfied that, for a particular purpose as described above, the live-link interpretation would not adversely affect or otherwise undermine or limit the suspect's ability to communicate confidently and effectively for that purpose, (s)he must so inform the suspect, his/her solicitor and (if applicable) the appropriate adult. At the same time, the operation of live-link interpretation must be explained and demonstrated to them, they must be advised of the chief officer's obligations concerning the security of live-link communications, and they must be asked if they wish to make representations that live link interpretation should not be used or if they require more information about the operation of the arrangements. They must also be told that at any time live-link interpretation is in use, they may make representations to the custody officer or the interviewer that its operation should cease and that the physical presence of an interpreter should be arranged.

When the authority of an inspector is required   If representations are made that live-link interpretation should not be used, or (where live-link interpretation is in use) that its operation should cease and the physical presence of an interpreter arranged, and the custody officer or interviewer (subject to the provisions at **5.81**(b) is unable to allay the concerns raised, live-link interpretation may not be used, or (as the case may be) continue to be used, unless authorised in writing by an inspector (or above). Such authority may be given if that officer is satisfied that for the purpose(s) in question at the time an interpreter is required, live-link interpretation is necessary and justified. In making this decision, the officer must have regard to the:

(a) circumstances of the suspect;
(b) nature and seriousness of the offence;
(c) requirements of the investigation, including its likely impact on both the suspect and any victim(s);
(d) representations made by the suspect, his/her solicitor and (if applicable) the appropriate adult that live-link interpretation should not be used;
(e) availability of a suitable interpreter to be physically present compared with the availability of a suitable interpreter for live-link interpretation; and
(f) risk if the interpreter is not physically present, evidence obtained using link interpretation might be excluded in subsequent criminal proceedings; and
(g) likely impact on the suspect and the investigation of any consequential delay to arrange for the interpreter to be physically present with the suspect.

For the purposes of Code E and live-link interpretation, there is no requirement to make a visual recording which shows the interpreter as viewed by the suspect and others present at the interview. The audio recording required by Code E is sufficient. However, the authorising officer, in consultation with the officer in charge of the investigation, may direct that the interview is conducted and recorded in accordance with Code F. This will require the visual record to show the live-link interpretation arrangements and the interpreter as seen and experienced by the suspect during the interview. This should be considered if it appears that the admissibility of interview evidence might be challenged because the interpreter was not physically present or if the suspect, solicitor or appropriate adult make representations that Code F should be applied.

Documentation   A record must be made of the actions, decisions, authorisations, and outcomes arising from the requirements of Annex N.

Modifications   Part 2 of Annex N makes the following modifications to Code C for the purposes of live-link interpretation:

1. For the third sentence of the third paragraph of **5.82**, there is substituted: 'A clear legible copy of the complete record must be sent without delay via the live-link to the interviewer. The interviewer, after confirming with the suspect that the copy is legible and complete, must allow the suspect to read the record, or have the record read to him/her by the interpreter and to sign the copy as correct or indicate the respects in which (s)he considers it inaccurate. The interviewer is responsible for ensuring that the signed copy and the original record made by the interpreter are retained with the case papers for use in evidence if required and must advise the interpreter of his/her obligation to keep the original record securely for that purpose.'

2. There is substituted for (b) in the last paragraph of **5.82**: '(b) a clear legible copy of the complete statement must be sent without delay via the live-link to the interviewer. The interviewer, after confirming with the suspect that the copy is legible

and complete, must invite the suspect to sign it. The interviewer is responsible for ensuring that the signed copy and the original record made by the interpreter are retained with the case papers for use in evidence if required and must advise the interpreter of his/her obligation to keep the original record securely for that purpose.'

3. There is inserted after the first sentence of the third paragraph of **5.83**: 'A clear legible copy of the certified record must be sent without delay via the live-link to the interviewer. The interviewer is responsible for ensuring that the original certified record and the copy are retained with the case papers for use as evidence if required and must advise the interpreter of his/her obligation to keep the original record securely for that purpose.'

CHAPTER 6

Treatment, Charging, and Bail of Detainees

GENERAL

Whenever PACE Code C (**5.1**) requires a person to be given certain information (s)he does not have to be given it if (s)he is incapable at the time of understanding what is said to him/her or if (s)he is violent or in urgent need of medical attention, but (s)he must be given it as soon as practicable. **6.1**

Where video cameras are installed in the custody area, suspects and other people entering should be informed by prominently placed notices that cameras are in use. Any request by a suspect or other person to have video cameras switched off should be refused.

A reference to a 'juvenile' is to someone under 18. If a person appears to be under 18, (s)he must, in the absence of clear evidence that (s)he is older, be treated as a juvenile for the purposes of Code C. In respect of a reference to 'appropriate adult' and 'appropriate consent', see **5.7** and **7.42** respectively.

Reception of arrested persons at designated police stations

PACE, s 35(1) requires that the chief officer of police designates the police stations in his/her area which are to be used for the purpose of detaining arrested persons; his/her duty is to designate police stations which appear to him/her to provide sufficient accommodation for detaining arrested persons. By s 35(2A) the Chief Constable of the British Transport Police may designate police stations which (in addition to those designated under s 35(1)) may be used for the purpose of detaining arrested persons. **6.2**

Custody officers

Where a station is designated under s 35(1) or (2A), s 36(1), (2), and (3) requires that one or more custody officers must be appointed by the relevant chief officer of police or his/her nominee. They must be of the rank of sergeant (or above), but s 36(4) allows another officer of any rank to perform the duties of a custody officer under PACE at a designated police station if such an officer is not readily available to perform those duties. Such an officer is only not 'readily available' if (s)he is not actually at the police station and cannot, without much difficulty, be fetched there. **6.3**

None of the functions of a custody officer in relation to a person may be performed by an officer who at the time when the function falls to be performed is involved in the investigation of an offence for which that person is in police detention at that time. However, this does not prevent a custody officer:

(a) performing any function assigned to custody officers by PACE, or by a PACE code of practice;
(b) carrying out the duty imposed on custody officers by PACE, s 39 (below);
(c) doing anything in connection with the identification of a suspect; or
(d) doing anything under the Road Traffic Act 1988, s 7 (breath, blood, or urine specimens).

Where an arrested person is taken to a non-designated police station any officer not concerned in the investigation of the offence may, by PACE, s 36(7), assume the responsibilities of a custody officer. If no such officer is available, it may be the arresting officer.

Custody officer's duties

6.4   A custody officer's duties are laid down by PACE, ss 37 to 39. Basically, and this is provided for by s 39, (s)he is responsible for persons in detention, for ensuring that they and their property are treated in accordance with PACE and Code C, and for maintaining a chronological and contemporaneous record of every aspect of a person's treatment whilst in detention. These responsibilities cannot be overstated. The provisions in relation to detention are quite complex and the requirements for such a log mean that omissions in the keeping of it cannot be corrected. If a duty is carried out but not recorded, the failure to record as required by the Codes is sufficient to raise the issue of disciplinary proceedings. Sections 37 and 38 are dealt with at **6.47** and **6.53–6.55** respectively.

Where PACE and its Codes require that certain things are done by a custody officer, the requirement also applies to a police officer other than a custody officer who is performing the functions of a custody officer.

Where a person has been arrested, a custody officer does not have to satisfy him/herself that the arrest is lawful before (s)he can hold the arrested person in lawful custody.

A custody officer is required to perform the functions specified in Code C as soon as is practicable. A custody officer is not in breach of the Code in the event of delay provided that the delay is justifiable and that every reasonable step is taken to prevent unnecessary delay.

This covers cases where, for example, a large number of suspects are brought into the police station simultaneously to be placed in custody, or interview rooms are all being used, or where there are difficulties in contacting the appropriate adult, solicitor, or interpreter. The custody record must indicate where a delay has occurred and the reason why.

Police detention

6.5   A person is stated by PACE, s 118 to be in police detention for the purposes of the Act if:

(a) (s)he has been taken to a police station after being arrested for an offence, or after being arrested by a police officer under the Terrorism Act 2000 (TA 2000), s 41 on reasonable suspicion of being a terrorist; or
(b) (s)he is arrested at a police station after attending voluntarily at the station or accompanying a constable to it,

and (s)he is detained there or is detained elsewhere in the charge of a constable.

In most circumstances, of course, a person's detention begins under (a) on arrival at a police station under arrest. If a person is subsequently removed from a police station, eg, to attend an identification parade or to visit the scene of a crime or on transfer to another station, (s)he is in detention whilst (s)he remains in the charge of a constable. This is important when periods of detention are to be considered by custody officers. By way of an exception, s 118 provides that a person who is at a court after being charged is not in police detention for the purposes of the Act.

As already stated, a separate custody record must be opened as soon as practicable for each person detained at a police station.

Limitations on police detention

PACE, s 34 provides that a person arrested for an offence (as opposed to a breach of the peace, which is not an offence) must not be kept in police detention in the cases outlined below. However, no person in police detention may be released except on the authority of a custody officer.   **6.6**

Section 34(2) requires that if at any time a custody officer:

(a) becomes aware, in relation to any person (D) in police detention at that station, that the grounds for D's detention have ceased to apply; and
(b) is not aware of any other grounds on which D's continued detention could be justified under the provisions of the Act,

the custody officer must order D's immediate release from custody. However, a person who was unlawfully at large when arrested must not be released under these provisions.

PACE, s 34(5) provides that a person (P) released in these circumstances must be released without bail unless PACE, s 34(5A) applies, or on bail if s 34(5A) applies. Section 34(5A) applies if:

(a) it appears to the custody officer—
  (i) that there is need for further investigation of any matter in connection with which P was detained at any time during the period of P's detention; or
  (ii) that, in respect of any such matter, proceedings may be taken against P or P may be given a youth caution (which is only possible if the offender is under 18); and
(b) the pre-conditions for bail are satisfied.

The pre-conditions for bail in relation to the release of a person by a custody officer are:

(i) that the custody officer is satisfied that releasing the person on bail is necessary and proportionate in all the circumstances (having regard, in particular, to any conditions of bail which would be imposed); and
(ii) that an inspector (or above) authorises the release on bail (having considered any representations made by the person or the person's legal representative).

Where P is released under s 34(5), and the custody officer determines that:

(a) there is not sufficient evidence to charge P with an offence; or
(b) there is sufficient evidence to charge P with an offence but P should not be charged with an offence or given a simple, youth, conditional, or youth conditional caution in respect of an offence,

the custody officer must give P notice in writing that P is not to be prosecuted. This does not prevent the prosecution of P for an offence if new evidence comes to light after the notice was given.

An investigating officer may bring facts to the notice of a custody officer but issues of further detention or release are to be decided by the custody officer. It will be appreciated that custody officers may find themselves in dispute with officers of senior rank in respect of detention issues. If this occurs the superintendent responsible for the police station must be consulted.

For the above purposes and for the purposes of the detention provisions of PACE (ss 34–51) in general, a person who:

(a) attends a police station to answer street bail under **s** 30A (**4.57**),

(b) returns to a police station to answer to bail granted under the detention provisions, or

(c) is arrested under PACE, ss 30D or 46A (**4.57** and **6.79**),

is to be treated as arrested for an offence and that offence is the offence in connection with which (s)he was granted bail. The above provision does not apply in relation to a person who is granted bail subject to the duty to attend at a police station appointed by the custody officer for proceedings in relation to a live link direction or any preliminary hearing in relation to which such a direction is given, and who either attends a police station to answer to such bail, or is arrested under s 46A for failing to do so. Provision as to the treatment of such persons is made by s 46ZA (**6.73**).

Conditions of detention

6.7   (1) So far as practicable, not more than one person shall be detained in each cell.

(2) Cells in use must be adequately heated, cleaned and ventilated. They must be adequately lit, subject to such dimming as is compatible with safety and security to allow people detained overnight to sleep. No additional restraints should be used within a locked cell unless absolutely necessary, and then only approved restraint equipment which is reasonable and necessary in the circumstances having regard to the detainee's demeanour and with a view to ensuring the detainee's safety and the safety of others. If the detainee is deaf or a vulnerable person (**5.4**), particular care must be taken when deciding whether to use any form of approved restraints.

(3) Blankets, mattresses, pillows, and other bedding supplied should be of a reasonable standard and in a clean and sanitary condition.

(4) Access to toilet and washing facilities must be provided. This must take account of the dignity of the detainee.

(5) If it is necessary to remove a person's clothes for the purpose of investigation, for hygiene or health reasons or for cleaning, removal must be conducted with proper regard to the dignity, sensitivity, and vulnerability of the detainee, and replacement clothing of a reasonable standard of comfort and cleanliness must be provided. A person must not be interviewed unless adequate clothing has been offered to him/her.

(6) At least two light meals and one main meal must be offered in any period of 24 hours. Drinks should be provided at meal times and on reasonable request between meal times. Meals should, so far as practicable, be offered at recognised meal times, or at other times that take account of when the detainee last had a meal. Whenever necessary, advice must be sought from the appropriate healthcare professional (clinically qualified person working within the scope of his/her practice) on medical or dietary matters. As far as practicable, meals provided must offer a varied diet and meet any special dietary needs or religious beliefs that

a person may have. At the custody officer's discretion, the detainee may also have meals supplied by family or friends at his/her or their own expense. However, especially in the case of a person detained under the Terrorism Act 2000 (TA 2000), immigration detainees and others likely to be detained for an extended period, a custody officer is entitled to take account of the risk of items being concealed in any food or package and of his/her duties and responsibilities under food handling arrangements.

(7) Brief outdoor exercise must be offered daily if practicable.

(8) A juvenile must not be placed in a police cell, unless no other secure accommodation is available and the custody officer considers that it is not practicable to supervise him/her if (s)he is not placed in a cell or the custody officer considers that a cell provides more comfortable accommodation than other secure accommodation in the police station.

(9) A juvenile must not be placed in a cell with a detained adult.

Documentation

A record must be kept of replacement clothing and meals offered. If a juvenile is placed in a cell, the reason must be recorded. The use of any restraints, the reason for such use and any arrangements for enhanced supervision must be recorded.   **6.8**

Care and treatment of detainees

Detainees should be visited at least every hour. The requirement to visit detainees at least every hour is subject, in the case of a person to whom the Mental Health Act 1983 (Places of Safety) Regulations 2017 apply, to the provisions at **6.10**(1). If no reasonably foreseeable risk was identified in a risk assessment, a sleeping detainee need not be awakened. A person suspected of being under the influence of drink or drugs or having swallowed drugs, or whose level of consciousness causes concern, must, subject to any clinical directions given by the appropriate healthcare professional, be visited and roused at least every half hour, have his/her condition assessed, and clinical treatment arranged if appropriate. The detainee's condition must be assessed by entering the cell, calling his/her name and shaking him/her gently, before asking his/her name and where (s)he lives as well as where (s)he thinks (s)he is. (S)he should be asked to open his/her eyes and to lift one arm and then the other. These provisions would apply to a person in police custody by order of a magistrates' court under the Criminal Justice Act 1988 (CJA 1988), s 152 to facilitate the recovery of evidence where that person has been charged with drug possession or drug trafficking and is suspected of having swallowed the drugs. In the case of the healthcare needs of a person who has swallowed drugs, the custody officer, subject to any clinical direction, should consider the necessity for rousing every half hour. This does not negate the need for regular visiting of the suspect in the cell.   **6.9**

It is important to remember that a person who appears to be drunk or behaving abnormally might be suffering from illness or the effects of drugs or may have sustained injury (particularly a head injury) which is not apparent, and that someone addicted to certain drugs may experience harmful effects within a short time of being deprived of their supply. Consequently, police officers should always err on the side of caution when in doubt about calling an appropriate healthcare professional, and act with all due speed. A detainee may be dependent upon certain drugs, including alcohol, and may experience harmful effects within a short time of being deprived of their supply.

In these circumstances an appropriate healthcare professional should be consulted or an ambulance should be called. A record must be made of any intoxicating liquor supplied. Annex G to Code C limits the circumstances in which persons under the influence of drink or drugs, or whose mental state is affected in some other way, may be interviewed.

Should the custody officer feel in any way concerned about a detainee's condition, the officer must arrange for medical treatment.

As soon as practicable after arrival at the police station, each detainee must be given an opportunity to speak in private with a member of the custody staff, who if the detainee wishes may be of the same sex as the detainee, about any matter concerning the detainee's personal needs relating to the detainee's health, hygiene, and welfare that might affect or concern the detainee whilst in custody. Matters concerning personal needs include any requirement for menstrual products, incontinence products, and colostomy appliances, where these needs have not previously been identified. It also enables adult women to speak in private to a female officer about their requirements for menstrual products if they decline to respond to the more direct enquiry envisaged under the next paragraph. This contact should be facilitated at any time, where possible. If the detainee wishes to take this opportunity, the necessary arrangements must be made as soon as practicable. In the case of a juvenile or vulnerable person, the appropriate adult must be involved in accordance with **5.9**.

Each female detainee aged 18 or over must be asked in private if possible and at the earliest opportunity, if she requires or is likely to require any menstrual products whilst in custody. She must be told that they will be provided free of charge and that replacement products are available. At the custody officer's discretion, detainees may have menstrual products supplied by their family or friends at their expense. For girls under 18, see **5.10**.

For detailed guidance on matters concerning detainee healthcare and treatment and associated forensic issues, see the *Detention and Custody Authorised Professional Practice* produced by the College of Policing.

Clinical treatment and attention

6.10   (1) The custody officer must ensure that a detainee receives appropriate clinical attention as soon as reasonably practicable if that person appears to be suffering from physical illness or mental disorder or to be in need of clinical attention or is injured. This applies even if no request for clinical attention is made or if such attention has already been received elsewhere. The custody officer must also consider the need for clinical attention in relation to those suffering from the effects of alcohol or drugs.

These rules about seeking medical attention are not meant to prevent or delay the transfer to a hospital if necessary of a person detained under the Mental Health Act 1983 (MHA 1983), ss 135 and 136 (**4.114–4.116**). MHA 1983 Code of Practice provides more guidance about arranging assessments under MHA 1983 and transferring detainees from police stations to other places of safety. Additional guidance is published at https://www.gov.uk/government/publications/mental-health-act-1983-implementing-changes-to-police-powers.

Under the Mental Health Act 1983 (Places of Safety) Regulations 2017, when an adult is detained at a police station under MHA 1983, s 135 or s 136, a custody officer there must ensure that—

(a) the welfare of the detained adult (DA) is checked by a healthcare professional at least once every 30 minutes, and any appropriate action is taken for the treatment and care of DA; and

(b) so far as is reasonably practicable, a healthcare professional is present and available to DA throughout the period in which DA is detained at the police station (reg 4).

Subject to reg 7 (below), where either or both of the requirements in (a) and (b) is not met, the custody officer must arrange for DA to be taken to another place of safety (reg 4).

By reg 5, when an adult is detained at a police station under MHA 1983, s 135 or s 136, a custody officer there must, subject to the italicised words below and to regs 6 and 7 below—

(a) review the behaviour of DA at least once an hour and determine whether the behaviour of DA poses an imminent risk of serious injury or death to DA, or to another person; and whether, because of that risk, no place of safety other than a police station in the relevant police area can reasonably be expected to detain DA; and

(b) where those circumstances are determined not to exist, arrange for D to be taken to a place of safety other than a police station.

Before making a determination under (a), the custody officer must, where reasonably practicable, consult the healthcare professional who carried out the most recent check under reg 4. By reg 6, the frequency of the reviews referred to in (a) may be reduced, to no less than once every three hours, where DA is sleeping, and a healthcare professional who has checked DA's welfare has not, in the most recent check, identified any risk that would require DA to be woken more frequently.

Regulation 7 provides that, the requirements in the cases specified in regs 4 and 5 to take DA to a place of safety do not apply where—

(a) arrangements have been made which would enable an assessment of DA for the purpose of MHA 1983, s 135 or (as the case may be) s 136 to be commenced sooner at the police station than at another place of safety; and

(b) to postpone the assessment would be likely to cause distress to DA.

(2) If it appears to the custody officer, or (s)he is told, that a person brought to the police station under arrest may be suffering from an infectious disease or condition, reasonable steps must be taken to safeguard the health of the detainee and others. Advice must be sought from an appropriate healthcare professional. The person and his/her property may be isolated pending clinical direction.

(3) If a detainee requests a clinical examination, an appropriate healthcare professional must be called as soon as practicable. If a safe and appropriate healthcare plan cannot be provided, the police surgeon's advice must be sought.

(4) If a detainee is required to take or apply any medication in compliance with clinical directions prescribed before his/her detention, the custody officer must consult the appropriate healthcare professional prior to the use of the medication; such consultation and its outcome must be noted in the custody record. The custody officer is responsible for the safekeeping of any medication and for ensuring that the detainee is given the opportunity to take or apply prescribed or approved medication. However, no police officer may administer medicines which are also controlled drugs subject to the Misuse of Drugs Regulations 2001, Schs 1, 2, or 3. A detainee may administer controlled drugs to him/herself only under the personal

supervision of the registered medical practitioner authorising their use or other appropriate healthcare professional. The custody officer may supervise the self-administration or authorise other custody staff to supervise the self-administration of drugs listed in Schs 4 or 5, if (s)he has consulted the appropriate healthcare professional authorising their use and both are satisfied self-administration will not endanger the detainee, police officers, or anyone else.

(5) Where appropriate healthcare professionals administer drugs or other medication, or supervise self-administration, it must be within the scope of their practice as defined by their professional body.

(6) If a detainee has in his/her possession, or claims to need, medication relating to a heart condition, diabetes, epilepsy, or a condition of comparable potential seriousness then, even though (1), above, may not apply, the advice of the appropriate healthcare professional must be obtained.

A record must be made in the custody record of any clinical attention received, and of any request for a clinical examination under (3) and arrangements made in response. A record must also be made of any injury, ailment or condition causing such arrangements to be made, together with clinical directions and advice or clarifications given by a healthcare professional. Where applicable, a record must be made of responses made when attempting to rouse a detainee. Should a healthcare professional not record his/her clinical findings in the custody record, a note must be made in it of where the findings are recorded. Information necessary to ensure ongoing care and well-being of a detainee must be recorded openly in the custody record. A custody record must include all medication in the possession of a detainee on arrival, together with a note of medication which (s)he claims to need.

Whenever an appropriate healthcare professional is consulted, the custody officer must ask for an opinion concerning risks or problems which need to be taken into account when making a decision about detention, when to carry out an interview if applicable and the need for safeguards. Any doubts concerning directions, particularly in relation to the frequency of visits, must be cleared up.

Nothing in the above provisions prevents the police from calling an appropriate healthcare professional to examine a detainee for the purpose of obtaining evidence relating to an offence in which (s)he is suspected of being involved.

Complaints about treatment

6.11   If a complaint is made by or on behalf of a detainee about his/her treatment since arrest, or it comes to the notice of any officer that (s)he may have been treated improperly, a report must be made as soon as practicable to an inspector (or above) who is not connected with the investigation. If the matter concerns a possible assault or the unreasonable use of force, the appropriate healthcare professional must also be called as soon as practicable; a record must be made of any arrangements made.

A record must be made of any complaint reported under the above provisions, together with any relevant remarks by the custody officer.

Independent custody visitors for places of detention

6.12   The Police Reform Act 2002, s 51 requires local policing bodies to make arrangements for detainees to be visited by independent custody visitors. The arrangements may provide rights of access to police stations; examination of records; meetings with detainees; and the inspection of facilities. Access to a detainee may be denied if:

(a) it appears to an inspector (or above) that there are grounds for doing so at the time that it is requested;

(b) those grounds are specified within the arrangements; and

(c) the procedural requirements imposed by the arrangements in relation to a denial of access are complied with.

Local policing bodies must ensure that:

(i) the arrangements require independent custody visitors who visit a suspected terrorist detainee (ie someone detained under TA 2000, s 41 (**21.46**) to submit a report on that visit; and

(ii) a copy of the report is sent to the independent reviewer of terrorism legislation.

In addition, the arrangements may empower independent custody visitors to listen to audio recordings or view video recordings of police interviews with a suspected terrorist detainee, but a visitor may be denied access to these recordings, whether in whole or part, on the same basis as above.

The Secretary of State has issued a code of practice as to the carrying out of functions under the arrangements. The notice of rights and entitlements given to detainees explains that 'visitors' are members of the community who are allowed access to police stations unannounced to ensure that detainees have access to their rights. It explains that a detainee does not have a right to see an independent custody visitor and that (s)he cannot request to see one; that such a visitor acts independently of the police to check that welfare rights are protected; and that the detainee does not have to speak to such a visitor if (s)he does not wish to do so.

Information to be given to detainees

When the custody officer authorises the detention of a person who has not been charged, (s)he is required by PACE, s 37 to make, as soon as practicable, a written entry on the custody record of the grounds which exist for detention. This must be done in the person's presence and (s)he must be informed at that time of those grounds. If (s)he is asleep (s)he must be woken and told. (However, if at that time a person is incapable of understanding what is said to him/her or is violent or likely to be so, or is in urgent need of medical treatment, the information may be given as soon as practicable, and in any case before (s)he is questioned for the offence.) At the same time the detainee must be informed clearly of the rights set out at **5.9** and the other functions referred to there must be carried out.

6.13

Communication with others

See Chapter 5.

6.14

Access to legal advice

See Chapter 5.

6.15

Searching and retention of property

6.16   PACE, s 54 charges the custody officer with a duty to ascertain the property which a
person has with him/her when (s)he is:

(a) brought to a police station after being arrested elsewhere or after being committed
to custody by an order or a sentence of a court; or
(b) arrested at a police station or detained there:
   (i)   under s 37 (**6.47**),
   (ii) when answering police bail,
   (iii) for failing to answer police bail, or
   (iv) as a person treated by s 46ZA (**6.73**) as arrested and charged for an offence for
        which (s)he has been granted live link bail.

A detainee may be searched to the extent considered necessary if the custody officer
considers it necessary to ascertain the property which that person has with him/her,
but an intimate search (**6.21**) may not be carried out for this purpose under s 54. A strip
search (ie one involving the removal of more than outer clothing) is only permissible
if the custody officer reasonably considers that the detainee might have concealed an
article which (s)he would not be allowed to keep under the provision below. A strip
search, however, may only be carried out if the necessary requirements set out at **6.18**
and **6.19** are satisfied.

The custody officer may record all or any of the items which (s)he finds. In the case
of an arrested person, this may be done in the custody record.

Articles other than those subject to legal privilege (eg letters from solicitors, etc) may
be seized and retained. However, clothes and personal effects (which do not include
cash) may only be seized if the custody officer believes that they may be used by the
person:

(a) to cause physical injury to him/herself or another;
(b) to damage property;
(c) to interfere with evidence;
(d) to assist him/her to escape,

or if the custody officer has *reasonable grounds* for believing that they may be evidence
of an offence. The reference above to 'clothes' and 'personal effects' includes menstrual
and any other health, hygiene, and welfare products needed by the detainee (see **6.9**)
and a decision to withhold any such products must be subject to a further specific risk
assessment.

Section 54 also provides that a constable may at any time search a person who is in
custody at a police station, or is in police detention otherwise than at a police station, in
order to ascertain whether (s)he has with him/her any articles which (s)he could use for
any of the purposes in (a)–(d) above. A constable may seize and retain anything found
in such a search, except that clothes and personal effects may only be seized in the same
circumstances as mentioned above.

Any search under s 54 must be by a constable of the same sex (**4.5**) as the detainee.

Where articles are seized under s 54, the person must be told the reason unless (s)
he is violent or likely to be violent or is incapable of understanding. Items which are
seized on the grounds that they may be used to cause injury or damage, interfere with
evidence or assist escape, must be returned when the person is released from police
detention. Code C states that the custody officer is responsible for the safekeeping of
property taken from a person.

Where a female police officer had removed a heavily intoxicated detainee's soiled clothing and provided her with clean clothing, a judge in the Administrative Court held that, although that removal was not related to a search under s 54 and therefore s 54 did not apply, it had been necessary to remove the detainee's clothing for hygiene and health reasons, and a failure to do so would have been a breach of Code C (see **6.7**(5)) and the duty under PACE, s 39 (**6.4**).

Search and examination to ascertain identity

A detainee (D) at a police station may be searched or examined to establish: **6.17**

(a) whether D has any marks, features, or injuries that would tend to identify D as a person involved in the commission of an offence; or
(b) D's identity.

Where the appropriate consent has been withheld, or it is not practicable to obtain it, PACE, s 54A permits an inspector (or above) to authorise the search and/or examination of a detainee to find marks under (a). If D has refused to identify him/herself or the officer has reasonable grounds for suspecting that D is not who (s)he claims to be, an inspector (or above) may authorise a non-consensual search and/or examination to establish identity under (b). Either authorisation may be given orally but must be confirmed in writing as soon as is practicable.

Any identifying mark (including features and injuries) which is found may be photographed with consent, or if such consent is withheld or it is not practicable to obtain it, without consent. 'Photographed' covers the use of any means by which a visual image may be produced.

Such searches, examinations, and taking of photographs may only be conducted or taken by constables of the same sex as D, who may use reasonable force. If there is doubt as to whether D should be treated, or continue to be treated, as being male or female in the case of:

(a) a search carried out or observed by a person of the same sex as D; or
(b) any other procedure which requires action to be taken or information to be given that depends on whether D is to be treated as being male or female,
the gender of D and other parties concerned should be established and recorded in line with **4.5**.

An intimate search may not be carried out under the authority of s 54A. An examination of the mouth is a non-intimate search and therefore may be undertaken by a constable. A dentist has no power to conduct such examination without consent under these provisions.

Photographs taken under s 54A may be used by, or disclosed to, any person for any purpose relating to the prevention or detection of crime, the investigation of an offence, or the conduct of a prosecution. They may be retained after use or disclosure but may not be used or disclosed except for a related purpose.

The references to 'crime' in s 54A include conduct which is an offence (whether under UK law or the law of a country or territory outside the UK) or which is or corresponds to conduct which would be an offence in the UK if it all took place in any part of the UK.

When D is searched, examined or photographed under the above provisions, D must be informed of the:

(a) purpose of the search;

(b) grounds on which the relevant authority, if applicable, has been given; and

(c) purposes for which the photograph may be used, disclosed, or retained.

This information must be given before the search or examination commences or the photograph is taken, except if the photograph is to be taken covertly.

A record must be made when D is searched, examined, or a photograph of D, or any identifying marks found on D, is taken. The record must include the:

(a) identity (subject to the usual exception: **5.3**) of the officer carrying out the search, examination, or taking the photograph;

(b) purpose of the search, examination, or photograph and the outcome;

(c) detainee's consent to the search, examination, or photograph, or the reason the person was searched, examined, or photographed without consent; and

(d) giving of any authority, the grounds for giving it and the authorising officer.

If force is used when searching, examining, or taking a photograph in accordance with s 54A, a record must be made of the circumstances and those present.

SEARCHES

Strip searches

6.18   Atrip search is any search involving the removal of more than outer clothing. Outer clothing includes shoes and socks. Code C states that, for a strip search to take place, the custody officer must believe it necessary to remove an article which the detainee would not be allowed to keep, which there is reasonable suspicion the person might have concealed. Strip searches should not be routinely carried out where there is no reason for suspicion that articles have been concealed.

Conduct

6.19   The following procedures must be observed when strip searches are conducted:

(1) A police officer carrying out a strip search must be the same sex (**4.5**) as the person searched.

(2) The search must take place in an area where the suspect cannot be seen by anyone who does not need to be present, or by a member of the opposite sex (except the appropriate adult who has specifically been requested by the person being searched).

(3) Except in cases of urgency, where there is a risk of serious harm to the person detained or to others, whenever a strip search involves exposure of intimate parts of the body, there must be at least two people present other than the person searched, and if the search is of a juvenile or vulnerable person (**5.4**), one of those must be the appropriate adult. Except in urgent cases, a search of a juvenile may take place in the absence of the appropriate adult only if (s)he signifies in that adult's presence that (s)he prefers to be searched in that adult's absence and that adult agrees. A record must be made of the signification and signed by the appropriate adult. The presence of more than two people, other than the appropriate adult, may be permitted only in the most exceptional circumstances.

(4) The search must be conducted with proper regard to the dignity, sensitivity, and vulnerability of the detainee in these circumstances, including in particular, the detainee's health, hygiene, and welfare needs to which the provision referred to in the anti-penultimate and penultimate paragraphs of **6.9** apply. Every reasonable

effort must be made to secure the detainee's co-operation, maintain their dignity and minimise embarrassment. Detainees who are searched should not normally be required to have all their clothes removed at the same time; for example, a person should be allowed to remove clothing above the waist and re-dress before removing further clothing.

(5) Where necessary to assist the search, the suspect may be required to hold his/her arms in the air or to stand with his/her legs apart and to bend forward so that a visual examination may be made of the genital and anal areas provided that no physical contact is made with any body orifice.

(6) If, during a search, articles are found, the person must be asked to hand them over. If articles are found within any body orifice other than the mouth, and the person refuses to hand them over, their removal would constitute an intimate search which must be carried out in accordance with the provisions of the Code set out below.

(7) A strip search should be conducted as quickly as possible, and the suspect allowed to dress as soon as the procedure is complete.

Documentation

6.20 A note should be made on the custody record of a strip search. It should record the parts of the body searched, who searched, the reason it was considered necessary to undertake it, those present and any result.

Intimate searches

6.21 An intimate search is a search which consists of the physical examination of a person's body orifices other than the mouth, ie the nose, ears, anus, and vagina. The intrusive nature of such searches means the actual and potential risks associated with intimate searches must never be underestimated. Intimate searches are governed by PACE, s 55. Before an intimate search is authorised, every effort must be made to persuade a detainee to hand over the article. Whenever possible, a registered medical practitioner or registered nurse should be asked to make a risk assessment. The authorising officer must consider whether the grounds for believing that an article is concealed are reasonable. In cases of doubt, advice should be sought from a superintendent.

Authorisation

6.22 An intimate search must be authorised by an inspector (or above). To authorise such a search that officer must have reasonable grounds for believing that the detainee (D) may have concealed on him/herself:

(a) an article which D could, and might, use to cause physical injury to him/herself or others at the police station has been concealed; or

(b) a Class A drug which D intended to supply to another or to export;

and that, in either case, an intimate search is the only practicable means of removing it.

The authorisation of an intimate search may be given orally or in writing, but if orally it must be confirmed in writing as soon as practicable.

An *intimate drug offence search* may not be carried out without appropriate written consent (**7.42**). Before D is asked to give such consent D must be warned that if (s)he refuses without good cause the refusal may harm his/her case if it comes to trial. The following form of words may be used to give such a warning:

'You do not have to allow yourself to be searched but I must warn you that if you refuse without good cause, your refusal may harm your case if it comes to trial.'

In appropriate cases the warning may be given in a specified Welsh version. The warning may be given by a police officer or member of police staff. In the case of a juvenile or vulnerable detainee (5.4), the seeking and giving of consent must take place in the presence of the appropriate adult. A juvenile's consent is only valid if his/her parent's or guardian's consent is also obtained, unless the juvenile is under 14 (in which case his/her parent's or guardian's consent is sufficient in its own right). If not legally represented, the detainee must be reminded of his/her entitlement to legal advice, and the reminder must be noted in the custody record. Before the search begins, a constable must inform the detainee of the authorisation and the grounds for giving it and for believing that the item cannot be removed without an intimate search.

Where appropriate consent to a drug offence search is refused without good cause, a court or jury in any proceedings which follow may draw such inferences from that refusal as appear proper.

Execution

6.23   Intimate searches may take place only at:

(a) a police station (but not if it is a drug offence search);
(b) a hospital;
(c) a surgery; or
(d) other medical premises.

The search must be conducted with proper regard to the dignity, sensitivity, and vulnerability of the detainee, including in particular, their health, hygiene, and welfare needs to which the provisions in the ante-penultimate and penultimate paragraphs of **6.9** refer.

Before an intimate search takes place, the reasons why it is considered necessary must be explained to D (the person to be searched) and D must be reminded of his/her entitlement to have legal advice and the reminder must be noted in the custody record.

An intimate drug offence search may only be carried out by a registered medical practitioner or registered nurse. The same is the case for other intimate searches, unless an inspector (or above) considers that it is not practicable and the search takes place because there are reasonable grounds for believing that an article which could cause physical injury to the detainee or others at the police station has been concealed, in which case it must be carried out by a constable of the same sex (**4.5**) and the reason for the impracticability must be recorded. A proposal for a search to be carried out by someone other than a registered medical practitioner or a registered nurse must only be considered as a last resort where the risks involved in retention of the item outweigh those associated with its removal. With the exception in the next paragraph, no one of the opposite sex, other than a doctor or nurse, may be present, nor anyone whose presence is unnecessary, but a minimum of two people, other than the person searched, must be present during the search.

An intimate search at a police station of a juvenile or vulnerable detainee must take place in the presence of the appropriate adult of the same sex (unless the person specifically requests the presence of a particular adult of the opposite sex who is readily available). The search of a juvenile may take place in the absence of the appropriate adult only if the juvenile signifies in the adult's presence that (s)he prefers it to be done in his/her absence and that adult agrees. A record should be made of the signification and signed by the appropriate adult.

Documentation

After an intimate search has been carried out, an entry must be made as soon as prac- **6.24** ticable on the custody record, stating the authorisation to carry out the search; the grounds for the authorisation and for believing that the article could not be removed without an intimate search; which parts of the detainee's body were searched; who carried out the search; who was present; and the result. Where the intimate search is a drug offence search, a record must be made of the necessary warning and the fact that appropriate consent was given or (as the case may be) refused, and if refused, the reason given for the refusal.

The powers of seizure in respect of articles found are the same as those which apply to other searches. If an intimate search is carried out by a police officer, the reason why it was impracticable for a registered medical practitioner or a registered nurse to conduct it must be recorded.

X-rays and ultrasound scans

PACE, s 55A provides that where an inspector (or above) has reasonable grounds for **6.25** believing that a person arrested for an offence and in police detention may have swallowed a Class A drug (**20.1**), and was in possession of it with the appropriate criminal intent before the arrest, he may authorise an X-ray or an ultrasound scan (or both) to be taken. An X-ray or ultrasound scan must not take place without the appropriate consent in writing. The detainee (D) must be told beforehand by a police officer of the authorisation and its grounds. Before being asked for his/her consent, D must be warned that an unjustified refusal may harm his/her case if it comes to trial. This warning may be given by a police officer or member of police staff. In the case of a juvenile or vulnerable detainee (**5.4**), the notification of the authorisation and its grounds, and the seeking and giving of consent must take place in the presence of the appropriate adult. A *juvenile's* consent is only valid if his/her parent's or guardian's consent is also obtained, unless the juvenile is under 14 (in which case his/her parent's or guardian's consent is sufficient in its own right). If D is not legally represented, D must be reminded of his/her entitlement to have free legal advice, and the reminder noted in the custody record. The provisions in relation to the inferences which may be drawn where consent to an intimate drug offence search is refused also apply to refusals to permit X-rays and ultrasound scans to be carried out.

An X-ray or ultrasound scan may only be carried out by a suitably qualified person at a hospital, a doctor's surgery, or at some place used for medical purposes.

If authority is given for an X-ray or an ultrasound scan (or both), consideration should be given to asking a registered medical practitioner or registered nurse to explain to the detainee what is involved and to allay any personal concerns about the effect which such examinations may have upon him/her. If appropriate consent is not given, evidence of the explanation may, if the case comes to trial, be relevant to determining whether the detainee had good cause for refusing.

A record must be made as soon as practicable in D's custody record of the authorisation, its grounds, the warning, the giving (or otherwise) of the appropriate consent, and of the details of the X-ray or scan (if carried out).

TESTING FOR PRESENCE OF CLASS A CONTROLLED DRUGS

When permitted

6.26   PACE, s 63B permits the taking of urine or non-intimate samples from someone for the purpose of ascertaining whether (s)he has any specified Class A drug (ie cocaine and diamorphine (heroin), their salts, and any preparation containing either drug or its salts) in his/her body where:

(a) either the arrest condition or the charge condition is met;

(b) both the age condition and the request condition are met; and

(c) the notification condition is met in relation to the arrest condition, the charge condition, or the age condition (as the case may be).

The *arrest condition* is that the person concerned (PC) has been arrested for an offence but has not been charged with that offence and either:

(a) that offence is a 'trigger offence' (theft, attempted theft, robbery, attempted robbery, burglary, attempted burglary, aggravated burglary, taking a motor vehicle or other conveyance without authority, aggravated vehicle-taking, fraud, possessing articles used in frauds and making or supplying articles used in frauds, attempted fraud, handling stolen goods or attempted handling, going equipped for stealing, etc, and an offence under the Misuse of Drugs Act 1971 (if committed in respect of a specified Class A drug) of producing or supplying a controlled drug, possessing a controlled drug, or possessing a controlled drug with intent to supply, and begging or persistent begging contrary to the Vagrancy Act 1824, s 3 or 4 respectively); or

(b) an inspector (or above), who has reasonable grounds for suspecting that the misuse by PC of any specified Class A drug caused or contributed to the offence, has authorised such a sample to be taken.

The *charge condition* is either:

(a) that PC has been charged with a trigger offence; or

(b) that PC has been charged with an offence and an inspector (or above), who has reasonable grounds for suspecting that the misuse by PC of any specified Class A drug caused or contributed to the offence, has authorised the sample to be taken.

The *age condition* is:

(a) if the arrest condition is met, that PC is 18 or over; or

(b) if the charge condition is met, that PC is 14 or over.

The *request condition* is that a police officer has requested PC to give the sample (which in the case of a juvenile must be done in the presence of the appropriate adult). The *notification condition* is that:

(a) the relevant chief officer has been notified by the Secretary of State that appropriate arrangements have been made for the police area as a whole, or for the particular police station, in which the person is in police detention; and

(b) the notice has not been withdrawn.

Where a sample is taken from a person who satisfies the arrest condition, no other sample may be taken during the same continuous period of detention, but if the charge condition is met during that period, the sample already taken must be treated as one taken after charge. This must be recorded in the custody record.

In circumstances in which a person (P) is arrested for a first offence which satisfies the arrest condition but not the charge condition, and P would normally be liable to be released from custody before a sample is taken but P remains in custody by reason of arrest for another offence not falling within the arrest condition, a sample may be taken before the end of a period of 24 hours following P's *initial* arrest.

A sample must not be taken from a person in custody unless (s)he is brought before the custody officer. It may only be taken by a person authorised by the Police and Criminal Evidence Act 1984 (Drug Testing of Persons in Police Detention) Prescribed Persons Regulations 2001, viz a police officer, a person employed by a local policing body, or chief officer of police, or by a contractor engaged by such a body or chief officer, whose duties include taking samples for testing for the presence of specified Class A drugs. A sample may be taken from a juvenile only in the presence of the appropriate adult.

Disclosure of information obtained

Information obtained from a sample taken under s 63B may be disclosed: **6.27**

(a) for the purpose of informing any decision about granting bail in criminal proceedings to the person concerned, or about the giving of a conditional caution or a youth conditional caution (YCC);

(b) where the person concerned is in police detention or is remanded in or committed to custody by an order of a court or has been granted such bail, for the purpose of informing any decision about his/her supervision;

(c) where the person concerned is convicted of an offence, for the purpose of informing any decision about the appropriate sentence to be passed by a court and any decision about his/her supervision or release;

(d) for the purpose of:
   (i) an initial or follow-up assessment which the person concerned is required to attend;
   (ii) proceedings against the person concerned for an offence of failing to attend such an assessment; or
   (iii) ensuring that appropriate advice and treatment is made available to the person concerned.

Procedure

A request for a sample must be preceded by an explanation (a) of its *purpose*; (b) that   **6.28** failure without good cause to provide a sample may make the person requested liable to prosecution; and (c) where the taking of the sample has been authorised by an inspector (or above) under (b) of the arrest condition or of the charge condition (**6.26**), of the grounds for the authorisation. In addition, the person requested must be reminded of the right to have someone informed of his/her arrest; of the right to consult privately with a solicitor; of the availability free of charge of independent legal advice; and of the right to consult the relevant code of practice. When warning a person who is asked to provide a urine or non-intimate sample in these circumstances, the following form of words may be used (an alternative warning in Welsh is provided):

'You do not have to provide a sample, but I must warn you that if you fail or refuse without good cause to do so, you will commit an offence for which you may be imprisoned, or fined, or both.'

Custody officers may authorise continued detention for up to six hours from the time of charge to enable a sample to be taken.

Where the warning and information are given to a juvenile, this must be done in the presence of the appropriate adult, as must the taking of the sample.

Where a sample is taken following authorisation by an inspector (or above), the authorisation and the grounds for suspicion must be recorded in the custody record. An authorisation given by an inspector may be given orally but must subsequently be confirmed in writing as soon as practicable. Details of authorisations and the giving of warnings must be recorded, together with the time of charge and the time at which a sample was given.

A person who fails without good cause to give any sample which may be taken from him/her is guilty of an offence under s 63B(8).

Assessment of misuse of drugs after positive drug test

Initial assessment

6.29   Where an analysis of a sample reveals the presence of a Class A drug a police officer may, at any time before the person's release, require a person of 18 or over to attend an 'initial assessment' and to remain for its duration. A qualified 'initial assessor' will seek to establish dependency or a tendency to misuse any specified Class A drug and whether the person will benefit from further assessment or advice. If an initial assessor finds that a follow-up assessment is appropriate, (s)he must inform the person of the time and place at which it is to take place, confirm this in writing, and warn the person of the consequence of failing to attend.

Follow-up assessment

6.30   The police officer may, at the same time as (s)he imposes a requirement on the detainee to attend an initial assessment, require the person to attend a follow-up assessment and to remain for its duration. The officer must inform the detainee that the second requirement will cease to have effect, if at the initial assessment (s)he is informed that a follow-up assessment is not necessary. A follow-up interview will be concerned with a 'care plan'.

Attendance at assessments

6.31   A constable must inform the detainee (to be confirmed in writing) that failure to attend either or both forms of assessment and to remain there for its duration without good cause, may result in prosecution. This must be done prior to release from custody and a record must be made in the custody record.

The Drugs Act 2005, s 12(2) provides that a person who is required to attend an initial assessment commits an offence if he fails, without good cause, to attend at the specified time and place, or attends but fails to remain for the duration of the assessment. Where such a failure occurs, any requirement imposed in relation to a follow-up assessment ceases to have effect. It is an offence against s 14(3) to fail, without good cause, to attend a follow-up assessment or to fail to attend for its duration. These offences are not committed if a subsequent analysis of a sample reveals that a specified Class A drug was not present in the person's body. This is because any requirement imposed ceases to have effect.

Any sample taken may not be used for any purpose other than to ascertain whether the person has a Class A drug in his/her body. It can be disposed of as clinical waste unless it is to be sent for further analysis in cases where the test result is disputed at the point when the result is known, eg on the basis that medication has been taken, or for quality assurance purposes.

REVIEWS AND MAXIMUM PERIODS OF POLICE DETENTION

The following rules do not apply where a person has been arrested as a suspected terrorist under the Terrorism Act 2000 (TA 2000, s 41. In such a case TA 2000, Sch 8 (21.53) applies.   **6.32**

Reviews

PACE, s 40 requires periodic reviews of the detention of each person in police deten-   **6.33** tion. The review will be carried out:

(a) in the case of a person who has been arrested *and charged*, by the custody officer; and

(b) in the case of a person who has been arrested *but not charged*, by an inspector (or above) who has not been directly involved in the investigation.

The officer by whom the review is carried out is called the 'review officer'.

There will be some designated police stations at which the custody officer will be an officer of the rank of inspector. Where this is so (s)he could carry out both of the above functions.

PACE, s 40 is precise in relation to when these reviews must be carried out:

(a) the first review must be not later than *six hours* after the detention was first authorised;

(b) the second review must be not later than *nine hours* after the first; and

(c) subsequent reviews must be at intervals of *not more than nine hours*.

There is, however, a limited power to postpone a review (**6.37**).

Failure to carry out a review in accordance with these rules renders the previously lawful detention unlawful and entitles the detainee to damages for false imprisonment.

The review officer is responsible under s 40 for determining whether or not a person's detention continues to be necessary. This requirement continues throughout the detention period and, except when a telephone or a live link is used, the review officer must be present at the police station holding the detainee.

The case of a person who has been arrested but not charged

Here the review officer must proceed as follows: **6.34**

(1) Where the person (P) was detained because P was not in a fit state to be charged or released without charge (with or without bail), eg because P was under the influence of drink or drugs, the review officer must consider whether P is now in a fit state to be charged or released. If P is, one or other of these courses must be adopted. If P is not, further detention may be authorised, but consideration should be given to whether there is sufficient evidence to charge P with an offence.

(2) Where, although there is then insufficient evidence to charge P, the detention of P has been authorised by the custody officer on the basis that there are reasonable grounds to believe that P's detention without charge is necessary to secure or preserve evidence relating to an offence for which P is under arrest or to obtain such evidence by questioning P, the review officer may authorise further detention if this is necessary on the same basis.

The review of the detention of a person who has been arrested but not charged may be conducted in person or (under s 40A) remotely by live link or telephone discussion with persons at the police station where the arrested person is detained. The use of a telephone is not permitted if facilities for live review exist and it is reasonably practicable to carry out the review by live link (s 40A(2)). The references to 'live link' are to an arrangement by means of which the review officer is able to see and hear, and to be seen and heard by, the detainee concerned and the detainee's solicitor, appropriate adult, and interpreter (as applicable).

Subject to this, the decision on which form of communication is to be used is a matter for the review officer, who must take full account of the detainee's needs. The benefits of carrying out a review in person should always be considered, based on the particular circumstances with specific additional consideration if the person is:

(a) a juvenile (in which case his/her age should also be considered);
(b) a vulnerable person (**5.4**);
(c) in need of medical attention for other than routine minor ailments; or
(d) subject to presentational or community issues around his/her detention.

Where a review is conducted under s 40A, the review officer must require another officer at the station to carry out the review officer's functions under s 40 and Code C by making any record connected with the review in the detainee's custody record in the presence of the detainee (if applicable) and giving the detainee information about the review.

A live link or telephone review may be terminated by the review officer at any stage in favour of a review in person and a record should be made of such a decision.

Before deciding whether to authorise continued detention the review officer must give an opportunity to make representations about the detention to: the detainee, unless (s)he is asleep; the detainee's solicitor, if available; and the appropriate adult. Other people with an interest in the detainee's welfare may also make representations at the authorising officer's discretion.

The case of a person who has been charged

6.35   The person must be released, either on bail or without bail, unless:

(a) his/her name and address cannot be ascertained or there are reasonable grounds to doubt the name and address given; or
(b) in the case of an imprisonable offence, there are reasonable grounds to believe that his/her detention is necessary to prevent him/her committing an offence; or
(c) in the case of a non-imprisonable offence, there are reasonable grounds to believe that his/her detention is necessary to prevent him/her causing physical injury to any person or loss of or damage to property; or
(d) in the case where a drug sample may be taken (**6.26**), there are reasonable grounds to believe that his/her detention is necessary for a sample to be taken; or

(e) there are reasonable grounds to believe that (s)he will fail to answer bail or that his/her detention is necessary to prevent him/her interfering with the administration of justice or with police investigations or for his/her own protection; or

(f) in the case of an arrested juvenile, his/her detention is necessary in his/her own interests.

Before deciding whether to authorise continued detention the review officer must give an opportunity to make representations about the detention to: the detainee (unless the detainee is asleep); the detainee's solicitor if available at the time; and the appropriate adult if available at the time.

Other people having an interest in the detainee's welfare may also make representations at the authorising officer's discretion.

General

Representations may be made orally or in writing (including email). The authorising officer, however, may refuse to hear oral representations from the detainee (D) if the officer considers D unfit to make representations because of D's condition or behaviour.   **6.36**

Any comment D may make if the decision is to keep D in detention must be recorded and, if applicable, the review officer must be informed of the comment as soon as practicable. No officer may put specific questions to D regarding D's involvement in any offence, nor in respect of any comments D may make when given the chance to make representations or in response to the decision to keep him/her in detention. Such an exchange is likely to constitute an 'interview' and would require the safeguard of the Code's provisions concerning interviews generally.

Before conducting a review the review officer must ensure D is reminded of his/her entitlement to free legal advice, unless D is asleep. This reminder must be noted in the custody record.

The review officer must determine what documents and materials are essential (5.9: p 126) to challenging effectively the lawfulness of the detention, and these must be made available to D.

D need not be woken for the review. If D is likely to be asleep at the latest time when a review or extension authorisation may take place, the review officer should bring it forward, if possible, so that D may make representations without being woken up.

If, after considering any representations, the review officer authorises the detention of a person (whether charged or not) to continue, the review officer must make a written record of the grounds for the detention as soon as practicable. A detainee who is asleep at a review and whose continued detention is authorised must be informed about the decision and reason as soon as practicable after waking.

Where a person is in police custody in circumstances which are not subject to statutory review, for example:

(a) someone arrested on warrant for failure to answer to bail or for breach of a condition of bail;

(b) someone in police custody under the Crime (Sentences) Act 1997 for a specific purpose and period;

(c) a convicted or remanded prisoner held on behalf of the Prison Service;

(d) someone detained:

   (i) to prevent him/her causing a breach of the peace;

   (ii) on behalf of Immigration Enforcement; or

(iii) under CJA 1988, s 152 by order of a magistrates' court to facilitate the recovery of evidence where that person has been charged with drug possession or trafficking and is suspected of having swallowed drugs,

it is advised that reviews take place periodically to ensure that the power to detain still applies and that detention conditions are being complied with. Such reviews may be conducted by a sergeant.

Postponement of review

6.37   PACE, s 40 allows for the postponement of a review:

(a) if, having regard to all the circumstances at the latest time for the review, it is impracticable to carry out the review at that time; or

(b) if at that time the detainee (D) is being questioned by a police officer and the review officer is satisfied that an interruption of the questioning for the purpose of carrying out the review would prejudice the particular investigation; or

(c) if at that time no review officer is readily available.

If a review is postponed it must be carried out as soon as practicable after the normal latest time for it. The reason for any postponement must be recorded in the custody record. The postponement of a review does not affect the time at which any subsequent review must be carried out. Thus, a second review must be carried out nine hours after the latest time at which the first review should have taken place, which time is six hours after the detention was first authorised.

Where D (the person whose detention is under review) has not been charged before the time of the review, the first three paragraphs under the heading 'Duties of custody officer before charge' at **6.47** apply with the substitution of 'D (a person whose detention is under review)' for 'D (the person arrested)', and of 'review officer' for 'custody officer', and with the addition of 'asleep' in the list referred to at s 37(6).

Documentation

6.38   A record must be made as soon as practicable of the outcome of each review: if a detainee was asleep when continued detention was authorised, a record must be made of when (s)he was informed and by whom. The grounds for, and the extent of, any delay in conducting a review must be recorded and reasons for there being a live link, etc review must also be recorded, together with the place where the review officer was, and the method by which representations were made.

Limits on period of detention without charge

6.39   PACE, s 41 provides that a person must not be kept in police detention for more than 24 hours without being charged, except that detention beyond that period may be authorised in certain circumstances by a superintendent (or above) (s 42) or by a magistrates' court (ss 43 and 44). The maximum period of detention without charge is 96 hours, but is only 24 hours where an offence is not indictable.

Calculation of period of detention

6.40   PACE, s 41 refers to the 'relevant time', which is the time from which the detention of a particular person is to be calculated. This may be:

(a) in the case where a person, whose arrest is sought in one police area in England and Wales, is arrested in another area, and is not questioned in the area in which (s)he is arrested about the offence for which (s)he has been arrested, *the time at which the person arrives at the first police station in the area in which his/her arrest is sought, or the time 24 hours after that person's arrest, whichever is the earlier;*

(b) in the case of a person arrested outside England and Wales, *the time at which that person arrives at the first police station to which (s)he is taken in the police area in England and Wales in which the offence for which (s)he is arrested is being investigated, or the time 24 hours after the time of that person's entry into England and Wales, whichever is the earlier;*

(c) in the case of a person who attends voluntarily at a police station, or who accompanies a constable to a police station without having been arrested, and is arrested at the police station, *the time of his/her arrest;*

(d) in the case of a person who attends a police station to answer bail under s 30A (**4.57**), *the time when (s)he arrives at a police station*; and

(e) in any other case, it is *the time at which the person arrested arrives at the first police station to which (s)he is taken after arrest, unless (s)he is in detention in an area* in England and Wales and his/her arrest for an offence is being sought *in some other police area* in England and Wales and (s)he is taken to that second area for the purpose of investigating that offence, without being questioned in the first area in order to obtain evidence in relation to it. In such a case the 'relevant time' will be the time 24 hours after (s)he leaves the place where (s)he is detained in the first area or the time at which (s)he arrives at the first police station to which (s)he is taken in the second area, whichever is the earlier.

Section 41 provides a safeguard in relation to a person in police detention who, whilst detained, is arrested for a second offence. The time does not start running again with his/her arrest for the second offence. It also provides for instances in which a person in police detention is removed to a hospital for medical treatment. Normally the period commencing with his/her journey to hospital and ending with his/her arrival back in police custody does not count towards his/her 24 hours in police custody; in effect, the clock may be stopped. However, any period, either during his/her journey or whilst in hospital, during which (s)he is questioned by a police officer for the purpose of obtaining evidence in relation to an offence, is included in his/her period of police detention.

Release from detention after 24 hours

6.41 PACE, s 41(7) provides that, subject to the following paragraph, if a person has not been charged after 24 hours in police detention he *must* be released without bail unless the pre-conditions for bail (**6.6**) are satisfied, or on bail if those pre-conditions are satisfied. A person so released must not be re-arrested without a warrant for the same offence unless, since his/her release, new evidence has come to light or an examination or analysis of existing evidence has been made which could not reasonably have been made before.

Release after 24 hours is not required if continued detention is authorised or a warrant of further detention is issued under the powers next discussed.

Authorisation of continued detention

6.42 The relevant provisions are set out by PACE, s 42. Under s 42(1), where a superintendent (or above) who is responsible for a police station at which a person (D) is detained has reasonable grounds for believing that:

(a) *the detention of D without charge is necessary to secure or preserve evidence relating to an offence for which D is under arrest or to obtain such evidence by questioning D;*

(b) *an offence for which D is under arrest is an indictable offence; and*

(c) *the investigation is being conducted diligently and expeditiously,*

(s)he may authorise the keeping of D in police detention for a period expiring at or before 36 hours after the relevant time.

This may not be done if D has been in detention for more than 24 hours from the relevant time when the authorisation of continued detention is sought, nor may such an authorisation be given before the second review of D's detention, that is the review at 15 hours after detention was first authorised. Where an officer authorises the keeping of D in police detention under s 42(1), (s)he must inform D of the grounds for D's continued detention, and record the grounds in D's custody record.

By s 42(2), if the first period of continued detention given does not take the detention time fully to 36 hours, a further period may be authorised by a superintendent (or above) if the conditions in the previous paragraph are satisfied, up to the maximum of 36 hours. This further period may be authorised at any time during the first extension, and even though more than 24 hours has elapsed from the relevant time.

By s 45ZA, the functions of a superintendent (or above) under s 42(1) or (2) may be performed, where D is at a police station, by such an officer who is not present at the police station where D is being held but who has access to the use of a live link (as defined at **6.34**, substituting 'superintendent (or above)' for 'review officer') may, using that live link, give authority to extend the maximum period of detention permitted before charge, if, and only if:

(a) a custody officer considers that the use of the live link is appropriate;

(b) D has requested and had advice from a solicitor on the use of the live link; and

(c) D has consented to the live link being used.

The reference to D's consent means:

— where D is 18 or over, the consent of D;

— where D is 14 to 17, the consent of D and of D's parent or guardian;

— where D is under 14, the consent of D's parent or guardian.

The consent of a person aged 14 to 17, or who is a vulnerable adult, may only be given in the presence of an appropriate adult. 'Appropriate adult' is defined in a similar way as under the PACE Codes (see **5.7**).

D's consent is only valid if:

(i) where D is 18 or over and is a vulnerable adult (defined for this purpose as an adult who, because of a mental disorder or for any other reason (eg drunkenness), may have difficulty understanding the purpose of an authorisation under s 42 or anything that occurs in connection with a decision whether to give it), information about how the live link is used and the reminder about D's right to legal advice, and D's consent, is given in the presence of the appropriate adult; and

(ii) in the case of a juvenile D—

(a) if information about how the live link is used and the reminder about D's right to legal advice is given in the presence of the appropriate adult; and

(b) if the juvenile is 14 or over, his/her consent is given in the presence of the appropriate adult.

When a live link is used:

(a) the authorising superintendent (or above) must, with regard to any requisite record connected with the authorisation, require an officer at the station holding D to make that record in D's custody record;

(b) the requirement to allow opportunity to make representations will be satisfied:

   (i) if facilities exist for the immediate transmission of written representations to the authorising officer, e.g. fax or email message, by allowing those who are given the opportunity to make representations, to make their representations in writing by means of those facilities or orally by means of the live link; or

   (ii) in all other cases, by allowing those who are given the opportunity to make representations, to make their representations orally by means of the live link.

The authorising officer can decide at any stage to terminate the live link and attend the police station where D is held to carry out the procedure in person. The reasons for doing so should be noted in the custody record.

D must be informed on all occasions of the grounds for his/her continued detention and the custody record must be endorsed with them.

Before deciding whether to authorise continued detention, the officer responsible must give an opportunity to make representations about the detention to D (unless asleep), or D's solicitor if available, and the appropriate adult if available; 'available' means available to attend in person or contactable by telephone or other electronic means at the time of the review. Representations may be made orally in person or by telephone or in writing. However, where the superintendent's functions are performed by live link under the above live link provisions, this provision is to be read as if it required the superintendent to give the persons mentioned an opportunity to make representations:

(a) if facilities exist for the immediate transmission of written representations to the officer, either in writing by means of those facilities or orally by means of the live link; or

(b) in any other case, orally by means of the live link.

During consideration of authorisations of continued detention, specific questions must not be put to D regarding D's involvement in an offence or in respect of any comments which D makes. Such an exchange could be considered to be an interview. Before conducting a review the officer responsible must ensure that D is reminded of his/her entitlement to free legal advice. This reminder must be noted in the custody record.

If, when an extension of detention is authorised, D has not yet exercised his/her right to have some person informed of his/her detention, or right of access to legal advice, the custody officer must remind D of these rights, decide whether D must be permitted to exercise them, and record his/her decision in the custody record. However, the requirement to inform D of his/her right of access to legal advice is disapplied where the superintendent's functions are performed by live link under the above live link provisions.

A detainee (D) who is asleep at a review and whose continued detention is authorised must be informed about the decision and reason as soon as practicable after waking.

Any comment made by D about an authorisation of continued detention must be recorded and, if applicable, the authorising superintendent must be informed of the comment as soon as practicable.

Detaining a juvenile or mentally vulnerable person for longer than 24 hours is dependent upon the circumstances of the case and with regard to the person's special vulnerability, the legal obligation to provide an opportunity for legal representations to be made prior to a decision about extending detention, the need to consult and consider the views of the appropriate adult, and any alternatives to police custody.

Under s 42(10), where D has been the subject of continued detention, D must be released not later than 36 hours after the relevant time, without bail unless the pre-conditions for bail (**6.6**) are satisfied, or on bail if those pre-conditions are satisfied. This does not apply if D has been charged with an offence, or D's continued detention is authorised or otherwise permitted by a warrant of further detention (below).

Warrant of further detention

6.43   Such a warrant is governed by PACE, s 43. A magistrates' court may issue a warrant of further detention following an application on oath by a constable which is supported by a written information. The court must be satisfied that there are reasonable grounds for believing that further detention is justified. A person's further detention is only justified for the purpose of s 43 (or s 44 (extension of warrant of further detentions below)) if the italicised conditions set out at **6.42** in relation to s 42 are satisfied. A magistrates' court may give a live link direction for the purpose of the hearing of an application under s 43 (or s 44) if, and only if:

(a)   a custody officer considers that the use of a live link for that purpose is appropriate;
(b)   the detainee (D) to whom the application relates has requested and received legal advice on the use of the live link;
(c)   D's consent (see **6.42** for the relevant provisions as to such consent, and its validity) to the use of the live link has been given; and
(d)   it is not contrary to the interests of justice to give the direction.

Where a live link direction is given, the requirement for D to be brought before the court for the hearing does not apply. For these purposes, 'live link' means an arrangement by means of which D, when not present in the court where the hearing is being held, is able to see and hear, and to be seen and heard by, the court during the hearing.

An application for a warrant of further detention may be made at any time before the expiry of 36 hours after the 'relevant time', or, if it is not practicable for the magistrates' court to sit at that time but it will sit during the six hours following that period, at any time before the expiry of those six hours. In the latter case, D may be kept in police detention until the application is heard, and the custody officer must record in the custody record the fact that (and the reason why) D was detained for more than 36 hours after the relevant time. It is not sufficient for the police to have the application on the court lists for hearing within the sitting of the court. The application must be brought to the notice of the court before the end of the relevant period.

If an application for a warrant of further detention is made after the expiry of 36 hours after the relevant time, and it would have been reasonable for the police to make it before the expiry of that period, the court must dismiss the application. Where on an application for a warrant of further detention a magistrates' court is not satisfied that there are reasonable grounds for believing that the further detention of the person to whom the application relates is justified, it must:

(a)   refuse the application; or
(b)   adjourn the hearing of it until not later than 36 hours after the relevant time (in which case D may be kept in police detention during the adjournment).

Where an application is refused, D must forthwith be charged or released without bail unless the pre-conditions for bail (**6.6**) are satisfied, or on bail if those pre-conditions are satisfied, but D need not be released before the expiry of:

(a) 24 hours after the relevant time; or

(b) any longer period for which D's continued detention is or has been authorised under s 42 (**6.42**).

Where an application is refused, no further application may be made for a warrant of further detention in respect of D (the person in question), unless supported by evidence which has come to light since the refusal.

A warrant of further detention must state the time of issue and the period of detention which it authorises, which must not be longer than 36 hours.

A 'magistrates' court' in the present context (and in that of s 44) is a court consisting of two or more justices, sitting otherwise than in open court for the purpose of these provisions.

At the expiry of the warrant of further detention, D must, unless the warrant is extended under s 44 (below), be charged or released from police detention without bail, unless the pre-conditions for bail are satisfied, or on bail if those pre-conditions are satisfied (s 43(18)).

Extension of warrant of further detention

This is dealt with by PACE, s 44. A magistrates' court may extend a warrant of further **6.44** detention, on an application made in the same way as under s 43, if satisfied that there are reasonable grounds for believing that further detention is justified. The extension may not exceed 36 hours or end later than 96 hours after the relevant time. If an extension ends earlier than 96 hours after that time, it may be further extended, provided that the extension ends no later than 96 hours after that time.

The provisions under s 43 ('Warrant of further detention') apply equally to an application under s 44.

Where an application for an extension is refused, the person to whom the application relates (D) must forthwith be charged or released without bail unless the pre-conditions for bail are satisfied, or on bail if those pre-conditions are satisfied. However, D need not be released before the expiry of any period for which a warrant of further detention has previously been extended or further extended.

General

When an application is made under PACE, s 43 or 44, D: **6.45**

(a) must, unless the court has made a live link direction, be brought to court for the hearing of the application;

(b) is entitled to be legally represented if D wishes, in which case the provisions about delay at **5.21** cannot apply; and

(c) must be given a copy of the information which supports the application and states:
  (i) the nature of the offence for which D has been arrested;
  (ii) the general nature of the evidence on which D was arrested;
  (iii) what inquiries about the offence have been made and what further inquiries are proposed;
  (iv) the reasons for believing continued detention is necessary for the purposes of the further inquiries.

A record must be made of the outcome of any determination whether to extend the maximum detention period without charge or an application for a warrant of further detention or its extension. Where an authorisation of continued detention has been given, the record must show the length of time by which the detention was extended or further extended. The same applies where a warrant (or extension) of further detention is granted.

Sections 41, 42, 43, and 44 each provide that, where D is released under their terms, and a custody officer determines that:

(a) there is not sufficient evidence to charge D with an offence; or
(b) there is sufficient evidence to charge D with an offence but D should not be charged with an offence or given a caution in respect of an offence,

the custody officer must give D written notice that D is not to be prosecuted.

This will not prevent the prosecution of D for an offence if new evidence comes to light after the notice was given.

Where a person has been released under s 41(7), 42(10), or 43(18) (**6.41–6.43**), (s) he may not be re-arrested without warrant for the same offence unless, since his/her re-lease, new evidence has come to light or an examination or analysis of existing evidence has been made which could not reasonably have been made before his/her release, but this does not apply to where (s)he fails to surrender to police bail.

Provision of essential documents

6.46 The officer reviewing the need for detention without charge (PACE, s 40), or (as the case may be) the officer considering the need to extend detention without charge from 24 to 36 hours (s 42), is responsible, in consultation with the investigating officer, for deciding which documents and materials are essential (**5.9**: p 1262) and must be made available to the detainee or his/her solicitor. The same applies when an application is made under s 43 or 44 for a warrant of further detention or an extension thereof.

DUTIES OF CUSTODY OFFICER BEFORE CHARGE

6.47 PACE, s 37 states that, where D (the person arrested) has been arrested without war-rant, or on a warrant which is not endorsed for bail, the custody officer must determine whether (s)he has sufficient evidence to charge D with the offence for which D was arrested. D may be detained at a police station for such period as is necessary for the custody officer to make that decision. The custody officer must record on the custody record the offence(s) for which D has been arrested (and the reasons). The custody of-ficer should also note on the custody record any comment D may make in relation to the arresting officer's account but should not invite comment. Section 37(2) provides that, if there is insufficient evidence of an offence, the custody officer must release D:

(a) without bail unless the pre-conditions for bail (**6.6**) are satisfied, or
(b) on bail if those pre-conditions are satisfied,

subject to s 37(3).

By s 37(3), if the custody officer has reasonable grounds for believing that D's deten-tion without being charged is necessary to secure or preserve evidence relating to an offence for which D is under arrest or to obtain such evidence by questioning D, (s)he may authorise D to be kept in police detention.

If the custody officer authorises the detention of an arrested person (D), (s)he must, as soon as practicable, record the grounds in the custody record (s 37(4)). This written record must be made in D's presence and at that time D must be informed by the custody officer of the grounds for detention (s 37(5)), unless D is incapable of understanding what is said to him/her, violent or likely to become violent or in urgent need of medical attention (s 37(6)). In any event, the custody officer must inform D of the grounds as soon as practicable and in any case before D is then questioned about any offence. The custody officer must note on the custody record any comment D may make in respect of the decision to detain him/her but may not invite comment, nor may (s)he put specific questions to D concerning his/her involvement in any offence, nor in respect of any comments D may make in response to the investigating officer's account or the decision to place D in detention. Such an exchange is likely to constitute an interview and would require the necessary safeguards.

Where:

(a) D is released under s 37 (2), and
(b) the custody officer determines that there is—
   (*i*) *insufficient evidence to charge D with an offence, or*
   (*ii*) *sufficient evidence to charge D with an offence but D should not be charged with an offence or given a caution in respect of an offence,*

the custody officer must give D notice in writing that D is not to be prosecuted (s 37(6A), (6B)). This does not prevent the prosecution of D for an offence if new evidence comes to light after the notice was given.

Section 37(7) provides that, if the custody officer determines that (s)he has sufficient evidence to charge D for the offence for which D was arrested, that person must be:

(a) released without charge and on bail, or kept in police detention, for the purpose of enabling a Crown Prosecutor to make a decision about charging or cautioning him/her;
(b) released without charge and without bail unless the pre-conditions for bail are satisfied;
(c) released without charge and on bail if those pre-conditions are satisfied but not for the purpose mentioned in (a); or
(d) charged.

By way of exception, if D is not in a fit state to be dealt with in this way (eg because D is under the influence of drink or drugs), D may be kept in police detention until in a fit state.

By s 37(7A), the decision as to how D is to be dealt with under s 37(7) is that of the custody officer, but in making that decision the custody officer must have regard to the DPP's *Guidance on Charging* below.

In relation to s 37(7)(a), it should be noted that Crown Prosecutors are often located at police stations to enable charge decisions to be taken relatively quickly. CPS Direct operates an out-of-hours service linked to the police by a high speed IT and telephony system.

If D is dealt with under s 37(7)(a), the custody officer must inform D that (s)he is being released, or as the case may be, detained, to enable a Crown Prosecutor to make a decision about charging or cautioning him/her. In such a case, an officer involved in the investigation must send to the CPS such information as is specified in the DPP's *Guidance on Charging*. A Crown Prosecutor must decide whether there is sufficient evidence to charge the person; must then decide whether (s)he should be charged or given a simple caution, conditional caution, youth caution or youth conditional caution; and, if so, state the offence. The Crown Prosecutor must notify the officer of the decision, at which stage the person must be charged (which constitutes the institution of criminal proceedings by the DPP), cautioned, or told that (s)he has been released from bail. If a person is arrested for a breach of bail granted under these circumstances (s)he must be charged, or released without charge, either on bail or without bail. The custody officer may subsequently appoint a different time, or an additional time, at which the person must attend a police station. A person who surrenders to such bail may be kept in police detention to enable him/her to be dealt with.

Where D is released under s 37(7)(b) or (c), and:

(i) a decision has not been taken about whether D should be prosecuted for the offence for which D was arrested, the custody officer must so inform D;

(ii) if the custody officer makes a determination as italicised above, (s)he must give D notice in writing that D is not to be prosecuted (s 37(8) and (8ZA)).

A person released without charge and without bail may be dealt with by one of the available alternatives to prosecution (**3.2**).

DPP's Guidance on Charging

6.48   Under the DPP's *Guidance on Charging*, the situation is as follows.

A custody officer may charge:

(a) any *summary only* offence (and criminal damage where the value of the loss or damage is less than £5,000) irrespective of plea;

(b) any offence of retail theft (shoplifting) or attempted retail theft irrespective of plea provided it is suitable for sentence in the magistrates' court; and

(c) any *either-way offence anticipated as a guilty plea and suitable for sentence in a magistrates' court*, provided it is not:

    (i)   a case requiring the consent of the DPP or Law Officer;

    (ii)   a case involving a death;

    (iii)   connected with terrorist activity or official secrets;

    (iv)   classified as hate crime or domestic violence under CPS Policies;

    (v)   a case of harassment or stalking;

    (vi)   an offence of violent disorder or affray;

    (vii) causing grievous bodily harm or wounding, or actual bodily harm;

    (viii) a Sexual Offences Act offence committed by or upon a person under 18;

    (ix)   an offence under the Licensing Act 2003.

A guilty plea may be anticipated where either:

- the suspect has made a clear and unambiguous admission to the offence and has said nothing that could be used as a defence;
- the suspect has made no such admission but has offered no explanation that is capable of being used as a defence and either—
  (i) the commission of the offence and the identification of the offender can be wholly established by police witnesses, or
  (ii) there is clear visually recorded evidence of the offence being committed and of the suspect being the offender; or
- the suspect has not been interviewed but either—
  (i) the commission of the offence and the identification of the offender can be wholly established by police witnesses; or
  (ii) there is clear visually recorded evidence of the offence being committed and of the suspect being the offender.

In relation to (c), an either-way case may be considered suitable for sentence in a magistrates' court unless the overall circumstances of the offence make it likely that the court will decide that a sentence in excess of six months' imprisonment is appropriate. The police will use this criterion to determine whether the police or CPS is responsible for making the charging decision in relation to a youth offender. The police must indicate whether or not a committal to the Crown Court should be sought in order to obtain a confiscation order, even if the offences charged are summary only offences; these cases are not suitable for sentence in a magistrates' court.

With reference to the exception to (c) in (c)(i) (cases requiring the consent of the DPP or a law officer):

— offences whose prosecution requires the DPP's consent appear at:
  4.92, 10.21, 10.31, 12.6, 12.19, 12.25, 12.26, 13.15, 18.2, 18.20, 18.21, 21.65, 22.1–22.3; and **22.15*, 22.16*, 22.18*** and **22.19*** (**the Director of the SFO's consent will do instead*); and
— offences whose prosecution requires a law officer's consent appear at:
  9.53, 9.59, 10.10, 11.77, 13.95, 18.40, 18.42, 18.43, 19.75, and **21.65.**

A Crown Prosecutor must make charging decisions in respect of any *offence triable only on indictment, any either-way offence not suitable for sentence in a magistrates' court or not anticipated as a guilty plea, and the offences specified in (c)(i)–(ix) above.* The offence must be referred to a Crown Prosecutor for this purpose.

In a case where any offences under consideration for charging by the police include any offence which must be referred to a Crown Prosecutor under the *Guidance* then all related offences in the case will be referred to a Crown Prosecutor to consider which should be charged.

By way of qualification to the last two paragraphs, an inspector may authorise the charging of an offence that should be referred to a Crown Prosecutor under the *Guidance* where the continued detention of the suspect after charge is justified and where it will not be possible to obtain a Crown Prosecutor's authority to charge before the expiry of any relevant PACE time limit applicable to the suspect. Any cases so charged under this provision must be referred to a Crown Prosecutor immediately for consideration of ratification of the offence charged.

Postponing release on bail for a drugs test

6.49 Where the offence for which the person is arrested is one in relation to which a sample could be taken for a drug test under PACE, s 63B (**6.26**), and the custody officer is required by s 37(2) to release that person and decides to release him/her on bail or decides in pursuance of s 37(7) to release that person without charge on bail, the detention of that person may be continued to enable such a sample to be taken, but this does not permit a person to be detained for more than 24 hours after the relevant time (as defined at **6.40**).

CHARGING DETAINEES

Procedures

6.50 When the officer in charge of the investigation reasonably believes that there is sufficient evidence to provide a realistic prospect of a detainee's conviction, (s)he must without delay inform the custody officer who will be responsible for considering whether or not the detainee should be charged. Where a person has been detained for more than one offence it is permissible to delay informing the custody officer until these conditions are satisfied in respect of all of the offences. If the detainee is a juvenile or vulnerable, any resulting action must be carried out in the presence of the appropriate adult (if present at the time). There is no power under PACE to detain a person and delay action solely to await the arrival of the appropriate adult. Reasonable efforts should therefore be made to give the appropriate adult sufficient notice of the time the decision (charge, etc) is to be implemented so that (s)he can be present. If the appropriate adult is not, or cannot be, present at that time, the detainee should be released on bail to return for the decision to be implemented when the adult is present, unless the custody officer determines that the absence of the appropriate adult makes the detainee unsuitable for bail for this purpose.

Where a custody officer determines in accordance with the DPP's *Guidance on Charging* that there is sufficient evidence to charge the detainee, (s)he may detain that person for no longer than is reasonably necessary to decide how that person is to be dealt with under PACE, s 37(7), (**6.47**) including, where appropriate, consultation with the duty prosecutor. The period is subject to the maximum period determined by ss 41 to 44 (**6.39-6.44**). Where a reference is made to the CPS a custody officer is responsible for ensuring that all specified information is sent with that reference.

Where a person is arrested under the provisions of the Criminal Justice Act 2003 (CJA 2003) allowing a person to be re-tried after being acquitted of a specified serious offence a superintendent (or above) who has not been directly involved in the investigation is responsible for determining whether the evidence is sufficient to charge.

Where a Crown Prosecutor is unable to make a charging decision based upon the information available at the time, a detainee may be released without charge or on bail. A detainee should be informed of the circumstances of such a decision.

Unless the restriction on drawing adverse inferences from silence (**5.29**) applies, a detainee who is charged or informed that (s)he may be prosecuted for an offence must be cautioned in the relevant terms set out at **5.27**. If the restriction does apply, the alternative terms of the caution set out at **5.29** must be used. The detainee must also be given a written notice showing particulars of the offence charged, which must include the name of the officer (with the usual exception: **5.3**) and the case reference number. The charge must be stated in simple terms but must show the precise offence in law. The notice must begin with the following words:

> 'You are charged with the offence(s) shown below' [and be followed by the appropriate caution in the particular circumstances].

Where applicable a copy of the notice must be given to the appropriate adult.

If, after these procedures have been carried out, a police officer wishes to tell a detainee about any written statement or interview with another person relating to the offence, the detainee must either be handed a true copy of the written statement or have the content of the interview record brought to his/her attention. Nothing must be done to invite a reply or comment except to caution the detainee that (s)he does not have to say anything, but that anything (s)he does say may be given in evidence, and to remind him/her of the right to legal advice. If the detainee cannot read, the document may be read to him/her. In relevant cases, the appropriate adult must be given a copy of the document, or the interview record must be brought to his/her attention.

When a *juvenile* is charged with an offence and the custody officer authorises his/her continued detention after charge, the custody officer must arrange for the juvenile to be taken into the care of a local authority to be detained pending appearance in court unless (s)he certifies that:

(a) for any juvenile, it is impracticable to do so. The reasons why it is impracticable must be set out in the certificate that must be produced to the court; or

(b) in the case of a juvenile at least 12 years old, no secure accommodation is available and other accommodation would not be adequate to protect the public from serious harm from that juvenile.

Chief officers should ensure that the operation of these provisions at police stations is subject to supervision and monitoring by an inspector (or above).

Questioning after charge

6.51 Further questions relating to the offence may not generally be asked of a person (D) after D has been charged with that offence, or informed that (s)he may be prosecuted for it. Exceptions are: where they are necessary to prevent or minimise harm or loss to some other person or to the public, or to clear up ambiguity in a previous answer or statement; or where it is in the interests of justice that D should have put to him/her (and have the opportunity to comment upon) new information concerning the offence. In such cases D must be cautioned before further questions are put that (s)he does not have to say anything, but that anything (s)he does say may be given in evidence. D must be reminded of the right to legal advice.

The giving of a warning or the service of a notice of intended prosecution required by the Road Traffic Offenders Act 1988, s 1 does not amount to informing a detainee that (s)he may be prosecuted for an offence and so does not preclude further questioning in relation to that offence.

For a special provision under the terrorism legislation, see **21.66**.

Documentation

6.52 A record must be made of anything a detainee says when charged. Questions put in an interview after a charge and answers given must be recorded in full during the interview on the forms provided and the record must be signed by the detainee or, if (s)he refuses, by the interviewer and any third parties present. If the questions are audibly recorded or visually recorded the arrangements set out in Code E or Code F apply.

DUTIES OF CUSTODY OFFICER AFTER CHARGE

6.53 PACE, s 38 requires that, when a person arrested otherwise than under a warrant endorsed for bail is charged with an offence, the custody officer, subject to the Criminal Justice and Public Order Act 1994 (CJPOA 1994), s 25 (**6.75**), must order his/her release from police detention, either on bail or without bail, *unless* the charge is murder or unless the following special rules apply. The Court of Appeal has held that the provisions of s 38 (described here and in the following paragraphs) do not apply where a person is in custody solely because (s)he has been arrested for breach of his/her bail conditions contrary to the Bail Act 1976, s 7.

Arrested person not a juvenile

6.54 PACE, s 38 provides that if the person charged (D) is not a juvenile the custody officer (CO) must order his release from police detention, with or without bail, unless:

(a) D's name and address cannot be ascertained or CO has reasonable grounds for doubting the truth of a name or address provided;

(b) CO has reasonable grounds for believing that D will fail to appear in court to answer bail;

(c) if D was arrested for an imprisonable offence, CO has reasonable grounds for believing that the detention of D is necessary to prevent him/her from committing an offence;

(d) in a case where a sample may be taken from D under PACE, s 63B (**6.26**), CO has reasonable grounds for believing that the detention of D is necessary to enable a sample to be taken from him/her;

(e) if D was arrested for a non-imprisonable offence, CO has reasonable grounds for believing that the detention of D is necessary to prevent him/her from causing physical injury to any other person or from causing loss of or damage to property; or

(f) CO has reasonable grounds for believing that the detention of D is necessary (i) to prevent D from interfering with the administration of justice or with the investigation of offences or of a particular offence; or (ii) for D's own protection.

Arrested person a juvenile

6.55 PACE, s 38 provides that if the person charged (JD) is a juvenile (ie a person appearing to be under 18) the custody officer (CO) must order JD's release from police detention, with or without bail, *unless* one of the above grounds ((a) to (f)) applies, or CO has reasonable grounds for believing that JD should be detained in JD's own interests. In the case of ground (d) this only applies if JD has reached the minimum age specified by s 63B (at present 14).

Section 38 also requires that, where CO authorises JD to be kept in police detention, CO must secure that JD is to be taken to local authority accommodation unless CO certifies:

(a) that by reason of such circumstances as are specified in the certificate, it is impracticable to do so. This requires a statement of the reasons, but detailed reasons are not required; or

(b) where JD is 12 or over, that no secure accommodation is available and that keeping him/her in other local authority accommodation would not be adequate to protect the public from serious harm from him/her.

Such a certificate must be produced to the court before which JD is first brought thereafter. The Court of Appeal has held that a failure to do so does not in itself render the detention unlawful.

Where JD is charged with a violent, sexual or terrorism offence, the reference in (b) to protecting the public from serious harm is to protection from death or serious personal injury, whether physical or psychological, occasioned by further such offences by him/her.

'*Sexual offence*' in this context means an offence under:

(a) the Protection of Children Act 1978 or the Criminal Justice Act 1988, s 160 (indecent photographs of children);

(b) any provision of the Sexual Offences Act 2003, Part 1, except s 71 (sexual activity in a public lavatory) or any provision repealed by that Act; or

(c) the Modern Slavery Act 2015, s 2 (human trafficking) where committed with a view to sexual exploitation.

'*Violent offence*' means an offence contained in a list of 86 offences against the person, firearms offences, explosives offences, public order offences and road traffic offences which lead, or are intended or likely to lead, to a person's death or to physical injury, and include an offence which is required to be charged as arson (whether or not it would otherwise fall within this definition). '*Terrorism offence*' means an offence under:

(a) the Terrorism Act 2000, s 11, 12, 54, 56, 58, 58, or 59;
(b) the Anti-terrorism, Crime and Security Act 2001, s 47, 50, or 113; or
(c) the Terrorism Act 2006, s 1, 2, 5, 6, 8, 9, 10, or 11.

Except as provided above, neither JD's behaviour nor the nature of the offence charged provides grounds for CO to decide that it is impracticable to arrange for the transfer to local authority care. A guidance note to Code C states that impracticability concerns the transport and travel requirements and the lack of secure accommodation which is provided for the purposes of restricting liberty does not make it impracticable to transfer the juvenile. Rather, 'impracticable' should be taken to mean that exceptional circumstances render movement of the child impossible or that the juvenile is due in court in such a short space of time that transfer would deprive him/her of rest or cause him/her to miss a court appearance. When the reason for not transferring the juvenile is an imminent court appearance, details of the travelling and court appearance times which justify the decision should be included in the certificate referred to above. The availability of secure accommodation is only a factor in relation to a juvenile aged 12 or over when the local authority accommodation would not be adequate to protect the public from serious harm from him/her.

Detention

If the release of a person arrested is not required by PACE, s 38, CO may authorise him/ **6.56** her to be kept in police detention, but CO may not authorise an adult to be so detained for a sample to be taken under PACE, s 63B after a six-hour period running from the charge.

Where CO authorises a person who has been charged to be kept in police detention CO must, as soon as practicable, record the grounds (ie reasons) in the custody record in that person's presence, unless the usual exceptions (**6.13**) apply.

PACE, s 46 requires that where a person (D) is charged with an offence; and after being charged, is kept in police detention or (in the case of a juvenile) is detained by a local authority, D must be brought before a magistrates' court as soon as practicable, and in any event not later than the first sitting after D is charged (or, if D is to be brought before a magistrates' court in another local justice area, not later than the first sitting of that court after D's arrival in that area). If D is to be brought before a magistrates' court in another area, D must be removed to that area as soon as practicable for the above purpose.

If no magistrates' court is due to sit on the day D is charged (or on the day D arrives in the other area) or on the next day, the custody officer must inform the designated officer for a local justice area that there is a person in the area who has been detained after charge, and that officer must arrange for a sitting of a magistrates' court not later than the day next following the day on which D is charged (or, if D has been transferred to another area, the day next following the day of D's arrival in that area). Christmas Day,

Good Friday, and any Sunday do not count as 'days next following' for this purpose; thus, for example, the day next following Saturday is Monday.

None of the above provisions requires a person who is in hospital to be brought before a court if (s)he is not well enough.

BAIL

Bail of arrested person (police bail)

6.57 When an investigating officer brings a person arrested for an offence (D) before the custody officer, the custody officer may decide that there is insufficient evidence to justify a charge. As seen at **6.47**, if this decision is reached, D must be released without bail unless the pre-conditions for bail (**6.6**) are satisfied, or on bail if those pre-conditions are satisfied, unless the custody officer has reasonable grounds for believing that D's detention without being charged is necessary to secure or preserve evidence relating to an offence for which D is under arrest or to obtain such evidence by questioning D, in which case (s)he may authorise D to be kept in police detention.

Alternatively, the custody officer may decide that there is enough evidence to charge at that stage. As seen at **6.47**, if this decision is reached, D may be (a) released without charge and on bail, or kept in police detention, pending a decision by a Crown Prosecutor; (b) released without charge and without bail unless the pre-conditions for bail are satisfied; (c) released without charge and on bail if those pre-conditions are satisfied but not for the purpose in (a); or (d) charged.

PACE, s 47 provides that, for the purposes of police bail, 'bail' refers to bail subject to a duty:

(a) to appear at a magistrates' court at such time and such place as the custody officer may appoint;

(b) to attend at such police station and at such time as the custody officer may appoint for the purposes of:
(i) proceedings relating to a live link direction, and
(ii) any preliminary hearing in relation to which any such direction is given; or

(c) to attend at such police station at such time, as the custody officer may appoint for purposes other than those mentioned at (b).

Release on police bail: general

6.58 PACE, s 47 states that a release on police bail must be a release on bail granted in accordance with the following provisions of the Bail Act 1976 (BA 1976). The normal powers to impose conditions of bail (ie the powers set out in the next paragraph) are available where a custody officer releases a person on bail (other than under ss 37C(2) (b) and 37CA(2)(b) (**6.79**)).

By BA 1976, ss 3 and 3A, a custody officer releasing a person on bail and without charge pending a decision of a Crown Prosecutor or after charge has power to require him/her to comply with such conditions on bail as appear to the custody officer to be necessary:

(a) to secure that the person:
(i) surrenders to custody;
(ii) does not commit an offence whilst on bail;

(iii) does not interfere with witnesses or otherwise obstruct the course of justice whether in relation to him/herself or any other person; or

(b) for the protection of the person concerned or, if (s)he is under 18, for his/her own welfare or in his/her own interests.

However, the custody officer does not have power to impose a requirement to reside in a bail hostel. A custody officer would not have sufficient time to make the necessary inquiries before such a condition might properly be imposed.

The combined effect of provisions in BA 1976, s 3 is to permit the following additional requirements to be made of a person before (s)he is released on bail:

(a) (s)he may be required:

(i) to provide a surety or sureties (see below) to ensure his/her surrender to custody;

(ii) to give security for his/her surrender which may be given by him/her or on his/her behalf;

(b) if a parent or guardian of a juvenile consents to be surety for him/her, the parent or guardian may be required to ensure that the juvenile complies with conditions imposed under (a)(i)–(iii) above, except that no such condition may be imposed where a juvenile will be 18 before the time appointed for surrender, and that a parent or guardian may not be required to secure compliance with any requirement to which his/her consent does not extend and may not, in respect of those requirements to which his/her consent does extend, be bound in a sum greater than £50.

Where a custody officer has granted bail in criminal proceedings, (s)he or another custody officer serving at the same police station may vary the condition of bail at the request of the person bailed, and in doing so (s)he may impose conditions or more onerous conditions.

BA 1976, s 5 provides that, where a custody officer grants bail or varies any conditions of bail, or imposes conditions in respect of bail, (s)he must make a record of the decision and, if requested by the person in question, give that person a copy of the record as soon as practicable; in practice this is always done. Sections 5 and 5A provide that, where a custody officer imposes conditions in granting bail or varies any condition of bail, (s)he must give reasons for doing so. This is to enable the person concerned to consider requesting the custody officer to vary those conditions. A note of the custody officer's reasons for imposing (or varying) conditions must be made in the custody record and a copy must be given to the person concerned.

A person who is bailed enters into a promise to appear as prescribed. That promise cannot be set against a recognisance from him/her that, if (s)he fails to appear, a specific sum of money shall be forfeit. Instead, BA 1976 provides its own penalties for non-appearance. In serious cases it may be necessary for a defendant to find one or more 'sureties', that is persons who *undertake to secure his/her attendance*. This is done by each surety entering a recognisance to forfeit a specified sum to the Crown in the event of the non-attendance of the accused. By the Magistrates' Courts Act 1980, s 43, the magistrates' court before whom the person is bailed to appear may enlarge the recognisances of any sureties.

Police bail of person not charged

The custody officer, when the investigating officer brings the person detained (D) before him/her, may decide under PACE, s 37 to release D on bail, such bail being **6.59** conditioned upon D's appearance at a police station at a given time, as opposed to appearing at a court. The custody officer may reach this decision because (s)he considers that there is not sufficient evidence at that stage to charge D, but that there probably will be when further inquiries have been made. Alternatively, (s)he may consider that there is sufficient evidence to charge D but decide under s 37 to release D on bail pending a decision by a Crown Prosecutor about charging or for some other purpose. The requirement to attend at a police station may be cancelled at any time by notice in writing from the custody officer.

Section 47(1B) provides that no application by the prosecutor may be made for a reconsideration of a custody officer's decision to grant bail where a person is released without charge on bail following arrest for breach of bail granted under s 37 to enable a Crown Prosecutor to make a decision about charging, or some other purpose.

Where a person is released on conditional bail in such circumstances, that person will not be entitled to apply to a magistrates' court for bail, but may seek variation of such conditions.

Where a person (P) has been granted bail subject to a duty to attend at a police station, a custody officer may subsequently appoint a different time, or an additional time, at which P is to attend at the police station to answer bail (s 47(4A)). The custody officer must give P written notice of this (s 47(4B)). The exercise of the power under s 47(4A) does not affect any conditions of bail (s 47(4C)). A time may not be appointed under s 47(4A) which is after the end of the *applicable bail period* (below) in relation to P (s 47(4D)). Section 47(4D) is subject to s 47ZL (**6.69**).

Limits on period of bail without charge

6.60   PACE, 47ZA to 47ZM apply in relation to a custody officer's power, when releasing a person (P) on bail under PACE, to appoint a time for P to attend at a police station in accordance with **6.57**(c).

Section 47ZA provides that the power must be exercised so as to appoint a time on the day on which *the applicable bail period* (below) in relation to P ends, unless:

(a) at the time of the exercise of the power P is on bail under PACE in relation to one or more offences other than the relevant offence (ie the offence in respect of which the power is exercised in relation to P) and the custody officer believes that it is appropriate to align P's attendance in relation to the relevant offence with Ps attendance in relation to the one or more other offences; or

(b) where the custody officer believes that a decision as to whether to charge P with the relevant offence would be made before the end of the applicable bail period in relation to P.

Where (a) or (b) applies, the power may be exercised so as to appoint a time on a day falling before the end of the applicable bail period (below) in relation to P.

Applicable bail period: initial limit

6.61   PACE, s 47ZB(1) provides that the 'applicable bail period' means, in an SFO case, the period of three months beginning with P's bail start date, or, in an FCA case or any other case, the period of 28 days beginning with P's bail start date.

For the purposes of the provisions about limits on the period of bail without charge, P's bail start date is the day after the day on which P was arrested for the relevant offence, and:

(a) an 'SFO case' is a case in which the relevant offence in relation to P is being investigated by the Director of the Serious Fraud Office, and a senior officer confirms this;

(b) an 'FCA case' is a case in which the relevant offence in relation to P is being investigated by the Financial Conduct Authority, and a senior officer confirms this.

'Senior officer' means a superintendent (or above).

By s 47ZB(2), the applicable bail period in relation to P may be extended under ss
47ZD to 47ZG (below) or treated as extended under s 47ZJ (below). Sections 47ZD to
47ZG are subject to ss 47ZL and 47ZM (below).

Extension of the applicable bail period

An extension of the period of bail beyond the initial bail period depends on the satis-   **6.62**
faction of some or all of the following conditions listed in PACE, s 47ZC:

Condition A is that the decision-maker has reasonable grounds for suspecting P to
be guilty of the relevant offence.

Condition B is that the decision-maker has reasonable grounds for believing:

(a) where P is or is to be released on bail under s 37(7)(c) (**6.47**) or after arrest for
breach of bail granted under s 37(7)(c), that further time is needed for making a
decision as to whether to charge P with the relevant offence; or

(b) otherwise, that further investigation is needed of any matter in connection with the
relevant offence.

Condition C is that the decision-maker has reasonable grounds for believing:

(a) where P is or is to be released on bail under s 37(7)(c) or after arrest for breach of
bail granted under s 37(7)(c), that the decision as to whether to charge P with the
relevant offence is being made diligently and expeditiously; or

(b) otherwise, that the investigation is being conducted diligently and expeditiously.

Condition D is that the decision-maker has reasonable grounds for believing that
the release on bail of P is necessary and proportionate in all the circumstances (having
regard, in particular, to any conditions of bail which are, or are to be, imposed).

In s 47ZC, 'decision-maker' means in relation to a condition which falls to be considered by virtue of:

– s 47ZD, the senior officer (**6.61**) in question;
– s 47ZE, the appropriate decision-maker in question;
– s 47ZF or 47ZG, the court in question.

Extension of applicable bail period in standard cases

By PACE, s 47ZD, where in a case other than an SFO case (**6.61**) the applicable bail   **6.63**
period in relation to P has not ended, and a senior officer is satisfied that conditions
A to D (above) are met in relation to P, the senior officer may authorise the applicable
bail period in relation to P to be extended so that it ends at the end of the three-month
period beginning with P's bail start date. Before determining whether to give such an
authorisation, the senior officer must arrange for P or P's legal representative to be
informed that a determination is to be made. In determining whether to give an authorisation, the senior officer must consider any representations made by P or P's legal
representative. The senior officer must arrange for P or P's legal representative to be
informed whether such an authorisation has been given.

Applicable bail period: extension of applicable bail period in designated cases

6.64   PACE, s 47ZE provides that, where:

(a) P's case is an SFO (**6.61**) case; or

(b) a senior officer (**6.61**) has authorised an extension of the applicable bail period in relation to P under s 47ZD, and

a qualifying prosecutor has designated P's case as being an exceptionally complex case, an appropriate decision-maker may extend the applicable bail period in relation to P until the end of the six-month period beginning with P's bail start date, provided that the decision maker is satisfied that conditions A to D (**6.62**) are met.

'Qualifying prosecutor' means a prosecutor of the description designated for these by the Chief Executive of the FCA, the Director of the SFO, or the DPP. An 'appropriate decision-maker' is a designated member of staff of the FCA (FCA case: **6.61**), a member of the SFO who is of the Senior Civil Service (SFO case), or a qualifying police officer (in any other case). 'Qualifying police officer' means an assistant chief constable or commander (or above).

Before determining whether to give such an authorisation:

(a) the appropriate decision-maker must arrange for P or P's legal representative to be informed that a determination is to be made, and

(b) if the appropriate decision-maker is a qualifying police officer, the officer must consult a qualifying prosecutor.

The appropriate decision-maker must consider any representations made by P or P's legal representative, and arrange for P or P's legal representative to be informed whether an authorisation has been given.

Any designation or authorisation under s 47ZE must be given before the applicable bail period in relation to P has ended.

Applicable bail period: first extension of limit by court

6.65   PACE, s 47ZF provides as follows where:

(a) P's case is an SFO (**6.61**) case;

(b) bail has been extended under s 47ZD; or

(c) bail has been extended under s 47ZE.

In such a case, before the applicable bail period in relation to P ends a qualifying applicant may apply to a magistrates' court for an extension of the applicable bail period. 'Qualifying applicant' means: a constable, a member of staff of the FCA of the description designated by its Chief Executive, a member of the SFO, or a Crown Prosecutor.

If the court is satisfied that conditions B to D (**6.62**) are met in relation to P, and the case does not fall within s 47ZF(7) (below), it may authorise the applicable bail period to be extended so as to end at the end of the following period beginning with P's bail start date: six months in a case within (a) or (b), and nine months in a case within (c).

However, if the court is satisfied that conditions B to D are met in relation to P, and the case falls within s 47ZF(7), it may authorise the applicable bail period to be extended so as to end at the end of the following period beginning with P's bail start date: nine months in a case within (a) or (b), and 12 months in a case within (c).

A case falls within s 47ZF(7) if the nature of the decision or further investigations mentioned in condition B means that that decision is unlikely to be made or those investigations completed if the applicable bail period in relation to P is not extended as specified in the previous paragraph.

Applicable bail period: subsequent extensions of limit by court

6.66 PACE, s 47ZG provides as follows where a court has authorised an extension of the applicable bail period in relation to P under s 47ZF.

Before the applicable bail period ends a qualifying applicant (**6.65**) may apply to a magistrates' court for it to extend that period.

The court may extend the bail period with effect from the end of the current applicable bail period in relation to P, if satisfied that conditions B to D (**6.62**) are met in relation to P:

(a) for a three-month period if satisfied that the decisions or investigations referred to in condition B are not unlikely to be made or completed within that period, or

(b) for six months if it is not likely that the decisions or investigations will be completed within the three-month period.

Where a court has authorised an extension under the above provisions, a qualifying applicant may make further applications for an extension under them.

Sections 47ZF and 47ZG: withholding sensitive information

6.67 By PACE, s 47ZH, where a qualifying applicant (**6.65**) makes an application to a magistrates' court under s 47ZF or 47ZG in relation to P, (s)he may apply to the court for it to authorise the specified information to be withheld from P and any legal representative of P. The court may grant such an application only if satisfied that there are reasonable grounds for believing that the specified information is sensitive information. For these purposes, information is sensitive information if its disclosure would have one or more of the following results:

(a) evidence connected with an indictable offence would be interfered with or harmed;

(b) a person would be interfered with or physically injured;

(c) a person suspected of having committed an indictable offence but not yet arrested for the offence would be alerted;

(d) the recovery of property obtained as a result of an indictable offence would be hindered.

PACE, ss 47ZF to 47ZH: proceedings in magistrates' court

6.68 An application made to a magistrates' court under PACE, ss 47ZF–47ZG is determined by a single justice on written evidence unless:

(a) the effect of the application would be to extend the applicable bail period in relation to P so that it ends at or before the end of the 12-month period beginning with P's bail start date, and a single justice considers that the interests of justice require an oral hearing; or

(b) the effect of the application would be to extend the applicable bail period in relation to P so that it ends after the end of the 12-month period beginning with P's bail start date, and P, or the person who made the application, requests an oral hearing.

In these two cases, the application is determined by two or more justices sitting otherwise than in open court.

An application under s 47ZH is determined by a single justice on written evidence unless the justice determines that the interests of justice require an oral hearing, in which case the application is determined by two or more justices sitting otherwise than in open court.

The above provisions are contained in s 47ZI.

Applicable bail period and bail return date: special case of release on bail under s 37(7)(a) or after arrest for breach of such bail

6.69   PACE, s 47ZL provides that, where P is released on bail under s 37(7)(a) in order for a Crown Prosecutor to make a charging decision, or after arrest for breach of bail so granted, the running of the applicable bail period in relation to P does not begin (in the case of a first release on bail), or is suspended (in any other case), subject to s 47ZL(6) (below). Accordingly, s 47ZA (**6.60**) does not apply to the exercise of the power to appoint a time for police station attendance when releasing P on bail.

If a 'DPP request' (a request by a Crown Prosecutor for the further information specified in the request to be provided before a decision whether there is sufficient evidence to charge P with the 'relevant offence') is made, a custody officer *must* under s 47(4A) appoint a different time for P to attend at the police station (and s 47(4B) to (4D) (**6.59**) apply). By s 47ZL(6), the applicable bail period in relation to P:

(a) begins on the day on which the request is made (in the case of a first release on bail), or

(b) resumes running on that day (in any other case).

Where a DPP request has been made in relation to P, and the applicable bail period would end before the end of the seven-day period beginning with the day on which the DPP request was made, the running of the applicable bail period is suspended for the number of days necessary to secure that the applicable bail period ends at the end of the seven-day period beginning with the day on which the DPP request was made.

If the DPP request made in relation to P is met, the running of the applicable bail period in relation to P is suspended.

The above references to the case of a first release on bail are to a case where P has not been released on bail in relation to the relevant offence under any other provision of the detention part of PACE, s 30A (**4.57**).

Applicable bail period: special cases of release on bail under s 30A and periods in hospital

6.70   Where P was released on street bail under PACE, s 30A, the 28-day period beginning with the day after the day on which P was arrested for the offence in relation to which bail is granted under s 30A is treated as being the 28-day period mentioned in s 47ZB (**6.61**), and any reference in s 47ZB to the relevant offence is read as a reference to the offence in respect of which the power in s 30A(1) was exercised.

If, at any time on the day on which the applicable bail period in relation to P would end, P is in hospital as an in-patient, the running of the applicable bail period treated as having been suspended for any day on which the patient was in hospital as an in-patient.

These provisions are contained in s 47ZM.

Comment

6.71   Consequent on the introduction of the restriction on the length of police bail, unless extended, the police now routinely release suspects under investigation. A person so released is not subject to bail and is not required to return to the police station. However, (s)he may be contacted by the police again and in appropriate circumstances may be arrested again. A divisional court has held that any suggestion that there was some sort of presumption that, whenever a suspect was released under investigation and then complied with the attendance requirements of a requisition, a court would subsequently grant him/her bail when (s)he was charged was misplaced.

Police bail after charge

By PACE, s 47(3A), where a custody officer grants bail subject to a duty to appear be- **6.72**
fore a magistrates' court, (s)he must appoint for the appearance:

(a) a date which is not later than the first sitting of the court after the person is charged with the offence; or
(b) where (s)he is informed by the designated officer for the relevant local justice area that the appearance cannot be accommodated until a later date, that later date.

The Magistrates' Courts Act 1980, s 43 enables a magistrates' court to fix a later time for appearance before it.

Alternatively, the custody officer may bail the person charged to attend a police station pursuant to a live link direction.

Live link police bail

Section 46ZA provides that an accused person answering live link bail is not to be **6.73**
treated as being in police detention, except where, at any time before the beginning of proceedings in relation to a live link direction at a preliminary hearing, (s)he is told by a constable that a live link will not be available or where the court determines not to give a live link direction. In such a circumstance, the accused is to be treated as arrested for and charged with the offence for which (s)he was granted bail, and as if (s)he had been so charged when that circumstance first applied to him/her. A person arrested under s 46A for failing to attend at a police station to answer to live link bail, and who is brought to such a station, is to be treated as if (s)he had been arrested for and charged with the offence for which (s)he was granted bail and had been so charged when brought to the station.

By s 54B, a constable may search at any time any person who is at a police station to answer to live link bail; and any article in the possession of such a person. If the constable reasonably believes a thing in the person's possession ought to be seized on any of the grounds that it:

(a) may jeopardise the maintenance of order in the police station;
(b) may put the safety of any person in the police station at risk; or
(c) may be evidence of, or in relation to, an offence,

the constable may seize and retain it or cause it to be seized and retained. Anything so seized and retained must be recorded. Such a search may only be conducted by a constable of the same sex as the person concerned. An intimate search may not be carried out under the authority of s 54B.

Section 54C empowers a constable to retain a thing seized under s 54B until the person from whom it was seized leaves the police station. However, retention beyond that point of time is permissible in two cases:

(a) a constable may retain something seized under s 54B in order to establish the thing's lawful owner, where there are reasonable grounds for believing that it has been obtained in consequence of the commission of an offence; or

(b) where the thing may be evidence of, or in relation to, an offence, a constable may retain it for use as evidence at a trial or for forensic examination or for investigation, if a photograph or copy would be insufficient for that purpose.

Arrest warrant endorsed for bail

6.74 The MCA 1980, s 117 allows a person arrested on a warrant endorsed for bail to be released on bail (ie admitted to bail 'on the spot') without being taken to a police station, provided that the endorsement for bail does not demand sureties. If sureties are required, the person must be taken to a police station.

Bail by a court

6.75 BA 1976, s 4 states that, when a person who is accused of an offence appears before a magistrates' court or the Crown Court in the course of, or in connection with, the proceedings for the offence, or when (s)he applies to a court for bail or for a variation of the conditions of bail in connection with the proceedings, (s)he must be granted bail if none of the exceptions specified in Sch 1 (below) applies. However, this is subject to the provisions of CJPOA 1994, s 25 (bail for defendant charged with or convicted of murder, attempted murder or manslaughter, or any offence under the Sexual Offences Act 2003, ss 1, 2, 4, 5, 6, 8, 30, and 31 (involving rape, non-consensual penetration, or sexual activity, the same activity with children under 13 and mentally disordered persons), or an attempt to commit any such offence, after a previous conviction for such an offence only to be granted if there are exceptional circumstances justifying it). Section 25 only operates where (a) a person has previously been convicted in a UK court of one of the above offences or culpable homicide (the Scottish equivalent of manslaughter), and (b) if that previous conviction is one of manslaughter or culpable homicide, resulted in a sentence of imprisonment or detention being imposed on an adult or long-term detention if not an adult. Section 25 also applies to a police officer considering the grant of bail in such circumstances.

A person charged with murder may be granted bail only by a Crown Court judge.

In taking decisions under BA 1976, Sch 1, a court must have regard, so far as it is relevant, to any misuse of a controlled drug by the defendant.

Exceptions to right to bail: imprisonable offences

6.76 An imprisonable offence is one which is punishable in the case of an adult with imprisonment.

The following exceptions to the right to bail apply under BA 1976, Sch 1, Part I, where the offence or one of them is imprisonable, except that they do not apply where the offence or each of them is triable only summarily or where the offence is criminal damage or aggravated vehicle-taking only involving damage where the value of the damage is less than £5,000.

(1) Bail need not be granted if the court is satisfied that there are *substantial grounds for believing that the defendant (D), if released on bail (whether subject to conditions or not), would:*
   (a) *fail to surrender to custody; or*
   (b) *commit an offence while on bail; or*
   (c) *interfere with witnesses or otherwise obstruct the course of justice, whether in relation to him/herself or any other person.*

This exception does not apply if D falls within (7) (below), unless the court is satisfied as mentioned there. In addition, this exception does not apply where:

(a) D has attained the age of 18,

(b) D has not been convicted of an offence in the proceedings, and

(c) it appears to the court that there is no real prospect that D will be sentenced to a custodial sentence in the proceedings.

(2) Bail need not be granted if the court is satisfied that there *are substantial grounds for believing that D, if released on bail (whether subject to conditions or not), would commit an offence while on bail by engaging in conduct that would, or would be likely to, cause (a) physical or mental injury to an associated person, or (b) an associated person to fear physical or mental injury.*

(3) Where the offence is an indictable offence (including an either-way offence), D need not be granted bail if it appears that D was on bail in criminal proceedings on the date of the offence. This provision does not apply in the circumstances shown in bold above in respect of (1).

(4) Bail need not be granted by the court:

    (a) if it is satisfied that D should be kept in custody for D's own protection or, if a child or young person (ie under 18), for D's own welfare; or

    (b) if D is in custody under a court order or in pursuance of any authority under the Armed Forces Act 2006; or

    (c) if the court is satisfied that there has not been time to obtain sufficient information upon which a decision about bail may be made.

(5) D need not be granted bail if, having previously been released on bail in, or in connection with, the proceedings, D has been arrested under BA 1976, s 7. This provision does not apply in the circumstances printed in bold above in respect of (1).

(6) A person charged with murder must not be granted bail unless the court is of the opinion that there is *no significant risk of him/her committing, while on bail, an offence that would, or would be likely to, cause physical or mental injury to any person other than him/herself.*

(7) An alleged drug offender aged 18 or over who is charged may not be granted bail (unless the court is of the opinion that there is *no significant risk of his/her committing an offence while on bail*) where:

    (a) a drug test indicates the presence of a Class A drug;

    (b) the offence is one of possession of a Class A drug *or* the court is satisfied that there are substantial grounds for believing that the misuse of a Class A drug caused or contributed to that offence or provided its motivation; and

    (c) the person concerned does not agree to a dependency/propensity to misuse assessment, or has undergone such assessment but does not agree to participate in any relevant follow-up offered.

However, (7) only applies in areas where facilities for relevant assessment or (as the case may be) follow-up are available.

In taking decisions about the words italicised in heads (1), (2), (6), and (7), the court must have regard to the nature and seriousness of the offence; the character, antecedents, associations, and community ties of D; D's record in respect of previous grants of bail; the strength of the evidence available (except where the case is merely being adjourned for inquiries or a report); and if there are substantial grounds for believing that D, if released on bail, would commit an offence while on bail, the risk that D may do so by engaging in conduct that would, or would be likely to, cause physical or mental injury to any person other than D.

BA 1976, Sch 1, Part IA applies where *the imprisonable offence (or each of them) is a summary offence or where the offence is criminal damage, or aggravated vehicle-taking only involving damage, and the value of the damage is less than £5,000.*

BA 1976, Sch 1, Part IA provides that D need not be granted bail if:

(a) there has been a previous failure to surrender to bail and the court believes that this will occur again;

(b) D was on bail when the offence was committed and the court believes that, if released on bail, D would commit an offence while on bail;

(c) the court is satisfied that there are substantial grounds for believing that, if released on bail, D would commit an offence by engaging in conduct likely to cause physical or mental injury, or fear of such injury to an associated person;

(d) the court is satisfied that D should be kept in custody for D's own protection, or, if under 18, for D's own welfare;

(e) D is in custody under the sentence of a court or of an officer under the Armed Forces Act 2006;

(f) having been released on bail in proceedings for the same offence, D has been arrested under BA 1976, s 7 for absconding or breaking conditions of bail, and the court is satisfied that there are substantial grounds for believing that if released on bail (with or without conditions) D would fail to surrender, commit an offence while on bail, interfere with witnesses or otherwise obstruct the course of justice;

(g) it has not been practical to obtain enough information to take the decisions required by Part 1A due to lack of time since the proceedings began; or

(h) the drug-users exception (see (7) above) applies.

Heads (a), (b), and (f) do not apply in relation to bail where:

(i) D has attained the age of 18,

(ii) D has not been convicted of an offence in those proceedings, and

(iii) it appears to the court that there is no real prospect that D will be sentenced to a custodial sentence in the proceedings.

Exceptions to right to bail: non-imprisonable offences

6.77   BA 1976, Sch 1, Pt II provides that D charged with a non-imprisonable offence need not be granted bail if (a), (d), or (e) immediately above applies. Head (a) only applies where D is under 18, or has been convicted in the proceedings of an offence.

D need not be granted bail if, having been released on bail in proceedings for the offence, D has been arrested under BA 1976, s 7 (arrest for absconding or breaking conditions of bail), and the court is satisfied that there are substantial grounds for believing that if released on bail (with or without conditions) D would, or would be likely to, cause physical or mental injury to an associated person, or to cause an associated person to fear physical or mental injury.

D need not be granted bail if:

(a) (s)he is under 18, or has been convicted in the proceedings of an offence;

(b) having been released on bail in proceedings for the offence, D has been arrested under BA 1976, s 7; and

(c) the court is satisfied that there are substantial grounds for believing that if released on bail (with or without conditions) D would fail to surrender to custody, commit an offence on bail or obstruct the course of justice (eg witness intimidation).

Conditions

In granting bail a court may impose conditions for a number of specified purposes, in-   **6.78**
cluding those specified for the purposes of conditional police bail.

Breach of requirements of bail subject to duty to attend police station: arrest, etc

Under PACE, s 46A, a constable may arrest without warrant any person who, having   **6.79**
been released on bail subject to a duty to attend at an appointed police station, fails to attend at that police station at the time appointed for him/her to do so. In addition, a person (P) who has been released on bail under PACE may be arrested without warrant under s 46A by a constable if (s)he has reasonable grounds for suspecting that P has broken any conditions of P's bail.

Such a person must be taken to the appointed police station as soon as practicable after his/her arrest.

Unlike the situation where the bailed person is under a duty to surrender to a court (**6.80**), there is no power of arrest in respect of an anticipated breach of conditions or failure to surrender. Breach of conditions is not an offence. There is one exception. This is where a person (P) arrested in respect of an offence to which the Counter-Terrorism Act 2008, s 41 applies is released without charge and on bail subject to a specified type of 'travel restriction condition'. (Examples of such an offence are under the Terrorism Act 2000, ss 11, 12, 15–18, 38B, or 57–58A, the Anti-terrorism, Crime and Security Act 2001, s 113, or the Terrorism Act 2006, ss 1, 2, 5, 6, or 8–11, all dealt with in Chapter 21; for a full list see Chapter 21 of the companion website.) In such a case, P commits an indictable (either-way) offence under the Policing and Crime Act 2017, s 68(3) if P's release on bail is subject to:

(a) a travel restriction condition not to leave the UK and P fails to comply with the condition; or

(b) any other specified type of travel restriction condition and P, without reasonable excuse, fails to comply with the condition.

The offence is punishable with up to 12 months' imprisonment.

Section 46A(1ZA) and (1ZB) has the effect of extending the power of arrest under s 46A to cover the case where a defendant attends a police station in response to a live link direction for a live link bail hearing, but (a) leaves before the beginning of the proceedings relating to the live link hearing, or (b) refuses to be searched under PACE, s 54B.

Section 47(2) provides that nothing in BA 1976 prevents a re-arrest without warrant of a person released on bail subject to a duty to appear at a police station if, since that release, new evidence has come to light or an examination or analysis of existing evidence has been made which could not reasonably have been made before the release. If such a person is re-arrested the detention provisions in PACE apply as if (s)he has been arrested for the first time, but this does not apply to a person arrested under s 46A for failure to surrender to police bail at a police station, or who has surrendered to that bail and who accordingly is deemed to have been arrested for that offence, or to a person who is treated under s 46ZA as if (s)he had been arrested.

Where a person has been released under s 37(7)(a) (**6.47**) and the Crown Prosecutor has not given notice of his/her decision, and the person is then arrested under s 46A for failing to answer police bail, the custody officer must, by s 37C(2):

(a) charge the person; or
(b) release him/her without charge, either on bail or without bail (s 37C).

The same provision applies where a person so released with bail is re-arrested for breach of that bail.

Where a person released on bail under s 37(7)(c) (**6.47**) is arrested under s 46A in respect of that bail, and is being detained at the appointed police station following that arrest, the custody officer must, by s 37CA(2):

(a) charge the person, or
(b) release him/her without charge, either without bail unless the pre-conditions for bail (**6.6**) are satisfied, or on bail if those pre-conditions are satisfied.

Where:

(i) a person is released under s 37CA(2), and
(ii) a custody officer determines that—
(a) there is not sufficient evidence to charge the person with an offence, or
(b) there is sufficient evidence to charge him/her with an offence but (s)he should not be charged with an offence or given a caution in respect of an offence,

the custody officer must, by s 37CA(6), give him/her notice in writing that (s)he is not to be prosecuted. Section 37CA(6) does not prevent the prosecution of the person for an offence if new evidence comes to light after the notice was given.

The decision as to how a person is to be dealt with under s 37C(2) or 37CA(2) is that of the custody officer.

A person released on bail under s 37C(2)(b) or 37CA(2)(b) must be released subject to the same conditions which applied immediately before his/her arrest.

MCA 1980, s 43 permits the enforcement of the recognisance of any surety for a person granted bail which is conditioned upon appearance at a police station in the same way as if conditioned on appearance at a magistrates' court.

PACE, s 47(6) provides that, where a person who has been granted police bail and either has attended at a police station or has been arrested under s 46A is detained at a police station, any previous time in custody must be included in any calculation of detention time and any time during which (s)he was on bail must not be so included. In practice, his/her old custody record will be continued. On the other hand, if the person has been re-arrested on the ground of new evidence the detention clock starts to turn again.

Breach of bail by person under duty to surrender into custody of court: arrest

6.80   BA 1976, s 7 provides a power to arrest without warrant. A person (D) who has been released on bail in criminal proceedings and is under a duty to surrender into the custody of a court may be arrested without warrant by a constable if:

(a) the constable has reasonable grounds for believing that D is not likely to surrender to custody, or

(b) the constable has reasonable grounds for believing that D is likely to break any of the conditions of D's bail or has reasonable grounds for suspecting that D has broken any of those conditions; or

(c) in a case where D was released on bail with one or more sureties, if a surety notifies a constable in writing that D is unlikely to surrender to custody, and that for that reason the surety wishes to be relieved of his/her obligations as a surety.

Unless D was arrested within 24 hours of the time appointed for surrender to custody, D must be brought before a justice as soon as practicable and in any event within 24 hours. Christmas Day, Good Friday, and any Sunday are excluded from the calculation of 24 hours. If arrested within 24 hours of the surrender time, D must be brought before the court at which D was to have surrendered to custody.

Failure to surrender: offence

By BA 1976, s 6(1), a person (D) who has been bailed commits an offence if D fails **6.81** without reasonable cause to surrender to custody as required. 'Surrender to custody' means surrendering into the custody of the court by entering the dock or of a constable (according to the requirements of the grant of bail) 'at the appointed time and place' and not 'at or about' the appointed time. Consequently, it is no defence that D was only slightly late. Moreover, if D had reasonable cause for failing to surrender at the appropriate time, D commits an offence, contrary to s 6(2), if (s)he fails to surrender to custody at the appointed place as soon after the appointed time as is reasonably practicable. It is for D to prove 'reasonable cause'.

An offence under s 6(1) or (2) is punishable on summary conviction with up to three months' imprisonment and/or an unlimited fine, or by the Crown Court as if it were a criminal contempt of court punishable with up to 12 months and/or an unlimited fine.

Where D has been released on bail by a court and subsequently fails to surrender to custody, D must be brought before the court at which proceedings in respect of which bail was granted are to be heard. No written charge should be issued to commence proceedings for such failure. The court in question should initiate proceedings for an offence of failing to surrender to bail on its own motion, following an express invitation by the prosecutor. On the other hand, where D has been bailed from a police station to appear before a magistrates' court, proceedings for an offence of failure to surrender to bail should be initiated by way of charging D.

CAUTIONS AS ALTERNATIVE TO PROSECUTIONS

The various types of cautions are among the alternatives to prosecution referred to at **6.82** 3.2. They are available where someone admits to committing an offence. They form part of an offender's criminal record. Offenders must be made aware of this before agreeing to accepting one of these cautions.

Simple cautions

A simple caution can be offered by a police officer to someone aged 18 or over.   **6.83**

Under the Criminal Justice and Courts Act 2015 (CJCA 2015), s 17(2), where the offence is *triable only on indictment*, a police officer may not give a simple caution except:

(a) in exceptional circumstances relating to the person or the offence, and
(b) with the consent of the DPP (given by a Crown Prosecutor).

In the case of *other offences*, the police may—subject to CJCA 2015, s 17(3) or (4)—make a decision to offer a simple caution without reference to a Crown Prosecutor, unless the DPP's *Guidance* (**6.48**) requires the case to be so referred. Whenever a case is referred to a Crown Prosecutor, the Crown Prosecutor's decision is binding on the police. Where the relevant requirements in CJCA 2015, s 17(3) or (4) (below) are not met, the offender must be charged with the offence.

By CJCA 2015, s 17(3), where the offence is a specified *either way* offence, a police officer may not give a simple caution except in exceptional circumstances relating to the person or the offence. The following offences are specified in the Criminal Justice and Courts Act 2015 (Simple Cautions) (Specification of Either-Way Offences) Order 2015, viz an offence under the:

(a) Children and Young Persons Act 1933, s 1(1);
(b) Prevention of Crime Act 1953, s 1(1) or 1A(1);
(c) Sexual Offences Act 1956, s 14(1) or 15(1) (indecent assault on woman/man respectively: both repealed), where the victim of the offence was under the age of 16 at the time of the offence;
(d) Indecency with Children Act 1960, s 1(1) (repealed);
(e) Firearms Act 1968, s 19, where the offence was committed in relation to a firearm within the meaning of s 57(1) and is triable either way;
(f) Misuse of Drugs Act 1971, s 4(2) or (3) or 5(3), where the offence was committed in relation to a Class A drug and is triable either way;
(g) Protection of Children Act 1978, s 1(1);
(h) Customs and Excise Management Act 1979, s 50(2) and (3), 68(2), 107(1), or 170(2) in connection with a prohibition having effect by virtue of the Misuse of Drugs Act 1971, s 3(1), where the offence was committed in relation to a Class A drug and is triable either way;
(i) Criminal Justice Act 1988, s 139(1), 139A(1) or (2), 139AA(1), or 160(1);
(j) Sexual Offences (Amendment) Act 2000, s 3(1) (abuse of position of trust) (repealed);
(k) Nationality, Immigration and Asylum Act 2002, s 145(1), (2) or (3) (traffic in prostitution) (repealed);
(l) Sexual Offences Act 2003, s 3(1) or 4(1) and (5), where the victim of the offence was under the age of 16 at the time of the offence;
(m) Sexual Offences Act 2003, s 7(1), 8(1) and (3), 9(1) and (3), 10(1) and (3), 11(1), 12(1), 14(1), 15(1), 16(1), 17(1), 18(1), 19(1), 25(1) and (4)(b), 25(1) and (5), 26(1) and (4)(b), 26(1) and (5), 47(1) and (4)(b), 47(1) and (5), 48(1), 49(1), 50(1), 57(1), 58(1), 59(1), or 59A(1) (ss 57–59A (sexual trafficking) repealed);
(n) Coroners and Justice Act 2009, s 62(1);
(o) Modern Slavery Act 2015, s 1 or 2.

By CJCA 2015, s 17(4), where:

(a) the offence is a summary offence or an either-way offence not specified above; and

(b) in the two years before the commission of the offence the person concerned has been convicted of, or cautioned for, a similar offence,

a police officer may not give him/her a simple caution except in exceptional circumstances relating to the person, the offence committed or the previous offence. (For the purposes of (b), a person has been 'cautioned for' an offence if (s)he has been given a simple caution, a conditional caution, a youth caution or a youth conditional caution.)

By CJCA 2015, s 17(5), it is for a police officer not below a rank specified in the Criminal Justice and Courts Act 2015 (Simple Cautions) (Specification of Police Ranks) Order 2015 to determine whether there are exceptional circumstances for the purposes of s 17(2), (3), or (4), and whether a previous offence is similar to the offence admitted for the purposes of s 17(4). Such a determination must be made in accordance with guidance to be issued by the Secretary of State. The ranks are as follows:

Determination	Minimum rank
Whether there are exceptional circumstances for the purposes of s 17(2)	Superintendent
Whether there are exceptional circumstances for the purposes of s 17(3) or s 17(4), or whether a previous offence is similar to the offence admitted for the purposes of s 17(4)	Inspector

Whatever the offence, a simple caution must not be offered to someone who has not admitted committing the offence, or who does not agree to accept it. In such a case, the offender must be charged with the offence. A simple caution may only be given if the decision-maker is satisfied that there is sufficient evidence to charge the offender (ie sufficient to provide a realistic prospect that the offender would be convicted). In addition, a simple caution must not be given if the decision-maker considers that it is in the public interest for the offender to be prosecuted. Although the victim should be consulted, (s)he cannot insist that the case is disposed of in a particular way.

Simple cautions do not have a statutory basis, although their existence is recognised by a number of statutes. The Ministry of Justice has issued guidance, *Simple Cautions for Adult Offenders*, which should be applied to all decisions relating to simple cautions. The guidance should be used in conjunction with the DPP's *Guidance on Charging*.

Youth cautions

A constable may give a child or young person (ie someone under 18) (Y) a youth caution under CDA 1998, s 66ZA if: **6.84**

(a) (s)he decides that there is sufficient evidence to charge Y with an offence,
(b) Y admits to the constable that Y committed the offence, and
(c) the constable does not consider that Y should be prosecuted or given a youth conditional caution (YCC) (below) in respect of the offence.

Unlike a simple (adult) caution, Y's agreement to the caution is not required. A youth caution may only be given if the decision-maker is satisfied that it would not be in the public interest to prosecute or offer a YCC.

Although the victim should be consulted, (s)he cannot insist that the case is disposed of in a particular way.

A youth caution must be given in the presence of an appropriate adult.

The constable must explain in ordinary language to Y and to the appropriate adult, the effect of a youth caution, and any guidance which has been given.

The Secretary of State has issued guidance (*Youth Cautions: Guidance for Police Officers and Youth Offending Teams*) as to the circumstances in which it is appropriate to give youth cautions, the places where youth cautions may be given, the form which youth cautions are to take, and the manner in which they are to be given and recorded.

Reference to youth offending team

6.85   CDA 1998, s 66ZB provides as follows.

If a constable gives a youth caution to a person (Y), (s)he must as soon as practicable refer Y to a youth offending team (YOT). On a referral, the YOT must assess Y, and unless they consider it inappropriate to do so, must arrange for Y to participate in a rehabilitation programme.

If Y has not previously been referred under the above provision and has not previously been given a YCC, the YOT may assess Y, and may arrange for Y to participate in a rehabilitation programme.

If a person who has received:

(a) two or more youth cautions is convicted of an offence committed within two years beginning with the date of the last of those cautions, or

(b) a YCC followed by a youth caution is convicted of an offence committed within two years beginning with the date of the youth caution,

the court:

(i) must not conditionally discharge him/her in respect of the offence unless it is of the opinion that there are exceptional circumstances relating to the offence or him/her that justify it doing so, and

(ii) where it does so, must state in open court that it is of that opinion and its reasons for that opinion.

A youth caution given to a person, and a report on a failure by a person to participate in a rehabilitation programme, may be cited in criminal proceedings in the same circumstances as a conviction of the person may be cited.

Conditional cautions

Conditional cautions: adults

6.86   CJA 2003, Part 3 (ss 22–27) makes provision for 'conditional cautions'. An '*authorised person*' (ie a constable, policing supporting officer or policing support volunteer designated under the Police Reform Act 2002, s 38, or person authorised by a relevant prosecutor (as defined below)) may give a conditional caution to a person aged 18 or over if:

(a) (s)he has evidence that the person has committed the offence;

(b) a relevant prosecutor (a Crown Prosecutor or the like) or the authorised person has decided that there is sufficient evidence to charge him/her and that a conditional caution should be given;

(c) the person admits to committing the offence;

(d) the authorised person has explained the effect of the conditional caution to the offender and has warned him/her that failure to comply with any of the conditions may result in a prosecution for the offence; and

(e) the offender has signed a document which contains details of the offence; an admission of guilt; his/her consent to being given a conditional caution; and the conditions attached to the caution.

Although the victim should be consulted, (s)he cannot insist that the case is disposed of in a particular way.

The conditions which may be attached to such a caution are those which have one or more of the following objects:

(a) facilitating the rehabilitation of the offender;
(b) ensuring that the offender makes reparation for his/her offence; and
(c) punishing the offender.

Typical conditions are payment of compensation, letters of apology, or drink or drugs referral interventions. CJA 2003, Part 3 provides that the conditions may include a condition that the offender pays a financial penalty (but only in respect of offences or descriptions of offences prescribed by the Secretary of State) which must not exceed one-quarter of the maximum fine to which the offender would have been liable on summary conviction or £250, whichever is the smaller.

The CJA 2003 (Conditional Cautions: Financial Penalties) Order 2013 provides that a financial penalty condition may be attached to the following descriptions of offence and that the amount listed against each description of offence is the maximum penalty which may be specified in such a condition:

Any summary offence	£50
Any either-way offence	£100
Any indictable only offence	£150

A conditional caution given to a 'relevant foreign offender' may have conditions attached to it that have the object of bringing about his/her departure from the UK and/or ensuring that (s)he does not return to the UK, whether or not these are in addition to conditions with objects specified above. A 'relevant foreign offender' means someone in respect of whom (a) directions for removal from the UK have been, or may be, given under the Immigration Act 1971, Sch 2 or the Immigration and Asylum Act 1999, s 10, or (b) a deportation order is in force under the Immigration Act 1971, s 5.

Before deciding what conditions to attach to a conditional caution, a relevant prosecutor or the authorised person must make reasonable efforts to obtain the views of the victim (if any) of the offence, and in particular the victim's views as to whether the offender should carry out any of the actions listed in the community remedy document. *If the victim expresses the view that the offender should carry out a particular action listed in the community remedy document, the prosecutor or authorised person must attach that as a condition unless it seems to the prosecutor or authorised person that it would be inappropriate to do so.* Where there is more than one victim and they express different views, or for any other reason the italicised words do not apply, the prosecutor or authorised person must nevertheless take account of any views expressed by the victim (or victims) in deciding what conditions to attach to the conditional caution.

The Secretary of State has issued a revised Conditional Cautioning Code of Practice and the DPP has issued a revised *Guidance on Adult Conditional Cautioning*. Provision is made for the National Probation Service to provide assistance in deciding whether conditional cautions should be given, the conditions which might be attached and the supervision and rehabilitation of persons so cautioned.

If an offender fails, without reasonable excuse, to comply with any of the conditions attached to the conditional caution, criminal proceedings may be instituted for the original offence and the document signed by the offender referred to above is admissible in evidence in those proceedings. The caution ceases to have effect with the institution of proceedings.

Youth conditional cautions

6.87   The Crime and Disorder Act 1998 (CDA 1998), s 66A provides for youth conditional cautions (YCCs). The Secretary of State has issued a Code of Practice for such cautions. The YCC provisions apply to those aged 10 or over but under 18. A YCC can be given by an 'authorised person' (**6.86**) to an offender aged 10 or over but under 18 where five requirements are satisfied. These requirements are essentially the same as those in respect of 'Conditional cautions: adults' as set out at **6.86**(a)-(e) substituting 'YCC' for 'conditional caution'. The requisite explanation and warning must be given in the presence of the appropriate adult.

The conditions which may be attached must have one or more of these objects:

(a) facilitating the offender's rehabilitation;
(b) ensuring that (s)he makes reparation;
(c) punishing him/her.

A condition may be attached that the offender attend at a specified place at a specified time (but not for more than 24 hours, excluding attendance required for the purpose of rehabilitation). In addition, a condition requiring the payment of a financial penalty may be attached to a YCC if the offence committed is a prescribed offence. The CDA 1998 (Youth Conditional Cautions: Financial Penalties) Order 2013 prescribes any summary offence, any offence triable either way, and any offence triable on indictment. The 2013 Order also prescribes the maximum penalties which may be attached to a YCC:

Description of offence	Offender 14 or over but under 18	Offender 10 or over but under 14
Any summary offence	£30	£15
Any either-way offence	£50	£25
Indictable only offence	£75	£35

There is a duty to consult victims; its terms are identical to those relating to adult conditional cautions, above.

If a YCC is given by an authorised officer, the officer must refer the offender to a youth offending team as soon as possible.

If the offender fails without reasonable excuse to comply with any of the conditions attached to a YCC (s)he may be prosecuted for the offence in question whereupon the YCC ceases to exist. The document signed by the young offender referred to above is admissible in evidence in proceedings for the offence.

Arrest for failure to comply

CJA 2003, s 24A(1), applied to YCCs by CDA 1998, s 66E, provides that, where a constable has reasonable grounds for believing that the offender has failed, without reasonable excuse, to comply with any of the conditions attached to a conditional caution or a YCC, the constable may arrest the offender without warrant. A person so arrested must be:

(a) charged with the offence in question;
(b) released without charge and without bail (with or without any variation in the conditions attached to the caution) unless (c)(i) and (ii) applies; or
(c) released without charge and on bail if—
  (i) the release is to enable a decision to be made as to whether the person should be charged with the offence; and
  (ii) the pre-conditions for bail (**6.6**) are satisfied (s 24A(2)).

Section 24A(2) also applies:

(i) where a person who, having been released on bail under (c) above, returns to a police station to answer bail or is otherwise in police detention at a police station;
(ii) where a person who, having been released on street-bail, attends to answer that bail or is otherwise in police detention; and
(iii) where a person has been arrested for failure to answer to police bail.

Section 24A(2) does not require a person who falls within (i) or (ii) above, and is in police detention in relation to another matter, to be so released if (s)he is liable to be kept in detention in relation to that other matter.

A person arrested under s 24A, or any other person to whom s 24A(2) applies, may be kept in police detention:

(a) to enable him/her to be dealt with in accordance with s 24A(2) (and this includes power to keep him/her in detention if necessary for the purpose of investigating whether (s)he has failed, without reasonable excuse, to comply with any of the conditions attached to his/her conditional caution); or
(b) where applicable, to enable a custody officer to appoint a different or additional time for answering to police bail.

Where a person is not in a fit state to be so dealt with, (s)he may be kept in police detention until (s)he is.

CJA 2003, s 24B provides that certain provisions in PACE (those dealing with limitations on police detention, duties of a custody officer, records kept by a custody officer, duties and responsibilities of custody officers, x-rays, and ultrasound scans) apply, with modifications, to persons arrested for suspected breach of a conditional caution or a YCC as they do to offenders arrested in respect of an offence.

PACE, ss 30 to 31, 34, 36, 37(4)–(6), 38, 39, and 55A apply (with any necessary modifications) to a person arrested under CJA 2003, s 24A.

CHAPTER 7

Identification Methods

7.1 Identification by witnesses who saw the crime committed may be made in a video identification, identification parade or similar procedure. There may also be an identification by fingerprints or footwear impressions, or by body samples and impressions. The relevant PACE Code of Practice (Code D: the Code of Practice for the Identification of Persons by Police Officers) is concerned with these methods. The Code is also concerned with the keeping of records and with the taking of photographs of arrested people.

GENERAL PRINCIPLES

7.2 PACE Code D provides certain general principles, in addition to those set out in the third paragraph of **4.1**, which apply to all methods by which identification can be made. It provides that:

(1) References to a police officer include a person designated under the Police Reform Act 2002, s 38 or 39 (**2.4**, **2.7**) acting in the exercise or performance of powers or duties under his/her designation.

(2) Where a record is made of any action requiring the authority of an officer of a specified rank, the name (except in the case of terrorism inquiries, or where it is reasonably believed that it might be a source of danger, in which cases his/her warrant or other identification number and his/her police station should be used) and rank of the officer must be included in the record.

(3) All records must be timed and signed by the maker. This is subject to the same exceptional provision as in (2).

(4) In the case of a detained person, records must be made in the custody record unless otherwise specified. References in Code D to written records, forms, and signatures include electronic records and forms and electronic confirmation that identifies the person completing the record or form.

(5) If an officer suspects, or is told, that a person may be mentally disordered or otherwise mentally vulnerable, that person must be treated as such, in the absence of clear evidence to dispel that suspicion.

(6) If a person appears to be under 18, (s)he must, in the absence of clear evidence that (s)he is older, be treated as a juvenile.

(7) Where the consent of the suspect to a procedure is required, the consent of a mentally disordered or mentally vulnerable suspect is only valid if given in the presence of the appropriate adult; and in the case of a juvenile his/her parent or guardian must consent in addition to the juvenile (unless (s)he is under 14, in which case his/her parent or guardian's consent suffices in its own right).

(8) Where a person is blind or seriously visually impaired or unable to read, the custody officer or identification officer must ensure that his/her solicitor, relative, the appropriate adult or some other person likely to take an interest in him/her (and not involved in the investigation) is available to help in checking any

documentation. Where Code D requires written consent or signification, the person assisting may be asked to sign if the detained person so wishes.

(9) If any information concerning the processes of an identification must be given to or sought from a suspect, it must be given or sought in the presence of the appropriate adult if the suspect is mentally disordered or mentally vulnerable, or a juvenile. If the appropriate adult is not present when the information is first given or sought, the procedure must be repeated in his/her presence when (s)he arrives. If the suspect is deaf or there is doubt about his/her hearing ability or his/her ability to understand English, the custody officer or identification officer must ensure that the necessary arrangements in accordance with PACE Code C are made for an interpreter to assist the suspect.

(10) Any procedure in Code D involving the participation of a witness who is or appears to be mentally disordered, otherwise mentally vulnerable or a juvenile should take place in the presence of a pre-trial support person unless the witness states that (s)he does not want a support person to be present. A support person must not be allowed to prompt any identification of a suspect by a witness. The support person should not be (or not be likely to be) a witness in the investigation.

The terms 'appropriate adult' and 'solicitor' where they appear above have the same meaning as in Code C (**5.7** and **5.23**).

Persons other than police officers, who are police staff but not 'designated persons', may be allowed to carry out procedures or tasks at the police station if the law allows. Such persons must be employees under the control of the chief officer of police or employed by a person contracted to provide services relating to persons arrested or otherwise in custody. The custody officer or the officer given custody must ensure that Code D is complied with.

IDENTIFICATION AND RECOGNITION OF SUSPECTS

In Code D, the material under this heading is divided into three parts: **7.3**

(a) Part A: Identification of a suspect by an eye-witness;
(b) Part B: Recognition by *controlled* showing of films, photographs and other images; and
(c) Part C: Recognition by *uncontrolled* viewing of films, photographs, and other images.

The eye-witness identification procedures in Part A should not be used to test whether a witness can recognise a person as someone (s)he knows and would be able to give evidence of recognition along the lines that 'On [the described date, time, location and circumstances] I saw an image of an individual who I recognised as AB.' In these cases, the procedures in Part B apply if the viewing is controlled and the procedure in Part C apply if the viewing is not controlled.

Part A: Identification of a suspect by an eye-witness

Part A applies when an eye-witness has seen someone committing the crime or has **7.4** seen someone in any other circumstances which tend to prove or disprove his/her involvement in the crime, for example, close to the scene of the crime, immediately before or immediately after it was committed. Part A sets out the procedures to be used to test the ability of that eye-witness to identify a person suspected of involvement in the offence as the person whom (s)he saw on the previous occasion.

(Code D states that, while it concentrates on visual identification procedures, it does not preclude the police making use of aural identification procedures such as a 'voice identification parade', where they judge that appropriate.)

A record of the description of the suspect (S) as first given by the eye-witness must be made and kept in a form which enables details of that description to be accurately produced from it, in a visible and legible form, which can be given to S or to his/her solicitor in accordance with Code D. Unless otherwise specified, the record must be made before the eye-witness takes part in any identification procedures. A copy must be provided to S or his/her solicitor before any procedures under the Code are carried out.

Methods where the suspect is known and available

7.5   References in Part A:

(1) to the identity of the suspect (S) being '*known*' mean that there is sufficient information known to the police to establish, in accordance with PACE Code G (Arrest), that there are reasonable grounds to suspect a particular person of involvement in the offence;

(2) to S being '*available*' mean that S is immediately available, or will be available within a reasonably short time, in order that (s)he can be invited to take part in at least one of the following *eye-witness identification procedures*:

Video identification (where the eye-witness is shown images of a known suspect, together with similar images of other people who resemble him/her: see **7.12**). Moving images must be used unless (a) or (b) apply:

(a) applies if:

(i) the identification officer (below: IO), in consultation with the officer in charge of the investigation, is satisfied that because of aging, or other physical changes or differences, the appearance of S has significantly changed since the occasion when the eye-witness claims to have seen S;

(ii) an image (moving or still) is available which IO and the officer in charge of the investigation reasonably believe shows the appearance of S as it was at the time S was seen by the eye-witness; and

(iii) having regard to the extent of change and the purpose of eye-witness identification procedures, IO believes that such an image should be shown to the eye-witness.

In such a case, IO may arrange a video identification procedure using the image described in (ii). S must first be given an opportunity to provide his/her own image(s) for use in the procedure but it is for IO and officer in charge of the investigation to decide whether, following (ii) and (iii), any image(s) provided by S should be used.

A video identification using an image described above may, at the discretion of IO, be arranged in addition, or as an alternative, to a video identification using *moving* images taken after S has been given the Notice to Suspect (**7.8**).

(b) applies if, in accordance with Annex A of Code D (**7.12**), IO does not consider that replication of a physical feature or concealment of the location of the feature can be achieved using a moving image. In these cases, still images may be used;

Identification parade (where the eye-witness sees S in a line of other people who resemble S: see **7.17**); and

Group identification (where the eye-witness sees S in an informal group of people: see **7.21**),
and it is practicable to arrange an effective procedure under the relevant provisions.

The arrangements for, and conduct of, the eye-witness identification procedures and circumstances in which any such procedure must be held are the responsibility of an officer *not below inspector rank* who must *not be involved in the investigation* ('the identification officer' (IO)). IO may direct another officer or police staff to make arrangements for, and to conduct, any of these identification procedures and references to IO include such an officer or police staff. In delegating these arrangements and procedures, IO must be able to supervise effectively and either intervene or be contacted for advice. Where any action referred to above is taken by another officer or police staff at the direction of IO, the outcome must, as soon as practicable, be reported to IO. For the purpose of these procedures, IO retains overall responsibility for ensuring that the procedure complies with Code D and in addition, in the case of a detained suspect, his/her care and treatment until returned to the custody officer.

Except as permitted by Code D, no officer or any other person involved with the investigation of the case against S may take any part in the eye-witness identification procedures or act as IO. There will be a breach of this prohibition, not only if an officer investigating an offence participates in the actual identification process, but also if (s)he takes the witness to the police station at which an identification is to be attempted. This does not prevent IO from consulting the officer in charge of the investigation to determine which procedure to use.

Circumstances in which an identification procedure must be held

If, before *any eye-witness identification procedure* has been held: **7.6**

(a) an eye-witness has identified a suspect or purported to have identified him/her; or
(b) there is an eye-witness available who expresses an ability to identify the suspect (S); or
(c) there is a reasonable chance of an eye-witness being able to identify S,

and the eye-witness in (a) to (c) has not been given an opportunity to identify S in any of the eye-witness identification procedures, an identification procedure must be held if S disputes being the person the eye-witness claims to have seen on a previous occasion, unless:

(i) it is not practicable to hold any such procedure; or
(ii) any such procedure would serve no useful purpose in proving or disproving whether S was involved in committing the offence, for example
  (a) where S admits being at the scene of the crime and gives an account of what took place and the eye-witness does not see anything which contradicts that; or
  (b) when it is not disputed that S is already known to the eye-witness who claims to have recognised S when seeing him/her commit the crime.

An eye-witness identification procedure may also be held if the officer in charge of an investigation considers it would be useful.

Selecting an identification procedure

7.7   If an identification procedure is to be held, S must initially be offered by the officer in charge a video identification, unless:

(a) this is impracticable, or
(b) an identification parade is practicable and more suitable, or
(c) a group identification is practicable and more suitable than the other two methods.

IO and the officer in charge of the investigation must discuss which option is to be offered. An identification parade may not be practicable because of factors such as the number of witnesses, their state of health, availability and travelling requirements. A video identification would normally be more suitable if, in a particular case, it could be arranged and completed sooner than an identification parade. Before an option is offered S must also be reminded of his/her entitlement to have free legal advice.

Where S refuses the first offer of one of these procedures S must be asked to state his/her reason and may obtain advice from his/her solicitor and appropriate adult if present. All such persons must be allowed to make representations as to why another identification procedure should be used. These matters must be recorded. After consideration of such reasons and representations IO must, if appropriate, arrange for an alternative which IO considers to be suitable and practicable to be offered to S. If IO decides that it is not suitable and practicable to offer an alternative, his/her reasons must be recorded.

Notice to Suspect

7.8   Before a video identification, an identification parade, or a group identification is arranged, the following must be explained to S:

(a) the purpose of the video identification, identification parade, or group identification;
(b) S's entitlement to free legal advice;
(c) the procedures for holding it (including S's right to have a solicitor or friend present);
(d) that S does not have to take part in a video identification, identification parade, or group identification;
(e) that, if S does not consent to and take part in a video identification, identification parade, or group identification, his/her refusal may be given in evidence in any subsequent trial and police may proceed covertly without his/her consent or make other arrangements to test whether an eye-witness can identify him/her;
(f) whether, for the purposes of a video identification procedure, images of S have previously been obtained either:
  (i) in accordance with the penultimate paragraph under the above heading (i.e. **7.8**), and if so, that S may co-operate in providing further, suitable images to be used instead; or
  (ii) in accordance with the provisions about when moving images do not have to be used (see **7.5**), and if so, that S may provide his/her own images for IO to consider using;
(g) where appropriate, the special arrangements for juveniles, or mentally disordered or mentally vulnerable persons;
(h) that should S significantly alter his/her appearance between being offered an identification procedure and any attempt to hold it, this may be given in evidence and IO may then consider other forms of identification;

(i) that a moving image or photograph may be taken of S when S attends for any iden-
    tification procedure;

(j) whether the eye-witness has been shown photographs, a computerised or artist's
    composite likeness or similar likeness or picture by the police;

(k) that if S changes his/her appearance before an identification parade it may not be
    practicable to arrange one on the day in question or subsequently and, because of
    his/her changed appearance, IO may consider alternative methods of identifica-
    tion; and

(l) that S or his/her solicitor will be provided with details of the description of S as first
    given by any eye-witnesses who are to attend the video identification, identification
    parade, group identification, or confrontation.

S must also be given a written 'Notice to Suspect' containing this information and
a reasonable opportunity to read it. S must then be asked to sign a second copy of the
notice to indicate whether (s)he is willing to participate in the making of a video or an
identification parade or group identification. IO must retain the signed copy.

In the case of a detained suspect, the above duties may be performed by the custody
officer or by another police officer or police staff not involved in the investigation as
directed by the custody officer if it is proposed to:

(a) release S in order that an identification procedure can be arranged and carried out
    (as where S is bailed to attend an identification parade) and an inspector is not
    available to act as IO before S leaves the station, or

(b) keep S in police detention while the procedure is arranged or carried out, and
    waiting for an inspector to act as IO would cause unreasonable delay.

Where IO and the officer in charge of the investigation have reasonable grounds to
suspect that, if given the above information and Notice, S would take steps to avoid
being seen by a witness in any identification procedure, IO may arrange for images of
S for use in a video identification procedure to be obtained before giving the information and Notice. If this is done, S may co-operate in providing new images to be used
instead, if suitable.

A witness must not be shown photographs, computerised or artist's composite likenesses or similar likenesses or pictures (including 'E-fit' images) if the identity of the
suspect (S) is known and S is available to take part in a video identification, identification parade or group identification.

Where the suspect is known but not available

Where a known suspect is not available or has ceased to be available, IO may make ar-   **7.9**
rangements for a video identification in accordance with Code D. If necessary, IO may
follow the video identification procedures but using *still* images. Any suitable moving
or still images may be used and these may be obtained covertly if necessary; covert
activity must be limited to that which is necessary. (A visual recording made in accordance with Code F may be used for eye-witness identification procedures to which the
previous three sentences apply.) Alternatively, IO may make arrangements for a group
identification. IO may arrange a confrontation where no other option is available. The
requirements for the giving of information to, and seeking it from, S, or for S to have
the opportunity to view the images before they are shown to an eye-witness, do not
apply if S's lack of co-operation prevents the necessary action.

These provisions would apply where a known suspect deliberately makes him/herself unavailable in order to delay or frustrate arrangements being made for obtaining evidence. They enable any suitable images of S (moving or still) which are available (eg from custody and other CCTV systems) or can be obtained to be used in a video identification.

Documentation

7.10   A record must be made of any video identification, identification parade, group identification, or confrontation. Where IO considers that it is impracticable to hold a video identification or identification parade requested by S, the reasons must be recorded and explained to S. So must a victim's failure or refusal to co-operate in a video identification, identification parade, or group identification. If applicable, the grounds for obtaining images must be recorded. Any records relating to these procedures must be made on the forms provided.

Where the identity of the suspect is not known

7.11   In such cases, an eye-witness may be taken to a neighbourhood or place to see whether (s)he can identify the person whom (s)he saw on the relevant occasion. Although there can be no control over the general mix of people, their age, sex, race and general description or manner of dress, the principles governing the formal identification procedures must be followed so far as practicable. For example:

(1) Where practicable, a record should be made of any description of S given by the eye-witness, before (s)he is asked to make an identification.

(2) Care must be taken not to provide the eye-witness with any information concerning the description of the suspect (if such information is available) and not to direct the eye-witness's attention to any individual unless this is unavoidable. This does not prevent an eye-witness being told to look carefully at people who are around, or to look towards a group or in a particular direction if this appears necessary to ensure that the witness does not overlook a possible suspect simply because the eye-witness is looking in the opposite direction and also to enable him/her to make comparisons between any suspect and others in the area.

(3) Where there is more than one eye-witness, every effort should be made to keep them separate and eye-witnesses should be taken to see whether they can identify a person independently.

(4) Once there is sufficient information to justify the arrest of a particular individual, eg, after the eye-witness makes a positive identification, the formal identification procedures must be adopted for any other eye-witnesses in relation to that individual.

(5) The officer or police staff accompanying the eye-witness must record in his/her report book the action taken as soon as practicable and in as much detail as possible; 'report book' includes any official report book or electronic recording device issued to police officers or police staff that enables the record to be made and dealt with in accordance with (5). That record should include the date, time, and place of the previous occasion when the eye-witness claims to have seen S; where any identification was made; how it was made and the conditions at the time (eg, the distance

which the witness was from S, the weather and light); if the witness's attention was drawn to S; the reason for this; and anything said by the eye-witness or S about the identification or the conduct of the procedure.

An eye witness must not be shown photographs, computerised or artist's composite likenesses, or similar likenesses or pictures (including 'E-fit' images) if, in accordance with **7.5**(1) and (2), the identity of S is known, and is available to take part in a video identification, identification parade or group identification. If S's identity is not known, the showing of any such images to a witness to see if (s)he can identify a person whose image (s)he is shown as the person (s)he saw on a previous occasion which is permissible, must be done in accordance with Annex E.

Annex E provides as follows.

A sergeant (or above) must be responsible for supervising and directing the showing of photographs, but the actual showing may be done by a constable or police staff. The supervising officer must confirm that the first description of S given by the eye-witness (W) has been recorded before W is shown the photographs. If the supervising officer is unable to confirm that the description has been recorded, (s)he must postpone the showing.

Only one eye-witness may be shown photographs at any one time. W must be given as much privacy as practicable and must not be allowed to communicate with any other witness in the case. Not less than 12 photographs must be shown at a time, which, as far as possible, must all be of a similar type.

When W is shown photographs, W must be told that the photograph of the person whom W has said that (s)he has previously seen might or might not be among them and that if (s)he cannot make an identification (s)he should say so. W must be told not to make a decision until (s)he has viewed at least 12 photographs. W must not be prompted or guided in any way but must be left to make any selection without help. If W makes an identification from photographs then, unless the person identified is otherwise eliminated from the inquiries or is not available, other eye-witnesses must not be shown photographs. However, both they and the eye-witness who has made the identification must be asked to attend a video identification, identification parade, or group identification unless there is no dispute about the identification of S. If W makes a selection but is unable to confirm the identification the person showing the photographs must ask W how sure W is that the photograph indicated is the person that W saw on a previous occasion.

Where the use of a computerised or artist's composite likeness or similar likeness has led to there being a known suspect who can be asked to participate in a video identification, an identification parade, or a group identification, that likeness may not be shown to other potential eye-witnesses.

None of the photographs used may be destroyed, whether or not an identification is made, since they may be required for production in court. The photographs should be numbered and a separate photograph taken of the frame or part of the album from which W made an identification as an aid to reconstituting it.

A visual recording made in accordance with Code F may be used for eye-witness identification procedures to which Annex E applies.

Whether or not an identification is made, a record must be kept of the showing of photographs. This must include anything said by the eye-witness about any identification or the conduct of the procedure, and any reasons it was not practicable to comply with any of the Code provisions governing the showing of photographs.

Execution: video identification

7.12   The following rules are laid down by Annex A to Code D as to how video identification should be carried out.

General

7.13   The arrangements for obtaining and ensuring the availability of a suitable set of images to be used in a video identification must be the responsibility of an investigating officer (IO) who has no direct involvement with the case.

The set of images must include the suspect (S) and at least eight other people who, so far as possible, resemble S in age, general appearance, and position in life. Only one suspect may appear on any set unless there are two suspects of roughly similar appearance, in which case they may be shown together with at least 12 other people.

If S has an unusual physical feature, eg a facial scar, tattoo or distinctive hairstyle or hair colour which does not appear on the images of the other people available to be used, steps may be taken, electronically or otherwise, to:

(a) conceal the location of the feature on the images of S and the other people; or
(b) replicate that feature on the images of the other people.

IO has discretion to choose whether to conceal or replicate the feature and the method to be used.

If a feature is concealed or replicated, the reason and whether the feature was concealed or replicated must be recorded.

If the eye-witness requests to view an image where an unusual physical feature has been concealed or replicated without that feature being concealed or replicated, the witness may be allowed to do so.

The images used to conduct a video identification must, so far as possible, show S and other people in the same positions or carrying out the same sequence of movements. They must also show S and other people under identical conditions unless IO reasonably believes that:

(a) because of S's failure or refusal to co-operate or other reasons, this is not practicable; or
(b) any difference in the conditions would not direct an eye-witness's attention to any individual image.

The reason why identical conditions were not practicable must be recorded.

Provision must be made for each person filmed to be identified by number. If police officers are filmed, any numerals or other identifying badges must be concealed. If a prison inmate is filmed, either as a suspect or not, either all or none of the persons filmed should be in prison clothing.

S or his/her solicitor, friend, or appropriate adult must be given a reasonable opportunity to see the complete set of images before it is shown to any eye-witness. If S has a reasonable objection to the set of images or any of its participants, S must be asked to state his/her reason. If practicable, steps must be taken to remove the grounds for objection. If this is not practicable, S and/or his/her representative must be told why his/her objections cannot be met. The objection, the reason given for it and why it cannot be met must be recorded.

Before the images are shown S or his/her solicitor must be provided with the details of the first description of S by any eye-witnesses who are to attend the video identification. S or his/her solicitor must also be allowed to view any material released to the media by the police for the purpose of recognising or tracing S, provided this is practicable and would not unreasonably delay the investigation.

No unauthorised people may be present when the video identification is conducted. S's solicitor, if practicable, must be given reasonable notification of the time and place it is to be conducted. S's solicitor may only be present at the video identification on request and with the prior agreement of IO, if IO is satisfied that the solicitor's presence will not deter or distract any eye-witness from viewing the images and making an identification. If IO is not satisfied and does not agree to the request, the reason must be recorded. The solicitor must be informed of the decision and its reason, and that (s)he may then make representations about why (s)he should be allowed to be present. The representations may be made orally or in writing, in person or remotely by electronic communication and must be recorded. These representations must be considered by an inspector (or above) who is not involved with the investigation and responsibility for this may not be delegated. If, after considering the representations, the officer is satisfied that the solicitor's presence will deter or distract the eye-witness, the officer must inform the solicitor of the decision and its reason and ensure that any response by the solicitor is also recorded. If allowed to be present, the solicitor is not entitled to communicate in any way with an eye-witness during the procedure but this does not prevent the solicitor from communicating with IO. S may not be present when the images are shown to any eye-witness and is not entitled to be informed of the time and place the video identification procedure is to be conducted. The video identification procedure itself must be recorded on video with sound. The recording must show all persons present within the sight or hearing of the eye-witness whilst the images are being viewed and must include what the eye-witness says and what is said to him/her by IO and by any other person present at the video identification procedure. A supervised viewing of the recording of the video identification procedure by S and/or S's solicitor may be arranged on request, at the discretion of the investigating officer (IO). Where the recording of the video identification procedure is to be shown to S and/or his/her solicitor, IO may arrange for anything in the recording that might allow the eye-witness to be identified to be concealed if IO considers that this is justified. IO may also arrange for anything in that recording that might allow any police officers or police staff to be identified to be concealed.

Conduct of video identification

IO must ensure that, before seeing the set of images, eye-witnesses cannot communicate with each other about the case, see any of the images, see, or be reminded of, any photograph or description of S or be given any other indication as to S's identity, or overhear an eye-witness who has seen the material. There must be no discussion with the eye-witness (W) about the composition of the set of images and W must not be told whether a previous eye-witness has made any identification.

7.14

Only one eye-witness (W) may see the set of images at a time. Immediately before the images are seen, W must be told that the person W saw might or might not appear in the images W is shown and that if W cannot make a positive identification should say so. W must be advised that at any point (s)he may ask to see a particular part of the set of images or to have a particular image frozen for him/her to study. Furthermore, it should be pointed out to W that there is no limit on how many times W can view the whole set of images or any part of them. However, W should be asked to refrain from making any decision until (s)he has seen the entire set at least twice.

Once W has seen the whole set of images at least twice and has indicated that (s)he does not want to view the images or any part of them again, W must be asked to say whether the individual (s)he saw in person on an earlier occasion has been shown and, if so, to identify him/her by number. W will then be shown that image to confirm the identification.

Care must be taken not to direct W's attention to any one individual image, or to give any other indication of S's identity. Where an eye-witness has previously made an identification by photographs, or a photofit, identikit, or similar picture, (s)he must not be reminded of such a photograph or picture once a suspect is available for identification by other means in accordance with Code D. Neither must W be reminded of any description of S.

If after the video identification procedure has ended, the eye-witness (W) informs any police officer or police staff involved in the post-viewing arrangements that (s)he wishes to change his/her decision about his/her identification, or (s)he has not made an identification when in fact (s)he could have made one, an accurate record of the words used by W, and of the circumstances immediately after the procedure ended, must be made. If W has not had an opportunity to communicate with other people about the procedure, IO has the discretion to allow W a second opportunity to make an identification by repeating the video identification procedure using the same images but in different positions.

Each eye-witness must be asked after the procedure whether (s)he has seen any broadcast or published films or photographs or any description of suspects relating to the offence; his/her reply must be recorded. The record must be in accordance with the provisions in **7.31** about a record insofar as they can be applied to the viewing in question.

Image security and destruction

7.15 IO must ensure that all relevant material containing sets of images used for a specific identification procedure is kept securely and its movement accounted for. In particular, no-one involved in the investigation against S may be permitted to view the material prior to its being shown to any witness.

Like any other photograph taken in accordance with PACE, s 64A (**7.83–7.86**), where a video film has been made all copies of it must be destroyed with the exceptions set out at **7.87**. An opportunity of witnessing the destruction must be given to S, if S so requests within five days of being cleared or informed that (s)he will not be prosecuted.

Documentation

7.16 A record must be made of all those participating in or seeing the set of images whose names are known to the police.

A record of the conduct of the video identification must be made on the forms provided. This must include anything said by the witness about the identification or the conduct of the procedure and any reasons why it was not practicable to comply with any provisions of Code D governing the conduct of a video identification.

Identification parades

7.17 Identification parades must be carried out in accordance with Annex B to Code D, which provides as follows.

A suspect (S) must be given a reasonable opportunity to have a solicitor or friend present, and S must be asked to indicate his/her wishes in this respect on a second copy of the 'Notice to Suspect'. A parade may take place in a normal room or in one equipped with a screen permitting eye-witnesses to see members of the parade without being seen.

Before the parade takes place, S or his/her solicitor must be provided with the details of the first description of S by any eye-witnesses who are to attend. S or his/her solicitor should be allowed to view any material released to the media by the police for the purpose of identifying and tracing S, provided it is practicable to do so and would not unreasonably delay the investigation.

Cases involving prison inmates

If a prison inmate is required for identification, and there are no security problems about his/her leaving the establishment, (s)he may be asked to participate in an identification parade or video identification.   **7.18**

A parade may be conducted in a Prison Department establishment. If it is, it must be conducted as far as practicable under normal parade rules. Members of the public must make up the parade unless there are serious security or control objections to their admission to the establishment. In such cases, or if a video or group identification is arranged within the establishment, other inmates may participate.

If S is an inmate, S should not be required to wear prison clothing for the parade unless the other persons taking part are other inmates in prison clothing or are members of the public who are prepared to wear prison clothing for the occasion.

Conduct of an identification parade

Immediately before the parade, S must be reminded of the procedure governing its conduct and given the appropriate caution (**5.27** and **5.29**). All unauthorised persons must be excluded from the place where the parade is held.   **7.19**

Once the parade has been formed, everything afterwards in respect of it must take place in the presence and hearing of S and of any interpreter, solicitor, friend, or appropriate adult who is present (unless the parade involves a screen, in which case everything said to or by any eye-witness must be said in the hearing and presence of S's solicitor, friend, or appropriate adult or be video recorded). No investigating officer should enter the room where the parade is being held.

The parade must consist of at least eight persons (other than S) who, so far as possible, resemble S in age, height, general appearance, and position in life. Where a suspect has an unusual physical feature, for example a facial scar or tattoo or distinctive hairstyle or hair colour which cannot be replicated on other members of the identification parade, steps may be taken to conceal the location of that feature on S and other members of the parade if S and his/her solicitor or appropriate adult agree. A plaster or a hat may achieve such an objective. It is also permissible to take reasonable steps in good faith to make non-suspects resemble S by the use of make-up, provided there is no objection.

One suspect only may be included in a parade unless there are two suspects of roughly similar appearance, in which case they may be paraded together with at least 12 other persons. In no circumstances may more than two suspects be included in one parade, and where there are separate parades they must be made up of different persons.

Where all members of a similar group are possible suspects, separate identification parades must be held for each member of the group unless there are two suspects of similar appearance. Where police officers in uniform form an identification parade, numerals or other identifying badges must be concealed.

When S arrives, S must be asked whether (s)he has any objection to the arrangements for the parade or to any of the other participants in it and to state reasons for any objection made. S may obtain advice from his/her solicitor or friend, if present, before the parade proceeds. If S has a reasonable objection to the arrangements or to any of the participants, steps must, where practicable, be taken to remove the grounds for objection. Where this is impracticable, the officer must explain to S why his/her objection cannot be met and the objection, the reason for it and why it cannot be met must be recorded on the forms provided.

S may select his/her own position in the line. Where there is more than one eye-witness, IO must tell S, after each witness has left the room, that S can if (s)he wishes change position. Each position must be clearly numbered, whether by means of a numeral laid on the floor in front of each parade member or by other means.

Appropriate arrangements must be made to ensure, before they attend the parade, that eye-witnesses are not able to:

(a) communicate with each other or overhear an eye-witness who has already seen the parade;
(b) see any member of the parade;
(c) see, or be reminded of, any photograph or description of S, nor are given any other indication of his/her identity; or
(d) see S, either before or after the parade.

The person conducting an eye-witness to the parade must not discuss with him/her its composition; in particular, (s)he must not disclose whether a previous witness has made any identification.

Eye-witnesses must be brought in one at a time. Immediately before a witness inspects the parade, (s)he must be told that the person (s)he saw might or might not be on the parade and that if (s)he cannot make a positive identification (s)he should say so. The eye-witness (W) must also be told that (s)he should not make any decision before looking at each member of the parade at least twice. When the officer or police staff member conducting the procedure is satisfied that W has properly looked at each member of the parade, (s)he must ask W whether the person (s)he saw in person on an earlier relevant occasion is on the parade and, if so, to indicate the number of the person concerned. Where this takes place behind a screen it is desirable for W to be asked to make a note of the number of the person identified so that (s)he may give direct evidence of that fact. However, if an eye-witness is unable to recall that number at a subsequent trial, evidence from the person who conducted the parade as to the number indicated by W is admissible as there is statutory authority for its admission. If W makes an identification after the parade has ended, S and, if present, his/her solicitor, interpreter, or friend must be informed. Where this occurs, consideration should be given to allowing W a second opportunity to identify S.

If an eye-witness wishes to hear any parade member speak, or to see him/her adopt any specified posture or move, W must first be asked whether (s)he can identify any persons on the parade on the basis of appearance only. When the request is to hear members of the parade speak, W must be reminded that the participants in the parade have been chosen on the basis of physical appearance only. Members of the parade may then be asked to comply with W's request to hear them speak or to see them move or adopt any specified posture.

If W requests that the person indicated by him/her remove anything used to conceal the location of an unusual physical feature, that person may be asked to remove it.

Each eye-witness must be asked after the parade whether (s)he has seen any broadcast or published films or photographs or any description of suspects relating to the offence; his/her reply must be recorded in the same way as at the end of **7.14.**

When the last eye-witness has left, S must be asked whether (s)he wishes to make any comments on the conduct of the parade.

Documentation

A video recording of the parade must normally be taken. Where this is impracticable, a **7.20** colour photograph must be taken. A copy of the video or photograph must be supplied, on request, to S or his/her solicitor within a reasonable time. The rules about the destruction and retention of such a video or photograph are the same as those described at **7.85-7.87.**

If any person is asked to leave the parade because (s)he is interfering with its conduct, the circumstances must be recorded. A record must be made of all those present at an identification parade whose names are known to the police. A record of the conduct of the parade must be made on the forms provided, including anything said by the witness or S about any identifications or the conduct of the procedure, and any reasons why it was not practicable to comply with any provision of Code D.

Group identification

Group identification must be carried out in accordance with Annex C to Code D, **7.21** which provides as follows.

General

A group identification may take place either with S's consent and co-operation or covertly without his/her consent.

The location is a matter for IO, although (s)he may take into account representations made by a suspect, appropriate adult, his/her solicitor, or friend. It should be somewhere where other people are passing by, or waiting around informally, in groups so that S is able to join them and is capable of being seen at the same time as others in the group, for example where people are leaving an escalator, walking through a shopping centre, or in queues. Where group identification is carried out covertly, its location depends on where S can be found (along with a number of other people). A suitable location might be a route regularly travelled by S, including buses, trains, and public places frequented by S.

In selecting the location, IO must consider the general appearance and number of persons likely to be present. In particular, (s)he must reasonably expect that persons broadly similar to S will appear from time to time during the period of the witness's observation. A group identification need not be held where IO believes that, because of S's unusual appearance, none of the practicable locations is likely to make the identification fair.

Immediately after a group identification (whether with or without S's consent) a colour photograph or a video should be taken of the scene, where practicable, so as to give a general impression of the scene and the number of people present. Alternatively, if practicable, the group identification may be video recorded. If such a photograph or video is impracticable, then a photograph or film of the scene may be made later when practicable.

If, at the time, S is on his/her own rather than in a group, the procedure remains a group identification.

Before the group identification takes place S or his/her solicitor must be provided with details of the first description of S by any eye-witness attending it, and should be allowed to view any material released to the media for the purpose of recognising or tracing S, provided that it is practicable to do so and it will not unreasonably delay the investigation.

After the procedure each eye-witness must be asked whether (s)he has seen any broadcast or published films or photographs or any descriptions of suspects relating to the offence; his/her reply must be recorded in the same way as at the end of **7.14**.

Group identification with the consent of the suspect

7.23   S must be given a reasonable opportunity to have a solicitor or friend present. (S)he must be asked to indicate his/her wishes on a second copy of the Notice to Suspect. The witness, person carrying out the procedure, S's solicitor, appropriate adult, friend, and any interpreter for the eye-witness (W) may be concealed from the sight of the persons in the group if this facilitates the identification. The person conducting a witness to the location must not discuss the forthcoming group identification nor disclose whether a previous witness has made an identification.

Anything said to or by an eye-witness during the procedure regarding the identification must be said in the presence and hearing of those at the procedure. Eye-witnesses who have not yet attended the identification must not be able to:

(a) communicate with each other about the case or overhear a witness who has already been given an opportunity to see S in the group;

(b) see S; or

(c) see or be reminded of any photograph or description of S or to be given any other indication of his/her identity.

Witnesses must be brought to the place singly and must be told that the person they saw might or might not be in the group and that if they cannot make an identification they should say so. W must then be asked to observe the group; the manner of doing so will depend upon whether the group is stationary or moving.

Moving groups   When the group in which S is to appear is moving, eg leaving an escalator, the following provisions apply.

If two or more suspects consent to a group identification, they should each be subject to different identification procedures, which may be conducted consecutively. The person conducting the procedure must ask W to observe the group and ask him/her to point out anyone (s)he thinks (s)he saw on the earlier occasion. S should then be allowed to take up whatever position in the group (s)he prefers. When an identification is made, W must, where practicable, be asked to take a closer look to confirm identification. If this is impracticable, or W is unable to confirm the identification, W must be asked how sure (s)he is that the person is the relevant person. The duration of the identification process must be such as the person conducting the procedure reasonably believes necessary for W to be able to make comparisons between S and other individuals of broadly similar appearance.

Stationary groups   When the group in which S is to appear is stationary, eg people in a queue, the following provisions apply.

If two or more suspects consent to a group identification, there should generally be two separate procedures. However, if they are of broadly similar appearance, they may appear in the same group. Separate stationary group identifications must consist of different people.

S may select his/her position. Where there is more than one eye-witness, S must be told, out of sight and hearing of any witness, that (s)he may change his/her position. W must be asked to pass along or amongst the group and to look at each person at least twice before making an identification. Once W has done so, (s)he must be asked if the person (s)he saw previously is in the group and to indicate that person by any means considered appropriate by the person conducting the identification. If this is not practicable, W will be asked to point out that person. W must, where practicable, be asked to take a closer look and confirm his/her identification. If this is impracticable, W must be asked how sure (s)he is that the person is the one seen on the earlier occasion.

All cases   An unreasonable delay by S in joining the group, or (having joined the group) his/her deliberate self-concealment from the sight of W, may be considered as a refusal to co-operate in the identification.

Where an eye-witness identifies someone other than S, that person should be asked if (s)he is prepared to give his/her name and address. (S)he is not obliged to do so. There is no duty to record persons present in the group or at the place where the procedure is conducted.

At the end of the procedure S must be asked to comment on the conduct of the procedure. Unless previously informed, S must be told of any identifications made by witnesses.

Group identifications without suspect's consent

These should, so far as possible, follow the rules set out above. As such an identification will take place without S's knowledge, no solicitor, etc will be present. Any number of suspects may be identified at the same time.   **7.24**

Group identifications in police stations

These must only take place for reasons of security or safety, or because it is impracticable to hold them elsewhere. The group identification may be in a room equipped with a one-way screen, or elsewhere in a police station. Safeguards applicable to identification parades must be followed where practicable.   **7.25**

Group identifications involving prison inmates

These may only take place in a prison or police station. They must follow the procedure which is applicable to group identifications in a police station. Where a group identification takes place in a prison, other inmates may participate. If S is in prison clothing, all persons taking part must be so dressed.   **7.26**

Documentation

Where a photograph or video is taken a copy must be supplied on request to S or his/her solicitor within a reasonable time. Such material must be destroyed in accordance with the rules described at **7.87**. A record of the conduct of the identification must be made on the forms provided, and must include anything said by the witness or suspect about any identification or the conduct of the procedure and any reason why it was impracticable to comply with any of the relevant provisions of the Code.   **7.27**

Confrontation by an eye-witness

A confrontation does not require S's consent.   **7.28**

The rules concerning confrontation, which are set out in Annex D to Code D, are simple.

Before the confrontation takes place:

(a) the eye-witness must be told that the person (s)he saw might or might not be the person (s)he is to confront and that if (s)he cannot make a positive identification (s)he should say so;

(b) S or S's solicitor must be provided with the details of the first description of S given by any eye-witness who is to attend the confrontation.

Where a broadcast or publication has been made for the purpose of recognition and tracing of suspects, S or S's solicitor should also be allowed to view any material released by the police to the media, provided that it is practicable to do so and would not unreasonably delay the investigation.

Force cannot be used to make S's face visible to an eye-witness.

S must be confronted independently by each eye-witness, who must be asked 'is this the person?'. If the eye-witness identifies the person but is unable to confirm the identification (s)he must be asked how sure (s)he is that the person is the person (s)he saw on the earlier relevant occasion. Confrontation must take place in the presence of S's solicitor, interpreter, or friend, unless this would cause unreasonable delay.

The confrontation should normally take place in the police station, either in a normal room or one equipped with a screen permitting an eye-witness to see S without being seen. In both cases, the procedures are the same, except that a room equipped with a screen may be used only when S's solicitor, friend, or appropriate adult is present or the confrontation is recorded on video.

After the confrontation each eye-witness must be asked whether (s)he has seen any broadcast or published films or photographs or any descriptions of suspects relating to the offence; his/her reply must be recorded in the same way as at the end of **7.14**.

Informal identification

7.29   Informal evidence of identification may be admitted provided that the evidence was obtained in good faith and has no adverse effect on the fairness of the proceedings.

Where a suspect and an eye-witness are well known to each other, there is less need for any formal out-of-court identification procedure to be used. However, where S asks for a parade, the procedure is governed by Code D and the failure to provide one may result in the exclusion of evidence, whether S and the witness were previously known to one another or not.

Part B: Recognition by controlled showing of films, photographs, and other images

7.30   Part B of Code D's section on the identification and recognition of suspects applies when, *for the purposes of obtaining evidence of recognition*, arrangements are made for a person, including a police officer, who is not an eye-witness:

(a) to view a film, photograph or any other visual medium; and

(b) on that occasion, to be asked whether (s)he recognises anyone whose image is shown in the material as someone who is known to him/her.

The arrangements for such viewings may be made by the officer in charge of the relevant investigation. Although there is no requirement for the identification officer to make the arrangements or to be consulted about them, nothing prevents this.

A guidance note states that the admissibility and value of evidence of recognition obtained when carrying out the procedures in Part B may be compromised if before the person is recognised, the witness who has claimed to know that person is given or is made, or becomes aware of, information about that person which was not previously known to him/her personally but which (s)he has purported to rely on to support his/her claim that the person is in fact known to him/her.

To provide safeguards against mistaken recognition and to avoid any possibility of collusion, on the occasion of the viewing, the arrangements should ensure that:

(a) the films, photographs, and other images are shown on an individual basis;
(b) any person who views the material:
   (i) is unable to communicate with any other individual to whom the material has been, or is to be, shown;
   (ii) is not reminded of any photograph or description of any individual whose image is shown or given any other indication as to the identity of any such individual;
   (iii) is not told whether a previous witness has recognised any one;
(c) immediately before a person views the material, (s)he is told that:
   (i) an individual who is known to him/her may, or may not, appear in the material (s)he is shown and that if (s)he does not recognise anyone, (s)he should say so;
   (ii) at any point, (s)he may ask to see a particular part of the material frozen for him/her to study and there is no limit on how many times (s)he can view the whole or any part or parts of the material; and
(d) the person who views the material is not asked to make any decision as to whether (s)he recognises anyone whose image (s)he has seen as someone known to him/her until (s)he has seen the whole of the material at least twice, unless the officer in charge of the viewing decides that because of the number of images the person has been invited to view, it would not be reasonable to ask him/her to view the whole of the material for a second time. A record of this decision must be included in the record that is made.

Documentation

Part B requires that a record of the circumstances and conditions under which the **7.31** person is given an opportunity to recognise the individual must be made and that the record must include a record of:

(a) whether the person knew or was given information concerning the name or identity of any suspect;
(b) what the person has been told before the viewing about the offence, the person(s) depicted in the images, or the offender and by whom;
(c) how and by whom the witness was asked to view the image or look at the individual;
(d) whether the viewing was alone or with others, and if with others the reason for it;
(e) the arrangements under which the person viewed the film or saw the individual and by whom those arrangements were made;
(f) the name and rank of the officer responsible for deciding that the viewing arrangements should be made in accordance with Part B;
(g) the date, time, and place images were viewed or further viewed or the individual was seen;

(h) the times between which the images were viewed or the individual was seen;

(i) how the viewing of images or sighting of the individual was controlled and by whom;

(j) whether the person was familiar with the location shown in any images or the place where (s)he saw the individual and if so, why; and

(k) whether or not on this occasion, the person claims to recognise any image shown, or any individual seen, as being someone known to him/her, and if (s)he does:

    (i)   the reason,

    (ii) the words of recognition,

    (iii) any expressions of doubt,

    (iv) what features of the image or the individual triggered the recognition.

The above record may be made:

(a) by the person who views the image or sees the individual and makes the recognition; and

(b) if applicable, by the officer or police staff in charge of showing the images to that person or in charge of the conditions under which that person sees the individual.

The person must be asked to read and check the completed record and as applicable, confirm that it correctly and accurately reflects the part (s)he played in the viewing.

It is important that this record (and the corresponding record under Part C below) is made as soon as practicable after the viewing and whilst it is fresh in the mind of the individual who makes the recognition.

Part C: Recognition by *uncontrolled* viewing of films, photographs, and images

7.32   Part C of Code D's section on the identification and recognition of suspects applies when, for the purpose of identifying and tracing suspects, films and photographs of incidents or other images are:

(a) shown to the public (which may include police officers and police staff as well as members of the public) through the national or local media or any social media networking site; or

(b) circulated through local or national police communication systems for viewing by police officers and police staff; and

the viewing is not formally controlled and supervised as set out in Part B.

A copy of the relevant material released to the national or local media for showing as described in (a), must be kept. The suspect or his/her solicitor must be allowed to view such material before any video identification, identification parade, group identification, or confrontation is carried out, provided it is practicable and would not unreasonably delay the investigation.

As noted above, each eye-witness involved in any of the eye-witness identification procedures must be asked, after (s)he has taken part, whether (s)he has seen any film, photograph, or image relating to the offence or any description of the suspect which has been broadcast or published as described in (a) and his/her reply recorded. *If (s)he has, (s)he should be asked to give details of the circumstances and subject to the eye-witness's recollection, the record described below should be completed.*

As soon as practicable after an individual (member of the public, police officer, or police staff) indicates in response to a viewing that (s)he may have information relating to the identity and whereabouts of anyone (s)he has seen in that viewing, arrangements should be made to ensure that (s)he is asked to give details of the circumstances and, subject to the individual's recollection, a record of the circumstances and conditions under which the viewing took place is made. This record must be made in accordance with the provisions of Part B insofar as they can be applied to the viewing in question.

IDENTIFICATION BY FINGERPRINTS AND FOOTWEAR IMPRESSIONS

PACE, ss 61 and 61A and Code D deal with the taking of fingerprints (which term in-   **7.33** cludes palmprints) and footwear impressions. These provisions do not apply in respect of someone arrested or detained as a suspected terrorist under the Terrorism Act 2000, s 41 (**21.46**) or Sch 7 (port and border controls) or the Counter-Terrorism and Border Security Act 2019, Sch 3, Pt 1 (border security); separate rules apply to such a person (see **21.52**).

Fingerprints: consent

A person's fingerprints may be taken only with the appropriate consent (which must be   **7.34** in writing if given at a police station) or (in the circumstances set out below) without the appropriate consent. As to 'appropriate consent', see **7.42**.

Where fingerprints are taken at a police station with the appropriate consent, the person must be informed that they may be subject to a speculative search against other fingerprints. A record must be made of these matters and that the person has been so informed.

Taking fingerprints without consent

Powers to take fingerprints from a person over the age of 10 without the appropriate   **7.35** consent are provided by PACE, s 61. This provides that fingerprints may be taken without the appropriate consent from a person:

(1) If (s)he is detained at a police station in consequence of his/her arrest for a record-able offence (**7.43**) or because (s)he has been charged with such an offence or in-formed that (s)he will be reported for it, and (s)he has not had his/her fingerprints taken in the course of the investigation of the offence by the police (s 61 (3) and (4)). (*By s 61(3A), where such a person has had his/her fingerprints taken in the course of the investigation by the police, this does not prevent a second set of prints from being taken if the first did not constitute a full set or are of unsatisfactory quality for their purpose.*)

(2) Where (s)he has answered to bail to appear at a court or police station, if the court, or an inspector (or above), authorises them to be taken (s 61(4A)). This may be done where the court or officer reasonably believes that a person who has sur-rendered to bail is not the person admitted to bail and the person bailed has been previously fingerprinted, or the person who has answered to bail claims to be a different person from the person who had his/her fingerprints taken on a previous occasion.

(3) If (s)he has been arrested for a recordable offence and released, and:

(a) (s)he has not had his/her fingerprints taken in the course of the investigation; or

(b) (s)he has had his/her fingerprints taken in the course of that investigation but (i) the first set did not constitute a full set or are of unsatisfactory quality for their purpose, or (ii) s 61(5C) applies. Section 61(5C) applies where the investigation was discontinued but subsequently resumed, and before that resumption the fingerprints were destroyed pursuant to s 63D(3) (**7.52**) (s 61(5A)).

(4) If, not detained at a police station, (s)he has been charged with a recordable offence or informed that (s)he will be reported for such an offence and:

(a) (s)he has not had his/her fingerprints taken in the course of the investigation; or

(b) (s)he has had his/her fingerprints taken in the course of that investigation but (i) the first set did not constitute a full set or are of unsatisfactory quality for their purpose; or (ii) s 61(5C) applies (s 61(5B)).

(5) If (s)he has been convicted of a recordable offence (or been found not guilty by reason of insanity or found under a disability but to have done the act charged) or (s)he has been cautioned for an admitted recordable offence (s 61(6)). Fingerprints may only be taken under this provision:

(a) if the person has not had his/her fingerprints taken since (s)he was convicted or cautioned, or

(b) if (s)he has had his/her fingerprints taken since then but s 61(3A) applies (s 61(6ZA)).

Fingerprints may only be taken under s 61(6) with the authorisation of an inspector (or above), who may only give such an authorisation if satisfied that taking the fingerprints is necessary to assist in the prevention or detection of crime.

(6) If a constable reasonably suspects that the person is committing or attempting to commit an offence, or has committed or attempted to commit an offence, and the name of the person is unknown to, and cannot be readily ascertained by the constable, or the constable has reasonable grounds for doubting whether a name furnished by the person as his/her name is his/her real name (s 61(6A)). This provision will apply to taking fingerprints at a police station or any other place. Before the power under s 61(6A) is exercised, the constable should:

(a) inform the person of the nature of the suspected offence and why (s)he is suspected of committing it;

(b) give him/her a reasonable opportunity to establish his/her real name before deciding that his/her name is unknown etc or that there are reasonable grounds to doubt that a name which (s)he has given is his/her real name;

(c) as applicable, inform the person of the reason why his/her name is not known etc or of the grounds for doubting that a name which (s)he has given is his/her real name, including, for example, the reason why a particular document the person has produced to verify his/her real name, is not sufficient.

The taking of fingerprints by virtue of s 61(6A) does not count for the purposes of PACE as taking them in the course of an investigation of an offence by the police. Fingerprints taken under this power will have to be destroyed as soon as the purpose for which they were taken has been achieved. They will not be retained or added to IDENT1 (the national automated fingerprint system). The reason for s 61(6A) is that the police may now use mobile digital fingerprint readers which are connected to IDENT1. This enables the taking of fingerprints (of two fingers) at a place other than a police station. These can immediately be checked to assist an officer in deciding upon a course of action.

(7) If:
- (a) the person has been 'convicted' of an offence under the law in force in a country or territory outside England and Wales;
- (b) the act constituting the offence would constitute a qualifying offence if done in England and Wales (whether or not it constituted such an offence when the person was convicted); and
- (c) either
  - (i) the person has not had his/her fingerprints taken on a previous occasion under the present provision, or
  - (ii) (s)he has had his/her fingerprints so taken on a previous occasion but s 61(3A) applies (s 61(6D) and (6E)).

A '*conviction*' for these purposes includes a finding equivalent to one of not guilty by reason of insanity or of disability but that the act charged was committed. Fingerprints may only be so taken with the authorisation of an inspector (or above), who must be satisfied that taking the fingerprints is necessary to assist in the prevention or detection of crime (s 61(6F) and (6G)).

A '*qualifying offence*' means the common law offences of murder, manslaughter, false imprisonment, kidnapping, or indecent exposure, an offence of indecent exposure under the Vagrancy Act 1824, s 4 or Town Police Clauses Act 1847, s 28 (both offences now repealed), the Offences Against the Person Act 1861, s 4, 16, 18, 20–24 or 47, the Explosive Substances Act 1883, s 2 or 3, the Infant Life Preservation Act 1929, s 1, the Children and Young Persons Act 1933, s 1, the Infanticide Act 1938, s 1, the Sexual Offences Act 1956, s 12 or 13 (repealed) (unless other party 16 or over and consenting), any other section of the Sexual Offences Act 1956 (largely repealed) other than ss 18 and 32, the Mental Health Act 1959, s 128 (repealed), the Indecency with Children Act 1960, s 1 (repealed), the Sexual Offences Act 1967, s 5 (repealed), the Criminal Law Act 1967, s 4 (in relation to murder), the Firearms Act 1968, ss 16–18, the Theft Act 1968, s 9, 10 or (where fatal accident involved) s 12A, the Genocide Act 1969, s 1(1) (repealed), the Criminal Damage Act 1971, s 1 (where required to be charged as arson), the Criminal Law Act 1977, s 54 (repealed), the Protection of Children Act 1978, s 1, the Aviation Security Act 1982, s 1, the Child Abduction Act 1984, s 2, the Prohibition of Female Circumcision Act 1985, s 1 (repealed), the Public Order Act 1986, s 1, the Aviation and Maritime Security Act 1990, s 9, the Sexual Offences (Amendment) Act 2000, s 3 (repealed), the International Criminal Court Act 2001, s 51, the Female Genital Mutilation Act 2003, ss 1–3, the Sexual Offences Act 2003, ss 1–19, 25, 26, 30–41, 47–50, 52, 53, 57–59A (repealed), 61–67, 69, and 70, the Domestic Violence, Crime and Victims Act 2004, s 5, or the Modern Slavery Act 2015, s 2, an offence listed under the Counter-Terrorism Act 2008 (C-TA 2008), s 41, aiding, abetting, counselling, or procuring one of the above offences, attempting or conspiring to commit any of them, or an offence of encouraging or assisting crime in relation to any of them.

Any power under PACE, s 61 to take fingerprints without the appropriate consent must be exercised by a constable. Reasonable force may be used.

Before any fingerprints are taken without the appropriate consent under any power under PACE, s 61, the person must be informed of:

- (a) the reason why his/her fingerprints are to be taken;
- (b) the power under which they are to be taken;
- (c) the fact that the relevant authority has been given where it is required;

(d) where the fingerprints are taken at a police station (or, in the case of (2) and (6), elsewhere), that his/her fingerprints may be the subject of a speculative search against other fingerprints; and

(e) that his/her fingerprints may be retained in accordance with the relevant provisions unless they were taken under the power in (6) when they must be destroyed after they have been checked.

A record must be made as soon as possible of these matters and of the fact that the person has been so informed. If a person is detained at a police station when the fingerprints are taken, the reason must be noted on his/her custody record. If force is used a record must be made of the circumstances and those present.

Taking footwear impressions

7.36   PACE, s 61A provides that, with the exceptions below, no impression of a person's footwear may be taken without the appropriate consent (**7.42**). Consent to the taking of an impression of a person's footwear must be in writing if it is given at a time when (s)he is at a police station.

Where a person is detained at a police station, a footwear impression may be taken without appropriate consent from a person over the age of 10 if:

(a) (s)he is detained in consequence of his/her arrest for a recordable offence (**7.43**), or has been charged with a recordable offence, or informed that (s)he will be reported for a recordable offence, and

(b) (s)he has not had an impression taken of his/her footwear in the course of the investigation of the offence by the police.

Reasonable force may be used. A record must be made as soon as possible of the reason for the non-consensual taking and, if force is used, of the circumstances and those present. If a person mentioned in (a) has already had an impression of his/her footwear taken in the course of the investigation of the offence, that fact must be disregarded if the impression taken previously is incomplete, or is not of sufficient quality to allow satisfactory analysis, comparison, or matching (whether in the case in question or generally). An impression taken without the appropriate consent must be taken by a constable.

In all cases where a footwear impression is taken, the person must be informed that the impression may be retained and may be the subject of a speculative search, and this must be recorded as soon as practicable. If the person is at a police station, the fact must be recorded in the custody record.

IDENTIFICATION BY INTIMATE SAMPLES AND NON-INTIMATE SAMPLES

7.37   PACE, ss 62 and 63 contain the basic provisions in this area, but (within the terms of these sections) it is Code D which sets out the detailed procedures. The powers set out below do not apply in respect of someone arrested or detained under the terrorism legislation referred to at **7.33**; separate rules apply to such a person. See **21.52**.

Intimate samples

An intimate sample means a sample of blood, semen, or any other tissue fluid, urine, or pubic hair, a dental impression, or a swab taken from any part of a person's genitals or from a person's body orifices other than the mouth. Intimate samples are governed by PACE, s 62.

An intimate sample may be taken from a person in police detention only with the appropriate consent (**7.42**) in writing and with the requisite authorisation (below) (s 62(1)).

An intimate sample may be taken from a person not in police detention if: (a) in the course of an investigation into an offence, two or more non-intimate samples have been taken which have proved unsuitable or insufficient for a particular form of analysis, (b) the requisite authorisation is given, and (c) the appropriate written consent is given (s 62(1A)).

In these two cases an intimate sample may only be taken if an inspector (or above) authorises it to be taken because (s)he has reasonable grounds for suspecting the involvement of the person in a recordable offence (**7.43**) and for believing that the sample will tend to confirm or disprove this involvement.

An intimate sample may also be taken from a person with the appropriate written consent where:

(a) two or more non-intimate samples suitable for the same means of analysis have been taken from the person under s 63(3E) and (3F) (persons convicted of offence outside England and Wales, etc) but have proved insufficient; and

(b) an inspector (or above), who must be satisfied that taking the sample is necessary to assist in the prevention or detection of crime, authorises it to be taken (s 62(2A)).

Before an intimate sample is taken from a person:

(1) (S)he must be informed:
    (a) of the reason;
    (b) that authorisation has been given and the provisions under which it has been given; and
    (c) that a sample taken at a police station may be the subject of a speculative search.

    The reason referred to in (a) must include, except in a case where the sample is taken under s 62(2A), a statement of the nature of the offence in which it is suspected that the person has been involved. After an intimate sample has been taken from a person, the following must be recorded as soon as practicable: the matters referred to in (a) and (b); if the sample was taken at a police station, the fact that the person has been informed as specified in (c); and the fact that the appropriate consent was given.

(2) (S)he must be warned that a refusal without good cause may harm his/her case in any proceedings against him/her. The Code suggests the use of the following words:

    'You do not have to provide this sample/allow this swab or impression to be taken, but I must warn you that if you refuse without good cause, your refusal may harm your case if it comes to trial.'

The warning must be recorded.

(3) (S)he must be reminded of his/her entitlement to free legal advice. The giving of this reminder must be recorded in the custody record if the person is in police custody.

In any proceedings for an offence, a court may draw such inferences as appear proper from a refusal, without good cause, to consent to the taking of an intimate sample.

An intimate sample, other than a sample of urine, may only be taken from a person by a registered medical practitioner or a registered nurse or registered paramedic. A dental impression may only be taken by a registered dentist.

Non-intimate samples

7.39   There are separate provisions, contained in PACE, s 63, for the taking of a *non-intimate sample*.

A non-intimate sample means hair, other than pubic hair, which includes hair plucked by the root; a sample taken from a nail or under a nail; a swab taken from any part of the body other than a part from which a swab taken would be an intimate sample; saliva; or a 'skin impression' which means any record, other than a fingerprint, which is a record, in any form and produced by any method, of the skin pattern and any other physical characteristics or features of the whole, or any part of, a person's foot or any other part of the body.

Where hair samples are taken for the purpose of DNA analysis (rather than for other purposes such as making a visual match) the suspect (S) should be permitted a reasonable choice as to where the hairs are to be taken from. When hairs are plucked they should be plucked individually unless S prefers otherwise; no more should be plucked than the plucker reasonably considers necessary for a sufficient sample.

Except in the following cases, a non-intimate sample may be taken from a suspect only with the appropriate consent (**7.42**) in writing. The exceptional cases are:

(1) A non-intimate sample may be taken from a person without the appropriate consent if:

    (a) (s)he is in detention as a result of an arrest for a recordable offence (**7.43**), and (s)he has not had a non-intimate sample of the same type and from the same part of the body taken in the course of the investigation of the offence by the police, or (s)he has had such a sample taken but it proved insufficient (s 63(2A)–(2C)); or

    (b) (s)he is being held in custody by the police on the authority of a court, and an inspector (or above) has authorised it to be taken without consent (s 63(3)). Such authorisation may only be given where that officer has reasonable grounds to suspect that the offence in question is a recordable offence and that the sample will tend to confirm or disprove S's involvement in it. An authorisation must not be given if the non-intimate sample concerned consists of a skin impression and such an impression has already been taken in the course of the investigation of the offence and that impression did not prove to be insufficient. Where such authorisation is given, S must be informed, before it is taken, of the grounds on which it has been given, including the nature of the suspected offence. (S)he must also be told that any sample taken may be the subject of a speculative search.

(2) A non-intimate sample may be taken from a person without the appropriate consent if (s)he has been arrested for a recordable offence and released and:

    (a) (s)he has not had a non-intimate sample of the same type and from the same part of the body taken from him/her in the course of the investigation of the offence by the police; or

    (b) (s)he has had a non-intimate sample taken from him/her in the course of that investigation but (i) it was not suitable for the same means of analysis, or (ii) it proved insufficient in quantity or quality for a particular form of analysis, or (iii) s 63(3AA) applies (s 63(3ZA)). Section 63(3AA) applies where the investigation was discontinued but subsequently resumed, and before that resumption any DNA profile derived from the sample was destroyed pursuant to s 63D(3) (**7.52**) and the sample itself was destroyed pursuant to s 63R(4), (5) or (12) (**7.68**).

(3) A non-intimate sample may be taken without the appropriate consent from any person (whether or not (s)he is in police detention or held in custody by the police on the authority of a court) if:

    (a) (s)he has been charged with a recordable offence or informed that (s)he will be reported for such an offence; and

    (b) (i) (s)he has not had a non-intimate sample taken from him/her in the course of the investigation; or

        (ii) (s)he has had a non-intimate sample taken from him/her, but it proved unsuitable or insufficient, or s 63(3AA) applies; or

        (iii) (s)he has had a non-intimate sample taken from him/her in the course of the investigation and the sample has been destroyed pursuant to s 63R (**7.68**) or any other enactment, and it is disputed, in relation to any proceedings relating to the offence, whether a DNA profile (ie information derived from a DNA sample) relevant to the proceedings is derived from the sample (s 63(3A)). A 'DNA sample' means any material that has come from a human body and consists of or includes human cells.

(4) A non-intimate sample may be taken from a person without the appropriate consent if:

    (a) (s)he has been convicted (see below) of a recordable offence, or (s)he has been given a caution in respect of an admitted recordable offence, and

    (b) either of the conditions mentioned in s 63(3BA) is met (s 63(3B)), viz:

        (i) a non-intimate sample has not been taken from him/her since the conviction or caution; or

        (ii) such a sample has been taken from him/her since then but it was not suitable for the same means of analysis, or it proved insufficient.

A non-intimate sample may only be so taken with the authorisation of an inspector (or above), who must be satisfied that taking the sample is necessary to assist in the prevention or detection of crime.

(5) A non-intimate sample may be taken without appropriate consent from a person detained following acquittal on the grounds of insanity or a finding of unfitness to plead (s 63(3C)).

(6) A non-intimate sample may be taken without the appropriate consent from a person if:

    (a) under the law in force in a country or territory outside England and Wales the person has been convicted of an offence under that law;

(b) the act constituting the offence would constitute a qualifying offence (**7.35**(7)) if done in England and Wales (whether or not it constituted such an offence when the person was convicted); and

(c) either:

  (i) the person has not had a non-intimate sample taken from him/her on a previous occasion under the present provision, or

  (ii) (s)he has had such a sample taken from him/her on a previous occasion under it but the sample was not suitable for the same means of analysis, or it proved insufficient (s 63(3E), (3F)).

A non-intimate sample may only be so taken with the authorisation of an inspector (or above), who must be satisfied that taking the sample is necessary to assist in the prevention or detection of crime.

A 'conviction' for these purposes includes a finding equivalent to one of not guilty by reason of insanity or of disability but that the act charged was committed.

Reasonable force may be used to take a non-intimate sample without S's consent under the above provisions.

Before any non-intimate sample is taken without the appropriate consent, the person must be informed of:

(a) the reason for taking it;

(b) the power under which it is taken;

(c) the fact that the relevant authority has been given if any power mentioned in s 63(3), 63(3B), or 63(3E) applies, including the nature of the suspected offence (except if taken (i) under PACE, s 63(3B) from a person convicted or cautioned, or (ii) under PACE, s 63(3E) from a person convicted outside England and Wales).

The 'reason' for taking the sample must include a statement of the nature of the suspected offence, except in cases under s 63(3B), (3E), or (3F).

Before a non-intimate sample is taken with or without consent at a police station, the person must be warned:

(i) that his/her sample or information derived from it may be the subject of a speculative search against other samples and information derived from them; and

(ii) that his/her sample and the information derived from it may be retained in accordance with the relevant provisions.

A record must be made as soon as practicable of the matters in (a) to (c) and that the person was informed of them and given the warnings in (i) and (ii). If force is used a record must be made of the circumstances and those present.

General

7.40   Where clothing needs to be removed in circumstances likely to cause embarrassment, no person of the opposite sex (**4.5**) who is not a medical practitioner or registered healthcare professional may be present (unless, in the case of a juvenile, or mentally disordered or mentally handicapped person, that person specifically requests the presence of a particular adult of the opposite sex who is readily available), nor shall anyone whose presence is unnecessary. However, in the case of a juvenile this is subject to the overriding proviso that such a removal of clothing may take place in the absence of the appropriate adult only if the juvenile signifies in that adult's presence that (s)he prefers the search to be done in his/her absence and the appropriate adult agrees.

As to the destruction of samples: see **7.68**.

IDENTIFICATION BY FINGERPRINTS, FOOTWEAR IMPRESSIONS, AND SAMPLES: GENERAL POINTS

The following general points apply to all these types of identification.   **7.41**

Appropriate consent

'Appropriate consent', in relation to someone, means: **7.42**

(a) if the person is aged 18 or over, the consent of that person;
(b) if the person is aged 14 to 17, the consent of that person and his/her parent or guardian; and
(c) if the person is aged under 14, the consent of his/her parent or guardian.

Recordable offence

Some provisions refer to a 'recordable offence'. The following offences are recordable: **7.43**

(a) any offence punishable with imprisonment (regardless of any statutory prohibition or restriction on the punishment of young offenders);
(b) persistently loitering or soliciting for the purposes of prostitution (Street Offences Act 1959, s 1);
(c) tampering with motor vehicle (RTA 1988, s 25);
(d) touting tickets for designated football match, and touting for car hire services (CJPOA 1994, ss 166(1) and 167);
(e) giving intoxicating liquor to a child under five, exposing children under 12 to risk of burning, and failing to provide for safety of children at entertainments (Children and Young Persons Act 1933, ss 5, 11, and 12);
(f) being drunk in public place (Criminal Justice Act 1967, s 91);
(g) failing to deliver up authority to possess prohibited weapon or ammunition, possession of assembled shotgun by unsupervised person under 15, and possession of air weapon or ammunition for air weapon by unsupervised person under 14 (Firearms Act 1968, ss 5(6), 22(3) and (4));
(h) trespassing in daytime in search of game, refusal by such trespasser to give name and address, and five or more found armed in daytime in search of game and using violence or refusing to give name and address (Game Act 1831, ss 30, 31, and 32);
(i) being drunk in highway or public place (Licensing Act 1872, s 12);
(j) obstructing authorised person exercising various powers under the Licensing Act 2003 (LA 2003, ss 59(5), 96(5), 108(3), and 179(4));
(k) failing to notify licensing authority of change of name or rules of club, or of convictions, and failing to notify court of the holding of a personal licence (LA 2003, ss 82(6), 123(2), and 128(6));
(l) keeping alcohol on premises for unauthorised sale, etc, allowing disorderly conduct on premises, selling alcohol to drunken person, obtaining it for such a person, failing to leave licensed premises, keeping smuggled goods, and allowing unaccompanied children on certain premises (LA 2003, ss 138(1), 140(1), 141(1), 142(1), 143(1), 144(1), and 145(1));
(m) selling alcohol to children, allowing sale to, purchase by or on behalf of children, allowing consumption by or delivery to children, sending a child to obtain

alcohol, delivering alcohol to children, and allowing unsupervised sale by children (LA 2003, ss 146(1) and (3), 147(1), 149(1), (3) and (4), 150(1) and (2), 151(1), (2) and (4), 152(1), and 153(1));

(n) making false statements in applications for licences, etc (LA 2003, s 158(1));

(o) allowing premises to remain open following closure order (LA 2003, s 160(4));

(p) making false statement in relation to an application for sex establishment licence (Local Government (Miscellaneous Provisions) Act 1982, Sch 3);

(q) falsely claiming professional qualification (Nursing and Midwifery Order 2001, art 44);

(r) taking or destroying game or rabbits by night (Night Poaching Act 1828, s 1);

(s) impersonating member of police force, and unlawful possession of police uniform (Police Act 1996, s 90(2) and (3));

(t) conduct likely to cause harassment, alarm, or distress, failing to give notice of public procession, non-compliance with condition imposed on public procession, taking part in prohibited public procession, non-compliance with condition imposed on public assembly, taking part in a trespassory assembly, and non-compliance with directions relating to such assembly (Public Order Act 1986, ss 5, 11, 12(5), 13(8), 14(5), 14B(2), and 14C(3));

(u) failing to co-operate with preliminary test (RTA 1988, s 6);

(v) in connection with designated sporting events, (i) allowing alcohol to be carried on public vehicle, (ii) being drunk on such vehicle, (iii) allowing alcohol to be carried in some other vehicles, (iv) being drunk at designated sports ground (Sporting Events (Control of Alcohol, etc) Act 1985, ss 1(2), 1(4), 1A(2), and 2(2));

(w) at designated football match, (i) throwing missiles, (ii) indecent or racialist chanting, and (iii) invading the playing area (Football (Offences) Act 1991, ss 2, 3, and 4);

(x) taking or riding a pedal cycle without the owner's consent (Theft Act 1968, s 12(5));

(y) purchasing or hiring a crossbow (or part) by person under 17, and unsupervised possession by person under 17 (Crossbows Act 1987, ss 2 and 3);

(z) begging, and persistent begging (Vagrancy Act 1824, ss 3 and 4);

(aa) individual subject to banning order (i) failing to comply with requirement to report etc, (ii) knowingly or recklessly providing false or misleading information in support of application for exemption from a reporting requirement (Football Spectators Act 1989, ss 19(6) and 20(10));

(bb) soliciting, and paying for sexual services of prostitute subjected to force (SOA 2003, ss 51A and 53A);

(cc) racially or religiously aggravated harassment (Crime and Disorder Act 1998, s 31(1)(c));

(dd) the offences under the Data Protection Act 2018.

Fingerprints and samples: requirement to attend

7.44   PACE, Sch 2A provides as follows.

Fingerprints

7.45   A constable may require a person to attend a police station for the purpose of taking his/her fingerprints under s 61(5A) (**7.35**(3)). This power may not be exercised in a case within s 61(5A)(b)(i) (fingerprints taken on previous occasion insufficient, etc) after six months from the day on which the appropriate officer was informed that s

61(3A) (**7.35**(1)) applied. The 'appropriate officer' means the officer investigating the offence for which the person was arrested. In addition, this power may not be exercised in a case falling within s 61(5A)(b)(ii) (fingerprints destroyed in a case where an investigation has been interrupted) after the end of a six-month period beginning with the day on which the investigation was resumed.

A constable may require a person to attend a police station for the purpose of taking his/her fingerprints under s 61(5B) (**7.35**(4)). This power may not be exercised after six months from:

(a) in a case within s 61(5B)(a) (fingerprints not taken previously in the course of investigation), the day on which the person was charged or informed that (s)he would be reported;

(b) in a case within s 61(5B)(b) (fingerprints taken on previous occasion insufficient, etc), the day on which the appropriate officer was informed that s 61(3A) applied. The 'appropriate officer' means the officer investigating the offence for which the person was charged or informed that (s)he would be reported; or

(c) in a case where fingerprints have been destroyed when the investigation was interrupted, the day on which the investigation was resumed.

A constable may require a person to attend a police station for the purpose of taking his/her fingerprints under s 61(6) (fingerprints of convicted or cautioned person) (**7.35**(5)). Where the condition in s 61(6ZA)(a) is satisfied (fingerprints not taken since conviction or caution), this power may not be exercised after two years from the day on which the person was convicted or cautioned. Where the condition in s 61(6ZA)(b) is satisfied (fingerprints taken on previous occasion insufficient, etc), this power may not be exercised after two years from the day on which an officer of the investigating police force was informed that s 61(3A) applied. These provisions do not apply where the offence is a qualifying offence (**7.35**(7)) (whether or not it was such an offence at the time of the conviction or caution).

A constable may require a person to attend a police station for the purpose of taking his/her fingerprints under s 61(6D), (6E) (**7.35**(7)).

Intimate samples

A constable may require a person to attend a police station for the purpose of taking **7.46** an intimate sample from him/her under PACE, s 62(1A) (**7.38**) if, in the course of the investigation of an offence, two or more non-intimate samples suitable for the same means of analysis have been taken from him/her but have proved insufficient.

A constable may require a person to attend a police station for the purpose of taking a sample from him/her under s 62(2A) (**7.38**) if two or more non-intimate samples suitable for the same means of analysis have been taken from him/her under s 63(3E), (3F) (**7.39**(6)) but have proved insufficient.

Non-intimate samples

A constable may require a person to attend a police station for the purpose of taking a **7.47** non-intimate sample from him/her under PACE, s 63(3ZA) (**7.39**(2)). This power may not be exercised in a case falling within s 63(3ZA)(b) (sample taken on a previous occasion not suitable, etc) after six months from the day on which the appropriate officer was informed of the matters specified in s 63(3ZA)(b)(i) or (ii). The 'appropriate officer' means the officer investigating the offence for which the person was arrested. Where a sample, and any DNA profile, have been destroyed in a case where an investigation has been interrupted, the power to require attendance for the taking of a non-intimate sample may not be exercised after the end of a six-month period beginning with the day on which the investigation was resumed.

A constable may require a person to attend a police station for the purpose of taking a non-intimate sample from him/her under s 63(3A) (**7.39**(3)). This power may not be exercised in a case falling within s 63(3A)(a) (sample has not been taken previously since conviction or caution) after six months from the day on which the person was charged or informed that (s)he would be reported. The above power may not be exercised in a case falling within s 63(3A)(b) (sample taken on a previous occasion not suitable, etc) after six months from the day on which the appropriate officer was informed of the matters specified in s 63(3A)(b)(i) or (ii). The 'appropriate officer' means the officer investigating the offence for which the person was charged or informed that (s)he would be reported.

A constable may require a person to attend a police station for the purpose of taking a non-intimate sample from him/her under s 63(3B) (**7.39**(4)). Where the first condition in s 63(3BA) (**7.39**(4)) is satisfied (sample has not been taken since conviction or caution), this power may not be exercised after two years from the day on which the person was convicted or cautioned. Where the second condition in s 63(3BA) is satisfied (sample taken on a previous occasion not suitable, etc), the above power may not be exercised after two years from the day on which an appropriate officer was informed that the sample was unsuitable or insufficient. The 'appropriate officer' means an officer of the police force which investigated the offence in question. These time limits do not apply where the offence is a qualifying offence (as defined at **7.35**(7)).

A constable may require a person to attend a police station for the purpose of taking a non-intimate sample from him/her under s 63(3E), (3F) (**7.39**(6)).

Multiple exercise of power in respect of fingerprints or non-intimate samples

7.48   Where fingerprints or a non-intimate sample have been taken from a person under PACE, s 61 or 63 (as the case may be) on two occasions in relation to any offence, (s)he may not be required to attend a police station to have another set of fingerprints or another such sample taken from him/her under the relevant section in relation to that offence on a subsequent occasion without the authorisation of an inspector (or above). The fact of the authorisation, and the reasons for giving it, must be recorded as soon as practicable after it has been given.

General

7.49   A power conferred by PACE, Sch 2A to require a person to attend a police station for taking fingerprints or a sample under any provision of PACE may be exercised only where the fingerprints or sample may be taken from the person under that provision (and, in particular, if any necessary authorisation for taking the fingerprints or sample under that provision has been obtained). A requirement under Sch 2A must (except in urgent cases) give at least seven days within which to attend the police station; and may require attendance at a specified time of day or between specified times of day. In urgent cases, a shorter period may be specified if authorised by an inspector (or above). The fact of such authorisation and the reasons for giving it must be recorded as soon as practicable. In specifying a period of time or times of day for attendance, the constable must consider whether the fingerprints or sample could reasonably be taken at a time when the person is for any other reason required to attend the police station. The time, etc at which the person must attend may be varied by written agreement between him/her and the constable.

The information and recording requirements referred to at the last paragraph of **7.35** (fingerprints), at **7.38**(1) (intimate samples), and the last three paragraphs at **7.39** (non-intimate samples) are applied by Code D to fingerprints taken under Sch 2A.

A constable may arrest without warrant a person who has failed to comply with a requirement under Sch 2A. A guidance note to Code D states that, to justify the arrest without warrant of a person for non-compliance with a requirement under Sch 2A, the officer making the requirement, or confirming in writing an agreed variation of it, should be prepared to explain how, when, and where the requirement was made or the variation was confirmed and what steps were taken to ensure that the person understood what to do and the consequences of not complying with the requirement.

Speculative searches

PACE, s 63A(1) provides that, where someone has been arrested on suspicion of being **7.50** involved in a recordable offence (**7.43**), or has been charged with a recordable offence, or has been informed that he will be reported for a recordable offence, fingerprints, footwear impressions, or samples or the information derived from samples taken under any power conferred by PACE from the person may be subjected to speculative searches, ie checked against:

(a) other fingerprints, footwear impressions, or samples to which the person seeking to check has access and which are held by or on behalf of any one or more relevant law enforcement agencies (eg a police force or NCA), or are held in connection with or as a result of an investigation of an offence;

(b) information derived from other samples if the information is contained in records to which the person seeking to check has access and which are held as described in (a).

Where:

(i) fingerprints, impressions of footwear, or samples have been taken from any person in connection with the investigation of an offence but otherwise than in circumstances to which s 63A(1) applies, and

(ii) that person has given his/her consent in writing to the use in a speculative search of the fingerprints, of the impressions of footwear, or of the samples and of information derived from them,

the fingerprints, etc and that information may be checked against any of the fingerprints, etc or information mentioned in (a) or (b) of s 63A(1) (s 63A(1C)). A consent given for the above purposes cannot be withdrawn (s 63A(1D)).

Section 63A(1E) and (1F) provide respectively that where fingerprints or samples have been taken under:

(a) s 61(6) or 63(3B) (persons convicted, etc), or

(b) s 61(6D), 62(2A), or 63(3E), (3F) (offences outside England and Wales, etc),

the fingerprints or samples, or information derived from the samples, may be subjected to a speculative search.

Fingerprints taken by virtue of s 61(6A) (**7.35**(6): reasonable suspicion that person whose name is not known is committing or has committed an offence, etc) may be checked against other fingerprints to which the person seeking to check has access and which are held by, or on behalf of, any one or more relevant enforcement authorities or which are held in connection with or as a result of an investigation of an offence.

DESTRUCTION, RETENTION, AND USE OF FINGERPRINTS, FOOTWEAR IMPRESSIONS AND DNA PROFILES SUBJECT TO PACE

7.51   PACE, ss 63D to 63U deal with this.

Destruction rule for fingerprints and DNA profiles

7.52   PACE, s 63D deals with the destruction of:

(a) fingerprints:
   (i) taken from a person under any power conferred by PACE (as described earlier in this chapter), or
   (ii) taken by the police, with the consent of the person from whom they were taken, in connection with the investigation of an offence by the police, and
(b) a DNA profile derived from a DNA sample taken as mentioned in (a)(i) or (ii).

Fingerprints and DNA profiles to which s 63D applies ('*s 63D material*') must be destroyed if it appears to the responsible chief officer of police that:

(a) the taking of the fingerprint or, in the case of a DNA profile, the taking of the sample from which the DNA profile was derived, was unlawful, or
(b) the fingerprint was taken, or, in the case of a DNA profile (as defined at **7.39**(3)(b) (iii)), was derived from a sample taken, from a person in connection with that person's arrest and the arrest was unlawful or based on mistaken identity (s 63D(2)).

The '*responsible chief officer of police*' is the chief officer of police for the police area:

(i) in which the material concerned was taken or derived, or
(ii) in the case of a DNA profile, in which the sample from which the DNA profile was derived was taken.

In any other case, s 63D material must be destroyed unless it is retained under any power conferred by ss 63E to 63O (including those sections as applied by s 63P) (s 63D(3)).

Nothing in s 63D prevents a speculative search, in relation to s 63D material, from being carried out within such time as may reasonably be required for the search if the responsible chief officer of police considers the search to be desirable.

Modification of s 63D destruction rules for particular circumstances

Retention of s 63D material pending investigation or proceedings

7.53   PACE, s 63E provides that s 63D (**7.52**) material taken (or, in the case of a DNA profile, derived from a sample taken) in connection with the investigation of an offence in which it is suspected that the person to whom the material relates has been involved may be retained until the conclusion of the investigation of the offence or, where the investigation gives rise to proceedings against the person for the offence, until the conclusion of those proceedings (for example, the time when charges are dropped or when a verdict is returned).

Further retention of s 63D material: persons arrested for or charged with a qualifying offence

PACE, s 63F deals with the case where s 63D (**7.52**) material: **7.54**

(a) relates to a person who is arrested for, or charged with, or informed that (s)he will be reported for, a qualifying offence (**7.35**(7)) but is not convicted (**7.72**) of that offence, and

(b) was taken (or, in the case of a DNA profile, derived from a sample taken) in connection with the investigation of the offence.

*If the person has previously been convicted of a recordable offence (**7.43**, **7.72**) which is not an excluded offence, or is so convicted before the material is required to be destroyed by virtue of s 63F, the material may be retained indefinitely (s 63F(2)).* The reference in s 63F(2) to a recordable offence includes an offence under the law of a country or territory outside England and Wales where the act constituting the offence would constitute a recordable offence if done in England and Wales (and, in the application of s 63F(2) where a person has previously been convicted, this applies whether or not the act constituted such an offence when the person was convicted).

An '*excluded offence*', in relation to a person, means a recordable offence:

(a) which:
   (i) is not a qualifying offence,
   (ii) is the only recordable offence of which the person has been convicted, and
   (iii) was committed when the person was under 18, and
(b) for which the person was not given a sentence of imprisonment or detention for five years or more.

If a person is convicted of more than one offence arising out of a single course of action, those convictions are treated as a single conviction for the above purposes.

For the purposes of the above definition of 'excluded offence':

(i) references to a recordable offence or a qualifying offence include an offence under the law of a country or territory outside England and Wales where the act constituting the offence would constitute a recordable offence or (as the case may be) a qualifying offence if done in England and Wales (whether or not it constituted such an offence when the person was convicted); and

(ii) in the application of (b) of the definition in relation to an offence under the law of a country or territory outside England and Wales, the reference to a relevant custodial sentence of five years or more is to be read as a reference to a sentence of imprisonment or other form of detention of five years or more.

Otherwise, by s 63F(3), material falling within s 63F(4), (5) or (5A) may be retained until the end of the retention period below.

Material falls within s 63F(4) if it:

(a) relates to a person who is charged with a qualifying offence but is not convicted of that offence, and

(b) was taken (or, in the case of a DNA profile, derived from a sample taken) in connection with the investigation of the offence.

Material falls within s 63F(5) if:

(a) it relates to a person who is arrested for a qualifying offence, other than a terrorism-related qualifying offence, but is not charged with that offence,
(b) it was taken (or, in the case of a DNA profile, derived from a sample taken) in connection with the investigation of the offence, and
(c) the Commissioner for the Retention and Use of Biometric Material has consented to the retention of the material concerned.

Material falls within s 63F(5A) if:

(a) it relates to a person who is arrested for a terrorism-related qualifying offence but is not charged with that offence; and
(b) it was taken (or, in the case of a DNA profile, derived from a sample taken) in connection with the investigation of the offence.

For these above purposes, a terrorism-related offence is an offence to which the terrorism notification requirements (**21.67**) apply.

The retention period is three years beginning with the date on which (as the case may be):

(a) the fingerprints were taken, and
(b) the DNA sample from which the profile was derived was taken (or, if the profile was derived from more than one DNA sample, the date on which the first of those samples was taken).

The responsible chief officer of police (**7.52**) or a specified chief officer of police may apply within a three-month period ending on the last day of the retention period to a District Judge (Magistrates' Courts) for an order extending the retention period for an additional two years. Appeal lies to the Crown Court against an extension order, or a refusal to make such an order. A 'specified chief officer of police' means:

(i) the chief officer of the police force of the area in which the person from whom the material was taken resides, or
(ii) a chief officer of police who believes that the person is in, or is intending to come to, the chief officer's police area.

Consent of Commissioner   Section 63G provides as follows in relation to the consent of the Commissioner for the Retention and Use of Biometric Material referred to above. The responsible chief officer of police may apply in writing to the Commissioner for consent to the retention of s 63D material which falls within s 63F(5)(a) and (b). An application may be made on one of two bases—that the material:

(a) was taken (or, in the case of a DNA profile, derived from a sample taken) in connection with the investigation of an offence where any alleged victim (or intended victim) of the offence was, at the time of the offence, under 18, a vulnerable adult, or associated with the person to whom the material relates; or
(b) is not material to which (a) relates, but the retention of the material is necessary to assist in the prevention or detection of crime.

The responsible chief officer of police must give the person to whom the material relates (P) written notice of an application under s 63G, and the right to make representations. The notice may, in particular, be given to P by leaving it at his/her usual or last known address (whether residential or otherwise), by sending it to P by post at that address, or by sending it to P by email or other electronic means. There is no need to give notice if the whereabouts of P is not known and cannot, after reasonable inquiry, be ascertained.

On such application the Commissioner may consent to the retention of material to which the application relates if (s)he considers that it is appropriate to retain the material. Before deciding whether or not to give consent, the Commissioner must consider any representations by P which are made within 28 days beginning with the day on which the notice is given.

Retention of s 63D material: persons arrested for or charged with a minor offence

PACE, s 63H deals with the retention of s 63D material (**7.52**) which: **7.55**

(a) relates to a person who:
   (i) is arrested for, or charged with, or informed that (s)he will be reported for, a recordable offence (**7.43**) other than a qualifying offence (**7.35**(7)), and
   (ii) if arrested for, or charged with, or informed that (s)he will be reported for, more than one offence arising out of a single course of action, is not also arrested for or charged with a qualifying offence, and
   (iii) is not convicted (**7.72**) of the offence or offences in respect of which (s)he is arrested or charged, and
(b) was taken (or, in the case of a DNA profile, derived from a sample taken) in connection with the investigation of the offence or offences in respect of which the person is arrested or charged.

If, in such a case, the person has previously been convicted of a recordable offence which is not an excluded offence (as defined at **7.54**), the material may be retained indefinitely.

The reference to a recordable offence includes an offence under the law of a country or territory outside England and Wales where the act constituting the offence would constitute a recordable offence if done in England and Wales (and, in relation to the previous paragraph, where a person has previously been convicted, this applies whether or not the act constituted such an offence when the person was convicted).

If a person is convicted of more than one offence arising out of a single course of action, the convictions are treated as a single conviction for the above purpose.

Retention of material: persons convicted of a recordable offence

PACE, s 63I provides that: **7.56**

(a) s 63D material (**7.52**) other than material to which s 63K (below) applies, which:
   (i) relates to a person who is convicted of a recordable offence (**7.43** and **7.72**), and
   (ii) was taken (or, in the case of a DNA profile, derived from a sample taken) in connection with the investigation of the offence or
(b) material taken under s 61(6) (**7.35**(5)) or 63(3B) (**7.39**(4)) which relates to a person who is convicted of a recordable offence,

may be retained indefinitely.

A decision in 2020 of the European Court of Human Rights is of interest in relation to s 63I. In that case, the Court was concerned with a policy in Northern Ireland (where PACE does not apply) to retain indefinitely the DNA profile, fingerprints, and photographs of a person convicted of a recordable offence. The Court held that such indefinite retention constituted a disproportionate interference with the person's right to private life and was not necessary in a democratic society; it was therefore a breach of the European Convention on Human Rights, art 8. It would seem very likely that the Court would take the same view of the indefinite retention of material under s 63I, and that it would also take the same view in respect of s 63F(2) referred to at **7.54**.

Retention of material: persons convicted of an offence outside England and Wales after taking of s 63D material

7.57   By PACE, s 63IA, where:

(a) s 63D material (**7.52**) is taken (or, in the case of a DNA profile, derived from a sample taken) in connection with the investigation of an offence;

(b) at any time before the material is required to be destroyed by virtue of PACE, the person is convicted (**7.72**) of an offence under the law of a country or territory outside England and Wales, and

(c) the act constituting the offence mentioned in (b) would constitute a recordable offence (**7.43**) if done in England and Wales,

the material may be retained indefinitely. Section 63IA does not apply where s 63KA (**7.60**) applies.

Retention of material: persons convicted of an offence outside England and Wales: other cases

7.58   PACE, s 63J provides that:

(a) fingerprints taken from the person under s 61(6D), (6E) (power to take fingerprints without consent in relation to offences outside England and Wales: see **7.35(7)**), or

(b) a DNA profile derived from a DNA sample taken from the person under s 62(2A) or 63(3E), (6F) (powers to take intimate and non-intimate samples in relation to offences outside England and Wales: see **7.38** and **7.39(6)**),

which relate to a person who is convicted (**7.72**) of an offence under the law of any country outside England and Wales may be retained indefinitely.

Retention of s 63D material: exception for persons under 18 convicted of first minor offence

7.59   PACE, s 63K deals with the case where s 63D material (**7.52**), which relates to a person (P) who:

(a) is convicted (**7.72**) of a recordable offence (**7.43**) other than a qualifying offence (**7.35(7)**),

(b) has not previously been convicted of a recordable offence, and

(c) is under 18 at the time of the offence,

was taken (or, in the case of a DNA profile, derived from a sample taken) in connection with the investigation of the offence.

In (b), the reference to a recordable offence includes an offence under the law of a country or territory outside England and Wales where the act constituting the offence would constitute a recordable offence if done in England and Wales (whether or not it constituted such an offence when the person was convicted).

Section 63K(2)–(4) provides as follows where P is given:

(a) a custodial sentence in respect of the offence, the material may be retained until the end of the period consisting of the term of the sentence plus five years (if sentence less than five years), or indefinitely (if the sentence is five years or more);

(b) a sentence other than a relevant custodial sentence in respect of the offence, the material may be retained until the end of the period of five years beginning with the date on which (as the case may be):

   (i) the fingerprints were taken, and

   (ii) the DNA sample from which the profile was derived was taken, or, if the profile was derived from more than one DNA sample, the date on which the first of those samples was taken.

However, if, before the end of the relevant period, P is again convicted of a recordable offence, the material may be retained indefinitely (s 63K(5)). In s 63K(5), a recordable offence includes an offence under the law of a country or territory outside England and Wales where the act constituting the offence would constitute a recordable offence if done in England and Wales.

Retention of s 63D material under s 63IA: exception for persons under 18 convicted of first minor offence outside England and Wales

Under PACE, s 63KA, the following provisions apply where: **7.60**

(a) s 63D material (**7.52**) is taken (or, in the case of a DNA profile, derived from a sample taken) in connection with the investigation of an offence;

(b) at any time before the material is required to be destroyed by virtue of PACE, the person is convicted (**7.72**) of an offence under the law of a country or territory outside England and Wales;

(c) the act constituting the offence mentioned in (b) would constitute a recordable offence (**7.43**) if done in England and Wales but would not constitute a qualifying offence (**7.35(7)**);

(d) the person is aged under 18 at the time of the offence mentioned in (b); and

(e) the person has not previously been convicted of a recordable offence (for these purposes a recordable offence includes an offence under the law of a country or territory outside England and Wales where the act constituting the offence would constitute a recordable offence if done in England and Wales (whether or not it constituted such an offence when the person was convicted)).

Where the person is sentenced to imprisonment or another form of detention for less than five years in respect of the offence mentioned in (b), the s 63D material may be retained until the end of the period consisting of the term of the sentence plus five years.

Where the person is sentenced to imprisonment or another form of detention for five years or more in respect of the offence mentioned in (b), the material may be retained indefinitely.

Where the person is given a sentence other than a sentence of imprisonment or other form of detention in respect of the offence mentioned in (b), the material may be retained until the end of the period of five years beginning with the date on which (s)he was arrested for the offence (or, if (s)he was not arrested for the offence, the date on which (s)he was charged with it).

However, if, before the end of the period within which material may be retained by virtue of s 63KA, the person is again convicted of a recordable offence, the material may be retained indefinitely (s 63KA(6)). In s 63KA(6), a recordable offence includes an offence under the law of a country or territory outside England and Wales where the act constituting the offence would constitute a recordable offence if done in England and Wales.

Retention of s 63D material: persons given a penalty notice

7.61 PACE, s 63L provides that, where s 63D material (**7.52**) relates to a person who is given a penalty notice under the Criminal Justice and Police Act 2001 (**3.40**) and in respect of whom no proceedings are brought for the offence to which the notice relates, and was taken (or, in the case of a DNA profile, derived from a sample taken) from the person in connection with the investigation of that offence, the material may be retained for two years beginning with the date on which (as the case may be):

(a) the fingerprints were taken,

(b) the DNA sample from which the profile was derived was taken, or, if the profile was derived from more than one DNA sample, the date on which the first of those samples was taken.

Retention of s 63D material for purposes of national security

7.62 PACE, s 63M provides that s 63D (**7.52**) material may be retained for as long as a national security determination made by a chief officer of police (ie a determination that the retention of any such material is necessary for the purposes of national security) has effect in relation to it. Such a determination must be made in writing, has effect for a maximum of five years, and may be renewed.

Retention of s 63D material given voluntarily

7.63 PACE, s 63N provides that s 63D material consisting of fingerprints, or a DNA profile derived from a DNA sample, taken from a person with his/her consent, may be retained until it has fulfilled the purpose for which taken or derived (s 63N(1) and (2))).
Such material which relates to a person who:

(a) is convicted (**7.72**) of a recordable offence (**7.43**), or

(b) has previously been convicted of a recordable offence (other than a person who has only one 'exempt conviction'),

may be retained indefinitely (s 63N(3)). A conviction is exempt if it is in respect of a recordable offence, other than a qualifying offence (**7.35**(7)), committed when the person is under 18 (s 63N(4)).

The reference to a recordable offence in s 63N(3)(a) includes an offence under the law of a country or territory outside England and Wales where the act constituting the offence would constitute a recordable offence if done in England and Wales; and the references to a recordable offence in s 63N(3)(b) and (4), and to a qualifying offence in s 63N(4), include an offence under the law of a country or territory outside England and Wales where the act constituting the offence would constitute a recordable offence or (as the case may be) a qualifying offence if done in England and Wales (whether or not it constituted such an offence when the person was convicted).

If a person is convicted of more than one offence arising out of a single course of action, the convictions are treated as a single conviction for the above purpose.

Retention of s 63D material with consent

7.64 By PACE, s 63O, fingerprints (other than fingerprints taken under s 61(6A) (**7.35**(6)) to which s 63D (**7.52**) applies), and a DNA profile to which s 63D applies, may be retained for so long as the person to whom the material relates consents to its retention in writing. Such consent can be withdrawn at any time.

Retention of s 63D material in connection with different offence

If s 63D material (**7.52**) is taken (or, in the case of a DNA profile, derived from a sample taken) from a person (X) in connection with the investigation of an offence, and X is subsequently arrested for, or charged with a different offence, or convicted (**7.72**) of or given a penalty notice for a different offence, ss 63E to 63O (above) and ss 63Q and 63T (below) have effect in relation to the material as if the material were also taken (or, in the case of a DNA profile, derived from a sample taken):   **7.65**

(a) in connection with the investigation of the different offence,
(b) on the date on which X was arrested for that offence (or charged with it, or given a penalty notice for it, if X was not arrested) (s 63P).

Retention of further sets of fingerprints

By PACE, s 63PA, the following provisions apply where s 63D material (**7.52**) is or includes a person's fingerprints ('the original fingerprints').   **7.66**

A constable may make a determination under s 63PA in respect of any further fingerprints taken from the same person ('the further fingerprints') if any of conditions 1 to 3 are met.

Condition 1 is met if the further fingerprints are s 63D material, and the further fingerprints or the original fingerprints were taken in connection with a terrorist investigation.

Condition 2 is met if the further fingerprints were taken from the person in England or Wales under the Terrorism Act 2000, Sch 8 (**21.52**), the Terrorism Prevention and Investigation Measures Act 2011, Sch 6 (**21.68**), or the Counter-Terrorism and Border Security Act 2019, Sch 3.

Condition 3 is met if the further fingerprints are material to which the Counter-Terrorism Act 2008, s 18 (**7.76**) applies, and are held under the law of England and Wales.

Where a determination under s 63PA is made in respect of the further fingerprints:

(a) the further fingerprints may be retained for as long as the original fingerprints are retained under a power conferred by ss 63E to 63O (above) (including those sections as applied by s 63P (above)); and
(b) a requirement under any enactment to destroy the further fingerprints does not apply for as long as their retention is authorised by (a).

However, (a) does not prevent the further fingerprints being retained after the original fingerprints fall to be destroyed if the continued retention of the further fingerprints is authorised under any enactment.

A written record must be made of a determination under s 63PA.

Destruction of copies of s 63D material

This is dealt with by PACE, s 63Q. If fingerprints are required by s 63D (**7.52**) to be destroyed, any copies of the fingerprints held by the police must also be destroyed. If a DNA profile is required by s 63D to be destroyed, no copy may be retained by the police except in a form which does not include information which identifies the person to whom the DNA profile relates.   **7.67**

Destruction rules for samples and footwear impressions

Destruction of samples

7.68   PACE, s 63R provides as follows. Samples taken under any power conferred by PACE, or taken by the police, with the person's consent, in connection with the investigation of an offence, must be destroyed if it appears to the responsible chief officer of police (**7.52**) that:

(a) the taking of the samples was unlawful, or
(b) the samples were taken from a person in connection with that person's arrest and the arrest was unlawful or based on mistaken identity (s 63R(2)).

Where neither (a) nor (b) applies, (i) a DNA sample so taken must be destroyed as soon as a DNA profile has been derived from the sample, or if sooner, before the end of six months beginning with the date on which the sample was taken; and (ii) any other sample so taken must be destroyed before the end of six months beginning with the date on which it was taken (s 63R(4) and (5)).

The responsible chief officer of police may apply to a District Judge (Magistrates' Courts) for an order to retain a sample beyond the date on which it would otherwise be required to be destroyed by virtue of (i) or (ii) if the sample was taken from a person in connection with the investigation of a qualifying offence (**7.35**(7)), and the responsible chief officer of police considers that the following condition is met. The condition is that, having regard to the nature and complexity of other material that is evidence in relation to the offence, the sample is likely to be needed in any proceedings for the offence for the purposes of disclosure to, or use by, a defendant, or responding to any challenge by a defendant in respect of the admissibility of material that is evidence on which the prosecution proposes to rely. Such application must be made before the date on which the sample would otherwise be required to be destroyed.

If, on such an application, the District Judge is satisfied that the above condition is met, (s)he may make an order which allows the sample to be retained for 12 months beginning with the date on which the sample would otherwise be required to be destroyed, and may be renewed (on one or more occasions) for a further period of not more than 12 months from the end of the period when the order would otherwise cease to have effect. An application for such an order (other than an application for renewal) may be made without notice of the application having been given to the person from whom the sample was taken, and may be heard and determined in private in the absence of that person. A sample retained by virtue of such an order must not be used other than for the purposes of any proceedings for the offence in connection with which the sample was taken (s 63R(11)). A sample that ceases to be retained by virtue of such an order must be destroyed (s 63R(12)).

Section 63R does not prevent a speculative search, in relation to samples to which it applies, from being carried out within such time as may reasonably be required for the search if the responsible chief officer of police considers the search to be desirable. Section 63R has not been brought into force for the purpose of its application to samples that are, or may become, disclosable under the Criminal Procedure and Investigations Act 1996 or its attendant Code of Practice.

Destruction of footwear impressions

7.69   PACE, s 63S provides that impressions of footwear taken from a person under any power conferred by PACE, or taken by the police, with the person's consent, in connection with the investigation of an offence, must be destroyed, with the proviso that they may be retained for as long as is necessary for purposes related to the prevention or detection of crime, the investigation of an offence or the conduct of a prosecution.

Supplementary provision for material subject to PACE

Use of retained material

PACE, s 63T deals with the use of material to which PACE, s 63D, 63R, or 63S applies. **7.70** 'Use of material' includes allowing any check to be made against it and to disclosing it to any person.

Any material to which s 63D, 63R, or 63S **(7.52, 7.68, 7.69)** applies must not be used other than:

(a) in the interests of national security,
(b) for the purposes of a terrorist investigation (**21.17**),
(c) for purposes related to the prevention or detection of crime, the investigation of an offence or the conduct of a prosecution, or
(d) for purposes related to the identification of a deceased person or of the person to whom the material relates.

For the purposes of s 63T, the reference to crime includes a reference to any conduct which constitutes one or more criminal offences (whether under the law of England and Wales or of any other country or territory), or is, or corresponds to, any conduct which, if it all took place in England and Wales, would constitute one or more criminal offences, and the references to an investigation and to a prosecution include references, respectively, to any investigation outside England and Wales of any crime or suspected crime and to a prosecution brought in respect of any crime in a country or territory outside England and Wales.

Material which is required by s 63D, 63R, or 63S **(7.52, 7.68, 7.69)** to be destroyed must not at any time after it is required to be destroyed be used in evidence against the person to whom the material relates, or for the purposes of the investigation of any offence.

Exclusions for certain regimes

PACE, ss 63D to 63T **(7.52-7.70)** do not apply to biometric data held under the **7.71** Terrorism Act 2000, Sch 8, the Terrorism Prevention and Investigation Measures Act 2011, Sch 6 or the Counter-Terrorism and Border Security Act 2019 (border security) (PACE, s 63U). A broadly equivalent regime for the destruction, retention, and use of such material operates under these statutes (see **21.52** and **21.68** in relation to the first two Acts).

Nor do PACE, ss 63D to 63T apply to material which is or may become disclosable under the Criminal Procedure and Investigations Act 1996 or its associated code of practice (**8.75-8.83** and **8.85-8.93**), or to material taken from someone which relates to another person (s 63U).

A sample which is or may become disclosable under the italicised provisions above, which would otherwise be required to be destroyed under s 63R, must not be used other than for the purposes of any proceedings for the offence in connection with which the sample was taken, and once the italicised provisions above no longer apply the sample must be destroyed (s 63U).

Other supplementary provisions

Cautions, etc treated as convictions

For the purposes of this chapter: **7.72**

(a) any reference to a person who is convicted of an offence includes a reference to a person who has been given a caution (or, if under 18 at the time, a warning or reprimand (now abolished)), or to a person who has been found not guilty by reason of insanity or found to be under a disability and to have done the act charged; and

(b) a person has been convicted of an offence under the law of a country or territory outside England and Wales if a court exercising jurisdiction under the law of that country or territory has made in respect of such an offence a finding equivalent
   (i) to a finding that the person is not guilty by reason of insanity, or
   (ii) to a finding that the person is under a disability and did the act charged.

Spent conviction rules do not apply

7.73 PACE, s 65B(2) provides that the provisions of PACE about fingerprints, samples etc apply irrespective of the Rehabilitation of Offenders Act 1974 (under which certain convictions and cautions are treated as being spent, and therefore disregarded for most purposes, after the end of a specified expiry period). However, a person is not to be treated as having been convicted of an offence if that conviction is a conviction or caution for the repealed offences of buggery or gross indecency between men and the Secretary of State has ordered it to be disregarded (s 65B(3)).

Inclusion of DNA profiles on National DNA Database

7.74 A DNA profile derived from a DNA sample which is retained under PACE, ss 63E to 63L (including those sections as applied by s 63P) will have to be recorded on the National DNA Database (s 63AA). This provision puts on a statutory footing the pre-existing National DNA Database, as does s 63AB which requires the Secretary of State to make arrangements in respect of the National DNA Database Strategy Board whose role is to oversee the National DNA Database. The Board is required to issue guidance about the destruction of DNA profiles taken or retained under the above provisions of PACE in accordance with which chief officers of police must act, and may issue guidance about applications under s 63G (**7.54**).

In order to bring clarity to the oversight for the operation of the non-statutory National Fingerprints Database (IDENT1), the governance structure of the National DNA Database Strategy Board has been extended to include the National Fingerprints Database which retains on a non-statutory basis fingerprint image. Although the Board's statutory name remains the National DNA Database Strategy Board, it is now called the Forensic Information Database (FIND) Strategy Board on a non-statutory basis.

Commissioner for the Retention and Use of Biometric Material

7.75 The Protection of Freedoms Act 2012, s 20 provides for the appointment of a Commissioner for the Retention and Use of Biometric Material, whose functions are to keep under review determinations under PACE, s 63M and corresponding terrorism provisions (see Chapter 21) made by chief officers of police and others that the fingerprints and DNA profiles derived from a DNA sample of a person are required to be retained for national security purposes and the use to which fingerprints and DNA profiles so retained are being put. The Secretary of State has issued guidance about the making or renewing of such determinations, to which regard must be had.

To enable the Commissioner to discharge his/her functions, s 20 requires persons making national security determinations to notify the Commissioner in writing of the making of a determination, including the reasons for it, and to provide such other information as the Commissioner may require.

The Commissioner has power under s 20 to order the destruction of the fingerprints and DNA profile held pursuant to it where satisfied that it is not necessary for any material retained under the determination to be so retained. There is no appeal against such a ruling by the Commissioner save by way of judicial review. The Commissioner may not order the destruction of material that could otherwise be retained pursuant to any other statutory provision.

The Commissioner must also keep under review the retention and use in accordance with PACE, ss 63A and 63D to 63T (or the corresponding terrorism provisions) of any material to which s 63D or 63R (or corresponding terrorism provisions) applies, and any copies of any material to which s 63D (or corresponding terrorism provision) applies.

DESTRUCTION, RETENTION, AND USE OF NATIONAL SECURITY MATERIAL NOT SUBJECT TO EXISTING STATUTORY RESTRICTIONS

This is dealt with by C-TA 2008, ss 18–18E. The basic provision is s 18 which applies to fingerprints (as defined at **7.33**), DNA samples, and DNA profiles derived from a DNA sample that: **7.76**

(a) are held for the purposes of national security by a law enforcement authority under the law of England and Wales or Northern Ireland, and
(b) are not held subject to existing statutory restrictions.

Such material is described as 's 18 material'.
'Law enforcement authority' means:

(a) a police force,
(b) NCA,
(c) HM Revenue and Customs Commissioners, or
(d) a foreign authority so far as exercising functions which correspond to those of a police force, or otherwise involve criminal investigation or prosecution.

The prime example of material not held subject to 'existing statutory restrictions' is material on the counter-terrorism DNA database maintained by the Metropolitan Police which comprises DNA material and fingerprints obtained from crime scenes, covert surveillance or from foreign police or intelligence services.

Destruction of material

By C-TA 2008, s 18, s 18 material must be destroyed if it appears to the responsible officer (the chief officer of the police force or other entity by whom the material was obtained or acquired) that the material has not been: **7.77**

(a) obtained by the law enforcement authority pursuant to an authorisation under the Police Act 1997, Pt 3 (**4.110**);
(b) obtained by the law enforcement authority in the course of surveillance, or use of a covert human intelligence source, authorised under the Regulation of Investigatory Powers Act 2000, Part II (**4.98**);
(c) supplied to the law enforcement authority by another law enforcement authority; or
(d) otherwise lawfully obtained or acquired by the law enforcement authority for any of the purposes mentioned in s 18D below.

In any other case, ie where the s 18 material has been so obtained, s 18 material must be destroyed unless it is retained by the law enforcement authority under any power conferred by s 18A or 18B (non-DNA samples: below), but this is subject to a rule that a DNA sample to which s 18 applies must be destroyed as soon as a DNA profile has been derived from the sample, or, if sooner, before the end of the period of six months beginning with the date on which it was taken.

Section 18 material which ceases to be retained under a power conferred by s 18A or s 18B may continue to be retained under any other such power which applies to it.

Section 18 does not prevent s 18 material from being the subject of a speculative search within such time as may reasonably be required for the check, if the responsible officer considers the check to be desirable.

Retention of material: general

7.78   By C-TA 2008, s 18A, s 18 material (**7.76**) which is *not a DNA sample* and relates to a person who has no previous convictions or only one exempt conviction may be retained by the law enforcement authority until the end of the specified retention period. The retention period is three years beginning with the date on which (as the case may be):

(a) the fingerprints were taken, and

(b) the DNA sample from which the profile was derived was taken (or, if the profile was derived from more than one DNA sample, the date on which the first of those samples was taken).

A person 'has no previous convictions' if (s)he has not previously been convicted (**7.72**) in England and Wales or Northern Ireland of a recordable offence (see **7.43** and the corresponding Northern Irish legislation). 'Spent convictions' are treated in the same way as under PACE, s 65B(2) and (3) (**7.73**). A 'conviction' is exempt if it is in respect of a recordable offence, other than a qualifying offence, committed when the person was under 18. A qualifying offence has the meaning as at **7.35**(7) and a corresponding meaning in Northern Ireland. Convictions of more than one offence arising out of a single course of action are treated as a single conviction for the purposes of calculating under s 18A whether the person has been convicted of only one offence.

Section 18 material which is *not a DNA sample* and relates to a person who has previously been convicted of a recordable offence (other than a single exempt conviction), or is so convicted before the material is required to be destroyed by virtue of s 18A, may be retained indefinitely.

Section 18 material which is *not a DNA sample* may be retained indefinitely if it is held by the law enforcement authority in a form which does not include information identifying the person (P) to whom the material relates, and the authority does not know, and has never known, the identity of P. Where s 18 material is being so retained by a law enforcement authority, and the authority comes to know the identity of P, and P has no previous convictions or only one exempt conviction, the material may be retained by the authority until the end of the period of three years beginning with the date on which the P's identity comes to be known by the authority.

Retention for purposes of national security

7.79   C-TA 2008, s 18B provides that s 18 material (**7.76**) which is *not a DNA sample* may be retained for as long as a national security determination made by the responsible officer (**7.77**) or a chief officer of police has effect in relation to it.

A national security determination is made if the responsible officer or a chief officer of police determines that it is necessary for any such s 18 material to be retained for the purposes of national security. A national security determination must be written, has effect for a maximum of five years, and may be renewed.

Retention of further fingerprints

By C-TA 2008, s 18BA, where s 18 material (**7.76**) is or includes a person's fingerprints ('the original fingerprints'), a constable may make a determination in respect of the retention of any further fingerprints taken from the same person ('the further fingerprints') if conditions 1 and 2 are met.   **7.80**

Condition 1 is met if the further fingerprints are:

(a) s 18 material;

(b) taken or provided under or by virtue of PACE or the corresponding Northern Irish provision, or the Terrorism Prevention and Investigation Measures Act 2011 (**21.68**), or the Counter-Terrorism and Border Security Act 2019; or

(c) material to which the Counter-Terrorism Act 2008, s 18 (**7.76**) applies.

Condition 2 is met if both the original fingerprints and the further fingerprints are held by a law enforcement authority under the law of England and Wales or the law of Northern Ireland.

Where such a determination is made in respect of the further fingerprints:

(i) the further fingerprints may be retained for as long as the original fingerprints are retained under s 18A or 18B (**7.78**, **7.79**); and

(ii) a requirement under any legislation, including that of Northern Ireland, to destroy the further fingerprints does not apply for as long as their retention is authorised by (i).

(i) does not prevent the further fingerprints being retained after the original fingerprints fall to be destroyed if the continued retention of the further fingerprints is authorised under any legislation, including that of Northern Ireland.

A written record must be made of a determination.

Destruction of copies

C-TA 2008, s 18C provides that, if fingerprints are required by s 18 to be destroyed, any copies of the fingerprints held by the law enforcement authority concerned must also be destroyed. If a DNA profile is required by s 18 to be destroyed, no copy may be retained by the law enforcement authority concerned except in a form which does not include information which identifies the person to whom the DNA profile relates.   **7.81**

Use of retained material

By C-TA 2008, s 18D, s 18 material must not be used other than for the same purposes as under PACE, s 63T (**7.70**).   **7.82**

Subject to this, s 18 material may be subjected to a speculative search if the responsible officer (**7.77**) considers the check to be desirable.

Material which is required by s 18 to be destroyed must not at any time after it is required to be destroyed be used in evidence against the person to whom the material relates, or for the purposes of the investigation of any offence.

The references above to using material have the same meaning as under PACE, s 63T, and so does the reference to crime (with the substitution of 'UK' for 'England and Wales').

PHOTOGRAPHS

(a) Persons detained at a police station

7.83   PACE, s 64A provides powers to take a photograph of a detainee with the appropriate consent (**7.42**), or without such consent where it is withheld or it is impracticable to obtain consent.

(b) Otherwise than at a police station

7.84   Section 64A also authorises the taking of a photograph elsewhere than at a police station with the appropriate consent, or without such consent where it is withheld or it is impracticable to obtain it, if the person has been:

(a) arrested by a constable for an offence;

(b) taken into custody by a constable after a citizen's arrest;

(c) required to wait by a designated community support officer or designated community support volunteer;

(d) given a fixed penalty notice by a constable, by a designated community support officer or designated community support volunteer, or by an accredited person by virtue of his/her accreditation; or

(e) given a direction to leave and not return to a specified location for up to 48 hours by a constable (under the Anti-social Behaviour, Crime and Policing Act 2014, s 35).

General to (a) and (b)

7.85   For these purposes, the term 'photograph' includes a moving image. A photograph obtained without the appropriate consent may be obtained by making a copy of an image taken on a camera system installed anywhere in a police station. In the event of non-co-operation, where it is not possible to take the photograph covertly, reasonable force may be used to take the photograph. The use of reasonable force to take the photograph of a suspect elsewhere than at a police station must be carefully considered.

Only a police officer may take a photograph under s 64A. The officer may require the person to remove anything worn on, or over, all or any part of the head or face. In the event of non-compliance, the officer may remove the item or substance. In order to obtain a suspect's consent and co-operation to remove an item of religious headwear to take a photograph, a constable should consider whether in the situation the removal of the headwear and the taking of the photograph should be by an officer of the same sex as the person. It would be appropriate for these actions to be conducted out of public view.

The suspect must be informed of the reason for taking the photograph and the purposes for which it may be used. This must be done beforehand, unless the photograph is taken covertly or by making a copy of an image (in which cases (s)he must be informed as soon as practicable thereafter).

A photograph so taken may be used or disclosed only for purposes related to the prevention or detection of crime (including a foreign crime corresponding to one under UK law), the investigation of an offence, the conduct of a prosecution, or the enforcement of a sentence.

After being so used or disclosed, a photograph may be retained but may only be used or disclosed for the same purposes. Its retention is subject to the guidance on retention contained in the section of the College of Policing's Authorised Professional Practice relating to the Management of Police Information (MOPI), with which chief officers are required to comply by the Code of Practice on MOPI, issued by the Secretary of State.

(c) Persons voluntarily at a police station

When there are reasonable grounds for suspecting the involvement of a person in a criminal offence, but that person is at a police station voluntarily and not detained, the above provisions apply, except that force may not be used to take the person's photograph.   **7.86**

Destruction of photographs

Code D provides that the photograph of a person which is not taken in accordance with the provisions under headings (a) or (b) above or under PACE, s 54A (**6.17**), must be destroyed (together with any negatives and copies) unless he: **7.87**

(a) is charged with, or informed that he may be prosecuted for, or is prosecuted for a recordable offence (**7.43**);
(b) is cautioned for a recordable offence; or
(c) gives informed, written consent to the photograph being retained.

Such a person must be given an opportunity to witness the destruction or to have a certificate confirming destruction, if (s)he so requests within five days of notification that the destruction is required.

Documentation

A record must be made of the identity of the photographer (with the usual exceptions), the purpose of the photograph and its outcome, and the detainee's consent (or the reason for taking the photograph without consent).   **7.88**

CHAPTER 8

The Law
of Criminal Evidence

8.1   The law of criminal evidence determines two things:

(a) the means by which the facts in issue are proved in a court; and

(b) those facts which may (or may not) be proved in a court.

CLASSIFICATION OF EVIDENCE

Direct evidence and circumstantial evidence

8.2   *Direct evidence* is evidence which (if believed) directly establishes a particular fact in issue itself. For example, the existence of a firearm alleged to have been possessed by the defendant (D) may be proved by its production in court and the fact that D was in possession of it may be proved by a statement from a person who claims to have discovered D in possession.

Circumstantial evidence is evidence of a fact or facts from which a fact in issue may be inferred. Suppose that D is charged with murder. If an eye-witness gives evidence that (s)he saw D fire a gun at the victim, this is direct evidence of a fact in issue. On the other hand, evidence that D was seen in possession of a gun near the scene of the crime shortly before it was committed is circumstantial evidence since it is evidence of a fact from which the fact in issue (that D fired the gun) may be inferred.

Oral evidence, documentary evidence, and real evidence

Oral evidence

8.3   Most evidence given in a court is oral evidence. This consists of statements made in court by witnesses concerning matters of which they have knowledge, such as something which they have seen, or heard, or smelt, or touched.

Normally, oral evidence must be given on oath (ie sworn). However, where it is not possible to administer the oath in the manner appropriate to the religious beliefs of the witness (W), or where W objects to being sworn, W may make a solemn affirmation.

Determination of whether a witness should be sworn   The question whether W may be sworn is determined by the court as follows in accordance with the Youth Justice and Criminal Evidence Act 1999 (YJCEA 1999), s 55.

W may not be sworn unless W has reached 14 and has sufficient appreciation of the solemnity of the occasion and of the particular responsibility to tell the truth which is involved in taking an oath. If W is able to give intelligible testimony, W is presumed to have sufficient appreciation if no evidence is adduced to the contrary. If such evidence is adduced, it is for the party seeking to have W sworn to prove on the balance of probabilities that W has reached 14 and has sufficient appreciation of the specified matters.

Reception of unsworn evidence   YJCEA 1999, s 56 deals with a person of any age who is competent to give evidence in criminal proceedings (**8.15**) but is not permitted to be sworn by s 55. It provides that the evidence of such a person may be given unsworn.

Unsworn evidence may also be given by a person called simply to produce a document.

Evidence by live link   Provision is made by CJA 2003, Part 8 (ss 51–56) under which a 'live link' is a live television link or other arrangement by which a witness, while at a place in the UK which is outside the building where the proceedings are being held, is able to see and hear a person at the place where the proceedings are being held and to be seen and heard by the defendants, judge and/or justices, jury (if any), legal representatives, and any interpreter or other person appointed by the court to assist the witness.

A witness (other than D) may, if the court so directs, give evidence through a live link in criminal proceedings. The court may give such a direction on its own motion or on the application of a party. The court must be satisfied that it is in the interests of the efficient or effective administration of justice for this to take place.

For a temporary modification of these provisions until the end of 24 March 2022 (or longer if that period is extended), see Chapter 8 of the companion website.

Evidence by video recording   For video recorded evidence under the 'special measures provisions', see **8.23**.

Refreshing memory   Witness statements are usually recorded well in advance of trial. Recognising the difficulties associated with refreshing memory from an audio or video recording, CJA 2003, s 139 states that a witness giving evidence in criminal proceedings may refresh his/her memory from a document made or verified by him/her at an earlier time provided that:

(a) (s)he states in his/her oral evidence that the document records his/her recollections of the matter at that earlier time; and
(b) his/her recollection of the matter is likely to have been significantly better at that time than it is at the time of his/her oral evidence.

Where two witnesses have acted together (and this is common in the case of police officers) they may refresh their memories from notes which they made together.

Documentary evidence

It is a general rule that the contents of a document (a term which includes not only a   **8.4** written document but also a map or drawing, photograph, disc, tape or the like, and any film or the like) may be proved only by production of the original or an admission as to its contents (primary evidence). However, there are now so many exceptions to this general rule that it has lost much of its importance. When one of these exceptions applies, secondary evidence (eg a copy authenticated in a way approved by the court or oral evidence) of the contents of the document may be given.

The principal exceptions whereby secondary evidence of the contents of a document may be given are as follows:

(1) Where the original is proved to have been lost or destroyed, secondary evidence of the contents may be given.
(2) The contents of public documents can always be proved by means of secondary evidence, although some statutes providing this exception for a particular type of public document limit the nature of this secondary evidence. A 'public document' is a document prepared for the purpose of the public making use of it and being

able to refer to it. Examples are Registers of Births, Deaths, and Marriages, the contents of which can be proved by a copy of an entry certified by a person who has lawful custody of the register. Judicial notice is taken of Acts of Parliament and their contents, ie they do not have to be proved by evidence. A statutory instrument is proved by production of the Queen's Printer's copy.

(3) By the Bankers' Books Evidence Act 1879, ss 3–5, an examined copy of any entry in a banker's book kept in the ordinary course of business is admissible as prima facie evidence of such entry.

Certain statutory provisions specially provide for evidence to be given by the production of a document. For examples, see **8.42** and **8.43**.

Real evidence

8.5 Real evidence is the production of an object for the inspection of the court or jury. Where a document is produced as evidence of the matter contained in it, as opposed to proof of its physical existence, it is, as we have just seen, documentary evidence. The firearm which was produced by a witness in the example quoted at **8.2** is real evidence. So, it has been held, is a computer printout recording attempts to enter a website.

Original evidence and hearsay evidence

8.6 Original evidence is evidence of a fact by 'first-hand' evidence of it. Hearsay evidence is 'second-hand' evidence, since it consists of any representation of fact or opinion made by whatever means (eg orally, in writing, or by conduct (such as a nod of the head)) by a person, otherwise than in oral evidence in the particular proceedings, which is tendered as evidence to establish the matter stated in it. It must be emphasised that not all evidence of what someone else stated is hearsay. If it is produced merely to prove the fact that it was stated, as opposed to being produced with the object of establishing the truth of what was stated, it is original evidence and not hearsay evidence. For examples of the operation of these rules see **8.32**.

The distinction between original evidence and hearsay evidence is important, because hearsay evidence is inadmissible in criminal cases unless its admission is permitted under a variety of rules (see **8.33–8.49**). The distinction is dealt with further at **8.32**.

PROOF

8.7 The general rule is that the prosecution must prove the existence of any fact on which it relies. There are, however, certain facts which do not need to be proved.

Facts which may be established by means other than proof

Judicial notice

8.8 The court may take 'judicial notice' of certain matters which are so notorious or well known that evidence of their existence need not be adduced. One does not have to prove that beer is intoxicating, for example, as this is a matter of general knowledge, whereas it may be necessary to prove that a less well-known drink is intoxicating.

Presumptions

Sometimes there is a presumption of *law* that if a particular fact is proved some other **8.9** fact *must* be presumed to exist. There are two kinds of presumption of law: irrebuttable and rebuttable.

When the presumption is irrebuttable, no evidence can be received to contradict the presumed fact. An example of an irrebuttable presumption of law is provided by the rule that a child under 10 is incapable of committing an offence.

Where there is a rebuttable presumption of law against D, the jury (or magistrates) must find that the presumed fact existed unless (depending on the presumption) D proves the contrary on the balance of probabilities, or there is evidence raising a doubt that the presumption is rebutted (in which case the prosecution will have to disprove that evidence beyond reasonable doubt). An example of a rebuttable presumption of law is that mechanical instruments are in proper working order, in the absence of evidence to the contrary.

Formal admissions

A formal admission of a fact *dispenses with the need* for proving that fact since it is **8.10** conclusive evidence of that fact as against the party admitting it. Provision is made for formal admissions, by or on behalf of either prosecution or defence, by the Criminal Justice Act 1967, s 10. Such a formal admission may be made before or at the proceedings. Unless made in court, it must be made in writing and signed by the person making it (or by an officer of the company if made by a company).

CJA 2003, s 118 preserves the common law rules under which in criminal proceedings:

(a) an admission made by D's agent (eg solicitor or counsel); or
(b) a statement made by a person to whom D refers a person for information,

is admissible as evidence against D of any matter stated.

The burden of proof

The burden of proof is often described as the 'persuasive' or 'legal' burden. The general **8.11** rule is that the prosecution has the burden of proving D's guilt beyond reasonable doubt. In more detail, the position is as follows. The prosecution must prove beyond reasonable doubt that D committed the actus reus of the offence charged with the requisite mens rea. In relation to defences, D normally has the burden of adducing sufficient evidence to raise a defence (an evidential burden); if D does so it is then for the prosecution to disprove the alleged defence beyond reasonable doubt.

Exceptionally, D has the burden of proving a defence on the balance of probabilities. The burden of proof is imposed on D in the following cases:

Defence of insanity   This was outlined in Chapter 1.

Express statutory provision   A statute sometimes provides that it is a defence if D proves certain facts. For example, under the Homicide Act 1957, s 2, D has the burden of proving the defence of diminished responsibility on a charge of murder.

Provisos and exemptions in statutory offences   Where a statute governing any offence provides any exception, exemption, proviso, excuse or qualification, as where it prohibits the doing of an act save in specified circumstances (or by persons of specified classes, or with specified qualifications, or with the licence or permission of specified authorities), the onus of proving such an exception, exemption, etc is impliedly cast on D. This rule is provided in the case of summary proceedings by the Magistrates' Courts Act 1980, s 101, and applies in the case of trial on indictment by virtue of the common law. In exceptional cases, however, the rule will not apply if the court construes the legislation as only imposing an evidential burden.

Rebuttable presumption of law against the defendant where the defendant bears the burden of proving that the presumption is rebutted   This was dealt with at **8.9**.

Placing a persuasive burden on D may be incompatible with the presumption of innocence guaranteed by the European Convention on Human Rights, art 6(2), depending on the nature of the provision which imposes that burden. In each case it depends on whether imposing a persuasive burden serves a legitimate aim and is a justifiable and proportionate response to it; if not it will be incompatible. The Human Rights Act 1998, s 3 gives the courts a liberal power of statutory interpretation which enables them to read an express or implied statutory requirement for D to prove something as simply imposing an evidential burden on D if this makes the provision compatible with the Convention when it would not otherwise be so.

Corroboration

8.12   'Corroboration' is evidence from a source (or sources) independent of the witness whose evidence is to be corroborated, which confirms or supports that evidence in some material particular. Corroboration is not generally required as a matter of law. This means that, generally, although corroboration may aid proof, it is possible for the prosecution to discharge its burden of proof by adducing only one item of evidence.

However, by way of exception, corroboration is *required* by statute in respect of:

(a) perjury and other offences under the Perjury Act 1911; and
(b) speeding contrary to the Road Traffic Regulation Act 1984, s 89.

Thus, in these cases the relevant fact is not proved on uncorroborated evidence given on behalf of the prosecution.

A judge has a *discretion* to give a warning to the jury about the danger of convicting on the uncorroborated evidence of a witness. For it to be appropriate for a warning to be given, there must be an evidential basis for suggesting that a witness is unreliable.

Proof of convictions and acquittals

8.13   The Police and Criminal Evidence Act 1984 (PACE), s 73 provides that, where the fact that a person has in the UK been convicted or acquitted of an offence, otherwise than by a Service court, is admissible in evidence, it may be proved by producing a certificate of conviction or acquittal relating to that offence, and proving that the person named in the certificate is the person whose conviction or acquittal for the offence is to be proved. The certificate must be signed by the 'proper officer' of the court of conviction or acquittal. Where such a certificate is relied on to prove a conviction against D, it must be proved by the prosecution beyond reasonable doubt that D is the person named in that certificate. In respect of any foreign conviction, there is a corresponding provision in the Evidence Act 1851, s 7.

PACE, s 73 supplements other previous provisions. The Criminal Procedure Act 1865, s 6 provides that if, on a witness being lawfully questioned about a conviction, there is a denial or refusal to answer, the cross-examining party may prove the conviction. To do so, a procedure corresponding to that under PACE, s 73 will have to be followed. The Road Traffic Offenders Act 1988, s 31 provides that, where a person is convicted of an offence involving obligatory or discretionary disqualification, any previous conviction endorsed on his/her driving record is prima facie evidence of that conviction.

In any proceedings where evidence is admissible of the fact that D has committed an offence, D is rebuttably presumed by PACE, s 74 to have committed it if D is proved to have been convicted of it by a UK court or by a Service court outside the UK.

Section 74 also provides that the fact that a person *other than* D has been convicted in the UK or by a Service court outside the UK of an offence by such a court is admissible in evidence for the purpose of proving that that person committed that offence, where evidence of his/her having done so is admissible. If that conviction is proved that person is rebuttably presumed to have committed it.

Although it is possible to prove previous convictions by adducing Police National Computer (PNC) records admitted under CJA 2003, s 117 (**8.35**), s 117 cannot be relied on so as to make admissible details on the PNC about the offences (as opposed to the dates of convictions, the offences charged, and the sentences).

COMPETENCE, COMPELLABILITY, AND PRIVILEGE

Who can give evidence? Who is obliged to? When is a witness entitled to refuse to answer questions? The answers to these questions are to be found respectively in the law relating to competence to give evidence, to compellability to give evidence, and to various types of privilege. **8.14**

Competence

YJCEA 1999, s 53 provides that anyone, of whatever age, is competent to give evidence in criminal proceedings unless (s)he is unable to understand questions put to him/her as a witness, or unable to answer them in a way which can be understood. The Court of Appeal has emphasised that this alone is the test of competence, and that a witness need not understand the special importance that the truth should be told in court, and the witness need not understand every single question or give a readily understandable answer to every question. A witness may need the assistance of 'special measures' (see **8.20–8.23**). **8.15**

A person charged in criminal proceedings is not competent to give evidence *for the prosecution* (whether or not (s)he is the only person, or is one of two or more persons, charged in the proceedings). This reference to a person charged in criminal proceedings does not include a person who is not, or is no longer, liable to be convicted of any offence in the proceedings (whether because of a guilty plea, or otherwise).

Under s 54, questions of competence are decided by the court and in the absence of the jury if there is one. The party calling the witness must satisfy the court that the witness, on the balance of probabilities, is competent to give evidence.

As a general rule a witness who is competent is also compellable.

Compellability of D's spouse or civil partner

PACE, s 80 deals with the 'compellability' of the *spouse or civil partner of a defendant* to give evidence. **8.16**

In any proceedings, the spouse or civil partner (W) of a person charged in the proceedings is compellable to give evidence *on behalf of that person*, unless (s)he (ie W) is also charged in those proceedings.

In addition, provided that the spouse or civil partner (W) of a person charged in the proceedings is not also charged in those proceedings, W is compellable to give evidence:

(a) *on behalf of any other person charged in those proceedings*, but only in respect of any specified offence with which that other person is charged, or;
(b) *for the prosecution*, but only in respect of any specified offence with which any person is charged in the proceedings.

In relation to W, an offence is a specified offence for these purposes if it:

(a) involves an assault on, or injury or a threat of injury to, W or a person who was at the material time under 16;
(b) is a sexual offence alleged to have been committed in respect of a person who was at the material time under 16; or
(c) consists of encouraging or assisting, attempting or conspiring to commit, or of aiding, abetting, counselling or procuring the commission of, an offence under (a) or (b).

For the purposes of (b), a 'sexual offence' is an offence under:

(i)   the Protection of Children Act 1978 (PCA 1978) (taking, etc indecent photographs, etc of a child);
(ii) the Sexual Offences Act 2003 (SOA 2003), Part 1 (ss 1–79);
(iii) the Sexual Offences Act 1956 or the Indecency with Children Act 1960, both of which have been repealed; or
(iv) an offence under the Modern Slavery Act 2015, s 2 (human trafficking) committed with a view to exploitation consisting of or including sexual exploitation.

The references above to a person charged in any proceedings do not include a person who is not, or is no longer, liable to be convicted of any offence in the proceedings (whether as a result of pleading guilty or for any other reason).

The failure of a spouse or civil partner of a person charged in any proceedings to give evidence in the proceedings may not be commented on by the prosecution.

An ex-spouse or ex-civil partner is compellable to give evidence as if (s)he had never been married to the defendant.

Privilege

Self-incrimination

8.17   A person required to answer questions or produce something may refuse to do so on the grounds that the evidence might incriminate him/her. Examples of exceptions to this rule are provided by statutes concerned with drink/drug-driving and with the care and protection of children.

Lawyer/client

8.18   Communications between a qualified lawyer and his/her client which are concerned with the giving of legal advice or as to the presentation of material, and those between a qualified lawyer, his/her client and a third party which are concerned with litigation and legal advice, are privileged. Privilege is not afforded in respect of such communications in the furtherance of crime or fraud. In addition, where such communications are disclosed from other sources (document coming into possession of police) or where privilege is waived by the lawyer's client, such evidence may be given.

Exclusion of evidence on grounds of public interest immunity

Evidence held by the prosecution may need to be excluded from disclosure at trial or in preliminary proceedings (and from pre-trial disclosure to the defendant) by the court on grounds that its disclosure is not in the public interest. Such material is said to be excluded on the basis of public interest immunity (PII).   **8.19**

PII may apply to many police matters, eg information relied upon for the issue of search warrants; reports to the DPP; the disclosure of the identity of informants and the siting of police observation posts.

PROTECTION OF WITNESSES

Special measures directions in case of vulnerable and intimidated witnesses (other than defendants)

Eligibility for special measures

Witnesses who are eligible for assistance on grounds of age or incapacity YJCEA 1999, s 16(1) provides that a witness (W) (other than the defendant (D)) is eligible for assistance by special measures to help him/her in giving his/her evidence: **8.20**

(a) if W is *under 18 at the time of the hearing* (s 16(1)(a)); or
(b) if the court considers that the quality of the evidence given by W is likely to be diminished by reason of W:
  (i) suffering from mental disorder within the meaning of the Mental Health Act 1983; or
  (ii) otherwise having a significant impairment of intelligence and social functioning; or
  (iii) having a physical disability or suffering from physical disorder (s 16(1)(b) and (2)).

References in (b) to the quality of W's evidence are to its quality in terms of completeness, coherence, and accuracy. 'Coherence' refers to W's ability in testifying to give answers which address the questions put to him/her and can be understood both individually and collectively (ie in relation to a particular question and to his/her evidence generally).

Where the age of a witness who is the complainant in respect of a 'relevant offence' is uncertain, and there are reasons to believe that the witness is under 18, that witness is presumed to be under 18. For these purposes, 'relevant offence' means an offence under SOA 2003, Part 1 (ss 1 to 79), PCA 1978, s 1, the Criminal Justice Act 1988, s 160 (possessing indecent photographs, etc of children), or the Modern Slavery Act 2015 (MSA 2015), s 1 or 2 (slavery, etc; human trafficking).

Witnesses eligible for assistance on ground of fear or distress about testifying YJCEA 1999, s 17 provides that a witness (W) in criminal proceedings (*other than* D) is eligible for assistance by special measures where the court is satisfied that the quality of evidence given by W is likely to be diminished by reason of fear or distress on W's part in connection with testifying in the proceedings. In determining whether it is so satisfied, the court must consider in particular:

(a) the nature and the alleged circumstances of the offence to which the proceedings relate;
(b) W's age;
(c) such of the following matters which appear relevant:
  (i)   W's social and cultural background and ethnic origins;
  (ii) W's domestic and employment circumstances; or
  (iii) W's religious beliefs or political opinions, if any;
(d) any behaviour towards W on the part of:
  (i)   D;
  (ii) members of the family or associates of D; or
  (iii) any other person who is likely to be a defendant or a witness in the proceedings.

Where the complainant in respect of a sexual offence or an offence under MSA 2015, s 1 or 2 is a witness in proceedings relating to that offence (or to that and any other offences), there is a presumption that the witness is eligible for assistance by special measures unless the witness waives that entitlement. *For the purposes of the provisions of YJCEA 1999 in this chapter, 'sexual offence' means an offence (a) under SOA 2003, Part 1 (ss 1 to 79) or any 'relevant superseded offence' (ie a corresponding offence superseded by SOA 2003, Part 1), or (b) under MSA 2015, s 2 (human trafficking) committed with a view to sexual exploitation.*

There is a presumption that a witness in proceedings relating to a relevant offence involving weapons (or to a relevant offence and any other offences) is eligible for assistance by special measures, unless the witness waives that entitlement. For these purposes, an offence is a relevant offence if it is an offence described in YJCEA 1999, Sch 1A, viz:

(a) murder, or manslaughter, where it is alleged that *a firearm or knife* was used to cause the death in question, or where it is alleged that:
  (i)   D was carrying such a weapon during the commission of the offence, and
  (ii) a person other than D knew or believed during the commission of the offence that D was carrying such a weapon;
(b) an offence under the Offences Against the Person Act 1861, ss 18, 20, 38, or 47 where it is alleged that *a firearm or knife* was used to commit the offence in question, or in a case where it is alleged that (i) or (ii) above applies;
(c) an offence under the Prevention of Crime Act 1953, s 1 or s 1A or the Criminal Justice Act 1988, s 139 or s 139A or s 139AA (which deal with having an offensive weapon or bladed or pointed article in public places or schools or threatening with such an article);
(d) an offence under the Firearms Act 1968, ss 1, 2(1), 3, 4, 5(1), 5(1A), 16, 16A, 17, 18, 19, 20, 21, 21A, or 24A;
(e) an offence under the Violent Crime Reduction Act 2006, ss 28, 32, or 36.

For the above offences see Chapters 9, 10 and 19.

The reference in (a) and (b) (above) to an offence ('offence A') includes a reference to:

(i) an attempt or conspiracy to commit offence A in a case where it is alleged that the attempt or conspiracy was to commit offence A in a manner or circumstances described in (a) or (b);

(ii) an offence under the Serious Crime Act 2007 (SCA 2007), Part 2 (encouraging or assisting crime) in relation to which offence A is the offence (or one of them) which the person intended or believed would be committed in a case where it is alleged that the person intended or believed offence A would be committed in the manner or circumstances described in (a) or (b); and

(iii) aiding, abetting, counselling or procuring the commission of offence A in a case where it is alleged that offence A was committed, or the act or omission charged in respect of offence A was done or made, in the manner or circumstances described in (a) or (b).

A reference in (c)–(e) to an offence ('offence A') includes a reference to:

(i) an attempt or conspiracy to commit offence A;

(ii) an offence under the SCA 2007, Part 2 in relation to which offence A is the offence (or one of the offences) which the person intended or believed would be committed; and

(iii) aiding, abetting, counselling or procuring the commission of offence A.

Special measures direction

By YJCEA 1999, s 18 a range of special measures is made potentially available to witnesses in criminal proceedings who are eligible for special measures under s 16 or s 17. The 'special measures' are set out in detail by ss 23 to 30 (below). **8.21**

Special measures direction relating to eligible witness YJCEA 1999, s 19(2) provides that, once a court has determined that a witness is eligible for special measures, it must then determine whether any of those measures (or a combination of them) would be likely to improve the quality of the witness's evidence and, if so, determine which of those measures would be likely to maximise so far as practicable the quality of his/her evidence. Having made this second determination, it must give a 'special measures' direction in relation to the measure(s) so determined. An exception to this approach is provided by s 21.

Special provisions relating to child witnesses YJCEA 1999, s 21 provides special protection for one type of person eligible for 'special measures' under s 16: 'a child witness' (ie someone under 18 at the time of the hearing). It provides that, where in making a determination under s 19, a court determines that a witness (W) is a 'child witness', it must first have regard to the following provisions. Under them the 'primary rule' is that in respect of W it must give a special measures direction providing for a video recording of W's evidence-in-chief to be admitted (unless this is contrary to the interests of justice) and providing for any evidence given by W which is not by means of video recording (whether in chief or otherwise) to be given by means of a live link. The primary rule does not apply to the extent that the court is satisfied that compliance with it would not be likely to maximise the quality of W's evidence so far as practicable (whether because the application to that evidence of one or more other special measures available in relation to the witness would have that result or for any other reason).

If W informs the court of his/her wish that the primary rule should not apply or should apply only in part, the primary rule does not apply to the extent that the court is *satisfied that not complying with the rule would not diminish the quality of W's evidence.* Where, as a consequence of all or part of the primary rule being disapplied under this provision, W's evidence or any part of it would fall to be given as testimony in court, the court must give a special measures direction for W to be screened from D when giving the evidence or that part of it. However, this screening requirement is subject to two limitations:

(a) if W informs the court of his/her wish that the requirement should not apply, the requirement should not apply to the extent that the court *is satisfied that not complying with it would not diminish the quality of W's evidence*; and
(b) the requirement does not apply to the extent that the court is satisfied that screening would not be likely to maximise the quality of W's evidence so far as practicable.

In making a decision under the italicised provisions, the court must take into account the following factors (and any others it considers relevant):

(a) W's age and maturity;
(b) W's ability to understand the consequences of giving evidence otherwise than in accordance with the requirements of the primary rule or a screening requirement (as the case may be);
(c) the relationship (if any) between W and D;
(d) W's social and cultural background and ethnic origins;
(e) the nature and alleged circumstances of the offence to which the proceedings relate.

Only after it has had regard to the above provisions must the court have regard to s 19(2) (above).

If W is not under 18 at the hearing but was under 18 when a video recording of W's evidence-in-chief was made, YJCEA 1999, s 22 provides that the primary rule applies, *so far as relating to the giving of a direction for the recording to be admitted.*

Special provisions relating to sexual offences

8.22   YJCEA 1999, s 22A makes special provisions relating to the trial of a sexual offence (**8.20**) (or a sexual offence and other offences) *in the Crown Court where the complainant is a witness (eligible for special measures) in the proceedings and is not under 18 at the time of the hearing.* (If the complainant is under 18, s 21 (above) applies.) *Where it applies*, s 22A establishes a rule in favour of admitting the complainant's video-recorded evidence-in-chief. Section 22A does not mean that such evidence is inadmissible in a magistrates' court, but merely that there is no rule in favour of admitting it.

The operation of s 22A is as follows:

(a) If a party to the proceedings applies for a special measures direction in relation to the complainant, that party may request that the direction provide for a video recording of an interview of the complainant to be admitted as evidence-in-chief of the complainant.
(b) If the court determines that the complainant is eligible for assistance by virtue of s 16 (1)(b) and (2) or s 17 (**8.20**), it must give such a special measures direction in relation to the complainant, unless:
    (i) it considers that in the interests of justice the recording should not be admitted; or
    (ii) it is satisfied that compliance with such a direction would not be likely to maximise the quality of the complainant's evidence so far as practicable (whether because the application to that evidence of one or more other special measures available in relation to the complainant would have that result or for any other reason).

Special measures available

The special measures are: **8.23**

(1) *Screening (s 23)* Screens may be authorised to shield W from D (but the judge, jury, justices, a legal representative from each side, any interpreter, and any person appointed to assist W must be able to see W).

(2) *Evidence by live link (s 24)* 'Live link' means a live television link or other arrangement whereby a witness, while absent from the place where the proceedings are being held, is able to see and hear a person there and to be seen by the judge or justices (or both) and the jury (if there is one), legal representatives, and any interpreter or other person appointed to assist the witness. Where a live link direction is given, evidence may not be given in any other way without the consent of the court.

(3) *Evidence given in private (s 25)* The court may be cleared of non-essential personnel but this measure is only available in relation to a sexual offence (**8.20**) or an offence under MSA 2015, s 1 (slavery, etc), or when the court reasonably believes that someone has tried to intimidate, or will try to intimidate, W.

(4) *Removal of wigs and gowns (s 26)* This applies to the judiciary as well as legal representatives.

(5) *Video-recorded evidence-in-chief (s 27)* Where this special measure is directed, it will provide for a video recording of an interview of W to be admitted in evidence-in-chief. However, the direction may not provide for a video recording, or part of it, to be admitted if its admission would not be in the interests of justice. If it is decided to permit only an edited version to be shown, the court must consider whether the exclusion of part of the recording is prejudicial. A court may later exclude a recording if its making is not properly proved, but may nevertheless admit it in such circumstances. The party tendering the evidence must call W, unless a special measures direction provides for cross-examination under s 28 (below) or the parties have agreed to non-attendance.

The *Criminal Practice Directions 2015* require that the party who made the application to admit video-recorded evidence-in-chief must edit the recording in accordance with the judge's directions and send a copy of the edited recording to the appropriate officer of the Crown Court and to every other party to the proceedings.

(6) *Video-recorded cross-examination or re-examination (s 28)* The provisions of s 28 have been brought into force in Crown Courts sitting at a large number of locations in relation to a witness (W) eligible for special measures under s 16 (**8.20**: W under 18 or suffering incapacity).

The provisions of s 28 have also been brought into force in Crown Courts sitting at Kingston-upon-Thames, Leeds or Liverpool where a complainant in respect of a sexual offence or modern slavery offence is a witness in proceedings relating to that offence (or that offence and any other offence) and is thereby eligible under s 17 (**8.20**) for special measures.

When s 28 is fully and nationally in force, the law will be as follows. Where a special measures direction provides for video recording of W's evidence-in-chief to be admitted, the special measures direction may also provide that W may be cross-examined before trial and that that cross-examination (and any re-examination) may be recorded for use at trial. This will not occur in the physical presence of D, although D will be able to see and hear it and to communicate with his/her legal adviser (live link). Nor need it take place in the physical presence of the judge or magistrates and the defence and prosecution legal representatives, although they must be able to see and hear the examination and to communicate with those present. However, a judge or magistrate must control the proceedings and this person will normally be the trial judge or magistrate. Where a recording has been made of the examination of W under the above power, W may not be subsequently cross-examined or re-examined in respect of his/her evidence unless the court makes a further direction to this effect. Such a further direction may only be given: (a) where the proposed cross-examination is sought by a party to the proceedings as a result of having become aware since the original recording of a matter which (s)he could not with reasonable diligence have ascertained by then, or (b) where it is in the interests of justice to do so.

The *Criminal Practice Directions* provide that witnesses eligible for special measures under s 28 should be identified by the police. The police and CPS should discuss, with the witness or with the witness' parent or carer, special measures available and the witness' needs, such that the most appropriate package of special measures can be identified. This may include use of a registered intermediary (below). The *Directions* also provide that for access to special measures under s 28, the witness' interview must be recorded in accordance with the *Achieving Best Evidence* guidance referred to at **8.24**.

The following two additional measures under ss 29 and 30 are available only in the case of someone eligible for special measures under s 16:

Examination of witness through intermediary (s 29)   An intermediary is an interpreter or someone else whom the court approves to communicate to W the questions asked in court, and then to communicate W's answers. The intermediary may also explain such questions or answers, if necessary to facilitate understanding. An intermediary will normally be a specialist. An intermediary can act however and wherever the examination is conducted. The Witness Intermediary Scheme operated by the NCA identifies intermediaries who may be used by the prosecution and defence. Such an intermediary is matched to the particular witness based on expertise, location, and availability. An intermediary appointed through, and accredited by, the WIS is a registered intermediary. Registered intermediaries are bound by Codes of Practice and Ethics issued by the Ministry of Justice (which oversees the WIS).

Aids to communication (s 30)   A special measures direction may require a witness to be provided with an appropriate device to assist communication.

General   Evidence given using any of the special measures set out above must be treated in the same way as oral evidence.

If a witness who would normally be sworn gives unsworn evidence by means of a video recording, that evidence is admissible. However, where a person authorised to administer an oath is present, the evidence could be taken under oath in appropriate cases.

Where a video recording is to be adduced during proceedings before a Crown Court, it must be produced and proved by the interviewer, or any other person who was present at the interview with the witness at which the recording was made. The parties may agree to accept a written statement in lieu of the attendance of such a person. The party adducing the video recording must arrange for the operation of the video-playing equipment.

Best practice in questioning child and vulnerable witnesses

8.24   The Ministry of Justice has published assistance for those dealing with witnesses subject to special measures and for those preparing video-recorded interviews, *Achieving Best Evidence in Criminal Proceedings: Guidance on Interviewing Victims and Witnesses, and Guidance on Using Special Measures.*

The guidance is primarily aimed at police officers conducting visually-recorded interviews with vulnerable, intimidated and significant witnesses, those tasked with preparing and supporting such witnesses during the criminal justice process and those involved at the trial, both in supporting and questioning the witness in court. While the guidance is advisory, compliance with *Achieving Best Evidence* and effective training is likely to maximise the quality of interviews with such witnesses and is likely to benefit the interviewer, the witness, practitioners, and the courts alike. Where there has been a breach of *Achieving Best Evidence* an interview will only be inadmissible if the court cannot be sure that the witness has given a credible and accurate account.

Evidence of vulnerable defendant

YJCEA 1999, s 33A allows a court, on the application of the defendant (D), to direct **8.25** that D's evidence should be given over a live link. The court must be satisfied that it would be in the interests of justice for D to give evidence through a live link, and:

(a) if D is under 18, that D's ability to participate effectively as a witness giving oral evidence is compromised by D's level of intellectual ability or social functioning, and that use of a live link would enable D to participate more effectively as a witness;

(b) if D is 18 or over, that D is unable to participate in the proceedings effectively as a witness in court giving oral evidence because D has a mental disorder or a significant impairment of intelligence and social function, and that use of a live link would enable D to participate more effectively as a witness.

Section 33A is intended to provide a structured approach to decision-making and to ensure that the giving of evidence in this way is reserved for exceptional cases. Where a juvenile is concerned the test is less strict as there is no reference to mental disorder or impairment.

Where a direction has been given under s 33A, D must give all his/her oral evidence through a live link and cross-examination should be carried out by the same means.

A divisional court has held that the courts have a common law duty to appoint an intermediary to help a child defendant (or, indeed, an adult defendant) follow the proceedings and give evidence where (s)he would otherwise not have a fair trial.

Protection of witnesses from cross-examination by defendant in person

Complainants in proceedings for sexual offences

YJCEA 1999, s 34 provides that no person charged with a sexual offence (as defined **8.26** at **8.20**) may in any criminal proceedings cross-examine in person a witness who is a complainant, either in connection with the offence, or in connection with any other offence (of whatever nature) with which that person is charged in the proceedings.

Child complainants and other witnesses who are children

In relation to an offence to which it applies, YJCEA 1999, s 35 makes similar provisions **8.27** concerning a 'protected witness'. A 'protected witness' is a witness who:

(a) either is a complainant or a witness to the offence; and

(b) either is a child or falls to be cross-examined after giving evidence-in-chief as a child.

Section 35 applies to any offence under SOA 1956, ss 33–36 (brothels), the Protection of Children Act 1978 (taking, etc indecent photographs, etc of a child), SOA 2003, Part 1 (ss 1–79) or any 'relevant superseded enactment', or MSA 2015, ss 1 and 2. For the purposes of YJCEA 1999, s 35, where an offence is under one of these pieces of legislation, a 'child' is someone under 18.

Section 35 also applies to kidnapping, false imprisonment, child abduction (Child Abduction Act 1984, ss 1 or 2), cruelty to children (Children and Young Persons Act 1933, s 1), or any offence involving an assault on, or injury (or a threat of injury) to, any person. For the purposes of these other specified offences, a 'child' is someone under 14.

Direction prohibiting cross-examination

8.28   YJCEA 1999, s 36, which is not limited to the sexual offences or other offences to which s 35 applies, permits a court to prohibit an unrepresented defendant from cross-examining witnesses in other cases, where the provisions of ss 34 and 35 do not apply. This may be done where the court is satisfied that the quality of the evidence given by the witness on cross-examination is likely to be diminished if cross-examination is undertaken or continued by the defendant in person, and would be likely to be improved if a direction was given, and that such a prohibition will not be contrary to the interests of justice.

The term 'witness' does not include any other person who is charged with an offence within the proceedings.

Representation of such persons for the purpose of cross-examination

8.29   Where D is prohibited from cross-examining a witness under the provisions of YJCEA 1999, ss 34, 35, or 36, the court must invite D to appoint a legal representative to cross-examine on D's behalf. If no such appointment is made within the prescribed time limits, but the court considers that it is necessary in the interests of justice it must appoint a legal representative.

Witness anonymity

8.30   The Coroners and Justice Act 2009, s 86 provides for the making of witness anonymity orders (WAOs) in criminal proceedings, on the application of the prosecutor or the defendant (D).

A WAO is an order that requires such specified measures to be taken in relation to a witness (W) in criminal proceedings as the court considers appropriate to ensure that the identity of W is not disclosed in or in connection with the proceedings.

By s 88, a court may make a WAO only if it is satisfied that conditions A to C below are all met.

Condition A is that the measures to be specified in the order are necessary in order:

(a) to protect the safety of W or another person or to prevent any serious damage to property; or

(b) to prevent real harm to the public interest (whether affecting the carrying on of any activities in the public interest or the safety of a person involved in carrying on such activities, or otherwise).

Condition B is that, having regard to all the circumstances, the taking of those measures would be consistent with D receiving a fair trial.

Condition C is that the importance of W's testimony is such that in the interests of justice the witness ought to testify and—

(a) the witness would not testify if the proposed order were not made, or

(b) there would be real harm to the public interest if the witness were to testify without the proposed order being made.

MEANS OF PROOF WHICH MAY BE INADMISSIBLE

Opinion

Opinion evidence is generally inadmissible because it is the function of the court or **8.31** jury, and not of a witness, to draw conclusions from the facts proved. If a witness alleges that a particular driver was at fault and caused an accident, that evidence is inadmissible because that is the issue which the court or jury must decide. A fine line can sometimes exist between 'opinion' and 'fact', and for this reason evidence as to the identification of a person or thing (which must always be an opinion to some extent) is admissible. In addition, by way of exception to the general rule, the following opinion evidence of experts is admissible:

(a) opinion evidence of experts on matters (outside the knowledge or experience of a jury) of a scientific, technical, or artistic nature, such as doctors of medicine, forensic scientists, metallurgists, or literary experts: a police officer can be an expert under this heading in appropriate circumstances; for example, the opinion evidence of a police officer that a quantity of drugs was too great for personal use was held admissible as expert evidence because the officer based his opinion on his 17 years' experience and on published and unpublished material;

(b) opinion evidence from persons who are experts in handwriting comparison; and

(c) opinion evidence by a lawyer shown to have knowledge of a particular system of foreign law.

An expert witness should provide independent, objective assistance to the court in relation to matters within his/her expertise. His/her duties are owed to the court and override any obligation to those instructing or paying him/her. Consequently, for example, if an expert instructed by the prosecution carries out a test which casts doubt on his/her opinion, (s)he must disclose this to his/her instructing solicitor, who must disclose it to the defence.

Hearsay

As said at **8.6**, hearsay evidence is 'second-hand' evidence since it consists of any repre- **8.32** sentation of fact or opinion made by whatever means (eg orally, in writing, or by conduct) by a person otherwise than in oral evidence in the particular proceedings which is tendered as evidence of any matter stated *in it*. Hearsay evidence is inadmissible, unless its admission is permitted under a variety of rules, because the law of evidence generally requires a fact to be proved by direct evidence of it.

Of course, as already indicated, not all evidence of what someone other than a witness said or wrote is hearsay. If it is produced merely to prove the fact that it was said or written, as opposed to establishing a matter contained in the oral or written statement, it is original evidence and not hearsay evidence, and therefore admissible if relevant to the facts in issue. This distinction can be illustrated as follows.

If a witness says that the deceased, while in hospital with injuries from which he unexpectedly died, told him, 'Fred did this to me', this is hearsay evidence since it is produced as evidence of the truth of what the deceased stated, that is, that Fred caused the injuries. Likewise, if a witness says that X had told him/her that Y had broken the windows of a greenhouse, this is hearsay evidence since it is produced as evidence of the truth of what X stated, ie that Y had broken the windows.

By way of contrast, it is not hearsay evidence for evidence to be called to show that D *said* that he had acted under a threat of death made by Jones, or for a witness to say 'Mrs Bird *complained* to me that Tate had fondled her breasts'. This is original evidence because it is produced as evidence of the fact that a threat or complaint had been made. The latter statement is, however, hearsay if it is intended to show the fact that the sexual offence has been committed.

The rule against hearsay evidence has gradually been whittled away, culminating in CJA 2003, Part 11, Chapter 2 (ss 114–136) which abolished the common law rules about hearsay evidence and created a set of rules about its admissibility. It should be noted that hearsay evidence does not have to be shown to be reliable or independently verified before it can be admitted under these rules.

The cases where a statement is admissible in criminal proceedings as evidence of the truth of its contents under an exception to the rule against hearsay are as follows.

Admissibility of hearsay evidence: the basic provisions

8.33   CJA 2003, Part 11, Chapter 2 commences by stating in s 114(1) that hearsay evidence is admissible in criminal proceedings as evidence of any matter stated if, but only if:

(a) a statutory provision makes it admissible (**8.34-8.43**);
(b) any rule of law preserved by s 118 (**8.44-8.48**) makes it admissible;
(c) all parties to the proceedings agree to it being admissible; or
(d) the court is satisfied that it is in the interests of justice for it to be admissible.

In deciding under s 114(1)(d) whether it is in the interests of justice, the court must have regard to the following factors (and any others it considers relevant):

(i)   how much probative value the statement has in relation to a matter in issue (ie how much value is it in determining the matter) or how valuable it is for understanding other evidence;
(ii)   what other evidence has been, or can be, given on the matter or evidence mentioned in (i);
(iii)   how important the matter or evidence mentioned in (i) is in the context of the case as a whole;
(iv)   the circumstances in which the statement was made;
(v)   how reliable the maker of the statement appears to be;
(vi)   how reliable the evidence of the making of the statement appears to be;
(vii)   whether oral evidence of the matter stated can be given and, if not, why it cannot;
(viii) the amount of difficulty involved in challenging the statement; and
(ix)   the extent to which the difficulty would be likely to prejudice the party facing it (s 114(2)).

A court does not have to reach a conclusion about all nine of these factors in order to admit hearsay evidence under (d). The Court of Appeal has held that s 114(1)(d) must be cautiously applied in the case of an absent witness who does not fall within s 116 ('Witness unavailable': **8.34**); otherwise the conditions in s 116 would be circumvented.

The Court of Appeal has held that an anonymous witness statement made by a person whose identity is not known at all cannot be admitted under the hearsay rules in CJA 2003, Part 11, Chapter 2.

CJA 2003, Part 11, Chapter 2 does not affect the exclusion of evidence of a statement on grounds other than the fact that it is hearsay. For example, a confession (a type of admissible hearsay) might be excluded under the provisions of PACE (**8.52**).

CJA 2003, s 115 provides that, for the purposes of CJA 2003, Part 11, Chapter 2, references to a 'statement' are to any representation of fact or opinion made by a person by whatever means, including a representation made in a sketch, photofit or other pictorial form (which the Court of Appeal held clearly includes an E-Fit). By s 115, a 'matter stated' is one to which CJA 2003, Part 11, Chapter 2 applies only if the purpose (or one of the purposes) of the maker of the statement appears to have been to cause another person to believe the matter or to cause another person to act or a machine to operate on the basis that the matter is as stated. As a result a statement which was not made with such a purpose, but from which a fact may be inferred, is original evidence and not hearsay evidence. Examples would be a statement made in a secret diary or in a text to another person mentioning a fact that the texter believes the other knows.

The exceptions to the hearsay rule to which s 114(1)(a) and (b) relate are as follows.

Witness unavailable

CJA 2003, s 116 provides that in criminal proceedings a statement which is not made in oral evidence in the proceedings is admissible in evidence as evidence of any matter stated if:

8.34

(a) oral evidence given by the person making the statement would be admissible as evidence of that matter;

(b) the person who made the statement is identified to the satisfaction of the court; and

(c) that person ('the relevant person'):

    (i) is dead;

    (ii) is unfit to be a witness because of his/her bodily or mental condition;

    (iii) is outside the UK and it is not reasonably practicable to secure his/her attendance;

    (iv) cannot be found although such steps as it is reasonably practicable to take to find him/her have been taken; or

    (v) through fear does not give (or continue to give) oral evidence in the proceedings, either at all, or in connection with the subject matter of the statement, and the court gives leave for the statement to be given in evidence.

'Unfit to be a witness' in (c)(ii) applies not only to a person's physical inability to attend a court but also to his/her mental capacity when there to give evidence.

'Cannot be found' in (c)(iv) includes 'cannot be contacted'.

'Fear' in (c)(v) is to be widely construed and (for example) includes fear of the death or serious injury of another person or of financial loss. There must, if possible, be direct evidence (of fear) from the witness (W). However, the evidence of fear may, for example, be given by a police officer. Nonetheless, it is important to establish that the fear existed at the time of the trial. Before a court can properly be satisfied as to the requirements, it should be informed of what steps have been taken to persuade W to attend or to alleviate W's fears. Leave to admit may only be given under (c)(v) where the court considers that the statement ought to be admitted in the interests of justice, having regard to its contents, to the risk of unfairness and, in appropriate cases, to the fact that a special measures direction could be made in relation to the person making the statement, and to any other relevant circumstances.

Any condition in (c)(i)–(v) which is in fact satisfied will be treated as not having been satisfied if any of the circumstances described in (c) have effectively been caused by the party in the trial in support of whose case it is sought to admit the statement, or by any person acting on his/her behalf, in order to prevent the relevant person referred to in (c) from giving oral evidence in the proceedings (whether at all or in connection with the subject matter of the statement). It is irrelevant whether the steps which had the relevant effect were taken before or after the commencement of the proceedings.

Business and other documents

8.35   CJA 2003, s 117 provides that in criminal proceedings a statement contained in a document is admissible as evidence of any matter stated if:

(a) oral evidence given in the proceedings would be admissible as evidence of that matter;

(b) the following requirements are satisfied:
   (i) the document or the part containing the statement was created or received by a person in the course of a trade, business, profession, or other occupation, or as the holder of a paid or unpaid office;
   (ii) the person who supplied the information contained in the statement (the relevant person) had, or may reasonably be supposed to have had, personal knowledge of the matters dealt with; and
   (iii) each person (if any) through whom the information was supplied from that relevant person to the person mentioned in (i) received the information in the course of a trade, business, profession, or other occupation, or as the holder of a paid or unpaid office; and

(c) additional requirements are satisfied where the statement was prepared for the purpose of pending or contemplated criminal proceedings, or for a criminal investigation, but was not obtained pursuant to a request under the Crime (International Co-operation) Act 2003, s 7, an order under the Criminal Justice Act 1988, Sch 13 or an overseas production order under the Crime (Overseas Production Order) Act 2019 (all of which relate to overseas evidence). The additional requirements referred to are any of the five conditions set out at **8.34**(c), or that the relevant person cannot reasonably be expected to have any recollection of the matters dealt with in the statement (having regard to the length of time since (s)he supplied the information and all other circumstances).

The persons mentioned in (b)(i) and (ii) may be the same person. Thus, for example, a note made by an operator working for a paging company that messages had been left for a customer would be admissible under s 117.

Section 117 is obviously sensible. It would be unnecessarily burdensome if oral evidence was always required from a person who was either the creator or keeper of the document, or the supplier of the information contained in the document.

As a result of s 117, entries in a police officer's notebook (or a computerised crime record produced by a police officer) are admissible if the terms of s 117 are satisfied. An example of a case where s 117 was satisfied is where a man presented a stolen debit card at a supermarket checkout and a supervisor saw him making off in a car, the registered number of which she noted; a record made by a second supervisor, at the dictation of the first, was held to be a document created or received in the course of a business, profession, etc for these purposes; the fact that she could recall other matters which occurred at the time (such as the colour of the offender) did not mean that she could be expected to remember the registered number of the car.

It is important to note the limitations of s 117. For example, where a shopper had left the offending vehicle's registration number on the damaged car's windscreen after it had been hit in a supermarket car park and that number had been reported in the police incident log after it had been given by the girlfriend of the owner of the damaged car, the Court of Appeal held that the information in the log should not have been admitted under s 117 because the girlfriend had not received it in the course of trade or business and therefore condition (b)(iii) was not satisfied. The Court of Appeal added that the police record would have been admissible under the rule relating to the admissibility of multiple hearsay (see condition **8.38**(c)).

A statement is not admissible under s 117 if the court makes a direction to that effect, which it can do if satisfied that the statement's reliability as evidence is doubtful, either in its contents, the source of its information, the way or circumstances in which the information was supplied or received, or the way in which the document was created or received.

Inconsistent statements

CJA 2003, s 119 provides that, where a witness (W) in criminal proceedings gives oral **8.36** evidence and admits making a previous inconsistent statement, or a previous inconsistent statement made by W is proved to have been made, the previous statement is admissible as evidence of any matter in respect of which oral evidence would be admissible. Thus, the fact that W made a previous inconsistent statement does not merely affect W's credibility but it is also some evidence of the truth of the facts which W had previously stated.

Other previous statements of witness

Where it is suggested that a witness (W) in criminal proceedings has fabricated oral **8.37** evidence, a previous statement by W is admissible not only on the issue of credibility but also to prove the truth of the matters previously stated. This is the effect of CJA 2003, s 120(2) which states that any previous statement which W has made is admissible in evidence as evidence of any matter which it contains.

In addition, s 120(3) provides that where W refreshes his/her memory from a written document and is cross-examined on a statement in that document, and the document is in consequence received in evidence, that statement in the document will become evidence of any matter stated in it. Section 120(3) does not make admissible as truth of the matters stated a statement made in a written document in circumstances where a re-reading of a previous out-of-court document has failed to refresh W's memory when giving oral evidence at the trial. The situation where W has made a previous statement when the matters were fresh in W's memory, but does not remember them at the time of the trial, even when W has attempted to refresh his/her memory, is dealt with specifically by (b) in the following provisions.

By s 120(4)–(7) a previous statement is also admissible as evidence of the facts contained in it if W states that (s)he made the statement and believes it to be true and any one of the following conditions applies:

(a) that the statement describes or identifies a person, object, or place;

(b) that it was made by W when the matters stated were fresh in W's memory, and W cannot reasonably be expected to remember the matters stated well enough to give oral evidence of them; or

(c) that:

    (i) W claims to be a victim of an offence to which the proceedings relate,

    (ii) the statement consists of a complaint by W about conduct which (if proved) would constitute the offence (or part of it),

    (iii) the complaint was not made as a result of a threat or promise, and

    (iv) W gives oral evidence in respect of the matter before the statement is adduced.

Additional requirements for admissibility of multiple hearsay

8.38   'Multiple hearsay' refers to the case where the witness has no personal knowledge of the matter stated to him/her, but repeats a hearsay statement made to him/her by another, as where X makes a statement to Y who repeats it to Z. Repetition of the statement by Z is multiple (or second-hand) hearsay, whereas if Y gave the statement in evidence it would be first-hand hearsay. CJA 2003, s 121 provides that a hearsay statement is not admissible to prove the fact that an earlier hearsay statement was made unless:

(a) either of the statements is admissible under s 117 (business and other documents), 119 (inconsistent statements) or 120 (other previous statements by a witness), or

(b) all parties agree to its admissibility, or

(c) the court is satisfied that the value of the evidence in question, taking into account how reliable the statement appears to be, is so high that the interests of justice require the later statement to be admissible for that purpose.

CJA 2003, ss 116, 117, 119, and 120: general provisions concerning capability to make statements

8.39   CJA 2003, s 123 provides as follows.

Nothing in ss 116, 119, or 120 (**8.34**, **8.36**, and **8.37**) makes a statement admissible as evidence if it was made by a person who did not have the 'required capability' at the time that (s)he made the statement.

Nothing in s 117 (business and other documents: **8.35**) makes a statement admissible if any person who, in order for the requirements at **8.35**(b) (created or received by the particular persons in the course of their business, etc) to be satisfied, must have supplied or received the information concerned (or created or received the document or part concerned):

(a) did not at the time have the required capability, or

(b) cannot be identified but cannot reasonably be assumed to have had the required capability at that time.

A person has the 'required capability' if (s)he is capable of understanding questions put to him/her about the matter and of giving answers which can be understood.

Evidence from computer records

8.40   The ordinary law on evidence applies to computer evidence. In the absence of evidence to the contrary, courts will presume that the computer system was working correctly. If there is evidence that it might not have been, the party seeking to introduce the evidence will need to prove that it was working. Such proof can be satisfied by the evidence of a person familiar with the operation of the computer, who need not be a computer expert. The House of Lords has held that evidence by a store detective that computerised cash tills were working satisfactorily was admissible where it was apparent from the nature of her evidence that she was thoroughly familiar with the operation of the tills and the central computer, even though she did not understand the technical operation of the computer.

CJA 2003, s 129 provides that, where a representation of fact is generated by a machine and depends for its accuracy on information supplied by a person, the representation will only be admissible as evidence of the fact where it is proved that the information was accurate. This is not directly related to proof of the reliability of the machine, but is associated, in that there is a requirement to show that accurate information was fed into the machine.

Procedural rules

A party who wants to introduce hearsay evidence under ss 114(1)(d) (evidence admissible in the interests of justice), 116 (witness unavailable), 117 (head (c) statement prepared for criminal proceedings) or 121 (multiple hearsay) (**8.33**, **8.34**, **8.35** and **8.38** respectively) must give notice in accordance with the Criminal Procedure Rules 2020, Part 20; see Chapter 8 on the companion website.   **8.41**

Evidence by certificate

The Criminal Justice Act 1948, s 41 provides that a certificate signed by a constable or qualified person certifying that a plan or drawing exhibited in criminal proceedings is a plan or drawing made by him/her of an identified place or object specified in the certificate, and that the plan or drawing is correctly drawn to a scale so specified, is evidence of the relative position of things shown on the plan or drawing. The plan, etc is admissible to the same extent that oral evidence would be admissible. A 'qualified person' means a registered architect or a chartered engineer of one of certain types.   **8.42**

In order for the certificate to be admissible, a copy of such a plan, etc must be served on D not less than seven days before the hearing. Moreover, if D serves the appropriate notice that D requires the attendance of the person who signed the certificate, the certificate will not be admissible.

Proof by written statements

The Criminal Justice Act 1967, s 9 provides for the admissibility of written statements in criminal proceedings. It states that a written statement signed by the person who made it is admissible as evidence to the like extent as oral evidence given by that person. There are certain conditions which must be satisfied: **8.43**

(a) the statement must contain a declaration by that person to the effect that it is true to the best of his/her knowledge or belief and that (s)he has made the statement knowing that, if it were tendered in evidence, (s)he would be liable to prosecution if (s)he wilfully stated in it anything which (s)he knew to be false or did not believe to be true;

(b) before the hearing at which the statement is tendered in evidence, a copy of the statement must be served, by or on behalf of the party proposing to tender it, on each of the other parties to the proceedings; and

(c) none of the other parties or their solicitors must, within the relevant period, have served a notice on the party so proposing objecting to the statement being tendered in evidence under the section.

(b) and (c) do not apply if the parties agree before or during the hearing that the document shall be so tendered.

In practice, statement forms carried by police officers incorporate the declaration at (a) so that the evidence of any witness can be offered in written form if it is accepted by the other side.

'The relevant period' in (c) is defined by the Criminal Procedure Rules 2020, Part 16. It is the period of five business days after service of the statement unless:

(i)   the court extends that time limit;
(ii) the 'pleading guilty by post' provisions (**3.27**) apply, in which case the time limit is the later of five business days after service of the statement or five business days before the hearing date; or
(iii) the trial by single justice on the papers procedure (**3.28**) applies, in which case the time limit is 15 business days after service of the statement.

It has been held that 'written statement' in s 9 means the written assertion of facts made by the person who signed the document, and that what is required by (a) is that the written assertion of facts is contained in a document in identical terms to those in the signed document. Thus, it was irrelevant that the copy of the statement which had been served on D had not been signed by the maker of the written statement; the important thing was that the assertions of facts in the copy were identical to those in the signed statement.

The Criminal Procedure Rules 2020, Part 16 additionally require the statement to contain:

(a) the witness's (W's) name and, if under 18, age; and
(b) if W cannot read the statement, a signed declaration by someone else that that person read it to W.

In addition, where the statement refers to a document or object as an exhibit, the statement must clearly identify that document or object.

Statements received as part of the res gestae

8.44   A statement is part of the res gestae if it is so closely associated in time, place, and circumstances with the fact in issue that it can be said to be part of the same transaction. At common law, under the 'res gestae rule', a statement which is part of the res gestae is admissible if the conditions set out below are satisfied.

The res gestae rule is specifically preserved by CJA 2003, s 118 which preserves any rule of law under which, in criminal proceedings, a statement is admissible as evidence of any matter stated if the statement:

(a) was made by a person so emotionally overpowered by an event that the possibility of concoction or distortion can be disregarded;
(b) accompanied an act which can be properly evaluated as evidence only if considered in conjunction with the statement; or
(c) relates to a physical sensation or mental state (such as intention or emotion).

Thus, an oral statement made by a person involved in an unusual, startling, or exciting event is admissible as part of the res gestae as evidence of the facts stated, provided it was so clearly made spontaneously that, in the light of the circumstances, the possibility of concoction or distortion can be disregarded. If the issue is one of murder it is probable that a victim's statement 'don't shoot, Sam', which was made as (or immediately before) the gun was fired, would be admissible as part of the res gestae, although it is hearsay, and that likewise the victim's agitated shout immediately afterwards, 'Look what you've done. Get a doctor, quick', would be admissible. Where a man was stabbed and, whilst being given first aid treatment by a constable, named his attackers, this evidence was admitted as a part of the res gestae at D's trial after the declarant had died.

Confessions

Although hearsay, a confession is generally admissible. This common law rule is specif- **8.45** ically preserved by CJA 2003, s 118. The law relating to confessions is dealt with below **(8.50-8.54)**.

Public information

CJA 2003, s 118 preserves the rule of law which permits, in criminal proceedings, the **8.46** admissibility of:

(a) facts of a public nature stated in published works dealing with matters of a public nature (eg histories, scientific works, dictionaries, maps), or
(b) facts stated in public documents (eg public registers, and returns under public authority with respect to matters of public interest), or in court records, treaties, etc, or
(c) evidence as to a person's age or date or place of birth in a birth certificate.

Expert reports

An expert report is admissible in evidence in criminal proceedings, whether or not **8.47** the person making it attends to give oral evidence. However, if it is proposed that the maker of the report will not give oral evidence, the report is only admissible with the leave of the court. In determining whether to give leave, the court must have regard to the contents of the report; the reasons why it is proposed that oral evidence will not be given; the likely risk to fairness in relation to D; and any other relevant circumstance. An 'expert report' is one written by a person dealing wholly or mainly with matters on which (s)he is (or if living would be) qualified to give expert evidence.

This common law rule is preserved by CJA 2003, s 118.

Reputation

CJA 2003, s 118 preserves the common law rules whereby, in criminal proceedings, **8.48** hearsay evidence of a person's reputation is admissible for the purpose of proving his/her good or bad character, but only in so far as it allows the court to treat such evidence as proving his/her good or bad character.

Section 118 also preserves the common law rules about the admissibility of evidence of reputation or family tradition for the purpose of proving pedigree or the existence of a marriage, the existence of any public or general right, or the identity of any person or thing. The use of these rules will be rare.

Supplementary points about hearsay

Credibility of hearsay evidence By CJA 2003, s 124, a challenge may be made to the **8.49** credibility of the maker of a statement admitted in criminal proceedings as hearsay evidence if (s)he does not give oral evidence in the proceedings. In certain circumstances, the person against whom the hearsay evidence has been admitted may produce evidence to discredit the maker of the statement or to show that (s)he has contradicted him/herself.

Stopping case where hearsay evidence unconvincing A judge has the duty under CJA 2003, s 125 to stop a case and either direct acquittal, or (if the judge considers that there ought to be a re-trial) discharge the jury, if the prosecution case is based wholly or partly on an out-of-court statement which is *so unconvincing that, considering its importance to the case against D, a conviction would be unsafe*. If any untested out-of-court statement is not shown to be reliable and it is a statement which is part of the central body of evidence without which the case cannot proceed, the italicised words will be satisfied.

A magistrates' court would have a similar duty to that under s 125 in the above circumstances.

General discretion to exclude hearsay evidence   CJA 2003, s 126 gives a court a discretion to exclude a statement as evidence of a matter stated if the statement was made otherwise than in oral evidence in the proceedings, and the court is satisfied that the case for excluding the statement, taking account of the danger that to admit it would result in undue waste of time, substantially outweighs the case for admitting it, taking account of the value of the evidence. This is in addition to any discretion under PACE, s 78 (**8.74**) or any other power of a court to exclude evidence.

CONFESSIONS

Confession by one defendant: evidence against co-defendant?

8.50   An out-of-court confession by *one defendant is not evidence against another co-defendant* unless the co-defendant expressly or impliedly adopts the statements contained in it. There are qualifications to this statement. First, the House of Lords has held that when, in a joint trial of two or more for an offence alleged to have been committed jointly, proof of the guilt of one of the defendants, A, is essential in proving the case against another of them, B, and the evidence against A consists solely of A's own out-of-court confession, then A's confession will be admissible as evidence, not only against A but also against B in so far as the fact of A's guilt of itself establishes B's guilt. Second, an out-of-court accusation against a co-defendant contained in a confession is *capable* of being admitted as hearsay evidence if it falls within CJA 2003, s 114(1)(d) (interests of justice for evidence to be admissible: **8.33**).

Confession by defendant as evidence against him/her

8.51   For the purposes of the rules which follow, a 'confession' is defined by PACE, s 82 as including 'any statement wholly or partly adverse to the person who made it, whether made to a person in authority or not and whether made in words or otherwise'. Because of the phrase 'in words or otherwise', a defendant's confession may simply consist of a gesture of acceptance of a statement adverse to him/her made by another. Statements made which are intended to vindicate a suspect, for example, where (s)he gives explanations intended to justify his/her possession of goods, and if taken to be true, would do so, do not amount to a confession, even if they are later shown to be false or inconsistent with the maker's evidence. In such a case, however, a record should be made as soon as possible; the reason for there being no contemporaneous notes should be recorded and the suspect should be given an opportunity to check the record.

PACE, s 76(1) allows a confession by D to be given in evidence against D in so far as that confession is relevant to a matter in issue in the proceedings and has not been excluded by the court under s 76(2) on the basis explained below.

Exclusion

If it is represented by D (or by D's counsel), on the basis of some credible, more than insignificant or trivial evidence, that a confession by D was, or may have been, obtained by oppression or in consequence of anything said or done which was likely to render the confession unreliable, the court must not allow the confession to be given in evidence against that person except in so far as the prosecution proves beyond reasonable doubt that the confession (notwithstanding that it might be true) was not obtained by these means.

8.52

PACE, s 76(2) provides that a *confession by oppression must always be excluded*. 'Oppression' includes torture, inhuman or degrading treatment, and the use of threats or violence. Apart from this, 'oppression' bears its ordinary dictionary meaning, namely, 'the exercise of power or authority in a burdensome, harsh, or wrongful manner; unjust or cruel treatment'. It will almost inevitably involve some impropriety on the part of the questioner. The fact that a confession has been obtained in circumstances involving a breach of a code of practice does not in itself constitute oppression. However, bullying questioning may exceptionally amount to oppression.

Section 76(2) also provides that a *confession must be excluded if it was obtained in consequence of anything said or done which was likely, in the circumstances existing at the time, to render unreliable any confession which might be made by D in consequence thereof.* An example of something which will be held to be likely to render a confession unreliable is making an untrue inducement to bring about a confession (eg 'if you confess, the matter will go no further'). Another example is the use of hostile and intimidating interview techniques. For such a likelihood of unreliability to be found there is no need for any hint of impropriety. Where a suspect was mentally handicapped and was interviewed without an adult person being present, it was held that once it had been established that there had been a breach of PACE Code C, the onus was on the prosecution to satisfy the judge beyond reasonable doubt that the confession was not obtained in breach of s 76(2). The circumstances existing at the time were all important in relation to reliability.

Where it is shown that there has been aggressive and hostile questioning it becomes a matter of degree as to whether the threshold is passed beyond which the behaviour of police officers has made the confession unreliable in all circumstances. However, all the circumstances must be examined. In the course of an interview concerning drug offences an officer interjected, implying that if the suspect (D) did not tell the truth (D) would be held in custody. Some 16 minutes later D confessed to an offence. The Court of Appeal endorsed the trial judge's refusal to exclude D's confession being satisfied that D was astute, had experience of being interviewed at a police station, and that his will had not been broken. He had continued to deny other offences.

Facts discovered as a result of inadmissible confession PACE, s 76(4) provides that, even if a confession, or a part of it, is excluded under the above provisions, this does not affect the admissibility in evidence:

(a) of any facts discovered as a result of the confession; or
(b) where the confession is relevant as showing that D speaks, writes, or expresses himself in a particular way, of so much of the confession as is necessary to show that D does so.

The effect of (a) is as follows. If evidence of a fact is discovered as a result of a confession, or part of a confession, and that confession, or a relevant part of it, is excluded under s 76(2), that evidence may nevertheless be adduced by the prosecution. However, no reference should be made to the fact that the discovery was the result of a confession; only D, or someone acting on D's behalf, may disclose this. Thus, if D is arrested for thefts of motor cars and makes a confession which includes details of the persons to whom D sold the vehicles, and that confession is excluded, the prosecution may give evidence of the recovery of the vehicles from those persons but it may not mention that they were discovered as a result of the excluded confession.

Confessions given in evidence for co-defendant

8.53   PACE, s 76A provides that a confession, as defined by s 82 (**8.51**) , made by a defendant may be given in evidence for a co-defendant in so far as it is relevant to a matter in issue unless excluded by the court on one of the two grounds under s 76. If it is alleged that one of them applies, the co-defendant seeking to rely on the confession must prove on the balance of probabilities that the confession was not obtained by oppression, etc. The Court of Appeal has held that s 76A does not apply so as to permit evidence of a confession to be given when D who gave the confession has pleaded guilty and is therefore not on trial with the co-defendant; s 76A it said, is designed to cater for the joint trial of the defendant and co-defendant.

The fact that a confession is wholly or partly excluded under s 76A does not affect the admissibility in evidence of the matters referred to in s 76(4) (above).

Confessions by the mentally handicapped

8.54   PACE, s 77 is concerned with confessions, as defined by s 82 (**8.51**), made by mentally handicapped defendants. Where the case against D depends wholly or substantially upon D's confession and the court is satisfied that D is mentally handicapped and that the confession was not made in the presence of an independent person, the jury must be warned of, or the magistrates' court must heed, the special need for caution before convicting D in reliance on the confession. A police officer or member of police staff is not an independent person in this context. Generally speaking, to question a suspect in the absence of an independent person amounts to a breach of PACE Code C.

Of course, if a mentally handicapped person's confession is obtained by oppression or by words or conduct likely to make it unreliable (as in a case where D confessed at a fifth interview after 36 hours' detention and without having received any legal advice), the confession will be inadmissible under s 76.

Procedure—video- or audio-recorded confessions: Crown Court

8.55   If an audio- or video-recording is to be adduced during proceedings before the Crown Court, the *Criminal Practice Directions 2015* requires that: (a) the recording should be produced and proved by the interviewing officer, or any other officer who was present at the interview; (b) the prosecution must ensure that someone is available to operate any video or audio equipment; and (c) counsel must indicate to the operator the parts of a recording which it may be necessary to play.

DEFENDANT'S RIGHT OF SILENCE

The Criminal Justice and Public Order Act 1994 (CJPOA 1994), ss 34 to 37 permit a   **8.56**
court or jury to draw such inferences from D's failure to mention facts as appear proper.
The circumstances, generally, in which such inferences may be drawn are where:

(a) D has failed, when questioned under caution or on being charged or officially informed that (s)he may be prosecuted (or after being charged when questioned
under the Counter-Terrorism Act 2008, s 22), to mention facts later relied upon as
part of his/her defence, and which it is reasonable to expect him/her to have mentioned (s 34);
(b) D fails, without good cause, to give evidence or answer questions at trial (s 35);
(c) having been arrested by a constable, D fails or refuses to account for possession of
objects, substances or marks when requested to do so (s 36);
(d) having been arrested by a constable, D fails or refuses to account for his/her presence at a particular place, when requested to do so (s 37).

By s 38, an adverse inference drawn under ss 34 to 37 cannot be the sole basis for a
finding of a case to answer, or for a finding of guilt. As a result, the importance of ss 34
to 37 is that an adverse inference under them may enable the judge, jury, or magistrates
to find that other evidence, when considered with the adverse inference, enables them
to be sure beyond reasonable doubt of the truth and accuracy of that other evidence
and in consequence of D's guilt.

Effect of defendant's silence when questioned under caution or on being charged, etc

The relevant provision is contained in CJPOA 1994, s 34, which applies in relation to   **8.57**
the determination of:

(a) applications for dismissal of charges made by a person sent to the Crown Court
for trial;
(b) the issue of whether a person accused of an offence (of whatever type) has a case to
answer; and
(c) whether a person is guilty of the offence charged.

By s 34, where, *in the circumstances* existing at the time, D could reasonably have
been expected:

(a) when questioned under caution to discover whether or by whom the offence was
committed; or
(b) when charged with it or officially informed that (s)he might be prosecuted for it; or
(c) when questioned under the Counter-Terrorism Act 2008, s 22 after being charged,

to have mentioned a fact (as opposed to theory, possibility, or speculation) on which
(s)he later relies in his/her defence, a judge, jury, or magistrates may draw such inferences as appear proper from D's failure. The Court of Appeal has held that 'in the circumstances' includes such matters as the time of day, and D's age, experience, mental
capacity, state of health, sobriety, tiredness, knowledge and personality, and legal advice, which might all be relevant. It has also held that s 34 does not permit an adverse
inference to be drawn from D's failure to leave his police cell to be interviewed because
this does not fall within the ambit of 'being questioned'. On the other hand, where an
interview of D, who had been cautioned, consisted of D being invited to respond to the

interviewing officer's summary of the allegation against him, the Court of Appeal held that a defendant is questioned under caution is expressly or by necessary implication invited to give his/her account of the matter which has given rise to the interview; it is not necessary that specific questions are put to him/her in the interview. As already implied, s 34 is not confined to questioning that amounts to an 'interview'.

The Court of Appeal has held that a person who, when questioned, gives the interviewing officer a prepared statement from which (s)he does not depart when giving evidence at his/her subsequent trial has 'mentioned facts' and therefore falls outside s 34, notwithstanding that the prepared statement was not given in response to questioning and that (s)he said 'no comment' to all subsequent police questions.

The rule that an inference may only be drawn from a failure to mention facts which D could have been reasonably expected to mention when questioned requires an assessment of the situation at the time of that questioning, not with the benefit of hindsight as at the date of trial. An inference cannot be drawn from a failure to mention a fact of which D was unaware when questioned.

Section 34 does not apply in a case where D was at a police station at the time of the failure if D had not been allowed an opportunity to consult a solicitor prior to being questioned, charged or informed that (s)he might be prosecuted.

It should be noted that there are significant areas of overlap between the rule under s 34 and the rules under ss 36 and 37 (dealt with below). However, unlike ss 36 and 37, s 34 is not confined to the questioning of persons who are under arrest. In addition, ss 36 and 37 operate irrespective of whether a fact is relied upon as part of the defence, or irrespective of whether any defence is in fact made. By contrast, it is this reliance which is at the heart of s 34. Lastly, under s 36 or s 37, a constable is under a duty to explain the effect of the requirement in ordinary language, while the key prerequisite in s 34 is the formal caution.

There will be 'reliance' on a fact, even though D does not give evidence at his/her trial, if D relies on a fact by evidence through a witness on his/her behalf or through cross-examination of a prosecution witness.

Where D follows a solicitor's advice to remain silent, the ultimate question for the jury remains under s 34 whether the facts relied on at the trial were facts which D could reasonably have been expected to mention at interview. If they were not, that is the end of the matter. If the jury consider that D genuinely depended on the advice, that is not necessarily the end of the matter. It might still not have been reasonable for D to depend on the advice, or the advice might not have been the true explanation for D's silence, the true reason being that D had no or no satisfactory explanation consistent with innocence to give.

By s 34(3), subject to any directions given by the court, evidence which tends to establish the particular failure may be given before or after the evidence which tends to establish the fact which D is alleged to have failed to mention. Thus, where an interview is concerned with D's possession of a stolen watch, evidence of D's failure to offer an explanation for his/her possession of it may be given before, or after, evidence of it being found in D's possession. In most circumstances, the most appropriate time will be after evidence of D's possession of the property has been given. Section 34(3) does not prejudice the admissibility of evidence of D's silence or other reaction which would otherwise be admissible (eg reaction to things said in D's presence and hearing which relate to D's involvement in an offence).

The Court of Appeal has discouraged prosecutors from too readily seeking to activate s 34. Its mischief, it said, was primarily directed at the positive defence following a 'no comment' and/or 'ambush' defence. Where that was not the case, the Court warned against the further complicating of trials and summings-up by invoking s 34.

Effect of defendant's silence at trial

CJPOA 1994, s 35 deals with the case where D fails to give evidence or refuses without good reason to answer a question. It does not apply where: **8.58**

(a) D's guilt is not in issue, or
(b) it appears to the court (from evidence, and not just from a submission by an advocate) that D's physical or mental condition makes it undesirable (and not just difficult) for D to be called upon to give evidence.

For the purposes of (a), D's guilt will not be in issue if D has pleaded guilty and the hearing is merely concerned to resolve matters relevant to sentencing. Nor will it be in issue in preliminary hearings or in issues concerning the admissibility of evidence. Where the defence wish to rely on (b) they must adduce evidence on the issue referred to in (b).

Where s 35 applies, the court must, at the conclusion of the evidence for the prosecution, satisfy itself (in the presence of the jury where applicable) that D is aware that:

(a) the stage has been reached at which evidence can be given for the defence;
(b) (s)he can, if (s)he wishes, give evidence; and
(c) if (s)he chooses not to give evidence, or, having been sworn, without good cause refuses to answer any question, it will be permissible for the court or jury to draw such inferences as appear proper from such failure or refusal.

This rule will not apply if, at the conclusion of the evidence for the prosecution, it is established that D will give evidence.

For the purposes of (c), a refusal will be taken to be without good cause where a person, having been sworn, refuses to answer any question unless:

(a) (s)he is entitled to refuse to answer the question by virtue of any enactment, whenever passed or made, or on the ground of privilege; or
(b) the court in the exercise of its general discretion excuses him/her from answering it.

The privilege against self-incrimination (**8.17**) and legal professional privilege (**8.18**) therefore apply. Outside such matters 'good cause' may be limited to relevance and propriety and perhaps where a question may be considered oppressive.

Where s 35 applies, the jury or magistrates in determining whether D is guilty of the offence charged may draw such inferences as appear proper from D's failure to give evidence or from D's refusal, without good cause, to answer any question.

The Court of Appeal has ruled that, apart from the two exceptions included in s 35 (ie D's guilt not in issue and D's physical or mental condition), it is open to a jury or magistrates to decline to draw an inference from silence where the circumstances of the case justified such a course. However, there must be some evidential basis, or exceptional factors, making that a fair course to take. The inferences permitted by s 35 were only such as 'appear proper'.

The Court of Appeal has highlighted the need for a jury to be told that:

(a) the burden of proof remains on the prosecution;
(b) D is entitled to remain silent;
(c) an inference cannot by itself prove guilt;

(d) the jury must be satisfied that the prosecution has established a case to answer before drawing such an inference; and

(e) if, despite any evidence relied upon to explain silence or in the absence of any such evidence, the jury concludes that silence can only sensibly be attributed to D's having no answer, or none that would stand up to cross-examination, they may draw an adverse inference.

Effect of defendant's failure or refusal to account for objects, marks, etc

8.59   CJPOA 1994, s 36 is the relevant provision. Where:

(a) a person (D) is arrested by a constable and there is:
  (i)   on his/her person; or
  (ii) in or on his/her clothing or footwear; or
  (iii) otherwise in his/her possession; or
  (iv) in any place in which (s)he is at the time of his/her arrest,
  any object, substance, or mark, or there is any mark on any such object; and
(b) that or another constable investigating the case reasonably believes that the presence of the object, substance or mark may be attributable to the participation of D in the commission of an offence specified by the constable; and
(c) the constable informs D that (s)he so believes, and requests D to account for the presence of the object, substance or mark; and
(d) D fails or refuses to do so,

then if, in any proceedings (of the type to which s 34 applies (**8.57**)) against D for the offence so specified, evidence of those matters is given, the judge, jury or magistrates, as appropriate, may draw such inferences from the failure or refusal as appear proper. These provisions of s 36 apply to the condition of clothing or footwear as they apply to a substance or mark thereon. They do not apply unless D was told in ordinary language by the constable when making the request in (c) what the effects of s 36 would be if (s)he failed or refused to comply with the request.

An inference may not be drawn under s 36 where D was at an authorised place of detention (a police station or other authorised place) at the time of the failure or refusal, if D had not been allowed an opportunity to consult a solicitor prior to the request being made.

The provisions of s 36 do not preclude the drawing of inferences from a failure or refusal to account for objects, marks, etc which could properly be drawn apart from s 36.

Effect of defendant's failure or refusal to account for presence at a particular place

8.60   CJPOA 1994, s 37 provides that, in the circumstances below, in proceedings of the type to which s 34 applies (**8.57**), a judge, jury, or magistrates may draw such inferences as appear proper from D's failure or refusal to account for certain matters. Such inferences are permitted where:

(a) a person (D) arrested by a constable was found by him/her at a place at or about the time the offence for which (s)he was arrested is alleged to have been committed; and

(b) that or another constable investigating the offence reasonably believes that the presence of D at that place and at that time may be attributable to his/her participation in the commission of the offence; and

(c) the constable informs D that (s)he so believes, and requests him/her to account for his/her presence; and

(d) D fails or refuses to do so.

Section 37 does not apply unless D was told in ordinary language by the constable when making the request in (c) what the effects of s 37 would be if D failed or refused to comply with the request.

An inference may not be drawn under s 37 where D was at an authorised place of detention (see above) if D had not been allowed an opportunity to consult a solicitor prior to the request being made.

Once again, the provisions of s 37 do not prevent the drawing of any other inference from a refusal or failure to account which could properly be drawn apart from s 37.

EVIDENCE AS TO CHARACTER

By the defendant of his/her good character

D may always give evidence of his/her good character.   **8.61**

D's good character may be relevant to D's credibility and to the likelihood that D would commit the offence (propensity). The judge must give a direction to the jury as to the relevance of good character to credibility and propensity if D has an absolute good character (no convictions or cautions recorded, or other reprehensible conduct alleged) or effective good character (conviction or cautions recorded but old, minor and irrelevant to the charge, and the judge decides that in all the circumstances it is fair to treat D as of good character).

If D has convictions or cautions which are not in the same category as the offence alleged, and is not judged to be of effective good character and adduces them under head (b) of CJA 2003, s 101 (**8.64**), eg to found an argument that D has no propensity for the type of offence alleged, D is not entitled to a good character direction in respect of credibility and propensity, but the judge has a discretion to give any part of the direction if fairness dictates.

If D has no previous convictions or cautions but other misconduct is relied on by the prosecution, the judge may weave comments about D's residual good character into a direction about the relevance of D's bad character.

Bad character

Evidence of bad character is only admissible if the relevant provisions of CJA 2003, Part   **8.62**
11, Ch 1 (below) are satisfied.

CJA 2003, s 98 provides that references in the relevant provisions to a person's 'bad character' are to evidence of, or of a disposition towards, misconduct on his/her part, *other than evidence which*:

(a) *has to do with the alleged facts of the offence with which D is charged, or*

(b) *is evidence of misconduct in connection with the investigation or prosecution of that offence.*

'Misconduct' refers to the commission of an offence or other reprehensible behaviour. 'Reprehensible behaviour' is clearly intended to include conduct which is scandalous, disgraceful, dishonest, or improper, without being criminal. Behaviour for which a person has been charged with an offence can clearly amount to misconduct even if the prosecution is later dropped or the person is acquitted.

In relation to exception (a), the Court of Appeal has held that for evidence 'to do' with the alleged facts of the offence there must be a nexus in time between that offence and the evidence of misconduct; the evidence must be reasonably contemporaneous and closely associated with the alleged facts. However, the Court has subsequently held that there is no 'nexus in time requirement' where the evidence relates to incidents alleged to have created the motive for the offence; such evidence therefore falls within the exception.

The Court of Appeal has held that exception (b) is not limited to misconduct by the prosecuting authorities and that it could be said to encompass evidence of intimidation or blackmail by a co-defendant which could be said to be connected with the prosecution or investigation of the offence.

Where the evidence falls within either of the italicised exceptions it is admissible without more ado; it does not have to pass through any of the gateways under ss 100 (non-defendants) or 101 (defendants).

Evidence of non-defendant's bad character

8.63   CJA 2003, s 100 provides that, in criminal proceedings, evidence of the bad character of any person other than D is admissible *only* if:

(a) it is important explanatory evidence;

(b) it has substantial probative value in relation to a matter which is in issue in the proceedings and is of substantial importance in the context of the case as a whole; or

(c) all parties in the proceedings agree to its admissibility.

Section 100 applies to a witness, victim, or any other person.

As to (a), 'important explanatory evidence' has the same meaning as it does in respect of s 101 (**8.64**).

In relation to (b), evidence of a non-defendant's bad character would be of probative value if it assisted in establishing an issue one way or another. In assessing the probative value of evidence for the purposes of (b) the court must have regard to the following factors (and any other relevant factors):

(i)   the nature and number of the events, or other things, to which the evidence relates;

(ii) when those events or things are alleged to have happened or existed;

(iii) where (*a*) the evidence is evidence of a person's misconduct, and (*b*) it is suggested that the evidence has probative value by reason of similarity between that misconduct and other alleged misconduct, the nature and extent of the similarities and the dissimilarities between each of the alleged instances of misconduct;

(iv) where (*a*) the evidence is evidence of a person's misconduct, (*b*) it is suggested that that person is also responsible for the misconduct charged, and (*c*) the identity of the person responsible for the misconduct charged is disputed, the extent to which the evidence shows or tends to show that the same person was responsible each time.

As a result of (b) and the above provisions, evidence of a non-defendant's bad character may be admitted, for example, as evidence of *his/her* propensity to commit the offence in question or of his/her credibility as a witness.

The Court of Appeal has held that a police report recording unproven allegations against a non-defendant of criminal behaviour is most unlikely to have substantial probative value; first, it is, at best, hearsay, and it would fall to be judged by reference to the rules for the admission of hearsay and (given the difficulties of assessing such evidence) it would be rare for it to be judged of substantial probative value; second, if there is no complainant prepared to support the allegation, that robs it of a great deal of probative value.

Except in the case of (c), evidence of the bad character of a non-defendant may only be given with the leave of the court.

A party who wants to introduce evidence of a non-defendant's bad character must make an application in accordance with the Criminal Procedure Rules 2020, Part 21; see Chapter 8 on the companion website.

Evidence of defendant's bad character

CJA 2003, s 101 deals with such matters.   **8.64**

By s 101(1), evidence of D's bad character is admissible in criminal proceedings only if one of the following seven gateways is satisfied:

(a) all parties agree to its being admissible;
(b) it is evidence adduced by D him/herself or is given in answer to a question asked by D during cross-examination and intended to elicit it;
(c) it is important explanatory evidence;
(d) it is relevant to an 'important matter' (ie a matter of substantial importance in the context of the case as a whole) in issue between D and the prosecution;
(e) it has 'substantial probative value' (ie an enhanced capability of proving or disproving) in relation to an 'important matter' in issue between D and a co-defendant;
(f) it is evidence to correct a false impression given by D; or
(g) D has made an attack on another person's character.

Evidence must not be admitted under gateways (d) or (g) if D has made an application to exclude it and the judge or magistrates' court considers that the admission of the evidence would have such an adverse effect on the fairness of the proceedings that it ought not to admit it. The test to be applied is the same as that set out in PACE, s 78 (**8.74**). On such an application, regard must be had, in particular, to the time which has passed between the matters to which the evidence relates and the matters forming the subject of the offence charged.

By CJA 2003, s 108, which applies in addition to s 101, in proceedings for an offence by D when aged 21 or over, evidence of D's conviction for an offence when under 14 is not admissible unless both offences are triable only on indictment and the interests of justice require the evidence to be admissible.

Where D's conviction under 14 was for an offence under the law of a country outside England and Wales ('the previous offence') and that offence would be an offence under the law of England and Wales ('the corresponding offence') if done in England and Wales at the time of the proceedings with which D is now charged, the previous offence is regarded as triable only on indictment if the corresponding offence is so triable.

Gateway (c): important explanatory evidence

8.65   For the purpose of gateway (c) under CJA 2003, s 101, evidence is important explanatory evidence if, in its absence, the jury or court would find it impossible or difficult properly to understand other evidence in the case, and its value for understanding the case as a whole is substantial. In an assault case, for example, evidence may be given that D (who claims that (s)he acted in self-defence) has previously made unprovoked attacks on the victim.

Gateway (d): important matter in issue between defendant and prosecution

8.66   CJA 2003, s 103(1) provides that, for the purposes of gateway (d), matters in issue between D and the prosecution *include*:

(a) the question whether D has a 'propensity' to commit offences of the kind with which D is charged, except where D's having that propensity makes it no more likely that D is guilty of the offence; and

(b) the question whether D has a propensity to be untruthful, except where it is not suggested that D's case is untruthful in any respect.

In respect of (b), the Court of Appeal has held that the question whether D has a propensity to be untruthful will not normally be describable as an 'important matter in issue between D and the prosecutor'; the only circumstance where there is likely to be an important issue as to whether D has a propensity to be untruthful is where telling lies is an important element of the offence, and even then the propensity is only likely to be significant if the lying was in the context of committing criminal offences (in which case evidence is likely to be admissible under (a)).

The Supreme Court has held that, where the prosecution relies on several incidents to establish propensity on the part of D, proving the propensity does not require proving beyond reasonable doubt that each incident happened in precisely the way that is alleged, nor does it require that the facts of each incident be considered in isolation from all the others; the proper approach is for the jury or court to consider the evidence in the round to determine whether propensity has been established beyond reasonable doubt.

Section 103(2) provides that, where (a) applies, D's propensity to commit offences of the kind with which D is charged may (without prejudice to any other way of doing so) be established by evidence of a previous conviction for an offence of the same description as that with which D is charged, or an offence of the same category as that offence. For these purposes, two offences are of the same description as each other if the statement of the offence in a written charge or indictment would, in each case, be in the same terms, and two offences are of the same category if they are of the same *category of offences as prescribed by order*. The CJA 2003 (Categories of Offences) Order 2004 prescribes two categories of offences: one category is specified offences against the Theft Acts 1968 and 1978, and the other is specified offences against the Sexual Offences Act (SOA) 2003 where the victim was under 16. They are offences under the Theft Act 1968, ss 1, 8, 9 (if theft intended, committed or attempted), 10 (if the above type of offence under s 9 involved), 12, 12A, 22, or 25; the Theft Act 1978, s 3; or, provided complainant under 16, the Sexual Offences Act 2003, s 1, 2, 3, 4, 5, 6, 7, 8, 9, 10, 14, 16, 17, 25, 26, 30, 31, 34, 35, 38, or 39; and aiding, abetting, counselling or procuring, encouraging or assisting, or attempting the commission of such an offence. The sexual offences category also includes corresponding offences superseded by the Sexual Offences Act 2003.

By CJA 2003, s 103(3), s 103(2) does not apply if the court is satisfied, by reason of the length of time since the conviction or for any other reason that it would be unjust for it to apply.

CJA 2003, s 103(7) provides that where:

(a) D has been convicted of an offence under the law of any country outside England and Wales ('the previous offence'), and

(b) the previous offence would constitute an offence under the law of England and Wales ('the corresponding offence') if it were done in England and Wales at the time of the trial for the offence with which D is now charged ('the current offence'),

then for the purpose of determining for the purposes of s 103(2) whether the previous offence and the current offence are of the same description or category:

(i) the previous offence is of the same description as the current offence if the corresponding offence is of that same description, as set out in the 2004 Order referred to above;

(ii) the previous offence is of the same category as the current offence if the current offence and the corresponding offence belong to the same category of offences as prescribed by the 2004 Order.

Similar provision is also made by s 103(10) in respect of a conviction of an offence under a foreign service law which would constitute an offence under the law of England and Wales or a service offence if done in England and Wales by a member of HM forces.

In referring to offences of the same description or category, CJA 2003, s 103(2) is not exhaustive of the types of conviction which may be relied on to show evidence of propensity to commit offences of the kind charged. Indeed, the provision is not limited to previous convictions; the fact that D has previously asked for offences to be taken into consideration can be admitted.

Gateway (e): substantial probative value in relation to an important matter in issue between the defendant and co-defendant

8.67

This is intended to deal with 'cut-throat' defences. CJA 2003, s 104 provides that evidence relevant to whether D has a propensity to be untruthful is admissible on that basis under (e) only if the nature or conduct of D's defence is such as to undermine the co-defendant's defence. Only evidence adduced by the co-defendant or by a witness in cross-examination by the co-defendant is admissible under (e).

Gateway (f): correcting a false impression

8.68

CJA 2003, s 105 provides that, for the purposes of gateway (f), D gives a false impression if D is responsible for the making of an express or implied assertion which is apt to give a false or misleading assertion about D, and that evidence to correct such an impression is evidence which has probative value in correcting it.

D is treated as responsible for the making of an assertion if it is made by D in the proceedings, or by a defence witness, or by any witness in response to a question by D which was intended or likely to elicit it, or if it was made by D when questioned under caution or on being charged with the offence, or if it was made out-of-court by any person and D adduces evidence of it.

The Court of Appeal has held that gateway (f) is concerned with attempting to mislead the court in a way that goes beyond denying the offence. The false impression has to be one which is given to the court.

Gateway (g): attack on another person's character

8.69 By CJA 2003, s 106, D attacks the character of another person if D adduces evidence to the effect that this person has committed any offence or has behaved in a reprehensible way, or if D asks questions in cross-examination which are intended or likely to elicit such evidence, or if evidence is given of an imputation about the other person by D on being questioned under caution before charge or on being charged or informed that (s) he might be prosecuted. Where evidence of D's bad character has been admitted as a result of D's attack upon the character of another person, it can be used, if relevant, to establish a propensity on the part of D to commit offences of the type with which D is charged.

Assumption of truth in assessment of relevance or probative value

8.70 CJA 2003, s 109 provides that any reference in the Act's provisions about bad character to the relevance or probative value of evidence is a reference to its relevance or probative value on the assumption that it is true. However, in assessing the relevance or probative value of an item of evidence, a court need not assume that evidence is true if it appears, on the basis of material before it, that no court or jury could reasonably find it to be true.

Notice of introduction of evidence of a defendant's bad character

8.71 A party who wants to introduce evidence of a defendant's bad character must serve notice in accordance with the Criminal Procedure Rules, 2020, Part 21; see Chapter 8 on the companion website.

Spent convictions etc

8.72 The provisions of the Rehabilitation of Offenders Act 1974 whereby an offender's convictions become 'spent' and the offender is thereafter treated as if (s)he had not been convicted do not prevent the admissibility in criminal proceedings of evidence relating to a spent conviction which is otherwise admissible (s 7). A similar rule applies in respect of spent cautions (Sch 2).

   A person (X) who has a conviction or caution for the abolished offence of buggery or the abolished offence of gross indecency between men which has been ordered to be disregarded by the Secretary of State under the Protection of Freedoms Act 2012, s 92 is to be treated as if X had not committed, or been charged with, or prosecuted for, or convicted of, or sentenced for, or cautioned for, that offence. In particular, no evidence is admissible to prove any of these matters and X is not, in any court proceedings, to be asked (and, if asked, is not to be required to answer) any question relating to X's past which cannot be answered without acknowledging or referring to the conviction or caution or any circumstances ancillary to it (s 96).

Proof of conviction

8.73 See **8.13**.

EXCLUSION OF UNFAIR EVIDENCE

8.74 The purpose of all the rules concerning the gathering of evidence and the manner of its presentation in court is to ensure absolute fairness to an accused person. PACE, s 78 provides that, in any proceedings, a judge or magistrates' court may refuse to allow evidence (including confessions) on which the prosecution proposes to rely to be given if it appears that, having regard to all the circumstances, including the circumstances in which the evidence was obtained, the admission of the evidence would have *such an adverse effect on the fairness of the proceedings that the court ought not to admit it.* 'Fairness of the proceedings' is directed primarily to fairness of the actual conduct of the proceedings but it is not strictly limited to this.

Evidence obtained by a breach by the police of PACE or one of the Codes of Practice does not necessarily render evidence inadmissible under s 78, but it will if it has such an adverse effect on the fairness of the proceedings that it ought not to be admitted. The issue is whether there has been a significant and substantial breach. There is no general requirement for the breach to have been committed deliberately or in bad faith before evidence is excluded under PACE, s 78. Bad faith on the part of police officers will usually lead to the exclusion of evidence (bad faith may make significant and substantial a breach which might not otherwise be so), but evidence may be excluded even though the police acted in good faith if there is a significant and substantial breach of police powers.

An example of the operation of s 78 is provided by a case where a person had been convicted on evidence based solely on a confession obtained after police had falsely pretended that his fingerprints had been found at the scene of the crime; the Court of Appeal ruled that such evidence should have been excluded under s 78 on the ground that it posed a threat to the fairness of proceedings. Another example is provided by various cases where evidence obtained after a significant and substantial breach of the provisions described in Chapter 5 relating to the right to legal advice or to the conduct of an interview has been excluded on the ground that in the circumstances of the case it would be unfair to admit it.

Section 78 does not affect the common law powers of a court to exclude evidence on the basis that its prejudicial effect is likely to be greater than its probative value, and to exclude self-incriminatory evidence unfairly obtained from the defendant after the commission of the alleged offence. Although s 78 goes much further than the common law power, particularly because it significantly widens the discretion to exclude evidence which has been unfairly or unlawfully obtained, there is one context in which the common law power is more extensive than s 78. This is where the evidence in question has already been adduced. Section 78 does not apply in this context, but if the evidence ought not to have been admitted the court may, under the common law power, take any necessary steps to prevent an injustice, whether by directing the jury to ignore the offending evidence or, if necessary, by discharging the jury.

Sometimes, in cases of unfairness to D, it is more appropriate for the court to stay proceedings on the grounds of an abuse of process on the basis that a fair trial would not be possible or, if it would, that a trial would be contrary to the public interest in the integrity of the criminal justice system, for example because the actions of the police threaten a basic human right. One such case recognised by the House of Lords is police entrapment, ie luring D into committing an offence and then seeking to prosecute him/her. Alleged entrapment does not prevent D seeking to have the evidence obtained by entrapment excluded as inadmissible, but it must be borne in mind that the tests are different. In staying proceedings the test is that just stated; under s 78 it is whether admitting the evidence would have such an adverse effect on the fairness of proceedings that the evidence ought not to be admitted. The House of Lords has held that in entrapment cases a stay of proceedings should normally be regarded as the appropriate response.

DISCLOSURE OF EVIDENCE

The disclosure provisions of the Criminal Procedure and Investigations Act 1996

8.75   The Criminal Procedure and Investigations Act 1996 (CPIA 1996), Part 1 (ss 1 to 21) contains provisions relating to the advance disclosure of material which apply where:

(a) a person charged with a summary offence pleads not guilty;

(b) a person of 18 or over charged with an offence triable either way, in respect of which a court proceeds to summary trial, pleads not guilty; or

(c) a person under 18 charged with an indictable offence, in respect of which a court proceeds to summary trial, pleads not guilty.

The provisions also apply where a person is charged with an indictable offence and is sent for trial in the Crown Court.

For the purpose of the disclosure provisions, 'material' refers to material of all kinds, and in particular refers to information, and to objects of all descriptions.

Initial duty of prosecutor to disclose

8.76   By CPIA 1996, s 3, the prosecutor must:

(a) disclose to the defendant (D) previously undisclosed prosecution material which might reasonably be considered capable of undermining the prosecution case, or of assisting the case for D; or

(b) give to D a written statement that there is no material of a description mentioned in (a).

Information must not be disclosed by the prosecutor to the extent that the court, on the prosecutor's application, concludes it is not in the public interest to disclose it and orders accordingly. Nor should it be disclosed if its disclosure is prohibited under the Investigatory Powers Act 2016.

At the same time as (s)he acts under s 3, the prosecutor must give D any schedule of non-sensitive material previously given to him/her by the disclosure officer (s 4).

The Court of Appeal has held that s 3 applies to documentation created or received by D (and now in the possession of the prosecution), including documents said to be required by D to refresh D's memory.

The Court of Appeal has held that the disclosure responsibilities imposed on the prosecution by the disclosure provisions cannot be sidestepped by not making an enquiry. It stated that a police officer who believes that a person may have information which might undermine the case for the prosecution or assist the case for the suspect or defendant cannot decline to make enquiries of that person in order to avoid the need to disclose what the person might say.

The Court of Appeal has stated by way of guidance that full disclosure, under s 3, of material which is not part of the formal case against D, and which either weakens its case or strengthens D's, is essential to avoid a miscarriage of justice. The prosecution is not, however, required to disclose material which is neutral or adverse to D. Especially in large and complex cases, investigating and prosecuting agencies must apply their respective case management and disclosure strategies and policies and be transparent with the defence and the courts about how the prosecution has approached complying with its disclosure obligations in the context of the individual case. The defence will be expected to play their part in defining the real issues in the case, and will be invited to participate in defining the scope of the reasonable searches that may be made of any digitally stored material by the investigator to identify material that may reasonably be expected to undermine the prosecution case or assist the defence. It is not the duty of the prosecution to comb through all the material in its possession on the look-out for anything which may conceivably or speculatively assist the defence.

Compulsory disclosure by defendant

The relevant provisions are in CPIA 1996, s 5. Where cases are to be tried on *indict-* **8.77** *ment*, and the prosecutor complies (or purports to comply) with his/her initial duty of disclosure, and provided that D has received documents containing the prosecution's case, D is required to provide the court and the prosecutor during the relevant period (**8.83**) with a defence statement.

By s 6A(1), a defence statement must be in writing and must set out the nature of D's defence, including any particular defences which are to be relied upon; indicate the matters of fact on which D takes issue with the prosecution, and why D takes issue; set out the matters of fact on which D intends to rely for the purposes of his/her defence; and indicate any point of law (including any point as to the admissibility of evidence or an abuse of process) which D wishes to take and any authority upon which D intends to rely. The Court of Appeal has held that if D raises no positive case at all in a defence statement but *simply* requires the prosecution to prove its case, there is no failure to comply with s 6A. It also held that both legal professional privilege (**8.18**) and D's privilege against self-incrimination (**8.17**) have survived s 6A. Section 6A requires D to disclose what is going to happen at the trial. It does not compel disclosure of D's confidential discussions with his/her advocate. Nor is D obliged to incriminate him/herself.

Any defence statement which includes an alibi is required by s 6A(2) to give *the name, address and date of birth of any witness D believes is able to give evidence in support of the alibi, or as many of those details as are known to D, and any other information in his/her possession which might assist in identifying or finding any such witness where such details are not known to D at that time.* Evidence in support of an alibi is evidence tending to show that, by reason of D's presence at a particular place or area at a particular time, D was not, or was unlikely to have been, at the place where the offence is alleged to have been committed at the time of its alleged commission.

Where a defence statement identifies an alibi witness under s 6A(2) or where D gives a notice under s 6C (**8.79**) indicating that D intends to call an identified witness at his/her trial, guidance is provided by a code of practice to police officers and others charged with the duty of investigating offences in relation to arranging and conducting interviews of persons notified. The Code of Practice for Arranging and Conducting Interviews of Witnesses Notified by the Accused provides as follows.

If an investigator wishes to interview a witness identified as above, the witness must be asked whether (s)he consents to being interviewed and informed that:

(a) an interview is being requested following his/her identification by D as a proposed witness under ss 6A(2) or 6C;
(b) (s)he is not obliged to attend the interview;
(c) (s)he is entitled to be accompanied by a solicitor; and
(d) a record will be made of the interview and (s)he will be sent a copy of it.

If the witness (W) consents to being interviewed, W must be asked whether (s)he:

(a) wishes to have a solicitor present;

(b) consents to a solicitor attending on behalf of D, as an observer; and

(c) consents to a copy of the record being sent to D. If W does not consent, W must be informed that the effect of disclosure requirements in criminal proceedings may nevertheless require the prosecution to disclose the record to D (and any co-defendant) in the course of the proceedings.

The investigator must notify D or D's legal representatives:

(a) that the investigator requested an interview with W;

(b) whether W consented to the interview; and

(c) if W consented to the interview, whether W also consented to a solicitor attending on behalf of D, as an observer.

If D is not legally represented in the proceedings, and if W consents to a solicitor attending the interview on behalf of D, D must be offered a reasonable opportunity to appoint a solicitor to attend it.

The investigator must nominate a reasonable date, time, and venue for the interview and notify W of them and any changes to them.

If W has consented to the presence of D's solicitor, D's solicitor must be notified that the interview is taking place, invited to observe, and provided with reasonable notice of the date, time, and venue of the interview and any changes.

The identity of the investigator conducting the interview must be recorded. (S)he must have sufficient skills and authority, commensurate with the complexity of the investigation, to discharge his/her functions effectively. (S)he must not conduct the interview if that is likely to result in a conflict of interest, for instance, if (s)he is the victim of the alleged crime which is the subject of the proceedings. The advice of a more senior officer must always be sought if there is doubt as to whether a conflict of interest precludes an individual conducting the interview. If thereafter the doubt remains, the advice of a prosecutor must be sought.

D's solicitor may only attend the interview as an observer if W has consented to his/her presence as an observer. Provided that D's solicitor was given reasonable notice of the date, time, and place of the interview, the fact that D's solicitor is not present will not prevent the interview from being conducted. If W at any time withdraws consent to D's solicitor being present at the interview, the interview may continue without the presence of D's solicitor.

Where W has indicated that W wishes to appoint a solicitor to be present, that solicitor must be permitted to attend the interview.

If W is under 18 or is mentally disordered or otherwise mentally vulnerable, W must be interviewed in the presence of an appropriate person.

A record must be made of the interview, wherever it takes place. It must be made, where practicable, by audio recording or by visual recording with sound, or otherwise in writing. Any written record must be made and completed during the interview, unless this would not be practicable or would interfere with the conduct of the interview, and must constitute either a verbatim record of what has been said or, failing this, an account of the interview which adequately and accurately summarises it. If a written record is not made during the interview it must be made as soon as practicable after its completion. Written interview records must be timed and signed by the maker.

The Code concludes by providing that a copy of the record must be given, within a reasonable time of the interview, to:

(a) W, and

(b) if W consents, to D or D's solicitor.

Any police officer or other person charged with the duty of investigating offences who arranges or conducts such an interview must have regard to the Code. Any provision of the Code or any failure to have due regard to the Code, either of which is relevant to any question arising in civil or criminal proceedings, can be taken into account by a court or tribunal conducting those proceedings in deciding that question.

Voluntary disclosure by defendant

By CPIA 1996, s 6, where the case is to be tried *summarily*, and the prosecutor complies (or purports to comply) with his/her initial duty of disclosure, D or D's solicitor *may* give a defence statement to the prosecutor during the relevant period. If (s)he does so, (s)he must also give the statement to the court. **8.78**

Notification of intent to call witnesses

By CPIA 1996, s 6C, referred to at **8.77**, D must also give the court and the prosecutor notification of his/her intention to call witnesses together with their identity or details which might lead to their identification. Changes must be notified. **8.79**

Continuing duty of disclosure by prosecutor

After the prosecutor has complied (or purported to comply) with his/her initial duty of disclosure, the prosecutor remains under a continuing duty under CPIA 1996, s 7A, up to the time of acquittal, conviction, or a decision not to proceed with the case, to keep under review whether there is any undisclosed prosecution material which might reasonably be considered capable of undermining the case for the prosecution against D or of assisting D's case. If there is, the prosecution must disclose it as soon as reasonably practicable. **8.80**

Where D gives a defence statement under s 5 or 6 (**8.77** and **8.78**), as a result of which the prosecutor is required by the above provision to make any disclosure, or further disclosure, the prosecutor must do so within the relevant period. If (s)he considers that (s)he is not so required (s)he must give D a written statement to that effect within that period.

Application by defendant for further disclosure

Following compliance (or purported compliance) by the prosecutor with the provision just mentioned, or a failure by the prosecutor to comply with a duty under that provision to disclose, and provided that D has given a defence statement under CPIA 1996, s 5 or 6, D may apply to the court under s 8 for the disclosure of material which D reasonably believes is required to be disclosed to him/her but has not been. An order for further disclosure can only be made in respect of material: (a) in the possession of the prosecution in connection with the case against D, or (b) which the prosecutor has inspected or is entitled to inspect in pursuance of the CPIA 1996 (s 23(1)) Code of Practice (**8.85**), or (c) of which in pursuance of that Code the prosecutor is entitled to have a copy. An order for further disclosure cannot be made against a third party. **8.81**

Failure in disclosure by defendant

CPIA 1996, s 11 deals with this. It applies in the three cases set out below. The first case is where s 5 (**8.77**) applies and D: **8.82**

(a) fails to give an initial defence statement;
(b) gives one out of time;
(c) sets out inconsistent defences in his/her defence statement; or

(d) at his/her trial: (i) puts forward a defence which differs from that set out in a defence statement, (ii) relies on a matter (or any particular of any matter of fact) which, in breach of s 6A (**8.77**), was not mentioned in his/her defence statement, (iii) adduces evidence in support of an undisclosed alibi, or (iv) calls a witness to support an alibi about whom (s)he made no disclosure in his/her defence statement.

The second case is where s 6 (**8.78**) applies, D gives an initial defence statement, and D gives that statement after the end of the relevant period for doing so, or does any of the things mentioned in (c) or (d) above.

The third case is where D gives a witness notice but does so after the relevant period for doing so, or at his/her trial calls a witness (other than him/herself) not included or adequately identified in a witness notice.

In these three cases, the court, or any other party (in some cases only with the judge's leave), may make such comment as appears appropriate. In addition, the court or jury may draw such inferences as appear proper in deciding whether D is guilty of the offence concerned, but a person may not be convicted solely on the basis of such an inference.

Time limits

8.83   The CPIA 1996 (Defence Disclosure Time Limits) Regulations 2011 provide that the relevant period for compulsory disclosure by D (CPIA 1996, s 5), voluntary disclosure by D (s 6), or notification of intent to call defence witnesses (s 6C) begins with the day on which the prosecutor complies or purports to comply with the prosecutor's initial duty of disclosure (s 3), and that in respect of:

(a) summary proceedings, the relevant period for s 6 and s 6C expires at the end of 14 days beginning with the first day of the relevant period;
(b) proceedings in the Crown Court, the relevant period for s 5 and s 6C expires at the end of 28 days beginning with the first day of the relevant period.

Where such a period expires on a weekend, Christmas Day, Good Friday, or a bank holiday, the relevant period is extended so as to expire on the next day which is not such a day.

The Regulations also provide that on an application by D the court may extend, or further extend, the relevant period by so many days as it specifies, if it is satisfied that it would be unreasonable to require D to give a defence statement under s 5 or s 6, or give notice under s 6C, as the case may be, within the relevant period. D's application must be made within the relevant period, specify the ground on which it is made, and state the number of days which D wishes the relevant period to be extended. There is no limit to the number of applications for an extension.

The Regulations do not prescribe a particular period of time in relation to disclosure by the prosecution. However, s 13 requires such disclosure as soon as is reasonably practicable after a person has pleaded not guilty in summary proceedings or copies of the evidence have been served on a person sent for trial in the Crown Court.

Common law disclosure rules

8.84   Common law rules about disclosure by the prosecution still apply in two types of case:

(a) where the investigation began before 1 April 1997 (when the statutory—CPIA 1996—disclosure provisions came into force); and

(b) at any stage in the process when the statutory provisions do not apply, ie after the institution of proceedings but before the time when D pleads guilty or not guilty in a magistrates' or youth court (eg an 'early stage' bail application), or is sent for trial, and after a conviction.

The prosecution's duty at common law is to disclose to the defence all relevant material, which might reasonably assist D in the early preparation of D's defence or at a bail hearing.

Police procedure

Police procedure is controlled by the CPIA 1996, (s 23(1)) Code of Practice. In summary: **8.85**

The officer in charge of an investigation

The officer in charge of an investigation is any police officer involved in the conduct of a **8.86** criminal investigation. All investigators are responsible for the recording and retention of materials and this includes negative material resulting from the interview of persons who could give no positive evidence. The officer in charge of an investigation must make such material available to the disclosure officer. While the officer in charge of an investigation may delegate tasks to other investigators, (s)he remains responsible for ensuring that the duties relating to disclosure are properly carried out. All reasonable enquiries, whether pointing towards or away from a suspect must be investigated. It is a matter for the investigator, with the assistance of the prosecutor if required, to decide what constitutes a reasonable line of inquiry in each case.

Where an investigating officer believes that other parties may be in possession of relevant material which has not been obtained (s)he may invite those persons to retain the material in case a request for disclosure is received. The disclosure officer should inform the prosecutor that those persons have the material. There is no requirement to make speculative enquiries; there must be some reason to believe that the persons concerned have relevant material.

Material which might be relevant to a criminal investigation but is not recorded must be held in a durable or retrievable form. Information must be recorded at the time at which it is obtained or as soon as is practicable thereafter. There is no requirement to take a statement where it would not otherwise be taken. The investigator has a duty to retain material obtained in a criminal investigation which might be relevant to the investigation. Where material which previously has been examined but not retained becomes relevant, the officer in charge of an investigation must take steps to obtain it or ensure that it is retained. The duty to retain material includes in particular the duty to retain material falling into the following categories:

(a) records derived from tapes or recordings of telephone messages containing descriptions of an alleged offence or offender;

(b) any incident logs relating to the allegation;

(c) contemporaneous records of the incident, such as: crime reports and crime report forms; an investigation log; any record or note made by an investigator (including police notebook entries and other handwritten notes) on which (s)he later makes a statement or which relates to contact with suspects, victims or witnesses; an account of an incident or information relevant to an incident noted by an investigator; records of actions carried out by officers (such as house-to-house interviews,

CCTV, or forensic enquiries) noted by a police officer; CCTV footage, or other imagery, of the incident in action.

(d) D's (the defendant's) custody record or voluntary attendance record;

(e) any previous accounts made by a complainant or any other witnesses;

(f) interview records (written records, or audio or video tapes, of interviews with actual or potential witnesses or suspects);

(g) any material casting doubt on the reliability of a witness;

(h) final versions of witness statements (and draft versions where their content differs from the final version), including any exhibits mentioned (unless these have been returned to their owner on the understanding that they will be produced in court if required);

(i) material relating to other suspects in the investigation;

(j) communications between the police and experts, reports of work carried out by experts, and schedules of scientific material prepared by the expert for the investigator, for the purposes of criminal proceedings;

(k) records of the first description of a suspect by each potential witness who purports to identify or describe the suspect.

In addition, the duty to retain material which might be relevant to the investigation includes in particular the duty to retain material which may satisfy the test for prosecution disclosure, such as information provided by D which indicates an explanation for the offence with which D is charged, and material casting doubt on the reliability of a confession or a prosecution witness.

All such material which may be relevant to the investigation must be retained until a decision has been taken whether to institute proceedings against a person for an offence. If proceedings are instituted the material must be retained until the proceedings result in conviction or acquittal or a decision not to proceed has been taken. Where there is a conviction, material must be retained until the person is released from custody, or, in other circumstances, six months from the date of conviction. If a convicted person is released from custody within six months of conviction the material must be retained until the six months has elapsed. Where an appeal is lodged, material must be retained until the appeal is determined.

The disclosure officer

8.87 The disclosure officer is the person responsible for examining material retained by the police during an investigation; revealing material to the prosecutor during the investigation and any criminal proceedings resulting from it, and certifying that (s)he has done this; and disclosing material to D at the request of the prosecutor. In any criminal investigation, one or more deputy disclosure officers may be appointed to assist the disclosure officer, and such a deputy may perform any of the above functions of the disclosure officer.

The functions of the disclosure officer (DO) may be carried out by the investigator or officer in charge of the investigation. Indeed, the functions of all three roles may be carried out by one person. Where this is not done there must be full consultation between both officers. DO is the link between the investigators and the Crown Prosecution Service and is responsible for providing the material for 'primary disclosure' and performing any other tasks required by the prosecutor. (S)he must ensure, by liaison with the officer in charge of the investigation where that is a different person, that all material is made available for examination.

Preparation of material for the prosecutor: (1) schedules of unused material

The officer in charge of the investigation, the disclosure officer (DO) or an investigator **8.88** may seek advice from the prosecutor about whether any particular item of material may be relevant to the investigation. Material which may be so relevant and has been retained in accordance with the Code, and which DO believes will not form part of the prosecution case, must be listed on the appropriate schedule of unused material.

DO must ensure that the appropriate schedule of unused material is prepared in the following circumstances: where D charged with an offence which is triable only on indictment; where D is charged with an offence which is triable either way, and it is considered that the case is likely to be tried on indictment; where D is charged with an either-way offence that is likely to remain in the magistrates' court, and it is considered that D is likely to plead not guilty; where D is charged with a summary offence and it is considered that D is likely to plead not guilty.

However, by way of *exception*, where D is charged with a summary offence or an either-way offence, and it is considered that D is likely to plead guilty, a schedule is not required unless a not guilty plea is subsequently entered or indicated.

Irrespective of the anticipated plea, the common law test (**8.84**) for disclosure requires material to be disclosed if there is material known to DO that might assist the defence with the early preparation of its case or at a bail hearing (eg, a key prosecution witness has relevant previous convictions or a witness has withdrawn his/her statement). A note must be made on the case summary for the prosecutor of any such material, which must be revealed to the prosecutor who will review it and consider whether it is disclosable. Where there is no such material, a certificate to that effect must be completed.

Material in the following list (which where it exists will have been retained or recorded in accordance with the duty to retain material) is likely to include information which meets the test for prosecution disclosure. This material must therefore, subject to the *exception* two paragraphs above, be scheduled and provided to the prosecutor. In reviewing this material, DOs and prosecutors must start with a presumption that it is likely to meet the disclosure test, although the material will need to be carefully considered and the disclosure test applied before a decision is made:

(a) records derived from tapes or recordings of telephone messages containing descriptions of an alleged offence or offender;

(b) any incident logs relating to the allegation;

(c) contemporaneous records of the incident, such as: crime reports and crime report forms; an investigation log; any record or note made by an investigator (including police notebook entries and other handwritten notes) on which (s)he later makes a statement or which relates to contact with suspects, victims or witnesses; an account of an incident or information relevant to an incident noted by an investigator; records of actions carried out by officers (such as house-to-house interviews, CCTV or forensic enquiries) noted by a police officer; CCTV footage, or other imagery, of the incident in action;

(d) D's custody record or voluntary attendance record;

(e) any previous accounts made by a complainant or by any other witnesses;

(f) interview records (written records, or audio or video tapes, of interviews with actual or potential witnesses or suspects);

(g) any material casting doubt on the reliability of a witness, eg relevant previous convictions of any prosecution witnesses and any co-defendant.

This material must be listed on the schedule by DO in addition to all other material which may be relevant to an investigation; it is likely that some of this material will need to be redacted (see **8.90**).

Preparation of material for the prosecutor: (2) ways in which material is to be listed on schedule

8.89 Material which DO does not believe is sensitive must be listed on a schedule of non-sensitive material, which must include a statement that DO does not believe the material is sensitive. Where there is sensitive unused material, see (**8.90**). DO should ensure, subject to what is said later in this paragraph, that each item of material is listed separately on the schedule, and is numbered consecutively. The description of each item should contain sufficient detail to enable the prosecutor to decide whether (s)he needs to inspect the material before deciding whether or not it should be disclosed. In some investigations it may be disproportionate to list each item of material separately. These may be listed in a block or blocks and described by quantity and generic title. Even if some material is listed in a block, DO must ensure that any items among that material which might satisfy the test for prosecution disclosure are listed and described individually.

Preparation of material for the prosecutor: (3) redaction of sensitive material

8.90 DO should redact any sensitive information contained in material that is likely to satisfy the test for prosecution disclosure. DO should also redact any personal, confidential information (eg a person's date of birth, address, and phone number) in material that is to be disclosed. Any material which is believed to be sensitive must be listed on a schedule of sensitive material. In *exceptional circumstances*, where its existence is so sensitive (listing likely to lead directly to loss of life or threat to national security) that material cannot be listed, it should be revealed to the prosecutor separately. If there is no sensitive material, DO must record this fact on a schedule of sensitive material, or otherwise so indicate. Subject to the *exceptional circumstances* just mentioned, DO must list on a sensitive schedule any material the disclosure of which (s)he believes would give rise to a real risk of serious prejudice to an important public interest, and the reason for that belief. The schedule must include a statement that DO believes the material is sensitive.

Examples of sensitive material are specified in the Code. The full list is of examples is material:

(a) relating to national security;

(b) received from the intelligence and security agencies;

(c) relating to intelligence from foreign sources which reveals sensitive intelligence-gathering methods;

(d) given in confidence;

(e) relating to the identity or activities of informants, or undercover police officers, or witnesses, or other persons supplying information to the police who may be in danger if their identities are revealed;

(f) revealing the location of any place used for police surveillance, or the identity of any person allowing a police officer to use them for surveillance;

(g) revealing, directly or indirectly, techniques and methods relied upon by a police officer in the course of a criminal investigation, eg covert surveillance techniques, or other methods of detecting crime;

(h) whose disclosure might facilitate the commission of other offences or hinder the prevention and detection of crime;

(i) upon the strength of which search warrants were obtained;

(j) containing details of persons taking part in identification parades;

(k) supplied to an investigator during a criminal investigation which has been generated by an official of a body concerned with the regulation or supervision of bodies corporate or of persons engaged in financial activities, or which has been generated by a person retained by such a body;

(l) supplied to an investigator during a criminal investigation which relates to someone under 18 and which has been generated by a local authority social services department, an Area Child Protection Committee, or other party contacted by an investigator during an investigation; or

(m) relating to the private life of a witness.

Revelation of material to prosecutor

8.91 Where cases have been charged on the Full Code Test in the Code for Crown Prosecutors and it is anticipated that D will plead not guilty, DO should provide the schedules concerning unused material to the prosecutor at the same time as submitting the case file. In all other cases DO must provide the schedules as soon as possible after a not guilty plea has been either indicated or entered.

DO should draw the attention of the prosecutor to any material retained by an investigator which it is considered may satisfy the test for prosecution disclosure in CPIA 1996, explaining the reasons for that view.

DO must give the prosecutor a copy of any such material (unless already supplied as part of the filed material for the prosecution case), together with (a) any information provided by a defendant which indicates an explanation for the offence with which (s) he has been charged; (b) any material casting doubt on the reliability of a confession, and (c) any material casting doubt on the reliability of a prosecution witness.

DO must give the prosecutor a copy of any material listed at **8.88**(a)-(g) (material likely to meet the test for disclosure) which has been scheduled in accordance with the provisions applying to that material, indicating whether it is, or is not, considered to satisfy the test for prosecution disclosure, and in either case explaining the reasons for coming to that view.

DO must comply with a request from the prosecutor to be allowed to inspect material which has not already been copied to him/her. If the prosecutor asks to be provided with a copy of such material it should be provided, except where (having consulted the officer in charge of the investigation) DO believes that the material is too sensitive to be copied and can only be inspected.

Where material is in a form other than in writing, it should be given to the prosecutor in a form agreed between DO and the prosecutor.

Subsequent action by disclosure officer

8.92 When a schedule of non-sensitive material is prepared, DO may not know exactly what material will form the case against D. In addition, the prosecutor may not have given advice about the likely relevance of particular items of material. Once these matters have been determined, DO must give the prosecutor, where necessary, an amended schedule listing any additional material: which may be relevant to the investigation, does not form part of the case against D, is not already listed on the schedule, and (DO believes) is not sensitive, unless DO is informed in writing by the prosecutor that the prosecutor intends to disclose the material to the defence.

As mentioned at **8.80**, a continuing duty is imposed on the prosecutor, for the duration of criminal proceedings against D, to disclose material which might reasonably be considered *to undermine the case for the prosecution, or which might assist D*. To enable this to be done, any new material coming to light should be treated in the same way as the earlier material. In particular, after a defence statement has been given, or details of the issues in dispute have been recorded on the Preparation for Effective Trial form or the Plea and Trial Preparation Hearing form, DO must look again at the material which has been retained and must draw the attention of the prosecutor to any material which might reasonably be considered capable of having the *italicised effect* just mentioned and must reveal it to the prosecutor in accordance with the third and fourth paragraphs of **8.91**.

DO must certify to the prosecutor that, to the best of his/her knowledge and belief, all relevant material which has been retained and made available to him/her has been revealed to the prosecutor in accordance with the Code. It will be necessary to certify not only at the time when the schedule and accompanying material is submitted to the prosecutor, and when relevant material which has been retained is reconsidered after D has given a defence statement, but also whenever a schedule is otherwise given or material is otherwise revealed to the prosecutor.

Disclosure of material to defendant

8.93   A prosecutor must review the schedules of unused material provided by DO and endorse the schedule to indicate whether each item of material does or does not meet the test for disclosure. If any of the material does meet the test for disclosure, the prosecutor should record the reason for this decision. A prosecutor must additionally review any material listed at **8.88**(a)–(g) (material likely to meet the test for disclosure). The prosecutor must endorse the schedule to indicate whether the material does or does not meet the test for disclosure, and must record the reason for the decision.

When a prosecutor provides material to the defence in accordance with the obligation under CPIA 1996, s 3 or 7A (**8.76** and **8.80**), (s)he must at the same time provide the schedule of non-sensitive material to the defence.

Other than early disclosure under the common law duty, in the magistrates' court the schedule (and any relevant unused material to be disclosed under it) must be disclosed to D either at the hearing where a not guilty plea is entered, or as soon as possible following a formal indication from D or his/her representative that a not guilty plea will be entered at the hearing.

In the Crown Court, initial disclosure should if possible, be served prior to the plea and trial preparation hearing (PTPH). Where this has not been done, it should be served as soon as possible after that hearing and in accordance with any direction made by the Court.

If material has been copied to the prosecutor, and it is to be disclosed, whether it is disclosed by the prosecutor or DO is a matter of agreement. If material has not already been copied to the prosecutor, and (s)he requests its disclosure to D on the ground that it satisfies the test for prosecution disclosure, or that the court has ordered its disclosure, DO must disclose it to D. Disclosure to D by DO must be made by the provision of a copy or an opportunity for D to inspect the material. Where D requests the provision of an inspected copy, DO must supply it, unless in DO's opinion that is either not practicable (eg because the material consists of an object which cannot be copied, or because the volume of material is so great), or not desirable (eg because the material is a statement by a child witness in relation to a sexual offence). Where material inspected by D consists of information which is recorded other than in writing, it is for DO to decide whether it should be given to D in its original form or in the form of a transcript. If the material is transcribed, DO must ensure that the transcript is certified to D as a true record of the material which has been transcribed.

If a court concludes that an item of sensitive material satisfies the prosecution disclosure test and that the interests of the defence outweigh the public interest in withholding disclosure, it will be necessary to disclose the material if the case is to proceed. This does not mean that sensitive documents must always be disclosed in their original form: for example, the court may agree that sensitive details still requiring protection should be blocked out, or that documents may be summarised, or that the prosecutor may make an admission about the substance of the material under CJA 1967, s 10 (**8.10**).

PROTECTING PEOPLE

CHAPTER 9

Non-Fatal Offences Against the Person

COMMON ASSAULT AND BATTERY

Common assault and battery are separate summary offences, punishable with up to six **9.1** months' imprisonment and/or an unlimited fine, contrary to the Criminal Justice Act 1988, s 39(1). Rather confusingly, the word 'assault' is used in some other statutes to refer to assault or battery.

Assault

A person is guilty of the separate offence of assault if (s)he intentionally or recklessly **9.2** causes another person to apprehend the immediate application to him/herself of unlawful force.

Any act, even mere words, can suffice if it has the requisite result. An example would be where, during an argument in a pub, a man holding a beer glass loses his temper and shouts out to his antagonist, 'I'll glass you for that'. Although words alone can constitute an assault, threatening words are more likely to be prosecuted as an offence under the Public Order Act 1986, s 4, 4A, or 5 (**18.12–18.19**).

The requirement that the immediate application of unlawful force must be apprehended means that it is an assault to aim a blow at V, whether or not that blow hits him, unless V is blind, or the blow is aimed from behind V, or for some other reason V does not apprehend force. The requirement of 'immediacy' has been given a liberal interpretation by the courts. In one case, where a woman (V) had been caused psychiatric harm after repeated telephone calls and letters from the defendant (D), the last two of which contained threats, the Court of Appeal held that the jury were entitled to find that the last letter had caused V fear of immediate force. It emphasised that D, who was known to V, lived near her and she thought that something could happen at any time. The Court of Appeal, albeit accepting the requirement of the apprehension of immediate force, said that it was enough for the prosecution to prove fear of force 'at some time not excluding the immediate future', which seems to require a fear of the application of force at some time in the future, including the immediate future. Cases involving repeated conduct such as that just described are now better dealt with by bringing a prosecution for an offence under the Protection from Harassment Act 1997, described later in this chapter.

If V is put in fear of immediate force, it is irrelevant that D could not in fact carry out his/her threat; for example, pointing an unloaded gun at someone who is unaware of its harmlessness can be an assault.

The force apprehended must be unlawful; see **9.19**.

The mens rea required for an assault is an intention to cause the other person to apprehend the immediate application of unlawful force or recklessness as to whether (s)he might so apprehend.

Battery

9.3 A person is guilty of battery if (s)he intentionally or recklessly applies unlawful force to another person. Most batteries are preceded by an assault, but this is not always so. If, for example, V is clubbed down from behind there is certainly a battery, but, as seen in **9.2**, if V was unaware that the blow was coming, there will not have been an assault.

A battery requires some conduct on D's part which results in unlawful force being applied to another (V). Technically, the slightest degree of force, even a mere touching, suffices, but a prosecution is most unlikely unless some harm has been caused. The force can be applied directly, as where D punches V, or sits on V, or hits V with an instrument, or indirectly, as where D puts a tripwire across an alley over which V trips or D puts acid in a hand drier which is blown onto the hands of the next user. The fact that a battery requires an application of force means that causing someone psychiatric harm by a threat does not involve a battery.

Normally, the force must be applied as a result of an act by D. However, liability can also be based on an omission (with the appropriate mens rea) to take such steps as are in D's power to counteract a dangerous situation created by him/her, even if inadvertently. This was held by a divisional court in a case where V, a police officer, approached D and told him that she intended to carry out a full body search. V asked D to turn out his pockets. D did so and produced some syringes without needles. V asked D if he had any needles on him and he replied 'No'. When V searched one of D's pockets her finger was pierced by a hypodermic needle, at which D smirked. Clearly, D had failed to counteract a danger, which his assurance had created, by not warning V not to put her hand in the pocket. As a result, force (the needle) had been applied to V's finger and there could be a conviction for battery since D had the mens rea for that offence.

The force applied must be unlawful; see **9.19**.

The mens rea required for a battery is an intention to apply unlawful force to the other or recklessness as to whether unlawful force might be so applied. It follows that, if D punches V mistakenly believing that V is attacking him/her, there is no battery because, as we shall see, D will not have intended, nor been reckless as to, the application of *unlawful* force.

Clearly, it is not a battery to hit someone accidentally (since there is no intention to apply force to another), unless D can be proved to have realised the risk that his/her act, eg of swinging an arm, might possibly result in unlawful force being applied to another and unreasonably run that risk (in which case D would be proved to have been reckless as to the risk).

Procedural matters

9.4 Where the person has been merely 'put in fear' (ie assault) D must be charged that (s)he 'did assault' that person. If force has been applied (ie battery), the charge should allege 'did assault by beating'.

AGGRAVATED ASSAULTS

9.5 There are a number of offences of aggravated assault. Among them are assault with intent to rob (**14.23**), and racially or religiously aggravated assault (**13.86**(c)) and (c)). Like the aggravated assaults discussed below, they require an assault or battery which is accompanied either by a particular intention or by a special circumstance or consequence.

Assault occasioning actual bodily harm

9.6 It is an indictable (either-way) offence, contrary to the Offences Against the Person Act 1861 (OAPA 1861), s 47, to assault any person, thereby occasioning him/her actual bodily harm (ABH). The offence is punishable with up to five years' imprisonment.

What is required is an assault or battery which has occasioned ABH. 'Harm' is a synonym for 'injury'. 'Actual' indicates that the injury should not be so trivial as to be wholly insignificant. Harm which is both transient and trifling is not ABH. 'Bodily harm' refers to injury to any part of the body (including a person's hair) or an identifiable psychiatric injury brought about by psychological factors (eg post-traumatic stress disorder, or battered wife syndrome, or reactive depression) but not mere emotions such as fear, distress, or panic which are not themselves evidence of an identifiable clinical condition. Consequently, to cause someone psychiatric injury by a threat of 'immediate' force can amount to an offence under s 47. Where V claims to have suffered psychiatric illness or injury as a result of a non-physical assault, there must be psychiatric evidence as to whether the symptoms alleged by V amount to a psychiatric illness or injury. It has been held that loss of consciousness falls within the meaning of 'harm', because it involves an injurious impairment to the victim's sensory functions.

There must be a causal connection between the 'assault' and the bodily harm occasioned and in most circumstances this will be apparent. If D punches V in the face and causes ABH, for example cuts or bruises, there has been a battery and the harm has been occasioned thereby. Cases where the assault is not the only cause of the ABH are governed by the same principles of causation as apply to murder and manslaughter (**10.7**), substituting 'ABH' for 'death'.

The mens rea required for this offence is the mens rea required for an assault or battery (as the case may be). It is not necessary to establish that D intended to cause some bodily harm or was reckless as to the risk of doing so.

Assault with intent to resist arrest

9.7 Under OAPA 1861, s 38, it is an indictable (either-way) offence, punishable with up to two years' imprisonment, for someone to assault (ie by an assault or by a battery) any person with intent to resist or prevent the lawful apprehension or detaining of him/herself, or any other person, for any offence. Section 38 deals with assaults on any person effecting an arrest, and therefore protects members of the public who are making 'citizens' arrests'. It does not apply, however, if the arrest is for a breach of the peace (which is not itself an offence) or in civil process; it is limited to an arrest for an offence.

An offence under s 38 is proved if it is established that the arrest was lawful; that D intended to resist it; and that D knew that the person whom D assaulted was seeking to arrest D. The issue of whether or not an offence which merited arrest had been committed is irrelevant, as is an alleged belief on D's part that (s)he had not committed the offence.

Assault on emergency worker

9.8 The Assaults on Emergency Workers (Offences) Act 2018, s 1 applies to an offence of assault or battery that is committed against an emergency worker acting in the exercise of functions as such a worker (s 1(1)). By s 1(2), such an offence is an indictable (either-way) offence punishable with up to 12 months' imprisonment. For the purposes of s 1(1), the circumstances in which an offence is to be taken as committed against a person acting in the exercise of functions as an emergency worker include circumstances where the offence takes place at a time when the person is not at work but is carrying out functions which, if done in work time, would have been in the exercise of functions as an emergency worker (s 1(3)). A divisional court has held that 'in the exercise of functions as an emergency worker' does not have the same meaning as 'in the execution of his/her duty' referred to in **9.11** and does not import a requirement that the emergency worker be acting lawfully.

In s 1, 'emergency worker' means:

(a) a constable;

(b) a person (other than a constable) who has the powers of a constable or is otherwise employed for police purposes or is engaged to provide services for police purposes;

(c) a NCA officer;

(d) a prison officer;

(e) a person (other than a prison officer) employed or engaged to carry out functions in a custodial institution of a corresponding kind to those carried out by a prison officer;

(f) a prisoner custody officer, so far as relating to the exercise of escort functions;

(g) a person certified as approved by the Secretary of State for the purpose of performing escort functions;

(h) a person employed for the purposes of providing, or engaged to provide—

 (i) fire services or fire and rescue services;

 (ii) search services or rescue services (or both);

(i) a person employed for the purposes of providing, or engaged to provide—

 (i) NHS health services; or

 (ii) services in the support of the provision of NHS health services,

 and whose general activities in doing so involve face-to-face interaction with individuals receiving the services or with other members of the public.

It is immaterial for the purposes of this list whether the employment or engagement is paid or unpaid.

Assault on, obstruction of, or resistance to a constable in the execution of his/her duty

Assault on a constable in the execution of his/her duty

9.9 The Police Act 1996 (PA 1996), s 89(1) makes it a summary offence punishable with up to six months' imprisonment and/or an unlimited fine for a person to assault (by an assault or by a battery) a constable in the execution of his/her duty. The more serious offence under the 2018 Act just mentioned overlaps with that under s 89(1). CPS guidance suggests that the former offence should usually be prosecuted.

While, of course, D must have the necessary mens rea for the assault or battery which D commits, it is irrelevant that D does not know that V is a constable acting in the execution of his/her duty. However, if D, ignorant that V is a constable, applies force to V who is exercising one of his/her powers, and that force would have been reasonable on the ground of self-defence if V had not been a constable, D does not commit an offence. D will not have intentionally or recklessly applied unlawful force (the mens rea for a battery) because of his/her ignorance of V's status.

Obstructing or resisting a constable in the execution of his/her duty

9.10 Under PA 1996, s 89(2), it is a summary offence punishable with up to one month's imprisonment and/or a fine not exceeding level 3, for a person to resist or wilfully obstruct a constable in the execution of his/her duty, or a person assisting him/her.

To '*obstruct*' is to do any conduct which prevents or makes it more difficult for a constable to carry out his/her duty, and in this sense those who give warning of police speed checks may obstruct the constables in the execution of their duty. It has been held that someone who warns motorists of a police speed check is only guilty of obstructing a constable in the execution of his/her duty if the prosecution proves that the motorists warned were in fact speeding or were likely to be speeding at that location. Someone who deliberately drinks alcohol after an accident to negate the breath-testing procedure is also guilty of this offence.

The obstruction must be 'wilful', which in this context means that:

(1) D's conduct which has resulted in the obstruction must have been deliberate and intended by him to bring about a state of affairs which, in fact, prevented or made it more difficult for the constable to carry out his/her duty, whether or not D realised that that state of affairs would have that effect.

(2) Also, D must have had no lawful excuse. Police officers often experience difficulty in obtaining names and addresses from offenders but a refusal to give such information will not amount to a wilful obstruction unless that person has a duty to give that information, because otherwise (s)he will have a lawful excuse for his/her refusal. Nor, for the same reason, is it a wilful obstruction to advise someone not to answer police questions which (s)he is not obliged to answer, even if the advice is given in an abusive way. Much of the traffic legislation imposes a duty to give particular types of information, but there is no such requirement in relation to most offences. Just as in the case of a failure to provide information, so in the case of other failures to assist the police (eg by failing to accord entry to a constable), there is only a wilful obstruction if the constable has the right to require the assistance in question, so that D is under a legal duty to provide it. An example would be where a constable has a statutory right of entry. A refusal to admit the constable in breach of the duty to do so would be a wilful obstruction.

It has been held that a defendant who believed that the person obstructed was not a constable could not be convicted of the present offence.

Like obstruction, *resistance* does not require an assault or battery. Probably, any resistance is also an obstruction, but resistance is a more appropriate word in certain cases (such as where a person arrested by a constable tears himself away).

Acting in the execution of duty

9.11 To be acting in the execution of his/her duty, a constable (C) must be acting within the general scope of a duty imposed on C by law (such as C's duties to protect life and property, to keep the peace, to prevent and investigate crimes and to prevent obstruction of the highway) and C must not be acting unlawfully at the time. Thus, even if C is acting within the general scope of one of his/her duties, C is not acting in the execution of his/her duty if C has no power to do the thing in question (and is, therefore, committing a trespass against a person or a person's property). The courts have held that, since a constable (C) does not have the power without arresting someone to physically detain him/her for questioning, C will not be acting in the execution of his/her duty if C does, even if the detention is only for a few seconds. It would be different if C is exercising a power of arrest.

It has been held in the Administrative Court that, as a constable has the power to ask someone to stop and answer questions if (s)he does not physically detain him/her or threaten to do so, a constable who so asks is acting in the execution of his/her duty at that time, even if (s)he intends unlawfully to detain later the person asked to stop.

A constable does not act outside the execution of his/her duty if what (s)he does involves no more than a trivial touching; so to act is lawful because the conduct would be generally acceptable in the ordinary conduct of daily life (**9.22**). Divisional courts have held in cases where a constable took hold of someone's arm or tapped him on the shoulder not in order to detain him, but in order to speak with him or draw something to his attention, that the constable had been acting in the execution of his duty. On the other hand, where a police officer restricted a person's movements (blocking him) in a doorway for a few seconds, without having intended or purported to arrest him, in order to question him, the Court of Appeal held that this went beyond conduct which was generally acceptable in the ordinary conduct of daily life and was unlawful. The constable was therefore not acting in the execution of his duty at the time.

The test whether a constable is acting in the execution of his/her duty is judged objectively. Thus, for example, an officer who makes an unlawful arrest is not acting in the execution of his/her duty even though (s)he believes that (s)he has the necessary reasonable grounds for making an arrest. Officers who arrive on the scene and assist a colleague in what turns out to be an unlawful arrest will also not be acting in the execution of their duty.

A person who has not been arrested is entitled to leave a police station at any time unless (s)he is detained under particular provisions which allow detention. A constable who attempts to prevent such a person from leaving a police station is not acting in the execution of his/her duty.

Constables are frequently asked to assist with the expulsion of persons from premises, where the owner of the premises considers them to be intruders or otherwise unwelcome. An officer may lawfully assist the owner of property in these circumstances, but (s)he is not bound to do so. Unless there are particular circumstances which demand such expulsion, for example, the removal of violent, quarrelsome, disorderly persons from various premises as required by law (see, eg, **20.119**), or where a breach of the peace is taking place or apprehended, it is unlikely that (s)he will be considered to have been acting in the execution of his/her duty.

On occasions constables are given the authority to enter premises, and if they enter under such an authority they are acting in the execution of their duty. For example, the common law authorises a constable to enter premises to deal with a breach or reasonably apprehended breach of the peace (and it also authorises him/her to remain for this purpose if (s)he is already on the premises); such a constable is acting in the execution of his/her duty. Response to a burglar alarm gives police an implied permission to enter premises for a reasonable time for the purpose of a search, but there is no legal right to enter premises found insecure at night. Where a constable is invited to enter premises by a member of the family and is later told to leave by the occupier, and (s)he is assaulted by the occupier while immediately complying with that request, it is an assault upon him/her in the execution of his/her duty. On the other hand, (s)he would no longer be in the execution of his/her duty if (s)he did not comply with the request within a reasonable time because (s)he would become a trespasser; if (s)he was assaulted after the expiry of such a time an offence under s 89(1) would not be committed.

A divisional court has held that, where a constable has acted unlawfully (eg by using unlawful force) during a sequence of events, this does not mean that the constable's previous or subsequent exercise of his/her powers is also outside the execution of his/her duty.

Where a constable (C) does something mistakenly believing that the subject of his/her action is committing a particular offence whereas that person is committing another offence, and C would have power to act as C did in either event, C will be acting in the execution of his/her duty if C has a reasonable basis for his/her mistaken belief and is acting in good faith.

Other points

By PA 1996, s 89(3) the offences under s 89 may also be committed against constables **9.12** of the Police Service of Scotland or of Northern Ireland who are executing warrants or acting in England and Wales by virtue of any enactment. Section 89 also applies to a constable of the British Transport Police Force or a member of the Civil Nuclear Constabulary (Railways and Transport Safety Act 2003, s 68; Energy Act 2004, s 68). A foreign police or customs officer carrying out surveillance in England and Wales as part of a foreign surveillance operation is treated as if (s)he were acting as a constable in the execution of his/her duty (Crime (International Co-operation) Act 2003, s 84).

Under s 89(1) and (2) it is also an offence to assault, obstruct, or resist a person assisting a constable in the execution of his/her duty. The reference in s 89 to 'a person assisting a constable in the execution of his duty' includes reference to any person who is neither a constable nor in the company of a constable but who is carrying out his/her functions as a member of an international joint investigation team led by a member of a police force.

The Police Reform Act 2002, s 46(1) and (2) creates offences identical to those in PA 1996, s 89(1) and (2) in relation to assaults on, or the obstruction of or resistance to, a designated or accredited person acting in the execution of his/her duty, or someone assisting such a person.

OFFENCES INVOLVING WOUNDING OR GRIEVOUS BODILY HARM

Malicious wounding or infliction of grievous bodily harm

OAPA 1861, s 20 provides two indictable (either-way) offences punishable with up to **9.13** five years' imprisonment: malicious wounding and malicious infliction of grievous bodily harm. Section 20 provides that a person who unlawfully and maliciously wounds or inflicts any grievous bodily harm on any other person, either with or without any weapon or instrument, is guilty of an offence.

Both offences have two elements in common: 'unlawfully' and 'maliciously'. The difference between them relates to their actus reus: 'wounding' and 'infliction of grievous bodily harm', and these terms will be discussed first.

Wounding or infliction of grievous bodily harm

The term 'wound' indicates a breaking of the continuity of all layers of the skin. **9.14** Consequently, all injuries involving broken bones are excluded unless the bone pierces the skin. However, such injuries will normally amount to grievous bodily harm.

'Grievous bodily harm' (GBH) means really serious harm, but 'really serious' appears to mean no more than 'actually serious'; 'bodily harm' can include psychiatric injury or loss of consciousness but, of course, such injury must be serious in order to be grievous. It is not necessary that the nature of the harm should be either permanent or dangerous.

A wound can be caused or grievous bodily harm can be 'inflicted' even though it does not result from the application of force; it is enough, instead, that it is caused by something done by D. The same principles of causation apply as apply to murder and manslaughter (**10.7**), substituting 'GBH' or 'wounding' for 'death'.

Unlawfully

9.15   This means 'without lawful justification' and is merely intended to except from the offence, in certain circumstances, acts done with a justification rendering the harm lawful (**9.19–9.27**), for example harm lawfully caused in self-defence.

Maliciously

9.16   The mens rea of an offence under OAPA 1861, s 20 is that D should have wounded or inflicted GBH 'maliciously'. This does not mean that D must have acted out of spite or ill-will. Instead, what is required is that D must have intended his/her act to cause some unlawful harm to another, or been reckless as to whether some unlawful harm might result from his/her act. It must be emphasised that it is not necessary that D should have intended or foreseen a wound or really serious harm; foresight that *some* harm, albeit of a minor character, might result is enough.

Because D must have been aware that his/her act might cause some *unlawful* bodily harm, someone who mistakenly believes that (s)he is acting in self-defence or has some other legal justification is not guilty under s 20 if (s)he wounds or inflicts grievous bodily harm on someone.

Wounding or causing grievous bodily harm with intent to do grievous bodily harm or to resist or prevent arrest

9.17   OAPA 1861, s 18 provides that a person who unlawfully and maliciously by any means whatsoever wounds or causes GBH to any person with intent to do GBH to any person, or with intent to resist or prevent the lawful apprehension or detainer of any person, is guilty of an offence triable only on indictment and punishable with a maximum of life imprisonment.

OAPA 1861, s 18 provides two offences: wounding with intent to do unlawful GBH or with intent to resist the lawful apprehension or detaining of any person, and causing GBH with one of these intents.

What was said in relation to OAPA 1861, s 20 in relation to 'unlawfully', 'wound', and 'GBH' is equally applicable to OAPA 1861, s 18. However, s 18 specifies that GBH must be 'caused' (as opposed to 'inflicted'). This difference in terminology between s 18 and s 20 raises the question of whether there is a difference of substance. GBH can be 'caused' by a deliberate and culpable omission to act; but opinions differ as to whether GBH can be 'inflicted' by such an omission.

Mens rea

9.18   If D is charged with wounding *with intent to do GBH*, or with causing GBH *with such intent*, 'maliciously', the meaning of which was explained at **9.16**, is rendered redundant by the stricter requirement of an intent to do GBH. The GBH intended need not be GBH to the particular victim. D may be convicted of wounding V with intent, even though D thought that V was X, or even though D fired at X and hit V by accident. If D fires a gun into a group of people without taking particular aim, but intending to harm someone, D may be charged with a s 18 offence against the person whom D hits.

The basic distinction between wounding or causing GBH with intent to do GBH and attempted murder is that, in the former offences, only an intent unlawfully to do GBH is required, while the latter requires an intent unlawfully to kill.

Where D is charged with wounding *with intent to resist or prevent the lawful apprehension or detaining of any person*, or with causing GBH *with such intent*, 'maliciously' is relevant. It must be proved that D intended his/her conduct to cause *some* unlawful harm to another (or was reckless as to this occurring), ie that (s)he was 'malicious', *and* that (s)he intended to resist or prevent the lawful apprehension or detaining of him/herself or another. The lawful arrest which D intends to resist or prevent need not be for an offence; it could be for a breach of the peace or in civil process.

UNLAWFUL FORCE OR HARM

It is an integral part of the offences referred to above that the requisite force (appre-   **9.19**
hended or applied) or harm (as the case may be) must be unlawful. In this context, the essential point is that, if the victim has given a valid consent to the force or harm, or if D has acted in self-defence, prevention of crime, or the like, it is lawful.

Consent

Case law establishes that consent can validly be given to an act which does not cause   **9.20**
actual bodily harm. 'Actual bodily harm' means any injury which is not so trivial as to be wholly insignificant. See, further, **9.6**.

Consent cannot validly be given if an act causes actual bodily harm which D intended or foresaw as a risk *unless the case falls within one of a number of recognised exceptions or a new exception is recognised on public interest grounds*. The result of all this is that if D intentionally causes V actual bodily harm or the risk of actual bodily harm caused was foreseen, it is generally irrelevant whether or not V has consented, since, generally, V cannot give a valid consent in such a case. Thus, assuming the other elements of the offence are proved, there can generally be a conviction for an assault occasioning actual bodily harm or some other offence involving harm to the person in such a case, despite the victim's apparent consent. For example, men who fight each other to 'settle a score' and cause actual bodily harm commit an assault occasioning actual bodily harm when they fight each other, despite their consent to such harm since actual bodily harm is clearly intended or foreseen.

Under the Domestic Abuse Act 2021, s 71(2), it is not a defence to an offence under OAPA 1861, s 18, 20 or 47 (**9.6** and **9.13–9.18**) that the victim consented to the infliction of GBH, wounding or ABH for the purposes of sexual gratification. For these purposes, it does not matter whether such harm was inflicted for the purposes of obtaining sexual gratification for D, for V or for some other person. Section 71(2) does not apply in the case of an offence under s 20 or s 47, where such harm consists of, or results from, the infliction of V with a sexually transmitted infection in the course of sexual activity, and V consented to that activity knowing or believing that D had that infection (s 71(4)).

The recognised exceptions

A person can give a valid consent to 'any actual bodily harm' caused by a reasonable   **9.21**
surgical operation or procedure carried out by someone medically qualified to perform

it; if (s)he could not, the surgeon would commit a battery or an assault occasioning actual bodily harm or some more serious offence against the person.

It has been held that a valid consent can be given to ear piercing, to being tattooed (see below for an exception) or, even, to being branded with one's spouse's initials, since the bodily harm caused by these activities is not contrary to public policy. It has been held, on the other hand, that a valid consent cannot be given to 'body modification', ie the removal of parts of the body (eg ear- or nipple-removal) or mutilation (eg tongue splitting) for aesthetic or other non-medical reasons, which is unregulated, requires no particular training or qualification and would never be done by a reputable medically qualified person. By statute the tattooing of someone under 18 is unlawful as (in Wales) is the intimate piercing of such a person.

Those who agree to take part in a lawful sport consent to the rules of that sport and, if those rules allow forms of physical contact, they validly consent to the risk of actual bodily harm which is likely to result from physical contact which is within the rules or is a minor infringement of them. For example, a blow struck in a boxing match under the Queensberry Rules (in which boxers wear approved gloves) is not a battery or any other offence, regardless of the injury caused, unless the blow is struck in circumstances far outside the rules (eg hitting an opponent when (s)he is lying unconscious on the floor, or hitting an opponent with a glove in which is concealed a heavy object). Likewise, in soccer and rugby, the participants consent to the risk of actual bodily harm resulting from something within the rules of the game or not too far removed from them, but not to the risk of such harm resulting from something which is far outside the rules, such as a head-butt or deliberately kicking a player who is on the ground.

Not all sports are lawful. For example, a prize-fight, where gloves are not worn and the fight continues until one of the participants can no longer continue, is an unlawful sport. Thus, the participants cannot give a valid consent to the actual bodily harm intended or foreseen, with the result that the force which they apply to each other is always unlawful.

Other points on consent

9.22   Sometimes when a person has consented to the application of force, his/her consent is invalid, even though actual bodily harm is not caused by the force. This occurs where (s)he is so young or mentally impaired as not to be able to make a rational decision whether or not to consent, or where his/her apparent consent has been procured by duress or by a deception as to the identity of the defendant or as to the nature or quality of the act or its purpose. Deception as to identity may include deception as to the defendant's professional qualification if that is integral to the defendant's identity and crucial to the victim's consent.

Assuming it is valid, a consent need not be express; it can be implied from the circumstances. Everyday living demands a certain amount of physical contact. People are often touched in order to attract their attention, and there are frequent collisions in shopping precincts, and the consent of people to such things can normally be implied. It is, of course, different if the person touched has indicated that (s)he does not want to be touched. If V tells D, who has been pestering him, to go away, V clearly does not impliedly consent to D touching him/her soon after in order to attract his/her attention. Of course, there is a limit to what a person impliedly consents to. For example, one does not impliedly consent to collisions in a shopping precinct caused by hooligans charging about.

Another way of expressing cases based on implied consent is that they fall within a general exception embracing all physical contact which is generally acceptable in the ordinary conduct of daily life.

Other factors which render force lawful

Disciplinary use of force and corporal punishment

Subject to the limitations below, parents and other people in loco parentis are entitled **9.23** as a disciplinary measure to apply a reasonable degree of force to their children or charges old enough to understand its purpose. In relation to the following offences:

(a) wounding or causing grievous bodily harm with intent, or unlawful wounding or infliction of grievous bodily harm;
(b) assault occasioning actual bodily harm; or
(c) cruelty to a person under 16,

battery of a child cannot be justified on the ground that it constituted reasonable punishment.

Teachers are no longer entitled by virtue of their position as such to apply reasonable corporal punishment as a disciplinary measure. However, the prohibition on corporal punishment by teachers does not affect the power of members of school staff to use reasonable force to restrain a pupil from:

(a) committing an offence or continuing to do so;
(b) causing or continuing to cause personal injury or damage to property; or
(c) behaving or continuing to behave in a way prejudicial to good order and discipline at school or among the pupils.

Prevention of crime or effecting arrest

The Criminal Law Act 1967 (CLA 1967), s 3 provides that it is lawful to use such force as **9.24** is reasonable in the circumstances in the prevention of crime or in effecting (or assisting in) the lawful arrest of offenders, suspected offenders, or persons unlawfully at large.

A divisional court has held that the defence of using force to prevent a crime requires an apprehension of a need to use force to prevent an actual, imminent or immediate crime. The court also stated that the defence under s 3(1) applied where there was a direct application of force by D, although the force did not necessarily have to be applied directly against a person.

It must be emphasised that, if the force used to prevent a crime or to make an arrest is unreasonable in the circumstances, it will be unlawful and the person using it will not have a defence to a charge of battery or of another offence against the person. Like anyone else, a police officer has no defence, even though (s)he uses reasonable force, if (s)he is acting in furtherance of an unlawful arrest. Nor does (s)he have a defence if (s)he uses force to restrain someone whom (s)he does not intend or purport to arrest, even though an arrest would have been justified.

The Court of Appeal has held that a mistake of law cannot found a defence under s 3(1); it is no answer for D to say that, because of a mistaken belief as to the law, (s)he mistakenly thought the victim was committing a crime, when the victim was not.

Self-defence and defence of property or of another

Self-defence and the defence of property or of another are common law defences. They **9.25** are limited to defence against an actual or imminent or immediate unlawful (ie criminal or tortious) act. However, a person who acts in defence of him/herself or another or of property is almost invariably acting in the prevention of crime, in which case (s)he also has the defence under CLA 1967, s 3. For practical purposes, the terms of both the common law and the statutory defences are identical in their requirements.

Use of force against an innocent person

9.26   The Court of Appeal has held that self-defence and the defence under CLA 1967, s 3 extends to the use of force against an innocent third party where such force is used to prevent a crime being committed by someone else. Two examples of cases where the defence is capable of arising include a police constable bundling a man out of the way to get to another man who is about to detonate an explosive device, and where a person knocks car keys out of the hands of a third party to prevent the keys being given to a drunken person who is attempting to drive.

Applicable principles

9.27   The Criminal Justice and Immigration Act 2008, s 76, which applies to the common law defences of self-defence, defence of another, and defence of property, and to the defences under CLA 1967, s 3(1) of the use of force in the prevention of crime or making of a lawful arrest, provides that the question whether the degree of force used by D was reasonable in the circumstances is to be decided by reference to the circumstances as D believed them to be, and that in connection with deciding that question:

(a) if D claims to have held a particular belief as regards the existence of any circumstances:
   (i) the reasonableness or otherwise of that belief is relevant to the question whether D genuinely held it, but
   (ii) if it is determined that D did genuinely hold it, D is entitled to rely on it, whether or not it was mistaken or (if it was mistaken) the mistake was a reasonable one to have made;

   but (ii) does not enable D to rely on any mistaken belief attributable to intoxication that was voluntarily induced;

(b) in a householder case, the degree of force used by D is not to be regarded as having been reasonable in the circumstances as D believed them to be if it was grossly disproportionate in those circumstances. This provision does not direct that any degree of force less than grossly disproportionate is reasonable. Whether it was or not reasonable will depend on the particular facts and circumstances of the case. 'A householder case' is a case where:
   (i) the defence concerned is the common law defence of self-defence or defence of another;
   (ii) the force concerned is force used by D while in or partly in a building, or part of a building, that is a dwelling or is 'forces accommodation' (armed services living accommodation), or is both;
   (iii) D is not a trespasser at the time the force is used; and
   (iv) at that time D believed the victim to be *in, or entering*, the building or part as a trespasser.

   Where a part of a building is a dwelling where D dwells, another part of it is a place of work for D or another person who dwells in the first part, and that other part is internally accessible from the first part, that other part, and any internal means of access between the two parts, are each treated for the above purposes as a part of a building that is a dwelling. There is a similar provision in respect of 'forces accommodation'.

   For the above purposes, 'building' includes a vehicle or vessel;

(c) in a case other than a householder case, the degree of force used by D is not to be regarded as having been reasonable in the circumstances as D believed them to be if it was disproportionate in those circumstances. Although this means that force which is disproportionate is automatically unreasonable, it does not mean that force which is proportionate is necessarily reasonable. In almost all cases if the degree of force is proportionate it will also be reasonable, but the two terms are not to be equated;

(d) in deciding whether the degree of force used by D was reasonable in the circumstances, a possibility that D could have retreated is to be considered (so far as relevant) as a factor to be taken into account, rather than as giving rise to a duty to retreat;

(e) in deciding whether the degree of force used by D was reasonable in the circumstances, the following considerations are to be taken into account (so far as relevant in the circumstances of the case):

(i) that a person acting for the purpose of crime prevention, effecting a lawful arrest, self-defence, defence of another, or defence of property ('a legitimate purpose') may not be able to weigh to a nicety the exact measure of any necessary action, and

(ii) that evidence of a person's having only done what the person honestly and instinctively thought was necessary for a legitimate purpose constitutes strong evidence that only reasonable action was taken by that person for that purpose.

For the avoidance of doubt, it must be stated that the mere fact that a person who has used force against another was provoked to lose self-control (as opposed to acting in self-defence, etc) is no defence.

PROTECTION FROM HARASSMENT ACT 1997

Prohibitions of harassment

9.28 The Protection from Harassment Act 1997 (PHA 1997) sets out two prohibitions of harassment. The first, contained in s 1(1), prohibits a person from pursuing a course of conduct which:

(a) amounts to harassment of another person; and
(b) D knows or ought to know amounts to harassment of the other person.

The second prohibition of harassment, contained in s 1(1A) prohibits a person from pursuing a course of conduct:

(a) which involves harassment of two or more persons; and
(b) which D knows or ought to know involves harassment of those persons; and
(c) by which D intends to persuade any person (whether or not one of those mentioned above),
(i) not to do something that (s)he is entitled or required to do, or
(ii) to do something that (s)he is not under any obligation to do.

Section 1(1A) was added to protect company employees from harassment by animal rights groups. Section 1(1) does not protect such employees if they are only harassed on one occasion.

Course of conduct

9.29 In the case of conduct relating to a single person (see PHA 1997, s 1(1)), a 'course of conduct' must involve conduct on at least two occasions in relation to that person. In the case of conduct in relation to two or more persons (see s 1(1A)), a 'course of conduct' must involve conduct on at least one occasion in relation to each of those persons. Publishing a series of articles in a newspaper can constitute a course of conduct.

The term 'conduct' in PHA 1997 includes speech. By s 7(3A), a person's (X's) conduct on any occasion which is aided, abetted, counselled, or procured by another (Y) is taken:

(a) also to be on that occasion the conduct of Y; and
(b) to be conduct in relation to which Y's knowledge and purpose (and what Y ought to have known) are the same as they were in relation to what was contemplated or reasonably foreseeable at the time of the aiding, abetting, etc.

This means in relation to the wording in s 1(1) that there can be a breach of the prohibition on harassment by Y if, for example, on one occasion the harassing conduct is by X aided and abetted by Y and on the second occasion is by Y himself, provided that the two pieces of conduct can be regarded as a 'course' and that part (b) of s 7(3A) is satisfied. The same principle will apply to conduct aimed at different company employees so far as s 1(1A) is concerned.

By s 1(2), for the purposes of ss 1 and 2A (**9.34**) the person whose course of conduct is in question ought to know that his/her conduct amounts to or involves harassment of another if a reasonable person in possession of the same information would think the course of conduct amounted to or involved harassment of the other. The reasonable person in this context is not imbued with any mental illness or other characteristic which the defendant has.

Although, a company cannot be harassed for the purposes of PHA 1997, it can be liable for harassment.

Harassment

9.30 By s 7(2), 'harassment' in PHA 1997 *includes* alarming another person or causing that person distress. As a divisional court has held, this is not an exhaustive definition of harassment. The Court of Appeal has held that, in addition, the conduct must be unacceptable and oppressive, either physically or mentally, such that it would sustain criminal liability; it must be targeted at an individual, and must result in fear or distress. The Court also stated that, when considering whether conduct amounts to or involves harassment, it is right to have regard to what the ordinary person would understand by harassment.

While s 1 was fashioned with 'stalkers' in mind (in this respect note s 2A at **9.34**), its terms are wide enough to embrace situations in which harassment is caused to neighbours or to persons living in a particular area in which unruly juveniles tend to congregate. Section 1 can also cover the activities of political protesters and members of the 'paparazzi', and more mundane things like sending a series of letters demanding payment of a debt, if the other elements are established.

Exceptions

9.31 By PHA 1997, s 1(3), the prohibitions of harassment in s 1(1) and (1A) do not apply to a course of conduct if the person who pursued it shows that it was pursued for the purpose of preventing or detecting crime; or that it was pursued under any enactment or rule of law or to comply with a condition or requirement lawfully imposed; or that, in the particular circumstances, the pursuit of the course of conduct was reasonable. It has been stated by a divisional court that whether conduct was reasonable involves balancing the interests of the victim against the purpose and nature of the course of conduct pursued, including the right to peaceful protest.

Offence of harassment

PHA 1997, s 2(1) makes it an offence to break the prohibition of harassment under s **9.32**
1(1) or s 1(1A). The offence is a summary one punishable with up to six months' imprisonment and/or an unlimited fine.

Civil remedy

PHA 1997, ss 3 and 3A make provision for an actual or apprehended breach of the pro- **9.33**
hibition of harassment under s 1(1) or s 1(1A) respectively to be the subject of a civil
claim in the High Court or county court by a person who is or may be the victim or a
victim, under which damages (s 3 only) or an injunction restraining a defendant from
pursuing a course of conduct amounting to harassment may be awarded.

Breach of an injunction under s 3 or s 3A without reasonable excuse is an indictable (either-way) offence punishable with up to five years' imprisonment (ss 3(6)
and 3A(3)).

Offence of stalking

PHA 1997, s 2A(1) provides that a person is guilty of an offence if (a) he pursues a **9.34**
course of conduct *in breach of the prohibition of harassment in s 1(1)*, and (b) the course
of conduct amounts to stalking. For the purposes of (b), a person's course of conduct
amounts to stalking of another person if:

(a) it amounts to harassment (**9.30**) of that person,
(b) the acts or omissions involved are ones associated with stalking, and
(c) the person whose course of conduct it is knows or ought to know that the course of
    conduct amounts to harassment of the other person (s 2A(2)).

The offence under s 2A is a summary one punishable with up to six months' imprisonment and/or an unlimited fine. It is a more appropriate offence than that under s 2 to
deal with stalkers. However, the offence of stalking may alternatively be dealt with by a
prosecution under PHA 1997, s 2.

Section 2A(3) gives the following examples of acts or omissions which, in particular
circumstances, are ones associated with stalking:

(a) following a person,
(b) contacting, or attempting to contact, a person by any means,
(c) publishing any statement or other material:
    (i) relating or purporting to relate to a person, or
    (ii) purporting to originate from a person,
(d) monitoring the use by a person of the internet, email or any other form of elec-
    tronic communication,
(e) loitering in any place (whether public or private),
(f) interfering with any property in the possession of a person,
(g) watching or spying on a person.

Power of entry in relation to offence of stalking

PHA 1997, s 2B provides that a justice of the peace may, on an application by a con- **9.35**
stable, issue a warrant authorising a constable to enter and search premises if the justice
of the peace is satisfied that there are reasonable grounds for believing that:

(a) an offence under s 2A has been, or is being, committed,

(b) there is material on the premises which is likely to be of substantial value (whether by itself or together with other material) to the investigation of the offence,

(c) the material is likely to be admissible in evidence at a trial for the offence, and does not consist of, or include, items subject to legal privilege, excluded material or special procedure material (as defined at **4.68**), and

(d) either entry to the premises will not be granted unless a warrant is produced, or the purpose of a search may be frustrated or seriously prejudiced unless a constable arriving at the premises can secure immediate entry to them.

A constable may seize and retain anything for which a search has been so authorised. The additional powers of seizure provided by the Criminal Justice and Police Act 2001, s 50 (**4.61**) apply to this power of seizure. A constable may use reasonable force, if necessary, in the exercise of any power conferred by virtue of PHA 1997, s 2B.

Putting people in fear of violence

9.36 PHA 1997, s 4 creates the indictable (either-way) offence, punishable with up to 10 years' imprisonment, of putting people in fear of violence. Section 4(1) provides that a person whose course of conduct causes another to fear, on at least two occasions, that violence will be used against him, is guilty of an offence if he knows or ought to know that his course of conduct will cause that other person so to fear on each of those occasions. The Court of Appeal has held that the course of conduct must amount to harassment of another. Direct evidence from the victim that (s)he was caused to fear such violence is not essential, because a court may infer such fear if there is other evidence entitling it to do so, but without direct evidence from the victim proof may be difficult.

For the purposes of s 4, the person whose course of conduct is in question ought to know that it will cause another to fear that violence will be used against him/her on any occasion if a reasonable person in possession of the same information would think the course of conduct would cause the other so to fear on that occasion.

Section 4 provides similar defences to those under s 1(3) (**9.31**) with the substitution of 'circumstances in which the pursuit of his/her course of conduct was reasonable for the protection of him/herself or another or for the protection of his/her or another's property', for 'the pursuit of the course of conduct being reasonable in the circumstances'.

Stalking involving fear of violence or serious alarm or distress

9.37 PHA 1997, s 4A(1) provides that a person (D) whose course of conduct:

(a) amounts to stalking, and

(b) either:

    (i) causes another (V) to fear, on at least two occasions, that violence will be used against V, or

    (ii) causes V serious alarm or distress which has a substantial adverse effect on V's usual day-to-day activities,

is guilty of an indictable (either-way) offence if D knows or ought to know that D's course of conduct will cause V so to fear on each of those occasions or (as the case may be) will cause such alarm or distress. The offence is punishable with up to 10 years' imprisonment.

In relation to (b)(i), the Court of Appeal has held that there can be a fear of violence sufficient for (b)(i) where the fear of violence is of violence on a separate and later occasion. The Court added that there is no requirement for the fear to be of violence on a particular date or time in the future, or at a particular place or in a particular manner, or for there to be a specific threat of violence. These points are equally applicable to the corresponding phrase in s 4.

For the purposes of s 4A, D ought to know that his/her course of conduct will cause V to fear that violence will be used against V on any occasion if a reasonable person in possession of the same information would think the course of conduct would cause V so to fear on that occasion; and D ought to know that his/her course of conduct will cause V serious alarm or distress which has a substantial adverse effect on V's usual day-to-day activities if a reasonable person in possession of the same information would think the course of conduct would cause V such alarm or distress.

The same defences are available as apply to an offence under s 4.

The offence under s 4A may alternatively be dealt with by a prosecution under s 4.

Breach of restraining order

The Sentencing Code, s 360 empowers a court to make a restraining order when sentencing a convicted person for *any* offence. Such an order is of particular importance in respect of cases under PHA 1997. The Code provides that, in addition to any other sentence imposed, the court may make an order protecting the victim(s) of the offence, or any other person mentioned in the order, from further conduct which amounts to harassment or will cause a fear of violence.   **9.38**

PHA 1997, s 5A permits a court before which D is acquitted of an offence under PHA 1997 or otherwise, if it considers that it is necessary to do so to protect a person from harassment by D, to make an order prohibiting D from doing anything specified in the order.

If, without reasonable excuse, a person does anything which is prohibited by a restraining order, (s)he commits an indictable (either-way) offence (Sentencing Act 2020, s 363(1) and PHA 1997, s 5A(2)). Such an offence is punishable with up to five years' imprisonment. Proof of a knowing breach of the order is not required.

Police directions stopping the harassment, etc of a person at his/her home

The Criminal Justice and Police Act 2001 (CJPA 2001), s 42 provides that a constable who is at the scene may give a direction to a person if: **9.39**

(a) this person is present outside, or in the vicinity of, premises used by someone (the resident) as a dwelling;

(b) that constable believes, on reasonable grounds, that this person is there for the purpose (by his presence or otherwise) of representing to the resident or another individual, or of persuading the resident or other individual that:

　(i) he should not do something which he is entitled or required to do, or

　(ii) he should do something which he is not under any obligation to do; and

(c) the constable also believes on reasonable grounds that the presence of this person (either alone or together with others who are likely to be present):

　(i) amounts to, or is likely to result in, the harassment of the resident, or

　(ii) is likely to cause alarm or distress to the resident.

A direction under s 42 is one which requires its addressee to do things which the constable considers to be necessary to prevent the harassment of the resident or the causing of alarm or distress to the resident. It may be given orally, either to an individual or a group. Requirements that may be made under s 42 include:

(a) a requirement to leave the vicinity of the premises in question; and
(b) a requirement to leave that vicinity and not to return to it within such period as the constable may specify, not being longer than three months.

A requirement to leave may be to do so immediately or after a specified period. A direction may include exceptions and may make exceptions subject to conditions, including conditions as to the distance from the premises in question at which, or the location where, those who do not leave must remain, and conditions as to the numbers, or identity, of the persons who are authorised by the exception to remain in the vicinity.

When there is more than one officer at the scene, only the senior officer present may give the direction. The power to give a direction under s 42 does not include the power to direct a person to refrain from activity which is lawful under the Trade Union and Labour Relations (Consolidation) Act 1992, s 220.

A person who knowingly fails to comply with a requirement in a direction (other than a requirement not to return under (b) above) commits an offence (s 42(7)). A person who in breach of a requirement under (b) returns to the vicinity of the premises within the period specified for the purpose of persuading the resident or another individual as described above also commits an offence (s 42(7A)). Both offences are summary and are punishable with up to six (s 42(7A)) or three (s 42(7)) months' imprisonment and/or a fine not exceeding level 4.

Offence of harassment, etc of a person in his/her home

9.40   CJPA 2001, s 42A provides that a person also commits an offence if:

(a) he is present outside or in the vicinity of any premises that are used by any individual (the resident) as a dwelling; and
(b) he is there to represent to the resident or another individual, or to persuade the resident or other person, that he should not do something which he is entitled or required to do, or should do something he is not obliged to do;
(c) he intends his presence to amount to harassment, alarm or distress to the resident, or knows or ought to know that his presence is likely to do so; and
(d) his presence amounts to, or is likely to result in, harassment of, or alarm or distress to, the resident, a person in the resident's building or a person in another dwelling in the vicinity of the resident's dwelling.

This is a summary offence punishable with up to six months' imprisonment and/or a fine not exceeding level 4.

Stalking protection orders

9.41   The Stalking Protection Act 2019 (SPA 2019) provides for a civil order, the stalking protection order. Guidance about the Act has been issued by the Secretary of State.

A stalking protection order (SPO) is an order which, for the purpose of preventing D from carrying out acts associated with stalking, prohibits D from doing anything described in it, or requires D to do anything described in it (SPA 2019, s 1(2)). In s 1(2) and the rest of SPA 2019 'acts' includes omissions.

A SPO has effect for a fixed period (of at least two years) specified in the order, or until a further order; different periods may be specified in relation to different prohibitions or requirements.

Application for order

By SPA 2019, s 1, a chief officer of police for a police area or the Chief Constable of the **9.42** British Transport Police or Ministry of Defence Police (hereafter 'chief officer of police') may apply to a magistrates' court (or youth court) for a SPO in respect of a person (D) if it appears to the chief officer that:

(a) D has carried out acts associated with stalking, examples of which are those given by PHA 1997, s 2A (**9.34**), wherever (in a part of the UK or elsewhere) they were carried out and even if carried out before the commencement of SPA 2019, s 1;

(b) D poses a risk associated with stalking to another person; and

(c) there is reasonable cause to believe the proposed order is necessary to protect another person from such a risk (whether or not the other person was the victim of the acts mentioned in (a)).

A risk associated with stalking may be in respect of physical or psychological harm to the other person, and may arise from acts which D knows or ought to know are unwelcome to the other person even if, in other circumstances, the acts would appear harmless in themselves.

A chief officer of police may apply for a SPO only in respect of a person who resides in the chief officer's police area, or who the chief officer believes is in that area or is intending to come to it.

Power to make SPOs

SPA 2019, s 2 deals with this. **9.43**

A magistrates' court may make a SPO if satisfied that:

(a) D has carried out acts associated with stalking, wherever (in a part of the UK or elsewhere) they were carried out and even if carried out before the commencement of s 2;

(b) D poses a risk associated with stalking to another person; and

(c) the proposed order is necessary to protect another person from such a risk (whether or not the other person was the victim of the acts mentioned in (a)).

A magistrates' court may include a prohibition or requirement in a SPO only if satisfied that the prohibition or requirement is necessary to protect the other person from a risk associated with stalking. Prohibitions or requirements must, so far as practicable, be such as to avoid conflict with D's religious beliefs, and such as to avoid interference with any times at which D normally works or attends an educational establishment. A prohibition or requirement has effect in all parts of the UK unless expressly limited to a particular locality.

Variations, renewals, and discharges

SPA 2019, s 4 provides that D or a relevant chief officer of police may apply to a magis- **9.44** trates' court for an order varying, renewing or discharging a SPO. For these purposes, 'relevant chief officer of police' means:

(a) the chief officer of police for the area in which D resides;

(b) a chief officer of police who believes that D is in, or is intending to come to, that chief officer's police area; and

(c) the chief officer of police who applied for the SPO to which the application relates.

The court may make any order varying, renewing or discharging the SPO that the court considers appropriate. However, the court may not:

(a) in renewing or varying an order, impose an additional prohibition or requirement unless satisfied that it is necessary to do so in order to protect a person from a risk associated with stalking;

(b) discharge an order before the end of two years beginning with the day of the order without D's consent and—

   (i) where the application under s 4 was made by a chief officer of police, that chief officer; or

   (ii) in any other case, the chief officer of police who applied for the SPO and (if different) the chief officer of police for the area in which D resides, if that area is in England or Wales.

Interim SPOs

9.45 Where an application for a SPO (the 'main application') has not been determined, a magistrates' court (or youth court) may, by SPA 2019, s 5, make an interim SPO in respect of D on an application made at the same time and by the same chief officer of police as the main application, or, if the main application has already been made, on an application made by the chief officer of police who made that application. If it considers it appropriate to do so, the court may make an interim SPO prohibiting D from doing anything described in the order, or requiring D to do anything described in it. Prohibitions or requirements must, so far as practicable, be such as to avoid conflict with D's religious beliefs, and such as to avoid interference with any times at which D normally works or attends an educational establishment. A prohibition or requirement has effect in all parts of the UK unless expressly limited to a particular locality.

An interim SPO has effect only for a fixed period specified in the order, and ceases to have effect, if it has not already done so, on the determination of the main application.

D or the chief officer of police who applied for an interim SPO may apply to a magistrates' court for an order varying, renewing or discharging the interim SPO. On such an application, the court may make any order varying, renewing or discharging the interim SPO that it considers appropriate.

Breach

9.46 A person who, without reasonable excuse, breaches a SPO or an interim SPO commits an indictable (either-way) offence under SPA 2019, s 8(1). The offence is punishable with up to five years' imprisonment.

Notification requirements

9.47 SPA 2019, ss 9–11 make provision about notification requirements. They are set out in Chapter 9 on the companion website.

ATTEMPTED CHOKING, AND POISONING, ETC

9.48 OAPA 1861, ss 21 to 24 include other offences which are associated with the causing, or attempted causing, of forms of bodily harm. These offences, which are all triable only on indictment, are punishable with a maximum of life imprisonment (ss 21 and 22), 10 years' imprisonment (s 23), and five years' imprisonment (s 24).

Attempting to choke, etc

This offence, provided by OAPA 1861, s 21, consists of: **9.49**

(a) an attempt by D, by any means whatsoever, to choke, suffocate, or strangle any other person, or
(b) an attempt by D, by any means calculated to choke, suffocate, or strangle, to render any other person insensible, unconscious, or incapable of resistance,

with intent thereby to enable D or another person to commit any indictable offence, or with intent to assist another to do so.

Using chloroform, etc to commit an indictable offence

By OAPA 1861, s 22, D commits an offence if (s)he unlawfully applies or administers **9.50** to, or causes to be taken by, any person a stupefying or overpowering drug, matter, or thing, with intent thereby to enable him/herself or another person to commit any indictable offence, or with intent to assist another to do so.

Administering poison, etc so as thereby to endanger life, etc or with intent to injure, etc

OAPA 1861, s 23 provides that a person is guilty of an offence if he unlawfully and **9.51** maliciously administers to, or causes to be administered to or taken by, any other person any poison, or other destructive or noxious thing, *so as thereby to endanger the life of such person, or so as thereby to inflict upon such person any grievous bodily harm.*

OAPA 1861, s 24 provides that a person is guilty of an offence if he unlawfully and maliciously administers to, or causes to be administered to or taken by, any other person any poison, or other destructive or noxious thing, *with intent to injure, aggrieve, or annoy such person.*

A 'poison' means a recognised poison, in whatever quantity it may be administered, etc. A 'noxious thing' is any other drug or thing which is harmful in the dosage and manner in which it was administered, etc; 'harmful' is not limited to meaning 'injurious to health' but also includes 'unwholesome'. Noting these points, the Court of Appeal has held in a case under s 24 that urine can be a noxious substance.

'Administer' is not limited to the direct administration of the thing, eg injecting the victim (V), holding a glass containing a noxious thing to V's, or spraying a noxious thing at V. Where D acts in concert with V who self-injects a drug (for example by holding a tourniquet around V's arm while V self-injects or by preparing and giving V the drug for immediate self-injection), D does not *thereby* administer the drug. If the act of self-injection by V was not free, or not deliberate or not informed, as where V was forced to self-inject, D would cause the thing to be administered by V.

A poison, or other destructive or noxious thing, is caused to be taken if it is left for a person who then drinks it in ignorance of what it is or under duress. To leave it for the purpose of it being taken would be to attempt to cause it to be taken.

The consent of the person to whom the poison is given is no defence to charges under s 23 or s 24. For example, heroin can be a noxious substance for the purposes of these offences. If a shot of heroin is administered to another with his/her consent, this will amount to an administration.

The distinction between the offences under ss 23 and 24 lies partly in the fact that the offence under s 23 requires the additional element that the administration must be such as thereby to endanger life or to inflict grievous bodily harm, and partly in the fact that the requirement of mens rea is not the same for each offence. The differences between the two offences are shown by the words italicised in the definitions given above.

The mens rea required is as follows. In the case of the offence under s 23, the Court of Appeal has held that where endangering of life or grievous bodily harm is caused directly (eg by injection) the requirement of malice is satisfied by its deliberate administration; no mens rea is required to be proved as to the risk of injury. The Court stated that where the endangering of life, etc was caused indirectly (eg by causing the victim to inhale gas) foresight of injury must be proved. In the case of an offence under s 24 not only must an intentional or reckless administration, etc of a poison or other destructive or noxious thing be proved but also an intention to injure, aggrieve, or annoy.

GUNPOWDER, ETC OFFENCES

9.52   The Offences Against the Person Act 1861 (OAPA 1861), ss 28, 29, 30, and 64 create a number of offences related to the use or possession of explosives. They are triable only on indictment, and punishable with a maximum of life imprisonment (ss 28 and 29), 14 years' imprisonment (s 30), and two years' imprisonment (s 64).

Section 28 provides that anyone, who unlawfully and maliciously, by the *explosion* of gunpowder or other explosive substance, *burns, maims, disfigures, disables, or does any grievous bodily harm to any person*, is guilty of an offence.

Section 29 provides that anyone, who unlawfully and maliciously *causes an explosion, or sends or delivers to any person an explosive substance (or any other dangerous or noxious thing), or places or throws at any person any corrosive fluid or any destructive or explosive substance, with intent to burn, maim, disfigure, or disable any person, or to do some grievous bodily harm to any person*, is guilty of an offence, whether or not any bodily injury is effected.

Section 30 provides that it is an offence unlawfully and maliciously to *place or throw* in, into, upon, against, or near any building or vessel an *explosive substance with intent to do any bodily injury to any person, whether or not any explosion occurs and whether or not anyone is injured.*

It has been held that a petrol bomb is an explosive substance under s 29. This decision is equally applicable to the meaning of 'explosive substance' in ss 28 and 30.

Section 64 makes it an offence to make, manufacture, or knowingly possess any explosive substance or machine, engine, or instrument with intent to commit any offence under OAPA 1861, or for the purpose of enabling others to do so.

TORTURE

9.53   The offence of torture, which is triable only on indictment and punishable with a maximum of life imprisonment, is governed by the Criminal Justice Act 1988, s 134 (**6.48**).

By s 134(1), a public official or person acting in an official capacity, whatever his/her nationality, commits the offence of torture if (s)he intentionally inflicts severe physical or mental pain or suffering on another in the performance or purported performance of his/her official duties.

By s 134(2), a person not acting in such an official capacity commits the offence of torture if (s)he intentionally inflicts severe physical or mental pain or suffering on another at the instigation or with the consent or acquiescence of a public official or person acting in that capacity, and the official or other person is performing or purporting to perform his/her official duties when (s)he instigates the commission of the offence or consents to or acquiesces in it.

It is immaterial whether the pain or suffering is physical or mental and whether it is caused by an act or omission. It is a defence for a person to prove that (s)he had lawful authority, justification or excuse for that conduct.

KIDNAPPING

This offence, which is triable only on indictment and punishable with a maximum of life imprisonment, involves four requirements: **9.54**

(1) The taking or carrying away of one person by another

This can include the case where a person is induced to walk or drive away him /herself, if the other elements of the offence are fulfilled (as where a woman's car is hijacked and she drives away under the instructions of the hijackers). Although the person taken or carried away is often secreted thereafter, this is not a requirement of the offence.   **9.55**

(2) The taking or carrying away must be by force or by fraud

'Force' is not limited to physical force or the threat of it. It encompasses any conduct which, coupled with the taking or carrying away (requirement (1)), overrides the true consent of the person taken or carried away (requirement (3)). Thus, the exercise of mental or moral power or influence to compel another to do something against his/her will can suffice if it overcomes his/her will. Any fraud which induces the victim to consent to being taken or carried away will suffice, and will invalidate the victim's apparent consent.   **9.56**

(3) The taking or carrying away must be without the consent of the person taken or carried away

This requirement must be satisfied whatever the age of that person. However, there can be a kidnapping, even though the person carried away consents at first, if (s)he changes his/her mind and ceases to consent while still being carried away. A very young child does not have the understanding or intelligence to give consent so that the absence of consent will be a necessary inference from its age. In the case of an older child, it is a question of fact for the jury whether the child had sufficient understanding or intelligence to give consent and, if so, whether absence of consent has been proved. Unlike child abduction (**13.14–13.17**), the presence or absence of consent on the part of the person having custody or care and control of a child victim is immaterial (except that such consent may support a defence of lawful excuse). Likewise the presence or absence of consent on the part of other people, such as the spouse of a person who is taken, is immaterial.   **9.57**

It must be proved that a person charged with kidnapping knew or was reckless that (s)he did not have the victim's consent.

(4) The taking or carrying away must be without lawful excuse

9.58   Clearly, for example, someone with custody of a child will often have a lawful excuse for taking or carrying away the child. An exception would be where this contravenes a court order in relation to the child.

HOSTAGE-TAKING

9.59   A person of any nationality commits an offence (**6.48**) against the Taking of Hostages Act 1982, s 1(1) if (s)he detains any other person ('the hostage') and, in order to compel any state, international governmental organisation, or person to do or abstain from doing any act, (s)he threatens to kill, injures, or continues to detain the hostage. The offence is triable only on indictment and punishable with a maximum of life imprisonment.

EVICTION AND HARASSMENT

9.60   The Protection from Eviction Act 1977 (PEA 1977), s 1 provides three indictable (either-way) offences, punishable with up to two years' imprisonment, whose aim is to protect tenants: the offence of eviction and two offences of harassment. The Act specifically authorises district councils to institute proceedings for these offences.

The offences use the term 'residential occupier'. In relation to any premises, the term means a person occupying the premises as a residence, whether under a contract or by virtue of any enactment or rule of law giving him/her the right to remain in occupation or restricting the right of any other person to recover possession of the premises. Therefore, a tenant, or even a lodger who is living in a furnished room under an agreement with the owner, is a residential occupier. However, someone like a lodger who simply had a contractual licence which has expired is not a residential occupier. A squatter is not a residential occupier of the premises.

Eviction

9.61   PEA 1977, s 1(2) provides that if any person unlawfully deprives the residential occupier of any premises of his occupation of the premises or any part thereof, or attempts to do so, he is guilty of an offence unless he proves that he believed, and had reasonable cause to believe, that the residential occupier had ceased to reside in the premises.

It is not necessary that any form of violence or intimidation is used; it would be sufficient if the residential occupier was tricked into leaving so that the owner could regain occupation, or that the owner entered by stealth during the residential occupier's absence. The deprivation of occupation need not be permanent. Consequently, a person who unlawfully excludes a residential occupier from his/her premises, intending the exclusion to be permanent, can be convicted under s 1(2) even though (s)he repents almost immediately and lets the occupier back in.

The commission of the offence is not restricted to the owner or any person having an interest in the property; it can be committed by anyone (eg an employee of the owner) who unlawfully deprives, or attempts to deprive, a residential occupier of his/her premises.

It is a defence to prove a reasonable belief that the residential occupier has ceased to reside in the premises.

Harassment

PEA 1997, s 1(3) provides that if any person with intent to cause the residential occupier of any premises:

(a) to give up the occupation of the premises or any part thereof; or
(b) to refrain from exercising any right or pursuing any remedy in respect of the premises or part thereof;

does any act likely to interfere with the peace or comfort of the residential occupier or members of his household, or persistently withdraws or withholds services reasonably required for the occupation of the premises as a residence, he is guilty of an offence.

Provided that it is likely to interfere with the peace or comfort of the residential occupier, any *act*, such as intimidation, threats, or even interference with the building, perhaps by removing window frames or doors on the pretence that they are to be replaced, will do; the House of Lords has held that it is irrelevant that the act in question was not wrongful in civil law. Alternatively the offence can consist of a *lack of action* which results in services reasonably required for the occupation of the premises being persistently withdrawn or withheld, as where the landlord fails to pay for essential services to the premises occupied by the tenant. The use of the word 'persistently' which is used as an adverb to both 'withdraws' and 'withholds' indicates that this must be done for some period of time.

The intention must be to cause the *residential occupier* to give up permanently occupation of the premises or to refrain from exercising any right or pursuing any remedy in respect of the premises or part of them.

Proof of the requisite intention under s 1(3) may not always be easy. Consequently, s 1(3A) provides that the landlord of a residential occupier or an agent of the landlord commits an offence if he does acts likely to interfere with the peace or comfort of the residential occupier or members of his household, or if he persistently withdraws or withholds services reasonably required for the occupation of the premises, and (in either case) *he knows or has reasonable cause to believe that this conduct is likely to cause* the residential occupier to give up occupation of the whole or part of the premises or to refrain from exercising any right or pursuing any remedy in respect of the whole or any part of the premises. A person is not guilty under s 1(3A) if (s)he proves that (s)he had reasonable grounds for doing the acts or withdrawing or withholding the services in question.

Caravans: eviction and harassment

The Caravan Sites Act 1968 (CSA 1968), s 3 provides similar indictable (either-way) offences to those under PEA 1977 in relation to a residential caravan on a protected site. The Act only applies in England. The Mobile Homes (Wales) Act 2013 makes similar provision in Wales. See Chapter 9 on the companion website.

CONTAMINATION OF GOODS, ETC WITH INTENT

The Public Order Act 1986 (POA 1986), s 38(1) provides that it is an indictable (either-way) offence for a person to contaminate or interfere with goods, or make it appear that goods have been contaminated or interfered with, or to place goods which have been contaminated or interfered with, or which have that appearance, in a place where goods of that description are consumed, used, sold or otherwise supplied, with the intention of causing:

(a) public alarm or anxiety;

(b) injury to members of the public consuming or using the goods; or

(c) economic loss to any person by reason of (i) the goods being shunned by members of the public, or (ii) steps taken to avoid any such alarm or anxiety, injury or loss.

Section 38(2) makes it an indictable (either-way) offence for a person to threaten that (s)he or another will do, or claim that (s)he or another has done, any of the acts referred to in s 38(1) with such intention as is mentioned in (a) or (c) above. This does not include someone who, in good faith, reports or warns that such acts have been, or appear to have been, committed. Thus, the broadcast of a warning received, if carried out in good faith, is excused.

By s 38(3), possession of contaminated goods, apparently contaminated goods, or materials with which to contaminate goods or to make it appear that goods have been contaminated, is also an indictable (either-way) offence.

Thus, the activities of groups, calculated to hit at businesses with which they are not in sympathy, are made punishable. However, it must be remembered that such activities may also amount to attempts to commit offences (for example, attempted murder, if poisons are placed in foodstuffs with an intention of killing). In addition, if someone actually consumes a contaminated product and suffers harm, there will be liability for the relevant 'full' offence against the person, depending on the degree of harm.

An offence under s 38 is punishable with up to 10 years' imprisonment.

SENDING THREATENING LETTERS OR OTHER COMMUNICATIONS

9.65   See **12.35–12.36**.

CHAPTER 10

Homicide and Related Offences

Murder, manslaughter, infanticide, and corporate manslaughter are among the offences dealt with in this chapter. The various 'death by driving' offences are dealt with in Chapters 24 and 26. **10.1**

MURDER

Murder is a common law offence triable only on indictment and punishable with a mandatory sentence of life imprisonment. It is committed where someone unlawfully kills another human being, who is under the Queen's peace, with intent unlawfully to kill or cause grievous bodily harm. **10.2**

In three types of exceptional case a person is not guilty of murder, but only of voluntary manslaughter, even though the definition of murder is satisfied. Those exceptions are where the defendant (D): was suffering from diminished responsibility, has the defence of loss of control or was acting in pursuance of a suicide pact.

With the exception of the reference to D's intent, the terms of the definition of murder also apply to involuntary manslaughter, which is discussed later in this chapter.

Unlawfully

A killing is unlawful unless it falls within one of the following categories: **10.3**

Prevention of crime or effecting arrest

It is lawful to use such force as is *reasonable* in the circumstances as D believed them to be in the prevention of a crime under the law of England and Wales or in effecting (or assisting in) the lawful arrest of offenders, suspected offenders or persons unlawfully at large. See, further, **9.24** and **9.27**. **10.4**

Self-defence and defence of another or of property

Self-defence and defence of another or of property render a killing lawful if the force used is *reasonable* in the circumstances as D believed them to be. See, further, **9.25** and **9.27**. **10.5**

Misadventure

Death is caused by misadventure where the killing is not murder, manslaughter, or any other offence of homicide. As an example, if a patient dies as a result of a lawful operation carried out by a surgeon with proper care, the killing is by misadventure, and so not unlawful, and therefore the surgeon is not guilty of any offence of homicide. **10.6**

Kills

10.7   Although a failure to act can suffice for homicide if D was under a legally recognised duty to act, some form of action which proves to be a substantial cause of death is generally required. 'Substantial' in this context simply means 'more than minimal'; D's conduct need not be the sole or major cause of death.

If D's original act was a substantial cause of death it does not matter that some other intervening event finally caused the death, provided that this intervening event (as opposed to its details) was reasonably foreseeable in the ordinary course of things. If D knocks V unconscious on the beach and V is killed by the incoming tide, D's act is regarded in law as the cause of V's death because the incoming tide was clearly reasonably foreseeable. On the other hand, if two men (D and V) fight in a park and D leaves V unconscious but not seriously hurt and V is later killed by a tree which falls upon him when it is blown down by a gale, the fact that the fall of the tree was not reasonably foreseeable means that D's act is not regarded in law as a cause of V's death. A similar rule applies where D threatens someone who takes evasive action and is killed in doing so, as where a woman jumps out of a first-floor window to avoid rape. Provided that the evasive action is likely, the threat will be a cause of death if it was a substantial cause (as almost inevitably it will be).

If the injuries inflicted by D are an operating cause of death it is irrelevant that an intervening act by a third party also contributed to the death, provided that D's act was a substantial contribution. Suppose that D causes such serious injuries to V that V can only be kept alive by a life-support machine. If, subsequently, because there is no hope of recovery, doctors switch off that machine, D has still killed that person as his act was a substantial and operating cause of death. Suppose, on the other hand, that D's act is not an operating cause of death but merely provides the setting in which an intervening act by a third party is the immediate cause of death. D's act will not be a legal cause of death unless the third party's act is not free or not deliberate or not informed. In one case D, in attempting to evade arrest, snatched a girl in front of him as he fired at police officers. The officers fired back instinctively, killing the girl. Although the officers' act resulted in the girl's death, it was held that D's act was in law a cause of the girl's death; the officers' reaction was instinctive (and therefore not deliberate) and was done in self-defence (and therefore not free). By way of qualification, the Court of Appeal has approved a statement that even an accidental or unintended intervention may break the chain of causation if it was not reasonably foreseeable in the circumstances. This does not mean that the exact form of any such intervention must have been foreseeable at the time of the original assault, etc in order for the chain of causation to remain unbroken. If the general form and risk of further harm was reasonably foreseeable, it may not then matter if the specific manner in which it occurred was entirely unpredictable.

Even where negligence in the treatment of a victim was the immediate cause of death, this does not exclude D's responsibility, unless the original injury had ceased to be an operative cause of death and the negligent treatment was so independent of D's acts and itself so potent in causing death that D's contribution was insignificant.

Sometimes a victim contributes to his/her death by doing something (other than taking evasive action, dealt with above) after D's act. The normal rule here is that if the victim's intervening act is not free, or is not deliberate, or is not informed, D's act will be a legal cause of death, whereas if the victim's act was free, deliberate and informed D's act will not be a legal cause of death.

If an act is committed which leads to death, it is legally immaterial that the person injured had a medical condition rendering him more susceptible to death (eg haemophilia or a weak heart) or refused medical treatment. A person who commits violent acts must take his/her victim as (s)he finds him/her, including the victim's mental condition or, even, the victim's religion. Where a girl, who was a Jehovah's witness, was stabbed by an assailant and refused a blood transfusion required before surgery and died, the act of her attacker was held to be a cause of her death.

In addition to the above rules, where D was under a legally recognised duty to V to act (as where a parent or similar person has the care of a child or helpless person), and fails to fulfil that duty (eg by not seeking medical attention for V when V becomes ill) and V dies, D can be found to have caused V's death only if V *would* (not just might) have survived if D had fulfilled his/her duty to V.

The victim

Any human being, however deformed or subnormal, is protected by the law of homicide.   **10.8**

This raises the question of the point of time at which a foetus becomes a human being. The law of homicide only protects those who have been born alive, and this requires that a child must have completely emerged into the world and have a separate existence from its mother. To have had that separate existence it is not essential that the cord has been severed or even that the afterbirth has been expelled, but it must have breathed.

The wilful destruction of a child capable of being born alive before it is born alive may amount to the offence of child destruction (**10.32**), while the intentional procuring of a miscarriage may constitute the offence of abortion (**10.33**). If D injures a pregnant woman, and as a result of the attack she goes into premature labour and her child, although born alive, subsequently dies owing to its prematurity, D is guilty of manslaughter but cannot be convicted of murder despite the fact that D intended to kill the woman or seriously harm her. It would make no difference that D intended also to destroy the foetus in the womb because such an intent does not suffice for murder.

Under the Queen's peace

For the purposes of homicide offences, all persons are under the Queen's peace whether   **10.9**
they are Her Majesty's subjects or not, with the exception of alien enemies (and possibly rebel subjects) in the actual heat and exercise of war.

Lapse of time between act and death

An offence of homicide may be committed by a person who carries out an appropriate   **10.10**
act with the necessary mens rea regardless of the time which has passed since the injury, etc was inflicted. However, where the injury alleged to have caused the death was sustained more than three years before the death occurred, or where the person whom it is intended to prosecute for an offence of homicide has already been convicted of an offence in circumstances alleged to be connected with the death, no prosecution for an offence of homicide may be brought without the consent of the Attorney General. These provisions also apply to the offences of encouraging or assisting suicide and of causing or allowing the death of a child or vulnerable adult.

Intent unlawfully to kill or cause grievous bodily harm

10.11   The mens rea required for murder is an intent unlawfully to kill another human being or unlawfully to cause grievous bodily harm to another human being. This is described as 'malice aforethought'. 'Malice' is a misleading term because an intentional killing motivated by compassion (a mercy killing) is murder unless reduced to manslaughter on grounds of diminished responsibility. 'Aforethought' is also misleading because it suggests that the killing must have been premeditated, which is certainly not a legal requirement. Provided that D's fatal act was done with intent unlawfully to kill or cause grievous bodily harm, it is irrelevant that D acted on the spur of the moment, the intention only being formed a brief second before the killing. A person who intends to kill or cause grievous bodily harm will not intend to do so unlawfully if on the facts, as (s)he believes them, his/her use of force is reasonable to prevent crime or in self-defence.

MANSLAUGHTER

10.12   Manslaughter is a term which covers a variety of unlawful homicides which do not amount to murder. Manslaughter continues to be a common law offence triable only on indictment and punishable with a maximum of life imprisonment. There are two varieties: voluntary and involuntary manslaughter.

Voluntary manslaughter

10.13   Voluntary manslaughters embody all the characteristics of murder including the necessary malice aforethought. It is the presence of particular circumstances acting upon the mind of the defendant, which has the effect of reducing the nature of the crime to voluntary manslaughter. These circumstances are: where the person was acting under 'loss of control' at the material time; where the person was suffering from 'diminished responsibility' at that time; and where a killing occurred in consequence of a suicide pact.

Loss of control

10.14   The Coroners and Justice Act 2009 (C&JA 2009), s 54(1) provides that where a person (D) kills or is a party to the killing of another (V), D is not to be convicted of murder if:

(a) D's acts and omissions in doing or being a party to the killing resulted from D's loss of self-control,

(b) the loss of self-control had a qualifying trigger, and

(c) a person of D's sex and age, with a normal degree of tolerance and self-restraint and in the circumstances of D, might have reacted in the same or in a similar way to D.

D does not have to prove the defence. If sufficient evidence is adduced to raise an issue with respect to the defence under s 54(1), the jury must assume that the defence is satisfied unless the prosecution proves beyond reasonable doubt that it is not.

The fact that one party to a killing is by virtue of the defence of loss of control not liable to be convicted of murder does not affect the question whether the killing amounted to murder in the case of any other party to it.

Section s 54(2) provides that it does not matter whether or not the loss of self-control was sudden. However, by s 54(4) the defence does not apply if, in doing or being a party to the killing, D acted in a considered desire for revenge.

Qualifying trigger   D's loss of self-control must have had a qualifying trigger. By s 55(2)–(5), a loss of self-control would have a qualifying trigger if D's loss of self-control was attributable to:

(a) D's fear of serious violence from V against D or another identified person; or
(b) a thing or things done or said (or both) which—
   (i)   constituted circumstances of an extremely grave character, and
   (ii) caused D to have a justifiable sense of being seriously wronged; or
(c) a combination of the above matters.

There are two limits to the first trigger (ie (a)):

- the fear of serious violence must be of violence from V, the person killed; and
- the fear must be of such violence to D or some other identified person.

In determining whether a loss of self-control had a qualifying trigger, D's fear of serious violence must be disregarded to the extent that it was caused by a thing which D incited to be done or said for the purpose of providing an excuse to use violence.

The second trigger (ie (b)) deals with cases of gross provocation. In determining whether a loss of self-control had a qualifying trigger:

- a sense of being seriously wronged by a thing done or said is not justifiable if D incited the thing to be done or said for the purpose of providing an excuse to use violence;
- the fact that a thing done or said constituted sexual infidelity must be disregarded. (However, the Court of Appeal has held that, although sexual infidelity must be disregarded where it is the only element in support of a qualifying trigger, sexual infidelity may be relied on when it is not the only element relied on in support and is integral to and forms an essential part of the context.)

Objective requirement   Assuming the other two requirements in s 54(1) are satisfied, D is not guilty of murder if a person of D's sex and age, with a normal degree of tolerance and self-restraint and in the circumstances of D, might have reacted in the same or in a similar way to D. As can be seen this requirement is not wholly objective. It is not simply concerned with whether someone with a normal degree of tolerance and self-restraint might have reacted as D did (ie might (s)he have reacted, and if so might (s)he have reacted as D did?) but also whether a person of D's sex and age (the latter in particular may be relevant to tolerance and self-restraint) and in D's circumstances might have reacted as D did. However, not all D's circumstances can be taken into account. This is because s 54(3) provides that the reference in s 54(1)(c) to 'the circumstances of D' is a reference to all of D's circumstances other than those whose only relevance to D's conduct is that they bear on D's general capacity for tolerance or self-restraint. The closing words of s 54(3) make it clear that characteristics or other circumstances whose only relevance to D's conduct is that they bear on D's general capacity for tolerance or self-restraint cannot be referred to in applying the present requirement. Thus, the fact that D was intoxicated or intolerant or irritable or excessively jealous or had problems in controlling his impulses or otherwise had impaired powers of self-control must be ignored.

Diminished responsibility

The Homicide Act 1957, s 2 governs this defence. Section 2(1) provides that a person (D) who kills or is a party to the killing of another is not to be convicted of murder if D proves that he was suffering from an abnormality of mental functioning which: **10.15**

(a) arose from a recognised medical condition;

(b) substantially impaired D's ability to do one or more of the things mentioned below; and

(c) provides an explanation for D's acts and omissions in doing or being a party to the killing.

The things referred to in (b) are set out in s 2(1A). They are:

(i) to understand the nature of D's conduct;

(ii) to form a rational judgment;

(iii) to exercise self-control.

As regards (c), by s 2(1B), an abnormality of mental functioning provides an explanation for D's conduct if it causes, or is a significant contributory factor in causing, D to carry out that conduct. This means that abnormality of mental functioning need not be the sole cause of D's conduct. Thus, for example, the fact that other explanations or causes were also operative does not in itself negate the defence of diminished responsibility.

Although acute intoxication is a recognised medical condition, the Court of Appeal has held that the defence of diminished responsibility cannot be based on it.

Suicide pacts

10.16   The Homicide Act 1957, s 4 declares that it is manslaughter, and not murder, for a person to kill another or be party to someone else killing another, if (s)he was acting in pursuance of a suicide pact between him/herself and the person killed. This is a matter for the defence to prove.

A 'suicide pact' is a common agreement between two or more persons, having for its object the death of them all, whether or not each is to take his/her own life. Nothing done by a person entering into a suicide pact may be treated as in pursuance of such a pact unless it is done while (s)he has the settled intention of dying in pursuance of the pact.

Involuntary manslaughter

10.17   This category covers cases where D, who has unlawfully killed another (ie has committed the actus reus of murder), is not guilty of murder because (s)he lacked malice aforethought but acted with some lesser degree of mens rea.

There are three types of involuntary manslaughter, between which there is a degree of overlap:

(a) killing by an unlawful and dangerous act;

(b) killing by gross negligence; and

(c) killing with recklessness as to death or serious bodily harm.

Killing by an unlawful and dangerous act

10.18   This mode of committing manslaughter is commonly known as 'constructive manslaughter'. It cannot be committed by an omission to act; an unlawful act by D is required. Three elements must be proved by the prosecution:

(1) *That D has committed the actus reus of an offence (other than homicide) with the mens rea required for that offence*; proof of this is proof of the 'unlawful act'. In most cases the offence will be a battery, but constructive manslaughter is certainly not limited to that offence; for example it is not unusual for a constructive manslaughter conviction to be based on an offence of administering a noxious thing

(9.51). However, the mere supplying of a noxious thing to another who then freely, deliberately and voluntarily injects himself with it will not amount to an offence of administration of a noxious thing and therefore is not an unlawful act and cannot make the supplier guilty of constructive manslaughter. Dangerous driving or careless driving which results in death can never constitute constructive manslaughter.

(2) *That the unlawful act was dangerous.* For an unlawful act to be dangerous it must be such that all sober and reasonable people would inevitably recognise that it must subject another person to the risk of some physical harm (including physical harm—eg a heart attack—resulting from shock), albeit not serious harm. This element is applied on the basis of the facts known to D at the time of his/her unlawful act and of any other facts of which a reasonable person would have been aware or, if the act continues over a period of time, of facts which became known during that period. The Court of Appeal has held that a fatal unlawful act which is not itself a crime of violence or dangerous can be dangerous for the purposes of underpinning an offence of constructive manslaughter because of the circumstances surrounding its commission.

(3) *That the unlawful and dangerous act was a cause of death.* It is not enough that some other act by D caused the death. As to the requirement of causation, see **10.7**.

Killing by gross negligence

This type of manslaughter may be committed by an act or by a failure to act (if D has failed in breach of a legal duty to do the act); it is irrelevant whether or not D's act or omission would have constituted an offence if death had not resulted. Manslaughter by gross negligence can no longer be committed by an organisation to which the offence of corporate manslaughter (**10.21**) applies. **10.19**

The requirements of manslaughter by gross negligence are:

(1) *The existence of a duty of care.* Such a duty arises when it is reasonably foreseeable that negligence will cause injury to the victim (V); there is a relationship of sufficient proximity between D and V, and it would be 'just and reasonable' to impose liability.

(2) *A gross breach of duty.* Normally, proof of negligence simply involves proof that, whether or not (s)he realised the risk (of which (s)he should have been aware), the person subject to the duty did something, or failed to do something, in a way which fell below the standard of conduct expected of a reasonable person in all the circumstances (including D's expertise and training, if they are relevant in the context). This is not enough in the case of gross negligence. For there to be gross negligence, D's conduct must:
   (a) have involved reasonable foreseeability of an obvious and serious risk of death to another, in assessing which account should not be taken of what D did not know but would have known but for his/her breach of duty unless the 'warning signs' were there for 'all to know'; and
   (b) in respect of that risk D's conduct must have fallen so far below the standard to be expected of a reasonable person, ie be so bad, that it should be judged criminal.

(3) *The breach must cause death.* This simply repeats the requirement of causation.

Killing with recklessness as to death or serious bodily harm

In this context, a person is reckless as to a risk of death or serious bodily harm if (s)he foresees that risk as a highly probable consequence of the fatal act or omission and (s)he takes that risk, and in all circumstances it is unreasonable for him/her to do so. **10.20**

The present type of involuntary manslaughter will often overlap with constructive manslaughter and manslaughter by gross negligence, but it will not do so where the fatal act is not otherwise unlawful and there is no risk of death.

CORPORATE MANSLAUGHTER

The offence

10.21   The Corporate Manslaughter and Corporate Homicide Act 2007 (CMCHA 2007) governs the offence of corporate manslaughter (**6.48**) which is triable only on indictment and punishable with an unlimited fine. CMCHA 2007, s 1(1) provides that an organisation to which s 1 applies is guilty of an offence if the way in which its activities are managed or organised:

(a) causes a person's death; and
(b) amounts to a gross breach of a relevant duty of care owed by the organisation to the deceased.

By s 1(3), an organisation is guilty of corporate manslaughter only if the way in which its activities are managed or organised by its senior management is a substantial element in the breach referred to in s 1(1). Because the offence requires a gross breach of a duty of care this offence is one requiring proof of gross negligence within its terms.

Organisations to which CMCHA 2007, s 1 applies

10.22   Section 1 applies to:

(a) a corporation;
(b) a government department or other public body listed in CMCHA 2007, Sch 1;
(c) a police force; and
(d) a partnership, or a trade union or employers' association, that is an employer (s 1(2)).

Senior management

10.23   'Senior management', in relation to an organisation, means the persons who play significant roles in: (a) the making of decisions about how the whole or a substantial part of its activities are to be managed or organised; or (b) the actual managing or organising of the whole or a substantial part of those activities.

Relevant duty of care

10.24   Section 2 states that, subject to the exclusions referred to below, a 'relevant duty of care', in relation to an organisation, means any of the following duties owed by it under the law of negligence, ie a duty owed:

(a) to employees to provide safety at work;
(b) as occupier of buildings or land to provide for the safety of persons on the premises;
(c) in connection with the use or keeping of any plant, vehicle or other thing;
(d) in connection with supplying goods or services;
(e) in connection with carrying on any other activity on a commercial basis;
(f) in connection with carrying on construction or maintenance operations; or
(g) to a person who is held in detention. 'Detention' means detention in a prison or similar establishment, or in a custody area at a court, police station or customs

premises, or detention in service custody premises, or detention in immigration detention facilities, or transportation under escort to such places, or placement in secure accommodation for children and young persons, or detention under mental health provisions.

Whether or not a duty of care was owed to a particular individual is a matter of law to be decided by a judge.

Exclusions

Sections 3 to 7 set out exclusions from the 'relevant duty of care'. **10.25**

Decisions relating to public policy taken by public authorities such as government departments and local authorities are excluded, as are activities involved in carrying out statutory inspections unless the duty is connected to the organisation's duty as an employer or occupier of premises.

Other exclusions include those relating to policing and law enforcement. Section 5 provides that any duty of care owed by the police and other enforcement agencies in respect of:

(a) operations for dealing with terrorism, civil unrest or serious disorder, which involve policing or law enforcement activities in the course of which officers or employees of the authority come under attack or face the threat of attack or violent resistance;
(b) activities carried on in preparation for, or directly in support of, such operations; or
(c) training of a hazardous nature, or training carried out in a hazardous way, which it is considered needs to be carried out, or carried out in that way, in order to improve or maintain the effectiveness of officers or employees of the public authority with respect to such operations,

is not a 'relevant duty of care'. Section 5 also provides that any duty of care owed by the police and other law enforcement agencies in respect of policing and law enforcement is not a 'relevant duty of care' unless it falls within (a), (b) or (g) at **10.24**.

Exclusions from the relevant duty of care exist for fire and rescue services, NHS bodies, ambulance services and the armed forces in respect of the way in which they respond to emergencies, and for child-protection and probation functions.

Gross breach of duty

By s 1(4)(b), a breach of a duty of care by an organisation is 'gross' if the conduct alleged **10.26** to amount to a breach of that duty falls far below what can reasonably be expected of the organisation in the circumstances.

INFANTICIDE

The Infanticide Act 1938, s 1 states that where a woman by any wilful act or omission **10.27** causes the death of her child, being a child under the age of 12 months, but at the time of that act or omission the balance of her mind was disturbed by reason of not having fully recovered from the effect of giving birth to the child, or by reason of the effect of lactation consequent on the birth of her child, then if the circumstances were such that, but for the Infanticide Act 1938, the offence would have amounted to murder or manslaughter she is guilty of infanticide, an offence triable only on indictment and punishable with a maximum of life imprisonment. The Court of Appeal has held that s 1 does not require the disturbance of the balance of the mind to be caused solely by a failure to fully recover from the effect of giving birth or lactation; so long as a failure to recover from that effect or lactation is an operative or substantial cause of the disturbance of the balance of the mind that will be sufficient, even if other underlying mental problems are part of the overall picture.

The basis of the law on infanticide is that depression after childbirth, or the effect of breast-feeding a child, are factors which can cause a mother to act out of character by committing some wilful act, or omitting to do something which a caring mother would do, that act or omission leading to the death of a child.

Infanticide may be charged in the first instance as an offence triable only on indictment or may be raised as a defence to a charge of murder or manslaughter.

CAUSING OR ALLOWING A CHILD OR VULNERABLE ADULT TO DIE OR SUFFER SERIOUS PHYSICAL HARM

10.28   In cases of child abuse, or of abuse of a vulnerable adult, where the victim has died or been injured as a result of the act or default of someone in the same household, it is sometimes difficult to prove who perpetrated the crime, or to prove that either A aided and abetted B to do so or B aided and abetted A to do so (so that both can be convicted of an offence of homicide described above or a non-fatal offence, as the case may be).

This problem is resolved by the Domestic Violence, Crime and Victims Act 2004, s 5. Section 5(1) provides that a person (D) is guilty of an offence if:

(a) a child under 16 or vulnerable adult (V) dies or suffers serious physical harm as a result of the unlawful act of a person who:
    (i) was a member of the same household as V, and
    (ii) had frequent contact with him;
(b) D was such a person at the time of that act;
(c) at that time there was a significant risk of serious physical harm being caused to V by the unlawful act of such a person; and
(d) either D was the person whose act caused the death or serious physical harm, or:
    (i) D was, or ought to have been, aware of the risk mentioned in (c) above,
    (ii) D failed to take such steps as (s)he could reasonably have been expected to take to protect V from the risk, and
    (iii) the act occurred in circumstances of the kind that D foresaw or ought to have foreseen.

The prosecution does not have to prove whether it is the first alternative in (d) or the second ((i)–(iii)) that applies.

There are two offences under s 5(1), one where death has occurred, and one where only serious physical harm has occurred.

If D was not the mother or father of V:

(a) D may not be charged with an offence under s 5 if (s)he was under 16 at the time of the act that caused the death or serious physical harm;
(b) for the purposes of (d)(ii), D could not have been expected to take any such step as is referred to there before attaining 16.

For the purposes of s 5:

(i) 'act' includes a course of conduct and also includes omission;

(ii) an 'unlawful' act is one that constitutes an offence, or (except where the act is by D) would constitute an offence but for being the act of a person under 10 or of a person who has the defence of insanity;

(iii) 'vulnerable adult' means a person (V) aged 16 or over whose ability to protect him/ herself from violence, abuse or neglect is significantly impaired through physical or mental disability or illness, through old age or otherwise; and

(iv) a person is to be regarded as a 'member' of a particular household, even if (s)he does not live in that household, if (s)he visits it so often and for such periods of time that it is reasonable to regard him/her as a member of it.

The Court of Appeal has held that 'or otherwise' at the end of (iii) refers to impairment of self-protection from some other cause which significantly impaired V's ability to protect him/herself from violence, abuse, or neglect. The cause might be physical, psychological, or arise from V's circumstance. The Court added that 'utter dependency' was not required; a victim of long-term sexual or domestic abuse or modern slavery, for example, might have been left scared, cowed, and with a significantly impaired ability to protect him/herself.

It is important to note that s 5 is not limited to the case where it is unclear which of two defendants killed or seriously harmed the vulnerable person. Thus, it can apply to make D guilty under it where it is clear that another person killed or seriously harmed a vulnerable person and that D failed to protect that person.

An offence under s 5(1) is triable only on indictment and is punishable with up to 14 years' imprisonment (if V died) or 10 years' (if V suffered serious physical harm).

THREATS TO KILL

10.29 By the Offences Against the Person Act 1861 (OAPA 1861), s 16 it is an indictable (either-way) offence punishable with up to 10 years' imprisonment for any person, without lawful excuse, to make to another person a threat to kill him/her or a third person, intending that that other person would fear that it would be carried out. An unborn child is not 'a third person' for the purposes of s 16.

Self-defence can amount to lawful excuse for a threat provided that the threatened force is reasonable in the circumstances as the maker of the threat believes them to be.

SOLICITATION OF MURDER

10.30 It is an offence contrary to OAPA 1861, s 4 for a person to solicit, encourage, persuade, or endeavour to persuade, or to propose to any person, to murder any other person. There must be some form of communication and this may be in any form. It is not essential that the person solicited, etc was affected by the communication. The offence is triable only on indictment and punishable with up to life imprisonment.

ENCOURAGING OR ASSISTING SUICIDE

10.31 The Suicide Act 1961 (SA 1961) provides as follows.

By SA 1961, s 2(1), a person (D) commits an offence (**6.48**) if:

(a) D does an act capable of encouraging or assisting the suicide or attempted suicide of another person; and

(b) D's act was intended to encourage or assist suicide or an attempt at suicide.

Section 2(1A) provides that the person referred to in (a) need not be a specific person (or class of persons) known to, or identified by, D. Thus, provided the necessary mens rea can be proved, a person who posts information about how to commit suicide to a suicide chat room may be convicted of the present offence. By s 2(1B), D may commit an offence under s 2 whether or not a suicide, or an attempt at suicide, occurs.

Section 2A elaborates on what constitutes an 'act capable of encouraging or assisting the suicide or attempted suicide of another person'. It provides that:

(a) if D arranges for D2 to do an act that is capable of encouraging or assisting the suicide or attempted suicide of another person and D2 does that act, D is also to be treated as having done it. Thus, if D tells D2 to assist in the commission of suicide by T, and D2 does so, D is treated as having done the act of assistance done by D2. Both of them can be convicted of an offence of encouraging or assisting suicide;

(b) where the facts are such that an act is not capable of encouraging or assisting suicide or attempted suicide, it is to be treated as so capable:

    (i) if the act would have been so capable had the facts been as D believed them to be at the time of the act (as where D wrongly believes that the harmless drugs which (s)he supplies to assist another's suicide are lethal); or

    (ii) had subsequent events happened in the manner D believed they would happen (as where D posts a lethal drug to another to assist the other's suicide but this is lost in the post); or

    (iii) both;

(c) a reference to a person (D) doing an act that is capable of encouraging the suicide or attempted suicide of another person includes a reference to D doing so by threatening another person or otherwise putting pressure on another person to commit or attempt suicide.

A reference to an act in SA 1961 includes a reference to a course of conduct, and a reference to doing an act is to be read accordingly.

The offence is triable only on indictment. The maximum punishment is 14 years' imprisonment.

CHILD DESTRUCTION

10.32   The Infant Life (Preservation) Act 1929 (ILPA 1929), s 1(1) provides that an offence triable only on indictment is committed by a person who, with intent to destroy the life of a child capable of being born alive, by any wilful act causes a child to die before it has an existence independent of its mother. The maximum sentence is life imprisonment.

A child is capable of being born alive when it has reached a state of development in the womb in which it is capable, if born then, of living and breathing through its own lungs without any connection with its mother. ILPA 1929 provides a presumption that a child is capable of being born alive at any time after the 28th week of pregnancy. However, the offence can be committed in relation to a younger child if it is proved that it was capable of being born alive. Provided that the child was capable of being born alive, it is irrelevant that it is not capable of sustained survival.

There is a proviso to the offence, namely that a person is not guilty of it unless it is proved by the prosecution that D did not act in good faith for the purpose only of preserving the life of the mother. This proviso has been construed by the judges as including acting to preserve the mother's physical or mental health. In addition, the Abortion Act 1967 (AA 1967), s 5(1) provides that no offence under ILPA 1929 is committed by a registered medical practitioner who terminates a pregnancy in accordance with the provisions of AA 1967 (**10.37**).

ABORTION

The OAPA 1861, s 58 makes it an offence to do one of a number of specified acts with intent to procure a miscarriage, ie the expulsion of an ovum implanted in the uterus (whether or not this occurs). Because there must be such an intent, someone who does something to *prevent* the implantation of a fertilised ovum in the uterus (as where the 'morning-after' pill is used) does not commit an offence under s 58 because a pregnancy only begins, and a miscarriage can only occur, after implantation. An offence under s 58 is triable only on indictment and punishable with up to life imprisonment. The common description of such an offence as abortion is somewhat misleading because the relevant offence does not require the abortion (miscarriage) of a foetus. **10.33**

The woman herself

Section 58 states that it is an offence for a woman, *being with child* and with intent to procure her own miscarriage, unlawfully to administer to herself any poison or other noxious thing, or unlawfully to use any instrument or other means whatsoever. **10.34**

Any other person

Section 58 goes on to provide that anyone (other than the woman herself) who, with intent to procure the miscarriage of any woman, unlawfully administers to her, or causes to be taken by her, any poison or other noxious thing, or who, with the same intent, unlawfully uses any instrument or other means whatsoever, commits an offence, *whether or not the woman is pregnant*. **10.35**

The means

'Poison' has been defined as a recognised poison. If such a thing is taken, etc, it is irrelevant that the quantity is too small to cause harm. The term 'noxious thing' means any substance, other than a recognised poison, which is harmful in the dosage in which it was administered even though it might be harmless in smaller quantities. Clearly 'noxious thing' is a wide term. If a dosage is insufficient to render a substance a noxious thing, although D believes it is, there can be a conviction for an attempt to commit an offence under s 58. The term 'instrument' covers the range of surgical instruments usually associated with medical operations and also instruments such as knitting needles. The phrase 'other means whatsoever' embraces any other way in which a person may seek to bring about an abortion, such as manual manipulation with the fingers. **10.36**

Unlawfully: the effect of the Abortion Act 1967

By AA 1967, s 5(2) anything done with intent to procure a woman's miscarriage (or, in the case of a woman carrying more than one foetus, her miscarriage of any foetus) is unlawfully done unless authorised by s 1. **10.37**

Section 1 legalises abortions (including abortion operations which are unsuccessful or not completed) carried out by a registered medical practitioner where two registered medical practitioners are of the opinion formed in good faith that:

(a) the pregnancy has not exceeded its 24th week and that the continuance of the pregnancy would involve risk, greater than if the pregnancy was terminated, of injury to the physical or mental health of the pregnant woman or any existing children of her family (a question in the determination of which account may be taken of the pregnant woman's actual or reasonably foreseeable environment); or

(b) the termination is necessary to prevent *grave* permanent injury to the physical or mental health of the pregnant woman (a question in the determination of which account may be taken of the woman's actual or reasonably foreseeable environment); or

(c) the continuance of the pregnancy would involve risk to the life of the pregnant woman, greater than if the pregnancy was terminated; or

(d) there is a substantial risk that if the child were born it would suffer from such physical or mental abnormalities as to be seriously handicapped.

In order to be lawful, any treatment for the termination of pregnancy must be carried out in a NHS hospital, NHS Trust hospital (in Wales), NHS Foundation Trust hospital, or other approved place.

In an emergency any treatment for the termination of pregnancy may be carried out by a registered medical practitioner without complying with the above requirements if it is necessary to do so immediately to save the life of a pregnant woman, or to prevent grave permanent injury to her physical or mental health. This might occur during an operation upon the woman or at the scene of, or immediately following, a serious road accident.

The person supplying or procuring the means

10.38   OAPA 1861, s 59 punishes those who unlawfully supply or procure any poison or other noxious thing, or any instrument or physical thing whatsoever, knowing that it is intended to be unlawfully used or employed with intent to procure the miscarriage of any woman, whether she be or be not with child. The offence is triable only on indictment and punishable with up to five years' imprisonment. 'Supply' should be given its ordinary meaning of transferring physical control of something from one person to another. 'Procure' in s 59 means to obtain possession of something for some other person. As to 'intent to procure miscarriage', see **10.33**.

CONCEALMENT OF BIRTH

10.39   OAPA 1861, s 60 provides that, if any woman is delivered of a child, every person who, by any secret disposition of the dead body of that child, endeavours to conceal its birth is guilty of an indictable (either-way) offence punishable with a maximum of two years' imprisonment. It is irrelevant whether the child died before, at or after its birth. However, in the case of a stillborn child, it must have reached a sufficient state of maturity that, but for some accidental circumstance, it might have been born alive.

The secret disposition may be done by anyone, but typically it is done by the mother following an unattended birth. In such a case, the offence is of importance where it cannot be proved how and/or when a child died, so that the mother cannot be convicted of an offence of homicide or of child destruction.

The abandonment of the child's body is not enough. There must be a secret disposition of it in an endeavour to conceal its birth from the world at large, but there can be such an endeavour even though some of the defendant's confidantes know of the birth. Clearly, a secret disposition in an endeavour to conceal the birth from a particular person does not constitute the offence.

CHAPTER 11

Sexual Offences

INTRODUCTION

The law in relation to sexual offences is principally governed by the Sexual Offences **11.1**
Act 2003 (SOA 2003).

Sexual

For the purposes of the various offences in SOA 2003, s 78 provides that, except in **11.2**
relation to the offences of sexual communication with a child and sexual activity in a
public lavatory (**11.30** and **11.75**), penetration, touching or any other activity is sexual
if a reasonable person would consider that:

(a) whatever its circumstances or any person's purpose in relation to it, it is because of
its nature sexual; or
(b) because of its nature it *may* be sexual and because of its circumstances or the purpose of any person in relation to it (or both) it *is* sexual.

The Court of Appeal has held that (b) contains two requirements. The first is whether
a reasonable person would consider that because of its nature the actual act could be
sexual. In relation to this requirement, the circumstances before or after the act took
place, or any evidence as to the purpose of any person in relation to it are irrelevant. If
the answer to the question posed by this requirement is 'No', the act is not sexual. If the
answer to the question is 'Yes', the second requirement comes into play, and requires
the jury or magistrates to ask themselves whether because of the circumstances of the
activity and/or the purpose of *any* person in relation to it (not just the person who does
the act, but—eg—someone who encourages the act to be done), the activity *is* sexual.

Consent

For the purposes of the offences under SOA 2003 involving the absence of consent, **11.3**
'consent' is defined by SOA 2003, s 74 as follows: a person consents if (s)he agrees by
choice, and has the freedom and capacity to make that choice.

An agreement can be express or implied, and may be evidenced by words or conduct, past or present.

Consent to sex can take many forms, ranging from willing enthusiasm to reluctant
acquiescence. An example of reluctant consent would be where someone who has the
capacity and freedom to make a choice nevertheless submits to a demand that (s)he
feels unable to resist.

Capacity to choose

11.4   SOA 2003, s 74 does not lay down a test of capacity. However, reference to the offences under ss 30 and 31 against persons with a mental disorder impeding choice shows that a person lacks capacity if that person lacks the capacity to choose whether to agree (whether because that person lacks sufficient understanding of the nature or reasonably foreseeable consequences of what is being done, or for any other reason). The words 'for any other reason' are clearly capable of encompassing a wide range of circumstances in which a person's mental disorder may rob that person of the ability to make an autonomous choice, even though (s)he may have sufficient understanding of the information relevant to making it. It has been held in the High Court that capacity is not only 'issue-specific' in relation to different types of transaction (so that someone may have capacity for one purpose but not for another), but it is also issue-specific in relation to different transactions of the same type (so that a vulnerable adult (B) may have the capacity to consent to one type of sexual activity whilst lacking the capacity to consent to some other (and to B unfamiliar) type of sexual activity). In addition, the House of Lords has held, capacity to choose can be 'person-specific' or 'situation-specific' as well as 'issue-specific'. Thus, for example, B can have capacity vis-à-vis C but not vis-à-vis A.

Speaking in the context of rape, the Court of Appeal has held that if through drink (or for any other reason) the complainant (B) has temporarily lost capacity to choose whether to have intercourse on the relevant occasion, B is not consenting and, subject to questions about whether the defendant had mens rea, if intercourse takes place, this will be rape. However, it said, where B has voluntarily consumed even substantial quantities of alcohol, but nevertheless remains capable of choosing whether or not to have intercourse, and in drink agrees to do so, this will not be rape.

Freedom to choose

11.5   A person may not have a freedom to make a choice whether or not to agree if violence is being used or threatened against him/her or another at the material time or immediately beforehand. A person may also lack freedom of choice for other reasons, as where someone agrees because (s)he is unlawfully detained or where there was the use, or threat, of violence to destroy property which was of special value, financially or emotionally, or threat of dismissal by an employer to an employee, or threat to remove children. It would all depend on the nature of the threat and the other circumstances, the perception of the complainant, and whether in the light of these factors the complainant was not in reality free to agree or disagree.

Another example where there may not be freedom of choice is provided by a Court of Appeal case where A's conviction for rape of B, a young adult, who had from childhood been sexually abused, dominated, and controlled by A was upheld. The Court held that from such conduct an apparent consent might be found not to be a real consent. Likewise the fact that B has been groomed may mean that an apparent consent is not a real consent. The Court of Appeal has stated that one of the consequences of grooming is that it has a tendency to limit or subvert the alleged victim's capacity to make free decisions.

Irrebuttable presumptions as to absence of consent

11.6   Apparent agreement may be invalidated by deception. Two kinds of deception are specifically addressed in SOA 2003, s 76, by means of irrebuttable presumptions, but only in the context of alleged offences under SOA 2003, ss 1–4 (rape, sexual assault, etc).

Section 76 provides that if in proceedings for an offence under ss 1 to 4 it is proved that the defendant did the *'relevant act'* and that either of the *specified circumstances* existed, it is to be conclusively (ie irrebuttably) presumed that:

(a) the complainant (B) did not consent to the relevant act; and
(b) the defendant did not believe that the complainant consented to the relevant act.

The *specified circumstances* are: that the defendant intentionally deceived the complainant as to the nature or purpose of the relevant act; and that the defendant intentionally induced the complainant to consent to the relevant act by impersonating a person known personally to B.

By the SOA 2003, s 77, in relation to an offence to which ss 75 (below) and 76 apply, references to the *relevant act* and to the complainant are to be read as follows:

Offence	Relevant act
Under s 1 (rape)	The defendant intentionally penetrating, with his penis, the vagina, anus or mouth of the complainant
Under s 2 (assault by penetration)	The defendant intentionally penetrating, with a part of his/her body or anything else, the vagina or anus of the complainant where the penetration is sexual
Under s 3 (sexual assault)	The defendant intentionally touching the complainant, where the touching is sexual
Under s 4 (causing person to engage in sexual activity without consent)	The defendant intentionally causing the complainant to engage in an activity, where the activity is sexual

Deception not covered by s 76

Whether or not there is an effective consent in cases of deception not covered by the **11.7** Sexual Offences Act 2003, s 76 must be determined by the jury by applying the terms of s 74 in the light of the following principles laid down by case law. The statement in s 74 that a person consents if (s)he agrees by choice and has the freedom to make a choice requires an informed choice. A deception which is closely connected with 'the nature or purpose of the act', because it relates to the sexual act itself rather than the broad circumstances surrounding it, has been held capable of negating a complainant's free exercise of choice for the purposes of s 74 of the 2003 Act. Thus, where A has induced B to consent to a sexual act by lying about whether a condom would be worn, or by falsely promising to withdraw from intercourse before ejaculation, or by misrepresenting A's gender, A's deception is capable of invalidating B's consent, whereas where A has induced B to consent to a sexual act by lying that A has had a vasectomy B's consent is not vitiated because A's lie will not relate to the physical performance of the sexual act but to the risks and consequences of it; it is insufficiently closely connected to the performance of the sexual act. As can be seen the type of case where a deception not covered by s 76 can vitiate consent is tightly defined. Inducing B to consent to sexual activity by lying about one's status or falsely promising marriage, for example, clearly cannot invalidate consent.

Non-disclosure

11.8 A failure to disclose, however material the undisclosed fact, falls outside the scope of s 74. Thus, where A engages in a sexual act with B without telling B that A has a sexually transmissible disease, B's consent to that act is not thereby invalidated. The sexual act remains consensual. However, A will not have any defence, in such a case, to any charge of an offence against the person which may result from harm created by the act, merely by virtue of that consent, because such consent does not include consent to infection by the disease.

Rebuttable presumptions as to absence of consent

11.9 Section 75 provides that if in proceedings for an offence to which s 75 applies (those under ss 1 to 4) it is proved that:

(a) A did the *relevant act* (see table at **11.6**);
(b) any of the *specified circumstances* below existed; and
(c) A *knew* that those circumstances existed,

B is to be taken (ie presumed) not to have consented to the relevant act *unless sufficient evidence is adduced to raise an issue as to whether B consented*, and A is to be taken not to have reasonably believed that B consented *unless sufficient evidence is adduced to raise an issue as to whether A reasonably believed it*. Evidence is not sufficient if it is fanciful or speculative.

The *specified circumstances* are that:

(a) any person was, at the time.of the relevant act or immediately before it began, using violence against B or causing B to fear that immediate violence would be used against him/her;
(b) any person was, at the time of the relevant act or immediately before it began, causing B to fear that violence was being used, or that immediate violence would be used, against another person;
(c) B was, and A was not, unlawfully detained at the time of the relevant act;
(d) B was asleep or otherwise unconscious at the time of the relevant act;
(e) because of B's physical disability, B would not have been able at the time of the relevant act to communicate to A whether B consented;
(f) any person had administered to or caused to be taken by B, without B's consent, a substance which, having regard to when it was administered or taken, was capable of causing or enabling B to be stupefied or overpowered at the time of the relevant act.

If sufficient evidence is adduced to raise an issue as to one (or both) of the presumed matters, the matter must be determined in the normal way.

Penetration, touching and parts of the body

11.10 By SOA 2003, s 79, 'penetration' is a continuing act from entry to withdrawal. It has long been established by the courts that the slightest degree of penetration is enough. SOA 2003, s 79 also provides that references to a part of the body include references to a part surgically constructed (in particular through gender re-assignment surgery); that 'vagina' includes vulva (external female genital organs); and that 'touching' includes touching with any part of the body, or with anything else, or through anything, and in particular includes touching through penetration. The Court of Appeal has held that 'touching' includes touching a person's clothing without applying any force to his/her body.

NON-CONSENSUAL SEXUAL OFFENCES

Rape

By SOA 2003, s 1(1), a person (A) commits an offence triable only on indictment if: **11.11**

(a) he intentionally penetrates the vagina, anus or mouth of another person (B) with his penis;
(b) B does not consent to the penetration; and
(c) A does not reasonably believe that B consents.

The absence of consent does not have to be demonstrated by offering resistance or by communicating it to A.

Reasonable belief in consent

SOA 2003, s 1(2) provides that whether a belief in consent is reasonable is to be deter- **11.12** mined having regard to all the circumstances, including any steps A has taken to ascertain whether B consents. The words 'all the circumstances' refer to those which might be relevant to the issue, including any characteristic of the defendant, permanent or transient, which might affect his ability to perceive or understand whether or not the victim is consenting. However, the Court of Appeal has held that account may not be taken of voluntary intoxication, or of mental disorder if it leads to an irrational belief in consent.

Proof of lack of consent and of the absence of reasonable belief is assisted by the rebuttable and irrebuttable presumptions in ss 75 and 76 (**11.9** and **11.6**).

Rape of a child under 13

By SOA 2003, s 5(1), a person (A) commits an offence triable only on indictment if: **11.13**

(a) he intentionally penetrates the vagina, anus or mouth of another person (B) with his penis; and
(b) B is under 13.

Whether or not the other person consents to the penetration is irrelevant; a child under 13 is legally incapable of giving a legally significant consent.

The House of Lords has confirmed not even a reasonable belief that B was 13 or over will excuse A.

Exceptions

The offence is one of the offences in respect of which there are exceptions from criminal **11.14** liability under SOA 2003, s 73, on the basis of aiding, abetting or counselling the commission of the offence if a person acts for the purpose of:

(a) protecting the child from sexually transmitted infection;
(b) protecting the physical safety of the child;
(c) preventing the child from becoming pregnant; or
(d) promoting the child's emotional well-being by the giving of advice,

and not for the purpose of obtaining sexual gratification or for the purpose of causing or encouraging the activity constituting the offence or the child's participation in it.

Assault by penetration

11.15   By SOA 2003, s 2(1), a person (A) commits an offence triable only on indictment if:

(a) A intentionally penetrates the vagina or anus of another person (B) with a part of his/her body (eg a finger or tongue) or anything else;
(b) the penetration is sexual;
(c) B does not consent to the penetration; and
(d) A does not reasonably believe that B consents.

Section 2(2) provides that whether a belief is reasonable must be determined having regard to all the circumstances, including whether A has taken reasonable steps to ascertain whether B consents.

Proof of lack of consent and of the absence of reasonable belief is assisted by the rebuttable and irrebuttable presumptions in ss 75 and 76 (**11.9** and **11.6**).

Assault of a child under 13 by penetration

11.16   By SOA 2003, s 6(1), a person (A) commits an offence triable only on indictment if:

(a) A intentionally penetrates the vagina or anus of another person with a part of his/ her body or anything else;
(b) the penetration is sexual; and
(c) the other person is under 13.

The absence of consent by the other person is not an element of this offence. The above exceptions under s 73 (**11.14**) from liability for aiding, abetting or counselling apply to this offence.

Sexual assault

11.17   By SOA 2003, s 3(1), a person (A) commits an indictable (either-way) offence if:

(a) A intentionally touches another person (B);
(b) the touching is sexual;
(c) B does not consent to the touching; and
(d) A does not reasonably believe that B consents.

Section 3(2) provides that whether a belief is reasonable must be determined with regard to all the circumstances, including any steps A has taken to ascertain whether B consents.

The Court of Appeal has held that the 'intentional touching' element of this offence simply requires a deliberate touching; A is not required to intend his/her touching to be sexual. It has also held that this intent is a basic intent, with the result that evidence of voluntary intoxication cannot be relied upon to negate it. These points are equally applicable to other offences under SOA 2003 involving intentional penetration or touching.

Proof of lack of consent and of the absence of reasonable belief is assisted by the rebuttable and irrebuttable presumptions in ss 75 and 76 (**11.9** and **11.6**).

Sexual assault of a child under 13

By SOA 2003, s 7(1), a person (A) commits an indictable (either-way) offence if: **11.18**

(a) A intentionally touches another person;
(b) the touching is sexual; and
(c) the other person is under 13.

The absence of the other person's consent is not an element of the offence. The exceptions under s 73 (**11.14**) from liability for aiding, abetting or counselling apply to this offence.

Causing a person to engage in sexual activity without consent

By SOA 2003, s 4(1) and (5), a person (A) commits an indictable (either-way) offence if: **11.19**

(a) A intentionally causes another person (B) to engage in an activity;
(b) the activity is sexual;
(c) B does not consent to engaging in the activity; and
(d) A does not reasonably believe that B consents.

A more serious offence triable only on indictment is committed under s 4(1) and (4) if the activity caused involves one of four circumstances:

(a) penetration of B's anus or vagina;
(b) penetration of B's mouth with a person's penis;
(c) penetration of a person's anus or vagina with a part of B's body or by B with anything else; or
(d) penetration of a person's mouth with B's penis.

Section 4(2) provides that whether a belief is reasonable must be determined with regard to all the circumstances, including any steps A has taken to ascertain whether B consents.

Proof of lack of consent and of the absence of reasonable belief is assisted by the rebuttable and irrebuttable presumptions in ss 75 and 76 (**11.9** and **11.6**).

Causing or inciting a child under 13 to engage in sexual activity

By SOA 2003, s 8(1) and (3), a person (A) commits an indictable (either-way) offence if: **11.20**

(a) A intentionally causes or incites another person (B) to engage in an activity;
(b) the activity is sexual; and
(c) B is under 13.

A more serious offence triable only on indictment is committed under s 8(1) and (2) if the activity caused or incited involved one of the four circumstances set out at **11.19** above in relation to s 4.

The absence of B's consent is not an element of an offence under s 8.

An offence of intentionally inciting a child under 13 to engage in sexual activity can be committed even though it is not possible to identify any specific or identifiable child to whom the incitement was addressed. It matters not whether the incitement was directed at a particular child or a very large group of children. These points are equally applicable to the other offences of incitement to sexual activity in SOA 2003.

The Court of Appeal has held that 'intentional' causing or inciting for the purposes of s 8 means deliberate causing or inciting; recklessness will not do. This is equally applicable to the other offences of intentionally causing or inciting under SOA 2003.

Punishment

11.21   An offence under SOA 2003, s 1(1), 2(1), 4(1), and (4), 5(1), 6(1), or 8(1) and (2) is punishable with up to life imprisonment; an offence under s 7(1) or 8(1) and (3) with up to 14 years' imprisonment, and an offence under s 3(1) or 4(1) and (5) with up to 10 years'.

CHILD SEX OFFENCES

11.22   In these offences, it is irrelevant that the child may have consented.

Sexual activity with a child

11.23   By SOA 2003, s 9(1), a person aged 18 or over (A) commits an offence if:

(a) A intentionally touches another person (B);
(b) the touching is sexual; and
(c) either:
  (i)   B is under 16 and A does not reasonably believe that B is 16 or over; or
  (ii) B is under 13.

An offence under s 9(1) is triable only on indictment if the touching involves one of the following four circumstances:

(a) penetration of B's anus or vagina with a part of A's body or anything else;
(b) penetration of B's mouth with A's penis;
(c) penetration of A's anus or vagina with a part of B's body; or
(d) penetration of A's mouth with B's penis.

If the touching occurs in other circumstance the offence is an indictable (either-way) offence.

The exceptions under s 73 (**11.14**) from liability for aiding, abetting or counselling apply to an offence under s 9.

Causing or inciting a child to engage in sexual activity

11.24   By SOA 2003, s 10(1), a person aged 18 or over (A) commits an offence if:

(a) A intentionally causes or incites another person (B) to engage in an activity;
(b) the activity is sexual; and
(c) either:
  (i)   B is under 16 and A does not reasonably believe that B is 16 or over; or
  (ii) B is under 13.

An offence under s 10(1) is triable only on indictment if the activity caused or incited involves:

(a) penetration of B's anus or vagina;
(b) penetration of B's mouth with a person's penis;

(c) penetration of a person's anus or vagina with a part of B's body or by B with anything else; or

(d) penetration of a person's mouth with B's penis.

In other circumstance the offence is an indictable (either-way) offence.

Engaging in sexual activity in the presence of a child

By SOA 2003, s 11(1), a person aged 18 or over (A) commits an indictable (either-way)   **11.25**
offence if:

(a) A intentionally engages in an activity;

(b) the activity is sexual;

(c) for the purpose of obtaining sexual gratification, A engages in it:
   (i)   when another person (B) is present or is in a place from which A can be observed; and
   (ii) knowing or believing that B is aware, or intending that B should be aware, that A is engaging in it; and

(d) either:
   (i)   B is under 16 and A does not reasonably believe that B is 16 or over; or
   (ii) B is under 13.

'Observed' in (c) and elsewhere in SOA 2003 means observation whether direct or by looking at an image produced by any means.

The Court of Appeal has held that A must engage in the sexual activity in the presence or under the observation of a child (B) for the purpose of A's obtaining sexual gratification *from B's presence or observation*. If A does not have this purpose, A does not commit an offence under s 11 if (s)he engages in sexual activity, for the purposes of sexual gratification *from the sexual activity itself*, knowing that B is present or observing. This decision is equally applicable to other offences under SOA 2003 in which the phrase 'for the purpose of obtaining sexual gratification' appears.

Causing a child to watch a sexual act

By SOA 2003, s 12(1), a person aged 18 or over (A) commits an indictable (either-way)   **11.26**
offence if:

(a) for the purpose of obtaining sexual gratification, A intentionally causes another person (B) to watch a third person engaging in an activity, or to look at an image of any person engaging in an activity;

(b) the activity is sexual; and

(c) either:
   (i)   B is under 16 and A does not reasonably believe that B is 16 or over, or
   (ii) B is under 13.

'Image' in (a), and elsewhere in the Act, includes a moving or still image, however produced, and, where—as here—the context permits, a three-dimensional image. It does not include written material. References to an image of a person include references to an image of an imaginary person.

The purpose of obtaining sexual gratification need not relate to immediate gratification (eg from seeing the victim watch the images). The Court of Appeal has held that the offence under s 12 can be committed where A's purpose involves immediate, or deferred, or immediate and deferred, gratification. Thus, the offence can be committed, for example, where A causes a child to watch a sexual act to put the child in the mood for future sexual abuse, as well as where A does so because A derives enjoyment at the time from seeing the child watch the sexual act. This decision is equally applicable to other offences under SOA 2003 in which the phrase 'for the purpose of obtaining sexual gratification' appears.

Child sex offences committed by children or young persons

11.27 By SOA 2003, s 13(1), a person under 18 (A) commits an indictable (either-way) offence if A does anything which would be an offence under SOA 2003, ss 9 to 12 if A were aged 18. It is, however, a less serious offence than those offences. The exceptions under s 73 (**11.14**) from liability for aiding, abetting or counselling apply to this offence if what is done would be an offence under s 9 if the offender were aged 18 or over.

Arranging or facilitating the commission of a child sex offence

11.28 By SOA 2003, s 14(1), a person (A) commits an indictable (either-way) offence if:

(a) A intentionally arranges or facilitates something that A intends to do, intends another person to do, or believes that another person will do, in any part of the world; and

(b) doing it will involve the commission of an offence under ss 9 to 13.

Section 14(2) and (3) provides exceptions for people who seek to protect a child from pregnancy or sexually transmitted disease, to protect a child's physical safety or to give it advice, where they do not intend an offence under ss 9 to 13 to be committed but believe that it will. They provide that A does not commit an offence under s 14 if:

(a) A arranges or facilitates something that A believes another person will do, but that A does not intend to do or intend another person to do; and

(b) any offence within ss 9 to 13 which the doing of that thing would involve would be an offence against a child for whose protection A acts.

In this context, a person acts for the protection of a child if (s)he acts for the purpose of:

(a) protecting the child from sexually transmitted infection;
(b) protecting the child's physical safety;
(c) preventing the child from becoming pregnant; or
(d) promoting the child's emotional well-being by the giving of advice,

and not for the purpose of obtaining sexual gratification or for the purpose of causing or encouraging the activity constituting an offence within ss 9 to 13 or the child's participation in it.

The Court of Appeal has stated that s 14 covers taking preparatory steps (with the necessary intent) to commit an offence under ss 9 to 13. It also stated that an arrangement may be made without the agreement or acquiescence of anyone else.

Meeting a child following sexual grooming etc with a view to engaging in sexual activity with it

SOA 2003, s 15(1) deals with sexual grooming by providing that a person aged 18 or **11.29** over (A) commits an indictable (either-way) offence if:

(a) having met or communicated with another person (B) anywhere in the world on one or more occasions:
  (i) A subsequently intentionally meets B, or
  (ii) A subsequently travels with the intention of meeting B in any part of the world or arranges to meet B in any part of the world, or
  (iii) B travels with the intention of meeting A in any part of the world;
(b) A intends to do anything to or in respect of B, during or after the meeting mentioned in (a)(i)–(iii) above, and in any part of the world, which if done will involve the commission by A of a relevant offence;
(c) B is under 16; and
(d) A does not reasonably believe that B is 16 or over.

A relevant offence is an offence referred to at **11.11–11.75** and **13.61–13.68**, or conduct outside England and Wales which would be such an offence if committed in England and Wales.

Sexual communication with a child

A person aged 18 or over (A) commits an indictable (either-way) offence under SOA **11.30** 2003, s 15A(1) if:

(a) for the purpose of obtaining sexual gratification, A intentionally communicates with another person (B);
(b) the communication is sexual or is intended to encourage B to make (whether to A or another) a communication that is sexual; and
(c) B is under 16 and A does not reasonably believe that B is 16 or over.

For these purposes, a communication is sexual if:

(i) any part of it relates to sexual activity, or
(ii) a reasonable person would, in all the circumstances but regardless of any person's purpose, consider any part of the communication to be sexual, and

in (i) 'sexual activity' means an activity that a reasonable person would, in all the circumstances but regardless of any person's purpose, consider to be sexual.

The offence is intended to criminalise a defendant at the very start of the grooming process.

Punishment

An offence under SOA 2003, s 9(1), 10(1), or 14(1) is punishable with up to 14 years' **11.31** imprisonment; an offence under s 11(1), 12(1), or 15(1) with up to 10 years' imprisonment; and an offence under s 13(1) with up to 5 years', and an offence under s 15A with up to two years'.

ABUSE OF POSITION OF TRUST

11.32   All the offences under this heading are indictable (either-way) offences punishable with up to five years' imprisonment.

Abuse of position of trust: sexual activity with a child

11.33   Provided that (s)he has any necessary mens rea, a person aged 18 or over (A) commits an offence under SOA 2003, s 16(1) if:

(a) A intentionally touches another person (B);
(b) the touching is sexual;
(c) A is in a position of trust in relation to B; and
(d) B is under 18.

B's consent is, of course, irrelevant in offences involving abuse of a position of trust. The exceptions under s 73 (**11.14**) from liability for aiding, abetting or counselling apply to this offence if B is under 16.

Abuse of position of trust: causing or inciting a child to engage in sexual activity

11.34   Provided that (s)he has any necessary mens rea, a person aged 18 or over (A) commits an offence under s 17(1) if:

(a) A intentionally causes or incites another person (B) (**11.20**) to engage in an activity;
(b) the activity is sexual;
(c) A is in a position of trust in relation to B; and
(d) B is under 18.

Abuse of position of trust: sexual activity in the presence of a child

11.35   Provided that (s)he has any necessary mens rea, a person aged 18 or over (A) commits an offence under SOA 2003, s 18(1) if:

(a) A intentionally engages in an activity;
(b) the activity is sexual;
(c) for the purpose of obtaining sexual gratification (**11.25**), A engages in it:
    (i)   when another person (B) is present or is in a place from which A can be observed, and
    (ii) knowing or believing that B is aware, or intending that B should be aware, that A is engaging in it;
(d) A is in a position of trust in relation to B; and
(e) B is under 18.

Abuse of position of trust: causing a child to watch a sexual act

11.36   Provided that (s)he has any necessary mens rea, a person aged 18 or over (A) commits an offence under SOA 2003, s 19(1) if:

(a) for the purpose of obtaining sexual gratification (**11.26**), A intentionally causes another person (B) to watch a third person engaging in an activity, or to look at an image of any person engaging in an activity;
(b) the activity is sexual;
(c) A is in a position of trust in relation to B; and
(d) B is under 18.

Position of trust

For the purposes of the above offences, a person (A) is, by SOA 2003, s 21, in a position of trust in relation to another person (B) if: **11.37**

(a) any of paras (1)–(11) below apply; or
(b) any condition specified in an order made by the Secretary of State is met. No order has yet been made.
(1) If A looks after persons under 18 who are detained in an institution by virtue of a court order or under an enactment, and B is so detained in that institution.
(2) If A looks after persons under 18 who are resident in a home or other place for children in residential care, and B is in residential care there.
(3) If A looks after persons under 18 who are accommodated and cared for in a hospital, an independent clinic (in Wales), a care home, residential care home or private hospital, a community home, voluntary home, children's home or the like, or a place in Wales at which a care home service or secure accommodation service is provided, and B is accommodated and cared for in that institution.
(4) If A looks after persons under 18 who are receiving education at an educational institution and B is receiving, and A is not receiving, education at that institution.
(5) If A is engaged in the provision of a careers service or similar service and, in that capacity, looks after B on an individual basis.
(6) If A regularly has unsupervised contact with B (whether face to face or by any other means) in the provision of accommodation for children in need thereof, or in police protection or detention, or on remand.
(7) If A, as a person who is to report to the court on matters relating to the welfare of B, regularly has unsupervised contact with B (whether face to face or by any other means).
(8) If A is a personal adviser appointed for B under the Children Act 1989, or under the Social Services and Well-being (Wales) Act 2014 in respect of category 1 or category 2 young persons, and, in that capacity, looks after B on an individual basis.
(9) If:
   (a) B is subject to a care order, a supervision order or an education supervision order, and
   (b) in the exercise of functions conferred by virtue of the order on an authorised person or the authority designated by the order, A looks after B on an individual basis.
(10) If A is an officer of the Children and Family Court Advisory Support Service (CAFCASS) or a Welsh family proceedings officer appointed for B, or is appointed children's guardian or guardian ad litem of B under the relevant rules, and, in that capacity, regularly has unsupervised contact with B (whether face to face or by any other means).

(11) If:

  (a) B is subject to requirements imposed by or under an enactment on his/her release from detention for a criminal offence, or is subject to requirements imposed by a court order made in criminal proceedings, and

  (b) A looks after B on an individual basis in pursuance of the requirements.

For the purposes of paras (1), (2), (3), and (4), a person looks after persons under 18 at an institution or the like if (s)he is regularly involved in caring for, training, supervising or being in sole charge of such persons there, not necessarily the child abused.

Paragraphs (5), (8), (9), and (11) refer to a person looking after another on an individual basis. A person (A) looks after another (B) on such a basis if:

  (a) A is regularly involved in caring for, training or supervising B; and

  (b) in the course of his/her involvement, A regularly has unsupervised contact with B (whether face to face or by any other means).

Mens rea

11.38 In addition to requiring A intentionally to do the thing specified by each individual section, ss 16 to 19 make further provision as to the mens rea required in respect of the child's age and the existence of a position of trust.

In terms of the mens rea as to the age of the child (B), A must not reasonably believe that B is 18 or over. However, this provision does not apply if B is under 13 at the material time; in such a case the offences are undoubtedly ones of strict liability as to age. Although the prosecution ultimately has the persuasive burden of proof of the absence of reasonable belief that B is 18 or over where B is aged 13 to 17, the prosecution is assisted by the provision that, where in proceedings for an offence under ss 16 to 19 it is proved that B was under 18, A is to be taken not to have reasonably believed that B was 18 or over unless sufficient evidence is adduced to raise an issue as to whether A reasonably believed it.

Where A is in a position of trust in relation to B by virtue of circumstances within paras (1), (2), (3), or (4) above, and, in each case, A is not also in a position of trust by virtue of other circumstances, it must be proved that A knew or could reasonably be expected to know of the circumstances by virtue of which A is in a position of trust in relation to B. Proof of this is aided by the provision that, where it is proved that A was in a position of trust in relation to B by virtue of the four circumstances just referred to, and it is not proved that A was in such a position of trust by virtue of other circumstances, it is to be presumed that A knew or could reasonably have been expected to know of the circumstances by virtue of which A was in such a position of trust unless sufficient evidence is adduced to raise an issue as to whether A knew or could reasonably have been expected to know of those circumstances.

Where a position of trust arises wholly or partly by virtue of the other categories of circumstance referred to above, the Act does not require proof of any mens rea as to the position of trust.

Exceptions: marriage, civil partnership and existing sexual relationships

11.39 Conduct by a person (A) which would otherwise be one of the offences under SOA 2003, ss 16 to 19 against another person (B) is not such an offence if:

(a) B was 16 or over, and the defendant proves that A and B were lawfully married (or civil partners) at the time of the conduct; or

(b) the defendant proves that, immediately before the position of trust arose, a sexual relationship existed between A and B, but this exception does not apply if at that time sexual intercourse between A and B would have been unlawful, eg because B was under 16.

FAMILIAL SEXUAL OFFENCES

Offences involving a child family member

Sexual activity with a child family member

A person (A) commits an indictable (either-way) offence under SOA 2003, s 25(1) if: **11.40**

(a) A intentionally touches another person (B);

(b) the touching is sexual;

(c) the relation of A to B is within the specified relationships (**11.42**);

(d) A knows or could reasonably be expected to know that his/her relation to B is of a description falling within those relationships; and

(e) either:
  (i) B is under 18 and A does not reasonably believe that B is 18 or over; or
  (ii) B is under 13.

SOA 2003 has the effect of distinguishing between two offences. The more serious is committed where A is 18 or over at the time of the offence. If the touching involves:

(a) penetration of B's anus or vagina with a part of A's body or anything else;

(b) penetration of B's mouth with A's penis;

(c) penetration of A's anus or vagina with part of B's body; or

(d) penetration of A's mouth with B's penis,

the offence is triable only on indictment; otherwise it is an indictable (either-way) offence (s 25(1) and (4)). In either case, it is punishable with up to 14 years' imprisonment.

The less serious offence is where A was not 18 or over at the material time, whatever type of activity was involved (s 25(1) and (5)). It is an indictable (either-way) offence punishable with up to five years' imprisonment.

Where it is proved that the relation of A to B was of a description falling within the specified relationships, it is to be taken that A knew or could reasonably have been expected to know that his/her relation to B was of that description unless sufficient evidence is adduced to raise an issue as to whether A knew or could reasonably have been expected to know that it was.

In respect of the mens rea in (e)(i) (no reasonable belief (where B is not under 13) that B is 18 or over), where it is proved that B was under 18, A is to be taken not to have reasonably believed that B was 18 or over unless sufficient evidence is adduced to raise an issue as to whether A reasonably believed it.

The exceptions under s 73 (**11.14**) from liability for aiding, abetting or counselling apply to an offence under s 25 if B is under 16.

Inciting a child family member to engage in sexual activity

A person (A) commits an offence under SOA 2003, s 26(1) if: **11.41**

(a) A intentionally incites another person (B) (**11.20**) to touch, or allow him/herself to be touched by, A;

(b) the touching is sexual;

(c) the relation of A to B is within the specified relationships;

(d) A knows or could reasonably be expected to know that A's relation to B is of a description falling within those relationships; and

(e) either:

    (i) B is under 18 and A does not reasonably believe that B is 18 or over; or

    (ii) B is under 13.

There are two offences under s 26: by s 26(4) and (5) they are distinguished (and triable and punishable) in a similar way as in s 25 (**11.40**).

The mens rea in respect of the fact that the relation of A to B is within the specified relationships is the same as that in the offences described in relation to s 25. Thus, where it is proved that the relation of A to B was of a description falling within the specified relationships, it is to be taken that A knew or could reasonably have been expected to know that his/her relation to B was of that description unless sufficient evidence is adduced to raise an issue as to whether A knew or could reasonably have been expected to know that it was.

Likewise, in respect of the mens rea in para (e)(i) (no reasonable belief (where B is not under 13) that B is 18 or over), where it is proved that B was under 18, A is to be taken not to have reasonably believed that B was 18 or over unless sufficient evidence is adduced to raise an issue as to whether A reasonably believed it.

Family relationships

11.42   For the purposes of SOA 2003, ss 25 and 26, the relation of A and B must be within s 27. The relation of A to B is within the specified relationships if it is within paras (a)–(c) below, including an adoptive relationship (as well as a biological one).

The relation of A and B is within the specified relationships if:

(a) one of them is the other's parent, grandparent, brother, sister, half-brother, half-sister, aunt or uncle, or A is or has been B's foster parent;

(b) A and B live or have lived in the same household, or A is or has been regularly involved in caring for, training, supervising or being in sole charge of B, and:

    (i) one of them is or has been the other's step-parent,

    (ii) A and B are cousins,

    (iii) one of them is or has been the other's stepbrother or stepsister, or

    (iv) the parent or present or former foster parent of one of them is or has been the other's foster parent;

(c) A and B live in the same household, and A is regularly involved in caring for, training, supervising or being in sole charge of B, eg a nanny or au pair. It will be noted that (c) only applies while A is living in the same household.

A 'step-parent' includes someone who is neither married to, nor the civil partner of, a parent, if that person is a parent's 'partner'. In (b)(iii), the reference to a stepbrother or stepsister is to be read as follows: X's stepbrother or stepsister includes someone who is the son/daughter of the spouse, civil partner or partner of X's parent (but not the son/daughter of either of X's parents). For these purposes, a person is another's partner if they live together as partners in an enduring family relationship.

Exceptions: marriage, civil partnership, and existing sexual relationships

11.43   By SOA 2003, ss 28 and 29, conduct by A which would otherwise be an offence under s 25 or s 26 against B is not an offence thereunder if:

(a) B is 16 or over, and A and B are lawfully married (or civil partners); or

(b) (i) the relation of A to B is not within **11.42**(a), either biologically or adoptively; and

　　(ii) immediately before the relation of A to B first became such as to fall within the specified relationships, a sexual relationship existed between A and B (as where, when A's mother marries B's father, A and B were already in a sexual relationship).

The defence in (b) does not apply if at the time referred to in (b)(ii) sexual intercourse between A and B would have been unlawful, eg because one of them was under 16.

It is for the defendant to prove the relationship mentioned in (a) or (b).

Sex with an adult relative

A person aged 16 or over (A) (subject to SOA 2003, s 64(3A)—see below) commits an **11.44** indictable (either-way) offence under s 64(1) if:

(a) A intentionally penetrates another person's vagina or anus with a part of his/her body or anything else, or penetrates another person's mouth with his penis;

(b) the penetration is sexual;

(c) the other person (B) is aged 18 or over;

(d) A is related to B in a specified way; and

(e) A knows or could reasonably be expected to know that (s)he is related to B in that way.

A person aged 16 or over (A) (subject to s 65(3A)—see below) commits an indictable (either-way) offence under s 65(1) if:

(a) another person (B) penetrates A's vagina or anus with a part of B's body or anything else, or penetrates A's mouth with B's penis;

(b) A consents to the penetration;

(c) the penetration is sexual;

(d) B is aged 18 or over;

(e) A is related to B in a specified way; and

(f) A knows or could reasonably be expected to know that A is related to B in that way.

The specified ways that A may be related to B are as a parent, grandparent, child, grandchild, brother, sister, half-brother, half-sister, uncle, aunt, nephew, or niece.

Where a child has been adopted, not only are the biological parent and child related to each other for the purposes of ss 64 and 65, whenever the child was adopted, but 'parent' also includes an adoptive parent and 'child' includes an adopted person. However, where s 64(1) or 65(1) apply in a case where A is related to B as B's child by virtue of being B's adopted child, A does not commit an offence under s 64(1) or 65(1), as the case may be, unless A is 18 or over. This is provided by ss 64(3A) and 65(3A).

It remains the case that other adoptive relationships, eg that of adoptive brother and sister, are excluded from the application of s 64 or 65.

These references to an adoptive relationship are to be read as including a corresponding relationship arising by virtue of a parental order under the Human Fertilisation and Embryology Act 2008 (the 'test tube babies' Act).

In terms of the requirement that the prosecution must prove that A knew or could reasonably be expected to know that A is related to B in a specified way, it is to be taken that A knew or could reasonably have been expected to know that A was related in that way unless sufficient evidence is adduced to raise an issue as to whether A knew or could reasonably have been expected to know that A was.

An offence under s 64(1) or 65(1) is punishable with up to two years' imprisonment.

OFFENCES AGAINST PEOPLE WITH A MENTAL DISORDER

Meaning of mental disorder

11.45   For the purposes of these offences, 'mental disorder' means any disorder or disability of the mind. A person with a learning disability falls within this definition.

Offences against persons with a mental disorder impeding choice

Sexual activity with a person with a mental disorder impeding choice

11.46   A person (A) commits an indictable (either-way) offence under SOA 2003, s 30(1) if:

(a) A intentionally touches another person (B);
(b) the touching is sexual;
(c) B is unable to refuse because of or for a reason related to a mental disorder; and
(d) A knows or could reasonably be expected to know that B has a mental disorder and that because of it or for a reason related to it B is likely to be unable to refuse.

Under s 30(1) and (3), A commits a more serious offence triable only on indictment if the touching involves one of the four types of penetration set out at **11.40**.

B is unable to refuse if:

(a) B lacks the capacity to choose (**11.4**) whether to agree to the touching (whether because B lacks sufficient understanding of the nature or reasonably foreseeable consequences of what is being done, or for any other reason); or
(b) B is unable to communicate such a choice to A.

In relation to the first type of inability, the House of Lords has held that the words 'for a reason related to mental disorder' in the definition of the offences were clearly capable of encompassing a wide range of circumstances in which a person's mental disorder might rob him/her of the ability to make an autonomous choice, even though (s) he might have sufficient understanding of the information relevant to making it. Those circumstances could include the kind of compulsion which drove a person with anorexia to refuse food, the delusions which drove a person with schizophrenia to believe that she had to do something, or the phobia or irrational fear which drove a person to refuse a life-saving injection. Irrational fear plainly was capable of depriving a person of capacity; the question was whether it did in a particular case.

The second type of inability to refuse covers people who might be able to make a choice but are unable to communicate it. The House of Lords has held that it is not limited to cases where B was physically unable to communicate because of his/her mental disorder, and that it also includes cases of an inability to communicate which was the result of or associated with a disorder of the mind. Thus, for example, inability to communicate includes someone with such a degree of learning difficulty that (s)he has never acquired the gift of speech, so that it is impossible to discover whether or not (s)he can understand or make a choice.

The exceptions under s 73 (**11.14**) from liability for aiding, abetting or counselling apply to an offence under s 30 if B is under 16.

Causing or inciting a person with a mental disorder impeding choice to engage in sexual activity

11.47 A person (A) commits an indictable (either-way) offence under SOA 2003, s 31(1) if:

(a) A intentionally causes or incites (**11.20**) another person (B) to engage in an activity;
(b) the activity is sexual;
(c) B is unable to refuse because of or for a reason related to a mental disorder; and
(d) A knows or could reasonably be expected to know that B has a mental disorder and that because of it or for a reason related to it B is likely to be unable to refuse.

Under s 31(1) and (3), A commits a more serious offence triable only on indictment if the activity caused or incited involves:

(a) penetration of B's anus or vagina;
(b) penetration of B's mouth with a person's penis;
(c) penetration of a person's anus or vagina with a part of B's body or by B with anything else; or
(d) penetration of a person's mouth with B's penis.

B is unable to refuse if:

(a) B lacks the capacity to choose whether to agree to engaging in the activity caused or incited (whether because B lacks sufficient understanding of the nature or reasonably foreseeable consequences of the activity, or for any other reason); or
(b) B is unable to communicate such a choice to A.

Engaging in sexual activity in the presence of a person with a mental disorder impeding choice

11.48 A person (A) commits an indictable (either-way) offence under SOA 2003, s 32(1) if:

(a) he intentionally engages in an activity;
(b) the activity is sexual;
(c) for the purpose of obtaining sexual gratification (**11.25**), A engages in it:
  (i) when another person (B) is present or is in a place from which A can be observed, and
  (ii) knowing or believing that B is aware, or intending that B should be aware, that A is engaging in it;
(d) B is unable to refuse because of or for a reason related to a mental disorder; and
(e) A knows or could reasonably be expected to know that B has a mental disorder and that because of it or for a reason related to it B is likely to be unable to refuse.

B is unable to refuse if:

(a) B lacks the capacity to choose whether to agree to being present (whether because B lacks sufficient understanding of the nature of the activity, or for any other reason); or
(b) B is unable to communicate such a choice to A.

Causing a person with a mental disorder impeding choice to watch a sexual act

11.49 A person (A) commits an indictable (either-way) offence under SOA 2003, s 33(1) if:

(a) for the purpose of obtaining sexual gratification (**11.26**), A intentionally causes another person (B) to watch a third person engaging in an activity, or to look at an image of any person engaging in an activity;

(b) the activity is sexual;

(c) B is unable to refuse because of or for a reason related to a mental disorder; and

(d) A knows or could reasonably be expected to know that B has a mental disorder and that because of it or for a reason related to it B is likely to be unable to refuse.

B is unable to refuse if:

(a) B lacks the capacity to choose whether to agree to watching or looking (whether because B lacks sufficient understanding of the nature of the activity, or for any other reason); or

(b) B is unable to communicate such a choice to A.

Punishment

11.50 An offence under s 30 or 31 is punishable with up to life imprisonment if penetration is proved. Otherwise the above offences are punishable with up to 14 years' imprisonment (s 30 or 31) or 10 years' (s 32 or 33).

Inducements, etc to persons with a mental disorder

Inducement, threat or deception to procure sexual activity with a person with a mental disorder

11.51 A person (A) commits an indictable (either-way) offence under SOA 2003, s 34(1) if:

(a) with the agreement of another person (B) A intentionally touches that person;

(b) the touching is sexual;

(c) A obtains B's agreement by means of an inducement offered or given, a threat made or a deception practised by A for that purpose;

(d) B has a mental disorder; and

(e) A knows or could reasonably be expected to know that B has a mental disorder.

Under s 34(1) and (2), A commits a more serious offence triable only on indictment if the touching involves one of the four types of penetration set out at **11.40**.

It is not necessary in either of these offences or in the case of the related offences under ss 35 to 37 that B's mental disorder made B unable to refuse. These offences are concerned with gaining the consent of a mentally vulnerable person by inducement (eg a reward), a threat (eg 'I won't tell on you if you agree') or a deception (eg 'This is what everyone does').

The exceptions under s 73 (**11.14**) from liability for aiding, abetting or counselling apply to an offence under s 34 if B is under 16.

Causing a person with a mental disorder to engage in, or agree to engage in, sexual activity by inducement, threat, or deception

11.52 A person (A) commits an indictable (either-way) offence under SOA 2003, s 35(1) if:

(a) by means of an inducement offered or given, a threat made or a deception practised by A for this purpose, A intentionally causes another person (B) to engage in, or to agree to engage in, an activity;

(b) the activity is sexual;

(c) B has a mental disorder; and

(d) A knows or could reasonably be expected to know that B has a mental disorder.

A more serious offence triable only on indictment is committed under s 35(1) and (2) if the activity caused or agreed involves one of the four types of penetration set out at **11.47**.

Engaging in sexual activity in the presence, procured by inducement, threat, or deception, of a person with a mental disorder

A person (A) commits an indictable (either-way) offence under SOA 2003, s 36(1) if: **11.53**

(a) A intentionally engages in an activity;
(b) the activity is sexual;
(c) for the purpose of obtaining sexual gratification (**11.25**), A engages in it:
   (i) when another person (B) is present or is in a place from which A can be observed, and
   (ii) knowing or believing that B is aware, or intending that B should be aware, that A is engaging in it;
(d) B agrees to be present or in the place referred to in (c)(i) because of an inducement offered or given, a threat made or a deception practised by A for the purpose of obtaining that agreement;
(e) B has a mental disorder; and
(f) A knows or could reasonably be expected to know that B has a mental disorder.

Causing a person with a mental disorder to watch a sexual act by inducement, threat, or deception

A person (A) commits an indictable (either-way) offence under SOA 2003, s 37(1) if: **11.54**

(a) for the purpose of obtaining sexual gratification (**11.26**), A intentionally causes another person (B) to watch a third person engaging in an activity, or to look at an image of any person engaging in an activity;
(b) the activity is sexual;
(c) B agrees to watch or look because of an inducement offered or given, a threat made or a deception practised by A for the purpose of obtaining that agreement;
(d) B has a mental disorder; and
(e) A knows or could reasonably be expected to know that B has a mental disorder.

Punishment

An offence under s 34 or 35 is punishable with up to life imprisonment if penetration   **11.55** is proved. Otherwise the above offences are punishable with up to 14 years' imprisonment (s 34 or 35) or 10 years' (s 36 or 37).

Offences by care workers with persons with mental disorder

Sexual activity by a care worker with a person with mental disorder

A person (A) commits an indictable (either-way) offence under SOA 2003, s 38(1) if: **11.56**

(a) A intentionally touches another person (B);
(b) the touching is sexual;
(c) B has a mental disorder;
(d) A knows or could reasonably be expected to know that B has a mental disorder; and
(e) A is involved in B's care in a way that falls within the provisions at **11.61**.

A more serious offence triable only on indictment is committed under s 38(1) and (3) if the touching involves one of the four types of penetration set out at **11.40**.

The exceptions under s 73 (**11.14**) from liability for aiding, abetting or counselling apply to an offence under s 38 if B is under 16.

Offences under s 38, and the other 'care worker offences' referred to below, are designed to protect a mentally disordered person with the capacity to refuse consent from exploitation by a care worker of his/her relationship with that person.

Care worker causing or inciting a person with mental disorder to engage in sexual activity

11.57   A person (A) commits an indictable (either-way) offence under SOA 2003, s 39(1) if:

(a) A intentionally causes or incites (**11.20**) another person (B) to engage in an activity;
(b) the activity is sexual;
(c) B has a mental disorder;
(d) A knows or could reasonably be expected to know that B has a mental disorder; and
(e) A is involved in B's care in a way that falls within the provisions at **11.61**.

A more serious offence triable only on indictment is committed under s 39(1) and (3) if the sexual activity caused or incited involves one of the four types of penetration set out at **11.47**.

Sexual activity by care worker in presence of person with mental disorder

11.58   A person (A) commits an indictable (either-way) offence under SOA 2003, s 40(1) if:

(a) A intentionally engages in an activity;
(b) the activity is sexual;
(c) for the purpose of obtaining sexual gratification (**11.25**), A engages in it:
    (i)   when another person (B) is present or is in a place from which A can be observed, and
    (ii) knowing or believing that B is aware, or intending that B should be aware, that A is engaging in it;
(d) B has a mental disorder;
(e) A knows or could reasonably be expected to know that B has a mental disorder; and
(f) A is involved in B's care in a way that falls within the provisions at **11.61**.

Care worker causing person with mental disorder to watch a sexual act

11.59   A person (A) commits an indictable (either-way) offence under SOA 2003, s 41(1) if:

(a) for the purpose of obtaining sexual gratification (**11.26**), A intentionally causes another person (B) to watch a third person engaging in an activity, or to look at an image of any person engaging in an activity;
(b) the activity is sexual;
(c) B has a mental disorder;
(d) A knows or could reasonably be expected to know that B has a mental disorder; and
(e) A is involved in B's care in a way that falls within the provisions at **11.61**.

Rebuttable presumption of mens rea as to mental disorder

11.60   If it is proved at a trial for an offence under ss 38 to 41 that B had a mental disorder, it is to be taken that A knew or could reasonably have been expected to know that B had a mental disorder unless sufficient evidence is adduced to raise an issue as to whether

A knew or could reasonably have been expected to know it (ss 38(2), 39(2), 40(2), and 41(2)).

Definition of care worker

For the purposes of the offences described in ss 38 to 41, a person (A) is involved in the   **11.61**
care of another (B) ('care worker' for short) if para (a), (b), or (c) below applies, that is:

(a) if:
  (i)   B is accommodated and cared for in a care home, community home, voluntary home, children's home, or premises in Wales at which a secure accommodation service is provided; and
  (ii) A has functions to perform in the course of employment in the home (or premises in Wales) which have brought A or are likely to bring A into regular face-to-face contact with B;
(b) if B is a patient for whom services are provided:
  (i)   by a National Health Service body or an independent medical agency; or
  (ii) in an independent clinic (in Wales) or an independent hospital, and A has functions to perform for the body or agency or in the clinic or hospital in the course of employment which have brought him/her or are likely to bring him/her into regular face-to-face contact with B;
(c) if:
  (i)   A is, whether or not in the course of employment, a provider of care, assistance or services to B in connection with B's mental disorder; and
  (ii) as such, has had or is likely to have regular face-to-face contact with B.

Exceptions: marriage, civil partnership, and existing sexual relationships

By SOA 2003, ss 43 and 44, conduct by a person (A) which would otherwise be an   **11.62**
offence under ss 38 to 41 against another person (B) is not such an offence if respectively:

(a) B was 16 or over at the time of the conduct and the defendant proves that A and B were lawfully married (or civil partners) at that time; or
(b) the defendant proves that immediately before A became involved in B's care in a way described above, a sexual relationship existed between A and B. This exception does not apply if at that time sexual intercourse between A and B would have been unlawful, eg because B was under 16.

Punishment

An offence under s 38 or 39 is punishable with up to 14 years' imprisonment if pene-   **11.63**
tration is proved. Otherwise the above offences are punishable with up to 10 years'
imprisonment (s 38 or 39) or seven years' (s 40 or 41).

PREPARATORY OFFENCES

Administering a substance with intent

A person (A) commits an indictable (either-way) offence, punishable with up to 10   **11.64**
years' imprisonment, under SOA 2003, s 61(1) if A intentionally administers a substance to, or causes a substance to be taken by, another person (B):

(a) knowing that B does not consent; and
(b) with the intention of stupefying or overpowering B, so as to enable any person to engage in a sexual activity that involves B.

Although the nature of the substance administered is not restricted provided that it is intended to have the specified effect, this offence is aimed at the activity which has become known as 'date rape' which is frequently carried out following the 'spiking' of a drink with a stupefying drug. However, it would also include circumstances where a person's drink was spiked with strong alcohol if that person believed that (s)he was drinking a non-alcoholic drink or where a person was given an injection or caused to inhale chloroform. It would not, of course, cover circumstances in which a person was merely encouraged to drink alcoholic drinks where that person was aware that (s)he was drinking alcohol.

A need not carry out the act so that *(s)he* may thereby be enabled to engage in sexual activity; it will be sufficient if A does so to enable some other person to so engage in that activity. It does not matter whether sexual activity actually takes place; it is sufficient that A administered the substance, knowing that there was no consent and with the appropriate intention.

Committing an offence with intent to commit a sexual offence

11.65 A person commits an offence under SOA 2003, s 62(1) if (s)he commits any offence with the intention of committing a relevant sexual offence. A 'relevant sexual offence' is any offence under SOA 2003, Part 1 (ss 1 to 79) (which offences are dealt with in this chapter and—in terms of the sexual exploitation of children and prostitution—in Chapter 13), including an offence of aiding, abetting, counselling or procuring such an offence. Generally, an offence under s 62 is an indictable (either-way) offence punishable with up to 10 years' imprisonment. A person who commits an offence under s 62 by kidnapping or false imprisonment is guilty of a more serious offence triable only on indictment and punishable with up to life imprisonment (s 62(1) and (3)).

Trespass with intent to commit a sexual offence

11.66 A person commits an indictable (either-way) offence, punishable with up 10 years' imprisonment, under SOA 2003, s 63(1) if:

(a) (s)he is a trespasser (whether or not (s)he entered as a trespasser) on any premises;
(b) (s)he intends to commit a relevant sexual offence (as defined in the previous paragraph) on the premises; and
(c) (s)he knows that, or is reckless as to whether, (s)he is a trespasser.

MISCELLANEOUS SEXUAL OFFENCES

11.67 All the statutory offences under this heading besides the last (**11.75**) are indictable (either-way) offences punishable with up to two years' imprisonment. The offence at **11.75** is a summary offence punishable with up to six months' imprisonment and/or an unlimited fine.

Exposure

11.68 A person commits an offence under SOA 2003, s 66(1) if:

(a) (s)he intentionally exposes his/her genitals (ie the sexual organs of reproduction); and
(b) (s)he intends that someone will see them and be caused alarm or distress.

Although this offence is framed in gender-neutral terms, its commission by a woman is, of course, extremely uncommon.

The indecent exposure of any part of the male or female body in a place to which the public has access though not necessarily as of right, or a place where what is done is capable of public view, can constitute the common law offence of outraging public decency, if it is capable of being seen by at least two people who are actually present (even if they do not actually see it). This offence is rarely charged and should not be invoked in a case covered by s 66 unless there is good reason for doing so. It is indictable (either-way) and punishable with imprisonment for a period fixed at the discretion of the judge in the Crown Court.

Voyeurism: observing

A person commits an offence under SOA 2003, s 67(1) if: **11.69**

(a) for the purpose of obtaining sexual gratification, (s)he observes another person doing a private act; and
(b) (s)he knows that the other person does not consent to being observed for his/her sexual gratification.

For the purposes of s 67, a person is doing a private act if the person is in a place which, in the circumstances, would reasonably be expected to provide privacy, and:

(a) the person's genitals, buttocks or breasts are exposed or covered only with underwear;
(b) the person is using a lavatory; or
(c) the person is doing a sexual act of a kind not ordinarily done in public.

The Court of Appeal has held that the reference to breasts in (a) is limited to female breasts. The Court has also held that in the context of communal showers or changing rooms there is no reasonable expectation of privacy from casual observation by other users of the communal showers or changing rooms, whereas there is a reasonable expectation of privacy from being spied on in such places by someone who has drilled a hole in the wall for this purpose. Thus, casual observation of other users of communal showers or changing rooms does not amount to the offence of voyeurism, even if the observer gains sexual gratification from what (s)he sees.

A person commits an offence under s 67(4) if (s)he installs equipment, or constructs or adapts a structure or part of a structure, with the intention of enabling him/herself or another person to commit the above offence. 'Structure' includes a tent, vehicle or vessel or other temporary or movable structure.

Voyeurism: operating equipment

A person commits an offence under SOA 2003, s 67(2) if: **11.70**

(a) (s)he operates equipment with the intention of enabling another person to observe, for the purpose of obtaining sexual gratification, a third person (B) doing a private act (as defined above); and
(b) (s)he knows that B does not consent to his/her operating equipment with that intention.

For the purposes of ss 67 and 67A (below), operating equipment includes enabling or securing its activation by another person without that person's knowledge. 'Observation' means any observation, whether direct or by looking at a moving or still image (eg by installing a webcam).

Voyeurism: recording

11.71   A person commits an offence under SOA 2003, s 67(3), if:

(a) (s)he records another person (B) doing a private act (as defined above);
(b) he does so with the intention that (s)he or a third person will, for the purpose of obtaining sexual gratification, look at a moving or still image of B doing the act; and
(c) (s)he knows that B does not consent to his/her recording the act with that intention.

The Court of Appeal has held that s 67(3) is not limited to protecting the privacy of B from secret filming by someone (A) who is not present during the private act in question. A participant in an activity could be guilty of a s 67(3) offence if (s)he secretly recorded what was otherwise a lawful event in which (s)he had participated.

Voyeurism: upskirting

11.72   SOA 2003, s 67A deals with what is currently described as upskirting.
Section 67A(1) provides that a person (A) commits an offence if:

(a) A operates equipment beneath the clothing of another person (B);
(b) A does so with the intention of enabling A or another person (C), for a purpose mentioned in s 67A(3), to observe—
  (i) B's genitals or buttocks (whether exposed or covered with underwear); or
  (ii) the underwear covering B's genitals or buttocks,
  in circumstances where the genitals, buttocks, or underwear would not otherwise be visible; and
(c) A does so—
  (i) without B's consent; and
  (ii) without reasonably believing that B consents.

By s 67A(2), a person (A) commits an offence if:

(a) A records an image beneath the clothing of another person (B);
(b) the image is of—
  (i) B's genitals or buttocks (whether exposed or covered with underwear); or
  (ii) the underwear covering B's genitals or buttock;
  in circumstances where the genitals, buttocks, or underwear would not otherwise be visible,
(c) A does so with the intention that A or another person (C) will look at the image for a purpose mentioned in s 67A(3); and
(d) A does so—
  (i) without B's consent, and
  (ii) without reasonably believing that B consents.

By s 67A(3), the purposes referred to in s 67A(1) and (2) are:

(a) obtaining sexual gratification (whether for A or C);
(b) humiliating, alarming, or distressing B.

Intercourse with an animal

A person commits an offence under SOA 2003, s 69(1) if: **11.73**

(a) he intentionally performs an act of penetration with his penis;
(b) what is penetrated is the vagina or anus of a living animal; and
(c) he knows that, or is reckless as to whether, that is what is penetrated.

In relation to an animal, references to the vagina or anus include references to any similar part.

A person (A) commits an offence under s 69(2) if:

(a) A intentionally causes, or allows, A's vagina or anus to be penetrated;
(b) the penetration is by the penis of a living animal; and
(c) A knows that, or is reckless as to whether, that is what A is being penetrated by.

Sexual penetration of a corpse

A person commits an offence under SOA 2003, s 70 if: **11.74**

(a) (s)he intentionally performs an act of penetration with a part of his/her body or anything else;
(b) what is penetrated is a part of the body of a dead person;
(c) (s)he knows that, or is reckless as to whether, that is what is penetrated;
(d) the penetration is sexual.

Sexual activity in a public lavatory

A person commits an offence under SOA 2003, s 71 if: **11.75**

(a) (s)he is in a lavatory to which the public or a section of the public has, or is permitted to have access, whether on payment or otherwise;
(b) (s)he intentionally engages in an activity; and
(c) the activity is sexual.

For the purposes of s 71, an activity is sexual if a reasonable person would, in all the circumstances but regardless of any person's purpose, consider it to be sexual.

RESTRICTIONS ON EVIDENCE OR QUESTIONS ABOUT COMPLAINANT'S SEXUAL HISTORY

The Youth Justice and Criminal Evidence Act 1999 (YJCEA 1999), s 41 imposes such **11.76** restrictions where a person is charged with (a) any sexual offence under SOA 2003, Part 1 (ss 1 to 79), ie all the offences discussed so far in this chapter, as well as those described in Chapter 13 under SOA 2003, Part 1 relating to sexual exploitation of children or to prostitution; (b) human trafficking committed with a view to sexual exploitation (**13.75-13.76**), or (c) aiding, abetting, counselling, or procuring the commission of such an offence, conspiracy or attempt to commit such an offence, or encouraging or assisting the commission of such an offence.

YJCEA 1999, s 41 requires that, where someone is charged with a sexual offence then, except with the leave of the court, no evidence may be adduced, and no questions may be asked in cross-examination, by or on behalf of any defendant about the 'sexual behaviour' of the complainant. 'Sexual behaviour' means any sexual behaviour or other sexual experience, whether or not involving any defendant or other persons, but excluding (except as specified in italics in (a)(iii) and (b) below) anything alleged to have taken place as part of the event which is the subject matter of the charge.

If the defence wishes to introduce evidence or ask questions about such matters, it must apply to the court for leave to do so. The court can only grant leave when two conditions are satisfied. The first is that a refusal of leave might have the result of rendering unsafe a conclusion of the jury or (as the case may be) of the court on any relevant issue in the case. The second condition is that:

(a) the evidence in question relates to a relevant issue in the case and either:
   (i) that issue is not an issue of consent (for example, if the defence is that the defendant *believed* the complainant was consenting, evidence of recent consensual activity between them is admissible in relation to the issue of the defendant's belief, but not as to whether the complainant had consented), or
   (ii) it is an issue of consent and the complainant's sexual behaviour to which the evidence or question relates is alleged to have occurred at or about the same time as the event which is the subject matter of the charge against the defendant (for example, that the complainant had consented to intercourse with the defendant a couple of hours earlier), or
   (iii) it is an issue of consent and the complainant's sexual behaviour to which the evidence or question relates is so similar either to *any sexual behaviour of the complainant which (according to the defence's version) took place as part of the event which is the subject matter of the charge,* or to any other sexual behaviour of the complainant which (according to the defence's version) took place at or about the same time as that event, that the similarity cannot reasonably be explained as a coincidence; or
(b) the evidence or question relates to *any evidence adduced by the prosecution about any sexual behaviour of the complainant,* and would go no further than necessary to enable the evidence adduced by the prosecution to be rebutted or explained by the defendant.

To be admitted under the above provisions, the evidence must relate to a specific instance, or instances, of sexual behaviour. If the court considers that the purpose (or main purpose) of the evidence which the defence seeks to have admitted is to undermine or diminish the complainant's credibility, it will not allow the evidence to be given.

The House of Lords has held that under (a)(iii) the test of admissibility is whether the evidence, and questioning relating to it, is so relevant to the issue of consent that to exclude it would endanger the fairness of the trial, contrary to the defendant's right to a fair trial under the European Convention on Human Rights, art 6. If that test is satisfied, the sexual behaviour evidence should not be excluded.

ANONYMITY OF COMPLAINANTS

11.77 The Sexual Offences (Amendment) Act 1992 (SO(A)A 1992) provides for the anonymity of complainants in the case of any offence under the provisions of SOA 2003, Part 1, except ss 64 and 65 (sex with an adult relative), 69 (intercourse with an animal), and 71 (sexual activity in a public lavatory). The Act also applies to the offence of human trafficking, contrary to the Modern Slavery Act 2015, s 2 (**13.75–13.76**).

Where an *allegation* is made that one of these offences has been committed against a person, no matter relating to that person may, during that person's lifetime, be included in any publication if it is likely to lead members of the public to identify that person as the person against whom the offence was committed. These matters include particularly (if their inclusion in any publication is likely to have such a result):

(a) the person's name or address;
(b) the identity of any educational establishment attended by the person or of any place of work; and
(c) any still or moving picture of the person.

For the purpose of the anonymity provisions, 'publication' is widely defined. It includes any communication, for example, a radio programme, a film, a written communication, or a speech, addressed to the public at large or a section of the public.

This prohibition ceases to apply once a person has been accused of one of the above offences. Thereafter, a corresponding provision applies whereby, where a person has been accused of one of the above offences, no matter whatsoever (including (a)–(c) above) which is likely to lead to the identification of the complainant may be included in any publication during the complainant's lifetime. Provision is made for this prohibition to be set aside or relaxed on application being made to the trial judge where it is considered necessary to induce persons to come forward as witnesses and that the applicant's defence will be substantially prejudiced if the direction is not given. The judge also has a 'public interest' discretion to set aside the prohibition.

If any matter is published in breach of the above rules, a summary offence (**6.48**), punishable with an unlimited fine, is committed under SO(A)A 1992, s 5(1):

(a) in the case of a newspaper or periodical, by any proprietor, editor or publisher of it;
(b) in the case of a publication in a programme, by any body corporate engaged in providing the service and by anyone involved in the programme corresponding to an editor of a newspaper; or
(c) in the case of any other publication, by its publisher.

INFORMATION ABOUT GUESTS AT HOTELS BELIEVED TO BE USED FOR CHILD SEXUAL EXPLOITATION

11.78   Under the Anti-social Behaviour, Crime and Policing Act 2014, s 116 an inspector (or above) may issue a written notice to the owner, operator, or manager of a hotel which (s)he reasonably believes has been or will be used for purposes of child sexual exploitation, or conduct that is preparatory to, or otherwise connected with, such sexual exploitation.

'Child sexual exploitation' means conduct that constitutes an offence listed in (a) or (b), or an offence in (c) or (d) against someone under 18, namely:

(a) an offence under any of SOA 2003, ss 5 to 13, 16 to 19, 25 and 26 (all dealt with above), and 47 to 50 (sexual exploitation of children: **13.64–13.69**);
(b) an offence under the Protection of Children Act 1978, s 1 (indecent photographs of children: **12.6–12.14**);
(c) an offence under any of SOA 2003, ss 1 to 4, 30 to 41, 61, 66 and 67 (all dealt with previously), and s 59A (trafficking people for sexual exploitation) (repealed), and
(d) an offence under the Modern Slavery Act 2015, s 2 (human trafficking) committed with a view to sexual exploitation (**13.75–13.76**).

The notice must specify the hotel to which it relates and the dates on which it comes into effect and expires (not more than six months after it comes into effect), and explain the effect of: the provisions in the following paragraph, the provisions about the right of appeal to a magistrates' court against a notice within 21 days of its issue, and the provisions of s 118 (below).

A constable may require someone issued with a notice to provide him/her with information about the names and addresses of guests at the hotel.

A requirement must be in writing, and specify the period to which it relates and the date(s) on or by which the required information must be provided. The period specified must end no later than the expiry of the notice.

'Guest' means someone who, for a charge payable by him/her or another, has the use of a guest room at the hotel in question; and 'hotel' includes any guest house or similar establishment at which accommodation is provided for a charge.

A person who fails, without reasonable excuse, to comply with a requirement imposed under s 116 commits an offence (s 118(1)). A person commits an offence if, in response to a requirement imposed under s 116, (s)he provides incorrect information which (s)he did not take reasonable steps to verify or to have verified, or knows to be incorrect (s 118(2)). A person does not commit an offence under s 118(2) on grounds of failure to take reasonable steps to verify it or to have it verified if there were no steps that (s)he could reasonably have taken to verify the information or to have it verified (s 118(3)). An offence under s 118(1) or (2) is a summary offence punishable with a fine not exceeding level 4.

NOTIFICATION REQUIREMENTS

11.79   SOA 2003, Part 2 requires a 'relevant offender' (a person subject to the notification requirements under the Act: hereafter RO) to make an initial notification to the police of his/her date of birth, National Insurance number, name and home address (and any other name and any other address which (s)he uses), and any other information which has been prescribed. The Sexual Offences Act 2003 (Notification Requirements) (England and Wales) Regulations 2012 prescribe the following additional information:

(a) where an RO resides, or stays for a period of at least 12 hours, at a relevant household (a place where a child resides or stays, and to which the public do not have access), (s)he must notify:
    (i)   the date on which (s)he begins to reside or stay at a relevant household,
    (ii) the address of the relevant household, and
    (iii) where (s)he holds such information, the period or periods for which (s)he intends to reside or stay at the relevant household;

(b) where an RO solely or jointly holds an account with a banking institution (defined as a bank, building society or any other institution providing banking services), a debit card in relation to such an account, a credit card account, or a credit card, (s)he must notify specified details relating to it;

(c) where an RO holds any passport, other identity document, or (in a case where (s)he does not hold any passport or other identity document) any other document in which his/her full name appears, (s)he must notify the number and his/her full name as it appears in the document and (except in the case of a passport) describe the document. If (s)he holds a passport the information must be given in relation to it. If (s)he does not, the information must be given in relation to another identity document if (s)he holds one. If (s)he does not, it must be given in relation to any other document.

Notification by the RO must be made within three days of the relevant date and must be made by attending at any prescribed police station in his/her local police area, and giving an oral notification to any police officer, or to any person authorised for the purpose by the officer in charge of the station. The relevant date is the date of conviction, finding or caution but the three-day period does not include any time when the RO is in custody by a court order or in prison, detained in hospital, or outside the UK. On giving notification, the RO may be required by the police officer or authorised person to have his/her fingerprints and a photograph taken.

After such an original notification, an RO must give notice, within a three-day period of the change, of a new (ie unnotified) name; of a change of home address; of a new (ie unnotified) place of residence in the UK where (s)he has resided for a qualifying period of seven days, or periods amounting to seven days in a 12-month period; of his/her release from custody, imprisonment or detention in hospital; or of any prescribed change of circumstances. The 2012 Regulations prescribe the following changes of circumstance:

(a) where the RO resides, or stays for a period of at least 12 hours, at a relevant household in relation to which there has been no initial notification, or ceases to reside or stay at a relevant household in relation to which there has been an initial notification. A notification of such a change must disclose the date from which (s)he resides or stays, or the date on which (s)he ceases to reside or stay, at a relevant household;

(b) where the RO's account is opened or closed, a debit or credit card is obtained, no longer held, or has expired and information previously notified by him/her has altered or become inaccurate or incomplete. A notification of such a change must give the same specified details about the account or card as an initial notification;

(c) where the RO obtains a passport, other identity document or other document in relation to which there has been no initial notification, or ceases to hold a passport, other identity document or other document in relation to which there has been an initial notification.

Although notification is permitted before use of the new name or the change of address, if the change subsequently takes place more than two days before the date notified that notification is invalid. Moreover, if the event to which an advance notification relates has not occurred within three days of the date notified, the offender must within six days of the notified date notify the police that it did not occur within that period.

An RO must notify the police of his/her intention to leave the UK, and of his/her return, in accordance with the Sexual Offences Act 2003 (Travel Notification Requirements) Regulations 2004.

An RO must also annually re-notify the police of the information required to be given on initial notification. By way of exception, where the home address initially notified, or notified as a change, was a location where the RO can regularly be found in the absence of his/her having a fixed abode in the UK, the duty to re-notify is not an annual one but is to re-notify every seven days. Where an RO is in custody by a court order, in prison, detained in hospital, or outside the UK at the relevant time, the re-notification period is three days from release, or return to the UK.

The persons subject to the notification requirements under SOA 2003, Part 2 (ie ROs) are those convicted of an offence listed in SOA 2003, Sch 3; those found not guilty of such an offence by reason of insanity; those found to be under a disability (ie unfit to plead) and to have done the act charged in respect of such an offence; and those cautioned in respect of such an offence. Persons previously convicted of offences which made them subject to the Sex Offenders Act 1997, Part 1 remain subject to the notification requirements then imposed, until the completion of the period for which notification was required.

Schedule 3 offences

11.80   The schedule includes offences under SOA 2003 of rape, assault by penetration, most sexual assaults, child sex offences, sexual grooming, sexual communication with a child, offences involving an abuse of trust in relation to children, familial child sexual offences, offences against persons with a mental disorder including offences by care workers, sexual exploitation of children offences, administering substances with intent, trespassing with intent to commit a sexual offence, sex with an adult relative, exposure, voyeurism offences, intercourse with an animal, and sexual penetration with a corpse. In brief, this list includes every offence under SOA 2003, Part 1 (ss 1 to 79) besides those relating to prostitution (other than prostitution involving sexual exploitation of a child) or to sexual activity in a public lavatory. The listed offences also include offences relating to an indecent photograph or pseudo-photograph of a child, possession of a paedophile manual, possession of an extreme pornographic image, and possession of a prohibited image of a child. Encouraging or assisting, attempting, conspiring to commit, and aiding and abetting, such offences are also Sch 3 offences. Reference should be made to the Schedule as the offences often only give rise to the notification requirements in particular circumstances or in the case of particular sentences.

Duration of notification requirement

11.81   Relevant offenders (ROs) sentenced to life imprisonment, or to 30 months' imprisonment or more, or persons sentenced to imprisonment for public protection under the Criminal Justice Act 2003, s 225 (repealed), or persons admitted to hospital under a restriction order, are indefinitely subject to the notification requirement; ROs sentenced to more than six months but less than 30 months are subject for a period of 10 years; ROs sentenced to less than six months or admitted to a hospital without a restriction order are subject for a period of seven years; and ROs of any other description, for a period of five years. However, where ROs are under 18, the periods of 10, seven, and five years are halved. ROs who are cautioned are subject to the requirements for two years. Those who are conditionally discharged are subject to the requirements during the period of conditional discharge.

The Court of Appeal has held that the whole of the term of an extended sentence (ie the aggregate of the appropriate custodial term and the extension period for which the offender is to be subject to a licence) constitutes the term for which a person is 'sentenced to imprisonment' for the purposes of determining the notification period.

In the case of a young offender the court may direct a person with parental responsibility for the offender to comply with the notification requirements until the offender reaches the age of 18 or until an earlier date specified in the order.

The Sexual Offences Act 2003 (Remedial) Order 2012 provides that an RO subject to an indefinite notification requirement, and not subject to a sexual harm prevention order (or an interim order), may apply in writing to the relevant chief officer of police for the area in which (s)he is residing or staying (according to his/her latest notification) for a determination that (s)he is no longer subject to the indefinite notification requirements. Such an application must be made on or after the qualifying date or, as the case may be, the further qualifying date. Where RO was 18 or over when the notification requirement began, the qualifying date is the day after the end of the 15-year period beginning with the day on which the RO gives the first notification after release from prison or other detention in relation to the conviction giving rise to the indefinite requirement. Where (s)he was under 18 when the requirement began, it is the day after the end of the eight-year period beginning with that day. As to the further qualifying date, see below.

For the purposes of the determination of an application for review, the RO must satisfy the chief officer of police on the balance of probabilities that it is not necessary for the purpose of protecting the public or any particular members of the public from sexual harm for him/her to remain subject to the indefinite notification requirements.

If the chief officer determines that the RO should not remain subject to the indefinite notification requirements, the RO ceases to be subject to them on the date of receipt of the notice of determination.

If the chief officer determines that the RO should remain subject to the indefinite notification requirements, the notice of the determination must:

(a) state the reasons for the determination; and
(b) inform the RO of his/her right to appeal within 21 days to a magistrates' court.

In the event of such a determination, the further qualifying date (ie for another review) will be eight years from the determination, or up to 15 years (as determined by the chief officer) if (s)he considers that the risk of sexual harm posed by the RO is sufficient to justify a continuation of those requirements after the end of the eight-year period beginning with the day on which the determination is made.

A divisional court has held that a chief officer's functions under SOA 2003 can be delegated to an officer of an appropriate level; a superintendent (or above) in the present context.

Offences relating to notification requirements

It is an indictable (either-way) offence under SOA 2003, s 91(1) for a person: **11.82**

(a) to fail without reasonable excuse to comply with the notification requirements;
(b) to notify to the police, in purported compliance with these requirements, any information which (s)he knows to be false.

Such an offence is punishable with up to five years' imprisonment.

PREVENTIVE ORDERS

Notification orders

Under SOA 2003, s 97, a chief officer of police (see last paragraph of **11.81**) may apply   **11.83**
to a magistrates' court for a civil order (a notification order) in respect of a defendant who resides in his/her police area, or whom (s)he believes is in, or is intending to come to, his/her police area, where it appears to him/her that the following conditions are satisfied with respect to the defendant (D):

(a) under the law in force in a country outside the UK:
   (i) D has been convicted of a relevant offence (whether or not D has been punished for it),
   (ii) a court exercising jurisdiction under that law has made in respect of a relevant offence a finding equivalent to a finding that D is not guilty by reason of insanity,

(iii) such a court has made in respect of a relevant offence a finding equivalent to a finding that D is under a disability and did the act charged against him/her in respect of that offence, or

(iv) D has been cautioned in respect of the relevant offence;

(b) condition (a) above is met because of a conviction, finding or caution which occurred on or after 1 September 1997 (or although D was convicted before that date D was dealt with after it) or because, although there was a conviction before that date, D was detained in the foreign country, or subject to supervision or a community sentence; and

(c) the notification requirement period specified (**11.81** applies with appropriate modification) in respect of the relevant offence has not expired.

If it is proved that these conditions exist the court must make a notification order. A 'relevant offence' means an act which constituted an offence under the law in force in the country concerned and would have constituted an offence listed in SOA 2003, Sch 3 if it had been done in any part of the UK.

The effect of a notification order is to subject the defendant to the notification requirements under SOA 2003, Part 2 for a period corresponding to that which would apply if (s)he had been convicted in the UK. The notification period runs from the date of the foreign conviction, etc. The initial notification must be made within three days of the service of the order.

Provision is made for an 'interim notification order' pending determination of an application for a notification order. Where a relevant offence has been committed abroad it may take some time to obtain the necessary documentation to put before the court. A person subject to an interim order is subject to the notification requirements while the order is in force and must make that notification within three days of service of the order.

Sexual harm prevention orders

11.84   Sexual harm prevention orders (SHPOs) replaced sexual offences prevention orders (SOPOs) and foreign travel orders in 2015. SHPOs are governed by SOA 2003, ss 103A to 103K and the Sentencing Code, ss 343 to 358 whose terms are almost wholly identical, except for the grounds for making orders under them.

Effect of SHPO

11.85   A SHPO prohibits the defendant (D) from doing anything described in the order. With one exception, a prohibition contained in a SHPO has effect for a fixed period, specified in the order, of at least five years, or until further order. The exception is that a prohibition on foreign travel contained in a SHPO must be for a fixed period of not more than five years, although it may be extended for a further such period.

A SHPO may specify:

(a) that some of its prohibitions have effect until further order and some for a fixed period;

(b) different periods for different prohibitions.

The only prohibitions that may be included in a SHPO are those necessary for the purpose of protecting:

(a) the public in the UK or any particular members of the public from sexual harm from D, or

(b) children (persons under 18) or vulnerable adults generally, or any particular children or vulnerable adults, from sexual harm from D outside the UK.

'Vulnerable adult' means someone aged 18 or over whose ability to protect him/herself from physical or psychological harm is significantly impaired through physical or mental disability or illness, through old age or otherwise.

A 'prohibition on foreign travel', referred to above, means a prohibition on travelling:

(a) to any country outside the UK named or described in the order;
(b) to any country outside the UK other than a country named or described in the order; or
(c) to any country outside the UK (in which case the order must require D to surrender all his/her passports at a police station specified in the order on or before the date when the prohibition takes effect, or within a period so specified).

Any passports surrendered under (c), other than a non-UK passport returned to its issuing authority, must be returned as soon as reasonably practicable after D ceases to be subject to a SHPO containing a prohibition within (c).

A person subject to a SHPO (or an interim order under **11.89**) is subject to the notification requirements of SOA 2003, Part 2 for the duration of the order.

Where a court makes a SHPO in relation to a person who is already subject to such an order (whether a SHPO under SOA 2003 or a SHPO under the Sentencing Code made on conviction: see below) the earlier order ceases to have effect.

When can a SHPO be made?

A SHPO may be made in three cases:

11.86

(a) by the Sentencing Code, s 345, a criminal court may make a SHPO when dealing with a person convicted of a 'listed offence';
(b) by SOA 2003, s 103A, a criminal court may make a SHPO when dealing with a person who has been found not guilty of a 'listed offence' by reason of insanity, or found to be under a disability and to have done the act charged in respect of such an offence; and
(c) by SOA 2003, s 103A, a magistrates' court (or youth court) may make a SHPO on application by the police or NCA (SHPO on application).

SHPO after conviction

The Sentencing Code, ss 345 and 346 provide that where:

11.87

(a) D is convicted of an offence listed in SOA 2003, Sch 3 or 5 (**11.80** and fourth paragraph of **11.89**); and
(b) the court dealing with D in respect of the offence is satisfied that it is necessary to make a SHPO for the purpose of protecting the public or any particular members of the public in the UK from sexual harm (below) from D, or protecting children (persons under 18) or vulnerable adults generally, or any particular children or vulnerable adults, from sexual harm from the offender outside the UK,

the court may make a SHPO.

'Sexual harm' means physical or psychological harm caused by D committing a Sch 3 offence or offences, or (in the case of harm outside the UK) by D doing, outside the UK, anything which would constitute a Sch 3 offence if done in any part of the UK. A 'vulnerable adult' has the same meaning as defined in **11.85**.

SHPO after a finding of insanity or of disability, etc in criminal proceedings

11.88   Where:

(a) a court deals with D after a finding that D is not guilty of an offence listed in SOA 2003, Sch 3 or 5 by reason of insanity, or a finding of disability and to have done the act charged in respect of such an offence; and

(b) the court is satisfied that it is necessary to make a SHPO for the purpose of protecting the public or any particular members of the public in the UK from sexual harm (**11.87**) from D, or protecting children or vulnerable adults (above) generally, or any particular children or vulnerable adults, from sexual harm from D outside the UK,

the court may make a SHPO.

SHPO on application

11.89   A chief officer of police or the Director General of the NCA (the Director General) may apply by complaint (ie in civil proceedings, as to which see **13.47**) to a magistrates' court (or youth court) for a SHPO in respect of D if it appears to him/her that D is a qualifying offender and has since the appropriate date acted in such a way as to give reasonable cause to believe that it is necessary for a SHPO to be made. The 'appropriate date' is the date (or first date) on which D was convicted, found, or cautioned as mentioned below. As in the case of applications for other preventive orders, a chief officer or the Director General can delegate this function to an officer of an appropriate level.

On such an application, the court may make a SHPO if it is proved that D is a 'qualifying offender', and the court is satisfied that D's behaviour since the 'appropriate date' makes it necessary to make a SHPO, for the purpose of protecting the public or any particular members of the public from sexual harm (**11.87**) from D, or protecting children or vulnerable adults generally, or any particular children or vulnerable adults, from sexual harm from D outside the UK.

Where the Director General has made such an application, (s)he must as soon as practicable notify the chief officer of police for a relevant police area of the application.

A person is a 'qualifying offender' for these purposes if (s)he has been:

(a) convicted of a Sch 3 or Sch 5 offence;

(b) found not guilty of such an offence by reason of insanity;

(c) found to be under a disability and to have done the act charged against him/her in respect of such an offence; or

(d) cautioned in respect of such an offence.

Schedule 3 offences have been described at **11.80**, but it should be noted that any restrictions referred to in Sch 3 in terms of particular circumstances or penalties do not apply where a Sch 3 offence is in issue for the purpose of a SHPO. *Schedule 5 offences* include offences of child cruelty or neglect, etc, child abduction, outraging public decency, as well as murder and other offences of serious violence, harassment, stalking, exploitation of prostitution offences, and human trafficking.

A person is also a 'qualifying offender' for these purposes if, under the law of a country outside the UK:

(a) (s)he has been convicted of a relevant offence;

(b) a court exercising jurisdiction under that law has made, in respect of a relevant offence, a finding equivalent to a finding of not guilty by reason of insanity, or a finding equivalent to a finding that (s)he is under a disability and did the act charged against him/her in respect of the offence; or

(c) (s)he has been cautioned in respect of a relevant offence.

A 'relevant offence' means an act which constituted an offence under the law in force in the country concerned, and would have constituted a Sch 3 or Sch 5 offence if it had been done in any part of the UK.

For these purposes, the 'appropriate date', in relation to a qualifying offender, means the date or (as the case may be) the first date on which (s)he was convicted, found or cautioned as mentioned above.

Under s 103F, *interim SHPOs* may be made in appropriate circumstances by the magistrates' court to which an application for a SHPO is or has been made pending determination of the application for a SHPO. The applicant or D may apply to the court that made the interim SHPO for the order to be varied, renewed or discharged. On an application for an interim SHPO made by a chief officer of police, the court may, if it considers it just to do so, make an interim notification order (either in addition to or instead of an interim SHPO).

On an application for a SHPO made by a chief officer of police, the court must make a notification order in respect of D (either in addition to or instead of a SHPO) if the applicant invites the court to do so, and it is proved that the three conditions ((a)–(c) at **11.83**) for such an order are met.

Variation, renewal, and discharge

A SHPO made under **11.87**, **11.88**, or **11.89** may be varied, renewed, or discharged but (unless its only prohibition is one on foreign travel) it may not be discharged before the end of five years beginning with the day on which the order was made, without the consent of D and (a) where the application is made by a chief officer of police, that chief officer, or (b) in any other case, the chief officer of police for the area in which D resides.

11.90

The Court of Appeal has held that it would be wrong for a court to vary a SHPO or other preventive order merely because the applicant thought that some new restriction should have been included in the order. However, it accepted that a variation is not precluded, even though there has been no change of circumstances, where the evidence which provides the basis for the variation was not known by the court who made the original order, but should have been.

Breach

Breach of a prohibition in a SHPO (or an interim SHPO) without reasonable excuse is an offence under the Sentencing Code, s 354(1) (SHPO under **11.87**) or under SOA 2003, s 103I(1) (SHPO under **11.88** or **11.89**). The offence under SOA 2003, s 103I(1) also applies to breach of a SOPO or interim SOPO.

11.91

Failure, without reasonable excuse, to comply with the passport-surrender condition in a prohibition on foreign travel is an offence under the Sentencing Code, s 354(3) (**11.87**: SHPO after conviction) or SOA 2003, s 103I(2) (**11.88** or **11.89**: SHPO in other cases).

D only has an evidential burden in respect of these defences of reasonable excuse.

The above offences are indictable (either-way) offences punishable with up to five years' imprisonment.

Sexual risk orders

Sexual risk orders (SROs) replaced risk of sexual harm orders (RSHOs) in 2015. SROs are governed by SOA 2003, s 122A to 122K.

11.92

Effect of order

By SOA 2003, s 122A(7), a SRO prohibits D from doing anything described in the order; it has effect for a fixed period (not less than two years) specified in the order or

11.93

until further order. A prohibition on foreign travel contained in a SRO must not be for a period of more than five years, although it may be extended for a further such period. It may specify different periods for different prohibitions. The only prohibitions that may be imposed are those necessary for the purpose of protecting the public in the UK or any particular members of the public from physical or psychological harm caused by D doing an act of a sexual nature, or protecting children or vulnerable adults generally, or any particular children or vulnerable adults, from such harm from D outside the UK (s 122A(9)). 'Child' and 'vulnerable adult' mean the same as for a SHPO (**11.85**).

By s 122C, the same rules apply to a 'prohibition on foreign travel', referred to earlier, as apply to such a prohibition in relation to a SHPO (**11.85**).

Section 122F provides that a person in respect of whom a court makes a SRO (other than one that replaces an interim SRO), or an interim SRO, must, within three days beginning with the date of service of the order, notify to the police his/her name (and, where (s)he uses more than one, each of his/her names), and his/her home address, unless (s)he is already subject to the notification requirements. If (s)he uses a name which has not been notified, or changes his/her address, (s)he must notify the police of this within three days beginning with the date on which this happens.

Where a court makes a SRO in relation to someone already subject to a SRO the earlier order ceases to have effect.

Applications

11.94 By SOA 2003, s 122A(1), a chief officer of police or the Director General of NCA may apply by complaint (ie in civil proceedings) to a magistrates' court (or youth court) for a SRO in respect of D if it appears to him/her that D has done an act of a sexual nature as a result of which there is reasonable cause to believe that it is necessary for a SRO to be made. As in the case of applications for other preventive orders, a chief officer or the Director General can delegate this function to an officer of an appropriate level.

The Director General must as soon as practicable notify the chief officer of police for a relevant police area of any application that (s)he has made.

On such an application, the court may make a SRO if satisfied that D has done an act of a sexual nature as a result of which it is necessary to make such an order for the purpose of protecting the public or any particular members of the public from harm from D, or protecting children or vulnerable adults generally, or any particular children or vulnerable adults, from harm from D outside the UK.

Under s 122E, interim SROs may be made. The rules about interim SROs are identical to those which apply to an interim SHPO (**11.89**).

Variation etc

11.95 A SRO may be varied, renewed or discharged but (unless its only prohibition is one on foreign travel) it may not be discharged before the end of two years beginning with the day on which the order was made, without the consent of D and (a) where the application is made by a chief officer of police, that chief officer, or (b) in any other case, the chief officer of police for the area in which D resides.

Breach

11.96 Breach of a prohibition in a SRO or interim SRO without reasonable excuse is an offence under SOA 2003, s 122H(1). This offence also applies to breach of a RSHO or interim RSHO.

Failure, without reasonable excuse, to comply with the passport-surrender condition in a prohibition on foreign travel is an offence under s 122H(2).

D only has an evidential burden in respect of these defences of reasonable excuse.

The above offences are indictable (either-way) offences punishable with up to five years' imprisonment.

Where:

(a) D is convicted of an offence under s 122H, or is found not guilty of such an offence by reason of insanity or to be under a disability and to have done the act charged, or is cautioned for such an offence; and

(b) D was subject to the notification requirements under SOA 2003, Part 2; and

(c) D would otherwise cease to be subject to those requirements while the relevant order has effect,

D remains subject to the notification requirements.

Where D was not subject to the notification requirements at the time of such conviction, 'finding' or caution, D becomes subject to the notification requirements and remains so until the relevant order ceases to have effect. In such a case, the 'relevant date' for the purposes of those requirements is the date when this provision first applies to D.

Application of orders throughout UK

SOA 2003, s 136ZA provides that, for the purposes of the offences under that Act relating to their breach, the prohibitions imposed by a SHPO (whether under SOA 2003 or the Sentencing Code), interim SHPO, SRO, or interim SRO (or one of the orders which they have replaced) made in one part of the UK apply (unless expressly confined to a particular location) throughout that and every other part of the UK.   **11.97**

GUIDANCE

The Secretary of State has issued guidance for police and legal practioners regarding the provisions about notification requirements and preventive orders under SOA 2003.   **11.98**

CHAPTER 12

Indecent Images, Material, and Communications

OBSCENE PUBLICATIONS

12.1 Under the Obscene Publications Act 1959 (OPA 1959), s 2(1) an indictable (either-way) offence, punishable with up to five years' imprisonment, is committed by a person who:

(a) publishes an obscene article, whether for gain or not; or
(b) has an obscene article for publication for gain, whether for him/herself or another.

Writing or otherwise creating an obscene article is not an offence under OPA 1959, nor is simply possessing it or possessing it to give as a gift.

Obscene publication via a broadcast or cable service is a rather specialised type of obscene publication, and is not dealt with further in this book.

What is an 'obscene article'?

12.2 An 'article' for present purposes means any article containing or embodying matter to be read or looked at or both, any sound record, and any film or other record of a picture or pictures (such as a photograph, video cassette, or computer disc). OPA 1959 also applies to things like negatives from which an article may be reproduced or manufactured.

An article is deemed to be obscene if its effect, or (where the article comprises two or more distinct items) the effect of any of its items is, taken as a whole, such as to tend to deprave and corrupt a significant proportion of persons who are likely, having regard to all the circumstances, to read, see or hear the matter contained in it.

Though a novel may be considered as a whole, a magazine must be considered item by item and, if any one of the items is obscene, this suffices. 'Deprave and corrupt' are strong words; to lead morally astray is not necessarily to deprave and corrupt. Obscenity is not confined to that which has a tendency to corrupt sexual morals; a book depicting the career of a drug addict has been held to be obscene because of its likely effect. On occasions, articles may be directed at persons who may already be considered to have become depraved and corrupted but this might still amount to an offence if the object is to maintain that state of depravity and corruption and to prevent escape from it. Whether an article is obscene because it is likely to deprave and corrupt is essentially a matter for the jury or magistrates.

Publication, etc

12.3 To 'publish' means to distribute, circulate, sell, let on hire, give, or lend, or to offer for sale or hire. 'Publishing' also includes 'making available', as where X gives Y a key to a library containing obscene articles. Additionally, in the case of a record, a film, etc 'publish' includes showing, playing, or projecting it. A person who enables another to access electronically stored obscene matter on a computer 'shows' the other that matter, and therefore 'publishes' it. Where the matter is data stored electronically, a person publishes it if (s)he transmits that data.

As implied above, there can be a publication for the purposes of OPA 1959 even if the obscene material has been communicated only to one person.

Defences

Under OPA 1959, s 2(5), it is a defence for D to prove (s)he had not examined the article and had no reasonable cause to suspect that it was such that his/her publication of it, or possession, as the case may be, would make him/her liable under OPA 1959.   **12.4**

Section 4 provides a defence of 'public good'. If it is proved that publication was justified as being for the public good on the grounds that it is in the interests of science, literature, art or learning, or of other objects of general concern, a person must not be convicted of the above offences. Expert evidence may be given to establish or negative such a defence.

Police powers

By OPA 1959, s 3, if an information on oath is laid before a justice that there are reasonable grounds for suspecting that obscene articles are kept on any premises, stall or vehicle for publication for gain, (s)he may issue a warrant authorising a constable to search for and seize any articles which the constable has reason to believe to be obscene and to be kept for publication for gain. The additional powers of seizure provided by the Criminal Justice and Police Act 2001, s 50 (**4.61**) apply where such a search warrant is executed. If the justice considers that any articles seized are obscene the justice may issue a summons to the occupier of the premises, etc to appear before a magistrates' court and show cause why the articles should not be forfeited. If the court is satisfied that the articles seized are obscene and kept for publication for gain, it must order their forfeiture, unless the 'public good defence' applies.   **12.5**

INDECENT PHOTOGRAPHS OR PSEUDO-PHOTOGRAPHS OF CHILDREN

The Protection of Children Act 1978 (PCA 1978), s 1 and the Criminal Justice Act 1988 (CJA 1988), s 160 provide a range of indictable (either-way) offences (**6.48**) relating to such photographs or pseudo-photographs. An offence under PCA 1978 is punishable with up to 10 years' imprisonment, and one under CJA 1988, s 160 five years'.   **12.6**

General definitions

Photographs or pseudo-photographs

References in the offences to an indecent photograph include an indecent film, a copy of an indecent photograph or film, and an indecent photograph comprised in a film. References to photographs also include the negative as well as the positive version, and data stored on a computer disc or by other electronic means which is capable of conversion into a photograph. References to a photograph also include (a) a tracing or other   **12.7** image whether made by electronic or other means (of whatever nature) which is not itself a photograph or pseudo-photograph, but which is derived from the whole or part of a photograph or pseudo-photograph (or a combination of either or both); and (b) data stored on a computer disc or by other electronic means which is capable of conversion into an image within (a). Such items could be described as 'derivative photographs'. Such derivatives will include line-traced and computer-traced images such as might be taken on a mobile telephone.

A 'pseudo-photograph' is an image, whether made by computer graphics or otherwise howsoever, which appears to be a photograph. The term includes:

(a) a copy of an indecent pseudo-photograph; and
(b) data stored on a computer disc or by other electronic means which is capable of conversion into an indecent pseudo-photograph.

An exhibit obviously consisting of parts of two different photographs taped together cannot be said to 'appear to be a photograph' and is therefore not a pseudo-photograph, although, if it was itself photocopied, it could be.

Indecent

12.8   The term 'indecent' is an objective one. The test is whether the image was indecent according to recognised standards of propriety, as judged by the ordinary person. A photograph or pseudo-photograph can be indecent even though it is not obscene. The child's age (or in the case of a pseudo-photograph apparent age) is a relevant factor as to whether or not the photograph or pseudo-photograph is indecent, but the circumstances in which it was taken or made, or the motivation of the photographer (or the maker of a pseudo-photograph), is not. Something abstracted from a decent set of images is capable of being indecent if the abstracted matter satisfies the test of indecency. This was held by the Court of Appeal in a case where D copied a television programme of a medical examination of a boy's genitals and subsequently removed the commentary and slowed down the filming. The Court of Appeal held that the jury were entitled to look at the images independently of the original decent television programme in order to decide whether they were indecent.

The Court of Appeal has emphasised that the categorisation of images into various levels for sentencing purposes is irrelevant to the question of whether an image is indecent.

Child

12.9   A photograph or pseudo-photograph must be of a child. In the case of a photograph, a child is someone under 18. A person in a photograph is taken to have been a child at the material time if it appears from the evidence as a whole that (s)he was then under 18. It is for a jury or magistrates to decide whether an unknown person in an indecent photograph was under 18. Such case law as there is on the point indicates that whether that person was under 18 must be decided without the assistance of expert evidence, except where it shows only the lower part of his/her body. Since a pseudo-photograph will not be an image of a real person with a real age, it is provided that, if the impression conveyed by the pseudo-photograph is that the person shown is a child under 18, the pseudo-photograph is to be treated for all purposes as showing such a child, and so is a pseudo-photograph where the predominant impression conveyed is that the person shown is a child under 18 notwithstanding that some of the physical characteristics shown are those of an adult.

Taking or making

PCA 1978, s 1(1)(a) provides that, subject to ss 1A and 1B, it is an offence for a person **12.10**
to take, or permit to be taken, or to make any indecent photograph or pseudo-photograph of a child.

A person (D) who intentionally downloads an indecent image of a child from a web
page or via email 'makes' a photograph or pseudo-photograph. If D does this knowing
that the image was, or was likely to be, an indecent one of a child, D commits an offence.
Where D takes an indecent photograph of a child deliberately, D commits an offence,
and such knowledge is not required.

It is irrelevant whether or not D's motive is sexual gratification.

Even if the images originated outside England and Wales, the downloading or
printing of them in England and Wales creates new material which has been *made* inside England and Wales.

Defences

PCA 1978, s 1A provides a limited defence of consent by the child or reasonable belief **12.11**
in consent. It states that, where in such proceedings, D proves that the photograph or
pseudo-photograph was of the child *aged 16 or over*, and that at the time of the offence
charged the child and D were married or civil partners of each other, or lived together
as partners in an enduring family relationship, and sufficient evidence is adduced to
raise an issue as to whether the child consented to the photograph or pseudo-photograph being taken or made, or as to whether D reasonably believed that the child so
consented, D is not guilty of the offence unless it is proved that the child did not so
consent and that D did not reasonably believe that the child so consented. Section 1A
applies whether the photograph or pseudo-photograph showed the child alone or with
D, but not if it showed any other person.

In addition, s 1B provides that in respect of a charge of *making an indecent photograph or pseudo-photograph*, for example downloading of an image from the internet,
or copying a photograph from a computer hard drive, a person will not be guilty of an
offence if (s)he proves that it was done for the purpose of the prevention, detection or
investigation of crime, or for the purpose of criminal proceedings. A similar defence
also applies to a member of the Security Service, the Secret Intelligence Service, or
GCHQ who is able to prove that it was necessary to make the photograph or pseudo-
photograph for the exercise of any functions of that Service or GCHQ.

Distributing or showing or possession with a view
to distribution or showing

By PCA 1978, s 1(1)(b) it is an offence to distribute or show indecent photographs **12.12**
or pseudo-photographs of a child. Section 1(1)(c) provides that it is an offence for a
person to have in his/her possession (**12.15**) such a photograph or pseudo-photograph
with a view to its distribution or showing by him/herself or others.

'Showing' includes 'making available', as where a person gives another access to a
computer or cupboard containing indecent photographs. The same would be so where a
person makes available a password which enables someone to download a 'photograph'
stored on a computer to his/her own computer. The offence of possession with a view
to showing' under s 1(1)(c), requires D to be in possession with a view to showing the
indecent photographs to a third party; a view to showing them to himself is insufficient.

A person possesses indecent photographs, etc with a view to their being distributed or shown by him/her to others only if one reason (not necessarily the primary reason) for possessing them is that they would be distributed or shown to others.

Defences

12.13   PCA 1978, s 1(4) provides a defence to charges *of distributing or showing or possession with a view to distribution or showing* if D proves either that:

(a) D had a legitimate reason for distributing or showing the photographs or pseudo-photographs or (as the case may be) having them in his/her possession; or

(b) D had not seen the photographs or pseudo-photographs and did not know, nor had any reason to suspect, them to be indecent.

In respect of proceedings for *distributing or showing an indecent photograph or pseudo-photograph*, PCA 1978, s 1A provides an additional defence for the case where D proves that the photograph or pseudo-photograph was of the child *aged 16 or over*, and that at the time of the offence charged (ie the distributing or showing), or at the time when D obtained the photograph or pseudo-photograph, the child and D were married or civil partners of each other, or lived together as partners in an enduring family relationship. In such a case, D is not guilty of the offence unless it is proved that the showing or distributing was to someone other than the child.

In respect of proceedings for *possessing an indecent photograph or pseudo-photograph of a child with a view to its being distributed or shown*, s 1A provides a defence where the child consents to D's possession of the photograph or pseudo-photograph (or that D reasonably believed in such consent) and D intended to distribute or show the photograph or pseudo-photograph only to the child. It states that where, in such proceedings, D proves the photograph or pseudo-photograph was of the child *aged 16 or over*, and that at the time of the offence (ie the possession) charged, or at the time when D obtained the photograph or pseudo-photograph, the child and D were married or civil partners of each other, or lived together as partners in an enduring family relationship, and sufficient evidence is adduced to raise an issue as to whether both:

(a) the child consented to the photograph or pseudo-photograph being in D's possession, or as to whether D reasonably believed that the child so consented; and

(b) D had the photograph or pseudo-photograph in his/her possession with a view to its being distributed or shown to anyone other than the child,

D is not guilty of the offence unless it is proved either that the child did not so consent and that D did not reasonably believe that the child so consented, or that D had the photograph or pseudo-photograph in his/her possession with a view to its being distributed or shown to a person other than the child.

The above defences under s 1A apply whether the photograph or pseudo-photograph showed the child alone or with D, but not if it showed any other person.

Publishing, or causing to be published, advertisements

12.14   PCA 1978, s 1(1)(d) provides that it is an offence for a person to publish or cause to be published any advertisement likely to be understood as conveying that the advertiser distributes or shows indecent photographs or pseudo-photographs of a child, or intends to do so.

Simple possession

CJA 1988, s 160(1) makes it an offence for a person to have any indecent photograph or pseudo-photograph of a child in his/her possession.

The Court of Appeal has held that, for the purposes of an offence under PCA 1978 or under CJA 1988, s 160, or an offence at **12.19** or **12.26**, the prosecution must establish (a) that the images were within D's custody or control so that D was capable of accessing them, and (b) that D had known that (s)he possessed the material (eg a digital file) containing the images; proof that D was aware of the contents of the material was not required. The Court stated that although, where unsolicited images were sent to D by the messaging application 'WhatsApp' and automatically downloaded to D's phone's memory, it was highly likely that (a) would be made out, whether (b) was made out would depend on whether D knew (s)he had received the material containing the images. Because 'possession' means custody or control, it follows, for example, that an assistant in a sex shop containing indecent photographs of children is, like his/her employer, in possession of them. Where the image concerned is a computer file, D is in possession of it at the relevant time if it is within his/her control (as where, for example, (s)he can produce it on his/her screen, make a hard copy of it or send it to someone else). Thus, images which have been emptied from the computer's recycle bin may be considered to be within the control of a defendant who is skilled in the use of computers and owns the software necessary to retrieve the images; whereas the images may not be considered to be within the control of a defendant who does not possess the requisite skill, and does not own the necessary software.

Defences

By CJA 1988, s 160(2), a person charged under s 160 has a defence if (s)he proves that:

(a) (s)he had a legitimate reason for having the photograph or pseudo-photograph in his/her possession; or

(b) (s)he had not him/herself seen the photograph or pseudo-photograph and did not know, nor had any reason to suspect, it to be indecent; or

(c) the photograph or pseudo-photograph was sent to him/her without any previous request made by him/her or on his/her behalf and that (s)he did not keep it for an unreasonable time.

The Court of Appeal has held that it is implicit in (b) that a defendant who had not seen the photograph or pseudo-photograph but had cause to suspect that it was indecent has a defence if (s)he proves that (s)he had no reason to suspect that it was an indecent photograph or pseudo-photograph *of a child*.

By s 160A, if D proves that the photograph or pseudo-photograph was of the child *aged 16 or over*, and that at the time of the offence charged, or at the time when D obtained the photograph or pseudo-photograph, the child and D were married or civil partners of each other, or lived together as partners in an enduring family relationship, and sufficient evidence is adduced to raise an issue as to whether the child consented to the photograph or pseudo-photograph being in D's possession, or as to whether D reasonably believed that the child so consented, D is not guilty of the offence unless it is proved either that the child did not so consent or that D did not reasonably believe that the child so consented. Section 160A applies whether the photograph or pseudo-photograph showed the child alone or with D, but not if it showed any other person.

Prosecution

12.17   The Court of Appeal has stated that, where photographs are contained in books available from reputable outlets, the proper course (if it is claimed that they are indecent) is to prosecute the publisher or retailer under PCA 1978, and not an individual purchaser under CJA 1988, s 160.

Police powers

12.18   Under PCA 1978, s 4, on information laid on oath by a constable or by or on behalf of the DPP, a justice may issue a warrant to authorise entry, search for, and seizure of indecent photographs or pseudo-photographs of a child. The additional powers of seizure provided by CJPA 2001, s 50 (**4.61**) apply where a search warrant under PCA 1978, s 4 is executed.

PCA 1978, Sch permits forfeiture of indecent photographs or pseudo-photographs of children, and the devices that hold them. Forfeiture is automatic (ie without the involvement of a court) unless the owner or some other person with an interest in the material objects. Computer hard drives which contain indecent photographs or pseudo-photographs of children may be forfeited where it is not technically possible to separate them. PCA 1978, Sch applies irrespective of the power under which material was seized. Thus, a computer seized during an investigation into fraud and found to contain indecent photographs of children may be forfeited.

POSSESSION OF PROHIBITED IMAGES OF CHILDREN

12.19   The Coroners and Justice Act 2009 (C&JA 2009), s 62(1) provides that it is an offence (**6.48**) for a person to be in possession (**12.15**) of a prohibited image of a child, other than an excluded image. An offence under s 62(1) is an indictable (either-way) offence punishable with up to three years' imprisonment.

Prohibited image

12.20   A prohibited image is an image which:

(a) is pornographic;
(b) focuses solely or principally on a child's genitals or anal region, or portrays any of the following acts:
   (i) the performance by a person of an act of intercourse or oral sex with or in the presence of a child;
   (ii) an act of masturbation by, of, involving or in the presence of a child;
   (iii) an act which involves penetration of the vagina or anus of a child with a part of a person's body or with anything else;
   (iv) an act of penetration, in the presence of a child, of the vagina or anus of a person with a part of a person's body or with anything else;
   (v) the performance by a child of an act of intercourse or oral sex with an animal (whether dead or alive or imaginary);
   (vi) the performance by a person of an act of intercourse or oral sex with an animal (whether dead or alive or imaginary) in the presence of a child; and
(c) is grossly offensive, disgusting, or otherwise of an obscene character.

An 'image' includes (a) a moving or still image (produced by any means), or (b) data (stored by any means) which is capable of conversion into an image within (a), but it does not include an indecent photograph, or indecent pseudo-photograph, of a child within the meaning of PCA 1978. References to an image of a person (or child) include references to an image of an imaginary person (or child).

'Child' means a person under 18. An image of a person is treated as an image of a child if the impression conveyed by the image is that the person shown is a child, or the predominant impression conveyed is that the person shown is a child.

Exclusion of classified film

C&JA 2009, s 63 provides that the offence under s 62 does not apply to an 'excluded   **12.21**
image', ie an image which forms part of a series of images contained in a recording of the whole or part of a classified work. But such an image is not an 'excluded image' if it is contained in a recording (including a device for storing data electronically and from which images may be produced) of an extract from a classified work and it is of *such a nature that it must reasonably be assumed to have been extracted (whether with or without other images) solely or principally for the purpose of sexual arousal.* 'Classified work' means a video work in respect of which a classification certificate has been issued under the Video Recordings Act 1984 by a designated authority.

Where an extracted image is one of a series of images contained in the recording, the question whether the image is of the nature italicised above is to be determined in the same way as the determination of whether an image which forms part of a series of images is pornographic (see below).

Pornographic

An image is 'pornographic' if it is of such a nature that it must reasonably be assumed to   **12.22**
have been produced solely or principally for the purpose of sexual arousal. The Court of Appeal has held that the assumed purpose (of production) of sexual arousal refers to the sexual arousal of anyone who came to have the image, whether (s)he was the producer of the image, a distributor, or the ultimate recipient.

Where an image forms part of a series of images, which as a whole might not be of a nature to be assumed to have been produced for the purpose of sexual arousal, the question whether it is pornographic is to be determined by reference to:

(a) the image itself; and
(b) (if the series of images is such as to be capable of providing a context for the image) the context in which it occurs in the series of images.

So, for example, where:

(i) an image forms an integral part of a narrative constituted by a series of images; and
(ii) having regard to those images as a whole, they are not of such a nature that they must reasonably be assumed to have been produced solely or principally for the purpose of sexual arousal,

the image may, by virtue of being part of that narrative, be found not to be pornographic, even though it might have been found to be pornographic if taken by itself. For the purposes of this provision it is irrelevant that the series was created by someone other than the creator of the individual image or images in issue.

Defences

12.23   C&JA 2009, s 64 provides three defences to an offence under s 62. The terms of s 64 are identical to those of CJA 1988, s 160(2) (**12.16**), with the substitution of 'image' for 'photograph or pseudo-photograph' and 'prohibited' for 'indecent'.

General

12.24   By C&JA 2009, s 67, the provisions of PCA 1978 (**12.18**) relating to entry, search and seizure, and to forfeiture apply in relation to prohibited images of children as they apply in relation to indecent photographs of children.

POSSESSION OF PAEDOPHILE MANUAL

12.25   It is an offence (**6.48**) under the Serious Crime Act 2015, s 69(1) to be in possession of any 'item' (which term includes anything in which information of any description is recorded) that contains advice or guidance about abusing children sexually. 'Abusing children sexually' means doing anything that constitutes an offence under:

(a)   the Sexual Offences Act 2003, Part 1 (ss 1–79) (**11.1–11.75** and **13.61–13.69**), or a corresponding Northern Irish offence, against a person under 16;

(b)   the Protection of Children Act 1978, s 1 (**12.6–12.14**), or a corresponding Northern Irish offence, involving indecent photographs (but not pseudo-photographs); or

(c)   the Modern Slavery Act 2015 (MSA 2015), s 2 (human trafficking) committed with a view to sexual exploitation within s 3(3) (**13.75–13.76**).

By s 69(2), it is a defence for D to prove:

(a)   that (s)he had a legitimate reason for being in possession of the item;

(b)   that—
  (i)   (s)he had not read, viewed, or (as appropriate) listened to the item, and
  (ii) (s)he did not know, and had no reason to suspect, that it contained advice or guidance about abusing children sexually; or

(c)   that the item was sent to him/her without any request made by him/her or on his/her behalf, and that (s)he did not keep it for an unreasonable time.

An offence under s 69 is an indictable (either-way) offence punishable with up to three years' imprisonment. The police powers set out at **12.18** in relation to indecent photographs of children also apply in relation to the offence.

POSSESSION OF EXTREME PORNOGRAPHIC IMAGES

12.26   Possession (**12.15**) of an extreme pornographic image is an offence (**6.48**) under the Criminal Justice and Immigration Act 2008 (CJIA 2008), s 63(1). The offence is an indictable (either-way) offence punishable with up to three years' imprisonment.

The definitions of 'image' and 'pornographic' are identical to those at **12.20** and **12.22**. An image is 'extreme' if:

(a)   it portrays, in an explicit and realistic way:
  (i)   an act which threatens a person's life,
  (ii) an act which results, or is likely to result, in serious injury to a person's anus, breast or genitals,

(iii) an act which involves sexual interference with a human corpse, or

(iv) a person performing an act of intercourse or oral sex with an animal (whether dead or alive),

and a reasonable person looking at the image would think that any such person or animal was real; and

(b) it is grossly offensive, disgusting or otherwise of an obscene character.

These references to parts of the body include references to any part which is surgically reconstructed (in particular through gender reassignment surgery).

In addition, an image is also 'extreme' if:

(a) it portrays, in an explicit and realistic way, either of the following:
   (i) an act which involves the non-consensual penetration of a person's vagina, anus, or mouth by another with the other person's penis, or
   (ii) an act which involves the non-consensual sexual penetration of a person's vagina or anus by another with a part of the other person's body or any-thing else,

   and a reasonable person looking at the image would think that the persons were real; and

(b) it is grossly offensive, disgusting or otherwise of an obscene character.

For these purposes, penetration is a continuing act from entry to withdrawal, and 'va-gina' includes vulva.

Exclusion of classified films, etc

CJIA 2008, s 64 excludes from s 63 an image forming part of a series of images con-tained in films which have been classified by a designated authority under the Video Recordings Act 1984. The terms of s 64 are identical to those under C&JA 2009, s 63 (**12.21**).   **12.27**

Defences

CJIA 2008, s 65 provides three defences which are identical to those under CJA 1988, s 160(2) (**12.16**), with the substitution of 'image' for 'photograph or pseudo-photograph' and 'extreme pornographic' for 'indecent'.   **12.28**

In addition, where:

(a) a person (D) is charged with an offence under s 63; and

(b) the offence relates to an image that portrays an act or acts constituting an extreme image within either definition of 'extreme image' (but does not portray a person performing intercourse or oral sex with an animal),

it is a defence under s 66 for D to prove:

(i) that D directly participated in the acts portrayed; and

(ii) that the act or acts did not involve the infliction of any non-consensual harm on any person; and

(iii) if the image portrays an act involving sexual intercourse with a human corpse, that what is portrayed as a human corpse was not in fact a corpse, and if the image por-trays an act within the second definition of 'extreme image', that what is portrayed as non-consensual penetration was in fact consensual.

For these purposes, harm inflicted on a person (P) is non-consensual harm if:

(a) the harm is of such a nature that P cannot, in law (**9.20**), consent to it being inflicted on himself; or
(b) where P can, in law, consent to it being so inflicted, P does not in fact consent to it being so inflicted.

DISCLOSING OR THREATENING TO DISCLOSE PRIVATE SEXUAL PHOTOGRAPHS AND FILMS WITH INTENT TO CAUSE DISTRESS ('REVENGE PORN')

12.29   A person (D) commits an indictable (either-way) offence under the Criminal Justice and Courts Act 2015, s 33(1) if:

(a) D discloses, or threatens to disclose, a private sexual photograph or film in which another individual (V) appears;
(b) by so doing, D intends to cause distress to V; and
(c) the disclosure is, or would be, made without the consent of V.

Disclosure, or threatened disclosure, to V is not an offence under s 33(1) (s 33(2)).

Where a person is charged with an offence under s 33(1) of threatening to disclose a private sexual photograph or film, it is not necessary for the prosecution to prove:

(a) that the photograph or film referred to in the threat exists; or
(b) if it does exist, that it is in fact a private sexual photograph or film (s 33(2A)).

A person charged with an offence under s 33(1) is not to be taken to have intended to cause distress by disclosing, or threatening to disclose, a photograph or film merely because that was a natural and probable consequence of the disclosure or threat (s 33(8)).

An offence under s 33(1) is punishable with up to two years' imprisonment.

Defences

12.30   It is a defence for D to prove that (s)he reasonably believed that the disclosure was necessary for the purposes of preventing, detecting or investigating crime (s 33(3)).

It is a defence for D to show that:

(a) the disclosure or threat to disclose was made in the course of, or with a view to, the publication of journalistic material; and
(b) D reasonably believed that, in the particular circumstances, the publication of the journalistic material was, or would be, in the public interest (s 33(4)).

It is a defence for D to show that:

(a) D reasonably believed that the photograph or film had previously been disclosed for reward, whether by V or another person; and
(b) D had no reason to believe that the previous disclosure for reward was made without the consent of V (s 33(5)).

D is taken to have shown the matters mentioned in s 33(4) or (5) if sufficient evidence of the matters is adduced to raise an issue with respect to it, and the contrary is not proved beyond reasonable doubt (s 33(6)).

General

The following definitions apply for the purposes of s 33.

A person *'discloses'* something to a person if, by any means, (s)he gives or shows it to the person or makes it available to the person. Something that is given, shown or made available to a person is disclosed whether or not it is given, shown or made available for reward, and whether or not it has previously been given, shown or made available to the person.

'Photograph or film' means a still or moving image in any form that:

(a) appears to consist of or include one or more photographed or filmed images; and

(b) in fact consists of or includes one or more photographed or filmed images.

References to a photograph or film include a negative version of, or stored data capable of conversion into, an image. The reference in (b) to photographed or filmed images includes photographed or filmed images that have been altered in any way.

'Photographed or filmed image' means a still or moving image that was originally captured by photography or filming, or is part of an image originally captured by photography or by filming on any medium, from which a moving image may be produced.

'Consent' to a disclosure includes general consent covering the disclosure, as well as consent to the particular disclosure.

A photograph or film is *'private'* if it shows something that is not of a kind ordinarily seen in public.

A photograph or film is *'sexual'* if:

(a) it shows all or part of an individual's exposed genitals or pubic area;

(b) it shows something that a reasonable person would consider to be sexual because of its nature; or

(c) its content, taken as a whole, is such that a reasonable person would consider it to be sexual.

In the case of a photograph or film that:

(a) consists of or includes a photographed or filmed image that has been altered in any way;

(b) combines two or more photographed or filmed images; and

(c) combines a photographed or filmed image with something else,

the photograph or film is not private and sexual if—

(i) it does not consist of or include a photographed or filmed image that is itself private and sexual;

(ii) it is only private or sexual by virtue of the alteration or combination mentioned in (a)–(c); or

(iii) it is only by virtue of the alteration or combination mentioned in (a)–(c) that V is shown as part of, or with, whatever makes the photograph or film private and sexual.

'Publication' of journalistic material means disclosure to the public at large or to a section of the public.

INDECENT DISPLAYS, ETC

12.32   The Indecent Displays (Control) Act 1981 (ID(C)A 1981), s 1(1) provides that it is an offence to make a public display of indecent matter, or cause or permit such a display. This is an indictable (either-way) offence punishable with up to two years' imprisonment.

Matter is deemed to be publicly displayed if displayed in, or visible from, a public place. For these purposes, 'public place' means any place to which the public have access (whether on payment or otherwise) while that matter is displayed, except:

(a) a place to which the public are permitted to have access only on payment which is for, or includes payment for, that display, or

(b) a shop, or any part of a shop, to which public access can only be gained by passing beyond an adequate warning notice,

but the exclusions in (a) and (b) only apply where persons under 18 are not allowed to enter while any such display is actually taking place. The 'adequate warning notice' should read as set out below and this notice should be looked for in premises where it is known that such displays are held:

> **'WARNING**
> Persons passing beyond this notice will find material on display which they may consider indecent. No admittance to persons under 18 years of age.'

'Matter', for the purposes of ID(C)A 1981, includes anything capable of being displayed, but does not include the actual human body or a part of it. Thus ID(C)A 1981 is not concerned with strippers. This is a useful Act to police officers as it can be applied to everything from the display of obscene graffiti scrawled on a wall to indecent film shows, etc which are open to the public.

Official TV programmes, art galleries, museums, Crown buildings, local authority buildings, theatres, and arenas controlled by other legislation are totally outside the provisions of the Act.

Police powers

12.33   ID(C)A 1981, s 2, a constable may seize articles which (s)he has reasonable grounds for believing to be indecent, or to contain indecent matter, or to have been used in the commission of an offence under the Act.

In addition, a justice may grant a search warrant, on information on oath, authorising entry within 14 days and seizure of material reasonably suspected to have been used in an offence under the Act. The additional powers of seizure provided by CJPA 2001, s 50 (**4.61**) apply where a search warrant under ID(C)A 1981 is executed.

SENDING INDECENT, ETC MATTER THROUGH POST

12.34   Under the Postal Services Act 2000, s 85(3) and (4) respectively it is an indictable (either-way) offence, punishable with up to 12 months' imprisonment, to send a postal packet which encloses any indecent or obscene matter, or which has on the packet indecent or obscene words, marks, or designs. So far as this offence is concerned the words 'indecent' and 'obscene' are directed at offending against the recognised standards of propriety. It is for a court or jury to decide, applying these standards, whether the material is indecent or obscene.

INDECENT, GROSSLY OFFENSIVE, THREATENING, OR FALSE COMMUNICATIONS

Malicious Communications Act 1988

The Malicious Communications Act 1988 (MCA 1988), s 1(1) makes it an offence for   **12.35**
any person to send to another:

(a) a letter, electronic communication, or article of any description which conveys:
   (i) a message which is indecent or grossly offensive,
   (ii) a threat, or
   (iii) information which is false and known or believed to be false by the sender; or
(b) any other article or electronic communication which is wholly or partly of an indecent or grossly offensive nature,

if his/her purpose or one of his/her purposes in sending it is that the message, threat, or information should cause distress or anxiety to the recipient or to any other person to whom (s)he intends that it, or its content or nature, should be communicated. Such an offence is an indictable (either-way) offence punishable with up to two years' imprisonment.

A divisional court has held that the fact that a communication is political or educational in nature has no bearing on whether it is indecent or grossly offensive. The court added that it is possible to interpret s 1 in a way that is compatible with the European Convention on Human Rights, arts 9 and 10 (freedom of thought, conscience, and religion and freedom of expression), by giving a heightened meaning to the terms 'grossly offensive' and 'indecent' or by reading into s 1 a provision to the effect that s 1 will not apply where to create an offence would be a breach of a person's Convention rights.

In relation to the sending of a threat, a defence exists if D shows that the threat is to reinforce a demand made by him/her on reasonable grounds and that (s)he believed and had reasonable grounds for believing that it was a proper means of reinforcing the demand.

For the purposes of MCA 1988, the term 'electronic communication' includes any oral or other communication by means of an electronic communication network, and any communication (however sent) that is in electronic form.

Communications Act 2003

The Communications Act 2003, s 127(1) provides that a person commits an offence   **12.36**
who:

(a) sends by means of a *public* electronic communications network, a message or other matter which is grossly offensive or of an indecent, obscene, or menacing character; or
(b) causes any such message or matter to be so sent.

A divisional court has held that a message does not have to be received by a human being for an offence under (a) to be committed, and that therefore (for example) the upload of a video to YouTube by D with the intention that people might view it, or sending it to an inanimate answering machine, was sufficient. Another divisional court has held that a tweet is a message sent by means of a public electronic communications network because, although the twitter social networking platform is owned by a private company, it operates through the internet, a public electronic network provided for the public and paid for by the public through internet service providers.

A divisional court has held that pasting a hyperlink which connected with the YouTube site and allowed immediate streaming of a grossly offensive video fell within s 127(1)(b); the hyperlink set in train the sending process, and its purpose was to cause the material to be sent.

The House of Lords has held that whether a message or other matter is grossly offensive for the purposes of s 127 must be judged by the standards of an open and just multiracial society; the words must be judged taking account of their context and all other relevant circumstances. To be guilty of a 'grossly offensive offence' under s 127, D must intend his/her words to be grossly offensive to those to whom they relate, or be aware that they might be taken to be so. If D has such a state of mind, it is irrelevant that a recipient of the message was not offended.

A divisional court has held that a message or other matter which does not create fear or apprehension in those to whom it is communicated, or who may reasonably be expected to see it, is not menacing. Thus, if the person or persons who receive or read a message containing a threat would brush it aside as a silly joke or empty bombastic banter or would regard it as no more than merely 'nasty', it is not menacing. In deciding whether a message is menacing, its precise terms, and inferences to be drawn from it, need to be examined in the context in which (and the means by which) it was sent. The court added that to be guilty of a 'menacing offence' under s 127 D must intend that the message should be of a menacing character, or be aware that it may create fear or apprehension in any reasonable member of the public who reads or sees it.

In addition, it is an offence against s 127(2) to send, *for the purpose of causing annoyance, inconvenience or needless anxiety to another,* by means of a *public* electronic communications network, a message which is known to be false, or to cause such a message to be sent, or persistently to make use of such a network. A divisional court has held that s 127(2) does not criminalise forms of expression, the contents of which are no worse than annoying or inconvenient in nature or such as to cause anxiety for which there is no need. Examples of the types of behaviour falling within s 127(2) were, it said, repeated instances of prank calls, silent calls, or 'heavy breathing calls'. The court added that to be regarded as 'persistent' a series of messages had to have an element of frequency and a connecting theme or factor.

An offence under s 127(1) or (2) is a summary offence punishable with up to six months' imprisonment and/or an unlimited fine. An offence under s 127(2) is a 'penalty offence' for the purposes of CJPA 2001: see **3.40**.

CHAPTER 13

Protection of Vulnerable People

OFFENCES OF CRUELTY TO PERSON UNDER 16

'Child abuse' is dealt with by the criminal law by the offences of cruelty towards chil- **13.1**
dren and young persons which are governed by CYPA 1933, s 1. Section 1(1) provides
that someone *of 16 or over who has the responsibility for any child or young person under
16* commits an indictable (either-way) offence punishable with up to 10 years' imprisonment, if (s)he wilfully:

(a) *assaults,*
(b) *ill-treats (whether physically or otherwise),*
(c) *neglects,*
(d) *abandons,* or
(e) *exposes him*; or
(f) *causes or procures him to be assaulted, ill-treated (whether physically or otherwise),
     neglected, abandoned, or exposed,*

*in a manner likely to cause him unnecessary suffering or injury to health (whether the
suffering or injury is of a physical or a psychological nature).*

 CYPA 1933, s 1(1) does not prevent a parent, or other person with the right to do so,
administering punishment to a child or young person. As explained at **9.23**, the right
to administer corporal punishment is now very limited.

Key elements

Responsibility for a child or young person

In this context a person is presumed to have responsibility for a child or young **13.2**
person if:

(a) (s)he has parental responsibility for him/her under ChA 1989 (which the mother
    and father will both have if they were married to each other when the child or
    young person was born, or which only the mother will have if the mother and
    father were not so married, or which another person (including the father in the
    instance just given or a local authority may acquire by operation of law); or
(b) (s)he is otherwise liable to maintain him/her; or
(c) (s)he has care of him/her.

Wilfully

13.3   'Wilfully' makes it clear that any offence under s 1 requires mens rea on the part of the defendant (D) as to the risk of unnecessary suffering or injury to health. For example, on a charge of wilfully neglecting, D must be aware of the risk of unnecessary suffering or injury to health resulting from the neglect, and if (to continue the example) that charge involves failure to provide adequate medical aid, the requirement of wilfulness can only be satisfied where D was aware that the child or young person's health might be at risk if (s)he was not provided with medical aid or where D's non-awareness of this risk was due to D's not caring whether the child's health was at risk or not.

Assault

13.4   'Assault' for the purpose of s 1 requires more than a mere common assault or battery. An assault or battery may amount to no more than frightening a person or giving him/her a light slap, but this could hardly be done 'in a manner likely to cause unnecessary suffering or injury to health'.

Ill-treat

13.5   This signifies a continuous course of conduct leading to unnecessary suffering. A series of assaults, each of which, if considered on its own, would not amount to an offence against s 1, might together form ill-treatment over a period of time. Likewise, persistent frightening or bullying will suffice if it is likely to cause unnecessary suffering.

Neglect

13.6   This signifies a want of adequate care. The likelihood of causing unnecessary suffering or injury to health can be caused by a deliberate omission to supply medical or surgical aid. Direct proof of such likelihood is not always strictly necessary as this may be inferred from the evidence of neglect and its actual effect.

Section 1(2) provides that a parent or other person legally liable to maintain a child or young person or his legal guardian is deemed to have neglected him in a manner likely to cause injury to health if he has failed to provide adequate food, clothing, medical aid, or lodging for him or if, having been unable to provide such food, clothing, medical aid, or lodging, he has failed to take steps to procure it to be provided under the enactments applicable in that behalf.

Section 1(2) also declares that if the death of an infant under three years of age is caused by suffocation (other than by disease or a foreign body in the throat or air passage) while the infant was in bed with a person of 16 or over (D), D, if D was, when D went to bed or at any later time before the suffocation, under the influence of drink or a prohibited drug, will be deemed to have neglected the infant in a manner likely to cause injury to health. The reference to the infant being 'in bed' with another ('the adult') includes a reference to the infant lying next to the adult in or on any kind of furniture or surface being used by the adult for the purpose of sleeping (and the reference to the time when the adult 'went to bed' is to be read accordingly). A drug is a prohibited drug for the above purposes in relation to a person if the person's possession of the drug immediately before taking it constituted an offence under the Misuse of Drugs Act 1971, s 5(2) (**20.11**).

Abandon or expose

13.7   It is helpful to consider these terms together since an abandonment frequently leads to exposure. To abandon a child means leaving it to its fate. If a woman separated from her husband takes her child to her husband's home and leaves the child at his door, she abandons that child. From the moment of abandonment the child is exposed and, if that exposure leads to suffering, the child has been abandoned and exposed in a manner likely to cause unnecessary suffering. If the husband becomes aware that the child has been left on his doorstep and allows it to remain there, he cannot then disclaim his custodial responsibilities. 'Exposure' does not cover exposure to risk. A father, who took his son and other boys on to a baulk of timber and floated the timber into deep water in London Docks, was not guilty of an offence under s 1, because this was not the type of exposure to which the section refers.

OTHER OFFENCES

There are a number of offences under CYPA 1933 which require little explanation. **13.8** These offences are committed in relation to children of a specified age, which varies from offence to offence.

Allowing person aged 4 to 15 to reside in or frequent a brothel

By CYPA 1933, s 3(1), it is a summary offence, punishable with up to six months' imprisonment and/or a fine not exceeding level 2, for any person who has responsibility for a child or young person aged four or above, but under 16, to allow him/her to reside in or frequent a brothel (as defined at **13.71**). It is important to remember that a woman who is a prostitute and receives men in her own room, but does not allow other women to use her room, is not keeping a brothel. **13.9**

Giving intoxicating liquor to child under five

It is a summary offence, contrary to CYPA 1933, s 5, to give, or cause to be given, intoxicating liquor to a child under five, otherwise than on the orders of a duly qualified medical practitioner or in an emergency. The offence is punishable with a fine not exceeding level 1. **13.10**

Exposing child under 12 to risk of burning

If a person of 16 or over, having responsibility for a child under 12, allows the child to be in a room containing an open fire grate or any heating appliance liable to cause injury to a person by contact with it, and it is not sufficiently guarded against the risk of the child being burnt or scalded without taking reasonable precautions against that risk and in consequence the child is killed or seriously injured, that person commits a summary offence, punishable with a fine not exceeding level 1, contrary to CYPA 1933, s 11. **13.11**

Section 11 adds that a prosecution under s 11 will not affect a person's liability to prosecution for any indictable offence.

PROTECTING CHILDREN IN EMERGENCIES

Emergency protection orders

The Children Act 1989 (ChA 1989), s 44 empowers a court, on the application of any person, to make an emergency order (EPO) for the protection of a child. It may do so if it is satisfied that: **13.12**

(a) there is reasonable cause to believe that the child is likely to suffer significant harm if:
  (i) (s)he is not removed to accommodation provided by or on behalf of the applicant, or
  (ii) (s)he does not remain in the place in which (s)he is then being accommodated;
(b) in the case of an application made by a local authority:
  (i) inquiries are being made with respect to the child under the authority's duty to investigate where it reasonably suspects that a child is suffering, or is likely to suffer, significant harm, and
  (ii) *these inquiries are being frustrated by access to the child being unreasonably refused to a person authorised to seek access and that the applicant has reasonable cause to believe that access to the child is required as a matter of urgency;* or
(c) in the case of an application made by an authorised person (NSPCC or one of its officers):
  (i) the applicant has reasonable cause to suspect that the child is suffering, or is likely to suffer, significant harm,
  (ii) the applicant is making inquiries with respect to the child's welfare, and
  (iii) the italicised words in (b)(ii).

An EPO directs any person in a position to do so to produce the child, and authorises the child's removal to accommodation provided by or on behalf of the applicant or the prevention of the removal of the child from a hospital or other place. It also gives the applicant parental responsibility for the child. An EPO's duration is such period, not exceeding eight days, specified by the court, but the court has power (on one occasion only) to extend it for up to a further seven days. An EPO can include an exclusion requirement in specified circumstances. Such a requirement enables the child to stay in its home by excluding someone else, such as a suspected child abuser, from it.

By s 44(15), it is an offence intentionally to obstruct a person exercising the power (under an EPO) to remove or prevent the removal of a child. The offence is a summary offence, punishable with a fine not exceeding level 3.

Removal and accommodation of children by police in emergencies

13.13   ChA 1989, s 46 empowers police officers to take a child into 'police protection' in prescribed circumstances.

Where a constable has reasonable cause to believe that a child would otherwise be likely to suffer significant harm, (s)he may:

(a) remove the child to suitable accommodation and keep the child there; or
(b) take such steps as are reasonable to ensure that the child's removal from any hospital, or other place, in which the child is then being accommodated is prevented.

The courts have held that, in cases of emergency too pressing to obtain an interim care order, the preferable procedure is to apply for an EPO, where this is practicable, rather than proceed under s 46; but the police must always have regard to the paramount need to protect children from significant harm. Where a police officer knows that an EPO is in force in respect of a child, (s)he should not exercise the power of removal under s 46, unless there are compelling reasons to do so.

As soon as is reasonably practicable after taking a child into police protection under s 46, the constable must:

(a) inform the local authority within whose area the child was found of the steps that have been, or are proposed to be, taken with respect to the child and the reasons for taking them;

(b) give details to the local authority within whose area the child is ordinarily resident ('the appropriate authority') of the place at which the child is being accommodated;

(c) inform the child (if (s)he appears capable of understanding):

    (i) of the steps that have been taken and of the reasons, and

    (ii) of the further steps which may be taken with respect to the child under s 46;

(d) take such steps as are reasonably practicable to discover the wishes and feelings of the child;

(e) secure that the case is inquired into by an officer designated for the purposes of s 46 by the chief officer of police; and

(f) where the child was taken into police protection by being removed to accommodation which is not provided:

    (i) by or on behalf of a local authority, or

    (ii) as a refuge (ie a voluntary home or registered children's home certified as a refuge),

secure that (s)he is moved to accommodation which is so provided.

The constable must also, as soon as reasonably practicable, inform:

(a) the child's parents;

(b) any other person who has parental responsibility for him/her; and

(c) any other person with whom the child was living immediately before being taken into police protection,

of the steps that (s)he has taken under s 46 with respect to the child, the reasons for them, and the further steps that may be taken with respect to the child under the section.

When the case has been inquired into by the designated officer, (s)he must release the child from police protection unless (s)he considers that there is still reasonable cause for believing that the child would be likely to suffer significant harm if released.

No child may be kept in police protection for more than 72 hours. However, at any time while the child is in police protection, the designated officer may apply *on behalf of the appropriate local authority* (whether or not it consents) for an EPO to be made with respect to the child.

Whilst a child is in police protection, the designated officer must do what is reasonable in all the circumstances of the case for the purpose of safeguarding or promoting the child's welfare (having regard in particular to the length of time the child will be so protected).

The designated officer must allow:

(a) parents;

(b) any other person with parental responsibility;

(c) any person with whom the child was living immediately before the child was taken into police protection;

(d) any person named in a child arrangements order as someone with whom the child is to have contact;

(e) where the child is in local authority care, any person allowed to have contact with the child by virtue of a court order; and

(f) any person acting on behalf of any of these persons,

to have such contact (if any) with the child as, in the opinion of the designated officer, is both reasonable and in the child's best interest. However, if a child taken into police protection is in accommodation provided by, or on behalf of, the appropriate authority, these contact responsibilities are those of the authority rather than the designated officer.

CHILD ABDUCTION

13.14 Two indictable (either-way) offences of child abduction are provided by the Child Abduction Act 1984 (CAA 1984). These offences, which are punishable with up to seven years' imprisonment, contain a number of common features dealt with after the separate offences have been outlined.

Abduction from the United Kingdom by parent, etc

13.15 By CAA 1984, s 1(1), a person connected with a child under 16 commits an offence (**6.48**) if (s)he takes or sends the child *out of the UK* without the appropriate consent. A person is regarded as 'sending' a child if (s)he causes the child to be sent. A person connected with a child does not commit an offence if the appropriate consent has been given for a child to be removed from the UK for a defined period but the child is kept out of the UK after the end of that period and the appropriate consent no longer exists.

A person (P) is 'connected with' a child if:

(a) P is the child's parent; or

(b) in the case of a child, whose parents were not married to each other at the time of his/her birth, there are reasonable grounds for believing that P is its father; or

(c) P is a guardian or special guardian of the child; or

(d) P is named in a child arrangements order as a person with whom the child is to live; or

(e) P has custody of the child.

Only persons within these categories can commit the present offence.

The reference to the 'appropriate consent' means:

(a) the consent of *each* of the following:

(i)   the child's mother,

(ii) the child's father, if he has parental responsibility for him,

(iii) any guardian or special guardian of the child,

(iv) any person named in a child arrangements order as a person with whom the child is to live, and

(v)   any person who has custody of the child; or

(b) the leave of the court granted under or by virtue of the Children Act 1989, Part II (orders with respect to children in family proceedings); or

(c) if any person has custody of the child, the leave of the court which granted custody to him.

It is irrelevant that the child consents to what occurs.

P does not commit the present offence by taking or sending a child out of the UK without the appropriate consent if P is:

(a) a person named in a child arrangements order as a person with whom the child is to live and P takes or sends the child out of the UK for a period of less than one month; or

(b) a special guardian of the child and takes or sends it out of the UK for a period of less than three months,

unless P is in breach of an order under the Children Act 1989, Part II.

P does not commit the present offence by doing anything without the consent of another person whose consent is required if:

(a) P does it in the belief that the other person has consented or would consent if (s)he was aware of all the relevant circumstances; or

(b) P has taken all reasonable steps to communicate with the other person but has been unable to communicate with him/her; or

(c) the other person has unreasonably refused to consent.

However, condition (c) does not apply where the person who refused consent is a person named in a child arrangements order as a person with whom the child is to live, or who has custody of the child; or where the taking or sending out of the UK is in breach of an order made by a court in the UK.

CAA 1984, s 1 is modified by Sch 1 for the purposes of its application where the child is in the care of a local authority or a voluntary organisation, detained in a place of safety, remanded to local authority accommodation, or the subject of proceedings (or an order) for adoption, or is in a place of safety.

Abduction of child by other persons

This offence is governed by CAA 1984, s 2. It can be committed by anyone other than: **13.16**

(a) the father or mother (if they were married when the child was born) or the mother (if they were not); or

(b) a person within (a)(iii), (iv), or (v) at **13.15**.

Other important differences from the offence under s 1 are that the child need not be abducted from the UK and that the offence can be committed by taking *or detaining* a child.

Section 2(1) provides that anyone, other than someone mentioned above, commits an offence if, without lawful authority or reasonable excuse, he takes or detains a child under 16 so as either to remove him from the lawful control of anyone having lawful control of him or to keep him out of the lawful control of any person entitled to it. Removal from control does not require any removal in a geographical sense; it suffices to deflect the child from what, with the consent of those having lawful control of him/her, (s)he would be otherwise doing into some activity induced by D, as where D finds a child in a park pursuing a particular activity (eg playing football) and induces it to go elsewhere in the park for another activity (eg to look for an alleged lost bicycle).

D's conduct need not be the sole cause of the abduction. If D's conduct is more than merely peripheral, ie is an effective cause of the child accompanying or staying with D, that is sufficient. It is no defence that another cause may be the child's own decision or state of mind. This was held by a divisional court in a case where, knowing of a child abduction warning notice that prohibited contact or communication with a named 14-year-old child, D allowed the child to enter his house and remain with him, thereby keeping her out of the lawful control of her mother. The court held that the fact that the child had visited D and remained with him willingly was no defence.

A person is regarded as detaining a child if (s)he causes the child to be detained or induces the child to remain with him/her or another.

Although it must be proved that there was an intentional or reckless taking by D, the objective consequence of which was to remove or keep the child from the lawful control of anyone having lawful control of the child, it does not have to be proved that D intended or was reckless as to this consequence occurring.

It is a defence for D to prove that, at the time of the alleged offence, D believed the child was 16 or over. Alternatively, in the case where the father and mother of 'the child in question' (ie the child taken or detained) were not married to each other at the time of birth, it is a defence for D to prove that he is the father of the child taken or detained or that he had reasonable grounds to believe he was that child's father. Lastly, D is not guilty of an offence under s 2 if D mistakenly believed, reasonably or not, in facts which—if they had been as D believed—would have given him/her a lawful authority or reasonable excuse. On this basis, a man who takes a child, thinking that it is his child whereas in truth it is another child, will not be guilty if, on the facts as he believes them to be, he would have a lawful authority or reasonable excuse for taking the child.

A point common to ss 1 and 2

13.17   For the purposes of both offences, a person is regarded as taking a child if (s)he causes or induces the child to accompany him/her or any other person, or causes the child to be taken.

Abduction of children in care, etc

13.18   It is a summary offence against ChA 1989, s 49(1) for any person, knowingly and without lawful authority or reasonable excuse, to take a child to whom s 49 applies from the responsible person, or to keep such a child away from the responsible person, or to induce or assist or incite such a child to run away or stay away from the responsible person. Section 49 applies to a child who is in care, the subject of an emergency protection order, or in police protection. A 'responsible person' means any person who for the time being has care of him/her by virtue of the care order, the emergency protection order, or ChA 1989, s 46 (**13.13**), as the case may be.

This offence is punishable with up to six months' imprisonment and/or an unlimited fine. Where such abduction has occurred, s 50 permits a court to issue a 'recovery order'.

PARENTING ORDERS

13.19   A parenting order can be made under the Crime and Disorder Act 1998 (CDA 1998), s 8, the Anti-social Behaviour Act 2003 (A-sBA 2003), s 20, 26, 26A, or 26B, or the Sentencing Code, s 366, 368, or 369.

A parenting order is a court order requiring a parent or guardian (ie the person for the time being with care), *referred to in the account below of the parenting order provisions simply as 'the parent'*:

(a) to comply, for a period not exceeding 12 months, with such requirements ('the compliance requirements') as are specified in the order; and

(b) to attend, for a concurrent period not exceeding three months, such counselling and guidance programme as may be specified in directions given by the responsible officer.

A parenting order need not include a counselling and guidance-session requirement under (b) if the parent has previously been made subject to a parenting order, although it may do so.

The 'responsible officer', eg a probation officer, social worker, or youth offending team member, who is to specify the counselling or guidance sessions to be attended will be specified in the order. The responsible officer's role is also to ensure that parents attend whatever sessions are specified in his/her directions and to ensure compliance with the order and any requirements under it.

The compliance requirements which may be specified are those which the court considers desirable in the interest of preventing any repetition of the kind of conduct which 'triggers' the making of a parenting order or (as the case may be) the commission of any further offence of the type which is a 'trigger'.

Parenting orders under CDA 1998, s 8

By CDA 1998, s 8, a parenting order may be made in any court proceedings where: **13.20**

(a) a child safety order is made in respect of a child or the court determines that a child has failed to comply with such an order;
(b) an injunction under the Anti-social Behaviour, Crime and Policing Act 2014 (A-sB,C&PA 2014), s 1 is granted, a criminal behaviour order is made, or a sexual harm prevention order is made, in respect of a child or young person; or
(c) currently only in 10 areas, a parental compensation order is made in relation to a child's behaviour.

If the court is satisfied that 'the relevant condition' is met, it may (or must, see below) make a parenting order in respect of a person who is the parent of the child or young person.

The 'relevant condition' is that the parenting order would be desirable in the interests of preventing any repetition of the kind of behaviour which led to the order or injunction referred to (a), (b), or (c) being made.

CDA 1998 provides that a court is not normally obliged to make a parenting order if one of the relevant conditions under s 8 is satisfied; it simply has the power to do so. There is an exception, provided by s 9, where a court grants an injunction under A-sB,C&PA 2014, s 1 or makes a criminal behaviour order in respect of a young person under 16. If it is satisfied that the relevant condition is satisfied, a court must make a parenting order. If the court is not so satisfied, it must say so in open court and give its reasons why it is not so satisfied.

Parenting orders under A-sBA 2003, s 20 in cases of exclusion or potential exclusion from school

A-sBA 2003, s 20 enables applications for a parenting order to be made to a magistrates'   **13.21**
court in relation to the parent of a pupil who has been excluded (or whose behaviour warrants his/her exclusion) from school on disciplinary grounds. A local authority (or, in England, the school governors) may apply for a parenting order as a first measure, or where a parent refuses to sign a 'parenting contract'.

The court may make such an order in respect of a pupil's parent if a relevant condition prescribed by the Education (Parenting Contracts and Parenting Orders) (England) Regulations 2007, or the Education (Parenting Contracts and Parenting Orders) (Wales) Regulations 2010, as the case may be is met and the court is satisfied that an order is desirable in the interests of improving the pupil's behaviour.

Section 20 complements a court's powers under the Sentencing Code, s 369 (**13.25**).

Parenting orders under A-sBA 2003, s 26 in respect of criminal conduct and anti-social behaviour: youth offending teams

13.22   Similar provision is made by A-sBA 2003, s 26 for a parenting order where a youth offending team applies to a magistrates' court for such an order in respect of a child or young person who has been referred to the team. If the court is satisfied that such a child or young person has engaged in criminal conduct or anti-social behaviour and that the making of the order is desirable in preventing further criminal conduct or anti-social behaviour, it may make such an order. Only a member of a youth offending team can be the responsible officer in respect of an order under s 26.

Parenting orders under A-sBA 2003, s 26A or 26B in respect of anti-social behaviour: local authorities and registered social landlords

13.23   A-sBA 2003, ss 26A and 26B respectively enable a local authority or registered social landlord to apply to a magistrates' court for a parenting order against a parent in respect of anti-social behaviour by his/her child or young person. Such an application may also be made to a county court as an adjunct to other proceedings in that court.

General point about parenting orders under CDA 1998, s 8 and A-sBA 2003, s 20, 26, 26A, or 26B

13.24   Provision is made for the court which made a parenting order of any of the above types to vary or discharge it.

As long as a parenting order of any of these types is in force, it is a summary offence, punishable with a fine not exceeding level 3, for a parent to fail without reasonable excuse to comply with any requirement included in a parenting order, or specified in directions given by the responsible officer (CDA 1998, s 9(7)). Proceedings may be brought by a relevant local authority.

Parenting orders under the Sentencing Code, s 366, 368, or 369

13.25   A parenting order under the Sentencing Code, s 366 is available to a court by or before which an offender aged under 18 is convicted of an offence. If an order under s 366 is available, and the offender is under 16 at the time of conviction, the court *must*:

(a)   make such a parenting order in respect of a parent of the offender if it is satisfied that the order would be desirable in the interests of preventing the commission of any further offence by the offender, or

(b)   state in open court that it is not so satisfied, and why not,

but this does not apply if the court makes a referral order in respect of the offender.

If the offender is aged 16 or 17 at the time of conviction, the court *may* make a parenting order under s 366 if it is satisfied that the order would be desirable in the interests of preventing the commission of any further offence by the offender.

A parenting order is available under s 368 to a youth court where an offender has been referred to a youth offender panel, and a parent of the offender is referred by the panel to the youth court in respect of a failure to comply with an order requiring attendance at meetings of the panel. In such a case, the youth court may make a parenting order if satisfied that the parent has failed without reasonable excuse to comply with the attendance order, and the parenting order would be desirable in the interests of preventing the commission of any further offence by the offender.

A parenting order is available under s 369 to the court by or before which an offender is convicted of an offence under the Education Act 1996 (EA 1996), s 443 (failure to comply with school attendance order), or EA 1996, s 444 (failure to secure regular attendance at school of registered pupil). In such a case, the court may make such an order in respect of the offender if satisfied that the order would be desirable in the interests of preventing the commission of any further offence under s 443 or 444.

Provision is made for the court which made the parenting order to vary or discharge it.

It is a summary offence, contrary to the Sentencing Code, s 375(1), for a person in respect of whom a parenting order made under the above provisions of the Sentencing Code to comply with any requirement included in the order, or specified in directions given by the responsible officer. The offence is punishable with a fine not exceeding level 3.

REMOVAL OF TRUANTS AND EXCLUDED PUPILS TO SCHOOLS OR DESIGNATED PREMISES

CDA 1998, s 16 provides that when a local authority has designated premises in a police **13.26** area to which children and young persons of compulsory school age may be removed under s 16, and has notified the chief constable of the designation, a police superintendent (or above) may direct that the powers set out in the next paragraph are to be exercisable as respects any area falling within the police area that is specified in the direction. The powers will only be exercisable during the period specified in the direction. A British Transport Police superintendent (or above) has power to make such a direction but in such a case the powers set out below are only exercisable in any area or in the vicinity of any premises policed by the British Transport Police or in the premises themselves.

Section 16 provides that, where a police officer has reasonable cause to believe that a child or young person found by him in a public place in a specified area during a specified period:

(a) is of compulsory school age; and
(b) (i) is absent from a school without lawful authority, or
    (ii) has been excluded from school on disciplinary grounds and has no justification for being in a public place,

he may remove the child or young person to designated premises, or to the school from which he is so absent. Condition (b)(ii) only applies where the excluded juvenile is so found during school hours.

The constable does not have to be in uniform. The child must be 'found in a public place' by a police officer. There is no power to enter private premises (eg the child's home) nor to remove a truant found on private premises on which the police officer is already lawfully present. 'Public place' means any highway and any place to which at the material time the public or any section of the public has access, on payment or otherwise, as of right or by virtue of express or implied permission, eg shopping centres, local authority parks, recreation grounds, and amusement arcades.

For the purpose of (b)(i) (absence from a school without lawful authority), a child or young person's absence is deemed to be without lawful authority unless the absence is with leave, or because attendance at school is prevented by sickness or any unavoidable cause, or because the absence is on a day exclusively set aside for religious observance by the religious body to which his/her parent belongs.

PENALTY NOTICE FOR AN OFFENCE OF FAILURE TO SECURE REGULAR ATTENDANCE AT SCHOOL

13.27   By the Education Act 1996, s 444A, the offence of failure to secure regular attendance at a relevant school of a registered pupil under the Education Act 1996, s 444 is one in respect of which a penalty notice may be given in England by an authorised officer. An 'authorised officer' for these purposes is a constable, a community support officer, a community support volunteer, an accredited person, an authorised local authority officer, or an authorised staff member (head teacher or staff member authorised by the head teacher). A relevant school is a maintained school, a pupil referral unit, an academy school, an alternative provision academy, a city technology college, or a city college for the technology of the arts. The Supreme Court has held that 'regular attendance' means attendance 'in accordance with the rules prescribed by the school'.

The Education (Penalty Notices) (England) Regulations 2007 prescribe the form of the notice and allow a period of 28 days during which no proceedings will be taken. The Regulations require that a code of conduct be drawn up by each local authority in consultation with other parties, to ensure consistency in the issuing of penalty notices. The penalty to be paid is £60 where payment is made within 21 days of receipt of the notice and £120 where paid within 28 days of receipt of the notice. The Education (Penalty Notices) (Wales) Regulations 2013 make similar provision in Wales but specify 28 and 42 days in place of 21 and 28 respectively.

POSSESSION OF ALCOHOL IN PUBLIC PLACE BY PERSON UNDER 18

Offence of persistently possessing alcohol in a public place

13.28   The Policing and Crime Act 2009, s 30(1) provides that a person under 18 is guilty of an offence if, without reasonable excuse, he is in possession of alcohol in any relevant place on three or more occasions within a period of 12 consecutive months.

'Relevant place' means any public place, other than excluded premises, or any place, other than a public place, to which the person in question has unlawfully gained access. 'Excluded premises' means premises which may by virtue of the Licensing Act 2003, Parts 3, 4, or 5 (**20.94–20.96**) be used for the supply of alcohol.

The s 30 offence is a summary offence punishable with a fine not exceeding level 2.

Confiscation of alcohol—young person

13.29   The Confiscation of Alcohol (Young Persons) Act 1997 (CA(YP)A 1997), s 1 provides that, where a constable reasonably suspects that P, a person in any public place other than licensed premises, or any place (other than a public place) to which P has unlawfully gained access, is in possession of alcohol and that either:

(a) P is under 18; or
(b) P intends that any of the alcohol should be consumed by a person under the age of 18 in that or another similar place; or
(c) a person under the age of 18 who is, or has recently been, with P, has recently consumed alcohol in that or another similar place,

the constable may require P to surrender anything which P possesses which is, or the constable reasonably believes to be, alcohol or a container for it. A constable must also require P to state his/her name and address and may, if the constable reasonably suspects that P is under 16, remove him/her to his/her place of residence or to a place of safety. If P, without reasonable excuse, fails to comply with a surrender of alcohol requirement or a name and address requirement P commits a summary offence under s 1(3), which is punishable with a fine not exceeding level 2. A constable who makes a surrender requirement must inform P of his/her suspicion and that failing without reasonable excuse to comply with either type of requirement is an offence.

A constable may dispose of anything surrendered to him/her under CA(YP)A 1997 in such a manner as (s)he considers appropriate.

BEING DRUNK IN CHARGE OF A CHILD

13.30 The Licensing Act 1902, s 2 provides a summary offence, punishable with up to one month's imprisonment or a fine not exceeding level 2, of being found drunk on any highway or other public place, whether a building or not, or on any licensed premises, while having the charge of a child apparently under the age of seven.

ILL-TREATMENT OR NEGLECT

13.31 In addition to the offences of cruelty under the Children and Young Persons Act 1933 (CYPA 1933), s 1 (13.1), there are the following offences involving ill-treatment or neglect.

Ill-treatment or neglect: persons lacking (or apparently lacking) capacity

13.32 The Mental Capacity Act 2005 (MCA 2005), s 44(1) and (2) provides that if D:

(a) has the care of a person (P) *who lacks, or whom D reasonably believes to lack, capacity*;
(b) is the donee of a lasting power of attorney, or an enduring power of attorney created by P; or
(c) is a deputy appointed by the court to make decisions for P,

D is guilty of an offence if D ill-treats or wilfully neglects P.

The Court of Appeal has held that the italicised words in (a) must also be proved in respect of P in a case within (b) or (c).

For the above purposes, P lacks capacity in relation to a matter if at the material time P is unable to make a decision for him/herself in relation to the matter because of an impairment of, or a disturbance in the functioning of, the mind or brain, whether permanent or temporary. P is unable to make a decision for him/herself if P is unable to understand the information relevant to the decision, to retain, use or weigh it as part of the decision-making process, or to communicate his/her decision (whether by talking, sign language or otherwise).

Ill-treatment or wilful neglect: care worker offence

13.33   By the Criminal Justice and Courts Act 2015 (CJCA 2015), s 20(1), it is an offence for an individual who has the care of another individual by virtue of being a care worker to ill-treat or wilfully to neglect that individual. 'Care worker' means an individual who, as paid work, provides:

(a) health care for an adult or child, other than excluded health care (ie health care at an educational institution, children's home or residential family centre); or
(b) social care for an adult,

including an individual who, as paid work, supervises or manages individuals providing such care or is a director or similar officer of an organisation which provides such care.

Ill-treatment or wilful neglect: care provider offence

13.34   A care provider commits an offence under CJCA 2015, s 21(1) if:

(a) an individual who has the care of another by virtue of being part of the care provider's arrangements ill-treats or wilfully neglects that individual;
(b) the care provider's activities are managed or organised in a way which amounts to a gross breach of a relevant duty of care owed by the care provider to the individual who is ill-treated or neglected; and
(c) in the absence of this breach, the ill-treatment or wilful neglect would not have occurred or would have been less likely to occur.

'Care provider' means:

(a) a body corporate or unincorporated association which provides or arranges for the provision of:
    (i)   health care for an adult or child, other than excluded health care, or
    (ii) social care for an adult, or
(b) an individual who provides such care and employs, or has otherwise made arrangements with, other persons to assist him/her in providing such care,

except that a local authority, or a person or body carrying out certain functions for a local authority, or a registered adoption society or adoption support agency, is not a care provider.

A 'relevant duty of care' means a duty owed under the law of negligence, but only to the extent that the duty is owed in connection with providing, or arranging for the provision of, health care or social care.

A breach of a duty of care by a care provider is a 'gross' breach if the conduct alleged to amount to the breach falls far below what can reasonably be expected of the care provider in the circumstances.

General

13.35   Ill-treatment and neglect have the same meaning as in CYPA 1933, s 1, but unlike that section the above provisions do not include the words 'in a manner likely to cause unnecessary suffering or injury to health'.

The above offences are indictable (either-way) offences. The offences at **13.32** and **13.33** are punishable with up to five years' imprisonment. The offence at **13.34** is punishable with an unlimited fine; in addition or instead, a person convicted of that offence may be given a remedial order and/or a publicity order (which requires the recipient to publish details about the offence and its outcome), breach of which is an indictable either-way offence punishable with an unlimited fine.

Causing or allowing a child or vulnerable adult to die or suffer serious physical harm

This is dealt with in **10.28**.   **13.36**

DOMESTIC VIOLENCE OR ABUSE

Domestic violence or abuse is defined by the Government as any incident of control-   **13.37**
ling, coercive, or threatening behaviour, violence, or abuse between those aged 16 or over who are or have been intimate partners or family members, regardless of their gender or sexuality. Anyone can be can be a victim. The abuse can encompass, but is not limited to psychological, physical, sexual, financial, or emotional abuse.

Reference is made below to the victim of domestic violence or abuse as a woman but—as the previous paragraph recognises—it must not be forgotten that sometimes the victim is a man. A woman who has been injured by her husband, civil partner, or cohabitant is competent and compellable to give evidence against that person in a criminal court in relation to the offence in question, and may wish to do so. Likewise, a family member who has been subject to domestic violence or abuse is competent and compellable. However, it will be appreciated that this is frequently an extremely difficult decision for a wife, partner, or family member to make. Many women, for example, feel trapped and helpless and, being unable to face up to life on their own, prefer to remain with a man who treats them badly. In such circumstances they will not wish to give evidence against him because of their fear that this will lead to a final breakdown of the relationship. The social services are experienced in the handling of these situations and will help if the woman will accept such aid.

Domestic violence or abuse, particularly domestic violence, can result in civil proceedings (dealt with in **13.39**, **13.40**, **13.46**, and **13.47**), or in liability for one of the various offences under the criminal law which are not limited to such conduct. In addition, it may constitute the following offence.

Controlling or coercive behaviour in an intimate or family relationship

Under the Serious Crime Act 2015, s 76(1), a person (D) commits an indictable (either-   **13.38**
way) offence, punishable with up to five years' imprisonment, if:

(a) D repeatedly or continuously engages in behaviour towards another person (V) that is controlling or coercive;
(b) at the time of the behaviour, D and V are personally connected;
(c) the behaviour has a serious effect on V; and
(d) D knows or ought to know that the behaviour will have a serious effect on V.

For the purposes of (d), D 'ought to know' that which a reasonable person in possession of the same information would know.

D and V are 'personally connected' if:

(a) D is in an intimate personal relationship with V; or
(b) D and V live together and they are members of the same family, or they have previously been in an intimate personal relationship with each other.

D and V are members of the same family if:

(a) D and V are, or have been, married to each other;
(b) they are or have been, civil partners of each other;
(c) they are relatives;
(d) they have agreed to marry one another (whether or not the agreement has been terminated);
(e) they have entered into a civil partnership agreement (whether or not the agreement has been terminated);
(f) they are both parents of the same child (ie someone under 18);
(g) they have, or have had, parental responsibility for the same child.

Examples of coercive or controlling behaviour can include violence, threats, humiliation, and intimidation or other abuse used to harm, punish, or frighten the victim (coercive behaviour); or acts designed to make a person subordinate and/or dependent, eg by limiting someone's access to family, friends, or finances, by controlling someone's access to social media accounts, by surveillance through apps, by dictating what someone wears, or by threats to reveal or publish private information.

D does not commit an offence under s 76(1) if at the time of the behaviour in question:

(a) D has parental responsibility for V, or is otherwise legally liable to maintain V, or has care of him/her; and
(b) V is under 16 (s 76(3)).

D's behaviour has a 'serious effect' on V if it causes V to fear, on at least two occasions, that violence will be used against V, or if it causes V serious alarm or distress which has a substantial adverse effect on V's usual day-to-day activities.

It is a defence for D to show that:

(a) in engaging in the behaviour in question, D believed that (s)he was acting in V's best interests; and
(b) the behaviour was in all the circumstances reasonable (s 76(8)).

D is to be taken to have shown these facts if sufficient evidence of the facts is adduced to raise an issue with respect to them, and the contrary is not proved beyond reasonable doubt (s 76(9)). This defence is not available to D in relation to behaviour that causes V to fear that violence will be used against V (s 76(10)).

The Secretary of State has issued guidance about the investigation of offences under s 76.

Non-molestation orders

13.39   A non-molestation order is made in civil family proceedings under the Family Law Act 1996 (FLA 1996), s 42.

A non-molestation order is an order containing either or both of the following provisions:

(a) prohibiting a person (the respondent, R) from molesting another person who is associated with R; and
(b) prohibiting R from molesting a relevant child.

The Court of Appeal has held that 'molestation' in the present context implies some deliberate conduct aimed at a high degree of 'harassment' (as defined at **9.30**) of the other party, sufficient to call for the intervention of the court.

A non-molestation order may be made:

(a) on the application (whether in other family proceedings or without any other family proceedings being instituted) of a person who is associated with R; or
(b) during family proceedings to which R is a party, if the court considers that such an order will benefit any other party or a relevant child even though no application has been made.

By FLA 1996, s 62, a person is '*associated with*' R if they are spouses (or ex-spouses), civil partners (or ex-civil partners), or cohabitants (or ex-cohabitants), or members (or ex-members) of the same household, or relatives, or are or have been engaged, or have or have had an intimate personal relationship which is or was of significant duration, or are parties to family proceedings other than proceedings for a non-molestation order. For these purposes, '*cohabitants*' are two persons who are neither married to each other nor civil partners of each other but are living together as if they were a married couple or civil partners.

In relation to a child (ie someone under 18), a person is associated with R if each of them is a parent of the child or has had parental responsibility for it. Where a child has been adopted or freed for adoption, two persons are associated with each other if one is its natural parent (or a parent of such a natural parent) and the other is the child or someone who is its adoptive parent (or an applicant for adoption) or with whom the child has been placed for adoption.

A '*relevant child*' in relation to such proceedings is any child who is living with or might reasonably be expected to live with either party to the proceedings, or any child in relation to whom an order under the Adoption Act 1976, the Adoption and Children Act 2002, or the Children Act 1989 is in question in the proceedings, or any other child whose interests the court considers relevant.

A non-molestation order may refer to molestation in general, to particular acts of molestation, or both. It may be made for a specified period, or until a further order is made. An order which is made in other family proceedings ceases to have effect if those proceedings are withdrawn or dismissed.

Application without notice to the other party

Where it appears just and convenient to do so a court may, under FLA 1996, s 45(1), **13.40** make a non-molestation order even though R has not been given notice of the proceedings. In such cases, the court must consider:

(a) the risk of significant harm if the order is not made immediately;
(b) whether if such an order is not made the applicant is likely to be deterred or prevented from pursuing the application; and
(c) whether there is reason to believe that R is aware of the proceedings but is deliberately evading service of the notice and the applicant, or a relevant child, will be seriously prejudiced by the delay involved in effecting service of the proceedings.

Where an order is made on an application without notice to R, the court must afford R an opportunity to make representations as soon as just and convenient at a full hearing.

Arrest warrant

13.41   Where a non-molestation order has been made, and the applicant considers that R has failed to comply with it, the applicant may apply to the court for a warrant to arrest R.

Offence of breaching non-molestation order

13.42   Under FLA 1996, s 42A(1), a person who without reasonable excuse does anything that (s)he is prohibited from doing by a non-molestation order is guilty of an indictable (either-way) offence punishable with up to five years' imprisonment. In the case of an order made without notice to the other party, a person can be guilty of the offence only in respect of conduct engaged in when (s)he was aware of the existence of the order. Where a defendant has adduced evidence which raises the issue of reasonable excuse, the prosecution must disprove reasonable excuse beyond reasonable doubt.

Undertakings

13.43   Where a non-molestation order could be made, a court may accept an undertaking from any party to the proceedings. Breach of such an undertaking is not an offence. However, the undertaking is enforceable in civil law in the same way as a court order, except that no power of arrest attaches to its breach.

Domestic violence protection notices and orders

13.44   The Crime and Security Act 2010 empowers a superintendent (or above) to issue a domestic violence protection notice (DVPN). A DVPN prohibits a suspected perpetrator (P) from molesting a victim of domestic violence (V) and, where they cohabit, may require P to leave those premises. Sections 24 to 26 deal with DVPNs.

The issue of a DVPN triggers an application for a domestic violence protection order (DVPO), which is an order lasting between 14 and 28 days. A DVPO prohibits P from molesting V and may also make provision about access to shared accommodation by P and V. Sections 27 to 30 deal with DVPOs.

Under the scheme the police and courts can protect V when V is most vulnerable, in the immediate aftermath of an attack, by preventing P from contacting V or returning home for up to 28 days. This helps victims who may otherwise have had to flee their home and gives them the space and time to access the support they need and to consider their options with the help of a support agency.

The scheme is intended to plug a gap in protection for victims of domestic violence due to either the police being unable to charge the perpetrator due to lack of evidence (so V cannot be protected through strict bail conditions) or the process for granting longer-term non-molestation orders taking several days or weeks to apply for.

DVPN

13.45   A DVPN may be issued to P, if P is aged 18 or over, where the authorising officer (superintendent (or above)) has reasonable grounds for believing that:

(a) P has been violent, or has threatened violence, towards an associated person, V, and

(b) the issue of a notice is necessary in order to protect V from violence or the threat of violence by P.

An 'associated person' means a person who is associated with P within the meaning of FLA 1996, s 62 (**13.39**).

Before issuing a DVPN, the authorising officer must consider the welfare of any child whose interests the officer considers relevant. (S)he must take reasonable steps to find out the opinion of V as to whether the DVPN should be issued. Consideration must also be given to any representation P makes in relation to the issuing of the DVPN. Where the DVPN is to include conditions in relation to the occupation of premises shared by P and V, reasonable steps must also be taken to find out the opinion of any other associated person who lives in the premises. The issue of the DVPN does not require V's consent, as the authorising officer may nevertheless have reason to believe that V requires protection from P.

A DVPN must contain provision to prohibit P from molesting V for the duration of the DVPN, which may be expressed so as to refer to molestation in general, particular acts of molestation, or both.

Where P and V share living premises, the DVPN may explicitly:

(a) require P to leave the premises; or
(b) prohibit P from evicting or excluding V from the premises, from entering the premises, or from coming within a certain distance of the premises (as specified in the DVPN) for the duration of the DVPN.

It does not matter for these purposes whether the premises are owned or rented in the name of P or V.

Where a DVPN is issued which prevents P from entering (or requires P to leave) premises, and the authorising officer believes that P is subject to service law and the premises are service living accommodation, (s)he must make reasonable efforts to inform P's commanding officer that the notice has been issued.

The DVPN must state the grounds for issuing a DVPN; that a constable may arrest without warrant if (s)he has reasonable grounds to believe that P is in breach of the DVPN; that the police will make an application for a DVPO which will be heard in court within a 48-hour period and a notice of the hearing will be given to P; that the DVPN will continue to be in effect until the DVPO application is determined; and the provision that may be included in a subsequent DVPO. A DVPN must be in writing and only be served on P personally by a constable. The constable serving the DVPN must ask P to supply an address in order to enable P to be given notice of the hearing for the DVPO.

If (s)he has reasonable grounds to believe that P is in breach of the DVPN, a constable may arrest P without warrant as set out above. P must then be held in custody and brought within a period of 24 hours before the magistrates' court that will hear the application for the DVPO. In calculating the end of that period, a Sunday, Christmas Day, Good Friday, or a Bank Holiday is disregarded. However, if the DVPO hearing has already been arranged to take place within that 24-hour period, then P must be brought before the court for *that* hearing. If P is brought before the court in advance of the DVPO hearing, the court may remand P; it may also remand P if it adjourns the hearing.

DVPO

Once a DVPN has been issued, a constable must apply by complaint to a magistrates' court for a DVPO. The magistrates' court hearing must be no later than 48 hours (calculated as for a DVPN) after the time when the DVPN was served. Notice of the hearing must be given to P. Where no address has been given by P, the court must be satisfied that reasonable efforts have been made to give P the notice of the hearing. **13.46**

If the hearing of the application is adjourned, the DVPN continues in effect until the application is determined by the court. V is not compelled to attend the hearing of an application for a DVPO, or to answer questions (unless V has given oral or written evidence at the hearing).

Two conditions must be met for a DVPO to be made; the court must:

(a) be satisfied on the balance of probabilities that P has been violent, or threatened violence, towards an associated person, V;

(b) think that the DVPO is necessary to secure the protection of V from violence, or the threat of violence, by P.

Before making a DVPO, the court must consider: the welfare of any person under 18 whose interests the court considers relevant to the DVPO; the opinion (if known) of V; and, where the DVPO is to include conditions in relation to the occupation of premises shared by P and V, the opinion (if known) of any other associated person who lives in the premises. The court may issue a DVPO regardless of whether or not V consents.

A DVPO must contain provision explicitly prohibiting P from molesting V for the duration of the DVPO. The prohibition may be expressed so as to refer to molestation in general, particular acts of molestation, or both. Where P and V share living premises, the DVPO may explicitly: prohibit P from evicting or excluding V from the premises; prohibit P from entering the premises; require P to leave the premises; or prohibit P from coming within a certain distance of the premises (as specified in the DVPO). The DVPO must state that a constable may arrest P without warrant if the constable has reasonable grounds for believing that P is in breach of the DVPO.

A DVPO may be in force for a minimum of 14 days, and a maximum of 28 days, from the day on which it is made. It must state the period for which it is to be in force.

If P is arrested for breach of a DVPO, P must be held in custody and brought before a magistrates' court within a 24-hour period (calculated as for a DVPN). If the matter is not disposed of when P is brought before the court, the court may remand P.

General

13.47   Like various other orders, a DVPO is a civil order. As the proceedings are civil in nature the law of civil evidence applies. Under that law, the court can also take into account evidence of bad character untrammelled by the conditions for doing so in criminal proceedings under the Criminal Justice Act 2003, s 101. In determining whether the conditions are met the court can also take into account hearsay evidence admissible under the Civil Evidence Act 1995 (CEA 1995).

Under CEA 1995, s 1, subject to certain safeguards in later sections, hearsay evidence is admitted in civil cases.

CEA 1995, s 1 does not give a party to a civil action an unfettered right to introduce hearsay evidence in proceedings. In particular, a statement is not admissible at all if the maker of it (not the witness but the person whose words it is sought to admit) would not have been a competent witness. In addition a number of sections provide specific safeguards:

CEA 1995, s 4 provides that in estimating the degree of weight which is given to the statement the court 'shall have regard to any circumstances from which any inference can reasonably be drawn as to the reliability or otherwise of the evidence'. Section 4(2) lists a number of factors to which the court should have regard in assessing weight but does not restrict the court to these factors. Factors listed include whether it would have been reasonable and practicable to call the maker of the statement, whether the statement was contemporaneous with events narrated in it, and whether the statement involved multiple hearsay.

CEA 1995, s 5 allows the credibility of the maker of the statement to be impeached and also allows the admission of inconsistent statements made by the maker of the statement. With the leave of the court the maker of the statement, if available, may be called as a witness.

Normally, CEA 1995, s 2 provides that a party proposing to adduce hearsay evidence must give notice to that effect and provide further details if requested to do so by the other party to the action. The Magistrates' Courts (Hearsay Evidence in Civil Proceedings) Rules 1999 require a party who desires to give hearsay evidence to serve a hearsay notice not less than 21 days before the date fixed for the hearing. Failure to give notice (or provide further details if requested) does not affect the admissibility of the evidence but may affect the weight that is given to it. CEA 1995, s 2 does not, however, apply to DVPO proceedings.

If what are relied on are oral statements to a police officer, the officer should give direct evidence of what was said and the circumstances in which it was said.

Rights associated with the matrimonial home

Where domestic disputes occur, regardless of whether or not a non-molestation order or a DVPO has been sought, there is the issue of rights of occupation in relation to the home. FLA 1996 deals with this. **13.48**

Where one spouse or civil partner has no estate, etc

The Family Law Act 1996 (FLA 1996), s 30 provides that, where one spouse or civil partner (X) is legally entitled to occupy a dwelling house and the other spouse or civil partner (Y) has no such legal entitlement, Y has 'matrimonial home rights' which means that: **13.49**

(a) if Y is in occupation, Y has a right not to be evicted or excluded from the dwelling house or any part of it by X without the leave of the court under FLA 1996, s 33 (occupation order provisions); and

(b) if Y is not in occupation, Y has a right with the leave of the court under FLA 1996, s 33 to enter and occupy the dwelling house.

There is, therefore, no lawful way by which one party to a marriage or civil partnership can be removed from the home without the circumstances being examined by a civil court.

Occupation orders

FLA 1996, s 33 provides for the making of an occupation order where the applicant (A) has an estate or interest, etc entitling A to occupy a dwelling house or has matrimonial home rights in it, and the dwelling house is or has been the home of A and of someone else (the respondent, R) with whom A is associated (or was intended by both such people to be their home). If an occupation order is made it may: **13.50**

(a) enforce A's occupation rights as against R;

(b) require R to permit A to enter and remain in that dwelling house or part of it;

(c) regulate the occupation by either or both parties;

(d) prohibit, restrict or suspend the exercise by R of R's occupation rights, if R is entitled via an estate or interest etc to occupy;

(e) if R has home rights and A is the other spouse or civil partner, restrict or terminate those rights;

(f) require R to leave the dwelling house or part of it; or

(g) exclude R from a defined area in which the dwelling house is included.

In deciding whether to make an occupation order and (if so) in what manner, the court must have regard to all of the circumstances including the housing needs and housing resources of both parties and any relevant child (**13.39**), the financial resources of both parties, the likely effect of any order (or of a failure to make an order) on the health, safety, or well-being of each party and any relevant child, and the conduct of the parties in relation to each other and otherwise.

Former spouse, etc not entitled to occupy, and situations in which neither spouse, former spouse, etc entitled to occupy

13.51 FLA 1996 makes similar provisions for occupation orders in relation to former spouses or civil partners, and cohabitants (**13.39**) or former cohabitants, where one of them is legally entitled to occupy a dwelling house, and the other is not. It also makes provision for occupation orders in circumstances in which neither spouse or civil partner or co-habitant (or ex-spouse or ex-civil partner or ex-cohabitant) has a legal entitlement to occupy a dwelling house which is (or was) the matrimonial home. These provisions are set out in ss 35 to 38.

Arrest for breach of occupation order and other points

13.52 Under FLA 1996, s 47, there is a power to arrest without warrant for breach of an occupation order where such a power has been attached to the order; otherwise an application must be made for an arrest warrant. 'Undertakings' and orders without notice to the other party may be made in respect of these orders in the same way as in the case of non-molestation orders.

VAGRANCY OFFENCES

Begging

13.53 The Vagrancy Act 1824 (VA 1824), s 3 makes it a summary offence, punishable with up to one month's imprisonment or a fine not exceeding level 3, where a person wanders abroad, or places him/herself in any public place, street, highway, court or passage, to beg or gather alms, or who cause, procure or encourage any child to do so. A divisional court has held that the offence does not cover those who collect alms in an orderly manner for a specific purpose, eg striking workers who seek assistance by asking for contributions towards their cause.

VA 1824, s 4 deals with the aggravated forms of begging by the exposure of wounds or deformities in public or by going about seeking charitable contributions of any kind by false pretences. An offence under s 4 is a summary offence punishable with up to three months' imprisonment or a fine not exceeding level 3.

Causing or allowing persons under 16 to be used for begging

13.54 Causing or procuring a child or young person under 16 to be in any street, premises, or place for the purpose of begging or receiving alms, or of inducing the giving of alms (whether or not there is any pretence of singing, playing, performing, offering anything for sale, or otherwise), is an offence contrary to CYPA 1933, s 4(1). This offence is in addition to the similar offence under the Vagrancy Act 1824, s 3 above.

It is also an offence under CYPA 1933, s 4(1) for the person having the responsibility for such a child or young person to allow him/her so to act. Such a person allows a child or young person to beg in the street if (s)he fails to prevent this when (s)he could and should have prevented it. If it is proved that a child or young person was in a street, etc for the purpose of begging, etc and that a person with responsibility for the child or young person allowed him/her to be in the street, etc, that person is presumed to have allowed the child or young person to be there for that purpose, until the contrary is proved.

An offence under s 4(1) is a summary offence punishable with up to three months' imprisonment and/or a fine not exceeding level 2.

Sleeping out, etc

It is an offence contrary to VA 1824, s 4 (**13.53**) for any person, wandering abroad and **13.55** lodging in any barn or outhouse, or in any deserted or unoccupied building, or in the open air, or under a tent, or in any cart or wagon, not to give a good account of him/herself.

Genuine hikers who, being tired and hungry, rest in a barn or outhouse are not guilty of this offence because they can easily give a good account of themselves by explaining their presence. In contrast, tramps found sleeping in a barn would find it much more difficult to give a good account of themselves, particularly if they have made a temporary home in that building.

Two important limits were imposed on the offence by the Vagrancy Act 1935.

First, the Act amended VA 1824, s 4 by providing that the reference to a person lodging under a tent or in a cart or waggon does not include a person lodging under a tent, cart, or waggon with or in which he travels. It was thereby made quite clear that the present offence was not concerned with gypsies travelling in their own waggons, nor with persons sleeping out in their own tents.

Second, the Vagrancy Act 1935 requires that, before a person (D) can be guilty of the present offence, it must be proved either that:

(a) on the occasion in question, D had been directed to a reasonably accessible place of free shelter and that D failed to apply for, or refused, accommodation there; or

(b) D is a person who persistently wanders abroad and, notwithstanding that a place of free shelter is reasonably accessible, lodges or attempts to lodge in a way described above; or

(c) by, or in the course of, lodging in a way described above D caused damage to property, infection with vermin, or other offensive consequence, or D so lodged in such circumstances as to appear to be likely to do so.

As a result, the present offence is of little practical significance to police officers. The reason is that there are few places of free shelter to which people may be directed, or which are reasonably accessible to the person who persistently sleeps out. The provision of greatest practical significance is that at (c), which can apply to the roadster who destroys hay or feed in a barn by his presence, or who causes the barn or outbuilding to become verminous by his presence.

PROSTITUTION

Prostitute

For the purposes of legislation other than the Sexual Offences Act 2003 (SOA 2003), the **13.56** term 'prostitute' is defined by case law. This establishes that a *prostitute* is a woman (or a man) who offers her/his body commonly for sexual intercourse or acts of a sexual nature, in return for payment. It has been held by the Court of Appeal that it is immaterial that the woman (or man) is dishonest and intends simply to pocket advance payment and not to provide sexual services. The reference to other acts of a sexual nature extends the definition of 'prostitute'. Masseuses, for example, who carry out acts of masturbation on request are engaging in such acts. So are women who engage in sado-masochistic sessions for the sexual pleasure of their partner. Since sexual intercourse is not required, even a virgin can be a prostitute. If a person offers intercourse for payment regularly to one person, the offeror is not a prostitute as services are restricted to one person. If the offer is made to a number of persons because of the promiscuous nature of the offeror and no payment is required for the services, the offeror is not a prostitute as no payment is received in return.

For the purposes of SOA 2003, 'prostitute' is defined as a person (A) who, on at least one occasion and whether or not compelled to do so, offers or provides sexual services to another person in return for payment or a promise of payment to A or a third person; 'prostitution' in SOA 2003 is interpreted accordingly. 'Payment' for the purposes of SOA 2003 means any financial advantage, including the discharge of an obligation to pay or the provision of goods or services (including sexual services) gratuitously or at a discount. Thus, any form of financial arrangement will be classified as payment. The settlement of a debt by the provision of sexual services, or the supply of drugs, or their provision at a discounted price will be sufficient. There is no doubt that a similar approach will be followed in relation to other statutes.

It is not an offence in itself to be a prostitute, but there are a number of offences which can only be committed by, or in respect of, prostitutes.

Persistent loitering or soliciting by a prostitute

13.57   The Street Offences Act 1959 (StOA 1959), s 1(1) creates the summary offence committed by a prostitute (whether male or female) who persistently loiters or solicits in a street or public place for the purposes of prostitution. The offence can only be committed by someone aged 18 or over. It is punishable with a fine not exceeding level 2 (level 3 if a previous conviction). In lieu of punishment an offender, an offender can be required to attend three meetings with a supervisor to address the causes of the offending conduct and ways and find ways to cease engaging in it. Failure to comply with such an order without reasonable excuse can result in a magistrates' court revoking the order and dealing with the offender, for the offence in respect of which the order was made, as if the offender had just been convicted of the offence

Persistently loitering or soliciting for the purposes of prostitution

13.58   Loitering or soliciting is persistent if it takes place on two or more occasions in any period of three months.

Loitering by a prostitute does not need to be for the purpose of making approaches to others; it is sufficient that there is loitering for the purpose of being approached by potential clients (as where a prostitute loiters in an area noted as a haunt of prostitutes). Loitering need not be on foot. It can, for example, be done in a slowly moving vehicle.

Soliciting need not be by words. Movements of the body, arms, hands, as well as facial expressions and gestures can be equally compelling forms of solicitation. Tapping on window panes, leaning out of windows with signals to indicate price, signalling the position of the entry door with the fingers are all forms of solicitation. The test to be applied should be, 'Is it clear to the reasonable person that sex for money is being offered?'

Street or public place

'Street' for the purpose of StOA 1959 includes any bridge, road, lane, footway, subway, **13.59** square, court, alley or passage, whether a thoroughfare or not, which is for the time being open to the public. In addition, the doorways and entrances to premises abutting on a street and the ground adjoining and open to a street are treated as forming part of the street.

The definition is quite wide. As the courts have confirmed, it includes places which are upon private property if they are open to a street, eg a balcony or a room whose window faces onto a street.

'Public place' is not defined but generally the courts have accepted that a public place is one where the public go, no matter whether they have a right to go or not. What is a public place may vary from time to time; the question is whether the place was public at the material time.

Placing of advertisements relating to prostitution

The Criminal Justice and Police Act 2001 (CJPA 2001), s 46(1) creates an offence of pla- **13.60** cing on, or in the immediate vicinity of, a public telephone, an advertisement relating to prostitution, with intent that the advertisement should come to the attention of any other person. An advertisement is covered by s 46 if it is for the services of a prostitute (**13.56**), whether male or female, or indicates that premises are premises in which such services are offered. Any advertisement which a reasonable person would consider to be an advertisement relating to prostitution is presumed to be so, unless shown not to be.

For these purposes, 'public telephone' means any telephone located in a public place for use by the public (or a section of the public) together with any structure in which it is housed. A 'public place' means any place to which the public have, or are permitted to have, access, whether on payment or otherwise, other than (a) any place to which children under the age of 16 are not permitted to have access, whether by law or otherwise, and (b) any premises which are wholly or mainly used for residential purposes.

The above is a summary offence punishable with up to six months' imprisonment and/or an unlimited fine.

Exploitation of prostitution

The two offences under this heading are indictable (either-way) offences punishable **13.61** with up to seven years' imprisonment.

SOA 2003, s 52(1) provides that it is an offence for a person intentionally to cause or incite another person to become a prostitute (**13.56**) in any part of the world if he does so for or in the expectation of gain for himself or for a third person. Section 53(1) makes it an offence for a person intentionally to control any of the activities of another person relating to that person's prostitution in any part of the world for or in the expectation of gain for himself or a third person. 'Control' is not limited to conduct which forces another to carry out the relevant activity; it can be exercised in various ways. It is enough if a defendant instructed or directed a prostitute to carry out the relevant activity or do it in a particular way.

For the purposes of these offences the term 'gain' means:

(a) any financial advantage, including the discharge of an obligation to pay or the provision of goods or services (including sexual services) gratuitously or at a discount; or

(b) the goodwill of any person which is or appears likely, in time, to bring financial advantage.

This definition makes it clear that there is no requirement that money is intended to change hands from the prostitute to the defendant (D). It may be that D wants no more than an occasional 'freebee' for himself or another, or intends that the prostitute should pay for his drug habit out of her earnings.

Sections 52 and 53 are of particular importance in relation to the protection of persons aged 18 or over, because ss 47 to 50 (**13.65-13.68**) specifically protect those under 18.

Soliciting to obtain a prostitute's sexual services

13.62   Under SOA 2003, s 51A(1), it is an offence for a person in a street (as defined by the StOA 1959 (**13.59**)) or public place to solicit another (B) for the purpose of obtaining B's sexual services as a prostitute (**13.56**). This is a summary offence punishable with a fine not exceeding level 3.

The reference to a person in a street or public place includes a person in a vehicle in a street or public place.

Paying for sexual services of a prostitute subjected to force, etc

13.63   By SOA 2003, s 53A, a person (A) commits a summary offence punishable with a fine not exceeding level 3 if:

(a) A makes or promises payment for the sexual services of B, a prostitute (**13.56**),
(b) a third person (C) has engaged in exploitative conduct of a kind likely to induce or encourage B to provide the sexual services for which A has made or promised payment, and
(c) C engaged in that conduct for or in the expectation of gain for C or another person (apart from A or B).

It is irrelevant where in the world the sexual services are to be provided or whether those services are provided. It is also irrelevant whether A is, or ought to be, aware that C has engaged in exploitative conduct.

For the purposes of s 53A:

(a) C engages in exploitative conduct if C uses force, threats (whether or not relating to violence), or any other form of coercion, or C practises any form of deception; and
(b) 'payment' and 'gain' have the meanings given at **13.56** and **13.61** respectively.

Sexual exploitation of children

13.64   All the offences under this heading are indictable (either-way) offences, besides those under s 47(1) and (3) or 47(1) and (4)(a) (which are triable only on indictment).

Paying for the sexual services of a child

13.65   SOA 2003, s 47(1) and (5) makes it an offence for a person (A) intentionally to obtain for himself the sexual services of another person (B) under 18 where, in advance, such services have been paid for by A or (to A's knowledge) another or where payment

(**13.56**) has been promised by A or (to A's knowledge) another. The maximum punishment is seven years' imprisonment.

A more serious offence (maximum 14 years' imprisonment) is committed under s 47(1) and (4)(a) if the offence involved one of the four types of penetration, set out below, or s 47(1) and (4)(b) if it does not, and in either case B is under 16.

An even more serious offence, punishable with a maximum of life imprisonment, is committed under s 47(1) and (3) if B is under 13 and the offence involved:

(a) penetration of B's anus or vagina with a part of A's body or anything else,
(b) penetration of B's mouth with A's penis,
(c) penetration of A's anus or vagina with a part of B's body or by B with anything else, or
(d) penetration of A's mouth with B's penis.

The payment or promise may be made to B or to someone else. Where B is under 13 an offence will be committed regardless of any belief which A might have concerning age. However, where B is 13 or over A is only guilty if A did not reasonably believe that the victim was 18 or over; it will be for the prosecution to prove that such a belief did not exist.

Causing or inciting sexual exploitation of a child

13.66

SOA 2003, s 48(1) makes it an offence, punishable with up to 14 years' imprisonment, for a person (A) intentionally to cause or incite another person (B) under 18 to be *sexually exploited* (**13.69**) in any part of the world. Where B is under 13, it is irrelevant what belief A (however reasonably) might have had as to B's age. If B was 13 or over, the prosecution must prove that A did not reasonably believe that B was under 18. Section 48 does not require proof that A acted for gain. Nor do ss 49 and 50 below.

Controlling a child in relation to sexual exploitation of a child

13.67

SOA 2003, s 49(1) makes it an offence, punishable with up to 14 years' imprisonment. if a person (A) intentionally controls any of the activities of another person (B) relating to B's *sexual exploitation* (**13.69**) in any part of the world, and either B is under 18 and A does not reasonably believe that B is 18 or over, or B is under 13. As to 'control' see **13.61**.

Arranging or facilitating sexual exploitation of a child

13.68

SOA 2003, s 50(1) makes it an offence, punishable with up to 14 years' imprisonment, for a person (A) intentionally to arrange or facilitate the *sexual exploitation* in any part of the world of another person (B), where either B is under 18 and A does not reasonably believe that B is 18 or over, or B is under 13.

This offence covers, for example, those who 'convey' young persons to places or premises where sexually exploitative activities are intended to take place and those who merely make the administrative arrangements for such activities.

Sexually exploited

13.69

For the purposes of ss 48 to 50, B is sexually exploited if:

(a) on at least one occasion and whether or not compelled to do so, B offers or provides sexual services to another person in return for payment (**13.56**) or a promise of payment to B or a third person; or
(b) an indecent image of B is recorded or streamed or otherwise transmitted,

and 'sexual exploitation' is to be interpreted accordingly.

Closure orders

13.70   SOA 2003, Part 2A gives the courts the power to make premises closure orders where there is evidence of the premises being used for activities relating to certain prostitution or child sex offences. See Chapter 13 on the companion website.

OFFENCES RELATING TO BROTHELS

Keeping a brothel

13.71   By the Sexual Offences Act 1956 (SOA 1956), s 33, it is an offence for a person to keep a brothel, or to manage it, or to act or assist in its management. Premises are a brothel if they are used by persons for illicit heterosexual or homosexual intercourse or other indecent behaviour. It is not necessary to show that some of the people resorting to the premises are prostitutes or that they received payment for their services, but there must be at least three people who use the premises in this way. It does not matter that one of them is the occupier, and it does not matter that only two people at a time ever use the premises for sexual activities.

The above is a summary offence punishable with up to three months' imprisonment and/or or a fine not exceeding level 3 (six and 4 respectively in the event of a previous conviction for an offence under s 33 (or s 34, 35 or 36 (**13.73**)).

Keeping a brothel for prostitution

13.72   By SOA 1956, s 33A(1), it is an indictable (either-way) offence punishable with up to seven years' imprisonment for a person to keep, or to manage, or to act or assist in the management of, a brothel *to which people resort for practices involving prostitution* (**13.56**) (whether or not also for other practices).

If separate and self-contained flats are separately let, each to one prostitute, it is likely that the building in its entirety will not be classed as a brothel as there is only one prostitute in each of the flats. However, if single rooms are let to prostitutes in one building, it may be sufficient if the rooms are sufficiently close to constitute what might be described as a nest of prostitutes.

Where a brothel is used for prostitution, SOA 1956, s 33A catches the keeper of the premises, who is most likely to be the residential landlord, a manager, who looks after the maintenance and day-to-day needs of the building and its tenants, the 'madame' of the trade, and those who act or assist in the management of the brothel. 'Assisting in the management of a brothel' covers any conduct which contributes to the management of the brothel. It does not require proof that the person actively exercised some control over the brothel or carried out some specific act of management. A person who takes advertisements to a post office and pays for them assists in the management of the brothel. So does a person who discusses with a potential customer the nature of the sexual activities on offer, or who negotiates the price. On the other hand, it has been held that a cleaner at a brothel does not assist in its management.

Related offences

13.73   SOA 1956, ss 34, 35 and 36 go further and add offences (triable and punishable in the same way as an offence under s 33: **13.71**) to cover other possibilities in relation to responsibility for the brothel. First, a lessor or landlord of premises (or his/her agent) who has knowledge of their intended or actual use as a brothel is guilty of an offence (s 34). In addition, the tenant or occupier, or person in charge, of premises who knowingly permits the whole or part of them to be so used is also guilty of an offence (s 35(1)). Lastly, by s 36, it is an offence for the tenant or occupier of any premises knowingly to permit the whole or part of the premises to be used for the purposes of habitual prostitution (whether any prostitute involved is male or female).

SLAVERY AND TRAFFICKING

Slavery, servitude, and forced or compulsory labour

A person commits an indictable (either-way) offence under the Modern Slavery Act **13.74** 2015 (MSA 2015), s 1(1) if:

(a) (s)he holds another person in slavery or servitude and the circumstances are such that (s)he knows or ought to know that the other person is held in slavery or servitude; or

(b) (s)he requires another person to perform forced or compulsory labour and the circumstances are such that (s)he knows or ought to know that the other person is being required to perform forced or compulsory labour.

In determining whether a person is being held in slavery or servitude or required to perform forced or compulsory labour, regard may be had to all the circumstances, including (i) any of his/her personal circumstances (such as his/her being a child, his/her family relationships, and any mental or physical illness) which may make him/her more vulnerable than other persons, and (ii) any work or services provided by the person, including work or services provided in circumstances which constitute exploitation within s 3(3) to (6) (**13.76**).

The consent of a person (whether an adult or a child) to any of the acts alleged to constitute holding the person in slavery or servitude, or requiring the person to perform forced or compulsory labour, does not preclude a determination that the person is being held in slavery or servitude, or required to perform forced or compulsory labour.

The maximum punishment for an offence under s 1 is life imprisonment.

Human trafficking

A person commits an indictable (either-way) offence under MSA 2015, s 2(1) if (s)he **13.75** arranges or facilitates the travel of another person (V) with a view to V being exploited. It is irrelevant whether V consents to the travel (whether V is an adult or a child). A person may in particular arrange or facilitate V's travel by recruiting V, transporting or transferring V, harbouring or receiving V, or transferring or exchanging control over V. The Court of Appeal has stated that 'arranging' and 'facilitating' should be given their ordinary meaning, so that 'arranging' includes, by way of example, such matters as transporting V, procuring a third person to transport V, or buying a ticket for V, and 'facilitating' includes 'making easier'.

A person arranges or facilitates V's travel with a view to V being exploited only if:

(a) (s)he intends to exploit V (in any part of the world) during or after the travel; or

(b) (s)he knows or ought to know that another person is likely to exploit V (in any part of the world) during or after the travel.

'Travel' means arriving in, or entering, any country, departing from any country, or travelling within any country. In relation to the repealed trafficking provisions of the Sexual Offences Act 2003, the Court of Appeal has held that the key factor was that travel, a journey, was contemplated. The fact that the journey might be short did not affect whether there had been travel for the purposes of the offence.

'Country' includes territory or other part of the world.

A person who is a UK national commits an offence under s 2(1) regardless of where the arranging or facilitating takes place, or where the travel takes place. 'UK national' means a British citizen, a person who is a British subject by virtue of the British Nationality Act 1981, and who has a right of abode in the UK, or a person who is a British overseas territories citizen by virtue of a connection with Gibraltar.

A person who is not a UK national commits an offence under s 2(1) if any part of the arranging or facilitating takes place in the UK, or the travel consists of arrival in or entry into, departure from, or travel within, the UK.

The maximum punishment for an offence under s 2 is life imprisonment.

Meaning of exploitation

13.76   For the purposes of MSA 2015, s 2, a person (V) is exploited only if one or more of the following provisions apply in relation to V.

Slavery, servitude, and forced or compulsory labour   V is the victim of behaviour which involves the commission of an offence under s 1, or which would involve the commission of an offence under s 1 if it took place in England and Wales (s 3(2)).

Sexual exploitation   Something is done to or in respect of V which involves the commission of an offence under the Protection of Children Act 1978, s 1(1)(a) (making/taking indecent photographs of children) or the Sexual Offences Act 2003, Part 1 (ss 1 to 79), or which would involve the commission of such an offence if it were done in England and Wales (s 3(3)).

Removal of organs etc   V is encouraged, required, or expected to do anything which involves the commission, by V or another person, of an offence under the Human Tissue Act 2004, s 32 or 33 (prohibition of commercial dealings in organs and restrictions on use of live donors), or which would involve the commission of such an offence, by V or another person, if it were done in England and Wales (s 3(4)).

Securing services etc by force, threats or deception   V is subjected to force, threats, or deception designed to induce V to provide services of any kind, to provide another person with benefits of any kind, or to enable another person to acquire benefits of any kind (s 3(5)).

Securing services etc from children and vulnerable persons   Another person uses or attempts to use V for a purpose within s 3(5), having chosen V for that purpose on the grounds that (a) V is a child (if under 18), is mentally or physically ill or disabled, or has a family relationship with a particular person, and (b) an adult, or a person without the illness, disability or family relationship, would be likely to refuse to be used for that purpose (s 3(6)). The Court of Appeal has held that, although V's youth must be one of the reasons for the choice of that person, it did not need to be the sole or main reason. In addition, the Court emphasised that there was no need to prove a lack of consent on the part of the young person, etc or any element of coercion.

Anonymity of complainant

An offence under MSA 2015, s 2 is subject to the anonymity rules set out at **11.77**.

Police powers

MSA 2015, s 12 provides that, if a person (P) has been arrested for an offence under s 2, **13.78** a constable or senior immigration officer may detain a relevant land vehicle, ship (any vessel, including a hovercraft, used in navigation) or aircraft:

(a) until a decision is taken as to whether or not to charge P with the offence,

(b) if P has been charged, until P is acquitted, the charge is dismissed or the proceedings are discontinued, or

(c) if P has been charged and convicted, until the court decides whether or not to order forfeiture of the vehicle, ship or aircraft.

A land vehicle, ship, or aircraft is relevant if the constable or officer has reasonable grounds to believe that an order for its forfeiture could be made under s 11 (forfeiture on conviction on indictment of an offence under s 2) if P were convicted of the offence.

Committing offence with intent to commit offence under MSA 2015, s 2

A person commits an offence under MSA 2015, s 4 if (s)he commits any offence ('the **13.79** index offence') with the intention of committing an offence under s 2 (including an index offence committed by aiding, abetting, counselling or procuring an offence under s 2). An offence under s 4 is an indictable (either-way) offence punishable with up to 10 years', unless the index offence is kidnapping or false imprisonment in which case it is triable only on indictment and carries a maximum sentence of life imprisonment.

Preventive orders

MSA 2015, ss 14 to 22 make provision for 'slavery and trafficking prevention orders' **13.80** (much like sexual harm prevention orders described at **11.84–11.91**). MSA 2015, ss 23 to 29 make provision for 'slavery and trafficking risk orders' (much like sexual risk orders described at **11.92–11.96**. Breach of an order made under one of these sets of provisions is an indictable (either-way) offence punishable with up to five years' imprisonment.

Special provisions about protection of complainants in investigations into a 'human trafficking offence'

The Trafficking People for Exploitation Regulations 2013 provide measures aimed at **13.81** the protection of complainants where there is a police investigation into a 'human trafficking offence', ie:

(a) an offence under the Sexual Offences Act 2003, ss 57 to 59A (repealed) (trafficking for sexual exploitation);

(b) an offence under the Asylum and Immigration (Treatment of Claimants, etc) Act 2004, s 4 (repealed) (trafficking people for exploitation);

(c) an offence under MSA 2015, s 2;

(d) attempting or conspiring to commit an offence under (a), (b) or (c), or aiding, abetting, counselling or procuring the commission of such an offence, or an offence of assisting or encouraging under such an offence under the Serious Crime Act 2007, Part 2.

Regulation 3 provides that, without prejudice to the rights of the defendant (D: someone who is alleged to have committed, or has committed, a 'human trafficking offence'), and in accordance with an individualised assessment of the personal circumstances of the complainant (C: someone against or in relation to whom a human trafficking offence is alleged to have been committed, or has been committed), the relevant chief officer of police (the chief officer of police for the police area in which an investigation of a human trafficking offence takes place) must ensure that C receives specific treatment aimed at preventing secondary victimisation by avoiding, as far as possible, during an investigation of a human trafficking offence:

(a) unnecessary repetition of interviews;
(b) visual contact between C and D, using appropriate means including communication technologies;
(c) unnecessary questioning concerning C's private life.

By reg 4, where C is under 18, and without prejudice to D's rights, during an investigation of a human trafficking offence, the relevant chief officer of police must ensure that:

(a) interviews with C take place without unjustified delay after the facts have been reported;
(b) interviews with C take place, where necessary, in premises designed or adapted for the purpose;
(c) interviews with C are carried out, where necessary, by or through professionals trained for the purpose;
(d) if possible and where appropriate, the same persons conduct all the interviews with C;
(e) the number of interviews with C is as limited as possible and interviews are carried out only where strictly necessary for the purposes of the investigation;
(f) C may be accompanied by an adult of C's choice, unless a reasoned decision has been made to the contrary in respect of that adult.

Duty to notify Secretary of State about suspected victims of slavery or human trafficking

13.82 If a 'public authority', eg a chief officer of police or the NCA, has reasonable grounds to believe that a person may be a victim of slavery or human trafficking it is required by MSA 2015, s 52 to notify:

(a) the Secretary of State; or
(b) if regulations made by the Secretary of State require it to notify a public authority other than the Secretary of State, that public authority.

The information to be included in a notification is specified in the Modern Slavery Act 2015 (Duty to Notify) Regulations 2015. No regulations have been made under (b).

A person is a victim of slavery if (s)he is a victim of conduct which constitutes an offence under s 1, or conduct which would have constituted an offence under s 1 if s 1 had been in force when the conduct occurred. A person is a victim of human trafficking if (s)he is the victim of conduct which:

(a) constitutes an offence under s 2, or would constitute an offence under s 2 if the person responsible for the conduct were a UK national, or

(b) would have been within (a) if s 2 had been in force when the conduct occurred.

National Referral Mechanism

13.83

The National Referral Mechanism has no legislative basis. It provides for the formal identification of a victims of trafficking. It operates as follows. A suspected victim is referred by 'first responders' (a term which includes the police) to the Single Competent Authority (SCA), part of the Home Office, for a decision to be made on whether (s)he is a victim of trafficking. The identification of an individual as a victim of trafficking is a two-stage process. The first stage is the 'reasonable grounds decision', acting as an initial filter, SCA deciding whether it was 'reasonable to believe' that the person was a victim of trafficking on the information available. At this stage, the standard of proof is whether SCA 'suspects but cannot prove' that a person was a potential victim. The second stage involves a further inquiry and leads to a 'conclusive grounds decision' as to whether the person is in fact a victim. At this stage, the requisite standard of proof is proof on the balance of probabilities.

Defence for slavery or trafficking victims who commit an offence

13.84

MSA 2015, s 45(1) provides that a person (D) is not guilty of an offence if:

(a) D is 18 or over when D does the act which constitutes the offence;

(b) D does that act because D is compelled to do it;

(c) the compulsion is attributable to slavery or to 'relevant exploitation' (defined below); and

(d) a reasonable person in the same situation and having D's 'relevant characteristics' (defined below) would have no realistic alternative to doing that act.

D may be compelled to do something by another person or by D's circumstances (s 45(2)). Compulsion is attributable to slavery or to relevant exploitation only if:

(i) it is, or is part of, conduct which constitutes an offence under s 1 or conduct which constitutes relevant exploitation; or

(ii) it is a direct consequence of a person being, or having been, a victim of slavery (**13.82**) or a victim of relevant exploitation (s 45(3)).

'Relevant exploitation' is exploitation (within the meaning of s 3, see **13.76**) that is attributable to the exploited person being, or having been, a victim of human trafficking (**13.82**).

A divisional court has accepted that the seriousness of the offence committed by D is a significant consideration when determining under s 45(1)(d) what an adult reasonable person would have done.

Section 45(4) provides that a person (D) is not guilty of an offence if:

(a) D is under 18 when D does the act which constitutes the offence;

(b) D does that act as a direct consequence of being, or having been, a victim of slavery or a victim of relevant exploitation; and

(c) a reasonable person in the same situation and having the person's relevant characteristics would do that act.

For the above purposes, 'relevant characteristics' means 'age, sex and any physical or mental illness or disability', and references to an act include an omission.

A divisional court has held with reference to s 45(4)(c) that age is only one factor to consider when considering the situation and D's relevant characteristics, and that the seriousness of the offence will be of less significance than it would in respect of the defence under s 45(1).

The Court of Appeal has held that s 45 does not implicitly require D to bear the 'persuasive' or 'legal' burden of proof of any element of a defence under s 45(1) or (4). It held that the burden on D is evidential. It is for D to raise evidence of each of those elements and then for the prosecution to disprove one or more of them beyond reasonable doubt.

Offences to which the defence does not apply

13.85   MSA 2015, s 45(1) and (4) do not apply in respect of any of the long list of serious offences in Schedule 4. These offences include murder; manslaughter; false imprisonment; kidnapping; perverting the course of justice; offences under: the Infanticide Act 1938, s 1; the Domestic Violence, Crime and Victims Act 2005, s 5; the Infant Life (Preservation) Act 1929, s 1; the Road Traffic Act 1968, s 1 or 3A; the Offences Against the Person Act 1861, ss 4, 16, 18, 20, 21–23, 28–30, 32, 33 and 38; the Crime and Disorder Act 1998, ss 29 or (if based on the Public Order Act 1986, s 4 or 4A) 31; the Female Genital Mutilation Act 2003, ss 1–3; the Taking of Hostages Act 1982, s 1; the Protection from Harassment Act 1997, s 4 or 4A; the Child Abduction Act 1984, s 1 or 2; the Children and Young Persons Act 1933, s 1; the Explosive Substances Act 1883, ss 2–4; the Firearms Act 1968, s 5, 16, 16A, 17(1) or (2) or 18; the Public Order Act 1936, s 2; the Public Order Act 1986, s 1 or 2; the Terrorism Act 2000, ss 54–57 or 59; the Terrorism Act 2006, ss 5, 6 or 9–11; the Anti-terrorism, Crime and Security Act 2001, s 47, 50 or 113; the Sexual Offences Act 2003, ss 1–10,13–15, 16–19, 25, 26, 30–41, 47–50, 61–67A or 70; the Modern Slavery Act 2015, s 1 or 2; the Theft Act 1968, s 8, 9, 10, 12A or 21; the Criminal Damage Act 1971, s 1 (if arson or offence under s 1(2) other than arson); Immigration Act 1971, s 25 (assisting unlawful immigration to an EU state or the UK); encouraging or assisting, attempting or conspiring to commit one of these offences, or aiding, abetting, counselling, or procuring one of them.

In the case of these offences, or indeed any offence, a prosecutor will still be able to decide not to prosecute a slavery or trafficking victim who may have committed an offence as a direct consequence of his/her slavery or trafficking, if it would not be in the public interest to do so.

HATE CRIMES

Racially or religiously aggravated 'assaults'

13.86   The Crime and Disorder Act 1998 (CDA 1998), s 29 provides that a person is guilty of an indictable (either-way) offence under this section if he commits:

(a) an offence under OAPA 1861, s 20 (**9.13–9.16**);

(b) an offence under OAPA 1861, s 47 (**9.6**);

(c) common assault (or battery) (**9.1–9.3**),

which is racially or religiously aggravated (see below).

An offence within (a) or (b) is punishable with up to seven years' imprisonment, and one within (c) two years'.

On a charge of an offence under CDA 1998, s 29, the prosecution must prove that D has committed one of the relevant specified basic offences and that it (the basic offence) was racially or religiously aggravated.

Racially or religiously aggravated

By CDA 1998, s 28, an offence is racially or religiously aggravated for the purposes of ss 29 to 32 if: **13.87**

(a) at the time of committing the offence, or immediately before or after doing so, the offender demonstrates towards the victim of the offence hostility based on the victim's membership (or presumed membership) of a racial or religious group (s 28(1)(a)); or

(b) the offence is motivated (wholly or partly) by hostility towards members of a racial or religious group based on their membership of that group (s 28(1)(b)).

In (a), 'membership', in relation to a racial or religious group, includes association with members of that group; 'presumed' means presumed by the offender.

Religious group

'Religious group' means a group of persons defined by reference to religious belief or lack of religious belief. Muslims and Rastafarians are religious groups, for instance.   **13.88**

Racial group

'Racial group' means a group of persons defined by reference to race, colour, nationality (including citizenship), or ethnic or national origins. These words are given a broad, non-technical interpretation.   **13.89**

A group of people will be a racial group defined by reference to race if, in ordinary speech, those people would be regarded as belonging to a named race. In one case, for example, it was held by a divisional court that 'African' described a racial group defined by reference to race because in ordinary language 'African' denotes a limited group of people regarded as common stock and as one of the major divisions of humankind having distinct physical features in common; it 'denotes a person characteristic of the blacks in Africa'. This was said to be so despite the fact that, strictly, 'African' is capable of covering Egyptians and white South Africans who would not fit the 'common stock' definition of 'Africans'. This broad, non-technical approach has been taken further where the victims of the words 'bloody foreigners' were Spanish. The House of Lords considered that these words were capable of satisfying the requirements of s 28(1)(a). It held that people who are not of British origin constitute a racial group for the purposes of the definition of 'racial group'. Thus, a racial group can be defined exclusively by reference to what its members are not in terms of race, colour, nationality, or ethnic or national origins, eg non-British or non-white, as well as inclusively by reference to what they are in such terms, eg Spanish or black.

Section 28(1)(a)

Because s 28(1)(a) does not require D's conduct to be motivated by racial or religious hatred, but simply requires the demonstration of such hatred (as defined), it is possible for D to fall within s 28(1)(a) even though D is of the same colour, etc as the victim of his/her conduct. This was held by a divisional court in a case where it upheld the conviction of D whose conduct had demonstrated hostility based on the victim's membership of a racial group which was the same as D's.   **13.90**

For the test under s 28(1)(a) to be satisfied, D must have formed the view that the victim was a member of a racial group (or religious group, as the case may be) and D must have done or said something which demonstrated hostility towards the victim (V) based on that membership. Words used need not expressly identify the racial or religious group to which the victim belongs. This was held by the Court of Appeal in a case where V was Indian and brown-skinned. He was called an 'immigrant doctor' by D immediately before D assaulted him. The Court of Appeal held that it was open to the jury to conclude that D had identified V as falling within the racial groups 'Indian' and 'brown-skinned' and that the use of 'immigrant' demonstrated hostility based on the victim's membership of such groups.

Section 28(1)(a) does not require the hostility demonstrated to be based only on V's membership of a racial or religious group, or even principally on it. However, the more incidental the words or other conduct with a racial or religious content, the more difficult it will be to prove that D has demonstrated racial or religious hostility.

The word 'immediately' in s 28(1)(a) qualifies 'after' as well as 'before'; s 28(1)(a) strikes at words uttered or acts done in the immediate context of the basic substantive offence. Thus, a divisional court held that a racially aggravated offence was not made out where D demonstrated racial hostility to the victim only 20 minutes after committing the basic offence, while being questioned by the police.

Section 28(1)(b)

13.91   Section 28(1)(b) requires that the offence is motivated (wholly or partly) by hostility towards members of a racial or religious group based on their membership of that group. Section 28(1)(b) does not require D to be motivated by racial or religious hostility towards the victim of the offence but 'merely' by hostility 'towards members of a racial or religious group'. Hostility towards one member of a racial or religious group based on his/her membership of that group is sufficient to qualify so long as it forms part of the motivation for the conduct. Normally the victim will be a member of that group (and thus included within the ambit of that hostility) or at least be associated with it but this is not a requirement of s 28(1)(b). For example, if D, motivated by hostility towards members of the Jewish religious community, attacked a bricklayer whom D knew was not Jewish who was building a synagogue, D would fall foul of s 28(1)(b).

Although s 28(1)(a) requires proof of what D did at the time of committing the offence, s 28(1)(b) can be established by evidence relating to what D might have said or done on other occasions (since such evidence may be relevant to D's motivation at the time of the offence).

Racially or religiously aggravated harassment or stalking offences

13.92   A person commits an indictable (either-way) offence under CDA 1998, s 32 if (s)he commits an offence under the Protection from Harassment Act 1997 (PHA 1997) which is racially or religiously aggravated (**13.87–13.91**). CDA 1998, s 32 does not create one offence which can be committed in more than one way; it creates separate aggravated offences based on the offences under the following sections of PHA 1997: s 2 or 2A (maximum punishment two years' imprisonment) or s 4 or 4A (maximum 14 years') (**9.28–9.37**).

Racially or religiously aggravated public order offences

13.93 A person commits an offence under the Crime and Disorder Act 1998 (CDA 1998), s 31 if (s)he commits an offence under POA 1986, s 4, 4A, or 5 (**18.12–18.19**) which is racially or religiously aggravated (**13.87–13.91**). There are three separate offences under s 31, each based on one of the three basic offences under POA 1986. Those based on s 4 or 4A are indictable (either-way) offences and punishable with up to two years' imprisonment and/or a fine. That based on s 5 is a summary offence punishable with a fine not exceeding level 4.

Racially or religiously aggravated criminal damage

13.94 A person commits an indictable (either-way) offence, punishable with up to 14 years' imprisonment, under the Crime and Disorder Act 1998, s 30(1) if (s)he commits an offence under CDA 1971, s 1(1) (**14.148**) which is racially or religiously aggravated (**13.87–13.91**).

For the purposes of s 30(1), s 28(1)(a) (**13.90**) has effect as if the person to whom the property belongs or is treated as belonging for the purposes of CDA 1971 was the victim of the offence.

Racial or religious hatred or hatred on the grounds of sexual orientation

13.95 POA 1986, Part III (ss 17 to 29) provides indictable (either-way) offences relating to racial hatred. POA 1986, Part IIIA (ss 29A to 29N) provides indictable (either-way) offences relating to religious hatred and to hatred on grounds of sexual orientation. Offences under Part II or IIIA (**6.48**) are punishable with up to seven years' imprisonment.

To a large extent, the offences of religious hatred and hatred on the grounds of sexual orientation mirror those under ss 17 to 29 which forbid threatening, abusive or insulting words, behaviour, or material intended or likely to cause racial hatred, *but they are limited to words, behaviour, or material which is 'threatening' and D must intend thereby to stir up religious hatred or hatred on grounds of sexual orientation.*

Section 17 provides that '*racial hatred*' means hatred against a group of persons defined by reference to colour, race, nationality (including citizenship), or ethnic or national origins (as to these words see **13.89**). Hereafter, such a group is described for convenience as a 'racial group'.

Section 29A provides that '*religious hatred*' means hatred against a group of persons defined by reference to religious belief or lack of religious belief.

Section 29AB defines '*hatred on the grounds of sexual orientation*' as meaning hatred against a group of persons defined by reference to sexual orientation (whether towards persons of the same sex, the opposite sex, or both).

Other introductory points

13.96 The comments about 'threatening' and (POA 1986, Part III only) 'abusive' and 'insulting' at **18.13** are equally applicable to the present offences.

POA 1986, s 29J provides that nothing in Part IIIA of the Act should be read or given effect in a way which prohibits or restricts discussion, criticism, or expressions of antipathy, dislike, ridicule, insult, or abuse of particular religions or the beliefs or practices of their adherents, or of any other belief system or the beliefs or practices of its adherents, or proselytising or urging adherents of a different religion or belief system to cease practising their religion or belief system. Section 29JA provides that, for the avoidance of doubt, discussion or criticism of:

(a) sexual conduct or practices or the urging of persons to refrain from or modify such conduct or practices; or
(b) marriage which concerns the sex of the parties to the marriage,

shall not be taken of itself to be threatening or intended to stir up hatred.

Use of words or behaviour or display of written material

13.97 POA 1986, s 18(1) provides that a person who uses threatening, abusive or insulting words or behaviour, or displays any written material which is threatening, abusive or insulting, is guilty of an offence if:

(a) he intends thereby to stir up racial hatred; or
(b) having regard to all the circumstances, racial hatred is likely to be stirred up thereby.

The parallel offence created by s 29B relating to religious hatred or hatred on grounds of sexual orientation requires the use of *threatening* words or behaviour, or the display of written material which is *threatening*, and an *intention* thereby to stir up religious hatred or hatred on the grounds of sexual orientation.

'Written material' includes any sign or other visible representation. POA 1986, ss 18 and 29B do not apply to words or behaviour used, or written material displayed, solely for the purpose of being included in a television or sound broadcasting or cable service. In such a case, however, an offence might be committed under s 22 (or s 29F) when the programme is transmitted.

These offences may be committed in a public place or a private place. There is one limit in relation to private places. As with similar offences under POA 1986, ss 4 to 5 (**18.12–18.19**) an offence is not committed by the use of words or behaviour, or the display of written material, by a person inside a dwelling which is not heard or seen except by other persons in that or another dwelling.

It is a defence for D to prove that (s)he was inside a dwelling and had no reason to believe that the words or behaviour used, or the written material displayed, would be heard or seen by a person outside that or any other dwelling.

Section 18(5) provides that a person who is not shown to have intended to stir up racial hatred is not guilty of an offence under s 18 if he did not intend his/her words or behaviour or the written material to be, and was not aware that it might be, threatening, abusive or insulting. Unlike comparable provisions in other offences in POA 1986, Part III, D does not have the burden of proving this lack of intent or awareness, although evidence of such lack must be adduced before the prosecution has to disprove it. There is no parallel provision to s 18(5) under s 29B as that section is limited to cases where D intended to stir up religious hatred or hatred on grounds of sexual orientation.

Racial or religious abuse or harassment unaccompanied by the mental element just described might, nevertheless, result in liability for an offence under CDA 1998, s 31 (**13.93**) or 32 (**13.92**).

Publishing or distributing

13.98 POA 1986, s 19(1) provides that a person who publishes or distributes written material which is threatening, abusive or insulting is guilty of an offence if:

(a) he intends thereby to stir up racial hatred; or

(b) having regard to all the circumstances, racial hatred is likely to be stirred up thereby.

'Written material' includes any sign or other visible representation. Articles in electronic form are written material.

There must be a publication or distribution to the public or to a section of the public. 'The public' and 'section of the public' are not defined by the Act. There is no minimum number of persons to whom publication or distribution must be made in order for it to be to 'the public'. Ultimately, the question must be solved by common sense, the question being whether the publication or distribution has been on a scale and on a basis such as to be describable as being to 'the public'.

The Court of Appeal has held that a family living together in one house is not a 'section of the public'. The Court stated that 'section of the public' refers to some identifiable group, 'in other words members of a club or association'.

Although the prosecution must prove that D intended to stir up racial hatred by the publication or distribution or that such hatred was likely to be stirred up thereby, it does not have to prove any knowledge on D's part in relation to the content of the written matter which D has published, or distributed, although D will almost inevitably have had such mens rea if D is proved to have intended to stir up racial hatred. However, under s 19(2), it is a defence for a defendant who is not proved to have intended to stir up racial hatred to prove that (s)he was not aware of the content of the matter and neither suspected nor had reason to suspect it of being threatening, abusive, or insulting. The defence under s 19 is of obvious importance to innocent publishers or distributors, like newsagents.

Section 29C provides that a person who publishes or distributes written material which is threatening is guilty of an offence if he intends thereby to stir up religious hatred, or hatred on the grounds of sexual orientation. The publication or distribution of written material must be to the public or a section of the public.

Possession of material

POA 1986, s 23(1) provides that a person who has in his/her possession written material which is threatening, abusive or insulting, with a view to its being displayed, published, distributed, or included in a television or radio broadcast or in a cable service (whether or not by him/herself) is guilty of an offence if: **13.99**

(a) he intends racial hatred to be stirred up thereby; or

(b) having regard to all the circumstances racial hatred is likely to be stirred up thereby.

Section 23(1) also makes similar provision in relation to a recording of visual images or sounds.

For the above purposes, regard must be had to such display, publication, distribution, showing, playing, or inclusion in a programme service, as D has, or it may reasonably be inferred that D has, in view.

As with s 19, the prosecution does not have to prove any mens rea on the part of a person charged with possession contrary to s 23 in relation to the content of the material possessed by him/her. Likewise, it need not necessarily be proved that D intended racial hatred to be stirred up by the publication or distribution, since it is enough that, if the material were published or distributed, racial hatred would be likely (having regard to all the circumstances) to be stirred up as a result of the publication or distribution. However, under s 23(3), it is a defence for a defendant who is not proved to have intended to stir up racial hatred to prove that (s)he was not aware of the content of the material, and neither suspected nor had reason to suspect it of being threatening, abusive, or insulting. This is of obvious importance to 'innocent' possessors of racialist material, such as warehousemen.

The person in possession of the material must have been in possession with a view to its publication or distribution. If there is a dispute about this, the magistrates or jury will have to draw such inferences as seem reasonable from the quantity and nature of the material possessed.

Section 29G parallels s 23 in relation to religious hatred or hatred on the grounds of sexual orientation. In this case, however, the material must be 'threatening' and intended to stir up religious hatred or hatred on the grounds of sexual orientation.

A justice of the peace, if satisfied by information on oath laid by a constable that there are reasonable grounds to suspect that a person has possession of written material or a recording in contravention of s 23 or s 29G, may issue a warrant authorising the entry and search of premises where it is suspected the material or recording is situated. A constable executing such a warrant may use reasonable force if necessary.

Other offences

13.100   POA 1986, ss 20, 21, and 22 respectively provide offences relating to threatening, abusive or insulting words, behaviour, or material in public plays, visual or sound recordings, and programmes in radio or television broadcasts or cable services, which are intended or likely to stir up racial hatred. A detailed explanation of these offences is outside the scope of this book.

Sections 29D, 29E, and 29F respectively provide offences corresponding to those under ss 20, 21, and 22 in relation to threatening words, behaviour, or material in public plays, visual or sound recordings, and programmes in broadcasts or cable services. In all these cases the play, recording, programme, or broadcast must be intended to stir up religious hatred or hatred on the grounds of sexual orientation.

General

13.101   None of the above offences applies to a fair and accurate report of proceedings:

(a) in Parliament, the Scottish Parliament, or Senedd Cymru/the Welsh Parliament; or
(b) publicly heard before a court or tribunal exercising judicial authority.

However, in the case of a report of the proceedings of a court or tribunal, the exemption *only* applies if the report is published *contemporaneously* with those proceedings or, if it is not reasonably practicable or would be unlawful to publish a report of them contemporaneously (because of the law of contempt of court), is published *as soon as publication is reasonably practicable and lawful*.

PART III

PROTECTING PROPERTY

CHAPTER 14

Offences against Property

THEFT

The Theft Act 1968 (TA 1968), s 1(1) provides that a person is guilty of theft if he dis- **14.1** honestly appropriates property belonging to another with the intention of permanently depriving the other of it. Theft is an indictable (either-way) offence punishable with up to seven years' imprisonment. By way of exception, low-value shoplifting (goods not in excess of £200 in value) is triable only summarily unless the defendant (D) is 18 or over and elects trial in the Crown Court. For these purposes, where D appears at a magistrates' court to answer charges of two or more offences of low-value shoplifting, the value of the goods is aggregated and the exception only applies if the total value does not exceed £200. A person guilty of low-level shoplifting is liable on summary conviction to six months' imprisonment and/or an unlimited fine. Any reference in PACE to an indictable offence has effect as if it included low-value shoplifting.

In the case of a conviction for stealing or attempting to steal a motor vehicle, discretionary disqualification (**24.40**) is available as an additional punishment

By TA 1968, s 30, the leave of the DPP is required for the institution of proceedings against one spouse or civil partner for the theft of the other's property, unless, by virtue of any judicial decree or order, the spouses were not obliged to cohabit at the material time or an order is in force providing for the separation of the civil partners, as the case may be.

Theft is a 'penalty offence' for the purposes of the Criminal Justice and Police Act 2001 (**3.40**). This may be an appropriate course of action in minor shoplifting cases.

The terms of the definition of theft in TA 1968, s 1(1) are defined, in whole or part, by ss 2 to 6.

For the purposes of exposition, it is best to start by noting that to be guilty of theft D must be proved to have:

(a) appropriated property belonging to another; and
(b) done so dishonestly and with the intention of permanently depriving the other of it.

Appropriation

TA 1968, s 3(1) describes appropriation as any assumption by a person of the rights of **14.2** an owner, and this includes, where (s)he has come by the property (innocently or not) without stealing it, any later assumption of a right to it by keeping or dealing with it as owner. It is immaterial whether the appropriation is made with a view to gain, or is made for the thief's own benefit. Thus, a postman who flushes postal packets down the lavatory to avoid delivering them, or who takes them to give to his son, is as guilty of theft as if he had taken them for his own benefit.

The essence of this definition is an 'assumption of the rights of an owner'. An owner of property has many rights in relation to it, including the rights to use it, to destroy it, to give it away, to sell it, and so on. The House of Lords has held that, despite the use of the words 'the rights' at the beginning of s 3(1), s 3 as a whole indicates that an appropriation does not require an assumption of all the rights of an owner and that it is enough that there has been an assumption of any of the rights of the owner.

The House of Lords has ruled that an act amounting to an assumption of a right of the owner done with the authority or consent of the owner can amount to an appropriation of goods for the purposes of TA 1968.

An appropriation can occur even though the assumption is only momentary. It has been held, for example, that there was an appropriation where a man wrested a bag from a woman's grasp, even though he then dropped it on the ground and did not make off with it.

A shopper who removes goods from a shelf in a supermarket and conceals them in his/her shopping bag thereby appropriates them (because this amounts to an assumption of one of the rights of the owner of the goods), and so does someone who simply puts goods in a supermarket basket without concealing them. In both cases, however, the person concerned would not be guilty of theft if (s)he intended to pay at the checkout because (s)he would not appropriate the goods dishonestly.

Switching the price labels on articles in a shop, so that a lesser price than the true price is paid at the cash desk, amounts to an assumption of the owner's right to fix the price, and therefore to an appropriation.

Someone can appropriate property even though (s)he never possesses it, as where, pretending to be the owner, (s)he points to another's car and offers to sell it (because (s)he has assumed the owner's right to sell). On the other hand, the Court of Appeal has held, a person who has never had possession or control of property but who deceives his/her victim into transferring it to a third party does not thereby appropriate it.

There may also be an appropriation through an innocent agent. If a person in authority signs a false invoice, intending that innocent people take further steps which result in money being debited and thus appropriated from a bank account, (s)he is guilty of theft.

Appropriation by those already in possession

14.3   A person can appropriate property even though (s)he is already in possession or control of it. This is made clear by the latter part of TA 1968, s 3(1), which provides that 'appropriation' includes, where D 'has come by the property (innocently or not) without stealing it, any later assumption of a right to it by keeping or dealing with it as owner'. It follows that a shop assistant who sells goods at less than the marked price thereby appropriates them because (s)he has assumed the owner's right to fix the price. A car-hirer who sells, or—even—offers to sell, the car thereby appropriates the car because (s)he assumes the right of the owner to sell it.

The latter part of s 3(1) can lead to the conviction of a person who originally came by the property dishonestly without stealing it. Suppose that D helps him/herself to V's umbrella in order to go out during a shower but intending to return it. D does not steal the umbrella at that stage because, although D has appropriated it, (s)he did not then intend permanently to deprive V. However, if D subsequently decides to keep the umbrella or to sell it, and does so, D is then guilty of theft because his/her later assumption of a right to it by keeping or dealing with it as owner constitutes an appropriation which is accompanied by an intent permanently to deprive V.

An express exception

TA 1968, s 3(2) excludes a particular type of case, which falls within the definition **14.4**
in s 3(1), from being an appropriation. It provides that, where property or a right or
interest in property is or purports to be *transferred for value* to a person *acting in good
faith*, no later assumption by him/her of rights which (s)he believed him/herself to be
acquiring shall, by reason of any defect in the transferor's title, amount to theft of the
property. The effect of s 3(2) is that, if X steals goods from V and sells them to D who
neither knows nor suspects that they are stolen, a refusal by D to restore the goods (or
D's actual disposal of them) after D's discovery of the theft by X is not theft by D from V.

Property

By TA 1968, s 4(1), 'property' includes money and all other property, real or personal, **14.5**
including things in action and other intangible property.

'Real property' means land and things forming part of the land, such as plants and
buildings, and rights in it. Although land and things forming part of the land are 'property' for the purposes of theft, there are special provisions (below) restricting the theft
of them.

'Personal property', in its tangible sense, means movable things which can be owned,
such as cars, computers, and television sets.

A 'thing in action' is intangible property. It is a personal right of property which
can only be enforced by a legal action, and not by taking possession. Its inclusion in
the definition of 'property' means that someone who dishonestly assumes a right of
ownership over a thing in action, such as a debt, copyright, or trade mark, with the
intention of permanently depriving the person entitled to it, is guilty of theft. Thus,
if D dishonestly assigns to X a debt owed to D and his partner, V, in order to defeat
V's rights, D is guilty of the theft of a thing in action belonging to V. Where a bank
account is in credit the bank owes a debt to its customer for the amount of that credit.
Consequently, if D dishonestly draws cheques on V's account and uses the proceeds
for his/her own purposes, D can be convicted of the theft of property belonging to
V because D will have appropriated a thing in action (the debt) owned by V when D
presents the cheque.

'Other intangible property' covers such things as gas stored in pipes, which is undoubtedly capable of being stolen, and patents.

Despite the wide terms of s 4(1), there are some things which do not, or may not,
come within the definition and hence cannot be stolen. A live human body is not property because it can never be owned. The same is true in relation to a human corpse.
However, where a corpse (or part of a corpse) has undergone the application of human
skill (such as embalming or dissecting) it becomes property. Thus, for example, an ana-
tomical or pathological specimen which has been embalmed or dissected for exhibition
or teaching purposes is property for the purposes of s 4.

It has been held that confidential information, such as a trade secret, is not property
for the purposes of theft, so that the mere abstraction of the information is not theft,
and it has also been held that electricity is not property for such purposes and cannot
be stolen. There is, however, a separate offence of abstracting electricity (**14.22**).

Land and things forming part of the land

TA 1968, s 4(2) provides that a person cannot steal land, or things forming part of land **14.6**
and severed from it by him or by his directions, except in the following cases:

(a) when he is a trustee or personal representative, or is authorised by power of attorney, or as liquidator of a company, or otherwise, to sell or dispose of land belonging to another, and he appropriates the land or anything forming part of it by dealing with it in breach of the confidence reposed in him; or

(b) when he is not in possession of the land and appropriates anything forming part of the land by severing it or causing it to be severed, or after it has been severed; or

(c) when, being in possession of the land under a tenancy, he appropriates the whole or any part of any fixture or structure let to be used with the land.

TA 1968, s 4(3) goes on to provide that a person who picks mushrooms growing wild on any land, or who picks flowers, fruit or foliage from a plant growing wild on any land, does not (although not in possession of the land) steal what he picks, unless he does it for reward or for sale or other commercial purposes. For these purposes, 'mushroom' includes any fungus, and 'plant' includes any shrub or tree.

These provisions can be explained as follows:

(1) Land as a whole cannot be stolen except where the appropriator is of a defined class and acts in a defined way. The class of appropriators comprises a trustee or personal representative, or a person authorised by power of attorney, or as a liquidator of a company, or otherwise, to sell or dispose of land belonging to another. The defined mode of appropriation is dealing with the land in breach of the confidence reposed in the appropriator. The result of the rule that land as a whole cannot be stolen except by a trustee, etc is that a person cannot steal land as a whole by moving a boundary fence or by occupying it as a squatter. The appropriation of land in such a way must be dealt with by civil process, or in the case of a residential building by a prosecution for the offence of squatting in such a building (**18.86**).

(2) Things forming part of the land, such as soil, houses, bricks in a wall, and fixtures, can only be stolen in the following cases:

(a) As for land as a whole, by the defined persons in the defined way.

(b) Where a person not in possession of the land appropriates the thing by severing it or causing it to be severed. If a trespasser digs up gravel, removes bricks from a building, digs up growing things, picks flowers from a cultivated plant, cuts hay, or cuts down trees or their branches, or causes such severance to be done, (s)he may be convicted of theft (although it will generally be more appropriate to charge him/her with, and convict him/her of, criminal damage).

The present provision (s 4(2)(b)) does not apply to the picking of *wild* mushrooms or fungi nor to picking *from wild* plants and the like, notwithstanding that there has been a severance. Such conduct is dealt with by s 4(3), whereby picking wild mushrooms or other fungi, or picking flowers, fruit or foliage *from* a plant, shrub or tree growing wild, by a person not in possession of the land cannot amount to theft unless it is done for reward or for sale or other commercial purpose.

Thus, wild mushroom gathering cannot amount to theft if it is done by a person who picks the mushrooms for his/her own use, but it can if done for sale. Likewise, picking a few sprigs of holly for use at home cannot be theft, but it can if done for reward (eg payment by a florist). The term 'pick from' does not include uprooting or sawing off the top of a Christmas tree; both are clear cases of severance covered by s 4(2)(b) and unaffected by s 4(3).

As Christmas approaches, police officers become increasingly involved in the protection of growing things which are a traditional part of Christmas decorations. Vehicles carrying Christmas trees should be accompanied by delivery notes issued by the Forestry Commission or the landowner in question. If they are not, there is reason to suspect that the trees have been stolen.

It should be noted that a person who gathers or plucks any part of a 'protected' wild plant without uprooting thereby commits an offence under the Wildlife and Countryside Act 1981, s 13, referred to in Chapter 16.

(c) Generally, a person in possession of land under a tenancy cannot steal things forming part of the land. Thus, (s)he cannot be convicted of theft if (s)he digs up a plant on the land, or picks blackberries from wild plants on the land in order to sell them. The only exception relates to the whole or part of any structure or fixture let to be used with the land; such is stealable by the tenant. The obvious example of a 'structure' is a building but the term also includes a wall or bridge. A 'fixture' is an article, such as a washbasin or fireplace, which is attached to the land or to a building so as to make a permanent improvement to the land or building; by law it becomes part of the land.

The result of all this is that a tenant may be convicted of theft if (s)he demolishes the garage on the land of which (s)he is a tenant, or if (s)he removes a fireplace there in order to sell it.

For the above purposes, 'tenancy' means any tenancy regardless of whether it is a long lease or a weekly tenancy, and includes an agreement for such a tenancy. A person who remains in possession as a statutory tenant or otherwise after the end of his/her tenancy is treated as having possession under a tenancy.

Of course, once a thing has been severed from the land it ceases to be part of the land and may thereafter be the subject of theft in the same way as any other piece of personal property, which it has become. In other words, the special provisions of TA 1968, s 4(2) and (3) no longer apply to it.

Wild creatures

TA 1968, s 4(4) states that wild creatures, whether tamed or untamed, are to be regarded **14.7** as property, but that a person cannot steal a wild creature, not tamed or ordinarily kept in captivity, or the carcase of any such creature, unless either it has been reduced into possession by or on behalf of another person and possession of it has not since been lost or abandoned or another person is in course of reducing it into possession.

Section 4(4) distinguishes two groups of wild creatures:

Wild creatures which have been tamed or are ordinarily kept in captivity   Such a creature can be stolen in the same ways as any other property. Thus a person may be guilty of theft by dishonestly appropriating a tamed fox or a bear from a zoo.

Wild creatures neither tamed nor ordinarily kept in captivity   Such a creature or its carcase cannot normally be stolen but becomes 'stealable':

(a) if reduced into possession by or on behalf of another (in which case it remains 'stealable' so long as possession has not subsequently been lost or abandoned); or
(b) if another person is in course of reducing it into possession.

Thus, it is not theft to poach game on another's land, unless for instance the game is taken from a sack into which another, even another poacher, has put the product of his/her own shooting (because there has been a reduction into possession by another) or is picked up from the ground where it is lying after it has been shot by another but not yet picked up by him/her (because another is in the course of reducing it into his/her possession). For the above purposes, possession of a live wild creature is abandoned if the possessor allows it to escape from his/her possession; it is lost if a wild creature not ordinarily kept in captivity escapes of its own volition. Possession of the carcase of a wild creature is not lost by a person who mislays it. Although s 4(4) means that poachers are not normally thieves, there are other offences dealt, with in Chapter 16, which they commit.

Belonging to another

14.8   The property appropriated must belong to another when appropriated.

The basic rule

14.9   TA 1968, s 5(1) states that property is regarded as belonging to any person having possession or control of it, or having in it any proprietary right or interest (not being an equitable interest arising only from an agreement to transfer or grant an interest). Possession or control need not be lawful. Thus, the appropriation of a drug in the unlawful possession of V can amount to theft.

Normally, it is obvious whether the property appropriated belonged to another person. If a wallet is taken from V's pocket it quite clearly belongs to V, since V will almost certainly be its owner (and complete ownership is the clearest example of a proprietary right) and, anyway, it will be in V's possession. If goods are taken from a shop they clearly belong to the proprietor of the shop for the same reasons. When there are joint owners of property, one of them will steal from the other if, with the necessary mens rea, (s)he assumes one of the rights of the owner, because the property will also belong to the other co-owner under s 5(1).

Someone who leaves his/her clock to be repaired still owns it and therefore it still belongs to him/her. The repairer now has possession of the clock, and, if it is then handed to one of his/her assistants to effect the repair, that assistant has control of it. The clock can now be stolen from either the owner, the repairer, or the assistant, and it can be stolen by one of these from the other. For example, if the assistant takes the clock to the pub at lunchtime and sells it, (s)he thereby appropriates property belonging to another (to the owner and to the repairer, since the clock is technically still in the repairer's possession). Likewise, if the owner sneaks into the repairer's shop and takes away the clock without paying for the repair (s)he appropriates property belonging to another.

A person who mislays property nevertheless still retains ownership of it, and (s)he also retains possession until the property comes into the possession of another. Thus, 'lost property' is still capable of being stolen. This must be contrasted with the situation where the property has been abandoned. When a person throws away his/her old bicycle, not caring what happens to it (ie (s)he abandons it), (s)he loses ownership and possession of it. Since the property has no owner or possessor, it cannot thereafter be stolen unless and until it comes into the possession or control of another. It would be different if a tyre punctured and the owner placed the cycle behind a hedge before setting off for a repair kit. The bicycle would not be abandoned because the owner cares about what may happen to it, and it therefore still belongs to him/her.

Property subject to a trust

14.10   Where property is subject to a trust, it is regarded as belonging to the beneficiaries (who have a proprietary interest in it) as well as to the trustees, with the result that trustees who appropriate trust property can be convicted of stealing it from the beneficiaries.

Charitable trusts and certain other types of trust do not, in law, have beneficiaries, with the result that under s 5(1) the trust property belongs only to the trustees. To prevent trust property being unprotected in such a case against appropriations by the trustees, s 5(2) provides that, where property is subject to a trust, the persons to whom it belongs shall be regarded as including any person having a right to enforce the trust, and that an intention to defeat the trust shall be regarded accordingly as an intention to deprive of the property any person having that right. In the case of a charitable trust, the Attorney General, although not a beneficiary, has the right to enforce the trust, so that appropriation of a charitable trust fund by the trustees is capable of amounting to theft since the fund belongs to the Attorney General under s 5(2).

Property received under an obligation to retain and deal with it in a particular way

TA 1968, s 5(3) provides that, where a person receives property from or on account of another and is under an *obligation* (ie a legal obligation) to the other to *retain* and *deal* with that property, or *its* proceeds (ie things into which it has been converted), *in a particular way*, the property or proceeds shall be regarded (*as against him*) as belonging to the other. This is important where ownership, possession and control of the property have been transferred to the recipient. **14.11**

The essence of s 5(3) is that property (usually money) or its proceeds is regarded (as against the defendant (D)) as belonging to another from or on whose account D has received the property if D is under a *legal obligation* to that person *to retain and deal* with the property or its proceeds *in a particular way*. Section 5(3) is clearly satisfied where D receives money from V which D is legally obliged to V to use in a particular way (eg to pay it into a Christmas Club which D runs), or where D is legally obliged to V to use in a particular way the proceeds of money received from V (eg to use the money to get some goods for V and bring them to V). In the latter case, both the money and the goods (its proceeds) will belong to another (V) under s 5(3). Section 5(3) is also satisfied if D, a shop assistant, receives money from a customer for some of D's employer's goods, since D has received the money on account of another (the employer) and is under a legal obligation to deal with it in a particular way (to put it in the till); as against D, the money will belong to the employer. On the other hand, s 5(3) is not satisfied where D, a decorator, is given a down-payment on a job by a customer. D is obliged to the customer to do the job, but not to retain and deal with the money in a particular way.

Property got by another's mistake

TA 1968, s 5(4) provides that where a person gets property by another's mistake, and is under an obligation to make restoration (in whole or in part) of the property or its proceeds or of the value thereof, then to the extent of that obligation the property or proceeds shall be regarded (*as against him*) as belonging to the person entitled to restoration, and an intention not to make restoration shall be regarded accordingly as an intention to deprive that person of the property or proceeds. **14.12**

The important point about this provision is that it only applies where the recipient of property transferred under a mistake is thereby under an immediate *legal obligation* to restore it (or its proceeds or value). For practical purposes, this provision is only of importance where the recipient was ignorant of the mistake when (s)he got the property and (s)he has acquired ownership, possession and control of the property to the exclusion of anyone else. If (s)he was aware of the mistake when (s)he got the property, (s)he could be convicted of theft on the basis that his/her appropriation at that time was accompanied by mens rea. On the other hand, if the recipient only discovers the mistake later, but decides not to return the property his/her appropriation with mens rea will be of property which is then in his/her ownership, possession and control. Under s 5(4), if the recipient is under an obligation to restore the property, its proceeds or value, the property is regarded (as against him/her) as belonging to the person entitled to restoration.

The best example of a case of a legal obligation to make restoration, where a person has received ownership, possession and control of property under a mistake, is where money is paid under a mistake which leads the transferor to believe that the transferee is legally entitled to the money. Thus, if V, by a mistake as to the number of hours of overtime worked, overpays D, and D, realising the mistake, appropriates the excess amount, D has appropriated money which by s 5(4) belongs to another. If D in such a case is paid by cheque and cashes the cheque and appropriates the cash received, D will appropriate property belonging to another under s 5(4) because the cash will be the proceeds of the cheque and D would be obliged to make restoration of it to V. On the other hand, if V is induced to give D some money as a gift by a self-induced mistaken belief that D is destitute, D cannot be convicted of theft if D appropriates the money on discovering V's mistake because, the requirements of s 5(4) not having been met (since D is not legally obliged to make restoration of the money), the money will not belong to another (ie other than D) when D appropriates it.

Property of a corporation sole

14.13   TA 1968, s 5 contains one other provision, s 5(5), which can be disposed of briefly. Section 5(5) provides that the property of a 'corporation sole', such as a bishop, a chief constable or a police and crime commissioner in his/her official capacity, shall be regarded as belonging to the corporation notwithstanding a vacancy in the corporation. Thus, the property of a police and crime commissioner 'belongs to another', and is therefore capable of being stolen, even though the commissioner has just died and not yet been replaced by a successor.

Dishonesty

14.14   The appropriation of property belonging to another must be committed dishonestly.

Theft Act 1968, s 2(1)

14.15   Section 2(1) expressly and as a matter of law excludes appropriations carried out with certain states of mind from being dishonest. Section 2(1) provides that a person's appropriation of property belonging to another is *not* to be regarded as dishonest:

(a) if he appropriates the property in the belief that he has in law the right to deprive the other of it, on behalf of himself or a third person; or

(b) if he appropriates the property in the belief that he would have had the other's consent if the other knew of the appropriation and the circumstances of it; or

(c) (except where the property came to him as a trustee or personal representative) if he appropriates the property in the belief that the person to whom the property belongs cannot be discovered by taking reasonable steps.

These provisions are concerned with the defendant's belief. *It is legally irrelevant that a belief in this context is unreasonable,* although, of course, justices or a jury are less inclined to accept an alleged belief as truly held if it is an unreasonable one.

Belief in legal right to deprive By s 2(1)(a), the element of dishonesty is excluded if a person appropriating property belonging to another believes that (s)he has a right in law to deprive the other of it, whether on behalf of him/herself or a third person. If D, who is owed money by V, takes some of V's property to the equivalent value, mistakenly believing that (s)he has a legal right to do so to recoup the debt, D is not guilty of theft.

Where a person acts under a belief in a *moral* right to deprive, the question of his/her dishonesty depends on the test at **14.16**.

Belief that the 'owner' would have consented if (s) he had known Section 2(1)(b) (belief that the person to whom the property belongs would have consented if (s)he had known of the appropriation and its circumstances) clearly covers the case, for example, of someone who takes a bottle of lager from a flatmate's room, leaving the price behind him/her and believing that the flatmate would have consented had (s)he known all the circumstances.

Belief that the 'owner' cannot be discovered by taking reasonable steps Section 2(1) (c) is primarily concerned with those who find property belonging to another. It exempts D from dishonesty if (s)he appropriates another's property under a belief that the person to whom the property belongs cannot be discovered by taking reasonable steps. The question is not whether the 'owner' could not be found by taking reasonable steps, but whether D believed this.

If D finds a £5 note in the street and appropriates it, it will be almost impossible to disprove a claim by D that D believed the owner could not be found by taking reasonable steps. It will be different if the note is contained in a purse bearing the owner's name and address (or containing other material identifying the owner); in that case it will be much easier to disprove a claimed belief that the owner could not be found by taking reasonable steps.

The exemption is not restricted to things which are found. Suppose that D's friend, on emigrating, left property under D's care until (s)he should return to this country. If, after many years, during which D has not heard from his/her friend, D sells the property, honestly believing that the friend will not return and cannot be traced by taking reasonable steps, D will not be guilty of theft.

Section 2(1)(c) expressly does not apply to a person who received the property as a trustee of property or a personal representative. This is sensible in view of the special obligations of such a person.

Dishonesty in cases outside s 2(1)

14.16 The negative definition of dishonesty in TA 1968, s 2(1) is only a partial definition. Consequently, a defendant's appropriation may not have been made dishonestly even though the case falls outside s 2(1). Whether or not a defendant who appropriated property with some alleged state of mind other than one of the three referred to in s 2(1) did so dishonestly is a question of fact for the jury. This means that, unlike the situation in which a belief of the type referred to in s 2(1) is pleaded (where the judge in the Crown Court must tell the jury that in law an appropriation with such a belief is not dishonest), if dishonesty is in issue for some other reason it is not for the judge to tell the jury whether or not an appropriation with the alleged state of mind is dishonest but for the jury to decide this according to the following test.

As the Court of Appeal, applying statements by the Supreme Court, has confirmed, the jury must first ascertain the actual state of D's knowledge or belief as to the facts. The reasonableness of that belief is a matter of evidence going to whether D held the belief at the material time, but it is not an additional requirement that the belief had to be reasonable. When D's state of mind is established, the jury must then determine whether, given that knowledge or belief, D's conduct was honest or dishonest by applying the objective standards of ordinary decent people.

Where a theft charge is tried in a magistrates' court this test is, of course, applied by the justices.

Theft Act 1968, s 2(2)

14.17   For the sake of completeness, it should be mentioned that TA 1968, s 2(2) says what has already been implied, by providing that an appropriation may be dishonest notwithstanding that D intends to pay for what (s)he took. It has been held that an appropriation can be dishonest even though the original owner of goods or money is not the poorer because of D's conduct.

Intention permanently to deprive

14.18   The dishonest appropriation of property belonging to another must be accompanied by an intention permanently to deprive the other of that property.

An intent permanently to deprive will usually be proved by evidence of what D did with the property appropriated. If D takes V's £20 note and spends it on drink this clearly indicates an intent permanently to deprive V of the note as D has passed the note into circulation. It is no use D alleging that (s)he intended to pay V back. Although this may prevent D being found to have been dishonest, and lead to an acquittal on that ground, D will nevertheless have intended permanently to deprive V of the thing (the actual £5 note) which D has appropriated. If a car is appropriated and a false registration book is produced, and the engine and chassis numbers are altered, this is clear evidence of an intent permanently to deprive.

In certain limited cases a person can be convicted of theft even though (s)he did not mean permanently to deprive, and even though (s)he positively intended to return the property at some future date (or did actually return it). A conviction in such a case is possible if the case falls within TA 1968, s 6, which extends the meaning of 'intention of permanently depriving'.

Section 6(1) provides that a person appropriating property belonging to another without meaning the other permanently to lose the thing itself is nevertheless to be regarded as having the intention of permanently depriving the other of it if his *intention is to treat the thing as his own to dispose of regardless of the other's rights;* and a borrowing or lending of it may amount to so treating it if, but only if, the borrowing or lending is for a period and in circumstances making it equivalent to an outright taking or disposal.

Treating as one's own to dispose of regardless of other's rights

14.19   The first part of s 6(1) is the key part. It operates to 'deem' a person to have intended permanent deprivation if (s)he intended to treat the thing as his/her own to dispose of regardless of the other's rights.

One type of case which may be caught by the first part of s 6(1) is that where D takes V's property, intending to return it to V only if a condition is fulfilled by V. In such a case, this does not automatically mean that D intends to treat the property as his/her own to dispose of regardless of V's rights, but depending on the facts D may do so. The position is that, if the condition attached by D to the return of the property is one which would not be fulfilled or would not be fulfilled in the foreseeable future, the circumstances may well amount to an intention to treat the property as D's own to dispose of regardless of V's rights, and hence to an intention permanently to deprive. On the other hand, if the condition can readily be fulfilled and may be fulfilled in the near future, this may well not amount to an intention to treat the property as D's own to dispose of regardless of V's rights, in which case there would not be an intention permanently to deprive.

Another type of case which may fall within s 6(1) is where D abandons the property and is indifferent as to whether it is recovered by the person to whom it belongs. If, by the circumstances of the abandonment and/or the nature of the property, it is (to D's knowledge) extremely unlikely that the property will be recovered, D can be said to intend to dispose of it regardless of the rights of the other. If V's car is taken and driven a distance of 200 miles by D then abandoned, the justices or jury are most unlikely to find that there was an intention to treat it as D's own to dispose of regardless of V's rights, as a car is easily identifiable and will certainly be returned to V. It would be different if what was taken and so abandoned was a watch.

Borrowing or lending

The second part of s 6(1) goes on to provide that a borrowing or lending may amount to treating property as one's own to dispose of regardless of the other's rights, if the borrowing or lending is for a period and in circumstances which make it equivalent to an outright taking or disposal. The Court of Appeal has ruled that this provision is only satisfied by a borrower if his/her intention is to return the thing only when 'all its goodness or virtue has gone'. **14.20**

Suppose that D takes V's monthly season ticket, intending to return it at the end of the month. D's borrowing is clearly for a period and in circumstances making it equivalent to an outright taking since, when it is returned, the season ticket will be a virtually worthless piece of paper. Because the borrowing is equivalent to an outright taking or disposal, D's intention so to act is regarded by s 6(1) as an intent to treat as his/her own to dispose of regardless of V's rights and, hence, as an intent permanently to deprive.

On the other hand, a cinema projectionist who borrows a film in order to make pirate copies does not intend to treat the film as his/her own to dispose of regardless of the owner's rights because on its return the film would not have lost all of its goodness or virtue.

An example of a case where a lending would satisfy the present provision would be where D, an assistant in a florists', lends some cut flowers to X for a week for a flower display, knowing that they will have died by the time of their return. D intends to treat the flowers as his/hers own to dispose of because, as D knows, the lending is for a period and in circumstances making it equivalent to an outright disposal.

Parting with property subject to a condition

Section 6(2) provides a further explanation of 'treating as one's own to dispose of regardless of the other's rights'. It provides that, where a person having possession or control (lawfully or not) of property belonging to another, parts with the property under a condition as to its return which he may not be able to perform, this (if done for purposes of his own and without the other's authority) amounts to treating the property as his own to dispose of regardless of the other's rights. Thus, where D, who is in possession of V's property, pawns it, intending to redeem it and return it if (s)he wins a bet, this amounts to treating as his own to dispose of regardless of the rights of the other (V) and his/her intention to do so is deemed by s 6(1) to be an intention permanently to deprive V of it. **14.21**

ABSTRACTING ELECTRICITY

14.22   As already indicated, it has been held that electricity is not property and therefore cannot be stolen. However, a special indictable (either-way) offence punishable with up to five years' imprisonment is provided by TA 1968, s 13, which states that a person who dishonestly uses without due authority, or dishonestly causes to be wasted or diverted, any electricity is guilty of an offence.

The offence will usually be encountered when a person, who has had his/her electricity supply disconnected, reconnects it, and thus uses electricity without authority; or when evidence is found that a consumer has by-passed his/her meter, since his/her authority to use electricity supplied by the electricity company is conditional upon that electricity having passed through the meter before use.

Section 13 covers wasting or diverting in addition to using, as where D diverts electricity for use by another. It seems that the person who by-passes his/her meter could be charged with either dishonest use or dishonest diversion, as (s)he does not have authority to use in that way and (s)he has certainly caused electricity to be diverted.

Section 13 is not restricted to mains electricity.

It is important not to forget that the element of dishonesty must be proved. The provisions of TA 1968, s 2 are limited to theft, and therefore do not apply to an offence under s 13, but what is said at **14.16** about dishonesty for the purposes of theft applies equally to an offence under s 13.

ROBBERY

14.23   Under TA 1968, s 8(1), a person is guilty of robbery (an offence triable only on indictment and punishable with a maximum of life imprisonment) if (s)he steals, and immediately before or at the time of doing so, and in order to do so, (s)he uses force on any person, or puts or seeks to put any person in fear of being then and there subjected to force. A person guilty of assault with intent to rob is similarly triable and punishable.

No theft: no robbery

14.24   Robbery is an aggravated form of theft, the theft being aggravated by the use of force or threat of force. If there is no theft, there is no robbery. Where a person uses or threatens force in order to steal but has not achieved the appropriation of any property, and is therefore not guilty of robbery, (s)he can be convicted of the offence of assault with intent to rob. There can be no conviction for robbery if it is found that the person who used or threatened force in order to appropriate the property believed that (s)he had a legal right to it, even though (s)he did not believe (s)he was entitled to use force to obtain it, because the essential ingredient for theft, 'dishonesty', is missing. Of course, there can be a conviction for some offence of assault (other than assault with intent to rob).

Use or threat of force immediately before or at time of theft

14.25   To constitute robbery, the force must be used or threatened 'immediately before or at the time of' the theft. There can be no robbery if force is only used or threatened after 'the time' of the theft. A person who pushes a woman to the ground in order to steal her handbag, and does so, commits robbery; so does someone who approaches her and steals her handbag by threatening to punch her in the face unless she hands it over. The first uses force at the time of the theft and the other threatens force immediately before the theft. However, a man who steals a woman's handbag without force but, being discovered by her in possession of the bag several minutes afterwards, assaults her, cannot be convicted of robbery. This is because the time of the theft will have elapsed when the force is used. The assailant could, of course, be convicted of theft and assault. The 'time' of the theft is not limited to the split second during which the appropriation with mens rea initially occurs, since an act of appropriation may be a continuing one. The 'time' of the theft lasts as long as the theft can be said still to be in progress in commonsense terms, ie so long as the defendant is 'on the job'. On this basis, there can be a robbery, for example, where a shoplifter takes goods from a supermarket shelf and, when approached by a store security guard, uses force on the guard in order to remove the goods.

Use of force on a person or threat then and there to subject to force

Where force is used, it must be used 'on' a person, but it has been held that the force **14.26** need not be used directly against the person. Consequently, it is enough to use force to obtain possession of property in the physical possession of another. For example, the use of force to wrest a handbag from a woman's grasp is robbery. Minimal force, such as that used to snatch a cigarette from another's fingers, will not suffice.

In the case of a threat of force, a threat of future force is insufficient; the threat must be 'then and there' to subject another to force. A threat of force to property will not suffice, nor will the actual use of force against it.

The references above to threats of force are a shorthand way of referring to the requirement that D 'puts or seeks to put any person in fear of being then and there subjected to force'. These words must not be forgotten. If D threatens violence to a month-old baby in a pram unless the mother parts with her money, this is not robbery because D cannot be said to have put or sought to put the mother in fear of immediate force to herself.

Although the force must be used or threatened on a person, it is not essential that the theft is carried out in his/her presence, provided that it is used or threatened for the purpose of stealing. It is robbery, for example, where a security guard is tied up so that property may be removed from some other part of the warehouse in which the guard is on duty.

The theft need not be from the person against whom the force is used or threatened. Thus, if D threatens a married couple that D will stab the wife unless the husband hands over his wallet, D can be convicted of robbery.

Use or threat of force in order to steal

The force used or threatened must be used in order to steal. If a man during an ar- **14.27** gument with another, threatens to give him a beating and the man threatened gives money to him not to do so, this cannot be robbery as the threat of force was not made in order to steal. Likewise, a man who knocks a woman to the ground to rape her, but who then changes his mind and instead takes the handbag which she has dropped, is not guilty of robbery or of assault with intent to rob.

REMOVAL OF ARTICLES FROM PLACES OPEN TO THE PUBLIC

14.28   Because theft requires an intention permanently to deprive another of his/her property, a charge of theft would fail against an art lover who took a painting from a public gallery, intending to enjoy its presence in his/her home for a year and thereafter to return it. For this reason, TA 1968, s 11 creates an indictable (either-way) offence, punishable with up to five years' imprisonment, to cover cases such as this. Section 11(1) provides that, where the public have access to a building in order to view the building or part of it, or a collection or part of a collection housed in it, any person who without lawful authority removes from the building or its grounds the whole or part of any article displayed or kept for display to the public in the building, or that part of it, or in its grounds, is guilty of an offence.

The removal of an article must be either from a building to which the public have access in order to view the building or part of it, or a collection or part of a collection housed in it, or from the grounds of such a building. The offence is therefore concerned with removals of articles from stately homes, galleries, historic buildings, etc which are open to the public in the above sense, or from their grounds. A person who removes the portrait of an Edwardian mayor from the entrance of the town hall does not commit the present offence because, although the public have access to the town hall, it is only for the purpose of paying council tax, making inquiries or seeing their councillors; they do not have access in order to view the building, or any collection in it, or any part of the building or collection. If someone removes a painting from a collection in a stately home, access to which is limited to members of the Women's Institute, (s)he does not commit the present offence because the public do not have access to the building in question.

The offence is not committed if the collection has been made or exhibited for the purposes of effecting sales or other commercial dealings. Thus, removals of paintings from commercial art galleries are not caught by s 11. Subject to this, it does not matter that the collection in question is one got together for a temporary purpose. The annual art exhibition in the village hall is therefore caught by s 11.

Generally, an offence under s 11 can be committed whether or not the building is open to the public at the time of the removal. There is one exception: if the thing removed is there otherwise than as forming part of, or being on loan for exhibition with, a collection intended for permanent exhibition to the public, it must be removed on a day when the public has access to the building or grounds. Thus, if a painting is removed from the collection at the National Gallery it is irrelevant that the Gallery is then closed over the Christmas period, whereas it is not an offence to remove a painting from the annual art exhibition in the village hall on a day when the hall is closed.

The offence under s 11 only applies where the thing removed is the whole or part of any article displayed or kept for display to the public in the building, or part of it, to which the public have access or in its grounds. If a visitor takes an old vase which forms a part of the display, (s)he commits this offence. If (s)he takes the attendant's coat on the way out (s)he does not, as the coat is not a part of the display.

There must be a removal from the building or its grounds. A visitor who moves the vase from one room to another does not commit the offence, but one who takes it out of the building does.

Section 11(1) makes it clear that no offence is committed where the person removing an article covered by it has lawful authority for doing so.

There is no need to prove dishonesty in respect of the act of removal but s 11(3) exempts someone who believes that (s)he has lawful authority, or that (s)he would have it if the person entitled to give it knew of the removal and its circumstances. Therefore, a furniture-remover who was removing articles on behalf of their owner would not commit an offence if (s)he took one of the exhibited articles, believing that it was one of the items to remove. Likewise, a picture-restorer who carried out work for the owner from time to time would not commit an offence if in the owner's absence (s)he took a painting for its five-yearly restoration, believing that, had the owner been present, (s)he would have consented.

TAKING VEHICLES OR OTHER CONVEYANCES

By TA 1968, s 12(1), a person is guilty of a summary offence, punishable with up to six months' imprisonment and/or an unlimited fine if, without having the consent of the owner or other lawful authority, (s)he takes any conveyance for his/her own or another's use, or, knowing that any conveyance has been taken without such authority, drives it or allows him/herself to be carried in or on it. If committed with reference to a motor vehicle, the offence is also punishable (see **24.40**) with discretionary disqualification for driving. **14.29**

There are two offences under s 12(1): taking a conveyance without authority, and driving or allowing oneself to be carried in or on a conveyance which one knows has been so taken. Neither offence requires proof of an intent permanently to deprive. Section 12 is aimed at 'joy-riders'.

'Conveyance' is defined to include any conveyance constructed or adapted for the carriage of a person or persons whether by land, water or air, but it does not include a conveyance constructed or adapted for use only under the control of a person not carried in or on it, and 'drive' is construed accordingly. 'Conveyance' therefore covers almost all motor vehicles, as well as aeroplanes, hovercraft, boats and ships, but it does not cover trailers or pedestrian-controlled vehicles, such as prams and trolleys. Nor does it cover pedal cycles. Pedal cycles are covered by a separate offence (**14.34**).

Taking without authority

Taking

The mere unauthorised assumption of possession or control is not enough to constitute a 'taking' of a conveyance; *there must be some movement of it, however small.* Consequently, a person who gets into a car and drives it a few feet takes a conveyance, as does a person who climbs into a boat on a boating lake and rows it to the other bank. On the other hand, people who unlawfully occupy a conveyance, either to shelter or to make love in it, do not take it. Of course, someone who gets into another's car and starts it is attempting to take a conveyance, but as the taking would amount to a summary offence there can be no charge of attempting to take it. In such a case there may be a conviction for the offence of interfering with vehicles (**14.36**). **14.30**

Unauthorised use of a conveyance by a person already in lawful possession or control of a conveyance may amount to a 'taking'. A lorry driver who uses his/her employer's lorry for his/her own purposes 'out of hours', or who appropriates it for his/her own use during the working day in a manner which is inconsistent with the rights of the employer and shows that (s)he has assumed control for his/her own purposes, thereby 'takes' it. A similar principle applies if D, a person who has borrowed a conveyance, uses it for a purpose other than that for which D has been given permission or after the time D is permitted to have the conveyance. By so using the conveyance, D takes it.

For the defendant's or another's use

14.31   Section 12(1) also requires that the taking be for D's own or another's use, and this means that either the conveyance must be used as a conveyance or it must be taken for later use as a conveyance. It follows that a person who cuts the mooring rope of a boat and allows it to drift away empty does not commit this offence, whereas (s)he would if (s)he was aboard or if (s)he towed it away for later use as a boat.

Without consent or other authority

14.32   The taking must be without the consent of the owner or other lawful authority. An apparent consent to a taking obtained by intimidation is not a true consent, so that the taking will be without consent. On the other hand, a consent which has been obtained by fraud is nevertheless valid and prevents the offence being committed, however fundamental (eg as to the identity of the deceiver) the mistake which is induced.

In relation to a conveyance subject to a hiring or hire-purchase agreement, 'owner' means the person in possession of it under that agreement. It follows that, during the currency of the agreement, such a person cannot commit the present offence in relation to that conveyance since (s)he can hardly be said to take it without the consent of the owner.

The addition of the words 'or other lawful authority' excuses lawful takings, such as the removal by a constable of a vehicle which is causing an obstruction. Finance companies on occasion reclaim vehicles which are on hire-purchase from them when the terms of the agreement have been broken by the hirer, and they will usually have lawful authority to do so under the terms of the agreement.

A mistaken belief in the existence of lawful authority or a mistaken belief that the owner would, if asked, have consented is also a defence.

Driving or allowing self to be carried

14.33   In relation to the taking of motor vehicles in particular, but not exclusively, it is frequently the case that, after a conveyance has been taken by one person, it is used to convey a number of persons, each of whom may take a turn at driving. The second offence in s 12(1) deals with this situation, by providing that an offence is committed by anyone who, *knowing* that a conveyance has been taken without the consent of the owner or other lawful authority, drives it or allows him/herself to be carried in or on it. In circumstances where a vehicle, known to have been 'taken' contrary to s 12(1), is seen to be moving (for it is essential that there is movement in order that a person can be said to be 'carried'), and all the occupants are seen to step out of the vehicle but all deny driving it, the issue of who was driving is unimportant because all have allowed themselves to be 'carried' in the vehicle (since that expression covers a person who was driving).

Where D takes a vehicle without consent and later picks up a friend and takes him/her for a drive, the friend commits no offence unless the friend knows that the vehicle has been so taken. Should the driver disclose that the vehicle has been unlawfully taken in the course of the journey, the friend is guilty if (s)he continues to allow him/herself to be carried.

Pedal cycles

Pedal cycles are separately dealt with by s 12(5). Section 12(5) provides that a person **14.34** commits a summary offence, punishable with a fine not exceeding level 3, if, without having the consent of the owner or other lawful authority, he takes a pedal cycle for his own or another's use, or rides a pedal cycle knowing it to have been taken without such authority. What was said earlier about the various elements of the offence under s 12(1) is equally applicable to the present offences.

Aggravated vehicle-taking

TA 1968, s 12A provides offences of aggravated vehicle-taking. An offence is committed **14.35** under s 12A(1) where a person has committed *either of the offences under TA 1968, s 12(1)* (the basic offence) in any way in relation to a *mechanically propelled vehicle* (**23.5**) and it is proved that, at any time after the vehicle was unlawfully taken (whether by him/herself or another) and before it was recovered, the vehicle was driven, or injury or damage was caused, in one or more of the following circumstances:

(a) the vehicle was driven in a dangerous manner on a road or other public place; the test of such driving in the Road Traffic Act 1988, s 2A(1) (**26.5**) applies here;

(b) owing to the driving of the vehicle, an accident occurred by which injury was caused to any person;

(c) owing to the driving of the vehicle, an accident occurred by which damage was caused to any property other than the vehicle; or

(d) damage was caused to the vehicle.

The prosecution does not have to prove that the dangerous driving was by D, or that the injury, damage or accident was caused by D's driving or by D at all. The Supreme Court has held that it is implied in (b) or (c), that it must be proved that there was some element of fault in the driving of the vehicle, whether or not amounting to careless or inconsiderate driving, which contributes in some more than minimal way to the injury or damage. 'Accident' in (b) and (c) includes a situation where a person has deliberately caused injury or damage. Once it is proved that the driving, injury or damage was caused during the period between the taking of the vehicle contrary to s 12(1) and its recovery, D is fixed with liability for an offence contrary to s 12A, unless D has one of the defences referred to in the next paragraph. A vehicle is 'recovered' when it is returned to its owner or other lawful possession or custody.

It is a defence for D to prove that the relevant driving, accident or damage occurred before D committed the basic offence, or that D was neither in, nor on, nor in the immediate vicinity of the vehicle when such driving, accident or damage occurred.

An offence under s 12A is an indictable (either-way) offence. By way of qualification, if the allegation is only that the vehicle was damaged and the value of the alleged damage is under £5,000 (or if the aggregate value is under £5,000 where D appears on charges of a series of two or more damage-only s 12A offences of the same or similar character), a magistrates' court must proceed *as if* the offence was triable only summarily.

The maximum imprisonment in a situation within (b) (above) where death is caused is 14 years. In all other situations within (a)–(d) it is two years. Since the maximum penalty is greater where death is caused in situation (b), s 12A(1) creates two offences: one relating to that situation, and the other for the other situations. Where a case is dealt with under the qualification in the previous paragraph the maximum sentence is three months' imprisonment and/or a fine not exceeding level 4. An offence under s 12A is (see **24.39**) additionally punishable with obligatory disqualification for driving.

It has been held that there is nothing to prevent both aggravated vehicle-taking (involving an allegation that the vehicle was driven dangerously) and dangerous driving being charged.

INTERFERENCE WITH VEHICLES

14.36   The Criminal Attempts Act 1981 (CAA 1981), s 9(1) provides that D is guilty of the summary offence of vehicle interference if he interferes with a motor vehicle (**23.6**) or trailer or with anything carried in or on a motor vehicle or trailer with the intention that an offence of:

(a) theft of the motor vehicle or trailer or part of it;
(b) theft of anything carried in or on the motor vehicle or trailer; or
(c) taking a conveyance, contrary to TA 1968, s 12(1),

shall be committed by himself or some other person. If it can be proved that D intended that one of these offences should be committed, it is immaterial that it cannot be shown which offence it was.

If a group of rowdy youths rock a vehicle in order to set off its intruder alarm, they do not commit an offence under s 9. They certainly interfere with the vehicle but they lack the intention to commit any of the specified offences. On the other hand, if they start to unscrew the aerial or a wing mirror with the intention of stealing it, they commit an offence under s 9. The removal of tarpaulins from goods vehicles carrying loads indicates an interference with the intention to steal part of the load.

If someone is seen trying to open the door of someone else's motor vehicle this is certainly an interference with that vehicle. However, it is unlikely to satisfy a court that (s)he intends to steal or take the vehicle, or to steal anything contained in it, unless there are further circumstances which point to such an intention. Such an intention becomes more apparent as attention is paid by the interferer to more than one vehicle, or (s)he is seen to examine the interior of the vehicle through its windows to check its contents, or even to try a number of doors on the same vehicle. The further the interference extends, the more likely it becomes that there is an appropriate criminal intent.

An offence under s 9 is punishable with up to three months' imprisonment and/or a fine not exceeding level 4.

GOING EQUIPPED TO STEAL

14.37   Under TA 1968, s 25(1), a person commits an indictable (either -way) offence if, when not at his/her place of abode, (s)he has with him/her any article for use in the course of or in connection with any burglary or theft.

The offence can be committed anywhere besides D's place of abode. It has been decided that, where a person lives in a motor vehicle, it is his/her place of abode whilst on the site where (s)he intends to abide, but that as soon as the vehicle leaves the site (s)he is no longer at his/her place of abode.

Although the offence is described in the marginal note to s 25 as 'going equipped for stealing, etc', and is commonly so described, it is not limited to conduct which could be described as 'going equipped'. The term 'has with him/her' has the same meaning as that referred to at **19.33**.

D must intend the article to be used in the future in the course of or in connection with any burglary or theft; it is not necessary to prove that the articles are to be used in connection with any particular burglary or theft. 'Possession' to enable someone else to use the article is sufficient for the purposes of s 25. For the purposes of the present offence, an offence under s 12(1) of taking a conveyance is treated as theft.

If D is found with an article made or adapted for use in committing burglary or theft, proof of the offence under s 25 is assisted by s 25(3), which states that where there is proof that D had with him such an article that is evidence that D had it with him for such use. Of course, this can be rebutted by evidence of a contrary intention. The result of s 25(3) is that, if D is found trespassing in the grounds of a dwelling house with a bunch of skeleton keys in his/her pocket, this is evidence that D had those articles with him/her for use in committing burglary.

An offence under s 25(1) is punishable with up to three years' imprisonment. Discretionary disqualification for driving (**24.40**) can also be imposed if the offence is committed with reference to the theft or taking of motor vehicles.

BLACKMAIL

Blackmail is triable only on indictment and is punishable with up to 14 years' imprisonment. It is dealt with by TA 1968, s 21(1) which provides that a person is guilty of blackmail if, with a view to gain for himself or another or with intent to cause loss to another, he makes any unwarranted demand with menaces; and for this purpose a demand with menaces is unwarranted unless the person making it does so in the belief that: **14.38**

(a) he has reasonable grounds for making the demand; and
(b) the use of the menaces is a proper means of reinforcing the demand.

It is irrelevant that D's demand did not result in anything being obtained.

Demand with menaces

The nature of the act or omission demanded is immaterial. An oral demand is made when the words are said; if it is made by letter, it is made when that letter is posted. A demand need not be made in terms of an express demand or requirement, since (taken together with the menaces) it may be implied by a suggestion or other conduct which is by no means aggressive or forceful. **14.39**

'Menaces' are not limited to threats of violence, since they include 'threats of action detrimental to or unpleasant to the person addressed'. It is immaterial whether the menaces do or do not relate to action to be taken by the person making the demand. The man who says, 'Pay me £5,000 or my daughter will tell the world that you seduced her', is as guilty of blackmail as the man who reinforces his demand with threats of action by himself.

Trivial threats will not suffice. The threat must be of 'such a nature and extent that the mind of an ordinary person of normal stability and courage might be influenced or made apprehensive so as to accede unwillingly to the demand'. By way of exception, if, although they would not have affected the mind of an ordinary person of normal stability, the threats affected the addressee's mind, they will amount to menaces if D was aware of the likely effect of his/her actions on the addressee, eg because D knew of an unusual susceptibility on the addressee's part.

If, on the facts known to D, his/her threats might have affected the mind of a person of ordinary stability, they will amount to menaces, even though they did not affect the addressee.

It is irrelevant whether D intends to carry out his/her menaces or is in any position to effect them.

With a view to gain or intent to cause loss

14.40   D must make his/her demand with menaces with a view to gain for him/herself or another, or with an intent to cause loss to another. The terms 'gain' and 'loss' are defined by s 34 as extending only to gain or loss in money or other property, whether temporary or permanent. Section 34 also provides that:

(a) 'gain' includes a gain by keeping what one has, as well as a gain by getting what one has not; and

(b) 'loss' includes not getting what one might get, as well as a loss by parting with what one has.

Unwarranted

14.41   TA 1968, s 21 provides that a demand with menaces *is unwarranted unless* the person making it does so in the *belief* that (s)he has reasonable grounds for making the demand *and* that the use of the menaces is a proper means of reinforcing the demand. If D has sex with a girl after promising payment of £500 and then refuses to pay, a demand by her for the money supported by a threat to inform D's wife will not be blackmail by the girl if she *believes* that she has reasonable grounds for making such a demand *and* that her threat is a proper means of reinforcing the demand.

D does not have to prove these beliefs, but this does not mean that the prosecution must negative the existence of a belief for which there is no evidence, since the prosecution need only negative the existence of one of the specified beliefs if there is evidence before the court in support of both types of belief; otherwise the jury is obliged to find that the demand with menaces was unwarranted.

BURGLARY

14.42   TA 1968, s 9(1) provides that a person is guilty of burglary if:

(a) he enters any building or part of a building as a trespasser and with intent to commit any such offence as is mentioned in s 9(2); or

(b) having entered any building or part of a building as a trespasser he steals or attempts to steal anything in the building or that part of it or inflicts or attempts to inflict upon any person therein any grievous bodily harm.

It has been held that TA 1968, s 9(1) creates two types of offence, the first being set out by s 9(1)(a) and the second by s 9(1)(b). Because in each type of offence there is a higher maximum term of imprisonment if the offence was committed in a dwelling in a dwelling (14 years) than otherwise (10 years), each of the two types of offence has in it two offences, one committed if it is alleged and proved that the building or part of it was a dwelling and the second if it is not alleged and proved that the building or part of it was a dwelling.

Burglary offences are triable only on indictment (a) where an offence triable only on indictment is committed or intended; or (b) where the burglary is in a dwelling and a person there is subjected to, or threatened with, violence; otherwise, they are triable either-way.

The definitions of the offences under s 9(1)(a) and (b) include many common terms which require explanation. These are given as s 9(1)(a) is described but what is said will be equally applicable to s 9(1)(b).

Burglary contrary to Theft Act 1969, s 9(1)(a): entry with intent

14.43 A person is guilty of burglary under s 9(1)(a) if (s)he enters any building or part of a building as a trespasser and with intent to commit one of the offences listed in s 9(2), namely to steal anything in the building or part of a building in question, or to inflict on any person therein grievous bodily harm, or to do unlawful damage to the building or anything therein.

Enters

14.44 The Court of Appeal has adopted the test of an 'effective entry'. This test excludes minimal intrusions, as where D's fingers are inserted through a gap between a window and its frame in order to open the window. However, where D was found stuck in a downstairs window of a house with his head and right arm inside but trapped by the window itself, which rested on his neck, the Court of Appeal held that there had been an entry. The issue of whether D was able, from that position, to steal anything was irrelevant.

Under pre-TA 1968 law, an entry could be effected merely by the insertion of an instrument without the intrusion of any part of the body *provided it was inserted to commit a relevant further offence*, but not if it was inserted merely to facilitate access by a person's body. Assuming that this remains the law, someone who, whilst remaining outside a building, inserts his walking stick in order to hook and steal a handbag inside enters the building for the purposes of burglary. On the other hand, someone who inserts a rod through a window which has been left ajar in order to release its catch does not enter the building because the insertion of the rod is merely to facilitate access by his body, not to commit an offence inside.

As a trespasser

14.45 A person enters a building or part of a building as a trespasser if it is in the possession of another and (s)he enters without a right by law or permission to do so; the entry need not involve any force at all.

Rights of entry are granted by statute to certain people, such as the police and public health inspectors, for an authorised purpose. Such a person is not a trespasser if (s)he enters for such a purpose. However, (s)he enters as a trespasser if (s)he enters premises for an unauthorised purpose, eg to steal something inside.

Likewise, a permission to enter will be given for a particular purpose or purposes. A person who enters for a purpose other than one for which (s)he has permission enters as a trespasser. For example, the Court of Appeal has held that a man, who had permission to enter his father's house, entered it as a trespasser when he entered to steal his father's television set because he entered in excess of his permission.

Permission to enter for a particular purpose or purposes may be given by the occupier or by someone with the express or implied authority of the occupier. For example, a member of the family has authority to invite someone into her parents' house for a cup of tea but does not have authority to invite that person in in order to steal her parents' property.

Permission to enter may be implied, instead of express. For instance, in the case of a shop there is an implied permission for members of the public to enter the public parts of the shop for the purposes of inspecting goods on display or making purchases, and a person who enters for such a purpose is not a trespasser.

A permission to enter might not necessarily extend to every part of the building. Thus, a person might lawfully enter a building such as a hotel or shop, but trespass in the manager's office or the stockroom; equally, (s)he might be a lawful guest at a meal in a private house, but enter a bedroom as a trespasser. In both of these cases the entry as a trespasser will be into a 'part of a building'.

The 'owner' of a building may trespass in part of it if that part is in the exclusive possession of someone else. For example, if a householder rents out the rooms in his attic to X, on terms whereby X obtains exclusive possession of them, the householder will enter them as a trespasser if (s)he enters without X's permission.

Building or part of a building

14.46 'Building' should be given its everyday meaning of a structure with walls and a roof and of a permanent or semi-permanent nature. Dwelling houses, warehouses, shops, and the like are all buildings; so are outhouses and greenhouses (provided they are at least semi-permanent) and substantial portable structures with most of the attributes normally found in buildings, provided that there is an element of permanence in their site.

TA 1968, s 9(4) states that references to a building also apply to an inhabited vehicle or vessel, and this is so whether or not the person having a habitation in it is there at the time. Clearly, a caravan or houseboat which is V's permanent home is an 'inhabited vehicle or vessel', even though V is not there at the time but, say, abroad on holiday; so is a caravan or boat which is used as a holiday home in the summer during those weeks or weekends in which it is being so used, but not at other times in the summer and not at all in the rest of the year when it is closed up.

'Part of a building' refers to a particular area of a building. An obvious example of 'part of a building' is a separate room but the term is not limited to this. It also includes a physically marked-out area in a room, such as the area behind a counter in a shop, from which D is plainly excluded, whether expressly or impliedly.

Mens rea

14.47 The mens rea required for burglary contrary to TA 1868, s 9(1)(a) is that, when entering the building or part of a building, D must satisfy two requirements.

First, D must know that (s)he is entering the building, or part, as a trespasser (ie D must know the facts which make him/her a trespasser), or at least be reckless as to whether (s)he is so entering.

Second, D must enter with intent to commit one of the offences listed in TA 1968, s 9(2):

(a) to 'steal' (ie to commit theft in respect of) anything in the building or, as the case may be, the part trespassed in;
(b) to inflict grievous (ie serious) bodily harm on any person in the building or, as the case may be, the part trespassed in; or
(c) to do unlawful (ie criminal) damage to the building or anything therein (whether or not D has trespassed in the part in which the damage is intended to occur).

It is, of course, irrelevant that it might be impossible for D to carry out his intent.

In most instances, unless there is an admission by D, the only way in which one of the requisite intentions will be apparent is if there has been some act almost amounting to an attempt to commit the intended offence.

Burglary contrary to Theft Act 1968, s 9(1)(b): having entered as a trespasser, stealing or inflicting grievous bodily harm, or attempting to do so

TA 1968, s 9(1)(b) provides that a person is guilty of burglary if having entered a building or part of a building as a trespasser he steals or attempts to steal anything in the building or that part of it or inflicts or attempts to inflict on any person therein any grievous bodily harm. **14.48**

Theft Act 1968, s 9(1)(a) distinguished

An important distinction between an offence of burglary under TA 1968, s 9(1)(a) and that under TA 1968, s 9(1)(b) is that the latter requires D, having entered a building or part of a building as a trespasser, actually to have committed or attempted to commit in the building or part trespassed in: **14.49**

(a) the offence of theft (contrary to TA 1968, s 1); or
(b) an offence involving the infliction of grievous bodily harm contrary to the Offences Against the Person Act 1861 (OAPA 1861), s 18, 20, or 23.

Another important distinction is that, unlike the offence under s 9(1)(a), the offence under s 9(1)(b) does not require D to have intended to commit one of the above offences when D entered as a trespasser. Consequently if D enters a building as a trespasser, but without one of the intents specified in s 9(2) (so that D is not guilty under s 9(1)(a)), D can only be guilty of burglary (under s 9(1)(b)) if D then steals or inflicts grievous bodily harm in the building or attempts to do either.

Mens rea

Apart, of course, from having the mens rea required for theft, or an offence involving the infliction of grievous bodily harm, or an attempt to commit such, as the case may be, D must know or be reckless that (s)he has entered as a trespasser when (s)he commits one of these offences. It is irrelevant whether or not D realised at the time of entry that (s)he was entering as a trespasser. Thus, if a man enters a building, thinking that he has permission, and later realises that he has not and then steals something inside or inflicts grievous bodily harm on someone inside (eg the occupier who is trying to eject him) he is guilty of burglary of the present type. **14.50**

Examples of Theft Act 1968, s 9(1)(b) offence

A person who enters an empty house as a trespasser to sleep in it for the night does not commit burglary contrary to s 9(1)(b), nor does (s)he if (s)he switches on the electric fire (because electricity is not property and cannot be stolen). But if (s)he turns on a gas fire, (s)he commits burglary contrary to s 9(1)(b) because gas is property and can be stolen. Likewise, (s)he would be guilty of burglary contrary to s 9(1)(b) if, being discovered there by V, (s)he deliberately strikes V with a piece of wood lying nearby and inflicts grievous bodily harm, whether (s)he intended to or not, since (s)he will have committed the offence of unlawfully and maliciously inflicting grievous bodily harm, contrary to OAPA 1861, s 20 (**9.13**). **14.51**

Of course, there is some overlap between the two types of offence of burglary. Where a person enters a building (or part) as a trespasser with intent to steal or to inflict grievous bodily harm therein, and (s)he commits the intended offence or attempts to do so, (s)he can be charged with either offence. It is normally preferable to charge burglary contrary to s 9(1)(b) in such a case since it is easier to prove.

AGGRAVATED BURGLARY

14.52   TA 1968, s 10(1) provides that a person is guilty of aggravated burglary if he commits any burglary and at the time has with him any firearm or imitation firearm, any weapon of offence, or any explosive. Offences of aggravated burglary are triable only on indictment and are punishable with a maximum of life imprisonment.

Firearm, imitation firearm, weapon of offence, or explosive

14.53   TA 1968, s 10(1) provides as follows.

'Firearm' includes an air gun or an air pistol. 'Firearm' is not otherwise defined. It is clear that the definition of a firearm given in the Firearms Act 1968 was not included in TA 1968 because it includes component parts. Aggravated burglary is aimed at burglars who are 'tooled up' and therefore pose a particular threat of injury. There is very little threat of injury posed by the possession of a magazine case for a rifle.

'Imitation firearm' means anything which has the appearance of being a 'firearm' whether or not capable of being discharged.

'Weapon of offence' means any article made or adapted for use for causing injury to or incapacitating a person, or intended by the person having it with him/her for such use. Thus, 'weapon of offence' covers not only articles which would be offensive weapons for the purposes of the Prevention of Crime Act 1953, such as coshes, knuckledusters, or even pickaxe handles, but also articles made, adapted, or intended to be used for incapacitating a person. Articles made for incapacitation include handcuffs, leg irons, or a straitjacket; those adapted would include a scarf or pair of tights adapted for use for tying-up; and those intended for such use would include drugs to induce sleep or handkerchiefs or pads for use with chloroform.

If the article in question is made or adapted for causing injury to, or incapacitating, a person, it will be necessary to prove no more than that a burglary was committed by D and that D was in possession of such a weapon of offence at the time. If it is alleged that D was in possession of an article which D intended to use to cause injury or to incapacitate, it will be necessary to prove D's intention so to use it. In the absence of an admission, this is only likely to occur where D threatens someone in the building in such a way that D indicates an intention to use that weapon. For the purposes of aggravated burglary, an article not made or adapted to injure or incapacitate can be a weapon of offence, even if the necessary intent is only formed an instant before it is used to injure or incapacitate (which makes an interesting contrast to the different rule which applies to other offences (see, eg, **19.48**). Thus, a trespasser who, hearing someone coming downstairs, takes a screwdriver from her bag, having decided to use it to injure, commits aggravated burglary if she causes serious injury.

'Explosive' means any article manufactured for the purpose of producing a practical effect by explosion, or intended by the person having it with him/her for that purpose. This definition makes clear that a thing which simply causes a 'pyrotechnic effect' (and

is therefore an 'explosive substance' under the Explosive Substances Act 1883) is not an 'explosive substance' for present purposes.

It is not essential that D should intend to use one of the above articles in the course of the burglary; it suffices that D had it with him/her for such use on another occasion (as where D intends to use a cosh to enable him/her to hijack a get-away car).

Has with him/her at time of burglary

In order to 'have with him/her' a firearm, etc, D must be armed with it; it is not enough that it is readily accessible to D. Thus, if the firearm, etc is carried not by the burglar but by an accomplice waiting outside the building, aggravated burglary is not committed. This is a stricter approach than is taken to the phrase 'has with him/her' in other offences where the term appears. In addition, D must know that (s)he has the thing with him/her: a burglar does not have with him a cosh which somebody has slipped into his swag bag, unknown to him, for example.

D must have had with him/her a relevant article *at the time of committing* the offence of burglary.

Where the burglary is alleged to have been committed by stealing or inflicting grievous bodily harm, or attempting to do so, after entry as a trespasser, the possession must be proved at the time at which the theft, etc was committed or attempted, because that is the time at which the offence of burglary will have been committed. Where the burglary is alleged to have been committed by an entry with intent to commit one of the offences specified for s 9(1)(a), D must be proved to have been in possession at the time of entry, because that is the time at which the offence of burglary will have been committed. For example, if D enters a workshop without a weapon, merely intending to trespass there (and is therefore not guilty of an offence of burglary under s 9(1)(a) at that stage), but when confronted by a security guard D picks up a hammer stored in the workshop and uses it to strike the guard, causing him grievous bodily harm, D will be guilty of aggravated burglary. This is because D had the hammer with him at the time of committing burglary under s 9(1)(b) and the hammer was clearly intended by him for use for causing injury (and was therefore a weapon of offence).

FRAUD

The Fraud Act 2006 (FA 2006), s 1(1) and (2) provides that a person is guilty of the offence of fraud if he is in breach of s 2, 3, or 4 (which provide different ways of committing the offence). Section 2 is concerned with fraud by false representation; s 3 with fraud by failing to disclose information; and s 4 with fraud by abuse of position.

The offence of fraud is an indictable (either-way) offence and is punishable with .up to 10 years' imprisonment.

Fraud by false representation

Under FA 2006, s 2, a person is in breach of s 1 if (s)he:

(a) dishonestly makes a false representation; and
(b) intends by making the representation to make a gain for him/herself or another, or to cause loss to another, or to expose another to a risk of loss.

This is the widest of the three ways of committing fraud.

14.54

14.55

14.56

Making false representation

14.57   Representation   FA 2006, s 2(3) provides that 'representation' means any representation as to fact or law, including a representation as to the state of mind of:

(a) the person making the representation, or
(b) any other person.

By s 2(4), a representation may be express or implied.

An obvious example of a representation of fact would be a statement that a worthless ring was a diamond ring. A statement as to the meaning of a statute is clearly a representation of law.

Because 'representation' includes a representation as to the state of mind of the representor or another, the making of a false statement of the present intentions of the person making the statement or another person (eg a false promise) can amount to a false representation. The intention may, of course, be implied from the nature of the transaction in which the defendant engages. A request for a loan of money implies an intention to repay, and the ordering of a meal in a restaurant implies a representation that the representor (or, possibly, another) intends to pay for it.

The requirement that the representation must be as to fact (ie a past or present verifiable thing) or law, or as to a state of mind, means that a statement which merely expresses an opinion does not in itself constitute a false representation if the opinion turns out to be unjustified. By way of example, if a statement that a picture is worth £1,000 turns out to be unjustified there is no false representation. However, if the person making the statement is aware that the opinion is unjustified, (s)he will at the same time by his/her conduct make a false representation as to his/her state of mind, because (s)he will impliedly represent that (s)he believes the opinion is justified when in fact (s)he does not so believe.

There is no limit on how an express representation may be made. Thus, it can be stated by words, whether oral or written. It can be communicated by conduct (as where a rogue dresses up in a security guard's uniform in order to convey the impression that (s)he is a security guard). The Court of Appeal has held that making a false representation is not limited to personal representations because it can be made by the unambiguous adoption of a false representation by a third party.

An implied representation may be made by words or by conduct, such as the implied representation by conduct, referred to above, made by someone who orders a meal in a restaurant that (s)he intends to pay for it, or such as that made by someone who sells property that (s)he has a right to sell.

A common example of an implied false representation by conduct concerns 'bouncing cheques'. The giver of a cheque normally impliedly represents that the state of facts existing at the date of delivery of the cheque is such that the cheque will be honoured in the ordinary course of events on presentation for payment on or after the date specified on the cheque. If the facts are not as represented, a false representation is made.

A person who uses a credit card or debit card impliedly represents by his/her conduct that (s)he has actual authority from the card company to use the card. If (s)he has no such authority the representation as to it is false.

By s 2(5), a representation may be regarded for the purposes of s 2 as made if it (or anything implying it) is submitted in any form to any system or device designed to receive, convey or respond to communications (with or without human intervention). Thus, s 2 can be breached where the false representation is submitted to a computer or other machine rather than being addressed to a human being, as where a person unauthorisedly enters someone else's number into a 'CHIP and PIN' machine or unauthorisedly 'swipes' another's card.

False representation By s 2(2), a representation is false if:

(a) it is untrue or misleading; and
(b) the person making it knows that it is, or might be, untrue or misleading.

A false representation may involve a half-truth. A representation is a half-truth where, although it is literally true, it omits a material matter, as where an applicant for a job correctly states that he does not have a criminal record but fails to state that he has been charged with theft to which he intends to plead guilty.

No one is required to have been deceived by a false representation.

The Court of Appeal has held that, because of the principle of innocent agency (**1.30**), if D, with the state of mind required by (b), procures another person innocently to make an objectively untrue or misleading representation, the terms of s 2(2) will be satisfied in respect of D.

Mens rea

In addition to knowing that his/her representation is, or might be, untrue or misleading: **14.58**

(a) D must make the false representation dishonestly. Where 'dishonesty' is in issue in a prosecution under s 2 or s 3 or 4 (below), it is for the jury to decide whether D's conduct was dishonest according to the following test.

The jury must first ascertain the actual state of D's knowledge or belief (reasonable or not) as to the facts. Having established D's state of mind, the jury must then determine whether, given that knowledge or belief, D's conduct was honest or dishonest by applying the objective standards of ordinary decent people.

Where a charge is tried in a magistrates' court this test is, of course, applied by the justices.

(b) D must intend, by making the representation, to make a gain for him/herself or another, or to cause loss to another or to expose another to a risk of loss. 'Gain' and 'loss' have the same meaning as in blackmail (**14.40**).

Fraud by failing to disclose information

FA 2006, s 3 provides that a person is in breach of s 1 if he: **14.59**

(a) dishonestly fails to disclose to another person information which he is under a legal duty to disclose; and
(b) intends, by failing to disclose the information, to make a gain for himself or another, or to cause loss to another or to expose another to a risk of loss.

A 'legal duty to disclose' means one in civil law. It may be one prescribed by statute; from the fact that the transaction is one of good faith (such as a contract of insurance); from the express or implied terms of a contract; from the custom of a trade or market; or from the existence of a fiduciary or similar relationship (such as that of a trustee and beneficiary, solicitor and client, or agent and principal) whereby one party (eg the trustee, solicitor or agent) is obliged to act in a disinterested way in the interests of the other. By way of example, a person in a position of trust has a duty to disclose material information when entering into a contract with a beneficiary.

Fraud by abuse of position

14.60   By FA 2006, s 4, a person is in breach of s 1 if (s)he:

(a) occupies a position in which (s)he is expected to safeguard, or not to act against, the financial interests of another person;

(b) dishonestly abuses that position; and

(c) intends, by means of the abuse of that position, to make a gain for him/herself or another, or to cause loss to another or to expose another to the risk of loss.

A person may be regarded as having abused his/her position even though his/her conduct consisted of an omission rather than an act.

The Court of Appeal has held that, to establish an abuse of position under FA 2006, s 4, the prosecution must demonstrate a breach of fiduciary duty or an obligation akin to a fiduciary duty. Accordingly, it held, s 4 did not apply in the general commercial arena where individuals or businesses competed in the market of one kind or another, including labour markets where participants were entitled to or expected to look after their own interests.

ARTICLES FOR USE IN FRAUD

Possession of articles for use in fraud

14.61   FA 2006, s 6(1) provides that a person commits an offence if he has in his possession or under his control any article for use in the course of or in connection with any fraud. An offence under s 6(1) is an indictable (either-way) offence punishable with up to five years' imprisonment.

The Court of Appeal has stated that D must have 'actual control or possession' of the article, and that an 'ability to control' is insufficient. A person would be in possession or control of data held on a computer in his/her possession or control. Two (or more) people can be in joint possession or control of an article if they each have possession or control of it.

D must intend the article to be used in the future in the course of or in connection with a fraud. A general intention to commit fraud is sufficient.

The Court of Appeal has held that, although the possession of an article which has already been used in the course of or in connection with any fraud is not within the scope of the offence under s 6(1), s 6(1) extends to the case where an article was created after an alleged fraud in order to mask or disguise that fraud (eg in order to provide a defence against prosecution).

The term 'articles' in s 6(1) and s 7(1) below includes any program or data held in electronic form. Computer programs can generate credit card numbers; computer templates can be used for producing blank bills; and computer files may contain credit card details of persons.

Making or supplying articles for use in fraud

14.62   A person commits an indictable (either-way) offence against FA 2006, s 7(1) if (s)he makes, adapts, supplies, or offers to supply any article:

(a) knowing that it is designed or adapted for use in the course of or in connection with fraud; or

(b) intending it to be used to commit, or assist in the commission of, fraud.

An offence under s 7(1) is an indictable (either-way) offence punishable with up to 10 years' imprisonment.

Where payment is made based upon energy consumed by a customer and consumption is measured by a meter, a person who makes or supplies a device intended to interfere with the record made by the device will intend it to be used to commit fraud.

FORGERY AND RELATED OFFENCES

General points

Meaning of 'instrument'

By the Forgery and Counterfeiting Act 1988 (FCA 1981), s 8, 'instrument' means: **14.63**

(a) any document, whether of a formal or informal character (other than a currency note);
(b) any stamp issued or sold by a postal operator (or a metered postage mark);
(c) any Inland Revenue stamp denoting any duty or fee; and
(d) any disc, tape, soundtrack or other device on or in which information is recorded or stored by mechanical, electronic, or other means. To be 'recorded' or 'stored' the information must be preserved for an appreciable time with the object of subsequent retrieval. Examples of items covered are microfilm records and information on computers or discs and tachograph record sheets, but not electronic impulses in a computer or its 'user segment' (which stores information momentarily while the computer searches its memory, eg to check a password). Thus, computer hacking is not forgery although it is an offence under the Computer Misuse Act 1990, s 1(**15.3**).

The Court of Appeal has held that a thing is only a document within (a) if it conveys two messages: a message about the thing itself (eg that it is a cheque) and a message to be found in the words or other symbols that is to be accepted and acted on (eg the message in a cheque to the banker to pay a specified sum). Thus, cheques, wills, and building society passbooks are examples of documents (and are therefore instruments) but paintings (even if falsely signed), false autographs, and any writing on manufactured articles or their wrappings indicating the name of the manufacturer or country of origin, are not.

Meaning of 'false'

FCA 1981, s 9 states that an instrument is false if it purports to have been: **14.64**

(a) *made in the form* in which it is made by *a person who did not in fact make it* in that form; or
(b) *made in the form* in which it is made on *the authority of a person who did not in fact authorise its making* in that form; or
(c) *made in the terms* in which it is made *by a person who did not in fact make it* in those terms; or
(d) *made in the terms* in which it is made *on the authority of a person who did not in fact authorise its making* in those terms; or
(e) *altered* in any respect *by a person who did not in fact alter it* in that respect; or
(f) *altered* in any respect *on the authority of a person who did not in fact authorise the alteration* in that respect; or

(g) *made* or *altered* on a *date* on which, or at a *place* at which, or *otherwise in circumstances* in which, it was *not in fact made or altered*; or

(h) *made* or *altered* by an *existing person* but he did not *in fact exist*.

The key point is that it is not enough that the document tells a lie (ie contains a false statement); what is required is that it should tell a lie about itself in relation to the person who made the instrument, or authorised its making, or altered it, or if it purports to have been made or altered on a date, or in a place, or otherwise in circumstances in which it was not in fact made or altered, or if it represents itself as having been made or altered by an existing person who in fact did not exist.

If D writes an application for a job and falsely alleges that (s)he has a degree, that instrument is not a forgery although its contents are untrue. It was made by the applicant, it has not been altered in any way or falsely dated, and it is made by an existing person. Should D make out a reference which is allegedly from his/her tutor that would be a false instrument, as it alleges that it was made by a person who did not make it or authorise its making.

If D finds a cheque book and signs the name of the person who holds the account, the cheque is a false instrument as it lies about itself by alleging that it was signed by another who did not make the instrument or authorise its making. On the other hand, someone who opens an account in a false name, in order to pay in a stolen cheque, does not make a false instrument when (s)he completes a withdrawal form in that name. The drawer's name is the same as the depositor's (although false) and the withdrawal form does not purport to be made by a person who did not make it.

A cheque, made out quite properly by the holder of the bank account for the sum of £100, becomes a false instrument when another person alters that cheque to show a sum of £1,000 by adding another zero to the figure shown on the face of the cheque and writing alongside it the initials of the account holder without that person's authority.

Trial and punishment

14.65   By FCA 1981, s 6, the offences dealt with below (**14.66-14.71**) are indictable (either-way) offences. They are punishable with a maximum of 10 years' imprisonment, with the exception of offences under s 5(2) or (4) (for which the maximum is two years').

Forgery

14.66   A person is guilty of forgery, contrary to FCA 1981, s 1(1), if (s)he makes a false instrument, with the intention that (s)he or another shall use it to induce somebody to accept it as genuine, and by reason of so accepting it to do or not to do some act to his/her own or any other person's prejudice.

D must make the false instrument with the intention that (s)he or another person shall use it to induce somebody to accept it as genuine, and with the intention to induce that person by reason of so accepting it to do or not to do some act to his/her own or any other person's prejudice. Consequently, it is not enough simply to intend to induce someone to believe that an instrument is genuine. For example, making a false birth certificate solely to induce a belief that one comes from a noble family is not forgery. If D acts with the necessary double intent, it is irrelevant whether the instrument is communicated to anyone, whether anyone is induced to accept the instrument as genuine or whether prejudice (within the meaning set out below) is caused. D need not intend to induce another human being, it suffices that D intends to induce a machine to respond to the instrument as if it were genuine.

The act or omission intended to be induced must be to the prejudice of the person induced or of someone else (besides D). FCA 1981, s 10(1) provides that, for the purposes of forgery and related offences, an act or omission intended to be induced is only to a person's prejudice if it is one which, if it occurs:

(a) will result:
   (i) in his temporary or permanent loss of property (including a loss by not getting what he might get as well as a loss by parting with what he has), as where a false cheque or will is made to cause him either to part with property or not to get property he might have got; or
   (ii) in his being deprived of an opportunity to earn remuneration or greater remuneration or to gain a financial advantage otherwise than by way of remuneration, as where a letter, falsely purporting to come from someone asked to give a reference for a job applicant or a person tendering for a contract, states that the applicant or tenderer is incompetent; or
(b) will result in somebody being given an opportunity:
   (i) to earn remuneration or greater remuneration from him (the person induced), as where a false degree certificate is made in order to obtain a job or better pay; or
   (ii) to gain a financial advantage from him otherwise than by way of remuneration, as where a false airline ticket is made in order to get a free flight; or
(c) will be the result of his having accepted a false instrument as genuine (or—and this is only relevant to offences under FCA 1981, ss 2 and 4 below—a copy of a false instrument as a copy of a genuine one) in connection with his performance of a duty, as where a false pass is made to induce a doorkeeper to admit an unauthorised person to premises. As this example shows, the intended prejudice under (c) need not be financial.

Where the intended inducement is of a machine (eg a cash dispenser), the act or omission intended to be induced by the machine responding to the instrument is treated as an act or omission to a person's prejudice.

By virtue of s 10(2), it is not forgery where the maker of a false instrument intends to induce someone to do something which (s)he is under an enforceable legal duty to do (or to induce someone not to do something (s)he is not legally entitled to do).

Provided that the requirements of s 10(1) are satisfied, an honest and reasonable belief in a legal or moral claim to the gain which D intended to make as a result of falsifying the instrument is no defence (unless s 10(2) applies). Indeed, it is not a defence in itself that D might actually have been entitled to have property transferred to him/her if D had made a true claim.

Copying a false instrument

FCA 1981, s 2 provides that it is an offence for a person to make a copy of an instrument **14.67** which is, and which he knows or believes to be, a false instrument, with the intention that he or another will use it to induce somebody to accept it as a copy of a genuine instrument, and with the intention to induce that person by reason of so accepting it to do some act to the prejudice of himself or some other person.

The essential points to prove are that the original instrument is a false instrument; that the person who made the copy knew or believed this to be so; and that the copy was made with intent to induce a belief that it is a copy of a genuine instrument and to induce someone to act, etc.

Using a false instrument or copy of it

14.68   FCA 1981, s 3 makes it an offence for a person (D) to use an instrument which is false, and which D knows or believes to be false, with the intention of inducing somebody to accept it as genuine, and with the intention of inducing that person by reason of so accepting it to do or not to do some act to his/her own or any other person's prejudice.

Section 4 provides a similarly worded offence committed where D uses a copy of a false instrument which is, and which D knows or believes to be, a false instrument, with the intention of inducing somebody to accept it as a copy of a genuine instrument, and by reason of so accepting it to do or not to do some act to his/her own or any other person's prejudice.

Any use of a false instrument (or, as the case may be, a copy of a false instrument) with the necessary intent suffices. The verb 'use' is wide in meaning and covers, for example, a person who offers, delivers, tenders in payment or exchange, or exposes for sale, a false instrument (or a copy of one).

Offences relating to false money orders, share certificates, etc

14.69   FCA 1981, s 5 provides a number of offences under this heading, which are concerned with the following instruments:

(a) money orders or postal orders;
(b) UK postage stamps;
(c) Inland Revenue stamps;
(d) share certificates;
(e) cheques or other bills of exchange;
(f) travellers' cheques;
(g) bankers' drafts;
(h) promissory notes;
(i) credit cards or debit cards;
(j) certified copies relating to an entry in a register of births, adoptions, marriages, civil partnerships, conversions (of a civil partnership into a marriage) or deaths and issued by the Registrar General, the Registrar General for Northern Ireland, a registration officer or a person lawfully authorised to issue such certified copies; and
(k) certificates relating to entries in such registers.

Any such instrument is hereafter referred to as a 'specified instrument'.

Custody or control of such instrument

14.70   By FCA 1981, s 5(1), it is an offence for a person to have in his/her custody or under his/her control a specified instrument which is, and which (s)he knows or believes to be, false, with the intention that (s)he or another shall use it to induce somebody to accept it as genuine, and by reason of so accepting it to do or not to do some act to his/her own or any other person's prejudice.

This 'possession' offence completes the cycle of offences which are likely to be committed if a false instrument is made for the purposes previously described. If D makes out a false cheque with intent to induce someone to accept it as genuine, and by reason of so accepting it to do something to his/her prejudice, D commits forgery contrary to s 1. If D walks through the streets to a bank, with the false cheque, in order to cash it, (s)he commits the present offence under s 5(1). If D then passes the cheque to a bank official in order to induce her to part with money, D commits the offence of 'using' contrary to s 3. Of course, it might be that different people will commit different offences in the cycle, as where the person who makes the false instrument gets other people to engage in the use of such false instruments.

If the intent required for an offence under s 5(1) cannot be proved, one can fall back on s 5(2). This makes it an offence for a person merely to have in his/her custody or control, without lawful authority or excuse, a specified instrument which is, and which (s)he knows or believes to be, false. 'Lawful authority or excuse' is likely to be limited to such matters as possession by a police officer after seizure of a specified instrument or possession by some other person who is in the course of handing over to the police such an instrument.

Making, custody, or control of equipment or materials

FCA 1981, s 5(3) and (4) is aimed at the tools of a forger's trade. Section 5(3) provides   **14.71**
that it is an offence for a person to make or to have in his custody or under his control a machine or implement, or paper or any other material, which to his knowledge is or has been specially designed or adapted for the making of a specified instrument, with the intention that he or another shall make a specified instrument which is false and that he or another shall induce somebody to accept it as genuine, and by reason of so accepting it to do or not to do some act to his own or another's prejudice. In this way the Act strikes at a would-be forger even before (s)he starts to make a false instrument. Whether or not the necessary intent for an offence under s 5(3) can be proved will often depend, in part, on the amount of forging equipment, etc in D's custody or control.

If the intent required under s 5(3) cannot be proved, one can fall back on s 5(4), which provides that it is an offence for a person to make or to have in his custody or control any such machine, implement, paper, or material, without lawful authority or excuse.

Forgery, etc: defence for refugees

It is a defence under the Immigration and Asylum Act 1999, s 31, for a refugee charged   **14.72**
with an offence under FCA 1981, Part I (**14.66-14.71**) or the Identity Documents Act 2010, s 4 or 6 (below), or an attempt to commit such an offence, to prove that, having come to the UK directly from a country where his/her life or freedom was threatened (within the meaning of the Refugee Convention), (s)he:

(a) presented him/herself to the authorities in the UK without delay;
(b) showed good cause for his/her illegal entry or presence; and
(c) made a claim for asylum as soon as was reasonably practicable after his/her arrival in the UK.

If, in coming from the country where his/her life or freedom was threatened, the refugee stopped in another country outside the UK, the defence applies only if (s)he proves that (s)he could not reasonably have expected to be given protection under the Refugee Convention in that other country.

Forgery, etc: police powers

14.73   See **14.92**.

FALSE IDENTITY DOCUMENT OFFENCES

General points

Identity document

14.74   'Identity document' means any document that is, or purports to be:

(a) an immigration document;

(b) a UK passport;

(c) a passport issued by or on behalf of the authorities of a country or territory outside the UK or by or on behalf of an international organisation;

(d) a document that can be used (in some or all circumstances) instead of a passport;

(e) a UK driving licence; or

(f) a driving licence issued by or on behalf of the authorities of a country or territory outside the UK (Identity Documents Act 2010 (IDA 2010), s 7).

In (a) above, 'immigration document' means:

(i) a document used for confirming—

    (a) the right of a person at a time before 31 December 2020 at 11.00 pm under the EU Treaties in respect of entry or residence in the UK; or

    (b) the right of a person under the Immigration (European Economic Area) Regulations 2016 (as they continue to have effect) in respect of entry or residence in the UK;

(ii) a document which is given in exercise of immigration functions and records information about leave granted to a person to enter or to remain in the UK; or

(iii) a registration card (within the meaning of the Immigration Act 1971, s 26A).

False identity document

14.75   An identity document is false only if it is false within the meaning given by FCA 1981 (**14.64**) (IDA 2010, s 9).

Improperly obtained identity document

14.76   An identity document is improperly obtained if false information was provided, in or in connection with, the application for its issue or an application for its modification, to the person who issued it or (as the case may be) to a person entitled to modify it. For these purposes:

(a) 'false' information includes information containing any inaccuracy or omission that results in a tendency to mislead;

(b) 'information' includes documents (including stamps and labels) and records; and

(c) the 'issue' of a document includes its renewal, replacement or re-issue (with or without modifications) (IDA 2010, s 9).

Personal information

14.77   'Personal information', in relation to an individual (I), means:

(a) I's full name;
(b) other names by which I is or has previously been known;
(c) I's gender;
(d) I's date and place of birth;
(e) I's external characteristics that are capable of being used for identifying I;
(f) the address of I's principal place of residence in the UK;
(g) the address of every other place in the UK or elsewhere where I has a place of residence;
(h) where in the UK and elsewhere I has previously been resident;
(i) the times at which I was resident at different places in the UK or elsewhere;
(j) I's current residential status;
(k) residential statuses previously held by I; and
(l) information about numbers allocated to I for identification purposes and about the documents (including stamps or labels) to which they relate.

'Residential status' means:

(i) an individual's nationality;
(ii) an individual's entitlement to remain in the UK; and
(iii) if that entitlement derives from a grant of leave to enter or remain in the UK, the terms and conditions of that leave (IDA 2010, s 8).

Possessing false identity documents, etc with improper intent

14.78 It is an offence, triable only on indictment and punishable with up to 10 years' imprisonment, against IDA 2010, s 4(1) for a person (D) with the improper intention to have in his/her possession or under his/her control:

(a) a false identity document D knows or believes to be false;
(b) an improperly obtained identity document which D knows or believes to have been improperly obtained; or
(c) an identity document relating to someone else.

The improper intention is:

(i) the intention of using the document for establishing personal information about D; or
(ii) the intention of allowing or inducing another to use it for establishing, ascertaining or verifying personal information about D or about any other person (with the exception, in the case of an identity document that relates to someone else, of the individual to whom it relates).

Possession of apparatus with prohibited intention

14.79 Under IDA 2010, s 5(1), it is an offence triable only on indictment and punishable with up to 10 years' imprisonment for D with the prohibited intention to make or to have in his/her possession or under his/her control:

(a) any apparatus which, to D's knowledge, is or has been specially designed or adapted for use in making false identity documents; or
(b) any article or material which, to D's knowledge, is or has been specially designed or adapted for use in making false identity documents.

The prohibited intention is the intention:

(i)   that D or another will make a false identity document; and
(ii) that the document will be used by somebody for establishing, ascertaining or verifying personal information about a person.

The references to the making of a false identity document include references to the modification of an identity document so that it becomes false (s 9).

Possession without reasonable excuse

14.80   Under IDA 2010, s 6(1), it is an indictable (either-way) offence, punishable with up to two years' imprisonment, for D to have in his/her possession or under his/her control, without reasonable excuse:

(a) a false identity document;
(b) an improperly obtained identity document;
(c) an identity document relating to someone else; or
(d) any apparatus, or article or material which, to D's knowledge, is or has been specially designed or adapted to make false identity documents, or to be used to make such documents.

The Court of Appeal has held that the mere fact that D did not know or believe that the document was false cannot of itself and without more amount to a reasonable excuse, but (it held) lack of such knowledge or belief can be relevant to a defence of reasonable excuse by way of explanation of D's possession—it might explain why the document had not been thrown away or handed to the police. Possession of a document for an innocent purpose cannot in itself amount to 'reasonable excuse'. Subject to these limits, 'reasonable excuse' is a matter for the jury or justices.

Defence for refugees

14.81   See **14.72**.

COUNTERFEITING AND RELATED OFFENCES

General points

Counterfeit

14.82   By FCA 1981, s 28, a thing is a counterfeit of a currency note or of a protected coin:

(a) if it is not a currency note or a protected coin, but *resembles a currency note or protected coin* (whether on one side only or on both) *to such an extent that it is reasonably capable of passing for a currency note or protected coin of that description*; or
(b) if it is a currency note or protected coin which has been *so altered that it is reasonably capable of passing for a currency note or protected coin of some other description*.

For the avoidance of any doubt on the matter, s 28 goes on to provide that a thing consisting of one side only of a currency note, with or without the addition of other material, is a counterfeit of such a note, and that a thing consisting of parts of two or more currency notes (or of parts of such note(s) and other material) is capable of being a counterfeit of a currency note. Thus, for example, a supposed currency note composed of parts of true currency notes and other materials is a counterfeit, as is a thing which consists of one side only of a currency note as a result of that note being 'split'.

For the purposes of FCA 1981, 'currency note' means:

(a) any note which has been lawfully issued in the UK, the Channel Islands, the Isle of Man or the Irish Republic, *is or has been* customarily used as money in the country of issue, and is payable on demand; or
(b) any note which has been lawfully issued in some other country, and *is* customarily used as money in that country.

Thus, a £20 note a €20 note and a 100 dinar note are all 'currency notes'; so is an old style £5 note, which is no longer legal tender, but not a note which is no longer valid currency in the foreign country where it was issued.

A 'protected coin' is defined as any coin which is customarily used as money in any country, ie any coin which *is still* valid tender at the time. It also includes a sovereign, half-sovereign, krugerrand (or a coin denominated as a fraction thereof), a Maria-Theresia thaler dated 1780, and any euro coin.

Passing or tendering

14.83 In the offences below, references to 'passing or tendering' a note or coin are not confined to passing or tendering it as legal tender, so that (for example) passing or tendering a note or coin to a coin dealer or as a collector's item is a 'passing or tendering' for the purpose of these offences. A person 'passes' a note or coin to somebody when the latter actually accepts it from him/her; a person 'tenders' a note or coin when (s)he offers to pass it to somebody. On the other hand, simply to produce a bundle of counterfeit notes to impress a lady friend does not constitute a tender of them, since it does not involve any offer to pass them to her or anyone else.

Trial and punishment

14.84 By FCA 1981, s 22, all the offences dealt with below (**14.85–14.91**) are indictable (either-way) offences, punishable as follows:

(a) up to 10 years' imprisonment in the case of offences under s 14(1), 15(1), 16(1), or 17(1);
(b) up to two years' imprisonment in the case of offences under s 14(2), 15(2), 16(2), 17(2), or 17(3);
(c) an unlimited fine in the case of an offence under s 18 or 19, neither of which is imprisonable.

Counterfeiting

14.85 FCA 1981, s 14(1) provides that it is an offence for a person to make a counterfeit of a currency note or protected coin, intending that he or another shall pass or tender it as genuine.

Section 14(2) makes it an offence for a person to make a counterfeit of a currency note or protected coin without lawful authority or excuse.

Both offences punish those who make a counterfeit currency note or protected coin; the difference between them is this. If the intention of the maker (D) is that D or another shall pass it into circulation, D commits an offence under s 14(1). If D did not so intend, or that intent cannot be proved, D can be convicted of the lesser offence under s 14(2) unless D had a lawful authority or excuse for making the thing (which will be rare).

Passing, etc counterfeit currency

Passing or tendering counterfeit as genuine

14.86   By FCA 1981, s 15(1)(a), it is an offence for a person to pass or tender as genuine anything which is, and which (s)he knows or believes to be, a counterfeit of a currency note or of a protected coin.

Someone who sells to another counterfeit notes or coins, having declared them to be such, does not pass or tender them as genuine.

Delivering counterfeit

14.87   FCA 1981, s 15(1)(b) makes it an offence for someone to deliver to another anything which is, and which (s)he knows or believes to be, a counterfeit of a currency note or protected coin, intending that the person to whom it is delivered or another shall pass or tender it as genuine. Section 15(1)(b) deals, for example, with the person who knowingly takes counterfeits from the maker and delivers them to a second person, intending that the second person should pass or tender them as genuine. It is irrelevant whether or not the second person is an innocent recipient.

Where it cannot be proved that the person who delivered the counterfeit intended that the recipient or another should pass or tender it as genuine, (s)he may be convicted of the offence of delivery without lawful authority or excuse, which is dealt with by s 15(2).

Custody or control of counterfeit currency, and making, custody, or control of counterfeiting materials and implements

14.88   The offences under FCA 1981, s 5 (**14.69-14.71**) have their counterparts in ss 16 and 17 in relation to people who have counterfeit currency or counterfeiting materials or implements in their custody or under their control.

Custody or control of counterfeit currency

14.89   FCA 1981, s 16(1) provides that it is an offence for a person to have in his custody or control a counterfeit of a currency note or of a protected coin, knowing or believing it to be so and intending either to pass or tender it as genuine or to deliver it to another with the intention that he or another shall pass or tender it as genuine. Section 16(2) makes it an offence for a person to have such custody or control, with such knowledge or belief, without lawful authority or excuse. On a charge under s 16(2) D's intention is irrelevant. Whilst a settled intention to hand in counterfeit currency may amount to a lawful excuse, the fact that a person has not yet decided what to do with it cannot amount to such an excuse.

The offences of 'possession' in respect of currency may be committed even though the coin or note is not in a fit state to be passed or tendered or even though its making or counterfeiting has not been finished or perfected. Consequently, these offences may be committed, for example, in relation to the part-finished efforts of a counterfeiter.

Making, custody, or control of counterfeiting materials and implements

FCA 1981, s 17(1) provides that it is an offence for a person to make, or to have in his **14.90** custody or under his control, anything *which he intends to use, or to permit any other person to use, for the purpose of making a counterfeit of a currency note or of a protected coin with the intention that it be passed or tendered as genuine.* The words 'anything which he intends to use' include *anything used as a part of the process of counterfeiting.* This includes chromolins (printers' proofs used to check the quality of an aluminium plate produced from a film) of a currency note, special papers and inks, and reprographic equipment.

Section 17(2) provides that it is an offence for a person without lawful authority or excuse to make or to have in his custody or under his control any thing which, to his knowledge, is or has been specially designed or adapted for the making of a counterfeit of a currency note. The intentions italicised above are not required.

By s 17(3), provides that it is an offence for a person to make or have in his custody or under his control, any implement which, to his knowledge, is capable of imparting to anything a resemblance to the whole or part of either side of a protected coin, or of the reverse of the image on either side of a protected coin. A person charged with this offence has a defence if (s)he proves that (s)he had the written consent of the Treasury or some other lawful authority or excuse. This is the only offence in the Act where the defendant has the burden of proving a lawful authority or excuse.

Reproducing British currency

FCA 1981, ss 18 and 19 deal respectively with the reproduction of British currency **14.91** notes without written consent to do so from the relevant authority, and with the making, sale, or distribution of imitation British coins in connection with a scheme intended to promote the sale of any product or service.

These offences are neither serious nor require any counterfeiting in its strict sense (ie notes and coins reasonably capable of passing as currency notes or protected coins). They set out to prevent the use of reproductions in any form, even though the copy reproduced would not fool any reasonable person. If there was no legislation to prevent the production of good quality 'stage money', this could lead to abuse. Colour supplement magazines could print a reproduction of one side of a currency note as a voucher entitling the holder to a price reduction in certain stores. This would no doubt be done innocently, but the reproduction might be used falsely by some other person. Section 18, goes to the extent of prohibiting copies of notes which are not reproduced on the correct scale, in order to prevent 'blow-ups' being produced which may be of assistance to a counterfeiter.

POLICE POWERS UNDER FCA 1981

FCA 1981, ss 7 and 24 authorise a justice who is satisfied upon information on oath **14.92** that there is reasonable cause to believe that a person has in his/her *custody* or under his/her *control*:

(a) anything which has been used, or is intended to be used, for the making of a false instrument or copy of a false instrument, contrary to s 1 or 2;

(b) any false instrument or copy which has been used, or is intended to be used, contrary to s 3 or 4;

(c) anything which it is unlawful to possess without authority, etc, under s 5 (false money orders, stamps, etc);

(d) anything which is a counterfeit currency note or protected coin, or a reproduction made in breach of s 18 or 19; or

(e) anything which has been used, or is intended to be used, for the making of such counterfeit or reproductions,

to issue a warrant authorising a constable to enter premises, search for and seize such objects. The additional powers of seizure provided by the Criminal Justice and Police Act 2001, s 50 apply where a search warrant under FCA 1981 is executed.

A constable may, at any time after seizure, apply to a magistrates' court for an order for the forfeiture and destruction or disposal of objects seized.

SUPPLYING SPECIALIST PRINTING EQUIPMENT

14.93 The Specialist Printing Equipment and Materials (Offences) Act 2015, s 1(1) provides an offence of supplying specialist printing equipment, knowing that it will be used for criminal purposes. The offence is triable only on indictment and punishable with up to 10 years' imprisonment.

Section 2 defines 'specialist printing equipment' as any equipment which is designed or adapted for, or is otherwise capable of being used for, the making of relevant documents (including any material or article that is used in the making of such documents). A 'relevant document' is anything that is or purports to be:

(a) an identity document (defined in the same way as at **14.74**, except that it does not include a driving licence);

(b) a travel document (ie a UK or other driving licence, a ticket or other document (or a concessionary permit) authorising travel on public transport services, or a 'blue badge');

(c) an entry document (ie a document authorising entry to premises, including a security pass, and a ticket or other document for a sporting or other event);

(d) a document used for verifying the holder's age or national insurance number;

(e) a currency note or protected coin, as defined at **14.82**;

(f) a debit or credit card;

(g) any other instrument to which the Forgery and Counterfeiting Act 1981, s 5 (**14.69**) applies.

OBTAINING SERVICES DISHONESTLY

14.94 A person commits an indictable (either-way) offence, punishable with up to five years' imprisonment, against FA 2006, s 11(1) if (s)he obtains services for him/herself or another:

(a) by a dishonest act; and

(b) in breach of the provisions of s 11(2).

By s 11(2), a person obtains services in breach of s 11 if:

(a) they are made available on the basis that payment has been, is being or will be made for or in respect of them;

(b) (s)he obtains them without any payment having been made for or in respect of them or without payment having been made in full; and

(c) when (s)he obtains them (s)he knows:
    (i)   that they are being made available on the basis described in (a), or
    (ii)   that they might be,
    but intends that payment will not be made, or will not be made in full.

The offence does not require a false representation. There must be an 'obtaining' of a service. An example would be where a film is exhibited in a cinema and D sneaks in to watch it. Obtaining services by the dishonest use of false credit card details or personal information would similarly amount to an offence against s 11. Trespassory entry into premises to view a free event is not an offence against s 11 as the service provided is not one for which a person is expected to pay.

Dishonesty bears the same meaning as described at **14.58**.

DISHONESTLY OBTAINING ELECTRONIC COMMUNICATIONS SERVICE, AND POSSESSION OR SUPPLY OF APPARATUS FOR DOING SO

It is an offence under the Communications Act 2003 (CA 2003), s 125 dishonestly to obtain an electronic communications service with the intention of avoiding payment of a charge in respect of that service. However, s 125 exempts the dishonest obtaining of a broadcasting service provided from a place in the UK.   **14.95**

It is an offence under CA 2003, s 126(1) to possess or have under one's control anything that may be used for obtaining an electronic communications service, or in connection with obtaining such a service, with certain specified intentions. The specified intentions are listed in s 126(3). They are an intent: to use the thing to obtain such a service dishonestly; to use it for a purpose connected with the dishonest obtaining of such a service; dishonestly to allow it to be so used; or to allow it to be used for a purpose connected with the dishonest obtaining of such a service. An intention does not fall within s 126(3) if it relates exclusively to obtaining a broadcasting service provided from a place in the UK.

Section 126(2) prohibits the supply, or offer to supply, of anything which may be used to commit the above offence where the supplier knows or believes that the person supplied, etc has one of the intentions specified in s 126(3).

These offences are indictable (either-way) offences punishable with up to five years' imprisonment.

MAKING OFF WITHOUT PAYMENT

The Theft Act 1978, s 3(1) provides that a person who, knowing that payment on the spot for any goods supplied or service done is required or expected from him, dishonestly makes off without having paid as required or expected and with intent to avoid payment of the amount due is guilty of an offence. The offence is an indictable (either-way) offence punishable with a maximum of two years' imprisonment. The offence is useful in that it covers cases, such as leaving a restaurant or self-service petrol station without paying, in which it may be difficult to prove that any particular intention existed at the time that the goods were supplied or the service was done.   **14.96**

Because D must make off 'without having paid as required or expected', an offence is not committed if the creditor (or his/her agent) has agreed that payment would be postponed (as where goods or services are supplied on credit), even if that postponement has been procured by a dishonest deception.

The Court of Appeal has held that 'making off without payment' involves a departure without paying from the place where payment would normally be made; for example, in the case of a taxi, payment might be made while sitting in it or standing at the window. Subsequently, the Court of Appeal has indicated that this statement about the taxi should not be applied too literally; if a taxi-passenger (D) honestly explained that she had to enter her house to get the fare, the moment of payment would be deferred and a decision not to return to the taxi would mean that, from that moment, D would make off without payment.

'Payment on the spot' includes payment at the time of collecting goods on which work has been done or in respect of which a service has been provided.

Section 3 does not apply if the payment required or expected is not legally due (eg because the person demanding it is seriously in breach of contract). Nor does the section apply to the supplying of goods or the doing of services which is contrary to the law, nor where the service done is such that payment is not enforceable. Consequently, for example, the offence is not committed by the man who makes off without paying a prostitute for services rendered.

The term 'knowing that payment on the spot is required or expected' relates to the knowledge D must have of when payment must be made (s 3(2)).

Dishonesty needs to be shown to have existed at the time of 'making off'. Dishonesty is a question of fact for the jury or justices and is approached in the same way as in the offence of fraud (**14.58**).

It has been held by the House of Lords that the requisite intention to evade payment is an intention permanently to evade payment; an intention to postpone (ie temporarily evade) payment is insufficient.

FALSE ACCOUNTING

14.97 The indictable (either-way) offences of false accounting contrary to the Theft Act 1968 (TA 1968), s 17 are not restricted to people who are employed for the purpose of bookkeeping in its traditional sense. The falsity need not necessarily be with a view to gain since it is sufficient that the falsity is accompanied by an intent to cause loss to another. The offences under s 17 are punishable with up to seven years' imprisonment.

TA 1968, s 17(1) provides that a person is guilty of an offence where he dishonestly, with a view to gain for himself or another or with intent to cause loss to another:

(a) destroys, defaces, conceals or falsifies any account or any record or document made or required for any accounting purpose, or

(b) in furnishing information for any purpose produces or makes use of any account, or any such record or document as aforesaid, which to his knowledge is or may be misleading, false or deceptive in a material particular.

Section 17(1) creates two quite separate and distinct offences which might be described as the falsification of accounts (s 17(1)(a)) and the use of false or deceptive accounts (s 17(1)(b)). Before examining these offences separately a number of points can be made which apply to both of them.

Mens rea

For the purposes of either type of offence, D's conduct must be carried out dishonestly **14.98** and with a view to gain for him/herself or another or with intent to cause loss to another. 'Dishonesty' in this context is understood in the same way as in the offence of fraud (**14.58**). 'With a view to gain … or with intent to cause loss …' bears the same meaning as in blackmail (**14.40**). There is no need for any gain or loss actually to be caused.

Account, record, or document made or required for any accounting purpose

The acts specified by TA 1968, s 17(1)(a) and (b) must be done in relation to 'any **14.99** account, record or document made or required for any accounting purpose'. These terms should be given their ordinary meaning. An accounts book, balance sheet, or a payroll record is made for an accounting purpose. That is the reason for its existence; it has no other real purpose as it can be used for little other than the keeping of accounts. Documents may be required for an accounting purpose, even though they have other purposes as well as an accounting purpose. For example, a housing benefit claim form which contained the only information used to calculate housing benefit has been held by a divisional court to be a document required for an accounting purpose, despite the fact that it is also used to determine entitlement to benefit.

The terms 'account' or 'record' are wide enough to cover an account or record produced by a mechanical device, such as a taxi meter or the turnstile at a soccer ground which records the number of persons admitted so that the entries may be related to the money collected.

The separate requirements of the two offences are as follows.

Falsification of accounts (s 17(1)(a))

What is required here is the destruction, defacement, concealment, or falsification of **14.100** any account, etc. If entry to a cinema is recorded by the issue of a ticket by a process which produces a duplicate copy on a roll, the destruction of any part of that record, if carried out with the necessary intent, would be an offence. If some of the duplicate copies were not handed over to the manager so that a false total figure could be shown to cover deficiencies, those copies would be 'concealed' for the purposes of TA 1968, s 17(1)(a). If, to cover deficiencies in the accounts, a corrosive liquid or other damaging agent, was applied to the duplicate records, this would amount to a defacement. However, the most common offence is always likely to be the falsification of figures in an account, record, or document. 'Falsification' covers the preparation of false accounts, as well as the falsification of existing ones.

Section 17(2) states that a person who makes or concurs in making in an account or other document an entry which is or may be misleading, false, or deceptive in a material particular, or who omits or concurs in omitting a material particular from an account or other document, is to be treated as falsifying the account or document. However, this is not an exclusive definition of falsification. In one case, it has been held that a turnstile operator, who allowed two people through the turnstile while only recording one of them, falsified the record. It might be that the falsification was within s

17(2) (by omitting a material particular) but if not there was a 'falsification' within the ordinary meaning of that term.

Somewhat surprisingly, the Court of Appeal has held that a person can falsify an accounting document by completely failing to fill in a blank form required for an accounting purpose, and even though the particular document so falsified cannot be identified. The Court of Appeal therefore upheld the conviction of an international telephone operator who had failed to log calls on the forms provided for this purpose. The court held that, as soon as a call was made, it was the operator's duty to fill in one of the forms in a pile in front of him and that thereby one of them became a document required for an accounting purpose and the fact that the particular form could not be identified (since the operator might not have chosen to use the first form) did not matter.

Use of false or deceptive account (s 17(1)(b))

14.101   Whereas TA 1968, s 17(1)(a) is concerned with those who falsify or destroy, etc any account, etc, s 17(1)(b) deals with those who, in furnishing information for any purpose, produce or make use of any account or any record or document made or required for any accounting purpose, *knowing* it is or might be misleading, false or deceptive in a material particular (ie liable to mislead in a significant way).

Of course, a person who has falsified an account contrary to s 17(1)(a) might go on to use it contrary to s 17(1)(b), in which case (s)he will have committed both offences. However, an offence under s 17(1)(b) may also be committed by a person who has not falsified the document in question, for it is not uncommon for persons with a view to gain to use in a dishonest way documents which have been made out erroneously by someone else. If a man, who is entitled to receive payment for money which he has spent on petrol for use in his firm's car, receives a receipt which he knows has been wrongly made out for a higher amount, he commits an offence against s 17(1)(b) if, when making out his claim (furnishing information), he makes use of that receipt since, to his knowledge, it is false in a material particular.

HANDLING STOLEN GOODS

14.102   The law concerning the indictable (either-way) offence of handling stolen goods is concerned with punishing not only the ultimate receiver but anyone who handles them en route to him/her or afterwards.

The Theft Act 1968 (TA 1968), s 22(1) provides that a person handles stolen goods if (otherwise than in the course of the stealing) knowing or believing them to be stolen goods he dishonestly receives the goods, or dishonestly undertakes or assists in their retention, removal, disposal, or realisation by or for the benefit of another person, or if he arranges to do so. Handling stolen goods is punishable with up to 14 years' imprisonment.

Goods

14.103   'Goods' are defined by TA 1968, s 34(2)(b), as including money and every other description of property except land, and includes things severed from the land by stealing.

Stolen

For the offence to be committed, the goods handled must be 'stolen goods' at that point **14.104**
of time. By TA 1968, s 24, 'stolen goods' are goods which have been:

(a) stolen, contrary to TA 1968, s 1;
(b) obtained as a result of blackmail, contrary to TA 1968, s 21;
(c) obtained by fraud, contrary to the Fraud Act 2006 (FA 2006), s 1; or
(d) 'stolen' abroad contrary to the law of that land, provided that had the 'stealing' occurred in England or Wales an offence contrary to TA 1968, s 1 or 21 or FA 2006, s 1 would have been committed.

In addition, references to stolen goods also include money which is dishonestly withdrawn from an account to which a 'wrongful credit' has been made, but only to the extent that the money derives from the credit. 'Wrongful credit' is defined at **14.146**.

If a child below the age of criminal responsibility appropriates property in circumstances which would amount to theft by someone over that age, a person who dishonestly handles that property is not guilty of this offence as the goods are not 'stolen goods'. However, that person would normally be guilty of theft.

The fact that goods were stolen goods when handled can be proved by evidence of the conviction of someone for their prior theft. Another way is by offering evidence of ownership, the owner identifying the goods and describing the circumstances of their loss. It is not necessary to prove that goods were stolen by any particular person; merely that they were stolen by someone. The acquittal of an alleged thief does not prevent the goods being found to be stolen.

When goods cease to be stolen

TA 1968, s 24(3) provides that no goods shall be regarded as having continued to be **14.105** stolen goods after they have been restored to the person from whom they were stolen or to other lawful possession or custody. Clearly, this goes further than repossession by the owner, since it includes cases in which goods have been restored to other lawful possession or custody. When a police officer in the course of his/her duties takes possession of stolen goods, they are thereby restored to 'other lawful possession or custody' and are no longer stolen; consequently, it is not an offence to handle them thereafter.

The issue is whether or not the stolen goods have been taken into possession or custody on the owner's account so that they may be returned to him/her. Where a security guard marked cartons of stolen cigarettes so that they would more easily be identified in the hands of a handler, they were not restored to lawful possession or custody. If P, a police officer, sees property inside a parked car which P suspects might be stolen and immobilises the car to ensure that P will have an opportunity of questioning the driver, P does not take possession of the stolen goods unless P immobilises the car with the intention of taking charge of those goods so that they cannot be removed. If P has retained an open mind as to whether (s)he should take possession, and merely immobilises the car to prevent the driver getting away without questioning, P does not reduce the goods into his/her possession or custody. It is P's state of mind which is the deciding factor in each case.

Section 24(3) also provides that goods cease to be stolen goods where the person from whom they were stolen and any other person claiming through him have ceased to have rights of restitution in respect of the theft. The right to restitution of property is a matter of civil law. There are various ways in which the right to restitution might have been lost. An example is where X obtains goods from Y by fraud. If Y, on discovering the fraud, nevertheless affirms the transaction Y thereby ceases to have any right to restitution of the goods, which therefore cease to be stolen goods at that point of time.

Goods representing those originally stolen

14.106   TA 1968, s 24(2) provides that references to stolen goods include, in addition to the goods originally stolen and parts of them (whether in their original state or not):

(a) any other goods which directly or indirectly represent or have at any time represented the stolen goods in the hands of *the thief* as being the proceeds of any disposal or realisation of the whole or part of the goods stolen or of goods so representing the stolen goods; and

(b) any other goods which directly or indirectly represent or have at any time represented the stolen goods in the hands of *a handler of the stolen goods or any part of them* as being the proceeds of any disposal or realisation of the whole or part of the stolen goods handled by him or of goods so representing them.

'The thief' means the person by whose conduct the goods were originally stolen. 'A handler' means any person who has committed the actus reus of handling with the appropriate mens rea. The Court of Appeal has held that goods are in the hands of the thief or of a handler if they are in his/her possession or under his/her control; physical custody is not required.

The rule can be illustrated as follows. A steals a car. A then sells it to B for £1,000 and receives that sum in cash. The car and the cash are now both stolen goods, the cash because it *directly represents the original goods (the car) in the hands of the thief—as the proceeds of the car's disposal or realisation.* Therefore, if A then gives the £1,000 (or part of it) to C, who receives it knowing it has represented the original stolen goods in A's hands and C buys a camera with the £1,000, the camera becomes stolen goods once it is in C's hands because *it indirectly represents the original stolen goods in the hands of a handler as the proceeds of the disposal or realisation of goods representing the original stolen goods.* Consequently if D receives the camera from C, knowing it has represented the stolen goods in C's hands, D can be convicted of handling stolen goods. However, if E receives the camera from D, unaware that it represents the original stolen goods, and then sells it for cash, the cash which E receives will not become stolen goods because, lacking mens rea, E is not a handler and therefore *that* cash does *not represent the original stolen goods in the hands of a handler as the proceeds of their disposal or realisation.*

Various forms of handling

14.107   TA 1968, s 22 defines 'handling' as receiving stolen goods, or undertaking or assisting in their retention, removal, disposal or realisation by or for the benefit of another person, or arranging to do one of these things.

It must not be forgotten that s 22 excludes from its provisions those who handle in one of these ways in the course of the original stealing. This is a commonsense approach to offences of handling. We are not concerned with thieves (and their accomplices) who handle stolen goods in the course of the original theft as they are punishable in respect of the theft, whatever form that theft may take.

The charge

14.108   Although an examination of the definition of 'handling' shows that there are 18 different ways in which handling can be committed, it is unnecessary to make exclusive selections of a particular variant of the offence. TA 1968, s 22 provides a single offence of handling stolen goods, which can be committed by receiving or by any of the other ways specified. Consequently, an information or written charge or indictment, which simply alleges handling of stolen goods, is unlikely to fail for duplicity.

Nevertheless, in order to be fair and clear to the defendant (D), the better practice is to particularise the form of handling relied on, and, if there is any uncertainty about the form of handling in question, it is advisable to have more than one written charge (or more than one count in the indictment). But only two written charges (or counts) should generally be used in such a case: one charging receiving, and the other charging the various forms of 'undertaking or assisting in' by or for the benefit of another (or such forms of these as are clearly the only ones relevant).

The forms of handling can be explained as follows.

Receiving

This requires proof that D obtained possession or control of stolen goods from someone   **14.109**
else; a finder of goods does not receive them. The presence of stolen goods on the premises of another does not necessarily mean that they are in the possession or control of the owner of the premises. To be a 'receiver' a person need not have contact with the goods in any physical sense, since a person can have control of goods without any physical contact with them, as where an employee or agent acting on his/her orders received them (in which case the employee or agent is also a receiver).

It is unnecessary that a receiver should act to gain any profit or even advantage from his/her possession of stolen goods.

Arranging to receive

This requires preparation by D, or a concluded agreement between D and another   **14.110**
(eg the thief), for the receiving of stolen goods by D. It is essential that the goods are stolen goods at the time the arrangement to receive is made. Thus, D does not commit handling where D agrees to receive goods which are to be stolen. However, in such a case, D and any other party to the agreement could be convicted of conspiracy to handle.

The other forms of handling set out below differ in a vital respect from receiving, or arranging to receive, in that they must be done 'by or for the benefit of' a person other than the alleged handler (or, the Court of Appeal has held, a co-defendant on the same charge of handling). The other person should be named in the charge if his/her identity is known. If it is not, the charge should indicate that the acts were done, as appropriate, 'by or for the benefit of' some person unknown and such a charge will have to be supported by strong evidence that that other person existed.

Undertaking the retention, removal, disposal, or realisation of stolen goods for the benefit of another

These four forms of handling cover the case where D, either alone or with another, re-   **14.111**
tains, removes, disposes, or realises the stolen goods for the benefit of another.

'Retention' means 'keeping possession of, not losing, continuing to have'. 'Removal' refers to the movement of stolen goods from one place to another, eg transporting stolen goods to a hideout for the benefit of the thief or another. 'Disposal' covers dumping, giving away, or destroying stolen goods, eg melting down candlesticks for the thief. 'Realisation' means the exchange of stolen goods for money or some other property. A person who sells stolen goods as agent for a third party (eg the thief) undertakes their realisation for the benefit of another. However, the House of Lords has held that a person who sells stolen goods on his/her own behalf does not undertake their realisation for the benefit of another because the buyer benefits from the purchase and not from the realisation (which benefits only the seller).

Of course, a person (D) who undertakes one of these four activities will often be in possession or control of the stolen goods. If D knew or believed they were stolen when (s)he acquired possession or control, D is guilty of handling anyway on the basis of receiving. However, if D lacked such a state of mind at that point of time but realised the goods were stolen when (s)he undertook one of the four activities, D can be convicted of handling on that basis, provided that his/her 'undertaking' was for the *benefit of another*.

Assisting in the retention, removal, disposal, or realisation of stolen goods by another

14.112   These four forms of handling are appropriate to cover cases where D provides assistance to another person, the thief or another handler, who is going to undertake the retention, removal, disposal, or realisation of stolen goods.

For there to be 'assistance', D must do something for the purpose of enabling the goods to be retained, etc, whether or not D succeeds.

There is some overlap with cases of 'undertaking'. For example, a person who joins with another in removing stolen goods not only undertakes their removal for the benefit of another but also assists in their removal by another.

A person *assists in the retention* of stolen goods by another if (s)he puts the thief in touch with a warehouse-keeper, or provides tarpaulins to conceal stolen goods, or tells lies so as to make it more difficult for the police to find stolen goods retained by the thief or a handler. On the other hand, the Court of Appeal has held, a refusal to answer police questions as to the whereabouts of stolen goods does not amount to handling, although it might well assist in their retention by another. A person, who innocently allows goods to be left on his/her premises but later discovers that they are stolen, assists in their retention by another if (s)he nevertheless permits them to continue to remain on the premises. Merely to use stolen goods does not suffice, since it does not in itself amount to assistance in their retention.

A person *assists in the removal* of stolen goods by another if (s)he lends a lorry for their removal. (S)he *assists in their disposal* by another if (s)he advises the thief as to how to get rid of the goods. (S)he *assists in their realisation* by another if (s)he puts a 'fence' in touch with the thief.

Arranging to undertake or assist in the retention, removal, disposal, or realisation of stolen goods by or for the benefit of another

14.113   The effect of TA 1968, s 22 is further extended by the addition of the words 'or arranging to do so'. Any arrangement to undertake or assist in one of the activities described at **14.111** and **14.112** above can suffice; it is irrelevant that nothing further is done. However, as mentioned in relation to the offence of 'arranging to receive', it must be proved that the arrangement in respect of one of these other forms of handling related to goods which were stolen goods at the time of the arrangement.

Mens rea

14.114   It must be proved that, when D handled the stolen goods, D knew or believed that they were stolen, and acted dishonestly.

Knowledge or belief

14.115   Wilful blindness, ie deliberately turning a blind eye to the question of whether or not goods are stolen, does not suffice for 'knowledge' or 'belief', but it has been held that, if wilful blindness as to the goods being stolen is proved, knowledge or belief *may* (not must) be inferred from it. This is important since, in the absence of an admission by the handler, knowledge or belief (at the time of the handling) that the goods were stolen can be difficult to prove, particularly in the common case of sales at bargain prices, of goods which 'fell off a lorry' and which turn out to have been stolen.

D need not be aware of the precise nature of the goods. If D receives boxes believing them to contain stolen whisky, it is immaterial that the boxes actually contained stolen wine.

Proof   Frequently, knowledge or belief is proved by reference to the surrounding circumstances. If stolen goods have been carefully concealed on private premises, this suggests not only that the occupier of the premises did not wish to be found in possession of them but also that (s)he knew that the goods were stolen.

In general, where D is found in possession of property which has been recently stolen, a judge may direct a jury that they may infer knowledge or belief if D fails to offer an explanation for his/her possession, or if they are satisfied beyond reasonable doubt that any explanation which is offered is false. This rule is merely an application of the ordinary rules of circumstantial evidence. In addition, TA 1968, s 27, which applies where D is being prosecuted at the trial in question only for handling stolen goods, recognises the difficulties which may exist in proving knowledge in handling offences by allowing special evidence to be offered in certain circumstances. These are that, if evidence has been given of an act of handling by D in relation to the goods in question, the following evidence may be given to assist proof of knowledge or belief that the property was stolen:

(a) evidence that D has had in his/her possession, or has undertaken or assisted in the retention, removal, disposal or realisation of, stolen goods from any theft taking place not earlier than 12 months before the offence charged; and

(b) provided that seven days' notice in writing have been given to D of the intention to prove the conviction, evidence that D has, within the five years preceding the date of the offence charged, been convicted of theft or of handling stolen goods.

Section 27 supplements the general rules about admissibility of 'bad character' evidence (**8.64–8.73**). It must be emphasised that it only permits the evidence described in (a) and (b) to be given to prove knowledge or belief that the goods were stolen. It does not enable such evidence to be used for any other purpose, for example to prove dishonesty on D's part or an act of handling by D.

The provisions in (b) must be read with PACE, s 73(2) which provides that where evidence of a previous conviction on indictment is admissible by means of a certificate of conviction the certificate must give 'the substance and effect of the indictment and conviction'. The House of Lords has held that, when (b) is so read, a certificate of previous conviction for theft or handling should:

(a) where the conviction was on indictment, state the substance and effect of the indictment and conviction, including the nature of the property concerned; and

(b) where the conviction was in summary proceedings, record the nature of the property concerned.

Dishonestly

Dishonesty is always a question of fact for the jury or justices, and the same approach is   **14.116**
to be taken by them, as it is in relation to the offence of fraud (**14.58**). It follows, for example, that if a person knowingly receives stolen goods, but does so in order to restore

them to their owner, or to hand them over to the police, his/her handling is unlikely to be found to be dishonest by the jury or justices.

ADVERTISING REWARDS FOR RETURN OF STOLEN OR LOST GOODS

14.117   TA 1968, s 23 punishes the public advertisement of a reward for the return of any lost or stolen goods which uses any words to the effect that no questions will be asked, or that the person producing the goods will be safe from apprehension or inquiry, or that any money paid for the purchase of the goods or advanced by way of loan on them will be repaid. The printer and publisher of such an advertisement are liable, as well as the advertiser. The offence under s 23 is a summary offence punishable with a fine not exceeding level 3.

MONEY LAUNDERING OFFENCES

14.118   Although many instances of handling stolen goods fall outside the popular conception of money laundering, the money laundering offences in the Proceeds of Crime Act 2002 (PCA 2002), ss 327 to 329 are, between them, capable of covering virtually anything which constitutes handling stolen goods. In addition, the mens rea requirement for these offences is less demanding since D is not required to know or believe that the property constitutes or represents the benefit from criminal conduct, but is simply required to know or suspect this. In addition, there is no requirement of 'dishonesty'.

Criminal property

14.119   The offences in PCA 2002, ss 327 to 329 relate to dealings with 'criminal property'. Section 340 provides that property is criminal property if:

(a) it constitutes a person's benefit from criminal conduct or it represents such a benefit (in whole or part and whether directly or indirectly), and

(b) the alleged offender knows or suspects that it constitutes or represents such a benefit.

For these purposes:

(i) property is all property wherever situated, including money, all forms of property, real or personal, heritable or moveable, and things in action and other intangible or incorporeal property;

(ii) criminal conduct is conduct which constitutes an offence in any part of the UK, or would constitute an offence in any part of the UK if it occurred there;

(iii) it is immaterial who carried out the conduct, and who benefited from it;

(iv) a person benefits from conduct if (s)he obtains property as a result of or in connection with the conduct;

(v) if a person obtains a pecuniary advantage as a result of or in connection with conduct, (s)he is taken to obtain as a result of or in connection with the conduct a sum of money equal to the value of the pecuniary advantage.

Concealing, etc

14.120   PCA 2002, s 327(1) provides that a person commits an offence if he:

(a) conceals criminal property;

(b) disguises criminal property;

(c) converts criminal property;

(d) transfers criminal property; or

(e) removes criminal property from England and Wales or from Scotland or from Northern Ireland.

Concealing or disguising criminal property includes concealing or disguising its nature, source, location, disposition, movement, or ownership or any rights with respect to it.

By s 327(2), a person does not commit such an offence under s 327(1) if—

(i) (s)he makes an authorised disclosure under s 338 (**14.123**) and (if the disclosure is made before (s)he does the act mentioned in s 327(1)) (s)he has the appropriate consent;

(ii) (s)he intended to make such a disclosure but had a reasonable excuse for not doing so;

(iii) the act (s)he does is done in carrying out a function (s)he has relating to the enforcement of any provision of PCA 2002 or of any other legislation relating to criminal conduct or benefit from criminal conduct.

Nor, by s 327(2A), does a person commit an offence under s 327(1) if—

(i) (s)he knows, or reasonably believes, that the relevant criminal conduct (the criminal conduct by reference to which the property concerned is criminal property) occurred in a particular country or territory outside the UK, and

(ii) the relevant criminal conduct was not, at the time it occurred, a criminal offence in that country, etc, and is not of a description prescribed in the Proceeds of Crime Act 2002 (Money Laundering: Exceptions to Overseas Conduct Defence) Order 2006.

A deposit-taking body or electronic money institution or payment institution that does an act mentioned in s 327(1) (c) or (d) does not commit an offence under s 327(1) if it does the act in operating an account maintained with it, and the value of the criminal property concerned is less than the threshold amount determined under s 339A ('the threshold amount') for the act (s 327(2C)).

Arrangements

PCA 2002, s 328(1) provides that it is an offence for a person to enter into or become **14.121** concerned in an arrangement which he knows or suspects facilitates (by whatever means) the acquisition, retention, use, or control of criminal property by or on behalf of another person.

Section 328(2) and (3) provides identical exceptions to those in s 327(2) and (2A) (**14.120**), save for the substitution of s 328(1) for the reference to s 327(1) in s 327(2). Section 328(5) provides a similar exception to that in s 327(2C), stating that a deposit-taking body, etc that does an act mentioned in s 328(1) does not commit an offence under s 328(1) if it does the act in operating an account maintained with it, and the arrangement facilitates the acquisition, retention, use or control of criminal property of a value that is less than the threshold amount for the act.

Acquisition, use and possession

By PCA 2002, s 329(1), it is an offence for a person to acquire, use or possess criminal **14.122** property.

By s 329(2), a person (P) does not commit an offence under s 329(1)if:

(a) P makes an authorised disclosure (**14.123**) and (if the disclosure is made before P does the act mentioned in s 329(1)) P has the appropriate consent of a constable, customs officer, or nominated person to whom that disclosure is made;

(b) P intended to make such a disclosure but had a reasonable excuse for not doing so;

(c) P acquired or used or had possession of the property for adequate consideration; or

(d) the act P does is done in carrying out a function P has relating to the enforcement of PCA 2002 or of any other legislation relating to criminal conduct or benefit from criminal conduct.

For these purposes—

(i) P acquires property for inadequate consideration if the value of the consideration is significantly less than the value of the property;

(ii) P uses or has possession of property for inadequate consideration if the value of the consideration is significantly less than the value of the use or possession;

(iii) the provision by P of goods or services which P knows or suspects may help another to carry out criminal conduct is not consideration.

Section 329(2A) and (2C) provides exceptions identical to those in s 327 (2A) and (2C) (**14.120**) with the substitution of references to s 329(1) for those to s 327(1).

Authorised disclosure

14.123 PCA 2003, s 338 provides that for the purposes of ss 327, 328, and 329 a disclosure is authorised if:

(a) it is a disclosure to a constable (including a NCA officer), a customs officer or a nominated officer (ie a disclosure in the course of the alleged offenders employment to a person nominated by his/her employer to receive disclosures) by the alleged offender (AO) that property is criminal property; and

(b) the first, second or third condition below is satisfied.

The first condition is that the disclosure is made before AO does the prohibited act (ie an act mentioned in s 327(1), 328(1), or 329(1), as the case may be).
The second condition is that:

(a) the disclosure is made while AO is doing the prohibited act;

(b) AO began to do the act at a time when, because AO did not know or suspect that the property constituted or represented a person's benefit from criminal conduct, the act was not a prohibited act; and

(c) the disclosure is made on AO's own initiative and as soon as is practicable after AO first knows or suspects that the property constitutes or represents a person's benefit from criminal conduct.

The third condition is that:

(a) the disclosure is made after AO does the prohibited act;

(b) AO has a reasonable excuse for failing to make the disclosure before doing the act; and

(c) the disclosure is made on AO's own initiative and as soon as it is practicable for AO to make it.

Mens rea

14.124 As already indicated, the mens rea requirement for the offences under PCA 2002, ss 327 to 329 is simply that D must know or suspect that the property constitutes or represents the benefit from criminal conduct.

Trial and punishment

The offences under s 327, 328 and 329 are indictable (either-way) offences punishable with up to 14 years' imprisonment.   **14.125**

Charging money laundering or handling: which?

The offences under PCA 2002, ss 327 to 329 clearly have the potential to be more at-   **14.126**
tractive to a prosecutor than that of handling stolen goods. However, they were created in the context of legislation directed primarily at money laundering and matters of serious criminality, and the courts have discouraged any practice of prosecuting in-appropriate cases under PCA 2002, ss 327 to 329 instead of charging handling stolen goods under TA 1968, s 22.

SEARCH WARRANTS

By TA 1968, s 26, a justice may grant a warrant, upon information being received on   **14.127**
oath that there is reasonable cause to believe that any person has in his/her custody or possession or on his/her premises any stolen goods, to search for and seize those goods.

A Crown Court judge has power to make a search and seizure warrant under PCA 2002, s 352 where someone is subject to a money laundering investigation.

The additional powers of seizure provided by the Criminal Justice and Police Act 2001, s 50 (**4.61**) apply where a search warrant under one of these two provisions is executed.

SCRAP METAL DEALERS

The relevant provisions are contained in the Scrap Metal Dealers Act 2013 (SMDA   **14.128**
2013).

Definitions

For the purposes of SMDA 2013:   **14.129**

(a) 'scrap metal dealer (SMD)' means a person who is for the time being carrying on business as a scrap metal dealer, whether or not authorised by a licence; and

(b) 'scrap metal' includes any old, waste or discarded metal or metallic material, and any product, article or assembly which is made from or contains metal and is broken, worn out or regarded by its last holder as having reached the end of its useful life. However, gold, silver, and any alloy of which 2 per cent or more by weight is attributable to gold or silver, are not scrap metal.

Licensing of SMDs

By SMDA 2013, s 1(1), no person may carry on business as a SMD unless authorised   **14.130**
by a scrap metal licence under the Act. It is a summary offence, punishable with an un-limited fine, to do so (s 1(3)).

A person carries on business as a SMD if (s)he:

(a) carries on a business which consists wholly or partly in buying or selling scrap metal, whether or not the metal is sold in the form in which it was bought; or

(b) carries on business as a motor salvage operator (so far as that does not fall within (a)).

For the purposes of (a), a person who manufactures articles is not to be regarded as selling scrap metal if (s)he sells scrap metal only as a by-product of manufacturing articles or as surplus materials not required for manufacturing them.

For the purposes of (b), a person carries on business as a motor salvage operator if (s)he carries on a business which consists:

(i) wholly or partly in recovering salvageable parts from motor vehicles for re-use or sale and subsequently selling or otherwise disposing of the rest of the vehicle for scrap; or

(ii) wholly or mainly in–

    (a) buying written-off vehicles and subsequently repairing and reselling them;

    (b) buying or selling motor vehicles which are to be the subject (whether immediately or on a subsequent re-sale) of any of the activities mentioned in (i) and (a), or

    (c) activities falling within (a) and (b).

Form and effect of licence

14.131   Local authorities are responsible for issuing scrap metal licences. Such a licence must be a site licence, or a collector's licence. For these purposes, 'local authority' means, (in England) a district council, a London borough council, or the Common Council of the City of London, and (in Wales) a county council or a county borough council.

A site licence authorises the licensee to carry on business at any site in the authority's area which is identified in the licence. It must:

(a) name the licensee,

(b) name the authority,

(c) identify all the sites in the authority's area at which the licensee is authorised to carry on business,

(d) name the site manager of each site, and

(e) state the expiry date of the licence.

'Site manager' means the individual who exercises day-to-day control and management of activities at the site, and 'site' means any premises used in the course of carrying on business as a SMD (whether or not metal is kept there).

A collector's licence authorises the licensee to carry on business as a mobile collector in the authority's area. It must:

(a) name the licensee,

(b) name the authority, and

(c) state the expiry date of the licence.

'Mobile collector' means someone who:

(i) carries on business as a SMD otherwise than at a site, and

(ii) regularly engages, in the course of that business, in collecting waste materials and old, broken, worn out or defaced articles by means of visits from door to door.

A person may hold more than one licence issued by different local authorities, but may not hold more than one licence issued by any one authority.

By SMDA 2013, s 8(1), an applicant for a scrap metal licence, or for the renewal or variation of a licence, must notify the authority to which the application was made of any changes which materially affect the accuracy of the information which the applicant has provided in connection with the application. A licensee who is not carrying on business as a SMD in the area of the authority which issued the licence must notify the authority of that fact within 28 days of the beginning of the period in which the licensee is not carrying on business in that area while licensed (s 8(2) and (3)). If a licensee carries on business under a trading name, ie a name other than that stated in the licence, the licensee must, within 28 days of the change, notify the authority which issued the licence of any change to that name (s 8(4) and (5)). An applicant or licensee who fails to comply with s 8 is guilty of a summary offence punishable with a fine nor exceeding level 3 (s 8(9)), but (s)he has a defence if (s)he proves that (s)he took all reasonable steps to avoid committing the offence (s 8(10)).

Closure of unlicensed sites

SMDA 2013, Sch 2 deals with this.   **14.132**

For the purposes of Sch 2, a person has an interest in premises if (s)he is the owner, leaseholder, or occupier of the premises. In the case of a local authority, the powers conferred by Sch 2 are exercisable only in relation to premises in the authority's area. In SMDA 2013, 'premises' includes any land or other place (whether enclosed or not).

• Closure notice

If a constable or the local authority is satisfied that: **14.133**

(a) premises (other than residential premises) are being used by a SMD in the course of business; and
(b) the premises are not a licensed site,

the constable or authority may issue a closure notice which:

(i)   states that the constable or authority is so satisfied;
(ii) gives the reasons for that;
(iii) states that the constable or authority may apply to the court for a closure order; and
(iv) specifies the steps which may be taken to ensure that the alleged use of the premises ceases.

The constable or authority must give the closure notice to:

(a) the person who appears to the constable or authority to be the site manager of the premises, and
(b) any person (other than the person in (a)) who appears to the constable or authority to be a director, manager, or other officer of the business in question, and may also give the notice to any person who has an interest in the premises.

Where:

(a) a person occupies another part of any building or structure of which the premises form part, and
(b) the constable or authority reasonably believes that the person's access to that other part would be impeded if a closure order were made in respect of the premises,

the constable or authority must give the notice to that person.

A closure notice may be cancelled by a notice issued by a constable or the local authority, which takes effect when it is given to any one of the persons to whom the closure notice was given. The cancellation notice must also be given to any other person to whom the closure notice was given.

Application for closure order

14.134   Where a closure notice has been given, a constable or the local authority may make a complaint to a justice of the peace for a closure order not less than seven days after the date on which the closure notice was given, nor more than six months after that date. On such a complaint, the justice may issue a summons to answer to the complaint; the summons must be directed to any person appearing to be the site manager or to be a director etc of the business, to whom the closure notice was given. If a summons is so issued, notice of the date, time, and place at which the complaint will be heard must be given to all other persons to whom the closure notice was given.

Closure order

14.135   If, on hearing the complaint, the court is satisfied that the closure notice was given to any person appearing to be the site manager or to be a director etc of the business and that:

(a) the premises continue to be used by a SMD in the course of business; or
(b) there is a reasonable likelihood that the premises will be so used in the future,

the court may make such order as it considers appropriate for the closure of the premises (a 'closure order').

A closure order may, in particular, require that:

(a) the premises be closed immediately to the public and remain closed until a constable or the local authority makes a certificate of termination (below);
(b) the use of the premises by a SMD in the course of business be discontinued immediately;
(c) any defendant pay into court such sum as the court determines and that the sum will not be released by the court to that person until the other requirements of the order are met.

A closure order including a requirement mentioned in (a) may, in particular, include such conditions as the court considers appropriate relating to:

(i)   the admission of persons onto the premises;
(ii) the access by persons to another part of any building or other structure of which the premises form part.

A closure order may include such provision as the court considers appropriate for dealing with the consequences if the order should cease to have effect under a certificate of termination.

As soon as practicable after a closure order is made, the complainant must fix a copy of it in a conspicuous position on the premises in respect of which it was made.

Termination of closure order by certificate of constable or authority

14.136   Where a closure order has been made, but a constable or the local authority is satisfied that the need for the order has ceased, the constable or authority may make a certificate to that effect, whereupon the closure order ceases to have effect. Any sum paid into court under the closure order must be released by the court to the defendant. As soon as practicable after making a certificate, the constable or authority must:

(a) give a copy of it to any person against whom the closure order was made;
(b) give a copy of it to the designated officer for the court which made the order; and
(c) fix a copy of it in a conspicuous position on the premises in respect of which the order was made.

The constable or authority must also give a copy of the certificate to any person who requests one.

Discharge of closure order by court

Any person to whom the closure notice was given, and any other person who has an interest in the premises, may make a complaint to a justice of the peace for an order that a closure order be discharged (a 'discharge order'). The court may not make a discharge order unless it is satisfied that there is no longer a need for the closure order.

14.137

Enforcement of closure order

A person is guilty of an offence under SMDA 2013, Sch 2, para 9(1) if, without reasonable excuse, (s)he permits premises to be open in contravention of a closure order, or otherwise fails to comply with, or does an act in contravention of, a closure order.

14.138

If a closure order has been made in respect of any premises, a constable or an authorised person may (if necessary using reasonable force) enter the premises at any reasonable time, and, having entered the premises, do anything reasonably necessary for the purpose of securing compliance with the order. A constable or an authorised person seeking to exercise these powers in relation to any premises must produce evidence of his/her identity, or of his/her authority to exercise those powers, if the owner, occupier, or other person in charge of the premises requires him/her to produce it. A person who intentionally obstructs a constable or an authorised person in the exercise of these powers is guilty of an offence under Sch 2, para 9(5). An 'authorised person' is a person authorised for the above purposes by the local authority.

An offence under Sch 2, para 9 is a summary offence punishable with an unlimited fine.

Conduct of business

Display of licence

A SMD who holds a site licence must display a copy of the licence at each site identified in the licence; the copy must be displayed in a prominent place in an area accessible to the public.

14.139

A SMD who holds a collector's licence must display a copy of the licence on any vehicle being used in the course of the dealer's business; the copy must be displayed in a manner which enables it easily to be read by a person outside the vehicle.

A SMD who fails to comply with these provisions commits a summary offence, punishable with a fine not exceeding level 3, under SMDA 2013, s 10(5).

Verification of supplier's identity

A SMD must not receive scrap metal from a person without verifying the person's full name and address (SMDA 2013, s 11(1)). That verification must be by reference to documents, data, or other information obtained from a reliable and independent source (s 11(2)). The SMDA 2013 (Prescribed Documents and Information for Verification of Name and Address) Regulations 2013 prescribe documents, data, or other information which are sufficient for these purposes.

14.140

The Regulations provide that, in order to verify a person's name and address, it is sufficient for the SMD to refer to either:

(a) a document listed below which bears the person's full name, photograph and residential address; or
(b) both of:
    (i) a document listed below which bears the person's full name, photograph, and date of birth, and
    (ii) a supporting document listed below which bears the person's full name and residential address.

The documents which apply for the purposes of (a) or (b)(i) are a valid:

(a) UK passport; or
(b) passport issued by an EEA state; or
(c) Great Britain or Northern Ireland photo-card driving licence; or
(d) UK biometric immigration document.

The supporting documents which apply for the purposes of (b)(ii) are:

(a) a bank or building society statement;
(b) a credit or debit card statement;
(c) a council tax demand letter or statement; or
(d) a utility bill, but not a mobile telephone bill,

provided that the date on which the document in question was issued is not more than three months before the date when the scrap metal is received by the SMD.

If a SMD receives scrap metal in breach of the above provision, each of the following is guilty of an offence under s 11(4):

(a) the SMD;
(b) if the metal is received at a site, the site manager;
(c) any person who, under arrangements made by a person within (a) or (b), has responsibility for verifying the name and address.

It is a defence for a person within (a) or (b) to prove that (s)he made arrangements to ensure that the metal was not received without the necessary verification, and that (s)he took all reasonable steps to ensure that those arrangements were complied with (s 11(5)).

A person who, on delivering scrap metal to a SMD, gives a false name or false address commits an offence (s 11(7)).

An offence under s 11 is a summary offence punishable with a fine not exceeding level 3.

Offence of buying scrap metal for cash, etc

14.141 By SMDA 2013, s 12(1), a SMD must not pay for scrap metal except by:

(a) a cheque which is not transferable, or
(b) an electronic transfer of funds (authorised by credit or debit card or otherwise).

'Paying' includes paying in kind (with goods or services).

If a SMD pays for scrap metal in breach of s 12(1), each of the following is guilty of an offence under s 12(4):

(a) the SMD;

(b) if the payment is made at a site, the site manager;

(c) any person who makes the payment acting for the dealer.

It is a defence for a person within (a) or (b) to prove that (s)he made arrangements to ensure that the payment was not made in breach of s 12(1), and that (s)he took all reasonable steps to ensure that those arrangements were complied with (s 12(5)).

An offence under s 12 is a summary offence punishable with an unlimited fine.

Records: receipt of metal

14.142 The following provisions set out in SMDA 2013, s 13 apply if a SMD receives any scrap metal in the course of his/her business.

The dealer must record the following information:

(a) the description of the metal, including its type (or types if mixed), form, condition, weight, any marks identifying previous owners, and any other distinguishing features;

(b) the date and time of its receipt;

(c) if the metal is delivered in or on a vehicle, the registration mark of the vehicle;

(d) if the metal is received from a person, the full name and address of that person;

(e) if the dealer pays for the metal, the full name of the person who makes the payment acting for the dealer (s 13(2)).

If the dealer receives the metal from a person, (s)he must keep a copy of any document which (s)he uses to verify the name or address of that person (s 13(3)). If the dealer pays for the metal by cheque, (s)he must keep a copy of the cheque (s 13(4)). If the dealer pays for the metal by electronic transfer (s)he must keep the receipt identifying the transfer, or (if no such receipt was obtained) (s)he must record particulars identifying the transfer (s 13(5)).

Records: disposal of metal

14.143 SMDA 2013, s 14 requires records to be kept in respect of the disposal of scrap metal by a SMD in the course of his/her business. For these purposes metal is disposed of whether or not:

(a) it is in the same form in which it was received;

(b) the disposal is to another person;

(c) the metal is dispatched from a site (s 14(2)).

Where the disposal is in the course of business under a site licence, the dealer must record the following information:

(a) the description of the metal, including its type (or types if mixed), form, and weight;

(b) the date and time of its disposal;

(c) if the disposal is to another person, the full name and address of that person;

(d) if the dealer receives payment for the metal (whether by way of sale or exchange), the price or other consideration received (s 14(3)).

Where the disposal is in the course of business under a collector's licence, the dealer must record the following information:

(a) the date and time of the disposal;

(b) if the disposal is to another person, the full name and address of that person (s 14(4)).

Records: supplementary

14.144   The information mentioned in SMDA 2013, ss 13(2) and (5) (**14.142**) and 14(3) and (4) (**14.143**) must be recorded in a manner which allows the information and the scrap metal to which it relates to be readily identified by reference to each other (s 15(1)).

The records mentioned in s 13(3) and (4) must be marked so as to identify the scrap metal to which they relate (s 15(2)). The dealer must keep the information and other records mentioned in ss 13(2) to (5) and 14(3) and (4) for a period of three years beginning with the day on which the metal is received or (as the case may be) disposed of (s 15(3)).

If a SMD fails to fulfil a requirement under s 13, 14, or 15, each of the following is guilty of a summary offence, punishable with an unlimited fine, under s 15(4)):

(a) the SMD;
(b) if the metal is received at or (as the case may be) dispatched from a site, the site manager;
(c) any person who, under arrangements made by a person within (a) or (b), has responsibility for fulfilling the requirement.

It is a defence for a person within (a) or (b) to prove that (s)he made arrangements to ensure that the requirement was fulfilled, and took all reasonable steps to ensure that those arrangements were complied with (s 15(5)).

Right to enter and inspect

14.145   SMDA 2013, s 16 provides the following rights of entry and inspection.

A constable or an officer of a local authority (hereafter 'officer') may enter and inspect a licensed site at any reasonable time on notice to the site manager (s 16(1)). A constable or an officer may enter and inspect a licensed site at any reasonable time, otherwise than on such notice, if:

(a) reasonable attempts to give such notice have been made and have failed, or
(b) entry to the site is reasonably required for the purpose of ascertaining whether the provisions of SMDA 2013 are being complied with or investigating offences under it and (in either case) the giving of notice would defeat that purpose (s 16(2)).

Section 16(1) and (2) do not apply to residential premises (s 16(3)). There is no entitlement to use force to enter premises in the exercise of the powers under s 16(1) and (2) (s 16(4)).

By s 16(5), a justice of the peace may issue a warrant authorising entry (in accordance with s 16(7) below) to any premises within s 16(6) if (s)he is satisfied by information on oath that there are reasonable grounds for believing that entry to the premises is reasonably required for the purpose of securing compliance with the provisions of SMDA 2013, or ascertaining whether those provisions are being complied with. Premises are within s 16(6) if they are a licensed site, or they are not a licensed site but there are reasonable grounds for believing that the premises are being used by a SMD in the course of business. A warrant is in accordance with s 16(7) if it is a warrant signed by the justice which specifies the premises concerned, and authorises a constable or an officer to enter and inspect the premises at any time within one month from the date of the warrant. A constable or an officer may, if necessary, use reasonable force in the exercise of the powers under a warrant under s 16(5) (s 16(8)).

A constable or an officer may:

(a) require production of, and inspect, any scrap metal kept at any premises mentioned in s 16(1) or (2) or in a warrant under s 16(5);
(b) require production of, and inspect, any records kept in accordance with s 13 or 14 and any other records relating to payment for scrap metal;
(c) take copies of or extracts from any such records (s 16(9)).

If a constable or an officer seeks to exercise powers under s 16 in relation to any premises, and the owner, occupier or other person in charge of the premises requires him/her to produce evidence of his/her identity, or of his/her authority to exercise those powers, (s)he must produce that evidence (s 16(10) and (11)).

It is a summary offence, punishable with a fine not exceeding level 3, to obstruct the exercise of a right of entry or inspection under s 16 or to fail to produce a record required to be produced under s 16 (s 16(13)).

DISHONESTLY RETAINING A WRONGFUL CREDIT

The offence of handling is not committed by someone into whose bank account a **14.146** 'wrongful credit', as defined below, has been received because that credit itself is not stolen goods.

This gap is filled by TA 1968, s 24A which provides that a person is guilty of an offence if:

(a) a wrongful credit has been made to an account kept by him or in respect of which he has any right or interest;
(b) he knows or believes that the credit is wrongful; and
(c) he dishonestly fails to take such steps as are reasonable in the circumstances to secure that the credit is cancelled.

The offence is an indictable (either-way) offence punishable with up to 10 years' imprisonment.

What is required is a failure by D to take such steps as are reasonable in the circumstances to secure the cancellation of a wrongful credit made to an account kept by D or in respect of which D has any right or interest. Nothing need be done by D.

A 'credit' refers to a credit of an amount of money in an account. 'Money' includes currencies other than sterling. A credit to an account is wrongful to the extent that it derives from theft, blackmail, fraud (contrary to FA 2006, s 1) or stolen goods. An 'account' is an account kept with:

(a) a bank;
(b) a person carrying on a business:
  (i) in the course of which money is received by way of deposit which is lent to others; or
  (ii) any other activity of which is financed, wholly or to any material extent, out of the capital of, or the interest on, money received by way of a deposit; or
(c) an issuer of electronic money (as defined by the Electronic Money Regulations 2011).

In determining whether a credit to an account is wrongful, it is immaterial whether the account is overdrawn before or after the credit is made.

D must know or believe that the credit is wrongful; D must know or believe the facts which make the credit 'wrongful' in law, although D need not know that they have this effect (since ignorance of the criminal law is no defence).

D must dishonestly fail to take such steps which are reasonable in the circumstances to secure that the wrongful credit is cancelled. The approach to the question of dishonesty is the same as in fraud (**14.58**).

Section 24A covers a range of situations. Suppose that X commits an offence of fraud and thereby causes a money transfer to be made to D's account, unknown to D. If D dishonestly fails to take reasonable steps to cancel the credit on discovering the truth, D commits an offence under s 24A because the credit is wrongful. Another example is where Y pays money which (s)he has stolen into his/her bank account, and then transfers the credit thereby created to D's bank account. If D dishonestly fails to take reasonable steps to cancel that credit D can be convicted of an offence under s 24A, because the credit is wrongful.

CRIMINAL DAMAGE

14.147 The Criminal Damage Act 1971 (CDA 1971) deals with most offences concerned with damage to property but other legislation still exists in this area. The Malicious Damage Act 1861 still deals with offences of obstructing railways or interfering with them and of concealing or removing navigation marks or buoys. If damage is caused by an explosion the offence under the Explosive Substances Act 1883, s 2 (**19.72**) should be considered.

Destroying or damaging another's property

14.148 CDA 1971, s 1(1) provides what is known as the simple offence of criminal damage. It states that a person who without lawful excuse destroys or damages any property belonging to another intending to destroy or damage any such property, or being reckless as to whether any such property would be destroyed or damaged, is guilty of an offence. The offence is an indictable (either-way) offence punishable with up to 10 years' imprisonment. By way of qualification, where the value of the property destroyed or of the alleged damage does not exceed £5,000, a magistrates' court must proceed *as if* the offence was triable only summarily. Where a defendant (D) appears at a magistrates' court to answer charges of two or more offences of criminal damage not in excess of £5,000 which form part of a series of two or more offences of the same or a similar character, the qualification has effect as if any reference in it to the values involved were a reference to the aggregate values involved. The same qualification applies where the offence charged consists of encouraging or assisting two or more such offences. Where the qualification applies to an offence of criminal damage the maximum punishment is three months imprisonment and/or a fine not exceeding level 4.

If the destruction or damage is by fire, there is a separate indictable (either-way) offence, punishable with up to life imprisonment, under s 1(1) and (3), required to be charged as arson.

Because of the difference in maximum sentence there are technically two offences: criminal damage otherwise than by fire contrary to s 1(1) and criminal damage committed by fire, contrary to, s 1(1) and (3).

By the Theft Act 1968 (TA 1968), s 30, the leave of the DPP is required for the institution of proceedings against one spouse or civil partner for criminal damage to the other's property, unless, by virtue of any judicial decree or order, the spouses were not obliged to cohabit at the material time or an order is in force providing for the separation of the civil partners.

Criminal damage contrary to CDA 1971, s 1(1) (but not s 1(1) and (3)) is a 'penalty offence' for the purposes of the Criminal Justice and Police Act 2001, and may be dealt with under a fixed penalty procedure (**3.40**). This may be an appropriate course of action in minor cases.

What is said below about the elements of an offence of criminal damage is equally applicable to the other offences under CDA 1971, unless the contrary is indicated.

Destroy or damage

Property may be damaged if it suffers physical harm which involves permanent or temporary impairment of the property's use or value. If a motor car is scratched when a coin is scraped along its side it is damaged thereby, because its value is impaired; likewise a wall is damaged if slogans are painted on it, as is beer if water is poured into it. If part of a machine is removed, without which it cannot work, the machine may be damaged, even though neither it nor the part suffers actual physical injury, because its use is impaired. Consequently, a car may be damaged if the rotor arm is removed from its engine. **14.149**

The fact that what is done is rectifiable does not prevent the property being damaged but the amount and cost of rectification are relevant factors in determining whether there has been damage; if they are minimal it may be found that the property has not been damaged.

By CDA 1971, s 10(5), any modification of the contents of a computer is not to be regarded as damaging any computer or computer storage medium (eg a hard disc), unless its effect on the computer or medium impairs its *physical* condition. Unauthorised acts with intent to impair the operation of a computer are dealt with by the Computer Misuse Act 1990, s 3 (**15.5**).

The destruction of property involves something which goes beyond damage, such as the demolition of a machine, the pulling down of a wall or other structure, or the killing of an animal.

If arson is charged, it must also be proved that the destruction or damage was caused by fire.

Property

CDA 1971, s 10(1) defines property as being property of a tangible nature, whether real or personal, including money and: **14.150**

(a) including wild creatures which have been tamed or are ordinarily kept in captivity, and any other wild creatures or their carcasses *if*, but only if, they have been reduced into possession which has not been lost or abandoned, or are in the course of being reduced into possession; but

(b) *not* including mushrooms growing wild on any land or flowers, fruit or foliage of a plant growing wild on any land.

The term 'mushroom' includes any fungus and 'plant' includes shrubs and trees.

Thus, 'property' is defined by s 10 in a similar way to the definition of that term in TA 1968 for the purposes of theft. One difference is that land itself, which generally cannot be stolen, is not subjected to any limits on when it can be the subject of criminal damage. A man who moves his fence to capture a little of his neighbour's lawn cannot be convicted of theft of that piece of lawn, but he would be guilty of criminal damage if he damaged it. Another difference is that intangible things, such as copyright, cannot be the subject of criminal damage although they may be stolen. A third difference is that wild mushrooms and wild flowers, etc can never be the subject of criminal damage, although they can be stolen in certain circumstances. The provisions of the Wildlife and Countryside Act 1981 (**16.64–16.71**) offer protection to some wild creatures and plants in circumstances which would include forms of damage.

Belonging to another

14.151 The class of persons to whom property is treated for the purposes of CDA 1971 as belonging by s 10(2) to (4) is by no means limited to the owners of property, but includes a range of other people with a connection with it.

Section 10(2) provides that property is to be treated as belonging to any person:

(a) having the custody or control of it;

(b) having in it any proprietary right or interest (not being an equitable interest arising only from an agreement to transfer or grant an interest); or

(c) having a charge on it.

Co-owners of property each have a proprietary right or interest in the property and if one damages the property (s)he may be convicted of criminal damage since the property 'belongs to another'. The adjective 'proprietary' excludes those, such as insurance companies, which, although they have an interest in the property, do not have a proprietary interest.

The best example of a charge on property is where, to buy a house, someone mortgages it by way of charge to a building society. As a result of s 10(2), the house will belong to the building society, as well as to the house-owner, and if (s)he intentionally or recklessly damages it without lawful excuse (s)he can be convicted of criminal damage, since the house will 'belong to another'.

Section 10(3) and (4) provides that, as in the case of theft, where property is subject to a trust, the person to whom it belongs shall include any person having the right to enforce the trust; and that property belonging to a corporation sole (**14.13**) is to be treated as belonging to the corporation notwithstanding a vacancy in the corporation.

Mens rea

14.152 To do an act deliberately, eg deliberately throwing a stone, is not enough; D must have intended his/her conduct to result in property belonging to another being destroyed or damaged, or been reckless as to the risk of such destruction or damage resulting from his/her conduct.

Without lawful excuse

14.153 In order to commit an offence under CDA 1971, s 1(1) or s 1(1) and (3), D must destroy or damage another's property 'without lawful excuse'. D does not have the burden of proof in respect of a lawful excuse, only an evidential burden. For these purposes, by s 5(2), a person is to be treated as having a lawful excuse if (s)he acted with one of two types of belief, neither of which need be reasonable (although the more reasonable an alleged belief the more likely that it will be found to have been held).

Belief in consent

Section 5(2)(a) provides that a person is to be treated as having a lawful excuse if, at the time of the act or acts alleged to constitute the offence, he believed that the person or persons whom he believed to be entitled to consent to the destruction of or damage to the property had so consented, or would have so consented if he or they had known of the destruction or damage and its circumstances. **14.154**

An employee who demolished a shed in his employer's yard, wrongly believing that his foreman had instructed him to do so by saying 'Clear that yard at the back', would have a lawful excuse. If no such direction had been given, but the worker honestly believed that he would have been given permission by the foreman had he asked, this would still amount to lawful excuse.

Belief in defence of property

Section 5(2)(b) provides that a person is to be treated as having a lawful excuse if he destroyed or damaged the property in question in order to protect property belonging to himself or another, or a right or interest (such as a right of way) in property which was or which he believed to be vested in himself or another, *and* at the time of the act he believed that the property, right or interest was in immediate need of protection, and at that time he believed that the means of protection adopted or proposed to be adopted were reasonable having regard to all the circumstances. It is immaterial whether or not the threatened harm which the defendant sought to prevent was unlawful or lawful. **14.155**

The property intended to be protected, unlike that damaged, need not be tangible; it can also consist of a right or privilege in or over land, whether created by grant, licence or otherwise. Just as a person is entitled in appropriate circumstances to shoot a dog attacking his/her sheep, so (s)he is entitled to demolish a wall barring a right of way which (s)he has (or believes (s)he has).

The defence under s 5(2)(b) involves the following four elements:

(1) *D must have destroyed or damaged the property in question in order to protect property or a right or interest in it.* Whether or not D did so depends on an objective test: 'Whatever D's state of mind and assuming an honest belief, can it be said that what D did was done in order to protect particular property?' This was stated by the Court of Appeal in a case where D had set fire to bedding in an isolated part of a sheltered accommodation in order, D said, to demonstrate that the fire alarm was not working and thereby to protect the building from the risk posed to it. Applying the above test, the Court of Appeal held that it admitted of only one answer: D's act was not done to protect the building; it was not an act which in itself protected or was capable of protecting property.

(2) *The property must belong to D or another.* D cannot rely successfully on s 5(2)(b) if what D seeks to protect is neither property belonging to him/herself or another nor a right or interest in property which is (or which D believes is) vested in him/ herself or another. Thus, where D went onto farmland and destroyed badger traps in order to protect wild badgers, a divisional court held that D could not rely on s 5(2)(b) because the badgers, being wild and not reduced into possession, did not belong to anyone.

(3) *D must believe that the property or right or interest is in need of immediate protection.* The Court of Appeal has held that whether D believes that the property, etc is in need of protection is a subjective question. On the other hand, it has been held that whether the property, etc is in need of 'immediate' protection is an objective

question, to be determined by the court or jury in the light of all the circumstances as D believed them to be.

(4) *D must believe that the means adopted were reasonable.* A divisional court has held that this is a purely subjective question. The question is not whether the means adopted by D were objectively reasonable having regard to the circumstances, but whether D believed them to be so.

Lawful excuse in a general sense

14.156 Quite apart from the statutory instances of lawful excuse provided by CDA 1971, s 5, any other defence recognised by law amounts to a lawful excuse. A police officer, in executing a search warrant, may, if denied entry, break a lock, thereby committing damage. (S)he would have a lawful excuse for his/her actions because the statutes under which search warrants may be granted authorise entry by force, if necessary. Other examples of lawful excuses besides those provided s 5 are self-defence and the defence of another.

On the other hand, a belief in a moral entitlement is not a lawful excuse, nor is a belief that the destruction or damage was reasonable in pursuit of some political objective.

Destroying or damaging property with intent to endanger life or recklessness as to life being endangered

14.157 CDA 1971, s 1(2) provides that it is an offence for a person without lawful excuse to destroy or damage any property, whether belonging to himself or another:

(a) intending to destroy or damage any property or being reckless as to whether any property would be destroyed or damaged; *and*

(b) intending the destruction or damage to endanger the life of another or being reckless as to whether the life of another would be thereby endangered.

An offence under s 1(2) is generally described as aggravated criminal damage.

By s 1(3), an offence committed under s 1(2) by destroying or damaging property by fire is required to be charged as arson and is charged as contrary to s 1(2) and (3).

There are four separate aggravated offences under s 1(2):

(a) damage with intent to endanger life;

(b) damage reckless as to whether life would be endangered;

(c) arson with intent to endanger life; and

(d) arson reckless as to whether life would be endangered.

It is, of course, open to the prosecution to charge more than one count, alleging different offences in the above list, for example one alleging offence (a) and another offence (b).

An offence under s 1(2) or under s 1(2) and (3) is triable only on indictment and punishable with up to life imprisonment.

An offence under s 1(2) or under s 1(2) and (3) can be committed in respect of property which 'belongs' only to the defendant (D), and the reason for this can easily be appreciated. If D sets fire to his house which 'belongs' only to him with his wife asleep inside it, D can be charged under s 1(2) and (3).

Except for the fact that the property need not belong to another, the actus reus of an offence under s 1(2) is the same as under s 1(1).

D must either have intended to destroy or damage the property or been reckless in that respect. D must additionally have intended to endanger the life of another, or been reckless as to whether the life of another would be endangered, *by the destruction or damaging* of property which (s)he intentionally or recklessly caused; it is not enough that (s)he merely intended to endanger life, or was reckless as to whether life would be endangered, by the act which caused the destruction or damage. Consequently, a man who fires a gun from outside a house at a person standing behind a window in it cannot be convicted under s 1(2), even though he intended to endanger that person's life, if he did not intend the damaging of the window to endanger life (and was not reckless as to that damage doing so).

The provisions in s 5(2) about lawful excuse do not apply to the present offences. 'Lawful excuse' for an offence under s 1(2), or s 1(2) and (3), is therefore restricted to those defences recognised under the general law, such as self-defence and defence of another.

Threats to destroy or damage property

Threats to destroy or damage property are dealt with under CDA 1971, s 2 which states **14.158** that a person who without lawful excuse makes to another a threat, intending that that other would fear that it *would be* carried out:

(a) to destroy or damage any property belonging to that other or a third person; or
(b) to destroy or damage his own property in a way which he knows is likely to endanger the life of that other or a third person,

is guilty of an offence. The Court of Appeal has held that the italicised words 'would be' include as an alternative the notion of a fear that the threat *might* be carried out.

The threat must be to do something which would be an offence against CDA 1971, s 1. It must be carried out with the intention of inducing a fear in the mind of the recipient that it would be carried out. It does not matter how the threat is communicated, nor that the person who offers the threat does not intend to carry it out (provided that (s)he intends the recipient to fear that (s)he would do so), nor that the recipient is not actually put in fear by the threat.

Someone who communicates the threat of an explosion to another person, intending that the recipient will fear that the threat will be carried out, commits the present offence if the threat is concerned with destruction or damage to property, etc. There is often confusion with the bomb hoax offences created by the Criminal Law Act 1977, s 51(2) (**19.80**) which deals with the offence of communicating false information to induce such a fear. That offence is distinguished from the one under CDA 1971, s 2 by its limitations to false information whereas the offence under s 2 can be committed whether or not the threat involves false information.

The provisions of s 5(2) (**14.153–14.155**) relating to 'lawful excuse' apply (with a minor modification) to threats to destroy or damage another's property but do not apply to threats to destroy or damage the threatener's own property in a way likely to endanger the life of some person.

An offence under s 2 is an indictable (either-way) offence punishable with up to 10 years' imprisonment.

Possessing thing with intent to destroy or damage property

The possession of things to be used for the purpose of causing damage is dealt with by **14.159** CDA 1971, s 3 which provides that a person who has anything in his custody or under his control intending without lawful excuse to use or cause or permit another to use it to destroy or damage:

(a) any property belonging to some other person; or

(b) his own or the user's property in a way which he knows is likely to endanger the life of some other person,

is guilty of an offence.

As can be seen, the possession of the 'thing' must be for the purpose of doing something (or causing or permitting something to be done) which would be an offence under s 1. There must be a clear intention to use the thing in such a way or to cause or permit another to do so; it is not sufficient to prove that D realised that it 'might' be so used. Provided that such an intention does exist it is immaterial that there is no immediate intention to use it but only a conditional one. A terrorist group which possesses explosives in a warehouse, intending to use them to destroy or damage property if, and when, the opportunity arises can therefore be convicted of the present offence.

'Possession' is referred to in s 3 by the use of 'in his custody or under his control'. The explosives contained in a warehouse might be in the custody of the keeper of that warehouse but they might also be under the control of the leaders of the terrorist group, who can order their removal and use at any time.

The offence is sufficient to include instances of possession of a terrorist arsenal, the possession of pickaxe handles by protection racketeers, or even the possession of paint by those who intend to endorse graffiti on the walls of buildings.

Where the offence involves an intent falling within (a), ie to destroy or damage another's property, 'without lawful excuse' is subject to s 5(2) (with a minor modification), but where the intent falls within (b), ie to destroy or damage property in a way known to be likely to endanger life, it is not.

An offence under s 3 is an indictable (either-way) offence punishable with up to 10 years' imprisonment.

Police powers

14.160   CDA 1971, s 6 empowers a justice, following information given on oath, to grant a search warrant if there is reasonable cause to believe that a person (P) has in P's custody or under P's control or on P's premises anything which there is reasonable cause to believe *has been* used, or is *intended for use*, without lawful excuse to destroy or damage property belonging to another or to destroy or damage property in a way likely to endanger the life of another.

If such a search warrant is granted, a constable may enter (if need be by force) any premises and search for the thing in question. (S)he may seize anything which (s)he believes to have been used or to be intended to be used as aforesaid.

SALE OF AEROSOL PAINT TO A CHILD

14.161   The Anti-social Behaviour Act 2003, s 54(1) makes it an offence for a person to sell an aerosol paint container to a person under the age of 16.

It is a defence for D to show that (s)he took all reasonable steps to determine the purchaser's age and that (s)he reasonably believed that the purchaser was not under 16 (s 1(4)). Where the sale was effected by another person, eg an assistant, it is a defence to prove that D took all reasonable steps to avoid the commission of the offence (s 1(5)).

The offence is a summary offence punishable with a fine not exceeding level 4.

CHAPTER 15

Cyber-Crime and Computer Misuse

TYPES OF CYBER-CRIME

There are two types of cyber-crime: the use of a computer and a network in the com- **15.1** mission of a wide range of offences which do not require the use of a computer for their commission ('cyber-enabled crime'), and those offences (dealt with below) where the misuse of a computer is the essence of the offence.

COMPUTER MISUSE

The misuse of computer hardware or software may involve one or more of four offences **15.2** under the Computer Misuse Act 1990 (CMA 1990), ss 1, 2, 3 and 3ZA set out below.

Unauthorised access to computer material

By CMA 1990, s 1(1), a person commits an offence if: **15.3**

- (a) (s)he causes a computer to perform any function with intent to secure access to any program or data held in any computer;
- (b) the access (s)he intends to secure or to enable to be secured is unauthorised; and
- (c) (s)he knows at the time when (s)he causes the computer to perform the function that this is the case.

This offence covers all forms of computer hacking, including hacking by someone who simply seeks access because of the challenge of breaking through a security system. On the other hand, the offence does not cover computer eavesdropping; mere surveillance of data displayed on a VDU screen does not trigger it, since D must cause the computer to perform a function if (s)he is to be guilty.

Actual access to any program or data held in a computer is not required, since it is enough that D simply causes a computer to perform a function with intent to secure access to a program or data held in it, such intended access being unauthorised to his/her knowledge.

A person intends to secure access to a program or data if (s)he intends, by causing a computer to perform any function, to:

- (a) alter or erase the program or data;
- (b) copy or move it to any storage medium other than that in which it is held or to a different location in the storage medium in which it is held;
- (c) use it; or

(d) have it output from the computer in which it is held (whether by having it displayed or in any other manner).

The intended access need not relate to any particular program or data, nor any particular type of program or data, nor to a program or data held in any particular computer. It is immaterial whether the program or data is unauthorisedly accessed directly from the computer containing it or indirectly via another computer.

Access of any kind by any person to a program or data held in a computer is unauthorised if:

(a) (s)he is not him/herself entitled to control access of the kind in question to the program or data; and
(b) (s)he does not have consent to access by him/her of the kind in question to the program or data from any person who is so entitled.

Because the definition refers to 'access of the kind' in question, a person who has authority to view data may nevertheless unauthorisedly access it if (s)he accesses it to alter or copy it if (s)he has no authority to access for such a purpose.

Unauthorised access with intent to commit or facilitate further offence

15.4   By CMA 1990, s 2(1), a person commits an offence if (s)he commits the unauthorised access offence under s 1 with intent:

(a) to commit an offence to which s 2 applies; or
(b) to facilitate the commission of such an offence (whether by him/herself or by any other person).

Section 2 applies to any offence whose sentence is fixed by law (essentially, murder), and to any offence for which an offender of 21 or over may be sentenced to imprisonment for five years (eg fraud, theft, and forgery).

It is immaterial whether the further offence is to be committed on the same occasion as the unauthorised access offence or on any future occasion, and it is also immaterial that the facts are such that the commission of the further offence is impossible.

Unauthorised acts with intent to impair, or with recklessness as to impairing, operation of computer, etc

15.5   By CMA 1990, s 3(1), a person is guilty of an offence if:

(a) (s)he does any unauthorised act in relation to a computer;
(b) at the time when (s)he does the act (s)he knows that it is unauthorised; and
(c) either s 3(2) or (3) (below) applies.

For the purposes of s 3:

(a) a reference to doing an act includes a reference to causing an act to be done;
(b) 'act' includes a series of acts.

An act is unauthorised if the person doing the act (or causing it):

(a) is not him/herself a person who has responsibility for the computer and is entitled to determine whether the act should be done; and
(b) does not have consent to the act from any such person.

Although the owner of a computer which is able to receive email is ordinarily to be taken to consent to the sending of emails to the computer, such implied consent is not without limits. For example, a divisional court has held, it does not extend to emails which are not sent for the purpose of communicating with the owner, but are sent as a 'mail bombing campaign' for the purpose of interrupting the proper operation and use of the system.

Section 3(2) applies if the person intends by doing the act:

(a) *to impair the operation of any computer;*
(b) *to prevent or hinder access to any program or data held in the computer; or*
(c) *to impair the operation of any such program or the reliability of any such data.*

'Impairing, preventing or hindering' something includes doing so temporarily.

An example of s 3(2) would be where someone, by misusing or bypassing a password, places in the files of a computer a bogus email pretending that the password holder was the author; such an addition would result in an unauthorised alteration of the contents of the computer and would clearly be done with intent to cause an alteration of the contents, and by so doing to impair the reliability of the data on the computer.

Section 3(3) applies if the person is reckless as to whether the act will do any of the things *italicised* above.

The intention or recklessness referred to in s 3(2) or (3) above need not relate to any particular computer or any particular program or data, or a program or data of any particular kind.

Unauthorised acts causing, or creating risk of, serious damage

A person is guilty of an offence under CMA 1990, s 3ZA(1) if: **15.6**

(a) (s)he does any unauthorised act in relation to a computer;
(b) at the time of doing the act (s)he knows that it is unauthorised;
(c) the act causes, or creates a significant risk of, serious damage of a material kind; and
(d) (s)he intends by doing the act to cause serious damage of a material kind or is reckless as to whether such damage is caused.

By s 3ZA(2), damage is of a 'material kind' if it is damage to human welfare in any place, to the environment of any place, to the economy of any country or to the national security of any country. For these purposes an act causes damage to human welfare only if it causes:

(i) loss to human life;
(ii) human illness or injury; or
(iii) disruption of a supply of money, food, water, energy, or fuel, of a system of communication, of facilities for transport or of services relating to health.

It is immaterial for the purposes of s 3ZA(2) whether or not an act causing damage does so directly, or is the only or main cause of the damage.

A reference above to doing an act includes a reference to causing an act to be done, 'act' includes a series of acts, and a reference to a country includes a reference to a territory, and to any place in (or part or region of) a country or territory.

Where an offence under s 3ZA(1) is committed as a result of causing or creating a significant risk of serious damage to human welfare or serious damage to national security, a more serious offence is committed under s 3ZA(1) and (7).

Making, supplying or obtaining articles for use in an offence against CMA 1990, s 1, 3 or 3ZA

15.7   CMA 1990, s 3A provides indictable (either-way) offences preparatory to an offence under s 1, 3 or 3ZA. It states that a person commits an offence in the following three cases if (s)he:

(a) makes, adapts, supplies or offers to supply any article intending it to be used to commit, or to assist in the commission of, an offence against s 1, 3 or 3ZA;

(b) supplies or offers to supply any article believing that it is likely to be so used; or

(c) obtains any article:

    (i) intending to use it to commit, or to assist in the commission of, an offence under s 1, 3, or 3ZA, or

    (ii) with a view to it being supplied for such use.

'Article' includes any program or data held in electronic form, and therefore includes a computer password or other means by which a computer system may be accessed.

These offences are aimed at 'hacker tools' which are increasingly being used in connection with organised crime.

Trial and punishment

15.8   The offences under CMA 1990, ss 1, 2, 3, and 3A are indictable (either-way) offences. An offence under s 3ZA is triable only on indictment. The maximum term of imprisonment is:

- ss 1 and 3A: two years;
- s 2: five years;
- s 3: ten years;
- s 3ZA: 14 years (or life, in the case of the more serious offence under s 3ZA(1) and (7)).

CHAPTER 16

Animal Welfare and Protection

ANIMAL WELFARE

The Animal Welfare Act 2006 (AWA 2006) seeks to prevent animal cruelty, and to pro- **16.1** mote animal welfare, in respect of animals to which it applies, ie vertebrates other than man.

AWA 2006 refers to a person responsible for an animal, ie a person responsible for an animal whether on a permanent or a temporary basis. 'Being responsible' includes being 'in charge'. A person who owns an animal is always regarded as being a person who is responsible for it. A person is treated as responsible for any animal for which a person under 16 of whom (s)he has actual care and control is responsible.

Prevention of harm

Causing unnecessary suffering

AWA 2006, s 4(1) provides that a person commits an offence if: **16.2**

(a) an act of his, or a failure of his to act, causes an animal to suffer;
(b) he knew, or ought reasonably to have known, that the act, or failure to act, would have that effect or be likely to do so;
(c) the animal is a protected animal; and
(d) the suffering is unnecessary.

In relation to (d), a divisional court has held that s 4(1) must be interpreted so as to require that the defendant knew, or ought to have known, that the suffering was unnecessary. Thus, as the court stated, a defendant would commit no offence under s 4(1) if (s)he acted in the honest and reasonable, but mistaken, belief that what (s)he was doing was necessary for the animal's welfare.

An animal is a 'protected animal' if it:

(a) is of a kind which is commonly domesticated in the British Islands (ie the UK, the Isle of Man, Jersey, and Guernsey, Alderney, Herm, and Sark);
(b) is under the control of man whether on a permanent or a temporary basis; or
(c) is not living in a wild state.

Section 4(2) provides that a person commits an offence if:

(a) he is responsible (**16.1**) for an animal;
(b) an act, or failure to act, of another person causes the animal to suffer;

(c) he permitted that to happen or failed to take such steps (whether by way of supervising the other person or otherwise) as were reasonable in all the circumstances to prevent that happening; and

(d) the suffering is unnecessary.

A divisional court has held that an offence under s 4(2) requires knowledge of the circumstances.

Section 4(3) states that the considerations to which it is relevant to have regard when determining for the purposes of s 4 whether suffering is unnecessary include whether:

(a) the suffering could reasonably have been avoided or reduced;

(b) the conduct which caused the suffering was in compliance with any relevant enactment or any relevant provisions of a licence or code of practice issued under an enactment;

(c) the conduct which caused the suffering was for a legitimate purpose, such as
    (i) benefiting the animal, or
    (ii) protecting a person, property, or another animal;

(d) the suffering was proportionate to the purpose of the conduct concerned; or

(e) the conduct concerned was in all the circumstances that of a reasonably competent and humane person.

By s 4(3A), in determining for the purposes of s 4(1) whether suffering is unnecessary in a case where it was caused by conduct for a purpose mentioned in (c)(ii) in the preceding paragraph (protecting a person, property or another animal), the fact that the conduct was for that purpose is to be disregarded if:

(a) the animal was under the control of a relevant officer at the time of the conduct,

(b) it was being used by that officer at that time, in the course of the officer's duties, in a way that was reasonable in all the circumstances, and

(c) that officer is not the defendant.

A 'relevant officer' means a constable, a person (other than a constable) who has the powers of a constable or is otherwise employed for police purposes or is engaged to provide services for police purposes, or a prisoner custody officer within the meaning of the Criminal Justice Act 1991, Part 4 (prisoner escorts).

Nothing in s 4 applies to the destruction of an animal in an appropriate and humane manner.

The responsibilities of persons responsible for animals are extended by AWA 2006, s 9 (**16.5**) which sets out responsibilities in respect of animal welfare.

Administration of poisons, etc

16.3   AWA 2006, s 7(1) provides that a person commits an offence if, without lawful authority or reasonable excuse, he administers to (or causes to be taken by) a protected animal (**16.2**) any poisonous or injurious drug or substance, knowing it to be poisonous or injurious.

It is an offence against s 7(2) for a person who is responsible (**16.1**) for an animal:

(a) to permit another person so to act without lawful authority or reasonable excuse; or

(b) knowing that the drug or substance is poisonous or injurious, to fail to take such steps (whether by way of supervising the other person or otherwise) as were reasonable in all of the circumstances to prevent the other person so acting.

A drug or substance may be poisonous or injurious by virtue of the manner or the quantity in which it is administered or taken.

Fighting

AWA 2006, s 8(1) provides that it is an offence for a person to: **16.4**

(a) cause an animal fight to take place, or attempt to do so;
(b) knowingly receive money for admission to an animal fight;
(c) knowingly publicise a proposed animal fight;
(d) provide information about an animal fight with the intention of enabling or encouraging attendance at the fight;
(e) make or accept a bet on the outcome of an animal fight or on the likelihood of anything occurring or not occurring in the course of an animal fight;
(f) participate in an animal fight;
(g) possess anything designed or adapted for use in connection with an animal fight with the intention of its being so used;
(h) keep or train an animal for use for or in connection with an animal fight; or
(i) keep any premises for use for an animal fight.

A divisional court has held that the involvement of money is not a necessary ingredient of an offence under s 8(1)(a), (c), (d), (f), (g), (h), or (i). A judge in the Administrative Court has held that 'keeping' in s 8(1)(h) covers someone who retains control of the animal whilst it is elsewhere in the home or place of another, as well as someone who has actual physical possession of the animal. Equally, the judge held, s 8(1)(h) covers a person who arranges for another to carry out the training of an animal to fight, as well as someone who him/herself trains an animal to fight.

For the purposes of s 8, an 'animal fight' means an occasion upon which a protected animal (as defined in the material above relating to s 4) is placed with another animal, or with a human, for the purpose of fighting, wrestling, or baiting (s 8(7)). A divisional court has held that there are two key elements in order for a protected animal to be placed with another animal for these purposes: physical proximity and control. 'Physical proximity' means that the other animal must be immediately present. 'Control' means that the other animal cannot escape. Therefore, the court held, letting dogs loose so that they can hunt and chase protected animals does not amount to a 'placing' within s 8(7).

It is an offence under s 8(2) for a person, without lawful authority or reasonable excuse, to be present at an animal fight.

Surveillance is essential to proof of an offence under s 8(1) or (2), and a record of what occurred and of the particular activities of those who attended must be made for evidential purposes.

Failure to take reasonable steps to ensure animal's welfare needs are met

AWA 2006, s 9(1) creates an offence which can be committed by a person in relation **16.5** to the welfare of animals for which (s)he is responsible (**16.1**). A person commits the offence if (s)he does not take such steps as are reasonable in all the circumstances to ensure that the needs of the animal for which (s)he is responsible are met to the extent required by good practice. A divisional court has held that s 9(1) sets a purely objective standard of care which a person responsible for an animal must provide. By s 9(2), the needs referred to in s 9(1) include the need for a suitable environment; the need for a suitable diet; the need to be able to exhibit normal behaviour patterns; any need to be housed with, or apart from, other animals; and the need to be protected from pain, suffering, injury, and disease.

In relation to this offence, the circumstances to which it is relevant to have regard include, in particular, any lawful purpose for which the animal is kept, and any lawful activity undertaken by the animal (s 9(3)). This will protect those who keep animals for use within various lawful activities against prosecution in the event of a complaint from an animal rights group to the effect that the manner of the keeping of animals, the environment within which they are kept, or their ability to display normal behaviour patterns, are unsatisfactory.

Section 9 does not apply to the destruction of an animal in an appropriate and humane manner.

A divisional court has held that if someone is convicted of an offence under AWA 2006, s 4 there should not be a separate conviction for an offence under s 9 where the neglect proved under s 9 is no wider than the conduct which caused the unnecessary suffering for which there is guilt under s 4, because the conduct would be entirely subsumed within the conduct giving rise to guilt of the more serious offence under s 4.

Improvement notices

16.6   AWA 2006, s 10 provides that an inspector (person appointed to be an inspector by a national authority or a local authority) may serve a notice on a person requiring him to take specified steps to improve if he is of the opinion that this person is not meeting the requirements of s 9. Where such a notice is served there can be no proceedings under s 9 before the end of the period for improvement specified in the notice. If the appropriate steps are taken, that is the end of the matter.

Powers

Powers in relation to animals in distress

16.7   Under AWA 2006, s 18(1), where an inspector (**16.6**) or constable reasonably believes that a protected animal (**16.2**) is suffering, (s)he may take, or arrange for the taking of, such steps as appear to be immediately necessary to alleviate the animal's suffering. This does not include the destruction of an animal (s 18(2)). However, if it appears that the condition of the animal is such that there is no reasonable alternative to destroying it and that the need for action is so pressing that it is not reasonably practicable to wait for a veterinary surgeon, an inspector or constable may destroy it, or arrange for its destruction (s 18(4)). In other circumstances a veterinary surgeon must certify that the condition of the protected animal necessitates destruction in its own interests. If (s)he does so, an inspector or constable may destroy the animal, or arrange for this to be done (s 18(3)).

An inspector or constable is empowered by s 18 to take a protected animal (and its dependent offspring) into his/her possession where a veterinary surgeon certifies that it is suffering or that it is likely to suffer if circumstances do not change (s 18(5) and (7)). An inspector or constable may so act without such certification where it appears that the need for action is so pressing that it is not reasonably practicable to wait for a veterinary surgeon to issue a certificate (s 18(6), as interpreted by a divisional court). Where such an animal is taken into possession an inspector or constable may remove it, or make arrangements for it to be removed, to a place of safety, or may care for it, or arrange for such care to be provided (s 18(8)). On the application of the owner, or any other person appearing to have an interest, a magistrates' court may make an order in relation to an animal so taken into possession (s 20).

Where actions are carried out under s 18 without the knowledge of a person responsible (**16.1**) for the animal, (s)he must be informed (by the person who acted) of the action taken as soon as reasonably practicable (s 18(11)).

It is an offence under s 18(12) intentionally to obstruct a person exercising his/her powers under s 18.

Powers of entry for the purposes of s 18

16.8 AWA 2006, s 19(1) empowers an inspector (**16.6**) or constable to enter premises for the purpose of searching for a protected animal and of exercising any power given by s 18 if (s)he reasonably believes that a protected animal is on the premises and that the animal is suffering or, if the circumstances of the animal do not change, is likely to suffer. 'Premises' includes any place, vehicle, vessel, aircraft, hovercraft, tent, or movable structure. Reasonable force may be used if it appears that entry is required before a warrant can be obtained. The power of entry does not extend to a private dwelling.

By s 19(4), a justice of the peace may, on the application of an inspector or constable, issue a warrant authorising an inspector or constable to enter premises, by force if necessary, for the purpose mentioned in s 19(1) if there are reasonable grounds for believing that a protected animal is on the premises and that the animal is suffering or is likely to suffer if its circumstances do not change. The justice must be satisfied that one of the following four conditions exists:

(a) the whole of the premises is used as a private dwelling and the occupier has been informed of the decision to apply for a warrant;

(b) no part of the premises is used as a dwelling and the occupier:
  (i) has been informed of the decision to seek entry to the premises and the reasons for that decision,
  (ii) has failed to allow entry to the premises on being requested to do so by an inspector or a constable, and
  (iii) has been informed of the decision to apply for a warrant;

(c) the premises are unoccupied or the occupier is absent and notice of intention to apply for a warrant has been left in a conspicuous place on the premises; or

(d) it is inappropriate to inform the occupier of the decision to apply for a warrant because this would defeat the object of entering the premises, or entry is required as a matter of urgency.

Seizure of animals involved in fighting offences

16.9 AWA 2006, s 22 empowers a constable to seize any animal if it appears to him/her that it is one in relation to which an offence under s 8(1) (causing, etc animal fighting) or s 8(2) (presence at animal fight) has been committed. (S)he may enter and search premises (**16.8**) for this purpose if (s)he reasonably believes that there is an animal on the premises and it is one in respect of which that power of seizure applies. Section 22 does not authorise such entry to a private dwelling.

Entry and search under warrant in connection with offences

16.10 By AWA 2006, s 23, a justice of the peace may issue a warrant authorising an inspector (**16.6**) or constable to enter premises (**16.8**), by force if necessary, in order to search for evidence of the commission of an offence under ss 4 to 9 (offences relating to prevention of harm) or 34(9) (breach of a disqualification order: (**16.15**)) if (s)he has reasonable grounds for believing that such an offence has been committed, or that evidence of the commission of such an offence is to be found on the premises.

Entry for purposes of arrest

16.11   PACE, s 17 (power of constable to enter and search premises for the purpose of arresting a person for offence under specified enactments: **4.63**) applies to offences against any of AWA 2006, ss 4, 7, and 8 (**16.2–16.4**).

Power to stop and detain vehicles

16.12   AWA 2006, s 54 empowers a constable in uniform, or an inspector (**16.6**) who is accompanied by such a constable, to stop and detain a vehicle for the purpose of entering and searching it in the exercise of his/her powers, or the exercise of powers under a warrant, under s 19 (**16.8**) or s 23 (**16.10**) or (constable in uniform only) under s 22 (**16.9**).

Prosecution, trial, and punishment

16.13   It is expressly provided by AWA 2006, s 30 that a local authority may prosecute proceedings for any offence under AWA 2006. Although there is no similar provision in respect of the RSPCA, it frequently brings a private prosecution for an offence under the Act.

Section 31 provides that a magistrates' court may try an offence under AWA 2006 if the written charge is issued or information is laid within:

(a)   three years of with the date of the commission of the offence; and

(b)   six months of the date on which evidence which the prosecutor thinks is sufficient to justify the proceedings comes to his/her knowledge.

An offence under s 4, 7, or 8 (**16.2–16.4**) is an indictable (either-way) offence punishable with up to five years' imprisonment. The other AWA 2006 offences dealt with in this chapter are summary offences (punishable with up to six months' imprisonment *and/or* a fine—either an unlimited fine (ss 9 (**16.5**) and 34(9) (**16.15**)) or one not exceeding level 4 (s 18(12) (**16.7**)).

Post-conviction powers

Deprivation

16.14   Where a person convicted of an offence under any of AWA 2006, s 4, 7, 8, 9, or s 34(9) is the owner of the animal, the court by or before which (s)he is convicted may, instead of or in addition to dealing with him/her in any other way, make an order depriving him/her of ownership of the animal and for its disposal (s 33).

Disqualification orders

16.15   Where a person has been convicted of an offence under AWA 2006, s 4, 7, 8, 9, or 34(9), a disqualification order may be made by the court in relation to animals generally, or in relation to animals of one or more kinds.

A disqualification order under s 34 may disqualify the offender under any one or more of s 34(2), 34(3), or 34(4) for such period as the court thinks fit. Disqualification under s 34(2) disqualifies the offender from owning animals, from keeping animals, from participation in the keeping of animals, and from being party to arrangements under which (s)he is entitled to control or influence the way in which animals are kept. Disqualification under s 34(3) disqualifies a person from dealing in animals. Disqualification under s 34(4) disqualifies a person from transporting animals and from arranging to do so. Breach of a disqualification order is an offence under s 34(9); see further **16.13**.

Section 35 permits the inclusion in an order of a power to seize animals which the person concerned owns or keeps.

A person who is disqualified under s 34 may apply to the court which made the order or, in the case of an order made by a magistrates' court, another magistrates' court in the same local justice area for the termination of the order. No application may be made within one year of the date on which the order was made or within any period specified by the court, nor subsequently within a one-year period after an unsuccessful application.

Destruction in the interests of an animal

Where conviction follows an offence under AWA 2006, s 4, 7, 8, or 9, the court may order the destruction of the harmed animal if it is satisfied, on the basis of evidence given by a veterinary surgeon, that it is appropriate to do so in the interest of the animal (s 37). It may also order the forfeiture and destruction of any equipment used in such offences (s 40).   **16.16**

BADGERS AND OTHER WILD ANIMALS

Badgers

Killing, injuring, or taking

Badgers are not domestic animals and are therefore not protected by AWA 2006, ss 4 and 7 (16.2 and 16.3). However, the Protection of Badgers Act 1992 (PBA 1992), s 1(1) gives special protection to badgers by creating offences of wilfully killing, injuring, or taking any badger, or attempting to do one of these things, except as permitted by the Act. Section 1(2) provides that if, in any proceedings for attempting to kill, take, or injure a badger, there is evidence from which it could reasonably be concluded that at the material time the defendant was doing so, (s)he must be presumed to have been doing so, unless the contrary is shown.   **16.17**

If a person is found in possession of a dead badger, or any part of it, (s)he is guilty of an offence under s 1(3), unless (s)he can show that the badger had not been killed in contravention of the provisions of the Act, or that the badger, or part, had been sold (whether to him/her or another) and, at the time of the purchase, the purchaser had no reason to believe that the badger had been killed in contravention of the Act.

Farmers and others who show that the action in question is necessary to prevent serious damage to land, crops, poultry, or other property are exempted by PBA 1992, s 7 from an offence of killing, taking or injuring under s 1. This exception is subject to a major qualification; if it was apparent before the time that the action was taken that it would prove necessary to prevent such damage (ie there is not a situation of urgency), a person is not exempted if (s)he had not applied for a licence under s 10 for this purpose as soon as reasonably practicable after the fact became apparent or if an application for such a licence had been determined.

Licences may be granted under s 10 by the Secretary of State (Welsh Ministers, in Wales) to permit killing or taking or interfering with a sett to prevent the spread of disease or serious damage to crops. Licences may also be so granted by the appropriate conservation body (Natural England and the Natural Resources Body for Wales) to allow taking for scientific purposes, zoological needs, or for ringing or marking. Acts authorised by such a licence are of course exempted from being an offence.

Cruelty

16.18   Offences of cruelly ill-treating, using badger tongs on, or digging for badgers are dealt with by PBA 1992, s 2(1). Section 2(2) provides, in relation to the offence of digging for a badger, that if there is evidence from which it could reasonably be concluded that the defendant was digging for a badger, he must be presumed to have been doing so, unless the contrary is shown.

In addition, s 2(1) also provides that the use, for the purpose of killing or taking a badger, of a firearm, other than a smooth-bore weapon of not less than 20 bore or a rifle using ammunition having a muzzle energy not less than 160 foot pounds and a bullet weighing less than 38 grains, is an offence.

Interfering with setts

16.19   PBA 1992, s 3 creates an offence of interfering with a badger sett by intentionally or recklessly damaging it, destroying it, obstructing access to it, causing a dog to enter it or disturbing a badger when it is occupying it. A 'badger sett' for the purposes of s 3 means the tunnel and chambers created by badgers and currently used by them, and their entrances. The term may also apply in other circumstances, as where badgers occupy coverts or disused sheds as their shelter or refuge. By s 8(1), a person is not guilty of an offence under s 3 if (s)he shows that his/her action is necessary to prevent serious damage to land, crops, poultry or other property. The same exception applies as under s 7 (**16.17**).

Under the provisions of the Regulatory Enforcement and Sanctions Act 2008 (RESA 2008), Part 3 (ss 36–65), as implemented in respect of PBA 1992, s 3 by the Environmental Civil Sanctions (England) Order 2010 and the Environmental Civil Sanctions (Wales) Order 2010, the Environment Agency, the Natural Resources Body for Wales and Natural England, as the specified regulators under RESA 2008, may deal with an offence under s 3 by imposing a civil sanction of the following types: a fixed monetary penalty; a variable monetary penalty; a compliance notice; a restoration notice; a stop notice; or an enforcement undertaking. Where official action is required in respect of an offence under s 3 it will be by way of a civil sanction, unless the breach of s 3 is a serious one (in which case a prosecution will be instituted).

Selling and possession of live badger

16.20   Under PBA 1992, s 4 it is an offence to sell, offer for sale, or have live badgers in one's possession or control.

A person is not guilty of an offence under s 4 by reason of having possession or control of a live badger if it is in his/her possession, in the course of his/her business as a carrier, or if it has been disabled otherwise than by his/her act and taken solely to give it the necessary tending until it has recovered.

General exceptions

16.21   There are general exceptions under PBA 1992, s 6 from liability for an offence under PBA 1992 for those who either take an injured badger to tend it or kill it as an act of mercy, and for those who unavoidably kill or injure a badger as an incidental result of lawful action.

Police powers

Where a constable has reasonable grounds for suspecting that a person is committing **16.22** an offence under PBA 1992, or has committed such an offence, and that evidence is to be found on that person, or in any vehicle or article (s)he has with him/her, the constable may (a) without warrant stop and search that person, vehicle, or article, and (b) seize and detain anything which may be evidence of the commission of such an offence (s 11).

Prosecution, trial, and punishment

The same prosecution time limit as in **16.13** applies to an offence under PBA 1992, **16.23** save that two years is substituted for three (s 12ZA(1) and (2)). The above offences are summary. Offences against s 1(1) or (3), 2, or 3 are punishable with up to six months' imprisonment and/or an unlimited fine. An offence under s 4 is punishable with an unlimited fine. The court must order the forfeiture of any badger or badger skin in respect of which the offence was committed and may, if it thinks fit, order the forfeiture of any weapon or article in respect of or by means of which the offence was committed.

Disqualification

Where a dog is used in the commission of an offence of killing, taking, etc, cruelty, or **16.24** interference with setts, a court may order its destruction, or may disqualify the offender from keeping or having the custody of a dog for such period as it thinks fit, or both (PBA 1992, s 13).

Cruelty to wild mammals

The Wild Mammals (Protection) Act 1996 (WM(P)A 1996), s 1 makes it a summary **16.25** offence for any person to mutilate, kick, beat, nail, or otherwise impale, stab, burn, stone, crush, drown, drag, or asphyxiate any wild mammal with intent to inflict unnecessary suffering upon it. A 'wild mammal' is any mammal which is not a protected animal within the meaning of AWA 2006 (**16.2**). WM(P)A 1996, s 2 provides exemptions from s 1 in respect of:

(a) the attempted mercy-killing of a wild mammal which has been so seriously disabled (otherwise than by his/her unlawful act) that there is no chance of recovery;
(b) the reasonably swift and humane killing of a wild mammal injured or taken in the course of either lawful shooting, hunting, coursing, or pest control;
(c) anything authorised under any enactment;
(d) any act made unlawful by s 1 if it was by means of a snare, trap, dog, or bird lawfully used for the purpose of killing or taking any wild mammal; or
(e) the lawful use of a poison or noxious substance.

For the purposes of WM(P)A 1996, s 2 hunting a wild mammal with a dog is treated as lawful if, and only if, it is exempt hunting within the meaning of the Hunting Act 2004 (below).

Some of these exemptions appear unnecessary. Those who fall within (a)–(c), for example, are unlikely to satisfy the terms of s 1.

WM(P)A 1996, s 4 provides that where a constable has reasonable grounds for suspecting that a person has committed an offence under WM(P)A 1996 and that evidence of the commission of the offence might be found on that person or in or on any vehicle which (s)he may have with him/her, the constable may:

(a) without warrant, stop and search that person and any vehicle or article (s)he may have with him/her; and

(b) seize and detain for the purposes of proceedings under any of those provisions anything which may be evidence of the commission of the offence or may be liable to be confiscated by a court under WM(P)A 1996, s 6.

An offence under s 1 is punishable with up to six months' imprisonment and/or an unlimited fine. In addition, the court may order the confiscation of any vehicle or equipment used in the commission of the offence.

Hunting

Hunting wild mammals with dogs

16.26   The Hunting Act 2004 (HA 2004), s 1 makes it an offence for a person to hunt a wild mammal with a dog, unless such hunting is exempt. A divisional court has held that 'hunt' does not include searching for a wild mammal for the purpose of stalking or flushing it out. A 'wild mammal' is any mammal living wild or a wild mammal which has been bred or tamed or is (or has been) captive. Exemptions ('exempt hunting') are specified in Sch 1.

Exempt hunting

16.27   *Stalking and flushing out*   Under Sch 1, para 1, stalking and flushing out a wild mammal is exempt if carried out in accordance with five prescribed conditions:

(1) The stalking or flushing out must be undertaken for the purpose of:
  (a) preventing or reducing serious damage which such a mammal would otherwise cause to livestock, game birds, food for livestock, crops, growing timber, fisheries or other property, or the biological diversity of an area;
  (b) obtaining meat for human or animal consumption; or
  (c) participation in a field trial (ie a competition in which dogs flush animals out of cover or retrieve shot animals so as to assess the dog's usefulness in connection with shooting).

(2) The stalking or flushing out must take place on land which belongs to the person carrying out the activity, or in respect of which permission has been granted for that purpose by the occupier or land owner.

(3) Not more than two dogs may be used for the purpose of stalking or flushing out.

(4) The stalking or flushing out must not involve the use of a dog below ground otherwise than in accordance with Sch 1, para 2 below.

(5) (a) (i) Reasonable steps must have been taken to ensure that the animal concerned is shot dead by a competent person; and
    (ii) each dog used in the stalking or flushing out must be kept under sufficient control to ensure that it does not prevent or obstruct the achievement of the objective in (a); or
  (b) In so far as stalking or flushing out is undertaken below ground in accordance with Sch 1, para 2 (below), the following is substituted for (a)(i) and (ii):
    (i) reasonable steps must be taken to ensure that as soon as possible after being found the animal is flushed out from below ground; and
    (ii) reasonable steps must be taken to ensure that as soon as possible after being flushed out, it is shot dead by a competent person; and
    (iii) in particular, the dog must be brought under control; and
    (iv) reasonable steps must be taken to prevent injury to the dog; and

(v) the manner in which it is used must comply with the approved Code of Practice for the Use of a Dog Below Ground in England and Wales.

Use of dogs below ground to protect birds for shooting   Schedule 1, para 2 provides that the use of a dog below ground in the course of stalking or flushing is exempt if:

(a) the stalking or flushing out is undertaken for the purpose of protecting game birds or wild birds which a person is keeping or preserving for the purpose of shooting;
(b) only one dog is used below ground at any one time; and
(c) the stalker or flusher out carries written evidence of ownership of the land or of permission by the occupier or owner of the land so to act, and makes the evidence immediately available for inspection by a constable on request.

Other exemptions   Exemptions are provided in respect of the hunting of rats, rabbits, and hares. There are also exemptions in respect of the flushing-out of a wild mammal for the purposes of falconry, in respect of the hunting of a wild mammal in order to recapture it or rescue it, and in respect of the observation or study of a wild mammal. These exemptions are made by Sch 1, paras 3–9.

General   A divisional court has ruled that D only has an evidential burden in respect of an exemption; ie D is only required to adduce evidence raising the issue of whether or not the hunting was exempt whereupon the prosecution must prove beyond reasonable doubt that the hunting was not exempt.

By s 4, it is a defence for D to show that (s)he reasonably believed that the hunting was exempt.

Hunting: assistance

16.28   HA 2004, s 3(1) provides that a person commits an offence if he knowingly permits land which belongs to him to be entered or used in the course of the commission of an offence against s 1. Section 3(2) provides that a person commits an offence if he knowingly permits a dog which belongs to him to be used in the course of the commission of such an offence. For the purposes of HA 2004, a dog belongs to a person if (s)he owns it, is in charge of it, or has control of it.

Hare coursing

16.29   A person commits an offence against HA 2004, s 5(1) if (s)he participates in a hare-coursing event (ie a competition in which dogs are, by the use of live hares, assessed as to skill in hunting hares), or attends, or knowingly facilitates such an event, or permits land which belongs to him/her to be used for such an event. By s 5(2), each of the following commits an offence if a dog participates in a hare coursing event: the person who enters the dog, the person who permits the dog to be entered, and the person who controls or handles the dog in the event.

Search and seizure

16.30   By HA 2004, s 8, a constable who reasonably suspects that a person is committing or has committed any of the offences under HA 2004 set out above has the following powers. If (s)he reasonably believes that evidence of the offence is likely to be found in or on a vehicle, animal or other thing of which the suspect appears to be in possession or control, (s)he may stop and search the vehicle, animal or other thing. (S)he may seize and detain such a vehicle, etc if (s)he reasonably believes that it might be used in evidence or made subject to a forfeiture order. A constable may without a warrant enter land, premises (other than a dwelling) or a vehicle for the purpose of exercising these powers.

Trial and punishment

16.31   Offences under HA 2004 are summary and punishable with an unlimited fine (HA 2004, s 6).

GAME

Day poaching

16.32   The Game Act 1831 (GA 1831), s 30 provides a summary offence, punishable with a fine not exceeding level 3, of poaching by day. It provides that anyone who trespasses by entering or being upon land in the daytime in search or pursuit of game (as defined at 18.34), woodcock, snipe or rabbits commits an offence. By s 32, an aggravated summary offence punishable with an unlimited fine is committed where five or more are involved and any of them is armed with a gun and uses violence, intimidation or menaces to prevent anyone exercising his/her powers under the Act. 'Daytime' begins one hour before sunrise and ends one hour after sunset.

Police powers

16.33   A constable has power to require any trespasser in search or pursuit of game to quit and give his/her name and address (GA 1831, s 31A). If a constable has reasonable cause for suspecting that a person is committing the offence of trespassing in pursuit of game in the daytime (s)he may enter land for the purpose of exercising this power (Game Laws (Amendment) Act 1960 (GL(A)A 1960), s 2). Where a person is lawfully arrested by a constable for an offence under GA 1831, s 30, a constable by or in whose presence (s) he was arrested may search him/her and may seize and detain any game or rabbits, or any gun, part of a gun, or cartridges or other ammunition, or any nets, traps, snares, or other devices of a kind used for the killing or taking of game or rabbits, which are found in his/her possession (GL(A)A 1960, s 4).

Night poaching

16.34   The Night Poaching Act 1828 (NPA 1828) is concerned with summary offences of poaching by night. Night commences one hour after sunset and continues until one hour before sunrise.

NPA 1828, s 1 creates two offences:

(a) by night, unlawfully taking or destroying any game or rabbits on land, open or enclosed (*including a public road, highway or path*); and

(b) by night, unlawfully entering or being on any land, open or enclosed, with any gun, net, engine or other instrument for the purpose of taking or destroying game there by night.

'Game' means hares, pheasants, partridges, grouse, heath or moor game, black game, or bustards. It will be noted that, unlike (a), (b) does not apply to rabbits. The italicised words in (a) do not apply to (b). Both offences are punishable with a fine not exceeding level 3. By s 2, anyone committing an offence under s 1 who assaults or offers violence with a gun or offensive weapon towards someone authorised to arrest him/her commits a summary offence punishable with up to six months' imprisonment and/or a fine not exceeding level 4.

By s 9, if any persons, to the number of three or more, by night unlawfully enter or are on/in any land whether open or enclosed, for the purpose of taking or destroying game or rabbits, any of them being armed with any gun, crossbow, firearms, bludgeon, or any other offensive weapon, each of them is guilty of a summary offence punishable with up to six months' imprisonment and/or a fine not exceeding level 4.

Police powers

A constable who has reasonable grounds for suspecting that an offence under NPA **16.35** 1828, s 1 or s 9 is being committed may enter land to deal with the offence (GL(A)A 1960, s 2). By GL(A)A 1960, s 4, where a person is lawfully arrested by a constable for an offence under NPA 1828, s 1 or s 9 a constable by or in whose presence (s)he was arrested may search him/her and may seize and detain any game or rabbits, or any gun, part of a gun, or cartridges or other ammunition, or any nets, traps, snares, or other devices of a kind used for the killing or taking of game or rabbits, which are found in his/her possession.

Powers of search and seizure

If a constable in any highway, street, or public place has good cause to suspect a person **16.36** of coming from land where (s)he has been unlawfully in search or pursuit of game, and of having in his/her possession any game unlawfully obtained, or any device of a kind used for the killing and taking of game, (s)he may, by the Poaching Prevention Act 1862, s 2, search that person or an accomplice of his/her. (S)he may also stop and search conveyances. The constable may seize and detain any game or other things connected with poaching which (s)he finds. Dogs and ferrets may not be seized.

If it is proved that such a person as just mentioned:

(a) has obtained the game by unlawfully going on any land in search or pursuit of game;
(b) has used any article or thing for unlawfully killing or taking game; or
(c) has been an accomplice to conduct of the type mentioned in (a) or (b),

(s)he is guilty under s 2 of a summary offence (maximum punishment a level 3 fine).

Rights to take game

In the first instance the right to take game rests with the landowner. However, if (s)he **16.37** leases the land, the right to take game on it automatically passes to the tenant, unless the landowner expressly reserves that right. Frequently the person with the right to take game may let the 'shooting rights' to another person or to a syndicate of persons.

The Ground Game Act 1880, s 1 provides that an occupier of land will always have the right to kill and take ground game (ie hares and rabbits) on the land which (s)he occupies, whether or not the landowner has contracted to some other person game rights generally. The occupier may also give written authorisation to other people to kill and take ground game, but (s)he may only authorise (again in writing) one person, in addition to him/herself, to kill ground game with firearms. Authorisations of these types may only be given by the occupier to his/her own resident household, people in his/her ordinary employment on the land, and one other person bona fide employed by him/her in the taking and destruction of ground game.

Close season for game

16.38 GA 1831, s 3 prohibits as an offence taking or killing game on Sundays or Christmas Day. Section 3 also prohibits as an offence the taking or killing of particular types of game birds during their close seasons, which are as follows:

(a) partridges—1 February to the following 1 September;
(b) pheasants—1 February to the following 1 October;
(c) black game—10 December to the following 20 August (1 September in Somerset, Devon, or New Forest);
(d) grouse—10 December to the following 12 August;
(e) bustard—1 March to the following 1 September.

An offence under s 3 is a summary offence punishable with a fine not exceeding level 1.

DEER

Deer poaching

16.39 By the Deer Act 1991 (DA 1991), s 1(1), entry onto land without the consent of the owner or occupier or other lawful authority in search or pursuit of any deer with the intention of taking, killing, or injuring it is an offence.

An offence is also committed by someone who:

(a) intentionally takes, kills, or injures any deer (or attempts to do so);
(b) searches for or pursues any deer with the intention of taking, killing, or injuring it; or
(c) removes the carcase of any deer (s 1(2)).

A person is not guilty of any of these offences if his/her act is done in the belief that:

(a) he would have had the consent of the owner or occupier of the land if such person knew of his/her doing it and the circumstances of it; or
(b) he has other lawful authority to do it (s 1(3)).

Taking or killing of deer: in close season or at night

16.40 It is an offence under DA 1991, s 2(1) to take or intentionally kill a deer during its close season. Deer farmers are permitted to take or kill their deer out of season, but their deer must be conspicuously marked. By s 7, authorised persons (occupiers of land, persons with shooting rights, and persons authorised by them) may shoot deer out of season in protection of crops, etc.

Schedule 1 provides close seasons for Chinese water, red, fallow, roe, and sika deer as follows (all dates are inclusive):

(a) Chinese water deer (*Hydropotes inermis*)—
Buck 1 April to 31 October;
Doe 1 April to 31 October;

(b) fallow deer (*Dama dama*)—
Buck 1 May to 31 July;
Doe 1 April to 31 October;

(c) red deer (*Cervus elaphus*)—
Stags 1 May to 31 July;
Hinds 1 April to 31 October;

(d) red/sika deer hybrids—
Stags 1 May to 31 July;
Hinds 1 April to 31 October;

(e) roe deer (*Capreolus capreolus*)—
Buck 1 November to 31 March;
Doe 1 April to 31 October;

(f) sika deer (*Cervus nippon*)—
Stags 1 May to 31 July;
Hinds 1 April to 31 October.

Other offences

DA 1991, s 4(1) makes it an offence to set a trap, snare, or poisoned or stupefying bait, calculated to inure any deer coming into contact with it, or to use such a thing for the purpose of taking or killing any deer (unless, in the case of a trap or net, this is done to prevent suffering to an injured or diseased deer). **16.41**

By s 4(2), it is an offence to use an arrow, spear, poisoned etc missile, smooth-bore gun (other than a specified type of slaughtering instrument), gun having a calibre of less than .240 inches, air gun, air pistol, air rifle, cartridge for a smooth-bore gun, or bullet other than one which is soft- or hollow-nosed, for the purpose of taking or killing or injuring any deer. This does not apply to the use as a slaughtering instrument of a smooth-bore gun of 12 bore or more, with a 24-inch (or more) barrel and loaded with a cartridge whose shot is .203 inches or more in diameter. The use of a rifle with a calibre of .220 inches or more and a muzzle energy of 1,356 joules, and a soft- or hollow-nosed bullet of 3.24 grammes or more, to take or kill a Chines water deer or muntjac is not an offence.

There is a mercy-killing defence to s 4(1) or (2).

Section 4(4) makes criminal the discharge at deer of firearms from a moving mechanically propelled vehicle (or one the engine of which is running) or the use of such vehicles for the purpose of driving deer.

Police powers

DA 1991, s 12 provides a constable, who suspects with reasonable cause that any person is committing or has committed any offence under the Act, with a power, without warrant, to search or examine persons, vehicles, animals, weapons, or other things on whom or on which (s)he reasonably suspects evidence of the offence is to be found, and to seize and detain anything which is evidence of an offence and any deer, venison, vehicle, animal, weapon, or other thing liable to be forfeited by a court under DA 1991, s 13 on conviction of an offence under DA 1991 (forfeiture of deer or venison in respect of which the offence committed, or of anything used to commit the offence (or capable of being so used) and found in offender's possession). For the purpose of exercising the above powers, or of exercising his/her power of arrest, a constable may enter any land other than a dwelling house. **16.42**

Prosecution, trial, and punishment

16.43 The time limit for a prosecution for any of the above offences under DA 1991 is the same as that in **16.13**, except for the substitution of two years for three (s 9(3) and (4)). Such an offence is a summary offence (maximum punishment three months' imprisonment and/or a level 4 fine). In addition, the court may:

(a) order the forfeiture of:
   (i) any deer or venison in respect of which the offence was committed or which was found in D's (the offender's) possession;
   (ii) any vehicle, animal, weapon, or other thing which was used to commit the offence or which was capable of being used to take, kill, or injure deer and was found in D's possession; and

(b) may cancel any firearm or shotgun certificate held by D.

DISEASES OF ANIMALS

16.44 The Animal Health Act 1981 (AHA 1981) sets out the duties of various persons in relation to notification of certain animal diseases and the action which must be taken. The administrative responsibilities lie with the Secretary of State (in Wales, the Welsh Ministers) but local authorities must appoint their own inspectors. Constables are frequently appointed as inspectors under the Act and when this occurs they have additional duties and responsibilities to those which all constables have under the Act.

Animals for the purposes of AHA 1981 are cattle, sheep, goats, and all other ruminating animals, swine (ie pigs), horses, asses, mules, and jennets (ie small Spanish horses). The Secretary of State (in Wales, the Welsh Ministers) has power, by order, to add to this list and, for the purposes of the special provisions relating to rabies mentioned at **16.46**, dogs and cats have been added to the definition of 'animal'.

The Act and orders thereunder specify the diseases with which they are concerned, the most important of which are foot-and-mouth disease, swine fever, and rabies. Persons having animals affected by a specified disease must so far as possible separate them from unaffected animals and speedily inform a constable. On receipt of such information a constable must forthwith inform a local authority inspector and the divisional veterinary inspector.

The action which follows is very much related to the particular disease, but action is always directed towards the prevention of the spread of disease.

Offences against AHA 1981

16.45 A person commits an offence against AHA 1981 if, without lawful authority or excuse, (s)he:

(a) does anything contravening AHA 1981, an order of the Secretary of State (in Wales, the Welsh Ministers) or a regulation of a local authority (s 73(a));
(b) fails to give, produce, observe or do any notice, licence, rule, or thing as required by AHA 1981 or by an order or regulation (s 73(b));
(c) does anything which AHA 1981 or an order makes or declares to be not lawful (s 72(b));
(d) does or omits anything the doing or omission of which is declared by AHA 1981 or an order to be an offence by him/her against AHA 1981 (s 72(a)).

It is an offence under s 66 to refuse entry, without lawful authority or excuse, to an inspector or other officer, or to obstruct or impede him/her in entering, or to obstruct or impede an inspector, constable, or other officer in the execution of his/her duty, or to assist another to do so.

In the case of each of these offences, the defendant has the onus of proving lawful authority or excuse.

The above offences are summary (maximum punishment six months' imprisonment and/or an unlimited fine). The time limit for a prosecution is the same as at **16.13**.

Duties and powers of a constable

AHA 1981, s 62(2), provides that, for the purpose of exercising any power to seize an animal or cause an animal to be seized, and (a) where an order is in force under the Act or (b) a power is given for the purpose of preventing the introduction of rabies into Great Britain, a constable may enter (if need be by force) and search any vessel, boat, aircraft, or vehicle in which there is, or in which (s)he with reasonable cause suspects that there is an animal to which that power applies. **16.46**

By PACE, s 17, a constable may enter and search premises for the purpose of arresting someone for one of the offences to which AHA 1981, s 61 applies, namely: the landing or attempted landing of any animal (including a dog or cat) in contravention of a Rabies Order; the failure by a person in charge of a vessel to discharge any obligation under such an order; or the unlawful movement of any such animal into, within, or out of a rabies-infected area.

DOGS

Control of dogs

Collars

The Control of Dogs Order 1992 requires that while on a highway or in a place of public resort every dog must wear a collar with the owner's name and address inscribed upon it, or on a plate or badge attached to it. This does not apply to: **16.47**

(a) any pack of hounds;
(b) any dog while being used:
    (i)   for sporting purposes;
    (ii) for the capture or destruction of vermin;
    (iii) for the driving or tending of cattle or sheep;
    (iv) on official duties by a member of the armed forces or Revenue and Customs or by the police force for any area; or
    (v)   in emergency rescue work; or
(c) any dog registered with the Guide Dogs for the Blind Association.

The owner or the person in charge of a dog who, without lawful authority or excuse, the proof whereof is on him/her, causes or permits the dog to be in a highway or place of public resort without the requisite collar is guilty of a summary offence against the Animal Health Act 1981, s 72(a). As to the prosecution time limit and punishment for this offence, see **16.45**.

Leads and muzzles

The Road Traffic Act 1988, s 27 empowers local authorities to make orders designating lengths of roads within their areas as roads upon which dogs must at all times be kept **16.48** on a lead. The chief officer of police must be consulted before such an order is made and the local authority is required to publish it and to place signs on the designated length of road subject to the order. When such an order is in force, it is a summary offence (maximum punishment a level 1 fine) under s 27(1) to cause or permit a dog to be on such a length of road if it is not held on a lead. The offence does not apply to dogs tending cattle or sheep in the course of a business nor to those being used for sporting purposes. Further exceptions may be made by the order.

Dangerous dogs: order to keep under control

16.49   If it appears to a magistrates' court that a dog is dangerous and not kept under proper control, it may, under the Dogs Act 1871, s 2, order the owner to keep it under control or order the dog's destruction. The proceedings must be by way of complaint. Complaints may be preferred by a police officer.

Such an order may be made whether or not the dog is shown to have injured any person. It may specify the measures to be taken for keeping the dog under control, whether by muzzling it, keeping it on a lead, or excluding it from specified places, or otherwise. Such an order may also require the neutering of a male dog.

Divisional courts have held that it is not necessary that the dog is dangerous to mankind; it is enough that it is dangerous to other animals of whatever kind, even another dog. In an old case, a dog which killed two pet rabbits was held by a divisional court not to be dangerous as it was within the natural instincts of a dog to chase, wound, or kill other small animals.

The Dangerous Dogs Act 1989 (DDA 1989), s 1(1) empowers a magistrates' court, when it makes a destruction order under the 1871 Act, to disqualify the owner from having custody for a specified period.

DDA 1989, s 1(3) provides summary offences of failing to keep a dog under proper control as ordered under the 1871 Act, and failing to deliver up a dog for destruction. These offences are punishable with a fine not exceeding level 3. In addition, the court may disqualify the offender from having custody of a dog for a specified period.

A person who has custody of a dog while disqualified under DDA 1989, s 1(1) or (3) commits a summary offence under s 1(6) (maximum punishment an unlimited fine).

Dangerous Dogs Act 1991

Possession or control of dogs bred for fighting

16.50   The Dangerous Dogs Act 1991 (DDA 1991), s 1(3) provides that no person may have in his possession or control a dog to which s 1 applies, except under a power of seizure or destruction or (by s 1(5)) under one of the two exemption schemes referred to at **16.51** and **16.53**. Section 1 applies to any dog of the type known as pit bull terrier, Japanese tosa, and any other type of dog designated by order of the Secretary of State, currently the Dogo Argentino and the Fila Braziliero. 'Type' has a wider meaning than 'breed'. Determining the limits of a type is a question of fact for determination by the court. The court is entitled to look at the American Dog Breeders' Association (ADBA) breed standard as a guide. The fact that a dog does not meet that standard in every respect is not conclusive that it is not one of the specified types. Thus, for example, it has been held that the fact that a dog is near to, or has a substantial number of, characteristics of a pit bull terrier as set out in the ADBA standard is sufficient for the dog to be found to be of the pit bull terrier type. It is relevant to consider whether the dog exhibited the behavioural characteristics of a pit bull terrier but that evidence would not be conclusive.

In any proceedings under s 1, it is presumed that a dog is of such a type unless D proves the contrary.

Breach of s 1(3) is a summary offence under s 1(7) (maximum punishment six months' imprisonment and/or an unlimited fine).

Court-ordered exemption scheme

16.51 The Dangerous Dogs Exemption Schemes (England and Wales) Order 2015 (DDES(E&W)O 2015), Part 2 (arts 4–11) deals with this. The prohibition in DDA 1991, s 1(3) does not apply to a dog provided that:

(a) a court has determined under s 4(1A) or 4B (**16.57**) that the dog is not a danger to public safety and has made the dog subject to a contingent destruction order under s 4A (**16.57**) or 4B;

(b) the following 'exemption conditions' are met in respect of that dog within the 'specified time period':

    (i) the dog is neutered and microchipped;

    (ii) third-party insurance in respect of the dog is in force; and

    (iii) a certificate of exemption is issued,

in each case in accordance with DDES(E&W)O 2015; and

(c) the requirements attached to the certificate of exemption are complied with throughout the dog's lifetime.

Subject to any extension granted by the court, the conditions in (b) must be complied with:

- in the case of an adult dog, within two months of the contingent destruction order; and
- in the case of a dog under the age of six months at the time of the contingent destruction order, within one month of the dog attaining six months.

A dog is not exempt from the prohibition in s 1(3):

(a) if the exemption conditions referred to in (b) above are not met within the time period just specified; or

(b) if the requirements attached to the certificate of exemption are not complied with at any time after the certificate is issued.

The Index of Exempted Dogs, the Agency designated to discharge functions under DDES(E&W)O 2015 (hereafter 'the Agency'), must issue a certificate of exemption in respect of the dog if it is satisfied that:

(a) the court, in determining that the dog is not a danger to public safety, has decided the person to whom the certificate is to be issued is a fit and proper person to be in charge of the dog and has made the dog subject to a contingent destruction order;

(b) the relevant fee plus VAT has been paid to the Agency; and

(c) conditions (i) and (ii) above have been met.

A certificate of exemption must contain requirements to:

(a) keep the dog at the same address as the person to whom the certificate is issued save for any 30 days in a 12-month period;

(b) notify the Agency: (i) of any proposed change of address (not to include any changes of address in the 30 days mentioned in (a)), and (ii) of the death or export of the dog;

(c) satisfy the Agency that a policy of third-party insurance is in force;

(d) keep the dog: (i) muzzled and on a lead when in a public place, and (ii) in sufficiently secure conditions to prevent its escape;

(e) provide access to the dog for the purpose of reading a microchip on request by a constable (or authorised local authority officer);

(f) produce to such a person the certificate of exemption and confirmation that third-party insurance is in force within five days of being requested to do so by him/her.

Substitution of person in charge of dog exempted

16.52   DDES(E&W)O 2015, Part 3 (arts 12–19) deal with this.

When a dog has been exempted from the prohibition in DDA 1991, s 1(3), a person may apply to a magistrates' court to be substituted as the person in charge of the dog only if the person determined by the court as being a fit and proper person to be in charge of the dog is unable to continue to be in charge of it by reason of the death of that person, or serious illness rendering him/her unable to be in charge. Subject to any extension by the court, the application must be made within six weeks of the death of the person in charge, or of an official letter from a medical practitioner confirming the serious illness of the person in charge. Non-compliance results in the dog no longer being exempted.

At least two weeks before making an application to the court, the applicant must provide to the chief officer of police for the area in which (s)he lives:

(a) his name, address, and date of birth; and

(b) details of the exempted dog and of the person to whom the certificate of exemption was issued.

The court may only substitute the applicant as the person in charge of the exempted dog if satisfied that the dog does not constitute a danger to public safety.

Until the application is determined by the court the dog may be kept at the applicant's address with the applicant on and after the date of the death of the person in charge, or of the official medical letter, as the case may be. Requirements (b)–(f) referred to at **16.51** which apply to a certificate of exemption also apply in relation to the applicant until a new certificate of exemption is issued to the applicant.

If the court is satisfied that the applicant is a fit and proper person to be substituted as the person in charge of the exempted dog the court must notify the Agency for the purposes of the issue of a new certificate of exemption.

The Agency must issue a new certificate of exemption to the applicant in respect of the dog if it is satisfied that—

(a) the court has determined the person to be substituted as the person in charge of the dog as a fit and proper person to be in charge of the dog;

(b) the relevant fee plus VAT has been paid to the Agency; and

(c) the third-party insurance required has been obtained.

The provisions attached to a certificate of exemption apply to a new certificate of exemption so issued.

A dog is not exempt from the prohibition in s 1(3) if there is a failure to comply with the requirements attached to the new certificate of exemption, or to comply with the requirements in (b)–(f) of the original certificate pending the issue of the new certificate.

Interim exemption scheme

DDES(E&W)O 2015, Part 4 (arts 20–26) provides as follows.

Where a dog suspected of being a dog to which DDA 1991, s 1 applies is seized under a statutory power conferred prior to the court's final determination in respect of whether the dog should be destroyed under s 4 or 4B (**16.57**), the chief officer of police for the area may release the dog to the person intending to apply for exemption of the dog under DDES(E&W) O 2015 ('the person in interim charge') only in accordance with Part 4.

The chief officer of police may only release the dog under Part 4 if satisfied that it does not constitute a danger to public safety. The same test applies to the determination of this as in relation to the substitution of the person in charge (above).

Where the chief officer of police is satisfied that the dog is not a danger to public safety, the dog is exempt on an interim basis from the prohibition in s 1(3) provided that:

(a) the chief officer of police is satisfied that exemption conditions (b)(i) and (ii) (**16.51**) have been met;
(b) the additional requirements referred to below are met throughout the period of interim exemption; and
(c) the person in interim charge confirms in writing that (s)he understands the conditions mentioned and the continuing additional requirements and the consequences of any failure to comply with those conditions or requirements.

If there is a failure to comply with the additional requirements, or if the chief officer of police is no longer satisfied that the danger to public safety test is met, the dog may be seized under DDA 1991, s 5 and the dog is not exempt under DDES(E&W)O 2015, Part 4 from the prohibition in DDA 1991, s 1(3).

The seized dog may only be released when the neutering and insurance exemption conditions have been met. Where a dog is under six months when the chief officer of police determines that the dog may be released the exemption conditions in (i) and (ii) must be met within one month of the dog attaining six months.

Where a dog is released to the person in interim charge, that person must comply with the following additional requirements:

(a) to keep the dog at the same address as that person;
(b) to notify the police of any proposed change of address;
(c) to satisfy the police on request that a policy of third-party insurance is in force;
(d) to provide access to the dog for the purpose of reading a microchip on request by a constable or authorised local authority officer;
(e) to keep the dog in sufficiently secure conditions to prevent its escape, and muzzled and on a lead when in a public place;
(f) any other requirement for the purpose of preventing the dog being a danger to public safety considered appropriate by the chief officer of police;
(g) to produce to a constable (or authorised local authority officer) the certificate of exemption and confirmation that third-party insurance is in force within five days of being requested to do so by that person.

When a dog has been released under the interim exemption scheme and is subsequently made subject to a contingent destruction order by the court under DDA 1991, s 4A or 4B (**16.57**):

(a) additional requirements under the interim exemption scheme continue to apply until a certificate of exemption is issued; and

(b) if the exemption conditions are not met within the specified time limits the dog may be seized under s 5 with a view to its destruction in accordance with the contingent destruction order.

Subject to the above, an interim exemption ceases when a court makes a final determination in respect of the dog under s 4 or 4B or determines that the dog is not one to which s 1 applies, or a decision is taken not to commence or continue with proceedings in respect of the dog.

Breeding, selling, etc of dogs bred for fighting

16.54 DDA 1991, s 1(2) and (7) makes it an offence to breed, sell, exchange, give, or offer to give, advertise, or expose for sale, exchange, or gift, a dog specified above in relation to s 1(3) (**16.50**).

Section 1(2) and (7) also makes it an offence for the owner (**16.58**) to abandon such a dog, or for the owner or person in charge of it to allow it to stray.

Finally, s 1(2) and (7) makes it an offence for the owner, or person for the time being in charge, of such a dog to allow it to be in a public place (**16.59**) without being muzzled and kept on a lead. This prohibition in s 1 is a strict one. If a dog of the requisite type is in a public place, it must not be allowed to be unmuzzled and must be kept on a lead. There are no circumstances in which the necessity of a situation, eg avoiding cruelty to the dog, can overtake the prohibition.

All the above offences are summary. They are punishable in the same way as a breach of s 1(3) (**16.50**), except that a person who publishes an advertisement in contravention of the first paragraph above is not liable to imprisonment if (s)he shows that (s)he published the advertisement to the order of someone else and did not him/herself devise it. In addition, such a person is not to be convicted if (s)he shows that (s)he did not know and had no reasonable cause to suspect that the advertisement related to a dog to which s 1 applies (s 1(7)).

Keeping dogs under proper control

16.55 Where any dog is dangerously out of control in any place (whether or not a public place), the owner or the person for the time being in charge of it is guilty of a summary offence under DDA 1991, s 3(1) (maximum punishment six months' imprisonment and/or an unlimited fine). There are three aggravated, indictable (either-way) offences under s 3(1) and (4) if the dog, whilst so out of control, injures any person (V) or 'assistance dog'. They are punishable with imprisonment as follows: up to 14 years' imprisonment if V dies as a result of being injured, up to five years' if V is injured, and up to three years' if an assistance dog is injured (whether or not it dies).

Under s 3(1A), a person (D) is not guilty of the offences just described in a case which is a 'householder case', ie a case where:

(a) the dog is dangerously out of control while in or partly in a building, or part of a building, that is a dwelling or is forces accommodation (or is both), and
(b) at that time (i) the person in relation to whom the dog is dangerously out of control (V) is in, or is entering, the building or part as a trespasser, or (ii) D (if present at that time) believed V to be in, or entering, the building or part as a trespasser.

The provisions set out in the last seven lines of **9.27**(b) apply equally here.

The Court of Appeal has held that a person who is the owner, or for the time being in charge, of a dog which is dangerously out of control is not guilty of an offence under s 3(1) unless the prosecution proves an act or omission on his/her part which to some more than minimal degree contributed to the dog being in the place in question, dangerously out of control. Thus, for example, if a dog escapes from D's house as a result of the unauthorised action of a third party, eg a burglar, and as a result is dangerously out of control in the place in question, D is not guilty of an offence under s 3(1).

Strict liability is imposed by s 3(1) on the owners or handlers of such dogs. However, an owner has a defence if (s)he can prove that at the time of the offence the dog was in the charge of a person whom (s)he reasonably believed to be a fit and proper person to be in charge of it. A divisional court has held that the defence only applies if there is plain evidence that 'charge' has been transferred to an identified person.

Police powers

By DDA 1991, s 5(1), a constable (or authorised local authority officer) may seize any **16.56** dog:

(a) which appears to be a dog to which s 1 applies and which is in a public place (**16.59**); or
(b) which is in a public place and which appears to be dangerously out of control.

Section s 5(1A) provides that a constable or authorised local authority officer may seize any dog in any place which is not a public place, if the dog appears to the constable or officer to be dangerously out of control.

If a justice of the peace is satisfied by information on oath that there are reasonable grounds to believe that one of the above offences has been committed, or that evidence of such an offence is to be found on any premises, (s)he may issue a warrant authorising a constable to enter and search them and to seize any dog or other thing which is evidence of such an offence (s 5(2)). The additional powers of seizure provided by the Criminal Justice and Police Act 2001, s 50 (**4.61**) apply to this power of seizure.

Destruction and disqualification orders

These are dealt with by DDA 1991, s 4, 4A, or 4B. **16.57**

Immediate destruction order Section 4(1)(a) provides that, where a person is convicted of an offence against s 1 or 3(1), the court may order destruction of the dog concerned. Indeed, it must do so in the case of an offence under s 1 or an aggravated offence contrary to s 3(1), unless (by s 4(1A)) it is satisfied that the dog would not constitute a danger to public safety.

Dogs may not be destroyed during the period allowed for notice of, and determination of, any appeal.

Where a court makes a destruction order under s 4(1)(a) or s 4B (below), it may appoint of a person to undertake the destruction of the dog and require it to be delivered up for that purpose.

Contingent destruction orders Section 4A, as interpreted by a judge in the Administrative Court, provides that, where:

(a) a person is convicted of an offence under s 1;
(b) the court does not order destruction of the dog under s 4; and
(c) in the case of an offence under s 1, the dog is subject to the prohibition under s 1(3),

the court must order the dog to be destroyed, unless the dog is exempted from that prohibition within the requisite period. That period, two months, may be extended by the court.

Section 4A also provides that, where a person is convicted of an offence under DDA 1991, s 3 the court may order that, unless its owner keeps it under proper control, the dog must be destroyed. Such an order may specify the measures to be taken to keep the dog under control whether by muzzling, keeping on a lead, excluding it from specified places, or otherwise, and, if it appears to the court that the dog is a male and would be less dangerous if neutered, may require that the dog be neutered.

Destruction order otherwise than on conviction   Section 4B makes provision for an order of destruction which may be made by a justice of the peace in respect of a dog which has been seized under s 5(1) or (2) or under any other statutory power (a) where there is no prosecution, or (b) where the dog cannot be released without contravention of s 1(3). A justice is not required to make such an order if (s)he is satisfied that the dog would not be a danger to public safety. Where, in a case falling within (b), a justice does not order the destruction of the dog, (s)he must order that, unless the dog is exempted from that prohibition in s 1(3) within the requisite period, it must be destroyed. That period, two months, may be extended by the court. It has been held in the Administrative Court that, if the dog is exempted within the requisite period, the police cannot subsequently summarily destroy the dog because they believe that the exemption has ceased to exist. It would seem that this decision, made in relation to an order under s 4B, could also apply to an order under s 4A.

Disqualification order   By s 4(1)(b), where a person is convicted of an offence under s 1 or 3 a court may also order him/her to be disqualified from keeping a dog for such a period as it thinks fit. A summary offence, punishable with a fine not exceeding level 3, is committed against s 4(8) where a person has custody of a dog while disqualified, or where (s)he fails to deliver up a dog for destruction as ordered.

Dogs owned by young persons

16.58 Where a dog is owned by a person who is under 16, the term 'owner' in DDA 1991, s 1 or 3 includes a reference to the head of the household, if any, of which that person is a member (s 6).

Public place

16.59 'Public place' in DDA 1991 means any street, road, or other place (whether or not enclosed) to which the public have or are permitted to have access whether for payment or otherwise, including the common parts of a building containing two or more separate dwellings (eg a block of flats) (s 10(2)). A front garden or driveway is not a public place in this context and neither is any other type of place which people enter by express or implied invitation. On the other hand, a dog in a private car on a public highway is in a public place.

DOGS WORRYING LIVESTOCK

16.60 The Dogs (Protection of Livestock) Act 1953 (D(PL)A 1953), s 1(1) declares that the owner of a dog, and, if it is in the charge of a person other than the owner, that person also, is guilty of an offence if the dog worries livestock on any agricultural land. An offence under s 1(1) is summary and punishable with a fine not exceeding level 3.

By s 1(2), 'worry livestock' covers attacking livestock, or chasing livestock in such a way as might reasonably be expected to cause injury or suffering to it, or, in the case of females, abortion, or loss of or diminution in their produce. It also covers a dog being at large (ie not on a lead or under close control) in a field or enclosure in which there are sheep, but not where the dog is owned by, or in the charge of, the occupier of the field or is a police dog, guide dog, trained sheepdog, working gun dog, or, perhaps unexpectedly, a pack of hounds.

'Livestock' means cattle, sheep, goats, swine, horses (including asses and mules), or poultry. Thus 'livestock' can generally be described as farm animals. 'Agricultural land' means land used as arable, meadow, or grazing land or for the purposes of poultry farming, pig farming, market gardens, allotments, nursery grounds, or orchards.

Defences

If the livestock are trespassing and the dog which attacks the livestock is owned by, or in the charge of, the occupier of that land or a person authorised by him/her, a defence is open to that person provided that (s)he did not cause the dog to attack the livestock (s 1(3)). **16.61**

The dog's owner has a defence if (s)he proves that at the time in question it was in the charge of some other person, whom (s)he reasonably believed to be a fit and proper person to be in charge (s 1(4)), as where (s) has allowed a reliable person to take the dog for a walk.

Police powers

By D(PL)A 1953, s 2(2), a constable is empowered to seize a dog, found anywhere, which (s)he reasonably believes to have been worrying livestock on agricultural land and to retain it until the owner is found and has paid the expenses of its detention. This power cannot be exercised if there is a person present who admits to being the owner of the dog or in charge of it. **16.62**

If, on an application by a constable, a justice is satisfied that there are grounds for believing that:

(a) an offence under the Act has been committed; and
(b) the dog in question is on premises specified in the application,

(s)he may issue a warrant under s 2A authorising a constable to enter and search the premises in order to identify the dog.

GUARD DOGS

Under the Guard Dogs Act 1975, ss 1(1) and 5(1), it is an offence to use or permit the use of a guard dog on any premises unless a person ('the handler') who is capable of controlling the dog is present on the premises and the dog is under the direct control of the handler at all times while it is being so used except while it is secured so that it is not at liberty to go freely about the premises. A 'guard dog' is one which is used either to protect the premises or property on them, or to protect a person guarding the premises or property. **16.63**

The handler of a guard dog must keep the dog under his/her control at all times while it is being used as a guard dog at any premises except:

(a) while another handler has control over the dog; or
(b) while the dog is secured so that it is not at liberty to go freely about the premises (s 1(2)).

Failure to do so is an offence under s 5(1).

The Act excludes from the term 'premises' agricultural land and land in the curtilage of a dwelling house, thus exempting farm dogs and those confined within a dwelling house or its yard or garden.

When guard dogs are kept upon 'premises' warning notices must be clearly displayed at all entrances to the premises (s 1(3)); failure to display such notices is an offence under s 5(1).

The above offences are summary offences, punishable with an unlimited fine,

WILD BIRDS, ANIMALS, AND PLANTS

Wild birds

Killing, destroying, damaging, or possessing

16.64 The relevant statute is the Wildlife and Countryside Act 1981 (WCA 1981), Part 1. WCA 1981, s 1 (1) creates the offences of:

(a) intentionally killing, injuring, or taking any wild bird (s 1(1)(a));
(b) intentionally taking, damaging, or destroying the nest of a wild golden eagle, white-tailed eagle, or osprey (species which re-use their nests) (s 1(1)(aa));
(c) intentionally taking, damaging, or destroying the nest of any wild bird while that nest is in use or being built (s 1(1)(b)); and
(d) intentionally taking or destroying an egg of such a bird (s 1(1)(c)).

It is equally an offence under s 1(2) to possess or control: a wild bird, whether alive or dead (including one which has been stuffed and mounted), or any part of it or thing derived from it (s 1(2)(a)), or an egg, or part of it, of such a bird (s 1(2)(b)). 'Knowledge' that the bird is wild is not required. Consequently, a defendant's belief that a wild bird was bred in captivity and was therefore not a 'wild bird' does not afford an excuse. It is a defence for the defendant to show that the bird or egg had not been killed or taken, or had been lawfully killed or taken (ie without contravention of WCA 1981, Part 1, or other wild bird legislation) otherwise than in contravention of 'the relevant provisions', or that the bird, egg, or other thing in his/her possession or control had been sold (whether to him/her or another) (s 1(3)). 'The relevant provisions' means the provisions of WCA 1981, Part 1, and orders thereunder, and (except in the case of an egg or part of it) the provisions of the Protection of Birds Acts 1954 to 1967 and orders thereunder. A person is not guilty of an offence under s 1(2)(b) if (s)he shows that the egg, or part of it, was in any person's possession or control before 28 September 1982 (the date when s 1(2) came into force) (s 1(3ZA)).

The purpose of the legislation is to provide complete protection to wild birds against the activities such as those of the falconer at one extreme and those of the youthful 'birds'-nester' at the other. The term 'wild bird' is widely defined to include all wild birds which are resident in, or are visitors to, the UK or the European territory of an EU state, with the exception of poultry (birds which could not readily be described as wild even if free-ranging) or game birds (which are protected by separate legislation discussed above). In addition, as already implied, 'wild bird' does not include a wild bird which has been bred in captivity. For a bird to be bred in captivity the parent birds must have been in captivity when the egg was laid. This restriction is to prevent the 'nest robbers' from legally rearing falcons and other rare birds in captivity. Captive birds must be ringed and registered.

It is an offence against WCA 1981, s 1(5) intentionally or recklessly to disturb any wild bird mentioned in Sch 1 to the Act whilst it is building a nest, or is in, on, or near such a nest containing eggs or young birds, or to disturb dependent young of such a bird. Schedule 1 contains all but the commonest of birds. It includes all birds of prey resident in the UK with the exception of the kestrel and sparrowhawk. Nest robbers target the nests of peregrine falcons, goshawks, and eagles. Frequent use is made of the provisions relating to 'disturbance' to combat such nest robbers. Where chicks are found in the possession of persons licensed to keep birds of prey, and the circumstances are suspicious, the parenthood of such chicks can be established by DNA fingerprinting. The RSPB will assist in this respect.

Exceptions

16.65 For the purpose of day-to-day enforcement, it is almost certain that persons found in the act of killing wild birds, or in possession of their eggs, will be guilty of an offence against this Act. However, there are some limitations on the offence of killing, taking, or injuring a wild bird. Some birds may be killed or taken outside their close season (generally their nesting season); these include the commoner species of duck and goose, plover, snipe, and woodcock, which are listed in WCA 1981, Sch 2, Part I. There are a number of other exceptions, relating to wild birds in general, where the action taken is done by an authorised person (ie a landowner or the occupier, or someone authorised by him/her or by an official body of a specified type) or officially required or done in relation to a disabled bird and in certain other cases.

Sale, etc

16.66 It is an offence under WCA 1981, s 6(1) to sell, or to expose or offer for sale, or to possess for sale, live wild birds or their eggs (s 6(1)(a)), or to publish, or cause to be published, advertisements to the effect that one buys or sells such birds or eggs or intends to do so (s 6(1)(b)). Section 6(2)(a) and (b) provide corresponding offences in relation dead wild birds or their eggs.

Wild animals

16.67 WCA 1981 contains similar offences to protect wild animals. These provisions are not always as easy for police officers to enforce as those relating to wild birds. The reason is that the protection afforded to wild animals by WCA 1981 is restricted to animals described in Sch 5 to the Act, some of which are not easily identifiable without specialist help. The animals listed in Sch 5 include most of the less common butterflies, moths, frogs, lizards and newts, porpoises, dolphins, red (but not grey) squirrels, and the common otter.

By WCA 1981, s 9, it is an offence intentionally to kill, injure, or take any wild animal listed in WCA 1981, Sch 5 (s 9(1)), or for a person to possess such animal alive or dead (in whole or part) (s 9(2)). By s 9(5), the same provisions as in the case of wild birds apply in relation to sale, exposing, or offering for sale, possessing for sale, or advertising for sale, any wild animal live or dead listed in WCA 1981, Sch 5.

Section 10 provides the following exceptions to s 9:

(1) Nothing in s 9 makes unlawful anything done in pursuance of a requirement by a Government Minister under the Agriculture Act 1947, s 98, or under, or in pursuance of an order made under, the Animal Health Act 1981.

(2) Notwithstanding anything in s 9, a person is not guilty of an offence by reason of:

(a) the taking of any such animal if (s)he shows that the animal had been disabled otherwise than by his/her unlawful act and was taken solely for the purpose of tending it and releasing it when no longer disabled;

(b) the killing of any such animal if (s)he shows that the animal had been so seriously disabled otherwise than by his/her unlawful act that there was no reasonable chance of its recovering; or

(c) any act made unlawful by s 9 if (s)he shows that the act was the incidental result of a lawful operation and could not reasonably have been avoided. A person cannot rely on the defence provided by (c) as respects anything done in relation to a bat otherwise than in the living area of a dwelling house unless (s)he had notified the conservation body for the area in which the act is to take place of the proposed action or operation and allowed them a reasonable time to advise him/her as to whether it should be carried out and, if so, the method to be used.

(3) Notwithstanding anything in s 9, an authorised person (**16.65**) shall not be guilty of an offence by reason of the killing or injuring of a wild animal included in Sch 5 if (s)he shows that his/her action was necessary for the purpose of preventing serious damage to livestock, foodstuffs for livestock, crops, vegetables, fruit, growing timber or any other form of property or to fisheries. An authorised person is not entitled to rely on this defence as respects any action taken at any time if it had become apparent, before that time, that that action would prove necessary for the above purpose and either a licence authorising that action had not been applied for as soon as reasonably practicable after that fact had become apparent, or an application for such a licence had been determined.

Wild plants

16.68   By WCA 1981, s 13(1)(a), it is an offence for any person intentionally to pick, uproot, or destroy any wild plant included in WCA 1981, Sch 8. Schedule 8 contains quite an extensive list of wild plants; all but the most common of our wild plants are listed.

It is an offence under s 13(1)(b) for someone who is not an authorised person intentionally to uproot any wild plant not included in Sch 8. An 'authorised person' is defined at **16.65**.

As can be seen, the intentional picking or destruction of wild plants, other than by uprooting, is only an offence if the plant concerned is one listed in Sch 8.

It is an offence under WCA 1981, s 13(2) to sell, offer, or expose for sale, possess for sale, or advertise for sale, etc any live or dead wild plant included in WCA 1981, Sch 8.

Licences

16.69   The above offences relating to wild birds, animals, and plants are not committed by a person acting within the terms of a licence granted by the appropriate authority.

Additional police powers

16.70   By WCA 1981, s 19(1) if a constable reasonably suspects that any person is committing or has committed any of the above offences under WCA 1981, (s)he may:

(a) stop and search that person if (s)he reasonably suspects that evidence of the commission of the offence may be found on him/her;

(b) search or examine anything which that person may then be using or have in his/her possession; and

(c) seize and detain for the purpose of proceedings under WCA 1981 the Act anything which might be evidence of the commission of the offence or might be liable to be forfeited.

These powers do not require a warrant.

For the purpose of exercising his/her powers as set out above, or for the purpose of arresting a person under PACE, s 24 for any of the offences under WCA 1981, a constable is empowered by WCA 1981, s 19(2) to enter onto any land other than a dwelling house. A justice may issue a search warrant under s 19(3) subject to the usual conditions in respect of the offence disclosed.

Prosecution, trial, and punishment

16.71 The time limit for prosecutions under WCA 1981 is the same as at **16.13**, except for the substitution of two years for three (s 20(2)). The above offences under WCA 1981 are summary and punishable with up to six months' imprisonment and/or an unlimited fine.

Civil sanctions

16.72 The provisions of the Regulatory Enforcement and Sanctions Act 2008, Part 3 (**16.19**) apply to the above offences under WCA 1981.

STRAYING HORSES, CATTLE, ETC

16.73 The Highways Act 1980, s 155(1) states that if horses, cattle, sheep, goats, or swine are found straying or lying on or at the side of a highway their keeper is guilty of a summary offence (maximum punishment a level 3 fine). A person in whose possession animals are is their keeper, whether or not (s)he derives any personal benefit from them. Highways which pass over common or unenclosed land are exempted from these provisions.

PART IV

PROTECTING COMMUNITIES

CHAPTER 17

Nuisances

COMMON LAW OFFENCE OF PUBLIC NUISANCE

It is a common law indictable (either-way) offence to cause a public nuisance. The sentence is at the discretion of the judge in the Crown Court. A public nuisance is an act not warranted by law, or an omission to discharge a legal duty, whose effect is to endanger the life, health, property or comfort of the public, or to obstruct a substantial section of the public in the exercise or enjoyment of rights common to all members of the public. An act which is specifically authorised by law cannot be a public nuisance if carried out as prescribed. Despite its width, this common law offence is little used as most 'nuisances' are covered by statutes.

17.1

DEPOSITING LITTER

The Environmental Protection Act 1990 (EPA 1990), s 87(1) provides that a person commits an offence if he *throws down, drops or otherwise deposits*, any litter in *any* place open to the air (a term including a covered place open to the air on at least one side) in the area of a principal litter authority, *and leaves it*. However, s 87(1) does not apply to a covered place open to the air on at least one side if the public does not have access to it, with or without payment. The offence of depositing litter is committed whether the litter is deposited on land or on water.

17.2

No offence is committed where the depositing of litter is authorised by law or is done with the consent of the owner or occupier or other person having control of the place where it is deposited. Consent may only be given to the depositing of litter in a lake, pond or watercourse by the owner, occupier or other person having control of all of the land adjoining that lake, pond or watercourse and all of the land into which water from those places directly or indirectly discharges, otherwise than by means of a public sewer.

The term 'litter' includes the discarded ends of cigarettes and like products and discarded chewing gum and the like.

A divisional court has held that 'leave' for these purposes does not mean abandon, and that an article deposited with no intention to remove it can be 'left' after only a short period of time. Someone who throws down fish wrappers, and refuses to pick them up immediately, commits an offence under s 87. (S)he has no intention of removing the wrappers and this is evidenced by his/her refusal.

The offence under s 87(1) is a summary offence punishable with a fine not exceeding level 4, as well as being a 'penalty offence' for the purposes of the Criminal Justice and Police Act 2001 (**3.40**).

Where the offence under s 87(1) is dealt with by a local authority 'litter warden' (who must be authorised in writing so to act on behalf of a litter authority), s 88 permits the use of a fixed penalty procedure under *that* section.

ABANDONMENT OF PROPERTY

Things other than motor vehicles

17.3   The Refuse Disposal (Amenity) Act 1978 (RD(A)A 1978), s 2(1)(b) provides the offence of abandoning, without lawful authority, on any land in the open air, or on any other land forming part of a highway, anything other than a motor vehicle which has been brought to the land for the purpose of being abandoned there. This deals with instances where someone takes refuse quite deliberately into the countryside and abandons it there.

Motor vehicles

17.4   RD(A)A 1978, s 2(1)(a) punishes the abandonment of motor vehicles in similar circumstances. The difference is that it is not necessary to prove that a motor vehicle was brought to the land for the purpose of abandonment. The offence is committed by someone who, without lawful authority, abandons on any land in the open air, or on any other land forming part of a highway, a motor vehicle or anything which formed part of a motor vehicle and was removed from it in the course of dismantling the vehicle on the land. Section 2A provides for the offence under s 2(1)(a) to be dealt with as a fixed penalty offence by an authorised local authority officer.

General

17.5   Section 2(2), which applies to s 2(1)(a) and (b), deals with a person who leaves anything on land in such circumstances or for such a period that (s)he may reasonably be assumed to have abandoned it or to have brought it to the land for the purpose of abandoning it there. It provides that such a person is deemed to have abandoned it there or, as the case may be, to have brought it to the land for that purpose, unless the contrary is shown.

An offence under s 2(1)(a) or (b) is a summary offence punishable with up to three months' imprisonment and/or a fine not exceeding level 4.

NUISANCE OR DISTURBANCE ON EDUCATIONAL PREMISES

17.6   The Education Act 1996, s 547(1) (which deals with schools) and the Further and Higher Education Act 1992, s 85A(1) (which deals with further education colleges and 16 to 19 Academies) each create a summary offence, punishable with a fine not exceeding level 2, which is committed by a person who is present without lawful authority on relevant educational premises and causes or permits nuisance or disturbance to the annoyance of persons who lawfully use those premises (whether or not any such persons are present at the time). 'Premises' includes playgrounds, playing fields and other premises for outdoor recreation. A constable or a person authorised by the appropriate authority may remove a person from the premises if (s)he has reasonable cause to suspect that that person is committing or has committed such an offence (s 547(3) and s 85A(3) respectively).

COMMUNITY PROTECTION NOTICES

17.7   These notices are governed by the Anti-social Behaviour, Crime and Policing Act 2014 (A-sB,C&PA 2014), Part 4, Ch 1 (ss 43–58). The community protection notice (CPN) is intended to deal with unreasonable, ongoing problems or nuisances which negatively affect the community's quality of life by targeting the person responsible.

A CPN is a notice imposing any of the following requirements on an individual or body issued with it, namely a requirement to:

(a) stop doing specified things;
(b) do specified things;
(c) take reasonable steps to achieve specified results.

A divisional court, having stated that CPNs constitute a significant interference with an individual's freedom and must therefore be clear in their terms and proportionate in their effect, went on (a) to comment that it would be best practice and consistent with legal certainty for any CPN to be limited in time, with that time clearly stated in the CPN, and (b) to emphasise the need for authorised persons (**17.8**) prior to issuing a CPN to consider with care the prohibitions and restrictions imposed to ensure that they go no further than is necessary and proportionate to address the behaviour which has led to the CPN being made.

Power to issue notices

By A-s B,C&PA 2014, s 43, an authorised person may issue a CPN to *an individual aged*   **17.8** *16 or over, or a body*, if satisfied on reasonable grounds that:

(a) the conduct ('conduct' includes a failure to act) of the individual or body is having a detrimental effect, of a persistent or continuing nature, on the quality of life of those in the locality, and
(b) the conduct is unreasonable.

The only requirements that may be imposed are ones that are reasonable to impose in order to prevent the detrimental effect referred to in (a) and (b) from continuing or recurring, or to reduce that detrimental effect or to reduce the risk of its continuance or recurrence.

An 'authorised person' means:

(a) a constable;
(b) the relevant local authority;
(c) a social housing provider if designated by the relevant local authority for these purposes.

For the purposes of CPNs, 'local authority' means, in relation to England, a district council, a county council for an area for which there is no district council, a London borough council, the Common Council of the City of London, or the Council of the Isles of Scilly; or in relation to Wales, a county council or a county borough council. 'The relevant local authority' means the local authority (or, as the case may be, any of the local authorities) within whose area the conduct specified in the CPN has, according to the notice, been taking place. For a fixed penalty notice (below), 'the relevant local authority' means the local authority (or, as the case may be, any of the local authorities) within whose area the offence in question is alleged to have taken place.

A person (A) may issue a CPN to an individual or body (B) only if:

(a) B has been given a written warning that the notice will be issued unless B's conduct ceases to have the detrimental effect referred to above; and
(b) A is satisfied that, despite B having had enough time to deal with the matter, B's conduct is still having that effect.

Before issuing a community protection notice, A must inform any body or individual (s)he thinks appropriate.

A CPN must identify the conduct referred to above, and explain the effect of ss 46 to 51 below. It may specify periods within which, or times by which, requirements to do things or take reasonable steps are to be complied with.

For the purposes of s 43, conduct on, or affecting, premises ('premises' includes any land) that a particular person owns, leases, occupies, controls, operates or maintains, is treated as conduct of that person, except that conduct on, or affecting, premises occupied for the purposes of a government department is treated as conduct of the Minister in charge of that department. An individual's conduct is not treated as that of another person if that person cannot reasonably be expected to control or affect it.

Where:

(a) an authorised person has power to issue a CPN;
(b) the detrimental effect referred to above arises from the condition of premises or the use to which premises have been put; and
(c) the authorised person has made reasonable enquiries to find out the name or proper address of the occupier of the premises (or, if the premises are unoccupied, the owner) but without success,

that person may post the CPN on the premises, and enter the premises, or other premises, to the extent reasonably necessary for that purpose. 'Owner' means a person (other than a mortgagee not in possession) entitled to dispose of the fee simple of the premises, whether in possession or in reversion, or a person who holds or is entitled to the rents and profits of the premises under a lease that (when granted) was for a term of not less than three years.

The CPN is treated as having been issued to the occupier of the premises (or, if the premises are unoccupied, the owner) at the time the notice is posted.

By s 46, a person issued with a CPN may appeal, within 21 days, to a magistrates' court against the notice on specified grounds. While an appeal is in progress, a requirement imposed by the notice to stop doing specified things remains in effect, unless the court orders otherwise, but any other requirement imposed by the notice is of no effect.

A divisional court has held that, leaving aside the possibility of a CPN being the subject of an appeal or judicial review, there is a power for an authorised person to revoke or vary a CPN, as well as issue one. It went on to state that, if the person or body (B) issued with the CPN sends written representations to an authorised person with a reasoned case that the CPN is inappropriate, the authorised person must consider those representations when considering the exercise of his/her discretion as to whether to retain, or revoke or vary, the notice. If (s)he fails to do so, B will be able to seek a judicial review of that failure. The court thought that, in going about their decision-making about a request for variation or discharge of CPNs, authorised persons should, as a minimum, operate a system for receiving and adjudicating such requests, and that relevant information should briefly be given with any CPN about how to seek a variation or discharge (eg on a change of circumstance), in addition to information about an appeal under s 46.

Failure to comply with notice

Remedial action by local authority

17.9   Where a person issued with a CPN ('the defaulter') fails to comply with a requirement of the notice, the relevant local authority (RLA) may take action as follows under A-s B,C&PA 2014, s 47.

The RLA may have work carried out to ensure that the failure is remedied, but only on land that is open to the air.

As regards premises other than land open to the air, if the RLA issues the defaulter with a notice:

(a) specifying work it intends to have carried out to ensure that the failure is remedied;
(b) specifying the estimated cost of the work; and
(c) inviting the defaulter to consent to the work being carried out,

the RLA may have the work carried out if the necessary consent is given. 'The necessary consent' means the consent of the defaulter, and (unless the RLA has made reasonable but unsuccessful efforts to contact the owner of the premises) of the owner of the premises on which the work is to be carried out (if that is not the defaulter). If work is carried out and the RLA issues a notice to the defaulter detailing that work, and specifying an amount that is no more than the cost to the authority, the defaulter is liable to the authority for that amount (subject to the outcome of any appeal to a magistrates' court, within 21 days, on the ground that the amount specified is excessive).

For the above purposes, the 'RLA' means—

(a) the local authority that issued the CPN;
(b) if the CPN was not issued by a local authority, the local authority (or, as the case may be, one of the local authorities) that could have issued it.

Offence of failing to comply with notice

17.10 A person (D) issued with a CPN who fails to comply with it commits an offence under A-sB,C&PA 2014, s 48(1). D does not commit the offence if (s)he took all reasonable steps to comply with the notice, or there is some other reasonable excuse for the failure to comply with it (s 48(3)). The offence under s 48(1) is a summary offence punishable with a fine not exceeding level 4 (if D is an individual) or an unlimited fine (if D is a body). On conviction for such an offence, the court may also make a remedial order to ensure that what the notice requires to be done is done, and may also make a forfeiture order in respect of any item used in the commission of the offence (ss 49 and 50).

Seizure of item used in commission of offence

17.11 A justice of the peace who is satisfied on information on oath that there are reasonable grounds for suspecting that an offence under A-sB,C&PA 2014, s 48 has been committed, and that there is an item used in the commission of the offence on premises specified in the information, may issue a warrant authorising any constable or person designated by the relevant local authority to enter the premises within 14 days from the date of issue of the warrant to seize the item (s 51). A constable or designated person may use reasonable force, if necessary, in executing such a warrant.

A constable or designated person who has seized an item under such a warrant:

(a) may retain the item until any criminal proceedings for the s 48 offence in whose commission the item was allegedly used have been finally determined, if such proceedings are started before the end of the period of 28 days following the day of the seizure;
(b) otherwise, must before the end of that period return the item to the person from whom it was seized.

Fixed penalty notices

17.12   A-sB,C&PA 2014, s 52 provides that an authorised person (**17.8**) may issue a fixed penalty notice to anyone who that person has reason to believe has committed an offence under s 48. A fixed penalty notice offers the person to whom it is issued the opportunity of discharging any liability to conviction for the offence by payment of a fixed penalty to a local authority specified in the notice (ie the local authority that issued the CPN to which the fixed penalty notice relates or, if that was not issued by a local authority, the local authority (or, as the case may be, one of the local authorities) that could have issued it).

Where a person is issued with a fixed penalty notice:

(a) no proceedings may be taken for the s 48 offence before the end of the period of 14 days following the date of the notice;

(b) the person may not be convicted of the offence if the person pays the fixed penalty before the end of that period.

The amount of fixed penalty specified in the notice must not exceed £100. The notice may specify two amounts and specify that, if the lower of those amounts is paid within a specified period (of less than 14 days), that is the amount of the fixed penalty.

Issuing of notices

17.13   A CPN or related fixed penalty notice may be issued by handing it to the person, leaving it at his/her proper address, or sending it by post to him/her at that address. A notice to a body corporate may be issued to the secretary or clerk of that body, and to a partnership to a partner or a person who has the control or management of the partnership business.

For these purposes, the proper address of a person is the person's last known address, except that in the case of:

(a) a body corporate or its secretary or clerk, it is the address of the body's registered or principal office;

(b) a partnership or person having the control or the management of the partnership business, it is the principal office of the partnership.

Guidance

17.14   Ministerial guidance has been issued to chief officers of police and local authorities about the exercise of functions under the above provisions.

PUBLIC SPACES PROTECTION ORDERS

17.15   These orders are governed by the A-s B,C&PA 2014, Part 4, Ch 2 (ss 59–76).

The public spaces protection order (PSPO) is intended to deal with a particular nuisance or problem in a particular area that is detrimental to the local community's quality of life, by imposing conditions on the use of that area. The order can also be used to deal with likely future problems.

Power to make a PSPO

17.16   A-s B,C&PA 2014, s 59 provides that a local authority may make a PSPO if satisfied on reasonable grounds that two conditions are met:

(a) that activities carried on in a public place within the authority's area have had a detrimental effect on the quality of life of those in the locality, or it is likely that activities will be carried on in a public place within that area and that they will have such an effect; and

(b) that the effect, or likely effect, of the activities:
    (i) is, or is likely to be, of a persistent or continuing nature,
    (ii) is, or is likely to be, such as to make the activities unreasonable, and
    (iii) justifies the restrictions imposed by the notice.

'Local authority' has the same meaning as for a CPN (**17.8**). 'Public place' is defined as meaning any place to which the public or any section of the public has access, on payment or otherwise, as of right or by virtue of express or implied permission.

A PSPO must identify the public place referred to in (a) ('the restricted area') to which it applies. It must also identify the activities referred to in (a), explain the effect of s 63 below (where it applies) and s 67 below, and specify the period for which it has effect.

A PSPO can prohibit specified things in respect of a specified public place (for example, drinking alcohol (other than on licensed premises), unauthorised busking, urinating (other than in a public toilet) or loitering in groups), require specified things to be done there (for example, keeping dogs on leashes), or both. The only prohibitions or requirements that may be imposed are ones that are reasonable to impose in order to prevent the detrimental effect referred to in (a) above from continuing, occurring or recurring, or to reduce that detrimental effect or to reduce the risk of its continuance, occurrence or recurrence.

A prohibition or requirement may be framed so as to apply:

(i) to all persons, or only to persons in specified categories, or to all persons except those in specified categories;

(ii) at all times, or only at specified times, or at all times except those specified;

(iii) in all circumstances, or only in specified circumstances, or in all circumstances except those specified.

There is power for a PSPO to be made by a body (other than a local authority) if it is designated by the Secretary of State in respect of land that it has a statutory power to regulate.

The following matters relating to a PSPO are of particular relevance to the police.

Failure to comply with a PSPO

Under A-s B,C&PA 2014, s 67(1), it is an offence for a person without reasonable excuse: **17.17**

(a) to do anything that (s)he is prohibited from doing by a PSPO, or

(b) to fail to comply with a requirement to which (s)he is subject under a PSPO.

A person does not commit the offence by failing to comply with a prohibition or requirement that the local authority did not have power to include in the PSPO (s 67(3)). The offence under s 67(1) is a summary offence punishable with a fine not exceeding level 3.

Consuming alcohol in breach of a PSPO is not an offence under s 67 (s 67(4)). As to such conduct, see s 63 below.

Prohibition on consuming alcohol

17.18   Special powers are given by A-s B,C&PA 2014, s 63 to a constable or an authorised person who reasonably believes that a person (P) is or has been consuming alcohol in breach of a prohibition in a PSPO, or intends to consume alcohol in circumstances in which doing so would be a breach of such a prohibition. 'Authorised person' means a person authorised for the purposes of s 63 by the local authority that made the PSPO.

The constable or authorised person may require P:

(a) not to consume, in breach of the order, alcohol or anything which the constable or authorised person reasonably believes to be alcohol;

(b) to surrender anything in P's possession which is, or which the constable or authorised person reasonably believes to be, alcohol or a container for alcohol.

A constable or an authorised person may dispose of anything surrendered under (b) in whatever way (s)he or she thinks appropriate.

As can be seen, it is not essential in order for a requirement to be made that a person is seen consuming alcohol. If a person has alcohol in a glass or open container this raises the inference that (s)he intends to drink it. Where alcohol is in a sealed container, notice must be taken of the circumstances. Possession of a 'four-pack' by someone walking home differs from possession by a group of persons assembled at that place for the purpose of drinking.

A constable or an authorised person who imposes such a requirement must tell P that failing without reasonable excuse to comply with the requirement is an offence.

A requirement imposed by an authorised person is not valid if (s)he or she is asked by P to show evidence of his or her authorisation, and fails to do so.

A person who fails without reasonable excuse to comply with a requirement commits a summary offence, punishable with a fine not exceeding level 2, contrary to s 63(6).

By s 62, a PSPO on consuming alcohol does not apply to:

(a) premises (other than council-operated licensed premises) authorised by a premises licence to be used for the supply of alcohol;

(b) premises authorised by a club premises certificate to be used by the club for the supply of alcohol;

(c) a place within the curtilage of premises within (a) or (b);

(d) premises which by virtue of the Licensing Act 2003, Part 5 (permitted temporary activities (**20.96**)) may at the relevant time be used for the supply of alcohol or which, by virtue of that Part, could have been so used within the 30 minutes before that time;

(e) a place where facilities or activities relating to the sale or consumption of alcohol are at the relevant time permitted by virtue of a permission granted under the Highways Act 1980, s 115E (council permission to set up objects and structures on a highway for income-producing purposes).

A prohibition in a PSPO on consuming alcohol does not apply to council-operated licensed premises (ie premises in respect of which a local authority holds the premises licence, or which are occupied by such an authority or managed by it or on its behalf) when the premises are being used for the supply of alcohol, or within 30 minutes thereafter.

Fixed penalty notices

17.19 Section 68 provides that a constable or an authorised person may issue a fixed penalty notice to anyone he or she has reason to believe has committed an offence under s 63 or 67 in relation to a PSPO. 'Authorised person' means a person authorised for the purposes of s 68 by the local authority that made the order.

Such a fixed penalty notice is a notice offering the person to whom it is issued the opportunity of discharging any liability to conviction for the offence by payment of a fixed penalty to a local authority specified in the notice, ie the one that made the PSPO. The provisions relating to a PSPO fixed penalty notice are the same as mentioned above in relation to those which apply to a fixed penalty notice under the community protection notice provisions (**17.12**).

Guidance

17.20 Ministerial guidance has been issued to chief officers of police and local authorities about the exercise of functions under the above provisions.

CHAPTER 18

Preserving the Peace

BREACH OF THE PEACE

18.1　There is a breach of the peace whenever and wherever (even on private premises):

(a) harm is *actually done*, or is *likely* to be done, to a person, whether by the conduct of the person against whom a breach of the peace is alleged or by someone whom it provokes; or

(b) harm is *actually* done, or is *likely* to be done, to a person's property in his/her presence; or

(c) a person is genuinely in fear of harm to him/herself or to his/her property in his/her presence as a result of an assault, affray, riot, or other disturbance.

Mere agitation or excitement does not amount to a breach of the peace where there is no question of harm or threat of harm.

A breach of the peace is not in itself a criminal offence, but it can result in an arrest without warrant being lawfully made or in other steps being lawfully taken or in a binding-over order being made.

A police officer (and, indeed, every citizen) has power to arrest without a warrant or to take other reasonable preventive steps where:

(a) a breach of the peace occurs in his/her presence; or

(b) (s)he reasonably believes that such a breach is about to occur, whether or not there has yet been a breach.

The House of Lords has confirmed that the power to prevent a breach of the peace does not arise when it is believed (however reasonably) that a breach of the peace is *likely to become imminent* and that it is reasonable to take action to prevent it; it must be believed that the breach of the peace is *actually imminent*.

An example of the exercise of the power to take preventive steps in relation to a reasonably apprehended imminent breach of the peace is 'kettling' (containment of people within a police cordon during a violent demonstration). The European Court of Human Rights, affirming a decision of the House of Lords, has held that 'kettling' does not amount to a deprivation of the right to liberty, guaranteed by art 5 of the European Convention on Human Rights, of those kettled so long as it is rendered unavoidable as a result of circumstances beyond the control of the authorities, is necessary to avert a real, imminent risk of serious injury or damage, and is kept to the minimum required for that purpose. The Court of Appeal has held that kettling of a group of relatively peaceful demonstrators to prevent another group of protesters, who were violent, taking over the former demonstration was legally permissible because there had been a reasonably believed risk of imminent and serious breaches of the peace. On the other hand, a divisional court has held that police who had lawfully kettled demonstrators had exceeded their powers by requiring them to be filmed and provide their personal details for identification purposes as the price for release.

Where a person's conduct is lawful but provocative, preventive steps cannot be taken against him/her if his/her behaviour is reasonable; if it is, preventive steps can be taken against opponents who are likely to react violently to the provocation if this reaction would be unreasonable. For example, where a person concerned in a lawful activity, eg a demonstration against neo-fascists, acts reasonably in making a reasoned but impassioned speech against neo-fascism, which is likely to produce violence from his/her opponents, preventive steps can be taken not against him/her but against them if their violent reaction would be unreasonable. It would be different if it is the speaker's conduct which is unreasonable.

A constable is entitled to enter private premises (or to remain on them if (s)he is already there), to deal with a breach of the peace or reasonably apprehended breach of the peace.

RIOT

The Public Order Act 1986 (POA 1986), s 1(1) provides that, where 12 or more persons who are present together use or threaten unlawful violence for a common purpose and the conduct of them (taken together) is such as would cause a person of reasonable firmness present at the scene to fear for his/her personal safety, each of the persons using unlawful violence for the common purpose is guilty of riot. A prosecution (**6.48**) for riot is triable only on indictment. Punishable with up to 10 years' imprisonment, it is the most serious offence against public order.   **18.2**

Riot may be committed in private as well as in public places.

Use of unlawful violence

A person does not perpetrate the offence of riot merely by threatening unlawful violence; (s)he must actually use violence in the prescribed circumstances. If 12 or more people simply threaten violence for a common purpose in a frightening way, but none of them uses violence, riot is *not* committed.   **18.3**

On the other hand, provided one of 12 or more people actually uses violence for the common purpose, the offence of riot is perpetrated by him/her (or by all those who so use violence if more than one does). Those who merely threaten violence for the common purpose can, however, be convicted as accomplices to riot if they assist or encourage the use of violence by another.

'Violence' is defined by POA 1986, s 8 as 'violent conduct'. It is not limited to violent conduct towards a person or persons since it includes violent conduct towards property (for example, smashing shop windows or overturning cars). Nor is it limited to conduct causing or intended to cause personal injury or damage to property, since it includes any other violent conduct (for example, throwing at or towards a person a missile of a kind capable of causing injury which does not hit or falls short).

The requirement that the violence be unlawful excludes from riot the use of violence which is justified by law (for example, under the rules relating to the use of reasonable force in self-defence, the prevention of crime or the effecting of an arrest).

Use or threat of unlawful violence for a common purpose by 12 or more present together

The 12 or more people need not form a cohesive group or be present pursuant to an agreement to come together.   **18.4**

The same comments apply to 'violence' as have been made above. If, during public disorder, the residents of a street use or threaten reasonable violence for the common purpose of defending themselves or their property from attack, their use or threat of violence does not constitute a riot because their violence is not unlawful.

In relation to threats, they may be by gestures alone (eg the brandishing of a weapon or the pointed display of it) or be by words alone, or be by a combination of both.

POA 1986, s 1(2) provides that it is immaterial whether or not the 12 or more use or threaten violence simultaneously. Equally, it is immaterial whether or not *any* of the others used or threatened violence at the time of the use of violence by the defendant (D). Provided that 12 or more persons who use or threaten violence are present *together* throughout, and that the violence is used or threatened by 12 or more for a common purpose, the offence of riot can be committed. Thus, it covers the situation where violence is used or threatened in one part of a crowd, then dies away, only to break out in another part at a later time.

It must be proved that the use or threat of violence by the 12 or more present together was for a purpose common to them (or at least to 12 of them). The question is not whether the 12 or more were present for a common purpose but whether they threatened or used violence for a common purpose. The common purpose need not be violence and it need not be an unlawful purpose (although, no doubt, it will normally be so). The common purpose may be inferred from the conduct of those involved.

Conduct such as would cause fear

18.5 The question is not whether the conduct of an individual defendant would cause fear but whether the conduct of the '12 or more present together ...' is such as would, *taken together,* cause fear (ie alarm or apprehension). The conduct of the 12 or more must be such as *would* cause a person (ie a third-party bystander) of reasonable firmness present at the scene to fear for his/her *personal safety*; it does not matter whether it actually caused fear to a person present at the scene or even *might* have.

No person of reasonable firmness need actually be, or be likely to be, present at the scene; in fact, no one else (beside the 12 or more) need be present or likely to be present at the scene. No doubt the case where there are no bystanders will be exceptional; where it occurs proof of the riot may be particularly difficult.

Mens rea

18.6 By POA 1986, s 6(1), a person is guilty of riot only if (s)he intends to use violence or is aware that his/her conduct may be violent. Section 6(5) provides that, for the purposes of the offence of riot, a person whose awareness is impaired by intoxication is to be taken to be aware of that of which (s)he would be aware if not intoxicated, unless (s)he shows either that his/her intoxication was not self-induced (as where his/her drink has been 'laced') or that it was caused solely by the taking or administration of a substance in the course of medical treatment. 'Intoxication' here means any intoxication, whether caused by drink, drugs, or other means (eg glue), or by a combination of means.

VIOLENT DISORDER

18.7 POA 1986, s 2(1) provides that, where three or more people who are present together use or threaten unlawful violence and their conduct (taken together) is such as would cause a person of reasonable firmness present at the scene to fear for his/her personal safety, each of the persons using or threatening unlawful violence is guilty of violent disorder, an indictable (either-way) offence punishable with up to five years' imprisonment. No person of reasonable firmness need actually be, or be likely to be, present at the scene.

As in the case of riot, violent disorder may be committed in private as well as in public places.

The prohibited conduct for this offence is substantially the same as that of riot (and the comments made when discussing the identical elements in riot are equally applicable here) with the exceptions that:

(a) D is guilty if (s)he uses or *threatens unlawful* violence;
(b) only three persons (including D) who are present together are required to use or threaten unlawful violence;
(c) neither D nor the other participants are required to use or threaten unlawful violence for a common purpose.

The operation of the above can be illustrated as follows. If a racist march or static demonstration takes place in the centre of an immigrant community, accompanied by threats of immediate violence which would make a person of reasonable firmness fear for his/her personal safety, the offence is committed; but not if the taunts are merely of a racist nature highly offensive to local inhabitants, although they might give rise to an offence under POA 1986, s 4A, 5, or 18 or the Crime and Disorder Act 1998, s 31 (**18.17–18.19, 13.97**, and **13.93**).

By POA 1986, s 6(2) a person is guilty of violent disorder only if (s)he intends to use or threaten violence or is aware that his/her conduct may be violent or threaten violence.

As in the case of riot, a person whose awareness is impaired by intoxication is to be taken to be aware of that of which (s)he would be aware if not intoxicated, unless (s)he shows either that his/her intoxication was not self-induced or that it was caused solely by the taking or administration of a substance in the course of medical treatment.

AFFRAY

A person is guilty of affray, contrary to POA 1986, s 3(1), if (s)he uses or threatens unlawful violence towards another and his/her conduct is such as would cause a person of reasonable firmness present at the scene to fear for his/her personal safety. Affray is an indictable (either-way) offence punishable with up to three years' imprisonment.   **18.8**

The prohibited conduct is that:

(a) D must use or threaten violence *towards* another; and
(b) his/her conduct must be such as would cause a person of reasonable firmness present at the scene to fear for his/her personal safety.

Like riot and violent disorder, affray may be committed in private as well as in public places. One result is that, if a fight breaks out at a party in someone's home, those who participate in it can be guilty of affray *if the terms of the offence are satisfied. Most* of the elements of the actus reus are common to riot and violent disorder.

Use or threat of unlawful violence towards another

Unlike the position in riot and violent disorder, 'violence' here does not include violent conduct towards property.   **18.9**

Another important difference between affray and riot and violent disorder is that a threat of violence cannot be made by the use of words alone. Of course, an affray can be committed where a threat of violence is made by a combination of words and gestures (such as shouting out, 'I'll get you for that', while brandishing a weapon or even shaking a fist) as well as by gestures alone. A threat or use of violence must be directed towards a person or persons actually present at the scene.

An affray, like riot and violent disorder, can only be committed if the violence is 'unlawful' as defined at **18.3**. Thus, a person who fights another in self-defence cannot be guilty of an affray, although his/her assailant can be if his/her use of violence would make a person of reasonable firmness fear for his/her personal safety.

Where two people use or threaten unlawful violence there is no need for them to do so for a common purpose.

Conduct such as would cause a person of reasonable firmness present at the scene to fear for his/her personal safety

18.10 This requirement provides an important limit on the offence, and excludes many fights from it. For example, it is most unlikely that a fight between two people, arising out of a personal quarrel but without any danger of the involvement of others, would constitute an affray.

Where two or more people use or threaten the unlawful violence, it is the conduct of them taken together that must be considered for the purpose of ascertaining whether the conduct would have the required effect.

As in the case of riot and violent disorder no one besides the participants (ie no bystander) need be present, or be likely to be present, at the scene, and it is expressly provided by POA 1986, s 3(4) that no person of reasonable firmness need actually be, or be likely to be, present at the scene. The offence of affray is, in reality, concerned with *three* persons: a person using or threatening unlawful violence (D), a person towards whom the violence or threat is directed (V), and a notional person of reasonable firmness. It is not enough that the victim of the violence or threat is put in fear for his/her personal safety. The question is whether, *if the notional person of reasonable firmness* had been so present, (s)he would have been caused to fear for *his/her* personal safety (as opposed to that of D or V or someone else).

Mens rea

18.11 By POA 1986, s 6(2), a person is guilty of affray only if (s)he intends to use or threaten violence or is *aware* that his/her conduct may be violent or threaten violence. As in the case of riot, a person whose awareness is impaired by intoxication—whether by drink, drugs, or other means (or a combination of these)—must be taken to be aware of what (s)he would have been aware of if not intoxicated, unless (s)he shows that his/her intoxication was not self-induced or that it was caused solely by the taking or administration of a substance in the course of medical treatment.

FEAR OR PROVOCATION OF VIOLENCE

18.12 A person commits an offence against POA 1986, s 4(1) if (s)he:

(a) uses towards another person threatening, abusive, or insulting words or behaviour; or

(b) distributes or displays to another person any writing, sign or other visible representation which is threatening, abusive, or insulting, with intent to cause that person to believe that immediate unlawful violence will be used against him/her or another by any person, or to provoke the immediate use of unlawful violence by that person or another, or whereby that person is likely to believe that such violence will be used or it is likely that such violence will be provoked. The offence under s 4(1) is a summary offence punishable with up to six months' imprisonment and/or an unlimited fine.

Threatening, abusive, or insulting

The words 'threatening, abusive or insulting' do not bear an unusual legal meaning. **18.13** Instead, the magistrates will decide as a question of fact whether D's conduct was threatening, abusive, or insulting in the ordinary meaning of those terms, and this is to be judged according to the impact which the conduct would have on a reasonable member of the public. Behaviour is not threatening, abusive, or insulting merely because it gives rise to a risk that immediate violence will be feared or provoked, nor simply because it gives rise to anger, disgust, or distress.

If conduct is threatening, abusive, or insulting, it does not matter whether or not anyone who witnessed it felt him/herself to be threatened, abused, or insulted.

Use, distribution, or display

The distribution or display of any writing, sign or other visible representation which is **18.14** threatening, abusive, or insulting covers handing out leaflets (distribution) or holding up a banner or placard (display).

POA1986, s 4 requires that threatening, abusive, or insulting words or behaviour must be used *towards another person* or that threatening, etc written material be distributed or displayed *to another*. In relation to the use of threatening, abusive, or insulting words or behaviour, 'towards another' imports a requirement that the words or behaviour in question must be directed towards (ie deliberately aimed at) another particular person or persons; if they are not, one must rely on the offence under s 4A or the lesser offence under s 5 (**18.17-18.19**). Conduct is not used towards another if (s)he is not present, in the sense that (s)he can perceive with his/her own senses the threatening words or behaviour, etc. Thus, a person who makes a threat against a person who is out of earshot and only learns of it through a third party who is not under the control or direction of the maker of the threat cannot be convicted of an offence under s 4. However, whilst the person towards whom the behaviour was aimed must have been present to perceive it, this does not mean that the only means of proving that the victim perceived the behaviour is by hearing evidence from that person. Justices may rely solely on evidence from a bystander and may draw the inference that the victim did perceive what was said and done by an accused.

The insertion of 'to another' after 'displays' requires that the display of written material be directed towards another particular person or persons (ie deliberately brought to his/her attention), rather than simply being displayed (eg as a poster).

Public or private place

With one exception, an offence under POA 1986, s 4 can be committed in private **18.15** places, as well as in public places. Thus, an offence can be committed by pickets who threaten working colleagues, whether the pickets are inside or outside factory premises, or by protesters who invade a military base.

The exception is that, in order to exclude domestic disputes, s 4 has the effect of providing that the use of words or behaviour inside a dwelling is only an offence if the addressee (ie another person towards whom the words or behaviour are used or the writing, etc is displayed) is not inside that dwelling (see below) or any other dwelling. Thus, to use threatening, abusive, or insulting words towards someone else in the same house cannot be an offence under s 4, and the same is true if such words are shouted to someone in the house next door. On the other hand, if such words are shouted in a house at a next-door neighbour who is in his/her back garden (see below), an offence under s 4 will be committed, provided that the other elements of the offence are satisfied.

Section 8 provides that, for the above purposes, 'dwelling' means any structure or part of a structure occupied as a person's home or as other living accommodation (whether the occupation is separate or shared with others). By s 8, 'structure' includes a tent, caravan, vehicle, vessel, or other temporary or movable structure. While accepting that a roof garden within a residential building would, as a matter of fact, be an integral part of that building and might rightly be deemed as 'part of a structure', a divisional court has held that a front or rear garden of a dwelling-house is not itself a 'structure', even though it might contain structures, such as a gazebo or greenhouse. Nor did the enclosure of a garden by a wall or fence around garden make it a structure, even though the wall or fence might itself be a 'structure'. The court added that, taking the definition of 'dwelling 'as a whole, even if a front or rear garden could be regarded in physical terms as a 'structure' or 'part of a structure', rather than merely being open space, it would not normally be regarded as being 'occupied as a person's home or as other living accommodation'. Section 8 provides that 'dwelling' does not include any part of a structure not occupied as a person's home or other living accommodation, eg as a garage, a shop with accommodation over, or the communal parts of a block of flats. Thus, if threatening words are shouted from a flat to a shop below, an offence under s 4 may be committed, and so may it if the words are shouted from the shop to the flat upstairs. The Court of Appeal has held that a police cell is not a place which a person occupies as a home or other living accommodation and is therefore not a dwelling for these purposes.

Mens rea

18.16   D must either intend the words, behaviour or writing, etc to be threatening, abusive, or insulting or be aware that they or it might be. The result of this requirement is that a person, who uses words which are seemingly innocuous but which are addressed to, or heard by, persons to whom (unknown to him/her) they are highly insulting, is not guilty of the present offence.

A person whose awareness is impaired by intoxication must be taken to be aware of that of which (s)he would be aware if not intoxicated, unless (s)he shows either that his/her intoxication was not self-induced or that it was caused solely by the taking or administration of a substance in the course of medical treatment.

POA 1986, s 4 also requires that D's use of the words or behaviour towards another (hereafter described as 'an addressee'), or D's distribution or display to an addressee of the writing, etc, must be intended by D or be likely (whether or not D realises this) either:

(a) to provoke the immediate use of unlawful violence by an addressee or another; or
(b) to cause an addressee to believe that immediate unlawful violence will be used against him/her or another.

It will be noted, under (a), that it is not necessarily an addressee who must be intended or likely to be provoked to immediate violence. It is sufficient that someone else present, towards whom the threatening, etc behaviour, etc was not directed, was intended or likely to be provoked. On the other hand, for the purposes of (b) (fear of immediate violence, intended or likely), the fear must be felt by an addressee (although it need not be fear of violence against him/herself, nor of violence by D).

It is important to note that the offence under s 4, unlike those under ss 1 to 3, is not concerned with the reactions of a hypothetical person of reasonable firmness. Instead, for example, a speaker must take his/her audience as (s)he finds it. If (s)he uses insulting words at a meeting, (s)he is guilty of the present offence if they are likely to provoke the immediate use of violence by the particular audience (s)he is addressing, even though (s)he does not intend to provoke this and even though his/her words would not be likely to cause a reasonable person so to react, provided that (s)he intends his/her words to be insulting or is aware that they might be.

It has been held that, since constables are under a common law duty to preserve the peace, they are unlikely to respond to threatening, abusive, or insulting conduct by using violence. Nevertheless, such conduct directed towards a constable will constitute an offence under s 4 if it is intended or likely to put him/her in fear of immediate unlawful violence or is intended to provoke him/her to such violence. If the conduct is so serious as to amount to a breach of the peace and make it likely that a constable to whom it is addressed will have to use violence in the exercise of his/her common law power to arrest for a breach of the peace, an offence under s 4 will not be committed because the violence likely to be provoked will not be unlawful.

The fact that the unlawful violence which is intended or likely to be feared or provoked must be *immediate* must be emphasised. However, that term has been given a liberal interpretation by a divisional court. 'Immediate' does not mean 'instantaneous'. Instead violence will be 'immediate' if it is likely to result in a relatively short period of time and without any intervening occurrence.

As indicated above, the immediate violence which is intended or likely to be feared or provoked will not be unlawful if it is reasonable force in self-defence, or is otherwise legally justified.

In the reference to an intention to cause, or the likelihood of causing, the apprehension of immediate unlawful violence or the provocation of it, 'violence' has the same meaning as in riot and related offences: see **18.3**.

HARASSMENT, ALARM, OR DISTRESS

POA 1986, s 4 does not deal with many minor acts of hooliganism or other anti-social **18.17** behaviour ('ASB') which are prevalent, particularly in inner city areas. It is at problems such as these, in particular, that the offences under POA 1986, ss 4A and 5 are aimed.

Under s 4A(1), a person is guilty of a summary offence, punishable with up to six months' imprisonment and/or an unlimited fine, if, *with intent to cause a person harassment, alarm or distress*, (s)he:

(a) uses *threatening, abusive, or insulting* words or behaviour, or *disorderly* behaviour; or

(b) displays any writing, sign or other visible representation which is *threatening, abusive, or insulting*,

thereby causing that or another person harassment, alarm or distress.

Under POA 1986, s 5(1), a person is guilty of a summary offence, punishable with a fine not exceeding level 3, if (s)he:

(a) uses *threatening or abusive* words or behaviour, or *disorderly* behaviour; or
(b) displays any writing, sign or other visible representation which is *threatening or abusive*,

within the hearing or sight of a person *likely to be caused harassment, alarm or distress thereby*. It will be noted that 'insulting' does not appear in s 5.

What is said about 'threatening', 'abusive' and 'insulting' at **18.13** is equally applicable here. A similar approach applies to 'disorderly'. Thus, it is a question of fact for the magistrates whether D's conduct was disorderly in the ordinary meaning of the term. A divisional court, in confirming this, has held that an element of 'violence' is not essential for there to be disorderly behaviour, and that neither is any feeling of insecurity, in an apprehensive sense, on the part of a member of the public.

In relation to the requirement that harassment, alarm or distress must be caused (s 4A) or be likely (s 5), 'harassment, alarm or distress' are not defined by POA 1986, but it has been held as follows for the purposes of ss 4A and 5.

Divisional courts have held that:

(1) 'Harassment' does not require any apprehension about one's personal safety, nor (probably) does 'distress'.
(2) A person can be harassed without experiencing emotional disturbance or upset.
(3) The words or behaviour in question must be likely to cause some real, as opposed to trivial, harassment.
(4) 'Distress' requires emotional disturbance or upset. That emotional disturbance does not have to be grave but the requirement should not be trivialised. There must be something amounting to real emotional disturbance or upset.

In relation to any threatening, abusive, or insulting writing, sign or other visible representation, the offences are limited to displaying and cannot (unlike an offence under s 4) also be committed by distribution. One result is that handing out threatening, abusive, or insulting leaflets is not caught by s 4A or s 5, unless the leaflets are so printed, and so held, that their contents can be said to be displayed in the sight of another.

Another distinction between the offences under s 4A and s 5 and that under s 4 is that the words or behaviour need not be used towards another person (nor need writing, etc be displayed to another). It follows that words or behaviour need not be directed towards another, nor need written material be deliberately brought to the attention of another. The display of graffiti or slogans can therefore suffice.

An offence under s 4A or s 5 may be committed in a public or a private place, except that no offence is committed where the words or behaviour are used, or the writing, sign or other visible representation is displayed, by a person inside a dwelling (**18.15**) and the other person who is harassed, alarmed or distressed thereby (s 4A), or within whose sight or hearing it occurs and who is likely to be harassed, alarmed or distressed thereby (s 5), is also inside that or another dwelling.

The separate elements: POA 1986, s 4A

18.18   While D's conduct need not be directed towards another (and written material need not be displayed by him/her to another), there must be a victim in the sense that someone else is actually caused harassment, alarm or distress.

D must either intend his/her words or behaviour, or the writing, etc, to be threatening, abusive, or insulting or be aware that they or it may be threatening, abusive, or insulting, or (as the case may be) intend his/her behaviour to be or be aware that it may be disorderly.

D must also intend his/her threatening, etc words or behaviour or his/her display of threatening, etc, writing, etc to cause a person harassment, alarm or distress; it is irrelevant that the person who was actually caused the harassment, etc was not the intended victim.

Section 4A(3) provides two defences. Section 4A(3)(a) provides a defence for a defendant who alleges that (s)he was inside a dwelling at the material time. It states that it is a defence for D to prove that (s)he was inside a dwelling and had no reason to believe that the words or behaviour used, or the writing, sign or other visible representation displayed, would be heard or seen by a person outside that or any other dwelling.

Section 4A(3)(b) provides that it is a defence for a defendant to prove that his/her conduct was reasonable, reasonableness being judged objectively. It would be reasonable, for example, to shout a threat at a pickpocket across the street to deter him/her. This is an exceptionally vague defence and proving it may often be difficult.

The separate elements: POA 1986, s 5

While D's conduct need not be directed towards another (and written material need not be displayed by him/her to another), there must be a victim in the sense that what D does must be within the hearing or sight of a person likely to be caused harassment, alarm or distress thereby, although no likelihood of violence being provoked or feared is required. Whilst there must be evidence that someone was able to see or hear the words or behaviour complained of, the prosecution does not have to call evidence that the words or behaviour were actually heard or seen. However, if it does call a witness who says nothing in giving evidence about experiencing harassment, alarm or distress it cannot be inferred that (s)he was likely to be caused harassment etc.

It is not necessary that a person who is likely to be alarmed should be alarmed for his/her own safety; it suffices that (s)he is likely to be alarmed about the safety of someone unconnected with him/her.

D must either intend his/her words or behaviour, or the writing, etc, to be threatening or abusive or be aware that it may be threatening or abusive, or (as the case may be) intend his/her behaviour to be or be aware that it may be disorderly. Consequently, a person who gives no thought to the nature of his/her conduct, or who honestly believes that there is no risk of it being threatening or abusive, does not commit this offence. The provisions of s 6(5) concerning intoxication, discussed at **18.6**, apply equally to an offence under s 5.

Although D's conduct must be in the hearing or sight of a person likely to be caused harassment, alarm or distress thereby, D is not required to intend this or to be aware that it might occur. On the other hand, s 5(3)(a) provides that it is a defence for D to prove that (s)he had no reason to believe that there was any person within hearing or sight who was likely to be caused harassment, alarm or distress.

Section 5(3)(b) provides a defence identical to that under s 4A(3)(a) (**18.18**).

Section 5(3)(c) provides a defence (conduct reasonable) identical to that under s 4A(3)(b) (**18.18**). A divisional court has held that if the prosecution proves, as it must, that D's conduct was threatening, abusive, or disorderly, and that (s)he intended it to be, or was aware that it might be, it would in most cases follow that his/her conduct was objectively unreasonable. The court also stated that, in considering whether

18.19

conduct was reasonable, the court must have regard to all the circumstances, and to the European Convention on Human Rights, art 10(2), which sets out the grounds on which the freedom of expression may be interfered with.

Offences against POA 1986, s 5 are 'penalty offences' for the purpose of the Criminal Justice and Police Act 2001 (**3.40**).

PROTECTION OF ACTIVITIES OF ANIMAL RESEARCH ORGANISATIONS

Interference with contractual relationships so as to harm animal research organisation

18.20   The Serious Organised Crime and Police Act 2005 (SOCPA 2005), s 145(1) provides that a person (A) commits an offence (**6.48**) if, with the intention of harming an animal research organisation, A:

(a) does a relevant act; or

(b) threatens that A or someone else will do a relevant act,

in circumstances in which that act or threat is intended or likely to cause a second person (B):

(i)   not to perform any contractual obligation owed by B to a third person (C) (whether or not such non-performance amounts to a breach of contract);

(ii) to terminate any contract B has with C; or

(iii) not to enter into a contract with C.

A 'relevant act' is an act amounting to a *criminal offence*, or to a *tortious act* (ie an act giving rise to liability for damages) causing B to suffer loss or damage of any description. For these purposes, a tortious act does not include an act which is actionable on the ground *only* that it induces another person to break a contract with B.

'Contract' includes any other arrangement, and to 'harm' an animal research organisation means to cause the organisation to suffer loss or damage *of any description*, or to prevent or hinder the carrying-out by the organisation of any of its activities.

Section 145 does not apply to any act done wholly or mainly in contemplation or furtherance of a trade dispute within the meaning of the Trade Union and Labour Relations (Consolidation) Act 1992, ie a dispute between workers and their employers relating wholly or mainly to employment-related issues.

Section 145 seeks to plug gaps in other legislation dealing with forms of harassment and disorder, which were exploited by those who engaged in activities of various kinds aimed at 'persuading' animal research organisations to abandon research involving the use of animals. Targeting was not restricted to the actual animal research establishments, since it was also directed at other organisations alleged to have business dealings with such establishments or with organisations which had such dealings.

Intimidation of persons connected with animal research organisation

18.21   Many animal research establishments, and the business premises of organisations with a connection with such establishments, are able to take effective measures to protect their property, to a large extent, but their employees and others associated with them present an easier target to protesters.

SOCPA 2005, s 146 therefore provides that a person (D) commits an offence (**6.48**) if, with the intention of causing a second person (V) to abstain from doing something which V is entitled to do (or do something which V is entitled to abstain from doing):

(a) D threatens V that D or somebody else will do a relevant act; and
(b) D does so wholly or mainly because V is:
- (i)   an employee or officer of an animal research organisation,
- (ii)   a student at an educational establishment that is an animal research organisation,
- (iii)   a lessor or licensor of premises occupied by an animal research organisation,
- (iv)   a person who has a financial interest in, or who provides financial assistance to, such an organisation,
- (v)   a customer or supplier of such an organisation,
- (vi)   a person contemplating becoming someone within (iii), (iv), or (v),
- (vii)   a person who is, or is contemplating becoming, a customer or supplier of someone within (iii), (iv), (v), or (vi),
- (viii) an employee or officer of someone within (iii), (iv), (v), (vi), or (vii),
- (ix)   a person with a financial interest in, or who provides financial assistance to, someone within (iii), (iv), (v), (vi), or (vii),
- (x)   a spouse, civil partner, friend, or relative of, or a person known personally to, someone within (i)–(ix),
- (xi)   a person who is, or is contemplating becoming, a customer or supplier of someone within (i), (ii), (viii), (ix), or (x), or
- (xii)   an employer of someone within (x).

The provisions in s 146 do not apply to trade disputes within the meaning of TULR(C)A 1992 (**18.20**).

'Relevant act' has the same meaning as in s 145, except that the limitation whereby a tortious act does not include an act actionable only because it induces a breach of contract does not apply.

General

18.22 An offence against SOCPA 2015, s 145 or 146 is an indictable (either-way) offence punishable with up to five years' imprisonment.

'Animal research organisation' means any person or organisation which or who:

(a) is the owner, lessee or licensee of premises constituting or including a place specified in a licence granted under the Animals (Scientific Procedures) Act 1986;
(b) is the employer of, or engages under a contract of service, a person who is the holder of a licence under that Act or who is specified by it; or
(c) holds a licence under the provisions for the licensing of undertakings involving the use of animals for scientific procedures.

PUBLIC PROCESSIONS

Advance notice

18.23 POA 1986, s 11 requires that, where it applies, written notice specifying the date a procession is intended to be held, the time when it is intended to start, its proposed route, and the name and address of the person (or one of the persons) proposing to organise it must be given to a police station in the police area in which it is proposed the procession will start. Section 11 applies if the procession is public and it is intended:

(a) to demonstrate support for or opposition to the views or actions of any person or body of persons;
(b) to publicise a cause or campaign; or
(c) to mark or commemorate an event.

If such a procession starts in Scotland, it is the first police area in England along the proposed route which must be given notice. The notice may be given by hand not less than six clear days before the date upon which the procession is intended to be held or, if that is not reasonably practicable (as where there is an unexpected event following which a group wishes to demonstrate immediately), as soon as delivery is reasonably practicable.

Section 11 permits delivery of the notice by recorded delivery service, if the notice is delivered not less than six clear days in advance. The provisions of the Interpretation Act 1978 under which a document sent by post is deemed to have been served when posted and to have been delivered in the ordinary course of post do not apply to the service of such notices.

Section 11 does not apply to processions commonly or customarily held (for example, a procession connected with an annual gala, a Remembrance Day parade, or a monthly campaigning mass cycle ride), or to funeral processions organised by funeral directors (as opposed to processions which suddenly appear in protest at a death). The House of Lords has held that a monthly campaigning cycle ride through central London which took a different route each time, although it set off from a fixed starting point, was a commonly or customarily held procession and therefore no prior notice needed to be given to the police.

Each of the persons organising a public procession is guilty of a summary offence, punishable with a fine not exceeding level 3, if notice has not been so given, or if the details given in the notice differ from the actuality of the procession (s 11(7)). It is a defence for D to prove that (s)he did not know of, and neither suspected nor had reason to suspect, the failure to give such notice, or that differences in the time, date, or route occurred due to circumstances beyond his/her control, or with the agreement of a police officer or by his/her direction (s 11(8) and (9)).

Conditions

18.24 By POA 1986, s 12, the senior police officer may impose conditions in relation to any public procession, having regard to its time, place, or circumstances, including its route or proposed route, if (s)he reasonably believes that:

(a) it may result in serious public disorder, serious damage to property, or serious disruption to the life of the community; or
(b) the purpose of the persons organising it is the intimidation of others with a view to compelling them not to do an act they have a right to do, or to do an act they have a right not to do.

These conditions may include any measures which appear necessary to prevent such disorder, damage, disruption, or intimidation, including conditions as to the route or proposed route of the procession or prohibiting it from entering any specified public place.

For these purposes, the 'proposed route' of a public procession is not limited to a specifically planned route; it can also cover one which is spontaneously chosen by the participants from time to time. Thus, conditions can be imposed in respect of what is reasonably believed might be the potential route.

In the case of a procession which is actually being held, the senior police officer is the police officer most senior in rank present at the scene; in such a case, it is not necessary for the notice to be in writing. In the case where people are assembling for a procession, the senior police officer is the chief officer of police; in such a case the chief officer must give the notice in writing.

Offences are committed by an organiser (s 12(4)) or participant (s 12(5)) who knowingly fails to comply with a condition. A divisional court has held that the participants in a procession (a protest march) where conditions under s 12 had been imposed designating its route were in breach of those conditions when they diverted from the designated route to join another protest, because they remained part of the original procession. The court added that there would not have been a breach if the participants in the procession had left it to go home or for some similar purpose, because they would no longer be part of the procession. It is a defence for D to prove that the failure arose from circumstances beyond his/her control. A person who incites another to participate in a procession and to fail to comply with a condition also commits an offence (s 12(6)).

The offences under s 12(4), (5), and (6) are summary offences punishable with up to three months' imprisonment and/or a fine not exceeding level 4 in the case of ss 12(4) and 12(6), and with a fine not exceeding level 3 in the case of s 12(5).

Prohibition

A chief officer of police may apply under POA 1986, s 13 to the council of the district **18.25** for an order prohibiting for a period, not exceeding three months, the holding of all public processions (or any class of procession specified) within that district. The chief officer must reasonably believe that, because of particular circumstances existing, his/her power to impose conditions will not be sufficient to prevent serious public disorder. Such an order may be made by the council with the consent of the Secretary of State. The Commissioners of the Metropolitan and City of London Police Forces may themselves make such an order in respect of their police areas with the consent of the Secretary of State.

A person who organises (s 13(7)), or takes part in (s 13(8)), a public procession commits an offence if (s)he knows that it has been prohibited. Those who incite others to participate in a prohibited procession are also guilty of an offence (s 13(9)).

The offences under s 13(7), (8), and (9) are summary offences punishable with up to three months' imprisonment and/or a fine not exceeding level 4 in the case of ss 13(7) and 13(9), and with a fine not exceeding level 3 in the case of s 13(8).

Delegation

A chief officer of police may delegate any of his/her functions under s 12 or 13 to an **18.26** assistant chief constable or an assistant commissioner of police (as the case may be).

PUBLIC ASSEMBLIES

Imposing conditions

POA 1986, s 14 authorises the senior police officer, on the basis of the same grounds of **18.27** reasonable belief as in the case of conditions on public processions (**18.24**), to impose such conditions in relation to the place at which any public assembly (an assembly of two or more persons in a public place which is wholly or partly open to the air) may be (or continue to be) held, its maximum duration, or the maximum number of persons who may constitute it, as appear to him/her to be necessary to prevent serious public disorder, serious damage to property, serious disruption to the life of the community, or intimidation. A divisional court has held that conditions may be imposed under this provision requiring those assembled to disperse by a specified route and to stay in a specified place as long as necessary to enable dispersal to take place safely and without disorder.

A divisional court has held that a 'public assembly' for the purposes of the Public Order Act 1086, s 14 has to be in a particular location to which the public or any section of the public had access, which was wholly or partly open to the air, and which location could be fairly described as a 'scene'. Thus, separate gatherings, separated both in time and by many miles, even if co-ordinated under the umbrella of one body, were not 'one public assembly' within the meaning of s 14. The court contrasted the provisions of POA 1986, s 14 with s 14A (trespassory assemblies (**18.28**)), noting that under s 14A there was the express power to impose a prospective area-wide ban on trespassory assemblies (plural), subject to temporal and geographical limits.

In the case of an assembly which is actually being held, the senior police officer is the police officer most senior in rank present at the scene; in such a case it is not necessary for the notice to be in writing. In the case of an assembly which is intended to be held, the senior police officer is the chief officer of police; in such a case the chief officer must give the notice in writing. Although extensive detail is not required, the chief officer's notice must provide sufficient detail of the reasons for his/her belief in the ground (or grounds) for a demonstrator to understand why a direction is being given.

A chief officer of police may delegate any of his/her functions under s 14 or 14A (below) to an assistant chief constable or an assistant commissioner of police (as the case may be).

Under s 14, summary offences, punishable in the same way, are committed by the same classes of persons as in relation to a breach of a public procession condition (s 14(4) (organiser), (5) (participant)), and (6) (inciter)): see **18.24**.

Prohibition of trespassory assemblies

18.28   These are dealt with by POA 1986, ss 14A, 14B, and 14C.

Section 14A is concerned with the prohibition of a trespassory assembly before it has taken place. Where a trespassory assembly has already commenced, police involvement is limited to the prohibition of its continuance if an officer believes that this is necessary to prevent a breach of the peace.

Section 14A provides that where a chief officer of police reasonably believes that an assembly of 20 or more people is intended to be held in any district at a place on land in the open air to which the public has no right of access or only a limited right of access and that the assembly:

(a) is likely to be held without the permission of the occupier of the land or to conduct itself in such a way as to exceed the limits of any permission of his/her or the limits of the public's right of access; and

(b) may result:

    (i) in serious disruption to the life of the community, or

    (ii) where the land, or a building or monument on it, is of historical, architectural, archaeological, or scientific importance, in significant damage to the land, building, or monument, (s)he may apply to the council of the district for an order prohibiting for a specified period the holding of all trespassory assemblies in the district, or a part of it as specified. An order may only be made by a council with the Secretary of State's consent. In London the relevant Commissioner may make an order with such consent. Where such an order is made, it will operate to prohibit any assembly of 20 or more which is held on land in the open air to which the public has no right of access or only a limited right of access and takes place without the permission of the occupier, or in excess of his/her permission or the public's right of access. Such an order may not prohibit such an assembly for a period exceeding four days or in an area exceeding a circle with a radius of five miles from a specified centre.

It is irrelevant that the anticipated assembly is open to the public or is a private ceremony, but it must be held on land in the open air. This can be contrasted with the power to impose conditions on a public assembly under s 14; there the assembly may be held wholly or partly in the open air. 'Land' in s 14A includes land forming part of the highway. It follows, for example, that an intended obstructive assembly of 20 or more on the highway (which is by definition trespassing, since the public only have a right to pass and repass on it and make other reasonable use of it) can be the trigger for an order under s 14A.

It is important to understand that these powers are restricted to assemblies which are reasonably believed to be trespassory in nature. Assemblies *with permission* are not covered by the legislation unless there is a reasonable belief that they will be conducted in such a way as to exceed that permission.

Offences

A person who organises an assembly which (s)he knows to be prohibited under s 14A **18.29** commits a summary offence (s 14B(1)). So does a person who takes part knowing that the assembly is prohibited (s 14B(2)), or who incites another to commit an offence under s 14B(2) (s 14B(3)). The offences are punishable with up to three months' imprisonment and/or a fine not exceeding level 4 in the case of a s 14B(1) or (3) offence, and with a fine not exceeding level 3 in the case of a s 14B(2) offence,

Central to these offences is that the assembly must be prohibited under s 14A, and an assembly is prohibited only if, within the area and duration of the prohibition order, it:

(a) is held on land in the open air to which the public has no right of access or only a limited right of access; and

(b) takes place in the prohibited circumstances, ie without the permission of the occupier of the land or so as to exceed the limits of any permission of his/her or the limits of the public's right of access.

In the case of assemblies on the highway, the public's right of access to the highway has been held by the House of Lords to include the right to hold a public assembly which is peaceful and non-obstructive of the rights of passage of other users of the highway provided the assembly is a reasonable use of the highway.

Police powers

By s 14C, if a constable in uniform reasonably believes that a person is on his/her way **18.30** to an assembly within the area to which such an order under s 14A applies, which the constable reasonably believes is likely to be an assembly which is prohibited by that order, (s)he may, within the area specified in the order:

(a) stop that person; and

(b) direct him/her not to proceed in the direction of the assembly.

A person who fails to comply with such a direction which (s)he knows has been given to him/her commits a summary offence, punishable with a fine not exceeding level 3, under s 14C(3).

Raves

18.31   CJPOA 1994, ss 63 to 65 provide the police with powers to deal with 'raves'. Section 63 is the main provision. It applies to a gathering *on land in the open air* of 20 or more persons *(whether or not trespassers)* at which amplified music is played during the night (with or without intermissions) and is such as, by reason of its loudness and the duration and the time at which it is played, is likely to cause serious distress to the inhabitants of the locality; and for this purpose:

(a)   such a gathering continues during intermissions in the music and, where the gathering extends over several days, throughout the period during which amplified music is played at night (with or without intermissions); and

(b)   'music' includes sounds wholly or predominantly characterised by the emission of a succession of repetitive beats.

'Land in the open air' includes a place partly open to the air. Thus, both an aircraft hangar without doors, or a Dutch barn, would be such a place. The Act does not define the term 'during the night'.

Section 63 also applies to a gathering if:

(a)   it is a gathering on *land* of 20 or more persons who are *trespassing* on the land; and

(b)   the gathering would be of a kind described above if it took place on land in the open air.

Section 63 does not apply to a gathering which is licensed by a local authority entertainments licence.

Where a superintendent (or above) reasonably believes, in respect of any land, that:

(a)   two or more persons are making preparations for the holding there of a gathering to which s 63 applies;

(b)   10 or more are waiting for such a gathering to begin there; or

(c)   10 or more are attending such a gathering which is then in progress,

(s)he may give a direction that those persons and any other persons who come to prepare or wait for or to attend the gathering are to leave the land and remove any vehicles or other property which they have with them on the land. The direction may be conveyed to the gathering by any constable. If reasonable steps have been taken to convey the direction, it is deemed to have been given.

Persons occupying or working on the land and their families are exempt from a direction under s 63.

A person who knows that such a direction has been given which applies to him/her who:

(a)   fails to leave the land as soon as reasonably practicable; or

(b)   having left, again enters the land within a period of seven days beginning with the day on which the direction is given,

commits a summary offence under s 63(6). It is a defence for D to show that (s)he had a reasonable excuse for failing to leave the land as soon as reasonably practicable or, as the case may be, for again entering the land (s 63(7)).

A person commits a summary offence under s 63(7A) if:

(a) (s)he knows that a s 63 direction has been given which applies to him/her; and
(b) (s)he makes preparations for or attends a gathering to which s 63 applies within the 24-hour period starting when the direction was given.

This means that those who have been given a direction to leave a rave cannot simply move to another site (and carry on raving) with impunity.

An offence under s 63(6) or (7A) is punishable with up to three months' imprisonment and/or a fine not exceeding level 4.

Police powers

Section 64 provides that, where a superintendent (or above) reasonably believes that **18.32** circumstances exist which would justify giving such a direction, (s)he may authorise a constable to enter the land without warrant to ascertain that these circumstances exist and to exercise powers conferred upon him/her. Where such a direction has been given, and a constable reasonably suspects that any person to whom the direction applies has, without reasonable excuse:

(a) failed to remove any vehicle or sound equipment on the land which appears to the constable to belong to him/her or to be in his/her possession or under his/her control; or
(b) entered the land as a trespasser with a vehicle or sound equipment within the period of seven days beginning with the day on which the direction was given,

the constable may seize and remove that vehicle or sound equipment.

Section 66 permits a court to order the forfeiture of sound equipment but also allows the owner of the equipment (if (s)he is not the person from whom it was seized) to claim it by applying to the court within six months. The Police (Disposal of Sound Equipment) Regulations 1995 provide for the disposal of the equipment where a forfeiture order has been made and any such claim has been unsuccessful.

Where a constable in uniform reasonably believes that a person is on his/her way to such a gathering in respect of which a direction has been given, (s)he may, by s 65, stop that person and direct him/her not to go in the direction of the gathering. This power may only be exercised within five miles of the boundary of the site of the gathering. It does not apply to occupiers of the land, etc. A person who, knowing that such a direction has been given, fails to comply with such a direction commits a summary offence under s 65(4). The offence is punishable with a fine not exceeding level 3.

PUBLIC MEETINGS

It is a summary offence, contrary to the Public Meeting Act 1908, s 1(1), for any person **18.33** at a lawful public meeting to act in a disorderly manner for the purpose of preventing the transaction of the business for which the meeting was called together. It is also a summary offence to incite others to do so (s 1(2)). Both offences are punishable with up to six months' imprisonment and/or an unlimited fine.

'Public meeting' means a meeting open to the public and not restricted to members of a particular organisation or club. Meetings which are open to the public but held on private premises are therefore public meetings. A police officer who reasonably suspects a breach of the peace may enter and/or remain on the private premises.

Section 1 only applies to a lawful public meeting. The fact that a meeting is held on a highway does not render it unlawful merely because it is so held. However, the circumstances in which the meeting is held may make the participants liable for other offences, such as obstructing the highway.

Section 1(3) empowers a constable who reasonably suspects a person of committing an offence under s 1(1) or (2) to require him/her immediately to give his/her name and address. However, a constable may only execute this power if requested to do so by the chairman of the meeting. The constable must decide whether or not the person to whom his/her attention is directed is merely asking questions and making points which the chairman and platform party do not like, or is acting in a disorderly manner for the purpose of preventing the transaction of the business. Refusal or failure to give a name or address, or the giving of a false name and address, is a summary offence, punishable with a fine not exceeding level 1, under s 1(3).

DISPERSAL POWERS

Authorisations to use powers

18.34 By the Anti-social Behaviour, Crime and Policing Act 2014 (A-sB,C&PA 2014), s 34, an inspector (or above) may authorise the use in a specified locality, during a period of not more than 48 hours specified in the authority, of the powers given by s 35 (**18.35**). 'Specified' means specified in the authorisation.

An officer may give such an authorisation only if satisfied on reasonable grounds that the use of those powers in the locality during that period may be necessary for the purpose of removing or reducing the likelihood of:

(a) members of the public in the locality being harassed, alarmed or distressed, or
(b) the occurrence in the locality of crime or disorder.

In deciding whether to give such an authorisation an officer must have particular regard to the rights of freedom of expression and freedom of assembly set out in the European Convention on Human Rights, arts 10 and 11.

An authorisation must be in writing, must be signed by the officer giving it, and must specify the grounds on which it is given.

Directions under s 35 excluding a person from an area

18.35 A-sB,C&PA 2014, s 35 provides that if the following conditions are met and an authorisation is in force under s 34, a constable in uniform may direct a person who is in a public place in the locality specified in the authorisation:

(a) to leave the locality (or part of the locality), and
(b) not to return to the locality (or part of the locality) for the period specified in the direction (the exclusion period).

'Public place' means a place to which at the material time the public or a section of the public has access, on payment or otherwise, as of right or by virtue of express or implied permission. The exclusion period may not exceed 48 hours. Provided it begins during the period specified in the authorisation, the exclusion period may expire after that period.

The first condition is that the constable has reasonable grounds to suspect that the behaviour of the person in the locality has contributed or is likely to contribute to:

(i) members of the public in the locality being harassed, alarmed, or distressed, or
(ii) the occurrence in the locality of crime or disorder.

The second condition is that the constable considers that giving a direction to the person is necessary for the purpose of removing or reducing the likelihood of the events mentioned in (i) or (ii).

A direction:

(a) must be given in writing, unless that is not reasonably practicable;
(b) must specify the area to which it relates;
(c) may impose requirements as to the time by which the person must leave the area and the manner in which the person must do so (including the route).

Unless it is not reasonably practicable, the constable must tell the person to whom the direction is given that failing without reasonable excuse to comply with the direction is an offence. If the constable reasonably believes that the person to whom the direction is given is under 16, (s)he may remove the person to a place where the person lives or a place of safety.

Any constable may withdraw or vary a direction, but a variation must not extend the duration of a direction beyond 48 hours from when it was first given. Notice of such a withdrawal or variation must be given to the person to whom the direction was given, unless that is not reasonably practicable, and, if given, must be given in writing unless that is not reasonably practicable.

By s 36, a constable may not give a direction:

(a) to a person who appears to him/her to be under 10;
(b) that prevents the person to whom it is given (P) having access to a place where P lives; or
(c) that prevents P attending at a place which P is required to attend for the purposes of his/her employment, or a contract of services, or by an obligation imposed by or under an enactment or by the order of a court or tribunal, or expected to attend for educational, training, or medical treatment purposes, at a time when P is required or expected (as the case may be) to attend there.

A constable may not give a direction to a person if the person is one of a group of persons who are:

(a) engaged in conduct that is lawful under the Trade Union and Labour Relations (Consolidation) Act 1992, s 220, or
(b) taking part in a public procession to which the Public Order Act 1986, s 11 applies (**18.23**) in respect of which written notice has been given in accordance with s 11 or is not required to be given under s 11.

In deciding whether to give a direction a constable must have particular regard to the rights of freedom of expression and freedom of assembly set out in the European Convention on Human Rights, arts 10 and 11.

Surrender of property

A-sB,C&PA 2014, s 37 provides that a constable who gives a person (P) a direction **18.36** under s 35 may also direct P to surrender to the constable any item in P's possession or control that the constable reasonably believes has been used or is likely to be used in behaviour that harasses, alarms or distresses members of the public. Such a direction (a s 37 direction) must be given in writing, unless that is not reasonably practicable. A constable giving a s 37 direction must (unless it is not reasonably practicable) tell P that failing without reasonable excuse to comply with it is an offence, and give P written information about when and how P may recover the surrendered item.

The surrendered item must not be returned to P before the end of the exclusion period. If after the end of that period P asks for the item to be returned, it must be returned (unless there is some other statutory power to retain it). However, if it appears to a constable that P is under 16 and is not accompanied by a parent or other responsible adult, the item may be retained until P is so accompanied. If P has not asked for the return of the item before the end of 28 days beginning with the day on which the direction was given, the item may be destroyed or otherwise disposed of.

Record-keeping

18.37   A constable who gives a direction under A-sB,C&PA 2014, s 35 is required by s 38 to make a record of:

(a) the individual to whom the direction is given,
(b) the time at which the direction is given, and
(c) the terms of the direction (including in particular the area to which it relates and the exclusion period).

A constable who withdraws or varies a direction under s 35 must make a record of:

(a) the time at which the direction is withdrawn or varied,
(b) whether notice of the withdrawal or variation is given to the person to whom the direction was given and if it is, at what time, and
(c) if the direction is varied, the terms of the variation.

A constable who gives a direction under s 37 must make a record of:

(a) the individual to whom the direction is given,
(b) the time at which the direction is given, and
(c) the item to which the direction relates.

Offences

18.38   A person given a direction under A-sB,C&PA 2014, s 35 who fails without reasonable excuse to comply with it commits a summary offence under s 39(1).

A person given a direction under s 37 who fails without reasonable excuse to comply with it commits a summary offence under s 39(3).

An offence under s 39(1) is punishable with up to three months' imprisonment or a fine not exceeding level 4. An offence under s 39(3) is punishable with a fine not exceeding level 2.

Guidance

18.39   Ministerial guidance has been issued to chief officers of police about the exercise by their officers of those officers' functions under the above provisions.

TRESPASSING ON PROTECTED SITE

18.40   A person commits a summary offence under the Serious Organised Crime and Police Act 2005 (SOCPA 2005), s 128(1) if (s)he enters, or is on, any protected site in England or Wales as a trespasser. The offence (**6.48**) is punishable with a maximum of six months' imprisonment and/or an unlimited fine.

A 'protected site' is either a nuclear site (ie the outer perimeter (fences, etc) of premises in respect of which a nuclear site licence exists and other premises within that perimeter) or a 'designated site'. A 'designated site' is one specified or described by order by the Secretary of State.

SOCPA 2005 (Designated Sites) Order 2005 designates HM Naval Base, Clyde; Northwood Headquarters; RAF Brize Norton; RAF Croughton; RAF Fairford; RAF Feltwell; RAF Fylingdales; RAF Lakenheath; RAF Menwith Hill; RAF Mildenhall; RAF Molesworth; RAF Waddington; RAF Welford; Royal Navy Armaments Depot Coulport; and Defence Science and Technology Laboratory, Porton Down as designated sites. SOCPA 2005 (Designated Sites under s 128) Order 2007 designates the following sites: 85 Albert Embankment; Buckingham Palace; the Ministry of Defence Main Building, Whitehall; Thames House, Millbank; St James's Palace Site; the Chequers Estate Site; the Downing Street Site; the GCHQ Hubble Road Site; the GCHQ Scarborough Site; the GCHQ Bude Site; the Highgrove House Site; the Palace of Westminster and Portcullis House Site; Senedd Cymru/Welsh Parliament site; the Sandringham House Site; the Windsor Castle Site; the Kensington Palace Site; and the Anmer Hall Site as designated sites.

It is a defence for a person charged with an offence under s 128 to prove that (s)he did not know, and had no reasonable cause to suspect, that the site was protected (s 128(4)).

PUBLIC ORDER ACT 1936

POA 1936 governs the wearing of political uniforms and participation in quasi-military organisations.   **18.41**

Political uniforms

It is a summary offence under POA 1936, s 1(1) for any person to wear uniform, in any **18.42** *public place or at any public meeting*, which signifies his/her association with any political organisation or with the promotion of any political object. By way of exception, a chief officer of police, with the consent of the Secretary of State, may by order permit the wearing of such uniforms at a ceremonial, anniversary, or other special occasion.

For these purposes, a 'meeting' is a meeting held for the purpose of discussing matters of public interest and a 'public meeting' includes any meeting in a public place and any meeting in a private place if the public (or a section of the public) are permitted to attend (whether on payment or otherwise). There are many organisations (whose objectives are quite harmless) whose members wear some form of identifying clothing within their organisation. This would be within the scope of s 1, but for the fact that it only deals with cases where the uniform signifies association with a political organisation or the promotion of a political object *and* it is worn in a public place or a public meeting.

The term 'uniform' includes any particular article of clothing which is worn by each member of a group and which is intended to indicate his/her association with such an organisation or object. The article does not have to cover any major part of the body and it is sufficient to prove that the article has been commonly used by members of a political organisation.

An offence (**6.48**) under s 1(1) is punishable with up to three months' imprisonment and/or a fine not exceeding level 4.

Quasi-military organisations

18.43   POA 1936, s 2 is concerned with the preparation of private quasi-military forces of any description. A person who takes part in the control or management of any association, or in its training, commits an indictable (either-way) offence under s 2(1) if its members or adherents are:

(a) organised, trained, or equipped for the purpose of enabling them to usurp the functions of the police or armed forces; or
(b) organised and trained or organised and equipped either to promote a political object by the use or display of physical force, or to arouse reasonable apprehension of that purpose.

In relation to the last words in (b), it has been held that the fact that there was no evidence of actual attacks on opponents, or of plans to attack them, did not necessarily remove grounds for 'reasonable apprehension of that purpose'.

Where a person is charged with taking part in the control or management of such an association, as opposed to training, (s)he has a defence if (s)he proves that (s)he neither consented to nor connived at the organisation, training or equipment in contravention of s 2.

The provision of a reasonable number of stewards to assist in the preservation of order at a public meeting held on private premises is permitted, as is the provision of badges and insignia for them. The instruction of such persons in their lawful duties is also permitted.

An offence under s 2(1) (**6.48**) is punishable with up to two years' imprisonment.

ANTI-SOCIAL BEHAVIOUR

Injunction

18.44   The Anti-social Behaviour, Crime and Policing Act 2014 (A-sB,C&PA 2014), Part 1 (ss 1–21) contains provisions for an injunction to deal with actual or threatened anti-social behaviour. Such an injunction, like injunctions in general, is a purely civil order.

Power to grant injunctions

18.45   A-sB,C&PA 2014, s 1(1) provides that a court may grant an injunction (commonly referred to as a 'Part 1 injunction') against a person aged 10 or over (the respondent, R) if two conditions are met:

(a) the court is satisfied, on the balance of probabilities, that R has engaged or threatens to engage in anti-social behaviour. In relation to this the civil rules of evidence (**13.47**) apply;
(b) the court considers it just and convenient to grant the injunction for the purpose of preventing R from engaging in anti-social behaviour. This is an exercise of judgment and does not involve a standard of proof.

For these purposes, 'anti-social behaviour' is defined by s 2(1) as meaning conduct:

(i) that has caused, or is likely to cause, harassment, alarm or distress to any person;
(ii) capable of causing nuisance or annoyance to a person in relation to that person's occupation of residential premise (only applies to applicant for injunction within **18.47**(a)–(d)); or
(iii) capable of causing housing-related nuisance or annoyance to any person ('housing-related' means directly or indirectly relating to the housing management functions of a housing provider or local authority).

Applications for injunctions

18.46 An application for a Part 1 injunction must be made to a youth court, in the case of a respondent under 18, and (subject to the following exception) to the High Court or the county court, in any other case. The exception, provided by the Magistrates' Courts (Injunctions: Anti-Social Behaviour) Rules 2015, is that a youth court may on application permit an application for a Part 1 injunction against a respondent aged 18 or over to be made to the youth court if:

(a) an application to the youth court has been, or is to be, made for a Part 1 injunction against a person under 18, and the application *for permission* includes details of the application for a s 1 injunction relating to the person aged under 18; and

(b) the youth court thinks that it would be in the interests of justice for the applications for a Part 1 injunction to be heard together, and for this purpose the application *for permission* must include a statement of the reasons why it is in the interests of justice for the applications for a Part 1 injunction to be heard together.

Where a respondent attains 18 after the commencement of proceedings for a Part 1 injunction, the proceedings must remain in a youth court, unless the court in which the proceedings were continuing when the respondent attained 18 directs that the proceedings be transferred to the High Court or county court.

The applicant

18.47 By A-sB,C&PA 2014, s 5, the injunction may be granted only on the application of specified bodies or persons, including:

(a) a local authority;
(b) a housing provider;
(c) the chief officer of police for a police area;
(d) the chief constable of the British Transport Police Force.

A chief officer of police can delegate his/her functions in respect of an application for an order to an officer or officers in his/her force judged suitable by him/her.

Ministerial guidance has been issued to the specified persons and bodies about the exercise of their functions under A-sB,C&PA 2014, Part 1.

Applications without notice

18.48 An application for a Part 1 injunction may be made without notice being given to R. If such an application is made the court must either adjourn the proceedings and grant an interim injunction (below), or adjourn the proceedings without granting an interim injunction, or dismiss the application (s 6).

Consultation, etc

18.49 By s 14, an applicant for a Part 1 injunction on notice must before doing so:

(a) consult the local youth offending team about the application, if R will be aged under 18 *when the application is made*;
(b) inform any other body or individual the applicant thinks appropriate of the application.

Where the court adjourns a without-notice application, before the date of the first on-notice hearing, the applicant must comply with (a) and (b) just mentioned, except that 'on that date' is substituted for the words italicised.

Prohibitions and requirements

18.50   A Part 1 injunction may, for the purpose of preventing R from engaging in anti-social behaviour, *prohibit* him/her from doing, or *require* him/her to do, anything described in the injunction (s 1(4)). Examples of requirements are attendance on an alcohol or drug awareness course, undergoing anger management training, or receiving help from a support (eg mental health) service.

Prohibitions and requirements in a Part 1 injunction must, so far as practicable, be such as to avoid:

(a) any interference with the times, if any, at which R normally works or attends school or any other educational establishment;

(b) any conflict with the requirements of any other court order or injunction to which R may be subject (s 1(5)).

Section 1(6) provides that a Part 1 injunction must specify the period for which it has effect, or state that it has effect until further order; in the case of an injunction granted before R has reached 18, a period must be specified and it must be no more than 12 months.

Under s 3, a Part 1 injunction that includes a requirement must specify the person (an individual or organisation) who is to be responsible for supervising compliance with the requirement.

A respondent (R) subject to a requirement included in a Part 1 injunction must, as a requirement of the injunction, keep in touch with P about that requirement, in accordance with any instructions given by P from time to time, and notify P of any change of address.

A Part 1 injunction may have the effect of excluding R from the place where (s)he normally lives ('the premises') only if:

(a) R is 18 or over,

(b) the injunction is granted on the application of—

(i) a local authority;

(ii) the chief officer of police for the police area that the premises are in; or

(iii) if the premises are owned or managed by a housing provider, that housing provider, and

(c) the court thinks that the anti-social behaviour in which R has engaged or threatens to engage consists of or includes the use or threatened use of violence against other persons, or that there is a significant risk of harm to other persons from R (s 13).

Interim injunctions

18.51   Section 7 provides as follows.

Where the court adjourns the hearing of an application (whether made with notice or without) for a Part 1 injunction, it may grant an interim Part 1 injunction lasting until the final hearing of the application or until further order if it thinks it just to do so. An interim injunction made at a 'without-notice hearing' may not have the effect of requiring participation in particular activities.

Variation or discharge

18.52   The court may vary or discharge a Part 1 injunction on the application of the person who applied for the injunction, or the respondent (s 8).

Arrest without warrant

18.53 By A-sB, C&PA 2014, s 4, a court granting a Part 1 injunction may attach a power of arrest to a prohibition or requirement (other than one requiring participation in particular activities) of the injunction if it thinks that the anti-social behaviour in which the respondent (R) has engaged or threatens to engage consists of or includes the use or threatened use of violence against other persons, or that there is a significant risk of harm to other persons from R. The court may specify a period for which the power of arrest is to have effect which is shorter than that of the prohibition or requirement to which it relates.

Where a power of arrest is attached to a provision of a Part 1 injunction, s 9 provides that a constable may arrest R without warrant if (s)he has reasonable cause to suspect that R is in breach of the provision. The constable must inform the applicant for the injunction. The arrested person must, within the period of 24 hours beginning with the time of the arrest, be brought before:

(a) a judge of the High Court or a judge of the county court, if the injunction was granted by the High Court;
(b) a judge of the county court, if the injunction was granted by the county court, or the injunction was granted by a youth court but R is aged 18 or over;
(c) a justice of the peace, if neither (a) nor (b) applies.

In calculating when the period of 24 hours ends, Christmas Day, Good Friday, and any Sunday are disregarded.

The judge before whom a person is brought under (a) or (b) may remand the person if the matter is not disposed of straight away. The justice of the peace before whom a person is brought under (c) must remand the person to appear before the youth court that granted the injunction. A remand under s 9 or under s 10 below may be in custody or on bail. In the case of someone under 18 it must normally be on bail.

Issue of arrest warrant

18.54 Section 10 deals with this.

If the applicant for a Part 1 injunction thinks that R is in breach of any of its provisions, the applicant may apply for the issue of a warrant for this arrest. The application must be made to:

(a) a judge of the High Court, if the injunction was granted by the High Court;
(b) as per **18.53**(b);
(c) as per **18.53**(c).

A judge or justice may issue such a warrant only if the judge or justice has reasonable grounds for believing that R is in breach of a provision of the injunction.

A warrant issued by a judge of the High Court, or of the county court, must require R to be brought before the High Court or county court, as the case may be. A warrant issued by a justice of the peace must require R to be brought before the youth court that granted the injunction, if the person is aged under 18, and before the county court, if (s)he is aged 18 or over.

A constable who arrests a person under a warrant issued under s 10 must inform the applicant for the injunction.

If R is brought before a court by virtue of such a warrant but the matter is not disposed of straight away, the court may remand R.

Contempt of court

18.55   Because an injunction is a civil order, breach of a Part 1 injunction is not a criminal offence. Where the person in breach of a Part 1 injunction is 18 or over, the breach is dealt with as a civil contempt of court (for which imprisonment for up to two years can be ordered). The breach is dealt with by the High Court or the county court, as appropriate. The breach must be proved beyond reasonable doubt.

Supervision orders and detention orders

18.56   A-sB,C&PA 2014, Sch 2 deals with cases where a person alleged to be in breach of a Part 1 injunction is under 18. It provides that, on the application of the person who applied for the injunction, a youth court can make one of the following orders, where it is satisfied beyond reasonable doubt that a person under 18 is in breach of that injunction:

(a) a supervision order;
(b) if the person is aged 14 to 17 inclusive, a detention order.

A person making an application for such an order must *before doing so*:

(a) consult the youth offending team specified in the injunction or, if none is specified, the local youth offending team; and
(b) inform any other body or individual the applicant thinks appropriate.

A supervision order imposes on the defaulter for a maximum of six months one or more of the following: a supervision requirement; an activity requirement; a curfew requirement. A supervision order containing a curfew requirement may also contain an electronic monitoring requirement.

A detention order is an order that the defaulter be detained for a period (not exceeding three months) specified in the order in youth detention accommodation.

Anti-social behaviour: requirement to give name and address

18.57   The Police Reform Act 2002, s 50(1) provides that, if a constable in uniform has reasonable grounds to believe that a person has been acting in an anti-social manner, (s)he may require that person to give his/her name and address.

For these purposes, 'anti-social behaviour' means conduct:

(a) that has caused, or is likely to cause, harassment, alarm, or distress to any person;
(b) capable of causing nuisance or annoyance to a person in relation to that person's occupation of residential premises; or
(c) capable of causing housing-related nuisance or annoyance to any person ('housing-related' means directly or indirectly relating to the housing management functions of a housing provider or local authority).

Failure to give a name and address when required do so under s 50(1), or the provision of a false or inaccurate name or address, is a summary offence, contrary to s 50(2), punishable with a fine not exceeding level 3.

Criminal behaviour orders

18.58   Under the Sentencing Code, ss 330–342, contained in the Sentencing Act 2020, following a conviction for any criminal offence in the Crown Court, a magistrates' court or a youth court, the court may make a criminal behaviour order (CBO). A CBO can be made where the court is satisfied that the offender (O) has engaged in behaviour that caused or was likely to cause harassment, alarm or distress to any person, and considers that the CBO will help in preventing O engaging in such behaviour. A CBO is an order which, for the purpose of preventing the offender from engaging in such behaviour, prohibits O from doing anything described in the order and requires O to do anything described in the order. The CBO must be additional to a sentence in respect of the offence. It may be made only on the application of the prosecution. A CBO may be varied or discharged by the court which made it on the application of the offender or the prosecution. An interim order may be made where a court adjourns the hearing of an application for a CBO. Where the CBO is made against someone under 18 the operation of the CBO is subject to a system of annual review as long as (s)he is under 18.

By the Sentencing Act 2020, s 339(1), a person who without reasonable excuse:

(a) does anything (s)he is prohibited from doing by a CBO, or

(b) fails to do anything (s)he is required to do by a CBO,

commits an indictable (either-way) offence. The offence is punishable with up to five years' imprisonment.

Anti-social behaviour, etc: the community remedy

A-sB,C&PA 2014, s 102 deals with the case where: **18.59**

(a) a constable, an investigating officer or a person authorised by a relevant prosecutor to give a conditional caution or youth conditional caution (P) has evidence that an individual (A) has engaged in anti-social behaviour (as defined below) or committed an offence,

(b) A admits to P that (s)he has done so,

(c) P thinks that the evidence is enough for taking proceedings against A for an injunction under A-sB,C&PA 2014, s 1, or taking other court proceedings, but decides that it would be appropriate for A to carry out action of some sort instead, and

(d) if the evidence is that A has committed an offence, P does not think that it would be more appropriate for A to be given a conditional caution, youth conditional caution or a fixed penalty notice.

For the purposes of s 102, anti-social behaviour is defined as conduct (i) that has caused, or is likely to cause, harassment, alarm or distress, (ii) capable of causing nuisance or annoyance to a person in relation to his/her occupation of residential premises, or (iii) capable of causing housing-related nuisance or annoyance to anyone. 'Housing related' means directly or indirectly relating to the management functions of a housing provider or local authority.

For the purposes of s 102, 'action' includes the making of a payment to the victim (but does not include the payment of a fixed penalty).

Section 102 provides that, before deciding what action to invite A to carry out, P must make reasonable efforts to obtain the views of the victim (if any) of the anti-social behaviour or the offence, and in particular the victim's views as to whether A should carry out any of the actions listed in the community remedy document prepared by the local policing body (eg the police and crime commissioner).

If the victim expresses the view that A should carry out a particular action listed in the community remedy document, P must invite A to carry out that action unless it seems to P that it would be inappropriate to do so. Where there is more than one victim and they express different views, or for any other reason the italicised words do not apply, P must nevertheless take account of any views expressed by the victim (or victims) in deciding what action to invite A to carry out.

If A does not agree to carry out the particular action which (s)he is invited to carry out, this can lead to a more formal sanction.

Review of response to complaints about anti-social behaviour

18.60   A-sB,C&PA 2014, s 104 provides for the 'community trigger'. The community trigger is a mechanism for victims of persistent anti-social behaviour to request that relevant bodies undertake a case review. An anti-social behaviour (ASB) case review entails the relevant bodies sharing information in relation to the case, discussing what action has previously been taken, and collectively deciding whether any further action could be taken. 'Relevant bodies' means *the relevant district council comprised within a county, or any type of unitary authority (or in Wales a county or county borough council); the chief officer of police for that body's area; each clinical commissioning group (Local Health Board in Wales);* and any local provider(s) of social housing co-opted by the italicised bodies pursuant to the requirements to do so. Any individual, community, or business can make an application for a case review, and the relevant bodies must carry out an ASB case review if the threshold is met. The threshold is set by the relevant bodies. The threshold may be set with reference to the persistence of the behaviour, the potential for harm to the victim, and the adequacy of response from agencies. The threshold should be set no higher than three complaints, but the relevant bodies may choose to set a lower threshold.

The ASB case review mechanism is intended as a backstop safety net for the victims of anti-social behaviour who consider that there has not been an appropriate response to their complaints about such behaviour.

The relevant bodies in each local government area must make and publish arrangements for ASB case review procedures. Joint arrangements may be made for a larger area such as the police force area. The bodies carrying out the ASB case review must inform the applicant of their decision on whether or not the threshold for review is met, the outcome of the review and any recommendations made as a result of the review. In addition, they may make recommendations to a person who carries out public functions, including any of the bodies that have taken part in the review, and that person must have regard to the recommendations.

Relevant bodies must publish information about the number of ASB case review applications received, the number of times the threshold was not met, the number of reviews carried out and the number of reviews that resulted in further action.

Schedule 4 makes additional provisions for ASB case reviews. Review procedures must include:

(a)   what happens when the applicant is dissatisfied with the way his or her application was dealt with or the review carried out;

(b)   an assessment of the effectiveness of the procedures and revising them.

In making and revising the procedures, the relevant bodies must consult the local policing body, and such local providers of social housing as they consider appropriate.

VIOLENT OFFENDER ORDERS

Nature of orders and applications for such orders

18.61   The Criminal Justice and Immigration Act (CJIA) 2008, Part 7 (ss 98 to 117) provides that a 'violent offender order' (VOO) may be made in respect of a qualifying offender imposing prohibitions, restrictions or conditions for the purpose of protecting the public from the risk of serious violent harm caused by the offender. The minimum term of a VOO is two years, and the maximum five. The reference to protecting the public includes the public in the UK, or any particular members of the public in the UK, from the risk of serious physical or psychological harm caused by that person committing one or more 'specified offences'.

'Specified offences' are:

(a) manslaughter;
(b) an offence under OAPA 1861, s 4, 18 or 20 (soliciting murder, wounding with intent to cause grievous bodily harm or malicious wounding);
(c) attempting to commit murder or conspiracy to commit murder; or
(d) a corresponding service offence.

A 'qualifying offender' is a person aged 18 or over who has been convicted of a specified offence and given a custodial sentence of at least 12 months or a hospital order, or a person found not guilty of such an offence by reason of insanity but made subject to a hospital order or a supervision order, or a person found to be under a disability to be tried and to have done the act charged who has been made subject to such an order. A person is also a qualifying offender if the above terms are satisfied in respect of proceedings outside England and Wales in respect of equivalent offences or murder under the corresponding foreign law.

A chief officer of police may apply by way of complaint to a magistrates' court for a VOO in respect of a person who resides in his/her area, or who (s)he believes is in, or intends to come to, that area, if it appears that: (a) that person is a qualifying offender; and (b) (s)he has acted in such a way as to give reasonable cause to believe that it is necessary for such an order to be made. That application may be made to any magistrates' court whose area includes any part of the chief officer's police area, or any place in which it is alleged that the person so acted. An order may be made if conditions (a) and (b) above are satisfied. It may not be made so as to come into force at any time when the offender is subject to a custodial sentence, on licence, or subject to a hospital order or supervision order, in respect of an offence.

A VOO may contain prohibitions, restrictions, or conditions preventing the offender from going to specified premises or places (whether at all or at specified times), attending specified events or having any (or specified) contact with a specified individual.

Provision is made for interim VOOs which apply before determination of an application.

Provision is also made for the variation, renewal and discharge of a VOO.

Notification requirements

Offenders who are for the time being subject to VOOs (or interim VOOs) must notify the police, within three days of the making of the order, of particulars specified in CJIA 2008, s 108 (personal particulars including date of birth, name(s) on the relevant date, National Insurance number, home address, and other addresses at which the offender may be found). Changes must be notified within three days of the change in circumstances occurring. In addition, the offender must renotify the police annually of the information required to be given on initial notification; in respect of renotification of addresses (other than a home address) at which the offender may be found. The periodic renotification requirement does not apply in the case of an interim order. In addition, the time periods referred to above are suspended while a person is remanded in

18.62

custody, in prison, or detained in hospital. The Criminal Justice and Immigration Act 2008 (Violent Offender Orders) (Notification Requirements) Regulations 2009 impose notification requirements in respect of travel outside the UK.

Offences

18.63   It is an indictable (either-way) offence against CJIA 2008, s 113 for a person to fail, without reasonable excuse, to:

(a) comply with any prohibition, restriction or condition contained in a VOO or interim VOO (s 113(1)); or

(b) fail to comply with a notification requirement or a requirement to provide fingerprints and/or a photograph if requested when making a notification (s 113(2)).

Knowingly providing false information when making a notification is also an indictable (either-way) offence under s 113(3).

These three offences are punishable with imprisonment with up to five years' imprisonment.

INJUNCTIONS TO PREVENT GANG-RELATED VIOLENCE AND DRUG-DEALING ACTIVITY

18.64   These injunctions are dealt with by the Policing and Crime Act 2009 (P&CA 2009), Part 4 (ss 34–50).

Those involved in gang-related violence or drug-dealing activity should be prosecuted for a criminal offence if there is sufficient evidence and it is in the public interest, but there may be instances where criminal proceedings have not yet been brought and applying for an injunction under the above provisions may be an appropriate response.

Power to grant injunctions to prevent gang-related violence and drug-dealing activity

18.65   P&CA 2009, s 34 empowers a youth court (if the respondent is under 18), and the High Court or a county court (in any other case), to grant an injunction in civil proceedings against a respondent aged 14 or over to prevent gang-related violence or gang-related drug-dealing activity if two conditions are met:

(a) that the court is satisfied on the balance of probabilities that the respondent has engaged in, or has encouraged or assisted, gang-related violence or gang-related drug-dealing activity; and

(b) that the court thinks it is necessary to grant the injunction for either or both of the following purposes:

   (i) to prevent the respondent from engaging in, or encouraging or assisting, gang-related violence or gang-related drug-dealing activity;

   (ii) to protect the respondent from gang-related violence or gang-related drug-dealing activity.

Something is 'gang-related' if it occurs in the course of, or is otherwise related to, the activities of a group that consists of at least three people, and has one or more characteristics that enable its members to be identified by others as a group. A judge in the Administrative Court has held, in reference to the requirement of a character-istic of a group that enables it to be identified by others as a group, that it is not suffi-cient that a police officer piecing together the available information is able to draw the conclusion of a link between members of the gang if there is no common character-istic that would enable others generally to be able to make that identification. Dealing with a gang composed of family members, he accepted that it is possible that such a gang might communicate to others that they have a cohesive characteristic of common family membership and they might indeed be known as a family gang but, he held, it is not sufficient if the group in question is an assortment of individuals who happen to have familial relationships if that is not an important badge of identification communi-cated to others and explains how others see them.

'Violence' includes a threat of violence.

'Drug-dealing activity' means the unlawful production, supply, importation, or ex-portation of a controlled drug or psychoactive substance, as defined in the Misuse of Drugs Act 1971 or Psychoactive Substances Act 2016.

An injunction under s 34 may (for either or both of purposes (i) and (ii) above) pro-hibit the respondent from doing anything described in the injunction, or require the respondent to do anything described in the injunction.

The prohibitions and requirements included in the injunction must, so far as prac-ticable, be such as to avoid any conflict with the respondent's religious beliefs and any interference with the times at which the respondent normally works or attends any educational establishment.

An injunction under s 34 may not include a prohibition or requirement that has effect after the end of the period of two years beginning with the day on which the in-junction is granted.

The court may attach a power of arrest in relation to any prohibition in the injunc-tion, or to any requirement in the injunction (other than a requirement which has the effect of requiring the respondent to participate in particular activities).

Applications for injunctions

18.66 By P&CA 2009, s 37, an application for an injunction under s 34 may be made by:

(a) the chief officer of police for a police area,
(b) the chief constable of the British Transport Police Force, or
(c) a local authority.

A chief officer of police can delegate the application for an injunction to an officer in his/her force.

An application can be made 'with notice' or 'without notice'.

Applications with notice

18.67 These require the applicant to notify the respondent of the application for the injunc-tion. In addition, before applying for an injunction, the applicant must consult:

(a) any local authority, and any chief police officer, that the applicant thinks it appro-priate to consult,
(b) the youth offending team(s) in the area(s) where it appears to the applicant that the respondent resides if the respondent is under 18 (and will be under 18 when the application is made), and
(c) any other body or individual that the applicant thinks it appropriate to consult.

If the court adjourns the hearing of a 'with notice' application for an injunction, it may grant an interim injunction if it thinks that it is just and convenient to do so.

Applications without notice

18.68   By s 39, an application under s 37 may be made without the respondent being given notice, in which case the above consultation requirement does not initially apply. If such an application is made the court must:

(a) dismiss it, or

(b) adjourn the proceedings until a full hearing, in which case the applicant must comply with the consultation requirement before the date of the first full hearing (ie a hearing of which notice has been given to the applicant and respondent).

If an application without notice is made, and the proceedings are adjourned, the court may grant an interim injunction if it thinks that it is necessary to do so (s 41). Such an interim injunction may not have the effect of requiring the respondent to participate in particular activities.

Applications without notice should not be routine but may be appropriate if an injunction is urgently required, if there is a risk that the respondent may flee if given prior notice of an injunction application, or if giving notice would endanger witnesses.

Variation and discharge of injunctions

18.69   An injunction may be varied or discharged after a review hearing, or on an application made by the applicant for the injunction or the respondent. If such an application is dismissed no further application may be made by anyone without the court's consent.

Breach of injunction and enforcement

Arrest without warrant

18.70   This is governed by P&CA 2009, s 43, which applies if a power of arrest is attached to a provision of an injunction. Its provisions mirror those which apply to a breach of a 'Part 1 injunction' (**18.53**).

Issue of warrant of arrest

18.71   If the applicant for the injunction considers that the respondent is in breach of any of its provisions, (s)he may apply under P&CA 2009, s 44 to a relevant judge for the issue of a warrant for the arrest of the respondent. A relevant judge may only issue a warrant if (s)he has reasonable grounds for believing that the respondent is in breach of any provision of the injunction.

Punishment

18.72   Breach of an injunction granted under P&CA 2009, s 34 is punishable by the High Court or county court as a civil contempt of court (**18.55**) where the person in breach is 18 or over. Where that person is under 18, special provision is made by Sch 5A (power to make a supervision order or a detention order). The provisions of Sch 5A correspond with those at **18.56**. If the respondent's behaviour constitutes a criminal offence, it should be dealt with as such.

Miscellaneous

The Secretary of State has published guidance on the use of gang-related injunc-   **18.73**
tions. A chief officer of police or a local authority must have regard to the guidance
published.

CLOSURE OF PREMISES ASSOCIATED WITH
NUISANCE OR DISORDER, ETC

A-sB,C&PA 2014, Part 4, Ch 3 (ss 76–93) sets out provisions for the closure of premises   **18.74**
associated with nuisance or disorder, etc. The provisions provide a fast, flexible power
that can be used to protect victims and communities by quickly closing premises which
are causing nuisance or disorder.

Closure notices

Power to issue closure notices

A-sB,C&PA 2014, s 76 provides as follows.   **18.75**

A police inspector (or above), or the local authority, may issue a closure notice if satisfied on reasonable grounds that:

(a) the use of particular premises has resulted, or (if the notice is not issued) is likely
    soon to result, in nuisance to members of the public; or
(b) there has been, or (if the notice is not issued) is likely soon to be, disorder near
    those premises associated with the use of those premises,

and that the notice is necessary to prevent the nuisance or disorder from continuing,
recurring or occurring.

A closure notice is a notice prohibiting access to the premises for a period specified
in the notice. A closure notice may prohibit access:

(a) by all persons except those specified, or by all persons except those of a specified
    description;
(b) at all times, or at all times except those specified;
(c) in all circumstances, or in all circumstances except those specified.

A closure notice may not prohibit access by people who habitually live on the premises, or by the owner of the premises, and accordingly they must be specified under list
entry (a) just mentioned.

A closure notice must:

(a) identify the premises;
(b) explain the effect of the notice;
(c) state that failure to comply with the notice is an offence;
(d) state that an application will be made under s 80 for a closure order;
(e) specify when and where the application will be heard;
(f) explain the effect of a closure order;
(g) give information about the names of, and means of contacting, persons and organ-
    isations in the area that provide advice about housing and legal matters.

A closure notice may be issued only if reasonable efforts have been made to inform:

(a) people who live on the premises (whether habitually or not), and
(b) any person who has control of or responsibility for the premises or who has an interest in them,

that the notice is going to be issued.

Before issuing a closure notice the police officer or local authority must ensure that any body or individual the officer or authority thinks appropriate has been consulted.

Duration

18.76   By A-sB,C&PA 2014, s 77, the maximum period that may be specified in a closure notice is 24 hours, except that the maximum period is 48 hours:

(a) if, in the case of a notice issued by a police officer, the officer is a superintendent (or above); or
(b) if, in the case of a notice issued by a local authority, the notice is signed by its chief executive officer (CEO) or a person designated by him/her for this purpose.

In calculating when the period of 48 hours ends, Christmas Day is disregarded.

A specified period of 24 hours (or less) may be extended by up to 24 hours if:

(a) in the case of a notice issued by a police officer, an extension notice is issued by a superintendent (or above), or
(b) in the case of a notice issued by a local authority, the authority issues an extension notice signed by its CEO or a person designated by him/her for this purpose.

An extension notice is a notice which identifies the closure notice to which it relates, and specifies the period of the extension.

Cancellation or variation

18.77   A-sB,C&PA 2014, s 78 deals with this. It applies where the relevant officer or authority is no longer satisfied as mentioned in s 76 (**18.75**), either (a) as regards the premises as a whole, or (b) as regards a particular part of the premises. Section 78 provides that, in a case within (a) the relevant officer or authority must issue a cancellation notice (a notice cancelling the closure notice), and that in a case within (b) the relevant officer or authority must issue a variation notice (a notice varying the closure notice so that it does not apply to the part of the premises referred to in (b)). Such a notice issued by a local authority must be signed by the person who signed the closure notice (or extension notice) (or, if unavailable, by another person qualified to sign it).

For these purposes, 'the relevant officer or authority' means:

(a) in the case of a closure notice issued by a police officer and not extended, that officer (or, if that officer is unavailable, another officer of the same or higher rank);
(b) in the case of a closure notice issued by a police officer and extended, the officer who issued the extension notice (or, if that officer is unavailable, another officer of the same or higher rank);
(c) in the case of a closure notice issued by a local authority, that authority.

Service

18.78   By A-sB,C&PA 2014, s 79, a closure notice, an extension notice, a cancellation notice, or a variation notice must be served by a constable (notice issued by a police officer), or by a representative of the authority (notice issued by a local authority). That person must, if possible, (a) fix a copy of the notice to at least one prominent place on the premises, to each normal means of access to the premises, and to any outbuildings apparently used with or as part of the premises; and (b) give a copy of the notice to at least one person apparently having control of or responsibility for the premises, to the residents on the premises, and to anyone not living there who was informed (under s 76) that the notice was going to be issued.

If the constable or local authority representative reasonably believes, at the time of serving the notice, that there are persons occupying another part of the building or other structure in which the premises are situated whose access to that part will be impeded if a closure order is made under s 80 (below), (s)he must also if possible serve the notice on those persons.

The constable or local authority representative may enter any premises, using reasonable force if necessary, for the purposes of fixing a notice on the premises.

Closure orders

Power of court to make closure orders

A-sB,C&PA 2014, s 80 provides that whenever a closure notice is issued an application **18.79** must be made to a magistrates' court for a closure order (unless the notice has been cancelled under s 78). Such an application must be made:

(a) by a constable, if the closure notice was issued by a police officer;
(b) by the authority that issued the closure notice, if the notice was issued by a local authority.

The application must be heard by the magistrates' court not later than 48 hours after service of the closure notice. In calculating that period, Christmas Day is disregarded.
The court may make a closure order if it is satisfied that:

(a) a person has engaged, or (if the order is not made) is likely to engage, in disorderly, offensive or criminal behaviour on the premises; or
(b) the use of the premises has resulted, or (if the order is not made) is likely to result, in serious nuisance to members of the public; or
(c) there has been, or (if the order is not made) is likely to be, disorder near those premises associated with the use of those premises,

and that the order is necessary to prevent the behaviour, nuisance or disorder from continuing, recurring or occurring.
A closure order is an order prohibiting access to the premises for a period specified in the order. The period may not exceed three months. A closure order may prohibit access:

(a) by all persons, or by all persons except those specified, or by all persons except those of a specified description;
(b) at all times, or at all times except those specified;
(c) in all circumstances, or in all circumstances except those specified.

A closure order may be made in respect of the whole or any part of the premises, and may include provision about access to a part of the building or structure of which the premises form part. The court must notify the relevant licensing authority if it makes a closure order in relation to premises in respect of which a premises licence is in force.

Proceedings for a closure order (or interim order, below) are civil in nature; for the implications of this, see **13.47**.

Temporary orders

18.80 Where an application has been made for a closure order, and the court does not make a closure order, it may (under A-sB,C&PA 2014, s 81(2)) order that the closure notice continues in force for a specified further period of not more than 48 hours, if satisfied that:

(a) the use of particular premises has resulted, or (if the notice is not continued) is likely soon to result, in nuisance to members of the public, or
(b) there has been, or (if the notice is not continued) is likely soon to be, disorder near those premises associated with the use of those premises,

and that the continuation of the notice is necessary to prevent the nuisance or disorder from continuing, recurring or occurring.

The court may adjourn the hearing of the application for a closure order for not more than 14 days to enable the occupier of the premises, the person with control of or responsibility for them, or any other person with an interest in them, to show why a closure order should not be made (s 81(3)). On an adjournment, the court may order that the closure notice continues in force until the end of the adjournment period (s 81(4)).

Extension of closure orders

18.81 At any time before the expiry of a closure order, an application may be made to a justice of the peace for an extension (or further extension) of the period for which the order is in force (A-sB,C&PA 2014, s 82). The application may be made by an inspector (or above) where the closure order was made on the application of a constable, and by the local authority where the closure order was made on its application. Such a police officer or local authority may apply only if satisfied on reasonable grounds that it is necessary for the period of the order to be extended to prevent the occurrence, recurrence or continuance of:

(a) disorderly, offensive or criminal behaviour on the premises;
(b) serious nuisance to members of the public resulting from the use of the premises, or
(c) disorder near the premises associated with the use of the premises,

and also satisfied that the appropriate consultee has been consulted about the intention to make the application. The 'appropriate consultee' means the local authority, in the case of an application by an inspector (or above); and the chief officer of police, in the case of an application by a local authority.

Where an application is made for an extension, the justice may issue a summons directed to any person on whom the closure notice was served, or any other person who appears to have an interest in the premises but on whom the closure notice was not served, requiring him/her to appear before the magistrates' court to respond to the application.

If the magistrates' court is satisfied that it is necessary for the period of the order to be extended to prevent the occurrence, recurrence or continuance of a matter mentioned in (a), (b), or (c) above, it may make an order extending (or further extending) the period of the closure order by a period not exceeding three months. The period of a closure order may not be extended so that the order lasts for more than six months.

Discharge of closure orders

A constable (where the order was made on the application of a constable), the local **18.82** authority (on whose application the order was made), a person on whom the closure notice was served, and anyone else who has an interest in the premises but on whom the closure notice was not served, may apply for the closure order to be discharged by a magistrates' court (A-sB,C&PA 2014, s 83). The court may not discharge the order unless satisfied that the closure order is no longer necessary to prevent the occurrence, recurrence or continuance of disorderly, offensive or criminal behaviour on the premises, serious nuisance to members of the public resulting from the use of the premises, or disorder near the premises associated with the use of the premises.

Enforcement

By A-sB,C&PA 2014, s 85, an authorised person may enter premises in respect of **18.83** which a closure order is in force, and do anything necessary to secure the premises against entry. An 'authorised person' is, in relation to an order made on the application of a constable, a constable or person authorised by the chief officer of police; and, in relation to an order made on the application of a local authority, a person authorised by the authority. A person so acting may use reasonable force. A person seeking to enter premises under s 85 must, if required to do so by or on behalf of the owner, occupier or other person in charge of the premises, produce evidence of his/her identity and authority before entering the premises. An authorised person may also enter the premises during the order to carry out essential maintenance or repairs to the premises.

Offences

A person who without reasonable excuse remains on or enters premises in contraven- **18.84** tion of a closure notice (including a notice continued in force under A-sB,C&PA 2014, s 81(**18.80**)) commits an offence contrary to s 86(1).

A person who without reasonable excuse remains on or enters premises in contravention of a closure order commits an offence contrary to s 86(2).

A person who without reasonable excuse obstructs a person acting under s 79 or under the power to enter and secure in s 85 commits an offence contrary to s 86(3).

The above offences are summary offences punishable with up to three months' imprisonment and/or an unlimited fine (s 86(1) and s 86(3)) or six months' imprisonment and/or an unlimited fine (s 86(2)).

Guidance

Ministerial guidance has been issued to chief officers of police and local authorities **18.85** about the exercise of the above functions under A-sB,C&PA 2014, Part 4, Ch 3.

SQUATTING IN A RESIDENTIAL BUILDING

A person commits a summary offence under the Legal Aid, Sentencing and Punishment **18.86** of Offenders Act 2012, s 144(1) if:

(a) (s)he is in a *residential building* as a trespasser having entered it as a trespasser,
(b) (s)he knows or ought to know that (s)he is a trespasser, and
(c) (s)he is living in the building or intends to live there for any period.

The offence is not committed by someone holding over after the end of a lease or licence (even if (s)he leaves and re-enters the building).

For the purposes of s 144:

(a) 'building' includes any structure or part of a structure (including a temporary or movable structure);

(b) a building is 'residential' if it is designed or adapted, before the time of entry, for use as a place to live; and

(c) the fact that a person derives title from a trespasser, or has the permission of a trespasser, does not prevent him/her from being a trespasser.

The offence under s 144 is punishable with up to six months' imprisonment and/or an unlimited fine.

ENTERING AND REMAINING ON PROPERTY

18.87   The Criminal Law Act 1977 (CLA 1977), Part 2 (ss 6–13) provides certain offences relating to entering and remaining on premises. Although these offences are often associated with 'squatters', the relevant provisions extend to situations beyond those involving squatting.

For the purposes of these offences, 'premises' is defined as any building, any part of a building under separate occupation (eg a flat), any land ancillary to a building, and the site comprising any building or buildings together with any land ancillary thereto, and 'building' includes any immovable structure, and any movable structure, vehicle, or vessel designed or adapted for use for residential purposes (eg a residential caravan or houseboat).

Violence for securing entry

18.88   By CLA 1977, s 6(1), it is an offence (a summary offence punishable with up to six months' imprisonment and/or an unlimited fine) for any person, without lawful authority, to use or threaten violence for the purpose of securing entry into *any premises* either for him/herself or for any other person, provided that:

(a) there is someone present on those premises at the time who is opposed to the entry which the violence is intended to secure; and

(b) the person using or threatening the violence knows that this is the case.

Use or threat of violence

18.89   The essence of this offence is the use or threat of violence against a person or property for the purpose of securing entry into premises on which a person opposed to the entry is present: actual entry is not required. It is immaterial whether the entry which the violence is intended to secure is for the purpose of acquiring possession of the premises or for some other purpose. People who use or threaten violence in order to secure entry to a nightclub are guilty of the present offence if they know that someone inside is opposed to their entry; so are would-be squatters who, with such knowledge, seek to enter a house by such means, and so are protesters who likewise seek to enter a public building.

The violence used or threatened may be against a person or property (whether (s)he or it is on or off the premises). It is important to recognise that the offence is to use 'violence', not 'force'. Although the difference may seem to be small, it is considerable in certain circumstances. It would no doubt amount to violence against the property to set fire to it to drive out those inside and thereby gain entry, but the degree of force necessary to insert a key and to secure entry does not amount to violence. It would be different if the lock was burst open by using violence against the door.

Someone on the premises opposed to the entry

Someone must be physically present on the premises who is opposed to the entry to them which the violence is intended to secure. (S)he might equally be the owner of the premises or a trespasser who is opposed to the owner's re-entry. Clearly, the present offence is not committed where someone breaks into an empty house. **18.90**

Mens rea

D must not only use or threaten violence for the purpose of securing entry into any premises but (s)he must also know that there is someone on the premises at the time who is opposed to the entry in question. **18.91**

Without lawful authority

An offence is not committed under CLA 1977, s 6(1) if the person using or threatening violence has lawful authority for acting in the prescribed way. This exemption is essential to protect the violent entries which might have to be made by police officers or bailiffs in executing warrants or orders of courts. However, the only entries so protected will be those where the form of violence used to secure entry is authorised by law. **18.92**

Section 6(2) states that the fact that a person has any interest or right to possession or occupation of any premises does not give him/her lawful authority to use or threaten the use of violence for the purpose of securing his/her entry onto those premises. This means that landlords and other non-residential occupiers must seek to recover possession of their premises by an action in the civil courts, unless it is possible for them to effect a peaceful re-entry. If an occupier does succeed in re-entering his/her premises, (s)he does not commit any offence by proceeding to eject any trespasser, whatever liability (s)he may incur by virtue of his/her entry.

Displaced residential occupiers and protected intending occupiers

Special provision is made for people falling within the definition of a 'displaced residential occupier' or 'protected intending occupier' of premises or any access to them. **18.93**

The offence under s 6(1) does not apply to a person who is a *displaced residential occupier* or a *protected intending occupier* of the premises in question or who is acting on behalf of such an occupier. This exemption does not have to be proved by D. Instead, if (s)he adduces sufficient evidence that (s)he was, or was acting on behalf of, such an occupier (s)he is presumed to be, or to be acting on behalf of, such an occupier unless the contrary is proved by the prosecution.

By s 12, any person who was occupying any premises as a residence immediately before being excluded from occupation by anyone who entered those premises, or any access to those premises, as a trespasser is *a displaced residential occupier* of the premises as long as (s)he continues to be excluded from occupation of the premises by the original trespasser or by any subsequent trespasser. A person who is a displaced residential occupier of premises by virtue of this provision is regarded as such an occupier also of any access to those premises. Section 12 also provides that a person who was him/herself occupying the premises as a trespasser before being excluded is not a displaced residential occupier. The obvious example of a displaced residential occupier is the householder who discovers squatters in his/her house when (s)he returns from work or from holiday.

An involved definition of '*protected intending occupier*' is provided by s 12A.

The first type is an individual who, at the time of the request to leave:

(a) has in the premises in question a freehold interest or leasehold interest with not less than two years still to run;

(b) requires the premises for his/her own occupation as a residence;

(c) is excluded from occupation of them by a person who entered them, or any access to them, as a trespasser; and

(d) holds, or a person acting on his/her behalf holds, a written statement, signed by him/her and witnessed by a magistrate or commissioner for oaths, which:

   (i) specifies his/her interest in the premises, and

   (ii) states that (s)he requires the premises for occupation as a residence for him/herself.

The purpose of the statement is to enable the police to identify the protected intending occupier and thereby prevent abuse of the protection given by the Act.

The second type of protected intending occupier is an individual who, at the time of the request to leave:

(a) has a tenancy of the premises (other than a tenancy falling within the other two definitions) or a licence to occupy them granted by a person with a freehold interest or a leasehold interest with not less than two years still to run in the premises;

(b) requires the premises for his/her own occupation as a residence;

(c) is excluded from occupation of the premises by a person who entered them, or access to them, as a trespasser; and

(d) holds, or a person acting on his/her behalf holds, a written statement which:

   (i) specifies that (s)he has been granted a tenancy of those premises or a licence to occupy them,

   (ii) specifies the interest in the premises of the person who granted that tenancy or licence to occupy (the landlord),

   (iii) states that (s)he requires the premises for occupation as a residence for him/herself, and

   (iv) is signed by the landlord and by the tenant or licensee and witnessed by a magistrate or commissioner for oaths.

The third type of protected intending occupier is an individual who, at the time of the request to leave:

(a) has a tenancy of the premises in question (other than a tenancy falling within the other two definitions) or a licence to occupy them granted by a local authority, the Regulator of Social Housing, or a registered housing association or certain other bodies;

(b) requires the premises for his/her own occupation as a residence;

(c) is excluded from occupation of them by a person who entered them, or any access to them, as a trespasser; and

(d) has been issued by or on behalf of the authority, Regulator or association with a certificate stating that the authority, etc is one to which these provisions apply and that (s)he has been granted a licence or tenancy to occupy the premises as a residence.

Adverse occupation of residential premises

CLA 1977, s 7(1) creates an offence of 'adverse occupation' which is triable and pun-   **18.94**
ishable in the same way as an offence under s 6(1) (**18.88**). Section 7(1) provides that
any person who is on any premises (including any access to them whether or not any
such access constitutes premises within the meaning of CLA 1977) as a trespasser, after
having entered as such, is guilty of an offence if (s)he fails to leave those premises on
being required to do so by or on behalf of a *displaced residential occupier*, or a *person
who is a protected intending occupier of the premises.*

Section 7 gives a displaced residential occupier or a protected intending residential
occupier of premises (eg a buyer or tenant who has not yet taken up occupation) who
has been excluded from them by trespassers a swifter remedy for recovering possession
of them than the available civil remedy. (S)he may require the trespassers to leave and
they commit an offence if they fail to do so.

There is no time set upon departure and it is submitted that this indicates that the
requirement is immediate.

Three defences are provided by s 7(3), the burden of proof in each case being on D:

(1) It is a defence that D believed that the person requiring him/her to leave was not a
displaced residential occupier or a protected intending occupier of the premises, or
someone acting on his/her behalf (s 7(2)). This plea will rarely succeed, particularly
in the light of the requirement in the case of a protected intending occupier of a
written statement or certificate to this effect.

(2) It is a defence that the premises in question are or form part of premises used
mainly for non-residential purposes and that D was not on any part of the premises
used wholly or mainly for residential purposes (s 7(3)). This means, for instance,
that people involved in a factory sit-in do not commit the present offence if they
fail to leave when required by a resident owner, so long as they are not in his/her
flat or in part of the premises used wholly or mainly for access to, or in connection
with, the flat.

(3) Where D was requested to leave by a person claiming to be (or to act on behalf of)
a protected intending occupier, it is a defence for D to prove that, although asked
to do so by D at the time that (s)he was requested to leave, the person requesting
him/her to leave failed at that time to produce a written statement, or certificate,
complying with CLA 1977 (s 12A(9)).

INTERIM POSSESSION ORDERS IN RELATION TO PREMISES

Obtaining a final possession order in respect of premises can take rather longer than   **18.95**
desirable. As a result an interim possession order, which can be obtained more speedily,
has been introduced by rules of court. By the Criminal Justice and Public Order Act
1994 (CJPOA 1994), s 76(2), it is an offence for a person to be present on premises as
a trespasser at any time during the currency of such an order. However, no offence is
committed if such a person leaves within 24 hours of the time of service of the order
and does not return, or if a copy of the order was not affixed to the premises in accordance with the rules of court. A person in occupation at the time of service, who leaves
the premises, commits an offence under s 76(4) if (s)he re-enters the premises as a
trespasser or attempts to do so after the expiry of the order but within a period of one
year from service of the order. An offence under s 76(2) or (4) is a summary offence
punishable with up to six months' imprisonment and/or an unlimited fine.

By s 75(1), a person commits an indictable (either-way) offence if, for the purpose of obtaining an interim possession order, (s)he makes a statement which (s)he knows to be false or is misleading in a material particular, or recklessly makes such a statement. Likewise, a person commits an indictable (either-way) offence under s 75(2) if (s)he knowingly or recklessly makes such a statement to resist the making of an interim possession order. An offence under s 75(1) or (2) is punishable with up to two years' imprisonment.

REMOVAL OF TRESPASSERS AND ASSOCIATED OFFENCES

18.96   Under the Criminal Justice and Public Order Act 1994 (CJPOA 1994), ss 61–62B there are two sets of provisions relating to:

(a) the removal of trespassers (ss 61 and 62); and
(b) the removal of trespassers where an alternative site is available (ss 62A, 62B, and 62C).

Removal of trespassers

Direction to leave

18.97   CJPOA 1994, s 61 provides that if the senior police officer (SPO) present at the scene reasonably believes that two or more persons are trespassing on land, that they are present with the common purpose of residing there for any period, that reasonable steps have been taken by or on behalf of the occupier to ask them to leave and that:

(a) any of those persons has caused damage to property on the land or used threatening, abusive, or insulting words or behaviour towards the occupier, a member of his/her family or an employee or agent of his/her; or
(b) those persons have between them brought six or more vehicles onto the land,

SPO may direct those persons, or any of them, to leave the land and to remove any vehicles or other property they have with them on the land. The direction must be to leave immediately or as soon as reasonably practicable, rather than at a future time.

Where the persons in question are reasonably believed by SPO to be persons who were not originally trespassers but have become trespassers on the land, SPO must reasonably believe that the other conditions above are satisfied after those persons became trespassers before SPO can exercise these powers.

SPO must reasonably believe that two or more persons are present on the land as trespassers with the common purpose of residence, no matter how brief that intended period of residence may be. A person can have a purpose of residing in a place notwithstanding that (s)he has a home elsewhere. 'Land' does not include buildings other than agricultural buildings, nor does it include scheduled monuments or land forming part of a highway unless it is a footpath, bridleway, byway open to all traffic, restricted byway, or cycle track. The term includes 'common land', whether public or privately owned common land.

SPO must also reasonably believe that reasonable steps have been taken by or on behalf of *the occupier* (the person entitled to possession of the land by virtue of an estate or interest held by him/her) to require the trespassers to leave.

SPO, reasonably believing these facts, may require such persons to leave without further reason if (s)he reasonably believes they have brought six or more vehicles onto the land. If they have not, (s)he must reasonably believe that any of those persons has caused damage to the land or to property on the land (eg crops or trees), or has used threatening, abusive, or insulting words or behaviour towards the persons specified (the occupier of the land, a member of his/her family or an employee or agent of his/her). Thus overnight campers are outside these provisions, provided that they have not caused damage to property on the land or used such words or behaviour. 'Damage' includes the deposit of any substance capable of polluting the land.

Where the land in question is *common land*, the references in s 61 to trespassing and trespassers include acts and persons doing acts which constitute a trespass as against the occupier or an infringement of the commoners' rights, and references to 'the occupier' include the commoners or any of them or, where the public has access to the common, the local authority as well as any commoner.

For the purposes of s 61, the term '*vehicle*' includes any vehicle, whether or not it is in a fit state for use on roads, and includes any chassis or body, with or without wheels, appearing to have formed part of such a vehicle, and any load carried by, and anything attached to, such a vehicle, and a caravan.

The purpose of s 61 is to give the occupier of the land a swifter remedy for recovering possession of it than the available civil remedy.

Offences

If a trespasser directed to leave complies with that direction, (s)he commits no offence. **18.98**
On the other hand, (s)he commits a summary offence, punishable with up to three months' imprisonment and/or a fine not exceeding level 4, under CJPOA 1994, s 61(4) if, knowing that a direction has been given which applies to him/her:

(a) (s)he fails to leave the land (with any vehicle or other property (s)he is required to remove), as soon as reasonably practicable; or

(b) having left, (s)he again enters the land as a trespasser within a three-month period beginning on the day on which the direction is given.

By s 61(6), it is a defence for D to prove that:

(i) (s)he was not trespassing on the land; or

(ii) (s)he had a reasonable excuse for failing to leave the land as soon as reasonably practicable or, as the case may be, for again entering the land as a trespasser.

The defence of 'not trespassing' refers to not trespassing at the time that the senior police officer forms his/her reasonable belief, and not the time of the prohibited conduct.

Police powers

By CJPOA 1994, s 62, where a direction under s 61 has been given, a constable may **18.99**
seize and remove a vehicle if (s)he reasonably suspects that a person to whom the direction applies has, without reasonable excuse, failed to remove the vehicle which appears to belong to him/her or be in his/her possession or control. The same powers apply where such a person has, without reasonable excuse, re-entered the land within a period of three months from the date of the direction.

Removal of trespassers where alternative site available

Direction to leave

CJPOA 1994, s 62A provides that, where the senior police officer present at the scene **18.100**
(SPO) reasonably believes that certain conditions are satisfied in relation to a person

and land, SPO may direct the person to leave the land and to remove any vehicle and other property which (s)he has with him/her on the land. The conditions are:

(a) that the person and one or more others ('the trespassers') are trespassing on land;
(b) that the trespassers have between them at least one vehicle on the land;
(c) that the trespassers are there for the common purpose of residing there for any period;
(d) if it appears to SPO that the person has one or more caravans in his/her possession or control on the land, that there is a suitable pitch on a relevant caravan site (ie a site in the same local authority area managed by that authority, a private registered provider of social housing, or a registered social landlord) for the caravan(s); and
(e) that the occupier of the land or someone acting on his/her behalf has asked the police to remove the trespassers.

Such a direction may be communicated to the person to whom it applies by any constable at the scene. 'Land' does not include buildings other than agricultural buildings or scheduled monuments. A person may be regarded as having a purpose of residing in a place even if (s)he has a home elsewhere, as would be the case where there is a 'sited' residential caravan. 'The occupier' and 'vehicle' have the same meaning as in s 61 (**18.97**), and the same provision is made in respect of common land as applies to s 61 (**18.97**).

Where a police officer proposes to give such a direction (s)he must consult every local authority within whose area the land is situated as to whether there is a suitable pitch for the caravan or each of the caravans on a relevant caravan site situated in their area.

Offences

18.101   By CJPOA 1994, s 62B(1), a person commits an offence triable and punishable in the same way as that under s 61(4) (**18.98**) if (s)he knows that a direction under s 62A has been given which applies to him/her and (a) (s)he fails to leave the land as soon as reasonably practicable, or (b) (s)he enters any land within the area of the relevant local authority as a trespasser within a period of three months with the intention of residing there.

Section 62B(5) provides defences where D can show that:

(i) (s)he was not trespassing on land in respect of which (s)he is alleged to have committed the offence; or
(ii) (s)he had a reasonable excuse for failing to leave as soon as reasonably practicable, or for entering land in the area of the relevant local authority as a trespasser with the intention of residing there; or
(iii) at the time the direction was given (s)he was under 18 and was residing with his/her parent or guardian.

Police powers

18.102   CJPOA 1994, s 62C provides that, where a constable reasonably suspects that a person subject to a direction under s 62A has, without reasonable excuse, failed to remove any vehicle which appears to belong to him/her or to be under his/her possession or control, or that such a person has entered any land in the area of the relevant local authority as a trespasser with a vehicle within the period of three months from the day of the direction, (s)he may seize and remove the vehicle.

UNAUTHORISED CAMPING WITH A VEHICLE

CJPOA 1994, s 77 empowers a county council, district council, London borough council or (in Wales) county borough council to direct persons residing in a vehicle or vehicles to leave: **18.103**

(a) any land in the open air forming part of a highway;
(b) any other unoccupied land in the open air; or
(c) any occupied land in the open air where they are camping without the consent of the occupier of the land.

Such a direction may be addressed to a particular person or persons or to all of the occupants of vehicles on the land. It is not necessary for the local authority to show that there has been any form of nuisance caused by their presence.

Where such notice of a direction has been served, any person who knows that the direction has been given, and that it applies to him/her, commits a summary offence under s 77(3), which is punishable with a fine not exceeding level 3:

(a) if (s)he fails, as soon as practicable, to leave the land or remove from the land any vehicle or any other property which is a subject of the direction; or
(b) if, having removed any such vehicle or property, (s)he again enters the land with a vehicle within a period of three months from the day upon which the direction was given.

Section 77(5) provides a defence where such failure to leave or remove a vehicle or other property as soon as practicable, or the re-entry with a vehicle, was due to illness, mechanical breakdown, or other immediate emergency. The onus is upon D to show such a defence.

If a direction under s 77 is not complied with, the local authority can make an application to a magistrates' court under s 78 for an order for the removal by local authority officers and employees of persons and vehicles on the land in contravention of the direction. Wilful obstruction of someone acting under such an order is a summary offence, punishable with a fine not exceeding level 3, under s 78(4). Although police officers do not actually execute an order under s 78, they will often be in attendance in view of the risk to public order.

AGGRAVATED TRESPASS

Offence of aggravated trespass

By CJPOA 1994, s 68(1), a person commits the summary offence of aggravated trespass, which is punishable with up to three months' imprisonment and/or a fine not exceeding level four, if (s)he trespasses on land and, in relation to any lawful activity which persons are engaged in, or are about to engage in, on that land or adjoining land, does there anything which is intended by him/her to have the effect of: **18.104**

(a) intimidating those persons or any of them so as to deter them or any of them from engaging in that activity;
(b) obstructing that activity; or
(c) disrupting that activity.

For this purpose, an activity on the part of a person or persons on land is 'lawful' if (s)he or they may engage in the activity on that land on that occasion without committing an offence against English law or trespassing on the land. The Supreme Court has held that an activity is not rendered unlawful by some unlawful act or event incidental or collateral to, or remote from, it; the unlawfulness must be integral to the activity in question.

The persons engaged, or about to engage, in the relevant lawful activity must physically be present on the land in question at the time of the alleged trespass.

Some act is required in addition to trespass. It must be specified in the charge and must be shown to be intended to have one of the effects set out at (a)–(c). Thus, a trespasser who gives drugged meat to gundogs commits such an act, as does someone who immobilises a bulldozer on a construction site. Indeed, a divisional court has held that a trespasser commits an offence under s 68 if (s)he does an act with intent to commit a further act, which is not committed, and thereby to intimidate, obstruct or disrupt a lawful activity, if the act done is sufficiently closely connected with the intended intimidation, etc as to be more than merely preparatory to it. On this basis, the court upheld the conviction under s 68 of a person who had trespassed on land and run after a hunt with the intention of getting close enough to do something to disrupt it.

The act must be committed on land. 'Land' does not include land forming part of a highway unless it is a footpath, bridleway, etc. A divisional court has held that 'land' in s 68 includes buildings.

A failure to do something is not enough. Trespassers who were already at a particular spot, who refused on an impulse to move to allow other persons to pass, would not commit an offence under s 68; whereas it would be an offence if they deliberately placed themselves there with the intention of disrupting the activity.

The offence is one requiring 'intent' to create one of the specified effects. The trespassing rambler who walks through grouse moors will not commit this offence even though (s)he disrupts a shoot, if that was not his/her intention. Those who protest at the building of a new road will commit the offence if they deliberately sit down on private land in front of the machines because they intend to obstruct or disrupt the lawful activity of the developers.

Since they are not acting to prevent an unlawful act, protesters who are otherwise guilty under s 68 cannot rely on the common law defence of defence of property as a defence to a charge of aggravated trespass, because that defence is limited to defence against an act which is criminal or tortious (**9.25**).

Direction to leave

18.105 By CJPOA 1994, s 69, where the senior police officer present at the scene reasonably believes that:

(a) a person is committing, has committed or intends to commit the offence of aggravated trespass on land; or

(b) two or more persons are trespassing on land and are present there with the common purpose of intimidating persons so as to deter them from engaging in a lawful activity or of obstructing or disrupting a lawful activity,

(s)he may direct that person, or those persons (or any of them), to leave the land. 'Land' has the same meaning as in s 68.

A person who, knowing that such a direction has been given which applies to him/her, fails to leave the land as soon as practicable or, having left, again enters the land as a trespasser within the period of three months beginning with the day on which the direction was given, commits a summary offence under s 69(3) which is punishable with up to three months' imprisonment and/or a fine not exceeding level four.

It is a defence to a charge of an offence under s 69(3) for D to prove that (s)he was not trespassing on the land, or that (s)he had a reasonable excuse for failing to leave the land as soon as practicable or, as the case may be, for again entering the land as a trespasser (s 69(4)).

PUBLIC ORDER AND ASSOCIATION FOOTBALL MATCHES

Alcohol on coaches, trains, etc

The Sporting Events (Control of Alcohol, etc) Act 1985 (SE(CA)A 1985), ss 1 and 1A   **18.106** provide offences designed to prevent drunken behaviour by football fans en route to or from matches, and to prevent them arriving at grounds drunk.

Specified vehicles

The offences under SE(CA)A 1985, s 1 apply to public service vehicles (ie coaches,   **18.107** buses, and the like) and passenger trains *which are being used for the principal purpose of carrying passengers for the whole or part of a journey to or from a 'designated sporting event'*. Section 1, therefore, does not apply to a bus or train on a normal scheduled service because it is not being used for the principal purpose of carrying passengers to or from a designated sporting event, even if the majority of the passengers are travelling to or from a match, since the words 'used' and 'principal purpose' must refer to use by, and the principal purpose of, the operator. On the other hand, s 1 does apply to a 'football special' or to a chartered coach or train, provided that it is travelling to or from a designated sporting event.

The offences under s 1A apply to a motor vehicle which:

(a) is not a public service vehicle but is adapted to carry more than eight passengers; and

(b) is being used for the principal purpose of carrying *two or more* passengers for the whole or part of a journey to or from a designated sporting event.

Designated sporting event

'Designated sporting event' in SE(CA)A 1985 means a sporting event or proposed   **18.108** sporting event designated by the Sports Grounds and Sporting Events (Designation) Order 2005.

The Sports Grounds and Sporting Events (Designation) Order 2005 designates as sporting events to which SE(CA)A 1985 applies:

(a) association football matches, at any sports ground in England or Wales, in which one or both of the participating teams represents a club which is for the time being a member (whether a full or associate member) of the Football League, the Football Association Premier League, the Football Conference National Division, the Scottish Professional Football League, or Welsh Premier League, or whose home ground is situated outside England and Wales, or represents a country or territory;

(b) association football matches, at any sports ground in England and Wales, in competition for the Football Association Cup (other than in a preliminary or qualifying round); and

(c) association football matches at a sports ground outside England and Wales in which one or both of the participating teams represents a club which is for the time being a member (whether a full or associate member) of the Football League, the Football Association Premier League, the Football Conference National Division, the Scottish Professional Football League, or Welsh Premier League, or represents the Football Association or the Football Association of Wales.

SE(CA)A 1985 does not apply to any sporting event or proposed sporting event where all competitors are to take part otherwise than for reward, and to which all spectators are to be admitted free of charge.

Causing or permitting carriage of alcohol on a vehicle

18.109   SE(CA)A 1985, s 1(2) provides that a person who knowingly causes or permits alcohol to be carried on a vehicle to which s 1 (**18.107**) applies is guilty of an offence if the vehicle is:

(a) a public service vehicle and (s)he is the operator of the vehicle or the employee or agent of the operator; or

(b) a hired vehicle and (s)he is the person to whom it is hired or the employee or agent of that person. Thus, an organiser of the supporters' club (or his/her agent) can be convicted if (s)he permits alcohol to be carried on a chartered train, but a train guard who fails to prevent this cannot because, although (s)he permits it, (s)he is not a person to whom the train is hired (nor the employee or agent of such a person).

Section 1A(2) provides that a person who knowingly causes or permits alcohol to be carried on a motor vehicle to which s 1A applies is guilty of an offence:

(a) if (s)he is its driver; or

(b) if (s)he is not its driver but its keeper, the employee, or agent of its keeper, a person to whom it is made available (by hire, loan or otherwise) by its keeper or the keeper's employee or agent, or the employee or agent of a person to whom it is so made available.

Possession of alcohol on a vehicle

18.110   Section s 1(3) makes it an offence for a person to have alcohol in his/her possession while on a vehicle to which s 1 applies. There is a corresponding offence under s 1A(3) in relation to motor vehicles to which s 1A applies.

Police powers

18.111   By s 7(3), a constable may stop a public service vehicle to which s 1 applies or a motor vehicle to which s 1A applies and may search *such a vehicle or a railway passenger vehicle* if (s)he has reasonable grounds to suspect that an offence under s 1 or s 1A is being or has been committed in respect of that vehicle.

As to search of a person see **18.116**.

Alcohol, containers, fireworks, etc at designated sports grounds

Possession of alcohol, etc at designated ground

18.112   SE(CA)A 1985, s 2(1) provides that a person who has alcohol or an article to which s 2 applies in his/her possession:

(a) at any time during the period of a designated sporting event (**18.114**) when (s)he is in any area of a designated sports ground (**18.115**) from which the event may be directly viewed (s 2(1)(a)); or

(b) while entering or trying to enter a designated sports ground at any time during the period of a designated sporting event at that ground (s 2(1)(b)),

is guilty of an offence.

Section 2(3) states that an article to which s 2 applies is any article capable of causing injury to a person struck by it, being:

(a) a bottle, can, or other portable container (including such an article when crushed or broken) which is for holding any drink and is of a kind which, when empty, is normally discarded or returned to, or left to be recovered by, the supplier; or
(b) part of an article falling within (a).

However, the definition expressly does not apply to anything that is for holding any medicinal product.

The container need not be made specifically to hold alcohol, and it is irrelevant that it has never contained alcohol or that it is broken. An empty soft drink bottle or can is caught. On the other hand, a re-usable plastic drinks container, a mug, a thermos flask, or a hip flask is not, since it is not of a kind normally discarded or returned to, or left to be recovered by, the supplier.

Possession of fireworks, etc

18.113 SE(CA)A 1985, s 2A(1), which is essentially aimed at reducing the risk of fire, makes identical provision in relation to the possession of a firework or of distress flares, fog signals, canisters of smoke, or visible gas, and similar articles. Matches and cigarette lighters are expressly excluded.

It is a defence for the defendant (D) to prove that (s)he had possession with lawful authority.

Period of a designated sporting event

18.114 Normally, the 'period of a designated sporting event' in SE(CA)A 1985 is the period beginning two hours before the start of the event or (if earlier, as where the start is delayed) two hours before the time at which it is advertised to start, and ending one hour after the end of the event. Where a match is postponed to a later day or cancelled, the period ends one hour after the advertised start time. In respect of a room in a designated sports ground from which the designated sporting event may be directly viewed and to which the public are not admitted (eg the directors' box), there is a different period in relation only to an offence of possession of alcohol, etc under s 2(1)(a) (above). This is a 'restricted period' beginning 15 minutes before the start of the event (or advertised start) and ending 15 minutes after the end of the event or 15 minutes after the advertised start (if the event is postponed to a later day or cancelled).

Designated sports ground

18.115 A 'designated sports ground' is any sports ground in England or Wales where accommodation is provided for spectators.

Sporting events: general police powers

18.116 By SE(CA)A 1985, s 7(1), a constable may, at any time during the period of a designated sporting event at any designated sports ground, enter *any* part of the ground for the purpose of enforcing SE(CA)A 1985.

By s 7(2), a constable may search a person (s)he has *reasonable grounds to suspect* is committing or has committed an offence under SE(CA)A 1985. The provisions of PACE, s 2 and PACE Code A apply to stop and search under s 7. See **4.11–4.18** and **4.24–4.30**.

Code A provides that its requirement that a person must not be searched, even with his/her consent, except under a specific power of search does not affect the searching of persons entering sports grounds or other premises with their consent given as a condition of entry.

Trial and punishment

18.117   The above offences under SE(CA)A 1985 are summary offences. They are punishable as follows:

s 1(2) or 1A(2)—a fine not exceeding level 4; and

s 1(3), 1A(3), 2(1), or 2A(1)—up to three months' imprisonment and/or a fine not exceeding level 3.

Misbehaviour at a designated football match

18.118   The Football (Offences) Act 1991 (F(O)A 1991) creates a number of summary offences punishable with a fine not exceeding level 3. They can only be committed at a 'designated football match'. Such matches are designated by the Football (Offences) (Designation of Football Matches) Order 2004. They are association football matches:

(a) in which one or both of the participating teams represents:
   (i) a club which is for the time being a member (whether a full or associate member) of the Football League, the Football Association Premier League, the Football Conference, the Welsh Premier League, or the Scottish Football League; or
   (ii) a club whose home ground is situated outside England and Wales; or
   (iii) a country or territory; or
(b) which is played in the FA Cup, other than a preliminary or qualifying round.

References to things done at a designated football match include anything done there in the period beginning two hours before the start of the match or (if earlier) two hours before the advertised start time and ending one hour after the end of the match. If the match does not take place, the period is two hours before the advertised start time until one hour after that time.

Throwing objects

18.119   By F(O)A 1991, s 2, it is an offence to throw anything at or towards:

(a) the playing area or any area adjacent to the playing area to which spectators are not generally admitted; or
(b) any area in which spectators or other persons are or may be present,

without lawful authority or excuse (which it is for D to prove).

Thus, those who throw objects onto the pitch, into the players' tunnel, etc, or into spectator areas will commit offences. Those who may prove 'lawful authority or excuse' would include vendors who throw packets of crisps, etc into the crowd, or spectators who throw money to such persons.

Chanting

F(O)A 1991, s 3(1) makes it an offence to engage or take part in chanting of an indecent **18.120** or racist nature at a designated football match. 'Chanting' means the repeated uttering of words or sounds (whether alone or in concert with one or more others), and 'racist nature' means consisting of or including matter which is threatening, abusive, or insulting to a person by reason of his/her colour, race, nationality (including citizenship), or ethnic or national origin.

Pitch invasion

By F(O)A 1991, s 4, it is an offence for a person to go onto the playing area, or any area **18.121** adjacent to the playing area to which spectators are not generally admitted, without lawful authority or lawful excuse (which it is for D to prove).

It is not easy to obtain a conviction where spectators surge forward onto the pitch. Most will contend that they were carried forward unwillingly by the momentum of the crowd. If this is not disproved, a conviction for an offence under s 4 will not be possible, since a person cannot generally be convicted if his/her conduct was beyond his/her control. Those who would have lawful excuse for going onto the pitch include trainers and official first aiders.

FOOTBALL BANNING ORDERS

The Football Spectators Act 1989 (FSA 1989), Part II deals with these orders. The rele- **18.122** vant provisions only apply in respect of a regulated football match.

Regulated football match

A 'regulated football match' is an association football match (whether in the UK or **18.123** elsewhere) which is a prescribed match or a match of a prescribed description. The Football Spectators (Prescription) Order 2004 prescribes 'regulated football matches in England and Wales' and 'regulated football matches outside England and Wales'.

Regulated football match in England and Wales

This is an association football match: **18.124**

(a) in which one or both of the participating teams represents:
  (i) a club which is for the time being a member (whether a full or associate member) of the Football League, the Football Association Premier League, the Football Conference, the Welsh Premier League, or the Scottish Professional Football League;
  (ii) a club whose home ground is situated outside England and Wales; or
  (iii) a country or territory; or
(b) which is played in the FA Cup, other than a preliminary or qualifying round.

Regulated football match outside England and Wales

This is an association football match involving: **18.125**

(a) a national team appointed by the Football Association to represent England or by the Football Association of Wales to represent Wales;
(b) a team representing a club which is for the time being a member (whether a full or associate member) of the Football League, the Football Association Premier

League, the Football Conference, the Welsh Premier League, or the Scottish Professional Football League (SPFL);

(c) a team representing any country or territory whose football association is for the time being a member of FIFA where the match is part of a competition or tournament organised by or under the authority of FIFA or UEFA and the competition or tournament is one in which a national team referred to in (a) is eligible to participate, or has participated; or

(d) a team representing a club which is for the time being a member (whether a full or associate member) of, or affiliated to, a national football association which is a member of FIFA, where the match is part of a competition or tournament organised by, or under the authority of, FIFA or UEFA and the competition or tournament is such that a club from the Football League, the Football Premier League, the Football Conference, the Welsh Premier League, or the SPFL is eligible to participate, or has participated.

When can a banning order be made?

18.126   A banning order may be made under FSA 1989, s 14A, s 14B, or s 22.

Banning orders following a conviction for a relevant offence

18.127   FSA 1989, s 14A is concerned with a situation in which an offender is convicted of a 'relevant offence'. The relevant offences are set out in FSA 1989, Sch 1. They are:

(a) any offence contrary to FSA 1989, s 14J(1) or 19(6) or 20(10) or 21C(2), (failure to comply with banning order requirement or requirement under s 14(2B) or (2C), or failure to comply with a requirement made at initial reporting at a police station or, being subject to a banning order, knowingly or recklessly providing false or misleading information in support of his/her application for exemption from requirements imposed under FSA 1989, or non-compliance with a notice under s 21B);

(b) an offence under the Police, Public Order and Criminal Justice (Scotland) Act 2006, s 68(1) or (5) (corresponding offences to those in (a));

(c) any offence contrary to SE(CA)A 1985, s 2 or 2A committed at a regulated football match or while entering or trying to enter the ground;

(d) any offence involving harassment, alarm or distress contrary to the Public Order Act 1986 (POA 1986), s 4A or 5, or racial or religious hatred or hatred on the ground of sexual orientation contrary to POA 1986, Part III or IIIA, or any offence involving the use of violence or threat of violence towards another person or property:

   (i) committed during a 'period relevant to a regulated football match' (the period from 24 hours before the start or advertised start of the match (whichever is earlier) to 24 hours after its end, or the period from 24 hours before the advertised start of a cancelled game or one postponed to another day to 24 hours after that time) while at, entering or leaving the ground (or trying to do so), or

   (ii) if the court makes a declaration of relevance (ie declares that the offence related to regulated football matches), while on a journey to or from a regulated football match;

(e) any offence involving harassment, alarm or distress (POA 1986, s 4A or 5) or racial or religious hatred or hatred on the ground of sexual orientation (POA 1986, Part III or IIIA), or any offence involving the use or threat of violence towards another person or property which does not fall within (d), which was committed during a period relevant to a regulated football match and as respects which the court declares that the offence related to that match or that match and other regulated football matches during that period;

(f) any offence of drunkenness (ie being found drunk, or being drunk and disorderly, in a public place), committed while on such a journey, as to which offence the court makes a declaration of relevance);

(g) any offence contrary to SE(CA)A 1985, s 1 or the Road Traffic Act 1988, s 4, 5, or 5A (drink or drugs and driving, etc) committed on a journey to or from a regulated match, as to which offence the court makes a declaration of relevance;

(h) any offence contrary to F(O)A 1991;

(i) any offence involving the use, carrying, or possession of an offensive weapon or firearm:
  (i) committed during a period relevant to a regulated football match while at, or entering or leaving the ground (or trying to do so), or
  (ii) if the court makes a declaration of relevance, committed while on a journey to or from a regulated football match;

(j) any offence involving the use, carrying, or possession of an offensive weapon or firearm which does not fall within (i)(i) or (i)(ii), which was committed during a period relevant to a regulated football match and as respects which the court declares that the offence related to that match or to that match and any other football offence during that period;

(k) ticket touting contrary to the Criminal Justice and Public Order Act 1994, s 166.

Attempting, conspiring, encouraging or assisting, or aiding, abetting, counselling, or procuring, the commission of any such offence is included within these provisions.

A person may be regarded as having been on a journey to or from a football match whether or not (s)he attended or intended to attend the match. A journey includes breaks (including overnight breaks).

If, upon conviction for such an offence, a court is satisfied that there are reasonable grounds to believe that making a banning order would help to prevent violence or disorder at, or in connection with, any regulated football match, it *must* make such an order in respect of the offender. If it is not so satisfied, it must state its reasons in open court.

Banning orders made on complaint

Under FSA 1989, s 14B, the chief officer of police of an area in which a person resides, or appears to reside, the chief constable of the British Transport Police or the DPP may make a complaint to a magistrates' court that the respondent has at some time contributed to violence or disorder in the UK or elsewhere.   **18.128**

If the court is satisfied as to the substance of the complaint and has reasonable grounds for believing that making a banning order would help to prevent violence or disorder at or in connection with any regulated football match it *must* make a banning order. In deciding whether to make an order a court may take into account (among other things):

(a) any decision of a court or tribunal outside the UK;
(b) deportation or exclusion from a country outside the UK;
(c) removal or exclusion from football grounds in the UK or elsewhere; or
(d) conduct recorded by video or by any other means.

However, a court may only take note of matters occurring within the 10 years preceding the application. It must also consider the reasons given by a court when, in relation to a relevant offence, it did not make a banning order.

The Court of Appeal has described the standard of proof as to the substance of the complaint as an exacting one which would be hard in practice to distinguish from the criminal standard of proof beyond reasonable doubt.

Banning order under FSA 1989, s 22

18.129 FSA 1989, s 22 governs the making of a banning order as a result of a conviction outside England and Wales for a 'corresponding offence'. A 'corresponding offence' is an offence under the law of a country specified outside England and Wales in an Order in Council. At the time of writing, there are no such orders in force. This is because FSA 1989, s 22 has fallen out of use, given that there is an alternative route to securing a banning order under FSA 1989 via s 14B.

The effect of a banning order

18.130 A 'banning order' means an order made by a court under FSA 1989, Part II which, in relation to regulated football matches:

(a) in the UK, prohibits the person who is subject to the order from entering premises for the purpose of attending such matches; and

(b) outside the UK, requires that person to report at a police station in accordance with Part II.

An order will apply to the whole range of matches which are prescribed, whatever the venue and whatever the club; a court cannot limit it to particular matches or teams. This is sensible. Clearly, a banning order would not be very effective if it barred a person only from the ground at which the offence was committed.

A banning order must require the person subject to the order to report to a police station specified in the order, within five days beginning with the day upon which the order is made.

A banning order must require the person subject to the order to notify the enforcing authority of the following events:

(a) a change of any of his/her names, of his/her home address, or of his/her temporary address (or of his/her ceasing to have one);

(b) his/her first use after the order of a name that was not disclosed at the time of the making of the order;

(c) his/her acquisition of a temporary address;

(d) his/her becoming aware of the loss of his/her passport;

(e) his/her receipt of a new passport;

(f) an appeal in relation to the order or against the making of a declaration of relevance in respect of an offence of which (s)he has been convicted; and

(g) an application for termination of the order.

The notification must be given before the end of the seven-day period beginning with the day on which the event in question occurs and:

(a) in the case of a change of a name or address or the acquisition of a temporary address, must specify the new name or address;

(b) in the case of a first use of a previously undisclosed name, must specify that name; and

(c) in the case of a receipt of a new passport, must give details of that.

A banning order must contain a requirement as to the surrender, in accordance with the Act (below), in connection with regulated football matches outside the UK, of the passport of the person subject to the order.

The court may impose additional requirements on a person subject to such an order and may subsequently vary an order so as to impose, replace or omit requirements on the application of the person subject to the order or the person who applied for the order or the prosecutor. In addition, the court has power to require a constable to photograph the person or cause him/her to be photographed.

Duration of banning order

Where a banning order is made under s 14A (**18.127**) and is in addition to a sentence **18.131** of imprisonment (any form of detention) taking immediate effect, the maximum duration is 10 years and the minimum six. In any other case following conviction the maximum is five years and the minimum three. Orders which are made under s 14B or s 22 may be for a maximum of five years and a minimum of three.

Termination of an order

After two-thirds of the period of the ban has passed, the person subject to it may apply **18.132** to the court by which the order was made to terminate it. The court, in considering whether or not to terminate the ban, must have regard to a person's character; conduct since the order was made; the nature of the offence or conduct concerned; and any other relevant circumstances.

Functions of the enforcing authority and local police forces

FSA 1989, s 19(2) provides that, when a banned person initially reports at a police sta- **18.133** tion, the officer responsible for that station may make such requirements of that person as are determined by the enforcing authority to be necessary or expedient for giving effect to the banning order, so far as matters are related to regulated football matches outside the UK. The 'enforcing authority' is the Football Banning Orders Authority established by the Secretary of State.

If, in connection with any regulated football match *outside the UK*, that authority considers that a requirement to report is necessary or expedient to reduce the likelihood of violence or disorder at or in connection with the match, the authority is required by s 19(2B) to give the person subject to the order a notice in writing to report as instructed and to surrender his/her passport as instructed. Under s 19(2B), the notice may also require compliance with additional requirements.

In the case of *any* regulated football match, the enforcing authority may by notice in writing require the person subject to the order to comply with any additional requirements of the order in the manner specified in the notice (s 19(2C)).

A notice under s 19 may not require the person subject to it to report except in the 'control period' in relation to a regulated football match outside the UK or an external tournament which includes such matches and may not require him/her to surrender a passport except in such a control period. The 'control period' in relation to such a regulated football match outside the UK is the period commencing five days before the day of the match and ending when the match is finished or cancelled. In relation to an external tournament the 'control period' means any period described in an order made by the Secretary of State beginning five days before the first match outside the UK and ending when the last match outside the UK has been finished or cancelled. However, qualifying matches do not count in determining the start of the tournament.

If the Secretary of State considers it necessary or expedient to do so in order to secure the effective enforcement of these provisions, the 'control period' may be extended by order in relation to any regulated football match to a maximum of not more than 10 days before the day of the match (or the day of the first match in the external tournament, as the case may be).

Failure to comply with a requirement

18.134   It is a summary offence, contrary to FSA 1989, s 19(6), to fail, without reasonable excuse, to comply with a requirement imposed under s 19(2). The offence is punishable with a fine not exceeding level 2.

It is a summary offence, contrary to s 14J(1), for a person subject to a banning order to fail to comply with any requirement imposed by the order or by s 19(2B) or s 19(2C). The offence is punishable with up to six months' imprisonment and/or an unlimited fine.

Summary measures

18.135   FSA 1989, ss 21A and 21B underpin the procedure for the application by way of complaint for a banning order. They provide for a constable in uniform to exercise powers of detention and of reference to a court in specified circumstances. The powers under ss 21A and 21B may be exercised only in relation to a British citizen.

Section 21A provides a constable in uniform with the power to detain a person in his/her custody if, during any 'control period' in relation to a regulated football match outside the UK or an external tournament:

(a) (s)he has reasonable grounds to suspect that the person has at any time caused or contributed to any violence or disorder in the UK or elsewhere; and

(b) (s)he has reasonable grounds to believe that making a banning order in his/her case would help to prevent violence or disorder at, or in connection with, any regulated football matches.

Such a person may be detained until the constable has decided whether or not to issue a notice under s 21B requiring him/her:

(a) to appear before a magistrates' court at a specified time;

(b) not to leave England and Wales before that time; and

(c) if the control period relates to a regulated football match outside the UK or to an external tournament which includes such matches, to surrender his/her passport to the constable,

and stating the grounds upon which his/her decision is based.

Such detention may not exceed four hours or, with the authority of an inspector (or above), six hours. A person so detained may not be further detained within the same control period unless new information becomes available.

The notice referred to above may be issued where the officer is authorised to do so by an inspector (or above). The time at which such a person must appear before a magistrates' court must be within 24 hours of receiving the notice or that person's detention, whichever is the earlier. Such a notice will be treated as an application for a banning order by way of complaint.

Where a person to whom such a notice has been given appears before a magistrates' court, the court may remand him/her. If (s)he is remanded on bail, (s)he may be required not to leave England and Wales before his/her appearance before the court and, if the control period relates to a regulated football match outside the UK or to an external tournament which includes such matches, (s)he may be required to surrender his/her passport to a constable.

It is a summary offence under s 21C to fail to comply with a notice under s 21B. It is punishable with up to six months' imprisonment and/or an unlimited fine,

TICKET TOUTS

18.136 The Criminal Justice and Public Order Act 1994, s 166 makes it a summary offence, punishable with an unlimited fine, for an unauthorised person to sell, or otherwise dispose of to another person, a ticket for a designated football match. 'Selling' includes offering or exposing a ticket for sale, making it available for sale by another, advertising it as available for purchase, and giving it to a person who pays or agrees to pay for some other goods or services or offers to do so.

The only persons who are 'authorised' are those authorised in writing by the organisers of the match. 'Ticket' includes anything which purports to be a ticket, so that false tickets are included. However, where a false ticket is involved a charge of fraud would be more appropriate where knowledge of such falsity can be proved. A 'designated football match' has the same meaning as 'regulated football match' (**18.123–18.125**), except that it does not cover the Scottish Professional Football League or matches in the FA Cup between clubs which are not members of the Leagues specified at **18.124**.

Police powers

18.137 The provisions of PACE 1984, s 32 (search of an arrested person and his/her vehicle) are extended to the case where the vehicle is reasonably believed to have been used for any purpose connected with an offence under s 166.

OFFENCES OF DRUNKENNESS

18.138 Drunkenness is not in itself an offence but becomes so in certain circumstances, which are described below. Drunkenness in this context is limited to intoxication through drink and does not include intoxication through drugs or through glue-sniffing. A divisional court has held that 'drunkenness' requires that the defendant has voluntarily consumed alcohol and that his/her has resulted in his/her becoming drunk. 'Drunk' bears its ordinary natural meaning; whether a defendant was drunk is a question of fact in each case.

Drunk on highway or other public place

18.139 The Licensing Act 1872, s 12 provides that a person is guilty of a summary offence, punishable with a fine not exceeding level 1, if (s)he is found drunk on any highway or other public place, whether a building or not, or on any licensed premises. The offence is often described as one of 'simple drunkenness' to distinguish it from other offences dealing with aggravated forms of drunkenness. It is particularly worth remembering that PACE, s 24 permits an arrest if it is necessary to prevent the person from suffering physical injury. Such injury would be likely if (s)he was left unconscious in the open.

The offence under LA 1872, s 12 is a 'penalty offence' for the purposes of CJPA 2001: see **3.40**.

Being drunk on a vehicle

18.140   The Sporting Events (Control of Alcohol) Act 1985 (SE(CA)A 1985), s 1(4) provides that a person who is drunk on a vehicle to which s 1 applies (**18.107**) is guilty of an offence. There is a corresponding offence under s 1A(4) in relation to motor vehicles to which s 1A applies (**18.107**). Both offences are summary offences punishable with a fine not exceeding level 2. As to police powers, see **18.111** and **18.116**.

Being drunk at a designated sports ground

18.141   By SE(CA)A 1985, s 2(2), a person who is drunk in a designated sports ground (**18.115**) at any time during the period of a designated sporting event (**18.108**) at that ground, or who is drunk while entering or trying to enter such a ground at any time during the period of a designated sporting event (**18.114**) at that ground, is guilty of a summary offence punishable with a fine not exceeding level 2. As to police powers, see **18.118**.

Drunk and disorderly

18.142   The Criminal Justice Act 1967, s 91 states that any person who in any public place is guilty, while drunk, of disorderly behaviour commits a summary offence punishable with a fine not exceeding level 3. A divisional court has held that whether the defendant's behaviour was disorderly is a simple question of fact. What is required is that his/her behaviour, viewed objectively, was disorderly. 'Disorderly behaviour' does not involve any element of mens rea. Where there is an intercom system and locks on the entrance to a block of flats, so that only those admitted by the occupiers are given access to the area, the landing area outside the flats is not a public place for the purposes of this/her offence. Nor would it be for the purposes of the offence of simple drunkenness.

An offence under s 91 is a 'penalty offence' for the purposes of CJPA 2001 (**3.40**).

Drunk in charge of a child

18.143   The Licensing Act 1902, s 2 creates the offence of being found drunk on any highway or other public place, whether a building or not, or on any licensed premises, while having the charge of a child apparently under the age of seven. The offence is punishable with up to one month's imprisonment or a fine not exceeding level 2.

CHAPTER 19

Preventing Harm from Firearms, Weapons, and Explosives

The Firearms Act 1968 (FiA 1968) controls the possession, use, manufacture, transfer **19.1** and acquisition of firearms or ammunition. Other Acts punish the possession or use of firearms in particular circumstances and deal with other forms of weapon and with explosives.

FIREARMS: DEFINITIONS

FiA 1968, s 57 defines 'firearm' for the purposes of that Act as: **19.2**

(a) a lethal barrelled weapon;
(b) a prohibited weapon;
(c) a relevant component part in relation to a lethal barrelled weapon or a prohibited weapon (see below);
(d) an accessory to a lethal barrelled weapon or a prohibited weapon where the accessory is designed or adapted to diminish the noise or flash caused by firing the weapon.

For the purposes of (a), 'lethal barrelled weapon' is defined as meaning a barrelled weapon (ie a weapon with a tube) from which a shot, bullet, or other missile, with kinetic energy of more than one joule at the muzzle of the weapon, can be discharged.

For the purposes of (b), see the definition of 'prohibited weapon' at **19.17**. Such a weapon includes weapons which are not lethal barrelled.

For the purposes of (c), each of the following is a relevant component part in relation to a lethal barrelled weapon or a prohibited weapon:

(i) a barrel, chamber, or cylinder;
(ii) a frame, body, or receiver;
(iii) a breech block, bolt, or other mechanism for containing the pressure of discharge at the rear of a chamber,

but only where the item is capable of being used as a part of a lethal barrelled weapon or a prohibited weapon.

By s 57A, an 'airsoft gun' is not a firearm for the purposes of FA 1968. An 'airsoft gun' is defined as a barrelled weapon of any description which:

(a) is designed to discharge only a small plastic missile (whether or not it is also capable of discharging any other kind of missile); and
(b) is not capable of discharging a missile (of any kind) with kinetic energy at the muzzle of the weapon that exceeds the permitted level.

'Small plastic missile' means a missile that:

(i)   is made wholly or partly from plastics;
(ii) is spherical; and
(iii) does not exceed 8 mm in diameter.

The permitted kinetic energy level is, in the case of a weapon capable of discharging two or more missiles successively without repeated pressure on the trigger, 1.3 joules. In any other case, it is 2.5 joules.

Antique firearms

19.3 With the exception of FiA 1968, ss 19, 20, and 21 (**19.33**, **19.34**, and **19.22**), nothing in the FiA 1968 relating to firearms applies to an antique firearm which is sold, transferred, purchased, acquired, or possessed as a curiosity or ornament. A firearm is an 'antique firearm' for the purposes of FiA 1968 if a firearm is an 'antique firearm' if:

(a) either the conditions in (i) below are met or the condition in (ii) below is met, and
(b) the firearm was manufactured before 1 September 1939.

The conditions referred to in (a) are that:

(i) (a) the firearm's chamber or, if the firearm has more than one chamber, each of its chambers is either a chamber that the firearm had when it was manufactured, or a replacement for such a chamber that is identical to it in all material respects;
     (b) the firearm's chamber or (as the case may be) each of the firearm's chambers is designed for use with a cartridge of a description specified in a lengthy list in the Antique Firearms Regulation 2021, Sch (whether or not it is also capable of being used with other cartridges). The list is in Chapter 19 of the companion website;
(ii) the firearm's propulsion system is of a description specified in the 2021 Regulations, namely any propulsion system:
     (a) which involves the use of a loose charge and a separate ball (or other missile) loaded at the muzzle end of the barrel, chamber, or cylinder of the firearm and which uses an independent source of ignition;
     (b) in a breech-loading cartridge firearm which uses an ignition system other than rim-fire or centre-fire;
     (c) which involves the use of rim-fire cartridges (other than .22 (5.58mm), .23 (5.8mm), 6mm, or 9mm rim-fire cartridges) in a breech-loading firearm, or
     (d) for an air weapon.

Where:

(a) immediately before the coming into force of the Antique Firearms Regulation 2021 on 22 March 2021, a person (P) had possession of a firearm that was an antique firearm for the purposes of FA 1968; and
(b) in consequence of the coming into force of the regulations, the firearm ceased to be an antique firearm for those purposes:
    (i) FA 1968, s 5 (**19.16**) does not apply in relation to P's possession of the firearm unless—
         (a) P carries on a business as a firearms dealer; and
         (b) the firearm is in P's possession for the purpose of the business;

(ii) an application by P for the grant or renewal of a firearm certificate or a shotgun certificate in respect of possession of the firearm may not be refused on the ground that P does not have a good reason for having possession of the firearm.

Changes in type

A weapon which at any time has been classified as a prohibited weapon (**19.17**) (or as a FiA 1968, s 1 firearm (**19.7**) or as a shotgun (**19.8**)) remains so classified notwithstanding anything done to convert it into a weapon of another type (eg prohibited weapon into a s 1 firearm). **19.4**

Imitation firearms readily convertible into firearms to which the Firearms Act 1968, s 1 applies

Under the Firearms Act 1982, s 1 the provisions of FiA 1968 concerned with a FiA 1968, s 1 firearm are widened so as to apply (with limited exceptions) to an imitation firearm which has the appearance of being a s 1 firearm and which is so constructed or adapted as to be readily convertible into such a firearm. The limited exceptions are indicated in the text. An imitation firearm is '*readily* convertible' if it can be converted into a firearm without special skill and without equipment or tools not in common use. '*Convertible*' refers to converting an item from which a missile could not be discharged into one from which a missile could be discharged. It does not matter whether that process would involve the permanent alteration of the item (eg by drilling) or some more temporary result; an item can be converted by altering its construction or by adapting the way it could be used (eg by making it capable of discharging a missile by the use of an external implement). Where an alleged offence involves an imitation firearm which is readily convertible into a s 1 firearm, it is a defence for the defendant to prove that (s)he did not know and had no reason to suspect that it was readily convertible. **19.5**

Ammunition

For the purposes of FiA 1968, 'ammunition' is defined by s 57 as meaning ammunition for any firearm; a blank cartridge is ammunition. Section 57 provides that the term also includes grenades, bombs and other missiles, whether capable of use with a firearm or not, and also includes prohibited ammunition (**19.18**). A 'bomb' is any explosive substance in a case, or a case containing poison gas, smoke or inflammable material which might be dropped from an aircraft, fired from a gun, or thrown or placed by hand. **19.6**

POSSESSION, PURCHASE, OR ACQUISITION OF FIREARMS OR AMMUNITION

Section 1 firearms and ammunition

FiA 1968, s 1(1) states that, subject to any exemption under FiA 1968, it is an offence for a person to have in his possession, or to purchase or acquire: **19.7**

(a) a firearm to which s 1 applies; or
(b) any ammunition to which s 1 applies,

without holding a firearm certificate, or otherwise than as authorised by such a certificate, or (ammunition only) in quantities in excess of those so authorised.

The offence is an indictable (either-way) offence punishable with up to five years' imprisonment or, if aggravated (weapon converted contrary to s 4(1) or (3): (**19.28**)), seven years'.

Section 1 applies to every firearm except:

(a) a shotgun, as defined at **19.8**; and
(b) an air weapon (ie an air rifle, air gun or air pistol which does not fall within FiA 1968, s 5(1) (**19.17**) and is not of a type declared by rules made by the Secretary of State under FiA 1968 to be specially dangerous (**19.9**)).

FiA 1968, s 1(1) applies to ammunition for a firearm except:

(a) cartridges containing five or more shot, none of which exceeds .36 inch in diameter;
(b) ammunition for any air gun, air rifle, or air pistol; and
(c) blank cartridges not more than one inch in diameter.

A person can be in possession of a firearm or ammunition even though (s)he does not have physical custody of it nor keeps it in his/her home; it is enough that (s)he knowingly has more than momentary control of it (as where (s)he stores a firearm at the home of a relative for safe-keeping). Indeed, someone can be in possession of a firearm or ammunition even though (s)he does not know or suspect that the thing under his/her control is a firearm or ammunition, as the case may be. For example, if a man has custody of a firearm in a holdall which he knows contains something without giving thought to the nature of its contents, he is in possession of the firearm. The fact that possession was brief, or that he did not know or could not reasonably have been expected to know that it contained a firearm, affords no defence. Because D need not know or suspect that the article possessed was a firearm or ammunition within the relevant meaning of those terms, an honest and reasonable mistaken belief that the article is not a firearm or is an antique firearm is no defence. *It has been held that in all the offences under FiA 1968 formulated in in terms of 'possession' that term is to be understood in the same way as under s 1.*

It is an indictable (either-way) offence under s 3(2), punishable with up to five years' imprisonment, for a person to sell or transfer a s 1 firearm or ammunition to a person other than a registered firearms dealer or person who has a firearm certificate authorising its purchase or acquisition.

Shotguns and ammunition

19.8   For the purposes of FiA 1968, a 'shotgun' is a smooth-bore gun (not being an airgun) which:

(a) has a barrel not less than 24 inches in length and does not have any barrel with a bore exceeding two inches in diameter;
(b) either has no magazine or has a non-detachable magazine incapable of holding more than two cartridges; and
(c) is not a revolver gun.

If the barrel of a shotgun is shortened to less than 24 inches it is no longer a shotgun, but becomes a s 1 firearm.

Shotguns are excepted from the provisions of FiA 1968, s 1 but FiA 1968, s 2(1) states that, subject to any exemption under FiA 1968, it is an indictable (either-way) offence, punishable with up to five years' imprisonment, for a person to have in his possession, or to purchase or acquire, a shotgun without holding a certificate under FiA 1968 authorising him/her to possess shotguns. Shotgun ammunition, other than s 1 ammunition, does not require a firearm certificate, shotgun certificate or any other form of certificate for its possession, acquisition or purchase.

It is an indictable (either-way) offence, punishable with up to five years' imprisonment, under s 3(2) for a person to sell or transfer a shotgun to a person other than a registered firearms dealer or person who has a shotgun certificate authorising its purchase or acquisition.

Air weapons and ammunition

Air weapons essentially operate by the release of compressed air. However, the Firearms   **19.9**
(Amendment) Act 1997, s 48 provides that any reference in the firearms legislation to an air rifle, air pistol, or air gun also includes a reference to a rifle, pistol, or gun powered by compressed carbon dioxide. An 'air weapon' (ie an air rifle, air gun, or air pistol) which is not a prohibited weapon and which has not been declared by the Secretary of State to be specially dangerous does not require a firearm certificate or any other form of certificate for its possession, acquisition or purchase. Nor does ammunition for any air weapon, even if declared especially dangerous. The Secretary of State has declared that an air weapon will be specially dangerous (and therefore a s 1 firearm) if:

(a) on discharge from the muzzle there is a kinetic energy in excess of 6 ft/lb in the case of an air pistol, or 12 ft/lb in the case of a weapon other than a pistol; or
(b) it is disguised as another object.

Firearm certificates and shotgun certificates

Firearm certificates

To obtain a firearm certificate an applicant must apply to the chief officer of police of   **19.10**
the area in which (s)he resides and must state such particulars as may be required by the application form. Information concerning previous names, residence, and convictions and written cautions (other than those for parking offences), must be given, and a personal health and medical declaration must be made. The applicant must sign a statement to the effect that the statements are true, rather than believed to be true. Under FiA 1968, s 28A(7), it is a summary offence, punishable with up to six months' imprisonment and/or an unlimited fine, knowingly or recklessly to make a statement false in a material particular. An applicant must provide a photograph and the names and addresses of two persons who have agreed to act as referees.

FiA 1968, s 27 states that a firearm certificate must be granted by the chief officer of police if (s)he is satisfied that the applicant:

(a) is fit to be entrusted with a firearm to which FiA 1968, s 1 applies and that (s)he is not prohibited by FiA 1968 (**19.22**) from possessing such a firearm;
(b) has good reason for having in his/her possession, or for purchasing or acquiring, the firearm or ammunition in respect of which the application is made; and
(c) can, in all the circumstances, be permitted to have the firearm or ammunition in his possession without danger to the public safety or peace.

The certificate must specify the nature and number of firearms to which it relates and certain other matters. A chief officer may impose conditions subject to which the firearm certificate is held. These conditions may refer to the nature of the storage of the weapons or their use. Conditions can be varied at any time by notice in writing to the holder, who may be required to return his certificate for variation. Failure to comply with a condition may have the effect of depriving the holder of the certificate of a statutory exemption that would otherwise be available to him/her (eg humane killing of animals) or even deprive him/her of legal authority to possess the weapon.

Police officers are required to carry out inquiries on behalf of the chief officer and, with these provisions in mind, they are required to check upon the intended usage of the weapon. The officer conducting the inquiry should always check the secure place in which the weapon will be stored.

Shotgun certificates

19.11   The rules relating to applications for shotgun certificates are not quite so strict. An application must be made to the chief officer of police on the application form. There are similar requirements in respect of the submission of a photograph and the name and address of *one* referee. The 'false statement offence' under FiA 1968, s 28A(7) (**19.10**) also applies to an applicant for a shotgun certificate.

By FiA 1968, s 28, such a certificate must be granted by the chief officer of police if (s)he is satisfied that the applicant can be permitted to possess a shotgun without danger to the public safety or to the peace, unless (s)he has reason to believe that the applicant is prohibited by FiA 1968 from possessing a shotgun (**19.22**), or is satisfied that the applicant does not have a good reason for possessing, purchasing or acquiring one. A sporting or competitive purpose is declared by FiA 1968 to be a good reason.

A shotgun certificate must specify the description of the shotguns to which it relates including, if known, the identification numbers of the guns.

General

19.12   The Firearms Rules 1998 provide that if a firearm or shotgun certificate is granted, the holder must on receipt sign it in ink. The holder must inform the chief officer:

(a) as soon as reasonably practicable but within seven days of the theft or loss of the certificate;

(b) without undue delay of a change of address.

Where the holder of the certificate is under 18, arrangements must be in place for ensuring that either (i) the holder's parent or guardian or (ii) a person aged 18 or over who is authorised under the Firearms Act 1968 to have possession of firearms and ammunition, or a shotgun, assumes responsibility for the secure storage of the firearms and ammunition, or the shotguns, to which the certificate relates.

Transfer of firearms, etc between authorised persons

19.13   The Firearms (Amendment) Act 1997 (Fi(A)A 1997), s 32 requires that where:

(a) a s 1 firearm or ammunition is sold, let on hire, lent or given; or

(b) a shotgun is sold, let on hire, given or lent for a period of more than 72 hours,

by any person, to a person who is not a firearms dealer nor exempt from holding a certificate, the following requirements must be complied with:

(i) the transferee must produce to the transferor the appropriate firearm or shotgun certificate or a visitor's permit (**19.20**);

(ii) the transferor must comply with any instructions contained in the certificate, etc; and

(iii) the transferor must hand the firearm to the transferee personally.

Failure to comply with these requirements is an offence under s 32(3).

Each party to the transfer (as described above) of a s 1 firearm or ammunition or of a shotgun who is the holder of a firearm or shotgun certificate must give notice of the transfer within seven days to the chief officer of police by registered post, the recorded delivery service or permitted electronic means. It is an offence under s 33(4) to fail to do so.

In the case of a s 1 firearm or ammunition, offences under s 32(3) or 33(4) are indictable (either-way) offences punishable with up to five years' imprisonment. In the case of a shotgun they are summary and punishable with up to six months' imprisonment and/or an unlimited fine.

De-activation, destruction, or loss of firearms or shotguns

19.14 Fi(A)A 1997, s 34 requires that where a firearm to which a firearm certificate or shotgun certificate relates is de-activated, destroyed or lost (by theft or otherwise), the certificate holder must give notice to the chief officer of police within seven days by registered post, the recorded delivery service or permitted electronic means. It is an offence under s 34(4) to fail to do so without reasonable excuse. In the case of a s 1 firearm, the offence is an indictable (either-way) offence, and in the case of a shotgun a summary offence. These offences are punishable in the same way those at **19.13**.

Revocation

19.15 By FiA 1968, s 30A, a firearm certificate may be revoked if the chief officer of police has reason to believe that the holder:

(a) is of intemperate habits or unsound mind or is otherwise unfitted to be entrusted with a firearm; or

(b) can no longer be permitted to have the firearm or ammunition to which the certificate relates without danger to the public safety or peace;

(c) is prohibited from possessing a s 1 firearm (**19.22**); or

(d) no longer has a good reason for possessing, purchasing or acquiring the firearm or ammunition which the certificate authorises him/her to have, etc.

A firearm certificate may also be revoked if the holder has failed to comply with a notice requiring him/her to deliver up his/her certificate (FiA 1968, s 30A).

In addition, a chief officer may partially revoke a firearm certificate in relation to any particular firearm or ammunition held under its authority if satisfied that the holder no longer has a good reason for possessing, purchasing or acquiring it (s 30B).

A shotgun certificate may be revoked only if the holder becomes a 'prohibited person', or cannot be permitted to possess a shotgun without danger to the public safety or to the peace (s 30C). It has been held that a chief officer, in deciding whether to revoke a shotgun licence, is entitled to take into account irresponsible conduct by the licence holder which does not involve the use of a shotgun. It is a matter for the chief officer's discretion to what extent (s)he should investigate a particular offence.

Where a certificate is revoked the holder must be notified in writing and required to surrender his/her certificate (s 30D(1) and (2)). It is a summary offence to fail to comply with such a notice within 21 days of the date of the notice (s 30D(3)).The offence is punishable with a fine not exceeding level 3.

POSSESSION, PURCHASE, OR ACQUISITION OF PROHIBITED WEAPONS OR AMMUNITION

19.16   Because prohibited weapons and ammunition can only properly be regarded as suitable for official use, FiA 1968, s 5(1) and (1A) provide that a person commits an offence if (s)he has in his/her possession, or purchases or acquires, any prohibited weapon or ammunition to which s 5(1) or (1A) respectively apply without the written authority of the Secretary of State. An offence under s 5(1) or s 5(1A) is one of strict liability as to the fact that the weapon or ammunition is prohibited, so that it is no defence that the possessor is reasonably unaware of the characteristic which makes the weapon or ammunition prohibited. The offence under s 5(1) (*except in relation to a prohibited weapon within (h) below*) and the offence under s 5(1A) (*in relation to a firearm disguised as another object*) are triable only on indictment; otherwise the above offences are indictable (either-way) offences. Offences under s 5(1) or 5(1A) are punishable with up to 10 years' imprisonment.

The requirement under s 5(1) and (1A) for the Secretary of State's authorisation applies over and above the need for a firearm certificate.

Prohibited weapons

19.17   For the purposes of an offence under FiA 1968, s 5(1), a prohibited weapon is:

(a) any firearm which is so designed or adapted that two or more missiles can be successively discharged without repeated pressure on the trigger;

(b) any self-loading or pump-action rifled gun other than one chambered for .22 inch rim-fire cartridges;

(c) any firearm which either has a barrel less than 30 cm in length or is less than 60 cm in length overall, other than an air weapon (as defined at **19.9**), a muzzle-loading gun, or a firearm designed as a signalling apparatus;

(d) any self-loading or pump-action smooth-bore gun which is not an air weapon or chambered for .22 inch rim-fire cartridges and either has a barrel of less than 24 inches in length or is less than 40 inches in length overall;

(e) any smooth-bore revolver gun other than one which is chambered for 9 mm rim-fire cartridges or a muzzle-loading gun;

(f) any rocket launcher, or any mortar, for projecting a stabilised missile, other than a launcher or mortar designed for line-throwing or pyrotechnic purposes or as signalling apparatus;

(g) any air rifle, air gun or air pistol which uses, or is designed or adapted for use with, a self-contained gas cartridge system; and

(h) any weapon of whatever description designed or adapted for the discharge of any noxious liquid, gas or other thing.

In addition, the following two weapons are prohibited weapons in relation to an offence of purchase or acquisition (and, in relation to an offence under s 5(2A) of manufacture, sale etc referred to at **19.40**). As from a day to be appointed they will be prohibited weapons in relation to an offence of possession:

(i) any rifle with a chamber from which empty cartridge cases are extracted using energy from propellant gas, or energy imparted to a spring or other energy storage device by propellant gas, other than a rifle which is chambered for .22 rim-fire cartridges;

(j) any device (commonly known as a bump stock) which is designed or adapted so that it is capable of forming part of or being added to a self-loading lethal barrelled weapon, and, if it forms part of or is added to such a weapon, it increases the rate of fire of the weapon by using the recoil from the weapon to generate repeated pressure on the trigger.

In relation to (a) above, the Court of Appeal has held that the words 'so designed or adapted that two or more missiles can be successively discharged without repeated pressure on the trigger' relate to what is objectively possible and not to the intention of the designer of the weapon. Thus, if that effect could be brought about, even if only in expert hands, the weapon is designed as a prohibited weapon.

For the purposes of (c) and (d), any detachable, folding, retractable or other movable butt-stock is disregarded in measuring the length of any firearm.

References to muzzle-loading guns in (c) and (e) are references to guns which are designed to be loaded at the muzzle end of the barrel or chamber with a loose charge and a separate ball (or other missile).

An air weapon referred to at (g) possessed on 30 April 2004 may be retained subject to the need to obtain a firearm certificate. A chief officer may not refuse an application for a firearm certificate in such circumstances, nor an application for renewal, on the ground that the person does not have a good reason for having the weapon.

List entry (h) covers a wide variety of weaponry including a flamethrower and a high-voltage electric stunning device. The fact that a stun-gun designed for an electrical charge is not working due to some unknown fault does not change its character as a prohibited weapon. The words 'designed or adapted' mean that any other type of weapon which is converted in any way for these purposes is a prohibited weapon as it has been 'adapted'. These weapons do not need to be either lethal or barrelled *provided* they are capable of discharging noxious liquid, gas or other thing. If the weapon is capable of discharging such a substance, and was either designed or adapted for that purpose, it is prohibited. A water pistol used to discharge gas would not be a prohibited weapon if it was used in an unaltered state, since it would not have been designed or adapted to discharge one of the prohibited substances. The same applies to a washing-up liquid bottle filled with hydrochloric acid. The container is neither designed nor adapted for the discharge of a noxious liquid. 'Noxious' before 'liquid, gas or other thing' must apply to all three things. A substance is noxious if it is harmful, hurtful, or injurious. A weapon which has not been designed for the purpose of discharging a noxious gas but is capable of doing so falls within list entry (j), at least (according to the Court of Appeal) if it has a 'deliberate design capability'.

Slaughtering instruments, humane killers, shot pistols used for killing vermin, starting pistols, trophies of war, firearms of historic interest, and weapons used for treating animals, are exempted from the prohibitions imposed upon weapons falling within (c) above (or, in the case of weapons used for treating animals, within (c) or (g)) subject to conditions set out in the Firearms (Amendment) Act 1997.

For the purposes of s 5(1A), the following are prohibited weapons:

(a) any firearm which is disguised as another object; and
(b) any launcher or other projecting apparatus, not a prohibited weapon under s 5(1), which is designed to be used with any rocket or ammunition designed to explode on or immediately before impact (ie a rocket or ammunition at **19.18**(i)) or with prohibited ammunition under s 5(1) (**19.18**(a)-(c)).

In relation to s 5(1A), there are a number of exemptions in terms of weapons and of ammunition. They relate principally to possession, purchase or acquisition for use for slaughtering animals, shooting vermin, or estate management purposes.

Prohibited ammunition

19.18   Prohibited ammunition for the purposes of FiA 1968, s 5(1) is:

(a) any cartridge with a bullet designed to explode on or immediately before impact (eg a 'dum-dum' bullet);

(b) any ammunition which contains, or is designed or adapted to contain, any noxious liquid, gas or other thing; and

(c) if capable of being used with a firearm of any description, any grenade, bomb (or other like missile), rocket or shell designed to explode on or immediately before impact,

other than ammunition used for treating animals.

Therefore containers designed or adapted to contain any noxious gas, etc for use as a missile or bomb are prohibited ammunition, whether filled or not. The other types of prohibited ammunition are explosive bullets and (if capable of being used with a firearm) grenades, bombs, rockets and shells.

Prohibited ammunition for the purposes of s 5(1A) is:

(i) any rocket or ammunition which is not prohibited ammunition (under (c) above) for the purposes of s 5(1) which consists in or incorporates a missile designed to explode on or immediately before impact and is for military use;

(ii) any ammunition for military use which consists in or incorporates a missile designed so that a substance contained in the missile will ignite on or immediately before impact;

(iii) any ammunition for military use which consists in or incorporates a missile designed, on account of its having a jacket and hard-core, to penetrate armour plating, armour screening or body armour;

(iv) any ammunition which is designed to be used with a pistol and incorporates a missile designed or adapted to expand on impact;

(v) anything which is designed to be projected as a missile from any weapon and is designed to be, or has been incorporated in:

(a) any ammunition falling within any of the above definitions; or

(b) any ammunition which would fall within any of those definitions but for its being specified in s 5(1).

LAWFUL POSSESSION WITHOUT A FIREARM CERTIFICATE OR SHOTGUN CERTIFICATE

Permits

19.19   Persons who hold a permit under FiA 1968, s 7 from the chief officer of police of the area in which they reside may, without holding a certificate, possess a firearm (including a shotgun) and ammunition in accordance with the terms of the permit. Such permits are frequently issued where the holder of a certificate dies and a relative requires some form of authority to possess the firearm pending its sale or disposal. It is unusual for such permits to be valid for more than one month.

Other exemptions

Firearms dealers (if registered: see **19.39**) and their employees may possess firearms **19.20** and ammunition without a certificate.

Other people may lawfully possess firearms and ammunition without holding a certificate. The principal exemptions provided by FiA 1968, ss 9 to 15 and 54 and the Firearms (Amendment) Act 1988 (Fi(A)A 1988), ss 16A, 16B, and 17 are set out in the following list and the rest of this numbered paragraph as follows:

(a) an auctioneer, carrier, warehouseman, or the employee of such a person, is frequently required to handle other people's firearms and ammunition in the course of his/her duties. Sensibly this is permitted, but it is necessary for such persons to take reasonable precautions for safe custody and to report loss or theft forthwith to the police;

(b) a slaughtering instrument and its ammunition may be possessed by a licensed slaughterman;

(c) a person carrying a firearm or ammunition belonging to another person who is the holder of a certificate may possess that firearm or ammunition under instructions from, and for the use of, that other person for sporting purposes only. The person carrying the firearm, etc can best be described as a 'gun bearer'. However, where the person carrying the firearm or ammunition is under 18, this exemption only applies if the other person is 18 or over;

(d) if aged 18 or over, a starter at an athletic meeting may possess a firearm for the purpose of starting races only. The exemption does not allow a starter to possess ammunition, so that (s)he is restricted to blanks not exceeding one inch in diameter (which do not, of course, require a certificate);

(e) subject to any exclusion of the club or restriction to specified types of rifle by the Secretary of State, a member of an approved rifle club (including a miniature rifle club) or of an approved muzzle-loading pistol club may possess a firearm and ammunition when engaged as such a member in connection with target shooting. An approval of a club may be limited to specified weapons;

(f) a person conducting or carrying on a miniature rifle range (whether for a rifle club or otherwise) or a shooting gallery at which no firearms are used other than air weapons or miniature rifles not exceeding .23 inch calibre may possess such miniature rifles and ammunition suitable for them;

(g) a person may use any such miniature rifle as is described in (f) above at such a range or gallery;

(h) a person who does not hold a shotgun certificate may borrow a shotgun from the occupier of private premises and use it on those premises, provided the occupier is present, but if the borrower is under 18 the occupier must be 18 or over;

(i) a person taking part in a theatrical performance, rehearsal or film may possess a firearm during the performance etc. This exception does not extend to ammunition, so that any ammunition used would have to be blanks not more than one inch in diameter;

(j) signalling apparatus or ammunition for it may be possessed on board an aircraft or at an aerodrome as a part of its equipment. It may also be transferred at an aerodrome from one aeroplane to another, or from or to an aeroplane at an aerodrome to or from an appointed place of storage there;

(k) a firearm and ammunition may be possessed on board a ship as a part of its equipment. However, if it is to be removed from or to a ship a permit to do so must be obtained from a constable (a similar permit is required if the signalling apparatus described at (j) is to be removed from or to an aeroplane or aerodrome);

(l) a Northern Ireland shotgun certificate authorises possession of a shotgun in Great Britain; and

(m) any 'person in the service of the Crown', a term which includes a member of a police force or a civilian officer, who is in possession of a firearm or ammunition in his/her capacity as such does not require a certificate. Nor does such a person require the Secretary of State's authority to possess a prohibited weapon or prohibited ammunition.

A person (the borrower: B) may, without holding a firearm certificate or shotgun certificate, borrow a rifle or shotgun from another person (the lender: L) on private premises and have possession of it on those premises if conditions (a) to (d) below are met, and, in the case of a rifle, B is aged 17 or over. The conditions are:

(a) that the borrowing and possession of the rifle or shotgun are for either or both of
   (i)   hunting animals or shooting game or vermin, or of
   (ii) shooting at artificial targets;
(b) L is 18 or over, holds a certificate in respect of the rifle or shotgun, and is either a person—
   (i)   who has a right to allow others to enter the premises for the purposes of hunting animals or shooting game or vermin; or
   (ii) who is authorised in writing by a person mentioned in (i) to lend the rifle or shotgun on the premises (whether generally or to persons specified in the authorisation who include B);
(c) B's possession and use of the rifle or shotgun complies with any conditions as to those matters specified in L's firearm certificate or shotgun certificate;
(d) during the period for which the rifle or shotgun is borrowed, B is in the presence of L or—
   (i)   where a rifle is borrowed, a person who, although not L, is 18 or over, holds a firearm certificate in respect of that rifle and is a person described in (b)(i) or (ii); or
   (ii) where a shotgun is borrowed, a person who, although not L, is 18 or over, holds a shotgun certificate in respect of that shotgun or another shotgun and is a person described in (b)(i) or (ii).

Where a rifle is borrowed on any premises in reliance on the exemption, B may, without holding a firearm certificate, purchase or acquire ammunition on the premises, and have the ammunition in his/her possession on those premises for the period for which the firearm is borrowed, if:

(a) the ammunition is for use with the firearm;
(b) L's firearm certificate authorises L to have in L's possession during that period ammunition of a quantity not less than that purchased or acquired by, and in the possession of, B; and
(c) B's possession and use of the ammunition complies with any conditions as to those matters specified in the certificate.

'Visitor's permits' may be issued by a chief officer of police in relation to both s 1 firearms and shotguns. A visitor's firearm permit permits a person to possess a firearm and ammunition and to acquire ammunition for it, and a visitor's shotgun permit permits a person to possess or acquire a shotgun (although there are exceptions in the case of a shotgun with a magazine). 'Group' applications may be made for not more than 20 permits. These will cover persons visiting Great Britain to take part in competitions.

Persons under the supervision of a member of the armed forces are authorised to possess a firearm and ammunition on service premises without holding a firearm certificate. Persons being trained or assessed in the use of firearms under the supervision of a member of the Ministry of Defence Police are authorised to possess a firearm or ammunition on premises used for any purpose of the Ministry of Defence Police. Such persons in such circumstances are also authorised to possess a prohibited weapon or prohibited ammunition without obtaining the Secretary of State's authority.

Museum licences

By virtue of Fi(A)A 1988, s 19 and Sch 1 specified museums do not need to have a   **19.21**
firearm certificate, shotgun certificate, or s 5 authority, as the case may be, in relation to exhibits displayed or stored at the museum if they have a museum firearm licence granted under those provisions. Specified museums are those registered with the Museums and Galleries Commission for the purpose of making them eligible for a museum firearm certificate.

PERSONS PROHIBITED FROM POSSESSING A FIREARM

FiA 1968, s 21 prohibits the possession of any firearm or ammunition by a person who   **19.22**
has been convicted of a crime and been sentenced to:

(a) custody for life or preventive detention, or to imprisonment, corrective training, youth custody, or detention in a young offender institution for three years or more. Such a person is banned for life;

(b) imprisonment or youth custody or detention in a young offender institution from three months to three years, or subject to a secure training order or a detention and training order. Such a person is banned for five years from the date of his/her release; or

(c) a suspended sentence of imprisonment for three months or more. Such a person is banned for five years beginning with the second day after the sentence was passed.

Air weapons are included in the prohibition. The exemption in respect of antique firearms referred to at **19.3** does not apply to an offence under s 21.

Contravention of these provisions is an offence under s 21(4). So is supplying a firearm or ammunition to someone (X) prohibited from possessing it, or repairing, testing or proving it for him, if D knows or has reasonable grounds to believe that X is so prohibited (s 21(5)). Both offences are indictable (either-way) ones and are punishable with up to five years' imprisonment.

A person subject to a prohibition under s 21 can apply to the Crown Court for its removal.

POSSESSION OR ACQUISITION BY YOUNG PERSONS

FiA 1968, ss 22 to 24 deal with the possession or acquisition of all types of firearms by   **19.23**
juveniles.

Section 1 firearms

19.24   If a person under 18 wishes to have a s 1 firearm, or ammunition for it, (s)he cannot buy or hire it him/herself; if (s)he does so (s)he commits an offence under FiA 1968, s 22(1). Under s 24(1), the seller or person letting it on hire also commits an offence, unless (by s 24(5)) (s)he can prove that (s)he reasonably believed that the other person was 18 or over.

It is an offence under s 24(2)(a) to give or lend a s 1 firearm or ammunition to a person under 14 years of age. Provided that someone is 14 or over (s)he may receive a firearm, together with ammunition, by way of a gift or loan. Where an adult wishes to buy a firearm as a gift for such a youth, (s)he must obtain a firearm certificate and so must the youth. The seller may then sell to the adult, who may then transfer the weapon to the youth, both notifying the chief officer of police of the transaction by registered post, recorded delivery, or permitted electronic means within 48 hours. Certificates granted to persons under 18 are endorsed to the effect that firearms or ammunition cannot be sold or hired to them until the specified date, which is the date of their 18th birthday.

It is an offence under ss 22(2) and 24(2)(b) respectively for a person under 14 to possess a s 1 firearm or ammunition, or for someone to part with possession of such a thing to such a person, with certain exceptions. In recalling these exceptions it is helpful to consider those persons who are permitted to possess s 1 firearms without a certificate and to identify the exceptions which might apply to persons under 14. The exceptions are when (s)he:

(a) is carrying the firearm or ammunition for another for sporting purposes, and that other person is 18 or over and the holder of a firearm certificate; or

(b) is using the firearm or ammunition at a shooting gallery or miniature rifle range where the only weapons used do not exceed .23 inch calibre; or

(c) as a member of an approved rifle club, is engaged in connection with target shooting.

If proved, a reasonable belief that the other person was 14 or over is a defence to a charge under s 24(2)(a) or (b) (s 24(5)).

The above offences are summary offences punishable with an unlimited fine and/or up to six months' imprisonment (three months in specified circumstance in respect of a s 22(1) or 24(1) offence).

Shotguns

19.25   It is an offence under FiA 1968, s 22(1) for a person under 18 to purchase or hire any shotgun or ammunition. Likewise, someone who sells or lets it on hire to such a person commits an offence under s 24(1). It is an offence under s 24(3) to make a gift of a shotgun to a person under 15. (These rules also apply to ammunition for a shotgun.) It is, of course, possible for a person of 15, 16, or 17 lawfully to acquire a shotgun by the same process as that outlined in relation to s 1 firearms. If proved, a reasonable belief that the other person was 18 or over, or 15 or over, respectively, is a defence to a charge under s 24(1) or (3) respectively (s 24(5)).

All young persons in possession of shotguns must have a shotgun certificate. Although FiA 1968 does not prescribe a minimum age at which a shotgun certificate may be granted, control is exercised by the chief officer of police who must consider the grant of such a certificate in the context of public safety.

Even if (s)he has a shotgun certificate, it is an offence under s 22(3) for a person under 15 to 'have with him/her' (**19.33**) an assembled shotgun, except while:

(a) under the supervision of a person of 21 or more; or
(b) it is so securely fastened with a gun cover that it cannot be fired.

The above offences are summary offences. As to the punishment of an offence under s 22(1) or 24(1), see **19.24**. An offence under s 22(3) or 24(3) is punishable with a fine not exceeding level 3.

Air weapons

The same general rule applies. By FiA 1968, s 22(1), it is an offence for a person under 18 to purchase or hire an air weapon or ammunition. The seller or person who lets on hire in such a case also commits an offence under s 24(1); if proved, a reasonable belief that the other person was 18 or over is a defence.

19.26

By s 22(4) a person under 18 commits an offence if (s)he has with him/her (**19.33**) an air weapon or ammunition for it except when one of the following exceptions in s 23 applies:

(a) (s)he is under the supervision of someone aged 21 or over, but the supervisor commits an offence under s 23(1) if the supervisor allows the juvenile to fire a missile beyond the premises. The supervisor will have a defence if (s)he shows that the only premises into or across which the missile was fired were premises occupied by a person who had consented to this;
(b) as a member of an approved club, (s)he is engaged in connection with target shooting;
(c) (s)he is using the weapon or ammunition at a shooting gallery or miniature range where the only firearms used are either air weapons or miniature rifles not exceeding .23 inch calibre; or
(d) (s)he is aged 14 or over and is on private premises with the consent of the occupier.

In relation to the exception under (a), it may be noted that a person of any age commits an offence under s 21A(1) if (s)he fires a missile from an air weapon beyond the premises where (s)he is, unless (s)he shows that the occupier of the premises into or across which the missile is fired had consented to this.

It is an offence under s 24(4)(a) for any person to make a gift of an air weapon or ammunition to a person under 18. By s 24(4)(b), it is an offence to part with the possession of an air weapon or ammunition to a person under 18 except where that person is not prohibited from having it with him/her by virtue of one of the exceptions in s 23. If proved, a reasonable belief that the person was 18 or over is a defence (s 24(5)).

By s 24ZA(1), it is an offence for a person in possession of an air weapon to fail to take reasonable precautions to prevent any person under the age of 18 from having the weapon with him, unless, by virtue of s 23, that person is not prohibited from having the weapon with him. *It is a defence to show that D reasonably believed that the other person was 18 or over. A person is taken to have shown these matters if sufficient evidence of them is adduced to raise an issue with respect to them, and the contrary is not proved beyond a reasonable doubt.*

The above are summary offences. An offence under s 22(1) or 24(1) in respect of an air weapon or ammunition is punishable with up to six months' imprisonment and/or an unlimited fine. An offence under s 21A(1), 22(4), 23(1), 24(4), or 24ZA(1) is punishable with a fine not exceeding level 3.

Purchase of imitation firearm by or sale to minors

19.27 FiA 1968, s 24A(1) and (2) respectively make it an offence for a person under 18 to purchase an imitation firearm, or for anyone to sell an imitation firearm to such a person. In relation to 'selling', the defence italicised above applies. For these purposes, 'imitation firearm' is defined by FiA 1968, s 57; see **19.30** for that definition.

The above offences are summary and punishable with up to six months' imprisonment and/or an unlimited fine.

FIREARMS OFFENCES RELATING TO PREVENTION OF CRIME AND PUBLIC SAFETY

Conversion of firearms

19.28 It is an indictable (either-way) offence under FiA 1968, s 4(1) to shorten the barrel of a shotgun to a length of less than 24 inches. It is not an offence for a registered firearms dealer to shorten a barrel for the sole purpose of replacing a defective part so as to produce a barrel not less than 24 inches in length.

It is an indictable (either-way) offence under s 4(3) for anyone other than a registered firearms dealer to convert into a firearm anything which, though having the appearance of a firearm, cannot discharge a missile through its barrel. Section 4(3) does not apply to an imitation firearm to which the definition in the Firearms Act 1982 (**19.5**) applies (Firearms Act 1982, s 2).

The above offences are punishable with up to seven years' imprisonment.

Possession with intent to endanger life

19.29 Under FiA 1968, s 16 it is an offence triable only on indictment and punishable with a maximum of life imprisonment to possess a firearm or ammunition with intent by means thereof to endanger life, or to enable another person by means thereof to endanger life, whether an injury has been caused or not.

Possession of a firearm, etc with intent to enable another person, by means thereof, to endanger life, means more than merely making it available to known criminals who might endanger life. An intention to endanger the life of a person abroad is sufficient for s 16; consequently, it covers possession by terrorist groups who might intend their mischief elsewhere. Possession with intent to commit suicide is not covered by s 16; an intent to endanger life must relate to the life of another.

The present offence is not committed by a person who intends to endanger life for a lawful purpose, as where a man whose house is besieged by an armed gang threatens them with his firearm in self-defence. This defence, however, is only available in circumstances where there is a specific risk of imminent attack: it does not apply where there is only a potential risk.

Possession with intent to cause fear of violence

19.30 By FiA 1968, s 16A, a person commits an offence triable only on indictment and punishable with up to 10 years' imprisonment if (s)he has in his/her possession any firearm or imitation firearm with intent:

(a) by means thereof to cause; or

(b) to enable any other person by means thereof to cause,

any person to believe that unlawful violence will be used against him/her or another.

For these purposes, 'imitation firearm' is defined by FiA 1968, s 57 (and not by the Firearms Act 1982 (**19.5**)) as meaning anything which at the material time has the appearance of being a firearm (other than a weapon for the discharge of a noxious liquid, gas or other thing) whether or not it is capable of discharging any shot, bullet or other missile. An automatic pistol with the firing pin removed has been held to be such an imitation firearm; clearly, it fell within the definition. So may things which have the appearance of being firearms in particular circumstances on specific occasions, but not in others. A piece of roughly fashioned wood held in the hand in a darkened room might certainly have the appearance of being a firearm. It would be a matter for the jury to decide whether something actually did have the appearance of being a firearm in the circumstances in question. The House of Lords has held that fingers positioned in a jacket so as to appear to be a gun are not in law capable of constituting a firearm.

The Court of Appeal has held that s 16A is not limited to cases where the intent specified in s 16A was formed before the material time. It held that, provided there is a coincidence of possession and intent, s 16A also embraces the situation where D forms an intention to cause fear of violence at or immediately before the time of his/her actions which were designed to cause such fear.

Use of firearms to resist arrest

19.31   It is an offence by FiA 1968, s 17(1) for a person to make, or attempt to make, any use whatsoever of a firearm or imitation firearm with intent to resist or prevent the lawful arrest or detention of him/herself or another person. If a firearm or imitation firearm is used in resisting lawful arrest or detention, whether of the person using the firearm, or some other person, the offence is complete. The important words are 'to make, or attempt to make, any use' of the firearm and the issue of in whose possession the firearm was before that moment does not arise. Someone who grabbed a gun from the person arresting him/her and made use of it in this way would be guilty. It must be proved that the firearm was used intentionally for such a purpose. The offence would not be made out, for example, if the firearm was used with the intention of preventing a search of premises and an arrest was not intended at that time.

Section 17(2) creates an offence of possessing a firearm or imitation firearm *at the time of committing or being arrested* for an offence specified in Sch 1, unless D can show that (s)he possessed it for a lawful object. The offences in Sch 1 include theft, robbery, burglary, blackmail, taking a conveyance, assaulting a constable in the execution of his/her duty, rape, assault by penetration, causing a person to engage in sexual activity without consent where the activity caused involved penetration, rape of a child under 13, assault of a child under 13 by penetration, causing or inciting a child under 13 to engage in a sexual activity where any activity involving penetration was caused, sexual activity with a person with a mental disorder impeding choice where the touching involved penetration, causing or inciting a person with a mental disorder impeding choice to engage in sexual activity where penetration was caused, criminal damage, malicious wounding, assault occasioning actual bodily harm, and assault with intent to resist arrest.

Where the case is one of possession *at the time of arrest*, there is no requirement that the prosecution prove that D actually committed the specified offence. It is only necessary to prove that D was in possession of the firearm when lawfully arrested for a specified offence.

For the purposes of both subsections of s 17, a 'firearm' does not include a component part or accessory. Subject to this, 'imitation firearm' in s 17(1) and (2) has the same meaning as in s 16A (**19.30**).

An offence under s 17(1) or (2) is triable only on indictment and punishable with a maximum of life imprisonment.

Having firearm with criminal intent

19.32   FiA 1968, s 18(1) deals with the offence committed by a person who has with him/her a firearm or imitation firearm with intent to commit an indictable offence, or to resist arrest or prevent the arrest of another, in either case while (s)he has the firearm or imitation firearm with him/her. The offence is triable only on indictment and punishable with a maximum of life imprisonment.

'Imitation firearm' has the same meaning as in FiA 1968 s 16A (**19.30**). As to 'has with him/her', see **19.33**.

The intent required by s 18 is a particular one. It is not sufficient that D intended to commit an indictable offence or to resist, etc the arrest. D must also intend to have a firearm or imitation firearm with him/her at the time of that commission or resistance. On the other hand, D does not have to intend to use or carry the gun in furtherance of the indictable offence. Section 18(2) provides that proof that D had a firearm or imitation firearm with him/her and intended to commit an offence, or to resist or prevent arrest, is evidence that D intended to have it with him/her while doing so.

Having firearm in a public place

19.33   A person commits an offence contrary to FiA 1968, s 19 if, without lawful authority or reasonable excuse (the proof whereof lies on him/her), (s)he has with him/her in a public place:

(a) a loaded shotgun;
(b) an air weapon (whether loaded or not);
(c) any other firearm (whether loaded or not) together with ammunition suitable for use in that firearm; or
(d) an imitation firearm (same meaning as in s 16A (**19.30**)).

Such an offence is an indictable (either-way) offence punishable with up to seven years' imprisonment (only 12 months' if the weapon is an imitation firearm) with the following exceptions. If it is committed in relation to a prohibited weapon under s 5(1) (**19.17**) (other than that at (h)), or a prohibited weapon of type (a) under s 5(1A) (**19.17**), the offence is triable only on indictment and punishable in the same way as the either-way offence. If the offence relates to an air weapon it is a summary offence punishable with up to six months' imprisonment and/or an unlimited fine.

A shotgun which has a loaded magazine is loaded, even though there is no round in the breach.

The exemption in respect of antique firearms referred to at **19.3** does not apply to an offence under s 19.

This is one of a number of other offences which refers to '*having with him/her*' rather than 'possessing'. The former term is a narrower one. As we have seen, a person can be in possession of a firearm if (s)he has control of it, even though it is not in his/her physical custody and is not immediately available to him/her, as where it is in his/her home in Norwich while (s)he is in London. In contrast, although a person can have a firearm with him/her, even though (s)he is not carrying it, (s)he must have a close physical link with it and it must have been readily accessible to him/her. A man who has a gun in his pocket clearly has it with him, and the same is true if it is in a bag which he is carrying or in the glove compartment of the car which he is driving. Provided that the firearm is readily accessible to him/her, a person may even have with him/her at the time a firearm which (s)he has left in his/her car which (s)he has parked down the street. A person does not 'have with him/her' something of whose presence (s)he is unaware, such as a firearm which has been slipped into his/her bag; but once (s)he knows of its presence it is no excuse that, at the material time, (s)he has forgotten about it, as where (s)he puts a firearm in the glove compartment of his/her car but has forgotten about it when (s)he is stopped by a patrol car a month later.

'Public place' has the same meaning as given at **19.49**.

The offence under s 19 is one of strict liability as to the nature of the item in question as defined above. Thus, knowledge or suspicion as to its nature is not required, and even a reasonably mistaken belief that it is not of that nature will not excuse.

It is difficult to imagine many circumstances in which someone could, with lawful authority or reasonable excuse, have with him/her a loaded shotgun in a public place. The possession of a shotgun certificate certainly does not authorise this. Depending on the circumstances, a gamekeeper, crossing a public highway in the course of his/her duties, might be considered to have a reasonable excuse for having a loaded shotgun with him/her. Someone going to his rifle club carrying a .22 rifle and ammunition in a rifle-case would doubtless be able to prove a reasonable excuse for having the rifle and ammunition with him, provided that (s)he possessed a firearm certificate.

Trespassing with a firearm

An indictable (either-way) offence is committed against FiA 1968, s 20(1) if a person, while (s)he has any firearm or imitation firearm with him/her, enters or is in any building or part of a building as a trespasser and without reasonable excuse (the proof whereof lies on him/her). Such an offence is triable only on indictment in the case of a prohibited weapon under s 5(1) listed at **19.17** (other than at (h)), or a prohibited weapon of type (b) under s 51A (**19.17**); otherwise it is an indictable (either-way) offence (except in the case of an air weapon, where it is only summary). On conviction on indictment a s 21(1) offence is punishable with up to seven years' imprisonment (12 months' if imitation weapon), and on summary conviction with up to six months' and/or an unlimited fine.

19.34

A summary offence punishable with up to three months' imprisonment and/or a level 4 fine is committed under s 20(2) where the trespass is on land; the other ingredients of this offence are identical to those in s 20(1).

'Imitation firearm' bears the same meaning as in s 16A (**19.30**). 'Enters' requires the bodily presence of D to some degree. As to 'trespasser', 'trespass' requires the absence of lawful authority or of permission (express or implied) from the occupier or someone on his/her behalf. 'Building' refers to a structure which has a roof and is of a reasonably permanent nature; the term 'part of a building' covers instances where a person might have a right to be in a building, for example a hotel, but is a trespasser in someone else's room, which would be a part of that building. For the purposes of s 20(2), the expression 'land' includes land covered with water, so that a person with a firearm trespassing in a boat on a lake is guilty of this offence.

The exemption in respect of antique firearms referred to at **19.3** does not apply to an offence under s 20.

POLICE POWERS

Stop and search in certain cases

19.35   FiA 1968, s 47(1) authorises a constable to require any person whom (s)he has reasonable cause to suspect:

(a) of having a firearm, with or without ammunition, with him/her in a public place; or
(b) to be committing or about to commit, elsewhere than in a public place, an offence of 'having with him/her' a firearm or imitation firearm with intent to commit an indictable offence or to resist arrest (contrary to s 18: **19.32**), or an offence of trespassing with a firearm (contrary to s 20: **19.34**),

to hand over the firearm or ammunition for examination. By s 47(2), failure to comply with such a demand is a summary offence punishable with up to three months' imprisonment and/or a fine not exceeding level 4. Imitation firearm bears the same meaning as at **19.30**.

Section 47(3) also provides that a constable who has reasonable cause to suspect the existence of one of the above circumstances (ie (a) or (b)) may search that person and may detain him for the purpose of doing so. This power extends to the search of vehicles and the constable may require a driver to stop for that purpose (s 47(4)).

A constable may enter any place to exercise his/her powers under s 47.

Production of certificates

19.36   FiA 1968, s 48(1) states that a constable may demand, from any person whom (s)he believes to be in possession of a firearm or ammunition to which s 1 applies, or of any shotgun, the production of his/her firearm certificate or (as the case may be) shotgun certificate.

If a person on whom a demand has been made under s 48 fails to produce the certificate or to permit the constable to read it, or to show that (s)he is exempt from the requirement to have a certificate, the constable may seize and detain the firearm, ammunition or shotgun and may require the person immediately to declare his/her name and address (s 48(2)). Under s 48(3), it is a summary offence, punishable with a fine not exceeding level 3, to refuse or fail to give a true name and address.

Search warrant

19.37   By FiA 1968, s 46(1), a justice, who is satisfied by information on oath that there is reasonable ground for suspecting that:

(a) an offence relevant for the purposes of s 46 has been, is being, or is about to be committed; or
(b) in connection with a firearm or ammunition, there is a danger to the public safety or to the peace,

may grant a search warrant.

The warrant will authorise a constable or civilian officer:

(a) to enter at any time any premises or place specified, if necessary by force, and to search them and every person found there;
(b) to seize and detain anything found on the premises or place, or on any such person, in respect of which or in connection with which (s)he has reasonable grounds for suspecting that:

(i) a relevant offence has been, is being or is about to be committed, or

(ii) in connection with a firearm, imitation firearm or ammunition there is a danger to the public safety or to the peace (s 46(2)).

A 'relevant offence' is any offence under FiA 1968 except that under s 22(3) (person under 15 having assembled shotgun otherwise than under supervision) or an offence relating specifically to air weapons.

The power of seizure and detention includes power to require information which is stored in any electronic form and is accessible from the premises or place to be produced in a form in which it is visible and legible (or from which it can readily be produced in such form) and can be taken away. The additional powers of seizure provided by the Criminal Justice and Police Act 2001, ss 50 and 51 (**4.61** and **4.60**) apply where a search warrant is executed under FiA 1968, s 46.

The Court of Appeal has held that the police may use reasonable force to restrain or detain the occupants of premises when executing a warrant under s 46.

It is a summary offence punishable with up to six months' imprisonment and/or an unlimited fine intentionally to obstruct a constable or civilian officer in the exercise of these powers (s 46(5)).

Entry into rifle clubs

By Fi(A)A 1988, s 15(7), a constable duly authorised in writing by a chief officer of police, on producing (if required) his/her authority, may enter any premises occupied or used by an approved rifle club and inspect them, and anything on them, for the purpose of ascertaining whether the provisions about such a club and any limitations in the approval are being complied with.   **19.38**

BUSINESS AND OTHER TRANSACTIONS RELATING TO FIREARMS AND AMMUNITION

Section 1 firearms and ammunition, shotguns and air weapons

Under FiA 1968, s 3(1), it is an indictable (either-way) offence punishable with up to five years' imprisonment for a person to: **19.39**

(a) manufacture, sell, transfer, repair, test or prove any s 1 firearm or ammunition, or a shotgun,

(b) expose for sale or transfer, or possess for sale, transfer, repair or proof, any such firearm or ammunition, or a shotgun, or

(c) sell or transfer, or expose for sale or transfer, an air weapon, or possess such a weapon for sale or transfer,

without being registered under FiA 1968 as a firearms dealer.

By way of exception, it is not an offence for an auctioneer to sell by auction a firearm or ammunition without being registered as a firearms dealer, provided the auctioneer holds a police permit for that purpose (s 9(2)).

A 'firearms dealer' is defined by FiA 1968, s 57 as a person who, by way of trade or business, *manufactures, sells, transfers, repairs, tests or proves firearms or ammunition to which FiA 1968, s 1 applies, or shotguns, or sells or transfers air weapons.*

The Violent Crime Reduction Act 2006 (VCRA 2006), s 32(1) and (2) makes it a summary offence punishable with up to six months' imprisonment and/or an unlimited fine to sell an air weapon by way of trade or business to a person who is not a registered firearms dealer when this is not done face to face.

The sale or transfer of a s 1 firearm or ammunition or of a shotgun to a person other than a registered firearms dealer is an offence under FiA 1968, s 3(2), unless that person produces the necessary certificate, or shows (s)he is legally entitled to purchase or acquire the firearm or ammunition without a certificate. This is an indictable (either-way) offence punishable with up to five years' imprisonment.

The chief officer of police must keep a register of firearms dealers. An applicant must (a) provide details of every place of business (including storage places) in the area, at which (s)he proposes to carry on business as a firearms dealer; (b) provide details of the precise nature of the business which (s)he intends to conduct; (c) complete a medical declaration, and (d) provide details of his/her 'servants' at each place of business for background check purposes. A registered firearms dealer or his/her employee is permitted to keep, purchase or acquire firearms and ammunition in the ordinary course of business without holding firearm certificates in respect of them, and this is so even though the place where the firearm or ammunition is possessed, purchased or acquired by the dealer or employee is not the dealer's place of business or has not been registered as his/her place of business.

Except on certain specified grounds, the chief officer of police must enter the applicant's name and place(s) of business in the register and grant him/her a certificate of registration. The chief officer of police may, however, impose conditions upon registration. These conditions are generally concerned with ensuring the safe-keeping of firearms. They usually include the following conditions; that:

(a) the dealer must, on being given reasonable notice, allow a police officer authorised in writing by the chief officer to enter and inspect his/her premises;
(b) hand-guns must be kept in a locked safe;
(c) other weapons must be chained together by the trigger guards and locked in a rack;
(d) ammunition is to be stored separately and locked up;
(e) rifle bolts must be removed and kept separately;
(f) the windows of cabinets for storage must be illuminated at night; and
(g) glass door panels and windows must be barred.

In addition, conditions are usually imposed concerning notification of dealings in various types of weapon.

All certificates of registration are renewable every three years. A new place of business must be notified to the chief officer and must be registered by him/her, unless the use of those premises for firearms dealing would endanger the public safety or the peace. A registered dealer may be removed from the register if (s)he ceases to deal in firearms, or to have a business place within the area, or if (s)he cannot be permitted to continue in business without danger to the public safety or the peace. Failure to comply with conditions also provides reason for removal from the register. Particular premises may be removed from the register on safety grounds.

Additional rules for prohibited weapons or ammunition

19.40   Under FiA 1968, s 5(2A), a person commits an offence, triable only on indictment and punishable with a maximum of life imprisonment, if without the written authority of the Secretary of State, (s)he:

(a) manufactures any prohibited weapon or ammunition specified in s 5(1) (**19.17** and **19.18**);

(b) sells or transfers *any* prohibited weapon or prohibited ammunition,

(c) has in his/her possession for sale or transfer *any* prohibited weapon or prohibited ammunition, or

(d) purchases or acquires for sale or transfer any prohibited weapon or prohibited ammunition.

Dealer to keep records

A registered dealer must keep a register of transactions. (S)he must within 24 hours **19.41** enter in his/her register the particulars of persons to whom firearms and ammunition (excluding ammunition to which FiA 1968, s 1 does not apply: see **19.7**) are sold or transferred.

Registered dealers must allow police officers or civilian officers, authorised in writing by the chief officer of police, to enter and inspect all stock in hand and must produce their registers for inspection. It is a summary offence punishable with up to six months' imprisonment and/or an unlimited fine to fail to do so, or knowingly or recklessly to make any false entry in a register (FiA 1968, s 40(5)). A police officer who is authorised in writing by his/her chief officer of police to carry out these duties must produce that written authority if required to do so.

Manufacture, import and sale of realistic imitation firearms

Under VCRA 2006, s 36(1), it is a summary offence to manufacture, import, or sell a **19.42** realistic imitation firearm (ie an imitation firearm which has an appearance that is *so realistic as to make it indistinguishable*, for all practical purposes, from a real firearm, and is neither a de-activated firearm nor itself an antique). Such an imitation firearm is not to be regarded as '*so distinguishable*' if it is distinguishable only by an expert, on close examination or by attempting to load or to fire it. Defences are provided by s 37 in relation to things done for specified legitimate purposes. The offence is punishable with up to six months' imprisonment and/or an unlimited fine.

VCRA 2006 (Realistic Imitation Firearms) Regulations 2007 specify the sizes and colours of imitation firearms which are to be regarded as unrealistic for the purposes of the definition of 'realistic imitation firearm'. The Regulations provide that an imitation firearm is not a 'realistic' one if:

(a) it is less than 38 mm high and 70 mm in length; or

(b) it is of one of the following colours: bright
    (i) red,
    (ii) orange,
    (iii) yellow,
    (iv) green,
    (v) pink,
    (vi) purple, or
    (vii) blue;
    or is made of transparent material.

The Regulations provide a defence to a charge under s 36 for D to show that his/her conduct was for the purpose only of making the imitation firearm available for the organisation and holding of permitted activities in respect of which there were insurance arrangements in respect of third-party liabilities, or for the purpose of display at a permitted event. 'Permitted activities' means the acting-out of military or law enforcement scenarios for the purpose of recreation (eg historical re-enactments of battles); and 'permitted events' means a commercial event at which firearms or realistic imitation firearms (or both) are offered for sale or displayed. D will have shown such a matter if there is sufficient evidence to raise an issue and the contrary is not proved beyond reasonable doubt.

FIREARMS OFFENCES UNDER OTHER ACTS

Drunk in possession of a loaded firearm

19.43   It is a summary offence contrary to the Licensing Act 1872, s 12 to be drunk when in possession on any highway or other public place of any loaded firearm (including a loaded air rifle). The Act does not define the term 'firearm' for the purpose of s 12 but that term bears its everyday meaning and includes an airgun. The offence is punishable with a fine not exceeding level one, or in the discretion of the court up to one month's imprisonment.

Discharge near the highway

19.44   By the Highways Act 1980, s 161(2)(b), a person commits a summary offence, punishable with a fine not exceeding level 3, if, without lawful authority or excuse, (s)he discharges any firearm or firework within 50 feet of the centre of any highway which consists of or comprises a carriageway, *and in consequence thereof* a user of the highway is injured, interrupted (eg by being forced to make a detour), or endangered. For the meaning of 'highway', see **23.3**.

Wanton discharge in a street

19.45   By the Town Police Clauses Act 1847, s 28, it is an offence wantonly to discharge a firearm or throw or set fire to a firework in any street to the obstruction, annoyance or danger of residents or passengers. The offence is punishable with a fine not exceeding level three, or in the discretion of the court up to 14 days' imprisonment.

OFFENSIVE WEAPONS

Offensive weapons in a public place

19.46   The Prevention of Crime Act 1953 (PCA 1953), s 1(1) provides that any person who without lawful authority or reasonable excuse, the proof whereof lies on him/her, has with him/her (**19.33**) in any public place any offensive weapon is guilty of an indictable (either-way) offence punishable with up to four years' imprisonment.

D need not know of the facts which render the article an offensive weapon within the meaning of the statute.

An offensive weapon is defined by s 1(4) as any article made or adapted for use for causing injury to the person, or intended by the person having it with him/her for such use by him/her or by some other person. This definition includes two classes of offensive weapon.

Articles made or adapted for causing injury to the person

19.47   Examples of articles *made* for causing injury to the person are petrol bombs, bayonets, coshes, knuckledusters, swords, flick knives, and butterfly knives. These weapons are described as offensive per se (ie in themselves). The Court of Appeal has held that an article which has all the characteristics of one made for causing injury to the person (in the instant case a flick knife) is a weapon offensive per se despite the fact that it has a harmless secondary characteristic, eg of being a lighter.

Examples of articles *adapted* for causing injury to the person are a piece of chain whose links have been sharpened and someone's cap in the peak of which a razor blade has been inserted with the cutting edge exposed. In such cases, articles which are harmless in themselves are made offensive weapons by their deliberate adaptation for use for causing personal injury.

Articles intended to be used for causing injury to the person

This class covers articles which are inoffensive per se (ie in themselves), because they **19.48** have not been made or adapted for causing personal injury, but which are rendered offensive by D's intention to use them for causing injury to the person. Such articles include belts, shoes, umbrellas, and dog leads, provided it can be shown that there was an intention to use the article for causing injury to the person. The difficulty of proving such an intention increases in proportion to the generally non-offensive character of the article in question.

It is not enough that someone, who is in innocent possession of an article in a public place, suddenly uses it for an offensive purpose. Thus, where a carpenter took a hammer from his tool bag during a fight at a railway station and used it, a divisional court held that he was not guilty of an offence under PCA 1953, s 1, as he did not have the hammer with him for causing injury to the person. This is a more restrictive rule than applies to similar wording in the offence of aggravated burglary (**14.53**).

Public place

D must have with him/her the offensive weapon in a public place; this can include **19.49** having it with him/her in a vehicle which is in a public place. 'Public place' includes any highway, and any other premises or place to which at the material time the public have or are permitted (ie invited or tolerated) to have access, whether on payment or otherwise. The issue is generally whether the public have access, or are permitted to have it if they wish. Public access is a matter of fact and if access is restricted to certain classes of person, it must be restricted to some considerable degree before it will be accepted that the public are not admitted.

The fact that a person is found with an offensive weapon in a private place may lead to a conviction for the present offence. For example, if a visitor to a dwelling house produces an offensive weapon, this gives rise to a strong inference that (s)he brought it with him/her through the streets and therefore had it with him/her in a public place.

Lawful authority or reasonable excuse

A person does not commit the offence if (s)he has lawful authority or reasonable excuse **19.50** for having with him/her the offensive weapon; D has the onus of proving such authority or excuse.

Those who may have offensive weapons with lawful authority include members of the armed services, or police forces, who may carry weapons which are offensive in themselves as part of their duty. On the other hand, it has been held that private security guards do not have lawful authority to carry truncheons or the like.

In terms of 'reasonable excuse', the appeal courts have held that whether or not D had a reasonable excuse depends on whether a reasonable person would think that D had a reasonable excuse for having the weapon with him/her in the circumstances at the time in question. On this basis, whether carrying a weapon for self-defence against imminent attack can amount to a reasonable excuse would depend on how imminent, how likely and how specific the perceived threat, and how serious the anticipated attack. However, as a matter of law, limitations have been placed on what a person might think in a context. For example, the courts have held that a person who has with him/her a weapon in order to commit suicide does not have a reasonable excuse, nor does a security guard who carries a truncheon 'as a deterrent' and 'as part of his/her uniform'. A claim by a person who has been proved to be in possession of a weapon which is offensive per se, that (s)he did not know that the article in question was an offensive weapon, has been held incapable of amounting to a reasonable excuse. The fact that D had forgotten that the article was in his/her possession does not in itself amount to a reasonable excuse.

Threatening with offensive weapon in public

19.51   Under PCA 1953, s 1A(1), a person is guilty of an indictable (either-way) offence punishable with up to four years' imprisonment if (s)he:

(a) has an offensive weapon (as defined at **19.46**) with him/her in a public place (**19.49**),

(b) unlawfully and intentionally threatens another person with the weapon, and

(c) does so in such a way that there is an immediate risk of serious physical harm to that other person.

For these purposes, physical harm is serious if it amounts to grievous bodily harm for the purposes of the Offences against the Person Act 1861 (**9.14**).

Ancillary offences to those under PCA 1953

Having article with blade or point in a public place

19.52   The Criminal Justice Act 1988 (CJA 1988), s 139 provides an ancillary offence which is useful where a knife or the like cannot be proved to have been made, adapted or intended to cause injury to the person. Section 139(1) makes it an indictable (either-way) offence, punishable with up to four years' imprisonment, for a person to have with him/her in a public place an article which is sharply pointed or with a blade, other than a folding pocket-knife with a blade not exceeding three inches. 'Has with him/her' has, and 'public place' essentially has, the same meaning as under PCA 1953 (**19.33**, **19.46**, and **19.49**).

A butter knife without a handle and with no cutting edge and no point has been held to be a bladed article within CJA 1988, s 139. The Court of Appeal has held that the 'folding pocket knife with a blade not exceeding three inches' exception only applies to an article which can properly be regarded as a pocket knife. It held that pocket knife was not an apt description of a cut-throat razor. Various judicial decisions indicate that, to be a folding pocket-knife, the blade must be readily and immediately foldable at all times, simply by the folding process. Thus, a knife whose blade is secured in an open position by a locking device and can only be released from the open position by the pressing of a release button is not a folding pocket knife within the meaning of s 139.

D has a defence if D proves lawful authority or good reason for having the article with him/her in a public place, or that (s)he had it with him/her there for use at work, for a religious reason or as part of any national costume (s 139(4) and (5)). Whether D was in possession of such a knife 'for use at work' is a matter for the court or jury as the statute uses ordinary, everyday language. Whether D had a good reason is a matter for the court or jury. A judge should not tell a jury what 'good reason' means, but in a clear case the judge can rule that D's explanation cannot amount to a good reason. The Court of Appeal has held that D's forgetfulness that (s)he has the article is not a good reason, but that forgetfulness combined with another reason may be. A fear of attack may be a good reason, depending on how imminent, how likely, how specific the perceived threat, and how serious the anticipated attack.

Having article with blade or point (or offensive weapon) on school premises

CJA 1988, s 139A(1) makes it an offence for a person to have with him/her on school premise any article to which s 139 applies. Section 139A(2) creates a similar offence in relation to an offensive weapon to which PCA 1953 applies. The term 'school premises' means land used for the purpose of a school excluding any land occupied solely as a dwelling by a person employed at the school. Similar defences exist; a person who proves that (s)he had good reason or lawful authority for having the article or weapon with him/her on the premises has a defence (s 139A(3)). So has a person who proves that (s)he had the article or weapon with him/her for the purposes set out above in relation to s 139 (s 139A(4)). In addition, (s)he may show possession 'for educational purposes' (s 139A(4)). The offences under s 139A(1) or (2) are indictable (either-way) offences punishable with up to four years' imprisonment.

19.53

Threatening with article with blade or point (or offensive weapon)

Under the Criminal Justice Act 1988, s 139AA(1), a person is guilty of an indictable (either-way) offence punishable with up to four years' imprisonment if (s)he:

19.54

(a) has an article to which s 139AA applies with him/her in a public place (as defined for s 139: **19.52**) or on school premises (as defined for s 139A: above),
(b) unlawfully and intentionally threatens another person with the article, and
(c) does so in such a way that there is an immediate risk of serious physical harm (as defined at **19.51**) to that other person.

In relation to a public place, s 139AA(1) applies to an article to which s 139 applies (**19.52**). In relation to school premises, s 139AA(1) applies to an article to which s 139 applies, and to an offensive weapon within the meaning of PCA 1953, s 1 (**19.46**).

Power of entry to search for articles with a blade or point and offensive weapons

CJA 1988, s 139B provides a constable with a power of entry, using reasonable force if necessary, into school premises and the power to search those premises and any person on those premises for articles to which CJA 1988, s 139 applies, or to which PCA 1953, s 1 applies, if the constable has reasonable grounds for suspecting that an offence under CJA 1988, s 139A or s 139AA (above) is being, or has been, committed. (S)he may seize any articles or weapons discovered in the course of such a search which (s)he reasonably suspects to be such articles or weapons.

19.55

Trespassing with offensive weapon

CLA 1977, s 8(1) provides that a person who is on any premises as a trespasser, after having entered as such, is guilty of an offence if, without lawful authority or reasonable

19.56

excuse (**19.50**), (s)he has with him/her (**19.33**) on the premises any weapon of offence. This is a summary offence punishable with up to three months' imprisonment and/or an unlimited fine.

The term 'weapon of offence' is defined as any article made or adapted for use for causing injury to or incapacitating a person, or intended by the person having it with him/her for such use. For a discussion of such a provision, see **14.53**.

RESTRICTION OF OFFENSIVE WEAPONS

Manufacture, sale, etc of flick knives and gravity knives

19.57   The Restriction of Offensive Weapons Act 1959 is concerned with the manufacture and distribution of flick knives and gravity knives. Under s 1(1), it is a summary offence, punishable with up to six months' imprisonment and/or an unlimited fine, for any person to manufacture, sell, hire, offer for sale or hire, or to expose or have in his/her possession for the purpose of sale or hire, or to lend or give to any person, either of these weapons.

A 'flick knife' is any knife which has a blade which opens automatically by hand pressure applied to a button, spring or other device in or attached to the handle of the knife. A 'gravity knife' is one which has a blade which is released from the handle or sheath thereof by the force of gravity or the application of centrifugal force and which, when released, is locked in place by means of a button, spring, lever or other device. The flick knife is therefore one with an up-and-over blade and a gravity knife is one which allows the blade to be shaken from the handle.

It is a defence for the defendant (D) to show that his/her conduct was only for the purposes of making the knife available to a public museum or gallery which does not distribute profits. D is to be taken to have shown this if sufficient evidence is adduced to raise an issue with respect to it, and the contrary is not proved beyond reasonable doubt.

Manufacture, sale, etc of specified weapons

19.58   By CJA 1988, s 141(1), a person who manufactures, sells or hires, or offers for sale or hire, exposes or has in his/her possession for the purpose of sale or hire, or lends or gives to any other person, any weapon specified by order by the Secretary of State, commits an indictable (either-way) offence punishable with up to four years' imprisonment.

The specified weapons

19.59   The Secretary of State has made the Criminal Justice Act 1988 (Offensive Weapons) Order 1988 which provides that s 141(1) applies to the following weapons, other than those which are antiques (ie ones manufactured more than 100 years before the alleged offence under s 141(1)). The specified weapons are: a knuckleduster, a swordstick, a handclaw, a belt buckle knife, a push dagger, a hollow kubotan (small truncheon with spikes), a footclaw, a death star, a butterfly knife, a telescopic truncheon, a blow-pipe, a kusari gama (sickle and chain), a kyoketsu shoge (hooked knife and chain), a manrikigusari or kusari (weights joined by chain), a disguised knife (any knife which has a concealed blade or concealed sharp point and is designed to appear to be an everyday object of a kind commonly carried on the person or in a handbag, briefcase or other hand luggage (such as a comb, brush, writing instrument, cigarette lighter, key, lipstick, or telephone)), a stealth knife (a knife or spike, which has a blade or a sharp point, made from a material that is not readily detectable by apparatus used for detecting metal, and which is not designed for domestic use or for use in the processing, preparation or consumption of food or as a toy), a straight, side-handled or friction-lock truncheon (sometimes known as a baton), a zombie knife, zombie killer knife, or zombie slayer knife (a blade with a cutting edge, a serrated edge, and images or words (whether on the blade or otherwise) suggesting that it is to be used for the purposes of violence), and a sword with a curved blade 50 cm or more in length (for which purpose the length of the blade is the straight line distance from the top of the handle to the tip of the blade). It is a defence to a charge under s 141 in relation to the last-named to show:

(a) that the sword was made before 1954 or was made at any other time according to traditional methods of making swords by hand;
(b) that D's conduct was for the purpose of making the weapon available for the purposes of the organisation and holding of a permitted activity (historical re-enactment or a sporting activity) in respect of which public liability insurance is held; or
(c) that D's conduct was for the purpose only of making the weapon available for the purposes of use in religious ceremonies.

D shows (a), (b), or (c) if sufficient evidence as to it is adduced to raise an issue with respect to it and the contrary is not proved beyond reasonable doubt.

Defences

19.60 By CJA 1988, s 141(8)-(11C), it is a defence to a charge under s 141(1) to show that the conduct in question was only for the purpose of making the weapon available: (a) to a public museum or gallery which does not distribute profits, or (b) for theatrical performances (or rehearsals) or the production of films or television programmes. D shows this if sufficient evidence of the matter is adduced to raise an issue with respect to it and the contrary is not proved beyond reasonable doubt.

Police powers

19.61 A justice may issue a warrant under CJA 1988, s 142 authorising entry and search on the application of a constable if the justice is satisfied that there are reasonable grounds for believing that there are on those premises knives such as are mentioned in the Restriction of Offensive Weapons Act 1959, s 1(1) or weapons to which CJA 1988, s 141 applies, that an offence under either of these provisions has been or is being committed in relation to them, and that one of the normal essential conditions for a search warrant exists.

Sale of knives, etc to persons under 18

19.62 By CJA 1988, s 141A(1), it is an offence to sell to a person under 18 a knife, knife blade or razor blade, any axe, and any other article which has a blade or which is sharply pointed and which is made or adapted for use for causing injury to the person. Section 141A does not apply to articles already controlled by the Restriction of Offensive Weapons Act 1959 or CJA 1988, s 141, or described in an order made by the Secretary of State under s 141A. The Secretary of State has made an order in respect of: (a) folding pocket knives with a blade not exceeding three inches, and (b) razor blades permanently enclosed in a cartridge or housing.

It is a defence if D proves that (s)he took all reasonable precautions and exercised all due diligence to avoid commission of the offence.

This is a summary offence punishable with up to six months' imprisonment and/or an unlimited fine.

USING SOMEONE TO MIND A WEAPON

19.63   VCRA 2006, s 28 provides that a person is guilty of an offence triable only on indictment if:

(a) (s)he uses another to look after, hide or transport a dangerous weapon for him/her; and

(b) (s)he does so under arrangements or in circumstances that facilitate, or are intended to facilitate, the weapon being available to him/her for an unlawful purpose.

For this purpose, the cases where a dangerous weapon is to be regarded as available to a person for an unlawful purpose include any case where it is available for him/her to take possession of it at a time and place at which his/her possession of the weapon would constitute, or be likely to involve or to lead to, the commission by him/her of an offence. A 'dangerous weapon' is a firearm (other than an air weapon or a component part of, or accessory to, an air weapon) or a weapon to which CJA 1988, s 141 or 141A applies (specified offensive weapons, knives, and bladed weapons: see **19.59** and **19.62**).

The maximum sentence for this offence is four or 10 years' imprisonment. The applicable maximum is determined by reference to specified circumstances.

CROSSBOWS

19.64   The Crossbows Act 1987 creates specific offences in respect of crossbows with a draw weight of at least 1.4 kg.

By s 1, it is an offence for any person to sell or let on hire such a crossbow or part of a crossbow to a person under the age of 18. No offence is committed if the seller or letter on hire believes the purchaser or hirer to be 18 or older, provided (s)he has reasonable ground for that belief. Similarly, by s 2, it is an offence for a person under 18 to purchase or hire a crossbow or part of a crossbow.

Section 3 provides that, unless (s)he is under the supervision of a person who is 21 or older, a person under 18 who has with him/her a crossbow which is capable of discharging a missile, or parts of a crossbow which together (and without any other parts) can be assembled to form a crossbow capable of discharging a missile, is guilty of an offence.

Where a constable suspects with reasonable cause that a person is committing or has committed an offence under s 3, (s)he may search:

(a) that person for a crossbow or part of a crossbow; or

(b) any vehicle, or anything in or on a vehicle, in or on which the constable reasonably suspects there is a crossbow or part of a crossbow connected with the offence.

A person or vehicle may be detained by the constable for the purpose of such a search and anything appearing to be a crossbow (or part) may be seized. The constable may enter any land other than a dwelling house to exercise these powers. These powers are provided by s 4.

The offences under ss 1 to 3 are summary and punishable with up to six months' imprisonment and/or an unlimited fine (s 1) or with a fine not exceeding level 3 (ss 2 and 3).

FIREWORKS

Offences under the Fireworks Act 2003

The Fireworks Regulations 2004 (FR 2004), made under the Fireworks Act 2003, contain a number of prohibitions. FR 2004, reg 12 requires chief officers of police to enforce the prohibitions (below) relating to possession or use. Enforcement of the prohibitions relating to unlicensed supply is a matter for the local weights and measures authority, except that in a metropolitan county licensing enforcement is a matter for that county's fire and rescue authority. **19.65**

The Fireworks Act 2003, s 11 provides that any person who contravenes a prohibition imposed by fireworks regulations (s 11(1)), or who fails to comply with a requirement imposed by or under such regulations to give or not give information (s 11(2)), or who knowingly or recklessly makes a false statement about such information (s 11(3)) is guilty of a summary offence punishable with up to six months' imprisonment and/or an unlimited fine. The defence of 'due diligence' provided by the Consumer Protection Act 1987, s 39 applies to these offences. The offence is a 'penalty offence' for the purposes of the Criminal Justice and Police Act 2001 (**3.40**).

Prohibition on possession

FR 2004, reg 4 provides that no person under 18 may possess an adult firework in a public place, ie any place to which at the material time the public have or are permitted access (whether on payment or otherwise). **19.66**

Regulation 5 prohibits anyone from possessing a 'category 4 firework'. This is a firework classified as category 4 under Part I of BS 7114.

However, reg 6 provides that nothing in regs 4 and 5 prohibits the possession of any firework by any person:

(a) employed by, or in business as, a professional organiser or operator of firework displays who possesses the firework for such purposes;

(b) employed in, or whose trade or business (wholly or partly) is, the manufacture of fireworks or assemblies containing fireworks who possesses the firework for the purposes of his trade, employment, or business;

(c) employed in, or whose trade or business (wholly or partly) is, the supply of fireworks or such assemblies in accordance with Pyrotechnic Articles (Safety) Regulations 2015;

(d) employed by a local authority, or a UK government department, or in a naval, military, or air force establishment, who, in the course of his employment, possesses the firework for the purpose of a firework display, in connection with a national public celebration or a national commemorative event, or (government employee only) for research or investigation.

(e) employed by or otherwise acting on behalf of a local authority, enforcement authority or other body, where the authority or body has enforcement powers applying to the firework in question, and the person possesses that firework for the purposes of the exercise of those powers;

(f) for use, in the course of a trade, business, or employment, for special effects purposes in the theatre, on film, or on television; or

(g) in business as or employed by a supplier of goods designed and intended for use in conjunction with fireworks or assemblies containing fireworks who possesses the firework for testing those goods in connection with safety.

Prohibition of use at night

19.67   FR 2004, reg 7 prohibits the use of an adult firework during 'night hours' (23.00 to 07.00) otherwise than on a permitted fireworks night or by a local authority employee in the course of a local authority display or at a national public celebration or commemorative event. Permitted fireworks nights are Chinese New Year (23.00 to 01.00); 5 November (23.00 to 24.00); the day of Diwali (23.00 to 01.00); and 31 December (23.00 to 01.00).

Throwing fireworks

19.68   It is a summary offence punishable with an unlimited fine to throw, cast or fire any fireworks in or into any highway, street, thoroughfare, or public place. This offence, provided by the Explosives Act 1875 (EA 1875), s 80, is extremely useful. Although there are other offences in relation to the use of fireworks in streets and public places, eg under the Highways Act 1980, s 161 and the Town Police Clauses Act 1847, s 28 (**19.44** and **19.45**), this offence is the most easily proved. There are no exceptions to the offence.

The offence under EA 1875, s 80 is a 'penalty offence' for the purposes of the Criminal Justice and Police Act 2001 (**3.40**).

Possession of pyrotechnic articles at musical events

19.69   Under the Policing and Crime Act 2017, s 134(1), it is a summary offence, punishable with up to three months' imprisonment and/or a fine not exceeding level 3, for a person to have a pyrotechnic article in his/her possession at any time when (s)he is:

(a) at a place where a qualifying musical event is being held; or
(b) at any other place that is being used by a person responsible for the organisation of a qualifying musical event for the purpose of—
    (i) regulating entry to, or departure from, the event; or
    (ii) providing sleeping or other facilities for those attending the event.

Section 134(1) does not apply to a person who is responsible for the organisation of the event, or who has the article in his/her possession with the consent of a person responsible for the organisation of the event (s 134(2)).

For the purposes of s 134:

(a) 'pyrotechnic article' means an article that contains explosive substances, or an explosive mixture of substances, designed to produce heat, light, sound, gas, or smoke, or a combination of such effects, through self-sustained exothermic chemical reactions, other than a match; and
(b) 'qualifying musical event' means an event at which one or more live musical performances take place and which is specified, or of a description specified, in the Policing and Crime Act 2017 (Possession of Pyrotechnic Articles at Musical Events) Regulations 2017, ie an event that—
    (i) is provided to any extent for members of the public or a section of the public; and
    (ii) takes place on premises in respect of which—
        (a) a premises licence has been granted; and
        (b) the licence authorises the premises to be used for the provision of regulated entertainment in the form of a performance of live music.

EXPLOSIVE SUBSTANCES ACT 1883

The Explosive Substances Act 1883 (ESA 1883) deals with offences which can be committed in relation to 'explosive substances'. **19.70**

Explosive substance

The term is defined by ESA 1883, s 9, which declares that an explosive substance is **19.71**
deemed to include:

(a) any material for making any explosive substance;
(b) any apparatus, machine, implement, or materials used, or intended to be used, or
adapted for causing, or aiding in causing, any explosion in or with any explosive
substance; and
(c) any part of any such apparatus, machine, or implement.

This wide definition covers, for example, an ingredient which would go into the making
of an explosive substance, an empty bomb case, and a detonator for a bomb.

ESA 1883 does not define the essential term 'explosive' but the Court of Appeal has
held that 'explosive' should be interpreted in the light of the definition of 'explosive' in
the Explosives Act 1875, s 3, which states that the term means:

(a) gunpowder, nitro-glycerine, dynamite, gun-cotton, blasting powder, fulminate of
mercury, or of other metals, coloured fires, and every other substance, whether
similar to those already mentioned or not, used or manufactured with a view to
producing a practical effect by explosion or a pyrotechnic effect; and
(b) includes fog-signals, fireworks, fuses, rockets, percussion caps, detonators, cartridges, ammunition of all descriptions, and every adaptation or preparation of an
explosive as above defined.

'Explosive substance' is, therefore, a wide term. A petrol bomb has been held to be
an 'explosive substance'.

Causing an explosion likely to endanger life

ESA 1883, s 2 makes it an offence for any person unlawfully and maliciously to cause **19.72**
by an explosive substance any explosion of a nature likely to endanger life or to cause
serious injury to property, whether such injury or damage occurs or not.

'Unlawfully' means without lawful justification, ie otherwise than in self-defence, prevention of crime, or the like, which will rarely be the case. 'Maliciously' simply means an
intention to cause some unlawful harm to another or to property, or recklessness as to
the risk of this resulting from the explosion. It does not mean or require spite or ill-will.

Attempt to cause explosion; making or keeping explosives with intent

ESA 1883, s 3(1) creates two offences. The first offence is concerned with acts done with **19.73**
the specified intent to cause an explosion, and the second with making or possessing an
explosive substance with the specified intent.

The offences are unlawfully and maliciously:

(a) to do any act with intent to cause by an explosive substance an explosion of a nature
likely to endanger life or cause serious injury to property; or

(b) to make or have in one's possession or under one's control an explosive substance with intent by means thereof to endanger life or cause serious injury to property, or to enable any other person to do so.

The intended consequences may be in the UK or elsewhere. 'Unlawfully' has the same meaning as in s 2. 'Maliciously' is redundant in this offence because of the specific nature of the required intent.

In examining circumstances which might lead to the identification of offences contrary to s 3, one should think in terms of those who have not yet caused an explosion (and are therefore not caught by ESA 1883, s 2) but are doing some act with that intention in mind. Section 3 extends to acts which are merely preparatory to causing an explosion and therefore could not amount to an attempt to commit a s 2 offence.

Making or possessing an explosive under suspicious circumstances

19.74 ESA 1883, s 4(1), as interpreted by the Court of Appeal, provides that any person who knowingly makes or knowingly has in his/her possession or under his/her control any explosive substance, under such circumstances as to give rise to a reasonable suspicion that (s)he is not making it or does not have it in his/her possession or under his/her control for a lawful object, commits an offence. If D claims that (s)he made, possessed, or controlled it for a lawful object, D must prove that this was so, in which case the offence is not committed.

Section 4 does not define what constitutes a 'lawful object' but it has been held that it requires a positive object which is lawful, and not simply the absence of a criminal purpose. Thus, it has been held, making a bomb with no intention of exploding it would not be making an explosive substance for a lawful object. On the other hand, it has been held that self-defence or the like against an imminent attack would be a lawful object if the defendant intended to use the explosive in a way which was no more than reasonably necessary to meet the imminent attack. The Supreme Court has held that experimentation and self-education being obviously lawful pursuits are 'lawful objects'. However, the Court of Appeal has subsequently qualified this by holding that any experimentation which gives rise to an obvious risk of harm to others or their property, or other unlawfulness such as causing a public nuisance, will taint D's object and prevent it being a lawful object.

General

19.75 An offence (**6.48**) under ESA 1883, s 2, 3, or 4 is triable only on indictment and punishable with a maximum of life imprisonment.

OTHER OFFENCES IN RELATION TO EXPLOSIVES

19.76 More general offences under OAPA 1861 and the Criminal Damage Act 1971 may also be relevant to the activities of bombers. These offences are described at **9.52** and **14.148–14.157**.

GENERAL POLICE POWERS IN RESPECT OF EXPLOSIVES

19.77 By the Explosives Act 1875, s 73, a constable may enter at any time, by force if necessary, any place (including a building, vehicle, or vessel) upon reasonable cause for believing that any offence has been or is being committed in that place with respect to an explosive if (s)he is in possession of:

(a) a justices' warrant granted following information on oath; or
(b) a written order from a superintendent (or above), which may be issued if the case is one of emergency and delay in obtaining a warrant would be likely to endanger life,

and to search for explosives, and take samples of any explosives and ingredients of an explosive.

BOMB HOAXES

The offences which generally attract the description of bomb hoaxes are those dealt with by the Criminal Law Act 1977 (CLA 1977), s 51. They are indictable (either-way) offences punishable with up to seven years' imprisonment.   **19.78**

Placing or dispatching an article

CLA 1977, s 51(1) creates two offences.   **19.79**

The first is committed by any person who places an article in any place whatsoever with the intention of inducing some other person to believe that it is likely to explode or ignite and thereby cause personal injury or damage to property. Section 51(1) states that the term 'article' includes any substance. Someone who deposits a fake bomb at a railway station, for example, commits the offence because (s)he clearly intends to induce others to believe that it is an explosive device.

The second offence is dispatching any article by post, rail, or any other means from one place to another with the intention to induce in some other person a belief that it is likely to explode or ignite and thereby cause personal injury or damage to property. If D sends through a post office sorting office a number of false devices with the intention of causing fear of an explosion to be aroused in the staff of that office, the offence is complete.

Neither offence requires D to have any particular person in mind as the person in whom D intends to induce the relevant belief.

False messages

CLA 1977, s 51(2) is concerned with false messages to the effect that an explosive device has been planted. It states that it is an offence for a person to communicate any information, which (s)he knows or believes to be false, to another person, with the intention of inducing in him/her or any other person a false belief that a bomb or other thing liable to explode or ignite is present in any place or location whatever. The use of the words 'there is a bomb' by a hoaxer is sufficient to give rise to the offence. It is not necessary that the person communicating the false information should identify a location.   **19.80**

When police officers receive such a call, it is essential to gain as much information as possible from the caller, or from the person who is passing on a message received from such a caller. The details should include sex, estimated age, urgency in voice, emotion, and accent, together with details of the time, date, duration of call, whether from a private telephone or a call box, and any background voices. All information given by the caller must be established; where, when, and why the bomb is likely to explode, description of the type of bomb and its appearance, and as many of the actual words used as can be recalled.

Although the offence is most commonly committed by means of a telephone call, it can be committed by any form of communication. It is not necessary that D had any particular person in mind in whom D intends to induce the relevant belief.

CHAPTER 20

Preventing Harm from Drugs and Alcohol

CONTROLLED DRUGS

Introduction

The Misuse of Drugs Act 1971 (MDA 1971) provides a number of offences intended **20.1** to control the misuse of drugs. Drugs which are subject to the Act are designated as 'controlled drugs'.

A 'controlled drug' is any substance or product for the time being specified:

(a) in MDA 1971, Sch 2, Part I, II, or III; or
(b) in a temporary class drug order as a drug subject to temporary control.

Part I of Sch 2 lists 'Class A drugs'; Part II, 'Class B drugs'; and Part III, 'Class C drugs'. Drugs can be added to the lists contained in Sch 2 by Order in Council, or moved from one class to another. One of the points of the classification of controlled drugs is that it affects the punishment of some of the offences under MDA 1971. As a result of a House of Lords decision, what may appear to be one offence (eg unlawful possession of a controlled drug) is in law divisible into distinct offences depending on the maximum penalty for the drug in question.

Class A drugs include narcotic drugs, such as cocaine, morphine, opium, pethidine, and heroin, and hallucinogenic drugs, such as mescaline, LSD, MDMA (ecstasy) however the MDMA is produced, or fungus of any kind which contains psilocin or an ester of psilocin. Class B drugs include amphetamines, cannabis, cannabis resin, cannabis derivatives, and ketamine. Class C drugs include benzyphetamine, GHB (often called the 'date rape' drug), khat, pipradol, temazepam, and anabolic steroids.

There are more than 100 drugs listed in Class C and a similar number are listed in Class A. There are fewer Class B drugs.

'Cannabis' and 'cannabis resin' are defined by MDA 1971. 'Cannabis' means any part of the genus *Cannabis* or any part of any such plant (by whatever name designated) except that it does not include cannabis resin, or any of the following products, after separation from the rest of the plant, namely:

(a) mature stalk of any such plant;
(b) fibre produced from mature stalk of any such plant; and
(c) seed of any such plant.

'Cannabis resin' means the separated resin, whether crude or purified, obtained from any plant of the genus *Cannabis*.

As indicated, a 'controlled drug' is a specified substance or *product*. In addition 'controlled drug' includes any 'preparation or other product' containing a specified substance or product. Some growing things, such as certain types of mushroom, contain a specified substance. In their natural state such things are not a specified substance or product, but if they are picked and subjected to some process to enable them to be used as a drug, they become a 'preparation' containing a specified substance and become a controlled drug. If they are picked, packed, and frozen they become a 'product' containing a specified substance and become a controlled drug.

'*Temporary class drug*' means any substance or product which is for the time being a controlled drug by virtue of a temporary class drug order. *Where the punishment depends on the class of the controlled drug a temporary class drug is treated in the same way as a Class B drug.* The Secretary of State may make a temporary class drug order specifying any substance or product (other than one specified in Sch 2) as a drug subject to temporary control. A temporary class drug order must be made by statutory instrument and ceases to have effect unless approved by each House of Parliament. A substance or product specified in a temporary class drug order ceases to be a controlled drug by virtue of the order:

(a) at the end of one year beginning with the commencement day of the order; or
(b) if earlier, upon the coming into force of an Order in Council by virtue of which the substance or product is specified as a controlled drug in Sch 2.

This is without prejudice to the power of the Secretary of State to vary or revoke a temporary class drug order.

A temporary class drug order may provide for the exception of a temporary class drug from the prohibitions on the import, export, production, or supply of controlled drugs, and for circumstances in which a person's possession of the drug is to be treated as excepted possession for the purposes of MDA 1971.

MDA 1971, s 27 provides that, where someone is convicted of an offence under the Act, the court may order anything relating to the offence to be forfeited and destroyed or otherwise dealt with. See, further, Chapter 20 on the companion website.

Unlawful import and export

20.2   MDA 1971, s 3 prohibits the import or export of a Class A, B, or C controlled drug, otherwise than as authorised by regulations made under the Act or by a licence issued by the Secretary of State. MDA 1971, s 3 also prohibits the import or export of a temporary class drug unless the drug has been excepted in the circumstances from s 3 by a provision in the temporary class drug order.

Breach of s 3 is not an offence under the Act but it is an indictable (either-way) offence under the Customs and Excise Management Act 1979, ss 50 (importation), 68 (exportation), and 170 (fraudulent evasion of prohibition). Such an offence is punishable with up to life imprisonment (Class A drug) or 14 years' imprisonment (Class B or C).

Unlawful production

20.3   MDA 1971, s 4(2) states that it is an offence for any person unlawfully to produce a controlled drug or to be concerned in the production of a controlled drug. An offence under s 4(2) is subject to the defence provided by s 28 (**20.19**). It is an indictable (either-way) offence punishable with up to life imprisonment (Class A drug) or 14 years' imprisonment (Class B or C).

The production of any Class A, B, or C controlled drug, otherwise than as authorised by regulations, is an unlawful production. The Misuse of Drugs Regulations 2001 authorise production by drug companies, by research establishments for experimental purposes, and by chemists in the course of their business. The production of a temporary class drug is an unlawful production unless a provision of a temporary class drug order excepts the drug from the provisions of s 4 or authorises its production in the circumstances in question.

Produce

'Produce' means to produce a controlled drug by manufacture, cultivation, or any other   **20.4** method, and 'production' has a corresponding meaning.

The Court of Appeal has held that the conversion of cocaine hydrochloride into freebase cocaine, in which form it would vaporise and be capable of being inhaled, by dissolving cocaine hydrochloride in water and either baking powder or household ammonia, amounts to production of a Class A drug 'by any other method' than manufacture or cultivation since the drug, in these two forms, is chemically different. It did not matter that the cocaine hydrochloride was already a Class A drug before the process began. By way of further example, where cannabis plants are stripped to take out those parts which could be used for smoking, it has been held that this amounts to production of a Class B drug (cannabis) as a controlled drug is produced by some 'other method'. 'Any other method' has also been held by the Court of Appeal to include the making of an infusion from a controlled drug which, although it does not alter the drug's chemical make-up, has been put into a form which enables it to be consumed.

Since the offence is to produce a controlled drug, it is essential that a controlled drug is actually produced before the offence can be committed.

Concerned in the production

The effect of these words is that criminal liability is not limited to those who actually   **20.5** participate in the production of a controlled drug, since those who arrange for the delivery of ingredients to the place of manufacture of such a drug, knowing the purpose for which they are required, are concerned in its production, as are those who knowingly allow their premises to be used.

Unlawful supply

MDA 1971, s 4(3) states that it is an offence for a person unlawfully to supply a con-   **20.6** trolled drug to another or to be concerned in the supplying of such a drug to another, or to offer to supply a controlled drug to another, or to be concerned in the making to another of an offer to supply such a drug. The Court of Appeal has held that there are separate and distinct offences under s 4(3):

(a) unlawfully supplying or offering to supply a controlled drug to another;
(b) being concerned in the unlawful supply of such a drug to another; and
(c) being concerned in the making to another of an offer unlawfully to supply such a drug.

The defence provided by s 28 (**20.19**) applies to an offence under s 4(3). Such an offence is an indictable (either-way) offence punishable in the same way as a s 4(2) offence (**20.3**).

Unlawful

20.7   The supply of a Class A, B, or C controlled drug is unlawful unless authorised by regulations. The circumstances in which a Class A, B, or C controlled drug may be supplied lawfully are set out in the Misuse of Drugs Regulations 2001 made under the Act. They are not difficult to imagine. For example, doctors may issue drugs direct from their own dispensaries; pharmacists may supply them upon prescription; nurses may supply patients in hospital; laboratory analysts, inspectors, and quality controllers may also handle drugs and pass them from one to another. The inclusion of the term 'unlawfully' ensures that all circumstances in which Class A, B, or C controlled drugs are supplied to a person outside any exception amount to offences under MDA 1971.

A 'supply' of a temporary class drug is not unlawful if a provision of a temporary class drug order excepts the drug from s 4 or authorises its supply in the circumstances in question.

Supply to another or offer to supply

20.8   'Supply' means more than the mere transfer of physical control from one person to another; it means furnishing to another the drug in order to enable the other to use it for his/her own purposes, whether this is done by an importer, producer, 'drug mule', pusher, or anyone else in the distribution chain. The person who distributes drugs at a party supplies the drugs to another. So does someone who returns a drug to a person who already owns it so that he can use it, or who hands a 'joint' to someone so that he can take a puff. On the other hand, a person who hands a drug to another (A) for safe-keeping or to an employee of a professional courier service does not supply it to A. Nor does a person who injects another if that drug is already in the other's control.

Although a person who makes a joint purchase of drugs for consumption by himself and another, paying with their joint funds, supplies the other when he hands over the latter's share of the drugs, the Court of Appeal has stated that a charge of supplying the latter is undesirable.

Before an offence involving supplying can be committed the substance supplied must be a controlled drug. It is not sufficient that the supplier believed that the substance was a drug, when in fact it was not, although (s)he could be convicted of an attempt to supply in such a case or, depending on the circumstances, of an offer to supply.

Direct evidence of supply is not required; circumstantial evidence can suffice.

In terms of an offer to supply, it is the making of an offer which is the important factor. If an offer is made to supply a controlled drug, an offence is committed even though the thing offered is not intended to be supplied, or is not in fact a controlled drug as the offeror knows. Likewise, the offence of offering to supply a controlled drug is committed if the offeror does not intend to supply anything.

Concerned in the supply or an offer to supply

20.9   The wide meaning given to the term 'concerned' when discussed in its application to offences of production (**20.5**) should be applied.

Proof that the defendant (D) was concerned in the supply of a controlled drug does not require proof that D physically supplied it to another. The Court of Appeal has stated that D's participation can take other forms, eg setting up a meeting, being a middleman, providing the finance, or arranging the contacts.

The Court of Appeal has held that the word 'supply' *in the present context* is not confined to an 'actual delivery', but refers to the entire process of supply. Thus, the Court upheld convictions of defendants who had been concerned in transporting controlled drugs from London to Portsmouth to be delivered to others in that area. That, said the Court, fell plainly within 'supply'.

Supply, etc of article usable for administration

MDA 1971, s 9A makes it an offence to supply or offer to supply an article which may be used or adapted to be used (whether by itself or in combination with another article) in the administration by a person of a controlled drug (s 9A(1)), or which may be used to prepare a controlled drug for administration (s 9A(3)), in either case believing it would be so used in circumstances which would be unlawful.   **20.10**

It is not an offence under s 9A(1) to supply or offer to supply a hypodermic syringe. In addition, under the Misuse of Drugs Regulations 2001, it is lawful for someone employed or engaged in the lawful provision of drug treatment services to supply or offer to supply aluminium foil as part of structured steps to engage a patient in a drug treatment plan, or which form part of such a plan. Aluminium foil can be used to warm some drugs and inhale them; inhaling is safer than injecting.

The offences under s 9A(1) and (3) are summary offences punishable with up to six months' imprisonment and/or an unlimited fine.

Unlawful possession

The offences of unlawful possession of controlled drugs are those with which police officers are most commonly involved. MDA 1971, s 5(2) states that it is an offence for a person unlawfully to have a controlled drug in his/her possession. The defence under s 28 (**20.19**) applies to these offences. An offence under s 5(2) is indictable (either-way and punishable with up to seven years' (Class A drug), five years' (Class B), or two years' (Class C) imprisonment. Possession of various forms of cannabis and cannabis derivatives or of khat (or any preparation or other product containing one of these things) is a 'penalty offence' for the purposes of the Criminal Justice and Police Act 2001 (**3.40**).   **20.11**

The offences under MDA 1971, s 5(2) do not apply in relation to a temporary class drug (s 5(2A)).

Unlawful

The Misuse of Drugs Regulations 2001 specify when possession of a controlled drug is lawful. They provide that such possession is lawful if it is under the authority of a licence issued by the Secretary of State or under a doctor's, first-level nurse's, nurse independent prescriber's, pharmacist's, pharmacist independent prescriber's, or registered midwife's prescription. In addition, a constable who comes into possession of controlled drugs in the course of his/her duties is in lawful possession of them, and so are carriers, postal workers, despatchers, workers in forensic laboratories who examine drugs on behalf of the police, medical personnel, ship's masters, etc, in circumstances properly connected with their duties.   **20.12**

Possession

Physical custody of the drug is not necessary for possession but physical control over it is. It follows that a person who has bought a controlled drug is not in possession of it if it is still hidden in the seller's car or stored at the seller's home. On the other hand, if D leaves a drug in his car or at home while he is away he remains in possession of the drug since he retains physical control over it.   **20.13**

Possession can be joint: for example, if two people share a car which they know contains cannabis, they are both in possession of the cannabis if each shares with the other the right to control what is done with it. Moreover, MDA 1971, s 37(3) states that for the purposes of the Act the things which a person has in his possession shall be taken to include anything subject to his control which is in the custody of another. Therefore there may be a number of persons in possession of a particular controlled drug. If a man imports cannabis and hands it to his business manager to store pending distribution, both are in possession as the drug is subject to their control. If the business manager then passes it on to the warehouseman to keep pending its distribution, all three are in possession of it.

Possession cannot begin until the person with control is aware that the thing is under his/her control; if a drug is slipped into X's pocket, unknown to X, X is not in possession of it. (As an exception, X is in possession of a drug delivered to X's home, even if X is unaware that it has arrived, provided it is delivered in response to a request by X.)

Knowledge of a thing's quality is not required. It follows that a mere mistake by D as to the quality of the thing under his/her control (as where D thinks some heroin tablets are aspirins) is not enough to prevent D being in possession.

In the case of drugs in a parcel, packet or other container in D's physical control, D is in possession of those drugs if D knows that (s)he is in control of that container and that it contains something, even though D thinks that the thing is something different in kind from a drug and even though D has no right to open the container to check its contents.

Possession, once begun, continues as long as the thing is in the person's control, even though (s)he has forgotten about it or mistakenly believes it has been destroyed or disposed of.

Although it is not necessary to prove that a minimum or usable quantity of a controlled drug was unlawfully in D's possession, D must have been in possession of a quantity of it which was visible, tangible, and measurable. Persons who are found under the influence of a drug are not then in possession of it for the purposes of this offence, even though traces of it are found in a blood or urine sample. This is because, once consumed, the thing changes its character and can no longer be considered a controlled drug. However, evidence of the presence of a drug in a blood or urine sample can be given to support an allegation of possession of the drug in its true state at some earlier time, ie before it was taken into the body.

Defence

20.14 A person is not criminally liable for the unlawful possession of a controlled drug contrary to MDA 1971, s 5(2) in the circumstances outlined by s 5(4).

Section 5(4) provides that if it is proved that D had a controlled drug in his/her possession it is a defence for D to prove that:

(a) knowing or suspecting it to be a controlled drug, (s)he took possession of it for the purpose of preventing another from committing or continuing to commit an offence in connection with that drug and that as soon as possible after taking possession (s)he took all such steps as were reasonably open to him/her to destroy the drug or to deliver it into the custody of a person lawfully entitled to take custody of it; or

(b) knowing or suspecting it to be a controlled drug, (s)he took possession of it for the purpose of delivering it into the custody of a person lawfully entitled to the custody of it and that as soon as possible after taking possession of it (s)he took all such steps as were reasonably open to him/her to deliver it into the custody of such a person.

Despite the fact that MDA 1971, s 5(4) requires D to 'prove' a defence under it, it would seem likely from a House of Lords' decision that all D has to do is to adduce sufficient evidence to raise the defence, whereupon it will be for the prosecution to prove beyond reasonable doubt that the defence is not made out.

Possession with intent unlawfully to supply

20.15 MDA 1971, s 5(3) provides offences, triable and punishable in the same way as an offence under s 4(2) (**20.3**), of having a controlled drug in one's possession, whether lawfully or not, with intent to supply it unlawfully to another. It need not be proved that D knew the identity of the substance. The defence under s 28 (**20.19**) applies to such an offence.

The Court of Appeal has held that the intended supply must be a supply in the UK.

The Court of Appeal has also held that, to come within s 5(3), the intention to supply (of someone in possession of a controlled drug) has to be an intention to supply *the* thing of which (s)he was then in possession. Thus, it held that where D intended to supply cannabis plants then in D's possession when they reached maturity in a few months' time there was no offence under s 5(3).

An offence of possession with intent to supply may be committed by persons who are lawfully in possession in the first instance. For example, it can be committed by a doctor who is in possession of drugs lawfully, but forms an intention to supply them unlawfully (eg merely for profit, as opposed to bona fide treatment). However, the offence is usually committed by drug pushers and the like.

In terms of proving that possession was with intent unlawfully to supply, evidence of drug-related paraphernalia, evidence of an extravagant lifestyle, and evidence of the possession of large amounts of cash *which are prima facie explicable only if derived from drug dealing* are relevant—but not conclusive—to the issue of *intent to supply* but not normally to the issue of possession. A case where such evidence might be relevant to possession, as well as to intent to supply, is where there is evidence of frequent brief visits by different young people, who then leave carrying small packages, and when the premises (whose occupant is long-term unemployed) are searched large sums of money and some drugs are found. Such evidence may be admitted as evidence that the occupant knew of the drugs and was in control of them (ie in possession) as well as of an intent to supply them.

Whilst evidence in the form of documents relating to transactions and cash in D's possession is relevant, the relevance of the cash in D's possession must be related to evidence of ongoing (and not merely previous) drug transactions in order for it to be admissible. Evidence which shows that D has supplied drugs in the past is potentially admissible under the 'bad character' provision in the Criminal Justice Act 2003, s 101(1)(d) (see condition (d) at **8.66**).

Cultivation of cannabis

20.16 MDA 1971, s 6(2) states that it is an offence unlawfully to cultivate any plant of the genus *Cannabis*. The only lawful cultivation of cannabis is that authorised by a licence issued by the Secretary of State. The defence under s 28 (**20.19**) applies to an offence under s 6(2). This is an indictable (either way) offence punishable with up to 14 years' imprisonment.

'Cultivate' indicates some form of attention to the plant during the process of its growth. A person who puts seeds in the ground cultivates, as does someone who hoes, waters, prunes or generally cares for a plant during the process of its growth. It is doubtful whether a person could be held to have cultivated plants merely because (s) he failed to remove those which were growing wild but this would depend upon the circumstances. If they were deliberately preserved, by caring for the ground in which they were growing, this would amount to the type of care which could be described as cultivation.

Medical necessity

20.17   The defence of duress of circumstances is not available to a defendant (D) in relation to offences of cultivation, production, possession, or possession with intent to supply cannabis or resin, where D's purpose is to alleviate pain arising from an illness.

Activities relating to opium

20.18   MDA 1971, s 9 prohibits a person (D) from:

(a)   smoking or otherwise using prepared opium; or
(b)   frequenting a place used for the purpose of opium smoking; or
(c)   having in D's possession:
   (i)   any pipes or other utensils made or adapted for use in connection with the smoking of opium, being pipes or utensils which have been used by D or with D's knowledge and permission in that connection or which D intends to use or permit others to use in that connection; or
   (ii)   any utensils which have been used by D or with D's knowledge or permission in connection with the preparation of opium for smoking.

The defence under s 28 (**20.19**) applies to an offence under s 9. The offence is an indictable (either-way) offence punishable with up to 14 years' imprisonment.

'Prepared opium' is opium prepared for smoking and includes dross and any other residue remaining after opium has been smoked.

'Frequenting a place' means to go there often. There is no requirement that a person who frequents a place used for opium smoking must have been shown to have been involved in opium smoking. The place does not need to be a building. Proof that it is used as an opium den will be necessary.

Statutory defence

20.19   MDA 1971, s 28 provides a defence in relation to charges contrary to s 4(2), 4(3), 5(2), 5(3), 6(2), or 9 (**20.3–20.18**). For convenience, the defence will be explained in relation to the offence of unlawful possession but what is said will be equally applicable (with the appropriate changes of words) to the other offences just mentioned.

Assuming that the prosecution has proved that D was in unlawful possession of a controlled drug, D can be convicted of that offence, even though it is not proved that D knew that what was in possession was a controlled drug. However, s 28 provides D with a defence in the circumstances outlined below. Although s 28 says that D has to prove the defence, the House of Lords has held that s 28 should be read as simply imposing an evidential burden on D so as to make it compatible with the presumption of innocence under the European Convention on Human Rights, art 6. The result is that, provided that sufficient evidence is adduced to raise a defence under s 28, the defence will succeed unless the prosecution proves beyond reasonable doubt that its terms are not satisfied.

The basic definition of the defence is contained in s 28(2), which states that it is a defence for D to 'prove' that *(s)he neither knew of, nor suspected, nor had reason to suspect* the existence of some fact alleged by the prosecution which it is necessary for the prosecution to prove if (s)he is to be convicted of the offence charged. (This does not affect the need for the prosecution to prove the element of knowledge required to establish 'possession'.)

This provision is subject to a qualification, provided by s 28(3), where D alleges, and 'proves', that (s)he did not know, suspect, or have reason to suspect that the thing in question was the controlled drug alleged, and proved, by the prosecution to have been involved. In this case, such 'proof' by D is not enough to give D a defence. In order to be acquitted D must also 'prove' one of two things:

(a) that (s)he neither believed nor suspected, nor had reason to suspect, that the thing in question was a controlled drug at all; or
(b) that (s)he believed that the thing in question was a controlled drug which (s)he was, in fact, legally entitled to possess (or supply or produce, etc as the case may be).

Occupier, etc of premises permitting drug activities

20.20 A person commits an offence under MDA 1971, s 8 if, being the occupier or concerned in the management of any premises, (s)he knowingly permits or suffers any of the following activities to take place on those premises:

(a) unlawfully producing or attempting to produce a controlled drug;
(b) unlawfully supplying or attempting to supply a controlled drug to another, or offering to supply a controlled drug unlawfully to another;
(c) preparing opium for smoking; or
(d) smoking cannabis, cannabis resin or prepared opium.

There must be proof that the conduct in (a), (b), (c), or (d) actually occurred; merely giving tacit approval in advance is insufficient. An offence under s 8 is an indictable (either-way) offence punishable with a maximum of 14 years' imprisonment.

The occupier

20.21 To be 'the occupier' a person does not have to be a tenant or have an estate in the premises. A person (P) is the occupier of premises if P is entitled to exclusive possession of them, in the sense that P has the requisite degree of control over them to exclude from them those who might otherwise carry on one of the forbidden activities there. Thus, a student who had a room in a college hostel was held to be the occupier of it because his contractual licence gave him such exclusivity of possession, whether or not he was entitled to exclude the college authorities. If there is drug-taking in a house, and it is knowingly permitted by the householder (ie 'the occupier', D), D commits an offence under s 8. However, D would not commit that offence if, in D's absence and unknown to D, D's teenage son knowingly permitted drug-taking on the premises. Nor would the son be guilty of an offence under s 8 on the basis of being 'the occupier' of the premises since he would not have that status. Nevertheless, depending on the circumstances, the son might be guilty of an offence under s 8 on the basis of 'being concerned in the management of the premises' (below).

Concerned in management

20.22   A person 'concerned in the management of premises' is anyone who is concerned in exercising control over the premises or in running or organising them. It is possible that such a person will have some control over who shall be permitted to enter the premises and who shall not, but this is not a prerequisite for a person to be concerned in their management. If drug-taking was generally permitted on the premises of a club, it is possible that only the general manager would have the right to permit entry but other officials of the club might control activities in different rooms. If drug use is generally and knowingly permitted, then all who are concerned in any way in the management of the premises (ie the general manager and other officials) would be guilty of this offence.

Persons who occupy premises as trespassers (and are therefore not 'occupiers' for the purpose of the offence) might nevertheless be concerned in the management of those premises.

Premises

20.23   MDA 1971 does not define the term 'premises'. The term should be given its normal, everyday meaning, and in this sense 'premises' includes any form of building and the grounds in which a building stands and also land without any building on it. Therefore, the organiser of an open-air pop festival who knowingly permitted one of the activities described above would (as a person concerned in the management of the premises) be guilty of the present offence, as would the occupier of the site if (s)he knowingly permitted one of these activities.

Knowingly permits or suffers

20.24   'Permit' and 'suffer' are synonymous. The Court of Appeal has held that 'permit' (and presumably 'suffer') require proof of unwillingness to prevent the prohibited activity, which can be inferred from failure to take reasonable steps readily available to prevent it. The fact that D believed that (s)he had taken reasonable steps to prevent the prohibited activity is irrelevant.

The inclusion of the word 'knowingly' is an important point. As mentioned at **1.20**, 'knowledge' may (not must) be inferred from wilful blindness, ie suspecting what is going on but deliberately refraining from making inquiries.

On a charge of permitting the premises to be used for producing or supplying a controlled drug, it is not necessary for the prosecution to prove more than that D knew of the production or supply of a controlled drug; it need not be proved that D knew that it was the particular type of controlled drug supplied. It remains to be seen whether, on a charge of permitting the smoking of cannabis, D could be convicted even if D thought that cannabis resin or opium was being smoked, and so on.

Police powers in respect of controlled drugs

Inspection

20.25   MDA 1971, s 23(1) empowers a constable, or other person authorised by the Secretary of State, to enter the premises of a business producing or supplying controlled drugs and to demand and inspect any books or documents relating to dealings in them and to inspect any stocks of them. By s 23(4), it is an indictable (either-way) offence, punishable with up to two years' imprisonment, to conceal from such a person any such books, documents, stocks, or drugs, or to fail without reasonable excuse to produce such books or documents when demanded.

Search, seize, and detain

MDA 1971, s 23(2) provides that, if a constable has reasonable grounds to suspect that any person (P) is in *possession of a controlled drug in contravention of MDA 1971 or of any regulations or temporary class drug orders thereunder*, the constable may: **20.26**

(a) search P, and detain P for the purpose of searching P;
(b) search any vehicle or vessel in which the constable suspects that the drug may be found, and for that purpose require the person in control to stop it;
(c) seize and detain, for the purpose of proceedings under MDA 1971, anything found in the course of the search which appears to the constable to be evidence of an offence under MDA 1971.

It must be emphasised that the exercise of these powers depends upon there being reasonable grounds to suspect possession of a controlled drug in contravention of MDA 1971 or regulations or orders. As to this requirement of reasonable suspicion, see **4.11**.

As already noted, simple possession of a temporary class drug is not a contravention of MDA 1971. Therefore, s 23(2) does not apply to it. For this reason, s 23A provides a further power to search, seize and detain in relation to temporary class drugs which applies where a constable has reasonable grounds to suspect that a person (P) is in *possession of a temporary class drug*, and it does not appear to the constable that a power under s 23(2) applies to the case. Section 23A provides that in such a case the constable may exercise the powers under (a) and (b) above and may seize and detain anything found in the course of the search *which appears to the constable to be a temporary class drug* or to be evidence of an offence under MDA 1971. If a constable reasonably believes that anything so detained is a temporary class drug but is not evidence of any offence under MDA 1971, the constable may dispose of the drug in such manner as the constable thinks appropriate.

A temporary class drug order may provide for circumstances where a person's possession of a temporary class drug is to be treated as 'excepted possession' for the purposes of MDA 1971. If any provision has been made about excepted possession by a temporary class drug order that applies to the temporary class drug in question, the powers in s 23A apply only if the constable has no reason to believe that P's possession of the drug is to be treated as excepted possession.

The above powers to search a person must extend to searching things in his/her immediate possession, for example a suitcase or a holdall.

In a case where there had been forcible use of emetics to induce vomiting by a man suspected of swallowing 'bubble wraps' of cocaine, the European Court of Human Rights held that forcible medical intervention to obtain evidence had to be *necessary* otherwise the European Convention on Human Rights, art 3 (prohibition of inhuman or degrading treatment) will be contravened. The court considered that in the circumstances it would have been possible to wait for the natural discharge of the 'bubbles'.

Search warrant

MDA 1971, s 23(3) authorises a justice to grant a search warrant if satisfied by information on oath that there is a reasonable ground for suspecting that controlled drugs are, in contravention of MDA 1971 or any regulations or temporary class drug orders   **20.27** thereunder, in the possession of a person on any premises, or that a document directly or indirectly relating to drug dealing is in the possession of persons on any premises. The warrant will name the particular premises, and it is only those premises (and persons in them) which may be searched on its authority. If necessary, force may be used to enter the premises.

If there is reasonable ground for suspecting that an offence has been committed in relation to any controlled drugs found on the premises or in the possession of anyone there, or that a document so found directly or indirectly relates to drug dealings, the drugs or document may be seized and detained.

A divisional court has held that, where a search warrant has been issued under both MDA 1971 and PACE, and the warrant refers to both persons and premises, MDA, s 23 provides the power to detain persons for the purpose of a search and PACE, s 117 provides the power to use reasonable force for the purpose of executing the warrant, including moving persons to one room while another is searched.

Additional powers of seizure

20.28   The additional powers of seizure provided by the Criminal Justice and Police Act 2001, ss 50 and 51 (**4.61** and **4.60**) apply where a search is made under the power in MDA 1971, s 23(2) or where a search warrant under s 23(3) is executed.

Obstruction

20.29   It is an indictable (either-way) offence intentionally to obstruct a constable (or, where relevant, a person) in the execution of his/her powers under MDA 1971, s 23 or s 23A, respectively (s 23(4), s 23A(6)). Both offences are punishable with up to two years' imprisonment.

Seizure and forfeiture of drug-cutting agents

20.30   The Serious Crime Act 2015, Part 4 (ss 52 to 65) and Sch 2 sets out provisions relating to the seizure and forfeiture of drug-cutting agents. A substance is a 'drug-cutting agent' if it is added to a controlled drug in connection with the supply or exportation of the drug, contrary to MDA 1971, s 3 or 4.

Applications for search and seizure warrants

20.31   Under SCA 2015, s 52, a justice of the peace may issue a warrant (a 'search and seizure warrant') authorising a police or customs officer:

(a) to enter premises; and

(b) to search them for substances that appear to be intended for use as drug-cutting agents,

if the justice is satisfied that there are reasonable grounds to suspect that a substance intended for such use is on the premises.

In SCA 2015, Part 4, 'police or customs officer' means a constable, a NCA officer, or a person designated as a general customs official (all referred to hereafter as an 'officer'). 'Premises' has the same meaning as in PACE (**4.59**) and therefore includes, for example, any place, and in particular includes any vehicle, vessel, tent, or movable structure.

A search and seizure warrant may be either an 'all-premises warrant' or a 'specific-premises warrant', defined in the same way as at **4.67**. The application may be made without notice being given to persons who might be affected by the warrant. Provisions corresponding to the search warrant provisions set out at **4.74** apply to the making of the application.

Section 53 sets out provisions relating to search and seizure warrants which correspond to the search warrant provisions set out at **4.75**.

Execution of search and seizure warrants

A search and seizure warrant may be executed by any officer. Reasonable force may **20.32** be used, if necessary, for the purpose of entering premises under the warrant. By SCA 2015, s 54(2), an entry on or search of premises under a search and seizure warrant is unlawful unless it complies with Sch 2 whose terms correspond with those of PACE, s 16 relating to the execution of a search warrant, set out at **4.76**. Where those provisions refer to an authorisation by an inspector (or above) an authorisation in the present context may also be given by a NCA officer of grade 3 (or above). Where an officer has power under s 55 (below) to seize a substance from premises, the officer or a person authorised by the warrant to accompany him/her may inspect or test the substance on the premises with a view to establishing whether or not it is a substance that is suitable for use as a drug-cutting agent. An officer who enters premises under a search and seizure warrant must take reasonable steps to ensure that when (s)he leaves the premises they are as secure as they were before (s)he entered.

By s 54(4), a summary offence, punishable with a fine not exceeding level 3, is committed by a person who without reasonable excuse obstructs an officer executing or seeking to execute a search and seizure warrant.

Seizure of substances under search and seizure warrant

SCA 2015, s 55 provides that an officer searching premises under a search and seizure **20.33** warrant may seize any substance on the premises that (s)he has reasonable grounds to suspect is intended for use as a drug-cutting agent.

Seizure of substances without search and seizure warrant

By SCA 2015, s 56, if an officer: **20.34**

(a) is lawfully on premises that are not subject to a search and seizure warrant; and
(b) finds a substance there that the officer has reasonable grounds to suspect is intended for use as a drug-cutting agent,

(s)he may seize the substance.

Notice to be given where substances seized

An officer who has seized a substance under SCA 2015, s 55 or 56 is required by s 57 **20.35** to make reasonable efforts to give written notice explaining the effect of ss 59, 60, 61, and 63 (below):

(a) to the person from whom the substance was seized; and
(b) if the officer thinks that the substance may belong to a different person, to that person.

Seizure of containers

Under SCA 2015, s 58, an officer who seizes a substance under s 55 or 56 may also **20.36** seize any container holding it. Reasonable efforts must be made to return a seized container to the person from whom it was seized, or (if different) a person to whom it belongs, unless the container appears to be of negligible value, it is not practicable for the container to be returned, or the container is or may be needed for use as evidence at a trial for an offence.

Initial retention of seized substances

20.37   This is dealt with by SCA 2015, s 59.

Where a substance has been seized under SCA 2015, s 55 or 56; and there continue to be reasonable grounds to suspect that the substance was intended for use as a drug-cutting agent, it may be retained until the end of the 30th day after the date of seizure.

Where:

(a) a substance has been seized under another legislative provision and is lawfully in the possession of an officer;

(b) the specified retention period during which the substance may lawfully be retained under that provision expires; and

(c) there are reasonable grounds to suspect that the substance was intended for use as a drug-cutting agent,

it may be retained until the end of the 30th day after the specified retention period referred to in (b).

Continued retention or return of seized substances

20.38   Under SCA 2015, s 60, on an application made by an officer, a magistrates' court or justice of the peace may extend the period for which a substance may be retained under s 59 if satisfied that condition (a) or (b) below is met.

Condition (a) is that the continued retention of the substance is justified while its intended use is further investigated, or while consideration is given to bringing proceedings against any person for an offence with which the substance is connected. If this condition is met, an order under s 60 may authorise the retention of the substance for a specified period ending no later than the 60th day after the date of seizure (in the case of seizure under s 55 or 56), or the end of the specified retention period referred to above (in any other case).

Condition (b) is that proceedings started against any person for an offence with which the substance is connected have not been concluded. If this condition is met, an order under s 60 may authorise the retention of the substance until the proceedings are concluded. Proceedings against a person for an offence are concluded when:

(i) (s)he is convicted or acquitted of the offence and either the time allowed for making an appeal, or applying for permission to do so, has expired, or, if an appeal is made, the appeal is determined or otherwise dealt with;

(ii) the charge is withdrawn; or

(iii) proceedings in respect of the charge are discontinued, or an order is made for the charge to lie on the file.

If the court or justice is not satisfied that the condition (a) or (b) is met, the court or justice must order the substance to be returned to a person entitled to it, ie the person from whom it was seized, and (if different) any person to whom it belongs.

Where an extension order is made under s 60, and no person entitled to the substance was present or represented at the hearing, an officer must make reasonable efforts to give written notice to the person from whom the substance was seized and, if the officer thinks that the substance may belong to a different person, to that person. The notice must explain the effect of the court's order, and the effect of s 63 (below).

Forfeiture and disposal, or return, of seized substances

Under SCA 2015, s 61, an officer may apply to a magistrates' court for the forfeiture of **20.39** a substance retained under s 59. The court must order the forfeiture of the substance if satisfied that it was intended for use as a drug-cutting agent. A substance ordered to be forfeited may be disposed of in whatever way the officer who applied for the order thinks is suitable. A substance must not be so disposed of before the end of the 30-day period within which an appeal to the Crown Court may be made, or, if an appeal is made, before it is determined or otherwise dealt with.

Return of substance to person entitled to it, or disposal if return impracticable

SCA 2015, s 63 provides as follows. **20.40**

Where the retention of a substance has been, but is no longer, authorised under SCA 2015, Part 4, the substance must (subject to s 63(4)) be returned to a person entitled to it. A person who claims to be entitled to a substance retained under Part 4 may apply to a magistrates' court for an order for return under s 60 or under s 63.

Where a court makes an order requiring a substance to be returned to a particular person, and reasonable efforts have been made, without success, to find that person, or it is for some other reason impracticable to return the substance to that person, the order has effect as if it required the substance to be returned to any person entitled to it.

Section 63(4) provides that, where:

(a) a substance is required by a provision of Part 4, or an order under Part 4, to be returned to a person entitled to it; and

(b) reasonable efforts have been made, without success, to find a person entitled to the substance, or it is for some other reason impracticable to return it to a person entitled to it,

an officer may dispose of the substance in whatever way the officer thinks is suitable.

Travel restrictions on drug trafficking offenders

The Criminal Justice and Police Act 2001, s 33 empowers a court to make a 'travel re- **20.41** striction order' where a person convicted of a drug-trafficking offence and has been sentenced to four years' or more imprisonment. The reference to a sentence of four years' or more imprisonment is to a single sentence of imprisonment, and does not include consecutive sentences of imprisonment totalling four years or more. Such an order prohibits the offender from leaving the UK at any time after his/her release from prison until such date as is set out in the order (not less than two years). An order may contain a direction that the offender's passport must be surrendered. It is a summary offence, punishable with up to six months' imprisonment and/or an unlimited fine, to fail to comply with a direction to do so (s 36(3)). It is an indictable (either-way) offence, contrary to s 36(1), to leave the UK while subject to a travel restriction. The offence is punishable with up to five years' imprisonment.

For the purposes of s 33 a 'drug-trafficking offence' means any of the following offences (including one committed by aiding, abetting, counselling or procuring):

(a) production and supply of controlled drugs, contrary to MDA 1971, s 4(2) or (3);

(b) assisting in or inducing commission outside the UK of an offence punishable under a corresponding law, contrary to MDA 1971, s 20;

(c) any offence designated by order of the Secretary of State;

(d) improper import, export or fraudulent evasion contrary to the Customs and Excise Management Act 1979 in connection with any prohibition or restriction on importation or exportation of a controlled drug;

(e) encouraging or assisting, or conspiring, or attempting to commit any of the offences in (a) to (d), contrary to the Serious Crime Act 2007, Pt 2; Criminal Law Act 1977, s 1; Criminal Attempts Act 1981, s 1, respectively; and

(f) inciting another person to commit any offence under MDA 1971, contrary to MDA 1971, s 19.

PSYCHOACTIVE SUBSTANCES

Introduction

20.42   The Psychoactive Substances Act 2016 (PSA 2016) creates a blanket ban on the production, distribution, sale and supply of psychoactive substances.

PSA 2016, s 2 defines a 'psychoactive substance' for the purposes of the Act as any substance which is:

(a) capable of producing a psychoactive effect in a person who consumes it; and

(b) not an exempted substance (see below).

For the purposes of s 2, a substance produces a psychoactive effect in a person if, by stimulating or depressing his/her central nervous system, it affects his/her mental functioning or emotional state. The Court of Appeal has held that, for these purposes, the words 'substance produces a psychoactive effect in a person if, by stimulating or depressing his/her central nervous system' encompass a substance which indirectly has the effect of stimulating or depressing his/her central nervous system, as well as one which directly has that effect.

Schedule 1 excludes the following 'exempted substances' from the definition: controlled drugs; medicinal products; *alcohol or alcoholic products*; nicotine; *tobacco products*; *caffeine or caffeine products*; any substance which is ordinarily consumed as food or drink, and does not contain a prohibited ingredient (ie any psychoactive substance which is not naturally occurring in the substance, and the use of which in or on food is not authorised by UK legislation). In the case of the *italicised* products mentioned, the exemption does not apply if they contain any psychoactive substance other than alcohol, nicotine or caffeine, as the case may be.

For these purposes, 'medicinal product', has the same meaning as in the Human Medicines Regulations 2012, reg 2(1), viz (a) any substance or combination of substances presented as having properties of preventing or treating disease in human beings, or (b) any substance or combination of substances that may be used by or administered to human beings with a view to (i) restoring, correcting, or modifying a physiological function by exerting a pharmacological, immunological or metabolic action, or (ii) making a medical diagnosis. The Court of Appeal has held that the regulatory regime under the 2012 Regulations (which imposes obligations, backed by criminal sanctions, on manufacturers and suppliers of medicinal products) contemplates that a substance may be a medicinal product for one purpose (and thus subject to the Regulations) but not another. It held that the manufacture and supply of nitrous oxide for medical purposed would clearly be covered by the Regulations because in that context it would fall within the definition of a medicinal product, whereas nitrous oxide possessed for non-medical purposes, eg for recreational use as laughing gas or making whipped cream would not be a medicinal product; the gas would modify the physiological functions of those who inhaled it but it would bring neither short-tern nor long-term beneficial effects to human health.

For the purposes of PSA 2016, a person (P) consumes a substance if P causes or allows the substance, or fumes given off by the substance, to enter P's body in any way.

Offences

PSA 2016, ss 4, 5, 7, and 8 contain the following provisions about offences relating to psychoactive substances. Exceptions to those offences are provided by s 11.   **20.43**

Production

A person commits an offence contrary to PSA 2016, s 4(1) if (s)he: **20.44**

(a) intentionally produces a psychoactive substance;
(b) knows or suspects that the substance is a psychoactive substance; and
(c) intends to consume the psychoactive substance for its psychoactive effects, or knows, or is reckless as to whether, it is likely to be consumed by some other person for its psychoactive effects.

For the purposes of PSA 2016, any reference to producing a substance is a reference to producing it by manufacture, cultivation or any other method.

Supply

A person commits an offence contrary to PSA 2016, s 5(1) if: **20.45**

(a) (s)he intentionally supplies a substance to another person;
(b) the substance is a psychoactive substance;
(c) (s)he knows or suspects, or ought to know or suspect, that the substance is a psychoactive substance; and
(d) (s)he knows, or is reckless as to whether, the psychoactive substance is likely to be consumed by the person to whom it is supplied, or by some other person, for its psychoactive effects.

A person (D) commits an offence contrary to s 5(2) if:

(a) D offers to supply a psychoactive substance to another person (R); and
(b) D knows or is reckless as to whether R, or some other person, would, if D supplied a substance to R in accordance with the offer, be likely to consume the substance for its psychoactive effects.

For the purposes of (b), the reference to a substance's psychoactive effects includes a reference to the psychoactive effects which the substance would have if it were the substance which D had offered to supply to R.

For the purposes of PSA 2016, any reference to supplying a substance includes a reference to distributing.

Possession with intent to supply

A person commits an offence contrary to PSA 2016, s 7(1) if (s)he: **20.46**

(a) is in possession of a psychoactive substance;
(b) knows or suspects that the substance is a psychoactive substance; and

(c) intends to supply the psychoactive substance to another person for its consumption, whether by any person to whom it is supplied or by some other person, for its psychoactive effects.

For the purposes of PSA 2016, the items which are in a person's possession include any items which are subject to that person's control, but in the custody of another person. 'Simple possession' of a psychoactive substance is not an offence.

Importing or exporting a psychoactive substance

20.47   A person commits an offence contrary to PSA 2016, s 8(1) if:

(a) (s)he intentionally imports a substance;
(b) the substance is a psychoactive substance;
(c) (s)he knows or suspects, or ought to know or suspect, that the substance is a psychoactive substance; and
(d) (s)he intends to consume the psychoactive substance for its psychoactive effects, or knows, or is reckless as to whether, the psychoactive substance is likely to be consumed by some other person for its psychoactive effects.

A person commits an offence contrary to s 8(2) if (s)he intentionally exports a substance, and (b), (c) and (d) just mentioned are satisfied.

Where a person imports or exports a controlled drug suspecting it to be a psychoactive substance, (s)he is treated for the purposes of s 8 as if (s)he had imported or exported a psychoactive substance suspecting it to be such a substance. 'Controlled drug' has the same meaning as in the Misuse of Drugs Act 1971 (**20.1**).

Exceptions to above offences

20.48   PSA 2016, s 11 provides that it is not an offence for a person to carry on any listed activity if, in the circumstances in which it is carried on by him, the activity is an exempted activity.

An 'exempted activity' means an activity listed in Sch 2, ie any activity carried on:

(a) by a person who is a health care professional and is acting in the course of his/her profession;
(b) for the purpose of, or in connection with the supply to, or the consumption by, any person of a substance prescribed for that person by a health care professional acting in the course of his/her profession or in accordance with the directions of a health care professional acting in the course of his/her profession;
(c) in respect of an active substance by a person who is registered in accordance with the Human Medicines Regulations 2012, reg 45N, or is exempt from any requirement to be so registered;
(d) in the course of, or in connection with, approved scientific research.

Trial and punishment

20.49   The offences under PSA 2016, ss 4, 5, 7, and 8 are indictable (either-way) offences, punishable with up to seven years' imprisonment.

Powers for dealing with prohibited activities

20.50   PSA 2016, ss 13 to 35 contain powers for dealing with prohibited activities in respect of psychoactive substances. These are powers for a 'senior officer' or a local authority (both defined at **20.52**) to give prohibition notices or premises notices and for a court to make prohibition orders or premises orders.

Meaning of 'prohibited activity'

The following are a 'prohibited activity':   **20.51**

(a) producing a psychoactive substance that is likely to be consumed by individuals for its psychoactive effects; or
(b) supplying, offering to supply, importing, or exporting such a substance; or
(c) assisting or encouraging the carrying on of a prohibited activity referred to in (a) or (b),

unless the carrying on of that activity would not be an offence under PSA 2016 by virtue of s 11 (**20.48**).

Prohibition notices (s 13)

A prohibition notice is a notice requiring the person (P) to whom it is given not to   **20.52**
carry on any prohibited activity or a prohibited activity of a description specified in the notice. A prohibition notice may not be given to someone under 10.

A senior officer or a local authority may give a prohibition notice if the senior officer or local authority reasonably believes that:

(a) P is carrying on, or is likely to carry on, a prohibited activity; or
(b) it is necessary and proportionate to give the notice to prevent P from carrying on any prohibited activity.

In PSA 2016, 'senior officer' means a constable who is an inspector (or above), a designated NCA officer of grade 3 (or above) or a general customs official of the grade of higher officer (or above), and 'local authority' means a county council, district council, or the like.

A notice given to someone under 18 must specify the period for which it has effect, and may not have effect for more than three years.

Premises notices (s 14)

A premises notice is a notice requiring the person (P) to whom it is given to take all rea-   **20.53**
sonable steps to prevent any prohibited activity, or a prohibited activity of a description specified in the notice, from being carried on at any premises specified in the notice that are owned, leased, occupied, controlled or operated by P.

For the purposes of s 14 (and ss 20 and 22 (**20.58** and **20.60**)), a person (other than a mortgagee not in possession) 'owns' premises if (s)he is the freeholder or the lessor of them under a lease of three years or more.

'Premises' in PSA 2016 has the same meaning as in PACE (**4.59**).

A premises notice may not be given to an individual who is under 18.

A senior officer or a local authority may give a premises notice to P if:

(a) (i) the senior officer or local authority reasonably believes that a prohibited activity is being, or is likely to be, carried on at particular premises, and
    (ii) P owns, leases, occupies, controls, or operates the premises; and
(b) that officer or authority reasonably believes that it is necessary and proportionate to give the premises notice to prevent any prohibited activity from being carried on at any premises owned, leased, occupied, controlled, or operated by P.

General points about prohibition notices and premises notices

20.54 A notice must set out the grounds for giving it, and explain the possible consequences of not complying with the notice.

A prohibition or premises notice, or notice that it has been withdrawn, takes effect when it is given. It may be given to a person (P) by handing it to him/her, leaving it at P's proper address, posting it to P at that address, or (in specified circumstances) sending it to P by electronic means.

Prohibition orders

20.55 A prohibition order is an order prohibiting the person (P) against whom it is made from carrying on any prohibited activity or a prohibited activity of a description specified in the order (s 17). A prohibition order may be made on application to a court, or by a court dealing with someone following conviction of an offence under any of PSA 2016, ss 4, 5, 7, and 8 (**20.44–20.47**) or a related offence.

Prohibition orders on application (s 18)

20.56 An adult magistrates' court (youth court if P is under 18) may make a prohibition order against P if:

(a) the court is satisfied on the balance of probabilities that P has failed to comply with a prohibition notice; *or*

(b) where no prohibition notice has been given (or one has been withdrawn), the court is satisfied on the balance of probabilities that P is carrying on, or is likely to carry on, a prohibited activity, and considers that P would fail to comply with a prohibition notice; *and in either case*

(c) the court considers it necessary and proportionate to make the order to prevent P from carrying on any prohibited activity.

If a prohibition order is based on (a) having been met, the prohibition notice is deemed withdrawn.

A prohibition order under s 18 may not be made against someone under 10. An order against someone under 18 at the time of the order must specify the period for which it has effect, and may not take effect for more than three years.

Prohibition orders following conviction (s 19)

20.57 Where a court is dealing with a person who has been convicted of a relevant offence, it may add a prohibition order to the sentence if it considers that it is necessary and proportionate to prevent him/her from carrying on any prohibited activity.

If a court makes a prohibition order under s 19, any prohibition notice given to the person against whom the order is made is deemed withdrawn. A s 19 prohibition order against someone under 18 must specify its period of effect (maximum three years).

In s 19, 'relevant offence' means an offence under any of ss 4, 5, 7, or 8 (**20.44–20.47**), an offence of being an accomplice to such an offence, or an offence of encouraging or assisting, conspiring, or attempting, to commit such an offence.

Premises orders (s 20)

20.58 A premises order is an order against an adult that requires the person (P) against whom it is made to take all reasonable steps to prevent any prohibited activity, or a prohibited activity of a description specified in the order, from being carried on at any premises specified in the order that are owned (**20.53**), leased, occupied, controlled, or operated by P. A magistrates' court may make a premises order against a person if:

(a) the court is satisfied on the balance of probabilities that P has failed to comply with a premises notice; *or*

(b) where no premises notice has been given (or one has been withdrawn):

    (i) the court is satisfied on the balance of probabilities that a prohibited activity is being, or is likely to be, carried on at particular premises,

    (ii) P owns, leases, occupies, controls, or operates the premises, and

    (iii) the court considers that P would fail to comply with a premises notice if given; and in either case

(c) the court considers it necessary and proportionate to make the premises order for the purpose of preventing any prohibited activity from being carried on at any premises owned, leased, occupied, controlled, or operated by P.

If a court makes a premises order based on (a) having been met, the premises notice is deemed withdrawn.

Applications for prohibition orders and premises orders

20.59 An application for a prohibition order under PSA 2016, s 18 or a premises order may be made only by the chief officer of police for a police area, by the chief constable of the British Transport Police Force, by the Director General of the NCA, by the Secretary of State by whom general customs functions are exercisable, or by a local authority. However, where an application is made based on a failure to comply with a prohibition notice or a premises notice (as the case may be) given by a senior officer, the application must be made by the chief officer etc, Director General, Secretary of State, or local authority of whose force, organisation, etc the senior officer is a member.

Provision that may be made by a prohibition order or premises order

20.60 Under PSA 2016, s 22, a court making a prohibition order or a premises order, or varying such an order, may impose any prohibitions, restrictions, or requirements that it considers appropriate (in addition to the prohibition referred to in s 17 or the requirement referred to in s 20, as the case may be).

The prohibitions that may be imposed on a person by a prohibition order or a premises order include, for example, an 'access prohibition', ie a prohibition prohibiting access to premises owned (**20.53**), occupied, leased, controlled, or operated by the person for a 'specified' period (ie specified in the order) not exceeding three months (or six months if subsequently extended by the court). An access prohibition may prohibit access:

(a) by all persons, or by all persons except those specified, or by all persons except those of a specified description;

(b) at all times, or at all times except those specified;

(c) in all circumstances, or in all circumstances except those specified.

By s 23, an authorised person (a constable, a person authorised by the chief officer of police or the BTP chief constable, a general customs official, or a person authorised by the Director General of the NCA, by a local authority or by the Secretary of State with a customs function) may enter premises in respect of which an access prohibition is in effect, and do anything necessary to secure the premises against entry; reasonable force may be used.

Someone seeking to enter premises under the above power must, if required to do so by the occupier of the premises or by another person appearing to be in charge of the premises in the occupier's absence, give his/her name, and (if not a uniformed constable) produce documentary evidence that (s)he is an authorised person. An authorised person may also enter premises in respect of which an access prohibition is in effect to carry out essential maintenance or repairs to the premises.

Offence of failing to comply with a prohibition order or premises order

20.61   A person (D) against whom a prohibition order or a premises order is made commits an indictable (either-way) offence contrary to PSA 2016, s 26(1) by failing to comply with the order. D does not commit the offence if D took all reasonable steps to comply with the order, or there is some other reasonable excuse for the failure to comply (s 26(3)). The offence is punishable with up to two years' imprisonment.

Offence of failing to comply with an access prohibition, etc

20.62   Where a prohibition order or a premises order imposes an access prohibition:

(a) a person, other than the person against whom the order was made, who without reasonable excuse remains on or enters premises in contravention of the prohibition commits an offence contrary to PSA 2016, s 27(1) and (2);

(b) a person who without reasonable excuse obstructs a person enforcing (under s 23) the prohibition commits an offence contrary to s 27(1) and (3),

An offence under s 27 is a summary offence, punishable with up to six months' imprisonment and/or an unlimited fine.

Variation and discharge

20.63   A prohibition order or a premises order may be varied or discharged by a court, normally the one which made the order.

Nature of proceedings

20.64   Proceedings before a court for a prohibition order or premises order (or its variation or discharge), including those relating to a prohibition order following conviction, are civil proceedings, and the standard of proof to be applied by the court in the proceedings is the balance of probabilities (s 32(1) and (2)). The court is not restricted to considering evidence that would have been admissible in the criminal proceedings in which the person concerned was convicted (s 32(3)). See further **13.47**.

Special measures

20.65   The Youth Justice and Criminal Evidence Act 1999 (YJCEA 1999), ss 16 to 33C (special measures directions in the case of vulnerable and intimidated witnesses: (**8.20–8.23**) applies with necessary modifications to proceedings referred to in 20.64 as it applies to criminal proceedings.

Enforcement powers

Power to stop and search persons

20.66   By PSA 2015, s 36, a constable, designated NCA officer or general customs official who has reasonable grounds to suspect that a person has committed, or is likely to commit, an offence under any of PSA 2016, ss 4, 5, 7, 8 (**20.44–20.47**), 9 (possession in custodial institution), or 26 (**20.61**) may search the person (P) for evidence that such an offence has been committed, and stop and detain P for that purpose. These powers may be exercised in any place to which the constable etc lawfully has access (even if the public does not).

Power to enter and search vehicles

PSA 2016, s 37 provides that a constable, general customs official or designated NCA **20.67** officer (C) who has reasonable grounds to suspect that there is evidence in a vehicle that an offence under any of PSA 2016, ss 4, 5, 7 to 9, or 26 has been committed may at any time enter the vehicle and search it for such evidence, and for this purpose may stop and detain the vehicle. This power does not apply if the vehicle is a dwelling.

Where a vehicle has been stopped under this power, and C considers that it would be impracticable to search the vehicle in the place where it has stopped, C may require the vehicle to be taken to such place as C directs to be searched. In addition, C may require anyone travelling in a vehicle, or its registered keeper, to afford such facilities and assistance with respect to matters under that person's control as C considers would facilitate the exercise of any of the above powers.

These powers under s 37 may be exercised in any place to which C lawfully has access (even if the public does not).

In s 37, 'vehicle' does not include any vessel or aircraft.

Section 38 contains similar powers in respect of boarding and searching vessels and aircraft.

Search warrant

PSA 2016, s 39 provides that, where a justice is satisfied that the requirements below are **20.68** met in relation to any premises (as defined at **4.59**), (s)he may issue a search warrant authorising *a relevant enforcement officer* (a constable, designated NCA officer, general customs official or local authority officer) to enter the premises, and to search them for 'relevant evidence' (evidence that an offence under any of PSA 2016, ss 4, 5, 7 to 9, or 26 has been committed).

As in the case of a warrant under PACE (**4.67**), such a warrant may be either a 'specific premises warrant' or an 'all-premises warrant'.

The requirements referred to above are met in relation to premises if there are reasonable grounds to suspect that: there are items on the premises that are relevant evidence, and, in a case where the premises are specified in the application, any of the following conditions is met:

(a) that it is not practicable to communicate with any person entitled to grant entry to the premises;
(b) that it is not practicable to communicate with any person entitled to grant access to the items;
(c) that entry to the premises is unlikely to be granted unless a warrant is produced;
(d) that the purpose of entry may be frustrated or seriously prejudiced unless a relevant enforcement officer arriving at the premises can secure immediate entry to them.

Section 40 and Sch 3 contain provisions about applications for a search warrant under PSA 2016 which correspond with the provisions of PACE, ss 15 and 16 (**4.74–4.76**). The only notable differences are that entry and search must be within one month from its date of issue, that it refers to a 'relevant enforcement officer' (ie a 'senior officer' as defined at **20.52**) wherever those sections refer to a constable, and that the reference to an inspector (or above) in PACE, s 16 is replaced by a reference to a 'relevant enforcement officer'. As a result of this correspondence, for example, s 40 provides that the warrant may authorise persons to accompany any relevant enforcement officer who executes it.

Powers of examination, etc

20.69   Where a relevant enforcement officer is exercising a power of search conferred by s 37, 38, or 39 in relation to any premises, (s)he has power under PSA 2016, s 41 to examine anything that is in or on the premises, and to measure or test (including taking a sample from a live plant) anything which (s)he has such power to examine.

For the purpose of exercising a power of search conferred by s 37, 38 or 39, or any power conferred by s 41, the officer may, so far as is reasonably necessary for that purpose, break open any container or other locked thing.

The officer may require any person in or on the premises to afford such facilities and assistance with respect to matters under that person's control as the officer considers would facilitate the exercise of a power of search conferred by s 37, 38, or 39, or any power conferred by s 41.

Power to require production of documents, etc

20.70   Where a relevant enforcement officer is exercising a power of search of a vehicle or premises conferred by PSA 2016, s 37, 38, or 39 in relation to any premises, (s)he has power under s 42 to require any person in or on the premises to produce any document or record that is in the person's possession or control. 'Produce any document' includes a reference to the production of a hard copy of information recorded otherwise than in hard copy form, or information in a form from which a hard copy can be readily obtained.

Powers of seizure, etc

20.71   These are dealt with by PSA 2016, s 43 as follows.

A constable, general customs official or designated NCA officer who is exercising the power of search of a person conferred by PSA 2016, s 36 may seize and detain anything found in the course of the search. Where a relevant enforcement officer is exercising a power of search under s 37, 38 or 39 in relation to any premises, or is otherwise lawfully on premises, (s)he may seize and detain or remove any item found on the premises, and take copies of or extracts from any document or record found on them.

A relevant enforcement officer to whom any document or record has been produced in accordance with a requirement under s 42 may seize and detain or remove that document or record, and take copies of or extracts from it.

These powers under s 43 may only be exercised for the purposes of determining whether an offence under any of ss 4, 5, 7, 8 (**20.44–20.47**), 9 (possession in a custodial institution), or 26 (**20.61**) has been committed, or in relation to an item which a relevant enforcement officer reasonably believes to be (a) evidence that such an offence has been committed, or (b) a psychoactive substance (whether or not it is relevant evidence).

There is no power under s 43 to seize an item which is an excluded item, ie an item subject to legal privilege, excluded material, and special procedure material (all of which are defined at **4.68**) (s 44).

Further provision about seizure under s 43

20.72   Where:

(a) any items which a relevant enforcement officer wishes to seize and remove are in a container; and

(b) the officer reasonably considers that it would facilitate the seizure and removal of the items if they remained in the container for that purpose, any power to seize and remove the items conferred by s 43 includes power under s 45 to seize and remove the container.

If a container is seized under s 45, reasonable efforts must be made to return it to the person from whom it was seized, or (if different) a person to whom it belongs, unless the container appears to be of negligible value, or if it is not practicable for the container to be returned, or while the container is or may be needed for use as evidence at a criminal trial.

If, in the opinion of a relevant enforcement officer (O), it is not for the time being practicable for O to seize and remove any item, O may require:

(a) the person from whom the item is being seized; or
(b) where O is exercising a power of search conferred by s 37, 38, or 39 in relation to any premises, any person in or on the premises,

to secure that the item is not removed or otherwise interfered with until such time as O may seize and remove it.

Where a relevant enforcement officer, or a person accompanying a relevant enforcement officer, seizes any item under s 43, s 46 provides as follows. When the item is seized, the officer (O) must make reasonable efforts to give written notice to each of the following persons:

(a) in the case of an item seized from a person, the person from whom it was seized;
(b) in the case of an item seized from premises, any person who appears to O to be their occupier or otherwise to be in charge of them;
(c) if O thinks that the item may belong to any person not falling within (a) or (b), that other person.

If the item is seized from premises, and at that time it is not reasonably practicable to give a notice to any affected person (ie a person falling within any of (a) to (c)), O must leave a copy of the notice in a prominent place on the premises. The notice must state what has been seized and the reason, specify any offence which O believes has been committed, and explain the effect of ss 49 to 51 and 53 (**20.75–20.78**). O must make a record of what has been seized. If a person appearing to a relevant enforcement officer to be an affected person asks for a copy of that record, the officer must, within a reasonable time, provide a copy of that record to that person.

Powers of entry, search, and seizure: supplementary provisions

A relevant enforcement officer (or a person authorised under s 40 to accompany him/her) may use reasonable force, if necessary, for the purpose of exercising any power conferred by ss 36 to 45 (**20.66–20.72**) (or ss 39 to 45) respectively (s 47). **20.73**

Offences in relation to enforcement officers

A person commits an offence contrary to PSA 2016, s 48(1) if, without reasonable excuse, (s)he intentionally obstructs a relevant enforcement officer in the performance of any of the officer's functions under ss 36 to 45. **20.74**

A person commits an offence contrary to s 48(2) if:

(a) (s)he fails without reasonable excuse to comply with a requirement reasonably made, or a direction reasonably given, by a relevant enforcement officer in the exercise of any power conferred by ss 37 to 45; or
(b) (s)he prevents any other person from complying with any such requirement or direction.

The above references to a relevant enforcement officer include a reference to a person authorised under s 40 (**20.68**) to accompany a relevant enforcement officer.

Both offences are summary offences punishable with up to six months' imprisonment and/or an unlimited fine.

Retention and disposal of items

Retention of seized items

20.75 By PSA 2016, s 49, an item seized under s 43 (**20.71**) may be retained so long as is necessary in all the circumstances. In particular, it may be retained for use as evidence at a trial for an offence under PSA 2016, or for forensic examination or for investigation in connection with an offence under PSA 2016, unless a photograph or a copy would suffice.

Power of police, etc to dispose of seized psychoactive substances

20.76 PSA 2016, s 50 provides that, if:

(a) a constable, general customs official or designated NCA officer (C) has seized an item found during the course of a search under s 36, 37, or 38;

(b) the search was carried out in a place to which C lawfully had access without a warrant;

(c) C reasonably believes that the item (i) is a psychoactive substance which, if it had not been seized, was likely to be consumed by an individual for its psychoactive effects, but (ii) is not evidence of any offence under this Act, and

(d) C has no reason to believe that, at the time of the seizure, the item was being used for the purposes of, or in connection with, an 'exempted activity' (see **20.48**) carried on by the person from whom it was seized or, (if different) any person to whom it belongs,

C may dispose of the item in whatever way C thinks suitable.

Forfeiture of seized items by court on application

20.77 A relevant enforcement officer may apply under PSA 2016, s 51 to an adult magistrates' court (or youth court where the individual to whom the application relates is under 18) for the forfeiture of an item retained under s 49. See, further, Chapter 20 on the companion website.

Return of item to person entitled to it, or disposal if return impracticable

20.78 PSA 2016, s 53 deals with this.

Where the retention of an item has been, but is no longer, authorised under PSA 2016, the item must normally be returned to a person entitled to it, and a magistrates' court (or youth court, as the case may be) must, if asked to do so by a person entitled to the item, order it to be returned to him/her. See, further, Chapter 20 on the companion website.

DRUG DEALING TELECOMMUNICATIONS RESTRICTION ORDERS

Introduction

20.79 The provisions about drug dealing telecommunications restriction orders (DDTRO) respond to an operational requirement of the police and NCA to support them in tackling the issue of county lines drug dealing and its related violence and criminal exploitation. 'County lines' refers to urban gangs supplying drugs to suburban areas and market and coastal towns, using dedicated anonymous mobile phone lines. These gangs exploit children and engage them in trafficking, and exploit vulnerable adults; such children and adults are then at high risk of extreme physical and sexual violence, and gang recriminations. The phone is the way drug dealers communicate with their addicts and run operations. The phone is kept well away from street-level drug dealing, in, as it were, the headquarters of the drug gang. Where possible, the police pursue criminal prosecutions, but they do not always have the evidence to do so. The provisions about DDTROs are targeted at those cases where there is insufficient evidence for prosecution but it is necessary to disrupt the criminal activity.

The Serious Crime Act 2015, s 80A empowers the making of regulations about DDTROs. It defines a DDTRO as an order requiring a communications provider to take whatever action the order specifies for the purpose of preventing or restricting the use of communication devices in connection with drug dealing offences. For these purposes:

A 'communications provider' means a provider of a telecommunications service.

A 'communications device' is a device capable of transmitting or receiving images, sounds, or information by electronic communications (including a mobile telephone), a component part of such a device, or an article designed or adapted for use with such a device (including any disk, film, or other separate article on which images, sounds, or information may be recorded).

'Drug dealing offence' means an offence under MDA 1971, s 4(3) (**20.6**) or PSA 2016, s 5 (**20.45**), and a communication device is used in connection with such an offence if it is used by a person ('the user') in the course of:

(a) the user committing a drug dealing offence;

(b) the user facilitating the commission by the user or another person of a drug dealing offence; or

(c) conduct of the user that is likely to facilitate the commission by the user or another person of a drug dealing offence (whether or not an offence is committed).

Drug Dealing Telecommunications Restriction Orders Regulations 2017

Drug Dealing Telecommunications Restriction Orders Regulations 2017 provide as follows. **20.80**

Application for a DDTRO

A DDTRO may be made only on an application to a county court by (a) the Director General or Deputy Director General of the NCA, or (b) a police superintendent (or above). The application must be made without notice of the application or hearing having been given to a person (P) who would be likely to be affected by the order (if made) or P's legal representative, and must be heard and determined in the absence of P or P's legal representative. The hearing of an application must be held and determined in private. **20.81**

Power to make a DDTRO

A county court may make a DDTRO if the court: **20.82**

(a) is satisfied that a relevant item identified in the order—
   (i) has been, or is likely to have been, used; or
   (ii) is likely to be used in the future,
   in connection with drug dealing offences; and

(b) it has reasonable grounds to believe that the order would prevent or restrict the use of a communication device in connection with drug dealing offences.

A 'relevant item' is:

(a) a communication device; or
(b) a phone number or something else that may be used with a communication device (including a subscriber identity module (SIM) or device ID). 'Device ID' includes an International Mobile Equipment Identity (IMEI) number, a Mobile Equipment Identifier (MEID) number, an Electronic Serial Number (ESN), a Mobile Station International Subscriber Directory Number (MSISDN), an Integrated Circuit Card Identifier (ICCID), an International Mobile Subscriber Identity (IMSI), a mobile identification number (MIN), or a mobile subscription identification number (MSIN), a media access control address (MAC), a Universal Device Identifier (UDID), an Android identifier number (Android_ID), or any other serial number or identification number, used with a communications device.

A DDTRO may provide for the order, or any specified requirements of it, not to apply in relation to any relevant item that the applicant (ie the applicant for the DDTRO) discovers is not being used in connection with drug dealing offences.

A DDTRO has effect until the date, or the end of the period, that the order specifies (if any), or, if no date or period is specified, until discharged by the court.

A DDTRO must specify the date and time on or before which the requirements of the order are to be complied with, which must be the end of the fifth working day (ie a day other than Saturday, Sunday, Christmas Day, Good Friday, or a bank holiday) after the date of the order unless the applicant requests that the court specify a different date and time and the court agrees that the request is reasonable.

Notice and information to be given by the applicant

20.83   The applicant must, as soon as reasonably practicable after a DDTRO has been made by the Court:

(a) serve a copy of the DDTRO on the communications provider; and
(b) take reasonable steps to bring the making of the DDTRO to the attention of any other affected person (ie a person who has been or is likely to be affected by the order) of whom the applicant is, or ought reasonably to be, aware.

Where:

(a) a DDTRO contains provision made by virtue of the italicised provision in **20.82**; and
(b) the applicant discovers that a relevant item in relation to which the order, or any particular requirement of it, would apply (but for that provision) is not being used in connection with drug dealing offences,

the applicant must comply with the following requirements:

(i) the applicant must as soon as reasonably practicable notify the communications provider of the discovery and inform the provider in writing that, accordingly, the order or provision does not apply in relation to that relevant item; and
(ii) where the communications provider has already complied with the terms of the DDTRO, the applicant must as soon as reasonably practicable notify any other affected person of the discovery and inform that person in writing that, accordingly, the order or provision does not apply in relation to that relevant item.

Information to be given by a communications provider

A communications provider required to comply with a DDTRO must notify the appli- **20.84**
cant and the Office of Communications (OFCOM) as soon as is reasonably practicable
once the requirements of the order have been complied with.

Costs of complying with a DDTRO

Where a communications provider is likely to incur (or has incurred) costs in com- **20.85**
plying with a DDTRO, the order (or a further order) may include a requirement for
the applicant to pay any or all of the communications provider's costs that the court
considers reasonable.

Variation or discharge

The court may discharge an order made under the Regulations, or extend or otherwise **20.86**
vary it, on the application of the applicant for the DDTRO or an affected person (ie a
person who has been or is likely to be affected by the order). Such an application must
set out the grounds on which it is made and (in the case of an application for extension
or variation) the terms of the extension or variation sought. Where an application is to
be made by an affected person who is a communications provider, notice in writing of
the intention to apply must be served on the applicant at least 24 hours before the application is made. Where an application is to be made by an affected person who is not
a communications provider:

(a) notice in writing of the intention to apply must be served on the applicant and any
communications provider required to comply with the DDTRO at least 24 hours
before the application is made; and

(b) both the applicant and any communications provider required to comply with the
DDTRO must be named as parties to the application.

ALCOHOL AND LICENSING

Licensing system

The Licensing Act 2003 (LA 2003) introduced a system of licensing operated by li- **20.87**
censing authorities, essentially local authorities. The 'licensable activities' to which the
provisions of the Act apply are not restricted to matters involving the sale or supply of
alcohol; they extend to the control of various forms of entertainment and to the provision of late-night refreshment.

The key to understanding the regime under LA 2003 is that a *licensable activity*
(**20.88**) must be carried on (a) under and in accordance with a premises licence, or
(b) under circumstances where it is carried out in accordance with a temporary event
notice (TEN). In addition, a personal licence is required by a person selling alcohol
by retail (or supplying it by, or on behalf of, a club to, or to the order of, a member of
the club) on premises which have a premises licence. A *qualifying club activity* (**20.89**)
must be carried out in accordance with a club premises certificate.

On certain occasions it is an offence under LA 2003 intentionally to obstruct an
authorised person (ie an officer of a licensing authority or a person with a health, safety
or environmental role appointed or authorised by certain bodies). Such a person shares
powers with a constable in relation to the enforcement of many provisions of the Act.
On each of these occasions it will also, of course, be an offence against the Police Act
1996, s 89(2) wilfully to obstruct a constable in the execution of his/her duties.

By LA 2003, s 4, a licensing authority must carry out its functions under LA 2003 with a view to promoting the licensing objectives under that section:

(a) the prevention of crime and disorder ('and' has been held to mean 'and/or');
(b) public safety;
(c) the prevention of public nuisance; and
(d) the protection of children from harm.

Terminology

Licensable activities

20.88 By LA 2003, s 1(1), the following are licensable activities:

(a) sale by retail of alcohol;
(b) supply of alcohol by or on behalf of a club to, or to the order of, a member of the club, otherwise than a sale by retail;
(c) provision of regulated entertainment; and
(d) provision of late-night refreshment.

The reference to 'sale by retail' in (a) means that 'business-to-business' wholesale sales are outside the provisions of the Act. The reference to 'otherwise than a sale by retail' in para (b) of this definition can be explained as follows. In the case of a members' club (as opposed to a proprietary one) the property of the club belongs to its members for the time being jointly in equal shares. Consequently if alcohol is supplied to a member at a price, this is not a sale by retail but a release by the other members of their interest in the drink supplied. This would not be covered by para (a), although a supply at a price to a guest of a member of the club would be.

Section 173 excludes activities which would otherwise be licensable activities from being so if they are carried on in particular places, for example aboard an aircraft, hovercraft or train on a journey, at an approved wharf at a designated port or hoverport, on board a vessel on an international journey, or at an examination station at a designated international airport (area beyond security check-in). Premises permanently or temporarily used for the purpose of the armed forces are also exempted.

Section 175 provides that the promotion of an incidental lottery will not constitute a licensable activity by reason only of one or more of the prizes in the lottery consisting of or including alcohol, provided that the alcohol is in a sealed container.

Qualifying club activities

20.89 By LA 2003, s 1(2), the following licensable activities are also qualifying club activities:

(a) supply, otherwise than a sale by way of retail, of alcohol by or on behalf of a club to, or to the order of, a member of the club;
(b) sale by retail of alcohol by or on behalf of a club to a guest of a member of the club for consumption on the premises where the sale takes place; and
(c) provision of regulated entertainment where that provision is by or on behalf of a club for members of the club or members of the club and their guests.

Alcohol

20.90 'Alcohol' in LA 2003 means spirits, beer, wine, cider, or any other fermented, distilled, or spirituous liquor in any state, except alcohol with a strength not exceeding 0.5 per cent and obvious exceptions like perfume, medicinal, or veterinary products, and liqueur confectionery.

Regulated entertainment

LA 2003, Sch 1 provides that 'the provision of regulated entertainment' means 'entertainment' of a type described below *where the following two conditions are satisfied*:

(a) that the entertainment is provided:
  (i) to any extent for members of the public or a section of the public;
  (ii) exclusively for members of a club (or members and their guests) which is a qualifying club in relation to the provision of regulated entertainment; or
  (iii) in any case not falling within (i) or (ii), for consideration and with a view to profit;
(b) that the premises on which the entertainment is provided are made available for the purpose, or for purposes which include the purpose, of enabling the entertainment concerned.

The *descriptions of 'entertainment'* are:

(a) a performance of a play;
(b) an exhibition of a film;
(c) an indoor sporting event ('sporting event' means any contest, exhibition, or display of any sport other than a boxing or wrestling entertainment);
(d) a boxing or wrestling entertainment (ie any contest, exhibition or display of boxing or wrestling, or which combines boxing or wrestling with one or more martial arts);
(e) a performance of live music;
(f) any playing of recorded music;
(g) a performance of dance; or
(h) entertainment of a similar description to that falling within (e), (f), or (g),
  where the following two conditions are satisfied (so far as relevant):

The first condition is that the entertainment takes place in the presence of an audience, and is provided for the purpose, or for purposes which include the purpose, of entertaining that audience.

The second condition is relevant only to a performance of a play or dance, or an indoor sporting event, and is that one or more of the following applies:

(i) the audience consists of more than 500 persons (play or dance) or 1,000 (indoor sporting event);
(ii) the entertainment takes place before 08.00 on any day;
(iii) the entertainment takes place after 23.00 on any day; or
(iv) (dance only) the entertainment is a *'relevant entertainment'* (defined as any live performance, or live display of nudity, whose sole or principal purpose, other than financial gain, is sexually stimulating any member of the audience).

So much of any entertainment of a description specified in paragraphs (a) to (h) above as does not satisfy the conditions in (i) to (iv) (so far as relevant) is not 'entertainment'.

Exemptions

The following are not regulated entertainments:

(a) playing of live TV and radio programmes;
(b) music or films incidental to an activity which is not itself 'entertainment', eg background music in a shop, or a piano in a restaurant;
(c) films used for product demonstration, advertisement, information, education, or instruction, or in an exhibition in a museum or art gallery;

(d) films at community premises if the following conditions (i) to (v) are satisfied:

  (i) prior written consent for the entertainment to take place at the community premises must have been obtained, by or on behalf of an organiser or manager of the above entertainment:

  (a) from the management committee of the community premises; or

  (b) where there is no management committee, from—

    (i) a person who has control of the community premises in connection with the carrying on by that person of a trade, business or other undertaking (for profit or not); or

    (ii) where there is no such person, an owner of the community premises;

  (ii) the entertainment must not be provided with a view to profit;

  (iii) the entertainment must take place in the presence of an audience of no more than 500;

  (iv) the entertainment must take place between 08.00 and 23.00 on the same day

  (v) the film classification body or the relevant licensing authority must have made a recommendation concerning the admission of children to an exhibition of the film and the admission of children is subject to such restrictions (if any) as are necessary to comply with the recommendation of that body or authority (or of that authority, if they have both made recommendations);

(e) entertainment incidental to a religious service or at a place of religious worship;

(f) entertainment at a garden fete;

(g) Morris dancing or dancing of similar type;

(h) entertainment provided on vehicles in motion;

(i) 'relevant entertainment' (**20.91**(iv)) at premises for which a sex establishment licence is required (or waived) by virtue of the Local Government (Miscellaneous Provisions) Act 1982, Sch 3, or would be required but for the infrequency of the relevant entertainment;

(j) any entertainment provided by or on behalf of health care providers, local authorities and school proprietors, if:

  (i) it takes place:

    (a) if it is provided by or on behalf of a health care provider, on any premises forming part of a hospital in which that provider has a relevant property interest (freeholder or lessor of a lease of three years or more), or which are lawfully occupied by that provider;

    (b) if it is provided by or on behalf of a local authority, on any premises in which that authority has a relevant property interest or which are lawfully occupied by that authority, and

    (c) if it is provided by or on behalf of a school proprietor, on the premises of the school;

  (ii) the premises are not domestic premises (ie premises occupied as a private dwelling, including any garden, yard, garage, outhouse, or other appurtenance of such premises whether or not used in common by the occupants of more than one such dwelling);

  (iii) the entertainment takes place between 08.00 and 23.00 on the same day (or, where a special occasion order (**20.100**) has effect, between the hours specified in that order); and

  (iv) the entertainment is not a 'relevant entertainment' for the purposes of the Local Government (Miscellaneous Provisions) Act 1982, Sch 3 (licensing of sex establishments);

(k) entertainment consisting of a performance of live music and/or the playing of recorded music if the following conditions are satisfied:

   (i) that the entertainment takes place at:

      (a) community premises that are not authorised, by a premises licence or club
premises certificate, to be used for the supply of alcohol for consumption
on the premises;

      (b) the premises of a hospital or of a school; or

      (c) premises in which a local authority has a relevant property interest (as
above) or which are lawfully occupied by a local authority;

   (ii) that the premises are not domestic premises (as above);

   (iii) that the entertainment takes place in the presence of an audience of no more
than 500 persons;

   (iv) that the entertainment takes place between 08.00 and 23.00 on the same day
(or, where a special occasion order has effect, between the hours specified in
that order); and

   (v) that a person concerned in the organisation or management of the entertainment has obtained the prior written consent of a relevant person for the entertainment to take place.

Where the entertainment takes place at community premises, a 'relevant person'
means the management committee of the premises, or, if there is none, a person
who has control of the premises in connection with the carrying on by that person
of a trade, business or other undertaking (for profit or not) or (in the absence of
such a person) a person with a relevant property interest in the premises. Where
the entertainment takes place at the premises of a hospital, it means a health care
provider which has a relevant property interest in or lawfully occupies those premises. Where the entertainment takes place at premises in which a local authority
has a relevant property interest or which are lawfully occupied by a local authority,
it means that authority. Where the entertainment takes place at school premises it
means the school proprietor;

(l) live music or recorded music on premises authorised to be used for the supply of
alcohol for consumption on the premises by a premises licence or club premises
certificate, provided:

   (i) the premises are then open for the supply of alcohol for consumption on the
premises,

   (ii) if the music is amplified, it takes place in the presence of an audience of no
more than 500 persons,

   (iii) the music takes place between 08.00 and 23.00 on the same day (or, where a
special occasion order has effect, between the hours specified in that order), and

   (iv) conditions have not been included in the licence or certificate by virtue of LA
2003, s 177A(3) or (4) (**20.126**);

(m) live music at a workplace provided:

   (i) the place is not licensed under LA 2003 (or is so licensed only for the provision
of late night refreshment),

   (ii) the audience does not exceed 500, and

   (iii) the live music takes place between 08.00 and 23.00 on the same day;

(n) subject to s 177A(3) and (4), live unamplified music which takes place between
08.00 and 23.00 on the same day;

(o) any entertainment that consists of or forms part of a performance by a travelling
circus, if:

(i)   the entertainment is not an exhibition of a film, or a boxing or wrestling entertainment (as defined at **20.91**(d));

(ii) it takes place between 08.00 and 23.00 on the same day;

(iii) it takes place wholly within a moveable structure, inside which the audience is also wholly accommodated; and

(iv) the travelling circus has not been located on the same site for more than 28 consecutive days;

(p) entertainment consisting of a boxing or wrestling entertainment, if it:

(i)   is a contest, exhibition or display of Greco-Roman wrestling, or of freestyle wrestling, between two participants (regardless of their sex);

(ii) takes place in the presence of no more than 1,000 spectators;

(iii) takes place between 08.00 and 23.00 on the same day; and

(iv) takes place wholly inside a building, inside which the spectators are wholly accommodated.

Late-night refreshment

20.93   LA 2003, Sch 2 provides that 'late-night refreshment' means the supply at any time between 23.00 and 05.00 of hot food or hot drink for consumption on or off the premises: (a) to members of the public (or a section of it), or (b) when members of the public (or a section of it) are admitted to any premises, to any persons (or to a person of a particular type) on or off those premises. 'Hot' for the purposes of the Act means food or drink which is heated on the premises or elsewhere or that which, after it is supplied, may be heated on the premises.

The following are not the provision of late-night refreshment for the purposes of LA 2003:

(a) supply to persons staying at hotels, camping sites or other premises supplying accommodation as their main purpose;

(b) supply to members of recognised clubs;

(c) supply to employees of particular employers (works canteens, etc);

(d) premises already licensed under other Acts, eg 'near beer' premises in London;

(e) supply free of charge provided no charge has been made for admission to the premises or some other item;

(f) supply by a registered charity;

(g) supply by means of a vending machine;

(h) supply on a vehicle not permanently or temporarily parked;

(i) supply on or from premises which are wholly situated in an area designated by the relevant licensing authority;

(j) supply on or from premises which are of a description designated by the relevant licensing authority;

(k) supply during a period (beginning no earlier than 11.00 and ending no later than 05.00) designated by the relevant licensing authority.

A licensing authority may designate a description of premises for the purposes of (j) only if the description is prescribed by the Licensing Act 2003 (Late Night Refreshment) Regulations 2015, ie motorway service areas; filling stations; licensed premises authorised to sell by retail alcohol for consumption on the premises between 23.00 and 05.00; and (with exceptions) local authority premises, school premises, hospital premises, or community premises.

Premises licences, club premises certificates, and temporary event notices

Premises licences

Premises licences are governed by LA 2003, Part 3 (ss 11 to 59).

20.94

A premises licence is granted by the licensing authority for the area and authorises its holder to use the premises to which the licence relates for licensable activities (**20.88**).

In addition to the general right to apply for a review of a premises licence, an involved process, a chief officer of police may apply for an expedited review of a premises licence. A chief officer may do so where a superintendent (or above) has certified that (s)he considers that premises which are licensed are associated with serious crime or disorder. A serious crime is one defined by the Regulation of Investigatory Powers Act 2000, s 81 (generally, those crimes for which a sentence of imprisonment for at least three years may be imposed and offences involving violence, resulting in substantial gain or involving a large number of persons in pursuit of a common purpose).

By s 59, a constable or authorised person (**20.87**) may enter the premises to which an application for a premises licence or provisional statement (about the effect of building works on the grant of a premises licence) relates, before that application has been determined, in order to assess the likely effect on the promotion of the licensing objectives of the grant of the application. An authorised person must produce evidence of his/her authority, if requested. It is a summary offence against s 59(5) intentionally to obstruct an authorised person exercising such a power. The offence is punishable with a fine not exceeding level 2.

A licence will be granted subject to conditions consistent with the operating schedule in the licensing application and with any mandatory conditions which are appropriate.

The premises licence-holder must ensure that the licence (or a certified copy) and a list of relevant mandatory conditions are kept on the premises in his/her custody or control or that of an employee there nominated in writing, and that the summary of the licence (or a certified copy) and the position of any nominee referred to above are prominently displayed on the premises. Non-compliance by the premises licence-holder, without reasonable excuse, is a summary offence under s 57(4). A constable or authorised person may require production of the premises licence (or a certified copy) or a list of relevant mandatory conditions. An authorised person must, on request, produce evidence of his/her authority. Non-compliance without reasonable excuse with such a request is a summary offence under s 57(7). An offence under s 57(4) or (7) is punishable with a fine not exceeding level 2.

Club premises certificates

These are dealt with by LA 2003, Part 4 (ss 60 to 97).

20.95

Club premises certificates will be granted to a qualifying club by the licensing authority for the area and will authorise the use of club premises for qualifying club activities (**20.89**) specified in the certificate.

A club may wish to provide entertainment to members of the public on certain occasions and, in consequence, it may hold a club premises certificate in respect of its day-to-day activities and a premises licence authorising the provision of entertainment.

A club premises certificate will be subject to conditions in the same way as a premises licence.

A club premises certificate may authorise the supply of alcohol to its *members* for consumption off the premises. In general, a club premises certificate:

(a) may not authorise the supply of alcohol for consumption off the premises unless it also authorises its supply to members for consumption on the premises;

(b) authorising the supply of alcohol for consumption off the premises must include three conditions as to such supply. The supply must be:

   (i) made at a time when the premises are open for the purposes of supplying alcohol, in accordance with the club premises certificate, to members of the club for consumption on the premises,

   (ii) in a sealed container, and

   (iii) to a member of the club in person.

Sale of alcohol by retail to a guest for such consumption is not authorised by the certificate.

No conditions attached to a club premises certificate may prevent the sale by retail of alcohol or the provision of regulated entertainment to members of an associated club or their guests if those are permitted activities.

A constable or authorised person (**20.87**) may require production of a certificate (or certified copy) or a list of relevant mandatory conditions. Failure to comply, without reasonable excuse, is a summary offence against s 94(9). The club secretary must ensure that the certificate (or certified copy) is held on the club's premises by a person nominated by him/her, and the nominated person must ensure that the summary of the certificate and a notice of the nominated individual responsible for it on the premises are prominently displayed. Failure to comply with one of these requirements, without reasonable excuse, is a summary offence against s 94(5) (failure by secretary to ensure holding) or (6) (failure by nominated person to ensure display). An offence under s 94(5), (6) or (9) is punishable with a fine not exceeding level 2.

By s 96, on production of his authority a constable authorised by the chief officer of police or an authorised person may enter and inspect premises in respect of which an application has been made for the grant of a certificate. 48 hours' notice must have been given to the club. It is a summary offence against s 96(5), punishable with a fine not exceeding level 2, to obstruct an authorised person exercising these powers.

By s 97, a constable may enter and search club premises (by force, if necessary) where (s)he has reasonable cause to believe that an offence in respect of the supply of controlled drugs or psychoactive substances has been, is being, or is about to be, committed or that there is likely to be a breach of the peace.

Temporary event notices

20.96 Sometimes, a person wishes to carry out licensable activities at premises which are not licensed, eg to provide bar facilities at a wedding reception on premises not licensed for the sale of alcohol, or to set up a disco constituting regulated entertainment in premises which, although licensed for the sale of alcohol, are not licensed for the provision of entertainment. This can be achieved under the provisions of LA 2003, Part 5 (ss 98 to 110) which deals with permitted temporary activities. Under Part 5, licensable activities can occur on premises which do not have a premises licence or club premises certificate on a temporary basis (for a period not exceeding 168 hours) subject to conditions and limitations.

A licensable activity is a 'permitted temporary activity' if:

(a) it is carried out in accordance with:

   (i) a temporary events notice (TEN) given, no later than 10 working days before the event starts and with the prescribed fee, to the relevant licensing authority; and

(ii) where relevant, any conditions imposed; and

(b) it satisfies the following conditions:

(i) the TEN has been duly acknowledged (if required: no requirement if licensing authority has issued counter-notice following police objection or if permitted limits already exceeded) by the licensing authority;

(ii) the TEN has not been withdrawn by the individual giving the notice; and

(iii) no counter-notice has been issued by the relevant licensing authority.

The individual who gives the TEN is called the 'premises user'.

The temporary event notice

A TEN must contain details including: **20.97**

(a) the licensable activities to be carried out;

(b) the date(s) when the licensable activities are to be carried out (and the total length of the event);

(c) the times during the event at which the licensable activities are to be carried out;

(d) the maximum number of people to be allowed on the premises at any one time, which must be less than 500;

(e) whether any alcohol sales are to be made for consumption on or off the premises (or both).

If the proposed licensable activities include the supply of alcohol, the notice must include the condition that all supplies will be made by, or under the authority of, the premises-user.

Temporary event notices: objections and counter-notices

Where a 'relevant person' (a chief officer of police or local authority with an envir-   **20.98** onmental or health function) considers that if the temporary event should proceed it would undermine a licensing objective, the premises-user, the relevant licensing authority, and every other relevant person must be informed by an objection notice, setting out the reasons, before the end of the second working day following the day on which the TEN was received. If there is such an objection there must be a hearing unless all parties agree that it is unnecessary. Where the licensing authority accepts the objection it must give the premises-user a counter-notice preventing the event from taking place; this must be done at least 24 hours before the proposed event. If the counter-notice is not given as required, the premises-user may proceed with the event.

Where there has been an objection by a relevant person, at any time before a hearing that person may, with the consent of the premises-user, modify the terms of the TEN. If this is done the notice takes effect with those modifications.

In addition to the case where an objection is accepted, a licensing authority must give a counter-notice in certain cases; the effect is to prevent the event taking place.

Temporary event notices: rights of entry, production of notice, etc

LA 2003, s 108 provides that a constable or an authorised officer of the licensing au-   **20.99** thority may, at any reasonable time, enter premises to assess the probable impact of the proposed event upon the crime prevention objective. An authorised officer must produce evidence of his/her authority, if requested. It is a summary offence intention- ally to obstruct such an officer (s 108(3)).

The premises-user must ensure that:

(a) a copy of the TEN together with any statement of conditions is prominently dis- played on the premises; or

(b) the TEN and such statement are kept there in his/her custody or that of a nominated person. Where the TEN and any statement of conditions are in the custody of a nominated person, the premises-user must ensure that a notice to that effect is prominently displayed on the premises.

It is a summary offence against s 109(4) for a premises-user, without reasonable excuse, to fail to comply with one or other of these requirements.

Where a TEN or any statement of conditions is not prominently displayed and there is no notice displayed as to its whereabouts, a constable or authorised officer may require the premises-user to produce the TEN or statement of conditions. Where a notice relating to a 'nominated person' is displayed on the premises a constable or authorised officer may require the nominated person to produce the actual notice or statement of conditions. It is a summary offence under s 109(8) for a person to fail, without reasonable excuse, to produce a TEN or statement of conditions when requested to do so.

An offence under s 108(3), 109(4) or (8) is punishable with a fine not exceeding level 2.

Opening hours of premises in respect of which there is a premises licence or club premises certificate

20.100   LA 2003 does not prescribe 'permitted hours' within which alcohol may be sold or supplied for consumption on or off the premises. In addition, there are no general restrictions placed upon any other licensable activity. An applicant for a premises licence or a club premises certificate may choose the hours within which it would like to be authorised to carry out the licensed activities. The licence will be granted on those terms unless, following representations, the authority considers it necessary to reject the application or alter its terms bearing in mind the licensing objectives set out at **20.87**.

The Secretary of State is empowered by LA 2003, s 172 to make a special occasion order to permit premises with a premises licence or a club premises certificate to open for specified, generally extended hours, on special occasions of international, national or local significance.

Early morning alcohol restriction orders

20.101   LA 2003, s 172A empowers a licensing authority to make an early morning alcohol restriction order if it considers it appropriate for the promotion of the licensing objectives.

An order will provide that premises licences and club premises certificates granted by the authority, and TENs given to the authority, do not have effect to the extent that they authorise the sale or supply of alcohol during the period specified in the order. The period that may be specified in the order must begin no earlier than 24.00 and end no later than 06.00. The order would, therefore, apply not only to pubs, bars, and nightclubs but also to non-profit clubs such as sports, political, and working men's clubs, supermarkets, and convenience stores, and temporary events.

The Licensing Act 2003 (Early Morning Restriction Orders) Regulations 2012 prescribe exceptions from the effect of an early morning alcohol restriction order. The prescribed exceptions are:

(a) premises which are a hotel, guest house, lodging house or hostel at which the supply of alcohol between 00.00 and 06.00 on any day may only be made to persons staying at the premises for consumption in their rooms (eg alcohol supplied by room service or by virtue of a mini-bar);

(b) premises which are authorised to supply alcohol for consumption on the premises between 00.00 and 06.00 only on 1 January in every year.

Thus, premises of these descriptions may continue to supply alcohol notwithstanding that they are situated in an area to which an early morning alcohol restriction order applies.

An order made under s 172A is subject to an order made under s 172 (special occasion order) unless the s 172 order provides otherwise.

Personal licences

Personal licences are dealt with by LA 2003, Part 6 (ss 111 to 135).

A personal licence is a licence granted by a licensing authority to an individual which authorises him/her to supply alcohol, or to authorise the supply of alcohol, in accordance with a premises licence. Supplying alcohol in this context means selling it by retail, or supplying it by or on behalf of a club to, or to the order of, a member of the club. On any premises with a premises licence authorising their use for the sale or supply of alcohol, the person nominated for the day-to-day running of the premises must hold a personal licence and is known as the 'designated premises supervisor'. There may be more than one personal licence-holder on the licensed premises, but it is not necessary for all members of staff to have personal licences. However, all supplies of alcohol under a premises licence must be made by or under the authority of a personal licence-holder.

A personal licence is one which relates solely to the supply of alcohol. A personal licence is not required in respect of the other activities regulated by LA 2003.

The licensing authority must grant a personal licence if it appears to it that:

(a) the applicant (A) is aged 18 or over;

(b) A is entitled to work in the UK;

(c) A possesses a recognised qualification or is a person of a prescribed description;

(d) no personal licence held by A has been forfeited in the five-year period ending with the date of application; and

(e) A has not been convicted of any 'relevant offence' (a long list of offences ranging from theft to possession of drugs) or any 'foreign offence' (an offence (other than a relevant offence) under the law of any country outside England and Wales) or required to pay an immigration penalty.

A licensing authority must reject an application if the applicant fails to meet the condition in (a), (b), (c), or (d).

If an applicant meets the conditions in (a), (b), (c), and (d), but fails to meet condition (e), the authority must notify the chief officer of police for its area. If the chief officer is satisfied, by reference to (i) any conviction for a relevant offence, or (ii) any conviction for a foreign offence which (s)he considers comparable to a relevant offence, or (iii) a requirement to pay an immigration penalty, that granting a personal licence to the applicant would undermine the crime prevention objective of LA 2003, (s)he must within 14 days give the authority an objection notice.

If the applicant meets conditions (a) to (d) but fails to meet condition (e) by virtue of having been convicted of a specified immigration offence or of a foreign offence comparable to an immigration offence, or required to pay an immigration penalty, the authority must inform the Secretary of State. Where, having regard to such a conviction or requirement, the Secretary of State is satisfied that granting the licence would be prejudicial to the prevention of illegal working in licensed premises, the Secretary of State must, within 14 days of being informed, give the authority the reasons for being so satisfied in an 'immigration objection notice'.

If an objection notice or immigration objection notice is not given in the requisite time, the authority must grant the personal licence. If such a notice is given in the requisite time, the authority must hold a hearing to consider the notice; having regard to the notice, the authority must:

(i) where the notice is an objection notice, reject the application if it considers it appropriate for the promotion of the crime prevention objective to do so; or

(ii) where the notice is an immigration objection notice, reject the application if it considers it appropriate to the prevention of illegal working in licensed premises to do so,

otherwise the application must be granted.

A personal licence specifies the holder's name and address, the authority which granted it, and convictions for relevant or foreign offences. The holder must notify the authority of any change in name or address. It is a summary offence, punishable with a fine not exceeding level 2, to fail, without reasonable excuse, to do so (s 127(4)).

The system of personal licences permits licensed persons to move from one set of premises to another, regardless of area.

Where the holder of a personal licence is on premises to sell or authorise the sale of alcohol, and such sales are (a) authorised by a premises licence, or (b) a permitted temporary activity on the premises by virtue of a TEN in respect of which (s)he is the premises user, (s)he may be required by a constable or an authorised officer of the licensing authority to produce his/her licence. Failure to comply with such a request, without reasonable excuse, is a summary offence against s 135(4). The offence is punishable with a fine not exceeding level 2.

Licensing offences

Unauthorised licensable activities

20.103 It is a summary offence against LA 2003, s 136(1)(a) for a person to carry on, or attempt to carry on, or against s 136(1)(b) for a person knowingly to allow to be carried on, a licensable activity on or from any premises otherwise than under or in accordance with an authorisation provided by a:

(a) premises licence;
(b) club premises certificate; or
(c) TEN.

An offence under s 136(1)(a) or (b) is punishable with imprisonment for up to six months and/or an unlimited fine.

A divisional court has held that the offence under s 136(1)(a) is directed at persons who *actually* carry on, or attempt to carry on, a licensable activity otherwise than in accordance with such authorisation. Thus, for example, where premises subject to a premises licence are used in breach of a condition under the premises licence the premises licence-holder is not automatically guilty of the offence. This is of obvious importance where (as in the case referred to) the premises licence-holder owns the licensed premises but lets them to someone else who actually carries out the licensable activity there and who breaches a condition of the licence unknown to the licence-holder.

Exposing alcohol for unauthorised sale

20.104 It is a summary offence against LA 2003, s 137(1) to expose, on any premises, alcohol for sale by retail in circumstances where the actual sale by retail of that alcohol would be an unauthorised licensable activity. As can be seen, a sale, or an attempted sale, is not necessary. The offence under s 137(1) is punishable in the same way as a s 136 offence.

Keeping alcohol on premises for unauthorised sale

LA 2003, s 138(1) creates the summary offence, punishable with a fine not exceeding **20.105** level 2, of possessing, or having under control, alcohol, with the intention of selling it by retail, or supplying it, where that sale or supply would be an unauthorised licensable activity.

Defence of due diligence

By LA 2003, s 139, it is a defence to an offence under s 136(1)(a), 137(1), or 138(1) that: **20.106**

(a) the defendant's act was due to a mistake, or to reliance on information given to him/her, or to an act or omission of another person, or to some cause beyond his/her control, and

(b) (s)he took all reasonable steps and exercised all due diligence to avoid committing the offence.

Offences in relation to children

Where the term is referred to in the various offences under LA 2003, including those **20.107** relating to children, 'relevant premises' are premises in respect of which there is a premises licence, or a club premises certificate, or premises which may be used for a permitted temporary activity.

Unaccompanied children prohibited from certain premises

Under LA 2003, s 145(1), it is an offence for a person, knowing that 'relevant premises' **20.108** (above) are within s 145(4), to allow a child under 16 to be on certain categories of relevant premises if (s)he is not accompanied by an adult (18 or over) and the premises are open for the supply of alcohol for consumption on the premises. 'Relevant premises' are within s 145(4) if they are:

(a) exclusively or primarily used for the supply of alcohol for consumption on the premises; or

(b) open for the purpose of being used for the supply of alcohol for consumption on the premises by virtue of LA 2003, Part 5 (permitted temporary activities) and, at the time the TEN in question has effect, they are exclusively or primarily used for such supplies.

It is also an offence against s 145(1) to allow an unaccompanied child under 16 to be on 'relevant premises' at a time between the hours of 00.00 and 05.00 when the premises are open for the supply of alcohol for consumption there.

By s 145(3), these offences may be committed by a person working on the premises in any capacity, paid or unpaid, with the authority to ask the child to leave; the premises licence-holder or designated premises supervisor; an officer or member of a club with that authority; or the premises-user who has given a TEN in respect of the premises.

No offence is committed where the unaccompanied child is merely passing through the premises, where this is the convenient route.

A person charged with an offence under s 145 by reason of his/her own conduct has a defence if (a) (s)he believed the child to be 16 or over or that an individual accompanying the child was 18 or over, and (b) either (s)he had either taken all reasonable steps to establish the individual's age, or nobody could reasonably have suspected from the individual's appearance that (s)he was aged under 16 or 18, as the case may be. A person will be treated as having taken all reasonable steps to establish an individual's age if (s)he asked for evidence of it and the evidence would have convinced a reasonable person. Where a person is charged because of the default of some other person, it is a defence to show that (s)he had exercised all due diligence to avoid committing the offence.

An offence under s 145 is a summary offence punishable with a fine not exceeding level 3.

Sale of alcohol to children

20.109   LA 2003, s 146(1) prohibits the sale of alcohol to an individual aged under 18 *anywhere*. Section 146(2) provides that a club commits an offence if alcohol is supplied by or on its behalf to, or to the order of, a member under 18, or to a person under 18 to the order of a member of a club. A person who supplies alcohol on behalf of a club in such circumstances commits an offence contrary to s 146(3).

Section 146(4) provides a defence where D (the person charged with an offence under s 146) believed that the purchaser was 18 or over and either:

(a) D took all reasonable steps to establish the purchaser's age; or
(b) no one could reasonably have suspected from the purchaser's appearance that he was under 18.

D will be deemed to have taken all reasonable steps if D asked the individual for evidence of his/her age and the evidence would have convinced a reasonable person. In the latter respect, the prosecution may show that the evidence of age produced was such that no reasonable person would have been convinced by it (it might be an obvious forgery or obviously belong to some other person).

Where a sale or supply was made by some other person than D (as where a barman supplies a drink on behalf of the holder of the premises licence) it is a defence to show that D exercised all due diligence to avoid committing the offence (s 146(5)).

An offence under s 146(1), (2) or (3) is a summary offence punishable with an unlimited fine. An offence under s 146(1) or (3) is a 'penalty offence' for the purposes of the Criminal Justice and Police Act 2001 (**3.40**).

It is an offence against s 147(1) for a person knowingly to allow the sale of alcohol to a person under 18 on 'relevant premises' as defined at **20.107**. The persons who can commit this offence are those workers, paid or unpaid, with authority to prevent sale or supply. An officer or member of a club, or a worker in it, with that authority commits an offence under s 147(3) if (s)he allows alcohol to be supplied to or to the order of a member under 18, or to someone under 18 to the order of such a member. An offence under s 147 is a summary offence punishable with an unlimited fine.

Persistently selling alcohol to children

20.110   LA 2003, s 147A(1) provides that an offence is committed if, on two or more different occasions in a period of three consecutive months, alcohol is unlawfully sold to an individual who is under 18 on the same premises, which *at the time of each sale* were licensed premises or premises authorised to be used for a permitted temporary activity. The offence, a summary one punishable with an unlimited fine, is committed by a person who was a 'responsible person' in relation to the premises *at each such time*. A person is a 'responsible person' for the purposes of s 147A(1) if, *at that time*, (s)he is the person (or one of the persons) who holds the premises licence or who is the premises-user in respect of a TEN. For the purposes of s 147A, alcohol is unlawfully sold to an individual under 18 if the person making the sale believed the individual to be under 18 or did not have reasonable grounds for believing him/her to be 18 or over. A person has reasonable grounds for so believing only if (s)he asked for evidence of age and the evidence produced was such that it would have convinced a reasonable person, or if no person could reasonably have suspected from the individual's appearance that (s)he was less than 18. The sales mentioned in s 147A(1) might, but need not, be to the same individual. The same sale may not be counted in respect of different offences for the purpose of enabling the same person to be convicted of more than one offence under s 147A, or for the purpose of enabling the same person to be convicted under s 147A and an offence under s 146 (sale of alcohol to children) or s 147 (allowing such a sale) referred to at **20.109**.

In determining whether an offence under s 147A has been committed, the following are admissible as evidence that there has been an unlawful sale of alcohol to an individual under 18 on any premises on any occasion:

(a) the conviction of a person for an offence under s 146 in respect of a sale to that individual on those premises on that occasion;
(b) the giving to a person of a caution in respect of such an offence; or
(c) the payment by a person of a fixed penalty in respect of such a sale.

On conviction of a premises licence-holder for an offence against s 147A, the court may suspend the premises licence for up to three months in respect of retail sales of alcohol.

As an alternative to a prosecution under LA 2003, s 147A, LA 2003, s 169A permits a 'relevant officer' (superintendent (or above) or an inspector of weights and measures) to issue a closure notice in the prescribed form in relation to any premises if:

(a) there is evidence of the commission of an offence against s 147A;
(b) the relevant officer considers that the evidence provides a realistic prospect of a conviction; and
(c) the offender is still the premises licence-holder (or one of them) in respect of the premises in question.

Such a notice proposes a prohibition of the sale of alcohol at the premises in question for a period specified in the notice and offers the offender the chance to discharge all criminal liability for the alleged offence against s 147A by the acceptance of the prohibition proposed by the notice. The length of the prohibition specified in the order must not exceed 336 hours. The period specified in the notice as the *start time* for closure must not be less than 14 days following service of the notice. Service of a notice must take place when licensable activities are taking place. The notice has the effect of temporarily suspending the premises licence. If sales take place during the period of prohibition, those sales will be unlawful in terms of s 136 referred to at **20.103**.

Purchase of alcohol by children or on their behalf

20.111 LA 2003, s 149(1) makes it an offence for D, a person under 18: (a) to buy or attempt to buy alcohol, or (b) if D is a member of a club, for D to have alcohol supplied to him/her by the club in circumstances in which D initiates the supply, or for D to attempt to do so. Test purchasing on behalf of a constable or trading standards officer is excepted from these provisions.

Section 149(3) makes it an offence for a person to act as an agent for a child in purchasing, or attempting to purchase, alcohol.

Section 149(4) provides an offence which is committed on 'relevant premises' (**20.107**) by prohibiting a person from buying or attempting to buy alcohol for consumption by a person under 18 on those premises. This would cover any relative or friend of a child who buys a drink for a child, or tries to do so, on 'relevant premises'. However, the offence is not committed by an adult who buys beer, wine, or cider for a person aged 16 or 17 to consume with a table meal (ie a meal eaten by a person seated at a table, counter or other structure used as a table by seated persons) taken on those premises in the company of an adult.

The offences under s 149(3) and (4) also apply in respect of the *supply* of alcohol in premises in respect of which there is in force a club premises certificate.

An offence under s 149 is a summary offence punishable with a fine not exceeding level 3 (s 149(1)) or an unlimited fine (s 149(3) or (4)). A s 149 offence is a 'penalty offence' for the purposes of the Criminal Justice and Police Act 2001 (**3.40**).

Consumption of alcohol by children

20.112 It is an offence against LA 2003, s 150(1) for an individual under 18 knowingly to consume alcohol on 'relevant premises' (**20.107**). Because of the presence of 'knowingly' the offence will not be committed where a child accidentally consumes alcohol, being unaware of the nature of his/her drink, or where something has been furtively added to it.

Section 150(2) makes it an offence for a person knowingly to allow such consumption to take place on 'relevant premises'. This offence can be committed by a worker at the premises who has authority to prevent the consumption or by a member or officer of a club in a capacity to do so.

These offences do not apply to a person of 16 or 17 who consumes beer, wine, or cider with a table meal on licensed, etc premises and is accompanied by a person of 18 or over.

An offence under s 150 is a summary offence punishable with a fine not exceeding level 3 (s 150(1)) or an unlimited fine (s 150(2)). It is a 'penalty offence' for the purposes of the Criminal Justice and Police Act 2001 (**3.40**).

Delivering alcohol to children

20.113 Under LA 2003, s 151(1), an offence is committed by a person working on 'relevant premises' (**20.107**), paid or unpaid, who knowingly delivers to a person under 18 (C), alcohol which is sold on the premises or supplied by a club. Delivery is a term which avoids 'sale or supply'. It may be that C's mother has bought and paid for a quantity of drink and has left it at the off-licence to be picked up. If C goes to pick up the alcohol and it is handed over to C, there has been a delivery. It is an offence (by s 151(2)) for a person working on the premises with the necessary authority to prevent the delivery to allow it to take place. A similar offence under s 151(4) applies to clubs.

None of the offences under s 151 is committed if alcohol is delivered to a home or workplace, nor does it apply where the job of a minor involves the delivery of alcohol, nor where alcohol is sold or supplied for consumption on the 'relevant premises' (as other offences are committed in this circumstance).

An offence under s 151 is a summary offence punishable with an unlimited fine. It is a 'penalty offence' for the purposes of the Criminal Justice and Police Act 2001 (**3.40**).

Sending child to obtain alcohol

20.114 In view of the offences involving delivery to a person under 18, it is not surprising to find that 'sending' such a person to obtain alcohol sold, or to be sold, on 'relevant premises' (**20.107**), or supplied or to be supplied by a club, is a summary offence, punishable with an unlimited fine, against LA 2003, s 152(1). Thus where a parent orders by telephone from an off-licence and sends a child to seek the order, an offence against the section is committed. Section 152 provides that the offence is committed whether the child is sent to the licensed premises, etc or to some other delivery point. Thus, if the 'order' has been sent to a distribution centre and it is collected from those premises, the offence is still committed. No offence is committed where the child works at the premises in a capacity involving the delivery of alcohol, nor if the child is assisting a constable or trading standards officer.

Unsupervised sales by children

Under LA 2003, s 153(1), a summary offence, punishable with a fine not exceeding level **20.115**
1, is committed where a responsible person knowingly allows on 'relevant premises' (**20.107**) a person under the age of 18 to sell or, in the case of a club, supply alcohol, unless each sale or supply has been specifically approved by a responsible person. A 'responsible person' means:

(a) in the case of licensed premises, the holder of the premises licence or the designated premises supervisor or an adult authorised by such a person to prevent the sale; or

(b) in the case of club certificate premises, any member or officer present in a capacity to prevent the supply; or

(c) at a temporary permitted activity, the premises-user or an adult authorised by the premises-user to prevent the sale,

as appropriate. Many of the checkout personnel in supermarkets which are authorised to sell for off-consumption are under 18, but provided that each sale is specifically approved no offence is committed. Section 153 exempts sales or supplies in a restaurant of alcohol to be taken with a table meal. Thus waiters and waitresses under 18 may serve drinks in such a part of the premises.

Making false statements in applications for licences and certificates and in notices

It is a summary offence against LA 2003, s 158(1) for a person knowingly or recklessly **20.116**
to make a false statement in or in connection with:

(a) an application for the grant, variation, transfer or review of a premises licence or club premises certificate;

(b) an application for a local authority provisional statement (about the effect of building works on the grant of a premises licence);

(c) a TEN or any other notice under the Act; or

(d) an application for the grant or renewal of a personal licence.

For the purposes of s 158, a person is treated as making a false statement if (s)he produces, furnishes, signs or otherwise makes use of a document that contains a false statement. The offence is punishable with an unlimited fine.

Drunkenness and disorderly conduct

Allowing disorderly conduct on licensed premises, etc

LA 2003, s 140(1) and (2) makes it an offence knowingly to allow disorderly conduct on **20.117**
'relevant premises' (**20.107**). The offence can be committed by any person who works

on the premises in a capacity, paid or unpaid, which gives him/her the authority to prevent such conduct. On premises with a premises licence, the offence can also be committed by the premises licence-holder or designated premises supervisor. In the case of a club, the offence can also be committed by an officer or member of the club who is present when the disorderly conduct takes place and who has the authority to prevent it. Members of a club committee would obviously be in such a position. In the case of a temporary event the offence can also be committed by the premises-user. An offence under s 140(1) and (2) is a summary offence punishable with a fine not exceeding level 3.

Sale of alcohol to a person who is drunk or obtaining alcohol for such a person

20.118   It is a summary offence, contrary to LA 2003, s 141, knowingly to sell, or attempt to sell, alcohol to a person who is drunk, or to allow alcohol to be sold to such a person, on 'relevant premises' (**20.107**). The offence also applies to the supply of alcohol by or on behalf of a club.

The offence may be committed by the same category of persons to which s 140 applies.

The s 141 offence is punishable with a fine not exceeding level 3. It is a 'penalty offence' for the purposes of the Criminal Justice and Police Act 2001 (**3.40**).

By s 142(1), it is a summary offence, punishable with a fine not exceeding level 3, for a person knowingly to obtain, or attempt to obtain, alcohol for consumption on the above types of premises by a person who is drunk.

Failure to leave licensed premises, etc

20.119   A person commits an offence against LA 2003, s 143(1)(a) if (s)he is drunk and disorderly and fails, without reasonable excuse, to leave 'relevant premises' (**20.107**) at the request of a police constable or:

(a) any person who works at the premises in a capacity, paid or unpaid, which gives him/her the authority to make that request;
(b) a premises licence-holder or designated premises supervisor;
(c) an officer or member of a club who is present at the time and has authority to make that request; and
(d) a premises-user who has given a TEN in respect of the event.

A person who is drunk and disorderly also commits an offence under s 143(1)(b) if, without reasonable excuse, (s)he enters, or attempts to enter, 'relevant premises' when requested not to do so by any such persons.

Both offences are summary and punishable with a fine not exceeding level 1.

Section 143(4) requires a constable to assist in the expulsion of persons who are drunk and disorderly from such premises, or to help prevent such a person from entering, if requested to do so by any of the persons described above.

Closure of premises due to disorder, actual or anticipated

20.120   LA 2003, s 160 deals with situations in which there is, or there is expected to be, disorder in a local justice area. In such cases, a magistrates' court acting for that area may make a closure order for a period not exceeding 24 hours, in relation to all premises situated at or near the place of the disorder or expected disorder and in respect of which a premises licence or a TEN has effect. Application for such an order must be made by a superintendent (or above) and the order must not be made unless the court is satisfied that it is necessary to prevent disorder.

It is a summary offence under s 160(4) for a manager of premises, the holder of a premises licence, the designated premises supervisor, or the premises-user in the case of a temporary event, knowingly to keep open premises to which the order relates, or to allow any such premises to be kept open, during the currency of that order. The offence is punishable with a fine not exceeding level 3.

By s 160(7), a constable may use such force as may be necessary to close such premises.

Closure of unlicensed premises under provisions of the Criminal Justice and Police Act 2001

Where a constable or a local authority is satisfied that premises are being, or within the last 24 hours have been, used for unauthorised sale of alcohol for consumption on, or in the vicinity of, the premises, CJPA 2001, s 19 provides that (s)he or it may serve a closure notice on a person having control of, or responsibility for, the activities carried on at the premises. A closure notice must also be served on any person occupying any other part of the building whose access would be impeded by the order. It may also be served on any other person with control of, or responsibility for, the activities on the premises, or any person with an interest in the premises. A closure notice must specify the nature of the alleged use of the premises, state the effect of s 20 (below), and set out the steps which may be taken to ensure that the alleged use of the premises ceases or (as the case may be) does not recur. Closure notices issued by a constable or a local authority may be cancelled at any time.

20.121

Section 20 provides that when a closure notice has been served the constable or local authority may seek an order from the justices, not less than seven days, and not more than six months, after the service of the notice. An order should not be sought where the constable or local authority is satisfied that the offending use of the premises has ceased and that there is no reasonable likelihood of resumption. Where the justices on hearing the case are satisfied that a notice was properly served under s 19, and that the premises and/or the vicinity of those premises continue to be used for the offending purpose or that there is a reasonable likelihood that there will be such use, they may make a closure order under s 21, requiring in particular:

(a) immediate closure of the premises to the public until the constable or local authority (as the case may be) has certified that the need for the order has ceased (such an order may include conditions relating to admission and access to other parts of the building);

(b) any offending use to be discontinued immediately; and

(c) any defendant to pay into court a sum of money which will not be released until the other requirements of the order have been complied with.

The constable or local authority (as the case may be) must fix a copy of the order to the premises.

Affected persons may seek discharge of the order. If the justices are satisfied that the need for the order has ceased, they may discharge it.

A person who, without reasonable excuse, permits premises to be open in contravention of a closure order, or who otherwise fails to comply with it, commits a summary offence against s 25(4) or (5) respectively. Such an offence is punishable with three months' imprisonment and/or an unlimited fine.

Police powers

20.122   By CJPA 2001, s 25(1) and (2), where a closure order has been made a constable or an authorised person (ie a person authorised for these purposes by the local authority for the area), who identifies himself if so required, may enter the premises at any reasonable time, if need be by reasonable force, and do anything reasonably necessary to secure compliance with the order.

It is a summary offence contrary to s 25(3) for a person intentionally to obstruct a constable or an authorised person of the local authority in the exercise of these powers. Where a constable is so obstructed, the offence is punishable with up to a month's imprisonment and/or an unlimited fine. An unlimited fine can be imposed in the case of an authorised person but not imprisonment.

Miscellaneous matters

Exclusion of persons convicted of offences of violence from licensed premises

20.123   The Licensed Premises (Exclusion of Certain Persons) Act 1980, s 1 permits courts to make exclusion orders in respect of someone convicted by or before them of offences of violence or threats of violence on premises in respect of which a premises licence authorising the supply of alcohol for consumption on the premises is in force, prohibiting such a person from entering specified licensed premises without the express consent of the licensee, his/her employee or agent. Entry in breach of an exclusion order is a summary offence under s 2(1), punishable with up to one month's imprisonment and/or a fine not exceeding level 3. By s 3, the licensee, his/her employee or agent may expel a person who has entered, or whom (s)he reasonably suspects of entering, in breach of an order. A constable must, on the demand of a licensee, his/her employee or agent, help to expel any person whom the constable reasonably suspects of being in breach of an exclusion order.

Prohibition of sales of alcohol at service areas, garages, etc

20.124   LA 2003, s 176 provides that no premises licence, club premises certificate, or TEN has effect to authorise the sale by retail or supply of alcohol on or from excluded premises. These are premises on the land of a motorway authority being used for the provision of facilities at motorway service stations or premises used primarily as a garage or which form part of such premises. The definition of excluded premises may be altered by the Secretary of State.

Dancing in certain small premises

20.125   LA 2003, s 177 is concerned with the situation where (a) a premises licence or a club premises certificate authorises dancing and the sale of alcohol for consumption on the premises, and (b) the premises are used primarily for the sale of alcohol for consumption on the premises and have a capacity limit of up to 200. Section 177 provides that, at any time when the premises are open for the supply of alcohol for consumption on the premises and are being used for dancing, any conditions imposed in respect of music entertainment (ie live music or dancing) by the licensing authority, other than those set out in the licence-holder's operating schedule, will be suspended unless they were imposed for public safety or the prevention of crime and disorder.

Section 177 may be disapplied in relation to conditions in respect of particular premises following a review of the licence or certificate.

Licence review for live and recorded music

LA 2003, s 177A provides that, where music (whether live or recorded) takes place on **20.126** premises authorised to be used for the supply of alcohol for consumption on the premises by a premises licence or club premises certificate, and:

(a) at the time of the music, the premises are open for the purposes of being used for the supply of alcohol for consumption on the premises;

(b) if the music is amplified, it takes place in the presence of an audience of no more than 500 persons;

(c) the music takes place between 08.00 and 23.00 on the same day (or, where a special occasion order (**20.100**) has effect, between the hours specified in that order),

any condition of the premises licence or club premises certificate which relates to music does not have effect in relation to the live music, recorded music, or both, unless, on a review of the licence or certificate, s 177A is disapplied in relation to that condition (s 177A(3)) or a condition is added relating to the music as if it were regulated entertainment and the licence or certificate licensed the music (s 177A(4)). For the purposes of s 177A, 'condition' means a condition included in a premises licence or club premises certificate, other than one which is a mandatory one, including one added by way of variation at the request of the licence-holder or the club, as the case may be.

Section 177A does not apply to music which, by virtue of a provision in Sch 1, other than the exemption relating to licensed etc venues (exemption (l) at **20.92**) or the exemption relating to live unamplified music (exemption (n) at **20.92**), is not regarded as the provision of regulated entertainment for the purposes of LA 2003.

Rights of entry to investigate licensable activities

LA 2003, s 179(1) provides for a police officer or authorised person (**20.87**) to enter **20.127** premises where (s)he has reason to believe that the premises are being, or are about to be, used for a licensable activity to ensure that the activities are being carried on under and in accordance with an authorisation. An immigration officer may enter premises where (s)he has reason to believe that they are being used for the retail sale of alcohol or the provision of late night refreshment to see whether an offence under the Immigration Acts is being committed in connection with the carrying on of such an activity. An authorised person or immigration officer must produce his/her authority if requested. Reasonable force may be used if necessary. By s 179(4), it is an offence intentionally to obstruct an authorised person or immigration officer acting under s 179. The offence is summary and punishable with a fine not exceeding level 3.

Section 180 provides that a police officer may enter and search premises where there is reason to believe an offence under the Act has been, is being or is about to be, committed, and may use reasonable force to gain entry.

CHAPTER 21

Terrorism Generally

21.1   This chapter refers to the offences and powers under the Terrorism Acts 2000 and 2006, the Anti-terrorism, Crime and Security Act 2001, the Counter-Terrorism Act 2008, and the Counter-Terrorism and Security Act 2015. Reference is also made to the power to impose terrorism prevention and investigation measures under the Terrorism Prevention and Investigation Measures Act 2011 to protect the public from terrorism. These Acts can be described as the 'terrorism legislation'.

MEANING OF 'TERRORISM' IN TERRORISM LEGISLATION

21.2   For the purposes of this legislation, 'terrorism' is defined by the Terrorism Act 2000 (TA 2000), s 1, as meaning the use or threat of action (inside or outside the UK) which:

(a) (i)   involves serious violence against a person wherever (s)he is,
   (ii)   involves serious damage to property wherever situated,
   (iii) endangers a person's life, other than that of the person committing the action,
   (iv) creates a serious risk to the health and safety of the public or a section of the public, or
   (v)   is designed seriously to interfere with or seriously to disrupt an electronic system;

(b) is designed to influence the government of the UK (or part thereof), or the government of any other country, or an international governmental organisation, or to intimidate the public or a section of the public anywhere; *and*

(c) is made for the purpose of advancing a political, religious, racial or ideological cause.

The Court of Appeal has held that it is not enough that the use or threat of action unwittingly or accidentally has an effect within (a)(i)–(v); it must have been done or made with intention or recklessness as to having such effect.

The use or threat of action falling within (a) which includes the use of firearms or explosives is terrorism whether or not (b) is satisfied.

The Court of Appeal has also held that the reference in (b) to the government of any other country is not limited to a representative or democratic government. The Supreme Court has held that, provided (b) and (c) are satisfied, military attacks by a non-state armed group against any state or inter-governmental armed forces in the context of a non-international conflict fall within the above definition.

There is no exemption from criminal liability for terrorist activities motivated by the alleged nobility of the terrorist cause.

PROSCRIBED ORGANISATIONS

Offences

21.3   TA 2000 contains a number of offences which can be committed in relation to a 'proscribed organisation' listed in TA 2000, Sch 2 (**21.4**):

(1) Belonging or professing to belong to a proscribed organisation, subject to a defence on 'proof' (interpreted by the House of Lords as only imposing an evidential burden) by the defendant (D) that D joined the proscribed organisation before it was proscribed and D has not taken part in the activities of the organisation since it became proscribed (TA 2000, s 11). The Court of Appeal has held that membership of a proscribed organisation depends on the nature of the organisation in question. Membership of a loose and unstructured organisation may not require an express process, whereas a more structured organisation may have an express process whereby a person becomes a member. The core elements of membership within s 11 will frequently be voluntary and knowing association with others with a view to furthering the aims of the proscribed organisation. Unilateral sympathy with the aims of an organisation, even coupled with acts designed to promote similar objectives, whilst being clear evidence of belonging, will not always be sufficient.

(2) Inviting support for a proscribed organisation, other than support with money or property (s 12(1)). The Court of Appeal has held that to secure a conviction under s 12(1), the prosecution must prove that: (a) the organisation was already proscribed for the purposes of TA 2000; (b) D used words which invited support for that organisation; and (c) D knew, when D did so, that (s)he was inviting support for that organisation. The Court noted that s 12 does not prohibit the mere holding of opinions or beliefs supportive of a proscribed organisation, or even the expression of those opinions or beliefs (although a profession of membership might, depending on the circumstances, amount to an offence under s 11). The Court rejected a submission that the word 'inviting' in s 12 should be construed as connoting or implying that the person issuing the invitation must already be engaged in providing such support or at least intend to do so, because the criminality lay in inviting support from third parties for the proscribed organisation, and not necessarily in inviting those third parties to join with D in providing it. The Court agreed with a statement that 'support' can encompass not only practical or tangible assistance, but also what was referred to as intellectual support (ie agreement with and approval, approbation, or endorsement of, that which is supported).

(3) Expressing an opinion or belief that is supportive of a proscribed organisation, where (in doing so) D is reckless as to whether a person to whom the expression is directed will be encouraged to support a proscribed organisation (s 12(1A)).

(4) Arranging, managing or assisting in arranging or managing a meeting of three or more people (whether or not the public are admitted) which D knows is:
    (a) to support a proscribed organisation,
    (b) to further the activities of a proscribed organisation, or
    (c) to be addressed by a person who belongs or professes to belong to a proscribed organisation (s 12(2)).
    Where an offence charged under (c) relates to a private meeting, D has a defence if D adduces sufficient evidence that (s)he had no reasonable cause to believe that such an address would support a proscribed organisation or further its activities, whereupon the defence is assumed to be satisfied unless the prosecution disprove this (ss 12(4) and 118).

(5) Addressing a meeting where the purpose is to encourage support for a proscribed organisation or to further its activities (s 12(3)).

(6) Wearing in a public place any item of clothing, or wearing, carrying, or displaying any article, in such a way or in such circumstances as to arouse reasonable apprehension that D is a member or supporter of a proscribed organisation (s 13(1)). A divisional court has held that s 13(1) does not require knowledge on D's part that

(s)he is wearing or displaying something which would give rise to a reasonable suspicion of membership or support for a proscribed organisation, or an intention to display support for such an organisation. By s 13(4), a constable may seize an item of clothing or any other article if (s)he reasonably suspects that it is evidence in relation to an offence under s 13(1), and is satisfied that it is necessary to seize it in order to prevent the evidence being concealed, lost, altered, or destroyed. In connection with exercising this power, a constable may require a person to remove the item of clothing or other article if the person is wearing it (s 13(5)). However, the powers conferred by s 13(4) and (5) may not be exercised so as to seize, or require a person to remove, an item of clothing being worn next to the skin or immediately over a garment being worn as underwear.

(7) Publishing an image (ie a still or moving image, however produced) of an item of clothing, or any other article, in such a way or in such circumstances as to arouse reasonable suspicion that D is a member or supporter of a proscribed organisation (s 13(1A)).

The offences listed at (1) to (5) are indictable either-way offences punishable with up to 14 years' imprisonment. Those at (6) and (7) are summary offences punishable with up to six months' imprisonment and/or an unlimited fine.

The organisations

21.4   The proscribed organisations associated with terrorism in Northern Ireland are at present:
the Irish Republican Army; Cumann na mBan; Fianna na hEireann; the Red Hand Commando; Saor Eire; the Ulster Freedom Fighters; the Ulster Volunteer Force; the Irish National Liberation Army; the Irish People's Liberation Organisation; the Ulster Defence Association; the Loyalist Volunteer Force; the Continuity Army Council; the Orange Volunteers; and the Red Hand Defenders.

Other proscribed organisations are:
Al-Qa'ida (also known as al-Nusrah Front, as Jabhat al Nusrah li-ahl al Sham, as Jabhat Fatah al-Sham, and as Hay'at Tahrir Al-Sham); Egyptian Islamic Jihad; Al-Gama'at al-Islamiya; Armed Islamic Group (Groupe Islamique Armée) (GIA); Salafist Group for Call and Combat (Groupe Salafiste pour la Prédication et le Combat) (GSPC); Babbar Khalsa; Harakat Mujahideen; Jaish e Mohammed; Lashkar e Tayyaba (also known as Jama'at ud Da'wa); Liberation Tigers of Tamil Eelam (LTTE); Islamic Movement of Uzbekistan; Jemaah Islamiyah; Al Ittihad Al Islamia; Ansar Al Islam; Ansar Al Sunna; Groupe Islamique Combattant Marocain; Harakat-ul-Jihad-ul-Islami; Harakat-ul-Jihad-ul-Islami (Bangladesh); Harakat-ul-Mujahadeen/Alami; Islamic Jihad Union; Jamaat ul-Furquan; Jundallah; Khuddam ul-IslamHizballah (Party of God); Hamas-Izz al-Din al-Qassem Brigades; Palestinian—Islamic Jihad—Shaqaqi; Abu Nidal Organisation; Islamic Army of Aden; Kurdistan Workers' Party (Partiya Karkeren Kurdistani) (PKK) including Teyrebazene Azadiye Kurdistan (TAK) and Hezen Parastina Gel (HPG); Revolutionary Peoples' Liberation Party—Front (Devrimci Halk Kurtulus Partisi-Cephesi) (DHKP-C) (also known as Revolutionary People's Liberation Front (Devrimci Halk Kurtulus Cephesi) (DHKC), as Revolutionary People's Liberation Party (Devrimci Halk Kurtulus Partisi) (DHKP), and as Revolutionary People's Liberation Front/Armed Propaganda Units (Devrimci Halk Kurtulus Cephesi/Silahli Propaganda Birlikleri) (DHKC/SPB); Basque Homeland and Liberty (Euskadi ta Askatasuna) (ETA); 17 November Revolutionary Organisation

(N17); Abu Sayyaf Group; Asbat Al-Ansar; Lashkar-e Jhangvi (also known as Ahle Sunnat wal Jamaa); Sipah-e Sahaba Pakistan (also known as Ahle Sunnat wal Jamaa); Baluchistan Liberation Army; Jammat-ul Mujahideen Bangladesh; Tehrik Nefaz-e Shari'at Muhammadi; Al-Shabaab; Tehrik-e Taliban Pakistan; Indian Mujahideen; Ansarul Muslimina Fi Biladis Sudan (Vanguard for the protection of Muslims in Black Africa) (Ansaru); Jama'atu Ahli Sunna Lidda Awati Wal Jihad (Boko Haram) and Minbar Ansar Deen (Ansar Al Sharia UK); Imarat Kavkaz (Caucasus Emirate); Ansar Bayt al-Maqdis (Ansar Jerusalem), Al Murabitun; Ansar al Sharia—Tunisia; Islamic State of Iraq and the Levant (Islamic State of Iraq and al-Sham) (Dawat al Islamiya fi Iraq wa al Sham (DAISh)) (also known as Islamic State (Dawlat al Islamiya), as Jaysh Khalid Bin Walid (JKbW) (JKW), as Jaysh Khalid bin al-Walid (KBW), and as Khalid ibn-Walid Army (KBWA)); Turkiye Halk Kurtulus Partisi-Cephesi (Turkish People's Liberation Party) (The Hasty Ones) (Mukavamet Suriye); Kateeba al-Kawthar (Ajnad al-sham) (Junud ar-Rahman al Muhajireen); Abdallah Azzam Brigades, including the Ziyad al-Jarrah Battalions; Popular Front for the Liberation of Palestine—General Command; Ansar al-Sharia-Benghazi (Partisans of Islamic Law); Ajnad Misr (Soldiers of Egypt); Jaysh al Khalifatu Islamiya (Army of the Islamic Caliphate) (Majahideen of the Caucasus and the Levant); Jund Al-Aqsa (Soldiers of Al-Aqsa); Jund al Khalifa—Algeria (Soldiers of the Caliphate in Algeria); Jamaat ul-Ahrar; The Haqqani Network; Global Islamic Media Front (including GIMF Bangla Team (Ansarullah Bangla Team) (Ansar-al Islam)); Mujahedeen Indonesia Timur (East Indonesia Mujahedeen); Turkestan Islamic Party (East Turkestan Islamic Party) (East Turkestan Islamic Movement) (East Turkestan Jihadist Movement) (Hizb al-Islami al-Turkistani); Jamaah Anshorut Daulah; National Action (also known as System Resistance Network, as Scottish Dawn and as NS131 (National Socialist Anti-Capitalist Action); al-Ashtar Brigades (Saraya al-Ashtar) (The Wa'ad Allah Brigades) (Islamic Allah Brigades) (Imam al-Mahdi Brigades) (al-Haydariyah Brigades); al-Mukhtar Brigades (Saraya al-Mukhtar); Hasam (Harakat Sawa'd Misr) (Harakat Hasm) (Hasm); Liwa al-Thawra; Jamaat Nusrat al-Islam Wal-Muslimin (JNIM) (Nusrat al-Islam) (Nusrat al-Islam wal Muslimeen) (NIM), including Ansar al-Dine (AAD), Macina Liberation Front (MLF), al-Murabitun, al-Qa'ida in the Maghreb and az-Zallaqa; Ansaroul Islam (Ansar ul Islam) (Ansaroul Islam Lil Irchad Wal Jihad); Sonnenkrieg Division; Feuerkrieg Division; Atomwaffen Division (AWD), including National Socialist Order (NSO); Al-Ghurabaa; and The Saved Sect. Al Muhajiroun (ALM); Call to Submission; Islam4UK; Islamic Path; London School of Sharia; Muslims against Crusades; Need4Khilafah; the Shariah Project; and Islamic Dawah Association, are treated as alternative names for both Al-Ghurabaa and The Saved Sect.

The Secretary of State has the power to add to, remove or amend a name included in TA 2000, Sch 2.

TERRORIST PROPERTY

Fund-raising

TA 2000, s 15 provides that a person is guilty of an offence if he:　　　　**21.5**

(a) invites any other person to provide money or other property, and intends that it should be used, or has reasonable cause to suspect that it may be used, for the purposes of terrorism (s 15(1)); or

(b) receives money or other property, and intends that it should be used, or has reasonable cause to suspect that it may be used, for the purposes of terrorism (s 15(2)); or

(c) provides money or other property, and knows or has reasonable cause to suspect that it will or may be used, for the purposes of terrorism (s 15(3)).

The references to the 'provision' of money or other property are references to its being given, lent or otherwise made available, whether or not for consideration. The definition of terrorism is at **21.2**.

The Supreme Court has rejected an argument that 'having reasonable cause to suspect' that the money or other property will or may be used for terrorism meant that the defendant (D) had actually to suspect, and for a reasonable cause, that it would or might be so used. The Court held that it was sufficient if, on the information known to D, a reasonable person would suspect that the money or property would or might be used for terrorism. This decision, in line with the general meaning of 'reasonable cause to suspect' in a statutory offence (as opposed to a police power), is clearly also of relevance to other offences under the Terrorism Act 2000 in which 'reasonable cause to suspect' or a corresponding phrase appears.

A person does not commit an offence under s 15 (or under ss 16 to 18 below) in the circumstances set out in s 21, 21ZA, 21ZB, or 21ZC (**21.12**).

Use and possession

21.6   TA 2000, s 16(1) prohibits as an offence the use of money or property for the purposes of terrorism. Section 16(2) creates a further offence of possessing money or other property intending that it should be used, or having reasonable cause to suspect that it might be used, for the purposes of terrorism.

Funding arrangements

21.7   TA 2000, s 17 makes it an offence for a person to enter into, or become concerned in, an arrangement as a result of which money or other property is made (or is to be made) available to another, knowing or having reasonable cause to suspect that it might be used for terrorism.

Insurance against payments made in respect to terrorist demands

21.8   TA 2000, s 17A(1) provides that the insurer under an insurance contract commits an offence if:

(a) the insurer makes a payment under the contract, or purportedly under it;
(b) the payment is made in respect of any money or other property that has been, or is to be, handed over in response to a demand made wholly or partly for the purposes of terrorism; and
(c) the insurer or the person authorising the payment on the insurer's behalf knows or has reasonable cause to suspect that the money or other property has been, or is to be, handed over in response to such a demand.

Money laundering

21.9   TA 2000, s 18(1) prohibits as an offence money laundering of terrorist property by concealment, removal from jurisdiction, transfer to nominees, or in any other way. Section

18(2) provides a defence if D proves lack of knowledge or reason to suspect that the arrangement was concerned with terrorist property.

'Terrorist property' means:

(a) money or other property which is likely to be used for the purposes of terrorism (including any resources of a proscribed organisation);
(b) proceeds of the commission of acts of terrorism; and
(c) proceeds of acts carried out for the purposes of terrorism.

The references to proceeds of an act include a reference to any property which wholly or partly, and directly or indirectly, represents the proceeds of the act (including payments or other rewards in connection with its commission).

Disclosure of information: duty

21.10 TA 2000, s 19(1) and (2) provides that a person commits an offence if he does not disclose to a constable, or a NCA officer authorised for the purposes of s 19 by the NCA's Director General, as soon as reasonably practicable, a belief or suspicion, and the information upon which it is based, where he:

(a) believes or suspects that another person has committed an offence under any of TA 2000, ss 15 to 18; and
(b) bases his belief or suspicion on information which came to his attention in the course of his employment (whether or not in a trade, profession or business).

However, s 19 does not apply if the information came to the person in the course of a business in the regulated sector. A business is in the regulated sector to the extent that it engages in accepting deposits, operating a bureau de change, dealing in, advising in relation to, or managing investments, accountancy business, estate agency or casino, or a range of other financially related activities listed in Sch 3A for the purposes of s 21A (**21.14**), which provides offences which can be committed by non-disclosure in those circumstances.

It is a defence to prove reasonable excuse for non-disclosure (s 19(3)). In the case of an employee, where there is an established procedure concerning disclosure, it is a defence for him/her to prove that disclosure was made within that procedure (s 19(4)). Section 19 does not require disclosure by a professional legal adviser of legally privileged information or of a belief or suspicion based on such information.

Co-operation with the police

21.11 TA 2000, s 21 provides that a person does not commit an offence under ss 15 to 18 if he is acting with the express consent of a constable.

Exceptions

21.12 TA 2000, ss 21ZA, 21ZB, and 21ZC provide exceptions which apply to the offences against ss 15 to 18. Section 21ZA provides an exception if D has made a disclosure to an authorised officer before becoming involved in a transaction or an arrangement and acts with the consent of the authorised officer. 'Authorised officer' means an NCA officer authorised for the purposes of s 21ZA by the NCA's Director General. Section 21ZB provides a further exception to cover those who become involved in a transaction or an arrangement and then make a disclosure, so long as there is a reasonable excuse for failure to make a disclosure in advance. Finally, s 21ZC provides a defence if D proves that he intended to make a disclosure under s 21ZA or 21ZB and has a reasonable excuse for failing to do so.

Trial and punishment

21.13   An offence under TA 2000, ss 15 to 18 is an indictable (either-way) offence and is punishable with up to 14 years' imprisonment. An offence under s 19 is similarly triable; it is punishable with up to five years' imprisonment.

Failure to disclose: regulated sector

21.14   TA 2000, s 21A makes further provision to prevent money laundering by terrorist organisations. Section 21A(1) to (4) creates an offence of failure by D to disclose certain information or other matter to a constable, a NCA officer authorised for the purposes of s 21A by the NCA's Director General or a nominated officer as soon as is practicable after that information comes into his/her possession. The offence is an indictable (either-way) offence punishable with up to five years' imprisonment. It occurs where:

(a) D knows or suspects or has reasonable grounds for knowing or suspecting that another person (X) has committed or attempted to commit an offence under any of TA 2000, ss 15 to 18; and

(b) that information or other matter, on which D's knowledge or suspicion is based or which gives reasonable grounds for such knowledge or suspicion, came to D in the course of a business in the regulated sector (**21.10**).

X will be taken to have committed an offence against ss 15 to 18 for these purposes if:

(a) X has taken action or been in possession of a thing; and

(b) X would have committed the offence if D had been in the UK at the time when (s)he took the action or was in possession of the thing.

D does not commit an offence under s 21A if:

(a) D has a reasonable excuse for not disclosing the information or other matter;

(b) D is a professional legal adviser or relevant professional adviser (a qualified accountant, auditor or tax adviser) and the information or other matter came to D in privileged circumstances; or

(c) (i) D is employed by, or in partnership with, such an adviser to provide the adviser with assistance or support,

(ii) the information or other matter comes to D in connection with the provision of such assistance or support, and

(iii) the information or other matter comes to D in privileged circumstances.

In deciding whether D has committed an offence under s 21A, a court must consider whether D had followed any relevant Treasury approved guidance issued by a supervisory authority and properly brought to the attention of affected persons.

A disclosure to a 'nominated officer' is one made to a person nominated by D's employer to receive such disclosures and is made in the course of D's employment in accordance with established procedures.

Protected disclosures

By TA 2000, s 21B a disclosure will not be taken to have breached any restriction upon **21.15** the disclosure of information (however imposed) if (a) the information came to the discloser in the course of a business in the regulated sector, (b) the information is such that it causes him/her to know or suspect, or gives him/her reasonable grounds for knowing or suspecting, that another person has committed an offence under ss 15 to 18, and (c) the disclosure is made to a constable, a NCA officer authorised for the purposes of s 21B by the NCA's Director General or a nominated officer as soon as is practicable.

Disclosures to the NCA

Where a disclosure is made under the above provisions to a constable, the constable **21.16** must disclose it in full as soon as practicable to a member of staff of the NCA authorised for this purpose by the Director General of the NCA; where a disclosure is made to a constable under TA 2000, s 21, this duty to disclose is a duty to disclose to an NCA officer specifically authorised to receive s 19 disclosures (s 21C).

TERRORIST INVESTIGATIONS

Police cordons

TA 2000, ss 33 to 36 provide powers to impose a police cordon for the purposes of a **21.17** 'terrorist investigation'. A 'terrorist investigation' means an investigation of:

(a) the commission, preparation, or instigation of acts of terrorism (**21.2**);
(b) an act which appears to have been done for the purposes of terrorism;
(c) the resources of a proscribed organisation (**21.4**);
(d) the possibility of making an order proscribing an organisation; or
(e) the commission, preparation, or instigation of an offence under TA 2000 or under the Terrorism Act 2006 (**21.57–21.63**) other than an offence under TA 2006, s 1 or s 2 (encouragement of terrorism and dissemination of terrorist publications).

TA 2000, s 34 permits a superintendent (or above), where (s)he considers it expedient for the purpose of a terrorist investigation, to authorise a cordon to be imposed on an area specified by him/her in that authorisation. (An officer below that rank may do so where there is great urgency. (S)he must, as soon as reasonably practicable, make a written record of the time of the designation of the area and ensure that an officer of at least the rank of superintendent is informed. Such an officer must confirm the designation or cancel it from a specified time, in which case his/her reason for doing so must be stated.)

The area on which a cordon is imposed must, so far as is reasonably practicable, be indicated by means of police tape or in such other manner as appears appropriate to the police officer responsible for carrying out the arrangements for applying the cordon. The period of time initially specified must not exceed 14 days, but may be extended by a superintendent (or above) by one or more written variations; however, the overall period must not exceed 28 days.

Persons must leave the area immediately when ordered to do so by a constable in uniform and persons must also leave premises which are wholly or partly in or adjacent to a cordoned area when so ordered. Drivers or persons in charge of vehicles must move them out of the area when ordered by such a constable (who also has power to remove such vehicles). A constable in uniform may prohibit or restrict vehicular or pedestrian access to a cordoned area. A summary offence is committed under s 36(2) by someone who does not comply with the constable's order, prohibition, or restriction. It is a defence to prove a reasonable excuse for the failure (s 36(3)). The offence is punishable with up to three months' imprisonment and/or a fine not exceeding level 4.

Offences relating to terrorist investigations

Information about acts of terrorism

21.18   By TA 2000, s 38B, a person who has information which (s)he knows or believes might be of material assistance in:

(a) preventing the commission by another person of an act of terrorism (**21.2**); or

(b) securing the apprehension, prosecution or conviction of another person, in the UK, for an offence involving the commission, preparation or instigation of an act of terrorism,

commits an indictable (either-way) offence if (s)he does not disclose the information as soon as reasonably practicable to a constable. Such an offence is punishable with up to 10 years' imprisonment.

This offence may be treated as having been committed in any place where the person to be charged is, or has at any time been, since (s)he first knew or believed that the information might be of material assistance, and proceedings may be taken in any such place.

Offences relating to disclosure

21.19   By TA 2000, s 39(2), an offence is committed by a person who knows or has reasonable cause to suspect that a constable is conducting, or proposes to conduct, a terrorist investigation (**21.17**) and who:

(a) discloses to another anything which is likely to prejudice the investigation; or

(b) interferes with material which is likely to be relevant to the investigation.

Where a person knows or has reasonable cause to believe that a disclosure has been or will be made under s 19 to 21B (**21.10–21.15**) or 38B (**21.18**), (s)he commits an offence under s 39(4) if (s)he:

(a) discloses to another anything which is likely to prejudice an investigation resulting from the disclosure under one of these sections; or

(b) interferes with material which is likely to be relevant to an investigation resulting from the disclosure under that section.

It is a defence to a charge under s 39(2) or (4) for a person (a) to adduce evidence that (s)he did not know and had no reasonable cause to suspect that the disclosure was likely to affect a terrorist investigation (whereupon the defence is assumed to be satisfied unless the prosecution disprove this) (ss 39(5)(a) and 118); or (b) to prove that (s)he had a reasonable excuse for the disclosure or interference (s 39(5)(b)).

The offences under s 39(2) and (4) are indictable (either-way) offences punishable with up to five years' imprisonment.

COUNTER-TERRORIST POWERS

Search of premises

By TA 2000, s 42, a justice of the peace may issue a warrant on the application of a **21.20** constable in relation to the search of specified premises if satisfied that there are reasonable grounds for suspecting that a person, whom the constable reasonably suspects to be a person who is or has been concerned in the commission, preparation or instigation of acts of terrorism, is to be found there. Such a warrant authorises entry and search for the purpose of the arrest of a person subject to arrest under the provisions of s 41 (**21.46**). This power of search includes a power to search a container on the premises.

Search warrants may also be granted under Sch 5 (**4.72**).

Code of practice relating to searches of persons and vehicles

Acting under TA 2000, s 47AB, the Secretary of State has issued a Code of Practice **21.21** for the Exercise of Stop and Search Powers under ss 43 and 43A of TA 2000 and the Authorisation and Exercise of Stop and Search Powers relating to s 47A of, and Sch 6B to, TA 2000 (hereafter 'the Terrorism Search Powers Code'). These statutory provisions are dealt with below. A constable must have regard to the Terrorism Search Powers Code when exercising any powers to which it relates. A failure on the part of a constable to act in accordance with any provision of the Code does not of itself make him/her liable to criminal or civil proceedings. However, the Code is admissible in evidence in any such proceedings, and a court may, in particular, take into account a failure by a constable to have regard to it in determining a question in any such proceedings.

The Code provides that chief officers and 'police authorities' (police and crime commissioners and the like) must have regard to it when discharging a function to which it relates. The Code must be followed by them unless there is good reason not to do so, in which case the decision not to follow it should be recorded in writing.

Search of persons

A constable is given power by TA 2000, s 43(1) to stop and search *a person* whom (s)he **21.22** *reasonably suspects* to be a terrorist to discover whether that person has in his/her possession anything which might constitute evidence that (s)he is a terrorist. A 'terrorist' is defined as a person who:

(a) has committed an offence under s 11 (membership of a proscribed organisation); s 12 (support for it); ss 15–18 (fund-raising, etc and money laundering); s 54 (weapons training); or ss 56–63 (directing terrorist organisation, possession for terrorist purposes, collecting information, eliciting, etc information about members of the armed forces, entering or remaining in a designated area, inciting terrorism overseas, and terrorist bombing and finance offences outside the UK); or

(b) is, or has been, concerned in the commission, preparation, or instigation of acts of terrorism (**21.2**).

This power to stop a person includes the power to stop a vehicle.

By s 43(2), a constable may also search anyone arrested under s 41 (**21.46**) to discover whether that person has anything in his/her possession which may constitute such evidence. Reasonable suspicion that such evidence may be found is not required.

The powers to search a person under s 43 include the power to search anything carried by the person, eg a bag. Anything found in a search under s 43(1) or (2) may be seized and retained if the constable reasonably suspects that it is evidence that the person is a terrorist (s 43(4)). The additional powers of seizure under the Criminal Justice and Police Act 2001, s 51 (**4.60**) are available.

Where a constable exercising the power under s 43(1) to stop a person stops a vehicle (s)he may:

(a) search the vehicle and anything in or on it to discover whether there is anything which may constitute evidence that the person concerned is a terrorist (but (s)he may not search someone in the same vehicle as the suspected terrorist solely on the basis that that person is with the suspected terrorist); and

(b) seize and retain anything which (s)he discovers in the course of such a search, and reasonably suspects may constitute evidence that the person is a terrorist (s 43(4A) and (4B)).

A 'vehicle' includes an aircraft, hovercraft, train, or vessel. The additional powers of seizure under the Criminal Justice and Police Act 2001, s 50 (**4.61**) are available. A constable stopping a vehicle must be in uniform.

The grounds for stopping and searching a person under s 43(1) are the same as the grounds for arrest under s 41: reasonable suspicion that the person is a terrorist. Stop and search is a less intrusive power than arrest and will be more appropriate in many situations, eg in encounters with individuals where a stop and search may help to allay suspicions. Stop and search should not be used in any situation where it is more appropriate to arrest or where an officer believes it may put him/her, or members of the public, in danger.

Powers to stop and search must be exercised fairly, responsibly, and in accordance with the Equality Act 2010 (referred to at **4.1**).

The same provisions apply to the removal of clothing in conducting a search as apply under PACE Code A: Stop and Search Code (**4.27**). Although there is no longer a statutory requirement for the constable conducting the search to be of the same sex as the person searched, where such a constable is readily available, that constable should carry out the search. Otherwise, searches may be carried out by a constable of the opposite sex unless the removal of clothes other than an outer coat or jacket is required. Particular regard should be paid to the sensitivities of some religious communities in respect of being searched by a member of the opposite sex.

Search of vehicles

21.23   By TA 2000, s 43A, if a constable *reasonably suspects* that a *vehicle* is being used for the purposes of terrorism (**21.2**), (s)he may stop and search:

(a) the vehicle,
(b) the driver of the vehicle,
(c) a passenger in the vehicle,
(d) anything in or on the vehicle or carried by the driver or a passenger,

to discover whether there is anything which may constitute evidence that the vehicle is being used for the purposes of terrorism (s 43A(1) and (2)). An unattended vehicle can be searched under s 43A.

'Driver', in relation to an aircraft, hovercraft, or vessel, means the captain, pilot or other person with control of the aircraft, hovercraft or vessel, or any member of its crew, and, in relation to a train, includes any member of its crew. A constable stopping a vehicle must be in uniform.

Searches may be undertaken of anything in or on the vehicle, but care should be taken not to damage a vehicle as part of a search or, in the case of an unattended vehicle, in order to gain entry into it. Vehicles stopped under s 43A and persons in those vehicles may be detained only for so long as is necessary to carry out the searches—at or near the place the vehicle was stopped.

A constable may seize and retain anything which (s)he:

(a) discovers in the course of such a search, and
(b) reasonably suspects may constitute evidence that the vehicle is being used for the purposes of terrorism (s 43A(3)).

A person who has the powers of a constable in one part of the UK may exercise a power under s 43A in any part of the UK.

The additional powers of seizure under the Criminal Justice and Police Act 2001, ss 50 and 51 (**4.61** and **4.60**) are available.

Power to give an authorisation to stop and search in specified areas or places

By TA 2000, s 47A(1), a senior police officer may give a stop and search authorisation under s 47A(2) or (3) (**21.27**) in relation to a specified area or place if (s)he: **21.24**

(a) reasonably suspects that an act of terrorism (**21.2**) will take place; and
(b) reasonably considers that the:
  (i) authorisation is necessary to prevent such an act;
  (ii) specified area or place is no greater than is necessary to prevent such an act; and
  (iii) duration of the authorisation is no longer than is necessary to prevent such an act.

Those parts of the Code of Practice (**21.21**) which relate to the exercise of a senior police officer's functions under s 47A are set out in Chapter 21 on the companion website.

Senior police officer

'Senior police officer' for the purposes of s 47A and Sch 6B (**21.24–21.37**) means an   **21.25** assistant chief constable (or above), or a commander (or above) in the case of the two London police forces. Authorising officers must be either substantive or on temporary promotion to the qualifying rank. Officers who are acting in the rank may not give authorisations.

Specified areas or places

An authorisation given by a senior police officer who is not a member of the British   **21.26** Transport Police, the Ministry of Defence Police or the Civil Nuclear Constabulary may specify an area or place in his/her police area together with the internal waters (ie waters such as a bay, estuary of a large river, or the sea near larger islands) which are not covered in any police area but are adjacent to that area or place, or a specified area of those internal waters. Police force areas do not cover the sea below the low water mark.

An authorisation by an assistant chief constable (or above) of the British Transport Police, the Ministry of Defence Police or the Civil Nuclear Constabulary may only be given in respect of specified areas or places policed by his/her force. An authorisation given by a senior police officer of the Civil Nuclear Constabulary does not have effect except in relation to times when the specified area or place is a place where members of that Constabulary have the powers and privileges of a constable.

Authorised conduct

21.27   An authorisation under TA 2000, s 47A(2) authorises any constable in uniform to stop a vehicle in the area or place specified in the authorisation and to search:

(a) the vehicle;
(b) the driver of the vehicle;
(c) a passenger in the vehicle;
(d) anything in or on the vehicle or carried by the driver or a passenger.

For these purposes, 'vehicle' includes an aircraft, hovercraft, train, or vessel, and 'driver', in relation to an aircraft, hovercraft, or vessel, means the captain, pilot, or other person with control of it or any member of its crew and, in relation to a train, includes any member of its crew.

An authorisation under s 47A(3) authorises any constable in uniform to stop a pedestrian in the area or place specified in the authorisation and to search:

(a) the pedestrian;
(b) anything carried by the pedestrian.

Written authorisation or confirmation

21.28   Authorisations should, where practicable, be given in writing. A senior police officer who gives an authorisation under s 47A orally must confirm it in writing as soon as reasonably practicable (TA 2000, Sch 6B).
  Where:

(a) a vehicle or pedestrian is stopped by virtue of s 47A(2) or (3), and
(b) the driver of the vehicle or the pedestrian applies for a written statement that the vehicle was stopped, or that the pedestrian was stopped, by virtue of s 47A(2) or (as the case may be) (3),

the written statement must be provided. Such an application must be made within the period of 12 months beginning with the date on which the vehicle or pedestrian was stopped.

Duration of authorisation

21.29   TA 2000, Sch 6B provides that an authorisation under s 47A has effect during the period:

(a) beginning at the time when the authorisation is given, and
(b) ending with the date or at the time specified in the authorisation.

The power under (b) includes power to specify different dates or times for different areas or places.
  The specified date or time in (b) must not occur after the end of the period of *14 days* beginning with the day on which the authorisation is given.

Confirmation of authorisation

21.30   The senior police officer who gives an authorisation must inform the Secretary of State of it as soon as reasonably practicable; the police should aim to have provided the written authorisation (or written confirmation of an oral authorisation) within two hours of an authorisation being given. Where practicable, notification of an intention to give an authorisation and (unless the authorisation is for less than 48 hours) a draft of that authorisation should be given before the authorisation is given. If the authorisation is not confirmed by the Secretary of State within 48 hours of being given it will automatically cease to have effect at the end of that period, but this will not affect the lawfulness of things done in reliance on the authorisation before then.

When confirming an authorisation, the Secretary of State may substitute an earlier date or time for the specified date or time, or substitute a more restricted area or place for the specified area or place, or both. Where an authorisation specifies more than one area or place, the latter power includes a power to remove areas or places from the authorisation.

The Secretary of State may cancel an authorisation which has been confirmed.

Cancellation or variation of authorisation by senior police officer

21.31 By TA 2000, Sch 6B, a senior police officer may cancel an authorisation, or substitute an earlier date or time, or a more restricted area or place, for that specified in the authorisation. Where an authorisation specifies more than one area or place this power of substitution includes a power to remove areas or places from the authorisation. Any cancellation or substitution by a senior officer in relation to an authorisation confirmed by the Secretary of State must be notified to the Secretary of State but does not require confirmation by the Secretary of State. If, during the currency of an authorisation, the authorising officer:

(a) no longer reasonably suspects that an act of terrorism of the description given in the authorisation will take place or no longer considers that the powers are necessary to prevent such an act, (s)he must cancel the authorisation immediately; or

(b) believes that the duration or geographical extent of the authorisation is no longer necessary for the prevention of such an act of terrorism, (s)he must substitute a shorter period, or more restricted geographical area.

New authorisation

21.32 The existence, expiry or cancellation of an authorisation does not prevent the giving of a new authorisation (Sch 6B).

Exercising the stop and search powers under s 47A

21.33 By TA 2000, s 47A(4), a constable in uniform may exercise the power conferred by an authorisation under s 47A(2) or (3) only for the purpose of discovering whether there is anything which may constitute evidence that the vehicle concerned is being used for the purposes of terrorism or (as the case may be) that the person concerned is or has been concerned in the commission, preparation or instigation of acts of terrorism. The search can therefore only be carried out to look for anything that would link the vehicle or the person to terrorism.

The power conferred by such an authorisation may be exercised whether or not the constable reasonably suspects that there is such evidence.

When exercising s 47A powers, officers should have a basis for selecting individuals or vehicles to be stopped and searched. This basis will be set by the tactical briefing which has been given. Constables should still consider whether powers requiring reasonable suspicion are more appropriate and should only use the powers conferred by a s 47A authorisation if they are satisfied that they cannot meet a threshold of reasonable suspicion sufficient to use other police powers.

When selecting individuals to be stopped and searched, officers should consider the following:

Deciding which power to use   If a s 47A authorisation is in place, the powers conferred by that authorisation may be used. However, if there is a reasonable suspicion that a person is a terrorist, then powers requiring reasonable suspicion in TA 2000, s 43 (**21.22**) or s 43A (**21.23**) should be used as appropriate instead.

Selecting an individual or vehicle using indicators   What are the geographical limits of the authorisation and what are the parameters within which the briefing allows stops and searches to be conducted? Behaviour: Is the person to be stopped and searched acting in a manner that gives cause for concern, or is a vehicle being used in such a manner? Could the person's clothing conceal an article of concern, which may constitute evidence that a person is a terrorist? Could an item being carried conceal an article that could constitute evidence that a person is a terrorist or a vehicle is being used for the purposes of terrorism?

Selecting individuals 'at random'   What are the geographical and other parameters of the operation as set out in the authorisation?

Explanation   Officers should be reminded of the need to explain to people why they or their vehicles are being searched.

Equality   The powers to stop and search must be used fairly, responsibly, and in accordance with the Equality Act 2010 (**4.1**). Officers should take care to avoid any form of racial or religious profiling when selecting people to search under s 47A powers. Racial or religious profiling is the use of racial, ethnic, religious, or other stereotypes, rather than individual behaviour or specific intelligence, as a basis for making operational or investigative decisions about who may be involved in criminal activity. Profiling in this way may amount to an act of unlawful discrimination, as would selecting individuals for a search on the grounds of any of the other protected characteristics. Profiling people from certain ethnicities or religious backgrounds may also lose the confidence of communities.

Great care should be taken to ensure that the selection of people is not based solely on ethnic background, perceived religion, or other protected characteristic. A person's appearance or ethnic background will sometimes form part of a potential suspect's description, but a decision to search a person under powers conferred by s 47A should be made only if such a description is available.

Removal of clothing

21.34   A constable may not require a person to remove any clothing in public except for headgear, footwear, an outer coat, a jacket, or gloves. Officers should be aware of the cultural sensitivities that may be involved in the removal of headgear.

Photography/film

21.35   It is important that police officers are aware, in exercising their counter-terrorism powers, that:

(a) members of the public and media do not need a permit to film or photograph in public places;

(b) it is not an offence for a member of the public or journalist to take photographs/film of a public building; and

(c) the police have no power to stop the filming or photographing of incidents or police personnel.

Under s 47A, police officers can stop and search someone taking photographs/film within an authorised area just as they can stop and search any other member of the public in the proper exercise of their discretion in accordance with the legislation and provisions of the Code. However, an authorisation itself does not prohibit the taking of photographs or digital images.

On the rare occasion that an officer reasonably suspects that photographs/film are being taken as part of hostile terrorist reconnaissance, a search under TA 2000, s 43 (**21.22**) or an arrest should be considered. It is important that police officers do not automatically consider photography/filming as suspicious behaviour. The size of the camera/video equipment should not be considered as a risk indicator.

Health and safety

When undertaking any search, officers should always consider their own safety and the health and safety of others. Officers should have an appropriate level of personal safety training and be in possession of personal protective equipment. Officers carrying out searches should use approved tactics to keep themselves and the public safe.   **21.36**

If, during the course of a stop and search there is a suspicion that a person is in possession of a hazardous device or substance, an officer should immediately request the assistance of officers appropriately trained and equipped to deal with the situation.

Seizure

Section 47A(6) provides that a constable may seize and retain anything which he: **21.37**

(a) discovers in the course of a search under an authorisation under s 47A(2) or (3); and

(b) reasonably suspects may constitute evidence that the vehicle concerned is being used for the purposes of terrorism or (as the case may be) that the person concerned is a person who is or has been concerned in the commission, preparation or instigation of acts of terrorism.

Film and memory cards may be seized as part of the search if the officer reasonably suspects they are evidence that the person is a terrorist, or a vehicle is being used for the purposes of terrorism, but officers do not have a legal power to delete images or destroy film. Cameras and other devices should be left in the state they were found and forwarded to appropriately trained staff for forensic examination. The person being searched should never be asked or allowed to turn the device on or off because of the danger of evidence being lost or damaged.

Seizures of cameras etc may only be made, following a stop and search, where the officer reasonably suspects that they constitute evidence that the person is a terrorist or that the vehicle is being used for the purposes of terrorism, as the case may be.

Anything seized may be retained for as long as necessary in all the circumstances. This includes retention for use as evidence at a trial for an offence.

A record should be made of any item seized or retained and made available with a copy of the record of the stop and search. If the reasonable suspicion referred to in (b) above ceases to apply, the item should be returned to the individual from whom it was seized, or the person in charge of the vehicle from which it was seized, unless there are other grounds for retaining it (eg in respect of the investigation of a separate offence). If there appears to be a dispute over the ownership of the article, it may be retained for as long as necessary to determine the lawful owner.

Points general to TA 2000, ss 43, 43A, and 47A

21.38   Given that they require reasonable suspicion in order to be exercised, the use of powers under ss 43(1) and 43A (**21.22** and **21.23**) should be prioritised for the purposes of stopping and searching individuals for the purposes of preventing or detecting terrorism. The authorisation of the no suspicion powers under s 47A should only be considered as a last resort, where reasonable suspicion powers are considered inadequate to respond to the threat. Use of the search powers under s 47A should only be made (in an authorised area) when the powers in ss 43 and 43A are not appropriate.

The Terrorism Search Powers Code provides guidance as to 'reasonable suspicion' which corresponds with that under PACE Code A: the Stop and Search Code (**4.11**).

Steps to be taken prior to a search

21.39   Before any search of a detained person or attended vehicle takes place the officer must take reasonable steps to give his/her identification number and name of police station (subject to the endangerment exception below) to the person to be searched or to the person in charge of the vehicle to be searched and to give that person the following information:

(a) that (s)he is being detained for the purposes of a search;

(b) the legal search power which is being exercised;

(c) a clear explanation of:

   (i) the object of the search (ie to search for evidence that the person is a terrorist or that a vehicle is being used for the purposes of terrorism); and

   (ii) in the case of s 47A, the nature of the powers conferred by s 47A, the fact an authorisation has been given and the reason why (s)he has been selected for a stop and search, or, in the case of s 43 or 43A, the grounds for suspicion;

(d) that (s)he is entitled to a copy of the search record if (s)he asks within three months of the search;

(e) that if (s)he is not arrested and taken to a police station as a result of the search and it is practicable to make the record on the spot, that immediately after the search is completed (s)he will be given (subject to the officer being called to an incident of higher priority), if (s)he requests, either:

   (i) a copy of the record, or

   (ii) a receipt which explains how (s)he can obtain a copy of the full record or access to an electronic copy of the record. A receipt may take the form of a simple business card which includes sufficient information to locate the record should the person ask for a copy, for example, the date and place of the search or a reference number;

(f) that if (s)he is arrested and taken to a police station as a result of the search, that the record will be made at the station as part of his/her custody record and (s)he will be given, if (s)he requests, a copy of his/her custody record which includes a record of the search as soon as practicable whilst (s)he is at the station.

A person who is not provided with an immediate copy of a search record may request a copy within three months of being stopped and searched. In addition a person is also entitled, on application, to a written statement that (s)he was stopped by virtue of the powers conferred by s 47A(2) or (3) (**21.27**), if requested within 12 months of the stop taking place.

If the person to be searched, or person in charge of a vehicle to be searched, does not appear to understand what is being said, or there is any doubt about the person's ability to understand English, the officer must take reasonable steps to bring information regarding that person's rights to his/her attention. If (s)he is deaf or cannot understand English and is accompanied by someone, then the officer may try to establish whether that person can interpret or otherwise help the officer to give the required information. This does not preclude an officer from conducting a search once (s)he has taken reasonable steps to explain the person's rights.

Conduct of stops and searches

The constable may detain the person or vehicle for such time as is reasonably required **21.40** to permit the search to be carried out at or near the place where the person or vehicle is stopped. The length of time must be kept to a minimum. A person or vehicle may be detained under the stop and search powers at a place other than where the person or vehicle was first stopped, only if that place, be it a police station or elsewhere, is nearby. Such a place should be located within a reasonable travelling distance using whatever mode of travel (on foot or by vehicle) is appropriate.

All stops and searches must be carried out with courtesy, consideration and respect for the person concerned. Every reasonable effort must be made to minimise the embarrassment that a person being searched may experience. The co-operation of the person to be searched must be sought in every case, even if the person initially objects to the search. A forcible search may be made only if it has been established that the person is unwilling to co-operate or resists. Reasonable force may be used as a last resort if necessary to conduct a search or to detain a person or vehicle for the purposes of a search.

Recording requirements

Searches which do not result in an arrest   When an officer carries out a search under TA **21.41** 2000, s 43(1) or 43A (**21.22** and **21.23**) or in the exercise of powers conferred by s 47A (**21.24**) and the search does not result in the person searched, or person in charge of the vehicle searched, being arrested and taken to a police station, a record must be made of it at the time, electronically or on paper, unless there are circumstances which make this wholly impracticable. If a record is not made at the time of the stop and search, the officer must make the record as soon as practicable after the search is completed. There may be situations in which it is not practicable to obtain the information necessary to complete a record, but the officer should make every reasonable effort to do so. If it is not possible to complete a record in full, an officer must make every reasonable effort to at least record details of the date, time and place where the stop and search took place, the power under which it was carried out, and the officer's identification number.

If the record is made at the time, the person who has been searched or who is in charge of the vehicle that has been searched must be asked if (s)he wants a copy of the record and if (s)he does, (s)he must be given immediately, either:

(a) a copy of the record, or
(b) a receipt which explains how (s)he can obtain a copy of the full record or access to an electronic copy of the record.

There is an exception: an officer is not required to provide a copy of the full record or a receipt at the time if (s)he is called to an incident of higher priority.

Where it is not practicable to provide a written copy of the record or immediate access to an electronic copy of the record or a receipt at the time, the officer should give the person details of the police station at which (s)he may request a copy of the record.

Searches which result in an arrest   If a search conducted under s 43, 43A, or 47A results in a person being arrested and taken to a police station, the officer who carried out the search is responsible for ensuring that a record of the search is made as part of that person's custody record. The custody officer must then ensure that the person is asked if (s)he wants a copy of the custody record and if (s)he does, that (s)he is given a copy as soon as practicable.

Record of search

21.42   The record of a search must always include the following information:

(a) a note of the self-defined ethnicity, and, if different, the ethnicity as perceived by the officer making the search, of the person searched or of the person in charge of the vehicle searched (as the case may be);

(b) the date, time, and place the person or vehicle was searched;

(c) the object of the search;

(d) in the case of the powers under:
   (i) s 47A, the nature of the powers, the fact an authorisation has been given and the reason the person or vehicle was selected for the search;
   (ii) s 43 or 43A, the grounds for suspicion;

(e) the officer's warrant number or other identification number (subject to the endangerment exception).

Officers should record the self-defined ethnicity of every person stopped according to the same categories as used in PACE Code A (the Stop and Search Code) (**4.30**). The procedure is identical to that set out at **4.30**.

For the purposes of completing the search record, there is no requirement to record the name, address, and date of birth of the person searched or the person in charge of a vehicle which is searched and the person is under no obligation to provide this information. An officer may remind a person that providing these details will ensure that the police force is able to provide information about the stop and search in future should the person request that information or if it is otherwise required.

The names of police officers are not required to be shown on the search record in the case of operations linked to the investigation of terrorism or otherwise where an officer reasonably believes that recording names might endanger the officers; this is the 'endangerment exception' referred to above. In such cases the record must show the officers' warrant or other identification number and duty station.

A record is required for each person and each vehicle searched. However, if a person is in a vehicle and both are searched, and the object and grounds of the search are the same, only one record need be completed. If more than one person in a vehicle is searched, separate records for each search of a person must be made. If only a vehicle is searched, the self-defined ethnic background of the person in charge of the vehicle must be recorded, unless the vehicle is unattended.

The record of the grounds for making a search must, briefly but informatively, explain the reason for suspecting the person concerned, by reference to the person's behaviour and/or other circumstances, or, in the case of searches under s 47A, the reason why a particular person or vehicle was selected.

After searching an unattended vehicle, or anything in or on it, an officer must leave a notice in it (or on it, if things on it have been searched without opening it) recording the fact that it has been searched. The notice must include the name of the police station to which the officer concerned is attached and state where a copy of the record of the search may be obtained and how (if applicable) an electronic copy may be accessed and where any application for compensation should be directed. The vehicle must if practicable be left secure.

Monitoring and supervising the use of stop and search powers

21.43 The Terrorism Search Powers Code contains provisions in this respect which correspond to those in PACE Code A (**4.32–4.33**). Statistical data on the use of powers should be provided quarterly to the Home Office.

Community engagement

21.44 Ongoing community engagement is essential in improving relationships with the community. Use may be made, for example, of existing community engagement arrangements. However, where stop and search powers affect sections of the community with whom channels of communication are difficult or non-existent, these should be identified and steps taken to engage. For example, if s 47A authorisations have primarily been made around transport hubs, effort should be made to engage with people using those hubs.

When planning a counter-terrorism search operation, police authorities and the local CONTEST Prevent strategic partnership should be involved at the earliest opportunity to provide advice and assistance in identifying mechanisms for engaging with communities.

If it is not possible to carry out community engagement prior to authorisation, police forces should carry out a retrospective review of the use of the powers, including stakeholders. Police forces should continue to monitor the use of s 47A powers for the duration of an authorisation, both in discussion with community representatives and by explaining how and why the powers are being used to individuals who are stopped and searched. Officers should be ready to explain to individuals why the powers are in place, insofar as this can be communicated without disclosing sensitive intelligence or causing undue alarm. Stop and search operations should form part of wider counter-terrorism policing, and public awareness of the powers should be considered as part of any wider communications strategy associated with an operation.

Prohibitions or restrictions on parking

21.45 TA 2000, s 48 permits a police officer of or above the rank of assistant chief constable (or commander), where it appears to him/her to be expedient in order to prevent acts of terrorism (**21.2**), to give an authorisation to any constable to prohibit or restrict the parking of vehicles on a specified road. This is to be done by the placing of traffic signs. A constable exercising his/her powers under s 48 may suspend a parking place and this will have the effect of permitting the removal of vehicles. Such an authorisation remains in force for a period, not exceeding 28 days, specified in it. It may be renewed subject to the same rules.

Failing to move a vehicle when ordered to do so by a constable is an offence under s 51(2). Parking in contravention of such a prohibition or restriction is an offence under s 51(1). It is a defence to either offence to prove a reasonable excuse. Current disabled persons' badge-holders are not exempt from these requirements and have no reasonable excuse for failure to comply on those grounds. The powers to impose these restrictions are additional to any other powers available to a constable.

The offences under s 51 are summary offences. An offence under s 51(1) is punishable with a fine not exceeding level 4, and an offence under s 51(2) with up to three months' imprisonment and/or a fine not exceeding level 4.

Arrest without warrant

21.46   TA 2000, s 41 provides that a constable may arrest without warrant a person whom he reasonably suspects to be a terrorist (as defined at **21.22**).

A person who has been so detained may be held for 48 hours. Continued detention beyond that period is subject to a warrant of further detention granted in accordance with the procedures prescribed by TA 2000, Sch 8 (**21.53**).

If a person detained under s 41, including by virtue of a warrant of further detention under Sch 8, is removed to hospital because (s)he needs medical treatment:

(a) any time during which (s)he is being questioned in hospital or on the way there or back for the purpose of obtaining relevant evidence is to be included in calculating any period which falls to be calculated for the purposes of s 42 or the warrant of further detention provision, but

(b) any other time when (s)he is in hospital or on the way there or back is not to be included.

For these purposes, 'relevant evidence' means, in relation to the detained person, evidence which:

(i) relates to his/her commission of an offence under any of the following sections of TA 2000: s 11 (membership of a proscribed organisation); 12 (support for it); ss 15–18 (fund-raising, etc and money laundering); s 54 (weapons training); or ss 56–63 (directing terrorist organisation, possession for terrorist purposes, collecting information, eliciting, etc information about members of the armed forces, entering or remaining in a designated area, inciting terrorism overseas, and terrorist bombing and finance offences outside the UK); or

(ii) indicates that (s)he is or has been concerned in the commission, preparation or instigation of acts of terrorism.

PACE Code H (**21.47**) provides that, if a person is arrested by one police force on behalf of another and the lawful period of detention in respect of that offence has not yet started in accordance with s 41, no questions may be put to him/her about the offence while (s)he is in transit between the forces except to clarify any voluntary statement which (s)he makes.

Treatment of persons detained under Terrorism Act 2000, s 41 (or Sch 7)

21.47   The following rules, set out in TA 2000, Sch 8, relate to someone detained under TA 2000, s 41 or Sch 7 (special powers of 'examining officers'—ie constables, immigration officers, and designated customs officers—to stop, question, and detain a person at seaports and airports to determine whether (s)he appears to be someone concerned in the commission, preparation, or instigation of acts of terrorism). The relevant code of practice in relation to those detained under s 41 is not PACE Code C (the Detention Code) but is PACE Code H: the Code of Practice in Connection with the Detention, Treatment and Questioning by Police Officers of Persons under s 41 of, and Sch 8 to, the Terrorism Act 2000 and the Treatment and Questioning of Detained Persons in

Respect of whom an Authorisation to Question after Charge has been given under C-TA 2008, s 22. In relation to the first part of its subject matter, Code H generally mirrors Code C (see Chapters 5 and 6). A separate Code of Practice for Examining Officers and Review Officers applies to those detained under Sch 7.

Except in relation to the special powers to question under Sch 7, TA 2000, s 114 empowers a constable to use reasonable force to exercise any power relating to the treatment of those detained under s 41 or Sch 7.

Place of detention

21.48 The Secretary of State is required to designate places at which persons may be detained under s 41 or Sch 7. A reference in Sch 8 to a police station includes a place so designated. Under Sch 8, a person arrested by a constable under s 41 must be taken, as soon as practicable, to the police station which the constable considers to be the most appropriate. If the person was arrested as a result of a search in which PACE Code A (the Stop and Search Code) or the Terrorism Search Powers Code applies, the officer carrying out the search is responsible for ensuring that the record of that stop and search is made as part of the person's custody record. The custody officer must then ensure that the person is asked if (s)he wants a copy of the search record and if (s)he does, that (s)he is given a copy as soon as practicable. The person's entitlement to a copy of the search record which is made as part of his/her custody record is in addition to, and does not affect, his/her entitlement to a copy of his/her custody record.

Identification

21.49 An 'authorised person' (which term includes a constable) may take any steps which are reasonably necessary for photographing, measuring, or identifying the detained person. However, these initial measures do not include the taking of fingerprints, non-intimate samples, or intimate samples.

Interviews

21.50 Following the arrest of a person under s 41, that person must not be interviewed about the relevant offence except at a place designated for detention (**21.48**), unless the consequent delay would be likely to: (a) lead to interference with, or harm to, evidence connected with an offence; interference with, or physical harm to, other people; or serious loss of, or damage to, property; (b) lead to alerting other people suspected of committing an offence but not yet arrested for it; or (c) hinder the recovery of property obtained in consequence of the commission of an offence. Interviewing in any of these circumstances must cease once the relevant risk has been averted or the necessary questions have been put in order to attempt to avert that risk.

Acting under a power in TA 2000, Sch 8, the Secretary of State has made an order requiring the video recording with sound of an interview at a police station by a constable of a person detained under TA 2000, s 41 or Sch 7. 'Police station' includes for these purposes any place designated as a place where a person may be detained under TA 2000, s 41. An interview with someone detained under s 41 must be conducted in accordance with the Code of Practice for the Video Recording with Sound of Interviews with Persons Detained under TA 2000, s 41 or Sch 7 and Post-Charge Questioning of Persons Authorised under C-TA 2008, s 22.

The Code provides in relation to such an interview that it may be necessary to delay an interview to make arrangements to overcome any difficulties that might otherwise prevent the record being made, eg non-availability of suitable recording equipment and interview facilities. If a person refuses to go into or remain in a suitable interview room, and the custody officer considers, on reasonable grounds, that the interview should not be delayed, the interview may, at the custody officer's discretion, be conducted in a cell using portable recording equipment. The reasons must be recorded.

Before any interview starts, the person and any appropriate adult (as defined at **5.7**) and interpreter must be given a written notice which explains that the interview must be video recorded with sound. At the same time, the person, the appropriate adult and interpreter must be informed verbally of the content of the notice.

If the person interviewed or the appropriate adult raises objections to the interview being recorded, the interviewing officer must explain that the interview is being recorded in order to protect both the person being interviewed and the interviewing officer and that there is no opt-out facility.

The provisions of the Code about the procedure relating to the recording mirror those under PACE Code F (**5.71-5.77**).

Rights

21.51   A person detained at a place under TA 2000, s 41 or Sch 7 is entitled, if (s)he so requests, to have a named person who is a friend, relative, or person known to him/her who is likely to take an interest in his/her welfare informed, as soon as is reasonably practicable, that (s)he is being detained there. If the detainee is transferred from one place to another, (s)he is entitled to exercise that right in respect of the place to which (s)he is transferred. A detainee must be informed of the above right on first being detained.

In addition, the detainee is entitled, if (s)he so requests, to consult a solicitor as soon as is reasonably practicable, privately and at any time. A detainee must be informed of this right on first being detained. A record must be made of such a request and the action taken.

Where a detainee under Sch 7 asks to consult a solicitor, the examining officer may not question him/her to determine whether (s)he appears to be concerned in the commission, preparation or instigation of acts of terrorism until the detainee has consulted a solicitor (or no longer wishes to do so). This does not apply if the officer reasonably believes that so postponing the questioning would be likely to prejudice determination of the relevant matters. The detainee is entitled to consult a solicitor in person, unless the examining officer reasonably believes that the time it would take to consult a solicitor in person would be likely to prejudice determination of the relevant matters in which case (s)he may require any consultation to take place in another way.

A superintendent (or above) may authorise a delay in informing the person named by the detainee or in permitting the detainee to consult a solicitor, but a detainee under s 41 must be permitted to exercise these rights within 48 hours of arrest. Such an officer may give an authorisation only if (s)he has reasonable grounds for believing that informing the named person or solicitor of the detainee's detention at the time when the detainee wishes this to be done, will lead to:

(a) interference with or harm to evidence of an indictable offence;

(b) interference with or physical injury to any person;

(c) alerting persons who are suspected of having committed an indictable offence but who have not been arrested for it;

(d) hindering the recovery of property obtained as a result of an indictable offence or in respect of which a forfeiture order in respect of the terrorist property offences in TA 2000, ss 15 to 18 or other terrorism offences could be made under the Act;

(e) interference with the gathering of information about the commission, preparation or instigation of acts of terrorism;

(f) alerting a person and thereby making it more difficult to prevent an act of terrorism; and

(g) alerting a person and thereby making it more difficult to secure a person's apprehension, prosecution or conviction in connection with the commission, preparation or instigation of an act of terrorism.

In addition, a similar authorisation may be given where a superintendent (or above) has reasonable grounds for believing that the detainee has benefited from his/her criminal conduct and that the recovery of the value of the property constituting the benefit will be hindered by informing the named person, or the exercise of the right to see a solicitor.

Where any delay is authorised orally under this provision, it must be confirmed in writing as soon as reasonably practicable. The detainee must be told of the reason for the delay as soon as reasonably practicable, and the reason must be recorded.

Where a detainee exercises the right to consult a solicitor, a superintendent (or above) may direct that the right may:

(a) not be exercised (or further exercised) by consulting the solicitor who attends for the purpose of the consultation or who would so attend but for the giving of the direction, but

(b) instead be exercised by consulting a different solicitor of the detainee's choosing.

Such a direction may be given before or after a detainee's consultation with a solicitor has started (and if given after it has started the right to further consult that solicitor ceases on the giving of the direction). A direction may only be given if the officer giving it has reasonable grounds for believing that:

(i) unless the direction is given, the exercise of the right of the detainee will have any of the consequences specified in (a)–(g) above; or

(ii) the detainee has benefited from his/her criminal conduct and that, unless the direction is given, the exercise of the right by the detainee will hinder the recovery of the value of the property constituting the benefit.

Fingerprinting, non-intimate, and intimate samples

Taking fingerprints or non-intimate samples   Fingerprints or non-intimate samples   **21.52**
(**7.39**) may be taken from a person detained under s 41 or Sch 7 by a constable:

(a) with the written appropriate consent (**7.42**) of the detainee; or

(b) without such consent where:

    (i) (s)he is detained at a police station and a superintendent (or above) gives an authorisation, or

    (ii) the detainee has been convicted of a recordable offence (and, in in the case of a non-intimate sample, (s)he was so convicted on or after 10 April 1995). As to 'recordable offence', see **7.43**.

A superintendent (or above) may make an authorisation under (b)(i) if a person is detained under TA 2000, s 41, where:

(i) the officer reasonably suspects that the person has been involved in an offence under s 11 (membership of a proscribed organisation); s 12 (support for it); ss 15–18 (fund-raising, etc and money laundering); s 54 (weapons training); or ss 56–63 (directing terrorist organisation, possession for terrorist purposes, collecting information, eliciting, etc information about members of the armed forces, entering or remaining in a designated area, inciting terrorism overseas, and terrorist bombing and finance offences outside the UK); and *the officer reasonably believes that the fingerprints or sample will tend to confirm or disprove his/her involvement, or,*

(ii) in any case, the officer is satisfied that taking the fingerprints or sample is necessary to assist in determining whether the person has been concerned in the preparation, commission or instigation of acts of terrorism.

A superintendent (or above) may also authorise the taking under (b)(i) of fingerprints from a person detained at a police station if (s)he is satisfied that this person's fingerprints will facilitate the ascertainment of his/her identity, and that person has refused to identify him/herself or the officer has reasonable grounds for suspecting that that person is not who (s)he claims to be. References to ascertaining a person's identity include references to showing that (s)he is not a particular person.

If non-intimate samples of hair (**7.39**) are taken, they may be taken by cutting or plucking provided that no more are plucked than is reasonably necessary to provide a sufficient sample.

Taking intimate samples   An intimate sample (**7.38**) may be taken by a constable from a person detained under TA 2000, s 41, but only if (s)he is detained at a police station, with the written appropriate consent and the authorisation of a superintendent (or above). However, an intimate sample other than a sample of urine or a dental impression may be taken only by a registered medical practitioner acting on the authority of a constable. A dental impression may be taken only by a registered dentist acting on the authority of a constable. An authorisation for the consensual taking of an intimate sample may only be given if the conditions italicised above are satisfied.

Where two or more non-intimate samples suitable for the same means of analysis have been taken from a person and these samples have proved insufficient, and the person concerned has been released from detention, an intimate sample may be taken by a constable with written appropriate consent and the authorisation of a superintendent (or above). Such an authorisation for the consensual taking of an intimate sample may only be given if the conditions italicised above are satisfied.

General   Before fingerprints or samples are taken, the person must be informed that they may be used for the purpose of checking against other fingerprints or samples held on behalf of police forces (speculative search) and, where taken with consent or in consequence of having been convicted of a recordable offence, of the reason for the fingerprints or samples being taken. All such matters must be recorded.

Destruction and retention of fingerprints and DNA profiles, etc   TA 2000, Sch 8, para 20A provides that paragraph 20A material, ie:

(a) fingerprints taken under Sch 8;
(b) a DNA profile derived from a DNA sample taken under Sch 8;
(c) relevant physical data taken or provided by virtue of provisions which deal with the retention, destruction and use of biometric material taken in Scotland;
(d) a DNA profile derived from a DNA sample taken by virtue of specified Scottish provision,

must be destroyed if it appears to the responsible chief officer of police that:

(i) the taking or providing of the material or, in the case of a DNA profile, the taking of the sample from which the DNA profile was derived, was unlawful, or
(ii) the material was taken or provided, or (in the case of a DNA profile) was derived from a sample taken, from a person in connection with that person's arrest under TA 2000, s 41 and the arrest was unlawful or based on mistaken identity.

'*Responsible chief officer of police*' means, in relation to fingerprints or samples taken in England or Wales, or a DNA profile derived from a sample so taken, the chief officer of police for the police area in which the material concerned was taken.

In any other case, para 20A material must be destroyed unless it is retained under any power conferred by paras 20B to 20E (below).

Para 20A material which ceases to be retained under a power mentioned below may continue to be retained under any other such power which applies to it.

Nothing in para 20A prevents a relevant search, in relation to para 20A material, from being carried out within such time as may reasonably be required for the search if the responsible chief officer of police considers the search to be desirable. A 'relevant search' is a search carried out for the purpose of checking the material against:

(a) other fingerprints or samples taken under the above provisions or a DNA profile derived from such a sample;
(b) any of the relevant physical data (as defined by the Criminal Procedure (Scotland) Act 1995, s 18(7A)), samples or information mentioned or held by virtue of specified corresponding Scottish legislation;
(c) material to which the Counter-Terrorism Act 2008, s 18 (**7.76**) applies;
(d) any of the fingerprints, data or samples obtained under the Terrorism Prevention and Investigation Measures Act 2011, Sch 6 or the Counter-Terrorism and Border Security Act 2019, Sch 3 or information derived therefrom;
(e) any of the fingerprints, samples and information mentioned in PACE, s 63A(1)(a) and (b) (**7.50**); and
(f) any of the fingerprints, samples and information mentioned in the corresponding Northern Irish legislation about checking of fingerprints and samples.

Any copies of the fingerprints or relevant physical data required to be destroyed by para 20A which are held by a police force must also be destroyed. No copy of a DNA profile required by para 20A to be destroyed may be retained by a police force except in a form which does not identify the person to whom the profile relates.

Retention of para 20A material   By TA 2000, Sch 8, paras 20B and 20C, the following provisions apply to para 20A material relating to a detainee under s 41 or Sch 7.

In the case of a person *who has previously been convicted* of a recordable offence (as defined at **7.43**) (other than a single exempt conviction (below)), or an imprisonable offence in Scotland, or is so convicted before the end of the period within which the material may be retained by virtue of para 20B, may be retained indefinitely.

The reference to a recordable offence in Sch 8, para 20B includes an offence under the law of a country or territory outside the UK where the act constituting the offence would constitute:

(a) a recordable offence under the law of England and Wales if done there;
(b) a recordable offence under the law of Northern Ireland if done there,

and, where a person has previously been convicted, this applies whether or not the act constituted such an offence when the person was convicted.

The reference to an imprisonable offence in Scotland includes an offence under the law of a country or territory outside the UK where the act constituted the offence would constitute an imprisonable offence under the law of Scotland if done there, and, where a person has previously been convicted, this applies whether or not the act constituted such an offence when the person was convicted.

In the case of a person *who has no previous convictions, or only one exempt conviction*, the material may be retained until the end of the relevant retention period. The retention period is, in the case of:

(a) fingerprints or relevant physical data, the period of three years (six months in the case of a Sch 7 detainee) beginning with the date on which the fingerprints or relevant physical data were taken or provided, and

(b) a DNA profile, the period of three years (six months in the case of a Sch 7 detainee) beginning with the date on which the DNA sample from which the profile was derived was taken (or, if the profile was derived from more than one DNA sample, the date on which the first of those samples was taken).

The responsible chief officer of police or a specified chief officer of police may apply to a District Judge (Magistrates' Court) for an order extending the retention period for two years. Such an application must be made within a three-month period ending on the last day of the retention period. A 'specified chief officer of police' means the chief officer of the police force of the area in which the person from whom the material was taken resides, or a chief officer of police who believes that the person is in, or is intending to come to, the chief officer's police area.

Nothing in para 20B or 20C prevents the start of a new retention period in relation to para 20A material if a person is detained again under TA 2000, s41 or Sch 7 when an existing retention period (whether or not extended) is still in force in relation to that material.

For the purposes of paras 20B and 20C, para 20D provides that a person is to be treated as having been convicted of an offence if:

(a) in relation to a recordable offence in England and Wales or Northern Ireland (s)he has been:
   (i) given a caution in respect of the offence (or has been warned or reprimanded under the Crime and Disorder Act 1998, s 65 (repealed)), or
   (ii) found not guilty by reason of insanity or to be under a disability and to have done the act charged,

(b) in relation to an imprisonable offence in Scotland:
   (i) (s)he has accepted or has been deemed to accept a conditional offer, a compensation offer, a combined offer, or a work offer,
   (ii) (s)he has been acquitted on account of insanity or committed to hospital for inquiry into his/her mental condition,
   (iii) having been given a fixed penalty notice, (s)he has paid the fixed penalty, or the sum which (s)he is liable to pay, or
   (iv) (s)he has been discharged absolutely.

For the purposes of 20B and 20C:

(a) 'offence', in relation to any country or territory outside the UK, includes an act punishable under the law of that country or territory, however it is described;

(b) a person has in particular been convicted of an offence under the law of a country or territory outside the UK if a court in that country or territory has made in respect of such an offence a finding equivalent to a finding of not guilty by reason of insanity, or of being under a disability and to have done the act charged.

For the purposes of the above provisions relating to the retention of para 20A material, 'spent convictions' count as convictions but a conviction in respect of the repealed offences of buggery and gross indecency with a man is not to be treated as a conviction if it is to be disregarded by virtue of a 'disregard decision' by the Secretary of State.

In addition:

(a) a person has *no previous convictions* if (s)he has not previously been convicted in England and Wales or Northern Ireland of a recordable offence, or, in Scotland, of an imprisonable offence, and

(b) a conviction for a *recordable* offence is *exempt* if it is in respect of such an offence, other than a qualifying offence (as defined at **7.35(7)**), committed when the person was under 18.

For the purposes of (a), a person (P) is to be treated as having previously been convicted in England and Wales (or Northern Ireland) of a recordable offence if P has previously been convicted of an offence under the law of a country or territory outside the UK, and the act constituting the offence would constitute a recordable offence under the law of England and Wales (or Northern Ireland) if done there (whether or not it constituted such an offence when P was convicted). A corresponding provision applies in respect of a Scottish imprisonable offence. Note in relation to (b) that the reference to a qualifying offence includes a reference to an offence under the law of a country or territory outside the UK where the act constituting the offence would constitute a qualifying offence under the law of England and Wales (or Northern Ireland) if done there (whether or not it constituted such an offence when the person was convicted).

If a person is convicted of more than one offence arising out of a single course of action, those convictions are treated as a single conviction for the purposes of calculating under para 20B whether the person has been convicted of only one offence.

Retention of para 20A material: national security determination Paragraph 20A material may be retained for as long as a national security determination made by a chief officer of police has effect in relation to it. A national security determination is made if a chief officer of police determines that it is necessary for any para 20A material to be retained for the purposes of national security.

A national security determination:

(a) must be made in writing,

(b) has effect for a maximum of five years beginning with the date on which it is made, and

(c) may be renewed.

Retention of further fingerprints Where paragraph 20A material is or includes a person's fingerprints ('the original fingerprints'), a constable may make a determination in respect of the retention of any further fingerprints taken from, or provided by, the same person ('the further fingerprints') if conditions 1 and 2 are met.

Condition 1 is met if the further fingerprints are:

(a) para 20A material;

(b) taken or provided under or by virtue of PACE or the corresponding Northern Irish or Scottish provisions, or the Terrorism Prevention and Investigation Measures Act 2011, or the Counter-Terrorism and Border Security Act 2019; or

(c) material to which the Counter-Terrorism Act 2008, s 18 (**7.76**) applies.

Condition 2 is met if:

(a) in a case where the further fingerprints are material to which the Counter-Terrorism Act 2008, s 18 applies, the original fingerprints and the further fingerprints are held under the law of the same part of the UK;

(b) in any other case, the original fingerprints and the further fingerprints were taken from or provided by the person in the same part of the UK.

Where such a determination is made in respect of the further fingerprints:

(i) the further fingerprints may be retained for as long as the original fingerprints are retained under the provisions above about retention of para 20A material, and

(ii) a requirement under any UK legislation to destroy the further fingerprints does not apply for as long as their retention is authorised by (i).

(i) does not prevent the further fingerprints being retained after the original fingerprints fall to be destroyed if the continued retention of the further fingerprints is authorised under any UK legislation.

A written record must be made of a determination.'

Destruction of samples   Paragraph 20G deals with para 20G material, ie samples taken under the provisions of TA 2000, Sch 8 referred to on pp 753-754, or by virtue of corresponding Scottish provisions.

Samples to which para 20G applies must be destroyed if it appears to the responsible chief officer of police that:

(a) the taking of the sample was unlawful, or

(b) the sample was taken from a person in connection with that person's arrest under TA 2000, s 41 and the arrest was unlawful or based on mistaken identity.

Subject to this, the following rules apply to a sample so taken.

A DNA sample must be destroyed:

(a) as soon as a DNA profile has been derived from the sample, or

(b) if sooner, before the end of the period of six months beginning with the date on which the sample was taken.

Any other sample must be destroyed before the end of the period of six months beginning with the date on which it was taken.

The responsible chief officer of police may apply to a District Judge (Magistrates' Courts) for an order to retain a sample beyond the date on which it would otherwise be required to be destroyed by the above two paragraphs if the sample was taken from a person detained under s 41 in connection with the investigation of a qualifying offence (7.35(7)), and the responsible chief officer considers that, *having regard to the nature and complexity of other material that is evidence in relation to the offence, the sample is likely to be needed in any proceedings for the offence for the purposes of*:

(a) *disclosure to, or use by, a defendant, or,*

(b) *responding to any challenge by a defendant in respect of the admissibility of material that is evidence on which the prosecution proposes to rely.*

Such an application must be made before the date on which the sample would otherwise be required to be destroyed.

If the District Judge is satisfied that the italicised condition is met, (s)he may make an order which:

(a) allows the sample to be retained for a period of 12 months beginning with the date on which the sample would otherwise be required to be destroyed, and

(b) may be renewed (on one or more occasions) for a further period of not more than 12 months from the end of the period when the order would otherwise cease to have effect.

An application for such an order (other than an application for renewal) may be:

(a) made without notice of the application having been given to the person from whom the sample was taken, and

(b) heard and determined in private in his/her absence.

A sample retained by virtue of such an order may only be used for the purposes of proceedings for the offence in connection with which it was taken. A sample that ceases to be retained by virtue of such an order must be destroyed. Nothing in para 20G prevents a relevant search (as defined above in relation to para 20A) in relation to the above types of samples from being carried out within such time as may reasonably be required for the search if the responsible chief officer of police considers the search to be desirable.

Use of para 20A or 20G material   Any para 20A or 20G material must not be used other than:

(a) in the interests of national security,

(b) for the purposes of a terrorist investigation (as defined at **21.17**),

(c) for purposes related to the prevention or detection of crime, the investigation of an offence, or the conduct of a prosecution, or

(d) for purposes related to the identification of a deceased person or of the person to whom the material relates.

The reference to crime includes a reference to any conduct which:

(i) constitutes one or more criminal offences (whether under the law of a part of the UK or of a country or territory outside the UK), or

(ii) is, or corresponds to, any conduct which, if it all took place in any one part of the UK, would constitute one or more criminal offences.

Subject to the above paragraph, a relevant search (as defined above) may be carried out in relation to material to which para 20A or 20G applies if the responsible chief officer of police considers the search to be desirable.

Material which is required by para 20A or 20G to be destroyed must not at any time after it is required to be destroyed be used:

(a) in evidence against the person to whom the material relates, or

(b) for the purposes of the investigation of any offence.

The references above to using material include a reference to allowing any check to be made against it and to disclosing it to any person. The references to an investigation and to a prosecution include references, respectively, to any investigation outside the UK of any crime or suspected crime and to a prosecution brought in respect of any crime in a country or territory outside the UK.

Limit   The above provisions about the destruction, use and retention of fingerprints, DNA profiles and samples do not apply to para 20A material relating to a person detained under s 41 which is, or may become, disclosable under CPIA 1996 or the CPIA 1996 (s 23(1)) Code of Practice (**8.75–8.93**).

Where a sample falls within the above paragraph, and would otherwise be required to be destroyed under para 20G, it must not be used for the purposes of any proceedings for the offence in connection with which the sample was taken. A sample that once fell within the above paragraph but no longer does, and so becomes a sample to which para 20G applies, must be destroyed immediately if the time specified for its destruction under para 20G has already passed.

Review of detention under s 41

21.53 The detention of a person arrested under TA 2000, s 41 must be periodically reviewed by a review officer. The first review must be as soon as reasonably practicable after the arrest; thereafter reviews must be carried out at intervals of no more than 12 hours. Code H provides that a review officer should carry out his/her duties at the police station where the detainee is held and be allowed such access to the detainee as is necessary to exercise those duties. A review officer may authorise continued detention only if (s)he does so:

(a) to obtain relevant evidence whether by questioning the detainee or otherwise;
(b) to preserve relevant evidence;
(c) while awaiting the result of an examination or analysis of relevant evidence;
(d) for the examination or analysis of anything with a view to obtaining relevant evidence;
(e) pending a decision to apply to the Secretary of State for a deportation notice to be served on the detainee, the making of any such application, or the consideration of any such application by the Secretary of State; and
(f) pending a decision to charge the detainee with an offence.

The review officer may not authorise continued detention unless satisfied that the relevant matter is being dealt with diligently and expeditiously. The detainee may not be held for more than 48 hours unless a warrant of further detention is granted.

Under Sch 8, a Crown Prosecutor or a superintendent (or above) may apply within the 48-hour period just mentioned or the six-hour period after its end to a designated District Judge (Magistrates' Court) for a warrant of further detention for a period of seven days from the time of the arrest under s 41, unless the person was being detained by an examining officer under Sch 7 (port and border controls) at the time of arrest (in which case time runs from the start of his examination under Sch 7).

An application for a warrant of further detention (or an extension, below) may be made orally or in writing depending upon the circumstances of the case and fairness to the detainee. It may not be heard unless notice of it has been given to the detainee. Code H provides that, when such an application is made, the detainee and his/her representative must be informed of the detainee's rights in respect of the application. These include the rights to:

(a) a written notice of the application;
(b) make oral or written representations to the designated District Judge (Magistrates' Court) about the application;
(c) be present and legally represented at the hearing of the application, unless specifically excluded by the judge;
(d) free legal advice.

Applications may make use of video conferencing facilities. A judge may require the physical presence of the detainee.

The judge may issue a warrant for a shorter period than seven days where the application is for a shorter period or (s)he is satisfied that there are circumstances that would make it inappropriate for the specified period to be as long as seven days. The period specified in the warrant may be extended or further extended, on application to a designated District Judge (Magistrates' Court), if the extension would not extend the period for more than 14 days after the person's arrest (or examination commencement) or a shorter period. Thus, the maximum period of detention of a person arrested under s 41 is 14 days. The grounds for a warrant of further detention or an extension are that there are reasonable grounds for believing that the further detention of the person concerned is necessary to obtain relevant evidence and the investigation is being conducted diligently and expeditiously. The Secretary of State has an emergency power to make an order temporarily extending the maximum period of detention of a person arrested under s 41 when Parliament is dissolved or when Parliament has met after a dissolution but before the first Queen's Speech. Such an order would extend the maximum period of detention to 28 days for a period of three months.

TA 2000, Sch 8 also makes provision about the review of a person's detention under Sch 7. See Chapter 21 on the companion website.

MISCELLANEOUS OFFENCES UNDER TA 2000

Possession of articles for terrorist purposes

21.54 Under TA 2000, s 57(1), a person (D) commits an offence if (s)he possesses an article in circumstances giving rise to a reasonable suspicion that the article is in his/her possession for a purpose connected with the commission, preparation or instigation of acts of terrorism. This is an indictable (either-way) offence punishable with up to 15 years' imprisonment. According to the Court of Appeal, the reasonable suspicion must be that the person intends the article to be used for the specified purpose. Where it is proved that at the time of the commission of such an alleged offence, D and the article were both present in any premises, or the article was in the premises of which D was the occupier, or which D habitually used other than as a member of the public, the court may assume that D possessed the article, unless D adduces sufficient evidence to raise the issue that (s)he did not know of its presence on the premises or that (s)he had no control over it, whereupon the prosecution must disprove this (ss 57(3) and 118).

A document or record in electronic or printed form is an article for the purposes of s 57.

The Court of Appeal has held that the offence can only be committed if there is a direct connection between the article and an intended act of terrorism.

It is a defence for D to adduce sufficient evidence to raise the issue that possession of an article was not for a purpose connected with the commission, preparation or instigation of an act of terrorism, whereupon the defence is assumed to be satisfied unless the prosecution disproves this (ss 57(2) and 118).

Unlawful collection, recording or possession of information

21.55 TA 2000, s 58(1) provides that a person commits an offence if he:

(a) collects or makes a record of information of a kind likely to be useful to a person committing or preparing an act of terrorism; or
(b) possesses a document or record containing information of that kind; or

(c) views, or otherwise accesses, by means of the internet a document or record containing information of that kind.

By s 58(1A), the cases in which a person collects or makes a record for the purposes of (a) include (but are not limited to) those in which the person does so by means of the internet (whether by downloading the record or otherwise). By s 58(2), 'record' includes a photographic or electronic record.

It has been held by the Court of Appeal that 'information of a kind likely to be useful to a person committing or preparing an act of terrorism' is broad enough to potentially include a series of suggestions as to the avoidance of detection or surveillance. On the other hand, the House of Lords has held, if a document containing information is one in everyday use (eg a published timetable or map) it cannot be treated as falling within s 58; the aim of s 58 is to catch information which would typically be of use to terrorists, as opposed to ordinary members of the public. The information collected or possessed must, of its very nature, be designed to provide practical assistance to someone committing or preparing an act of terrorism.

The House of Lords has also held that, in order for D to be convicted of an offence of possession contrary to s 58, the prosecution must prove beyond reasonable doubt that (s)he:

(a) had control of a record which contained information that was likely to provide practical assistance to a person committing or preparing an act of terrorism;
(b) knew (s)he had the record;
(c) knew the kind of information which it contained.

If these elements are established, D falls to be convicted subject to the defence of reasonable excuse (below). While s 57 focuses on the circumstances of D's possession of the article, s 58 focuses on the nature of the information which D collected, recorded or possessed in a document or record. Section 58 does not require proof that D had a terrorist purpose for doing what D did. Unless it amounts to a reasonable excuse, D's purpose in doing what D did is irrelevant. There is an overlap between ss 57 and 58. The possession of a document can, in an appropriate case, fall within both sections.

It is a defence for D to adduce sufficient evidence to raise the issue that (s)he had a reasonable excuse for his/her action or possession, whereupon the court must assume the defence is satisfied unless the prosecution disprove this (ss 58(3) and 118). By s 58(3A), the cases in which D has a reasonable excuse for these purposes include (but are not limited to) those in which:

(a) at the time of D's action or possession D did not know, and had no reason to believe, that the document or record in question contained, or was likely to contain, information of a kind likely to be useful to a person committing or preparing an act of terrorism; or
(b) D's action or possession was for the purposes of—
    (i) carrying out work as a journalist; or
    (ii) academic research.

The courts have provided the following guidance in respect of cases outside s 58(3A) about 'reasonable excuse' in s 58. For these purposes 'reasonable excuse' means an objectively reasonable excuse; possessing a document for purposes of carrying out a bank raid is a purpose not connected with terrorism, but it is not a reasonable excuse. The Court of Appeal has held that it cannot be a reasonable excuse to a charge of possessing documents, etc contrary to s 58 that the documents, etc originated as part of an effort to change an illegal and undemocratic regime.

The House of Lords has held that, if D adduces sufficient evidence of something which could be a reasonable excuse, the defence succeeds unless the prosecution disproves the explanation advanced or proves that, even if D's explanation is true, it does not amount to a reasonable excuse. If it does disprove that thing, the defence fails and D is guilty under s 58. The prosecution does not have to prove that D did not have a reasonable excuse at all.

The provisions of s 58 are wide enough to embrace the failure of close relatives of terrorists to disclose information in their possession as soon as reasonably practicable. It is recommended that relatives of a terrorist, who are not themselves involved in terrorism, should not be investigated with a view to obtaining evidence of offences by them against this section, unless particular extreme circumstances make this desirable. An example of such circumstances would be where the withholding of information could lead to death, serious injury or the escape of a terrorist offender.

An offence under s 58(1) is an indictable (either-way) offence punishable with up to 15 years' imprisonment.

Eliciting, publishing or communicating information about members of the armed forces, etc

A person commits an offence under TA 2000, s 58A(1) who: **21.56**

(a) elicits or attempts to elicit information about an individual who is or has been:
   (i) a member of Her Majesty's forces,
   (ii) a member of any of the intelligence services, or
   (iii) a constable,
   which is of a kind likely to be useful to a person committing or preparing an act of terrorism; or
(b) publishes or communicates any such information.

It is a defence for D to adduce sufficient evidence to raise the issue that (s)he had a reasonable excuse for his/her action, whereupon the defence is assumed to be satisfied unless the prosecution disproves it (ss 58A(2) and 118).

An offence under s 58A(1) is an indictable (either-way) offence punishable with up to 15 years' imprisonment.

ENCOURAGEMENT, ETC OF TERRORISM

Encouragement of terrorism

TA 2006, s 1(1) and (2) provides that a person (D) commits an offence if: **21.57**

(a) D publishes a statement to which s 1 applies or causes another to publish such a statement; and
(b) at that time, D intends members of the public to be directly or indirectly encouraged or otherwise induced by the statement to commit, prepare or instigate acts of terrorism or Convention offences, or is reckless as to whether they will be.

Section 1 applies to a statement that is likely to be understood by a reasonable person as a direct or indirect encouragement or other inducement, to some or all of the members of the public to whom it is published, to commit, prepare, or instigate acts of terrorism or Convention offences. In TA 2006, an 'act of terrorism' includes anything constituting an action taken for the purposes of terrorism, within the meaning of TA 2000 (**21.2**) (including a reference to action taken for the benefit of a proscribed organisation (**21.4**)). A 'Convention offence' means an offence listed in TA 2006, Sch 1 (eg specified explosives offences, biological, chemical or nuclear weapons offences, hijacking, hostage taking, offences relating to terrorist funds or directing terrorist organisations, and offences directed at a nuclear facility or interference with the operation of such a facility) or an equivalent offence under the law of a country or territory outside the UK.

For the purposes of s 1, the statements that are likely to be understood by *a* reasonable person as indirectly encouraging the commission of acts of terrorism or Convention offences include every statement which glorifies the commission or preparation (whether in the past, future or generally) of such acts or offences (whether or not in particular), and is a statement from which members of the public could reasonably be expected to infer that what is being glorified is conduct that should be emulated by them. 'Glorification' includes any form of praise or celebration.

Where it is not proved that D intended the statement directly or indirectly to bring about the commission, preparation or instigation of acts of terrorism or Convention offences, D has a defence if (s)he proves that the statement neither expressed his/her view nor had his/her actual endorsement, and that it was clear that this was so. In the case of electronic material, the defence will not be available where there has been a failure to comply with a notice under s 3 (**21.59**).

An offence under s 1(1) and (2) is an indictable (either-way) offence punishable with up to 15 years' imprisonment.

Dissemination of terrorist publications

21.58   TA 2006, s 2(1) makes it an offence to sell or otherwise disseminate terrorist publications, including information on the internet, if (a) D intends his/her conduct to encourage or induce people to engage in the commission, preparation or instigation of acts of terrorism, (b) D intends it to assist in the commission or preparation of such acts, or (c) D is reckless as to whether his/her conduct will have either effect.

By s 2(2), the offence can be committed by distributing or circulating a terrorist publication; giving, selling or lending a terrorist publication; offering such for sale or loan; providing a service which enables a person to obtain, read, listen to or look at such material; transmitting it by electronic means, or possessing it with a view to it being dealt with in one of these ways. Where D within the UK possesses terrorist material with the relevant mens rea, it is irrelevant that the dissemination of the material is to take place abroad.

A publication is a terrorist publication in relation to conduct within s 2(2) if it contains matter likely:

(a) to be understood by a reasonable person as a direct or indirect encouragement or other inducement, to some or all of the persons to whom it is or may become available as a result of that conduct, to the commission, preparation or instigation of acts of terrorism, or

(b) to be useful in the commission or preparation of such acts or offences and to be understood (by some or all of those persons) as contained in the publication, or made available, for the purpose of being so useful.

For the purposes of (a), matter that is likely to be understood by a reasonable person as indirectly encouraging the commission or preparation of acts of terrorism includes any matter which:

(i) glorifies the commission or preparation (whether in the past, in the future or generally) of such acts; and

(ii) is matter from which a person could reasonably be expected to infer that what is being glorified is being glorified as conduct that should be emulated by him/her in existing circumstances.

Where the publication is a terrorist publication by virtue of (a) and it is not proved that D acted with the intention of encouraging or inducing the commission, preparation or instigation of acts of terrorism, it is a defence for D to show that the material which encourages terrorism did not express his/her views, nor had his/her actual endorsement, and that it was clear, in all of the circumstances, that this was so. In the case of electronic material, the defence will not be available where there has been a failure to comply with a notice under s 3.

TA 2006, s 28 gives a justice of the peace power to issue a warrant authorising a constable to enter and search premises and seize any article on those premises if it is likely to be subject of conduct falling within s 2(2) and it would be treated as a terrorist publication, where (s)he is satisfied that there are reasonable grounds for suspecting that such an article is on the premises. Reasonable force may be used in effecting entry.

Bulk material may be removed in such circumstances under CJPA 2001, s 50 (**4.61**) for later examination. Notice must be given by the constable responsible for the seizure to every person whom (s)he believes to be the owner of any material. In the event of there being no such person, notice must be given to the person believed to be the occupier of the premises.

An offence under s 2(1) is an indictable (either-way) offence punishable with up to 15 years' imprisonment.

Internet activity

TA 2006, s 3 applies for the purposes of ss 1 and 2 in relation to cases where a statement is published or caused to be published in the course of, or in connection with, the provision or use of a service provided electronically, or where conduct falling within s 2(2) (**21.58**) was in the course of, or in connection with, the provision or use of such a service. Section 3 provides that the cases where a statement, or the article or record to which the conduct is related, is to be regarded as having the endorsement of a person at any time, include a case where: **21.59**

(a) that person has been given a notice by a constable which

(i) declares that in the constable's opinion the statement, article or record concerned is unlawfully terrorism-related,

(ii) requires him/her to secure that the statement, etc (so far as so related) is not available to the public,

(iii) warns him/her that a failure to comply with the notice within two working days will result in the statement, etc being regarded as having his/her endorsement, and

(iv) explains how (s)he may become liable by virtue of the notice if the statement, etc becomes available to the public after (s)he has complied with the notice;

(b) that time falls more than two working days after the notice was given; and

(c) that person has failed without reasonable excuse to comply with the notice.

The procedures relating to the giving of such notices are set out in s 4.

PREPARATION OF TERRORIST ACTS AND TERRORIST TRAINING

Preparation of terrorist acts

21.60   TA 2006, s 5(1) bites at an earlier stage than the offences of conspiracy or attempt by providing that a person commits an offence if, with the intention of:

(a) committing acts of terrorism (**21.2**), or
(b) assisting another to commit such acts,

he engages in any conduct in preparation for giving effect to his intention.

It is irrelevant whether the intention and preparations relate to one or more particular acts of terrorism, acts of terrorism of a particular description, or acts of terrorism generally. The offence is triable only on indictment and punishable with a maximum of life imprisonment.

Training for terrorism

21.61   TA 2006, s 6(1) makes it an offence for a person knowingly to provide instruction or training to a would-be terrorist in the making or handling of noxious substances or the use of terrorist methods or techniques (including design or adaptation of methods or techniques for the purposes of terrorism (**21.2**)). For these purposes, a 'noxious substance' means a pathogen or toxin, or any other substance which is hazardous or noxious or which may be or become hazardous or noxious only in certain circumstances. Section 6(2) makes it an offence for a would-be terrorist to receive such training.

It is irrelevant for the purposes of s 6(1) and (2) whether:

(a) any instruction or training that is provided is provided to one or more particular persons or generally;
(b) the acts or offences in relation to which a person intends to use skills in which (s) he is instructed or trained consist of one or more particular acts of terrorism or Convention offences (**21.57**), such acts, etc of a particular description or such acts, etc generally; and
(c) assistance that a person intends to provide to others is intended to be provided to one or more particular persons or to one or more persons whose identities are not yet known

A person who gives or receives training in the making or use of firearms, radioactive material, or weapons designed or adapted for the discharge of any radioactive material, explosives, or chemical, biological, or nuclear weapons, commits an offence under TA 2000, s 54(1) or (2) respectively. It is a defence for D to adduce sufficient evidence to raise the issue that his/her action or involvement was wholly for a purpose other than assisting, preparing for, or participating in terrorism, whereupon the defence is assumed to be satisfied unless the prosecution disprove it (ss 54(5) and 118).

An offence under TA 2006, s 6 or TA 2000, s 54 is an indictable (either-way) offence punishable with a maximum of life imprisonment.

Attendance at a place used for terrorist training

21.62   A person (D) commits an offence under TA 2006, s 8(1) if:

(a) D attends at any place, whether in the UK or elsewhere;

(b) while D is at that place, instruction or training of the type mentioned in TA 2006, s 6(1) or TA 2000, s 54(1) is provided there;

(c) that instruction or training is provided there wholly or partly for purposes connected with the commission or preparation of acts of terrorism (**21.2**) or Convention offences (**21.57**); and

(d) the following requirements of s 8(2) are satisfied in relation to D —

   (i) D knows or believes that instruction or training is being provided there wholly or partly for purposes connected with the commission or preparation of acts of terrorism or Convention offences; or

   (ii) a person attending at that place throughout the period of D's attendance could not reasonably have failed to understand that instruction or training was being provided there wholly or partly for such purposes.

It is immaterial whether D receives the instruction or training him/herself, and whether the instruction or training is provided for purposes connected with one or more particular *acts of terrorism or Convention offences*, such *acts, etc* of a particular description or *such acts, etc* generally.

The above offence is an indictable (either-way) offence punishable with up to 14 years' imprisonment.

OFFENCES INVOLVING RADIOACTIVE DEVICES AND MATERIALS AND NUCLEAR FACILITIES AND SITES

21.63 TA 2006, s 9(1) creates offences related to the making or possession of radioactive devices or the possession of radioactive material, for use in the commission or preparation of an act of terrorism or for purposes of terrorism.

Section 10(1) prohibits the use of such items in the course of or in connection with the commission of an act of terrorism, or for the purposes of terrorism. Section 10(2) creates an offence of using or damaging a nuclear facility in the course of or in connection with acts of terrorism, in such a manner that radioactive material is released, or that the risk that such material will be released is created or increased.

Section 11(1) provides that it is an offence, in the course of or in connection with the commission of an act of terrorism or for the purposes of terrorism, to demand the supply of a radioactive device or radioactive material, or that a nuclear facility, or access to a nuclear facility, be made available, if such a demand is supported by a threat of action if the demand is not met. The threat must be a credible one.

Section 11(2) makes it an offence to make a credible threat to use a radioactive device or material, or to use or damage a radioactive facility, in the course of or in connection with the commission of an act of terrorism or for the purposes of terrorism.

The offences under ss 9 to 11 are triable only on indictment and have a maximum punishment of life imprisonment.

OFFENCES INVOLVING THE USE OF NOXIOUS SUBSTANCES OR THINGS TO CAUSE HARM AND INTIMIDATE

21.64 The Anti-terrorism, Crime and Security Act 2001, s 113(1) creates the offence committed by any person who takes any action which:

(a) involves the use of a noxious substance or other noxious thing;

    (b) has or is likely to have an effect which:
- (i) causes serious violence against a person, or serious damage to property, anywhere in the world,
- (ii) endangers human life or creates a serious risk to the health or safety of the public or a section of the public, or
- (iii) induces in members of the public the fear that the action is likely to have the effect in (ii); and

    (c) is designed to influence the government or an international governmental organisation or to intimidate the public or a section of the public.

By s 113(3), an offence is committed by someone who:

(a) threatens that (s)he or another will take any action constituting an offence under s 113(1); and
(b) intends thereby to induce anyone anywhere in the world to fear that the threat is likely to be carried out.

Under s 114(1), a person commits an offence if he:

(a) places any substance or other thing in any place; or
(b) sends any substance or other thing from one place to another,

with the intention of inducing in a person anywhere in the world a belief that it is likely to be (or contain) a noxious substance or other noxious thing and thereby endanger human life or create a serious risk to human health.

A person is guilty of an offence under s 114(2) if (s)he communicates any information which (s)he knows or believes to be false with the intention of inducing in a person anywhere in the world a belief that a noxious substance or other noxious thing is likely to be present in any place and thereby endanger human life or create a serious risk to human health.

For these purposes, 'substance' includes any biological agent and any other natural or artificial substance (whatever its form, origin or method of production). In the case of s 113(3) or 114 the person concerned need not have any particular person in mind as the person in whom (s)he intends to induce the belief in question.

The above are indictable (either-way) offences. An offence under s 113 is punishable with up to 14 years' imprisonment, and an offence under s 114 with up to seven years'.

GENERAL

Consent of DPP or Attorney General

21.65   The consent of the DPP is required for any proceedings in relation to any of the above offences under the Terrorism Acts 2000 and 2006. Where an offence appears to have been committed outside the UK or for a purpose wholly or partly connected with the affairs of a country other than the UK, the DPP's consent may only be given with the consent of the Attorney General. Proceedings for an offence committed under the Anti-terrorism, Crime and Security Act 2001, s 113 outside the UK require the Attorney General's consent.

Post-charge questioning

21.66   The Counter-Terrorism Act 2008 (C-TA 2008), s 22 permits a judge of the Crown Court to authorise questioning of a person concerning an offence where (s)he has been charged with the offence or informed that (s)he may be prosecuted for it, or has been sent for trial for it, where that offence is a 'terrorism offence' or where the judge considers the offence to have a terrorist connection. A judge must not authorise such questioning unless (s)he is satisfied that:

(a) further questioning is necessary in the interests of justice;
(b) the investigation for the purposes of which the further questioning is being proposed is being conducted diligently and expeditiously; and
(c) the questioning would not interfere unduly with the preparation of the person's defence to the charge or any other criminal charge that (s)he may be facing.

'Terrorism offence' includes specified offences, mainly against TA 2000 (eg ss 11–13, 15–19, 21A, 21D, 38B, 39, 54, and 56–58B) and TA 2006 (ss 1, 2, 5, 6, and 8 to 11) and the Anti-terrorism, Crime and Security Act 2001, s 113. An offence has a 'terrorist connection' if the offence is, or takes place in the course of, an act of terrorism, or is committed for the purpose of terrorism.

PACE Code H applies to the treatment and questioning of the detainee.

Access to a solicitor may not be delayed, nor may the person be required to consult a solicitor within the sight and hearing of a police officer.

Unless the restriction on drawing adverse inferences from silence applies (5.29), before post-charge questioning begins the caution must be given in the terms set out at 5.27. The only restriction on drawing adverse inferences from silence applies where a person has asked for legal advice and is questioned before receiving such advice on grounds of urgency.

An interview by virtue of C-TA 2008, s 22 must be video recorded with sound in accordance with the Code of Practice for the Video Recording with Sound of Interviews with Persons Detained under TA 2000, s 41 or Sch 7, and Post-Charge Questioning of Persons Authorised under C-TA 2008, s 22.

NOTIFICATION REQUIREMENTS

21.67

C-TA 2008, Part 4 and associated Schedules and regulations:

(a) set out a system of notification requirements (initial notification, notification of change, periodic re-notification, and notification of return after absence from UK) for persons convicted of a terrorism offence or of a non-terrorism offence with a terrorist connection;
(b) make provision to apply notification requirements to persons convicted of equivalent offences outside the UK; and
(c) make provision for the making of foreign travel restriction orders against persons subject to the above notification requirements.

See Chapter 21 on the companion website for details.

IMPOSITION OF TERRORISM PREVENTION AND INVESTIGATION MEASURES

21.68

The Terrorism Prevention and Investigation Measures Act 2011 (TPIMA 2011) provides for the making of terrorism prevention and investigation measures against individuals suspected of being involved in terrorism-related activity.

Such measures may be imposed by the Secretary of State, normally with the prior permission of the High Court. The details about the regime under TPIMA 2011 are in Chapter 21 on the companion website.

TEMPORARY EXCLUSION FROM THE UNITED KINGDOM

21.69   The Counter-Terrorism and Security Act 2015 (C-TSA 2015), Part 1, Ch 2 (ss 2–15) and Schs 2–4 provide for the imposition of temporary exclusion orders.

A 'temporary exclusion order' is an order which requires an individual who has the right of abode in the UK and is suspected of terrorism-related activity outside the UK not to return to the UK unless the return is in accordance with a permit to return issued by the Secretary of State before the individual began the return, or the return is the result of the individual's deportation to the UK. 'Return to the UK' includes, in the case of an individual who has never been in the UK, a reference to coming to the UK for the first time.

Such an order may be imposed by the Secretary of State, normally with the prior permission of the High Court. The details about the regime under C-TSA 2015 are in Chapter 21 on the companion website.

CHAPTER 22

Preserving the Administration of Justice and Probity

ASSISTING OFFENDERS

By the Criminal Law Act 1967 (CLA 1967), s 4(1), where a person (P) has committed **22.1** a relevant offence, another person (D) is guilty of an offence (**6.48**) if (knowing or believing P to be guilty of the offence or of some other relevant offence) without lawful authority or reasonable excuse D does any act with intent to impede P's apprehension or prosecution. A 'relevant offence' is an offence for which:

(a) the sentence is fixed by law (essentially, murder); or
(b) a person of 18 years or over (not previously convicted) may be sentenced to a term of imprisonment for a term of five years.

D is not guilty, as an accomplice, of the relevant offence which has been committed by P, because D's conduct occurs after its commission; the offence under s 4(1) is a separate offence.

Although the *commission* of the 'relevant offence' by the principal offender (P) must be proved, it is not necessary that P should have been convicted of it. Indeed, the acquittal of P does not prevent a conviction of D for assisting that person; the commission of the relevant offence by P may be established in the case of D, even though it is not established against P, eg because evidence admissible against D was not admissible against P, or because, after P's acquittal, but before D's trial, the discovery of further evidence puts the commission of the offence by P beyond doubt.

An offence under s 4(1) is an indictable (either-way) offence. The maximum imprisonment varies depending on the relevant offence. It is 10 years (relevant offence murder or any other offence whose sentence is fixed by law), seven years (prison sentence of 14 years possible for relevant offence), five years (prison sentence of 10 years possible for relevant offence), and three years in all other cases. The effect of these different maxima is that there are four separate offences under s 4(1), punishable with 10, seven, five, and three years' imprisonment respectively.

CONCEALING OFFENCES

By CLA 1967, s 5(1), where a person (P) has committed a relevant offence (as defined **22.2** above), another person (D) is guilty of an offence (**6.48**) if D, knowing or believing that the offence or some other relevant offence has been committed, and that (s)he has information which might be of assistance in securing the prosecution or conviction of an offender for it, accepts or agrees to accept for not disclosing it any consideration other than the making good of loss or injury caused by the offence, or the making of reasonable compensation for it.

The offence under s 5(1) is an indictable (either-way) offence punishable with up to two years' imprisonment.

WASTEFUL EMPLOYMENT OF POLICE

22.3   The CLA 1967, s 5(2) states that it is a summary offence (**6.48**) for a person to cause any wasteful employment of the police by knowingly making to any person a false report tending to show that an offence has been committed, or to give rise to apprehension for the safety of any persons or property, or tending to show that the informant has any information material to any police inquiry. The maximum punishment is six months' imprisonment and/or a level 4 fine.

The Criminal Justice and Police Act 2001 provides that the offence is a 'penalty offence' in respect of which a 'penalty notice' may be given: see **3.40**.

PERJURY

22.4   Under the Perjury Act 1911 (PA 1911), s 1(1), it is an offence for a person, lawfully sworn as a witness or interpreter in a judicial proceeding, to wilfully make a statement material in that proceeding, which (s)*he knows to be false or does not believe to be true*. Perjury is triable only on indictment and punishable with up to seven years' imprisonment. The italicised words mean that the offence covers not only the cases where the maker of the statement knows it to be false or positively does not believe it to be true, but also the case where (s)he makes it recklessly, without thought at all, or not caring whether or not it is true.

A 'person lawfully sworn' includes a person who has made an affirmation or declaration that (s)he will tell the truth. A 'judicial proceeding' is a proceeding before any court, tribunal, or person having by law power to hear, receive, and examine evidence on oath. The presence of the word 'wilfully' means that it must be proved that the person concerned made that statement deliberately. A 'material statement' is one which might affect the decision of the court.

By the Youth Justice and Criminal Evidence Act 1999 (YJCEA 1999), s 29, this offence applies to an intermediary (**8.23**) as it applies to an interpreter lawfully sworn in a judicial proceeding.

PA 1911, s 13 states that a person shall not be convicted of perjury or any other offence under PA 1911 solely on the evidence of one witness as to the falsity of any statement alleged to be false, so that there is a requirement of corroboration in this respect.

Subornation of perjury

22.5   PA 1911, s 7(1) provides that a person who aids, abets, counsels, procures or suborns (ie procures by bribery or other corrupt means another to commit perjury) is liable for perjury 'as if he were a principal offender'. This provision is redundant since it adds nothing to the criminal liability which would otherwise be imposed under the general principle relating to accomplices.

OFFENCES AKIN TO PERJURY

False unsworn statement

YJCEA 1999, s 57(1) and (2) provides that an unsworn witness commits an offence if **22.6** he wilfully gives false evidence in such circumstances that, had the evidence been given on oath, he would have been guilty of perjury. The offence is a summary offence punishable with up to six months' imprisonment and/or a fine not exceeding £1,000. If the offender is under 14, the maximum punishment is a fine not exceeding £250.

False written statements tendered in evidence in criminal proceedings

By the Criminal Justice Act 1967, s 89(1), a person is guilty of an offence if in a written **22.7** statement tendered in evidence in criminal proceedings by virtue of s 9 of that Act (**8.43**) (s)he wilfully makes a statement material in those proceedings which (s)he knows to be false or does not believe to be true. The offence is an indictable (either-way) offence and is punishable with up to two years' imprisonment.

False statements on oath made otherwise than in a judicial proceeding

If any person: **22.8**

(a) being required or authorised by law to make any statement on oath for any purpose, and being lawfully sworn (otherwise than in a judicial proceeding), wilfully makes a statement which is material for that purpose and which (s)he knows to be false or does not believe to be true; or

(b) wilfully uses any false affidavit for the purposes of the Bills of Sale Act 1878,

(s)he is guilty of an offence under PA 1911, s 2. The offence is an indictable (either-way) one and is punishable with up to seven years' imprisonment.

False statutory declaration and other false statements without oath

Under PA 1911, s 5, a person commits an offence if (s)he knowingly and wilfully makes, **22.9** otherwise than on oath, a statement false in a material particular, and the statement is made:

(a) in a statutory declaration; or

(b) in an abstract, account, balance sheet, book, certificate, declaration, entry, estimate, inventory, notice, report, return, or other document which (s)he is authorised or required to make, attest, or verify by any statute for the time being in force; or

(c) in any oral declaration or oral answer which (s)he is required to make by, under, or in pursuance of any public general Act of Parliament.

The offence is an indictable (either-way) one and is punishable with up to two years' imprisonment.

PERVERTING THE COURSE OF JUSTICE

22.10   The common law offence of perverting the course of justice may be committed by any act or series of acts which has a tendency to pervert the course of justice and is intended to do so. The offence is triable only on indictment and is punishable with up to life imprisonment.

For the purposes of the offence, it has been held that the course of justice may be perverted by, for example: discontinuing a criminal prosecution in return for payment; making false claims about an offence; making false statements to police officers investigating an offence; making a false complaint to the police capable of being taken seriously, whether or not it identifies particular individuals; retracting a truthful allegation of an offence, or truthful evidence; interfering with a witness or a juror; producing fabricated evidence; and destroying evidence of an offence to hamper an investigation (actual or potential). The Court of Appeal has stated that there is no closed list of acts which may give rise to the offence, but that any extension by the judges in the range of recognised acts should take place with caution. The requisite acts are not limited to those giving rise to some other independent criminal wrongdoing.

A divisional court has held that the course of justice extends until the completion of any sentence imposed at the end of criminal proceedings.

INTIMIDATION OF WITNESSES, JURORS, AND OTHERS

Intimidation, etc before or during a criminal trial

22.11   By the Criminal Justice and Public Order Act 1994 (CJPOA 1994), s 51(1), a person commits an offence if (s)he:

(a) (s)he does an act which intimidates, and is intended to intimidate, another person (the victim);

(b) (s)he does the act knowing or believing that the victim is assisting in the investigation of an offence or is a witness or potential witness, or a juror or potential juror, in proceedings for an offence; and

(c) (s)he does the act intending thereby to cause the investigation or the course of justice to be obstructed, perverted or interfered with.

In respect of the requirement of the doing of an act which intentionally results in another person being intimidated, it is immaterial that the act is done otherwise than in the presence of the victim, or to a person other than the victim.

The Court of Appeal has held that 'intimidation' requires that a person is actually put in fear by violence or a threat. Threatened harm may be physical (to the person or a person's property) or financial.

Section 51(1) does not require that the person who is intimidated be actually assisting in the investigation of an offence at the time, or be a witness or potential witness, or a juror or potential juror, in proceedings for an offence. It is sufficient that D believes that the person so intimidated is involved in that way. There must, however, be an investigation in progress; it is insufficient that D believed that such an investigation was taking place. An 'investigation into an offence' in s 51 means an investigation by the police or other persons charged with the duty of investigating offences or charging offenders.

In respect of (c), CJPOA 1994, s 51(7) provides that if it is proved that the intimidating act was done with the required knowledge or belief, it must be presumed, unless the contrary is proved, that D did the act with the required intention. Thus, in such circumstances, the onus is upon D to show that (s)he did not have such an intention.

Reprisals against witnesses, jurors, and others in criminal cases

A person commits an offence under CJPOA 1994, s 51(2) if: **22.12**

(a) (s)he does an act which harms, and is intended to harm, another person or, intending to cause another person to fear harm, (s)he threatens to do an act which would harm that other person;

(b) (s)he does or threatens to do the act knowing or believing that the person harmed or threatened to be harmed (the victim), or some other person, has assisted in the investigation into an offence or has given evidence or particular evidence in proceedings for an offence, or has acted as a juror or concurred in a particular verdict in proceedings for an offence; and

(c) (s)he does or threatens to do it because of that knowledge or belief.

For the purposes of (a), it is immaterial that the act is or would be done, or that the threat is made, otherwise than in the presence of the victim, or to a person other than the victim. There is no requirement that the other person is actually intimidated thereby.

The harm which may be done or threatened may be physical (whether to the person or a person's property) or financial.

In respect of (c), s 51(8) provides that if it is proved that within the 'relevant period':

(i) D did an act which harmed, and was intended to harm, another person; or

(ii) intending to cause another person fear of harm, D threatened to do an act which would harm that other person,

and that he did the act, or (as the case may be) threatened to do the act, with the knowledge or belief required by s 51(2)(b), then D is presumed, unless the contrary is proved, to have done the act, or (as the case may be) threatened to do the act, because of that knowledge or belief.

For the purposes of this presumption, the 'relevant period' in relation to:

(a) *a witness or juror*, begins with the institution of proceedings and ends with the first anniversary of the conclusion of the trial, or of any appeal;

(b) *a person who has assisted in the investigation of an offence (or is believed by D to have done so) but who was not a witness in proceedings for an offence*, is the period of one year beginning with the act (or believed act) which assisted the investigation; and

(c) *a person who has assisted in the investigation of an offence (or is believed by D to have done so) and was also a witness in proceedings for an offence*, is the period beginning with the act (or believed act) which assisted in the investigation and ending with the first anniversary of the conclusion of the trial, or of any appeal.

General

CJPOA 1994, s 51(5) provides that the intention/motive required (in either case) need   **22.13**
not be the only or predominating intention/motive with which the act is done or

threatened. This excludes defences based upon the fact that the primary intention of the perpetrator was to avoid a miscarriage of justice.

An offence under s 51 is an indictable (either-way) offence punishable with up to five years' imprisonment.

Corresponding Provisions in Respect of Civil Proceedings

22.14   The Criminal Justice and Police Act 2001, ss 39 and 40 provide offences, corresponding to those in CJPOA 1994, s 51, for the protection of witnesses in civil proceedings before the Court of Appeal, the High Court, the Crown Court, the county court, or a magistrates' court.

BRIBERY

General bribery offences

Offences of bribing another person

22.15   The Bribery Act 2010 (BA 2010), s 1(1) provides that a person (P) is guilty of an offence (**6.48**) if either of the following cases applies.

Case 1 is where:

(a) P offers, promises, or gives a financial or other advantage to another person; and

(b) P intends the advantage to induce a person to perform improperly a relevant function or activity, or to reward a person for the improper performance of such a function or activity.

It does not matter whether the person to whom the advantage is offered, promised, or given is the same person as the person who is to perform, or has performed, the function or activity concerned.

Case 2 is where P offers, promises, or gives a financial or other advantage to another person, and P knows or believes that the acceptance of the advantage would itself constitute the improper performance of a relevant function or activity.

It does not matter in either case whether the advantage is offered, promised, or given by P directly or through a third party.

Offences relating to being bribed

22.16   By BA 2010, s 2(1), person (R) is guilty of an offence (**6.48**) if any of the following four cases applies.

Case 3 is where R requests, agrees to receive, or accepts a financial or other advantage intending that, in consequence, a relevant function or activity should be performed improperly (whether by R or another person).

Case 4 is where R requests, agrees to receive, or accepts a financial or other advantage, and the request, agreement, or acceptance itself constitutes the improper performance by R of a relevant function or activity.

Case 5 is where R requests, agrees to receive, or accepts a financial or other advantage as a reward for the improper performance (whether by R or another person) of a relevant function or activity.

Case 6 is where, in anticipation of or in consequence of R requesting, agreeing to receive, or accepting a financial or other advantage, a relevant function or activity is performed improperly by R, or by another person at R's request, or with R's assent or acquiescence.

In Cases 3 to 6 it does not matter whether:

(a) R requests, agrees to receive, or accepts (or is to request, agree to receive or accept) the advantage directly or through a third party,
(b) the advantage is (or is to be) for the benefit of R or another person.

In Cases 4 to 6 it does not matter whether R knows or believes that the performance of the function or activity is improper.

In Case 6, where a person other than R is performing the function or activity, it also does not matter whether that person knows or believes that the performance of the function or activity is improper.

General

Relevant function or activity   BA 2010, s 3 provides that a function or activity is a relevant function or activity if it is any: **22.17**

(a) function of a public nature;
(b) activity connected with a business, trade, or profession;
(c) activity performed in the course of a person's employment; or
(d) activity performed by or on behalf of a body of persons (whether corporate or unincorporate); and

it meets one or more of conditions A to C.

Condition A is that a person performing the function or activity is expected to perform it in good faith.

Condition B is that a person performing the function or activity is expected to perform it impartially.

Condition C is that a person performing the function or activity is in a position of trust by virtue of performing it.

A function or activity is a relevant function or activity even if it has no connection with the UK, and is performed in a country outside the UK.

Improper performance to which bribe relates   By BA 2010, s 4, a relevant function or activity is performed improperly if it is performed in breach of a relevant expectation, and is to be treated as being performed improperly if there is a failure to perform the function or activity and that failure is itself a breach of a relevant expectation.

A 'relevant expectation' in relation to a function or activity:

(a) which meets Condition A or B, means the expectation mentioned in the condition concerned; and
(b) which meets Condition C, means any expectation as to the manner in which, or the reasons for which, the function or activity will be performed that arises from the position of trust mentioned in that condition.

Anything that a person does (or omits to do) arising from or in connection with that person's past performance of a relevant function or activity is to be treated as being done (or omitted) by that person in the performance of that function or activity.

Expectation test   BA 2010, s 5 provides that, for the purposes of ss 3 and 4, the test of what is expected is a test of what a reasonable person in the UK would expect in relation to the performance of the type of function or activity concerned.

In deciding what such a person would expect in relation to the performance of a function or activity where the performance is not subject to the law of any part of the UK, any local custom or practice is to be disregarded unless it is permitted or required by legislation or a reported judicial decision applicable to the country concerned.

Defence   It is a defence under BA 2010, s 13 for a person charged with an offence:

(a) under s 1 which would not also be an offence under s 6 (below);
(b) under s 2;
(c) of aiding, abetting, counselling or procuring, attempting or conspiring, or encouraging or assisting, the commission of such an offence,

to prove that the person's conduct was necessary for the proper exercise of any function of an intelligence service, or the proper exercise of any function of the armed forces when engaged on active service.

Bribery of foreign public officials

22.18   This is dealt with by BA 2010, s 6(1) and (2), which provides that a person (P) who bribes a foreign public official (F) is guilty of an offence (**6.48**) if P's intention is:

(a) to influence F in F's capacity as a foreign public official; and
(b) to obtain or retain business, or an advantage in the conduct of business.

In this context, a trade or profession is a business.
    For the purposes of s 6, P bribes F if, and only if:

(a) directly or through a third party, P offers, promises or gives any financial or other advantage to F, or to another person at F's request or with F's assent or acquiescence, and
(b) F is neither permitted nor required by the legislation or a reported judicial decision applicable to F to be influenced in F's capacity as a foreign public official by the offer, promise or gift.

Failure of commercial organisations to prevent bribery

22.19   BA 2010, s 7 provides that (C), a body corporate or partnership carrying on a business, trade, or profession, is guilty of an offence (**6.48**) if a person (A) associated with C bribes another person intending to obtain or retain (a) business for C, or (b) an advantage in the conduct of business for C. For the purposes of s 7, A bribes another person if, and only if, A is, or would be, guilty of an offence under s 1 or 6 (whether or not A has been prosecuted for such an offence).

    It is a defence for C to prove that C had in place adequate procedures designed to prevent persons associated with C from undertaking such conduct. The Secretary of State has published guidance about procedures that relevant commercial organisations can put in place to prevent persons associated with them from bribing.

Trial and punishment

22.20   An offence under BA, s 1, 2 or 6 is an indictable (either-way) offence. An individual guilty of such an offence is punishable with imprisonment for up to 10 years. Where the offender is not an individual, it is liable to an unlimited fine. An offence under s 7 is punishable with an unlimited fine.

MISCONDUCT IN A PUBLIC OFFICE

22.21 A constable or other holder of a public office will commit the common law offence of misconduct in a public office if, while acting as such, (s)he wilfully neglects to perform his/her duty and/or wilfully misconducts him/herself to such a degree as to amount to an abuse of the public's trust in the office holder, without reasonable excuse or justification. The offence is triable only on indictment and is punishable with up to life imprisonment.

CORRUPT OR OTHER IMPROPER EXERCISE OF POLICE POWERS AND PRIVILEGES

22.22 A 'police constable' listed below commits an offence contrary to the Criminal Justice and Courts Act 2015, s 26(1) if (s)he exercises the powers and privileges of a constable improperly, and knows or ought to know that the exercise is improper. The offence is triable only on indictment and punishable with up to 14 years' imprisonment.

Section 26 applies to a constable or special constable of a police force in England and Wales; a British Transport Police Force (BTP) constable or special constable; a Civil Nuclear Constabulary or Ministry of Defence Police constable; and a National Crime Agency officer designated as having the powers and privileges of a constable.

For the purposes of s 26, a police constable exercises the powers and privileges of a constable improperly if:

(a) (s)he exercises a power or privilege of a constable for the purpose of achieving:
  (i) a benefit for him/herself, or
  (ii) a benefit or a detriment for another person, and
(b) a reasonable person would not expect the power or privilege to be exercised for the purpose of achieving that benefit or detriment.

'Benefit' and 'detriment' mean any benefit or detriment, whether or not in money or other property and whether temporary or permanent.

A police constable is to be treated as exercising the powers and privileges of a constable improperly where:

(i) (s)he fails to exercise a power or privilege of a constable, the purpose of the failure is to achieve a benefit or detriment described in (a) above, and a reasonable person would not expect a constable to fail to exercise the power or privilege for the purpose of achieving that benefit or detriment; or
(ii) (s)he threatens to exercise, or not to exercise, a power or privilege of a constable, the threat is made for the purpose of achieving a benefit or detriment described in (a) above, and a reasonable person would not expect a constable to threaten to exercise, or not to exercise, the power or privilege for the purpose of achieving that benefit or detriment.

The above references to exercising, or not exercising, the powers and privileges of a constable include performing, or not performing, the duties of a constable.

PART V

POLICING ROADS AND VEHICLES

CHAPTER 23

Roads and Vehicles: General Provisions

TERMINOLOGY

'Road', 'mechanically propelled vehicle', and 'motor vehicle' are recurring terms in this **23.1** and the next two chapters, so their definition is an obvious starting point.

Road

This is defined by RTA 1988, s 192 as meaning any highway and any other road to **23.2** which the public has access and including bridges over which a road passes.

Highway

A 'highway' is defined by the common law. It is a way over which there exists a public **23.3** right of passage on foot, riding, accompanied by a beast of burden, or with vehicles and cattle. The term is therefore wide enough to embrace public footpaths, bridleways, and carriageways, defined below. It is presumed that the boundary of the highway is defined by the buildings, hedges, fences, walls, or the like along its route. Consequently, grass verges are generally a part of the highway. If, for example, a fence behind a grass verge was erected by the highway authority it would clearly mark the limits of the highway. If it was erected by an adjoining landowner this would not necessarily be so.

A footpath is a highway over which the public has a right of way on foot alone.

A bridleway is a highway over which the public has a right of way while on foot, on horseback or leading a horse. By the Countryside Act 1968, s 30, the public also has a right to ride a pedal bicycle on a bridleway (provided that the cyclist gives way to pedestrians and persons on horseback). The right to ride a pedal cycle on a bridleway may be controlled by a local authority order. There may be a right on some bridleways to drive animals; such bridleways may be referred to locally as 'droves' or 'driftways'.

A carriageway is a way constructed or comprised in a highway, being a way (other than a cycle track) over which the public has a right of way on foot, on horseback, or with vehicles or cattle.

Other road

Apart from a highway, 'road' in the road traffic legislation means any road (ie a defin- **23.4** able way for passage between two points) to which the public has access. The essential factor is whether or not, as a question of fact, the public in general has access to the road. A private road leading to a farmhouse, which was maintained by the farmer, has been held to be a 'road' on evidence being offered that there was no gate and that it was regularly used by persons who had no business at the farm. Likewise, a divisional court has held that justices were entitled to find that tarmacked roadways across a caravan park to a beach, which had road markings and were easily definable as routes leading from one point to another on a map and to which the public had unrestricted access, were 'roads'.

It has been decided that any road may be regarded as a road to which the public has access, if members of the public are to be found on it who have not obtained access either by overcoming a physical obstruction or in defiance of a prohibition, express or implied. A pavement which is partly publicly owned and partly privately owned is a road if the public has access to the whole of it. It is essential to show that the public in general has access. Access which is restricted to certain classes of person is not usually sufficient to make a road 'public'.

Where a road within a housing estate has not been adopted by the local authority, the determining factor is not whether the road is repairable at public expense but whether the public has access to it. If members of the public are seen there, and their presence is tolerated, it is a road.

The House of Lords has ruled that a road is provided for the purpose of moving from one place to another as opposed to parking a car. It said that even though a part of a car park might be made up of routes giving access to parking bays this, by itself, was not enough if those routes merely gave access to parking bays and were not for the purpose of travelling from one place to another.

Mechanically propelled vehicle

23.5   The means of propulsion may be petrol, diesel, gas, steam, or electricity. Whether or not a vehicle is mechanically propelled is a question of fact.

A broken-down motor car remains a mechanically propelled vehicle; it is irrelevant that it might not be driven at that particular time. Before a vehicle can cease to be a mechanically propelled vehicle it must be in such a condition or in such circumstances that there is no reasonable prospect of it ever being driven again. Where a vehicle has been immobilised the extent of the immobilisation is the critical factor when considering whether it remains a mechanically propelled vehicle. A motor car in a scrap yard, stripped of all of its mechanical means of propulsion and without any reasonable prospect of the restoration of motive power, is obviously not a mechanically propelled vehicle. However, a motor car similarly stripped down in a garage to effect repairs remains a mechanically propelled vehicle as there is a reasonable prospect of it being restored to its former mobility.

The Road Traffic Act 1988 (RTA 1988), ss 1–4, 34, 163, and 170, referred to later, do not apply to a mechanically propelled single-person invalid carriage *whose unladen weight does not exceed 254 kgs, as long as it is being used by someone with a physical defect or physical disability, by someone taking it to or from repair, by a manufacturer testing or demonstrating it, by a seller demonstrating it, or by someone giving training in its use.*

Motor vehicle

23.6   This is defined by RTA 1988, s 185 as a mechanically propelled vehicle intended or adapted for use on a road. The words '*intended or adapted for use on a road*' are important. 'Adapted' does not mean 'altered'; it simply means 'fit and apt'. Some vehicles are quite obviously 'mechanically propelled' but they are not 'intended or adapted for use on a road'.

Whether or not a mechanically propelled vehicle is intended or adapted for use on a road depends on whether a reasonable person looking at the vehicle would say that one of its uses is general use on a road. On this basis, for example, it has been held that two types of motorised scooter known as a 'Segway' and a 'Go-ped' were motor vehicles within s 185. A 'Segway' consisted of a small gyroscopically stabilised platform mounted on two wheels, on which the traveller stood, powered by a battery driven electric motor. A vertical joy-stick was used to steer. Speed was controlled by leaning forward (to go faster) or standing up straight (to slow down). A divisional court held that, despite the intentions of the manufacturer that it should not be used on a road, the judge had not been wrong in concluding that a Segway was a motor vehicle within s 185. A 'Go-ped' consisted of a small foot platform attached to a sub-frame on which the person using the Go-ped would stand. Its engine gave it a maximum speed of 20 mph. The braking system was such that it could not stop the vehicle if it was travelling at any great speed, or in an emergency situation. Severe braking caused the rear wheel to lift from the road surface. A divisional court said that the roadworthiness of a conveyance or its capability to be used safely on a road were not conclusive in relation to whether or not its use on a road was contemplated. There was no obvious place in which a Go-ped could be used, other than on a road. Regardless of the fact that the manufacturers said that it was not to be used on a road, it would be and the reasonable person would recognise that to be so.

The above test does not depend on the owner's or manufacturer's intention. Unless there is evidence of regular use on a road, the particular use to which the vehicle is put at the time is irrelevant to the above test. In one case, a dumper truck, which was used on a site for the transport of material around that site and was occasionally driven on adjoining roads for short distances, was held not to be a motor vehicle because there was no proof of general (as opposed to occasional) use on roads. On the other hand, in another case, a Euclid earth scraper, which was primarily used to dig up earth on a building site and carry that earth under its own power to other places, was held to be intended for use on a road when evidence of its general use on roads and of its capability of reaching a speed of 45 mph was given.

Electric scooters

Like Segways and Go-peds (and other so-called 'powered transporters' such as hover-boards), electric scooters are motor vehicles, and therefore their use is illegal: **23.7**

(a) on a road without complying with a number of legal requirements, which potential users will find very difficult; and

(b) on pavements, cycle lanes, and other spaces set aside for use by pedestrians, cyclists, and horse-riders.

In order to assess the suitability of electric scooters for use on roads, the Electric Scooter Trials and Traffic Signs (Coronavirus) Regulations and General Directions 2020 make a number of amendments to road traffic law, so as to enable electric scooters being used in a trial to be used on roads. It must be emphasised that the amendments only apply to the use of an electric scooter in a trial, ie an assessment as to the suitability of electric scooters on roads conducted by virtue of:

(a) an order under the Road Traffic Act 1988, s 44 or 63 (no order has yet been made); or

(b) an arrangement made between two or more public authorities and a person who *hires out* electric scooters.

To be used lawfully on roads in the areas of these authorities, various of the requirements which apply to motor vehicles apply: the driver must be licensed and insured, a cycle helmet need not be worn, but other relevant safety provisions apply (including a pro-hibition on the use of a mobile phone) and the scooter must not be used on a pavement. On the other hand, the scooter does not need to be registered nor is vehicle excise duty payable. Speed must be limited by design to no more than 15.5 mph.

For the purposes of the 2020 Regulations, an electric scooter being used in a trial is defined as a vehicle which:

(a) is fitted with an electric motor with a maximum continuous power rating not ex-ceeding 500 watts;

(b) is not fitted with pedals that are capable of propelling the vehicle;

(c) has two wheels, one front and one rear, aligned along the direction of travel;

(d) is designed to carry no more than one person;

(e) has a maximum weight, excluding the driver, not exceeding 55 kgs;

(f) has a maximum design speed not exceeding 15.5 miles per hour;

(g) has a means of directional control through the use of handlebars which are mech-anically linked to the steered wheel;

(h) has a means of controlling the speed through hand controls; and

(i) has a power control that defaults to the 'off' position.

The 2020 Regulations also permit electric scooters being used in a trial (i) to use cycle lanes, and (ii) to use controlled crossings on the same basis as a pedal cycle.

Vehicles not treated as motor vehicles

23.8 RTA 1988, s 189 expressly provides that pedestrian-controlled mechanically propelled lawnmowers and 'electrically assisted pedal cycles' (**25.93**) are not motor vehicles. Mechanically propelled single-person invalid carriages are expressly stated not to be motor vehicles for the purposes of the Road Traffic Regulation Act 1984 and RTA 1988, except in relation to RTA 1988, s 22A (**24.83**), if the conditions italicised at **23.5** are satisfied.

A mechanically propelled vehicle originally manufactured for use on a road may cease to be a 'motor vehicle' for the purposes of s 185 if it is subsequently altered, but only if such alterations are very substantial.

REGISTRATION: MECHANICALLY PROPELLED VEHICLES

23.9 With the exceptions of electric scooters being used in a trial, electrically assisted pedal cycles and the other vehicles which do not require a vehicle licence or a nil licence (**23.41**), all vehicles used or kept on public roads (**23.38**) in the UK must be registered with the Secretary of State (via the Driver and Vehicle Licensing Agency (DVLA)). A 'vehicle' for this purpose is a mechanically propelled vehicle, or anything (whether or not a vehicle) that has been, but has ceased to be, a mechanically propelled vehicle. All records relating to vehicles are retained by the DVLA.

The Road Safety Act 2006 (RSA 2006), s 49A contains a provision for the disclosure by the Secretary of State to chief officers of police and certain other persons and bodies of information in foreign registers of vehicles which has been obtained.

The Road Vehicles (Registration and Licensing) Regulations (RV(R&L)R) 2002 pro-vide a single set of Regulations for the whole of the UK.

The Vehicle Excise and Registration Act 1994 (VERA 1994), s 21 provides that it is the duty of the Secretary of State to register a vehicle on the first issue of a vehicle licence or a nil licence (ie a licence for a vehicle exempt from excise duty), or where particulars in respect of the vehicle are received by the Secretary of State from a motor dealer before the first licence is issued. The registration mark assigned to a vehicle must be fitted in the prescribed manner. Since it is the Secretary of State's responsibility to register vehicles, the use of an unregistered vehicle on a road simply constitutes the offence (**23.34**) of use without a vehicle licence.

RV(R&L)R 2002, reg 10 authorises the Secretary of State to issue a registration document. (S)he may require the keeper of a vehicle to produce it for inspection or to produce other evidence that the vehicle accords with the particulars furnished with the application. If (s)he is not satisfied that the vehicle accords with the particulars, (s)he may refuse to issue a registration document or replacement registration document.

Registration document

When the Secretary of State registers a vehicle a registration document is issued and, **23.10** unless a registration mark has already been assigned by a motor dealer who has received a 'block' of marks, a registration mark is assigned to that vehicle. Even if a vehicle is exempt from the requirement to be licensed, it must still be registered. A registration document contains the registered particulars of the vehicle and the name and address of the person shown in the register as the owner or keeper of the vehicle.

The Secretary of State may issue a registration document in a microprocessor smart card format and may issue that type of registration document without charging a fee where the old type of document is surrendered. The Secretary of State may issue a replacement registration document where a registration document has, or may have been, lost, stolen, destroyed, or damaged, or it contains any particulars that have become illegible.

The registered owner of a vehicle is not necessarily its legal owner, and possession of a registration document is not in itself proof of ownership.

By reg 12, the keeper of a vehicle must produce a registration document for inspection if required to do so at any reasonable time by a constable or person acting on the Secretary of State's behalf. Failure to comply is an offence under VERA 1994, s 59(1).

VERA 1994, s 28A requires a person *using* a vehicle in respect of which a registration document has been issued to produce it for inspection by a constable or a person authorised by the Secretary of State (who must produce his/her authority). Failure to do so is an offence (s 28A(3)). However, no offence is committed if the person produces the document personally at a police station specified by him/her within seven days or as soon as is reasonably practicable. Nor is an offence committed if a vehicle is on lease or hire, and the vehicle is not registered (nor required to be registered) in the name of the lessee or hirer, and (s)he personally produces appropriate evidence of the lease or hire agreement in the same way as just described. However, (s)he must reasonably believe (or it must be reasonable for him/her to expect) that the lessor, etc is able to produce, or require production of, the registration document.

Notification of change of ownership

RV(R&L)R 2002, regs 22–24 are concerned with changes of ownership affecting **23.11** vehicles.

Where there is a 'private' sale or transfer of a vehicle, ie to a person other than a vehicle trader, the registered keeper is required by reg 22 to give to the new keeper the part of the registration document which provides for particulars of a new keeper, and must forthwith deliver to the Secretary of State on the remainder of the registration document, or otherwise in writing, or, if the Secretary of State thinks fit, orally by telephone or by electronic means, the following information:

(a) the name and address of the new keeper;

(b) the date of transfer; and

(c) a declaration by him/her that this information is correct to the best of his/her knowledge (reg 22).

A 'vehicle trader' is a person who: (i) holds a trade licence; (ii) carries on business as a dealer in, or auctioneer or dismantler of, motor vehicles; or (iii) in relation to a particular vehicle is a finance company which has repossessed it under a repossession order, or an insurer which has acquired it in satisfaction of a total loss claim.

Where the new keeper is a vehicle trader, the *registered keeper* must forthwith provide the Secretary of State with:

(i)   the name and address of the vehicle trader;

(ii) the date of the transfer to the vehicle trader; and

(iii) a declaration by him/herself that the transfer occurred on the specified date.

Where the new keeper is a vehicle trader, the *trader* must, on or before the appropriate date, notify the Secretary of State of the transfer of the vehicle to him/her, and the date on which (s)he became its keeper. The 'appropriate date' is the earliest of the following days: (a) the day of the trader's first use of the vehicle on a public road (**23.38**) otherwise than under a trade licence; (b) the day on which the trader first keeps it on such a road; or (c) the day following the expiry of a three-month period beginning with the day on which the vehicle was last kept by someone who was not a vehicle trader. If there is a transfer to another vehicle trader within the three-month period the registration document must be transferred with the vehicle (reg 24).

Where a vehicle trader transfers the vehicle to a person *other than another vehicle trader or transfers the vehicle to another trader outside the three-month period,* (s)he must forthwith deliver to the Secretary of State the following:

(a) the name and address of the new keeper;

(b) the date of transfer; and

(c) a declaration as to the correctness of the details under (a) and (b) (reg 24).

Where all parts of the registration document have been, or may have been, lost, stolen, or destroyed, the new keeper *whether or not a trader* must submit an application to the Secretary of State for a new registration document and send the prescribed fee. If the new keeper can produce that part of the registration document which is to be given to the new keeper, no fee need be sent.

Failure to comply with any of the above requirements is an offence under VERA 1994, s 59(1). It is not a continuing offence and therefore the time limitation upon proceedings runs from the day of the transfer of the vehicle. If, therefore, a period of six months has passed since the transfer of a vehicle no proceedings may be taken against either of the parties to the transaction if they have failed to comply with the requirement to notify the Secretary of State forthwith.

Notification of other changes

23.12   An owner of a registered vehicle who changes his/her name or address is required by RV(R&L)R 2002, reg 18(1) forthwith to notify the Secretary of State. Where the registration document has been, or may have been, lost, stolen or destroyed, notification must be accompanied by an application for the issue of a registration document, and a fee. A notification may be made on the registered keeper's behalf by a vehicle trader by whom the vehicle is sold or to whom the vehicle is disposed of.

In the event of a vehicle being sent permanently out of the UK the registered keeper is required by reg 17 immediately to notify the Secretary of State of that fact. The Secretary of State may, if (s)he thinks fit, accept such a notification by electronic means. A notification may be made on the registered keeper's behalf by a vehicle trader by, or to whom, the vehicle is sold or disposed of.

Regulation 17A requires that where an 'end-of-life' vehicle is transferred to an authorised treatment facility (facility operating under a waste disposal site licence) the facility must, in addition to issuing a 'certificate of destruction' to the last holder/owner of the vehicle, inform the Secretary of State of the issue of that certificate. No further records may then be made which are related to that vehicle.

Regulation 16 is concerned with procedures which must be followed when any alteration to a vehicle renders the particulars in the registration document incorrect, as where it is resprayed a different colour or has a different engine fitted. Notification of the alteration must be given to the Secretary of State:

(a) by the registered keeper; or
(b) on behalf of the registered keeper, by a vehicle trader to whom the vehicle is sold or by whom the vehicle is disposed of.

Non-compliance with any of the above requirements is an offence under VERA 1994, s 59(1).

Offence of using incorrectly registered vehicle

23.13 VERA 1994, s 43C(1) and (2) provides that it is an offence to use a vehicle (**23.36**) on a public road (**23.38**) or in a public place (**26.3**) if excise duty is chargeable in respect of it, or it is an exempt vehicle which requires a nil licence (**23.40**), where the name and address of its keeper are not recorded in the register, or any particulars recorded in the register are incorrect.

It is a defence under s 43C(3) for the defendant (D) to show that there was no reasonable opportunity, before the material time, to furnish:

(a) the name and address of the keeper of the vehicle; or
(b) particulars correcting the incorrect particulars.

It is a defence under s 43C(4) for D to show:

(a) that (s)he had reasonable grounds for believing, or it was reasonable for him/her to expect, that the name and address of the keeper or the other particulars of registration (as the case may be) were correctly recorded in the register; or
(b) that any prescribed exception is met. No such exception has been prescribed.

Trial and punishment

23.14 All the offences described in **23.10–23.13** are summary offences. With the exception of those at **23.10** (which are punishable with a fine not exceeding level 2), they are punishable with a fine not exceeding level 3.

Registration marks

Generally

23.15 The Secretary of State has power to assign to a vehicle:

(a) a registration mark on initial registration;
(b) a new registration mark to a vehicle in place of its existing one; and

(c) whether on first registration or not, a registration mark previously assigned to another vehicle.

The Secretary of State may also grant to a person the right to retain a registration number by transferring it from one vehicle registered in that person's name to another such vehicle.

The Road Vehicles (Display of Registration Marks) Regulations 2001 (RV(DRM)R 2001) deal with the forms of registration marks and the manner in which they are fixed to vehicles.

Regulation 10 and Sch 2 introduce a mandatory requirement for the use of registration plates made of retro-reflecting material conforming to British Standard specification BS AU 145a or an equivalent standard laid down by a European Economic Area (EEA) state on all vehicles first registered on or after 1 September 2001, and on all vehicles registered before that date but on or after 1 January 1973 if an existing plate is replaced. Regulation 10 and Sch 2 also set out the requirements for vehicles registered on or after 1 January 1973 and before 1 September 2001 in other circumstances (reflex-reflecting plates conforming to British Standard specification BS AU 145a or an equivalent standard laid down by an EEA state) but such vehicles *may* carry the forms of plates prescribed for vehicles first registered on or after 1 September 2001. In either case they must have black characters on a white background on a front plate; and black characters on a yellow background on a rear plate. The EEA comprises all EU member states plus Iceland, Liechtenstein and Norway.

Schedule 2 also deals with vehicles registered before 1 January 1973. It provides as follows. Where a plate is constructed so that it may be illuminated from behind by virtue of the translucency of its characters they must be white translucent characters on a black background, and when illuminated the characters must appear white against a black background. Where a plate is not so constructed, it must *either* comply with BS AU 145 carrying black characters on a white background at the front and black characters on a yellow background at the rear, *or* be white, silver, or light grey letters and numbers on a black surface which are indelibly inscribed on the plate and cannot readily be detached.

Permitted layouts

23.16   RV(DRM)R 2001, reg 13 and Sch 3 govern this. Registration marks in any non-permitted format are unlawful. The marks may be, for example, a group consisting of two letters and two numbers followed by a group of three letters (eg DE51 ABC); a group consisting of a single letter and not more than three numbers followed by a group of three letters (eg A123 ABC); a group of three letters followed by a group consisting of not more than three numbers and a single letter (eg ABC 123A); a group of four numbers followed by a single letter or a group of two letters (eg 1234 A, 1234 AB); a group of not more than three numbers followed by a group of not more than three letters (eg 123 ABC, 123 AB, 12 A); a group of not more than three letters followed by a group of not more than three numbers (eg ABC 123, AB 123, A 12); a single letter or group of two letters followed by a group of four numbers (eg A 1234, AB 1234); and in Northern Ireland, a group of three letters followed by a group of four numbers (ABZ 1234) or a group of four numbers followed by a group of three letters (eg 1234 ABZ).

It is not necessary that these letters and numbers all follow one another. The plates may be square permitting the letters and characters to be placed in two or three rows, except that plates containing three rows of characters are not permitted on vehicles first registered on or after 1 September 2001 or a replacement plate fixed to a vehicle first registered before then but on or after 1 January 1973. Plates containing all of the letters and numbers in a straight line are not permitted on motor cycles.

Character sizes and fonts

RV(DRM)R 2001, reg 14 and Sch 3 deal with the size and spacing of characters. Regulation 14 provides that the registration marks of vehicles must be 79 mm high. In respect of vehicles first registered before 1 September 2001 the registration marks may be 78 mm high instead of 79 mm, except where the vehicle was first registered on or after 1 January 1973 and the mark is displayed on a new registration plate. By way of further exception, the required height is 64 mm in relation to a motor cycle, motor tricycle, quadricycle, agricultural machine, works truck, or road roller. The width of characters, spacing, and margins are all precisely prescribed by Table B in Sch 3.

23.17

Regulation 14A applies special rules to a vehicle imported into the UK which does not have EC Whole Vehicle Type Approval and is so constructed that the area available for the fixing of the registration plate precludes the display on the plate of a registration mark in conformity with the requirements of reg 14. In such a case the prescribed height of characters is 64 mm; the width (except for the letter 'I' or figure '1') must be 44 mm; the width of every part of a stroke and the spacing of characters within a group must be 10 mm, the vertical spacing between groups of characters and the width of a margin between the mark and the top and lateral sides of the registration plate must not be less than 5 mm; and the space between the bottom of the mark and the bottom of the registration plate must not be less than 13 mm.

Regulation 12(2) permits deviation from the prescribed height of characters provided that the deviation is not more than 1 mm either way. In the case of other dimensions, including spaces, the permitted deviation is 0.5 mm either way.

Regulation 15 and Sch 4 deal with the font of characters displayed on a registration plate fixed to a vehicle first registered on or after 1 September 2001, or on a new registration plate fixed to any other vehicle (except where the vehicle was first registered before 1 January 1973). They require that each of the characters must be in the prescribed font. In relation to other cases, characters must be in the prescribed font or in a style which is substantially similar to the prescribed font so that the character is easily distinguishable but, in the latter case, characters must not be formed in italic script (or other script which is not vertical), or in script in which the curvature or alignment of the lines of the strokes is substantially different from the prescribed font, or in script using multiple or a broken stroke or in such a way that a character, or characters, appear like a different character or characters. A character will not be treated as substantially different solely on the grounds that it has, or does not have, serifs (small lines at the extremity of a main stroke).

Fixing of rear registration plates: post-1938 vehicles

RV(DRM)R 2001, reg 5 (which applies to all vehicles first registered on or after 1 October 1938 other than works trucks, road rollers, and agricultural machines) requires a rear registration plate to be fixed on the rear of the vehicle or (where it is towing a trailer) to the rear of the trailer or the rearmost trailer.

23.18

Where a vehicle or trailer has been constructed so as to satisfy the requirements of EU law, or retained EU law, in force at the time of construction as regards the space provided in accordance with those requirements, the plate may be fixed in the space provided in accordance with those requirements (reg 5(3)).

Except as provided in reg 5(3), reg 5(5) provides that, a rear registration plate, must be fitted vertically (or, if that is not reasonably practicable, as close to vertical as is reasonably practicable) in such a position that the characters are easily distinguishable from a distance of 22 m (where the characters are of a width of 57 mm); 21.5 m (where the characters are of a width of 50 mm); and 18 m (where the characters are of a width of 44 mm).

Regulation 9 requires a plate of the present type to be lit in accordance with the regulation when used on a road between sunset and sunrise. It provides that, except where it is fitted and lit in accordance with EU law or retained EU law in force at the time of construction as regards the rear registration plate lamp, a rear plate must be lit so that it is easily distinguishable from a distance of 18 m, but 15 m is substituted where the characters are of a width of only 44 mm.

Fixing of front registration plates: post-1938 vehicles

23.19   RV(DRM)R 2001, reg 6 deals with registration plates on a vehicle first registered on or after 1 October 1938, with the same exceptions as above in relation to rear plates. It requires that a front registration plate must be fixed vertically (or if that is not reasonably practicable, as close to vertical as is reasonably practicable) so that its characters are easily distinguishable from the rest of the plate in normal daylight. Regulation 6 requires that, in the case of motor cycles or motor tricycles which do not have a body of a type which is characteristic of the body of a four-wheeled vehicle, there must not be a front registration plate if the vehicle was first registered on or after 1 September 2001. A front registration plate need not be fitted if such a vehicle was registered before that date.

Fixing of registration plates: pre-1938 vehicles

23.20   RV(DRM)R 2001, reg 7 deals with registration plates fitted to the front of vehicles registered before 1 October 1938 and to their rear (and the rear of any trailer or rearmost trailer). It requires vertically fitted plates which are easily distinguishable.

There are similar provisions in respect of reasonable practicability and distinguishability of the characters on a plate as under regs 5 and 6. Likewise, there is no need for a front plate to be fitted to a motor bicycle or motor tricycle. The lighting requirements of reg 9 apply to the rear plates of these vehicles.

Works trucks, road rollers, and agricultural machines

23.21   Regulation 8 is concerned with works trucks, road rollers, and agricultural machines. Registration plates must be fitted vertically on both sides of the vehicle and on its rear, so that the characters on the mark are easily distinguishable from the sides and rear respectively. When the vehicle is towing a trailer and the plate is not fixed to the sides of the vehicle, a plate must be fixed on the trailer (or rearmost trailer) so that the characters of the mark are easily distinguishable from behind the trailer. In the case of a towing machine which is an agricultural machine (agricultural tractor, off-road tractor, light agricultural vehicle, agricultural engine, or mowing machine) the plate displayed on the trailer may be that of any agricultural machine kept by the keeper of the towing vehicle. The lighting requirements of reg 9 do not apply to these vehicles.

Use of reflex-reflecting material and other impediments to true photographs

23.22   RV(DRM)R 2001, reg 11 prohibits the application of reflex-reflecting material to any part of the registration plate or the treatment of the plate in such a way as to cause the registration mark to become retroreflective. In addition, it requires that the surface of a registration plate must not comprise nor incorporate any design, pattern, or texture, or be treated in any way which gives to any part of the plate the appearance of a design, pattern, or texture. It also prohibits any treatment of a registration plate which has the effect of making it less distinguishable or would prevent or impair the making of a true photograph. The use of a screw or bolt or other fixing device in a manner which has the effect of changing the appearance or legibility of any characters of a registration mark, or would impair the making of a true photograph, is also prohibited.

Plates with national words/letters and emblems, and green plates etc

23.23 RV(DRM)R 2001, reg 16 provides that no material other than a registration mark and the following material may be displayed on a registration plate. 'The following material', which, if displayed, must be positioned on the left-hand side of the plate, refers to the displayed letters denoting in full or by a specified abbreviation one of the constituent countries of Great Britain, Great Britain (as a whole), or the United Kingdom (as a whole) positioned below the corresponding emblem of those countries (an image of the Union Flag, the national flag of England or Scotland, or the Red Dragon of Wales). Thus, registration marks may identify vehicles as belonging to the UNITED KINGDOM (UK), GREAT BRITAIN (GB), ENGLAND (ENG), SCOTLAND (SCO), or WALES/CYMRU (CYM) by capital letters or (except GB) a mixture of initial capital letters followed by lower case together with the corresponding emblem. The display of the letters and the emblem must be not more than 50mm wide.

Regulation 16 also provides that an eligible vehicle (ie a vehicle which cannot produce any tailpipe emissions) and a trailer which it is towing may display a plate which has a panel which is green in colour. 'Green' means colour Pantone 7481c or a colour match as close as possible to this colour. Such a panel may display the words or letters and emblems referred to at the end of the previous paragraph. Such a panel must be not less than 40 mm and no more than 50 mm in width. The green panel must be positioned on the left-hand side of the registration plate.

Because only vehicles which have no emissions are permitted to display the green panels on their registration plates, hybrids, and plug-in hybrids are not eligible to display them. The aim of the amendment is to make it simpler for councils to offer cheaper parking to eligible vehicles.

Exemptions

23.24 RV(DRM)R 2001, reg 3 exempt s small purpose-built invalid carriages from the requirement to carry registration marks. The use of old-style number plates on 'classic' vehicles is also exempt from the Regulations.

Breach of the Regulations

23.25 VERA 1994, s 59(1) provides that a contravention of, or a failure to comply with, RV(DRM)R 2001 is a summary offence punishable with a fine not exceeding level 2. An offence under s 59(1) is a fixed penalty offence if it relates to a failure to fix prescribed registration marks to a vehicle in accordance with the regulations relating to the size, shape, and character of registration marks.

Offences relating to registration marks

23.26 VERA 1994, ss 42(1) and 43(1) provide, respectively, that, if a registration mark is not fixed on a vehicle as required by s 23, or if a registration mark is obscured or has become not easily distinguishable, the driver (or, if it is not being driven, its keeper) is guilty of a summary offence punishable with a fine not exceeding level 3. D has a defence to a charge under s 42(1) if (s)he proves that (s)he had no reasonable opportunity to register the vehicle and that it was being driven for the purpose of being registered (s 42(4)). D has a defence to a charge under s 43(1) if (s)he proves that (s)he took all reasonably practicable steps to prevent the mark being obscured etc (s 43(4)).

Forgery, fraud, or falsity

23.27   VERA 1994, s 44(1) and (2) makes it an indictable (either-way) offence, punishable with up to two years' imprisonment, to forge, fraudulently alter, fraudulently use, fraudulently lend, or fraudulently allow another to use a registration mark.

REGISTRATION: TRAILERS

23.28   In addition to the provisions referred to above relating to the registration of a mechanically propelled vehicles, or anything (whether or not a vehicle) that has been, but has ceased to be, a mechanically propelled vehicle, and registration marks and documents for them, there are provisions about the registration of certain trailers and registration marks and documents in relation to them *before they can be towed to and from many European countries*.

To address the requirements of the 1968 Vienna Convention on Road Traffic ('the 1968 Convention'), the Haulage Permits and Trailer Registration Act 2018 (HPTRA 2018), Part 2 provides for a UK trailer registration scheme for commercial trailers with a permissible maximum mass of more than 750 kg (in practice, almost exclusively, HGV trailers) and all trailers with such a mass of more than 3,500 kg. The 'permissible maximum mass' means the weight which the trailer is designed or adapted not to exceed when in normal use, travelling on a road and laden. Unless used only in the UK or only for journeys between the UK and Ireland (or only through a country not party to the 1968 Convention), such trailers need to be registered with the Secretary of State (via DVLA) by the person who keeps the trailer and is resident in the UK, a UK registered company, or a holder of a valid operator's licence granted under the Public Passenger Vehicles Act 1981 or the Goods Vehicles (Licensing of Operators) Act 1995 (or the equivalent Northern Irish licence). Registered trailers must display their own registration plate (separate from the vehicle towing them).

The Trailer Registration Regulations 2018 (TRR 2018), made under HPTRA 2018, set out the details of the registration scheme. The scheme is enforced domestically by the Driver and Vehicle Standards Agency (DVSA), primarily through the application of fixed penalty notices.

For further details, see Chapter 23 of the companion website.

REGULATION OF SUPPLIERS OF REGISTRATION PLATES

23.29   The Vehicles (Crime) Act 2001 (V(C)A 2001), Part 2 is concerned with the regulation of suppliers of registration plates. For these purposes, 'registration plate' means a vehicle registration plate or a trailer registration plate.

By s 17(1), a person commits an offence (**23.32**) if (s)he carries on business as a registration plate supplier without being registered with the Secretary of State. A person carries on a business as a registration plate supplier if his/her business consists wholly or partly in selling registration plates and (s)he is not exempt from the provisions of the Act. The Vehicles Crime (Registration of Registration Plate Suppliers) Regulations 2008, reg 3 exempts a dealer in vehicles who has arranged a first registration in the UK on behalf of the intended purchaser or keeper, or where the registration plate was not fixed to the vehicle by the dealer or on his/her behalf.

Under s 18, on payment of any prescribed fee, the Secretary of State must supply information from the register of registration plate suppliers on request by any person (subject to exceptions provided by regulations). The information supplied must be in the form of a certified copy of the register or of an extract from it. Any such certified copy is evidence of the matters mentioned in it. The Secretary of State may make all the information contained in the register, or prescribed parts of it, available for use by constables or police civilian staff in the investigation of offences against Part 2. Regulations limit the further disclosure by constables of information to which they have been given access.

By s 19(3), a person who, in applying for registration, makes a statement which (s)he knows to be false in a material particular, or recklessly makes a statement which is so false, commits an offence (**23.32**). On conviction, a court may make an order under s 20 providing for the removal of the entry relating to him/her in the register, and/or prohibiting him/her from making an application for registration within a period not exceeding five years specified by the court.

Under s 21, the Secretary of State may cancel a person's registration if satisfied that the person is not carrying on the business of a registration plate dealer and has not, while registered, been doing so for the past 28 days, but may not do so without serving notice under s 22 and allowing time for representations.

Information to be obtained from prospective purchasers of plates

The Vehicles Crime (Registration of Registration Plate Suppliers) Regulations 2008 require registered persons who are in the course of selling registration plates to obtain prescribed information from the prospective purchasers before the completion of the sale. They also require registered persons to keep records of prescribed matters.   **23.30**

The information to be obtained by a registered supplier from a prospective purchaser is prescribed by reg 6. (S)he is required to obtain:

(a) where the prospective purchaser is a partnership or is one or more partners of a partnership (or a limited liability partnership (LLP) or registered company) purchasing on behalf of such an organisation, the firm's name and the address of the principal place of business (the registered name and office, and principal place of business if different, in the case of a LLP or company);

(b) where the prospective purchaser is not as set out above at (a), the purchaser's name and residential or other address;

(c) where a person is acting as an agent for a prospective purchaser, the name and address of that agent;

(d) the registration mark to be displayed on the plate;

(e) the connection of the prospective purchaser with the registration mark or the vehicle on which the registration plate is intended to be fixed.

The information in (a)–(c) must be verified by the registered person. This must be done by reference to one of the following documents listed in Part 1 of the Schedule to the Regulations:

(a) a valid driving licence containing a photograph;

(b) a valid driving licence (whether or not issued in the UK);

(c) a registration document or registration certificate provided it is also used to verify the prospective purchaser's connection with the registration mark or vehicle (see below);

(d) a valid passport whether or not it is issued in the UK;

(e) a valid national identity card (whether or not issued in the UK);

(f) a valid debit card or credit card issued by a bank or building society;

(g) a police warrant card;

(h) a valid armed forces identity card;

(i) a dated bill or statement of account issued in respect of the supply of gas, electricity, water, or telecommunications services to premises at a specified address;

(j) a dated council tax bill or statement of account;

(k) a dated bill or statement of account issued in respect of rates payable in Northern Ireland; or

(l) a dated statement relating to an account held at a bank or building society.

If not dated, documents referred to in (i) to (l) must relate to a period ending no earlier than six months before that time.

In addition, the registered person must verify the purchaser's connection with the registration mark or the vehicle by one of the following documents listed in Part 2 of the Schedule to the 2008 Regulations: the registration document or registration certificate, or that part of such document or certificate relevant to the transfer of a vehicle; a certificate of entitlement to a registration mark; a retention document relating to the right of retention of a registration mark; a vehicle licensing reminder issued to the registered keeper; a temporary registration certificate; an authorisation issued by the Secretary of State for the purchase of a number plate; an authorisation for the purchase of the number plate issued by a company owning more than one vehicle stating that it holds the registration document or the registration certificate and giving the reference number of that document or certificate; or a notice of registration issued by the Secretary of State in accordance with the Trailer Registration Regulations 2018.

By V(C)A 2001, s 25(3), a person who contravenes any provision in reg 6 commits an offence (**23.32**) unless (s)he shows that (s)he took all reasonable steps and exercised all due diligence to avoid committing the offence.

Regulation 7 of the 2008 Regulations requires a registered person to keep records at his/her principal place of business or at any other premises at which (s)he carries on the business of a registration plate supplier. Records must be retained for a period of three years from the date of registration. Such records must contain:

(a) the information required by reg 6;

(b) the registration mark displayed on the registration plate (where not recorded under (a)); and

(c) details of all documents used for verification in accordance with reg 6 and those details must include:

   (i) in the case of a document listed in Part 1 of the Schedule which is used for verification, such particulars or numbers (if any) appearing on the document as purport to make it, or those particulars or numbers (or both), unique to the purchaser and which, in the case of a driving licence, must be the driver number, or

   (ii) in the case of a registration document or certificate the reference number of that document, and in the case of an authorisation to purchase the number plate, issued by a company owning more than one vehicle, the reference number referred to in that document.

Failure to comply with these provisions as to records is an offence (**23.32**) against V(C)A 2001, s 24(4) unless the defendant shows that (s)he took all reasonable steps and exercised all due diligence to avoid committing the offence.

Supplementary provisions and offences

23.31 V(C)A 2001, s 26 empowers a constable, or person authorised by the Secretary of State or the local authority for the area, to enter and inspect the registered premises of registered persons at any reasonable time. The constable or authorised person has power to require production of, and to inspect, any plates kept at the premises and any records which are required to be kept under V(C)A 2001, Part 2, and to take copies or extracts. Provision is also made for the issue to a constable or authorised person of a warrant. While force may not be used to obtain entry in normal circumstances, it may be used in the exercise of the constable or authorised person's powers under the authority of a warrant. When effecting entry without a warrant, a constable or authorised person must, if required by or on behalf of the owner or occupier or person in charge of the premises, produce evidence of identity, and of his/her authority for entering, before doing so. This also applies to an authorised person when executing a warrant. Section 26(7) also creates an offence (**23.32**) of obstruction of an authorised person in the exercise of his/her powers under the section. The Secretary of State is authorised to prosecute offenders.

A person registered under V(C)A 2001, Part 2 is required by s 27 to give notice of changes of circumstances. Section 27(4) creates the offence (**23.32**) of failure to do so. The same due diligence defence as is set out above in relation to ss 24 and 25 is available to a person accused of this offence.

Section 28 is concerned with counterfeit plates and those which do not conform with regulations. It creates five offences (**23.32**):

(a) selling a plate or other device which is not a vehicle registration plate as a vehicle registration plate, knowing that it is not such a plate or being reckless as to whether it is a vehicle registration plate (s 28(1));

(b) selling a plate or other device which is not a trailer registration plate as a trailer registration plate, knowing that it is not such a plate or being reckless as to whether it is a trailer registration plate (s 28(1ZA));

(c) selling a plate or other device which is not a vehicle registration plate only because the mark does not comply with the Road Vehicles (Display of Registration Marks) Regulations 2001 or is displayed otherwise than as permitted by them (s 28(1A));

(d) selling a plate or other device which is not a trailer registration plate only because the mark (i) does not comply with the Trailer Registration Regulations 2018 or (ii) is displayed otherwise than as permitted by them (s 28(1AA)); and

(e) supplying a plate, device or other object to a person who is carrying on a business which consists wholly or partly in activities which are unlawful contrary to (a), (b), (c), or (d), knowing or reasonably suspecting that the plate, device or other object will be used for the purposes of that other person's unlawful activities (s 28(2)).

Section 29(1) creates the offence of supplying a plate, device, or other object to an unregistered person (other than an exempt person to be defined by regulations) who is carrying on a business which consists wholly or partly in selling registration plates, knowing or reasonably suspecting that the plate, device, or other object will be used for the purpose of that other person's business as a registration plate or as part of a registration plate.

Trial and punishment

23.32 All the above offences under V(C)A 2001 are summary offences punishable with a fine as follows:

– s 17(1): unlimited fine;
– s 28(1), (1ZA), (1A), 1(AA), and (2): fine not exceeding level 4;
– ss 19(3), 24(4), 25(3), and 27(4): fine not exceeding level 3;
– s 26(7): fine not exceeding level 2.

LICENSING OF MECHANICALLY PROPELLED VEHICLES

23.33   The Vehicle Excise and Registration Act 1994 (VERA 1994), s 1 provides that vehicle excise duty must be charged in respect of:

(a) every mechanically propelled vehicle that:
    (i)   is registered under the Act, or
    (ii) (if not so registered) is used, or kept, on a public road (**23.38**).
As already noted, it is not necessary to prove that such a vehicle was intended or adapted for use on roads. Any mechanically propelled vehicle which is actually used (**23.36**) or kept (**23.37**) on a public road (**23.38**) requires a vehicle licence regardless of whether or not it was intended or adapted for use on a road; and
(b) everything (whether or not it is a vehicle) that has been, but has ceased to be, a mechanically propelled vehicle and:
    (i)   is registered under the Act, or
    (ii) (if not so registered) is used, or kept, on a public road.

Vehicle excise duty charged in respect of a vehicle is paid on a vehicle licence.

Although, by s 5, certain vehicles (**23.40** and **23.41**) are exempt from duty, they must generally have a nil licence. A vehicle licence issued in Northern Ireland is valid in Great Britain.

Where a vehicle for which a vehicle licence is in force is transferred by the holder of the licence to another person, the licence is to be treated for the purposes of ss 29 and 31A below as no longer in force.

In the rest of VERA 1994 'vehicle' means a mechanically propelled vehicle or any thing (whether or not it is a vehicle) that has been, but has ceased to be, a mechanically propelled vehicle.

Offences of using or keeping unlicensed vehicle

23.34   VERA 1994, s 29(1) provides that if any person *uses* (**23.36**) or *keeps* (**23.37**) a vehicle for which no vehicle licence or trade licence (**23.48**) is in force, he is guilty of an offence. The offence is a summary one punishable with a fine not exceeding level 3 or five times the duty chargeable whichever is greater. If the offence is committed within a prescribed time in breach of a statutory off-road declaration (SORN) the fine level is 4 (s 29(3A)).

Section 29(2A) provides that the offence under s 29(1) does not apply if the vehicle is an exempt vehicle requiring a nil licence (**23.40**) and such a licence is in force, or if it is an exempt vehicle not required to have a nil licence (**23.41**).

Section 29(1) does not apply if (a) the vehicle is neither being used or kept on a public road (**23.38**), (b) the requisite particulars and statutory declaration (the statutory off-road notification (SORN)) have been furnished and provided, and (c) the terms of that declaration have not been breached (s 29(2B)). In relation to the SORN, RV(R&L)R 2002, reg 26 and Sch 4 require that, where a person does not renew a vehicle licence on expiry, or keeps an unlicensed vehicle, (s)he must make a required declaration to the Secretary of State (in writing, orally, or by electronic means) not later than the day upon which the licence ceases to be in force (or three months after such expiry in the case of motor traders). The declaration is to the effect that (except for use under a trade licence) (i) the vehicle is not intended for the time being to be used or kept on a public road and (ii) a licence will be taken out before any such use. It is a summary offence, punishable with a fine not exceeding level 3, under VERA 1994, s 59(1) to fail to make such a declaration. If there is a transfer of such a vehicle during an unlicensed period and a licence is not taken out, a required declaration must be made by the new owner.

In addition, s 29(1) does not apply if the vehicle is kept by a motor trader or vehicle tester at his/her business premises (s 29(2C)).

Section 29(2E) provides that a defendant is not entitled to the benefit of any of these exceptions from s 29(1) unless evidence is adduced which is sufficient to raise an issue with respect to that exception, but where evidence is so adduced it is for the prosecution to prove beyond reasonable doubt that the exception does not apply.

Offence of being registered keeper of unlicensed vehicle

If a registered vehicle does not have a vehicle licence or trade licence in force the registered keeper is guilty of a summary offence contrary to VERA 1994, s 31A(1), unless it is an exempt vehicle of one of the types referred to in s 29(2A) above. However, s 31B provides a number of exceptions. The registered keeper does not commit an offence under s 31A if *at the relevant time*:   **23.35**

(a) (s)he is not the person (**23.37**) keeping the vehicle and, if previously (s)he was the person keeping it, (s)he has by the relevant time complied with *any requirements to furnish particulars which apply on not renewing a licence, or when keeping an unlicensed vehicle*;
(b) (s)he is keeping the vehicle, it is neither kept nor used on a public road (**23.38**) and (s)he has complied with any requirement italicised in (a);
(c) the vehicle has been stolen and has not been recovered, and within 14 days (s)he has notified a member of a police force with prescribed details about the theft; or
(d) the period of 'grace days' has not expired since the expiry of the last licence and a licence is taken out within the 14 'grace days'.

RV(R&L)R 2002, reg 26A deals with the 'requirements to furnish particulars', etc referred to in (a). They are the requirements in relation to the surrender or destruction of a registration document; delivery of a registration document to the Secretary of State; notification of transfer to a vehicle trader; and notification by a vehicle trader of a sale or transfer.

An offence under s 31A(1) is punishable in the same way as that under s 29(1) (**23.34**), save for the obvious omission of a reference to a SORN-breach.

Definitions and other general points

Uses

This term is described at **25.9**. An employer is liable for use by his/her employee even though blamelessly unaware of what the employee is doing. It is not good practice to proceed against the employee in normal circumstances.   **23.36**

Keeps

VERA 1994, s 62 states that a person keeps a vehicle on a public road if he causes it to be on such a road for any period, no matter how short, when it is not in use there. If the description 'use' cannot be applied to the vehicle's presence on the road at any particular   **23.37** time, it is 'kept' on that road by any person who causes it to be there. It is a question of fact in each case who that person is. It may be a driver who has parked it there whilst the vehicle still remains under his/her control. It may be the owner who allows it to remain in a back street unlicensed, or it may be some person in temporary possession. If an unlicensed vehicle is repaired at a garage and placed outside on the road by the proprietor when the repairs have been effected, the garage proprietor is keeping the vehicle on a road without there being a vehicle licence in force.

Public road

23.38   A public road is one which is repairable at public expense. Whether or not a road is repairable at public expense can be established by the local authority who will be able to say whether it has been 'adopted' in the sense that a highway authority is responsible for its maintenance under the Highways Act 1980 or another enactment. In most instances, particular reference to authorities is unnecessary as the road concerned is commonly known to be a public road and justices are entitled to apply their knowledge to such matters. Inquiries are advisable when use on roads within new housing estates is alleged, as the roads may still be the responsibility of the builder if development work is still continuing or has recently finished.

Licence is in force

23.39   The prosecutor invariably offers evidence of the lack of a vehicle licence but this does not strictly need to be proven. Once (a)(i) or (ii) or (b)(i) or (ii) at **23.33** is established, the defendant must prove that it was licensed. A 'licence' in this context includes a trade licence (**23.48**). Where a licence is obtained by means of a cheque which is dishonoured the licence is void from the time of issue on notice from the Secretary of State.

Exempt vehicles requiring nil licence

23.40   Vehicles which are exempt from excise duty are listed in VERA 1994, Sch 2. The following exempt vehicles do, however, require a nil licence if used or kept on a public road:
(a)   police vehicles;
(b)   fire engines;
(c)   ambulances, and health service vehicles;
(d)   veterinary ambulances;
(e)   invalid carriages not exceeding 508kg;
(f)   vehicles for export;
(g)   vehicles imported by members of foreign armed forces;
(h)   vehicles which are used only for purposes related to agriculture, horticulture, or forestry and are on public roads only in passing between different areas of land occupied by the same person and the distance so travelled on public roads in doing so does not exceed 1.5 km;
(i)   agricultural engines and off-road tractors;
(j)   mowing machines;
(k)   hedge and verge-cutting tractors;
(l)   electrically propelled vehicles (except a light passenger vehicle first registered on or after 1 April 2017, whose price exceeds £40,000);
(m)   snow ploughs;
(n)   gritters;
(o)   lifeboat vehicles;
(p)   road rollers; and
(q)   steam powered vehicles.

A vehicle used by or for persons with particular disabilities may, in certain circumstances, be exempt.

A vehicle constructed during the 12-month period beginning with 1 April in any year if it was constructed more than 40 years before 1 January in that year is exempt, except:

(a) a vehicle for which an annual rate is specified in Sch 1, Parts III, V, VI, VII and VIII (buses, recovery vehicles, vehicles used for exceptional loads, haulage vehicles, and goods vehicles); or

(b) a special vehicle (other than digging machine, works truck, mobile crane, mobile pumping vehicle, or road roller) where:
   (i) it is designed or adapted for use for the conveyance of goods or burden of any description (or for use with a semi-trailer attached);
   (ii) it is put to a commercial use on a public road; and
   (iii) that use is not a use for the conveyance of goods or burden of any description (or in a case where that use is a use with a semi-trailer attached, the semi-trailer is not used for the conveyance of goods or burden of any description).

A '*recovery vehicle*' is a vehicle which is constructed or permanently adapted primarily for any one or more of the purposes of lifting, towing, and transporting a disabled vehicle. A vehicle is not a recovery vehicle if at any time it is used for a purpose other than:

(a) the recovery of a disabled vehicle;

(b) the removal of a disabled vehicle from where it became disabled to premises where it is to be repaired or scrapped;

(c) the removal of a disabled vehicle from premises where it was taken for repair, to other premises where it is to be repaired or scrapped;

(d) carrying fuel and other liquids required for its propulsion and tools and other articles required for the operation of or in connection with integral or permanently mounted apparatus designed to lift, tow or transport a disabled vehicle;

(e) carrying any person who, immediately before a vehicle became disabled, was the driver of or a passenger in that vehicle, together with his/her personal effects, from the premises at which the vehicle is to be repaired or scrapped to his/her original intended destination;

(f) at the request of either a constable or a local authority empowered to remove a vehicle from a road, removing such vehicle to a place nominated by him/her or it;

(g) proceeding to a place at which the vehicle will be available for use for either of the purposes specified in (a) and (b) and remaining temporarily at such a place so as to be available for such use;

(h) proceeding from:
   (i) a place where the vehicle has remained temporarily so as to be available for such use; or
   (ii) a place where the vehicle has recovered a disabled vehicle; or
   (iii) any premises mentioned in (b) or (c).

Where a recovery vehicle is being used to recover a disabled vehicle or to remove such a vehicle from the place where it has become disabled to a place for repair or for being scrapped, certain uses are permitted which would otherwise take the vehicle outside the definition of a recovery vehicle:

(a) the carriage of a person who, immediately before the vehicle became disabled, was the driver of or a passenger in the vehicle;

(b) the carriage of any goods which, immediately before the vehicle became disabled, were being carried in the vehicle; or

(c) repairing a disabled vehicle where it became disabled or at a place to which it had been removed for safety reasons after becoming disabled; or

(d) drawing or carrying one trailer if the trailer was, immediately before a vehicle became disabled, being drawn or carried by it (VERA 1994, Sch 1 and RV(R&L)R 2002, Sch 7).

A vehicle is not a recovery vehicle at any time when it is used to recover more than two vehicles.

A '*special vehicle*' is:

(a) a vehicle which has a revenue weight exceeding 3,500 kg and is a digging machine, mobile crane, mobile pumping vehicle, works truck, or road roller;

(b) a vehicle designed or adapted for use for the conveyance of goods or burden of any description, but not so used, or not so used for hire or reward or in connection with a trade or business; or

(c) a vehicle designed or adapted for use with a semi-trailer attached but which is not so used or, if it is used, the semi-trailer is not used for the conveyance of goods or burden of any description.

'*Digging machines*' are machines designed for use for trench digging, excavating, or shovelling which are used on public roads for such a purpose or for getting to or from the place where they will so operate.

'*Mobile cranes*' are cranes used on public roads only in connection with work in the immediate vicinity of that road or in travelling to or from that place.

A '*works truck*' is a goods vehicle (other than a straddle carrier) designed for use on private premises and used on public roads only for carrying goods between such premises and a vehicle on a road in the immediate vicinity, or in passing from one part of any such premises to another or to other private premises in the immediate vicinity, or in connection with roadworks while at or in the immediate vicinity of the site of such works. 'Immediate vicinity' connotes a very considerable degree of closeness. Divisional courts have held that the use of a works truck six-tenths of a mile from premises, or more than one and a half miles away from premises, was not in the immediate vicinity of them.

RV(R&L)R 2002, Sch 5 exempts vehicles imported by members of visiting forces, members of a headquarters or other organisation, or a dependant of such a person. This exemption lasts for a period of 12 months only.

Exempt vehicles not requiring nil licence

23.41   A nil licence is not required in respect of the following exempt vehicles listed in VERA 1994, Sch 2.

An electrically assisted pedal cycle (**25.93**), an electric scooter being used in a trial (**23.7**), a registered Crown vehicle, a tram, a vehicle not constructed or adapted, or used, for the carriage of a driver or passenger, or a vehicle for export do not require a nil licence.

Nor does a vehicle when it is used solely for the purpose of submitting it by prior arrangement for examination or re-examination for a test certificate (or a vehicle weight test or vehicle identity check) or of taking it from such an examination or re-examination. This also applies to a charge of 'keeping'. Thus, if a driver parks whilst on his/her way to the testing station simply for the purpose of buying something in a shop, such 'keeping' of the vehicle on a public road is exempt because it is still possible to say that the vehicle was on the road *solely* for the purpose of going to the testing station. A vehicle used in the course of such an examination also does not require a nil licence.

A foreign vehicle is an exempt vehicle, and does not require a nil licence, if brought temporarily for up to one year into the UK by a non-UK resident is. The term 'foreign' includes vehicles from the Isle of Man and the Channel Islands.

Documentary evidence of commission of offence of using or keeping

23.42 The Road Traffic Offenders Act 1988 (RTOA 1988), s 20 provides that evidence of a fact relevant to proceedings for an offence to which it applies may be given by the production of a record produced by a prescribed device, accompanied by a certificate as to the circumstances in which it was produced, signed by a constable or a person authorised by or on behalf of a chief officer of police. An offence under VERA 1994, s 29 is prescribed as an offence to which RTOA 1988, s 20 applies and in respect of it a prescribed device is one designed or adapted to register:

(a) an image of a vehicle and its registration mark; and
(b) the time at which the image is registered,

and to record that information if, according to data stored by or otherwise accessible by the device, that vehicle is unlicensed. See **25.100** for more about s 20.

Duration and issue of licence

23.43 Vehicle licences may be issued for a period of 12 months or for six months in the case of (a) vehicles in respect of which the annual rate of duty exceeds £50, and (b) tractive units for which the 'tractive unit: special case' annual rate of duty is £10. In addition, where an application is made for a vehicle licence, the Secretary of State may issue a temporary licence for 14 days or such other period as is prescribed by regulations. Other than temporary licences, licences are valid from the first day of the month on which they are taken out. An application may be made out by any person who must make a declaration and furnish particulars which are prescribed. This is intended to counter the practice of 'ringing' stolen vehicles to assume the identity of legitimate vehicles.

Reliance is placed on DVLA's electronic vehicle register and use of tools like Automatic Number Plate Recognition to ensure that vehicles are correctly licensed and that vehicle excise duty has been paid.

The licence is issued to the vehicle, not the applicant, and it does not authorise that person to use or keep any other vehicle. On the sale of a vehicle an application must be made to DVLA for a rebate of duty; the licence cannot be transferred with the vehicle.

Annual rates of duty

23.44 Vehicle licences cost differing amounts depending upon the nature of the vehicle and its particular use. VERA 1994, Sch 1 sets out the various rates of duty but the actual amounts of the rates are set out in the Finance Acts and are regularly changed.

Immobilisation of unlicensed vehicles

23.45   The Vehicle Excise Duty (Immobilisation, Removal and Disposal of Vehicles) Regulations 1997 authorise the immobilisation and removal of stationary unlicensed vehicles in places other than the curtilage of a dwelling in circumstances in which an 'authorised person' (a person, such as a police officer or local authority employee, authorised by the Secretary of State) has reason to believe that an offence is being committed under VERA 1994, s 29(1) (**23.34**). Such persons are given power to enter any such place for the purpose of enforcement.

The vehicle may be wheelclamped where it stands, or moved to another place and wheelclamped there. This may be done by the authorised person or a person acting under his/her direction. An immobilisation notice must be fixed to the vehicle indicating that the device has been fitted, warning that no attempt should be made to move the vehicle until it has been released from the device, and providing other information, including the charge for release, removal, and disposal. Before an authorised person can release such a vehicle:

(a) the prescribed 'release charge' must be paid;

(b) (i) evidence must be produced that no offence contrary to VERA 1984, s 29(1) was being committed at the time of clamping or removal;

(ii) a vehicle licence for the vehicle must be in force;

(iii) the prescribed 'surety payment' (a sum payable where no such licence is in force) must be made; or

(iv) a declaration must be made to the effect that a licence was in force at the specified time, that a statutory off-road notification (SORN) was in force, or that the vehicle was an exempt vehicle which did not require a nil licence.

A voucher will be issued to a person who makes a 'surety payment'. Where a licence cannot be obtained immediately the 'surety payment' permits the vehicle to be used unlicensed for a period of 24 hours. The payment will be refunded if a claim to a refund is made within 15 days and a vehicle licence is then in force.

The Regulations authorise the removal of a vehicle which has been clamped for a period of 24 hours without release; it may be removed to the custody of an authorised 'custodian'. Where this has been done, a removal fee will be charged additionally, together with charges for storage.

The 1997 Regulations create the following offences:

(a) unauthorised removal of or interference with an immobilisation notice: reg 7(1) and (2);

(b) unauthorised removal, or attempted removal, of an immobilisation device: reg 7(3);

(c) false declaration with a view to securing release of vehicle from an immobilisation device: reg 8(2);

(d) false declaration with a view to securing possession of impounded vehicle: reg 13(1);

(e) making declaration in connection with obtaining voucher or refund relating to a surety payment, knowing that it is false or in a material respect misleading: reg 16(1); and

(f) forgery, fraudulent alteration, fraudulent use, or the fraudulent lending, of a voucher relating to surety payment: reg 16(2).

The offences under (a) and (b) are summary offences punishable with a fine not exceeding level 2 and 3 respectively. The offences under (c), (d), (e), and (f) are indictable (either-way) offences punishable with up to two years' imprisonment.

Fraudulent offences related to vehicle licences

The following offences are the most common fraudulent offences under VERA 1994:   **23.46**

(a) making, in connection with an application for the issue of a vehicle licence, a declaration which to the knowledge of the defendant (D) is false or in a material respect misleading: s 45(1);
(b) supplying or producing false information or documents which D knows (or is reckless as to) are false or in a material respect misleading in relation to the design weight of a vehicle (for the purposes of establishing the revenue weight): s 45(3A);
(c) forgery, alteration, or use of a certificate (design weight certificate) with intent to deceive, or lending or allowing such a certificate knowing or believing that it will be used for deception, or without reasonable excuse making or possessing a document so resembling such a certificate as to be calculated to deceive: s 45(3B).

These are indictable (either-way) offences punishable with up to two years' imprisonment.

Proceedings for offences and admissibility of evidence

VERA 1994, s 47 provides that proceedings for offences of using, etc without a licence   **23.47**
(23.34), of being a registered keeper of an unlicensed vehicle (23.35), of using a trade licence outside permitted uses (23.51), and of using a licensed vehicle for a purpose which attracts a higher rate of duty than that paid, can only be instituted by the Secretary of State or a constable; such a person is known as the authorised prosecutor. Moreover, no prosecution may be instituted by a constable for these offences without the approval of the Secretary of State; proof of such approval is required at the outset of proceedings in court.

Section 52 allows certified extracts from DVLA records to be admissible to the same extent as oral evidence. Evidence may therefore be offered of the last date upon which a vehicle was licensed by means of such a certified extract.

Section 46 requires that, where an offence under s 29 (23.34 and 23.35) or 34 (23.51) is alleged to have been committed in relation to a particular vehicle, the person keeping the vehicle must give such information as (s)he may be required by or on behalf of a chief officer of police or the Secretary of State to give as to the identity of any person concerned in the offence. Failure to do so is an offence under s 46(4), unless the keeper proves that (s)he did not know, and could not with reasonable diligence have ascertained, the identity of the person(s) concerned (s 46(6)). A divisional court has held that where there is evidence of using or keeping a vehicle on a road and a notice has been sent in accordance with s 46 to which the keeper does not respond, the justices should draw an adverse inference from the failure to respond and that this, coupled with the other evidence of using or keeping, is sufficient to support a conviction under s 46. The requirement to respond also extends to any person other than a 'keeper', and (in the case of a s 29 offence) to the person alleged to have been using the vehicle. Both types of person commit an offence under s 46(4) if they fail to give such information as to identity as it is in their power to give. An offence under s 46(4) is a summary offence punishable with a fine not exceeding level 3.

TRADE LICENCES

Without modification, the requirement that all vehicles which are registered, or which   **23.48**
are used or kept on a road, should be individually licensed under VERA 1994 would cause considerable problems for motor traders, through whose hands many vehicles

pass, most of which are retained for a short period of time. The purpose of trade licences is to permit traders temporarily to use vehicles for restricted purposes without the necessity for licensing in the manner described above. Trade licences are inexpensive and in consequence their permitted uses are carefully defined.

Who may apply for a trade licence?

23.49 By VERA 1994, s 11, a motor trader may apply to the Secretary of State for a licence to cover all vehicles which are from time to time temporarily in his/her possession in the course of his/her business as a motor trader. Any motor trader who is a manufacturer may also be granted a licence for the purpose of allowing him/her to carry out research and development work in the course of his/her business as a manufacturer, and for all other vehicles which are from time to time submitted to him/her by other manufacturers for testing on roads in the course of that business.

The term 'motor trader' means a manufacturer or repairer of, or dealer in, vehicles. A person is treated as a dealer in vehicles if (s)he carries on a business consisting wholly or mainly of collecting and delivering vehicles.

A person whose business is that of modifying vehicles (by fitting accessories or otherwise) or of 'valeting' vehicles (which means the thorough cleaning of a vehicle prior to first registration or in order to prepare it for sale, and includes removing wax and grease from the exterior, engine, and interior) is also a 'motor trader'.

A vehicle tester may also apply for a trade licence to cover his/her use of vehicles submitted to him/her for testing in the course of his/her business. A vehicle tester is a person, other than a motor trader, who regularly in the course of his/her business engages in the testing on roads of vehicles belonging to other persons.

Persons who satisfy the Secretary of State that they intend to commence business as motor traders or vehicle testers may also take out trade licences.

Trade licences may be taken out for a period of 12 months. Shorter-term licences are also available but are seldom used. There is nothing to prevent a motor trader from holding more than one licence to permit use of more than one vehicle at any one time.

Trade plates

23.50 The holder of a trade licence is issued with a set of two plates (which are generally referred to as 'trade plates') in respect of each licence held. These plates consist of red letters on a white background and show the registration mark assigned to the holder of the licence. One plate must be displayed to the front of the vehicle, and the other to the rear, in the same way as a registration plate. If the trader satisfies the Secretary of State that the vehicles which (s)he will use in the course of his/her business will include motor cycles as well as other vehicles, (s)he may be issued with a single additional plate for the motor cycles only.

Where a vehicle is used under a trade licence, its trade plate(s) must be fixed and displayed in the specified manner (reg 42). The plates remain the property of the Secretary of State and must be returned when the trader ceases to be the holder of the licence (reg 40(5)). Failure to comply with reg 40(5) or 42 is a summary offence punishable with a fine not exceeding level 3 (VERA 1994, s 59(1)).

Non-permitted uses of trade licences

23.51 VERA 1994, s 12 provides that the holder of a trade licence is not entitled by virtue of *that* licence:

(a) to use more than one vehicle at any one time; or
(b) to use any vehicle for any purpose other than a permitted one (**23.53**); or
(c) to keep any vehicle on a road if it is not being used thereon.

It is an offence, contrary to s 34(1), for the holder of a trade licence to use (**23.36**) on a public road (**23.38**) a greater number of vehicles than permitted by his/her licence (or licences), or to use a vehicle for a non-permitted purpose. Section 34(1) also punishes using a trade licence for the purposes of keeping on a public road a vehicle which is not being used at that time. These are summary offences punishable with a fine not exceeding level 3 or five times the duty chargeable whichever is greater.

Restrictions on use of trade licences

RV(R&L)R 2002, Sch 6 provides that the holder of a trade licence must not permit any   **23.52**
person to display trade plates on a vehicle other than one which that person is using for the purpose of the licence-holder's business (Sch 6, para 5), or other than when the vehicle is being used for one or more of the prescribed permitted purposes (Sch 6, para 6). Breach of para 5 or 6 is a summary offence, punishable with a fine not exceeding level 2, under VERA 1994, s 59(1). However, this does not prevent a person driving a vehicle on a road with the consent of the licence-holder, when the vehicle is being used for the licence-holder's business. Thus, an employee may drive vehicles in the course of his/her employer's (the licence-holder's) business.

Permitted purposes

The use of vehicles by a motor trader under a trade licence is controlled by RV(R&L)   **23.53**
R 2002, Sch 6, paras 10 to 15, breach of which is a summary offence, punishable with a fine not exceeding level 2, under VERA 1994, s 59(1).

Regulation 38 and Sch 6 prescribe the purposes for which the holder of a trade licence may use a vehicle by virtue of a trade licence. Those purposes do not include the carrying of any person on the vehicle or any trailer drawn by it except a person carried in connection with such a purpose. The prescribed purposes are without prejudice to the provisions of VERA 1994, s 11(2) to (4) which specify classes of vehicle for which a trade licence is a licence, in relation respectively to a motor trader who is a manufacturer of vehicles, any other motor trader and a vehicle tester.

Schedule 6, para 10 authorises a motor trader who is the holder of a trade licence to use a vehicle on a public road for purposes which meet each of the following requirements:

(a) business purposes, as specified by para 11;
(b) purposes specified by para 12; and
(c) purposes that do not include the conveyance of goods or burden of any description except specified loads (as defined by para 13).

Business purposes, para 11

A vehicle is used for 'business purposes' if used for purposes connected with the motor   **23.54**
trader's business:

(a) as a manufacturer or repairer of or dealer in:
  (i) vehicles; or
  (ii) trailers carried on in conjunction with his/her business as a motor trader;

(b) of modifying vehicles (whether by the fitting of accessories or otherwise); or

(c) of valeting vehicles.

Paragraph 12 purposes

23.55   A vehicle is used for a para 12 purpose if it is used for:

(a) the test or trial of the vehicle, or its accessories or equipment, in the ordinary course of construction, modification or repair, or after completion;

(b) proceeding to or from a public weighbridge to ascertain its unladen weight, or to or from any place for its registration or inspection by someone acting on the Secretary of State's behalf;

(c) its test or trial for the benefit of a prospective purchaser, including going either to or from a place of such test or trial at the instance of the prospective purchaser;

(d) its test or trial for the benefit of a person interested in promoting publicity in regard to the vehicle, including going either to or from a place of such test or trial at the instance of such a person;

(e) delivering it to a purchaser;

(f) demonstrating the operation of the vehicle or its accessories or equipment when handed over to a purchaser;

(g) delivering it between parts of the motor trader's own premises or to the premises of another manufacturer, dealer or repairer or bringing it back from there directly to his/her own premises;

(h) proceeding to or from a workshop where a body, or a special type of equipment or accessories, is to be or has been fitted to it or where it is to be or has been painted, valeted or repaired;

(i) proceeding from the premises of a manufacturer, repairer or dealer to a railway station, airfield or shipping dock for the purpose of transportation, or for proceeding to such premises from a railway station, etc to which it has been transported;

(j) proceeding to or from any garage, auction room or storage place where vehicles are usually stored or offered for sale and at which the vehicle is to be or has been stored or offered for sale as the case may be;

(k) proceeding to or from a place of testing; or

(l) proceeding to a place to be broken up or otherwise dismantled.

The use of a vehicle with a trailer is regarded as the use of a single vehicle under the licence.

Specified loads

23.56   RV(R&L)R 2002, Sch 6, para 13 defines the 'specified loads' referred to in **23.53**(c) as follows:

(a) a load which is carried by a vehicle being used for the purpose of testing or demonstrating the vehicle, or its accessories or equipment, within the terms at **23.55** (b), (d), (e), or (g), is carried solely for that purpose, and is returned to the place of loading without having been removed from the vehicle (except in the case of an accident, or for demonstrating its operation to a purchaser when handed over to him/her, or when the load consists of water, fertiliser, or refuse);

(b) in the case of a vehicle which is being delivered or collected and is being used for a relevant purpose (as described in (**23.55**) (f) to (k)), a load consisting of another vehicle used or to be used for travel to or from the place of delivery or collection;

(c) any load built in as a permanent part of the vehicle or permanently attached to it;

(d) in the case of a vehicle which is being used for a purpose falling within **23.55** (h), (i), or (j), a load which consists of a trailer or of parts, accessories or equipment designed to be fitted to the vehicle and of tools for fitting them.

Research and vehicle testing purposes

RV(R&L)R 2002, Sch 6, paras 14 and 15 deal respectively with manufacturers' research vehicles and use by vehicle testers as permitted purposes. The first restricts use to manufacturers for research and development purposes and the second restricts testing to vehicles and trailers drawn thereby, or any accessory or equipment on the vehicle or trailer, in the course of a business as a vehicle tester. **23.57**

No other permitted purpose

The above purposes are the only permitted purposes for which vehicles may be used under a trade licence, and even then only in the course of the business of the holder. If an employee used the trade plates and licence to remove his/her own private car to a paint shop for spraying, not in the course of his/her employer's business, the use would be unlawful. If the holder of the licence used under trade plates a vehicle which was in his/her possession in the course of his/her business to visit a cinema in the evening, that use of the motor vehicle would be unlawful as it would not be in the course of his/her business as a motor trader. **23.58**

Exhibition of altered, defaced, etc trade plates

RV(R&L)R 2002, Sch 6, para 3 provides that the holder of a licence must not, or must not permit any person to, exhibit on any vehicle any trade plate which has been altered, defaced, mutilated, or added to, upon which the figures or particulars have become illegible, or upon which the colour has been altered by fading or otherwise. **23.59**

Breach of Sch 6, para 3 is a summary offence, punishable with a fine not exceeding level 2, contrary to VERA 1994, s 59(1).

Forgery and fraud

By VERA 1994, s 44(1) and (2), forgery, fraudulent alteration or use, or fraudulent lending or allowing to be used by any other person, of a trade plate is an indictable (either-way) offence punishable with up to two years' imprisonment. So is the false declaration under s 45(1) (**23.46**) which also applies in respect of an application for a trade licence. **23.60**

FIXED PENALTY OFFENCES

The offences under VERA 1994, ss 34, 42, 43, 43C (**23.13**, **23.25**, **23.26**, and **23.50**, **23.51** and, in the instance at **23.25**, s 59, are fixed penalty offences, the fixed penalty being £100 (or £50 in the case of s 34 or 43C). As to fixed penalty offences, see **25.117–25.133**. **23.61**

Regulation and Responsibilities of Drivers

DRIVING LICENCES, PENALTY POINTS, AND DISQUALIFICATION

Driving licences

24.1 The licensing of drivers of motor vehicles is dealt with by the Road Traffic Act 1988 (RTA 1988), Part III (ss 87 to 109C) and the Motor Vehicles (Driving Licences) Regulations 1999 (MV(DL)R 1999).

For driving licence purposes, motor vehicles are divided by MV(DL)R 1999 into various categories and sub-categories and it is essential that licences are checked, not only to establish the identity of the driver, but also to ascertain that (s)he is authorised to drive the particular vehicle in which (s)he is found. Amendments made to MV(DL) R 1999 with effect from 19 January 2013 provide for the introduction of new categories and sub-categories of vehicles (principally mopeds and motor cycles). The categories and sub-categories of vehicles for licensing purposes are set out in MV(DL)R 1999, Sch 2 as follows:

Category or sub-category	Class of vehicle included	Additional categories and sub-categories covered
	Part 1	
	A licence authorising the driving of motor vehicles of a class included in a category or sub-category shown in Part 1 may be granted to a person who is entitled thereto by virtue of:	
	(a) *holding or having held a full licence, full Northern Ireland licence, full British external licence granted in the Isle of Man, Jersey or Guernsey, full British Forces licence, exchangeable licence (**24.25**) or Community licence (**24.23**) authorising the driving of vehicles of that class; or*	

(continued)

Category or sub-category	Class of vehicle included	Additional categories and sub-categories covered
	(b) having passed a test for a licence authorising the driving of motor vehicles of that class or a Northern Ireland or Gibraltar test corresponding to such a test.	
AM	Mopeds: two- or three-wheel vehicles with a maximum design speed (MDS) not exceeding 45km/h (excluding those with MDS less than 25km/h); electric scooters being used in a trial; and light quadricycles: quadricycles with unladen mass not exceeding 350kg (excluding batteries in case of electric vehicles), MDS not exceeding 45km/h, and engine capacity not exceeding 50cc (spark ignition) or maximum net power output not exceeding 4kw (other internal combustion engine) or maximum continuous rated power not exceeding 4kw (electric motor)	Q
A	Motor bicycles; electric scooters being used in a trial; and motor tricycles: motor vehicles with three symmetrically arranged wheels with MDS exceeding 45km/h and engine capacity exceeding 50cc (if internal combustion engine)	Q, AM, A1, A2, and K
A1	A sub-category of A comprising (a) A1 motor cycles: motor bicycles with engine capacity not exceeding 125cc, power not exceeding 11kw and power to weight ratio (PWR) not exceeding 0.1kw/kg; (b) A1 motor tricycles: motor tricycles with power not exceeding 15kw; and (c) electric scooters being used in a trial	Q and AM
A2	A sub-category of A comprising (a) A2 motor cycles: motor bicycles of power not exceeding 35kw, PWR not exceeding 0.2kw/kg and not being derived from a vehicle of more than double its power; and (b) electric scooters being used in a trial	Q, AM. and A1
A3	A sub-category of A comprising (a) A3 motor cycles: motor bicycles (i) of power exceeding 35kw or PWR exceeding 0.2kw/kg, or (ii) of power not exceeding 35kw with PWR not exceeding 0.2kw/kg and derived from a vehicle of more than double the particular cycle's power; (b) A3 motor tricycles: motor tricycles with power exceeding 15kw; and (c) electric scooters being used in a trial.	Q, AM, A1, A2, and K

(continued)

Category or sub-category	Class of vehicle included	Additional categories and sub-categories covered
B	Motor vehicles, other than vehicles included in category Q which are not electric scooters being used in a trial, AM, A, F, or K, having maximum authorised mass (MAM) not exceeding 3.5 tonnes and designed and constructed for not more than 8 passengers in addition to the driver, including: (a) combination of any such vehicle and trailer where trailer has MAM not exceeding 750kg, and (b) combination of any such vehicle and trailer where MAM of combination does not exceed 3.5 tonnes	Q, AM F, and K
B + E	Combinations of motor vehicle (being tractor vehicle in category B) and trailer or semi-trailer where (a) combination does not fall within category B or B96; and (b) MAM of trailer or semi-trailer does not exceed 3.5 tonnes	None
C	Motor vehicles having MAM exceeding 3.5 tonnes, other than vehicle falling within category D, F, G, or H, designed and constructed for the carriage of no more than 8 passengers in addition to the driver, and including such vehicle drawing trailer having MAM not exceeding 750kg	None
C1	A sub-category of C comprising motor vehicles having MAM exceeding 3.5 tonnes but not exceeding 7.5 tonnes, designed and constructed for the carriage of no more than 8 passengers in addition to the driver, and including such vehicle drawing trailer having MAM not exceeding 750kg	None
D	Motor vehicles designed and constructed for the carriage of more than eight passengers in addition to the driver, including such vehicle drawing trailer having MAM not exceeding 750kg	None
D1	A sub-category of D comprising motor vehicle designed and constructed for the carriage of not more than 16 passengers not including the driver with maximum length not exceeding 8m, and including such vehicle drawing a trailer with MAM not exceeding 750kg	None
C + E	Combination of motor vehicle and trailer where the tractor vehicle is in category C but the combination does not fall within that category	B + E

(continued)

Category or sub-category	Class of vehicle included	Additional categories and sub-categories covered
C1 + E	A sub-category of C + E comprising combinations of motor vehicle in category B, or in sub-category C1, and trailer or semi-trailer where combination's MAM is no more than 12 tonnes and: (a) if tractor vehicle is in category B, MAM of trailer or semi-trailer exceeds 3.5 tonnes, or (b) if tractor vehicle is in sub-category C1, MAM of trailer or semi-trailer exceeds 750kg	B + E
D + E	Combination of motor vehicle and trailer where tractor vehicle is in category D but the combination does not fall into that category	B + E
D1 + E	A sub-category of D + E comprising any combination of motor vehicle and trailer where: (a) tractor vehicle is in sub-category D1, and (b) MAM of trailer exceeds 750kg but not the un-laden weight of the tractor vehicle	B + E
F	Agricultural or forestry tractors, including any such vehicle drawing trailer but excluding any motor vehicle included in category H	K
G	Road rollers	None
H	Track-laying vehicles steered by their tracks	None
K	Mowing machines, not falling within category A, or vehicle controlled by pedestrian	None

Part 2

A licence authorising the driving of motor vehicles of a class included in any category or sub-category shown in Part 2 may not be granted to a person unless, at a time before 1 January 1997:

(a) in the case of a person applying for a full licence,
(i) (s)he held a full licence authorising the driving of motor vehicles of that class or a class which by virtue of MV(DL)R 1999 corresponds to a class included in that category or sub-category, or (ii) (s)he passed a test which at the time it was passed authorised the driving of motor vehicles of such a class or a Northern Ireland test corresponding to such a test;

(b) in the case of a person applying for a provisional licence, (s)he held a provisional licence authorising the driving of vehicles of that class or a class which by virtue of MV(DL)R 1999 corresponds to a class included in that category or sub-category

(continued)

Category or sub-category	Class of vehicle included	Additional categories and sub-categories covered
C1 + E (8.25 tonnes)	A sub-category of category C + E comprising combinations of motor vehicle and trailer in sub-category C1 + E where MAM of trailer exceeds 750kg and may exceed unladen weight of tractor vehicle, and MAM of combination does not exceed 8.25 tonnes	None
D1 (not hire or reward)	A sub-category of category D comprising motor vehicles in sub-category D1 driven otherwise than for hire or reward	None
D1 + E (not hire or reward)	A sub-category of D + E comprising motor vehicles in sub-category D1 + E where: (a) motor vehicle driven otherwise than for hire or reward, and (b) MAM of trailer exceeds 750kg and may exceed unladen weight of tractor vehicle	None
L	Vehicle propelled by electrical power	None
	Part 3	
B1 (invalid carriage)	A sub-category of category B comprising motor vehicles which are invalid carriages	None
	Part 4	
	A licence authorising the driving of motor vehicles of a class shown in column 2 of Part 4 opposite a former category or former sub-category (as the case may be) shown in column 1 may not be granted to a person unless, before 19 January 2013, that person held a licence authorising the driving of motor vehicles of that class or passed a test for a licence authorising the driving of motor vehicles of that class	
Former sub-category B1 (vehicle with three or four wheels and unladen weight not exceeding 550kg)	Motor vehicles having four wheels and unladen weight not exceeding 550kg, except light quadricycles	None
Former category B + E (as column 2)	Combination of motor vehicle and trailer where tractor vehicle is in category B and MAM of trailer exceeds 3.5 tonnes	None

(continued)

Category or sub-category	Class of vehicle included	Additional categories and sub-categories covered
Former sub-category D1 (as column 2)	Motor vehicles having more than 8 but not more than 16 seats in addition to driver's seat with maximum length exceeding 8m and including any such vehicle drawing trailer with MAM not exceeding 750kg	None
Former sub-category D1 + E (as column 2)	Combination of motor vehicle and trailer where: (a) tractor vehicle is in former sub-category D1, (b) MAM of trailer exceeds 750kg but not unladen weight of tractor vehicle (c) MAM of combination does not exceed 12 tonnes, and (d) trailer is not used for carriage of passengers	None
Former category P (vehicles having fewer than four wheels, MDS not exceeding 50km/hr and, if internal combustion engine, not exceeding 50cc)	Motor vehicle with fewer than four wheels, MDS exceeding 45km/hr but not exceeding 50km/hr and, if internal combustion engine, maximum 50cc	None

Part 5

Q These would be mopeds but for maximum speed limit of 25km/hr	Category Q vehicles: motor vehicle with less than four wheels which: (a) if propelled by internal combustion engine, has a cylinder capacity not exceeding 50cc and, if not equipped with pedals capable of propelling vehicle, MDS not exceeding 25km/hr; and (b) if not propelled by internal combustion engine, has MDS not exceeding 25km/hr A licence authorising the driving of category Q vehicles may not be granted to a person unless (s)he is entitled to be granted a full licence authorising the driving of motor vehicles of a class included in category AM, A, or B, or category P vehicles.	None

(continued)

Category or sub-category	Class of vehicle included	Additional categories and sub-categories covered
	Part 6	
B96	Combinations of motor vehicle and trailer where:	None
	(a) tractor vehicle is in category B,	
	(b) MAM of trailer exceeds 750kg,	
	(c) MAM of combination exceeds 3.5 tonnes but not 4.25 tonnes	
	A B96 licence is only granted if person holds or has held a Community licence to drive B96 vehicle	

The term 'maximum authorised mass' has the same meaning:

(a) in relation to goods vehicles as 'permissible maximum weight' (**24.16**); and
(b) in relation to any other vehicle or trailer as 'maximum gross weight' in the Road Vehicles (Construction and Use) Regulations 1986, reg 3(2), namely the weight which the vehicle is designed or adapted not to exceed when on a road.

MV(DL)R 1999 parallel categories of vehicle with the previous categories under earlier regulations. The 1999 Regulations provide that licences whether full or provisional granted before 1 January 1997 or 19 January 2013, as the case may be, are valid in respect of the new categories of vehicle, as set out in the table to the regulations.

Where a licence (whether full or provisional) granted before 19 January 2013 authorises the driving of:

(a) standard motor bicycles only, or
(b) standard motor bicycles and side-car combinations only,

any reference in that licence to motor vehicles in category A (save for those in sub-category A1) is a reference:

(i) where the standard access period (two years commencing on the day when the test to drive a standard motor bicycle was passed excluding any period of disqualification or any period when the licence was not in force) has not expired, to motor vehicles in sub-category A2; and
(ii) where that period has expired, to motor vehicles in category A.

Entitlement on passing a test

24.2   MV(DL)R 1999, reg 43 provides that, where a person passes a test prescribed in respect of any category for a licence which authorises the driving of motor vehicles included in that category or in a sub-category of that category, the Secretary of State must grant him a licence authorising him to drive vehicles of all classes included in that category or sub-category unless his licence is restricted to such vehicles fitted with automatic transmission, or vehicles specially adapted for the disabled, in which cases his entitlement will be so restricted. Such holders are also authorised to drive those vehicles shown in column 3 above as additional categories or sub-categories with the same limitations as set out above, should the test have been taken on those types of vehicle. However, where the additional category is Q, AM, F, or K the limitation in respect of 'automatics' will not apply.

Where a person has passed a test of competence for a licence to drive vehicles included in category B, that licence authorises the driving of vehicles within category Q or AM if and only if:

(a) the test was passed before 1 February 2001;
(b) the person concerned held at the date on which (s)he passed the test a valid certificate of successful completion of an approved training course for motor cyclists;
(c) the person concerned holds a valid certificate of successful completion of an approved training course for motor cyclists furnished after the date on which (s)he passed the test;
(d) the vehicles in category Q are electric scooters being used in a trial.

Where a person has passed the test of competence to drive a vehicle in category AM in a three-wheeled moped or light quadricycle, reg 44A provides for the grant of a licence restricted to three-or four-wheeled vehicles in category AM.

When licence-holder reaches the prescribed age

RTA 1988, s 99 provides as follows about the duration of driving licences.   **24.3**

Full driving licences issued to those who have passed the appropriate test are granted until the holder reaches 70 or for a period of three years whichever is the longer. After that, the licence can be renewed for three-year periods.

By way of exception, licences which authorise the holder to drive any prescribed class of goods or passenger-carrying vehicle are renewable on the holder's 45th birthday, or after five years, whichever is the longer, or *where the licence is issued to a person between 45 and 65 for the period ending on his/her 66th birthday or after five years, whichever is the shorter.* A licence granted after the age of 65 will remain in force for one year only. The italicised provisions have been temporarily amended by the Business and Planning Act 2020, s 14, so as to provide that, where a licence to drive any class of goods or passenger-carrying vehicle:

(a) was granted in the period 17 April 2020–24 March 2022;
(b) is the first licence to drive that class of vehicle granted after 16 April 2020 to the applicant; and
(c) was granted without the requirement of a medical report in support,

the reference to the period for which the licence is to remain in force has effect as a reference to one year. However, this does not apply if, in the 12-month period ending with the application, (i) the applicant was granted a driving licence under RTA 1988, Part III (**24.1**), and (ii) a supporting medical report was submitted.

Other licences

RTA 1988, s 88 provides that it is lawful for a person to drive at any time provided that   **24.4**
the driver has held a driving licence, a Community licence (**24.23**), a Northern Ireland licence, a British external licence granted in the Isle of Man, Jersey, or Guernsey or British forces licence, or an exchangeable licence (**24.25**), for the class of vehicle in question or a corresponding licence, and a 'qualifying application' for a licence under Part III of the Act which includes that time has been received. This provision does not apply where the application received was made as a result of, or in anticipation of, the expiry of a disqualification on or after 26 April 2013 which is relevant to the licence applied for, was made by a 'high-risk offender' (**24.15**), and the Secretary of State has

notified the applicant that (s)he must be medically assessed. The effect is that a high-risk driver will not be able to obtain a new licence before it is granted following a successful medical assessment.

Grant of licences

24.5   RTA 1988, s 97(1) requires the Secretary of State to issue a licence to a person who: (a) meets the relevant residence requirement, (b) applies in the prescribed manner, (c) pays the appropriate fee, (d) supplies necessary evidence to support his/her application, and (e) is not disqualified or otherwise barred from obtaining the licence. The Secretary of State may revoke the licence of someone if it appears that (s)he is not lawfully resident in the UK.

By s 97A, a person meets the relevant residence requirements if, on the date the application is made:

(a)   in the case of an application that is made by the holder of a Community licence, the applicant is lawfully resident in the UK and:
  (i)   is also normally resident in the UK, or
  (ii) has been attending a course of study in the UK during the period of six months ending on that date;

(b)   in the case of an application that is made by the holder of an exchangeable licence, the applicant is normally and lawfully resident in Great Britain but has not been so resident for more than the prescribed period;

(c)   in the case of an application for a provisional licence, the applicant is lawfully resident in Great Britain and the Secretary of State is satisfied that the applicant will remain so for not less than 185 days; and

(d)   in any other case, the applicant is lawfully resident in the UK and is also normally resident in Great Britain or has been attending a course of study in Great Britain during the six-month period ending with that date.

For the purposes of s 97A, a person is not lawfully resident in Great Britain or the UK if the person requires leave to enter or remain in the UK but does not have it.

Section 98 requires the licence to be in the form of a photocard of a description specified by the Secretary of State, or in such other form as may be specified. Photocard licences were introduced in 1998, but existing licences remain in force until they expire or the holder's details expire. Where a photocard licence is issued an applicant must supply the Secretary of State with a photograph which is a current likeness of him/her and with a specimen signature which can be electronically recorded and reproduced on the licence. If any other form of licence is granted, the holder must forthwith sign the licence in ink. Defaced or lost licences may be replaced by the Secretary of State. If a lost licence is subsequently found it must be returned to the authority.

By s 99, where a photocard licence remains in force after the end of the administrative validity period, its holder must nevertheless surrender the licence to the Secretary of State not later than the end of that period. The administrative validity period is 10 years for a licence other than one to drive any prescribed class of goods vehicles or passenger-carrying vehicle (vehicle constructed or adapted to carry eight or more passengers for hire or reward), which licence has a five-year validity period except that it is 10 years if issued before 19 January 2013. If a licence expires before the end of the standard five- or ten-year period (for instance in the case of licences issued for shorter periods on medical grounds), its administrative validity period will end on the expiry date of the licence. Subject to exceptions, the surrender of a photocard will be followed by the grant of a new one.

By s 89(1), a person applying for the first time for a full driving licence authorising the driving of a motor vehicle must:

(a) have passed a UK driving test within the previous two years;
(b) hold (or have held) a Community licence; or
(c) hold an exchangeable licence,

authorising the driving of vehicles of that or a corresponding class.

Disqualification of persons under age

A person may be disqualified for driving a particular class of vehicle by reason of age. **24.6**
A person is disqualified for holding or obtaining a licence to drive a motor vehicle of a particular class if (s)he is under the age applicable to that class of vehicle. The minimum ages at which persons may drive particular classes of vehicles are listed by RTA 1988, s 101, which provides that a person below the minimum age to drive a particular class of vehicle is disqualified for holding a licence to drive that class of vehicle. The requirements of s 101 are amplified by MV(DL)R 1999, reg 9. The minimum ages are as follows:

(a) 16 years:
    (i) invalid carriage,
    (ii) vehicle in category AM or category Q,
    (iii) agricultural or forestry tractor, provided it is a wheeled vehicle not exceeding 2.45 m in width and driven without a trailer (other than a two-wheeled or close-coupled four-wheeled trailer not exceeding 2.45 m in width); the driver must have passed a test for category F or be proceeding to or from such a test, or
    (iv) a small vehicle (as defined in (b)(ii) below) driven without a trailer attached where the driver of the vehicle is a person who:
        (a) has received an award of the higher rate component of the disability living allowance, which was in force immediately before his/her 16th birthday, provided (s)he has made a claim for personal independence payment under the Welfare Reform Act 2012, Pt 4; or
        (b) is in receipt of such a personal independence payment which includes the mobility component at the enhanced rate;

(b) 17 years:
    (i) A1 motor cycle or A1 motor tricycle,
    (ii) small vehicle, which means a motor vehicle (other than an invalid carriage, or a vehicle in category AM or category Q, or a motor tricycle) which:
        (a) is not constructed or adapted to carry more than nine persons inclusive of the driver, and
        (b) has a maximum gross weight not exceeding 3.5 tonnes, and includes a combination of such a vehicle and a trailer, but does not include a motor tricycle or light quadricycle,
    (iii) incomplete large vehicle not exceeding 3.5 tonnes, or
    (iv) road roller which is not steam propelled, whose unladen weight does not exceed 11.69 tonnes, which has no pneumatic, soft or elastic tyres, and which is not constructed or adapted to carry a load other than equipment of the vehicle;

(c) 18 years:
    (i) medium-sized goods vehicle, which means a motor vehicle constructed or adapted to carry or haul goods and not adapted to carry more than nine

persons inclusive of the driver, with a permissible maximum weight exceeding 3.5 tonnes but not 7.5 tonnes, and includes a combination of such a vehicle and a trailer where the relevant maximum weight of the trailer does not exceed 750 kg. However, the age of 21 applies if such a vehicle is drawing a trailer and the maximum authorised mass of the combination exceeds 7.25 tonnes,

(ii) motor-vehicle-and-trailer combination of sub-category C1 + E the maximum authorised mass of which does not exceed 7.5 tonnes,

(iii) motor vehicle of a class included in category C or C + E where that person:

    (a) has an initial qualification (ie initial Certificate of Professional Competence (CPC)) to drive a vehicle of that class;

    (b) has been issued by the Secretary of State with a document authorising him/her to drive the relevant vehicle for a specified period of up to 12 months while undertaking a vocational training course; or

    (c) is using the vehicle in the course of a lesson or driving test for the purpose of obtaining a driving licence or a CPC test;

(iv) motor vehicle of a class included in category C or C + E where being used by the fire service or for maintaining public order or is undergoing road tests for repair or maintenance purposes,

(v) motor-vehicle-and-trailer combination which is in sub-category C1 + E and the maximum authorised mass of the combination does not exceed 12 tonnes,

(vi) motor vehicle of a class included in category D or D + E, other than sub-category D1 or D1 + E, where the person driving the vehicle:

    (a) has an initial qualification authorising him/her to drive motor vehicles of that class and is either (*i*) engaged in the carriage of passengers on a regular service over a route which does not exceed 50 km, or (*ii*) is not engaged in the carriage of passengers;

    (b) is using the vehicle in the course of a driving lesson or driving test for the purpose of obtaining a driving licence or CPC; or

    (c) has been issued by the Secretary of State with a document authorising him/her to drive the relevant vehicle for a specified period of up to 12 months while undertaking a vocational training course;

(vii) motor vehicle of a class included in sub-category D1 (more than eight but not more than 16 seats in addition to driver's seat) or D1 + E, where one of the same conditions as at (iii) is satisfied; one of (b) or (c) in (vi) applies; or

(viii) incomplete large vehicle exceeding 3.5 tonnes but not exceeding 7.5 tonnes;

(d) 19 years:

    A2 motorcycle;

(e) 20 years:

    motor vehicle of a class included in category D or D + E, other than sub-category D1 or D1 + E, where the driver has an initial qualification authorising him/her to drive motor vehicles of that class and is engaged in the carriage of passengers otherwise than on a regular service restricted to 50 km;

(f) 21 years:

    (i) motor vehicle of a class included in category D or D + E, other than sub-category D1 or D1 + E, where the vehicle is being used by the fire service or for maintaining public order or is undergoing road tests for repair and maintenance purposes, and

    (ii) all other motor vehicles not referred to in (a) to (g);

(g) 24 years:

    (i) A3 motorcycle or A3 motor tricycle,

(ii) subject to (c)(vi) and (e), a motor vehicle of a class included in category D or D + E, other than sub-category D1 or D1 + E, save where the vehicle is being used by the fire service or for maintaining public order or is undergoing road tests for repair and maintenance purposes, or in respect of a person under the age of 24 who was entitled to a licence to drive a vehicle of that class before 19 January 2013, in which case the minimum age is 21.

The regulations permit members of the armed services, aged 17 or over, to drive A2 or A3 motorcycles, A3 motor tricycles and medium and large goods vehicles which are owned by the Secretary of State for Defence and are being used subject to his/her orders. Clearly this exemption is not a general one and the nature of the use of the vehicle at the time must be taken into account.

A1 motorcycles or A1 motor tricycles, A2 motorcycles, and A3 motorcycles can be ridden at 16, 18, and 20 respectively in the case of a person holding a Community licence authorising the driving of vehicles of that particular sub-category. An A3 motorcycle or A3 motor tricycle can be driven by a person at 21 where, for two or more years, (s)he has held a full licence for A2 motorcycles. An A2 motorcycle or A3 motorcycle can be driven at 17 by someone who passed a test before 19 January 2013 in respect of a motorcycle (other than one with an engine with maximum net power output of 11kw or less) and the standard access period (**24.1**: p 816) has elapsed. An A3 motorcycle can be driven by a person who passed an appropriate driving test after 18 January 2013 on a vehicle of a class in category B or sub-category A2 or A3, and was entitled, before 19 January 2013, to a licence to drive a motor bicycle with an engine having a maximum net power output exceeding 25kw or a power to weight ratio exceeding 0.16kw per kg. An A3 motor tricycle can be driven at 17 by someone who was, before 19 January 2013, entitled to drive vehicles having three or four wheels and an unladen weight not exceeding 550kg.

Although these provisions appear to be complex at first sight, the vehicles with which police officers are generally concerned are motor cycles, private saloon cars, goods vehicles, and public service vehicles and the age limits which generally apply are as follows: mopeds may be ridden at 16, motor cycles (unless they are A2 or A3 motorcycles) and private cars at 17, vehicles between 3.5 tonnes and 7.5 tonnes at 18, and those in excess of 7.5 tonnes at 21 (18 in the circumstances described above in relation to initial qualifications) in most circumstances. Small passenger vehicles, which are generally private cars (as the seating of such a vehicle must not exceed nine) may be driven at 17. Passenger vehicles with more than nine seats, with the exceptions outlined, may only be driven by a person of 21; most vehicles of the 'Transit' type fall within this category.

Grant of licences

24.7
A person may apply for a driving licence at any time within two months of the date from which the licence will take effect. Where the application is for a large goods or passenger-carrying vehicle driver's licence, it must be made within three months of such date.

Full licences

24.8
Where the application is for a full licence, an applicant must satisfy the Secretary of State that (s)he has passed a test at the time of his/her application. In support of his/her application (s)he must produce a valid test certificate.

MV(DL)R 1999, Part III deals with the constituent parts of driving tests, and the certificates to be issued to those who take tests. It requires that a test for a licence to drive a category B vehicle (eg the typical private motor car) must be in accordance with reg 40(2A) or (2B).

Regulation 40(2A) provides that a test thereunder must consist of two parts: (a) the standard test of driving theory and the standard test of hazard perception; and (b) the practical test of driving skills and behaviour.

Regulation 40(2B) provides that a test thereunder must consist of three parts: (1) the safe road use test; (2) the abridged standard test of driving theory and the standard test of hazard perception; and (3) the practical test.

In either type of test the theory part must be passed before the practical part is taken.

Where a person has been awarded the Safe Road User Award before 1 May 2010, the test for a category B licence is conducted in two parts, parts (2) and (3) as mentioned above.

As to driving tests for licences for motor bicycles, see **24.11**, and for buses and lorries, see Chapter 24 on the companion website.

MV(DL)R 1999 include a requirement that, where a person produces to an examiner an appropriate licence which does not include a photograph, (s)he must satisfy the person conducting the test as to his/her identity by producing a document to establish his/her identity as prescribed by Sch 6 (all of which have a photograph) or a document of a like nature. However, if the person's identity is clearly apparent from the facts known to, or other evidence in the possession of, the person conducting the test, this will be satisfactory.

Provisional driving licences

24.9   Full licences to drive a class of vehicle may only be granted to those who have passed the relevant driving test to drive that class of vehicle. Provisional licences are issued to those who wish to learn to drive motor vehicles and are issued subject to conditions set out in MV(DL)R 1999, reg 16. These conditions are concerned with the need for supervision, distinguishing marks, the drawing of trailers, and the carriage of passengers. However, these conditions do not apply where a provisional licence-holder has passed a test by virtue of which (s)he is entitled to be granted a licence authorising him/her to drive a vehicle of the class then being driven.

Supervision   With the exceptions listed below, a provisional licence-holder must not drive or ride a motor vehicle otherwise than under the supervision of a qualified driver who is present with him/her in or on the vehicle (reg 16).

A person is a qualified driver if (s)he:

(a) is 21 or over;
(b) holds a relevant licence (see below);
(c) has the relevant driving experience (see below); and
(d) in the case of a disabled driver, is supervising a provisional licence-holder who is driving a vehicle of a class included in categories B, C, D, C + E, or D + E and would in an emergency be able to take control of the steering and braking functions of the vehicle in which (s)he is a passenger.

However, for the purposes of supervising the holder of a provisional licence driving a vehicle of a class included in sub-category C1, C1 + E, D1, or D1 + E ('the learner vehicle') which the holder is authorised to drive by that licence, a person is not a qualified driver unless that person has, in addition to meeting the above requirements, passed a test in which the vehicle used in the practical test fell within the same sub-category as that of the learner vehicle.

Subject to provisions below relating to a disabled driver, a *'relevant licence'* means a *full licence authorising the driving of vehicles of the same class as the vehicle being driven by the provisional licence-holder.* However, subject to the provisions relating to a disabled driver, where a person holds a full licence authorising the driving of vehicles of the same class as that being driven by the provisional licence-holder, which class is included in a category or sub-category specified in column 1 of the table below, and that person has held the licence for less than three years, 'relevant licence' has a special meaning. It means a full licence authorising:

(a) where that class of vehicle is included within any sub-category specified in column 1 of the table below, the driving of vehicles in the sub-category specified in column 2 which is opposite that sub-category; or

(b) where (a) above does not apply, the driving of vehicles in the category specified in column 2 of that table which is opposite the category specified in column 1 that includes the class of vehicle being driven by the provisional licence-holder.

For the above purposes, the term 'full licence' includes a Northern Ireland licence and a Community licence.

The table referred to above is as follows:

Column (1)	Column (2)
Categories and sub-categories which include the vehicle being driven by the provisional licence-holder	Categories and sub-categories authorised by the relevant licence
C	D
C1	D1
C + E	D + E
C1 + E	D1 + E
D	C
D1	C1
D + E	C + E
D1 + E	C1 + E

In the case of a (supervising) disabled driver who holds a licence authorising the driving of vehicles in category B, a relevant licence must authorise the driving of vehicles other than a quadricycle or vehicles in sub-category B1 (invalid carriages). A 'disabled driver' is a person who holds a relevant licence which is limited by virtue of a declaration made with his/her application for the licence or a notice under RTA 1988, s 92(5)(b) to vehicles of a particular class.

Where the above italicised definition of 'relevant licence' only applies, a person has *'relevant driving experience'* if (s)he has held the relevant licence for a minimum period of three years. Where the definitions in (a) and (b) above apply, a person has *relevant driving experience* if (s)he has held the relevant licence authorising the driving of vehicles:

(a) of the same class as the vehicle being driven by the provisional licence-holder for a minimum period of one year; and

(b) in the category or sub-category specified in column 2 of the table above for a minimum period of three years.

For the purpose of meeting the requirements in this paragraph the minimum period of time for holding a full licence may be met either by holding that licence continuously for that period or for periods amounting in aggregate to not less than that period.

A period before 1 May 2010 during which a person ('the supervising driver') has held a licence authorising the driving of vehicles included in sub-category C1, C1 + E, D1, or D1 + E may only be taken into account in assessing whether the supervising driver has the relevant driving experience to supervise the holder of a provisional licence driving such a vehicle if the supervising driver has passed a test before 1 May 2010 in which the vehicle used for the practical test fell within one of those sub-categories.

The conditions requiring that a qualified driver is over 21 and that (s)he has relevant driving experience do not apply to a member of the armed forces of the Crown acting in the course of his/her duties. In addition, reg 8 provides that such a person may drive a dual-purpose vehicle (**25.1**) when it is being used to carry passengers for naval, military, or air force purposes:

(a) where it does not exceed 3.5 tonnes, with a category B (not former sub-category B1 or sub-category B1) licence;

(b) where it exceeds 3.5 tonnes but does not exceed 7.5 tonnes, a C1 licence; and

(c) in any other case, a C licence (other than C1).

The supervisor's duty is to make up for any deficiencies in the skill of the learner; as part of his/her duty (s)he must participate in the driving to such extent as could reasonably be expected to prevent danger to other persons or property. Because (s)he has a right of control over the learner driver, a supervisor can be convicted as an accomplice to a driving offence committed by the learner driver if (s)he deliberately fails to prevent it when (s)he could reasonably have done so.

A provisional licence-holder is not required to be supervised while undergoing a test. Although an examiner will be with him/her, the examiner is there for the purpose of assessing his/her competence, not for the purpose of supervision (and therefore not for the purpose of interference when a lack of skill is evident). The examiner is not, therefore, similarly exposed to charges of aiding and abetting offences by the learner.

A provisional licence-holder is not required to be supervised if (s)he is:

(a) driving a motor vehicle constructed to carry only one person which is not adapted to carry more than one person and which is a vehicle in sub-category B1 (invalid carriages), a motor tricycle, a motor vehicle having four wheels and an unladen weight not exceeding 550 kg, or a motor vehicle of a class included in category F, G, H or K;

(b) driving a motor vehicle of a class included in former sub-category B1 which is adapted to carry only one person, and the licence-holder has at any time between 1 August 2002 and 1 March 2003 had the use of an NHS invalid carriage that was issued to him/her by reason of his/her having a relevant disability. Former sub-category B1 vehicles may have three or four wheels and an unladen weight not exceeding 550 kg;

(c) riding a moped, or motor bicycle with or without a sidecar, a former category P vehicle, or a category Q vehicle; or

(d) driving a motor vehicle, other than a vehicle of a class in category C, C + E, D or D + E, on a road in an exempted island (a term which covers small islands like Lundy and the Isles of Scilly other than St Mary's).

It follows, for example, that a small three- or four-wheeler with two seats requires a supervisor in order to be driven by a provisional licence-holder, whereas if it is constructed with only one seat it does not (unless it has since been adapted to carry more than one person). The removal of a seat from a two-seater does not alter the position since it will have been constructed with two. It must be emphasised that there is a total exemption for two-wheeled motor bicycles, whether fitted with a sidecar or not.

Other licence conditions Motor vehicles which are driven or ridden by persons holding a provisional driving licence must display on the front and on the back of the vehicle the letter 'L' (or 'D' in Wales) in such a manner that it is clearly visible from a reasonable distance to other persons using the road. This provision does not apply where the holder of the provisional licence holder is driving an electric scooter being used in a trial.

The drawing of trailers by a motor vehicle driven by the holder of a provisional licence is prohibited, except that the holder of a provisional licence may do so in relation to the category of vehicles specified if his/her provisional licence authorises the driving of a vehicle of a class included in category B + E, C + E, D + E, or F (combination vehicles where the tractor falls in category B, C, or D respectively but the combination does not, or agricultural or forestry tractors).

Holders of provisional licences authorising the driving of a moped, a motor bicycle with or without a sidecar, or a former category P vehicle, or a category Q vehicle, must not drive such a vehicle while carrying on it another person.

The holder of a provisional licence authorising the driving of a motor bicycle other than an A1 motor cycle must drive under the supervision of a 'direct access instructor' who is accompanying him/her on another motor bicycle or, if the instructor is suffering from a relevant disability of such a nature that (s)he is unable to ride a two-wheeled vehicle, a three-wheeled vehicle in category AM or A; is able to communicate with him/her by radio other than a hand-held radio; is supervising only that person or, at the most, one additional provisional licence-holder; and is carrying a valid certificate issued by the Secretary of State. The 'learner' and the supervisor must both wear fluorescent or (during hours of darkness) luminous apparel.

The requirements concerning communication by radio do not apply to a person who has impaired hearing provided that a suitable means of communication with the instructor is arranged in advance. Direct access instructors hold additional qualifications in respect of larger motor cycles.

It is a condition of a provisional licence to drive a moped or A1 motor cycle that, when undergoing 'relevant training' (receiving professional tuition from a paid instructor in driving on a road after compulsory basic training), the holder of the licence cannot be in a group of more than three other such learners. Such an instructor must be present with him/her and riding a moped, A1 motor bicycle, or category P vehicle.

Offences

By RTA 1988, s 87(1), it is an offence for any person to drive on a road a motor vehicle **24.10** of any class otherwise than in accordance with a licence authorising him/her to drive a motor vehicle of that class. Once the prosecution has proved that D has driven a motor vehicle on a road, the onus is upon D to prove that (s)he is licensed to drive as this is a fact peculiarly within his/her own knowledge. It is desirable that there should be, where possible, a statutory demand for production of the relevant licence, but there is no obligation to do so.

It is an offence, under s 87(2), for a person to cause or permit another person to drive on a road a motor vehicle of any class otherwise than in accordance with a licence authorising that other person to drive a motor vehicle of that class. For the meaning of 'permit' and 'cause' see **25.10** and **25.11**.

An offence under s 87(1) or (2) is a summary offence punishable with a fine not exceeding level 3. The s 87(1) offence is also punishable (see **24.40**) with discretionary disqualification where the offender's driving would not have been in accordance with any licence that could have been granted to him/her.

Section 3ZB provides the indictable (either-way) offence, punishable with up to two years' imprisonment, of causing death by driving on a road when unlicensed. The offence is also punishable with obligatory disqualification (**24.39**). It is irrelevant to an offence under s 3ZB that the risk of death resulting from the driving was unforeseen or unforeseeable. The Supreme Court has held that, to be guilty of an offence under s 3ZB, D must be shown to have done something other than simply putting his/her vehicle on the road so that it is there to be struck. It must be proved that there was something which (s)he did or omitted to do by way of driving it which contributed in a more than minimal way to the death, and that thing must involve some element of fault (whether it amounts to careless or inconsiderate driving or is otherwise open to proper criticism). Thus, an unlicensed D is not guilty of an offence under s 3ZB if a car erratically driven by V crashes into D's carefully driven car and V is killed or if V jumps onto D's car as D drives under a bridge.

What constitutes 'driving' in relation to a motor vehicle is explained at **26.2** and it is sufficient to bear in mind that the essence of 'driving' is the use of the driver's controls (or, at least, one of them) in order to control the movements of the vehicle, however that movement is produced, provided that what occurs can in any sense be described as driving.

It should be noted that, when a police officer discovers an offence of driving without a driving licence in circumstances in which the offender's driving would not have been in accordance with any licence that could have been granted to him/her, the police officer should include in the report reference to whether or not the conditions applicable to that licence were being complied with. The reason is that this is important to the court in relation to the penalty points awarded for the offence.

Motor bicycles: some special rules

24.11   Restrictions on provisional licenses: two-wheeled vehicles. These are set out in MV(DL) R 1999, reg 15A. A provisional licence granted in respect of a vehicle in category AM or A is restricted to such vehicles as have no more than two wheels save where the applicant declares that (s)he is suffering from a relevant disability of such a nature that (s)he is unable to ride a vehicle which has two wheels. A provisional licence granted in respect of a vehicle with two wheels is restricted to vehicles of a class within category AM or A.

Approved training course   Motor cyclists and moped riders must undertake the two-part practical training which is applicable to them, within an approved training course for motor cyclists and moped riders. The first part is concerned with the basic handling and control of machines. The training for the first part can be undertaken without the necessity to ride on a road. The second part is the normal 'on the road' training to drive which takes place under the supervision of instructors.

A provisional licence does not authorise a person, before (s)he has passed a test of competence to drive, to drive on a road a motor bicycle or moped, except where (s)he has successfully completed an approved training course for motorcyclists or is undergoing training on such a course and is driving a motor bicycle or moped on a road as part of the training. Certificates will be issued to those who have successfully completed such courses.

A certificate is not valid:

(a) if the person to whom it is issued is at the time of issue ineligible to undertake the training course; and
(b) after whichever is the earliest of the following dates, namely in a case where:
  (i) the person to whom the certificate was furnished is subsequently disqualified by order of a court under the RTOA 1988, s 36, the date on which the order is made,
  (ii) the licence of the person to whom the certificate was furnished is subsequently revoked by the Secretary of State under the Road Traffic (New Drivers) Act 1995 (**24.53**), the date on which the revocation has effect,
  (iii) the certificate was issued before 1 February 2001, the last day of the period of three years beginning with the date of the certificate, or
  (iv) the certificate was issued on or after 1 February 2001, the last day of the period of two years beginning with the date of the certificate.

MV(DL)R 1999, reg 69 provides that the requirement that a person is not authorised, unless (s)he has passed a driving test of competence, to drive a motor bicycle on a road without having successfully completed an approved training course, does not apply to a person who is a provisional entitlement holder by virtue of having passed a test in respect of former category P (mopeds) on or after 1 December 1990 and before 19 January 2013, or passes the test in respect of category AM (mopeds and light quadricycles), and such a person is also exempt from the need to produce a certificate of such a pass on applying for a driving test to drive a motor bicycle. Similar exemptions exist in favour of persons resident on exempted islands (referred to at **24.9**:p 824).

The general proposition, that no person may be permitted to take a test of competence to drive a vehicle in category A (motor bicycles and motor tricycles) unless (s)he produces the prescribed certificate of completion of an approved training course, is subject to the following qualifications contained in reg 69. It does not apply to a person who is for the time being the holder of a full licence for a class of vehicle included in category A in respect of a test of competence to drive a vehicle of any other class included in that category. Such a holder is also exempt from the restrictions imposed by RTA 1988, s 97(3) (which prevent the driving of a vehicle in category AM or category A on a road, by the holder of a provisional licence, before (s)he has successfully completed an approved training course for motorcyclists) in respect of his/her driving of a vehicle of any class included in category AM or A. However, these qualifications do not apply in relation to the holder of a full licence authorising him/her only to drive a vehicle in category A having automatic transmission in respect of a test to drive vehicles with manual transmission or the driving of such a vehicle.

Driving test   By MV(DL)R 1999, reg 40(1A), the driving test for a licence for a category AM or a category A vehicle must be conducted in three parts: (a) the standard test of driving theory and the standard test of hazard perception; (b) the manoeuvres test; and (c) the practical test. All three parts must be passed and must be passed in the same category or sub-category of vehicle. Test (a) must be passed before test (b) is taken and test (b) must be passed before test (c) is taken.

Entitlement upon passing a test to drive a category AM vehicle   MV(DL)R 1999, reg 44A provides that, where a person has passed a test for a licence authorising the driving of vehicles included in category AM, the Secretary of State must grant:

(a) in a case where the test was passed on a three-wheeled moped or a light quadri-cycle, a licence authorising the driving of all vehicles having three or four wheels included in category AM;

(b) in any other case, a licence authorising the driving of all vehicles included in category AM.

Full licence as provisional licence

24.12   RTA 1988, ss 98(2) (and 99A(5)) provide that a full licence (or Community licence) may act as a provisional licence for any other classes of vehicle, unless the holder is below the minimum age at which the other class of vehicle may be driven. However, by MV(DL)R 1999, reg 19, this does not apply to a full licence which is restricted to a specially adapted vehicle for a person with a physical disability, nor (except in the case of an electric scooter being used in a trial) does s 98(2) authorise a person who has not passed a moped or motor bicycle test to drive a former category P or a category Q vehicle.

MV(DL)R 1999, reg 19 provides that RTA 1988, ss 98(2) and 99A(5) apply as follows. The holder of a full licence which authorises the driving of motor vehicles of a class included in a category or sub-category specified in column (1) of the table below may drive motor vehicles:

(a) of other classes included in that category or sub-category, and

(b) of a class included in each category or sub-category specified, in relation to that category or sub-category, in column (2) of the table,

as if he were authorised by a provisional licence to do so.

Column (1)	Column (2)
Full licence held	Provisional entitlement included
AM	A, B, F and K
A1	A, B, F and K
A2	A, B, F and K
A3	B, F and K
B	A, B + E, G and H
C1	C1 + E
C	C1 + E, C + E
D1	D1 + E
D	D1 + E, D + E
F	Q, AM and B
G	H
H	G

In the case of a full licence granted before 19 January 2013 which authorises the driving of a class of standard motor bicycles, other than motor bicycles with an engine the maximum net power output of which is 11 kilowatts or less, such licences do not authorise the driving of an A3 motorcycle before the expiration of the standard access period (**24.1**:p 816). Nor do they authorise the driving, as if authorised by a provisional licence, of vehicles of any class included in category B96.

Section 98(2) does not apply in so far as it authorises the full licence-holder to drive vehicles of any class included in category B + E, C + E, D + E, or K or in sub-category B1 (invalid carriages), C1 or D1 (not for hire or reward). By way of exception, holders of full licences restricted to vehicles with automatic transmissions may use those licences as provisional licences to drive manually controlled vehicles of a category or sub-category as specified in the table above. A corresponding provision applies to holders of Community licences.

Physical fitness of drivers

RTA 1988, s 92 provides that an applicant for a driving licence must declare his phys-   **24.13**
ical fitness to drive, stating whether or not he is suffering, or has in the past suffered, from any relevant or prospective disability. The term 'disability' for the purposes of the section includes disease and the persistent use of drugs or alcohol, whether or not such misuse amounts to dependency, and:

(a) a relevant disability means (i) any prescribed disability, and (ii) any other disability likely to cause the driving of a vehicle by him/her to be a source of danger to the public; and
(b) prospective disability means any other disability which, by virtue of its intermittent or progressive nature, may become a disability of the type specified in (a) in course of time.

The Secretary of State must not refuse a licence on account of a relevant disability if the applicant has had a previous licence and his/her disability is one of absence of a limb, deformity, or loss of use of a limb provided that the condition has not become more acute, or if the application is for a provisional licence.

The Secretary of State must refuse an applicant a licence if satisfied that the applicant is suffering from a prescribed relevant disability, unless (with most such disabilities) specified minimum standards are satisfied.

The prescribed disabilities are set out in MV(DL)R 1999, Part VI. For the purposes of the prescribed disabilities, licences are divided into Group 1 and 2 licences. Group 1 licences are those for vehicles in categories AM, A, B, B + E, F, G, H, K, and L (**24.1**) and the former category N (vehicles travelling six or less miles on public roads per week between two pieces of land in same occupation). Group 2 licences cover the other categories of licence; the rules relating to them are more stringent.

In respect of Groups 1 and 2 licences, the disabilities *prescribed* for the purposes of (a) above are:

(i)   severe mental disorder;
(ii) liability to sudden attacks of giddiness or fainting due to a heart condition for which a 'pacemaker' has been fitted. (A licence must not be refused if the person's driving is unlikely to endanger the public and (s)he has adequately arranged regular medical supervision and is conforming to those arrangements);
(iii) liability to sudden attacks of disabling giddiness or fainting for some other reason; or
(iv) persistent misuse of drugs or alcohol, whether or not such misuse amounts to dependency.

In respect of Group 1 licences, the following are also prescribed as relevant disabilities:

(a) impairment of vision;
(b) diabetes mellitus;

(c) epilepsy;

(d) isolated seizure (one or more unprovoked seizures occurring either over a single period not exceeding 24 hours, or over a period not exceeding 24 hours, where that period of seizure has occurred more than five years after the last unprovoked seizure).

In respect of Group 2 licences, the following are *also prescribed as relevant disabilities*:

(a) impairment of vision;

(b) diabetes mellitus;

(c) liability to seizure other than epilepsy;

(d) epilepsy;

(e) isolated seizure. The definition of 'isolated seizure' in respect of a Group 2 licence is different from that given above in relation to a Group 1 licence in that ten years is substituted for five years in that definition.

The details about the specified minimum standards are set out in Chapter 24 of the companion website.

Further points about disability

24.14   By RTA 1988, s 93, existing licences may be revoked totally if a relevant disability (**24.13**) actually arises and an application for the licence would have had to be refused, or may be revoked and replaced with shorter-term licences if a prospective disability (**24.13**) arises. Where a licence is revoked under s 93 it must be delivered forthwith to the Secretary of State. Failure without reasonable excuse to do so is an offence (s 93(3)). Similar provision is made in respect of the authorisation to drive conferred by (a) Community licences held by persons normally resident in the UK, and (b) holders of Northern Ireland licences, by ss 99C and 109B respectively. Offences under ss 99C(4) and 109B(4) correspond to that under s 93(3).

Licences may be issued which allow the driving of vehicles of a special construction—for example, a vehicle fitted with hand controls for a person who has lost both legs—or they may allow driving subject to certain specified conditions.

Persons holding driving GB licences who suffer an actual or prescribed disability must notify the Secretary of State forthwith. Failure without reasonable excuse to do so is an offence under RTA 1988, s 94(3). Driving after such a failure is an offence under s 94(3A).

A person who holds a licence and drives a motor vehicle of the class authorised on a road commits an offence under s 92(10) if (s)he knowingly made a false declaration in relation to a relevant disability to obtain the licence. A person who drives a motor vehicle on a road, otherwise than in accordance with a licence, commits an additional offence under s 94A(1) if a licence has earlier been refused or revoked on account of such a disability. This offence may also be committed by holders of Community licences normally resident in Great Britain and holders of Northern Ireland licences. As to discretionary disqualification for these offences, see **24.40**.

Section 96(1) makes it an offence for a person to drive a motor vehicle on a road while his/her eyesight is such (whether through a defect which cannot be or one which is not for the time being sufficiently corrected, eg by the use of spectacles) that (s)he cannot comply with any prescribed requirement as to eyesight.

A constable having reason to suspect a person driving a motor vehicle of committing a s 96(1) offence may require him/her to submit to an eyesight test; it is an offence under s 96(3) to refuse to submit to such a test.

All the above offences are summary ones. The offences under ss 93(3), 94(3), 94(3A), 96(1), 96(3), 99C(4), and 109B(4) are punishable with a fine not exceeding level 3, the offence under s 92(10) is punishable with a fine not exceeding level 4, and an offence under s 94A(1) is punishable with up to six months' imprisonment and/or an unlimited fine. An offence under s 92(10), 94(3A), 94A(1), 96(1), or (3) is punishable with discretionary disqualification for driving (**24.40**).

Disabilities requiring medical investigation: high-risk offenders

RTA 1988, s 94 provides that, if prescribed circumstances obtain in relation to a person **24.15** who is an applicant for, or the holder of, a driving licence or if the Secretary of State has reasonable grounds for believing that a person who is an applicant for, or the holder of, a licence may be suffering from a relevant or prospective disability (**24.13**), the Secretary of State may by notice in writing served on the applicant or holder require him/her to be medically assessed for the purpose of determining whether (s)he suffers or has suffered from such a disability, or (except where the licence in question is a provisional one) require him/her to take such driving test as the Secretary of State directs.

The prescribed circumstances are that the person has been disqualified for driving by a court:

(a) by reason that the proportion of alcohol in his/her body equalled or exceeded two-and-a-half times the prescribed limit;

(b) by reason of failure, without reasonable excuse, to provide a specimen when required to do so pursuant to RTA 1988, s 7;

(c) by reason of failure, without reasonable excuse, to give permission for a laboratory test of a specimen of blood taken pursuant to RTA 1988, s 7A; or

(d) on two or more occasions within any period of ten years for drink-driving or driving while unfit through drink.

For the purposes of (a) and (b) a disqualification must not be taken into account unless it was ordered on or after 1 June 1990 and (d) does not apply to a person unless the last disqualification was ordered on or after that date. For the purposes of (c) a disqualification must not be taken into account unless it was ordered on or after 1 June 2013.

Licences to drive passenger-carrying vehicles or large goods vehicles

The vehicles

There are two types of passenger-carrying vehicle (PCV). **24.16**

A vehicle is a '*large PCV*' if it is used to carry passengers and is constructed or adapted to carry more than 16 passengers, whether or not for hire or reward. The various forms of large PCVs are included in variations of category D types of vehicles under MV(DL) R 1999 (**24.1**).

A vehicle is a '*small PCV*' if it is used to carry passengers *for hire or reward* and constructed or adapted to carry more than eight but not more than 16 passengers. Small PCVs are category D1 type vehicles.

A large goods vehicle (LGV) is a motor vehicle (not being a medium-sized goods vehicle as defined at **24.6**(c)(i) which is constructed or adapted to carry or to haul goods and the permissible maximum weight of which exceeds 7.5 tonnes.

The *permissible maximum weights* of goods vehicles are as follows:

(a) in the case of a motor vehicle which neither is an articulated goods vehicle nor is drawing a trailer, the relevant maximum weight of the vehicle;

(b) in the case of an articulated goods vehicle:
  (i) when drawing only a semi-trailer, the relevant maximum train weight of the articulated goods vehicle combination;
  (ii) when drawing a trailer as well as a semi-trailer, the aggregate of the relevant maximum train weight of the articulated goods vehicle combination and the relevant maximum weight of the trailer;
  (iii) when drawing a trailer but not a semi-trailer, the aggregate of the relevant maximum weight of the articulated goods vehicle and the relevant maximum weight of the trailer;
  (iv) when drawing neither a semi-trailer nor a trailer, the relevant maximum weight of the vehicle;

(c) in the case of a motor vehicle (not being an articulated goods vehicle) which is drawing a trailer, the aggregate of the relevant maximum weight of the motor vehicle and the relevant maximum weight of the trailer.

The 'relevant maximum weight' of a goods vehicle is the maximum gross weight which is shown on the relevant plate of the vehicle. The relevant plate is always the Ministry plate if the vehicle has then been fitted with one. If it has not, then the relevant maximum weight will be shown on the vehicle.

The 'relevant maximum train weight' of an articulated goods vehicle combination is the gross weight of the unladen motor vehicle and its unladen trailer plus the maximum weight which the combination is permitted to carry. This is shown under the column marked 'gross train weight' on the plate attached to the vehicle.

The licensing requirements

24.17   A driving licence issued under RTA 1988, Part III (ss 87 to 109C) (**24.1**) is required to cover the driving of passenger-carrying vehicles (PCVs) or large goods vehicles (LGVs) or any class thereof on a road. However, by RTA 1988, s 110, licences under Part III to drive such vehicles must be granted by the Secretary of State in accordance with Part IV (ss 110 to 122) and, in so far as they authorise the driving of PCVs or LGVs, are subject to Part IV as well as Part III.

RTA 1988, s 111 requires a traffic commissioner to exercise the functions conferred by Part IV relating to the conduct of:

(a) applicants for and holders of PCV drivers' licences or LGV drivers' licences; and
(b) holders of PCV Community licences or LGV Community licences.

By RTA 1988, s 112, the Secretary of State may not grant a PCV driver's licence or LGV driver's licence unless satisfied, having regard to his/her conduct, that (s)he is a fit and proper person to hold the licence applied for. Any issue relating to an applicant's conduct may be referred to a traffic commissioner for a determination as to whether the applicant is, having regard to his/her conduct, a fit person to hold a PCV licence or LGV licence (s 113).

Someone who drives a PCV or LGV without holding a driving licence authorising the driving of a PCV or LGV of the class in question commits a summary offence, punishable with a fine not exceeding level 3, under s 87(1), and so does a person (under s 87(2)) who carries or permits him/her to do so. . As to disqualification in relation to s 87(1), see **24.40**.

Exemptions

MV(DL)R 1999, reg 50(1) provides that the provisions of RTA 1988, Part IV (ss 110– **24.18** 122) and the provisions of MV(DL)R 1999, regs 54 (below) and 55, 56, and 57 (special provisions relating only to LGV and PCV drivers' licences) do not apply to LGVs of a class included in categories F, G, or H (some tractors, road rollers, and track-laying vehicles) or sub-category C1 + E (8.25 tonnes), or to exempted vehicles. The list of other exempted vehicles is extensive, covering military vehicles, a variety of vehicles used in public works, industry, agriculture, articulated vehicles the unladen weight of which does not exceed 3.5 tonnes, unladen goods vehicles manufactured before 1 January 1960, and emergency vehicles. In addition, by reg 50(2) and (3), RTA 1988, Part IV and the provisions of regs 54 to 57 of the Regulations do not apply to a PCV:

(a) manufactured more than 30 years before the date upon which it is driven and not used for hire or reward or for the carriage of more than eight passengers;
(b) driven by a constable for the purpose of removing or avoiding obstruction to other road users or to other members of the public, for the purpose of protecting life or property (including the PCV and its passengers) or for other similar purposes.

LGV licences granted to persons under 21

These may be of two types. To permit those aged 18 or over but under 21 to be taught **24.19** to drive a LGV within a training agreement, provisional trainee drivers' licences may be issued. Other learner drivers will be issued with a standard provisional driving licence.

MV(DL)R 1999, reg 54(1) provides that a LGV driver's licence granted to a person under 21 is subject to the following additional conditions.

A full LGV driver's licence authorising the driving of a vehicle of category C is subject to the condition that its holder may not drive LGVs in category C + E (other than vehicles included in sub-category C1 + E whose maximum authorised mass does not exceed 7.5 tonnes) as if (s)he were authorised to do so by a provisional licence before the expiry of six months commencing on the date on which (s)he passed a test for category C (reg 54(4)).

It is an offence to fail without reasonable excuse to comply with any of these conditions (RTA 1988, s 114(1)), or knowingly to cause or permit someone under 21 to do so (s 114(2)). These summary offences are punishable with a fine not exceeding level 3.

Other provisions

In addition to the provisions at **24.9** MV(DL)R 1999, reg 16(8) prohibits a provisional **24.20** licence holder from driving a PCV while carrying any passenger. However, exceptions exist in favour of a supervisor and of a holder of a PCV drivers' licence who is receiving or giving instruction, has given or received it, or is to give or receive it.

MV(DL)R 1999, reg 7(6) provides that a full category B licence holder (but not former sub-category B1 or sub-category B1 (invalid carriages)) who has held that licence for at least two years, is over 21 and receives no consideration other than out-of-pocket expenses may drive, on behalf of a non-commercial body, for social purposes but not for hire or reward, a vehicle of a class in sub-category D1 (more than eight but not more than 16 seats in addition to the driver's seat) which is not drawing a trailer and has a maximum authorised mass not exceeding 3.5 tonnes, excluding weight attributable to special equipment for the carriage of disabled passengers, and 4.25 tonnes otherwise. The usual proviso applies in relation to 'automatics'. Where such a driver is aged 70 or over, (s)he must not be suffering from a relevant disability in respect of which the licensing authority would be bound to refuse him/her a Group 2 licence (within the meaning of reg 73—higher medical standards for large vehicles). MV(DL) R 1999, reg 76 provides that a person who held a licence authorising, on 31 December 1996, the driving of vehicles in category D otherwise than for hire or reward may drive the classes of vehicles included in category D which are driven under a permit granted under TA 1985, s 19 or, if not being so used, driven otherwise than for hire or reward.

By RTA 1988, s 114 it is an offence for the holder of a provisional PCV or LGV licence to fail, without reasonable excuse, to comply with any prescribed conditions. This summary offence is punishable with a fine not exceeding level 3.

Certificates of professional competence (CPC)

24.21   The Vehicle Drivers (Certificates of Professional Competence) Regulations 2007 (VD(CPC)R 2007) apply to a person who drives on a road a relevant vehicle, ie a vehicle in driving licence category D or D + E (vehicle with more than eight passenger seats, in addition to the driver, hereafter described as a 'bus' and a 'coach') or C or C + E (goods vehicle exceeding 3.5 tonnes maximum authorised mass, hereafter 'lorry'), and who is a national of:

(a) the UK;
(b) a third country employed or used by an undertaking established in the UK;
(c) a member state; or
(d) a third country employed or used by an undertaking established in a member state
     to any person who drives a relevant vehicle.

'Member state' means an EU state or another EEA state.

VD(CPC)R 2007 provide for an initial practical and theoretical driving test which is more extensive than the 'ordinary driving test'. There is also a requirement of periodic training of drivers every five years. Driving a bus, coach or lorry without having passed the requisite initial test, or driving such a vehicle when five or more years have elapsed since the completion of 35 hours of periodic training is a summary offence under reg 10(1); it is punishable with a fine not exceeding level 3.

Exemptions from the CPC Regulations

24.22   VD(CPC)R 2007 do not apply to the driving of a vehicle:

(a) which it is an offence for the person concerned to drive on any road at more than
     45 kph;
(b) which is being used by, or is under the control of, any of the following—
     (i)   the armed forces;
     (ii) a police force;
     (iii) a fire and rescue authority;
     (iv) an NHS ambulance service;
     (v)   the prison service;
     (vi) a local authority in the discharge of any function to deal with a major
           contingency;
     when the vehicle is being used as a consequence of a task assigned to such an au-
     thority, force or service;
(c) which is—
     (i)   undergoing road tests;
     (ii) being used to or from a previously arranged road test; or
     (iii) a new or rebuilt vehicle not yet put into service;

(d) which is being used in a state of emergency or is assigned to a rescue mission, including the non-commercial transport of humanitarian aid;

(f) which is being used for the non-commercial carriage of passengers or goods;

(g) which is carrying material or equipment, including machinery, to be used by that person in the course of his/her work, provided that driving that vehicle is not his/her principal activity;

(h) to which (a) to (g) do not apply, but which satisfies all of the following conditions –

   (i)   the vehicle must be being driven by a person whose principal activity in the course of his/her work is not driving relevant vehicles;

   (ii) the vehicle must be being driven within a 100 km radius of the driver's base;

   (iii) the driver must be the only person being carried on the vehicle;

   (iv) in so far as the vehicle may be carrying goods or burden, the goods or burden must only be equipment, including machinery, that is permanently fixed to the vehicle;

   (v)   the vehicle must be being used by an agricultural, horticultural, forestry, farming or fishery undertaking;

   (vi) the vehicle must be being used only for carrying goods in the course of that undertaking's business;

   (vii) the vehicle must be being driven by a person whose principal activity in the course of their work is not driving relevant vehicles.

Further details about the CPC are in Chapter 24 on the companion website.

Drivers from abroad

Community licence-holders

By RTA 1988, s 99A, a Community licence-holder (a holder of a licence issued within the EEA) may drive in Great Britain a motor vehicle of any class which (s)he is authorised by his/her Community licence to drive, provided that (s)he is not disqualified for holding or obtaining a licence under RTA 1988.   **24.23**

Thus, unlike other drivers from abroad, a Community licence-holder who becomes resident in Great Britain is authorised to drive here without the need to exchange his/her licence for a British licence within 12 months of becoming resident. (S)he does, however, have a right to exchange his/her licence. Where matters of validity, standards of health and fitness, and disqualification are concerned, the exchange of licences is mandatory. A resident Community licence-holder is subject to the same medical requirements as holders of a British licence. A Community licence-holder resident in Great Britain who is authorised to drive medium-sized or large goods vehicles or passenger-carrying vehicles of any class must, within a period of 12 months' residence, deliver his/her licence to the Secretary of State and provide details prescribed by s 99B.

Drivers from abroad other than Community licence-holders

The Motor Vehicles (International Circulation) Order 1975, art 2 states that it is lawful for *a person resident outside the UK who is temporarily in Great Britain* to drive for a period of 12 months from the date of his last entry into the UK if he holds a Convention driving permit (see below) or a domestic driving permit of a country outside the UK authorising him to drive the vehicle in question, provided he is not disqualified (eg by age) for holding or obtaining a British licence.   **24.24**

What has just been said is subject to qualification in respect of large passenger-carrying vehicles, privately operated passenger-carrying vehicles (vehicles not used for carrying passengers for hire or reward which are constructed or adapted to carry more than eight but not more than 16 passengers), large goods vehicles and medium-sized goods vehicles (vehicles constructed or adapted to carry or haul a load which is not adapted to carry more than nine persons including the driver and exceeds 3.5 tonnes but does not exceed 7.5 tonnes). A holder of either type of permit may drive such a vehicle if (s)he is resident in an EEA state, the Isle of Man, Jersey, or Guernsey. Other holders of such a permit may only drive such a vehicle if it has been temporarily brought into Great Britain. In the cases set out in this paragraph no other licence is required.

Generally, any person may cause or permit holders of such Convention or domestic driving permits to drive vehicles which they are authorised by their permit to drive. In the phrase 'temporarily in Great Britain', 'temporarily' is an element other than simply a time element. It involves a presence for casual purposes, for example a holiday, as contrasted with regular habits. For example, an overseas student studying here is not 'temporarily' resident here.

A Convention driving permit is usually referred to as an 'international driving licence'. There are now three types of Convention driving permits issued under the authority of a state, each dependent on which convention the state of issue operates under: the International Convention relative to Motor Traffic of 1926, the Convention on Road Traffic of 1949, and the Convention on Road Traffic of 1968. Some states are not party to any of these conventions; the USA and China are the most notable of these.

The term 'domestic driving permit' refers to a foreign driving licence issued to a person who has passed a driving test.

Article 2, above, merely recognises that persons visiting from abroad would experience difficulty in undertaking a test in Great Britain while they are temporarily in Great Britain. These permits are treated as Great Britain driving licences in every respect and a constable's powers to demand production of a licence and to demand the holder's date of birth apply equally to them.

MV(DL)R 1999, reg 80 makes the same provision for a person from abroad who becomes resident in Great Britain, the one-year period in this case running from when (s)he became a resident. Someone from abroad who is in this country but who falls outside the term 'temporarily in Great Britain', for example an overseas student, will be resident in Great Britain.

The holder of a relevant permit who becomes resident in Great Britain and is not disqualified for holding or obtaining a licence in Great Britain is, during the one-year period after (s)he becomes so resident, treated as the holder of a licence authorising him/her to drive all classes of small vehicles (generally vehicles not constructed or adapted to carry more than nine persons inclusive of the driver and vehicles not exceeding 3.5 tonnes, including a combination of such a vehicle and a trailer), motor cycles or mopeds which (s)he is authorised to drive by that permit. A 'relevant permit' is a 'domestic driving permit', or a 'Convention driving permit'. Where a question arises as to whether a person is normally resident in Great Britain or the UK, a person is deemed to be normally resident if (s)he shows that (s)he will have lived there for not less than 185 days preceding a test appointment.

Persons who become resident in Great Britain and who hold British external licences granted in the Isle of Man, Guernsey, or Jersey authorising the driving of large goods vehicles or passenger-carrying vehicles, and who are not disqualified for holding or obtaining a licence in Great Britain, may drive such vehicles under the authority of those licences for a one-year period from the date upon which they became resident.

People permitted to drive in this country under a domestic driving permit, etc who take out a provisional licence during the one-year period in order to take a driving test need not comply with the normal conditions applicable to a learner driver if they are still driving under the authority of their domestic permit, etc at the time (ie within the one-year period).

Exchangeable licences

Licences issued in Gibraltar and certain non-EEA countries designated by order under RTA 1988, s 108(2) by the Secretary of State are exchangeable by a holder resident in Great Britain for a British driving licence. The Secretary of State may designate a non-EEA country for the purpose of 'exchangeable licences' where (s)he is satisfied that the driving test in that country is satisfactory. (S)he may, however, restrict approval to the grant of exchangeable licences to particular circumstances, impose conditions to which they are subject, and limit the exchanged licence to particular classes of vehicles. An up-to-date list of the countries specified by order is contained in the Driving Licences (Exchangeable Licences) (Amendment) Order 2013. **24.25**

Production of driving licences

By virtue of RTA 1988, s 164(1), a constable or vehicle examiner (ie an examiner appointed under RTA 1988, s 66A by the Secretary of State (see **25.80**)) may demand the production of a driving licence by the following people: **24.26**

(a) a person driving a motor vehicle on a road; or
(b) a person whom a constable or vehicle examiner has reasonable cause to believe has been the driver of a vehicle at the time when an accident occurred owing to its presence on a road; or
(c) a person whom a constable or vehicle examiner has reasonable cause to believe has committed an offence in relation to the use of a motor vehicle on a road; or
(d) a person supervising the holder of a provisional licence while the holder is driving a motor vehicle on a road; or
(e) a person whom the constable or vehicle examiner has reasonable cause to believe was supervising the holder of such a licence when an accident occurred owing to the presence of the vehicle on a road or when an offence is suspected of having been committed by the holder in relation to the use of a vehicle on a road.

When a licence is produced to a constable or vehicle examiner pursuant to a request under RTA 1988, s 164(1), (s)he is entitled to ascertain the name and address of the holder of the licence, its date of issue, and the authority by which it was issued. Section 164(2) also provides a power whereby a constable may require a person (A) in circumstances prescribed by MV(DL)R 1999 to state his/her date of birth. These circumstances are where:

(a) A fails to produce forthwith for examination his/her driving licence on being required to do so by a constable; or
(b) on being so required, A produces a licence which the constable has reason to suspect:
    (i) was not granted to A, or
    (ii) was granted to him/her in error, or
    (iii) contains an alteration in its particulars made with intent to deceive; or
(c) on being so required, A produces a licence in which the driver number has been altered, erased, or defaced; or
(d) A is a person supervising the holder of a provisional licence while (s)he is driving a motor vehicle on a road, or is someone whom the constable reasonably suspects

to have been supervising such a person when an accident occurred or an offence was committed, and the constable has reasonable cause to suspect that A is under 21.

Section 164(4A) provides that, where a constable to whom a provisional licence has been produced by a person driving a motor bicycle has reasonable cause to believe that the holder was not driving it as a part of the training being provided on a training course for motorcyclists, the constable may require him to produce the prescribed certificate of completion of such a course.

The term 'licence' means a licence under RTA 1988, Part III, a Northern Ireland licence or a Community licence.

Police officers have direct access to details of driving licences and their holders through the Police National Computer.

Section 164(3) and (5) provides that where:

(a) a licence has been revoked by the Secretary of State but the holder has not surrendered it; or

(b) the holder has failed to produce his licence to a court when lawfully required to do so; or

(c) the Secretary of State has served notice on a Community licence-holder in pursuance of s 99C (relevant disability) or s 115A (conduct of holder of LGV or PCV Community licence) requiring delivery of the licence to him,

a constable or vehicle examiner may require the holder to produce them and on production may seize them.

A person who fails to produce his/her licence, or a certificate of completion of a training course for motorcyclists, or fails to state a date of birth, when required to do so under any of the provisions of s 164 commits a summary offence, punishable with a fine not exceeding level 3, under s 164(6). (There is a similar, similarly punishable, offence in respect of failure to produce a CPC (Vehicle Drivers (Certificates of Professional Competence) Regulations 2007, reg 11(8))).

An offence under s 164 of failing to produce a licence is not committed if the person produces a current receipt for the surrender of the licence issued under the fixed penalty procedure (**25.117**) and, if required, produces the licence in person immediately on its return at a police station specified at the time of the request, or if within seven days of the request (s)he produces the receipt in person at that police station and, if requested, produces the licence there in person immediately on its return. In addition, in proceedings for an offence of failing to produce a licence it is a defence for D to show that:

(a) within seven days after the production of his/her licence was required (s)he produced it in person at such police station as may have been specified by him/her at the time its production was required; or

(b) (s)he produced it in person there as soon as was reasonably practicable; or

(c) it was not reasonably practicable for him/her to produce it there before the day on which proceedings commenced.

If a licence is not produced at the time its production was required, it is the usual practice of the police to make out a form HO/RT1 which will be produced at the police station nominated by the driver, together with the licence. This is merely a practice followed by the police and there is no statutory requirement that this be done.

As to the power of seizure where a licence is not produced when required, see **24.70**.

Records

System of recording penalty points

24.27 Consequent on the abolition of the driving licence counterpart in Great Britain with effect from 8 June 2015, endorsements are no longer entered onto counterparts. They are instead entered solely onto an individual's electronic driving record, just as they had been previously in the case of unlicensed drivers and foreign drivers who did not have counterparts to their driving licences. Under the Road Traffic Offenders Act 1988 (RTOA 1988), s 97A, a 'driving record', in relation to a person, is defined as a record in relation to the person maintained by the Secretary of State and designed to be endorsed with particulars relating to offences committed by the person under the Traffic Acts (Road Traffic Act 1988, RTOA 1988 and Road Traffic Regulation Act 1984). An individual's driving record is maintained for the Secretary of State by the Driver and Vehicles Licensing Agency (DVLA) which is an executive agency of the Department for Transport. Endorsements placed on counterparts before 8 June 2015 are treated as endorsements placed on the driving record.

Under RTOA 1988, s 57A, the driving record of a person, whether or not (s)he is a licence-holder, who has been given a fixed penalty notice for an offence carrying obligatory endorsement (**25.117**), is endorsed, without the need for a court hearing, at the end of a suspended enforcement period unless (s)he gives notice requesting a hearing and has not paid the fixed penalty.

Access to driver licensing records

24.28 The Motor Vehicles (Access to Driver Licensing Records) Regulations 2001, reg 2 provides that the purposes for which constables may be given access to information held by the Secretary of State and made available are:

(a) the prevention, investigation or prosecution of a contravention of any provision of the following enactments:
  (i) RTA 1988;
  (ii) RTOA 1988;
  (iii) Road Traffic (Northern Ireland) Orders 1981 and 1995;
  (iv) Road Traffic Offenders (Northern Ireland) Order 1996;
  (v) Vehicle Drivers (Certificates of Professional Competence) Regulations 2007; and
(b) ascertaining whether a person has had an order made in relation to him under:
  (i) the Child Support Act 1991, s 40B(1) or (5) (disqualification for driving: further provisions);
  (ii) the Criminal Procedure (Scotland) Act 1995, s 248A(1) (general power to disqualify offenders) or 248B(2) (power to disqualify fine defaulters); or
  (iii) the Crime (Sentences) Act 1997, s 39(1) (offenders) or 40(2) (fine defaulters).

Regulation 3 provides that information to which constables have been given access may be further disclosed to an employee of a police authority, local policing body, or chief officer of police for any purpose ancillary to, or connected with, the use of the information by constables.

In addition, RV(R&L)R 2002, reg 27 authorises the Secretary of State to make available particulars contained in the register to:

(a) a local authority for the purpose of an investigation of an offence or for any purpose connected with its activities as a civil enforcement authority under the Traffic

Management Act 2004, Part 6 (parking and, in a few areas, some other road traffic contraventions);

(b) a chief officer of police;

(c) the Motor Insurers' Bureau in connection with its functions relating to the enforcement of the offence of keeping a vehicle which does not meet insurance requirements; or

(d) any person who can show satisfactory cause for requiring the information.

Regulation 28 authorises the Secretary of State to sell such particulars to such persons as (s)he considers fit, provided that the information does not identify any person or contain anything enabling such identification.

LICENCES TO OPERATE A PSV OR GOODS VEHICLE

PSV operator's licence

24.29 The Public Passenger Vehicle Act 1981 (PPVA 1981), s 12(1) provides that, subject to the exceptions in PPVA 1981, s 46 (fare-paying passengers on school buses), and TA 1985, s 18, a passenger service vehicle (PSV) must not be used on a road for carrying passengers for hire or reward except under a PSV operator's licence granted by s traffic commissioner. The exceptions in TA 1985, s 18 relate to the use on a 'not-for-profit' basis by an exempt body, ie a body concerned with education, religion, social welfare, or other activities of community benefit, under a 's 19 permit' (members of the public not to be carried) or a 's 22 permit' (community bus service). These exceptions are dealt with in Chapter 24 on the companion website.

A PSV operator's licence may be either a *standard licence* or *a restricted licence*. A *standard licence* authorises the use of any description of PSV and may authorise use either:

(a) on both national and international operations; or

(b) on national operations (UK only).

A *restricted licence* authorises the use (whether on national or international operations) of:

(a) PSVs not adapted to carry more than eight passengers; and

(b) PSVs not adapted to carry more than 16 passengers (ie basically, minibuses) when used:
   (i) otherwise then in the course of a business of carrying passengers, or
   (ii) by a person whose main occupation is not the operation of PSVs adapted to carry more than eight passengers.

Definition of public service vehicle

24.30 For the purposes of PPVA 1981, a PSV is defined as a motor vehicle (**23.6**) (other than a tramcar) which if:

(a) adapted to carry more than eight passengers, is used for carrying passengers for hire or reward (a 'large bus'); or

(b) not so adapted, is used for carrying passengers for hire or reward at separate fares in the course of a business of carrying passengers (a 'small bus') (PPVA 1981, s 1(1)).

'*Hire or reward*' in a general sense means a monetary reward legally due under a contract. (However, as explained below, there are some cases of carriage for hire or reward, even though no contract exists.) Money paid or promised under a purely so-cial arrangement (such as a car-sharing scheme) does not make the carriage for hire or reward. A contract to carry a passenger for hire or reward may be made before a pas-senger boards the vehicle (as in the case of a coach trip) or it may be made aboard the vehicle (as in the case of a bus journey).

The meaning of 'hire or reward' has been extended by PPVA 1981, s 1(5) as follows:

(a) a vehicle is to be treated as carrying passengers for hire or reward if payment is made for, or for matters which include, carrying passengers, irrespective of the person to whom the payment is made and, in the case of a transaction effected by or on behalf of (i) a member of any association of persons (whether incorporated or not) and (ii) the association or another member thereof, notwithstanding any rule of law as to such transactions;

(b) a payment made for carrying a passenger is to be treated as a fare notwithstanding that it is made in consideration of other matters in addition to the journey and ir-respective of the person by or to whom it is made;

(c) a payment is to be treated as made for carrying a passenger if made in consider-ation of a person being given a right to be carried, whether for one or more jour-neys and whether or not the right is exercised.

These provisions are not an exclusive definition of the circumstances in which a vehicle is used for carrying passengers for hire or reward. For example, in a ruling approved by the House of Lords it was held that the use of a vehicle for the systematic carriage of passengers under an informal arrangement going beyond the bounds of 'social kindness', whereby it was the passengers who paid for the cost of the fuel, no money being paid to the driver, was a use for hire or reward. As this case shows, there is no need to prove a legally binding contract. Thus, it has been held by a divisional court that, where school children were regularly carried to school in a private vehicle on the basis of 'petrol money' being paid, the vehicle was used for carriage for reward because there was a systematic carriage for hire or reward going beyond the bounds of social kindness and amounting to a business activity, whether or not direct demands had been made for payment. Likewise, where hotels regularly operate 'courtesy coaches' to be used free of charge by residents and visitors, there is a carriage for hire or reward as the service relates to the business activities of the hotel and the charges made for a room, or for a meal, can be taken to include such amenities.

The extended meaning of 'hire or reward' in s 1(5) is restricted in one special circum-stance by s 1(6) which provides that, where a fare is paid in respect of a journey by air and, due to mechanical failure, bad weather, or other circumstances outside the opera-tor's control, part of the journey has to be made by road, no part of the fare is to be treated for the purposes of s 1(5) as paid in consideration of the carriage of passengers by road.

Use of taxis and hire cars at separate fares

TA 1985, ss 10 to 12 make provision for the use of taxis for the carriage of passengers at **24.31** separate fares in three types of case.

Immediate hiring of taxis under conditions prescribed in a scheme TA 1985, s 10 provides that a licensed taxi may be hired for use for the carriage of passengers for hire or reward at separate fares without thereby becoming a PSV if:

(a) the taxi is hired in an area where a scheme made under TA 1985, s 10 is in operation;

(b) the taxi is licensed by the licensing authority for that area; and

(c) the hiring falls within the terms of the scheme.

For the purposes of TA 1985 'licensed taxi' means a vehicle licensed under the Town Police Clauses Act 1847, s 37, the Metropolitan Public Carriage Act 1869, s 6 or any similar enactment.

Such schemes *must* be concerned with the designation of places in the area from which taxis may be hired (authorised places), and must specify the requirements to be met in relation to the hiring at separate fares and other factors from time to time prescribed. They *may* deal with fares, display of documents, plates, marks or signs for indicating an 'authorised place', the manner in which arrangements are to be made for the carriage of passengers or the hiring, and the conditions to be applied to the hiring. The licensing authority in London is Transport for London and in other areas of England and Wales it is the authority responsible for licensing taxis.

For the purposes of s 10, the hiring of a taxi only falls within the terms of a scheme if it is hired from an authorised place and the hiring meets the licensing authority's requirements. A taxi is hired from an authorised place if it is standing at that place when it is hired and the persons hiring it are all present there. In other words, there must be an immediate hiring.

Advance bookings of taxis and hire cars   TA 1985, s 11 states that a licensed taxi or licensed hire car may be used for the carriage of passengers for hire or reward at separate fares without becoming a PSV provided that all the passengers carried on the occasion in question booked their journey in advance and each of them consented, when booking his journey, to sharing the use of the vehicle on that occasion with others, on the basis that a separate fare would be payable by each passenger for his own journey. For the purposes of TA 1968, s 11, 'licensed hire car' means a vehicle licensed under the Local Government (Miscellaneous Provisions) Act 1976, s 48 or the Private Hire Vehicles (London) Act 1998, s 7.

Use of taxis or hire cars in providing local services with restricted PSV operator's licence TA 1985, s 12 permits the holder of a taxi licence or a private hire vehicle licence to apply for a restricted PSV operator's licence. If the holder of the licence states in his/ her application that (s)he proposes to use one or more licensed taxis or licensed hire vehicles to provide a local service, a traffic commissioner must grant the application. However, the commissioner must attach conditions to the effect that any vehicle used under the restricted licence must have a taxi licence or private hire car licence and that it must not be used under the restricted licence otherwise than for providing a local service, although of course it may still be used as a taxi or hire vehicle under its separate licence. The term 'local service' does not include an excursion or tour. Such a vehicle is not treated as a PSV in every case. For example, it need not carry operator's discs and a driver does not require a PSV driver's licence. These restricted licences are referred to as 'special licences'. Vehicles operating under them must display a notice 'Bus' on the front plus the destination or route or nature of service, and must carry a fare table. The taxi code or hire car code (as appropriate) must be complied with in each case. Section 13 provides that, for the purposes of s 12, 'licensed hire vehicle' means a vehicle licensed under the Local Government (Miscellaneous Provisions) Act 1976, s 48 (which does not apply to the London area). However, by s 13A, Transport for London can make an order extending the definition of 'licensed hire vehicle' in s 12 to a vehicle licensed under the Private Hire Vehicles (London) Act 1998, s 7.

Use without operator's licence

By PPVA 1981, s 12(5), if a vehicle is used on a road as a PSV without the requisite PSV **24.32** operator's licence, its operator is guilty of a summary offence punishable with a fine not exceeding level 4, unless (s)he proves that (s)he took all reasonable precautions and exercised all due diligence to avoid the commission of that offence.

Operator's disc

Where a vehicle is used in circumstances requiring a PSV operator's licence, an opera- **24.33** tor's disc must, by PPVA 1981, s 18(1), be fixed and exhibited on the vehicle in the place previously occupied by a vehicle excise licence so that it does not interfere unduly with the driver's view and can easily be read in daylight from outside the vehicle. If a vehicle is used without the requisite operator's disc in the prescribed form exhibited in the prescribed way, its operator is guilty of a summary offence, punishable with a fine not exceeding level 3, under s 18(4) unless (s)he proves that (s)he took all reasonable care and exercised all due diligence to avoid the commission of that offence.

Other provisions

Chapter 24 of the companion website deals with other provisions relating to PSVs. **24.34**

Goods vehicle operator's licence

Definition of goods vehicle

For the purposes of the Goods Vehicles (Licensing of Operators) Act 1995 (GV(LO)A **24.35** 1995), 'goods vehicle' is a motor vehicle or trailer constructed or adapted for the carriage of goods; for this purpose the carriage of goods includes the haulage of goods. 'Goods vehicle' does not include a tramcar or trolley. 'Adapted' means 'altered physically so as to make fit for the purpose'. If the seating is stripped from a public service vehicle and the vehicle is then fitted out as a mobile shop, it is thereby adapted to carry goods and becomes a goods vehicle. Of course, the opposite can apply. For example, where a goods van was adapted for the carriage of passengers and its only use for the carriage of goods was when it carried samples to the owner's place of business, it was held not to be a goods vehicle. When a vehicle is originally constructed as a goods vehicle, any alterations made to it must be *substantial and dramatic* if its initial classification is to be changed. The test to be applied where a passenger vehicle has been converted to permit the carriage of goods is to examine its existing form and to ask whether, if it had been constructed in that form originally, it would have been classed as a passenger vehicle or as a goods vehicle.

Obligation to hold operator's licence

GV(LO)A 1995, s 2(1) states that, subject to the exceptions listed below and to any pro- **24.36** visions about temporary exemptions, no person may use a goods vehicle on a road for the carriage of goods for:

(a) hire or reward; or
(b) in connection with any trade or business carried on by him,

except under the authority of a goods vehicle operator's licence granted by a traffic commissioner.

By s 2(1B) and (1C) and Sch 1, (a) does not apply to a goods vehicle which (including any vehicle with which it is in combination) has a permissible laden mass not exceeding

3.5 tonnes (3,500 kg)), and (b) does not apply to the following types of 'small goods vehicle', ie a goods vehicle which:

(i)   does not form part of a vehicle combination (ie there is no trailer) and has a relevant plated (ie laden) weight not exceeding 3.5 tonnes or (not having a relevant plated weight) has an unladen weight not exceeding 1,525 kg; or

(ii)   forms part of a vehicle combination (not being an articulated combination) which is such that the total plated weight of the combination (except any small trailer (unladen weight not exceeding 1,020 kg)) does not exceed 3.5 tonnes (or, if one or more of the vehicles in the combination, other than a small trailer, is not plated, if the total unladen weight of those vehicles, excluding such a trailer does not exceed 1,525 kg); or

(iii)   forms part of an articulated combination which is such that the total of the unladen weight of the tractive unit and the plated weight of the trailer does not exceed 3.5 tonnes (or if the trailer does not have a plated weight, the total of their unladen weights does not exceed 1,525 kg).

'Plated weight' means a weight required to be marked on a vehicle by means of a plate in pursuance of regulations or required to be so marked by RTA 1988, s 57 or 58. 'Relevant plated weight' means a plated weight of the description specified in relation to that provision by regulations.

By s 2(2), s 2(1) does not apply to the use of a goods vehicle by a haulier established in Northern Ireland (and not in Great Britain) for international carriage.

The list of vehicles exempted from s 2(1) by the Goods Vehicles (Licensing of Operators) Regulations 1995, Sch 3 is extensive but, for ease of understanding, it is more practical to think in terms of categories of vehicle. Generally, the vehicles exempted are those used by public services. Thus, defence force, police, NCA, fire-fighting, rescue, and ambulance vehicles are exempt; so are vehicles used for purposes such as snow clearance, gritting and refuse disposal.

Also exempt under Sch 3 are vehicles used as farm and forestry tractors or machines, dual-purpose vehicles and their trailers, showmen's goods vehicles and their trailers drawn by them, recovery vehicles, cement-mixer lorries, vehicles used for funerals, and vehicles being held ready for use in an emergency by an undertaking for the supply of water, electricity, gas, or telephone services. Vehicles proceeding to or from a motor vehicle testing (MoT) station for the purposes of an examination of that vehicle are exempt, provided that the only load being carried is a load required for the purposes of the examination.

Motor vehicles constructed or adapted primarily to carry passengers or their effects, together with any trailer, are exempt under Sch 3 whilst being so used. Of course, the operation of public service vehicles must be licensed under the provisions referred to at **24.29–24.30**.

Lastly, s 2(1) does not apply to the temporary use of foreign goods vehicles in specified circumstances set out in Chapter 24 of the companion website.

By s 2(5), a person who uses a vehicle in contravention of s 2(1) commits an offence, a summary offence punishable with an unlimited fine.

If the vehicle belongs to the driver, or if it is in his/her possession under an agreement for hire, hire-purchase or loan, the driver is deemed by s 58(2) to be the user for licensing purposes; in any other case, the user will be the person whose employee or agent the driver is at the time. It is essential to prove that the goods vehicle was carrying goods at the time of the alleged offence. It must also be proved that the goods vehicle was being used *for hire or reward or for or in connection with any trade or business carried out by the user.*

Hire or reward   Generally speaking, a person uses a goods vehicle for hire or reward if (s)he hires out the use of the vehicle, together with its driver, to another person. For example, a furniture remover who uses his employees and his vehicles to assist a customer to move his household effects, an agreed fee being paid by the customer, uses the vehicle for the carriage of goods for hire or reward.

For or in connection with any trade or business carried on by him/her   A large number of goods vehicles are used by firms solely for, or in connection with, their own trade or business, as opposed to hauling goods for hire or reward. Supermarket chains operate fleets of vehicles solely for the purpose of effecting deliveries to their retail outlets, with the result that the vehicles are used for or in connection with their own trade or business, and require operator's licences. Likewise, the builder who uses his goods vehicle to carry building materials to the sites where his building operations are being carried out uses the vehicle in connection with his own business and requires an operator's licence. On the other hand, case law indicates that a person whose hobby is stock car racing and who transports the stock cars in a vehicle does not use that vehicle for or in connection with any trade or business, even though (s)he competes for prize money and receives sponsorship money.

For the purpose of operator's licensing, a local or public authority is deemed to be carrying on a business. Such an authority must therefore have operator's licences in respect of its goods vehicles.

Types of operator's licence

Under GV(LO)A 1995, s 3, there are two types of operator's licence: standard operator's **24.37** licences and restricted operator's licences.

Standard operator's licence   Standard operator's licences are all embracing in that they authorise the holder to carry goods for hire or reward or in connection with the holder's trade or business.

Standard licences may authorise transport operations both nationally and internationally (standard international vehicle operator's licence), or may be in respect of national operations only.

Restricted operator's licence   A restricted operator's licence authorises the use of goods vehicles solely for, or in connection with, a trade or business carried on by the holder of the licence, other than that of carrying goods for hire or reward. However, by s 3(4) a company (C) may use a goods vehicle on a road for the carriage of goods for hire or reward under a restricted licence instead of a standard licence if (but only if) the goods concerned are the property of a company which is:

(a) a subsidiary of C,
(b) a holding company for C, or
(c) a subsidiary of a company which is a holding company both for that subsidiary and for C.

Points applicable to both types of licence   An operator's licence will normally authorise an operator to use several vehicles. It is issued to the operator and is generally kept at an operating centre. The licence will specify the registration numbers of the vehicles which are authorised to be used. It will also specify, by type, the maximum number of trailers which may be used under the licence and also the maximum number of subsequently acquired motor vehicles which may be so used.

Except as provided in s 3(4) (above) and subject to temporary exemptions granted by the Secretary of State or a traffic commissioner, a person who uses a goods vehicle under a restricted licence for carrying goods for hire or reward is guilty of an offence (s 3(6)). A person who uses a goods vehicle for carrying goods for hire or reward on international transport operations under a standard licence which covers the carriage of goods on national transport operations only is guilty of an offence (s 3(7)). Both are summary offences punishable with a fine not exceeding £500.

Identity discs

24.38   So that vehicles specified on an operator's licence may bear evidence of the fact that they are being used under the authority of a licence, an identity disc is issued under the Goods Vehicles (Licensing of Operators) Regulations 1995, reg 23, in respect of each vehicle which is specified. The disc must be fixed to that vehicle in a waterproof container in the place previously occupied by a vehicle excise licence.

The holder of the licence must cause the identity disc to be displayed at all times when the vehicle is specified in his/her licence, regardless of whether or not the vehicle is being used at the time for a purpose for which a licence is required. Non-compliance is a summary offence punishable with a fine not exceeding level 1, contrary to GV(LOA) 1995, s 57(9).

DISQUALIFICATION BY A COURT AND PENALTY POINTS

Obligatory disqualification

24.39   RTOA 1988, s 34(1) deals with disqualification for driving following a conviction for an offence listed in Sch 2 involving obligatory disqualification. On such a conviction, a court must order disqualification for at least a specified minimum period, unless the court for special reasons thinks fit to order a shorter period of disqualification or not to order disqualification at all. The list of offences carrying obligatory disqualification is as follows:

Common law offence of manslaughter	
Road Traffic Act 1988	
Causing death by dangerous driving	s 1
Causing serious injury by dangerous driving	s 1A
Dangerous driving	s 2
Causing death by careless or inconsiderate driving	s 2B
Causing death by driving: unlicensed or uninsured driver	s 3ZB
Causing death by driving: disqualified driver	s 3ZC
Causing serious injury by driving: disqualified driver	s 3ZD
Causing death by careless driving when under the influence of drink or drugs	s 3A
Driving or attempting to drive vehicle when under the influence of drink or drugs	s 4(1)

(continued)

Common law offence of manslaughter	
Driving or attempting to drive with excess alcohol in breath, blood or urine	s 5(1)(a)
Driving with excess of a specified controlled drug in blood or urine	s 5A(1)(a) and (2)
Failing to provide, or allow to be taken, a specimen for analysis, where specimen required to assess ability, or alcohol or controlled drug level, at time of driving/attempting to drive	ss 7 and 7A
Motor racing and speed trials on a public highway	s 12
Using vehicle in a dangerous condition within three years of conviction under s 40A	s 40A
Theft Act 1968	
Aggravated vehicle-taking	s 12A

In the case of most of these offences, the specified minimum period of disqualification is 12 months. However, in the case of the offence of manslaughter, or an offence under RTA 1988, s 1, 1A, 3ZC, 3ZD or 3A, it is two years, and in a case within s 40A the specified minimum period is six months.

A two-year minimum period also applies in relation to a person on whom more than one disqualification for a fixed period of 56 days or more has been imposed within the three years immediately preceding the commission of the offence. (This does not apply if the period would have been less than 56 days but for an extension under s 35A or the Sentencing Code, s 166 (**24.41** and **24.47**).) A disqualification imposed as a result of an offence committed by using vehicles in the course of crime or in respect of a conviction for stealing or attempting to steal a motor vehicle, joyriding, or going equipped to steal, etc a motor vehicle, or attempting to commit any such offence, is disregarded for this purpose, as is an interim disqualification on committal for sentence.

If the conviction is for any of the above offences which are connected with the drink-driving laws, that is an offence under RTA 1988, s 3A, 4(1), 5(1)(a), or 5A(1)(a) and (2), referred to above, and there has been a previous conviction for such an offence within the preceding 10 years, a court *must* order disqualification for a period of not less than three years, unless there are special reasons for not doing so. The same applies to an offence under s 7(6) or 7A(6) where it involves obligatory disqualification.

Particulars of any disqualification must be endorsed on the driving record.

'*Special reasons*' are reasons special to the circumstances of the offence, as opposed to special to the offender. It is of no consequence that the loss of a driving licence will lose the offender his/her job, that (s)he is a person of previous good character and/or that (s)he has driven for many years without having been convicted of any motoring offence. The courts' approach is strict in this respect. They have refused to accept as special reasons the hardship to a country doctor and his patients or the problems caused for a disabled man. An example of a case where there would be reasons special to the offence is where someone has his/her supposed non-alcoholic drink 'laced' without his/her knowledge. In such a case, it is open to the court in its discretion to mitigate the period of disqualification, or not to disqualify at all, because of special reasons. Special reasons might have been found in the case of the country doctor if the country doctor had been called out to a man suffering a heart attack in circumstances in which no other doctor could reasonably have been summoned to attend.

Discretionary disqualification

24.40   RTOA 1988, s 34(2) deals with discretionary disqualification. Where a person is convicted of an offence listed in Sch 2 to the Act which is one which carries discretionary disqualification and either:

(a) the penalty points to be taken into account on that occasion number fewer than 12; or

(b) the offence is not one involving obligatory endorsement (and therefore does not carry penalty points),

the court may disqualify for any period which it thinks fit. Discretionary disqualification under s 34(2) may not be for an indefinite period, since the court must state the period of disqualification. Particulars of any disqualification must be endorsed on the driving record. If the penalty points to be taken into account are 12 or more the offender must be dealt with under s 35 (**24.44**: obligatory disqualification for repeated offences) and not under s 34(2).

The list of offences in Sch 2 carrying discretionary disqualification is as follows.

Road Traffic Regulation Act 1984	
Contravention of temporary speed restriction, if committed in respect of speed restriction	s 16
Use of motorway contrary to regulations (otherwise than by unlawful stopping or allowing vehicle to remain at rest on part where vehicles permitted so to remain in certain circumstances)	s 17
Breach of pedestrian crossing regulations by motor vehicle (MV)	s 25
MV not stopping at school crossing	s 28
MV exceeding speed limit	s 89
Road Traffic Act 1988	
Careless, or inconsiderate, driving	s 3
In charge of mechanically propelled vehicle when unfit	s 4
In charge of MV: (a) with excess alcohol in breath, etc	s 5(1)(b)
(b) with controlled drug concentration above limit	s 5A(1)(b) and (2)
Failing to co-operate with preliminary test	s 6
Failing to provide, or to allow to be taken, a specimen for analysis (otherwise than where specimen required to assess ability, or alcohol or controlled drug level, at time of driving/attempting to drive)	ss 7 and 7A
Dangerous parking, if MV	s 22
Unlawful carriage of passenger on motor cycle	s 23
MV failing to comply with traffic directions given by a constable or traffic officer	s 35
MV failing to comply with traffic signs at red (or at red with green arrow), double white lines, 'Stop' sign, 'No entry' sign, mandatory height restriction, requirement for large or slow vehicles to obtain level crossing permission	s 36

(continued)

Using vehicle in dangerous condition, where disqualification not obligatory	s 40A
Breach of requirements as to:	
(a) brakes, steering gear or tyres, unless offender proves no knowledge, and no reasonable cause to suspect, facts such that offence would be committed	s 41A
(b) control of vehicle and mobile phone	s 41D
Driving otherwise than in accordance with licence - only where offender's driving would not have been in accordance with any licence that could have been granted to him/her	s 87(1)
Driving after:	
(a) making false declaration as to physical fitness	s 92
(b) failing to notify disability, or	s 94
(c) refusal or revocation of licence driving disability	s 94A
Driving with uncorrected defective eyesight/refusing eyesight test	s 96
Driving while disqualified	s 103(1)(b)
Using vehicle while uninsured	s 143
Failing to stop and give particulars after accident or report accident	s 170(4)
Failing to give information as to driver's identity	s 172
Theft Act 1968	
Theft/attempted theft of MV	s 1
Taking MV without consent, etc/driving, etc MV so taken	s 12
Going equipped for stealing with reference to theft or taking of MV	s 25
Offences against the Person Act 1861	
Furious driving, if committed in respect of MPV	s 35

General Extension of disqualification period where custodial sentence also imposed

24.41 Under RTOA 1988, s 35A, where the offender has been sentenced to an immediate custodial sentence for the offence for which (s)he is ordered to be disqualified *under s 34 or s 35* (**24.39, 24.40,** and **24.44**), the court must add to the period of disqualification which it would otherwise have imposed under s 34 or s 35 an extension period (broadly, the custodial period which must be served). This is designed to require that the former period of disqualification takes effect after the offender's release from imprisonment.

Endorsement and penalty points

24.42 By RTOA 1988, s 44, unless there are special reasons not to do so, where a person is convicted of an offence specified in Sch 2 as involving obligatory endorsement the court must order particulars of the conviction to be endorsed on his/her driving record and the endorsement must also include:

(a) if the court orders disqualification, particulars of the disqualification; or

(b) if the court does not order disqualification, particulars of the offence (including its date) and the number of penalty points to be attributed as shown in respect of the offence in RTOA 1988, Sch 2 (or, where a range of numbers is so shown, a number falling within the range).

It follows that, where an offender is convicted and the court imposes disqualification under RTOA 1988, s 34(1), no penalty points are to be attributed for that offence in respect of which (s)he is convicted on that occasion. This prevents the possibility of 'double disqualification'.

Where a person is convicted of two or more offences included in Sch 2, all of which are committed on the same occasion, the number of points awarded is generally the highest single figure applicable to one of those offences.

Where a person is convicted for aiding and abetting offences involving obligatory disqualification the number of penalty points to be attributed to the offence is 10. The offences of manslaughter, causing death by dangerous driving, dangerous driving; causing death by driving: disqualified drivers; causing death by driving: unlicensed or uninsured driver; causing death by careless driving when under the influence of drink or drugs; causing serious injury by dangerous driving; causing serious injury by driving: disqualified driver; driving or attempting to drive vehicles when under the influence of drink or drugs or with excess alcohol in breath, blood or urine or with excess of a specified controlled drug in blood or urine; failing to provide a specimen for analysis; and an offence under the Theft Act 1968, s 12A, all carry obligatory endorsement of 3–11 penalty points, in addition to obligatory disqualification, unless there are special reasons. However, as indicated above, if disqualification is ordered the licence will not be endorsed with penalty points.

The offence of failing to stop after an accident carries 5–10 points, as does that of failing to report an accident. Uninsured use carries 6–8 points. Careless, or inconsiderate, driving, and furious driving carry 3–9. Failing to co-operate with a preliminary test (eg of breath) carries 4. Failing to identify the driver carries 6. Being in charge when unfit to drive through drink or drugs or with an excess alcohol or controlled drug level carries 10. Contravention of the construction and use regulations which constitutes an endorsable offence carries 3 points (but 6 in the case of mobile phone use). Dangerous parking, unlawful carriage of passengers on a motor cycle, failing to comply with traffic directions or signs, with the directions of school crossing patrols, or with pedestrian crossing regulations carry 3 points. All offences of exceeding a speed limit carry 3–6 penalty points or 3 (fixed penalty). Offences concerned with driving licences carry 3–6 penalty points, except driving whilst disqualified by a court order which carries 6 points, and driving with uncorrected eyesight/refusing eyesight test which carries 3 points.

Recognition of disqualification imposed in Ireland, Northern Ireland, etc

24.43 The Republic of Ireland is generally obliged to give notice to the UK about the disqualification by its courts of someone who is a UK resident or who (if not normally so resident) is the holder of a Great Britain or Northern Ireland licence.

The Crime (International Co-operation) Act 2003, Part 3, Chapter 1, as it applies in Great Britain, provides that, where an individual who is normally resident in the UK or who, although not normally so resident, is the holder of a Great Britain licence, is convicted in the Republic of Ireland:

(a) of an offence of reckless or dangerous driving (whether or not it results in death, injury or serious risk), drink or drug-driving, refusing to submit to a drink or drug test, failing to stop or report after an accident, speeding, or driving while disqualified and is disqualified in the Republic for holding or obtaining a driving licence; or

(b) for any other offence and disqualified in the Republic for six months or more, and provided certain conditions are satisfied, the appropriate Minister in Great Britain (the Secretary of State) must give a notice disqualifying that individual for holding or obtaining a licence for the unexpired period (see below) of the Irish disqualification if the unexpired period is not less than three months. Where the unexpired period is less than three months the Minister is not required to give such a notice but may do so.

RTA 1988, s 102A provides that a person disqualified for holding or obtaining a driving licence by a court in Northern Ireland, the Isle of Man, any of the Channel Islands, or Gibraltar, is disqualified for holding or obtaining a licence in Great Britain for the period of the disqualification.

Obligatory disqualification for repeated offences

RTOA 1988, s 35 applies to an offence involving discretionary disqualification and ob- **24.44** ligatory endorsement and to an offence involving obligatory disqualification in respect of which no order is made under s 34. Section 35 requires that where, on conviction for such an offence, the award of the penalty points for the offence, together with those already endorsed on the licence, brings the total to 12, the offender must be disqualified for a minimum period of six months if the points have been accumulated within three years of the commission (and not the conviction) of the offence for which penalty points are then awarded, unless the court is satisfied that there are grounds for mitigating the normal consequences of the conviction and thinks fit to order a shorter period of disqualification or not to order disqualification. If the total is in excess of 12 this may be recognised by disqualification for a longer period. The minimum period is one year's disqualification if there is a previous disqualification in the above period, or two if there is more than one. No account is to be taken, as a ground for mitigating the normal consequences, of any circumstances which are alleged to make the offence or any of the offences not a serious one; nor of any hardship other than exceptional hardship; nor of any circumstances which, within the preceding three years, have already been taken into account in ordering the offender to be disqualified for a shorter period or not to be ordered to be disqualified at all. Exceptional hardship which would be caused to a person other than the offender (for example a dependent, invalid wife) may be taken into account.

It follows from the above rules that, even if a court decides not to order disqualification when it is discretionary in relation to the offence in question, because of mitigating circumstances, it may be required to disqualify under s 35 as a result of the 'totting-up' of the penalty points attributed for the offence with previous penalty points.

A person who is already the holder of a driving licence authorising him/her to drive a vehicle of a particular class is disqualified by RTA 1988, s 102 for holding another licence for that class of vehicle. This is to prevent a person from holding more than one licence. If (s)he was able to do so (s)he could share out his/her penalty points by holding two licences indicating different addresses.

Once the disqualification has been imposed the driving record is in effect wiped clean. However, 'wiping clean' is restricted to disqualification under s 35 for repeated offences. Where a disqualification is imposed under s 34 for a specific offence, the penalty points previously accumulated remain effective at the end of the period of disqualification until the expiry of three years from the date of the offence for which they were imposed.

Use of vehicle in commission of crime: discretion to disqualify

The Sentencing Code, s 164(1) provides that a driving disqualification order is available **24.45** where an offender is convicted on indictment of an offence which is punishable with imprisonment for two years or more; and the Crown Court is satisfied that a motor vehicle was used (by the convicted person or anyone else) for the purpose of committing, or facilitating the commission of, the offence. For the purposes of s 164(1), facilitating the commission of an offence includes taking any steps after it has been committed for the purpose of disposing of any property to which the offence relates, or avoiding apprehension or detection.

Under s 164(3), a driving disqualification order is also available to the court by or before which an offender is convicted of an offence where:

(a) the offence is common assault, or any other offence involving an assault;
(b) the offence was committed on or after 1 July 1992; and
(c) the court is satisfied that the assault was committed by driving a motor vehicle.

Disqualification: offenders in general

24.46   The Sentencing Code, s 163 empowers a court on conviction of *any* offence to order the offender to be disqualified.

Legislation also empowers a magistrates' court to order disqualification in a case of default in paying a fine.

In these two cases, the disqualification is not endorsed on the offender's driving record, and RTOA 1988, s 35 (**24.44**) does not apply.

Disqualification under Sentencing Code: extension of period where custodial sentence also imposed

24.47   The Sentencing Code, s 166(1) mirrors RTOA 1988, s 35A (**24.41**). It obliges a court to extend the period of disqualification under a driving disqualification order under s 163 or 164 where it imposes an immediate custodial sentence on the offender for the same offence. In such a case, the court must add to the period of disqualification which it would otherwise have imposed an extension period (broadly the, the custodial period which must be served).

Effect of an order of disqualification

24.48   RTOA 1988, s 37(1) provides that, where a licence-holder is disqualified by order of a court, the licence is to be treated as revoked with effect from the beginning of the period of disqualification. However, where:

(a) the disqualification is for a fixed period shorter than 56 days disregarding any extension period added where an immediate custodial sentence is also imposed (as to which see **24.41** and **24.47**) in respect of an offence involving obligatory endorsement or
(b) where the order is made under s 26 (interim disqualification),

this does not prevent the licence from again having effect at the end of the period of disqualification (including any extension period).

Removal of disqualification

24.49   RTOA 1988, s 42 provides that a person who has been disqualified by a court may apply to have the disqualification removed as follows:

(a) if the disqualification is for less than four years, after two years;
(b) if it is for less than ten but more than four years, when half the disqualification has expired; or
(c) in any other case (life or ten or more years), when five years have passed.

Disqualification until passing of driving test

RTOA 1988, s 36 deals with this.   **24.50**

Where a person is disqualified under s 34 (**24.39**) on conviction for manslaughter by the driving of a motor vehicle, or for causing death by dangerous driving, or for causing serious injury by dangerous driving, or for dangerous driving, or for causing death by careless driving while under the influence of drink or drugs, or for causing death by driving: disqualified drivers, or for causing serious injury by driving: disqualified drivers, the court must order him/her to be disqualified until (s)he has passed the appropriate driving test.

Where a person is disqualified under s 34 on conviction for any other offence carrying obligatory endorsement, the court may order him/her to be disqualified until (s)he has passed the appropriate driving test.

An order of disqualification until the passing of a driving test ends when a certificate of competence is produced to the Secretary of State.

An extended driving test is required where a person is convicted of an offence involving obligatory disqualification or is disqualified under the totting-up provisions (RTOA 1988, s 35), and the ordinary driving test is required in any other circumstances.

A person disqualified until (s)he has passed a driving test is permitted by s 37(3) to take out a provisional licence, once any fixed period of disqualification has expired, in which case (s)he may drive under the conditions applicable to such a licence. A person who drives under a provisional licence granted by virtue of s 37(3), but who does not comply with the conditions attached to such a licence, commits an offence contrary to RTA 1988, s 103(1)(b) (**24.52**), of driving whilst disqualified. A person charged with driving while disqualified who relies on RTOA 1988, s 37(3) has the burden of proving that (s)he is the holder of a provisional licence and that (s)he was driving in accordance with the provisions of such a licence.

Offender escaping consequences of endorsable offence by deception

RTOA 1988, s 49 covers the case where, in dealing with a person (D) convicted of   **24.51**
an offence involving obligatory endorsement, a court was deceived regarding any circumstances that were or might have been taken into account in deciding whether, or for how long, to disqualify D. It provides that, if the deception constituted or was due to an offence committed by that person and he is convicted of *that* offence, the court has the same powers of disqualification as the court which was deceived. However, the court must take account of the order made by the first court on D's conviction for the endorsable offence.

Obtaining licence, or driving, while disqualified by a court

RTA 1988, s 103 creates two summary offences which can be committed by a person   **24.52**
disqualified *by a court*: obtaining a licence while so disqualified (s 103(1)(a)), and driving a motor vehicle on a road whilst so disqualified (s 103(1)(b)). The offence under s 103(1)(a) is punishable with a fine not exceeding level 3, and that under s 103(1)(b) with up to six months' imprisonment and/or an unlimited fine and also with discretionary disqualification for driving (**24.40**).

RTA 1988, s 3ZC provides a more serious offence, triable only on indictment and punishable with up to 10 years' imprisonment, as well as with obligatory disqualification

(**24.39**), of causing death by driving: disqualified driver. Section 3ZC provides that a person is guilty of an offence if he:

(a) causes the death of another person by driving a motor vehicle on a road; and
(b) at that time, is committing an offence under s 103(1)(b) of driving while disqualified.

Section 3ZD provides a similarly worded indictable (either-way) offence, the only difference being that it relates to causing serious injury rather than death. 'Serious injury' means physical harm which amounts to grievous bodily harm for the purposes of the Offences Against the Person Act 1861 (**9.14**). An offence under s 3ZD is punishable with up to four years' imprisonment and also with obligatory disqualification for driving.

What is said at **24.10** about the driving and death elements of an offence under s 3ZB is equally applicable to an offence under s 3ZC or 3ZD.

An order of disqualification remains valid unless it is suspended pending an appeal or is subsequently revoked. A subsequent revocation does not excuse driving whilst the order was in force. The prosecution must prove not only the order of disqualification (by means of certificate of conviction or entry in a court register), but that the person before the court is the person so disqualified. There is no prescribed way that the identification of D as the person disqualified must be proved; it can be proved by any admissible means. Proof of identity may be given, for example, by an admission under the Criminal Justice Act (CJA) 1967, s 10, or by evidence of a person who was present in court when the disqualification was imposed. In addition, a written statement made under CJA 1967, s 9, which refers to a person whom the deponent knew and stating that the deponent knew him/her under a particular name, may provide sufficient evidence of identification. It will normally be possible to establish a prima facie case on the basis of a match between the personal details of D and those recorded on the certificate of conviction. Indeed, where D has a highly unusual name it may not be necessary for the date of birth on the certificate to correspond with D's. If D calls no evidence to contradict a prima facie case, it is open to the court to be satisfied that it has been proved that D was the person disqualified. In addition, if it is proper and fair to do so and warning has been given, a failure to give contradictory evidence can additionally give rise to an adverse inference under the Criminal Justice and Public Order Act 1994, s 35, dealt with at **8.58**.

Revocation of licence of 'new driver'

24.53 Under the Road Traffic (New Drivers) Act 1995, where a qualified driver commits an offence involving obligatory endorsement during his/her 'probationary period' (two years from becoming a qualified driver) and the penalty points to be taken into account on that occasion number six or more, the court must send the Secretary of State a notice containing the particulars to be endorsed on the person's driving record together with the licence. A similar requirement is made of a fixed penalty clerk, where a fixed penalty offence is involved. The Secretary of State must, by notice, revoke the new driver's licence. Such a licence may not be granted (restored) until (s)he has passed a relevant driving test (ie re-test); the re-test must be within two years of that revocation. There are similar provisions about the revocation of a provisional licence and test certificate where the driver has not applied for a full licence.

The prescribed probationary period ends early where:

(a) an order is made under RTOA 1988, s 36 (disqualified until a test is passed); or

(b) after revocation of the licence (or provisional licence plus test certificate) by the Secretary of State (see above), the full licence is granted after the person has passed a re-test.

Revocation or suspension of passenger-carrying vehicle or large goods vehicle drivers' licences

RTA 1988, s 115(1) requires that a passenger-carrying vehicle (PCV) or large goods **24.54** vehicle (LGV) driver's licence (**24.16**) must be:

(a) revoked by the Secretary of State if, in relation to its holder, prescribed circumstances occur; or

(b) revoked or suspended by the Secretary of State if the holder's conduct is such as to make him/her unfit to hold such a licence. A divisional court has held that 'conduct' such as to make a driver 'unfit to hold such a licence' refers to conduct 'as a driver of a motor vehicle'.

The 'prescribed circumstances' referred to in (a) are set out in respect of a LGV licence MV(DL)R 1999, reg 55(1) and (2). They are that, in the case of the holder of such a licence who is *under 21*, the holder has been convicted (or is under fixed penalty rules treated as if (s)he had been convicted) of an offence as a result of which more than three penalty points are to be taken into account. No corresponding provision has been made in respect of PCVs.

Under s 115A there are similar provisions to those in s 115 in respect of the revocation of an authorisation to drive under a PCV or LGV Community licence. MV(DL)R 1999 reg 55 also applies here.

By RTA 1988, s 117:

(a) where a person's licence or authorisation is revoked under s 115(1)(a) the Secretary of State must order that that person be disqualified for driving indefinitely or for a fixed period. Where the Secretary of State determines that the disqualification is to be for a fixed period, it must until the driver reaches 21 or such longer period as the Secretary of State determines (reg 55(3)–(4));

(b) where a person's licence or authorisation is revoked under s 115(1)(b)—

    (i) *the Secretary of State may order that person to be disqualified either indefinitely or for such fixed period as (s)he thinks fit, or*

    (ii) *except where the licence is a provisional licence, if it appears to the Secretary of State that, owing to that person's conduct, it is expedient to require him/her to comply with the prescribed conditions applicable to provisional licences until (s)he passes a test to drive LGVs or PCVs, the Secretary of State may order him/her to be disqualified for holding or obtaining a full licence until (s)he passes a test.*

Section 117 is modified by MV(DL)R 1999, reg 56, which applies where PCV or LGV driver's licences are treated as revoked by RTOA 1988, s 37 (**24.48**) *consequent on disqualification by order of a court*, so as to provide that, where the licence to be treated as revoked is a LGV driver's licence held by a person under 21 (D), the Secretary of State must order that D be disqualified either indefinitely or for a fixed period, and where the Secretary of State determines that it shall be for a fixed period, D must be disqualified until D reaches 21 or for such longer period as the Secretary of State determines. Where the licence treated as revoked is a LGV driver's licence held by any other person, or is a PCV driver's licence, the provisions italicised above apply.

MOTOR VEHICLE INSURANCE

Using, or causing or permitting use of, a motor vehicle on a road without insurance

24.55   RTA 1988, s 143(1) states that a person must not use, or cause or permit any other person to use, a motor vehicle on a road or other public place unless there is in force in relation to the use of that vehicle by that person or that other person, as the case may be, such a policy of insurance in respect of third-party risks as complies with the requirements of RTA 1988, Part VI. By s 143(2), breach of s 143(1) is an offence. The offence is a summary offence punishable with an unlimited fine and discretionary disqualification for driving (**24.40**). Section 143(4) states that Part VI does not apply to invalid carriages not exceeding 254 kg unladen weight; a mobility scooter is a type of invalid carriage. Since electrically assisted pedal cycles (**25.93**) are not motor vehicles (unless they fall outside the relevant regulations), the provisions dealing with insurance do not apply to them.

As in the case of driving without a driving licence, the onus is upon the defendant (D), if proved to have used a motor vehicle on a road or other public place, to prove the existence of insurance as this is a fact peculiarly within D's knowledge. However, a divisional court has said that if the prosecution contends that D used the vehicle *in a way not permitted by that insurance* the prosecution must prove *this fact*.

For the meaning of 'road' and 'public place' see **23.2–23.4** and **26.3** respectively. The terms 'use', 'cause', and 'permit' are discussed in Chapter 25. Basically, a person who drives a vehicle 'uses' it, but 'using' is not limited to driving. 'Use' means 'have the use of', with the result that, for example, if someone parks a motor vehicle on a road (s)he can be said to be using it while it is parked, and it has been held that this is so even though the vehicle has been totally immobilised. Where an employee drives a motor vehicle owned by his/her employer in the course of his/her employment, the employer is regarded in law as also using it. A person 'causes' a vehicle to be used when, being in a position to do so, (s)he expressly orders or authorises the vehicle to be used; and (s)he 'permits' a vehicle to be used when (s)he allows another person to use it. In each case, of course, the use must be on a road or other public place without there being in force in relation to the use of the person using the vehicle a requisite third-party policy of insurance.

The prohibition in s 143(1) against using, or causing or permitting the use of, an uninsured vehicle is strict in the sense that it is no defence that D reasonably believed that (s)he, or the person allowed, to drive, was covered by an insurance policy. There is one exception: under s 143(3) it is a defence for a person (D) charged with *using* an uninsured vehicle to prove that it did not belong to D and was not in D's possession under a contract of hiring or loan, and that D was using it in the course of D's employment and neither knew nor had reason to believe that it was not properly insured.

Can a person be said to permit his/her vehicle to be used on a road or other public place without insurance when (s)he has lent it out on the express condition that it should only be used if its use is covered by insurance and its subsequent use is not so covered? The law provides a rather odd answer. If the condition is directly communicated to the person who uses the vehicle without insurance, the person imposing the condition does not permit his/her vehicle to be used without insurance. This seems sensible. How can you permit something which you have expressly forbidden? Thus, it is somewhat surprising that a divisional court has held that, if the condition as to use being covered by insurance has not been directly communicated to the user but has been communicated to someone else, for example to someone who borrows the vehicle, with a view to it being driven by the user, and it is used in breach of that condition, the person imposing the condition does in law permit uninsured use.

A person cannot permit another's use if (s)he was not in a position to forbid it. Thus A, who supervises B, a learner driver, in B's own car cannot be convicted of permitting its uninsured use.

Causing death by driving: uninsured drivers

24.56 RTA 1988, s 3ZB creates an indictable (either-way) offence, punishable with up to two years' imprisonment and obligatory disqualification for driving (**24.39**), of causing death by driving when uninsured. The offence is committed when it is shown that D was using a motor vehicle on a road while uninsured against third-party risks, contrary to s 143, and that the death of another person was caused by the driving of that vehicle. What is said at **24.10** about the driving and death elements of another offence under s 3ZB is equally applicable here.

Keeping vehicle not meeting insurance requirements

24.57 RTA 1988, s 144A creates a summary offence, punishable with a fine not exceeding level 3, of keeping (by the registered keeper) a motor vehicle which does not meet the insurance requirements. No use on a road or other public place is required. A vehicle will meet the insurance requirements if it is covered by a policy of insurance complying with RTA 1988 and either the policy, or a related certificate, identifies the vehicle covered by its registration mark or the policy covers any vehicle, or any particular type of vehicle, owned by the person named in it or a related certificate.

Section 144B specifies exceptions which are generally the same as those set out in s 144 (vehicles of local authorities, the police, and National Health Service, etc: **24.64**).

Section 144B provides further exceptions where the vehicle is no longer kept by the registered keeper, it is not kept for use on a road or other public place or it has been stolen. However, these exceptions only apply if a 'prior statement' has been made in respect of the vehicle. The Motor Vehicles (Insurance Requirements) Regulations (MV(IR)R) 2011, regs 2–4 set out the following requirements for the 'prior statement':

(1) In the case of the exception relating to a vehicle no longer kept by the registered keeper (RK):
   (a) where the vehicle has been sold or transferred by the RK (other than to be destroyed or sent permanently out of Great Britain) the RK must, by the relevant time and in relation to that vehicle, have furnished the information and documents and made the declarations in accordance with the appropriate requirements;
   (b) where the vehicle has been sold or transferred by the RK for the purpose of its destruction, that keeper must, by the relevant time and in relation to that vehicle, have furnished the information and documents and made the prescribed declarations, or been given a certificate of destruction; and
   (c) where the vehicle has been sent permanently out of Great Britain, the RK must, by the relevant time and in relation to that vehicle, have notified the Secretary of State of that fact (reg 2).

(2) In the case of the exception relating to a vehicle not kept for use on a road or other public place, the RK must, by the relevant time and in relation to that vehicle, have made a statutory off-road notification (SORN) (reg 3).

(3) In the case of the exception relating to a vehicle which has been stolen, the RK must, by the relevant time and in relation to that vehicle, have given notification of the theft to a member (or employee) of a police force (reg 4).

A further exception under RTA 1988, s 144B, to an offence under s 144A is that:

(a) the RK is at the relevant time keeping the vehicle;

(b) neither a licence nor a nil licence under VERA 1994 was in force on 31 January 1998;

(c) neither a licence nor a nil licence has been taken out for the vehicle for a period starting after that date; and

(d) the vehicle has not been used or kept on a public road after that date.

RTA 1988, s 144C authorises the Secretary of State to serve a fixed penalty notice on a person whom (s)he believes to have committed an offence against s 144A. That penalty is £100 but it is reduced to £50 if that amount is paid within 21 days of notice being given that a s 144A offence has been committed.

Immobilisation, removal, and disposal

24.58 The Motor Vehicles (Insurance Requirements) (Immobilisation, Removal and Disposal) Regulations 2011 (MV(IR)(I, R and D)R 2011) provide for the clamping of vehicles by an authorised person (AP) who has reason to believe that a s 144A offence (**24.57**) has been committed, and their removal and disposal. An AP is a person authorised by the Secretary of State for the purposes of MV(IR)(I, R and D)R 2011. An AP may be a local authority, an employee of a local authority, a member of a police force, or any other person. Different persons may be authorised for different purposes, but a person who is an AP for the purposes of appeals must not act as an AP for any other purpose.

By reg 4, MV(IR)(I, R and D)R 2011 do not apply where:

(a) a current disabled person's badge ('blue badge') is displayed on the vehicle;

(b) the vehicle appears to an AP to have been abandoned;

(c) the vehicle is a PSV being used for the carriage of passengers;

(d) the vehicle is being used for the removal of any obstruction to traffic, the maintenance, improvement or reconstruction of a public road, or the laying, erection, alteration, repair, or cleaning in or near a road of any traffic sign or sewer or of any main, pipe or apparatus for the supply of gas, water, or electricity, or of any telegraph or telephone wires, cables, posts, or supports; or

(e) the vehicle is being used by a universal service provider in the provision of a universal postal service and each side of the vehicle is clearly marked with the name of the provider concerned.

Power to immobilise vehicles   Where an AP has reason to believe that a s 144A offence is being committed as regards a stationary vehicle on a road or other public place, (s)he (or someone acting under his/her direction) may:

(a) fix an immobilisation device to the vehicle while it remains in the place where it is stationary, or

(b) move it from that place to another place on the same or another road or public place and fix an immobilisation device to it in that other place.

The exercise of this power does not prevent a prosecution for a motor vehicle insurance offence under RTA 1988 (reg 5).

Where a device is fixed to a vehicle, its fixer must also fix to the vehicle an immobilisation notice which:

(a) indicates that the device has been fixed to the vehicle and warns that no attempt should be made to move it until it has been released from the device;
(b) states why the device has been fixed;
(c) specifies the steps to be taken to secure its release, including the charges payable under MV(IR)(I, R and D)R 2011 and the person to whom and the means by which those charges may be paid; and
(d) states the right of the owner, or person in charge of the vehicle when it was immobilised, to appeal, and specifies the steps to be taken and the address to which representations to an AP should be sent.

Except in relation to a vehicle which is the subject of a hiring agreement or a hire-purchase agreement, references to the 'owner' of a vehicle at a particular time in MV(IR)(I, R and D)R 2011 are to the person by whom it was then kept and the registered keeper (RK) at a particular time is to be taken, unless the contrary is shown, to be the person by whom the vehicle was kept at that time.

For the above purposes and those of regs 11 and 12 (below), 'owner', in relation to a vehicle which is the subject of a hiring agreement or hire-purchase agreement, means a person who is a party to such an agreement.

Release of immobilised vehicle (reg 6)   A vehicle to which an immobilisation device has been fixed in accordance with reg 5:

(a) may only be released by, or under the direction of, an AP; and
(b) subject to (a), is to be released:
    (i) where an AP is satisfied that the vehicle was immobilised in any of the circumstances specified in reg 4; or
    (ii) if the following two requirements are met, viz.
        (a) that the prescribed charge (£100) for the release of the vehicle from the immobilisation device is paid in any manner specified in the immobilisation notice, and
        (b) that, in accordance with any instructions specified in the immobilisation notice, prescribed evidence is produced which establishes that any person who proposes to drive the vehicle away will not in doing so be guilty of an offence under RTA 1988, s 143, and that the RK is not, at the point of release, guilty of a s 144A offence as regards the vehicle.

For the purposes of MV(IR)(I, R and D)R 2011, 'prescribed evidence' means a policy or certificate of insurance.

Removal of or interference with immobilisation notice or device   An immobilisation notice must not be removed or interfered with except by or under the authority of an AP; it is an offence to do so (reg 7(1) and (2)).

Any person who, without being authorised to do so in accordance with reg 6, removes or attempts to remove an immobilisation device fixed to a vehicle is guilty of an offence (reg 7(3)).

The above are summary offences punishable with a fine not exceeding level 2 (reg 7(1) and (2)) or level 3 (reg 7(3)).

Other offences connected with immobilisation (reg 8)   Where MV(IR)(I, R and D)R 2011 would apply to a vehicle but for the fact that a current disabled person's 'blue badge is displayed on it and the vehicle was not, at the time it was stationary, being used:

(a) in accordance with the 'Blue Badges Regulations'; and

(b) in circumstances where a disabled person's concession would be available,

the person in charge of the vehicle at that time is guilty of a summary offence punishable with a fine not exceeding level 3.

Removal of vehicles (reg 9)   Where an AP has reason to believe that a s 144A offence:

(a) is being committed as regards a stationary vehicle on a road or other public place, or

(b) was being committed as regards a vehicle at the time when an immobilisation device fixed to the vehicle was fixed to it and:

   (i)   24 hours have elapsed since the device was fixed to the vehicle, and

   (ii) the vehicle has not been released in accordance with MV(IR)(I, R and D)R 2011,

the AP, or a person acting under his/her direction, may remove the vehicle and deliver it to a person authorised by the Secretary of State to keep vehicles so removed ('the custodian').

Disposal of removed vehicle (reg 10)   The custodian of a vehicle delivered in accordance with reg 9 may dispose of it by selling it or dealing with it as scrap, as (s)he thinks fit, in accordance with reg 10.

Recovery of prescribed charges (reg 11)   Where a vehicle has been removed and delivered into the custody of a custodian in accordance with reg 9, the Secretary of State or the custodian may (whether or not any claim is made under reg 12 or 14 (below)) recover from the person who was the owner of the vehicle when the vehicle was removed the prescribed charges for:

(a) its removal and storage; and

(b) its disposal, if the vehicle has been disposed of.

The relevant prescribed charges are:

Removal of a vehicle where possession is taken within the first 24 hours of such removal	£100
Removal of vehicle where the above does not apply	£200
Storage of vehicle for each period of 24 hours or part thereof	£21
Disposal of vehicle	£50

For the above purposes, 'owner' has the same meaning as for the purposes of reg 5 (above).
   Where, by virtue of (a) above, any sum is recoverable in respect of a vehicle by a custodian, the custodian is entitled to retain custody of it until that sum is paid.
   Where:

(a) it appears to a custodian of a vehicle that more than one person is the owner; and

(b) one of those owners, or a person authorised by one of those owners, has gained possession of the vehicle under reg 12 (below),

then the owner who gained possession of the vehicle under reg 12 must be treated as the owner from whom the sum is recoverable (reg 11(3)).
   Where:

(a) it appears to a custodian of a vehicle that more than one person is the owner; and

(b) one of those owners has made a claim under reg 13 that satisfies conditions (a), (b), and (c) in reg 13, set out below, then the owner who made the claim under reg 13 must be treated as the owner from whom the sum is recoverable (reg (11(4)).

Where:

(a) it appears to a custodian of a vehicle that more than one person is the owner; and
(b) neither reg 11(3) nor reg 11(4) applies,

then those persons must be treated as jointly and severally liable for the prescribed charges (reg 11(5)).

Taking possession of a vehicle (reg 12)   A person ('the claimant': C) may take possession of a vehicle which has been removed and delivered to a custodian and has not been disposed of under reg 10, if:

(a) C satisfies the custodian that C is the owner of the vehicle (above), or a person authorised by the owner to take possession of the vehicle;
(b) except where C produces evidence that no s 144A offence was committed or the custodian is satisfied that MV(IR)(I, R and D)R 2011 did not apply to the vehicle at the time it was immobilised or removed, C pays to the custodian:
  (i) the prescribed charge in respect of the removal of the vehicle; and
  (ii) the prescribed charge for the storage of the vehicle during the period it was in the custody of the custodian; and
(c) prescribed evidence is produced which establishes that—
  (i) any person who proposes to drive the vehicle away will not in doing so be guilty of an offence under RTA 1988, s 143; and
  (ii) is not, at the point of release, guilty of a s 144A offence as regards the vehicle.

On giving C possession of the vehicle, the custodian must give C a statement of the right of the owner, or person in charge of the vehicle at the time it was immobilised or (where it was not immobilised) it was removed, to appeal, of the steps to be taken in order to appeal and of the address to which representations to an AP made as mentioned in reg 14 should be sent.

Where it appears to a custodian of a vehicle that more than one person is the owner, or person authorised by the owner, the custodian must give possession of the vehicle to the first claimant who satisfies the conditions set out in (a), (b), and (c).

Claim by owner of a vehicle after its disposal (reg 13)   If, after a vehicle has been disposed of by a custodian pursuant to reg 10, a person claims to have been the 'owner' of the vehicle at the time when it was disposed of and:

(a) the person making the claim satisfies the custodian that that person was the owner of the vehicle at the time it was disposed of; and
(b) the claim is made before the end of the period of one year beginning with the date on which the vehicle was disposed of; and
(c) no previous claim in respect of the vehicle has been paid,

the custodian must pay that person an amount calculated by deducting from the proceeds of sale the prescribed charges for the removal, storage and disposal of the vehicle. For these purposes, where a vehicle is the subject of a hiring agreement or hire-purchase agreement, 'owner' means the person who, being a party to such agreement, has given up possession of the vehicle in return for payment under the agreement.

Where it appears to a custodian of a vehicle that more than one person is the owner, the custodian must treat the first person who makes a claim that satisfies the conditions set out in (a), (b), and (c) above as the owner for the purposes of reg 13.

Appeal (reg 14)   Where a dispute has arisen because:

(a) the claimant (C) has paid a charge in accordance with reg 6 or 12 in order to se-
    cure the release or to obtain possession of a vehicle and alleges that the charge ('the
    disputed charge') should be refunded to him/her on the ground that, at the time
    the vehicle was immobilised or, where it was not immobilised, at the time it was
    removed:
    (i) a s 144A offence was not being committed (*ground 1*); or
    (ii) any of the circumstances specified in reg 4 applied to the vehicle (*ground 2*);
    and
(b) the person to whom the disputed charge was paid refuses to refund the charge,

C may appeal against the refusal of a refund by sending, to the AP whose name is given
for this purpose in the immobilisation notice under reg 5 or the statement under reg
12 of the right of the owner etc, written representations stating the grounds on which
a refund is claimed.

The AP must uphold the appeal if satisfied that:

(a) ground 1 or ground 2 has been established; and
(b) the disputed charge has not been refunded.

Where the AP upholds an appeal, (s)he must inform C of the decision and the
Secretary of State must refund the disputed charge.

Where the AP rejects the appeal, (s)he must inform C of the decision and of the right
to make a further appeal to a magistrates' court.

Provision of information

24.59   MV(IR)R 2011, reg 7 provides that the Motor Insurers' Information Centre must make
available to the Secretary of State, for the purpose of enforcing the s 144A offence, information relating to vehicle insurance policies in the UK and to vehicles which are
permitted to be used on the road without insurance.

Third-party insurance policies

Certificate of insurance

24.60   The requirement made by RTA 1988, s 143 is that there must be a policy of insurance
in respect of third-party risks in accordance with Part VI (ie ss 143 to 162) of that Act
in relation to the use of the vehicle by the person using it. By s 147, an insurer issuing a
policy of insurance must deliver to the person insured a certificate of insurance in the
prescribed form. Delivery of the certificate may be by electronic transmission, or via a
website, with the insured's consent. 'Telephone insurance' is not valid; no certificate has
been delivered at that time giving evidence of a contract between insurer and insured.
Police officers frequently exercise discretion in circumstances in which a company has
clearly accepted liability for third-party risks but some technicality has delayed delivery
of a certificate. A cover note is a certificate.

What must be covered

24.61   RTA 1988, s 145 provides that, in order to comply with the requirements of Part VI, a
policy must be issued by an 'authorised insurer' and *must insure such person, persons or
classes of persons as may be specified in it in respect of any liability which may be incurred
by him/her or them in respect of the death of or bodily injury to any person (other than the*

driver) or damage to property up to a maximum value of £1,200,000 caused by, or arising out of, the use of the vehicle on a road or other public place in Great Britain.

The policy must also cover him/her or them:

(a) in respect of any liability for the emergency treatment of persons injured;

(b) in the case of a vehicle normally based in the territory of an EU state, insure him/her/them in respect of any civil liability which may be incurred by him/her or them as a result of an event related to the use of the vehicle in Great Britain if, according to the law of that territory, (s)he or they would be required to be insured in respect of a civil liability which would arise under that law as a result of that event if the place where the vehicle was used when the event occurred were in that territory, and the cover required by that law would be higher than that required by the provision italicised above; and

(c) in the case of a vehicle normally based in Great Britain, insure him/her/them in respect of any liability which may be incurred by him/her/them in respect of the use of the vehicle and of any trailer, whether or not coupled, in the territory of Northern Ireland and of each of the EU states according to the law on compulsory insurance against civil liability in respect of the use of vehicles applicable in the territory in which the event giving rise to the liability occurred; or if it would give higher cover, the law which would be applicable under RTA 1988 if the place where the vehicle was used when that event occurred were in Great Britain.

In the case of an automated vehicle (ie a vehicle listed as an automated vehicle under the Automated and Electric Vehicles Act 2018 (AEVA 2018), s 1) the policy must also provide for the insurer's obligations to an insured person under AEVA 2018, s 2(1) (liability of insurers etc where accident caused by automated vehicle) to be obligations under the policy. The reference to an 'insured person' is to a person who is covered under the policy for using the vehicle on a road or public place in Great Britain.

Extent of cover

The nature of the policy held will be specified in the certificate. A policy covering use **24.62** for social, domestic and pleasure purposes does not cover business use. However, to give a lift to a friend who is on business at the time is not a business use of the vehicle. To help move household goods for a friend without payment is not a business use, but it might be if payment was made. It is usual for non-business policies to exclude use for hire or reward, but RTA 1988, s 150 makes specific provisions for 'car-sharing' schemes. Provided that arrangements are made before the journey, that payment is in respect of running costs and depreciation only, and that the vehicle used is not adapted to carry more than eight passengers, the use is not for hire or reward.

The extent of cover is the cover specified in the policy of insurance. In disputed cases this depends on the construction of the insurance policy rather than the views of the insurance company or its willingness to meet liabilities.

Some policies allow persons to drive with the permission of the insured person if such a person 'holds or has held a driving licence' and is not disqualified. If a person has at any time held a licence, whether full or provisional, it is sufficient in law to satisfy such a requirement. The terms of the policy will have to be examined in circumstances in which the class of vehicle which the person is licensed to drive is different to the class of vehicle which is being driven. If the policy demands that such a person holds or has held a licence to cover the same class of vehicle as is covered by the policy, then use by a person who does not hold (and has not held) such a licence will amount to uninsured use, unless the licence which (s)he holds may act as a provisional one to cover use of the vehicle in question. Similarly, where a person has borrowed a vehicle subject to an implied limitation, of which (s)he was aware, concerning the purpose for which the vehicle was to be driven, (s)he does not have the consent of the owner for a purpose outside that limitation. Where, therefore, insurance cover was dependent upon the driver having 'the consent of the owner thereof', use outside that limitation was uninsured use.

The use of trailers in Great Britain does not expressly need to be covered in respect of third-party risks, whereas this is essential in EU states. However, the use of trailers in Great Britain is generally covered in all policies of insurance and would seem to be within the use of the motor vehicle in any case. Injury caused to a person by a trailer is certainly caused due to the use of the motor vehicle on a road or other public place.

Validity

24.63   When a policy has been taken out it remains in force until it expires or is validly set aside. If someone obtains motor insurance by failing to disclose his/her disqualification or driving convictions or some other material fact, then, although the insurance policy can be set aside by the insurance company, the insurance certificate remains valid until the company takes steps to set aside the policy. If the company does set it aside, this only invalidates the certificate for the future.

It is the practice of some insurance companies to insert restrictive clauses in their policies which can affect their validity in particular circumstances. However, RTA 1988, s 148 states that the third-party requirement of s 143 must be covered by companies regardless of restrictions which might be inserted covering the age, mental or physical condition of a driver; the condition of the vehicle; the number of persons carried; the load; the times or areas of use; the horsepower cylinder capacity or value of the vehicle; the carrying of any special apparatus, or the carrying on the vehicle of special identification marks.

When a person (D) is apprehended for taking motor vehicles without consent, it is good practice to check any certificate of insurance held by D in respect of motor vehicles. It is the terms of the policy which are important, not the moral implications. If D's policy covers the use by D of another vehicle not owned by D or hired to D by a hire-purchase agreement, that policy will cover such use even though the vehicle has been taken illegally. It would be different if the use covered was similar but 'with the consent of the owner of such vehicle'.

Exemptions from insurance requirements

24.64   The offences relating to use without third-party insurance, which are provided by RTA 1988, s 143, are subject to exemptions provided by s 144. Section 144 exempts from the requirement for third-party insurance a vehicle:

(a)   owned by a local authority, a National Parks Authority or the Broads Authority at a time when it is being driven under the owner's control;

(b)   owned by a local policing body or police authority being driven under its control, or a vehicle driven by a person for police purposes by or under the direction of a police constable, or by a person employed by a local policing body or police authority, or by a police volunteer or a member of the civilian staff of the police force; a police officer on duty using his/her own car for police purposes falls within this exemption;

(c)   being driven for salvage purposes pursuant to the Merchant Shipping Act 1995, Part IX;

(d) owned by a health service body, or an ambulance owned by an NHS Trust, or NHS foundation trust, at a time when it is being driven under the owner's control;

(e) made available by the Secretary of State or the Welsh Ministers to any person, body or local authority under the National Health Service Act 1977 at a time when it is being used within the terms on which it is being made available; or

(f) used by the Care Quality Commission, at a time when the vehicle is being used under the owner's control.

The reasoning behind these exemptions is clear: the vehicles are either owned by or being used by undertakings which are in a position to meet liabilities which might be incurred. Crown vehicles appear to be similarly exempt, as RTA 1988, s 183, which deals with the application of the provisions of the Act to the Crown, does not mention s 143.

Fraud, forgery, etc

By RTA 1988, s 174(5), it is an indictable (either-way) offence, punishable with up to   **24.65**
two years' imprisonment, to make a false statement, or to withhold any material information, for the purpose of obtaining the issue of a certificate of insurance, or any other document which may be produced in lieu of such a certificate of insurance. Any person who issues such a document which to his/her knowledge is false in any material particular commits a summary offence against s 175(1), punishable with a fine not exceeding level 4.

The alternative documents to which s 174(5) refers are specified by the Motor Vehicles (Third-Party Risks) Regulations 1972, reg 7 as follows:

(a) a certificate signed by a specified body, or by a local policing body or the like; or

(b) in the case of a vehicle normally based in an EU state, or Norway or Switzerland, a document issued by the insurer in the prescribed form.

The making of false statements to obtain insurance is not uncommon. The disclosure of previous convictions for motoring offences, particularly those involving endorsement, can lead to high premiums which can be avoided by non-disclosure. In this context, the Rehabilitation of Offenders Act 1974 is important. The Act provides for convictions to become 'spent'. It also provides that where a question seeking information with respect to a person's previous convictions is put to him, otherwise than in proceedings before a judicial authority, the question must be treated as not relating to spent convictions and answers may be made accordingly. Most road traffic offences will have resulted in a fine and as a result will become spent after a period of 12 months.

It is not necessary to prove that the person who made the statement was aware of its falsity (or even of the risk that it might be false).

RTA 1988, s 173(1) makes it an indictable (either-way) offence, punishable with up to two years' imprisonment, for a person with intent to deceive:

(a) to forge, alter, or use a certificate of insurance or a document which may be used in lieu thereof;

(b) to lend to, or allow to be used by, any other person such a certificate or document;

(c) to make or possess anything so closely resembling such a certificate or document as to be calculated (ie likely) to deceive. A pad of blank and bogus insurance certificates was held by the Court of Appeal to constitute documents for the purpose of (b) because they so closely resembled any certificate of insurance as to be likely to deceive.

Other requirements in respect of motor vehicle insurance

24.66   RTA 1988, s 154 requires any person against whom a claim is made in respect of any such liability as is required to be insured against by s 145 (**24.61**) to provide particulars of his/her insurance cover in respect of that liability. Under s 154(2), it is a summary offence, punishable with a fine not exceeding level 4, to fail to do so without reasonable excuse, or wilfully to make a false statement in reply to such a demand. Section 154 allows the administrative officers of hospitals, etc to recover the cost of emergency treatment from vehicle insurers.

Persons temporarily in Great Britain

24.67   The Motor Vehicles (International Motor Insurance Card) Regulations 1971, reg 5 authorises a visitor to Great Britain to use his/her motor vehicle under the internationally recognised international motor insurance card provisions. Such cards are referred to in the regulations as 'valid insurance cards'. A peculiarity of the provisions is that cards remain valid after the expiry date shown. This is to prevent the complications which would otherwise arise if holiday visitors decided to extend their stay. A constable's powers in respect of insurance certificates, etc apply to such cards. Cards issued under the 1971 Regulations have traditionally been referred to as 'green cards' but they may now be white, as well as green, in colour. They may now be in electronic or printed form.

It is not essential for visitors from the EU to hold a card because the necessary cover must be provided in accordance with RTA 1988, s 145 (**24.61**).

Certificates of insurance issued in Northern Ireland are, of course, valid in Great Britain.

Disclosure of vehicle insurance information to Secretary of State

24.68   The Disclosure of Vehicle Insurance Information Regulations 2005 require the Motor Insurers' Information Centre to make available 'relevant motor vehicle insurance information' to the Secretary of State for processing with a view to making the processed information available for use by police officers. 'Relevant motor vehicle insurance information' means information relating to vehicles whose use was, but is no longer, insured. Such information must be provided at such intervals as reasonably required. It must then be processed in a form which either (a) will be of assistance to a police officer in determining whether or not an offence of uninsured use of a vehicle has been committed, or (b) will enable chief officers of police to further process it to assist a police officer in determining whether such an offence has been committed. A constable may use the processed information to determine whether to demand production of evidence of insurance under RTA 1988, s 165. Processed information must not be disclosed other than for the purposes of proceedings.

Production of evidence of insurance

24.69   RTA 1988, s 165 states that a person:

(a) driving a motor vehicle, other than an invalid carriage, on a road; or

(b) whom a constable or vehicle examiner reasonably believes to have:

(i) been the driver of such a motor vehicle when an accident occurred owing to its presence on a road or other public place; or

(ii) committed an offence in relation to the use of such a motor vehicle on a road,

must, on being required by a constable or vehicle examiner, give his name and address and the name and address of the owner of the vehicle, and produce for examination the relevant *certificate of insurance (or electronic access to a copy of it where it was transmitted electronically or made available via a website by the insurer), or other evidence prescribed by the Motor Vehicles (Third Party Risks) Regulations 1972, reg 7 referred to at* **24.65**. The Court of Appeal has held that the 'relevant certificate of insurance' is a certificate covering the driving of a vehicle, whether or not the vehicle is identified in the certificate (as where a borrower of the vehicle is driving it on his/her insurance). Failure to do any of these things is a summary offence, punishable with a fine not exceeding level 3, under RTA 1988, s 165(3). The usual provisions concerning production of documents apply, allowing seven days for production; or production as soon as reasonably practicable; or proof that production was not reasonably practicable before the day on which proceedings were instituted. The documents (unlike a driving licence) need not be produced in person.

The provisions of s 165 requiring a person to give his/her name and address and the name and address of the owner of the vehicle also apply to a supervisor of a provisional licence-holder in each of the circumstances set out above.

Police powers of seizure of vehicles being driven without a driving licence or insurance

RTA 1988, s 165A provides that: **24.70**

(a) where a constable in uniform requires, under RTA 1988, s 164, the production of a licence (**24.26**), or, under s 165, the production of evidence of insurance; and

(b) it is not produced; and

(c) the constable has reasonable grounds for believing that the vehicle was being driven in contravention of s 87(1) (requirement to have a driving licence) or s 143 (requirement to have insurance),

the constable may seize the vehicle and remove it.

Section 165A also provides that:

(a) where a constable in uniform has required, under s 163 (**25.00/443**), a person driving a motor vehicle to stop it; and

(b) there is a failure to do so, or a failure to do so for a sufficient time to allow the constable to make appropriate enquiries, and the constable has reasonable grounds for believing that the vehicle is being driven in contravention of s 87(1) (driving without licence) or 143 (uninsured driving),

he may seize and remove it.

Section 165A empowers the constable to enter any premises (other than a private dwelling house) on which (s)he has reasonable grounds for believing the vehicle to be in order to seize it. The constable may use reasonable force to seize and remove the vehicle or to enter premises to seize and remove it.

Before seizing a vehicle under s 165A, and provided it is practicable to do so, a constable must warn the person by whom it appears that the vehicle is, or was being, driven in contravention of either s 87(1) or s 143 that (s)he will seize the vehicle if the relevant documents are not produced. On seizing a vehicle under s 165A, a constable must give a seizure notice (below) to the driver unless the circumstances make it impracticable to do so.

Where there has been a failure to stop, the power of seizure is valid for a period of 24 hours.

Retention and disposal of vehicles seized under powers provided by RTA 1988, s 165A

24.71   The Road Traffic Act 1988 (Retention and Disposal of Seized Motor Vehicles) Regulations 2005 provide for the retention, safe-keeping, and disposal by an authorised person (AP), ie a constable or other person authorised by the chief officer of police, of vehicles seized under RTA 1988, s 165A.

Such a vehicle must be passed into and remain in the custody of an AP until the AP permits it to be removed from his/her custody by a person appearing to him/her to be the registered keeper or owner of the vehicle, or until it has been disposed of in accordance with the Regulations.

An AP must, as soon as is reasonably practicable after (s)he has taken a vehicle into his/her custody, take such steps as are reasonably practicable to give a seizure notice to the person who is the registered keeper of the vehicle and to the owner, where that person appears to be someone different, unless (a) the AP is satisfied that a seizure notice has already been given, by a constable on seizing the vehicle, to the registered keeper and to the owner, where it appears that the owner is a different person; or (b) the vehicle has been released from his/her custody in accordance with the Regulations.

A seizure notice must contain information which can be or could have been ascertained from an inspection of the vehicle, or has been ascertained from another source, as to the registration mark and make of the vehicle. It also must include a statement as to the place of seizure and the place where the vehicle is now being kept; a requirement that the keeper or owner claim the vehicle from an AP on or before a date specified in the notice (not less than seven working days from the date upon which the notice is given); a statement that unless the vehicle is claimed on or before that date, an AP intends to dispose of it; a notice that charges are payable and that the vehicle may be retained until they are paid and a notice that the registered keeper or owner must either:

(a) produce at a specified police station a valid insurance certificate covering his/her use of that vehicle and a valid licence authorising him/her to drive that vehicle; or

(b) nominate for this purpose a third person who produces at a specified police station such a valid certificate and such a valid licence.

The notice must also make clear that the vehicle may be retained until (a) or (b) is satisfied.

A seizure notice may be given:

(a) by delivering it to the person to whom it is directed;

(b) in the case of the registered keeper, by leaving it at the registered address or by sending it by registered post to that address;

(c) in the case of the owner, by leaving it at his/her usual or last known address, or by sending it by registered post to that address;

(d) in the case of a body corporate, by delivering it to the secretary or clerk of the body at its registered or principal office, or by sending it by registered post addressed to that secretary or clerk at that office.

Where, before a relevant motor vehicle is disposed of, a person satisfies an AP that (s)he is the registered keeper or owner of that vehicle, pays the appropriate charge and produces at a police station specified in the notice a valid insurance certificate covering his/her use of the vehicle and a valid driving licence authorising him/her to drive the vehicle (or such documents are produced by a nominated person), the AP must permit him/her to remove the vehicle from custody. An AP may consider any documentary evidence produced to AP in determining whether a person who claims to be the owner of a motor vehicle is in fact the owner.

A person, who would otherwise be liable to pay the appropriate charge, is not liable to pay it if (s)he was not driving the vehicle at the time when it was seized under s 165A; did not know that the vehicle was being driven at that time; had not consented to it being driven; and could not, by taking reasonable steps, have prevented it from being driven.

An AP is empowered to dispose of a vehicle if:

(a) where the registered keeper and the owner appear to be the same person, that person fails to comply with any requirement in the seizure notice, or an AP was not able, after taking reasonably practicable steps, to give a seizure notice to that person;

(b) where the owner and registered keeper appear to be different:

    (i) where the seizure notice was given to both of those persons, neither the registered keeper nor the owner complies with all the requirements of the seizure notice;

    (ii) where, having taken such steps as were reasonably practicable, the AP was only able to give a seizure notice to one of them, that person fails to comply with any requirement in that seizure notice; or

    (iii) where, after taking reasonably practicable steps, the AP was not able to give a seizure notice to either the registered keeper or the owner.

An AP may not dispose of a vehicle:

(a) during the period of 14 days starting with the date of seizure; or

(b) if that date has expired, until after a date specified in the seizure notice; or

(c) if not otherwise covered by (a) or (b), during the period of seven working days starting with the date on which the vehicle is claimed under the above provisions.

Where there is a disposal by way of sale, the net proceeds of that sale must be paid to any person who, within a one-year period beginning with the date of sale, satisfies an AP that at the time of the sale (s)he was the owner of the vehicle. Where it appears that there may be more than one owner, the proceeds must be paid to such one of them as the AP thinks fit. The term 'net proceeds' refers to that part of the money remaining after the deduction of charges under the legislation.

INFORMATION AS TO IDENTITY OF DRIVER IN SPECIFIED CASES

The requirement to give information

RTA 1988, s 172(2) provides that where the driver of a vehicle is alleged to be guilty of an offence to which the section applies: **24.72**

(a) the person keeping the vehicle must give such information as to the identity of the driver as he may be required to give by or on behalf of a chief officer of police. A requirement made with the express or implied authority of the chief officer is given on his behalf; and

(b) any other person must if required as stated above give any information which is in his power to give and may lead to the identification of the driver.

The expression 'any other person' includes the driver himself.

Section 172 applies to:

(a) all offences against RTA 1988, except an offence under Part V (driving instruction), or s 13 (promoting motoring events on highway), s 16 (wearing of protective headgear), s 51(2) (certain test conditions—goods vehicles), s 61(4) (certain type approval requirements), s 67(9) (obstructing a vehicle examiner), s 68(4) (failing to proceed to place of vehicle inspection), s 96 (driving with uncorrected defective eyesight), or s 120 (offence against regulations dealing with licensing of drivers of large goods vehicles and passenger-carrying vehicles);

(b) an offence under RTOA 1988, s 25 (information as to date of birth and sex after conviction), s 26 (interim disqualification), or s 27 (production of a licence on conviction);

(c) any offence against any other enactment relating to the use of vehicles on roads; and

(d) manslaughter by the driver of a motor vehicle.

A requirement may be made by way of written notice and served by post. Where it is so made it has effect as a requirement to give the information within the 28-day period beginning on the day on which the notice is served.

Where more than one individual is registered as the keeper of a vehicle (eg where husband and wife are joint keepers), a notice under s 172 may be sent to them jointly; if there is a failure to comply and no defence is shown under s 172(4) or (7)(b) (below) each is guilty of the offence under s 172(3) (below).

The offence

24.73   By RTA 1988, s 172(3), a person who fails to comply with a requirement under s 172(2) commits a offence. However, by s 172(4), a person who *keeps* the vehicle is not guilty of the offence if (s)he shows that (s)he did not know, and could not with reasonable diligence have ascertained, who the driver of the vehicle was. A divisional court has held that the answer to the question of whether the driver's identity was known or could have been ascertained with reasonable diligence is to be assessed from the date the requirement is made and not earlier; the keeper is not required to keep a written record of those to whom (s)he lends the vehicle. Section 172(4) is subject to s 172(6) in the case of a corporate body; s 172(6) provides that s 172(4) does not apply unless the alleged offender also shows that no record of those who drove the vehicle was kept and that the failure to do so was reasonable. Any other person must be proved to have had the information within his/her power to give.

The fact that D does not know the driver's identity does not excuse D if D simply fails to respond to a requirement for information (as opposed to responding and saying that (s)he does not know). There must be evidence of the notice and that a requirement under s 172 has been made. This can be done by annexing a copy of the notice and requirement to a written statement under CJA 1967, s 9. By s 172(7)(b), where the requirement is posted, D is not guilty of the offence if D shows that (s)he gave the information as soon as reasonably practicable after the end of the 28-day period or that it had not been reasonably practicable for him/her to give it. A divisional court has held that it is not necessary to prove that D knew that (s)he was required to provide the specified information, but that in an appropriate case D might be able to show that it was not reasonably practicable for him/her to be aware of the notice, in which case s 172(7)(b) would apply. A judge in the Administrative Court has held that s 172(7)(b) imposes a personal responsibility on the recipient of a requirement under s 172(2) to give information about the driver's identity, and that D had not discharged the burden of proving this by simply leaving a signed notice in an appropriately addressed envelope in the post-room at his place of work in the expectation that it would be posted by the member of staff who was responsible for posting items left there; he had not placed the notice in the hands of the Post Office by posting it or handed it to one of the other private posting companies which provide a comparable service.

The offence under s 172(3) is a summary offence punishable with a fine not exceeding level 3 and with discretionary disqualification for driving (**24.40**).

Procedure in respect of the notice

It is not necessary to specify the nature of the offence alleged to have been committed. **24.74** Where a notice has been produced from an official source and has every appearance of authenticity, justices may draw an inference as to the validity of the notice.

A divisional court has held that the obligation to respond is an obligation to respond in the manner specified in the notice. Thus, for example, if the notice requires (as usual) that information be given in writing and signed, the obligation to respond is not satisfied by providing the required information by telephone.

Where (as usual) a requirement under RTA 1988, s 172(2) has been served by post, RTOA 1988, s 12 permits a court to accept a signed statement as evidence that the signatory was the driver of the vehicle at the time in question. A divisional court has held that where the name of the driver had been inserted in block capitals on the form sent to him, but there was no signature, this could not be a statement in writing purporting to be signed by D for the purposes of s 12 in the circumstances. On the other hand, where a registered keeper replied to a s 172(2) request by endorsing the form sent by the police, 'please see covering letter enclosed', and the letter was signed by him, a divisional court allowed an appeal by the keeper against his conviction for failing to provide the required information.

TRAFFIC ACCIDENTS

Duty to stop, etc

Where, owing to the presence of a mechanically propelled vehicle (MPV) on a road or **24.75** other public place, an accident occurs by which:

(a) personal injury is caused to someone other than the driver of that MPV; or
(b) damage is caused to a vehicle other than that MPV or a trailer drawn by it, or to an animal other than an animal in or on that vehicle or a trailer drawn by it, or to any other property constructed on, fixed to, growing in or otherwise forming part of the land on which the road or other public place in question is situated or land adjacent to such land (RTA 1988, s 170(1)),

the driver of the MPV is required by RTA 1988, s 170(2) to stop and, if required to do so by any person having reasonable grounds for so requiring, give his/her name and address, and also the name and address of the owner and the identification marks of the vehicle. If an accident might occur owing to the presence of two or more vehicles on a road, the driver of each vehicle must stop, etc if the consequences set out above occur.

The driver of a MPV, owing to whose presence on a road or other public place an accident occurs, is not obliged by s 170 (or any other provision) to stop, etc if:

(a) no one (besides that driver) is injured; and
(b) no vehicle (besides that vehicle or its trailer) is damaged; and
(c) no animal (other than one in that vehicle or its trailer) is injured; and
(d) no property forming part, etc of the road or land adjacent to it is damaged.

Section 170(2) places duties only upon the *drivers of MPVs*; the rider of a pedal cycle has no obligations to stop, etc under it (or any other provision).

An accident must have occurred owing to the presence of a mechanically propelled vehicle on a road or other public place

24.76 'MPV' and 'road' were defined at **23.5** and **23.2–23.4** respectively. As to 'other public place', see **26.3**.

For an *accident* to have occurred *owing to the presence* of a MPV on a road or other public place, there must be a direct causal connection between the presence of the MPV and the occurrence of the accident. If the driver of a vehicle brakes sharply on approaching a pedestrian who is crossing the road and the noise of that braking so startles the pedestrian that (s)he falls and is injured, there is a direct causal connection between the vehicle and the accident. Likewise, if a cyclist collides with a motor car waiting at traffic lights, there is a direct causal connection between the presence of the vehicle and the accident in that the cyclist was in contact with it.

A divisional court has held that 'accident' bears its ordinary popular meaning (as opposed to any technical meaning) and that the question is whether an ordinary person would say that an accident has occurred. It also held that there can be an accident even though part of the chain of events leading to the injury or damage was a deliberate act on someone's part. For the obligations in s 170 to apply to the driver of a MPV, it is not necessary that his/her vehicle is involved in an accident in the sense that it is in collision. For example, if D, the driver of a motor car, carelessly drives it across a junction with a major road, causing vehicles on the major road to take avoiding action which causes them to collide with each other, D certainly causes that accident although (s)he is not involved in the collision. In such a case, the obligations in s 170 will have to be discharged by D, as well as by the drivers of the other motor vehicles.

The accident must have caused one of the specified consequences

24.77 As already noted, these specified consequences are:

(1) Personal injury to a person other than the driver of the MPV owing to whose presence on a road or other public place the accident occurred.
(2) Damage to a vehicle other than the MPV, or its trailer, owing to whose presence on a road or other public place the accident occurred.
(3) Damage to any animal other than an animal in or on the MPV, or its trailer, owing to whose presence on a road or other public place the accident occurred. 'Animal' means any horse, cattle, ass, mule, sheep, pig, goat, or dog. Poultry are not included, nor are cats (no matter how valuable they may be).
(4) Damage to any other property constructed on, fixed to, growing in or otherwise forming part of the land on which the road or other public place in question is situated or land adjacent thereto.

It is a question of fact to be determined by the court whether or not property which is damaged forms part of, or is adjacent to, land on which the road or other public place in question is situated. If a car collides with the petrol pumps in a filling station adjacent to the road there is little doubt that a court would consider the petrol pumps to be on land adjacent to land on which a road is situated. Should damage be occasioned to a house which is immediately at the roadside, that house is undoubtedly constructed on land which is at least adjacent to land on which the road is situated. It is probable that, even if a garden exists at the front of the house, a court would consider that the house was constructed on land adjacent to the road. If, because of the presence of a car on a road, an accident occurs and the car crashes through a fence into a field of corn, the accident certainly causes damage to both the fence and the crops which are on land adjacent to the road.

The driver of the motor vehicle must stop and, if required by a person having reasonable grounds for so requiring, must give certain information

'*Driver*' means the person who takes the vehicle out on a road; (s)he remains its 'driver'   **24.78**
until that particular journey is over or someone replaces him/her as 'driver'.

The obligation to stop arises immediately the accident occurs. Thus, in one case, where a bus driver had injured a passenger by braking sharply and had driven on to a rendezvous with an ambulance which he had arranged by radio, a divisional court held that he had failed to stop as required by s 170(2). Because the duty to stop is an immediate one, a driver who leaves the scene and later returns to it does not 'stop' for the purpose of s 170(2).

The term 'stop' means that the driver should stop and remain where (s)he has stopped for a period of time sufficient to enable persons who have reasonable grounds for so doing to require of him/her directly and personally the information which (s)he may be required to supply. In determining what that period should be, particular regard should be paid to the character of the road or the place where the accident occurred. The duty is that of the driver and his/her responsibilities cannot be discharged by someone whom (s)he leaves at the scene. The driver must remain sufficiently near to the vehicle to allow any person to make the requests permitted by the section.

As already implied, the duty to stop does not cast on the driver the duty to go and seek persons to whom to give the above information.

Section 170(2) also requires *the driver to provide certain information if required so to do by 'any person having reasonable grounds for so requiring'*. Such a person will usually be the driver of any other vehicle involved in the accident, a person injured, or the owner of property or of an animal. However, there are other persons who might quite reasonably demand information from the driver of a MPV; for example, a friend of someone who has been injured. It will be a matter for the court to determine in cases of doubt. It may well look differently upon a demand made by a 'nosey parker' as opposed to one made by some person who quite genuinely had the other party's interests at heart.

The information which the driver is required by s 170(2) to give is:

(a) his name and address;
(b) the name and address of the owner of the vehicle; and
(c) the identification mark of the vehicle.

Section 170(2) can be satisfied by giving an address other than a home address, for example the address of the driver's solicitor or the driver's business address, provided the driver can be contacted via that address reasonably swiftly and easily.

This obligation is absolute in the sense that, if the requirement is made by a person having reasonable grounds to make it, the driver must supply this information.

However, if this information is supplied at the scene the obligations of the driver cease and there is no need to report the accident to the police. (If injury to some person is occasioned there is also a need to produce evidence of insurance (**24.81**)).

Duty to report to the police

24.79   RTA 1988, s 170(3) deals with a driver's obligation to report an accident to the police. Its wording is important: if for any reason the driver of the MPV does not give his/her *name and address* as required by s 170(2), (s)he must report the accident. If a driver gives his/her name and address to a person reasonably requiring information, but refuses to give the name and address of the owner or the identification mark of the vehicle, (s)he will not be obliged to report the accident to the police, although (s)he will commit an offence (**24.80**) in consequence of his/her refusal to give the other information.

The obligation to report an accident to the police arises whenever the driver has not given *his/her* name and address at the scene of the accident, whether or not (s)he was requested to do so by anyone there (and indeed, even though (s)he was known personally to any person with reasonable grounds to request his/her name and address). Thus, a divisional court has held, the obligation to report arises even though the driver could not have given his/her name and address at the scene of the accident because he was unconscious throughout the time he was there. The obligation is not negated by the fact that police are in attendance at the scene of an accident from which an unconscious driver is taken to hospital.

The obligation to report is to do so at a police station or to a constable as soon as reasonably practicable and, in any case, within 24 hours of the occurrence of the accident. 24 hours is, therefore, the maximum period in which the report may be made, so that if the driver does not report until 24 hours or more have elapsed since the accident (s)he is necessarily in breach of his/her obligation under s 170(3). Even if the report is made within 24 hours, the driver will nevertheless be in breach of his/her obligation if (s)he has not made it 'as soon as reasonably practicable'. What is 'reasonably practicable' is a matter for the court to decide on in the particular circumstances. If a driver who is not seeking urgent medical attention drives from the scene of an accident and past a police station, in order to go to his home, and only reports the accident 20 hours later, a court is unlikely to find that (s)he reported the accident as soon as reasonably practicable. On the other hand, in a case where a motorist's car left the road and collided, causing damage, at 23.00 and the driver left his address in his car at the scene and was interviewed by the police at 08.30 the next day (at which point of time he reported the accident), a divisional court held that the justices were entitled to conclude that in the circumstances there had been no failure to report as soon as reasonably practicable.

The obligation to report can only be performed officially and personally; telling a friend who is a police officer will not do, nor will a telephone message to a police station.

Offences

24.80   A person who fails to comply with any requirement of RTA 1988, s 170(2) or (3) commits a summary offence against s 170(4). In fact, if (s)he fails to stop (or to give specified information), contrary to s 170(2), and also fails to report the accident to the police, contrary to s 170(3), (s)he commits two offences. The first offence can be committed in two ways, either by failing to stop *or* having stopped, failing to provide the information which is required to be given by a person with reasonable grounds; neither failure is excused by a subsequent report to the police.

A divisional court has held that a person charged with an offence under s 170(4) has a defence if (s)he proves that (s)he did not know that an accident had occurred. However, in the case of the second offence, if, although unaware of the accident at the time of it, (s)he subsequently becomes aware of it within the 24-hour period and fails to report it to the police, (s)he can be convicted of that offence since (s)he will be obliged to report the accident and his/her failure at that time will be with knowledge of the accident. In addition, a divisional court has held that the defence of 'ignorance of accident' is not available where that ignorance is due to the defendant's voluntary intoxication.

An offence against s 170(4) is punishable with up to six months' imprisonment and/or an unlimited fine, and also with discretionary disqualification for driving (**24.40**).

Injury accidents: duty to produce insurance

24.81 RTA 1988, s 170(5) provides that if, in the case of a *personal injury* accident to which s 170 applies by virtue of **24.77**(1), the driver of a *motor vehicle* (other than an invalid carriage) does not at the time of the accident produce a certificate of insurance, or other evidence (as italicised at **24.69**):

(a) to a constable; or
(b) to someone who, having reasonable grounds for so doing, has requested him to produce it,

the driver must report the accident and produce such a certificate or other evidence. It is, of course, insufficient to wave a certificate of insurance in front of the eyes of a person requesting such information; it must be produced in the sense that that person is able to satisfy him/herself that an effective insurance is in force.

As in the case of the duty to report an accident under s 170(3), the driver must report the accident (and produce his/her insurance certificate or other evidence of insurance) at a police station or to a constable and must do so as soon as reasonably practicable, within 24 hours (s 170(6)). It is a summary offence, contrary to s 170(7), to fail to do so. However, the offence is subject to the proviso that a person is not to be convicted of it by reason *only* of a failure to produce a certificate or other evidence if, within seven days after the accident, the certificate is produced at the police station specified by him/her when the accident was reported. An offence under s 170(7) is punishable with a fine not exceeding level 3.

DANGEROUS ACTIVITIES WITH VEHICLES OR ON ROADS

Dangerous parking

24.82 RTA 1988, s 22 makes it an offence to cause or permit a vehicle or trailer drawn by it to remain at rest on a road in such a position, or in such condition or in such circumstances, as to involve danger of injury to other persons. The danger may be caused by the manner of the parking, the condition of the bodywork, etc which may come into contact with persons, or other circumstances. It has been held that, where a vehicle has been parked without the brake being properly set and the vehicle moves and injures a pedestrian, the offence is committed. An offence under s 22 is a summary offence punishable with a fine not exceeding level 3 and also with discretionary disqualification for driving (**24.40**) if committed in respect of a motor vehicle.

Causing danger to road users

24.83 By RTA 1988, s 22A(1), a person commits an indictable (either-way) offence, punishable with up to seven years' imprisonment, if (s)he intentionally and without lawful authority or reasonable cause:

(a) causes anything to be on or over a road; or
(b) interferes with a motor vehicle, trailer or cycle; or
(c) interferes (directly or indirectly) with traffic equipment,

in such circumstances that it would be obvious to a reasonable person that to do so would be dangerous. If the danger would have been obvious to a *reasonable* person, it is irrelevant that the defendant was unaware of it. In relation to (b), it has been held that 'interferes' suggests 'tampering' with the vehicle and that a person who stands in front of a vehicle does not interfere with it.

'*Dangerous*' refers to danger either of injury to any person while on or near a road, or of serious damage to property on or near a road. In determining what would be 'obvious' to a reasonable person in a particular case, regard must be had not only to the circumstances of which (s)he could be expected to be aware but also to any circumstances shown to have been within the defendant's knowledge.

With regard to (a) above, where persons are seen on a bridge over a motorway in possession of large pieces of concrete, and then are seen to *balance* these objects on the parapet, they certainly cause those objects to be over a road and it is submitted that it would be obvious to a reasonable person that what they are doing is dangerous. Protesters who erect a barrier across a road will not normally be guilty of an offence under s 22A if the barrier is solid and easily seen. Although their actions are otherwise unlawful, the essential element of 'danger' is normally missing in such circumstances. However, the stretching of a thin rope or wire across a road would cause the type of danger which the section seeks to outlaw.

A person who interferes with a vehicle's brakes or steering clearly interferes with it for the purposes of (b), above, in such circumstances that it would be obvious to a reasonable person that to do so would be dangerous. The Court of Appeal has held that interference with a vehicle etc does not have to take place prior to its being driven etc, nor does it have to have altered its physical state. Thus, conduct which interferes with a mechanical part of a vehicle, as where a passenger applies the handbrake of a moving car, interferes with the vehicle even though the interfering person is inside it. Interfering with the driver (eg grabbing him), as opposed to the vehicle, is not enough.

For the purposes of (c) above, 'traffic equipment' means:

(a) anything lawfully placed on or near a road by a highway authority;
(b) a traffic sign lawfully placed on or near a road by a person other than a highway authority;
(c) any fence, barrier or light lawfully placed on or near a road (i) to protect street works or undertakings, or (ii) by a constable or person acting on the instructions (whether general or specific) of a chief officer of police.

Any such thing placed on or near a road is, unless the contrary is proved, deemed to have been lawfully placed.

Tampering with motor vehicles

24.84 RTA 1988, s 25 provides that if, while a motor vehicle is on a road or on a parking place provided by a local authority, a person:

(a) gets on to the vehicle, or

(b) tampers with the brake or other part of its mechanism,

without lawful authority or reasonable cause he is guilty of an offence. The offence is a summary offence punishable with a fine not exceeding level 3.

Holding or getting on to vehicle in order to be towed or carried

If, for the purpose of being carried, a person without lawful authority or reasonable **24.85** cause takes or retains hold of, or gets on to, a motor vehicle or trailer while in motion on a road (s)he is guilty of under RTA 1988, s 26(1).

If, for the purpose of being drawn, a person takes or retains hold of a motor vehicle or trailer while in motion on a road (s)he is guilty of an offence under s 26(2).

Both offences are summary offences punishable with a fine not exceeding level 1.

Laser misuse

The Laser Misuse (Vehicles) Act 2018 (LM(V)A 2018), s 1(1) provides that a person **24.86** (D) commits an offence if:

(a) D shines or directs a laser beam towards a vehicle which is moving or ready to move; and

(b) the laser beam dazzles or distracts, or is likely to dazzle or distract, a person with control of the vehicle.

By s 1(2), it is a defence to show that D:

(a) had a reasonable excuse for shining or directing the laser beam towards the vehicle; or

(b) (i) did not intend to shine or direct the laser beam towards the vehicle, and

   (ii) exercised all due diligence and took all reasonable precautions to avoid doing so.

D is taken to have shown a fact mentioned in s 1(2) if sufficient evidence is adduced to raise an issue with respect to it, and the contrary is not proved beyond reasonable doubt (s 1(3)).

For the above purposes, 'vehicle' means any vehicle used for travel by land, water or air. A mechanically propelled vehicle which is not moving or ready to move but whose engine or motor is running is treated for the purposes of s 1(1)(a) as ready to move. In relation to an aircraft, the reference in s 1(1)(b) to 'a person with control of the vehicle' is a reference to any person on the aircraft who is engaged in controlling it, or in monitoring the controlling of it.

Under LM(V)A 2018, s 2, there are corresponding provisions in respect of the protection of air traffic control from laser misuse.

Offences under the LM(V)A 2018 are indictable (either-way) offences punishable with up to five years' imprisonment.

HIGHWAYS LAW

Highways

The meaning of the term 'highway' for the purposes of the Highways Act 1980 (HA **24.87** 1980) is given in HA 1980, s 328. Section 328 states that the term includes the whole or part of a highway other than a ferry or waterway. Where a highway passes over a bridge or through a tunnel, the bridge or tunnel is, for the purposes of the Act, a part of the highway. Section 328 does not define 'highway'; instead, that term is defined by common law (**23.3**).

Obstruction of highway

24.88   By HA 1980, s 137(1), it is a summary offence, punishable with a fine not exceeding level 3, for a person, without lawful authority or excuse, in any way wilfully to obstruct the free passage along the highway.

Whether or not there was an *obstruction* is a question of degree for a court to decide; a complete blockage of the highway is not required. Whether or not a use of the highway amounts to an obstruction depends upon whether or not it was unreasonable having regard to all the circumstances of the case, including where it occurs, its duration, its nature, its extent and its purpose, and whether there is an actual, as opposed to a potential, obstruction. Where a supermarket left trolleys in a pedestrian precinct for the convenience of shoppers there was an obstruction even though no complaint had been made by a member of the public. In another case there was held to be an obstruction where for 15 years a bridleway had been completely blocked by farm gates, tied by twine to hedges and held closed by a loop of twine, which could be opened easily. Simply to cause fear to users of a highway cannot amount to an obstruction. A divisional court so held in a case where D allowed his Rottweiler dogs to act in a menacing way behind a fence separating his land from a path constituting a highway. Pedestrians on the path were put in fear but, said the court, the highway was not obstructed.

The term '*wilfully*' in this context means that the particular obstruction was occasioned by some deliberate act which was freely carried out by D; consequently, a motorist who stops at a traffic light showing red does not wilfully obstruct the highway. In one case, a man who addressed a crowd, whose assembly interfered with traffic although traffic movement was not completely stopped, was held to have caused a wilful obstruction because, by the exercise of free will, he caused that obstruction to take place. On the other hand, if a queue forms outside a shop because many customers are attracted to it, it cannot be said that the shopkeeper has committed a deliberate and wilful act which caused the obstruction, as (s)he is trading in a normal way. It would be different if the obstruction was caused because (s)he was trading in an unusual way, for example through the window, as that unusual act would be wilful and would lead to an obstruction.

The question of whether or not a person has '*lawful authority or excuse*' will always be a question for the court to decide. Clearly, a road block established in accordance with PACE would have lawful authority. 'Lawful excuse' embraces activities lawful in themselves which are a reasonable use of the highway or incidental to the right of passage. The rights to freedom of expression and assembly under the European Convention on Human Rights, arts 10 and 11 do not give an absolute right to obstruct the highway; the rights are qualified ones and whether they provide lawful excuse for such obstruction depends on whether or not interference with those rights is proportionate. The Supreme Court has held that deliberate obstructive conduct by protestors is *capable* of constituting a lawful excuse where the impact of that obstruction on other highway users is more than minimal, and prevents them from passing along the highway (or is capable of doing so). However, the Court stated, whether or not there *is* a lawful excuse in such a case depends on whether or not interference with those rights is proportionate. The Court held that the factors applicable to evaluate proportionality included, among other things, the extent to which the continuation of the protest would breach domestic law, the importance of the precise location to the protesters, the duration of the protest, the degree to which the protesters occupy the land, and the extent of the actual interference the protest causes to the rights of others, including the property rights of the owners of the land, and the rights of any members of the public.

Danger or annoyance on highway

HA 1980, s 161 makes it an offence for any person, without lawful authority or excuse: **24.89**

(a) to deposit anything on a highway in consequence of which a user is injured or endangered (s 161(1)); or

(b) to light any fire on or over a highway which consists of or comprises a carriageway, or to discharge any firearm or firework within 50 feet of the centre of such a highway, if in consequence a user of the highway is injured, interrupted or endangered (s 161(2)).

Section 161A(1) prohibits a person from lighting a fire on any land not forming part of a highway which consists of or comprises a carriageway, or from directing or permitting such a fire to be lit, when in consequence a user of any highway which consists of or comprises a carriageway is injured, interrupted or endangered by, or by smoke from, that fire or any other fire caused by that fire. However, s 161A(2) provides a defence if it can be proved that at the time the fire was lit the person was satisfied on reasonable grounds that it was unlikely that users of any such highway would be injured, interrupted or endangered by the fire or by smoke from it, or from any other fire caused by it, and *either* that before or after the fire was lit (s)he did all that (s)he reasonably could to prevent the consequences *or* that (s)he had reasonable excuse for not doing so. Section 161A is aimed at preventing smoke clouds, caused by straw burning in surrounding fields, blowing over motorways and main roads.

Section 161(4) punishes those who, without lawful authority or excuse, allow filth, lime or dirt, or other offensive matter or thing, to run on to a highway from any adjoining premises.

Section 161(3) prohibits the playing of football or any other game on a highway to the annoyance of a user of the highway. This offence creates many problems for police officers. Usually a complaint is received by telephone from a householder who wishes to complain that (s)he is being annoyed by children playing football in the street. The offence is only committed when the game is 'to the annoyance of a user of the highway'; a person sitting in his/her house is not such a 'user'. A person who was not using the highway when annoyed cannot be subsequently annoyed by walking into the street unless the game continues to his/her annoyance when on the highway.

It is an offence contrary to s 162 to place, for any purpose, a rope, wire, or other apparatus across a highway in such a manner as to be likely to cause danger to persons using the highway, unless D proves that (s)he had taken necessary means to give adequate warning. A washing line in a back street with white sheets suspended from it certainly involves the stretching of a rope across the highway, but in the circumstances it can easily be proved that adequate warning was given. However, if the washing is taken in and only the rope is left, it will be harder to prove that the necessary steps were taken to give adequate warning.

The above offences are summary offences. An offence under s 161A(1) is punishable with an unlimited fine, and the others with fines not exceeding the following levels: 1 (s 161(3) and (4)) and 3 (ss 161(1) and (2) and 162).

Builders' skips

24.90   HA 1980, s 139 provides that a builder's skip must not be deposited on the highway without the permission of the highway authority; otherwise an offence is committed under s 139(3).

Such permission must be in writing; it is granted to a named person and the highway where the skip is to be located is specified. Each skip must be authorised on each occasion; a general permission to place skips is not possible. The authority may impose conditions on its permission. These may refer to the siting of the skip, its dimensions, the painting of the skip to make it visible, the care and disposal of its contents, the manner in which it is to be lighted or guarded and its removal at the end of the authorised period. Skips must be fitted with two oblong plates of diagonal red and yellow fluorescent or reflective material, similar to those fitted to heavy goods vehicles.

Section 139 also provides that, whatever the conditions imposed, an owner must secure that the skip is properly lit during the hours of darkness, that it is clearly and indelibly marked with the owner's name and with his telephone number or address, that it is removed as soon as possible after it has been filled, and that all conditions of the permission are complied with. Failure to do so is an offence under s 139(4) on the part of the owner. The term 'owner' in relation to a skip which is hired for a month or more, or one which is the subject of a hire-purchase agreement, means the person in possession of the skip under the hiring agreement.

Where the commission by any person of an offence under s 139 is due to the act or default of some other person, that other person is also guilty and may be convicted whether or not proceedings are taken against the first-mentioned person.

It is a defence to a charge under s 139 for D to prove that the commission of the offence was due to the act or default of another and that D took all reasonable precautions and exercised all due diligence to avoid the commission of the offence by him/herself or any other person under his/her control.

Section 140 provides that, regardless of whether or not permission has been obtained, the highway authority or a constable in uniform may require the owner of a skip to remove it, or reposition it, or cause it to be removed or repositioned. To fail to do so as soon as possible is an offence against s 140(3). A request by a constable must be made in person; a request by telephone, for example, is not sufficient.

The above offences are summary offences punishable with a fine not exceeding level 3.

Use of loudspeakers

24.91   The Control of Pollution Act 1974, s 62 bans the operation of loudspeakers in a street between 9 pm and 8 am the following morning for any purpose. It also bans their use at any other time for the purpose of advertising any entertainment, trade or business. A person who operates or permits a loudspeaker to be operated in breach of either ban commits a summary offence under s 62(1). Such an offence is punishable with an unlimited fine plus a further fine not exceeding £50 for each day on which the offence continues after the conviction.

The term 'street' means any highway and any other road, footway, square, or court which is for the time being open to the public. The ban is therefore effective both in built-up areas and in the country, provided, in either case, the loudspeaker is operated in a 'street'.

The exemptions to these provisions are predictable. They are: use for police, fire and rescue or ambulance purposes; by the Environment Agency (or the Natural Resources

Body for Wales) or a water undertaking in the exercise of its functions, or by a local authority; 'in-vehicle' entertainments or announcements; telephones; use by showmen at a pleasure fair; and use in an emergency provided that the loudspeaker is operated so as not to give reasonable cause for annoyance to persons in the vicinity.

The most common exception to the general rule against the use of loudspeakers for advertising is in favour of those who have a loudspeaker fixed to a vehicle used for the conveyance of a perishable commodity for human consumption, which is used solely for the purpose of informing the public (by means other than words) that the commodity is on sale from the vehicle. Such a loudspeaker must be operated so as not to give reasonable cause for annoyance to persons in the vicinity and it may only be used between noon and 7 pm of the same day. Thus 'ice-cream chimes' can be used between these times, but at no other.

SPECIAL EVENTS ON ROADS

The Road Traffic Regulation Act 1984, ss 16A and 16B permit traffic authorities to make temporary orders restricting/prohibiting traffic in connection with sporting events, social events, or entertainments on a road. By s 16C(1), contravention of such an order is an offence. The offence is a summary offence punishable with a fine not exceeding level 3. **24.92**

An order may not be made in relation to motor racing unless a motor race order (**26.35**) is made in relation to the 'race or trial', and may only be made in respect of a motor 'competition or trial' or a bicycle race or time trial if authorised by or under RTA 1988, s 13 or s 31 (**26.36** and **25.92**).

OFF-ROAD DRIVING

It is a summary offence, punishable with a fine not exceeding level 3, against RTA 1988, s 34(1) for a person to drive a mechanically propelled motor vehicle, without lawful authority, on to or on any common land, moorland or land of any other description which is not a part of the road, or on any road which is a footpath, bridleway, or restricted byway. However, s 34 provides that driving within fifteen yards of a road upon which a motor vehicle may be driven is not an offence if the driving is only for the purpose of parking. Nor is it an offence under s 34 for a person with an interest in the land, or a visitor to it, to drive to it along a road which before 2 May 2006 (England) or 16 November 2006 (Wales) was shown in a definitive map and statement as a road used as a public path. It is a defence to prove that the vehicle was driven for saving life, or extinguishing fire, or for dealing with any other similar emergency. **24.93**

The Police Reform Act 2002, s 59 provides powers for the seizure of vehicles in prescribed circumstances: **26.14**.

TOUTING FOR HIRE CAR SERVICES

The Criminal Justice and Public Order Act 1994, s 167(1) makes it an offence to solicit persons, in a public place, to hire vehicles, whether licensed taxis or otherwise, to carry them as passengers in circumstances other than those described at **24.31** in which a taxi would be operating lawfully. It is a defence under s 167(4) for a defendant to show that (s)he was soliciting for passengers to be carried at separate fares by PSVs on behalf of the holder of a PSV operator's licence for those vehicles whose authority (s)he had. **24.94**

The soliciting need not refer to any particular vehicle. The mere display of a sign is not soliciting.

An offence against s 167(1) is a summary offence punishable with a fine not exceeding level 4.

FIXED PENALTY OFFENCES

24.95   The offences under the following provisions dealt with in this chapter are fixed penalty offences under the Road Traffic Offenders Act 1988 (and their fixed financial penalties are):

(a) Highways Act 1980, s 137 (obstruction of highway by vehicle: **24.88**): £30;

(b) RTA 1988, ss 22 and 34 (dangerous parking of trailer and off-road driving: **24.82** and **24.93**): £50;

(c) RTA 1988, ss 22 and 87(1) (dangerous parking of motor vehicle and driving otherwise than in accordance with licence: **24.10** and **24.82**): £100;

(d) RTA 1988, s 172 (failure to identify driver: **24.73**): £200;

(e) Public Passenger Vehicles Act 1981, s 12, Goods Vehicles (Licensing of Operators) Act 1995, s 2 and RTA 1988, s 143 (use of PSV or goods vehicle on a road without operator's licence and uninsured use of motor vehicle: **24.32**, **24.36**, and **24.55**): £300.

As to fixed penalty offences, see **25.117–25.133**.

CHAPTER 25

Regulation of Vehicles

TERMINOLOGY

'Motor vehicle', 'mechanically propelled vehicle' (MPV), and 'road' have already been **25.1** defined (**23.1–23.8**). The following terms also used in this chapter are defined as set out below by the Road Traffic Act (RTA) 1988 and/or Road Vehicles (Construction and Use) Regulations 1986 (RV(C&U)R 1986).

A '*motor car*' is an MPV for the carriage of a load or passengers whose unladen weight does not exceed: (a) 3,050 kg (3 tonnes), if it is constructed solely for the carriage of passengers and adapted to carry no more than seven of them exclusive of the driver, or if it is constructed for the conveyance of goods; or (b) 2,540 kg (2.5 tonnes) in any other case.

A '*heavy motor car*' is an MPV (not being a motor car) constructed to carry a load, whether goods or passengers, and whose unladen weight exceeds 2,540 kg (2.5 tonnes). Although this definition covers buses and coaches, there are some requirements of the regulations which are specific to them. For this reason 'bus', 'large bus', and 'coach' are separately defined.

A '*bus*' is a motor vehicle constructed or adapted to carry more than eight seated passengers in addition to the driver, and a '*large bus*' is a vehicle constructed or adapted to carry more than 16 such passengers. A '*coach*' is a large bus with a maximum gross weight of more than 7.5 tonnes and a maximum speed exceeding 60 mph.

A '*dual-purpose vehicle*' is a vehicle constructed or adapted for the carriage both of passengers and of goods or burden, being a vehicle whose unladen weight does not exceed 2,040 kg, and which either:

(a) is so constructed or adapted that the driving power of the engine is, or can be, transmitted to all the wheels of the vehicle; or

(b) (i) is permanently fitted with a rigid roof;

    (ii) has at least one row of transverse, upholstered seats for two or more passengers permanently fitted;

    (iii) has on each side and at the rear a window or windows having an overall area of not less than 1,850 square cms on each side and not less than 770 square cms at the rear; and

    (iv) is such that the distance between the rearmost part of the steering wheel and the back-rests of the row of transverse seats in (ii) (or, if there is more than one such row of seats, the distance between the rearmost part of the steering wheel and the back-rests of the rearmost such row) is not less than one-third of the distance between the rearmost part of the steering wheel and the rearmost part of the floor of the vehicle.

A '*motor tractor*' is defined as a MPV which is not constructed itself to carry a load (other than one concerned with its own propulsion or maintenance) and the unladen weight of which does not exceed 7,370 kg. A '*locomotive*' is a similar vehicle, the unladen weight of which exceeds 7,370 kg (7.25 tonnes); there are two types: 'light locomotive'—unladen weight not exceeding 11,690 kg (11.5 tonnes), 'heavy locomotive'—unladen weight in excess of this.

ROAD VEHICLES (CONSTRUCTION AND USE) REGULATIONS 1986

25.2   RV(C&U)R 1986 are made under RTA 1988, s 41. They are divided into two main parts: Part II dealing with the construction, equipment, and maintenance of motor vehicles and trailers, and Part IV dealing with the various uses of motor vehicles and trailers on roads.

The 'construction elements' of the regulations are aimed at manufacturers in the main, as they amount to specifications for vehicle production. Many of these provisions are being replaced in relation to most modern vehicles by requirements relating to 'type approved' vehicles. The date of manufacture or first registration of vehicles will normally be the conclusive factor in deciding whether the 'construction elements' of RV(C&U)R 1986 apply, or those directed towards type approval. A vehicle which complies with the relevant type approval and has the relevant certificate of type approval or manufacturer's certificate of conformity to the type approval is exempt from certain of the 'construction elements'.

Certain special types of vehicle are authorised under the Road Vehicles (Authorisation of Special Types) General Order 2003 and do not have to comply with all of the requirements of the RV(C&U)R 1986. The clue to the nature of such vehicles lies within the term 'special'. The vehicles concerned are likely to be track-laying vehicles; those used for special engineering and maintenance purposes; military vehicles; and vehicles used for life-saving operations.

The type approval system is backed up by two summary offences under RTA 1988, ss 63 and 65:

(a) using, or causing or permitting to be used, on a road, a vehicle subject in whole or part to compulsory type approval, without a certificate; and

(b) selling, supplying, offering to sell or supply, or exposing for sale a vehicle which does not have a certificate.

These offences are punishable respectively with a fine not exceeding level 4 and an unlimited fine.

It is the provisions of RV(C&U)R 1986 about the equipment and maintenance, and use, of vehicles with which police officers are much more frequently concerned. They are dealt with at **25.12–25.33**.

CONSTRUCTION AND USE OFFENCES: GENERAL

25.3   RTA 1988, ss 40A, 41A, 41B, 41D, and 42 provide various offences which may be committed in relation to the use of a vehicle in a dangerous condition and/or in contravention of the RV(C&U)R 1986.

Using vehicle in dangerous condition

25.4   RTA 1988, s 40A makes it an offence for a person to use, cause or permit another to use a motor vehicle or a trailer on a road when:

(a) the condition of the motor vehicle or trailer, or of its accessories or equipment; or

(b) the purpose for which it is used; or

(c) the number of passengers carried by it, or the manner in which they are carried; or

(d) the weight, position or distribution of its load, or the manner in which it is secured,

is such that the use of the motor vehicle or trailer involves danger of injury to any person. The question of whether there was a danger for the purposes of s 40A is a matter which the court must consider by reference to the matters set out above.

Offences against s 40A are summary offences punishable with an unlimited fine in the case of a goods vehicle or a vehicle adapted to carry more than eight passengers, and otherwise with a fine not exceeding level 4. They are also punishable with discretionary disqualification for driving (**24.40**). However, if a s 40A offence is committed within three years of a previous conviction for such offence, disqualification is obligatory (**24.39**).

Breach of requirements as to brakes, steering-gear, or tyres

RTA 1988, s 41A makes it an offence for a person to: **25.5**

(a) contravene or fail to comply with a construction and use requirement as to brakes, steering gear or tyres; or

(b) use on a road a motor vehicle or trailer which does not comply with such a requirement, or cause or permit a motor vehicle or trailer to be so used.

Offences against s 41A are triable and punishable with a fine in the same way as an offence under s 40A (**25.4**). They are also punishable (see **24.40**) with discretionary disqualification for driving.

Breach of requirement as to weight: goods and passenger vehicles

RTA 1988, s 41B provides offences in relation to 'weight'. By s 41B(1), a person who: **25.6**

(a) contravenes or fails to comply with a construction and use requirement as to any description of weight applicable to:

(i) a goods vehicle; or

(ii) a motor vehicle or trailer adapted to carry more than eight passengers; or

(b) uses on a road a vehicle which does not comply with such a requirement, or causes or permits a vehicle to be so used,

commits an offence. Where the alleged contravention relates to any description of weight applicable to a goods vehicle, it is a defence to prove either that the vehicle was going to or coming from a weighbridge, or, where the relevant weight limit is not exceeded by more than 5 per cent, that this weight was not exceeded at the time of loading and this load had not been added to (s 41B(2)).

Offences against s 41B are summary and punishable with an unlimited fine. They do not carry disqualification.

Breach of requirements as to control of vehicle, mobile telephones, etc

RTA 1988, s 41D makes it an offence for a person to contravene or fail to comply with   **25.7** construction and use requirements (a) as to a driver being in such a position that (s) he can retain proper control of the vehicle, or a full view (**25.25** (s 41D(a)), or (b) as to not using a mobile telephone (**25.31**) (s 41D(b)). Offences against s 41D are summary offences punishable with a fine not exceeding level 4 in the case of a goods vehicle or a vehicle adapted to carry more than eight passengers, and otherwise with a fine not exceeding level 3. They are also punishable (see **24.40**) with discretionary disqualification for driving.

Breach of other construction and use requirements

25.8   RTA 1988, s 42 is concerned with contraventions of a construction and use requirement other than one dealt with by ss 41A, 41B, and 41D. It provides that a person who:

(a) contravenes or fails to comply with such a requirement; or
(b) uses on a road a motor vehicle or trailer which does not comply with such a requirement, or causes or permits a motor vehicle or trailer to be so used,

commits an offence. An offence under s 42 is a summary offence punishable with a fine not exceeding level 4 if committed in respect of a goods vehicle or vehicle adapted to carry more than eight passengers or, in any other case, level 3. It does not carry disqualification for driving.

An offence of using on a road a trailer which does not comply with regulations is an offence distinct from that of using a defective motor vehicle.

Use

25.9   This term should be given its ordinary meaning. A person uses a vehicle if (s)he controls, manages, operates it, or otherwise has the use of it as a vehicle.

RTA 1988, ss 40A to 42 impose strict liability upon those who use motor vehicles in contravention of RV(C&U)R 1986; consequently, it is irrelevant that the defendant (D) did not know of the defect, etc, nor ought to have known of it.

In law, an employer (E) can 'use' a vehicle in breach of RV(C&U)R 1986, even though E does not do so personally. This is because use of a vehicle by an employee in the course of employment is regarded in law as use by E as well (and E can therefore be held vicariously liable for it, regardless of whether E caused or permitted it).

The appeal courts have consistently refused to impose vicarious liability for 'using' outside the employer/employee relationship. Where such a strict employer/employee relationship cannot be established, as where the use is by a contractor, a charge of 'causing' or 'permitting' must be considered.

Where a vehicle is hired out with a driver for use by another firm, the driver will usually remain the employee of the owner of the vehicle, so that a 'using' by the driver will be a 'using' by the owner. Thus, where a haulier (H) hired a vehicle together with its driver to another haulier for a period of five years, and that other haulier operated the vehicle under its own livery, H was held to be using that vehicle when it was found to be overloaded. H had been in the business of hiring out vehicles with drivers who remained their employees and the vehicles were being used in the course of that business. Similarly, where the driver of a vehicle who was employed by company A was ordered by telephone to pick up a return load on behalf of company B, company B being responsible for loading the vehicle, its documentation, and the selection of route, a divisional court held that company A was guilty of using the vehicle when it was found to be overloaded.

As is apparent from the above, someone who lends his/her private motor car to his/her friend is not liable for 'using' that vehicle if RV(C&U)R 1986 are contravened by the friend.

Where a trailer is towed by another vehicle and the offence concerned relates to the trailer (eg a defective tyre or insecure load), the information or written charge should specify that it was the trailer which was defective or improperly used. This is essential where the tractor unit and the trailer are owned by different persons or companies.

Cause the use

This term, together with 'permitting the use', covers circumstances in which charges of **25.10** aiding and abetting would otherwise normally be appropriate. To be guilty of causing the use of a vehicle in contravention of RV(C&U)R 1986, D must actually have known of the contravention in question, and have been in a position to exercise some control over the driver. To 'cause' involves some express or positive command or direction from the person 'causing' to the person 'using'. If the foreman of a haulage depot, knowing that a vehicle has a defective tyre, orders an employee to drive the vehicle, the foreman 'causes' the driver to commit the offence. He is in a position to give orders to the employee in relation to the use of the vehicle and does so, and he knows of the defect. In such a case the haulage company would also be guilty of using the vehicle as it was used on their business and knowledge is not an essential ingredient of that offence.

A person who tows another vehicle is causing it to be used on a road.

Permit the use

Like 'causing the use' of a vehicle, 'permitting' a vehicle to be used in contravention of **25.11** RV(C&U)R 1986 requires proof that D actually knew of, or was wilfully blind as to, the contravention in question. However, the meaning of 'permitting' is much wider than the meaning of 'causing' because there does not need to be any express or positive mandate to use a motor vehicle by a person in a position to exercise control over the driver, and a general or particular permission to use a vehicle will be sufficient. However, for a person to be guilty of 'permitting' (s)he must be in a position to forbid another person to use the vehicle.

Many employees are given general permission to use their employers' vehicles in any way they choose in their own time. Having given such a general permission, it is unlikely that such an employer will be liable for contraventions of the RV(C&U)R 1986 in respect of that vehicle, but in some circumstances (s)he may. There is a distinction between knowledge that the vehicle is being used and knowledge that it is being used unlawfully. A sales representative (A) may be given total use of a motor car. As a driver A will be responsible for its use in such contravention but A's employers will not normally be liable for *permitting* such use, as they are unlikely to have knowledge of the particular unlawful use. This can be compared with the case where the driver is unlawfully using the vehicle in the course of employment; there a charge of using will be appropriate.

Someone who lends his/her car to his/her friend permits him/her to use it. That permission does not extend to unlawful use of a warning instrument as no permission has been given in that respect. On the other hand, if the tyres were defective, it would be a question of fact as to whether the owner of the vehicle knew of the defect when permission was given to use it.

PARTICULAR REGULATIONS UNDER RV(C&U)R 1986, PART II

Brakes

The differing braking systems with which different types of vehicles must be fitted are **25.12** dealt with in reg 15 (which applies to almost all vehicles with three or more wheels) and reg 16 and Sch 3 (which apply to motor bicycles and other vehicles). These provisions include exemptions from RV(C&U)R 1986 if type approval has been given. However, the fitting of various systems is not a matter of prime concern to police officers. On the other hand, the maintenance of brakes is. Regulation 18 deals with the maintenance of brakes and applies quite generally to braking systems (where fitted under these, or other, regulations).

Regulation 18(1) requires every braking system and the means of operation thereof fitted to a vehicle to be maintained in good and efficient working order, and properly adjusted. This applies to any vehicle to which a braking system is fitted, including a trailer which is not required to have a braking system fitted.

By reg 18(3) and (6) respectively, braking systems must be so maintained that service braking systems and secondary braking systems, and parking brakes, respectively have prescribed braking efficiencies set out in considerable detail in two tables in the regulations. The tables appear in Chapter 25 on the companion website.

Speedometer

25.13 By RV(C&U)R 1986, reg 35 every motor vehicle (with very limited exceptions) must be fitted with a speedometer. Regulation 35 also provides that speedometers fitted to vehicles first used after 1 April 1984 must indicate speed in both miles per hour and kilometres per hour; otherwise the indication given by a speedometer may be either in miles per hour or kilometres per hour. Instead of complying with reg 35, a vehicle may comply with the relevant Community Directive or ECE Regulation.

The following vehicles do not need to have speedometers fitted:

(a) invalid carriages, motor cycles not exceeding 100 cc, and works trucks (provided in each case they were first used before 1 April 1984);

(b) agricultural vehicles, not being a category T tractor (**25.16**), which are not driven at more than 20 mph;

(c) category T tractors which are not driven at more than 40 km/h;

(d) vehicles which legally or physically cannot exceed 25 mph; and

(e) vehicles first used before 1 October 1937.

Regulation 36 requires that a speedometer must be kept free from any obstruction which might prevent its being easily read and must at all times be maintained in good working order. On a charge involving a breach of this requirement, it is a defence to prove that the defect occurred in the course of the journey being then undertaken or that steps had already been taken to have repairs or replacements effected as soon as possible. These defences cover so many possibilities, and if alleged are so hard to disprove, that enforcement is almost limited to admitted long-standing defects in respect of which no repairs have been arranged or when the instrument registers incorrectly.

Mirrors

25.14 RV(C&U)R 1986, reg 33 sets out detailed requirements about the fitting of mirrors. Instead of complying with the above requirements about mirrors, a vehicle may comply with the relevant Community Directive or ECE Regulation. The various requirements are of particular relevance to the manufacturers and others in the motor trade.

Windscreen wipers and washers

By RV(C&U)R 1986, reg 34(1), all vehicles fitted with windscreens must be fitted with **25.15** one or more efficient automatic windscreen wipers, unless the driver can obtain an adequate view to the front without looking through the windscreen, for example by opening the windscreen or looking over it (as is the case with many cars produced before the 1950s). Regulation 34(1) also requires that the wipers fitted must be capable of clearing the windscreen so that the driver has an adequate view of the road in front of both sides of the vehicle and to the front of the vehicle.

Regulation 34(6), which deals with maintenance, requires that every windscreen wiper required by the Regulations to be fitted is, at all times while the vehicle is used on a road, maintained in good and efficient working order and is properly adjusted. As stated, reg 34(1) demands one or more wipers. If the driver's wiper provides the view ahead and to both sides, as required, any other wiper is not subject to the maintenance requirement.

By reg 34(2), a windscreen washer must be fitted to any vehicle which requires one or more automatic wipers, except:

(a) an agricultural motor vehicle, not being a category T tractor (**25.16**), unless it is first used on or after 1 June 1986 and is driven at more than 20 mph;
(b) a category T tractor, unless it is first used on or after 1 June 1986 and is driven at more than 40 km/h;
(c) a vehicle incapable of exceeding 20 mph;
(d) a track-laying vehicle; or
(e) a vehicle being used to provide a local bus service.

The washer must be such that, in conjunction with the wipers, it is capable of cleaning the area of the windscreen swept by the blades of mud or similar deposits.

In the alternative to the above requirement, a vehicle may comply with the relevant Community Directive.

Audible warning instrument

The following provisions are made by RV(C&U)R 1986, reg 37. **25.16**

All motor vehicles with a maximum speed of more than 20 mph must be fitted with a horn (but, in the case of an agricultural motor vehicle, only if it is being driven at more than 20 mph, and in the case of a category T tractor, only if it is being driven at more than 40 km/h). A category T tractor is a vehicle defined in EU Regulation 167/2013 and includes wheeled agricultural and forestry vehicles.

In the case of motor vehicles first used on or after 1 August 1973 the sound emitted by any horn must be continuous and uniform and not strident. This is to ensure that the horns of such vehicles give a sound which is not offensive to the ear.

With the exception of police vehicles, NCA vehicles, fire and rescue vehicles, vehicles used for ambulance purposes or for the purpose of providing a response to an emergency at the request of an NHS ambulance service, vehicles of HM Revenue & Customs used in the investigation of serious crime, and vehicles owned and operated by the Secretary of State for Defence and used by the special forces in response (or training or practising in responding) to a national security emergency, no motor vehicle may be fitted with a bell, gong, siren, or two-tone horn. However, vehicles may be fitted with an instrument or apparatus (not being a two-toned horn) to give a sound informing the public that goods are for sale from the vehicle; thus, ice-cream chimes, etc are not prohibited (but see also **24.91**). The rule against fitting gongs, etc does not apply to an anti-theft device nor to an alarm to summon help to a public service vehicle. Any anti-theft device fitted must have a cut-out limiting its use to five minutes. Where a horn can operate as an anti-theft device, such a cut-out must be fitted if the motor vehicle was first used on or after 1 October 1982.

A 'reversing alarm' fitted to a vehicle to warn persons that the vehicle is reversing or about to reverse, or a 'boarding aid alarm' fitted to a bus to warn that a power-operated ramp or lift is in operation is exempted from the requirement that the sound of a warning instrument shall be continuous and uniform but it must not be strident. The sound emitted must be such that it is not likely to be confused with the sound emitted from a pedestrian crossing (reg 99). A reversing alarm may only be used on a stationary vehicle if the vehicle's engine is running and it is about to move backwards or the vehicle is in danger from another moving vehicle.

Horns and other audible warning instruments must be maintained in good and efficient working order at all times.

Instead of complying with the provisions mentioned above concerning the nature and maintenance of audible warning instruments, a vehicle may comply with the relevant Community Directive or ECE Regulation.

Regulation 99 provides that no person shall sound, or cause or permit to be sounded, the horn or other audible warning device of a motor vehicle when it is stationary on a road at any time. The only exception is when there is danger due to another moving vehicle on or near the road. In addition, it provides that no person shall sound, or cause or permit to be sounded, in a vehicle in motion on a restricted road, a horn, etc, between 23.30 and 07.00 on the following day. This does not, of course, apply to anti-theft alarms, reversing devices, or boarding aid alarms.

Silencer

25.17   By RV(C&U)R 1986, reg 54, every vehicle propelled by an internal combustion engine must be fitted with an exhaust system including a silencer, and the exhaust gases from the engine must not escape into the atmosphere without first passing through its silencer. Specified noise levels are prescribed but they are directed at the manufacturers. Instead of complying with the above provisions, a vehicle may comply with the relevant Community Directive or ECE Regulation. Regulation 54 also requires that every exhaust system and silencer must be maintained in good and efficient working order and must not after the date of manufacture be altered so as to increase the noise made by the escape of exhaust gases.

Although not directly concerned with silencers, regs 97 and 98 also deal with noise. Regulation 97 prohibits the use, etc of a motor vehicle in a manner so as to cause excessive noise which could reasonably have been avoided by the driver. Regulation 98 requires the driver of a stationary vehicle to stop the action of machinery attached to it, or forming part of the vehicle, so far as may be necessary for the prevention of noise or of exhaust emissions.

Wings

25.18   RV(C&U)R 1986, reg 63 requires the following to be fitted with wings or similar fittings to catch, as far as practicable, mud or water thrown up by the wheels:

(a) invalid carriages;
(b) heavy motor cars, motor cars and motor cycles, not being agricultural motor vehicles or pedestrian-controlled vehicles;
(c) agricultural motor vehicles, not being category T tractors, driven at more than 20 mph;
(d) category T tractors (**25.16**) driven at more than 40 km/h; and
(e) trailers.

The requirements in relation to trailers are restricted to the rear wheels, if the trailer has more than two wheels.

Works trucks, unfinished vehicles, broken down vehicles, living vans, agricultural trailers not drawn in excess of 40 km/h and agricultural trailed appliances are exempt from the above requirements. Trailers used for or in connection with the carriage of round timber and the rear wheels of the motor vehicle parts of articulated vehicles whose semi-trailer is so used are also exempt.

Instead of complying with these requirements, a vehicle may comply with the relevant Community Directive.

Dangerous vehicles, etc

RV(C&U)R 1986, reg 100(1) requires that a motor vehicle, every trailer and all parts **25.19** and accessories of such vehicle or trailer are at all times in such a condition that no danger is caused or is likely to be caused to any person in or on the vehicle or trailer or on a road. The provision is concerned with actual or likely danger; a mere possibility of danger is not enough.

Regulation 100(1) applies to a vehicle which in its manufactured condition is inherently likely to cause danger to other road-users, as well as to a vehicle which has become dangerous through lack of maintenance. In relation to the requirement that all parts and accessories must be in such condition that no danger is caused, it is immaterial that parts may be in reasonable condition if they are dangerous because they are not in working order. Separate and distinct offences are involved in using a defective motor vehicle and using a defective trailer. This equally applies to articulated vehicles.

Regulation 100(1) also requires that the number of passengers carried, or the manner of their carriage, is such that no danger is caused or is likely to be caused to any person in or on the vehicle or trailer or on a road.

Regulation 100(1) also requires that the weight, packing, distribution, and adjustment of a load is such that no such danger is caused, or likely to be caused. This prohibition is really aimed at a badly loaded vehicle; insecure loads are dealt with by reg 100(2).

Regulation 100(2) requires that the load carried by a motor vehicle or trailer must at all times be so secured, if necessary by physical restraint other than its own weight, and be in such a position, that neither danger nor nuisance is likely to be caused to any person or property by reason of the load or any part of it falling or being blown from the vehicle or by reason of any other movement of the load or a part of it. In effect, this deals with insecure loads which for any reason cause danger or nuisance. It is necessary to specify which of these alternatives resulted from the insecurity of the load. Where a substantial part of a solid load falls onto a busy road there is little doubt that danger is caused. In circumstances where a goods vehicle, loaded with gravel, sheds part of its load steadily into the path of following traffic, there is no doubt that there is a breach of the requirement so to secure the load that no nuisance to persons or property is likely.

In considering whether a load has been adequately secured so that neither danger nor nuisance is likely to be caused one has to consider four things:

(a) the nature of the journey;
(b) the way in which the load was secured;
(c) the way in which the load was positioned; and
(d) the journey to be taken.

What might be secure for one journey in fine weather and on good roads might not be secure for another journey in poor weather and on less good roads. If the load is a high one, and too high to go under a bridge on the route, so that it is inevitable that the load will be knocked off, the load will not have been so secured that neither danger nor nuisance was likely.

Regulation 100(3) prohibits the use of a motor vehicle or trailer for any purpose for which it is so unsuitable as to cause danger or nuisance to any person on the vehicle or trailer, or on a road. The crucial question is whether the vehicle or trailer was unsuitable. If it is not properly loaded and it is the load which causes the danger or nuisance, this does not affect the suitability of the vehicle itself. In a case where an excavator which was loaded onto a trailer struck a bridge, it was held that the trailer did not become unsuitable when it was so loaded. It was the manner of the loading which was at fault; reg 100(1) was the relevant provision.

Unlike a conviction for the dangerous vehicle offence under RTA 1988, s 40A (**25.4**), for a conviction under s 42 (**25.8**) for a breach of reg 100(1) there can be no disqualification.

Maintenance of glass

25.20   All glass or other transparent material fitted to motor vehicles is required, by RV(C&U) R 1986, reg 30(3), to be maintained in such condition that it does not obscure the vision of the driver while the vehicle is being driven on a road. This regulation is not limited to the windscreen but applies to all glass affecting the driver's vision, such as the rear window and side windows through which the driver needs to look.

Maintenance of fuel tank

25.21   By RV(C&U)R 1986, reg 39 a fuel tank on a motor vehicle must be maintained so that any leakage of liquid or vapour from it is adequately prevented (except that this does not prevent a tank being fitted with a pressure release valve). Instead of complying with these requirements, a vehicle may comply with the corresponding requirements contained in the relevant Community Directive or ECE Regulation.

Maintenance of tyres

25.22   Except as provided below, RV(C&U)R 1986, reg 27(1) prohibits the use, on a road, of a motor vehicle or trailer, a wheel of which is fitted with a pneumatic tyre, if:

(a) the tyre is unsuitable for the use to which the vehicle or trailer is being put or to the type of tyres fitted to its other wheels;
(b) the tyre is wrongly inflated for such use;
(c) the tyre has a cut in excess of 25 mm or 10 per cent of the section width of the tyre, whichever is the greater, measured in any direction on the outside of the tyre and deep enough to reach the ply or cords;

(d) the tyre has a lump, bulge, or tear caused by separation or partial failure of its structure;

(e) the tyre has a portion of the ply or cord exposed;

(f) the base of any groove which showed in the original tread pattern is not clearly visible;

(g) either—

    (i) the grooves of the tread pattern of the tyre do not have a depth of at least 1 mm throughout a continuous band measuring at least three-quarters of the breadth of the tread and round the entire outer-circumference of the tyre; or

    (ii) if the grooves of the original tread pattern of the tyre did not extend beyond three-quarters of the breadth of the tread, any groove which showed in the original tread pattern does not have a depth of at least 1 mm;

(h) the tyre is not maintained in such condition as to be fit for the use to which the vehicle or trailer is being put or has a defect which might in any way cause damage to the surface of the road or damage to persons on or in the vehicle or to other persons using the road;

(i) the tyre is not a retreaded tyre and—

    (i) the week of manufacture marked on its sidewall in accordance with the relevant ECE Regulation falls more than 10 years before the date on which the motor vehicle is used on the road; or

    (ii) it does not have a week of manufacture marking which complies with that Regulation;

(j) the tyre is a retreaded tyre and—

    (i) the week of retreading marked on its sidewall in accordance with the relevant ECE Regulation falls more than 10 years before the date on which the motor vehicle is used on the road; or

    (ii) it does not have a week of retreading marking which complies with that Regulation; or

(k) a date marked on the tyre sidewall is illegible.

There are general exemptions from (a) to (g) or (i) to (k) in favour of agricultural trailers or trailed appliances, agricultural motor vehicles (not being category T tractors) which are not driven at more than 20 mph, category T tractors (**25.16**) which are not driven at more than 40 km/h, and broken-down vehicles (or vehicles en route to a breakers) being towed at a speed not exceeding 20 mph.

In addition, (f) and (g) do not apply to:

(i) a passenger vehicle, other than a motor cycle, constructed or adapted to carry no more than eight seated passengers in addition to the driver;

(ii) a goods vehicle with a maximum gross weight which does not exceed 3,500 kg; or

(iii) a light trailer not falling within (ii),

first used on or after 3 January 1933. Instead, the grooves of the tread pattern must be of a depth of at least 1.6 mm throughout a continuous band comprising the central three-quarters of the breadth of the tyre and round the entire outer-circumference of the tyre. Nor do (f) and (g) apply to three-wheelers not exceeding 102 kg unladen weight with a maximum possible speed of 12 mph, nor to pedestrian-controlled works trucks. In addition, (g) does not apply to motor cycles not exceeding 50 cc.

Nothing in (i) to (k) applies to a vehicle of historical interest used for non-commercial purposes; 'vehicle of historical interest' means a vehicle which is considered to be of historical interest to Great Britain and which was manufactured or registered for the first time at least 40 years previously, is of a type no longer in production, and has been historically preserved or maintained in its original state without substantial changes in the technical characteristics of its main components.

Subject to the general exemptions from (a) to (g) and (i) to (k) and to the exemption above for historic vehicles of historic interest:

(a) list entries (i) and (j) only apply to tyres fitted—
    (i) to a front steered axle of a bus other than a minibus;
    (ii) in single configuration on any axle of a minibus. 'Single configuration' means where no more than one wheel and tyre assembly is mounted at the end of the axle concerned;
    (iii) to a front steered axle of a goods vehicle with a maximum gross weight exceeding 3,500 kg.
(b) list entry (k) only applies to—
    (i) buses (including minibuses);
    (ii) goods vehicles with a maximum gross weight exceeding 3,500 kg.

Regulation 27(5) prohibits the fitting on a motor vehicle or trailer of a recut pneumatic tyre if its ply or cord has been cut or exposed in the process of recutting. Heavy-duty tyres fitted to goods vehicles can usually be recut without causing danger, but tyres fitted to saloon cars do not usually have sufficient thickness of rubber to accept the process.

Regulation 27(2) makes special provision to allow for the use of tyres designed to be used safely when deflated.

Special provision is made by reg 27(3) to allow passenger vehicles (not buses) to fit a temporary-use spare tyre, provided the vehicle does not exceed 50 mph.

Mixture of tyres

25.23   RV(C&U)R 1986, reg 26(1) provides that pneumatic tyres of different types of construction must not be fitted to the same axle of a vehicle. The regulation mentions three types of tyre:

(a) diagonal-ply tyres, which are commonly known as cross-ply tyres;
(b) bias-belted tyres, which are as above but have a reinforcing band around the outer-circumference of the tyre under the tread, but on top of the cords; and
(c) radial-ply tyres.

Regulation 26(2) provides that a motor vehicle having only two axles must not be fitted with the following combinations of tyres:

(a) cross-plys or bias-belted tyres on the rear axle and radials on the front; or
(b) cross-ply on the rear axle and bias-belted on the front.

This regulation is designed to prevent a gripping tyre from being fitted to the 'steering' wheels, whilst a tyre which is not so stable is fitted to the rear.

Not to emit smoke, etc

25.24   RV(C&U)R 1986, reg 61(5) provides that no person shall use, or cause or permit to be used, on a road a motor vehicle from which is emitted any smoke, visible vapour, grit, sparks or oily substance which causes, or is likely to cause, damage to property or injury or danger to any person who is, or who may reasonably be expected to be, on the road. This could occur where oil spillage made the road unsafe or fumes badly affected visibility.

Regulation 61(1) provides that a vehicle must be maintained so as not to emit any avoidable smoke or avoidable visible vapour.

CONDITIONS UNDER R(C&U)R 1986, PART IV RELATING TO USE

Position to retain proper view and control

RV(C&U)R 1986, reg 104 provides that a driver must at all times be in such a position **25.25** that he retains full control and has a full view of the road and traffic ahead. Someone who drove with a sheepdog on his lap has been held to be in breach of this provision.

Reversing

By RV(C&U)R 1986, reg 106, no person shall drive, or cause or permit to be driven, a **25.26** motor vehicle backwards on a road further than may be requisite for the safety or reasonable convenience of the occupants of the vehicle or other traffic.

Unnecessary obstruction

RV(C&U)R 1986, reg 103 provides that a person in charge of a motor vehicle or trailer **25.27** must not cause or permit it to stand on a road so as to cause any unnecessary obstruction. The various exemptions within local no-waiting orders in favour of goods vehicles which are loading or unloading has led to a belief by some van drivers that they may double park and even completely block roads for these purposes. This is not so. Any unreasonable use of the road which leads to the obstruction of other road users is an offence against reg 103. An obstruction can be caused by taking up so much of the road that normal two-way traffic is reduced to a one-way flow. In considering whether a particular use was reasonable or unreasonable, a court will apply its mind to the facts, including the duration of the obstruction, the nature of the place where it occurred, the purpose for which the vehicle was there and the *actual* (as opposed to potential) obstruction caused.

Parking facing the wrong way at night

RV(C&U)R 1986, reg 101 provides that, except with the permission of a uniformed **25.28** police officer, a person must not cause or permit a motor vehicle to stand on a road between sunset and sunrise otherwise than with the left or nearside of the vehicle as close as may be to the edge of the carriageway. If lights are not displayed in a place where parking lights are required to be shown, two offences are committed.

The exemptions from the prohibition under reg 101 are predictable: vehicles being used for fire and rescue or police purposes; vehicles being used for ambulance purposes or for the purpose of providing a response to an emergency at the request of an NHS ambulance service; taxi stands and bus stops; one-way streets; emergency operations, etc.

Stopping engine and setting parking brake

RV(C&U)R 1986, reg 107 provides that no person shall leave, or cause or permit to **25.29** be left, on a road any motor vehicle which is not attended by a person duly licensed to drive it, unless the engine has been stopped and the parking brake effectively set.

The exceptions include vehicles being used for fire and rescue authority or police purposes; vehicles being used for ambulance purposes or for the purpose of providing a response to an emergency at the request of an NHS ambulance service; and special vehicles which use their engines for particular operations other than the driving of the vehicle. Although there is a requirement to do two things, that is stop the engine and set the handbrake, the failure to carry out either of these operations completes the offence under RTA 1988, s 42.

Opening of doors

25.30 RV(C&U)R 1986, reg 105 provides that *no person* may open, or cause or permit to be opened, the door of a motor vehicle on a road so as to cause injury or danger to any person. It is not necessary to prove negligence: if the action causes injury or danger that is sufficient.

Mobile telephones

25.31 RV(C&U)R 1986, reg 110(1) prohibits the driving of a motor vehicle on a road if the driver is using a hand-held mobile telephone, or a hand-held device of any kind, other than a two-way radio, which performs an interactive communication function by transmitting and receiving data.

For the purposes of reg 110, a mobile telephone or device is classified as 'handheld' if it is, or must be, held at some point during the course of making or receiving a call or performing any other interactive communication function. The term 'interactive communication function' includes:

(a) sending or receiving oral or written messages, or
(b) facsimile documents, or
(c) still or moving images; and
(d) providing access to the internet.

The term 'two-way radio' refers to wireless telegraphy apparatus which is designed or adapted for the purpose of transmitting or receiving spoken messages and to operate on any frequency except a number of ranges. Such equipment is fitted to vehicles used by the emergency services and is therefore exempted from these provisions.

Regulation 110(2) prohibits causing or permitting any other person to drive a motor vehicle on a road while using a hand-held mobile telephone or hand-held device of the specified type. Regulation 110(3) forbids a person supervising a provisional licence holder to use a hand-held mobile telephone or such a hand-held device at a time when the provisional licence holder is driving a motor vehicle on a road.

Regulation 110(5) provides that there is no contravention of reg 110 where:

(a) the use is to call the emergency service on 112 or 999;
(b) it is in response to a genuine emergency; and
(c) it is unsafe or impracticable to cease driving in order to make the call, or for the provisional licence holder to cease driving while the call is made.

In addition, reg 110(5A) provides that a person does not contravene reg 110 if, at the time of the alleged contravention:

(a) the mobile telephone or other device is used only to perform a remote-controlled parking function of the motor vehicle; and

(b) that mobile telephone or device only enables the vehicle to move where the following conditions are satisfied—

   (i) there is continuous activation of the remote-control application of the telephone or device by the driver;

   (ii) the signal between the vehicle and the telephone or device is maintained; and

   (iii) the distance between the vehicle and the telephone or device is not more than six metres.

A divisional court has held that a driver who, while driving, had held a mobile phone in order to film the scene of an accident did not contravene RV(C&U)R 1986, reg 110. Regulation 110, the court stated, does not prohibit all use of a mobile phone held while driving; it only prohibits driving while using a mobile phone or other device for calls and other interactive communication and holding it at some stage during that process.

Trailers

25.32 RV(C&U)R 1986, reg 86 provides that, where a motor vehicle is drawing a trailer by means of a rope or chain, the length of the rope or chain must not exceed 4.5 m, and must not exceed 1.5 m unless the rope or chain is made clearly visible to other road users within a reasonable distance from either side.

Regulation 86A provides that trailers to which reg 15 (**25.12**) applies, which are not fitted with an automatic stopping device operative in the event of separation, must not be used on a road unless a secondary coupling is attached to the drawing vehicle and its trailer in such a way that, in the event of separation, the drawbar of the trailer would be prevented from touching the ground and there would be some residual steering of the trailer.

Trailers to which reg 15 applies which are fitted with a device which is designed to stop the trailer automatically in the event of separation but is dependent on a secondary coupling must not be used on a road unless that coupling is properly attached to the drawing vehicle and trailer.

Trailers which are living vans and have less than four wheels (or have two close-coupled wheels on each side) are prohibited by reg 90 from being used for the carriage of passengers. The only exception is in favour of testing, when a repairer, etc may be carried in the van for that purpose.

Regulation 83 limits the number of trailers which may be drawn by a motor vehicle on a road. It limits locomotives to a maximum of three trailers, motor tractors to one laden trailer or two unladen trailers, and heavy motor cars and motor cars to one trailer. For the purposes of reg 83, a vehicle drawn by a steam-powered vehicle and which is used solely to carry water for the purpose of the drawing vehicle is not a trailer.

Miscellaneous

25.33 RV(C&U)R 1986, reg 53 provides that mascots, emblems, or other ornamental objects fitted to vehicles first used on or after 1 October 1937 must not be in such a position that they are likely to strike persons with whom the vehicle may collide unless the nature of the mascot is such that injury is not liable to be caused.

Regulation 109 provides that no person shall drive, or cause or permit to be driven, a motor vehicle on a road if the driver is in such a position as to be able to see, whether directly or by reflection, a television receiving apparatus or other cinematographic apparatus used to display anything other than information concerning the vehicle or its journey.

WEARING OF SEAT BELTS

25.34   The circumstances in which seat belts must be worn by passengers in motor vehicles are now prescribed in the Motor Vehicles (Wearing of Seat Belts) Regulations (MV(WSB) R) 1993 and the Motor Vehicles (Wearing of Seat Belts by Children in Front Seats) Regulations (MV(WSBCFS)R) 1993 and not by RV(C&U)R 1986.

The offences committed by breaches of the Seat Belt Regulations are punishable under RTA 1988, ss 14 and 15 and not by RTA 1988, s 42. They are summary offences punishable with a fine not exceeding level 2. They do not carry disqualification for driving.

Adult

25.35   Subject to the exemptions below, MV(WSB)R 1993, reg 5 requires that every person:

(a) driving a motor vehicle (other than a two-wheeled motor cycle with or without a sidecar); or

(b) riding in a front or rear seat of a motor vehicle (other than a two-wheeled motor cycle with or without a sidecar),

must wear an adult seat belt. The regulation does not apply to a person under the age of 14, or if no adult belt is provided (driver) or available (passenger: **25.38**).

Failure to comply with reg 5 is an offence under RTA 1988, s 14(3) (**25.34**).

Regulation 6 provides the following exemptions from the requirement to wear a seat belt made by reg 5:

(a) a person holding a medical certificate to the effect that it is inadvisable for him/her to wear a seat belt;

(b) the driver of or a passenger in a motor vehicle constructed or adapted for carrying goods, while on a journey which does not exceed 50 metres and which is undertaken for the purpose of delivering or collecting anything;

(c) a person performing a manoeuvre which includes reversing;

(d) a qualified driver supervising a learner carrying out such a manoeuvre;

(e) a person by whom a test of competence to drive is being conducted where wearing a seat belt would endanger himself or another person;

(f) a person driving or riding in a vehicle being used for fire and rescue, or police or NCA purposes, or for carrying a person in lawful custody (who is also included in this exemption);

(g) a person riding in a motor ambulance while the person is providing medical attention or treatment to a patient which due to its nature or the medical situation of the patient cannot be delayed;

(h) the driver of a licensed taxi whilst being used to seek hire or answer a call, or when carrying a passenger for hire, or a private hire car whilst it is being used to carry a passenger for hire;

(i) a person in a vehicle being used under a trade licence for investigating or remedying a fault;

(j) a disabled person who is wearing a disabled person's belt; or

(k) a person riding in a vehicle whilst in a procession organised by or on behalf of the Crown.

There are two other exemptions under reg 6.

First, a person who is riding in a vehicle which is participating in a procession which is held to mark or commemorate an event is exempt from the requirement to wear a seat belt if the procession is one commonly or customarily held in a police area or areas, or if notice in respect of the procession has been given in accordance with the Public Order Act 1986, s 11 (**18.23**).

Second, the second seat belt requirement in reg 5, which relates to a person riding in a front or rear seat of a motor vehicle other than a two-wheeled motor cycle with or without a sidecar, does not apply to a person riding in a small or large bus which is being used to provide a 'local service' in a built-up area (ie entirely on restricted roads) or which is constructed or adapted for the carriage of standing passengers and on which the operator permits standing. For the purposes of the seat belt regulations, '*large buses*' and '*small buses*' are 'buses' as defined at **25.36**. The difference between them is that a large bus has a maximum laden weight exceeding 3.5 tonnes and a small bus has a maximum laden weight not exceeding 3.5 tonnes.

Front-seat child passengers

It is an offence, contrary to RTA 1988, s 15(1) and (2), for a person, without reason- **25.36** able excuse, to drive a motor vehicle on a road if a child under 14 is in the front of that vehicle, unless the child is wearing a seat belt in conformity with MV(WSBCFS) R 1993, reg 5. As to trial and punishment see **25.34**. The Regulations set out the rules with which a driver must conform. A 'front seat' is one which is wholly or partially in the front of the vehicle. The Regulations do not apply to two-wheeled motor cycles, with or without sidecars.

The description of the seat belts to be worn by front-seat child passengers set out in MV(WSBCFS)R 1993, reg 5 is as follows:

(a) in the case of a 'small child', the seat belt must be a child restraint (i) with a marking required under RV(C&U)R 1986, reg 47(7), or (ii) approved by an EU member state, or (iii) with markings required under UN Economic Commission for Europe (ECE) Regulation 129 about uniform provisions concerning the approval of enhanced child restraint systems if the markings indicate that it is suitable for his/her height, weight, and age;

(b) in the case of a 'large child', the seat belt must be a child restraint as described in (a) (i) or (iii), or an adult belt.

For the purposes of the Regulations, a '*small child*' is a child who is under 12 and under 135 cm in height, and a '*large child*' is a child under 14 who is not a 'small child'.

MV(WSBCFS)R, reg 7 exempts from these provisions:

(a) a 'small child' aged three or more who is riding in a bus and is wearing an adult belt if an appropriate seat belt is not available for him/her in the front or rear of the vehicle;

(b) a child for whom there is a medical certificate; or

(c) a disabled child who is wearing a disabled person's belt.

In addition, RTA 1988, s 15(1) does not apply in relation to a child riding in a bus which is being used to provide a local service in a built up area (**25.35**), or which is constructed or adapted for the carriage of standing passengers and on which the operator permits standing. The prohibition does not apply in relation to a 'large child' if no appropriate seat belt is available for him/her in the front of the vehicle.

RTA 1988, s 15(1A) requires that where:

(a) a child is in the front seat of a motor vehicle other than a bus;

(b) that child is in a rear-facing child restraining device; and

(c) the passenger seat where the child is placed is protected by a front air bag, a person must not, without reasonable excuse, drive the vehicle on a road unless the air bag is deactivated. 'Deactivation' includes the case where a bag is designed or adapted in such a way that it cannot inflate enough to pose a risk of injury to a child travelling in a rear-facing child restraining device in the seat in question. It is an offence contrary to s 15(2) for the person to drive a motor vehicle in contravention of s 15(1A) (**25.34**).

For seat belt purposes, '*bus*' means a motor vehicle with at least four wheels, constructed or adapted for the carriage of passengers, with more than eight seats in addition to the driver's seat, and with a maximum design speed exceeding 25 kph.

Rear-seat child passengers

25.37   RTA 1988, s 15(3) provides that a person must not, without reasonable excuse, drive a motor vehicle on a road if a child:

(a) under the age of three is in the rear of that motor vehicle; or
(b) of or over that age but under the age of 14 is in the rear of the motor vehicle and any seat belt is fitted in the rear of that vehicle,

unless the child is wearing a seat belt in conformity with MV(WSB)R 1993, reg 8. Regulation 8 is identical to the requirements which apply to children in front seats, ie MV(WSBCFS)R 1993, reg 5 (**25.36**).

RTA 1988, s 15(3A) provides that a person must not, without reasonable excuse, drive a passenger car on a road where a child under the age of 12 and less than 150 cm in height is in the rear of the car and no seat belt is fitted in the rear of that car, but a seat in the front of the passenger car is provided with a seat belt and is not occupied by any person.

Exemption from these prohibitions is provided by MV(WSB)R 1993, regs 9 and 10. Regulation 9 provides that they do not apply where the motor vehicle is a large bus (**25.35**), or where the vehicle is a licensed taxi or licensed hire car in which (in each case) the rear seats are separated from the driver by a fixed partition. Regulation 10 provides that the prohibitions in s 15(3) and (3A) do not apply in relation to:

(a) a child for whom there is a medical certificate;
(b) a small child (**25.36**) aged under three who is riding in a licensed taxi or a licensed hire car if no appropriate seat belt is available (**25.38**) for him/her in the front or rear of the vehicle;
(c) a small child aged three or more who is riding in a licensed taxi, a licensed hire car, or a small bus (**25.35**) and wearing an adult belt if an appropriate seat belt is not available for him/her in the front or rear of the vehicle;
(d) a small child aged three or more who is wearing an adult belt and riding in a passenger car or light goods vehicle where the use of child restraints by the child occupants of two seats in the rear of the vehicle prevents the use of an appropriate seat belt for that child and no appropriate seat belt is available for him/her in the front of the vehicle;
(e) a small child who is riding in a vehicle being used for the purposes of the police, security or emergency services to enable the proper performance of their duty;
(f) a small child aged three or more who is wearing an adult belt and who, because of an unexpected necessity, is travelling a short distance in a passenger car or light goods vehicle in which no appropriate seat belt is available for him; or
(g) a disabled child who is wearing a disabled person's belt or whose disability makes it impracticable to wear a seat belt where a disabled person's belt is unavailable to him.

By reg 10, the prohibition in s 15(3) does not apply in relation to a child under three riding in a rear seat of a small bus. Nor does it apply to a small child aged three or more riding in the rear of a small bus if neither an appropriate seat belt nor an adult seat belt is available for him/her in the front or rear of the vehicle. In addition, it does not apply in relation to a 'large child' in any vehicle, if no appropriate seat belt is available for him/her in the rear of the vehicle. Lastly, the prohibition in s 15(3) does not apply to a child riding in a small bus which is being used to provide a local service in a built-up area (**25.35**) or which is constructed or adapted for the carriage of standing passengers and on which the operator permits standing.

In addition, by reg 10, the prohibition in s 15(3A) does not apply to a child if no appropriate seat belt is available for him/her in the front of the vehicle.

A '*light goods vehicle*' is a motor vehicle with at least four wheels, a maximum design speed of more than 25 kph, and a maximum laden weight not exceeding 3.5 tonnes.

Breach of RTA 1988, s 15(3) or (3A) is an offence under s 15(4) (**25.34**).

'Availability' of seat belts

MV(WSB)R 1993, Sch 2, and MV(WSBCFS)R 1993, Sch 2, describe in the same terms **25.38** the circumstances in which a seat belt will be regarded as not being available. They are where:

(a) another person is wearing the relevant belt;
(b) a child is occupying the relevant seat and wearing a child restraint which is an appropriate child restraint for that child and this renders use of the seat belt impracticable;
(c) a person holding a medical certificate is occupying the relevant seat;
(d) a disabled person (not being the person in question) is occupying the relevant seat and wearing a disabled person's belt and this renders use of the seat belt impracticable;
(e) by reason of his/her disability, it would not be practicable for the person in question to wear the relevant belt;
(f) the person in question is prevented from occupying the relevant seat by the presence of a child restraint which could not readily be removed without the aid of tools; or
(g) the relevant seat is so designed that it can be adjusted to increase the space available for goods and effects and when it is so adjusted the seat cannot be used.

Wearing of seat belts in buses: additional rule

RTA 1988, s 15B(1) provides that, with the exception below, the operator of a bus **25.39** (**25.36**) in which any of the passenger seats are equipped with seat belts must take reasonable steps to ensure that every passenger is notified of the requirement to wear a seat belt fitted for his/her seat at all times when the bus is in motion. 'Operator' means the owner of the bus or other person in possession of it under a hire, or hire-purchase, conditional sale, loan or other agreement. Notification may be made by an official announcement, or an audio-visual presentation, made when the passenger joins the bus or within a reasonable time thereafter, or by a sign prominently displayed at each passenger seat equipped with a seat belt.

Section 15B(1) does not apply in relation to a bus which is being used to provide a local service in a built-up area (**25.35**) or which is constructed or adapted for the carriage of standing passengers and on which the operator permits standing.

If the operator fails to comply with the requirements of s 15B(1), (s)he commits a summary offence, punishable with a fine not exceeding level 4, against s 15B(4).

LIGHTS AND VEHICLES: INTRODUCTION

25.40 The lighting requirements for vehicles are set out in the Road Vehicles Lighting Regulations 1989 (RVLR 1989).

Terminology

25.41 RVLR 1989 define certain terms as follows for their purposes:

Daytime hours The time between half an hour before sunrise and half an hour after sunset.

Hours of darkness The time between half an hour after sunset and half an hour before sunrise.

Obligatory lamp, reflector, rear marking or device A lamp, reflector, rear marking, or device with which a vehicle, its load, or equipment, is required by the Regulations to be fitted.

Optional lamp, reflector, rear marking, or device A lamp, reflector, rear marking, or device with which a vehicle, its load, or equipment, is not required by the Regulations to be fitted.

Offences

25.42 Contravention of, or failure to comply with, any regulation contained in RVLR 1989 is a summary offence contrary to RTA 1988, s 42, except that where it relates to a pedal cycle it is a summary offence, contrary to the Road Traffic Offenders Act 1988 (RTOA 1988), s 91. Using on a road a motor vehicle or trailer which does not comply with RVLR 1989 or causing or permitting a motor vehicle or trailer to be so used is also an offence under RTA 1988, s 42.

For trial and punishment where the offence relates to a motor vehicle or trailer, see **25.8**. An offence under RTOA 1988, s 91 is punishable with a fine not exceeding level 3.

OBLIGATORY LAMPS, REFLECTORS, REAR MARKINGS, AND DEVICES

Required equipment

25.43 By RVLR 1989, reg 18, a person must not use a vehicle on a road, or cause or permit it to be so used, unless it is equipped with obligatory lamps, reflectors, devices, or markings as specified by Sch 1 to the Regulations. Column 1 of Sch 1 lists the type of lamp, reflector, rear marking, or device required; column 2 the installation (including height and lateral positioning) and performance requirement (by reference to other Schedules); and column 3 any exceptions to the general rule.

Each of the lamps, reflectors, devices, and markings listed below is obligatory for motor vehicles (with exceptions mentioned later) and some of them are obligatory for trailers drawn by motor vehicles and for other vehicles. Such lamps, etc must be fitted in the manner described by the Regulations.

As the requirements have increased, the position of vehicles manufactured before new requirements were made has had to be safeguarded. The various regulations therefore specify the date of application by applying the requirements to vehicles first used on or after a given date. However, reg 4 provides that, even if a motor vehicle is first used on or after that date, it will be exempt from the requirements of the particular regulation if manufactured more than six months in advance of that date. In some instances this provision is expressly repeated in a particular regulation but generally it must be implied into a particular regulation from reg 4.

For the purpose of the Regulations, a lamp is not treated as being a lamp if it is so painted over or masked that it is not capable of being immediately used or readily put to use, or if it is an electric lamp which is not provided with any system of wiring by means of which that lamp is (or can readily be) connected with a source of electricity.

Front lights

Front position lamps

Vehicles with three or more wheels (other than invalid carriages and pedal tricycles) **25.44** require two front position lamps, ie front side lamps.

Pedal cycles, solo motor bicycles, hand-propelled vehicles (see exemption at **25.64** where width does not exceed 800 mm), and invalid carriages require only one front position lamp, which must be fitted on the centre line or offside of the vehicle. Motorcycle combinations with a headlamp on the motor bicycle require a front position lamp on the centre line of the sidecar, or on the side of the sidecar furthest from the motor bicycle. A solo motor bicycle fitted with a headlamp need not be fitted with a front position lamp.

All front position lamps must be white in colour, unless they are incorporated in a yellow headlamp, in which case they may be yellow. They must be visible from a reasonable distance.

Dipped beam headlamps

Two dipped beam headlamps are required by motor vehicles with three or more wheels **25.45** in most circumstances. Solo motor bicycles or combinations, and three-wheelers first used before 1 January 1972, or with an unladen weight of not more than 400 kg and an overall width of not more than 1,300 mm, require only one headlamp on the centre line of the motor vehicle itself. A bus first used before 1 October 1969 need only have one dipped beam headlamp.

The light emitted by a dipped beam headlamp must be white or yellow. The lamp itself must be so constructed that the direction of the beam of light can be adjusted whilst the vehicle is stationary. Where two dipped beam headlamps are required to be fitted, they must form a matched pair and be capable of being switched on and off simultaneously and not otherwise.

Main beam headlamps

The provisions concerning the number of obligatory headlamps to be fitted are the **25.46** same as those relating to dipped beam headlamps, except that a bus must have two main beam headlamps, even if first used before 1 October 1969. The outer edges of the illuminated areas must not be outside those of the dipped beam headlamps; there is no maximum distance which should separate a pair of such lamps.

Main beam headlamps must emit a white or yellow light and must be constructed so that they can be deflected at the will of the driver to become a dipped beam, or so that they can be extinguished by the operation of a device which at the same time switches on a dipped beam or causes another lamp to emit a dipped beam. Thus headlamps of the 'long-range' type are invariably wired so that as they are extinguished a dipped beam is emitted from the normal headlamps. Main beam headlamps must be constructed so that the direction of the beam can be adjusted whilst the vehicle is stationary.

Dim-dip devices

25.47   These are devices capable of causing a dipped beam headlamp to operate at reduced intensity. Dim-dip devices must be provided on motor vehicles with three or more wheels first used on or after 1 April 1987. Vehicles having a maximum speed of 40 mph or less and home forces vehicles are exempt, as are vehicles which comply with Community Directive 76/756/EEC. Running lamps in the form of a matched pair of front lamps may be fitted as an alternative. A running lamp is defined as a lamp (not being a front position lamp, an end-outline marker lamp, headlamp, or front fog lamp) used to make the presence of a moving motor vehicle readily visible from the front; in effect, a higher intensity front position lamp.

Rear lights

Rear position lamps

25.48   Most vehicles with three or more wheels require two rear position lamps, ie rear side lamps. The following vehicles, some of which have three or more wheels, require only one rear position lamp: buses first used before 1 April 1955; solo motor bicycles; pedal cycles with less than four wheels; trailers drawn by pedal cycles; trailers (the overall width of which does not exceed 800 mm) drawn by solo or motorcycle combinations; invalid carriages having a maximum speed not exceeding 4 mph; and vehicles propelled by hand. (See exemption at **25.64** where width does not exceed 800 mm.) Some motor vehicles of maximum speed not exceeding 25 mph, and their trailers, require four rear position lamps; the details are in Sch 10 to the Regulations.

In cases in which only one rear position lamp is required it must be fitted on the centre line of the vehicle or on its offside.

All rear position lamps must be red.

Rear fog lamps

25.49   If it was first used on or after 1 April 1980, any motor vehicle having three or more wheels, and any trailer drawn by a motor vehicle, must, unless specifically dealt with elsewhere in the Regulations, have at least one rear fog lamp fitted, at or near the rear, on the centre line or offside of the vehicle. If two lamps are fitted there is no requirement concerning the distance these lights are placed from the sides of the vehicle. No more than two lamps may be fitted.

Vehicles first used before 1 April 1980 do not need to be fitted with rear fog lamps. Nor do motor vehicles whose maximum speed is 25 mph or less, nor motor vehicles or trailers which are no more than 1,300 mm in width, nor agricultural vehicles or works trucks first used before 1 April 1986. If one of these exempt vehicles is fitted with such lamps there are no restrictions in relation to the number of lamps which may be fitted, but any which are fitted must not be capable of illumination by a braking system.

A rear fog lamp must show a red light. It must not be fitted so that it can be illumin-ated by the application of any braking system of the vehicle. A tell-tale must be fitted to show that the lights are in operation. If two lamps are fitted to a motor vehicle first used on or after 1 April 1986, or on a trailer manufactured on or after 1 October 1985, they must form a matched pair. If two rear fog lamps are fitted, the rules for obligatory fog lamps apply to both of them.

Stop lamps

Stop lamps must show a red light. Motor vehicles having three or more wheels, and **25.50** trailers drawn by a motor vehicle, must, unless otherwise stated, be fitted with two stop lamps. Solo motor bicycles, combinations, invalid carriages, and trailers drawn by motor cycles, together with motor vehicles or trailers first used before 1 January 1971, need only be fitted with one stop lamp. Motor bicycles of less than 50 cc first used be-fore 1 April 1986 do not require any stop lamp, nor does any type of motor vehicle first used before 1 January 1936 or a motor vehicle whose maximum speed is 25 mph or less or an agricultural vehicle or works truck first used before 1 April 1986.

If two stop lamps are fitted they should be on each side of the longitudinal axis of the vehicle. If only one is fitted it should be on the centre line or offside.

The lamps must be operated by the application of the service braking system of the motor vehicle, and this also applies to the stop lamps of any trailer attached to that vehicle. Where two stop lamps are required to be fitted, they must form a pair.

Rear registration plate lamp

All vehicles which are required to be fitted with a rear registration plate must have **25.51** lighting which is capable of adequately illuminating the rear registration plate. This does not apply to a rear registration plate displaying a trailer registration mark (**23.28**) assigned to the trailer.

Other obligatory lamps

Direction indicators

The requirements for direction indicators are as follows: **25.52**

(1) Motor vehicles first used before 1 April 1936 and trailers manufactured before that date *may* have any arrangement of indicators but they are not required to have any at all.
(2) Motor vehicles first used on or after that date and before 1 April 1986, and trailers manufactured between 1 January 1936 and 1 October 1985, *must* be provided with any arrangement of indicators so as to satisfy the requirement of visibility to front and rear (or rear only in the case of a trailer).
(3) Motor vehicles first used on or after 1 April 1986, and trailers manufactured on or after 1 October 1985, are subject to detailed restrictions. Motor vehicles with three or more wheels, other than motorcycle combinations, must have a single front indicator, one side repeater indicator and one rear indicator on each side. One additional optional rear indicator may be fitted and any number of side repeater indicators may be added. Trailers manufactured after the specified date must have a rear indicator on each side. Additional optional indicators may be fitted as above. Motor bicycles and combinations must have a single front and a single rear indi-cator on each side of the vehicle.

The colour of the light shown by direction indicators is amber in most instances. Motor vehicles first used before 1 September 1965, and the trailer of such a vehicle, may show white or amber to the front or red or amber to the rear. However, if such an indicator is visible from both the front and the back it must show an amber light regardless of the date of first use. All indicators on any side of a vehicle or its trailer must be operated by one switch. There must be a tell-tale to show that the indicators are in operation. Flashing indicators must flash constantly at a rate of not less than 60 and not more than 120 times per minute.

Vehicles whose maximum speed does not exceed 15 mph or invalid carriages having a maximum speed not exceeding 4 mph are not required to have direction indicators fitted, nor are vehicles first used before 1 August 1986 which are agricultural vehicles, industrial tractors, or works vehicles.

Agricultural vehicles having an unladen weight not exceeding 255 kg do not require direction indicators.

Daytime running lamps

25.53 These are required on motor vehicles with three or more wheels, other than vehicles first registered before 1 March 2018. Such a lamp is a lamp emitting a white light and facing in a forward direction used to make the vehicle more easily visible when driven during daytime. Daytime running lamps fitted on or after 20 May 2018 must be two in number, fitted on each side of the vehicle's median longitudinal line. They must go off automatically when the headlamps are switched on or the engine is switched off.

Reversing lamps

25.54 The following obligations do not apply to vehicles first registered before 1 March 2018. One reversing lamp at the rear of the vehicle is obligatory on three or more-wheeled motor vehicles designed and constructed for the carriage of passengers and comprising no more than eight seats in addition to the driver's seat, and on all other vehicles with a length not exceeding 6,000 mm. Two are required on all three or more-wheeled motor vehicles with a length exceeding 6,000 mm, except the passenger-carrying vehicles just mentioned. Vehicles which are not required to be fitted with reversing lamps by the Road Vehicles (Approval) Regulations 2020 or retained EU Regulation 2018/858 are, however, exempt.

There are no special requirements as to the arrangement of obligatory reversing lamp(s). An obligatory reversing lamp must have electrical connections such that it can light up only if reverse gear is engaged and if the device which controls the starting and stopping of the engine is in such a position that operation of the engine is possible. It must not light up or remain lit if either of these conditions is not satisfied. A tell-tale is optional.

End-outline marker lamps

25.55 These are lamps fitted near the edge of a vehicle in addition to the front and rear position lamps to indicate the presence of a wide vehicle. They are required on motor vehicles first used on or after 1 April 1991, except those with a maximum speed not exceeding 25 mph; those having an overall width not exceeding 2,100 mm; and incomplete vehicles proceeding to works, etc.

There must be two white lights fitted to the front and two red lights to the rear, not more than 400 mm from the side of the vehicle. Each set must be a matched pair. Any number may be fitted.

Hazard warning signals

These signals are obligatory on motor vehicles having three or more wheels and first **25.56** used on or after 1 April 1986, except vehicles which are not required to be fitted with direction indicators. Hazard warning signals which are optionally fitted to other vehicles must comply with the same provisions. Each device must be operated by one switch which causes all direction indicators with which the vehicle or combination of vehicles is equipped to flash in phase. There must be a tell-tale, and the device must be capable of operation without the ignition being switched on.

Side marker lamps

Such a lamp is a lamp fitted to the side of a vehicle or its load and used to render **25.57** the vehicle more visible to other road users. Some motor vehicles with three or more wheels, and trailers drawn by motor vehicles are required to have side marker lamps on each side. Most are not, since motor vehicles first used before 1 April 1991 and trailers, the overall length of which does not exceed 6m (or 9.15m if manufactured before 1 October 1990), are not required to have side marker lamps, and nor are the following first used on or after that date:

(a) motor vehicles whose maximum speed does not exceed 25 mph;
(b) passenger vehicles;
(c) incomplete motor vehicles or trailers proceeding to a works for completion or to a place to be stored or displayed for sale;
(d) those not exceeding 6m in length;
(e) those first used before 1 April 1996 complying with Community Directive 76/756/ EEC (and trailers manufactured before 1 October 1985 which comply with this Directive);
(f) trailers, whose overall length, excluding any drawbar and any fittings for its attach-ment, exceeds 6m (9.15m if manufactured before 1 October 1990);
(g) agricultural and works trailers; and
(h) caravans and boat trailers.

A side marker lamp is a lamp which will show an amber light, if fitted to a vehicle first used on or after 1 October 1990, unless it is placed within 1m of the rear of the vehicle, when it may be red. Trailers manufactured before that date may have lamps which show a white light to the front and a red light to the rear.

The requirements are to have two, on each side, and as many more as are sufficient to ensure that the maximum distance from the front of the vehicle (including any drawbar) to the first lamp is 4m and the maximum distance from the rear in respect of the rearmost side marker lamp is 1m. The maximum separation distance of adjacent obligatory lamps on the same side of the vehicle is 3m or, if this is not practicable, 4m.

A vehicle, or a combination of vehicles whose overall length (including any load) exceeds 18.3m, must have additional side marker lamps, one lamp being no more than 9.5m from the foremost part of the vehicle or vehicles and one lamp no more than 3.05m from the rear (including loads in both circumstances). Other lamps must be placed to ensure that no more than 3.05m separates the lamps.

Where the length exceeds 12.2m but not 18.3m and the load is supported by any two vehicles, lamps must be placed behind the rearmost part of the drawing vehicle, but not more than 1,530 mm to the rear of that point. If the supported load extends more than 9.15m to the rear of the drawing vehicle, the lamp must not be forward of, or more than 1,530 mm to the rear of, the centre of the length of the load. These provisions do not apply to articulated vehicles.

Obligatory reflectors, etc

Rear retro reflectors

25.58   Generally, all vehicles must be equipped with two rear retro reflectors, ie reflectors used to indicate the presence and width of a vehicle when viewed from the rear. By way of exception, the following only require one retro reflector: solo motor bicycles; pedal cycles with less than four wheels (with or without a sidecar); trailers drawn by pedal cycles; trailers not exceeding 800 mm drawn by solo motorcycles or combinations; and invalid carriages having a maximum speed not exceeding 4 mph. In relation to hand-propelled vehicles there is an exemption from the requirement where their width does not exceed 800mm (see **25.64**). Some restricted vehicles (maximum speed 25 mph) require four reflectors.

Reflectors must be fitted at or near the rear of the vehicle. Where there is only one reflector it must be on the centre line or offside of the vehicle.

The colour of all rear reflectors must be red. A triangular-shaped rear reflector may only be fitted to a trailer or broken-down vehicle under tow.

Side retro reflectors

25.59   Side reflectors, ie reflectors fitted to the side of a vehicle or its load and used to render the vehicle more visible from the side, are obligatory on certain motor vehicles with three or more wheels and their trailers. The relevant requirements do not apply to a passenger vehicle (including private cars), nor to an incomplete vehicle travelling to a works for completion, etc, mobile cranes, plant and certain earth-removal vehicles, nor to a vehicle having a maximum speed not exceeding 25 mph. Nor do they apply to a goods vehicle whose overall length does not exceed 6m (if first used on or after 1 April 1986) or 8m (if first used before that date).

The requirements therefore are restricted to long goods vehicles. In the case of such vehicles, first used before 1 April 1986, and trailers manufactured before 1 October 1985, there must be two side retro reflectors on each side of the vehicle. In the case of those first used or manufactured on or after the relevant date, there must be two on each side of the vehicle and as many more as are required by Sch 17 to the Regulations. Side retro reflectors must be amber or, within 1m of the rear of the vehicle, they may be red. They must not be triangular in shape.

Rear markings

25.60   The rear markings provisions relate to 'long vehicle' markings, and therefore most vehicles are exempted from them. In the case of those motor vehicles first used before 1 April 1996 which are not exempted, those which do not exceed 13m in length need only carry the marker boards with diagonal lines, whereas those which exceed 13m must carry marker boards with the words 'long vehicle' in black on a yellow background surrounded by a red border or, as an alternative, boards of yellow retro reflective material surrounded by a red fluorescent border. Those used on or after 1 April 1996, which are not exempted, must carry boards of red and yellow diagonal stripes if they do not exceed 13m in length; if they exceed 13m, they must carry boards of yellow retro reflective material surrounded by a red fluorescent border. The same markings are required on certain trailers forming part of a combination of vehicles. In the case of a trailer manufactured on or after 1 October 1995 which does not, in combination, exceed 11m, it must carry a board of red and yellow diagonal stripes; if it exceeds 11m but not 13m it must carry boards of such red and yellow stripes or of yellow surrounded by red; and if it exceeds 13m, the boards must be of yellow surrounded by red. In the case of a trailer manufactured before 1 October 1995, the overall length does not exceed 11m, the marking must be of the diagonal line variety; if between 11m and 13m the marking may be of any approved variety; and if it exceeds 13m it must be the 'long vehicle' type, or a board of yellow reflective material surrounded by a red fluorescent border.

A vehicle fitted with conspicuity markings to the rear where the fitting complies with the relevant UN Economic Commission for Europe (ECE) conspicuity requirements is exempt from the rear marking requirements.

Conspicuity markings

A goods vehicle which is—   **25.61**

(a) a motor vehicle first used on or after 10 July 2011, or
(b) a trailer manufactured on or after 10 July 2011,

must be fitted with conspicuity markings (devices intended to increase the conspicuity of a vehicle, when viewed from the side or rear, by the reflection of light emanating from a light source not connected to the vehicle, the observer being situated near the source, ie high visibility 'conspicuity marking tape') complying with the relevant ECE conspicuity requirements.

This requirement does not apply to a goods vehicle which is—

(a) a motor vehicle with a maximum gross weight not exceeding 7,500kg;
(b) a trailer with a maximum gross weight not exceeding 3,500kg;
(c) an incomplete vehicle proceeding to a works for completion or to a place where it is to be stored or displayed for sale; or
(d) a motor car or heavy motor car intended to form part of an articulated vehicle.

A goods vehicle to which the above requirement applies is known as a 'relevant goods vehicle'.

If the overall length of a relevant goods vehicle does not exceed 6m, conspicuity markings need not be fitted to the side of that vehicle.

If the overall width of a relevant goods vehicle does not exceed 2.1m, conspicuity markings need not be fitted to the rear of that vehicle.

Pedal retro reflectors

Pedal cycles manufactured on or after 1 October 1985 must be provided with two   **25.62**
amber reflectors on each pedal.

Headlamp cleaning device

A headlamp cleaning device is a complete device with which all or part of the light   **25.63**
emitting surface of a headlamp or an adaptive front lighting system can be cleaned. An adaptive front lighting system is a type-approved lighting device providing beams with differing characteristics for automatic adaptation to varying conditions of use of the dipped-beam (passing beam) and, if it applies, the main-beam (driving-beam).

Such devices are required on motor vehicles with three or more wheels, other than vehicles which are not required to be so fitted by the Road Vehicles (Approval) Regulations 2020 or EU Regulation 2018/858. However, the requirements relating to such devices do not apply in the case of a headlamp cleaning device fitted to a lamp, other than a gas discharge headlamp or an adaptive front lighting system.

Fitting and performance

25.64   RVLR 1989 require that obligatory lamps, reflectors, markings, and devices are fitted and performing satisfactorily at all times. However, there are certain understandable exceptions to this rule. The Regulations do not require any lamp or reflector to be fitted between sunrise and sunset to:

(a) an incomplete vehicle proceeding to a works for completion,

(b) a pedal cycle,

(c) a pedestrian-controlled vehicle,

(d) a horse-drawn vehicle,

(e) a combat vehicle (ie a tank, armoured car, field gun, and the like, and vehicles constructed for the carriage of such a thing or of other weapons or ammunition), or

(f) a vehicle being taken to a testing station by an authorised vehicle examiner in order to submit the vehicle for an examination there in order to ensure that the examination carried out at the station is correctly carried out, or a vehicle being removed by an examiner following that examination..

To that list is added a vehicle which is not fitted with any front or rear position lamps, so that if a person is building a kit car and has not yet installed such lights (s)he will not commit an offence by using the vehicle on a road during daylight hours. In addition, the regulations relating to fitting do not apply to a vehicle based outside Great Britain which is on a journey in this country, provided that it has not been here for more than 12 months and that it complies with international requirements. Vehicles going to a port for export are similarly exempt. Hand-propelled vehicles whose overall width does not exceed 800 mm do not require lamps or reflectors between sunset and sunrise if used close to the nearside of the carriageway, or to cross the carriageway.

OPTIONAL LAMPS, REFLECTORS, REAR MARKINGS, AND DEVICES

25.65   The Regulations do not generally prevent the fitting of optional lamps, etc, ie lamps, etc which are not obligatory under the Regulations. Regulation 20 provides that every optional lamp, etc fitted to a vehicle must comply with provisions set out in the respective Schedules to the Regulations. A table included in the Regulations describes the types of optional lamps, etc and requires compliance with certain parts of the directions given in the relevant Schedule.

Optional lamps, etc can be divided into two categories:

(a) those which are additional to obligatory lamps, etc; and

(b) those which are not additional in that sense; there is no absolute requirement to have any lamp, etc of the particular type.

Additional lamps, etc

25.66   The position is:

(1) Any number of optional front position lamps may be fitted (except that a solo motor bicycle first used on or after 1 April 1991 must not be fitted with a total of more than two front position lamps). They must be white or, if incorporated in a yellow headlamp, yellow.

(2) Any number of optional dim-dip devices, or (where fitted on or after 20 May 2018) daytime running lamps may be fitted.

(3) Two optional dipped beam headlamps may be fitted to a vehicle with three or more wheels and a maximum speed exceeding 25 mph first used on or after 1 April 1991; any number may be fitted to other vehicles. These lamps must be white or yellow, comply with maximum and minimum height requirements, and be capable of adjustment while the vehicle is stationary.

(4) Any number of optional main beam headlamps may be fitted. They must be white or yellow, electrically connected so that they deflect or extinguish by the dip switch, and be capable of adjustment while the vehicle is stationary.

(5) Any number of optional rear position lamps may be fitted. They must be red.

(6) The number of optional rear fog lamps is controlled in the case of motor vehicles first used on or after 1 April 1980 (or, in the case of trailers, manufactured on or after 1 October 1979). No more than two such lamps are permitted in such a case. There is no limit in respect of other vehicles. Rear fog lamps must be red.

(7) Any number of stop lamps can be fitted. If additional lamps are fitted they must comply with the provisions for obligatory stop lamps, except those relating to position and angles of visibility. Motor vehicles first used on or after 1 April 1991 are subject to control in relation to the intensity of the light projected through the rear windows. Rear stop lamps must be red.

(8) Any number of side retro reflectors may be fitted. They must not be triangular. They must be amber except that, in respect of a vehicle used for the following purposes, the following colours are permitted:
- Police—amber, yellow, blue, white*
- Fire and Rescue Authority—amber, yellow, red
- Ambulance, or a vehicle providing emergency response at request of an NHS ambulance service (but only if vehicle owned, leased or hired by service)—amber, yellow, green, white*
- DVSA (Driver and Vehicle Standards Agency)—amber, yellow, silver, white*
- Traffic officer— amber, yellow, white.

(*Or, if within 1m of the rear of the vehicle, red.)
They must not be triangular.

(9) Any number of rear retro reflectors may be fitted. Optional rear retro reflectors must be red, except that, in respect of a vehicle used for police, ambulance, fire and rescue authority, DVSA or traffic officer purposes, or in respect of a vehicle providing emergency response at request of an NHS ambulance service (but only if vehicle owned, leased or hired by service), the following colours are permitted: red, yellow or orange (or any combination). Optional rear retro reflectors must not be triangular, except in the cases referred to at **25.58**.

(10) Reversing lamps. Such a lamp must show a white light. Not more than two optional lamps, facing the rear, may be fitted. In the case of a motor vehicle first used on or after 1 July 1954, they must be such that they cannot be illuminated other than automatically by the selection of the reverse gear or otherwise there must be a tell-tale. If a vehicle has an overall length in excess of 6m and is a bus or a vehicle which is not a passenger vehicle four reversing lamps may be fitted. If four are fitted to such a vehicle the configuration must be four to the rear or two to the rear and one on each side. If such side lamps are fitted:
(a) it must not be possible for them to be illuminated unless the front and rear position lamps of the vehicle are illuminated; and
(b) they must be switched on and off by the manual operation of one switch which has no other function; but
(c) they must switch off automatically if the vehicle is moving forward at a speed of 10km/h or more.

Non-additional optional lamps, etc

25.67   If such lamps, etc are fitted they must comply with the following requirements.

Front fog lamps

25.68   These must be white or yellow lamps. Where a pair of front fog lamps is used in conditions of seriously reduced visibility in place of the obligatory dipped beam headlamps, they must not be more than 400 mm from the side of the vehicle.

Motor vehicles (other than motor bicycles) first used on or after 1 April 1991 must not have more than two front fog lamps.

Warning beacons

25.69   The lights of such warning beacons must flash between 60 and 240 times per minute at constant intervals. Warning beacons, with the colours below, are permitted to be fitted in accordance with reg 11:

(1) Blue—a blue-and-white chequered light is permitted from a chequered domed lamp fitted to a police control vehicle and intended for use at the scene of an emergency; a blue light is permitted from a warning beacon or rear special warning lamp on an emergency vehicle, or on a special forces vehicle used in response to, or for training or practising to respond to, a national security emergency. No other type of vehicle may be fitted with a beacon of these types (or a device resembling one). Where an emergency or special forces vehicle fitted with a warning lamp is used for a purpose other than that to which the permission relates, as where an ambulance is used to take children to school, there is no requirement that the lamp should be covered or removed.

(2) Amber—permitted on road-clearance, refuse, or breakdown vehicles; those with an overall width exceeding 2.9m; road-service vehicles; special vehicles carrying abnormal loads; vehicles escorting an abnormal load; vehicles used for escort purposes other than escorting an abnormal load, while they are travelling at a speed not exceeding 25 mph; and vehicles of HMRC (fuel-testing vehicles). Regulation 17 requires motor vehicles with four or more wheels, other than those first used before 1 January 1947, which have a maximum speed which does not exceed 25 mph, or any trailer which they are drawing, to be fitted with, and display, at least one amber warning beacon when being driven on an unrestricted dual carriageway road. The Regulations do not apply to such vehicles when merely crossing such a road in the quickest possible manner.

(3) Green—permitted to be used by registered medical practitioners.

(4) Yellow—permitted to be used by airport vehicles.

Conspicuity markings

25.70   To the extent that a relevant goods vehicle (**25.61**) is exempt from the requirement to be fitted with conspicuity markings, conspicuity markings may be fitted to the side or rear (as the case may be).

A vehicle, other than:

(a) a passenger vehicle (except a bus);

(b) a trailer with a maximum gross weight not exceeding 750 kg,

which is not a relevant goods vehicle may be fitted with conspicuity markings.

USE OF LAMPS, REFLECTORS, ETC

No red lights to the front

By reg 11(1), no vehicle may be fitted with a lamp or retro reflective material which is **25.71** capable of showing a red light to the front. The important word is 'capable': the provision is not confined to the hours of darkness, nor is it necessary that the light is actually shown. The exceptions to the provision are obvious; they are (1) red-and-white chequered lamps or beacons fitted to a fire service vehicle for emergency purposes, (2) a red side marker lamp or a red side retro reflector, (3) red retro reflective material or a red side retro reflector fitted to any wheel or tyre of a pedal cycle (or its trailer or sidecar), a motor bicycle or motor-bicycle combination or an invalid carriage, and (4) a traffic sign attached to a vehicle.

Red lights to the rear

Regulation 11(2) provides that no vehicle may be fitted with a lamp or retro reflective **25.72** material which is capable of showing a light to the rear other than a red light. The same comments apply to 'capable' as made above. The exceptions are plentiful and include direction indicators, reversing lamps, conspicuity markings, interior illumination, rear number plates, public service vehicle route indicators, retro reflective material or side retro reflectors fitted to bicycles (or their sidecar or trailer), motor bicycles (or motor-bicycle combinations) and invalid carriages, and the various emergency lights described above. Pedal cycles may show an amber light to the rear from lamps fitted to their pedals. Pedal cycles, their trailers or sidecars may show a white or amber light to the rear from lamps fitted to their wheels or tyres if the lamp is designed to emit primarily to the side.

Maintenance of lamps, reflectors, rear markings, and devices

Regulation 23 prohibits using, or causing or permitting the use of, a motor vehicle on a **25.73** road in breach of the following requirements.

Every front position lamp, rear position lamp, headlamp, rear registration plate lamp, side marker lamp, end-outline marker lamp, rear fog lamp, retro reflector, rear marking, daytime running lamp, headlamp cleaning device, and reversing lamp with which the vehicle it is required to be fitted must be in good working order and, in the case of a lamp, clean. However, reg 23 does not apply to a rear fog lamp on a vehicle (a) which is a part of a combination of vehicles, if any part of the combination is not required to have a rear fog lamp, or (b) which is drawing a trailer.

Every stop lamp, running lamp, dim-dip device, headlamp levelling device, hazard warning signalling device, direction indicators, and front fog lamp even if it is in excess of one required by law, must be maintained at all times.

Regulation 23 does not apply to any defective lamp, reflector, dim-dip device or headlamp levelling device where the defect arose in the course of the journey on a vehicle in use between sunrise and sunset between sunrise and sunset, or if arrangements have been made to remedy the defect with all reasonable expedition; nor, to a lamp, reflector, headlamp levelling device or rear marking fitted to a combat vehicle. In addition, reg 23 does not apply to a front fog lamp or a daytime running lamp on a vehicle first registered before 1 March 2018.

Driving or parking without lights

25.74 Regulation 24 prohibits a person using, or causing or permitting to be used, a vehicle on a road between sunset and sunrise or (while the vehicle is in motion) during daytime hours in seriously reduced visibility unless every obligatory front position lamp, rear position lamp, rear registration plate lamp, side marker lamp and end-outline marker lamp is kept lit and unobscured. Except as indicated below, it also prohibits allowing (or causing or permitting to be allowed) a vehicle to remain at rest (ie parked) in similar circumstances between sunset and sunrise. There are variations to these provisions in respect of motorcycles and trailers not required to be fitted with front position lamps, since the prohibitions are not breached if there are fitted to these vehicles 'ad hoc' front position lamps.

Regulation 24 also provides that vehicles of certain classes may lawfully park between sunset and sunrise on roads subject to speed limits of 30 mph or less, without showing such lights. These are goods vehicles with gross vehicle weight not exceeding 2,500 kg, passenger vehicles other than buses, invalid carriages, and motorcycle, or pedal cycles (in either case with or without a sidecar). The exemption does not apply if the vehicle has a trailer attached or is carrying a load which requires lamps.

The exemption only applies to vehicles parked in:

(a) designated parking places on roads; or

(b) a lay-by which is clearly shown to be such; or

(c) elsewhere, provided, if the vehicle is parked on a one-way road, it is facing in the correct direction on either side of the road as close as possible to the kerb, or, if it is parked on an ordinary road, it is properly parked and facing the correct way, and in either case no part of the vehicle is less than 10m from a junction with the road upon which it is parked, whether the junction is on the same side of the road or not. Where a curving kerb exists, the 'junction' begins where the kerb begins to curve.

The lighting requirements under reg 24 do not apply to a solo motorcycle or pedal cycle, which is being pushed close to the nearside kerb, a pedal cycle which is halted and waiting to proceed (eg at traffic lights) if it is kept to the nearside, or a vehicle which is parked at roadworks and properly outlined by lamps or traffic signs.

Use of headlamps

25.75 RVLR 1989, reg 25 provides that a person must not use, or cause or permit to be used, on a road a vehicle which is fitted with obligatory dipped beam headlamps unless every such lamp is kept lit during:

(a) the hours of darkness, except on a road which is a restricted one by virtue of a system of street lighting, when those lights are lit; and

(b) daylight hours in seriously reduced visibility.

There are certain permitted variations; for example, motor vehicles with one obligatory dipped beam headlamp are exempt from the above requirement if a main beam or fog lamp is kept lit. Vehicles which are being towed, those which are parked and those propelling snow ploughs are exempt from the above requirement. In addition, in the case of a motor vehicle other than a motor tricycle or motor-bicycle combination, a pair of main beam lamps may be used as an alternative or, in seriously reduced visibility, a pair of front fog lamps may be used, provided that they are not more than 400 mm from the outer edges of the vehicle.

Prohibition of particular use of lamps or devices

Regulation 27 provides that a person must not use, or cause or permit to be used, on a road any vehicle on which any lamp, hazard signal warning device or warning beacon of a specified type is used in the manner listed below.

25.76

Headlamps must not be used so as to cause undue dazzle or discomfort to other persons using the road and shall not be lit when a vehicle is parked. These prohibitions also apply to front fog lamps, with the addition of a prohibition upon use at any time other than in conditions of seriously reduced visibility.

The use of rear fog lamps is similarly restricted but the reference to undue dazzle or discomfort is predictably restricted to following drivers. They must also not be used when a vehicle, other than an emergency vehicle, is parked.

Reversing lamps must not be used for any purpose other than that of reversing. Similarly the use of hazard warning signal devices is restricted to warning road users of a temporary obstruction (or the presence of a school bus which is stationary and loading or unloading school children under 16) when the vehicle is at rest, or, on a motorway or unrestricted dual carriageway, to warn of a temporary obstruction ahead, or, in the case of a bus, to summon assistance for the driver or conductor or an inspector who is on the vehicle.

Blue lamps and special warning lamps may only be used at the scene of an emergency, or to indicate the urgency of the journey or the presence of a hazard on the road. The use of amber lights is similarly restricted but they may also be used in connection with breakdowns or slow-moving vehicles on dual carriageways. Green lights may only be used whilst the vehicle is occupied by a registered medical practitioner and used in an emergency. Yellow light beacons must not be lit on a road. Work lamps must not dazzle, etc and must not be used other than for illuminating a working area, accident, breakdown, or works in the vicinity of the vehicle. No other lamp which is fitted to a vehicle may ever be used so as to cause undue dazzle or discomfort to other persons using the road.

Movement of lamps or reflectors and nature of light

Regulation 12 provides that no person shall use, or cause or permit to be used, on a road a vehicle fitted with a lamp or reflector capable of being moved (by swivelling, deflecting, or otherwise) while the vehicle is in motion. There are some obvious exceptions, eg dipping headlights, lamps which may be adjusted to compensate for loads, retracting headlamps, etc, amber pedal reflectors, and retro reflective material fitted to any wheel or tyre of pedal cycles or their trailer or sidecar.

25.77

Steady light

By regulation 13, the light shown by lamps covered by the Regulations must be a steady light, in that flashing lights are not permitted. There are exceptions for warning beacons prescribed for emergency vehicles, etc, and flashing front or rear position lamps fitted to pedal cycles (or their trailer or sidecar).

25.78

Projecting trailers and vehicles carrying overhanging or projecting loads

25.79   Regulation 21 prohibits a person using, or causing or permitting to be used, on a road:

(a) any trailer projecting laterally beyond the preceding vehicle in the combination; or
(b) any vehicle or combination of vehicles carrying a load or equipment,

which (in either case) does not comply with the following specifications:

(i) a trailer, which (or whose load) projects laterally more than 400 mm from the outermost part of the obligatory front position light on that side of the vehicle in front of it, must have white lights to the front which are not more than 400 mm from the outermost projection of the trailer (or, as the case may be, of the load);
(ii) a vehicle whose load projects laterally more than 400 mm must have lights at the front and rear not more than 400 mm from the outermost projection of the load;
(iii) a vehicle whose load projects more than 1m to the front or to the rear must have a front or rear lamp not more than 1m from the foremost or rearmost projection of the load (except that that distance is 2m in the case of an agricultural vehicle or a vehicle carrying a fire escape); or
(iv) a vehicle carrying a load which obscures any obligatory lamps, reflector or rear markings must show a lamp, etc in the prescribed position.

These requirements only apply when the vehicle/trailer is being used between sunset and sunrise or in circumstances of reduced visibility, except that in relation to obligatory stop lights and direction indicators in (iv) the requirement applies in all circumstances.

TESTING AND INSPECTION OF MOTOR VEHICLES

Testing vehicles on roads: RTA 1988, s 67

25.80   Authorised examiners may test a motor vehicle on a *road* to ascertain whether the requirements relating to its construction and use are being complied with, and also whether there is compliance with the requirement that the condition of the vehicle is not such that its use on a road would involve danger of injury to any person. There are two principal types of authorised examiners for the purposes of s 67: (a) police officers specially authorised by their chief constables for this purpose, and (b) vehicle examiners (ie examiners appointed under RTA 1988, s 66A by the Secretary of State in respect of statutory functions conferred on them in respect of public passenger vehicles, goods vehicles and other vehicles). For the purpose of testing a vehicle an examiner may require the driver to comply with his/her reasonable instructions and may drive the vehicle. All authorised examiners must produce their authority if required to do so.

Miscellaneous points

25.81   Where an examination of the vehicle under s 67 is required by an authorised examiner, the driver may elect that the test shall be deferred to a time and place to be arranged within the next 30 days.

The right to defer does not apply when:

(a) it appears to a constable that, following an accident on a road, the test must be taken forthwith. If this is so and the constable is not an authorised examiner, the constable may require that the vehicle is not taken away until the test has been carried out (s 67(7)); or

(b) the vehicle appears to a constable to be so defective that it ought not to be allowed to proceed (s 67(8)).

The provisions of s 67 do not mean that the requirement of RTA 1988 and RV(C&U) R 1986 may only be enforced on a road by an authorised examiner. Particular offences may still be reported by a constable who is not an authorised examiner. In addition, it has been held by a divisional court that any constable may test a vehicle with the owner's consent. In that case, a constable who was not an authorised examiner tested the efficiency of a handbrake with the driver's consent; evidence of his findings was admitted even though he was not an authorised examiner.

Testing vehicles off roads: RV(C&U)R 1986

Regulation 74 empowers any police constable in uniform or vehicle examiner to test and inspect the brakes, silencers, steering gear, and tyres, of any motor vehicle or trailer on any *premises* where that motor vehicle or trailer is, *subject to the consent of the owner of the premises*. This represents the basic difference between these powers and those set out above. RVLR 1989, reg 28 applies the provisions of reg 74 of the Road Vehicles (Construction and Use) Regulations 1986 to lighting equipment and reflectors with which a vehicle is *required* by the Regulations to be fitted. Regulation 74 empowers a constable in uniform to test and inspect lighting equipment on motor vehicles and trailers on any premises, subject to the consent of the owner of the premises.

Because the consent of the *owner of the premises* is necessary, there is no power of entry onto premises for the purpose of inspection under reg 74. In addition, the test and inspection cannot be carried out without the consent of the *owner of the vehicle*, unless notice of it has been served personally upon the owner at least 48 hours before the test, or 72 hours before it if service is effected by recorded delivery. The consent of the *owner of the vehicle* is not required if it has been involved in an accident within the preceding 48 hours.

Prohibition of unfit vehicles

RTA 1988, s 69 empowers an authorised constable to issue an immediate prohibition notice if, on an inspection under the various powers of testing and inspection under the Act, it appears to him/her that, owing to any defects in the vehicle, driving it (or driving it for any particular purpose or purposes or for any except one or more particular purposes) would involve a danger of injury to any person. The sections referred to deal with powers in relation to construction and use requirements; test certificates generally; inspections related to plated weights of goods vehicles, public service vehicles, and vehicles adapted to carry more than eight passengers which are not public service vehicles; and the testing of the condition of used vehicles at sale rooms, etc. A notice under s 69 may prohibit use absolutely, or for one or more specified purposes, or except for one or more specified purposes.

Vehicle examiners have a similar power.

The Road Vehicles (Prohibition) Regulations 1992 make provision for a vehicle which is subject to a prohibition issued by an authorised constable, or vehicle examiner, to be driven on a road solely for the purpose of submitting it, by prior arrangement, for test with a view to removal of that prohibition, or for it to be driven in the course of any test for the purpose of the removal of the prohibition, or (within three miles from where it is being, or has been, repaired) to be driven solely for the purpose of its test or trial with a view to the removal of the prohibition.

TEST CERTIFICATES

25.84   RTA 1988, ss 45 to 47 are concerned with tests of the satisfactory condition of motor vehicles, other than goods vehicles to which these sections do not apply and category T tractors (**25.16**) to which s 49 applies. The relevant provisions with regard to such goods vehicles are discussed in Chapter 25 on the companion website. The Secretary of State is authorised by RTA 1988, s 45 to make regulations for the examination of all vehicles to which ss 45 to 47 apply and the Secretary of State has made the Motor Vehicles (Tests) Regulations 1981.

By RTA 1988, s 47(1), a person commits an offence if (s)he uses, or causes or permits to be used (**25.9-25.11**), at any time on a road a motor vehicle to which s 47 applies, and as respects which no test certificate has been issued within the appropriate period before the said time. Once it is proved that D has used, etc a vehicle on a road, the burden shifts to D to prove that there was in force a test certificate for the vehicle. An offence under s 47(1) is a summary offence punishable with a fine not exceeding level 4 (vehicle adapted to carry more than eight passengers) or level 3 in any other case.

Vehicles to which s 47 applies

25.85   The answer to the question which vehicles must be tested depends on the *type of vehicle* and its *age*.

The following must be examined and receive a test certificate annually, *after the period of three years* defined in the next paragraph:

(a) passenger vehicles with not more than eight seats, excluding the driver's seat;
(b) goods vehicles up to 3,500 kg design gross weight;
(c) dual-purpose vehicles;
(d) motor cycles (including three-wheelers and mopeds);
(e) quadricycles, and
(f) motor caravans.

Section 47 applies to motor vehicles of the above types which were first registered not less than three years before the time in question (ie the time of the alleged use without a test certificate). If, for any reason, a vehicle is used on a road in Great Britain or elsewhere before being registered, a test certificate must be obtained three years from the date of manufacture. Thus, a serviceman returning from duty abroad who brings into this country a vehicle which is more than three years old must have it tested immediately.

RTA 1988 requires annual tests for the following motor vehicles *after one year from first registration* or three years from manufacture if used pre-registration:

(a) motor vehicles used for the carriage of passengers and with more than eight seats, exclusive of the driver's (mainly public service vehicles);
(b) taxis licensed to ply for hire;
(c) ambulances; and
(d) playbuses.

Exemptions

25.86   The Motor Vehicles (Tests) Regulations 1981 exempt a range of vehicles including:

(a) goods vehicles with a design gross weight exceeding 3,500 kg;
(b) articulated vehicles other than articulated buses;

(c) invalid carriages not exceeding 306 kg unladen weight (or 510 kg if supplied by the Department of Health);

(d) vehicles temporarily in Great Britain for a 12-month period;

(e) police vehicles maintained in police workshops, or vehicles provided for the purposes of NCA;

(f) electrically propelled goods vehicles with a design gross weight not exceeding 3,500 kg first registered before 1 March 2015;

(g) some licensed hackney carriages and private hire cars;

(h) any vehicle at a time when it is being used on a public road during any calendar week if it is being used only in passing from land in the occupation of its keeper to other land in his/her occupation, and it has not been used on public roads for more than an aggregate of six miles in that calendar week;

(i) heavy locomotives, light locomotives, motor tractors, track laying vehicles, agricultural motor vehicles and works vehicles;

(j) vehicles which have a Northern Ireland test certificate;

(k) a public service vehicle having a manufacture date before 1 January 1960 and which has not undergone substantial changes in the technical characteristics of its main components;

(l) a vehicle which is incapable, by reason of its construction, of exceeding a speed of 15.5 miles per hour on the level under its own power; and

(m) a vehicle of historical interest other than a public service vehicle.

For these purposes, 'vehicle of historical interest' means a vehicle which is considered to be of historical interest to Great Britain and which—

(i) was manufactured or registered for the first time at least 40 years previously;

(ii) is of a type no longer in production; and

(iii) has been historically preserved or maintained in its original state and has not undergone substantial changes in the technical characteristics of its main components.

The Regulations also permit use without a test certificate when a vehicle is being taken to a testing station by previous arrangement for a test, or when it is being brought away from such a test, or while it is being so tested by an authorised person or under his/her direction. In circumstances in which a test certificate is refused, the vehicle may be moved without a test certificate for the purpose of work being done (once again, by previous arrangement), or for the purpose of delivering it, by towing, to a place where it is to be broken up. Imported vehicles which need to be tested may be driven on arrival in this country to the place of residence of the owner of the motor vehicle.

Under the Road Vehicles (Certificate of Temporary Exemption) Regulations 2020, a public service vehicle adapted to carry more than eight passengers or a goods vehicle is not subject to RTA 1988, s 47 while a certificate of temporary exemption, issued by the Secretary of State, on grounds of exceptional circumstances, such as an accident, a fire, an epidemic, or severe weather, is in force in respect of it.

Test certificate

The contents of a test certificate are not prescribed. A certificate must be in a form sup- **25.87**
plied by the Secretary of State. A certificate has a serial number; indicates that the vehicle complied with the requirements on the date of the examination; shows the registration mark of the vehicle, the vehicle testing station number, the date of issue of the certificate, the date of expiry, and bears the signature of the person issuing the certificate. A

signature may be a facsimile of the signature of the examiner or of a person so author-ised by the Secretary of State. If the previous certificate had expired at the time of the examination, the serial number of the previous certificate may also be shown. Tests can only be conducted at testing stations authorised by the Secretary of State.

Requirement to produce test certificate

25.88   RTA 1988, s 165 provides that a person:

(a) driving a motor vehicle, other than an invalid carriage, on a road; or

(b) whom a constable or a vehicle examiner reasonably believes to have:

   (i)   been the driver of such a motor   vehicle when an accident occurred owing to its presence on a road or other public place; or

   (ii) committed an offence in relation to the use of such a motor vehicle on a road,

must, on being required by a constable or vehicle examiner, give his name and address and the name and address of the owner of the vehicle, and produce for examination a test certificate (if one is required for the vehicle)..

Failure to comply

25.89   Under RTA 1988, s 165(3), failure to comply with a requirement under s 165 is a sum-mary offence with a fine not exceeding level 3.

As with other types of document, there is a proviso which permits production within seven days at a specified police station; or production as soon as reasonably practic-able; or proof that production was not reasonably practicable before the day on which proceedings were commenced. Production in person is not required. If the proviso is satisfied, a person is not to be convicted under s 165(3).

CYCLES

Use of motor cycles

25.90   RTA 1988, s 23 restricts the carriage of passengers on a motor bicycle to one person who must be carried sitting astride the cycle on a proper seat securely fixed to the vehicle behind the driver's seat. Interestingly, a pillion passenger need not be carried facing the front. The driver commits a summary offence, punishable with a fine not exceeding level 3, if a passenger is carried in contravention of s 23. The offence is also punishable (see **24.40**) with discretionary disqualification for driving.

RV(C&U)R 1986, reg 102 prohibits the carriage of a passenger on a motor bicycle (whether it has a sidecar attached to it or not) on which there are not available suitable supports or rests for the feet for him/her. As to an offence of this type involving a breach of RV(C&U)R 1986, see **25.8-25.11**.

RTA 1988, s 16 empowers the Secretary of State to make regulations requiring per-sons driving or riding on a specified type of motor cycle (other than in a sidecar) which is on a road to wear prtective headgear. The type of headgear is prescribed by the Motor Cycles (Protective Helmets) Regulations 1998. For the purpose of these Regulations, 'motor cycle' means a motor bicycle, including one with a sidecar, and motor tricycles where the paired wheels are less than 460 mm apart. A divisional court has held that the BMW C1 motor cycle (which has a rigid rider cell in the form of a cab) is a motor cycle for these purposes. The Regulations do not apply to mowing machines, motor cycles being propelled by a person on foot, persons driving an electric scooter being used in a trial, or followers of the Sikh religion while wearing a turban. Helmets must comply with various British Standard or EEA specifications. Breach of the Regulations is a summary offence, punishable with a fine not exceeding level 2, contrary to RTA 1988, s 16(4). However, it is not an offence to aid and abet the failure to wear a crash helmet, unless the other party is under 16.

By virtue of s 18(3), it is a similarly triable and punishable offence to drive or ride on a motor cycle on a road without wearing eye protectors of the type prescribed by the Motor Cycles (Eye Protectors) Regulations 1999. For the purposes of s 18, the Regulations define 'motor cycle' in the same way as in relation to protective helmets. The present requirement does not apply to a person driving or riding on a mowing machine nor does it apply to a person driving or riding on a motor cycle temporarily brought into Great Britain by a foreign resident, if it has been here for less than one year. In addition, the requirement does not apply to a member of the armed forces of the Crown who is driving or riding on a motor cycle on duty, if (s)he is wearing an eye protector which is part of his/her service equipment. A person propelling a motor cycle on foot is not required to wear an eye protector.

Use of pedal cycles

By RTA 1988, s 24, the carriage of more than one person on a road on a bicycle not propelled by mechanical power is an offence by each of the persons carried, unless it is *constructed* or *adapted* for the carriage of more than one person. The offence is summary and punishable with a fine not exceeding level 1. A tandem is a cycle *constructed* for the carriage of more than one person, and a cycle which has a seat fitted to permit the carriage of a child has been *adapted* for the carriage of more than one person. **25.91**

The wilful riding of a pedal cycle (or the driving of any other form of vehicle) on a footpath or causeway by the side of any road made or set apart for the use or accommodation of foot passengers remains a summary offence, punishable with a fine not exceeding level 2, contrary to the Highways Act 1835, s 72. A person rides a cycle, even if (s)he merely sits astride it and propels him/her self with his/her feet. The offence under s 72 does not apply to an invalid carriage (eg a mobility scooter) if the requirements of the disabled persons 'Blue Badge' scheme are satisfied (Chronically Sick and Disabled Persons Act 1970, s 20).

Races or speed trials

RTA 1988, s 31 provides that a person who promotes or takes part in a race or speed trial for pedal cycles on a highway commits an offence, unless that race or speed trial is authorised under, and conducted in accordance with, the Cycle Racing on Highways Regulations 1960. This is a summary offence punishable with a fine not exceeding level 1. Time trials and bicycle races are dealt with separately by the Regulations. **25.92**

Regulation 2 defines a 'time trial' as a cycle race or trial of speed between bicycles or tricycles involving either single competitors, or groups of not more than four competitors, starting at intervals of not less than one minute. In the case of groups they must not compete against one another. The result must depend upon the time taken by a competitor, or if a group any member of it, to reach a finishing point; or on the distance covered in a fixed time. A time trial is authorised if its promoter gives the specified written notice 28 days in advance.

A 'bicycle race' is defined by reg 2 as a race or trial of speed which is not a time trial.

In order for a bicycle race to be authorised *in England*, the following additional requirements must be satisfied:

(a) the numbers taking part are restricted to, in not more than two races a year selected by the British Cycling Federation, 100 competitors, and in any other approved race, 80;

(b) races must not take place during the hours of darkness;

(c) if races follow a circuitous route, they must travel at least 10 miles before passing the same point on a highway twice (in whichever direction);

(d) no continuous part of the race may be on more than one-and-a-half miles of any road subject to a speed limit of 40 mph or less, and no further continuous part in a 40 mph-restricted section may be within three miles of the previous one; and

(e) its promoter must give the specified written notice 28 days in advance.

A chief officer of police is empowered to impose conditions on the holding or conduct of a race.

In order for a bicycle race to be authorised *in Wales*, the following requirements must be satisfied:

(a) the promoter of the race must complete a written risk assessment, which must include:

(i) a suitable and sufficient assessment of the risks to the health and safety of competitors, race officials, spectators and other road users arising out of the holding of the race, having regard to the maximum number of competitors eligible to participate in the race; and

(ii) details of appropriate actions that will be taken by the promoter of the race to ensure the safe operation of the race;

(b) races must not take place during the hours of darkness;

(c) if races follow a circuitous route, they must travel at least five miles before passing the same point on a highway twice (in whichever direction); and

(d) its promoter must give the specified written notice 28 days in advance.

A chief officer of police is empowered to impose conditions on the holding or conduct of a race. A chief officer who decides to do so must provide written reasons for his/her decision to the promoter of the race and the British Cycling Federation within five working days of making that decision

Electrically assisted pedal cycles

25.93 An 'electrically assisted pedal cycle' is not a motor vehicle for the purposes of the Road Traffic Acts (Road Traffic Regulation Act 1984, s 140); such a cycle is a pedal cycle with two or more wheels, is fitted with pedals by means of which it is capable of being propelled, and has an electric motor of a rated power not exceeding 250 watts and incapable of propelling the vehicle when it is travelling at more than 15.5 mph. There is no longer a maximum kerbside weight requirement. In other words, an electrically assisted pedal cycle is basically a pedal cycle with an auxiliary electric motor which cuts out when the vehicle is travelling in excess of 15.5 mph. A divisional court has held that, to satisfy the definition of an electrically assisted pedal cycle, the nature of the pedals fitted to such a cycle must be such that they are reasonably capable of propelling the vehicle in a safe manner in normal day-to-day use.

By RTA 1988, s 32(2), it is an offence for a person under 14 to drive an electrically assisted pedal cycle, or to cause or permit him/her to do so knowing that (s)he is under 14. This summary offence is punishable with a fine not exceeding level 2.

Use of an electrically assisted pedal cycle does not require a vehicle licence (or nil licence) or insurance. Nor do the requirements of testing, driving licences and protective headgear apply.

Brakes on pedal cycles

The Pedal Cycles (Construction and Use) Regulations 1983, reg 4A provides that no person may ride, or cause or permit to be ridden, on a road an electrically assisted pedal cycle unless it is fitted with braking systems which comply with para 4.6.8 of the 2014 BS EN Standard or with the standards required when the pedal cycle vehicle was type approved (if applicable). Regulation 5 prohibits a person from riding, or causing or permitting to be ridden, on a road such a cycle when the brakes have not been maintained in efficient working order. **25.94**

Regulation 6 provides that no person shall ride, or cause or permit to be ridden, a pedal cycle which is not electrically assisted if it does not comply with such of the braking-system requirements in regs 7 or 8 as apply to it.

Regulation 7 requires that at least one braking system be fitted to pedal cycles.

Pedal cycles manufactured on or after 1 August 1984 are subjected by reg 7 to additional braking requirements if the saddle height (fully raised and tyre fully inflated) is 635 mm (about 25 inches). The additional requirements are that fixed-wheel cycles must have a braking system acting on the front wheel or on at least two front wheels if there are more than one. Free-wheel cycles must have two independent systems, one acting upon the front wheel, or at least two front wheels if more than one is fitted, and the other acting upon the rear wheel, or at least two rear wheels if more than one is fitted. In the case of pedal cycles manufactured before 1 August 1984, these additional requirements apply under reg 8 to pedal cycles, any wheel of which exceeds 460 mm (approx 18 inches) in diameter including the fully inflated tyre, except that the braking system acting upon the rear need only act upon one rear wheel even if there is more than one rear wheel fitted. A non-goods tricycle, whenever manufactured, may instead have two independent systems both acting upon the front wheel if it has two rear wheels (and vice versa).

Regulation 9 provides that nothing in regs 7 or 8 applies to pedal cycles on which the pedals act directly on the wheel rather than through a gearing system (eg a penny-farthing) and those temporarily in Great Britain when ridden by a visitor, such a cycle having a braking system or system complying with provisions of the International Convention on Road Traffic.

Regulation 10 prohibits riding, or causing or permitting to be ridden, on a road a pedal cycle the braking system of which, fitted in accordance with these Regulations, is not in efficient working order.

Breach of regs 4A, 5, 6, or 10 is a summary offence, punishable with a fine not exceeding level 3, contrary to the Road Traffic Offenders Act 1988 (RTOA 1988), s 91. A constable in uniform is empowered by reg 11 to test and inspect a pedal cycle for the purpose of ascertaining that the braking requirements are satisfied. This may be done on a road, or on any premises where the cycle is if the cycle has been involved in an accident (provided that the test and inspection are carried out within 48 hours of the accident with the consent of the owner of the premises).

GOODS OR PASSENGER VEHICLE DRIVERS' HOURS AND RECORDS

25.95   The detailed and complex rules about these matters are dealt with in Chapter 25 on the companion website.

TRAFFIC SIGNS AND DIRECTIONS

25.96   These are governed by the Road Traffic Regulation Act 1984 (RTRA 1984) and the Traffic Signs Regulations and General Directions 2016 (TSR&GD 2016).

For the purposes of RTRA 1984, the term 'traffic sign' is defined by s 64 as any object or device (fixed or portable) for conveying, to traffic on roads or any specified class of traffic, warnings, information, requirements, restrictions or prohibitions of any description:

(a) specified by regulations made by the relevant Ministers acting jointly; or
(b) authorised by the Secretary of State,

and any line or mark on a road for so conveying such warnings, information, requirements, restrictions, or prohibitions. Traffic signs must be of a prescribed size, colour and type, or of any authorised character.

Powers to place traffic signs

25.97   RTRA 1984, s 65 empowers *highway authorities* to cause or permit traffic signs to be placed on or near any road in their area.

Sections 66 and 67 empower *a constable, or a traffic officer in uniform (in the case of s 67), or a person acting under the instructions (whether general or specific) of the chief officer of police*, to place on a road such authorised traffic signs indicating prohibitions, restrictions or requirements relating to vehicular traffic as may be required:

(a) to prevent an obstruction on public occasions or near public buildings or at an authorised cycle race (s 66). Section 66 permits police officers to place emergency signs on the road at special events such as air displays;
(b) to prevent or mitigate congestion or obstruction of traffic, or danger to or from traffic, in consequence of extraordinary circumstances (s 67). Among other things, s 67 empowers the police to place emergency signs on the road at the scene of an accident. Such signs may be maintained for a maximum period of seven days. If the need is a permanent one the matter must be dealt with by the highway authority.

Failure to comply with a sign placed under the provisions of RTRA 1984, s 65, 66, or 67 is a summary offence contrary to the Road Traffic Act 1988 (RTA 1988), s 36(1) (**25.98**).

Quite apart from the above, the TSR&GD 2016, Sch 13, Part 7, para 11 authorises the *crew of an emergency or breakdown vehicle* which is causing a temporary obstruction to place a 'keep right sign'. Non-compliance with such a sign is an offence against RTA 1988, s 36.

In addition, TSR&GD 2016, Sch 13, Part 7, para 12 authorises *any person* to place a 'road vehicle sign' (temporary obstruction by stationary vehicle) on a vehicle, and to place on any road:

(a) a 'flat traffic delineator', 'traffic cone', or 'traffic pyramid';
(b) a 'traffic triangle'; or
(c) a 'warning light',

for the purpose of *warning* traffic of a temporary obstruction, other than roadworks. Non-compliance with such a sign is not in itself an offence. A 'road vehicle sign' consists of a flexible sheet on which there is a triangle the outer edges of which are red, the inner part being either reflectorised white or fluorescent yellow. It is designed to be fixed to a stationary vehicle, facing approaching traffic. A 'traffic delineator', 'traffic cone', or 'traffic pyramid' is of the 'lane marker' type of device frequently used by emergency services. A 'traffic triangle' is the type of device frequently used by individual motorists, and a 'warning light' is a flashing amber signal.

The indications given by the signs and the form which they take is prescribed by TSR&GD 2016, Sch 13, Part 6.

The placing of signs otherwise than as prescribed is of no legal effect.

Drivers to comply with traffic signs

RTA 1988, s 36(1) states that an offence is committed by a person driving or propelling a vehicle who fails to comply with a lawfully placed traffic sign of the prescribed size, colour and type, or of another character authorised by the Secretary of State. It is no defence for a driver to allege that (s)he did not see a traffic sign. Such an offence is a summary offence punishable with a fine not exceeding level 3 and in the instances specified at **24.40** with discretionary disqualification for driving if committed in respect of a motor vehicle.   **25.98**

A traffic sign is not to be treated as lawfully placed unless (a) it indicates a statutory prohibition, restriction, or requirement, or (b) it is expressly provided by or under RTA 1988 or RTRA 1984 that RTA 1988, s 36 applies to the sign; such express provision has been made by the relevant parts of TSR&GD 2016. Examples of signs to which s 36 applies thereunder are described in the next three paragraphs (which deal in turn with signs indicating a prohibition, a restriction, and a requirement) and at **25.99** (particular signs).

Examples of signs which indicate a *prohibition* are 'No entry', 'No right turn', 'No U turn', 'Lorries prohibited', and 'No overtaking' signs. These signs frequently involve the use of symbols only. Except in the case of a 'No Entry' sign, these consist of a red circle with the enclosed symbol in black on a white background.

Signs dealing with *restrictions* include those dealing with waiting restrictions. These signs have a red outer circle with a blue background and a red band cutting across the face from the top left to bottom right of the sign as seen by approaching drivers. Information concerning the nature of the restrictions in force is provided on a plate, usually attached to the post supporting the sign. Restrictions on waiting may be additionally indicated by signs placed on the roadway but such a sign is only effective to create the restriction if used in conjunction with a restriction sign or plate of the type referred to above. A single continuous yellow line running parallel with the kerb restricts waiting (other than for loading or unloading) for a period of at least eight hours between 7 am and 7 pm on a minimum of four days a week. A double continuous yellow line indicates that no waiting at any time is permitted, unless there are signs specifically indicating seasonal restrictions. Broken yellow lines indicate other waiting restrictions, usually unilateral parking schemes, restricting waiting on particular sides of the road on particular days of the week.

Those signs which deal with *requirements* under the section are those which require a driver, etc to do something. 'Stop', 'Give Way', 'Keep Right', 'Keep Left', 'Traffic Lights' are all signs which indicate a requirement. These signs vary in appearance. The modern 'Stop' sign is octagonal and coloured red with 'Stop' in white capitals on it, although variations still exist. Other statutory requirements relating to stopping are red traffic lights and manually operated 'stop' signs at road works. 'Give way' signs consist of a red inverted triangle on a white, circular background with the words 'give way' inside the triangle. White arrows on a blue background indicate the route which a driver must follow.

For the purposes of RTA 1988, s 36, traffic signs placed on or near a road are deemed to be of the prescribed size, colour and type, or of another authorised character, and to have been lawfully placed, unless the contrary is proved. Consequently, the prosecution does not normally have to offer evidence of the nature of the sign beyond that which satisfies the court of the nature of the prohibition, restriction or requirement in question. However, if the sign is proved not to be of the prescribed size, etc, or not to have been lawfully placed, it will not be a valid sign and no offence can be committed in relation to it, unless the breach of the requirement is trivial. There is a presumption that automatic traffic lights are in working order, unless the contrary is proved.

Where the provisions of s 36 do not apply to a sign but it is governed by regulations concerning traffic control made by the Secretary of State, the Road Traffic Offenders Act 1998 (RTOA 1988), s 91 states that contravention or failure to comply is an offence under that section and punishable with a fine not exceeding level 3.

Particular signs

25.99   Certain signs indicate particular requirements. A 'Stop' sign requires that:

(a) every vehicle must, before entering the major road, stop at the transverse lines at the junction, or, if the lines are not visible, at the major road; and

(b) no vehicle may proceed past the transverse line painted nearest to the major road, or, if the lines are not visible, may not enter the major road, in such a manner or at such a time as is likely to cause danger to the driver of any other vehicle on the major road, or such as to necessitate the driver of any such other vehicle to change its speed or course in order to avoid an accident.

The latter provision (ie (b)) also applies to 'Give Way' signs. Thus, it is not sufficient to stop, or give way, and then to proceed, if the results of so proceeding are as described. There should be no interference with traffic on the major road. To emerge in such circumstances also indicates an element of carelessness which might amount to careless driving.

TSR&GD 2016, Sch 9, Part 7, para 9 deals with double white lines, which consist of either two continuous white lines or one continuous white line together with a broken white line painted along the middle of the carriageway itself. Any white unbroken line must be immediately preceded by a white warning arrow painted on the road; otherwise the line is not valid as a sign. If there are two continuous white lines vehicles travelling in both directions are required to keep to the nearside of the nearest continuous white line. A broken line with a continuous line requires vehicles to keep to the nearside of the continuous white line when that line is the nearer of the two to the vehicle. Neither of these rules applies if it is both safe to do so and necessary to do so:

(a) to enable the vehicle to enter, from the side of the road on which it is proceeding, roadside property or a side road joining that road;

(b) to pass a stationary vehicle;

(c) owing to circumstances outside the control of the driver;

(d) to avoid an accident;

(e) to pass a road maintenance vehicle which is in use, is moving at a speed not exceeding 10 mph, and is displaying to the rear a specified sign (angled white arrow on blue background within white edged circle, which may be mounted on a square or rectangular background);

(f) to pass a pedal cycle or an electric scooter being used in a trial moving, or a horse being ridden or led, at a speed not exceeding 10 mph; or

(g) to comply with any direction of a constable in uniform, or a traffic officer in uniform.

No vehicle may stop on either side of the road with only one traffic lane in each direction where there is a double white line marking of either description unless it is done:

(a) to pick up or set down passengers, to load or unload;

(b) in connection with building operations, the removal of an obstruction to traffic, or road or public utility works;

(c) by a vehicle used for police, traffic officer, fire and rescue, special forces, NCA, bomb disposal, or ambulance purposes or for the purpose of emergency-response at the request of an NHS ambulance service;

(d) by a pedal cycle or an electric scooter being used in a trial;

(e) to avoid an accident or because it is impossible to proceed; or

(f) with the permission or at the direction of a constable in uniform or a traffic officer in uniform.

This prohibition on stopping does not apply to a lay-by.

Records produced as evidence

25.100 RTOA 1988, s 20 provides that evidence of a fact related to an offence to which it applies may be given by the production of a record produced by a device of a description prescribed by the Secretary of State. Section 20 applies to failure by a person driving or propelling a vehicle to comply with a light signal which is at red, contrary to RTA 1988, s 36(1).

Evidence may only be so given if (in the same or another document) there is a certificate as to the circumstances in which it was produced, signed by a constable, or by a person authorised by (or on behalf of) the chief police officer for the area. The effect of s 20 is that the record and certificate can be tendered in evidence without the need for a witness to be called to prove them. The device must be of a type approved by the Secretary of State; it has been held that this requirement has been satisfied by the Secretary of State's approval of 'a device designed or adapted for recording by photographic or other image recording means the position of vehicles in relation to light signals' as a prescribed device. The device must be used in accordance with conditions subject to which the approval was given.

For a document of the type referred to above to be admissible under s 20, a copy of it must be served on the defendant (D) not less than seven days before the hearing or trial. If D, not less than three days before the hearing or trial, requires the attendance of the person who signed the document, the evidence of the circumstances in which the record was produced will not be admissible, although the record produced by the device will.

Where the prosecution does not serve the record and certificate on D within the seven days before the trial, a police officer may nevertheless attend as a witness to produce and prove those items as real evidence.

Drivers to comply with traffic directions

25.101   RTA 1988, s 35(1) provides that, where a constable or a traffic officer in uniform is for the time being engaged in the regulation of traffic in a road, a person driving or propelling a vehicle who neglects (ie fails) or refuses to:

(a) stop the vehicle; or
(b) make it proceed in, or keep to, a particular line of traffic,

when directed to do so by a constable in the execution of his duty, is guilty of an offence. This is a summary offence punishable with a fine not exceeding level 3 and (see **24.40**) with discretionary disqualification for driving if committed in respect of a motor vehicle.

Traffic surveys

25.102   RTA 1988, s 35(2) provides that, where a traffic survey is being conducted, a constable or a traffic officer in uniform may direct a person driving or propelling a vehicle to stop or to proceed in a line of traffic or to a particular point, but no requirement to give information may be made; non-compliance is an offence triable and punishable in the same way as an offence under s 35(1). Section 36(4) provides that signs such as 'Stop at Census Point' and 'Census Point—Stop if required' are traffic signs to which s 36 (**25.97**) applies.

Other powers to stop drivers and pedestrians

25.103   Apart from the powers given to the police by RTA 1988, s 35, RTA 1988, s 163 requires that a person driving a mechanically propelled vehicle or riding a cycle on a road must stop his/her vehicle on being required to do so by a constable in uniform or a traffic officer in uniform. Failure to do so is a summary offence against s 163(3). The offence is punishable with an unlimited fine, except in the case of a pedal cyclist where the fine must not exceed level 3.

In addition, by s 37, a pedestrian who fails to comply with the signals of a constable in uniform or traffic officer who is on traffic control commits an offence. Failure to give a name and address to a constable in such circumstances is an offence against s 169. Both offences are summary and punishable with a fine not exceeding level 3 or 1 respectively.

Under the Fire and Rescue Services Act 2004, s 44 an authorised employee of a fire or rescue authority may close off streets, or stop or regulate traffic in any street, whenever (s)he reasonably believes this is necessary for rescue or fire-fighting purposes. It is a summary offence, contrary to the Emergency Workers (Obstruction) Act 2006, s 1(1), without reasonable excuse to obstruct or hinder a fire and rescue authority employee taking action under s 44. The offence is punishable with an unlimited fine.

CROSSINGS

25.104   TSR&GD 2016 govern 'Zebra', 'Puffin' *pedestrian* crossings (below), and the other crossings mentioned at **25.111**. Zebra crossings are uncontrolled; Puffin crossings are light controlled. TSR&GD 2016 revoked the 'Zebra', 'Pelican' and 'Puffin' Pedestrian Crossing Regulations and General Directions 1997. However, the TSR&GD 2016 provides that the provisions of the 1997 Regulations are to be treated as remaining in force (with some modifications) in relation to Pelican crossings established before the TSR&GD 2016 came into force (22 April 2016) or within the six-month period beginning with that date.

By RTRA 1984, s 25(5), a person who contravenes the relevant regulations relating to *pedestrian* crossings is guilty of an offence. The offence is summary and punishable with a fine not exceeding level 3 and (**24.40**) with discretionary disqualification for driving if committed in respect of a motor vehicle.

'Zebra' crossings

25.105

TSR&GD 2016 prescribe the nature of uncontrolled crossings and the rules governing them. The *limits* of the crossing itself (the walking area for the pedestrians) are indicated by lines of square or circular marks or studs between which is an area marked by alternate black and white stripes. Flashing yellow globes indicate the presence of such a crossing to approaching drivers. The globes are mounted on poles which are also marked by black and white stripes. There must be a globe at either side of the crossing; if there is a central reservation or street refuge on the crossing, one or more globes may be placed there. The failure of the lamps does not invalidate a Zebra crossing; the regulations must still be complied with.

The approach to the crossing from each direction is marked out as a 'controlled area'. It is in effect a defensive area and drivers are made aware that they are entering it by white lines situated at each kerb and in the crown of the road on the approaching driver's side. This is known as the 'terminal line' and from it three zig-zag lines stretch to a distance of one metre from the crossing itself where they join a broken white line which is the 'give way' line.

These regulations only apply to Zebra crossings when traffic is not being controlled by a uniformed police constable or traffic officer. Immediately such a person takes control of traffic movements at such a crossing, the crossing ceases to be an uncontrolled one and the regulations cease to apply until that control ceases.

Precedence at a Zebra crossing

25.106

TSR&GD 2016, Sch 14, Part 5, para 7 states that every pedestrian on the carriageway within the limits of a Zebra crossing has precedence within these limits over any vehicle. That precedence must be afforded by a driver before any part of his/her vehicle enters the pedestrian limits of the crossing itself. The vehicle must be stopped at or before the broken white line give-way marking on the approach side of the crossing. Where there is a central refuge each side of it is treated as a separate crossing. A pedestrian only has precedence when (s)he is within the limits of the crossing before the vehicle enters the limits of the crossing. A pedestrian waiting at the kerbside is awaiting the courtesy of drivers who care to stop; such a person is not entitled to any precedence until (s)he steps on to the crossing.

The driver of a vehicle must always approach a Zebra crossing in such a manner that (s)he will be able to stop before reaching it unless (s)he can see that there is no one in the vicinity. Evidence of a failure to take reasonable care is unnecessary; the duty to accord precedence is strict and its discharge requires an approach to such crossings that will allow precedence to be given in any circumstances. The only exceptions are where there is a sudden defect in the vehicle or it is pushed on to the crossing by the vehicle behind. It is important to recognise that the issue of precedence does not arise until there is the question of who goes first. Clear evidence should be offered that the passage of a pedestrian was interfered with by the passage of a motor vehicle over the limits of the crossing. Taken to extremes, where a crossing spans a very wide street, a vehicle might pass over the limits while a pedestrian was on the crossing without the progress of that pedestrian being in any way interfered with. Alternatively, where a vehicle is stopped at a crossing and the pedestrians have passed the point at which it is waiting, the driver is free to proceed. If another pedestrian then steps on the crossing as the vehicle moves off, the driver commits no offence (provided no issue of 'who goes first' is raised).

'Puffin' and 'Pelican' crossings

25.107 By TSR&GD 2016, Sch 14, Part 5, para 5, when a red light is being displayed at a Puffin crossing a driver of a vehicle must not cause it to 'proceed beyond the stop line' or, if that line is not visible, the post on which the light signal is mounted. A red with amber signal denotes an impending change to green, but it conveys the same prohibition as a red signal. A steady amber signal when shown alone conveys the same prohibition as a red signal (with the usual proviso applicable to traffic lights), and there may be a green arrow showing, which signals that traffic may cross the stop line to proceed in a particular direction.

A Puffin crossing will have 'controlled areas' on each side of the crossing (or on one side in the case of one-way traffic), indicated by zig-zag lines stretching from the terminal line to the stop line (a line which is met before the limits of the crossing itself—identified by square or circular marks or studs—are reached). The primary signals shown to both drivers and pedestrians are by synchronised light signals. Whilst a steady green light is shown to drivers, a steady red light is shown to pedestrians, and vice versa. The light signals to pedestrians are by means of red and green figures.

Similar provision is made in relation to Pelican crossings by reg 23 of the 1997 Regulations which are treated as remaining in force for these purposes, as explained above, except that the light signals to pedestrians are, in the case of Pelican crossings, steady red, steady green, and flashing green figures reinforced by the illumination of a sign which reads 'WAIT'. Audible signals may be used at Pelican crossings to indicate to pedestrians when it is safe to cross.

Precedence at a Puffin or Pelican crossing

25.108 This is determined to a major extent by the light signals which are showing at the time.

However, at Pelican crossings (but not Puffin crossings) a driver may encounter a flashing amber signal which indicates that pedestrians, who are on the carriageway or a central reservation within the limits of the crossing before any part of the vehicle has entered those limits, must be accorded precedence by the driver of a vehicle.

Prohibition on waiting

25.109 TSR&GD 2016, Sch 14, Part 5, para 1 prohibits a driver from stopping his/her vehicle in the limits of a Zebra or Puffin crossing unless (s)he is prevented from proceeding by circumstances beyond his/her control or it is necessary to avoid an accident.

Moreover, para 3 prohibits the driver of a vehicle from causing the vehicle or any part of it to stop in a 'controlled area' (the zig-zag area). However, by para 4, this does not apply to a pedal cycle which is not mechanically propelled (unless it is an electrically assisted pedal cycle), nor to an electric scooter being used in a trial. Paragraphs 4 and 5 also exempt from para 3 vehicles:

(a) whose drivers have stopped to comply with the requirement to accord precedence, or to avoid an accident; or
(b) whose drivers have stopped to comply with a light signal or the direction of a constable in uniform or a traffic officer in uniform;
(c) whose drivers have been prevented from proceeding by circumstances beyond their control; or
(d) which have stopped for fire and rescue, ambulance, blood service, police, NCA, defence or special forces purposes, or for the purpose of emergency-response at the request of an NHS ambulance service, or in connection with the removal of obstructions, building works, road works or repairs to public utilities.

Vehicles may also halt to make right or left turns. Public service vehicles may stop to pick up or set down passengers but only on the far side of the crossing itself, not on its approach.

Similar provision is made in relation to Pelican crossings by regs 18 to 22 of the 1997 Regulations which are treated as remaining in force for these purposes, as explained above.

Prohibition on overtaking

By TSR&GD 2016, Sch 14, Part 5, para 6, a driver within a 'controlled area' of a Zebra **25.110** or Puffin crossing must not overtake *on the approach side* to a crossing a moving motor vehicle or a stationary motor vehicle according precedence to a pedestrian on a Zebra crossing or obeying a red light signal at a Puffin crossing. For these purposes, a vehicle overtakes another if any part of its foremost part passes ahead of the foremost part of another vehicle. Paragraph 6 limits the prohibition on overtaking to cases where the vehicle overtaken is the only other vehicle in the controlled area, or is the foremost vehicle. If this was not so, where there were two lines of traffic approaching a crossing, it would not be possible for the traffic on the offside to close up. In relation to the prohibition on overtaking a vehicle which has stopped to accord precedence, such a vehicle includes one which has stopped for this purpose even though a pedestrian intending to cross has not yet stepped on to the crossing.

Because of the drafting of para 6, the ban on overtaking does not apply where the overtaking vehicle was actually on the crossing when it passed ahead of the foremost part of the overtaken vehicle.

Similar provision is made in relation to Pelican crossings by reg 24 of the 1997 Regulations which are treated as remaining in force for these purposes, as explained above.

Parallel, equestrian, and 'Toucan' crossings

A parallel crossing is one where provision is made for pedestrians, drivers of elec- **25.111** tric scooters being used in a trial, and cyclists to cross the carriageway. Like a Zebra crossing it does not involve light signals. Its appearance in terms of yellow globes, zig-zag lines, black and white stripes, and so on is similar to that of a Zebra crossing but with a parallel crossing route for cyclists, or drivers of electric scooters being used in a trial, marked within the controlled area on one side of the striped crossing for pedestrians. The *limits* of the crossing are indicated by lines of square or round marks or studs. By TSR&GD 2016, Sch 14, Part 1, paras 21 to 24 corresponding rules to those which apply to Zebra crossings are provided in relation to parallel crossings in respect of according precedence, waiting and overtaking.

An equestrian crossing is one where provision is made for equestrian traffic to cross the carriageway. Its presence is indicated by a combination of traffic light signals, light signals for users of the crossing and road markings, including zig-zag lines.

A 'Toucan' crossing is one where provision is made for pedestrians, drivers of electric scooters being used in a trial, and pedal cyclists to cross the carriageway. Its presence is indicated in a similar way to an equestrian crossing.

By TSR&GD 2016, Sch 14, Part 1, para 14, corresponding rules to those which apply to Puffin crossings apply to equestrian crossings and Toucan crossings in respect of stopping, waiting and overtaking.

Breach of a requirement imposed in respect of one of the crossings under the above heading is an offence under RTA 1988, s 36 (**25.98**) (TSR&GD 2016, Sch 14, Part 4, para 1).

School crossing patrols

25.112   The stopping of vehicles at school crossings is dealt with by RTRA 1984, s 28.

When a vehicle is approaching a place in a road where a person is crossing or seeking to cross the road, a school crossing patrol wearing a uniform approved by the Secretary of State is empowered, by exhibiting a prescribed sign, to require the person driving or propelling the vehicle to stop it. When such a person has been required to stop:

(a) (s)he must cause the vehicle to stop before reaching the place where the person is crossing or seeking to cross and so as not to stop or impede his/her crossing; and

(b) the vehicle must not be put in motion again so as to reach the place in question so long as the sign continues to be exhibited.

As long as a prescribed sign has been properly exhibited, a driver must stop even if the persons have cleared the road and there are no others seeking to cross. A divisional court has held that the words 'so as not to stop or impede his/her crossing' in (a), above, merely describe the manner in which a driver should stop (ie (s)he must not halt across the path which those children would take) and are not meant to indicate that drivers can ignore a properly displayed sign if, in doing so, they would not directly impede pedestrians. The duty to stop is absolute.

It is an offence under s 28(3) to fail to comply with (a) or to cause a vehicle to be put in motion contrary to (b). These are separate summary offences punishable with a fine not exceeding level 3 and (**24.40**) with discretionary disqualification for driving if committed in respect of a motor vehicle.

The approved uniform is a cap, beret, or yellow turban and a coat with or without a fluorescent overgarment. Alternatively, a high visibility coat may be worn. The stopping of traffic may only be effected by exhibiting a sign prescribed by the TSR&GD 2016, Sch 14, Part 2 sign table. That sign will display the word 'Stop' in black letters with a black bar on a yellow fluorescent background surrounded by a red fluorescent border. The uniform worn, or a sign exhibited, by a patrol is rebuttably presumed to be approved (uniform) or in accordance with the regulations (sign).

A sign is exhibited if an approaching driver can see the words on it; but it need not be proved that it was full face to oncoming traffic.

REMOVAL OF VEHICLES

The police have powers to require the removal of vehicles and to remove them them- **25.113**
selves. Local authorities are legally obliged to remove abandoned vehicles.

Police powers

The Removal and Disposal of Vehicles Regulations 1986 (RDVR 1986), reg 3 allows a **25.114**
constable to require the owner, driver or other person in charge or control of a vehicle
which has:

(a) broken down, or been permitted to remain at rest, on a road or other land in such
a position, condition or circumstances as to cause obstruction to other persons
using the road or land concerned, or as to be likely to cause danger to such other
persons; or
(b) been permitted to remain at rest or has broken down and remained at rest on a
road or other land in contravention of a prohibition or restriction in or under any
enactment mentioned in Sch 1 to the Regulations,

to move or cause it to be moved. 'Other land' is not limited to land in the open air; the
land could, for example, include an underground car park.

'Obstruction' in (a) has been held by the Court of Appeal to have a different meaning
from that accorded to it in the Highways Act 1980, s 137 and the Road Vehicles
(Construction and Use) Regulations 1986, reg 103 (**24.88** and **25.27**). 'Obstruction' of
the highway in (a), the Court held, means more than simply impeding the free access
of members of the public to every part of the highway; what is required is obstructing
their passage by hindering or preventing them getting past. That 'obstruction', said the
Court, need not be an actual one; it includes obstructing people who might be expected
to be using the highway. On the other hand, the Court held, mere unreasonable use of
a highway does not make it an obstruction. No doubt a corresponding meaning applies
in respect of 'other land'.

The statutory prohibitions or restrictions included in Sch 1, referred to in (b) above,
are concerned with parking in 'No waiting' areas, in the controlled areas of pedestrian
crossings or in contravention of traffic signs—including police 'No waiting' signs. The
constable's requirement may include a requirement to move the vehicle to some other
place which may be a road or other land, or a requirement that it shall not be moved to
such road or other land as may be specified.

It is a summary offence, punishable with a fine not exceeding level 3, against RTOA
1988, s 91 to fail to move or cause a vehicle to be moved as soon as practicable when
required under the provisions of reg 3.

RDVR 1986, reg 4 allows a constable to remove, or arrange for the removal of, a
vehicle which is on a road or other land in the circumstances outlined in (a) or (b),
above. A constable may also do so if the vehicle, having broken down on a road or other
land, appears to have been abandoned without lawful authority, or if the vehicle has
been permitted to remain at rest on a road or other land in such a position or circumstances as to appear to have been abandoned without lawful authority.

The power to remove vehicles from land occupied by any person is subject to giving
notice as prescribed by RTRA 1984, s 99 and RDVR 1986, reg 8.

RDVR 1986, reg 4A empowers a *community support officer or volunteer or a person
accredited for this purpose under a community safety accreditation scheme* to remove,
or arrange for the removal of, a vehicle parked or broken down on a road and causing

an obstruction, where it is likely to cause danger to road-users, or where an offence is being committed in relation to a statutory prohibition or restriction included in Sch 1 to the Regulations, referred to above.

Associated with the powers relating to abandoned vehicles are the provisions of RTRA 1984, ss 100 to 102 (interim and absolute disposal of vehicles and charges for removal, storage, and disposal).

Duties of local authorities

25.115   The Refuse Disposal (Amenity) Act 1978, s 3(1) obliges local authorities to remove motor vehicles abandoned without lawful authority on any land in the open air or on any other land forming part of the highway within their area. Before removing vehicles from land other than a road the local authority must serve notice on the occupier of the land which is involved; if the occupier objects the authority is not entitled to remove the vehicle.

TRAFFIC OFFICERS APPOINTED UNDER THE TRAFFIC MANAGEMENT ACT 2004

25.116   Traffic officers are employed by Highways England (in England) or the Welsh Government (in Wales) to perform duties connected with the management of traffic on the network of 'relevant roads' in England or in Wales, depending on which body employs them. A uniformed traffic officer can exercise his/her powers over any 'relevant road', ie a road for which one of the following is the relevant traffic authority, unless his/her designation is limited or subject to conditions which will be specified. In England, Highways England is the relevant traffic authority for all highways for which the Secretary of State was previously the traffic authority, except the M6 toll road and the roads that form part of the Severn Crossings up to the border with Wales (for which the Secretary of State remains the relevant traffic authority). In Wales, the relevant traffic authority is the Welsh Government. In short, most motorways and major trunk roads are 'relevant roads'.

Traffic officers are required to comply with any directions given by a constable. Subject to that limitation, they must comply with any direction given by the relevant traffic authority.

FIXED PENALTY OFFENCES

25.117   RTOA 1988, Sch 3 lists the offences which are 'fixed penalty offences'. Schedule 2 indicates the offences which carry obligatory endorsement, a matter of considerable importance. To assist the memory of police officers in this respect, aides memoire have been prepared listing the various offences which are non-endorsable fixed penalty offences and those which are endorsable. The aide memoire in relation to endorsable fixed penalty offences also indicates the number of penalty points which apply, and gives a code number for each offence. The aide memoire for non-endorsable fixed penalty offences includes advice concerning enforcement policies existing within a police force, particularly where a vehicle rectification scheme is in operation.

The non-endorsable fixed penalty offences are generally concerned with those involving minor defects in the vehicle or its parts; vehicle registration and trade licence offences; failure to display goods vehicle plates; unlawfully stopping on the carriageway or hard shoulder or (England only) emergency refuge area of a motorway;

contravention of certain traffic directions (eg 'Give way', 'No entry', 'No U turn', 'One-way' signs, and temporary traffic signs); obstruction; waiting and parking offences; lighting and noise offences; offences relating to loads (other than dangerous loads); use of a vehicle without a test certificate; trailer offences; offences peculiar to motorcycles; offences of carrying more than one person on a pedal cycle and of cycling on a footway; various offences relating to the registration of trailers; and other miscellaneous motoring offences (eg failing to wear a seat belt).

The endorsable fixed penalty offences include those in the Road Vehicles (Construction and Use) Regulations 1986 which are concerned with danger (parts, loads, tyres, etc); driving without due care or attention, or without reasonable consideration; failure to comply with 'Stop' signs, double white lines, traffic lights, directions of a constable; illegal waiting at pedestrian crossings, overtaking a moving or stationary vehicle at such a crossing and failure to accord precedence at a pedestrian (or similar) crossing; speeding; stopping or reversing on the carriageway of a motorway; driving on a hard shoulder or central reservation; driving on a motorway by a provisional licence-holder; use of offside lane of three-lane motorway by a large goods vehicle or passenger-carrying vehicle; breaches of conditions of provisional licences; use, etc of a vehicle without insurance; failure to identify the driver; offences relating to motorcycle passengers; leaving a vehicle in a dangerous position; use of a vehicle in a designated play street; and driving with uncorrected defective eyesight.

Up to three penalty notices may be issued for one occurrence but only one may be for an endorsable offence.

Fixed penalty notices should not be given to juveniles as there are other recommended procedures for dealing with juveniles which recognise all the circumstances of each case.

The provisions discussed under this heading are contained in RTOA 1988, unless otherwise indicated.

Issue of a fixed penalty notice on-the-spot: person present

25.118 RTOA 1988, s 54 applies where a constable in uniform, or a vehicle examiner (ie an examiner appointed under RTA 1988, s 66A by the Secretary of State) who produces his/her authority, has reason to believe that a person (s)he finds:

(a) is committing or has on that occasion committed a fixed penalty offence (endorsable or non-endorsable); or

(b) has, within the period of 28 days before the day of that occasion, committed a Community drivers' hours offence (**25.95**) (RTOA 1988, s 54(1)).

In the above circumstances, a fixed penalty notice may be issued on-the-spot by the constable or vehicle examiner (s 54(2)). However, where s 54 applies by virtue of (b), a constable or vehicle examiner may not give a person (P) a fixed penalty notice for the Community drivers' hours offence if (s)he has reason to believe that:

(i) a fixed penalty notice has already been given to P under s 54 in relation to the offence;

(ii) a conditional offer has already been issued to P under s 75 (**25.132**) in relation to the offence;

(iii) proceedings have already been initiated against P for the offence; or

(iv) any other penalty has already been imposed on, or other proceedings have already been initiated against, P in respect of the breach of the applicable Community rules which constitutes the Community drivers' hours offence concerned.

Issue of a fixed penalty notice for endorsable offence: person present

25.119   RTOA 1988, s 54(2) states that, subject to the provisions described below, a fixed penalty notice for an endorsable offence *may* be issued to the offender on-the-spot by a constable in uniform, or a vehicle examiner who produces his/her authority, who has reason to believe that a person *is committing or has on that occasion committed* a fixed penalty offence.

By s 54(3), where it appears to the constable or the vehicle examiner that the fixed penalty offence involves obligatory endorsement (ie endorsable), the constable or vehicle examiner may only give the offender (O) a fixed penalty notice on-the-spot in respect of the offence if:

(a) the constable or vehicle examiner is satisfied, on accessing information held on O's driving record, that O would not be liable to be disqualified under the 'totting-up' provisions because the total number of penalty points would number 12 or more if O was convicted of the fixed penalty offence; and

(b) in the case of a person who is the holder of a licence, (s)he produces it for inspection by the constable or vehicle examiner and surrenders it to him/her to be retained and dealt with in accordance with RTOA 1988.

Section 56 requires a receipt to be given for a driving licence so surrendered. The receipt given is incorporated in the fixed penalty notice. It is valid for one month from the date of issue (or such longer period as may be prescribed). However, a person to whom the penalty is paid may issue a new receipt on the application of the licence-holder, which will expire on such date as may be specified in it. In any event, a receipt ceases to have effect on the return of the licence to the holder. Section 56 enables the receipt to be produced in place of a licence, subject to the same conditions concerning later production, on a request being made by a constable or examiner under RTA 1988, s 164, provided that, if required to do so, the licence-holder subsequently produces his/her driving licence at the specified police station immediately it is returned.

Further points

25.120   *Provisional fixed penalty notice*   Section 54(4) and (5A) provides that, where the offence appears to the constable or vehicle examiner to be endorsable, and:

(a) the constable or vehicle examiner is unable to satisfy him/herself, by accessing information held on O's driving record, that O would not be liable to a 'totting-up' disqualification because the total number of penalty points would number 12 or more if O was convicted of the fixed penalty offence; or

(b) where O is the holder of a licence, (s)he does not produce it for inspection by the constable or vehicle examiner,

the constable or vehicle examiner may give O a provisional fixed penalty notice.

Such a notice given by a constable must state that *if, within seven days after the notice is given, O personally delivers the notice together with his/her driving licence to a constable or an authorised person at the police station specified in the notice (being a police station chosen by O)* and the requirements in the next paragraph are met O will then be given a fixed penalty notice in respect of the offence (s 54(5) and (5A)(a)). (An 'authorised person' is a person at a police station authorised for this purpose by the chief officer of police.) In the case of a notice given by a vehicle examiner the italicised words are replaced by 'if, within 14 days O (either by post or in person) gives the notice etc to the Secretary of State at the place specified in the notice' (s 54(5A)(b)).

If O delivers the notice and (if O is the holder of a licence) his/her licence as required, and the following requirements are met:

(i)   the person to whom the notice is delivered is satisfied, on accessing information
      held on O's driving record, that O would not be liable to a 'totting-up' disqualifica-
      tion (as above); and

(ii) if O is the holder of a licence, it is delivered to be retained and dealt with in accord-
      ance with RTOA 1988,

the person to whom the notice is delivered must give O a fixed penalty notice in respect
of the offence (s 54(5B)).

If O is the holder of a licence and fails to produce it under the provisions of RTOA
1988, or it is found that O's penalty points are such that the issue of disqualification
arises, O will be prosecuted for the offence in the normal way and the provisional fixed
penalty notice is of no further effect.

General   In determining under s 54 whether O would be liable on conviction to disqualification it must be assumed, where the offence carries a range of penalty points,
that the points to be attributed to the offence would be the lowest in the range.

Endorsements of licences without hearing   RTOA 1988, s 57A provides that where a
fixed penalty notice under s 54 is given to a person in respect of an offence involving
obligatory endorsement, the Secretary of State may endorse the person's driving record
accordingly without any order of a court. If the fixed penalty is paid before the end of
the suspended enforcement period and the person to whom the payment is made is the
fixed penalty clerk, the clerk must send to the Secretary of State notice of the relevant
particulars which are to be endorsed on the person's driving record.

A person's driving record may not be so endorsed if at the end of the suspended enforcement period (as to which see **25.125**):

(a)   (s)he has given notice, in the manner specified in the fixed penalty notice, re-
      questing a hearing in respect of the offence to which the fixed penalty notice re-
      lates; and

(b)   the fixed penalty has not been paid.

Section 83 deals with the case where the appropriate person is deceived into endorsing a driving record under s 57A in circumstances in which the licence-holder
should have been disqualified under the totting-up procedure and the deception constituted, or was due to, an offence committed by the licence-holder. It provides that if
the licence-holder is convicted of that offence, the court before which (s)he is convicted
will have the same powers to disqualify as it would have had if it was convicting him of
the endorsable offence.

Driver absent: fixing notices to vehicles

RTOA 1988, s 62(1) provides that where a constable or vehicle examiner has reason   **25.121**
to believe that a fixed penalty offence is being or has been committed in respect of a
stationary vehicle, he *may* affix to that vehicle a fixed penalty notice in respect of the
offence unless the offence appears to him to involve obligatory endorsement.

Notice to owner

In cases in which a fixed penalty notice is affixed to a vehicle under RTOA 1988, s 62   **25.122**
and the fixed penalty has not been paid within the suspended enforcement period, *a
notice to owner* may be served under s 63 by or on behalf of the chief officer of police

(Secretary of State, if notice fixed by vehicle examiner) on *any person who appears to him/her* to be the owner of the vehicle (or a person authorised to act on such person's behalf). Such a notice must give particulars of the offence in question, of the fixed penalty concerned, and of the period allowed for response to the notice. It must also indicate that, if the fixed penalty is not paid before the end of that period, the person on whom the notice is served must furnish, before the end of that period, a statutory statement of ownership. The period allowed for response to such a notice is 21 days from the service of the notice. However, a notice need not be served when a request for a hearing has been made with an admission by the person making the request that (s)he was the driver of the vehicle on that occasion. If a person on whom a notice has been served was not the owner of the vehicle at the relevant time, and (s)he furnishes in time a statutory statement to that effect, (s)he is not liable for the offence in question. Otherwise, where a notice is served within six months of the commission of that offence and the fixed penalty has not been paid within the period allowed, a sum equal to the fixed penalty, plus one-half of that penalty, may be registered against the person on whom the notice to owner was served as a fine.

Where person receiving was not the driver

25.123   There will be occasions when the person who receives a notice to owner will not have been the driver at the relevant time. RTOA 1988, s 63 therefore requires that the notice must indicate that, before the end of the period permitted for response, the person on whom it is served may either:

(a) request a hearing; or
(b) if (s)he was not the driver and someone purporting to have been the driver wishes to give notice requesting a hearing, furnish, together with a statutory statement of ownership, a statutory statement of facts which has the effect of the actual driver requesting a hearing.

Once a statutory statement of facts has been supplied, the person named in the notice to owner is relieved of further responsibility and any sums due by way of fixed penalty may not be registered for enforcement.

Once a fixed penalty has been registered for enforcement as a fine against a person on whom a notice to owner has been served, however, no proceedings can be brought against any other person in respect of that offence.

Special provisions are applied to hire vehicles, ie a vehicle hired for temporary use, by RTOA 1988, s 66. Provided a form of agreement has been drawn up between the hire firm and the hirer under the Road Traffic (Owner Liability) Regulations 2000, and provided the hire firm produces a copy of this to the police (together with a statement of hirer liability signed by the hirer under that hiring agreement), liability is removed from the hire firm; and the hirer of the vehicle becomes its owner for the purpose of the fixed penalty provisions.

Fixed penalty offences generally

Number of penalty points

25.124   By RTOA 1988, s 28(3), the number of penalty points to be attributed to an endorsable offence where a fixed penalty notice has been issued is:

(a) where both a range of numbers and a number followed by the words '(fixed penalty)' is indicated in Sch 2, that number;

(b) where a range of numbers followed by the words 'or appropriate penalty points (fixed penalty)', the appropriate number of penalty points for the offence (such number imposed by order under s 28(3A) and (3B));

(c) where only a range of numbers is shown, the lower number in that range.

Proceedings

No proceedings may be brought for an offence to which a fixed penalty notice relates **25.125** until 21 days have passed since the day of the notice (or such longer period as may be specified in the notice). This period of time is referred to as the suspended enforcement period.

Where a fixed penalty notice has been given to an offender (O), no proceedings may be brought unless O has given notice requesting a hearing before the end of the suspended enforcement period.

If no such notice has been received and the fixed penalty has not been paid before the end of the suspended enforcement period, a sum equal to the fixed penalty plus one-half of that amount may be registered against O as a fine.

Evidence in court proceedings in fixed penalty notice cases

RTOA 1988, s 79 permits the service of the statement of evidence of a constable or a **25.126** vehicle examiner together with a fixed penalty notice or notice to owner. Such a statement will be deemed to have been served for the purposes of the Criminal Justice Act 1967, s 9 (**8.43**).

RTOA 1988, s 68 and Sch 4 permit 'statutory statements of ownership or of facts' to be used in evidence. A statutory statement of ownership is one where the declarant states whether or not (s)he was the owner of the vehicle at the relevant time. If (s)he was not it will state whether (s)he was ever the owner and, if so, when. A statutory statement of facts is one where the declarant states that (s)he was not the driver of the vehicle at the relevant time and states the name and address of the person who was.

Payment of penalties

Fixed penalties must be paid to the designated officer for a magistrates' court in a **25.127** manner specified within the notice. Payment may be made by way of a properly addressed, pre-paid letter. The payment is regarded as having been made at the time that the letter would be delivered in the course of normal post. If a fixed penalty is paid before the end of the suspended enforcement period, no further proceedings may be brought.

By RTOA 1988, s 53, the amount of the fixed penalty to be paid in respect of an offence is such amount as the Secretary of State may by order prescribe, or one-half of the maximum fine for it on summary conviction, whichever is less.

The amounts prescribed by the Secretary of State are currently contained in the Fixed Penalty Order 2000. Schedule 1 makes specific provision in respect of particular specified offences. In addition it provides as follows. The fixed penalty for offences involving obligatory endorsement is £100. Where the offence is non-endorsable the penalty is normally £50. However, where the non-endorsable offence consists of a fixed penalty parking offence it is £30 (or, in London, £60 (if red route) or £40 (otherwise)).

The following may be noted in relation to fixed penalties which are allocated to specific offences dealt with in this chapter.

The fixed penalty for an offence against RTA 1988, s 42 (**25.8**) is £30 where it relates to unnecessary obstruction contrary to RV(C&U) R 1986, reg 103.

The fixed penalty for offences against the Highways Act 1835, s 72 and RTA 1988, ss 24, 42, and 163 (cycling/driving on footway, pedal cycle offences, breach of miscellaneous construction, and use regulations (but see **25.128**), and non-compliance with requirement to stop: **25.8, 25.91,** and **25.103**) is £50.

The fixed penalty for offences against RTRA 1984, s 25 (pedestrian crossings: **25.104**), RTA 1988, ss 14, 15, 16, 18, 23, 35, 36, 40A, 41A, 41D(a), 42, and 47 (seat belt offences, motor cycles, traffic directions and signs, using vehicle in dangerous condition, requirements as to steering-gear, brakes, lack of control or (subject to **25.128**) tyres and other construction and use requirements, and use without a MOT certificate: **25.4, 25.5, 25.7, 25.34, 25.84, 25.90, 25.98, 25.101**) is £100.

The fixed penalty for an offence under RTA 1988, s 41D(b) (use of hand-held mobile telephones, etc while driving: **25.7**) is £200.

The fixed penalty for an offence under RTOA 1988, s 90D(6) (driving, etc in contravention of prohibition for failure to pay financial penalty deposit, etc: **25.133**) is £300.

For the amounts of fixed penalties payable in respect of offence referred to in Chapters 23, 24, and 26, see **23.61, 24.95, 26.10,** and **26.30**.

Graduated fixed penalties

25.128   RTOA 1988, s 53 permits fixed penalties to be different depending upon the circumstances including (in particular):

(a) the nature of the contravention or failure constituting the offence;
(b) how serious it is;
(c) the area, or sort of place, where it takes place; and
(d) whether the offender appears to have committed any offence or offences of a specified description, during the period so specified.

The Fixed Penalty Order 2000, Sch 2 specifies the fixed penalty to be paid for offences by reference to the matters set out above. In terms of offences dealt with in this chapter, graduated fixed penalties apply to offences under RTA 1988:

(a) s 41A (where offence consists of failure to have a tread pattern of at least 1 mm and the offence does not involve a motor cycle)—the penalty is £200;
(b) s 41B (weight limit offences: **25.6**) and vary from £100 to £300 depending upon the percentage of the excess weight; and
(c) s 42 (where offence relates to speed limiters and danger in respect of passengers or load) and vary from £100 to £200.

Fixed penalty notices given by vehicle examiners

25.129   Where a fixed penalty notice is issued by a vehicle examiner, in respect of offences, the system is operated independently of the police by the Secretary of State who is responsible for administering the system, collecting the penalties, and the endorsement of driving records.

Penalty registered against person who has no knowledge of the offence

25.130   RTOA 1988, ss 72 and 73 deal with the issue where a penalty is registered against a person who, for some reason, has no knowledge of the offence, as where a driver of someone else's vehicle does not inform the owner of the notice, or a fixed penalty notice is affixed to a vehicle which has changed ownership without DVLA records being amended.

Notices given to offender By s 72, if registration has followed non-receipt of a request for a hearing and the fixed penalty has not been paid, a statutory declaration may be made to the effect that the declarant:

(a) was not the person to whom the fixed penalty notice was given or
(b) had requested a hearing before the end of the suspended enforcement period.

Notices fixed to vehicles Section 73 provides that, where the registration has followed service of 'notice to owner', and the penalty has not been paid, the statutory declaration must state either that the declarant:

(i) did not know of the penalty, or fixed penalty notice, or notice to owner, until he received notice of the registration; or
(ii) was not the owner at the time of the offence alleged and that he has reasonable excuse for failing to comply with the notice to owner; or
(iii) requested a hearing as permitted by the notice to owner.

A declaration under s 72 or 73 must be served on the proper officer of the relevant court, within 21 days of receipt of the notification of registration. With the exception of (b) and (iii) respectively, the effect of the declaration is that the relevant notice, registration, or endorsement (as the case may be) is void; in case (b), the declarant will be treated as if (s)he had given notice requesting a hearing. In the case of (iii), the effect is to re-activate the 21-day response period referred to in the notice.

Miscellaneous points

RTOA 1988, s 30 requires that, if a person is convicted of an endorsable offence and **25.131** the court is satisfied that (s)he is liable to have penalty points endorsed upon his/her driving record within the fixed penalty procedure for an offence committed on the same occasion as those for which (s)he is convicted, the court must reduce the number of penalty points endorsed on the licence by the number which will be attached in consequence of the fixed penalty offence.

It is a summary offence contrary to RTOA 1988:

(a) s 62(2), to remove or interfere with a notice affixed to a vehicle under s 62(1) (**25.121**) unless it is done by or under the authority of the driver or person in charge of the vehicle or the person liable for the offence in question;
(b) s 67, recklessly to furnish a statement which is false in a material particular, in response to a notice to owner, or to furnish such a statement knowing it is false in that particular.

Such an offence is punishable with a fine not exceeding level 2 and 5 respectively.

Conditional offer of fixed penalty

RTOA 1988, s 75 provides as follows. Where a constable or vehicle examiner has reason **25.132** to believe that a fixed penalty offence has been committed and *no fixed penalty notice has been given at the time or fixed to the vehicle concerned*, a notice of 'conditional offer' may be sent to the alleged offender (O) by, or on behalf of, the chief officer of police (by the Secretary of State in the case of a vehicle examiner). Section 75 is of particular importance where speeding or traffic light offences have been detected by automatic devices.

A conditional offer must:

(a) give particulars of the circumstances of the alleged offence;
(b) state the amount of the fixed penalty; and
(c) state that proceedings cannot be commenced until the end of 28 days following the date of issue of the offer (or such longer period as may be specified therein).

It must also indicate that if (i) payment is made within the period specified under (c), and (ii) (where the offence is endorsable) O at the same time delivers his/her licence to the appropriate person, and that person is satisfied that O would not be liable to disqualification under RTOA 1988, s 35 (totting up), any liability to conviction will be discharged. For the purposes of (ii), it is assumed (where penalty points awardable for the offence are within a range) that the number to be attributed for the offence would be the lowest in the range. Where the conditional offer was made by a constable, 'the appropriate person' means the fixed penalty clerk; where it was issued by a vehicle examiner the term means the Secretary of State.

By RTOA 1988, s 76, if payment is made in accordance with the offer, no proceedings will be taken for the offence to which the conditional offer relates, unless:

(a) it appears to the appropriate person that O would be liable to be disqualified under RTOA 1988, s 35 if O were convicted of the offence;
(b) the appropriate person returns the payment and licence to O; and
(c) where the appropriate person is not the Secretary of State, the appropriate person notifies the chief officer of police that (s)he has done so.

Where O pays the fixed penalty and (if O is the holder of a licence) delivers his/her licence to the appropriate person, and proceedings against O are excluded by s 76, the Secretary of State will endorse O's driving record and O's licence will be returned to O (s 77A).

Financial penalty deposits

25.133   RTOA 1988, Part 3A (ss 90A–90F) empowers a constable or vehicle examiner to require from a person (P) the payment of a financial penalty deposit if P does not provide a satisfactory address in the UK, where the constable or vehicle examiner has reason to believe

(a) that P—
    (i) is committing or has on that occasion committed an offence relating to a motor vehicle or trailer; or
    (ii) has, within the period of 28 days before the day of that occasion, committed an offence relating to a motor vehicle which is a Community drivers' hours offence; and
(b) that P, the offence and the circumstances in which it is committed are of a description specified in the Road Safety (Financial Penalty Deposit) Order 2009.

However, a constable or vehicle examiner may not impose a financial penalty deposit requirement on P in relation to a Community drivers' hours offence where the constable or vehicle examiner has reason to believe that:

(a) a financial penalty deposit requirement has already been imposed on P in relation to the offence;
(b) proceedings have already been initiated against P for the offence; or

(c) any other penalty has already been imposed on, or other proceedings have already been initiated against P in respect of the relevant breach in Northern Ireland, an EU state or a contracting third country.

Under the Road Safety (Financial Penalty Deposit) Order 2009, P (the *specified person*) is the person whom the constable or vehicle examiner has reason to believe is committing, or has on the occasion concerned, committed a specified motor vehicle or trailer offence, or has, within the 28-day period before that occasion, committed a Community drivers' hours offence relating to a motor vehicle. The *specified offences* are set out in a lengthy list in the Schedule to the Order; almost all those offences are those described in this chapter and Chapters 23, 24, and 26. The *specified circumstances* are that the offence is being committed, or has on that occasion been committed, on a road or other public place.

A financial penalty deposit requirement is a requirement to make a payment of the appropriate amount to the Secretary of State and either immediately or within the relevant period. The Schedules to the Road Safety (Financial Penalty Deposit) (Appropriate Amount) Order 2009 specify the appropriate amount in respect of each of the specified offences referred to above. The 'relevant period' means:

(a) if P was given a fixed penalty notice and proceedings are not brought in respect of the offence before the end of the suspended enforcement period, the suspended enforcement period;
(b) if P was handed a conditional offer and proceedings are not brought in respect of the offence before the end of the 28-day period following the date on which the conditional offer was given or any longer period specified in the conditional offer, that period; and
(c) otherwise, the period ending with the person being charged with the offence.

Police and vehicle examiners are empowered to prohibit the moving of the vehicle if the deposit is not paid immediately, although a written direction may be given to remove it to a specified place. The Road Safety (Immobilisation, Removal and Disposal of Vehicles) Regulations 2009 (see Chapter 25 on companion website) provide for the detention, etc of vehicles prohibited from being driven following the imposition of a prohibition. The prohibition continues until the deposit is paid, or (where the offender has received a fixed penalty notice or conditional offer) the offender has paid the fixed penalty, or the offender is convicted or acquitted of the offence or informed that (s)he will not be prosecuted for it, or a 12-month period has elapsed since the deposit requirement. It is a summary offence, punishable with an unlimited fine, to fail to comply with a prohibition, to cause or permit a vehicle to be driven in contravention of it, or to fail to comply within a reasonable time with a direction requiring the person in charge of the vehicle to remove it to a specified place (RTOA 1988, s 90D(6)).

Thus, a means is provided to secure a deposit where an enforcement officer is not satisfied that a penalty or fine could be enforced in the UK. If a person who has paid a financial penalty deposit is subsequently acquitted of the offence, the deposit, with interest, will be returned; otherwise it will be set against any penalty imposed.

DRIVING INSTRUCTION

25.134   RTA 1988, s 123 prohibits the giving of paid driving instruction (ie for money or money's worth) in the driving of a motor car unless the instructor is a registered approved instructor or the holder of a licence authorising him/her to give such instruction (s 123(1)). The register of approved instructors is in two Parts:

(a) Part 1 contains the names of persons who satisfy the requirements for registration under Part 1 (including those European driving instructors who move to Great Britain and seek to provide driving instruction services on an established basis); and

(b) Part 2 contains the names of European driving instructors who are entitled to provide driving instruction on a temporary and occasional basis.

Paid instruction may not be given unless there is fixed to, and exhibited on, the motor car the current licence or certificate of registration, in a similar position to that formerly occupied by the vehicle licence (s 123(2)). Free instruction given as a perk when buying a car from a motor trader is deemed to be given for payment. By s 124, s 123 does not apply to police instructors.

Licensed instructor

25.135 To enable persons to gain experience with a view to undergoing the practical test of ability and fitness to instruct, which is part of the official driving instructor's examination for registration purposes, RTA 1988, s 129 allows him/her to be granted a licence to give driving instruction if (s)he has passed the written part, and the practical test of driving ability and fitness to drive, of the official driving instructors' examination. Licence-holders may only give instruction from premises named in the licence. If the premises are a driving school there cannot be more than one licence-holder to each registered instructor.

For the first three months of any licence (other than the second of two consecutive ones) the holder either must be under the direct personal supervision of a registered instructor for at least one-fifth of that time and keep a log to that effect, or must undertake a minimum of 20 hours' supplementary training. In the case of the latter alternative (s)he must undertake five further hours of training in a second consecutive period.

Offences

25.136 By RTA 1988, s 123(4), a person giving instruction in breach of s 123(1) commits an offence, as does his/her employer if (s)he is employed for that purpose. It is a defence for a defendant to prove that (s)he did not know and had no reasonable cause to believe that his/her name, or that of his/her employee, was not in the register.

If instruction is given in contravention of s 123(2), the person who gives it commits an offence, contrary to s 123(6), unless (a) (s)he proves that the instruction was given in accordance with a right conferred on him/her by the European Union (Recognition of Professional Qualifications) Regulations 2015, Part 2, and (b) (s)he had not received the certificate mentioned in s 123(2)

A person to whom a certificate of registration or a licence has been granted must produce it to a constable or authorised person on being required to do so. Failure to do so is an offence, contrary to s 137(3). However, it is a defence to show: production within seven days at a police station specified by the constable or at a place specified by the authorised person, or production as soon as reasonably practicable, or production was not reasonably practicable before the day on which proceedings commenced.

By s 135(2), it is an offence for any person who is not registered in Part 1 of the register to take or use a title prescribed for use by persons registered in Part 1, to wear or display a badge or certificate so prescribed, or to take or use any name, title or description which implies that (s)he is registered. Section 135(2A) provides a corresponding offence in respect of purporting to be registered under Part 2 It is a defence to an offence under s 135(2) or (2A) for the defendant to prove that (s)he did not know, and had no reasonable cause to believe, that his/her name was not in the relevant Part of the register at the material time.

Section 135(3) provides that a person (P) carrying on a driving instruction business who uses any such title or description in relation to an employee who is not registered under Part 1, is guilty of an offence under s 135(3) unless P proves that P did not know, and had no reasonable cause to believe, that the person's name was not in Part 1 of the register at the material time. There is a corresponding offence under s 135(4) where P uses such a title or description in relation to an employee not registered under Part 2.

All the above are summary offences. Those under ss 123(4) and 135(2)–(4) are punishable with a fine not exceeding level 4, and those under s 123 (6) and 137(3) level 3.

RESTRICTIONS ON THE USE OF MOTORWAYS

The Motorways Traffic (England and Wales) Regulations 1982 impose various restrictions on drivers on motorways of the classes of vehicles permitted to use motorways.   **25.137**

Vehicles which may use motorways

Under the 1982 Regulations, Class I and Class II vehicles are, in normal circumstances, permitted to use motorways.   **25.138**

Class I includes heavy and light locomotives, motor tractors, heavy motor cars, motor cars and motorcycles of not less than 50 cc and trailers drawn by such vehicles. (**25.1**). Track-laying vehicles are not within Class I. To be in Class I vehicles must be fitted with pneumatic tyres, must not be agricultural motor vehicles or machines and must not be pedestrian-controlled. All such vehicles must be capable of attaining a speed of 25 mph when unladen.

Class II vehicles are those specially made to transport abnormal, indivisible loads, and large vehicles (eg tank transporters) used by the armed services. In addition, earth movers and similar engineering plant are Class II vehicles if they are capable of attaining 25 mph when unladen.

Other provisions

The 1982 Regulations also contain provisions about: **25.139**

- the direction of driving;
- stopping;
- reversing;
- the general use, including by provisional licence-holders; and
- exclusion of certain vehicles from the outside lane,

on a motorway.

These provisions, and exceptions and relaxations from the effect of the 1982 Regulations, are set out in Chapter 25 on the companion website.

Offence

25.140   A person who uses a motorway in contravention of the 1982 regulations commits a summary offence under RTRA 1984, s 17(4). The offence is punishable with a fine not exceeding level 4, and with discretionary disqualification if committed in respect of a motor vehicle otherwise than by unlawfully stopping or allowing it to remain at rest on a part of the motorway on which vehicles are permitted in certain circumstances to remain at rest. It is a fixed penalty offence, the fixed financial penalty being £100 (or £50 in the exception just mentioned). As to fixed penalty offences, see **25.117-25.133**.

CHAPTER 26

Responsible Use of Vehicles

INTRODUCTION

Statutory offences were created in the nineteenth century to deal with the responsible **26.1** use of carriages and the like. Although these provisions were enacted before the days of mechanically propelled vehicles, they can be applied to such vehicles. Town Police Clauses Act 1847, s 28 provides that the furious driving in a street of any horse or carriage is a summary offence (maximum punishment a level 3 fine or 14 days' imprisonment). The Offences Against the Person Act 1861, s 35 punishes as an offence triable only on indictment the causing of bodily harm by wanton or furious driving on a road or otherwise by a person having charge of a carriage or vehicle; this charge can be useful in circumstances in which the negligent driving of a mechanically propelled vehicle did not take place on a road or other public place. This offence is punishable with up to two years' imprisonment and (see **24.40**) with discretionary disqualification for driving where the vehicle is a mechanically propelled vehicle. However, for all practical purposes, driving offences that are committed in relation to mechanically propelled vehicles are governed by the Road Traffic Act 1988 (RTA 1988). Those offences contain a number of general terms, which are defined below or elsewhere in this book.

'Driving'

The essence of 'driving' is the use of the driver's controls (or at least one of them) in **26.2** order to control the movement of the vehicle, however that movement is produced, provided that what occurs can in any sense be regarded as 'driving'. Thus, a person who releases the handbrake and 'coasts' downhill in a car is 'driving' it (even if the steering is locked). Someone who propels a moped by 'paddling' with his/her feet, sitting astride it and steering it, is driving it. A person in the driving seat of a car which is being towed is driving if (s)he has the ability to control its movements by means of the brakes or steering. This is so even if the vehicle is attached by means of a rigid tow bar which is attached to the towing vehicle by means of a ball hitch, and to the vehicle being drawn by means of a shackle. Such a combination leaves the person holding the steering wheel of the towed vehicle with a substantial potential for directional control because (s)he is able (indeed (s)he is required) to keep the towed vehicle in line with the towing vehicle by means of the steering wheel. It would be irrelevant that the towed vehicle does not have any brakes. On the other hand, a person in the driving seat of a vehicle on a fixed tow is not driving it, since (s)he cannot control its movements by the use of any of the driver's controls. Nor is a person who is pushing a car and steering it with his/her hand through the window driving it because this cannot in any sense be described as driving.

A divisional court has held that the defendant (D) was 'driving' in circumstances where the vehicle was stationary. D was sitting in the driver's seat of a vehicle on a grass verge engaging in an activity which he described as 'wheel spinning'. This involved the use of the engine, accelerator, clutch, and steering wheel with the vehicle in gear but the handbrake applied preventing any form of movement. The court held that the ordinary meaning of the word 'driving' included a situation where someone was ensuring that there was no movement but the wheels were spinning.

Two people may be driving a vehicle at the same time. For example, if A who is in the driving seat operates the clutch and brakes and gear shift but allows his/her passenger, B, to steer the vehicle, both A and B are driving.

The Road Traffic Act 1988 (RTA 1988), s 192 provides that, except for the purposes of the offence of causing death by dangerous driving, where a separate person acts as steersman of a motor vehicle he is driving the vehicle (as well as any other person engaged in driving it). This covers the case where one person is primarily concerned with the propulsion of the vehicle and another acts as its steersman. It is relevant only in the case of traction engines and the like.

An employer cannot be convicted of a driving offence physically committed by an employee.

Other terms

26.3   *Mechanically propelled vehicle* See **23.5**.
Motor vehicle See **23.6**.
Road See **23.4**.

Other public place   This term means a place (other than a 'road') to which the public, whether on payment or otherwise, have access with at least the tolerance of the landowner or proprietor. It is irrelevant whether the public could have access; the question is whether they actually have access to the place in question at the time in question. For example, the car park of a public house is open to the public at large during licensing hours and is therefore a public place during those hours, but the car park of a members' club is restricted to members of that club and their guests and therefore it is never a 'public place'. A car park situated within the business premises of a motor dealer for use by customers has been held to be a public place as members of the public using the car park did not cease to be members of the public and become a special class of persons merely because they used the car park as customers.

A place can be a 'public place' even though access to it is only permitted after a person has been screened provided that the screening does not relate to a special characteristic or reason personal to the person screened. Thus, if anyone can pass through the screening process and gain access simply by paying an entry fee (eg a site fee at a caravan park), or by satisfying conditions imposed by the landowner (eg access only for private vehicles or the possession of a boarding pass), the place is a public place. It would be different if the screening related to membership of the caravan club or status as a worker at a factory to whose car park access is sought.

The absence of a physical obstruction or of a notice forbidding entry does not of itself mean that the public have access.

DANGEROUS DRIVING, AND CAUSING DEATH OR SERIOUS INJURY BY DANGEROUS DRIVING

Dangerous driving

26.4   RTA 1988, s 2 provides that it is an offence for a person to drive a mechanically propelled vehicle dangerously on a road or other public place. The offence is an indictable (either-way) offence punishable with up to two years' imprisonment and (see **24.39**) with obligatory disqualification for driving.

The ways in which driving may be dangerous are set out in RTA 1988, s 2A.

Dangerous manner

RTA 1988, s 2A(1) provides that a person is to be regarded as driving dangerously if: **26.5**

(a) the way that he drives falls far below what would be expected of a competent and careful driver; and

(b) it would be obvious to a competent and careful driver that driving in that way would be dangerous.

This test in RTA 1988, s 2A(1) is concerned with whether the manner of D's driving falls *far* below what would be expected of a careful and competent driver in circumstances where it would be obvious to a competent and careful driver that driving in that way would be dangerous. The presence of the word '*far*' separates offences of dangerous driving from those which should more appropriately be charged as careless driving.

It would be no defence that D was doing his/her incompetent best, nor that (s)he did not intend to drive dangerously. This was held by the Court of Appeal in a case where a bus driver unintentionally pressed the accelerator when he meant to press the brake; as a result the bus travelled across a pedestrian island and pedestrians were killed. The court held that the driver's lack of an intention to drive dangerously was no defence.

The objective standards to be applied will vary according to the prevailing conditions and any exceptional circumstances. In fog, the competent and careful driver would drive in a manner quite different from that which (s)he would adopt in dry, clear conditions. Whilst driving at a speed of 70 mph in an area restricted to 30 mph might be described as dangerous, it would not often be so on a dual carriageway. If it occurred in the middle of the night when the road was not being used by any other person or vehicle, such driving in a restricted area might not be 'dangerous', whilst driving at that speed on a dual carriageway which is restricted by road works might be.

The opinion of witnesses as to whether or not D drove dangerously is, of course, inadmissible.

Once it has been established that the actual driving fell far below standards of a competent and careful driver, it becomes necessary to consider whether or not it would be 'obvious' to a competent and careful driver that such driving would be dangerous. Section 2A(3) states that 'dangerous' refers to danger either of injury to any person or of serious damage to property. It also states that, for the purposes of s 2A(1) and s 2A(2) (**26.6**), in determining what would be expected of, or obvious to, a competent and careful driver in a particular case, regard must be had not only to the circumstances of which (s)he could be expected to be aware but also to any circumstances shown to have been within the defendant's knowledge. Regard must not be had to any circumstance which D wrongly believed to exist. Consequently, it has been held that, where a police officer pursued a stolen car at speed through traffic lights at red, his mistaken belief that the junction was being controlled by other officers (so that he could cross safely) was irrelevant to the issue of dangerous driving.

The Court of Appeal has held that D's special skill as a driver (eg D had taken a police advanced driving course) is an irrelevant consideration when considering whether driving was dangerous: to take it into account would be inconsistent with the objective test of the competent and careful driver.

A breach of the Highway Code does not necessarily mean that an offence has been committed, but the Code's provisions can be considered by a jury or magistrates, provided that it is clearly explained that they do not provide a 'standard'.

Driving in a dangerously defective state Evidence of consumption of alcohol or drugs cannot *by itself* establish the fact that a driver has committed an offence of dangerous driving. Section 2A(1) is concerned with the dangerous manner of the driving, not with the dangerous state of the driver due to drink or drugs. The Court of Appeal has held that, while evidence of alcohol consumption is admissible in respect of an offence of dangerous driving if the amount of alcohol is proved or admitted, even though the driver's blood alcohol level is below the legal limit, it cannot be conclusive. The consumption of a drug, even if the amount is unquantified, has been held by the Court of Appeal to be admissible in respect of such an offence, although it cannot be conclusive. In either case, there must be evidence that the vehicle was driven dangerously within the definition in RTA 1988, s 2A.

Dangerous state of the vehicle

26.6   The second way in which driving may be dangerous is provided by RTA 1988, s 2A(2) which provides that a person is also to be regarded as driving dangerously if it would be obvious to a competent and careful driver that driving the vehicle in its current state would be 'dangerous' (in the sense defined by s 2A(3)). In determining the state of the vehicle, regard may be had to anything carried on or in it, and the manner in which it is attached or carried.

Because the danger involved in driving the vehicle in its current state must be obvious, the offence will not be committed if the defect involved is a latent one. A danger is 'obvious' in this context only if it could be seen or realised at first glance by a competent and careful driver. It is not enough that such a driver would have taken steps to check out whether the vehicle was in a condition which was not dangerous, and by doing so would have discovered the defect, perhaps by examining the underside of the vehicle.

'Current state' in s 2A(2) implies a state different from the original or manufactured state. Thus, where the alleged danger relates solely to something inherent in the original design of the vehicle (eg spikes forming part of a grab unit at the front of an agricultural vehicle) dangerous driving is not committed. It would be different if the original, unaltered state made the *manner* of the driving dangerous.

As with s 2A(1), in determining what would be obvious to a competent and careful driver, regard must be had not only to the circumstances of which (s)he could be expected to be aware but also to any circumstances shown to have been within D's knowledge. Apart from such knowledge, the state of D's mind is irrelevant.

Dangerous cycling

26.7   It is an offence contrary to RTA 1988, s 28(1) to ride dangerously on a road a bicycle, tricyle, or four-or-more wheeled cycle, not being in any case a motor vehicle. A person is to be regarded as riding dangerously if (s)he rides in a *manner* which equates to the manner of driving described at **26.5** (s 28(2) and (3)). The offence under s 28(1) is a summary offence punishable with a fine not exceeding level 4.

Causing death by dangerous driving

26.8   RTA 1988, s 1 provides that a person who causes the death of another person by driving a mechanically propelled vehicle dangerously (as defined in s 2A, see **26.5** and **26.6**) on a road or other public place is guilty of an offence. The offence is triable only on indictment and punishable with up to 14 years' imprisonment and (see **24.39**) with obligatory disqualification for driving. The victim may be someone else in D's vehicle.

It is irrelevant that the risk of death resulting from the dangerous driving in question was unforeseen or unforeseeable, as would be the case where the only obvious risk was of damage to property.

It is not necessary to show that the dangerous driving was the sole cause of death, since it is sufficient that it is a more than minimal cause. The rules about causation referred to at **10.7** apply to s 1 and to ss 1A, 2B, and 3A (below).

Since the offence relates to causing death *by* driving, and not causing death *while* driving, it is not necessary that the driver should have been driving the vehicle (ie it need not have been in motion) at the time that the fatal injury was caused. It is enough that there has been dangerous driving by the driver and that this driving is more than a minimal cause of death. This is shown by a case where a driver took his lorry on to a motorway after having been warned that his brake air-pressure gauges were not working (which constituted dangerous driving), and the handbrake system was activated due to loss of air pressure. The trailer unit blocked the nearside lane of the motorway. Some 12 minutes later a lorry collided with it and the driver of that vehicle was killed. The Court of Appeal held that the consequences of dangerous driving are capable of outlasting the time the driver spends at the wheel. It held that the dangerous driving must have played a part, not simply in creating the occasion of the fatal accident, but in bringing it about. On the facts the consequences of the dangerous driving were not 'spent' and too remote from the accident, and were more than a minimal cause of the death. As a result the driver's conviction for causing death by dangerous driving was upheld. The Court of Appeal has held that, where death is caused by another vehicle colliding with D's previously driven stationary vehicle, the jury (if sure that that D drove dangerously and thereby was a more than minimal cause of death) can convict D if they are sure that it could sensibly have been anticipated that a fatal collision might occur in the circumstances in which the collision did occur. In another case, the Court of Appeal added that the jury would not need to be sure that the particular circumstances of the collision were reasonably foreseeable.

The ingredients of the offences of manslaughter by gross negligence and of causing death by dangerous driving (a less serious offence in terms of punishment) are similar. CPS policy is to charge manslaughter only if the evidence shows a very high degree of risk, making the case one of the utmost gravity.

Causing serious injury by dangerous driving

A person commits an offence under RTA 1988, s 1A if (s)he causes serious injury to another person by driving a mechanically propelled vehicle dangerously (as defined in s 2A (see **26.5** and **26.6**)) on a road or other public place. 'Serious' injury means physical harm which amounts to grievous bodily harm for the purposes of the Offences against the Person Act 1861 (**9.14**). It is irrelevant that serious injury—or indeed any injury—was unforeseen or unforeseeable. The offence under s 1A is an indictable (either-way) offence punishable with up to five years' imprisonment and (see **24.39**) with obligatory disqualification for driving. **26.9**

CARELESS OR INCONSIDERATE DRIVING, AND CAUSING DEATH BY CARELESS OR INCONSIDERATE DRIVING

Careless, or inconsiderate, driving

RTA 1988, s 3 creates two separate offences by providing that if a person drives a mechanically propelled vehicle on a road or other public place without due care and **26.10** attention, or without reasonable consideration for other persons using the road or place, (s)he is guilty of an offence. The offences are summary offences punishable with an unlimited fine and (see **24.40**) with discretionary disqualification for driving. They are fixed penalty offences, the fixed financial penalty being £100. As to fixed penalty offences, see **25.117–25.133**.

Driving without due care and attention

26.11   RTA 1988, s 3ZA (which applies to s 2B and 3A (below) as well as to s 3) provides that:

(a) a person is to be regarded as driving 'without due care and attention' if (and only if) the way he drives falls below what would be expected of a competent and careful driver; and

(b) in determining what would be expected of a competent and careful driver in a particular case, regard must be had not only to the circumstances of which (s)he could be expected to be aware but also to any circumstances shown to have been within the knowledge of the defendant (D).

The definition of driving without due care and attention differs from that relating to dangerous driving in relation to the extent to which D's driving falls below the standard of a competent and careful driver. While dangerous driving requires a fall 'far' below what would be expected of a competent and careful driver, simply 'falling below' that standard, however marginally, suffices for careless driving.

The standard is an objective standard, impersonal and universal, fixed in relation to the safety of the other users of the highway. It is in no way related to the degree of proficiency or degree of experience attained by the individual driver, and it is unaffected by the fact that the driver was driving to an emergency.

Bearing in mind the objective approach to be applied, it is no defence that, as a learner, the driver was doing his/her best, if that best falls short of the standard which might be expected from a competent and careful driver. There is only one standard of competence and his/her actions fall short of it. In circumstances where a competent driver wears shoes with smooth leather soles and this causes his/her foot to slip from the brake pedal, the court will have to consider whether in the circumstances a competent and careful driver would have been aware of the possibility of this happening and would not have driven the car with such shoes. In the same way, it is essential to keep in mind that experienced drivers or specially trained drivers do not in consequence of that experience or training owe some higher standard of care. For example, the driver of a police vehicle owes the ordinary standard of care to other persons on the road.

In applying the appropriate test as to whether D has driven without due care and attention, all the circumstances of the case can be considered, including evidence that D had been affected by drink or that D had taken such an amount of drink as would be likely to affect a driver, since such evidence may indicate a lack of due care on D's part in driving in the way (s)he did. It must, however, be emphasised that it is not sufficient merely to rely on D's defective mental state in order to prove the offence of careless driving. D's condition is relevant and admissible but it does not determine whether the *way* in which D drove was careless.

Failure to observe a provision of the Highway Code does not of itself establish driving without due care and attention, but any such failure may be relied on as evidence of it.

In some cases the facts are such that, unless D offers some explanation consistent with him/her having taken due care, which is not disproved, the only proper inference is careless driving, in which case the justices must convict him/her.

Driving without reasonable consideration

RTA 1988, s 3ZA provides that a person is to be regarded as driving without reasonable **26.12** consideration for other persons only if those persons are inconvenienced by his driving. Those persons must be shown to have been inconvenienced. Thus, a driver may deliberately drive at speed through a pool of water, but will only commit this offence if other persons were either dampened or forced to flee.

Once again, an objective test should be applied as to whether particular forms of driving are carried out without reasonable consideration for other persons using the road or place. Thus, the question is whether D drove without showing the consideration to other road users which would be shown by a considerate driver. The driver who occupies the fast lane of a motorway, or even the outside lane of a dual carriageway, in circumstances in which there is nothing to prevent him/her from regaining the nearside lane, is unreasonably interfering with the progress of other drivers. The irresponsible use of full-beam headlamps may also amount to driving without reasonable consideration. A person may drive without reasonable consideration for other persons using the road even where his/her lack of consideration is to his/her own passenger.

A point about both offences

Because RTA 1988, s 3 creates two offences, an information or written charge alleging **26.13** both alternatives is likely to fail for breach of the rule against duplicity. It is sufficient to allege that a person drove without due care and attention at a particular time or place, or, as the case may be, to allege that (s)he drove without reasonable consideration at a time and place, since it is not necessary to specify the nature of his/her negligence in the information or written charge.

Vehicle used in manner causing alarm, distress, or annoyance

The Police Reform Act 2002 (PRA 2002), s 59(1) provides that, where a uniformed **26.14** constable has reasonable grounds for believing that a mechanically propelled vehicle (MPV) *is being used* in a manner which contravenes RTA 1988, s 3 and which is causing, or is likely to cause, alarm, distress, or annoyance to members of the public, he has the powers under PRA 2002, s 59(3) set out below.

By s 59(2), a uniformed constable has the powers under s 59(3) where (s)he has reasonable grounds for believing that a MPV *has been used* on any occasion in a manner falling within s 59(1).

The powers set out in s 59(3) are:

(a) if the vehicle is moving, to order the person driving it to stop it;
(b) to seize and remove the vehicle;
(c) for the purposes of exercising a power falling within (a) or (b), to enter any premises other than a 'dwelling house', which term does not include any garage or other structure occupied with the dwelling house, or any land appurtenant to the dwelling house, on which (s)he has reasonable grounds for believing the vehicle to be;
(d) to use reasonable force, if necessary, in the exercise of any power conferred by (a)–(c).

Section 59(4) provides that the power of seizure may not be exercised unless:

(a) the constable has warned the person appearing to be the person whose use falls within s 59(1) that he will seize the vehicle if that use continues or is repeated; and
(b) it appears that the use has continued or been repeated after the warning.

However, a warning need not be given if:

(i)   the circumstances make it impracticable;

(ii) the constable has already on that occasion given a warning under s 59(4) in respect of any use of that vehicle or of another MPV by that person or any other person;

(iii) the constable has reasonable grounds for believing that such a warning has been given on that occasion otherwise than by him/her; or

(iv) the constable has reasonable grounds for believing that the person whose use of that MPV on that occasion would justify the seizure is a person to whom a warning has been given (whether or not by that constable or in respect of the same vehicle or the same or a similar use) on a previous occasion in the previous 12 months.

Under s 59(6), it is a summary offence, punishable with a fine not exceeding level 3, to fail to comply with an order under s 59(3) to stop a MPV.

The Police (Retention and Disposal of Motor Vehicles) Regulations 2002 provide for the safe keeping, retention, release, and disposal of a vehicle seized under s 59(3). They provide that a person who would otherwise be liable to charges is not liable to pay if the use which caused the seizure was not a use by him/her and (s)he did not know of the use of the vehicle in the manner which led to its seizure, had not consented to its use in that manner and could not, by the taking of reasonable steps, have prevented its use in that manner.

These powers also apply where a uniformed constable reasonably believes that an offence against RTA 1988, s 34 (prohibition of off-road driving: **24.93**) is being or has been committed in a manner causing, or likely to cause, alarm, distress, or annoyance to members of the public.

Causing death by careless, or inconsiderate, driving

26.15   RTA 1988, s 2B provides that a person who causes the death of another person by driving a mechanically propelled vehicle on a road or other public place without due care and attention, or without reasonable consideration for other persons using the road or place, is guilty of an offence. The offence is an indictable (either-way) offence punishable with up to five years' imprisonment and (see **24.39**) with obligatory disqualification for driving. In essence, to be guilty of an offence under s 2B, D must have committed an offence under s 3 (**26.10**), and thereby killed another. It is not necessary to show that the careless driving was the sole cause, since it is sufficient that it is a more than minimal cause. It is irrelevant that the risk of death resulting from the careless driving was unforeseen or unforeseeable. These points also apply to **26.16**.

Causing death by careless, or inconsiderate, driving when under the influence of drink or drugs

26.16   RTA 1988, s 3A(1) provides that it is an offence for a person to cause the death of another person by driving a mechanically propelled vehicle on a road or other public place without due care and attention, or without reasonable consideration for other persons using the road or place, where he:

(a) is, at the time when he is driving, unfit to drive through drink or drugs; or

(b) has consumed so much alcohol that the proportion of it in his breath, blood or urine at that time exceeds the prescribed limit; or

(c) has in his body a specified controlled drug and the proportion of it in his blood or urine at that time exceeds the specified limit for that drug; or

(d) is, within 18 hours after that time, required under s 7 to provide a specimen, but without reasonable cause fails to provide it; or

(e) is required by a constable to give his permission for a laboratory test of a specimen of blood taken from him under s 7A, but without reasonable excuse, fails to do so.

The offence is triable only on indictment. It is punishable with up to 14 years' imprisonment and (see **24.39**) with obligatory disqualification for driving.

In essence, to be guilty of an offence under s 3A(1), D must have committed an offence under s 3 and thereby killed another, and have fallen within one of the drink (or drug)/drive elements in (a)–(e) above.

The term 'unfit to drive through drink or drugs' is explained at **26.51**. As for (b) and (c), see **26.48** and **26.49**. As to ss 7 and 7A, see **26.66** and **26.78**.

Despite the wording of s 3A(1), the offences at (b) to (e) apply only to persons driving motor vehicles as opposed to mechanically propelled vehicles.

Careless, or inconsiderate, cycling

There are separate offences, contrary to RTA 1988, s 29, of riding a bicycle, tricycle **26.17** or four-wheeled vehicle, not being in any case a motor vehicle, on a road without due care and attention, or without reasonable consideration for other persons using the road.

These are summary offences punishable with a fine not exceeding level 3.

DEFENCES TO OFFENCES INVOLVING DANGEROUS OR CARELESS DRIVING

The defences of duress of circumstances, duress by threats, and automatism, referred to **26.18** in Chapter 1, apply to the above offences. As seen, the terms of these defences are strict.

It should be noted that a defendant who, for example, falls asleep at the wheel or goes into a hypoglycaemic coma at the wheel can be convicted of dangerous or careless driving under the ordinary principles of liability, not in relation to the time when (s)he was an automaton but in relation to the time when (s)he realised or should have realised that (s)he was about to become unconscious and should have stopped. The reason is clear: to drive in such a case falls below the standard of a reasonable and prudent driver (who would have stopped).

The fact that the driving complained of was due to the sudden mechanical failure of an essential part of a vehicle can provide a defence to charges of dangerous or careless driving, provided that the defect was not known by D to exist prior to the occurrence which forms the basis of the charge, and was not such that it should have been discovered by a reasonably prudent driver. Thus, where a driver knew that the brakes of his vehicle pulled to the off-side but nevertheless drove it, the defence of mechanical defect was held not to be open to him.

Like the defence of automatism, the essence of this defence is that the dangerous situation was caused by a sudden loss of control which was in no way due to the fault of the driver.

SPEEDING

Speeding offences can be divided into those which relate (a) to restricted roads, (b) to **26.19** roads subject to a speed limit order, (c) to motorways, (d) to roads subject to maximum or minimum speed limits introduced by temporary orders, and (e) to speed limits which affect particular vehicles in particular places.

Restricted roads

26.20   By RTRA 1984, s 81 it is not lawful for a person to drive a motor vehicle on a restricted road at a speed exceeding 30 mph. It is an offence contrary to s 89(1) to do so.

A 'restricted road' is basically defined by s 82 as one upon which there is provided a system of street lighting furnished by means of lamps placed not more than 200 yards apart. A divisional court has held that an error of 12 yards between two lamps in a system of 24 does not prevent the road from being restricted. In addition to this basic definition, s 82 provides that a direction may be given by the traffic authority that a specified road with street lighting as described above shall cease to be a restricted road for the purpose of s 81 or that a road which is not provided with such lighting shall be a restricted road for the purpose of s 81. Such directions are made by the traffic authority, ie Highways England (in England) or the Welsh Ministers (in Wales) in the case of trunk roads, and by the local authority in the case of other roads (except that Transport for London is the traffic authority for a GLA road).

Signage

26.21   A road with street lighting which has been de-restricted by a direction must show the prescribed de-restriction signs (including repeater signs). Likewise, a road without such lighting which has been made a restricted road (or subjected to some other speed limit) must be provided with the prescribed restriction signs (including repeater signs). If these are not in place on the relevant part of the road, a driver cannot be convicted of exceeding the indicated speed limit. Moreover, even if they are, it is a requirement that the relevant signs can reasonably be expected to indicate the limit to an approaching driver in sufficient time to reduce from the previous lawful limit to a speed within the new limit. Thus, where a driver proved that the restriction sign, because of an overgrowing hedgerow, was not visible until the sign itself was reached, a divisional court allowed his appeal against conviction for speeding. Subject to this qualification, where signs are correctly placed, it is no defence to such a charge that the driver simply did not see the signs. The de-restriction or restriction signs which must be used are prescribed by the Traffic Signs Regulations and General Directions 2016.

The requirement for restriction signs does not apply to a road which is restricted by virtue of having a system of street lighting with lamps not more than 200 yards apart. Consequently, it is no defence for a driver accused of exceeding 30 mph on such a road that no signs were provided.

Roads subject to speed limit orders

26.22   Quite apart from 'restricted roads', many other roads are subject to speed limits. RTRA 1984, s 84 empowers a traffic authority to make an order prohibiting the driving of motor vehicles on a specified road at a speed exceeding that specified in the order (either at any time or during specified periods) or at a speed exceeding that indicated by traffic signs in accordance with the order. By s 89(1), a person who drives a motor vehicle at a speed in excess of that specified for the particular road is guilty of an offence. If a speed limit order is made under s 84, the specified road must bear the prescribed restriction signs (including repeater signs). If it does not at the point of enforcement, a person cannot be convicted of driving in excess of the specified limit.

Speed limits on motorways

26.23 The Motorways Traffic (Speed Limit) Regulations 1974 impose an overall speed limit of 70 mph on any motor vehicle using a motorway. If a section of motorway is subject to a lesser limit, this is expressly specified in the 1974 Regulations or subsequent regulations and that section must bear the prescribed signs indicating the lower limit (otherwise a driver cannot be convicted of exceeding that limit). Contravention of the Regulations is an offence contrary to RTRA 1984, s 17(4).

Temporary maximum speed limits

26.24 RTRA 1984, s 88, which does not apply to motorways, permits temporary speed limit orders to be made by the Secretary of State or the Welsh Ministers when it is desirable to do so in the interests of safety or for the purpose of facilitating the movement of traffic. Such orders may impose temporary maximum speed limits.

The 70 mph, 60 mph, and 50 mph (Temporary Speed Limit) Order 1977, which was made under the predecessor to s 88 and was continued in force indefinitely in 1978, imposes a maximum limit of 60 mph on single carriageways and of 70 mph on dual carriageways. These are general limits which apply to all roads (other than motorways) unless some other speed limit operates by virtue of the provisions described in this chapter.

The 1977 Order also imposes a limit of 60 mph on certain specified lengths of dual carriageway and of 50 mph on certain specified lengths of single carriageway. Such lengths must be provided with the prescribed restriction signs (including repeaters).

By RTRA 1984, s 89(1), it is an offence for a person to drive a motor vehicle on a road in excess of an applicable temporary speed imposed under the Order.

Temporary minimum speed limits

26.25 RTRA 1984, s 88 also permits temporary orders to be made imposing minimum speed limits, subject to such exceptions as may be specified. Signs must be displayed if a minimum speed limit is in force in respect of any road. By s 88(7), breach of such an order is an offence. Such a minimum speed limit only applies, of course, to motor vehicles.

Temporary speed restrictions in special cases

26.26 RTRA 1984, s 14 permits traffic authorities to impose temporary speed limits because of roadworks or work which is being undertaken near a road, or because of the likelihood of danger to the public or of serious damage to the road, or for the purpose of cleaning or clearing litter. Such an order cannot continue in force for more than 18 months unless the Secretary of State or Welsh Ministers consents to its further continuance. A divisional court has held that variable speed limits may be imposed under s 14.

A person who contravenes a speed restriction imposed under s 14 commits an offence contrary to RTRA 1984, s 16(1) but a divisional court has held that persons committing this offence should be charged under s 89(1).

Maximum speed limits on particular vehicles

26.27   Quite apart from the various speed limits which apply to roads, speed limits are also imposed on various types of vehicle. Thus, the driver of such a vehicle must not only observe the *speed limit applying to the road* in question but also the *speed limit applying to his/her vehicle on that road*. RTRA 1984, s 86 provides that it is unlawful for a person to drive a motor vehicle on a road at a speed in excess of a speed limit applicable to that vehicle. By s 89(1), a person who does so is guilty of an offence.

RTRA 1984, Sch 6 sets out the speed limits applying to particular classes of vehicles as follows:

Class of vehicle		M/ways	Dual carr	Other roads
1	Invalid carriage	n/a	20	20
2	Passenger vehicle, motor caravan, car-derived van or dual purpose vehicle drawing one trailer	60	60	50
3	Vehicle of the type mentioned in 2, drawing more than one trailer	40	20	20
4	Goods vehicle (except car-derived van) up to 7.5 tonnes mlw (maximum laden weight) not drawing a trailer	70	60	50
5	Articulated goods vehicle up to 7.5 tonnes mlw, and goods vehicles drawing one trailer where the combined mlw does not exceed 7.5 tonnes	60	60	50
6	Articulated goods vehicle over 7.5 tonnes mlw, goods vehicle over 7.5 tonnes mlw, and goods vehicles drawing one trailer where the combined mlw exceeds 7.5 tonnes	60	60	50
7	Goods vehicle, other than a car-derived van, drawing more than one trailer	40	20	20
8	Motor tractor, light loco, heavy loco	20	20	20
9	Motor tractor, light loco, heavy loco with certain requirements as to springs and wings being met	40	30	30
10	Works truck	18	18	18
11	Passenger vehicle exceeding 3.05 tonnes unladen weight or adapted to carry more than eight passengers:			
	Not exceeding 12 m length	70	60	50
	Exceeding 12 m length	60	60	50
12	Agricultural motor vehicle	40	40	40

The term 'car-derived van' means a goods vehicle which is constructed or adapted as a derivative of a passenger vehicle and which has a maximum laden weight not exceeding

2 tonnes. It is treated as if it was a passenger vehicle of the type from which it was derived.

A recovery vehicle equipped with a special boom for lifting vehicles is not a motor tractor. It is constructed to carry a load and is not, therefore, restricted to 40 mph on a motorway.

Exemptions for emergency services

RTRA 1984, s 87 exempts motor vehicles which are being used for fire and rescue, **26.28** ambulance or police purposes from the statutory provisions imposing speed limits on motor vehicles, if the observance of any such provision would be likely to hinder the use of the vehicle for the purpose for which it is being used on that occasion. The exemption relating to vehicles being used for ambulance purposes has been extended to cover vehicles which, although not being used for ambulance purposes, are being used for the purpose of providing a response to an emergency at the request of an NHS ambulance service. Section 87 also applies in relation to vehicles being used for NCA purposes, or for training drivers for use for NCA purposes, as it applies to a vehicle being driven for police purposes. However, other than in the case of training, s 87 does not apply in relation to a vehicle being driven for NCA purposes unless it is being driven by a person trained to drive vehicles at high speeds.

Exemptions for special forces

The Road Traffic Exemptions (Special Forces) (Variation and Amendment) Regulations **26.29** 2011 vary statutory provisions which impose speed limits to provide exemptions for drivers who are members of the special forces using vehicles in response to a national security emergency or whilst being trained or practising to do so.

It is a condition of the exemptions that drivers have been trained in the driving of vehicles at high speeds.

Regulation 2 provides that speed limits imposed by or under:

(a) RTRA 1984, s 81 (**26.20**);
(b) existing orders or notices under RTRA 1984, s 14 (**26.26**);
(c) existing orders under RTRA 1984, s 16A (**24.92**: prohibitions or restrictions on road in connection with sporting event, social event, or entertainment held on road), 84 (**26.22**) or 88 (**26.24** and **26.25**);
(d) motorway speed limits regulations (**26.23**);
(e) any local Act,

are varied so as to exempt special forces drivers using vehicles in response to a national security emergency or whilst being trained or practising to do so.

Speed limit offences: trial and punishment

The speed limit offences described at **26.20–26.27** are summary offences. With the **26.30** exception of the offence under RTRA 1984, s 17(4) (**26.23**), which is punishable with a fine not exceeding level 4, they are punishable with a fine not exceeding level 3, and, with the exception of the offence under s 88(7) (**26.25**), they are (see **24.40**) punishable with discretionary disqualification for driving. All are fixed financial penalty offences, the fixed penalty being £100 (£50 in the case of s 88(7)). As to fixed penalties, see **25.117–25.133**.

Procuring, etc speeding: a special provision

26.31   RTRA 1984, s 89(4) provides that, if a person who employs others to drive motor vehicles on roads publishes or issues any timetable or schedule, or gives any directions, under which any journey or part of a journey is required to be completed within a specified time, and it is not practicable for that journey (or part) to be completed in the specified time without the commission of an offence under s 89(1), that publication or issue, or giving of directions, may be produced as evidence that the employer procured or (as the case may be) encouraged or assisted his employees to commit such an offence.

Proof

26.32   RTRA 1984, s 89(2) states that a person must not be convicted of an offence of speeding *contrary to s 89(1)*, solely on the evidence of one witness to the effect that, in the opinion of the witness, the person prosecuted was driving the vehicle at a speed exceeding the specified limit. A similar provision applies to an offence under s 88(7) of non-compliance with a temporary minimum speed limit.

Section 89(2) requires corroborative evidence; it does not require that there must necessarily be more than one witness. Although, technically, there is no reason why a driver cannot be convicted on the evidence of two witnesses, where one corroborates the other by stating that in his/her opinion the vehicle was, for example, exceeding a limit of 30 mph, the courts would generally not be satisfied with two such opinions unless the speed was estimated to be far in excess of the limit imposed. Corroboration is therefore usually provided by some mechanical device, which is read by an operator (normally a police officer) who may then give evidence of its reading. The reading will provide corroboration of the operator's opinion of the speed of the vehicle in question.

Corroboration of speed may be obtained by the use of stopwatches to assess speeds over a measured distance, but it is usually provided by a speedometer, fixed camera, radar speed meter, VASCAR, or similar device. Where a speedometer is used, it is important to establish that the police vehicle containing the speedometer maintained an even distance from the vehicle being checked as this is relevant to the issue of speed.

A police officer's opinion of a vehicle's speed may also be corroborated from scientific calculations made by him/her (or another) based, for example, on damage to the vehicle and skid marks where it has crashed. It has been held by a judge in the Administrative Court that a police officer is entitled to corroborate his/her own opinion of whether a vehicle was exceeding a speed limit by reference to a prescribed device even if that device is not of a *type* approved by the Secretary of State (so that RTOA 1988, s 20 (below) cannot be relied on).

In instances where two police officers are involved in detecting speeding offences, they may keep one record of the transaction provided that both check and acknowledge the accuracy of that record at the time. This frequently occurs where one officer is engaged in checking the speed of the vehicle, whilst another stops and deals with the offending driver.

Radar meters are extremely accurate devices, and the courts will accept evidence of speed which is based upon meter readings provided that the operator can satisfy the court that (s)he is a trained operator and that there is clear evidence of the identity of the particular motor vehicle alleged to have exceeded the speed limit.

Hand-held radar guns are accurate, provided the batteries are fully charged or, if they are operating from another source, provided they are properly connected. However, their accuracy is affected if they are operated within a quarter of a mile of powerful VHF or UHF transmissions or within 100 yards of high-voltage cables. Care must also be taken to ensure that the beam is not bounced off metal objects and that it could not have picked up a reading from another moving object.

VASCAR and similar devices are also in common use within police forces. They record speeds averaged by a vehicle over a specified distance. Operators must be carefully trained as the device is operated by a series of switches. It is essential that the operator is able to satisfy a court that the switches were operated at the right time to ensure that correct distances are recorded, and that the vehicle was properly identified, together with the precise moment when it passes the object which marks the limit of the distance over which it is checked.

The above devices are prescribed devices (see below).

Evidence produced by prescribed devices

RTOA 1988, s 20 provides that a record produced by a prescribed device (ie a device **26.33** of a description prescribed by the Secretary of State) in an order made by statutory instrument is evidence of a fact related to any offence of speeding. The Road Traffic Offenders (Prescribed Devices) Order 1992 prescribes devices designed or adapted for measuring by radar the speed of motor vehicles. An identically named Order in 1993 prescribes devices designed or adapted for measuring speed by means of sensors or cables on or near the surface of a highway and those activated by means of a light beam or beams. Another identically named Order in 1999 prescribes cameras designed or adapted to record speed by correlating the image of the motor vehicle as it passes two positions, digitally recording each image and the time it is photographed, and calculating the average speed between the two positions. The cameras do not require film but transmit computer images to a computer centre. They are capable of measuring the speed of vehicles in different lanes. Another order with the same title in 2008 approved the use of manually operated devices that record the speed of a vehicle (vehicle A) from another vehicle (vehicle B) by (a) measuring the time it takes for vehicle A to travel between two points, (b) recording the distance between those points by means of the odometer pulses of vehicle B as it travels between the two points, and (c) calculating vehicle A's average speed between the two points by reference to the time in (a) and the distance calculated under (b) or measured manually.

The record must provide (in the same or another document) a certificate as to the circumstances in which the record was produced, signed by a constable or a person authorised by, or on behalf of, a chief police officer for the area in which the offence is alleged to have been committed.

A record produced by a prescribed device is inadmissible unless the device is of a *type* proved to have been approved by the Secretary of State and was used in accordance with conditions subject to which the approval was given.

The Orders referred to above show that a statutory instrument specifying a 'prescribed device' is drafted in general terms (as permitted by the legislation). On the other hand, approval of a *type* of device is given specifically in relation to a particular make and model. Evidence of the approval of the particular type of device used may be required and should be offered in the absence of judicial notice having been taken of its existence. This may be done by production of a valid copy of the instrument of approval or oral evidence may be given from a police officer who has knowledge of the approval, but approval may not be inferred merely from the fact that the device is in general use within police forces. A divisional court has held that the fact that a device is not properly used, in particular that a pre-operative test has not been performed, does not mean that the device is not of an approved type. The court went on to uphold a conviction for speeding where the expert evidence indicated that the failure properly to calibrate a laser speed device did not give rise to any material error.

The purpose and effect of RTOA 1988, s 20, is to enable a record relying on an accompanying certificate to be tendered in evidence without the necessity of a witness being called to prove them. A copy of evidence obtained by means of an approved device must be served on the person charged with the offence not less than seven days before the hearing or trial. If, not less than three days before the hearing or trial, the defendant requires the attendance of the person who signed the document, the evidence of the circumstances in which the record was produced will not be admissible, although the record produced by the device will. Where the record and certificate are not served on the defendant seven days before the trial, the magistrates should allow them to be produced by a police officer as real evidence.

Notices under RTA 1988, s 172

26.34   Notices may be sent under s 172, requiring the owner of the vehicle to identify the driver within 28 days. See further **24.72-24.74**.

Motor racing and competitions

Motor racing and speed trials on highway

26.35   It is a summary offence contrary to RTA 1988, s 12(1) to promote or take part in a race or trial of speed between motor vehicles on a highway. The offence is punishable with a fine not exceeding level 4 and (see **24.39**) with obligatory disqualification for driving.

RTA 1988, s 12(1) is subject to RTA 1988, ss 12A to 12F, which allow highway authority/authorities for the areas in question to make orders relating to the holding of a race or trial of speed between motor vehicles on a highway in England and Wales ('motor race orders'). Such an order is made on the application of the person proposing to promote the event, with the permission of a motor sport governing body authorised by regulations: currently The Royal Automobile Club Motor Sports Association Limited, and The Auto-Cycle Union Limited. Before issuing a permit an authorised body must consult with the local policing body for the area(s) concerned and other authorities and persons listed in s 12B.

By s 12E(2), a motor race order has the effect of disapplying s 12(1) in relation to the promoter of the event if that person:

(a) promotes the event in accordance with any conditions imposed on the promoter by the order, and

(b) takes reasonable steps to ensure that any other conditions specified in the order are met.

Section 12E(3) provides that the offences and provisions set out below do not apply in relation to a participant or an official or (as the case may be) in relation to a vehicle used by a participant or an official provided that:

(a) the participant has been approved by the motor sport governing body that issued a permit in respect of the event or (as the case may be) the official has been authorised by the promoter;

(b) the participant or official complies with any conditions specified in the motor race order that apply to participants or (as the case may be) officials; and

(c) the participant or official also complies with any conditions imposed on him or her by the promoter.

The offences disapplied by s 12E(3) are those under the Road Traffic Regulation Act 1984, ss 18, 20, 81, 84 (order thereunder), 86, 88 (order thereunder), 89, and 104; RTA 1988, ss 1, 1A, 2, 2B, 3, 3ZB, 12, 21, 22, 22A, 36, 40A, 41A, 41C, 42, 47, 87, 103 (in respect of driving), 143, 164, 165, 165A, and 170; the Vehicle Excise and Registration Act 1994, s 29; and the Removal and Disposal of Vehicles Regulations 1986; the Road Vehicles (Construction and Use) Regulations 1986, and the Road Vehicles Lighting Regulations 1989. The provisions disapplied are RTA 1988, s 38 (failure to observe Highway Code not an offence but may be relied on in legal proceedings as tending to establish or negative any liability in question) and the Vehicle Excise and Registration Act 1994, s 1(1)(b) (circumstances in which vehicle excise duty chargeable on unregistered vehicles).

Other competitions

RTA 1988, s 13(1) creates the summary offence, punishable with a fine not exceeding **26.36** level 3, of promoting or taking part in a 'competition or trial' (other than a race or trial of speed) which involves the use of motor vehicles on a highway unless the event is authorised and conducted in accordance with any conditions imposed. The Motor Vehicles (Competitions and Trials) Regulations 1969 deal with such events. Competitions or trials are authorised without the need for an application if:

(a) there are no more than 12 vehicles involved and the event does not take place within eight days of another similar event promoted by the same person or club;

(b) no merit is attached to completing the event with the lowest mileage and, in respect of any part of the event on a public highway, there are no performance tests and no route, competitors are not timed or required to visit the same places, although they may be required to finish at the same place by a specified time;

(c) in respect of any part of the event on the public highway, merit is attached to a competitor's performance only in relation to good road behaviour and compliance with the Highway Code; or

(d) all competitors are members of the armed forces of the Crown and the event is designed solely for the purpose of service training.

Regulation 6 permits the authorisation of other events by the Royal Automobile Club Motor Sports Association Ltd. Applications for such events must be made not less than two months before the event is due to be held (or, if it is to be held on more than one date, the date the event is due to begin). Except in the case of a specified event, which is an event which is run not more than once a year, for example the 'Veteran Car Run', and is specified in the schedule to the Regulations, the application must not be made more than six months before the event.

RTA 1988, s 13A states that a person is not guilty of an offence under s 1, 1A, 2, 2B or 3 (**26.4–26.15**) by virtue of driving a vehicle *in a public place other than a road* if (s)he shows that(s)he was driving in accordance with an authorisation for a motoring event given under regulations. The Motor Vehicles (Off Road Events) Regulations 1995 authorise a number of bodies to issue such authorisations.

Section 33(1) prohibits the promoting or taking part in a trial of any description between motor vehicles on a footpath, bridleway or restricted byway unless the holding of it has been authorised by the local authority. Contravention of s 33(1), or breach of a condition of an authorisation, is a summary offence, punishable with a fine not exceeding level 3, under s 33(3).

NOTICE OF INTENDED PROSECUTION

26.37   If not interviewed by the police at the time of a driving offence, a driver may experience considerable difficulty in recalling the circumstances some weeks after the event. For this reason, RTOA 1988, s 1(1) states that, in relation to certain named offences, a person (D) may not be convicted unless:

(a) D was warned at the time of the possibility of prosecution for the offence; or

(b) D was served with a requisition, single justice procedure notice or summons for the offence within 14 days of its commission; or

(c) a notice of intended prosecution specifying the nature of the alleged offence and the time and place where it is alleged to have been committed was served within 14 days on D or the person, if any, registered as the keeper of the vehicle at the time of the commission of the offence. (In the case of dangerous or careless cycling the notice must be served on the rider.)

The requirements of s 1(1) are deemed to have been complied with unless and until the contrary is proved.

The following offences require notice in one of these forms:

(a) RTA 1988, ss:

   2   Dangerous driving

   3   Careless, or inconsiderate, driving

  22   Leaving vehicle in a dangerous position

  28   Dangerous cycling

  29   Careless, or inconsiderate, cycling

  35   Failing to conform with traffic directions

  36   Failing to comply with traffic signs

(b) RTRA 1984, ss:

  16   Exceeding temporary speed restrictions imposed under s 14

  17(4)   Exceeding speed restriction on special road (ie a motorway)

  88(7)   Breach of temporary minimum speed limit

  89(1)   Speeding offences generally

(c) Aiding and abetting any of the above offences.

However, such notice need not be given in relation to an offence in respect of which a full or provisional fixed penalty notice has been given or fixed under RTOA 1988.

Warning at the time of the offence

26.38   The words 'at the time of the offence' mean at the *time of the incident and not the moment of the offence*. In one case, a warning, given by a police officer at the scene of an incident 35 minutes after its occurrence and while it was still being dealt with, was held to have been given 'at the time'. In other circumstances, where the police immediately traced the driver but the process took two-and-a-half hours, the notice was held to have been given 'at the time'. However, there is obviously a limit, and the matter should be judged in the context of the warning being given as a part of the continuous process of initial investigation, a question of fact and degree. If there has been any delay at the scene, it is best to send a written notice.

The warning must be heard and understood by the person concerned, and it must be to the effect that the question of prosecuting him/her for one or other of the above offences will be taken into consideration. There is no prescribed form of words and any clear statement to the above effect will do. It is appreciated that it is not always possible to decide whether the prosecution might be for dangerous or merely careless driving but a warning that it might be for either will suffice as the evidence upon which the charge will be based will be similar.

Service of requisition, single justice procedure notice, summons, or notice of intended prosecution

Where a warning is not given at the time, either a requisition, single justice procedure **26.39** notice, summons, or written notice of intended prosecution must be served within 14 days. The day on which the offence was committed is ignored.

A notice of intended prosecution may be served by delivering it to D or addressing it to D and leaving it at D's last known address. In addition, a notice may be served by sending it to D by registered post, recorded delivery service or first class post addressed to D at D's last known address. There is a difference between the first two postal methods and the third. A notice sent by registered post or recorded delivery service is irrebuttably deemed to have been served on D if addressed to D at D's last known address, and this includes being irrebuttably deemed to have been served within the period of 14 days. On the other hand, there is no such irrebuttable presumption in respect of a notice sent by first class post; consequently, if it is proved that a letter so sent was delivered after the end of the 14-day period it will not have been effectively served.

As to service of a requisition, single justice procedure notice or summons, see **3.10-3.14**.

It has been held that service of a notice on D's wife by handing her the notice was valid, since she was authorised to accept and deal with D's mail. It would be different if the notice was left with a hall porter, as he would not be so authorised.

Circumstances where non-compliance is no bar to conviction

By RTOA 1988, s 2(3), a failure to comply with the above requirements is no bar to **26.40** conviction if the court is satisfied that the defendant's or the vehicle's registered keeper's name and address could not with reasonable diligence have been ascertained in time for service of a requisition, single justice procedure notice, summons or notice of intended prosecution or that the defendant by his/her own conduct contributed to the failure.

Section 2(1) waives the requirements of s 1 in relation to any offence if, at the time of the offence or immediately thereafter, an *accident occurs owing to the presence on a road of the vehicle* in respect of which the offence was committed. This exemption exists because, in such circumstances, the offender will normally be aware of the circumstances surrounding the offence and the risk of prosecution. It does not apply, therefore, if the accident was so trivial that the driver was unaware that the accident had occurred. Here, the requirements of s 1 apply. On the other hand, the waiver of the requirements of s 1 does apply if the driver was unaware of the circumstances of an accident owing to the severity of his/her injuries, because subsequently the driver will be only too aware that the accident has occurred.

'Accident' in s 2(1) is given a commonsense meaning and is not restricted to unintended consequences having an adverse physical effect. It is for the prosecution to prove beyond reasonable doubt that an 'accident' has occurred. The words 'owing to the presence on a road of the vehicle' require there to be a direct causal link between the vehicle on the road and the accident.

DRINKING, OR DRUG-TAKING, AND DRIVING: INTRODUCTION

26.41 The Road Traffic Act 1988 (RTA 1988), ss 4 to 11 contain various offences relating to drinking, or drug-taking, and driving. The principal types of offence are driving or attempting to drive a motor vehicle with an alcohol concentration in excess of the prescribed limit (s 5), driving or attempting to drive a motor vehicle with a concentration of a specified controlled drug in excess of the specified limit (s 5A), and driving or attempting to drive a mechanically propelled vehicle when unfit to drive through drink or drugs (s 4).

Where a case could be prosecuted under s 5, or under s 4, it is normally preferable to prosecute for an offence under s 5, as appropriate, rather than one under s 4. This is because s 5 lays down an objective test; whereas under s 4 it is necessary for the prosecution to prove that the ability of the defendant (D) to drive properly was for the time being impaired by drink or drugs, which is a question of fact for the justices. The same comments apply in respect of s 5A as in respect of s 5.

DRIVING, ETC WITH EXCESS ALCOHOL

26.42 Under RTA 1988, s 5(1) it is an offence for a person:

(a) to drive or attempt to drive a motor vehicle on a road or other public place; or
(b) to be in charge of a motor vehicle on a road or other public place,

after consuming so much alcohol that the proportion of it in his/her breath, blood or urine exceeds the prescribed limit. The offences under s 5(1)(a) and (b) are summary offences. The s 5(1)(a) offence is punishable with up to six months' imprisonment and/or an unlimited fine, and the s 5(1)(b) offence with up to three months' imprisonment and/or a fine not exceeding level 4. The offences are also punishable with obligatory (s 5(1)(a)) or discretionary (s 5(1)(b)) disqualification for driving (see **24.39-24.40**).

The charge must state whether the person was driving, attempting to drive, or in charge of the vehicle and the nature of the specimen in which the prescribed limit was exceeded. If it does not it is likely to fail for breach of the rules against duplicity.

Driving, attempting to drive, or in charge of motor vehicle on a road or other public place

26.43 'Driving' (**26.2**), 'motor vehicle' (**23.6–23.8**), 'road' (**23.2–23.4**), and 'other public place' (**26.3**) are defined above.

Attempting to drive

26.44 As explained in Chapter 1, what is required is the doing of an act which is more than merely preparatory to the commission of the full offence, coupled with the intention to commit the full offence. For example, if a man gets no further than sitting in the driving seat of a motor car on a road and searching his pockets for the ignition key, the justices would find that his acts were merely preparatory and that he was not attempting to drive (although he could be convicted of the offence of being in charge with excess alcohol). On the other hand, if he gets as far as placing the key in the ignition switch and turning it is almost certain that the justices would find that the act was more than merely preparatory, and that therefore he was attempting to drive.

The offence of 'attempting to drive' is convenient where a driver with excess alcohol is stopped in a private place, eg a private car park, as (s)he drives towards a road. In such circumstances (s)he can be convicted of an attempt to drive on a road with excess alcohol.

Being in charge

A divisional court has held that there are broadly two distinct classes of cases: **26.45**

(a) If the defendant is the owner or lawful possessor or has recently driven the vehicle, (s)he will have been 'in charge' of it and the question will be whether (s)he is still in charge or has relinquished his/her charge. Usually such a defendant will be prima facie in charge unless (s)he has put the vehicle in someone else's charge. However, (s)he would not be so if in all the circumstances (s)he has ceased to be in actual control and there is no realistic possibility of his/her resuming actual control while unfit, eg if (s)he was home in bed for the night or a great distance from the car or if it was taken by another.

(b) If the defendant is not the owner, lawful possessor or recent driver, but is sitting in the vehicle or otherwise involved with it, the question will be whether (s)he has assumed being in charge of it. (S)he will have if (s)he is voluntarily in actual control of the vehicle or if, in the circumstances, including his/her position, his/her intentions, and his/her actions, (s)he may be expected imminently to assume control. Usually this will involve his/her having gained entry to the car and having indicated an intention to take control. But gaining entry may not be necessary if (s)he manifests that intention in some other way, eg by stealing the keys of a car in circumstances which show that (s)he means to drive it.

Because more than one person can be in charge of a vehicle, someone sitting in a passenger seat supervising a learner driver is in charge of the vehicle.

The defence under s 5(2)   RTA 1988, s 5(2) provides that it is a defence for a person charged with 'being in charge of a motor vehicle', etc to prove that, at the time (s)he is alleged to have committed the offence, the circumstances were such that there was no likelihood of his/her driving the vehicle whilst the proportion of alcohol in his/her breath, blood, or urine remained likely to exceed the prescribed limit. In determining whether there was such a likelihood the court may disregard any injury to him/her and any damage to the vehicle (s 5(3)). Normally, the defendant (D) will only succeed in proving this defence if there is expert evidence as to the rate of alcohol destruction by the body which indicates that D's alcohol level would not have been above the limit at the time D was likely to drive.

A divisional court has held that the defence is not primarily concerned with D's intentions, although they might be a factor to be considered. D's subjective intentions cannot be decisive in circumstances where D was affected by alcohol at a level well over the prescribed limit and has no means of knowing or proving when his/her alcohol levels had fallen below that limit.

The House of Lords has ruled that D bears a persuasive burden of proof of the s 5(2) defence (ie on the balance of probabilities).

Driving motor vehicle on a road or other public place

A divisional court has held that for a vehicle to be driven 'on' a road or other public **26.46** place it is not necessary that its wheels should be on the road etc or for the entirety of the vehicle to be on the road etc. It is enough that the vehicle materially encroaches on the road. The court, therefore, upheld a conviction where the boot of a car being man-oeuvred in a private driveway had projected into a road and collided with a car parked there.

After consuming so much alcohol that the proportion of it in breath, blood, or urine exceeds the prescribed limit

26.47   'Consuming' is a wide enough term to include other methods of ingestion than through the mouth. Thus, for example, an alcohol level which may have, in part, resulted from the injection of a substance, comes about through 'consumption'.

The prescribed limits

26.48   These are defined by RTA 1988, s 11 which states that the 'prescribed limit' means, as the case may require:

(a)   35 microgrammes of alcohol in 100 millilitres of breath;
(b)   80 milligrammes of alcohol in 100 millilitres of blood; or
(c)   107 milligrammes of alcohol in 100 millilitres of urine,

or such other proportion as may be prescribed by regulations made by the Secretary of State. So far, the Secretary of State has not prescribed any other proportion.

The 35 microgrammes is roughly equivalent to the 80 and 107 milligrammes of the two other levels.

DRIVING, ETC WITH CONCENTRATION OF SPECIFIED CONTROLLED DRUG ABOVE SPECIFIED LIMIT

26.49   The relevant provision is RTA 1988, s 5A. By s 5A(1), s 5A applies where a person (D):

(a)   drives or attempts to drive a motor vehicle on a road or other public place (s 5A(1)(a)); or
(b)   is in charge of a motor vehicle on a road or other public place (s 5A(1)(b)),

and there is in D's body a specified controlled drug. Section 5A(2) provides that D is guilty of an offence if the proportion of which in D's blood or urine exceeds the speci-fied limit for that drug. 'Specified' means specified by regulation. The following con-trolled drugs and limits (expressed in microgrammes (μg) per litre (l) of blood) have been specified:

(i)   Illicit drugs:
   Amphetamine, 250 μg/l
   Benzoylecgonine, 50 μg/l
   Cocaine, 10 μg/l
   Delta–9–Tetrahydrocannabinol (Cannabis and Cannabinol), 2 μg/l
   Ketamine, 20 μg/l
   Lysergic Acid Diethylamide (LSD), 1 μg/l
   Methylamphetamine, 10 μg/l
   Methylenedioxymethaphetamine (MDMA—Ecstasy), 10 μg/l
   6-Monoacetylmorphine (6-MAM—Heroin and Diamorphine), 5 μg/l
(ii)   Generally prescription drugs:
   Clonazepam, 50 μg/l
   Diazepam, 550 μg/l

Flunitrazepam, 300 µg/l
Lorazepam, 100 µg/l
Methadone, 500 µg/l
Morphine, 80 µg/l
Oxazepam, 300 µg/l
Temazepam, 1000 µg/l.

It is a defence to a charge under s 5A for D to show that:

(a) the specified controlled drug had been prescribed or supplied to D for medical or dental purposes;
(b) (s)he took the drug in accordance with any directions given by the person by whom the drug was prescribed or supplied, and with any accompanying instructions (so far as consistent with any such directions) given by the manufacturer or distributor of the drug; and
(c) his/her possession of the drug immediately before taking it was not unlawful under the Misuse of Drugs Act 1971, s 5 (restriction of possession of controlled drugs: **20.11**) because of an exemption under the Misuse of Drugs Regulations 2001 (s 5A(3)).

This defence is not available if D's actions were contrary to any:

(i) advice given by the person by whom the drug was prescribed or supplied, about the amount of time that should elapse between taking the drug and driving a motor vehicle; or
(ii) accompanying instructions about that matter (so far as consistent with any such advice) given by the manufacturer or distributor of the drug (s 5A(4)).

If evidence is adduced that is sufficient to raise an issue with respect to the above defence, the court must assume that the defence is satisfied unless the prosecution proves beyond reasonable doubt that it is not (s 5A(5)).

It is a defence to a charge of 'being in charge with an excess drug level' to prove that at the time that D is alleged to have committed the offence the circumstances were such that there was no likelihood of D driving the vehicle whilst the proportion of the specified controlled drug in D's blood or urine remained likely to exceed the specified limit for that drug (s 5A(6)). In determining whether there was such a likelihood, the court may disregard any injury to D and any damage to the vehicle (s 5A(7)).

The offences under s 5A are triable and punishable in the same way as their corresponding offence under s 5 (**26.42**).

DRIVING, ETC UNDER INFLUENCE OF DRINK OR DRUGS

RTA 1988, s 4(1) and (2) provides that: **26.50**

(1) A person who, when driving or attempting to drive a mechanically propelled vehicle on a road or other public place, is unfit to drive through drink or drugs is guilty of an offence.
(2) Without prejudice to subsection (1), a person who, when in charge of a mechanically propelled vehicle which is on a road or other public place, is unfit to drive through drink or drugs is guilty of an offence.

The charge must state whether the person was driving, attempting to drive, or in charge. The offences under s 4 are triable and punishable in the same way as their corresponding offence under s 5 (**26.42**).

Many of the elements of these offences have been dealt with already in relation to s 5. It is worth noting that s 4 applies to a mechanically propelled vehicle (**23.5**), whereas ss 5 and 5A are limited to a motor vehicle, and that s 4(3) provides a similar defence for persons charged with 'being in charge' to that provided by ss 5(2) and 5A(7) for those charged with the corresponding offences under ss 5 and 5A: namely, that it is a defence for the defendant to prove that there was no likelihood of his/her driving the vehicle while (s)he remained unfit through drink or drugs.

It remains to be added that 'drink' means alcoholic drink and 'drug' means any intoxicant other than alcohol, including medicines and glue.

Unfitness to drive

26.51   Section 4(5) provides that a person is to be taken to be unfit to drive if his ability to drive properly is for the time being impaired. The evidence before the court on this point will normally include evidence of the defendant's driving before his/her vehicle was stopped; any evidence of driving apparently outside the pattern of normal driving is relevant in that it may show some impairment of the ability to drive properly. In addition, the evidence should include the observation of the suspect's physical state prior to and during participation in a preliminary impairment test.

PROCEDURE

26.52   The procedures which precede a person being charged with an offence under RTA 1988, s 5 or 5A are regulated by RTA 1988. These procedures may begin with a constable requiring a preliminary test, usually at the roadside, followed by an arrest (whether as a result of a positive test or of a failure to co-operate) which then must be followed: (a) where a s 5 offence is suspected, by a requirement for specimens of breath (for an evidential breath test) or, in limited cases, a specimen of blood or urine (for analysis) in order to get an indication of the proportion of alcohol in the person's breath, blood or urine; or (b) where a s 5A offence is suspected by a requirement for a specimen of blood or urine (for analysis). The requirement for a specimen must be made at a police station or hospital or elsewhere (in the case of breath specimens) after an accident, except that a constable in uniform may move directly to an evidential breath test at the roadside *when a portable breath-testing device becomes available* (as to which see **26.71**).

A charge under s 3A (causing death by careless driving when unfit through drink or drugs or with excess alcohol (**26.16**)) or s 4 will normally be preceded by a preliminary test requirement.

PRELIMINARY TESTS

26.53   Under RTA 1988, s 6, there are three types of preliminary test whose purpose is to give an indication as to a defendant's 'impairment' in consequence of consumption of alcohol and/or of taking drugs, by means of a breath test, an impairment test or a drugs test.

Power to require preliminary tests

26.54   RTA 1988, s 6(1) provides that where any of s 6(2) to (5) applies, a constable may require a person to co-operate with one or more preliminary tests administered to that person by that constable or another.

Section 6(2) provides that such a test may be required *if a constable reasonably suspects that the person*:

(a) *is driving, is attempting to drive or is in charge* of a motor vehicle on a road or other public place; and

(b) *has alcohol or a drug in his body or is under the influence of a drug.*

Section 6(3) authorises a test where a constable *reasonably suspects* that the person:

(a) *has been driving, attempting to drive or in charge of* a motor vehicle on a road or other public place *while having alcohol or a drug in his/her body or while unfit to drive because of a drug*; and

(b) *still has alcohol or a drug in his/her body or is still under the influence of a drug.*

Section 6(4) authorises a test where a constable *reasonably suspects* that the person:

(a) *is or has been driving or attempting to drive or in charge* of a motor vehicle on a road or other public place; and

(b) *has committed a traffic offence while the vehicle was in motion.*

The type of traffic offence which the constable must reasonably suspect to have been committed is defined by s 6(8), which provides that 'traffic offence' in this context means an offence under any provision of the Public Passenger Vehicles Act 1981, Part II (fitness of vehicles and PSV operator's licence), the Road Traffic Regulation Act 1984, the Road Traffic Offenders Act 1988 except Part III (offences relating to fixed penalty procedures), or any provision of RTA 1988 except Part V (driving instruction for payment without being licensed or registered and other driving instructor offences). This effectively covers the range of road traffic offences which could be described as 'moving traffic offences'. All offences which are set out in regulations made under those Acts are also included within this description; for example, offences contrary to the Road Vehicles (Construction and Use) Regulations 1986 or the Road Vehicles Lighting Regulations 1989.

RTA 1988, s 6(5) authorises a test if:

(a) an *accident occurs* owing to the presence of a motor vehicle on a road or other public place; and

(b) a constable *reasonably believes* that the person *was driving, attempting to drive or in charge of the vehicle at the time* of the accident.

General The tests required (other than in the case of a test under s 6(5)) may only be administered by a constable in uniform. Whether or not a constable was in uniform is a question of fact in each case. A constable wearing his uniform except for his cap has been held to be in uniform so long as he is easily identifiable as a constable. A court is entitled to assume that a constable was in uniform unless this point is disproved.

The necessary suspicion or belief need not arise while the vehicle is in motion. Indeed, a uniformed constable may stop motorists, under his/her common law or statutory powers to do so, in order to see whether there is a reasonable suspicion that they have consumed alcohol, and if a reasonable suspicion then emerges of such consumption, go on to require a preliminary test. Thus, random stopping of motorists is not prohibited, although random preliminary tests are.

A constable does not have to administer a caution before requiring a preliminary test. It is only where the motorist has failed such a test (or failed to take it) that a caution needs to be administered.

Power where an accident has occurred

26.55   As already stated, there is a power, under RTA 1988, s 6(5), to require a 'preliminary' test if an accident occurs owing to the presence of a motor vehicle on a road or other public place. A constable may require any person whom (s)he has *reasonable cause to believe* was driving or attempting to drive or in charge of the vehicle at the time of the accident to provide a preliminary test, subject to s 9 (hospital procedure: **26.80**).

'Accident' in this context bears its ordinary meaning, and it has been held that a crash which is deliberately caused falls within that meaning. The person's vehicle need not have been physically involved but there must have been a direct causal connection between his/her vehicle being on the road and the accident occurring. Where a sequence of events begins on a road and leads to a vehicle leaving the road and colliding with an object some distance from the road, there is an accident for the purposes of s 6(5).

Of course, the fact that an accident has occurred owing to the presence of a motor vehicle on a road or other public place does not entitle a constable to require a specimen of breath under s 6(5) from anyone; that provision only empowers him/her to make such a requirement of any person whom (s)he has *reasonable cause to believe was driving or attempting to drive or in charge of* the vehicle at the time of the accident.

In contrast to the power to require a breath specimen under the other provisions of s 6, which is framed in terms of a constable having *reasonable cause to suspect* the specified things, s 6(5) uses the term *'has reasonable cause to believe'*, which requires more than 'reasonable suspicion' since 'believe' requires more than mere suspicion and refers to being virtually certain (ie having no substantial doubt) about the facts.

The nature of preliminary tests

Preliminary breath test

26.56   RTA 1988, s 6A provides that a preliminary breath test is a procedure whereby the person to whom the test is administered provides a specimen of breath to be used for the purpose of obtaining, by means of a device approved by the Secretary of State, an indication whether the proportion of alcohol in the person's breath or blood is likely to exceed the prescribed limit.

A preliminary breath test under s 6(2) to (4) may be administered only at or near the place where the requirement to co-operate with the test is imposed; where it is administered under s 6(5) (ie following an accident), it may be administered at or near that place or, if the constable who imposes the requirement thinks it expedient, at a police station specified by him.

Preliminary breath test devices used by police officers must be approved by the Secretary of State. Currently, the Secretary of State has approved two types of device, one which might be described as the 'blow in the bag' type of device and the other 'electronic'.

Approved devices of the 'blow in the bag' type are the Draeger Alcotest 80, the Alcotest R80A, and the Lion Alcolyser.

A number of electronic devices have been approved by the Secretary of State. There are various types of some of these devices (because new models have been approved as a particular device has been developed). In the following list the latest model of devices currently available is given: the Lion Alcolmeter 500 B, the Draeger Alcotest 6810 GB, the Intoximeter Alcosensor FST, and the Envitec AlcoQuant 6020.

The preliminary impairment test

Such a test is designed to indicate whether a person is unfit to drive, and, if (s)he is, **26.57** whether or not this is due to drink or drugs.

RTA 1988, s 6B provides that a preliminary impairment test is a procedure whereby the constable administering the test observes the person to whom the test is administered in his/her performance of tests specified by the constable, and makes such other observations of the person's physical state as the constable thinks expedient. Such tests may only be administered by a constable approved for the purpose by his/her chief officer of police. It may be administered at or near the place where the requirement to co-operate with the test is imposed or, if the constable who imposes the requirement thinks it expedient, at a police station specified by him/her.

The Secretary of State has issued the Code of Practice for Preliminary Impairment Tests for the use of constables trained and authorised to carry out such tests, to which a constable administering a preliminary impairment test must have regard. The Code sets the kind of task that may be specified for the purpose of such a preliminary impairment test, the kind of observation of physical state that may be made in the course of it, the manner in which it should be administered, and the inferences that may be drawn from observations made in the course of it.

The preliminary test procedure set out in the Code comprises:

(a) a pupillary examination;
(b) the modified Romberg balance test;
(c) the walk and turn test;
(d) the one-leg stand test;
(e) the finger-to-nose test.

The Code provides that in each of the preliminary impairment tests the officer must give a full explanation of what is required, demonstrating actions which a subject is required to perform where this is essential to understanding and establishing whether or not a disability, injury or illness exists, whether physical or mental, which might affect performance during the test. Obesity and age may be factors. The details about these tests and their execution are as follows:

A pupillary examination which involves the subject looking straight ahead with eyes wide open. The constable should establish whether or not (s)he is wearing contact lenses. A gauge is held adjacent to the appropriate side of the subject's face to enable, by a process of comparison, the size of the pupils of the eyes to be estimated. Conditions in which a subject's eyes are 'watering' or 'reddening' should also be noted.

The modified romberg balance test is concerned with a person's internal clock and ability to balance. (S)he should be required to stand with heels and toes together and arms by his/her side and retain that position while tilting the head back slightly and closing the eyes and then to bring the head forward, open the eyes and say 'stop' when (s)he thinks that 30 seconds have passed. A record will be made of ability to balance while being so instructed, whether (s)he steps, sways or raises his/her arms and whether the eyes were opened, or the head straightened, during the test, together with the actual number of seconds which elapsed.

The walk and turn test enables an assessment to be made of a person's ability to divide attention between walking, balancing, and processing instructions. The constable should identify a line (other than a kerb or other place where a person might fall) and instruct the person to place his/her left foot on the line, his/her right foot on the line in front of the left foot touching heel to toe, put arms down by the side and keep them there throughout the test and to maintain that position while further instructions are given. On the instruction 'start', the subject should take nine heel-to-toe steps along the line; after these nine steps, leave the front foot on the line and turn around using a series of small steps with the other foot (these manoeuvres can be demonstrated by the constable). After turning, the subject should take another nine heel-to-toe steps back along the line. The officer should count each step out loud.

A record must be made of whether a subject was able to stand still while being instructed, whether (s)he started too soon, turned correctly, occasions when (s)he stopped walking, missed heel-to-toe connection, stepped off the line, or raised his/her arms. A record must also be made of whether the steps were correctly counted; the point of any deviation from the straight line should be marked on a diagram on the appropriate form.

The one-leg stand test   tests balance and ability to count out loud. The subject should be instructed to stand with his/her heels and toes together and arms by his/her sides; to maintain that position while receiving instructions and not to begin until told to do so. On the instruction 'start' (s)he should raise his/her right foot 6 to 8 inches or 15 to 20 cms off the ground; keep the elevated leg straight with the toes pointing forward and the foot parallel with the ground; keep his/her arms by his/her side; keep looking at the elevated foot throughout the test; and while doing so to count out loud 'one thousand and one, one thousand and two, one thousand and three' and so on progressively until told to stop. The test should be carried out using each foot in turn.

A record must be made over a timed period of 30 seconds, for each foot, of any instances in which a subject sways, hops, puts a foot down, or raises the arms, together with the point in the test where that occurred.

The finger-to-nose test   is designed to test depth of perception and balance. A subject must stand with both feet together and extend both arms in front, palms uppermost with the fists closed and the index finger of each hand extended (this must be carefully demonstrated) and (s)he must maintain this position while waiting to be told to perform the remainder of the test. On being told to start (s)he should tilt his/her head back slightly and then close his/her eyes and, when told which hand to use, should attempt to touch the tip of his/her nose with the tip of the index finger then, having done so, to lower the hand, as the constable calls out the order in which each hand should be used—Left, Right, Left, Right, Right, Left.

A record should be made where the subject steps, sways or raises an arm, and whether the correct hand was used, or touched a part of the face other than the tip of his/her nose and whereabouts.

Safety and site conditions   The safety of the subject must be considered. A hard, level, non-slippery surface should be chosen wherever possible in a well-lit unobstructed area out of the public view and in appropriate weather conditions. Where this is not possible, consideration must be given to the possibility of the test being conducted at another nearby location or at a police station. If not, appropriate allowance must be made in interpreting the tests. If the nature of footwear may be an impediment, the subject should be given the opportunity to remove it.

A constable should not close his/her eyes when demonstrating a test. (S)he should always stand away from the subject and remain still.

Disabilities, injuries or illness   Any such factors must be recorded, whether physical or mental, if they might affect the performance of the subject during a test. Obesity or age must be recorded if likely to affect the test.

Where such factors are evident or claimed, an officer may continue to require co-operation with the test. However, (s)he must be particularly mindful of the possibility of the effect which this might have on performance when interpreting test results.

General   A constable may decide, in consequence of these tests, that there is sufficient evidence to justify an arrest on reasonable suspicion of an offence under RTA 1988, s 3A or 4 (**26.26** and **26.50**). There is no pass/fail standard, nor a scoring system. A decision must be based upon overall observation during the tests and the records made of those performances.

The preliminary drugs test

This is dealt with by RTA 1988, s 6C. A preliminary drugs test is a procedure by which   **26.58** a specimen of sweat or saliva is obtained and used for the purpose of obtaining, by means of a device of a type approved by the Secretary of State, an indication whether the person to whom the test is administered has a drug in his/her body and if so:

(a) whether it is a specified controlled drug;
(b) if it is, whether the proportion of it in the person's blood or urine is likely to exceed the specified limit for that drug.

It may be administered at or near the place where the requirement to co-operate with the test is imposed or, if the constable who imposes the requirement thinks it expedient, at a police station specified by him.

Up to three preliminary drug tests may be administered.

The Secretary of State has approved two devices for the purposes of s 6C: the Draeger DrugTest 5000 and Securetec DrugWipe 3S S303G, both of which test a saliva sample. Both types of device have been approved for use to administer a mobile preliminary drug test, either in a police station or at or near the place where the test-requirement is made, for the purpose of obtaining an indication whether a person has THC (the active ingredient in cannabis) or cocaine in his/her body.

General

PACE Code C: The Code of Practice for the Detention, Treatment and Questioning of   **26.59** Persons by Police Officers, referred to in Chapter 5, excludes the 'specimens for analysis' procedures under RTA 1988, s 7 (**26.66**) from the provisions of the Code which relate to 'interviews'; no similar exclusion is made in relation to the preliminary test procedure under s 6A. It is therefore advisable that, where there is conversation between an officer and a suspect which relates to his/her condition, that conversation is recorded and read and signed by the suspect. A divisional court, however, refused to interfere with a decision of the justices not to exclude evidence subsequently obtained following an officer's inquiry as to whether a suspect had been drinking, to which the suspect replied, 'Yes, I've had a couple of pints'. It said that the justices, although recognising that there had been a breach of Code C, were entitled to find that this breach had not been sufficiently 'significant and substantial' to merit exclusion of the evidence.

Nevertheless, care must be taken when engaging in conversation with drivers suspected of drink/drive offences. In one case police officers spoke to a man on three occasions: at his home prior to arrest and before caution; after his arrest and caution at his home; and in a police car, in the course of which he admitted that he had been to a funeral and had taken alcohol afterwards, that his last drink ('a couple of cans') had been taken at about 4 pm and that he had last driven his car at 5 pm and had not since taken a drink. It was held that these three conversations were 'interviews' for the purpose of Code C.

Arrest

Positive preliminary breath test or drug test

26.60   RTA 1988, s 6D(1) provides that a constable may arrest a person without warrant if as a result of a preliminary breath test or preliminary drug test the constable reasonably suspects that:

(a) the proportion of alcohol in the person's breath or blood exceeds the prescribed limit; or

(b) the person has a specified controlled drug in his body and the proportion of it in the person's blood or urine exceeds the specified limit for that drug.

The fact that evidential specimens of breath have been provided under s 7 (**26.66**) by the person concerned does not prevent s 6D(1) having effect if the constable who imposed on him/her the requirement to provide the specimens has reasonable cause to believe that the device used to analyse the specimens has not produced a reliable indication of the proportion of alcohol in the breath of that person (s 6D(1A)).

Failure to co-operate

26.61   By RTA 1988, s 6D(2), a constable may arrest a person without warrant if:

(a) the person fails to co-operate with a *preliminary test* in pursuance of a requirement imposed under s 6, and

(b) *the constable reasonably suspects that the person has alcohol or a drug in his/her body or is under the influence of a drug.*

Section 11(2) provides that 'fails' includes 'refusal'. If a person is given an opportunity to do something and does not do it, there is a failure to comply with that request. Where a person refuses to reply to such a clear request, there has been a failure and it is no defence to allege that the refusal to reply was consequent upon a previous caution. It is essential that the constable makes it clear to the person concerned that (s)he is required to co-operate in the test.

In a case where a driver refused to wait until a preliminary breath test device arrived at the scene, having been properly required to provide a specimen, it was held that he had failed to provide a specimen of breath. A demand that a test be deferred until the arrival of a solicitor amounts to a refusal. There can be no acceptance subject to conditions.

Section 11(3) provides that a person does not co-operate with a preliminary test unless his co-operation is sufficient to enable the test to be carried out, and is provided in such a way as to enable the objective of the test to be satisfactorily achieved. The importance of this can be shown by reference to the preliminary breath test. The various preliminary breath test devices require breath to be supplied in various ways to allow a sufficient sample to be obtained. If sufficient breath has not been supplied there has been a failure. Where a preliminary breath test device requires the illumination of two lights before a satisfactory specimen has been obtained, there is a 'failure' where only one of those lights is illuminated and a police officer is not required to 'read' such a specimen.

An important point to note about the power to arrest after a failure to co-operate is that, before effecting an arrest, the constable must have reasonable cause to suspect that the person has alcohol or a drug in his/her body or is under the influence of a drug.

General

Section 6D(2A) provides that a person arrested under s 6D, instead of being taken to a police station, be detained at or near the place where the preliminary test was, or would have been administered, with a view to imposing on him there a requirement for an evidential test under RTA 1988, s 7. Thus, a means is provided to carry out an evidential test at the roadside as an alternative to taking the person concerned to a police station.   **26.62**

Clear words must be used in all cases to indicate the reason for arrest and that the person concerned is being compulsorily taken to a police station in consequence of such a failure.

Normally, a constable will exercise his/her power of arrest in the above two cases but this is not a pre-condition of further steps in the procedure being adopted and is unnecessary if the person concerned is quite happy to proceed to the next stage in the procedure.

No arrest of hospital patient

A person may *not* be arrested under s 6D *while at hospital as a patient* (RTA 1988, s 6D(3)).   **26.63**

Power of entry

By RTA 1988, s 6E a constable may enter any place (using reasonable force if necessary) for the purpose of: **26.64**

(a) imposing a preliminary test requirement by virtue of s 6(5) following an accident in a case where the constable reasonably suspects that the accident involved injury of any person; or

(b) arresting a person under s 6D following an accident in a case where the constable reasonably suspects that the accident involved injury of any person.

It must be emphasised that, with the one exception in (a), police officers do not have power to enter a place without the express or implied consent of the occupier for the purpose of requiring a preliminary breath, impairment or drug test. Likewise, police officers, with the one exception in (b), do not have power to enter a place without the express or implied consent of the occupier for the purpose of making an arrest under s 6D.

Offence of failure to co-operate with a preliminary test under s 6

RTA 1988, s 6(6) provides that a person commits a summary offence, punishable with a fine not exceeding level 3 and (see **24.40**) with discretionary disqualification for driving, if without reasonable excuse he fails to co-operate with a preliminary test in   **26.65** pursuance of a requirement imposed under s 6. This offence can be committed notwithstanding that the defendant has not been warned that refusal to co-operate will be an offence. Such warning is not required.

As to what constitutes a 'failure to co-operate', see **26.61**. There cannot be a conviction for this offence unless the co-operation has been required under s 6, and it will not have been so required if the requirement made is invalid for some reason. For example, if the constable is a trespasser at the time of requiring a preliminary breath test (s)he will be behaving unlawfully and his/her requirement will not be valid.

In addition, where there is a preliminary breath test, there cannot be a conviction for failing to provide a specimen if the requisite procedure is not validly administered, as will be the case if the device used is not an approved one, or if it is defective, or if the constable fails to comply with the manufacturer's instructions as to *assembly*. However, if the constable realises the defect or mistake (s)he may validly require another breath test to be taken on another device. Provided the constable acts in good faith and not negligently, his/her non-compliance with the manufacturer's instructions as to the use of the device does not invalidate the test unless the non-compliance is prejudicial to the defendant.

A person cannot be convicted of failing to co-operate with a preliminary test unless his/her failure was without reasonable excuse. It is not a reasonable excuse that the defendant (D) did not think that (s)he had consumed any alcohol, nor that (s)he mistakenly believed that the test requirement was invalid, nor that (s)he had consumed alcohol after driving.

It has been stated in a number of cases that no excuse can be adjudged reasonable unless D was physically or mentally unable to provide the specimen or its provision would entail a substantial risk to D's health, as where D was unable to supply a sufficient specimen of breath because of a medical condition, or where D was concussed and unable to appreciate the requirement made of him/her. However, it has also been held that a foreigner who is unable to understand the purpose of the requirement and the penal consequence of a failure to comply has a reasonable excuse. Not surprisingly, someone who has made him/herself too drunk to understand these matters does not have a reasonable excuse, nor does someone who fails to provide a specimen because (s)he is in a state of self-induced agitation. See also **26.86**.

PROVISION OF SPECIMENS FOR ANALYSIS

26.66   RTA 1988, s 7(1) provides that, in the course of an investigation into whether a person has committed an offence under s 3A, 4, or 5, a constable may, subject to the other provisions of s 7 (below) and of s 9 (**26.80**), require him to provide:

(a) two specimens of breath for analysis by means of a device of an approved type, known as an 'evidential breath test' (s 7(1)(a)); or
(b) a specimen of blood or urine for a laboratory test (s 7(1)(b)).

Section 7(1A) provides that, in the course of an investigation into whether a person has committed an offence under s 5A, a constable may, subject to ss 7(3)–(7) and 9, require the person to provide a specimen of blood or urine for a laboratory test.

Introductory points

26.67   The fact that a requirement under s 7(1) or 7(1A) can be made 'in the course of an investigation' into whether a person has committed an offence under s 3A, 4, 5, or 5A indicates that it is not necessary that a preliminary test should have been required (indeed, as indicated at **26.52**, it may well not have been required where the investigation relates to drink-driving and a portable evidential breath-testing device is available (as to which see **26.71**)) and that, if there has been such a test, it is irrelevant that there has been some breach in the procedure relating to it.

For the procedure under s 7 to apply it is irrelevant that the person concerned has not been arrested, or that (s)he has been arrested on reasonable suspicion of driving etc while unfit through drink or drugs or on reasonable suspicion of a non-drink/drive offence, or that (s)he has been unlawfully arrested. Thus, if a driver reports an accident at a police station, and it is suspected that he has drugs in his body, he may be required to provide specimens for analysis. If someone is the subject of an investigation into whether an offence under s 3A, 4, 5, or 5A has been committed by him, a specimen may be required under s 7. In addition, a person who alleges that she was a passenger, not the driver of a car, at the time in question, may lawfully be required to provide a specimen as a person under an investigation for an offence under either s 3A, 4, 5, or 5A. It is not necessary to show that she was driving or in charge. Although, of course, liability for an offence under s 3A, 4, 5, or 5A will depend on proof that she was driving, attempting to drive, or in charge at the material time, she can be convicted under s 7 of failing to provide a specimen even if she was not driving, etc.

As already stated, the procedure under s 7 does not constitute an 'interview' for the purposes of PACE Code C (**5.33**). However, under PACE, s 78, the evidence obtained from the analysis of a specimen can be excluded if it appears to the court that, having regard to all the circumstances, including non-compliance with the preliminary test procedure under s 6, the admission of that evidence would have such an adverse effect on the fairness of the proceedings that the court ought not to admit it. Section 78 was so applied in a case where a preliminary test was improperly required (because the officer did not have one of the requisite reasonable suspicions specified by RTA 1988, s 6). A divisional court held that, since D had been denied the protection afforded by s 6, the prosecutor had obtained evidence which he would not otherwise have obtained and, as a result, D was significantly prejudiced in resisting the charge. The magistrates, it held, were therefore entitled to exclude the evidence.

Where the person in question is a patient at a hospital a special procedure, governed by s 9 (**26.80**), must be followed.

Because time is of the essence, it is only where the delay would be very short that the procedure can be delayed to enable the person to obtain legal advice. The courts have held that, except in such a case: (a) evidence obtained under the procedure will not have been obtained in breach of PACE, s 58, and (b) there will be no reasonable excuse should the person fail to provide a specimen.

Breath specimens

In terms of an investigation into whether an alcohol-related offence under RTA 1988, s 3A or 4, or an offence under s 5, has been committed, the important part of the procedure under s 7(1) is s 7(1)(a) whereby a constable may require two specimens of *breath*. It is only in exceptional circumstances that blood or urine samples may be required instead under s 7(1)(b). **26.68**

Where breath specimens can be required

Section 7(2) provides that a constable may make a requirement under s 7 to provide specimens of breath only if: **26.69**

(a) the requirement is *made* at a police station or a hospital;

(b) the requirement is imposed in circumstances where s 6(5) (**26.54**: after an accident) applies; or

(c) the constable is in uniform.

By s 7(2C), where a constable has imposed a requirement on the person concerned to co-operate with a preliminary breath test *at any place,* (s)he is entitled to remain at or near that place in order to impose on him/her there a requirement under s 7.

Section 7(2D) provides that if a requirement under s 7(1)(a) has been made at a place other than a police station, such a requirement may subsequently be made at a police station if (but only if):

(a) a device or a reliable device of the type mentioned in s 7(1)(a) was not available at that place or it was for any other reason not practicable to use such a device there, or

(b) the constable who made the previous requirement has reasonable cause to believe that the device used there has not produced a reliable indication of the proportion of alcohol in the breath of the person concerned.

The result of the analysis of a roadside evidential breath specimen is admissible in evidence to the same extent as the analysis of a specimen at a police station.

By s 7(5A), a constable may arrest a person (D) without warrant if:

(a) D fails to provide a specimen of breath when required to do so in pursuance of s 7; and

(b) the constable reasonably suspects that D has alcohol in his/her body.

This is of obvious importance where there has not been a preliminary test followed by an arrest.

A constable requiring a person to provide a breath specimen under s 7 is required by s 7(7) to warn him/her that a failure to supply it may result in his/her prosecution. The results of a breath test under s 7 taken in the absence of a warning are inadmissible, even if no prejudice to the person results from the lack of a warning.

Specimens of breath

26.70   RTA 1988, s 7 requires that the two specimens of breath which have been required under it be analysed by means of a device approved by the Secretary of State. Someone who has provided one specimen of breath which exceeded the prescribed limit but has failed to provide a second specimen cannot, on the basis of that specimen, be convicted of driving with excess alcohol. (S)he can, however, be convicted of failing to provide a specimen of breath (**26.84**).

Section 11(3) provides that a person does not provide a specimen of breath for analysis unless his co-operation is sufficient to enable the analysis to be carried out, and is provided in such a way as to enable the objective of the analysis to be satisfactorily achieved. This is an important provision, as it is necessary for the person providing the specimen to continue blowing until the indicator lights signify that sufficient breath has been obtained for analysis. If sufficient is not provided, there has been a 'failure'.

For the purposes of the legislation relating to the provision of 'specimens of breath', 'breath' does not mean deep-lung air but should be given its ordinary dictionary definition, ie 'air exhaled from anywhere'; therefore there is no need for the prosecution to prove that the reading from an approved device related solely to deep-lung alcohol and was not related to mouth alcohol.

Analysis of breath by approved devices

Procedure The following are approved evidential breath testing devices: Camic **26.71**
Datamaster; Lion Intoxilyzer 6000 UK; Intoximeter EC/IR. *At the time of writing no*
portable evidential breath testing device for roadside use had been approved.

The devices have been designed to overcome the problems likely to arise from an
evidential viewpoint. At the outset the devices are correctly calibrated but in actual use
the devices check themselves for accuracy. They check their correct calibration both
before and after each of the two breath samples and a record is made of those calibration checks on the eventual print-out slip. The slip therefore shows two separate readings of alcohol levels sandwiched between records of calibration checks to ensure that
the devices are operating correctly. The defendant is present throughout the procedure
and has the opportunity to see the device at work. The officer carrying out the test is not
obliged to explain to the defendant that the second specimen must be provided within
three minutes of the first, or the test will abort.

The devices provide a timed and dated print-out which gives evidence of two separate readings of alcohol content in the breath of the defendant. The normal procedure
thereafter is that the constable who has operated the device certifies all copies of the
print-out, which shows, in addition to the readings, the particulars of the person from
whom the sample is obtained, the signature of the officer, and the signature of the person
(or a record that such a signature was refused). The constable's statement declares the
lower of the two readings given by the device to be at the specified level and certifies
that copies of the statement were signed by him and by the defendant, or that the defendant refused. Where a print-out is not produced in evidence and no oral evidence
is given in relation to correct calibration, it is open to a court to find that calibration
was correct where there is evidence that the machine was used by a trained operator.

The devices are computers. In the absence of evidence to the contrary, courts will
presume that the computer system was working correctly. If there is evidence that it
might not have been, the party seeking to introduce the evidence will need to prove
that it was working. It is therefore desirable that police officers record on the appropriate procedural forms the fact that the computer was operating correctly, together
with any appropriate observations.

The officer also certifies that (s)he handed a copy of the statement to the defendant
who accepted, or declined to accept it. The mere fact that the copy handed to the defendant is not signed by the officer does not affect the validity of the original. RTOA
1988, s 16(3) requires that a copy is either 'handed to' a defendant at the time, or is
served upon him/her not later than seven days before the hearing. Where a defendant
signed all the copies of the statement but refused to accept one, a divisional court held
that s 16(3) had been complied with when the defendant was offered a copy although
there had been no physical transfer of possession of the document. Nevertheless, it will
be good practice in such circumstances subsequently to serve a copy of the statement
upon the defendant in accordance with s 16(3).

Evidence, however, of the test is not restricted to documentary evidence. Oral evidence may be given of the results of a test should the prosecutor, for some reason,
choose not to use the simplified procedure. If this is done, oral evidence of calibration
should also be given. It will be advisable, in such cases, to serve a copy of the evidence
of the police operator on the defendant in accordance with the Criminal Justice Act
1967, s 9. A divisional court has said that, where an officer is giving oral evidence of the
result of an analysis, there is no difference between those results having been seen on
the screen, and seen on a print-out.

Defence lawyers have no right to obtain documents kept in relation to breath-analysis devices, such as the log, repair reports, and memory roll, with a view to searching for material which might support a submission that the device was defective. They must rely upon the prosecution to carry out its duty to disclose material evidence which might be of assistance to the defence. The reliability of the approved device which has been used can be challenged either by direct evidence of some malfunctioning or by evidence from which the inference of unreliability can reasonably be drawn. On the other hand, the reliability of the device used cannot be challenged on grounds relating to all devices of the prescribed type because the type of device has been approved by the Secretary of State.

Only the lower reading is evidence   RTA 1988, s 8 provides that it is only the lower of the two readings given by the machine which may be used as evidence; the other must be disregarded. Where one of the two required readings is not obtained within the same operating cycle of the machine, a second cycle must be commenced. In such a case it is the lower specimen in the first cycle taken and the first of the second cycle which should be recognised. However, where neither of the specimens in the first cycle is valid, a second cycle is undertaken and third and fourth specimens are, effectively, the first and second specimens recognised by the machine.

Where only one satisfactory specimen of breath has been obtained due to a failure of the breath-analysis machine and, in consequence, an alternative blood or urine specimen has been obtained and a charge has been preferred alleging an excess of alcohol in that alternative specimen, evidence of the proportion of alcohol found in the one specimen of breath is inadmissible, as it is not relevant to a charge of excess alcohol in another specimen. However, where the accuracy of the analysis of the alternative specimen is challenged, the prosecution is required to prove beyond reasonable doubt that the blood/urine analysis was reliable and if the breath test result is broadly equivalent to the analysis of the alternative specimen then, notwithstanding the reason for requiring the alternative specimen, such evidence is at least capable of tending to support the reliability of the analysis of the alternative specimen. Provided that the significance of the relationship between the two analyses is explained by an expert, the evidence may be relevant.

Blood or urine specimens

26.72   RTA 1988, s 7(3) provides that a requirement under s 7 (**26.66**) to provide a specimen of *blood* or *urine* can only be made at a *police station* or at a *hospital*; and that it cannot be made at a police station unless:

(a) the constable making the requirement has reasonable cause to believe that for medical reasons a specimen of breath cannot be provided (because of inability) or should not be required (for some other reason, such as the taking of a drug which affects blood/alcohol levels). Intoxication rendering a person incapable of providing breath specimens can constitute a medical condition and, therefore, a medical reason for requiring blood or urine. However, if the person concerned (P) is too intoxicated to provide breath specimens when requested it does not follow that the constable is obliged to require an alternative specimen (although (s)he may do so). Nor does it follow that P has a 'reasonable excuse' if charged with failing to supply a breath specimen, contrary to s 7(6) (**26.84–26.87**). Provided that a reasonable cause to believe that there are medical reasons exists, it is irrelevant that the constable him/herself does not believe that for medical reasons a breath specimen cannot be provided or should not be required. The constable is not required to seek medical advice; the question is whether on the facts before the constable (s) he had reasonable cause to believe that medical reasons exist;

(b) specimens of breath have not been provided elsewhere and, at the time the requirement is made, an approved device or a reliable approved evidential breath testing device is not available at the police station or it is then for any other reason not practicable to use such a device there;

(c) an approved evidential breath testing device has been used (at the police station or elsewhere) but the constable who required the specimen of breath has reasonable cause to believe that the device has not produced a reliable indication of the proportion of alcohol in the breath of the person concerned;

(d) as a result of the administration of a preliminary drug test, the constable making the requirement has reasonable cause to believe that the person required to provide a specimen of blood or urine has a drug in his body; or

(e) the suspected offence is one under s 3A, 4, or 5A and the constable making the requirement has been advised by a medical practitioner or registered health care professional (**26.75**) thereafter 'practitioner or health care professional', that the condition of the person required to provide the specimen might be due to some drug;

but may then be made notwithstanding that the person required to provide the specimen has already provided or been required to provide two specimens of breath.

The fact that the constable has been advised by a practitioner or health care professional that the person's condition may be due to some drug may be proved by oral evidence from the practitioner or health care professional or by the police officer testifying to what the practitioner or health care professional said to him/her. In addition, where unchallenged evidence has been given by the custody officer as to what the practitioner or health care professional said and did and as to his/her completion of a procedural form recording the signed observations of the practitioner or health care professional, justices are entitled to find that this advice had been given. While an endorsement signed by the practitioner or health care professional is, in its contents, hearsay, the fact that a practitioner had signed the endorsement and said things which led the officer to complete the remainder of the form in a particular way, indicating that it was concerned with impairment through drugs after medical advice, has been held to be a matter to which the justices are entitled to have regard.

Reliable approved device not available or not practicable to use it

A '*reliable device*' referred to in RTA 1988, s 7(3)(b) (**26.72**) is one which the police **26.73** officer concerned reasonably believes to be reliable. The police officer must reasonably believe that no approved, reliable device *is* available. Where a police officer thought that the available device might be unreliable, because the defendant did not appear to be as badly affected as the device indicated, and required an alternative specimen, the conviction was quashed on the grounds that a belief that the device might be unreliable is insufficient. It is 'not practicable to use a device' if there is no officer available at the station who has been trained to use the device.

Where a suspect has provided two specimens of breath on a machine which is then found to be defective, (s)he may lawfully be required to provide two further specimens of breath for analysis by another device instead of being required to provide blood or urine. Where this involves a suspect being taken to another police station, a divisional court has said that it might be wise, *as a matter of an abundance of caution*, to repeat, at the second police station, the statutory warning that failure may render him/her liable to prosecution.

Unreliable indication reasonably believed

26.74   Under RTA 1988, s 7(3)(c) (**26.72**) an alternative specimen may be required where a breath-analysis device has been used and the constable who required the specimens has reasonable cause to believe that the device has not produced a *reliable indication* of the proportion of alcohol in the breath of the person concerned. Thus, if a device produces two readings which indicate significant differences in the levels of alcohol, it is open to the constable to require an alternative specimen if (s)he reasonably believes that the device has not produced a reliable indication of alcohol in the breath. The latest breath-analysis devices incorporate new software which enable them to identify and flag up automatically where it is suspected an interfering substance may be present, or the alleged offender produces mouth alcohol, or the difference between the readings of two specimens is greater than 15 per cent. In such situations a constable will be able to require blood or urine as an alternative.

Reliance on the guidance contained in procedural instructions about the use of breath-analysis devices, which leads to the conclusion that an indication is unreliable, will mean that the constable has reasonable grounds to believe that the device has not produced a reliable reading.

It has been held that, where the constable has reasonable cause to suspect that the unreliable indication is due not to the unreliability of the device but to the way in which the individual provided the breath specimens, the officer may offer the individual the opportunity to provide further breath specimens (instead of requiring him/her to provide a blood or urine specimen).

Decision as to blood or urine

26.75   RTA 1988, s 7(4) provides that, if the provision of a specimen other than a specimen of breath may be required in pursuance of s 7, the question whether it is to be a specimen of blood or a specimen of urine and, in the case of a specimen of blood, the question who is to be asked to take it shall be decided (subject to s 7(4A)) by the constable making the requirement. The limitation in s 7(4A) is that there must be no requirement to provide a blood specimen if:

(a) the medical practitioner who is asked to take the specimen is of the opinion that, for medical reasons, it cannot or should not be taken; or

(b) the registered health care professional who is asked to take it is of that opinion and there is no contrary opinion from a medical practitioner.

A 'registered health care professional' is a registered nurse, or a registered member of a health care profession designated by an order made by the Secretary of State. A registered paramedic is so designated. If the constable knows of a medical reason relevant to s 7(4A) affecting this issue (s)he must pass on that knowledge to the practitioner or professional concerned.

Section 7(4A) also provides that where, by virtue of s 7(4A), there can be no requirement to provide a specimen of blood, the constable may require a specimen of urine instead.

Where a constable requires blood as an alternative and the person required to give blood refuses but offers urine as an alternative, there is a 'failure' to provide a specimen as required under s 7 unless a medical reason exists.

Information to be given to suspect

On requiring a person to provide a specimen of blood or urine, a constable: **26.76**

(a) must warn him/her that a failure to provide it may render him/her liable to prosecution; and

(b) must state the reason why a breath specimen cannot be taken or used.

In addition, in order to ensure that the person is aware of the role of the medical practitioner or registered health care professional, the constable should tell him/her that it is for the constable to decide whether the specimen is to be of blood or urine and, if (as is usually the case) the constable decides to require blood, ask the person if there are any medical reasons why a blood specimen cannot or should not be taken from him/her by a medical practitioner or registered health care professional. Failure to do so will not lead to a dismissal of the charge unless the court has a reasonable doubt as to whether or not the person was prejudiced by not being told about the role of the practitioner or professional. Thus, for example, if (s)he was aware that a practitioner or professional would take the blood specimen, the charge will not be dismissed.

The House of Lords has held that there is no statutory requirement, nor any consideration of fairness, which requires a police officer to ask a suspect if any non-medical reason exists in consequence of which a specimen of blood should not be taken. Any such matter might support 'a reasonable excuse for failure to provide a specimen' but that is a matter for a court.

There is no need to give the person a chance to indicate his/her preference as to the alternatives of blood or urine.

Where a medical reason is given (for example that the suspect is taking tablets) there must be evidence that the police officer took this into account or had considered whether it could be a medical reason. Although medical possibilities might appear to be far-fetched, it is impossible to *know* that this is so. There must be evidence that the officer has asked questions. Where a 'medical reason' offered is capable of being valid the officer must refer the matter to a medical practitioner or registered health care professional. If (s)he does not, a requirement will not have been made pursuant to the Act and evidence of the resulting analysis will be inadmissible.

Taking of blood specimen

RTA 1988, s 11(4) provides that a person supplies a specimen of blood if and only if: **26.77**

(a) he consents to the taking of such a specimen from him; and

(b) the specimen is taken from him by a medical practitioner or by a registered health care professional.

Specimens of blood taken from a person incapable of consenting

RTA 1988, s 7A deals with the taking of blood specimens from a person who has been **26.78**
involved in an accident and is, for some reason, incapable of giving a valid consent to the taking of a specimen of blood.

Section 7A(1) provides that a constable may request a medical practitioner or registered health care professional (hereafter 'practitioner') to take a specimen of blood from a person ('the person concerned') irrespective of whether that person consents if:

(a) that person is a person from whom the constable would (in the absence of any incapacity of that person and of any objection under s 9 (**26.80**) be entitled under s 7 to require the provision of a specimen of blood for a laboratory test;

(b) it appears to that constable that this person has been involved in an accident which constitutes or is comprised in the matter that is under investigation or the circumstances of that matter;

(c) it appears to that constable that this person is or may be incapable (whether or not he has purported to do so) of giving a valid consent to the taking of a specimen of blood; and

(d) it appears to that constable that this person's incapacity is attributable to medical reasons.

Section 7A(2) provides that a request under s 7A:

(a) must not be made to a practitioner who for the time being has any responsibility (apart from the request) for the clinical care of the person concerned; and

(b) must not be made to a practitioner other than a police practitioner unless it is not reasonably practicable:

(i) for the request to be made to a police practitioner; or

(ii) for such a practitioner (assuming him to be willing to do so) to take the specimen.

A 'police practitioner' is a practitioner (as defined above) who is engaged under any agreement to provide medical or health care services for purposes connected with the activities of a police force.

By s 7A(3), it is lawful for a practitioner to whom a request is made:

(a) to take a specimen of blood from the person concerned irrespective of whether that person consents; and

(b) to provide the sample to a constable.

Section 7A(4) states that, if a specimen is taken pursuant to a request under s 7A, the specimen must not be subjected to a laboratory test unless the person from whom it was taken:

(a) has been informed that it was taken; and

(b) has been required by a constable to give permission for a laboratory test of the specimen; and

(c) has given permission.

On requiring a person to give permission for the purposes of s 7A for a laboratory test of a specimen, a constable is required by s 7A(5) to warn him/her that a failure to give the permission may render him/her liable to prosecution.

Other provisions about blood or urine specimens

26.79   *Blood*   If a specimen of blood is taken, RTOA 1988, s 15(4) provides that a specimen of blood must be disregarded unless:

(a) it was taken from the defendant (D) with D's consent by a medical practitioner or a registered health care professional (hereafter a 'practitioner'); or

(b) it was taken from D by a practitioner under RTA 1988, s 7A and D subsequently gave permission for a laboratory test of the specimen.

Evidence that a specimen of blood was taken from D with D's consent by a practitioner may be given by the production of a document purporting to certify that fact and to be signed by a practitioner.

Urine   A urine specimen must, by RTA 1988, s 7(5), be provided within one hour of the requirement and after the provision of a previous specimen. This means that a specimen must be taken and discarded, and another provided (for analysis) within one hour of the requirement. A police officer is not obliged to extend the time but, if (s)he does accept a second specimen of urine outside the time limit, the result of the analysis of that specimen is nevertheless admissible. It has been held that a constable is entitled to require a specimen of urine where, the breath-analysis device being inoperable, (s)he had required blood but a doctor was unable to obtain such a specimen when the defendant's vein collapsed. The rule is that an unproductive request for one specimen (ie the specimen is not provided) does not prevent a subsequent request for another specimen from being valid.

A divisional court has held that, where a urine sample is taken from someone who is catheterised, a specimen is provided for the purposes of s 7(5) each time the catheter bag is emptied, rather than at the point the urine leaves the body.

Providing D with part of specimen　RTOA 1988, s 15(5) provides that, where (at the time a blood or urine specimen was provided by D) D asked to be provided with such a specimen, evidence of the proportion of alcohol or any drug found in the specimen is inadmissible on behalf of the prosecution unless:

(a) the specimen is one of two parts into which the specimen was divided at the time it was provided; and
(b) the other part was supplied to D.

'At the time' does not mean 'then and there' nor 'in the presence of D'. It suffices that the division is closely linked in time and part of the same event as the taking of the sample. It is, however, desirable that the division takes place in D's presence. There is no obligation to inform D that D may request part of the specimen. Provided that at D's request the specimen has been divided and one part has been offered to D, it is irrelevant that D does not take up the offer; the statutory requirement is satisfied, and evidence of the alcohol level in the part retained by the police is admissible. Where someone is not in a position to take receipt of the second part of a sample, it may be provided to him/her for present purposes by handing it to someone else for safe-keeping.

RTOA 1988, s 15(5A) provides that, where a specimen was taken under RTA 1988, s 7A, evidence of the proportion of alcohol or any drug found in the specimen is not admissible on behalf of the prosecution unless:

(a) the specimen in which the alcohol or drug was found is one of two parts into which it was divided at the time it was taken; and
(b) any request to be supplied with the other part of the specimen taken from D, which was made by D when he gave permission for a laboratory test of the specimen, was complied with.

Labelling　When a specimen of blood or urine has been obtained, the labels identifying it with that person must be carefully made out and attached securely to the specimen which must then be sent to the forensic science laboratory.

HOSPITAL PATIENTS

RTA 1988, s 9 lays down a special procedure which applies while a person is at a hospital as a patient.　**26.80**

Section 9(1) provides that, while a person is at a hospital as a patient, he must not be required to co-operate with a preliminary test or to provide a specimen under RTA 1988, s 7 unless the medical practitioner in immediate charge of his case has been notified of the proposal to make the requirement, and:

(a) if the requirement is made, it must be for co-operation with a test administered, or for the provision of a specimen, at the hospital; but

(b) if the medical practitioner objects on the ground specified in s 9(2) (below), the requirement must not be made.

This procedure is concerned with whether the practitioner objects to a preliminary test of, or a specimen being provided by, a patient, and not whether the practitioner objects to a particular type of test or specimen.

Section 9(1A) provides that, while a person is at a hospital as a patient, no specimen of blood may be taken from him under s 7A (person incapable of consenting) and he must not be required to give permission for a laboratory test of a specimen taken under s 7A unless the medical practitioner in immediate charge of his case:

(a) has been notified of the proposal to take the specimen or to make the requirement; and

(b) has not objected on the ground specified in s 9(2).

By s 9(2), the ground on which the medical practitioner may object is:

(a) in a case within s 9(1), that the requirement or the provision of the specimen or (if one is required) the warning required by s 7(7) (**26.69**) would be prejudicial to the proper care and treatment of the patient; and

(b) in a case within s 9(1A), that the taking of the specimen, the requirement or the warning required by s 7A(5) (**26.78**) would be so prejudicial.

If the medical practitioner objects on the ground that the requirement, or the provision of a specimen, or the warning about the consequences of failing to provide it would be prejudicial to the proper care and treatment of the patient, the requirement in question must not be made.

It should be noted that if a preliminary test undergone by a patient proves positive, or (s)he fails to undergo it, (s)he may not be arrested, but this does not prejudice the rest of the procedure being followed (s 6D(3)).

It is not strictly necessary to obtain a preliminary test before requiring a specimen for analysis but an explanatory circular states that it is assumed that this will be done. It is good practice to give a person the opportunity quickly to clear him/herself of suspicion.

Where a requirement to provide a specimen of blood has been made at a hospital but the patient is discharged before the specimen can be taken, that requirement is not varied or discharged by the mere fact that the person to whom the requirement was made is then taken to a police station. The specimen of blood may be taken there. The only exception is if a police officer abrogates the procedure started at the hospital by asking for breath specimens and thereby setting in train the procedure under s 7.

EVIDENCE IN PROSECUTIONS UNDER ROAD TRAFFIC ACT 1988, S 3A, 4, 5, OR 5A

26.81   By RTOA 1988, s 15(2), evidence of the proportion of alcohol or any drug in a specimen of breath, blood or urine provided by or taken from D must, in all cases, be taken into account and it *must be assumed*:

(a) that the proportion of alcohol in D's breath, blood or urine at the time of the alleged offence was not less than in the specimen;

(b) that the proportion of a drug in D's blood or urine at the time of the alleged offence was not less than in the specimen.

Evidence of a breath, blood or urine specimen is not the only admissible evidence. The result is, for example, that, if the analysis or test reveals a proportion of alcohol below the prescribed limit, but the magistrates are sure from expert evidence that, given the lapse of time between the alleged offence and the provision of the specimen, the proportion of alcohol at the time of the alleged offence was over the prescribed limit, they may convict D of an offence under s 3A, 4, or 5. On the other hand, since D's alcohol level *must be assumed to be not less* than that found in the sample, D is not permitted to adduce evidence at trial that, although above the limit when the specimen was taken, D was below it when actually driving.

The assumption made by s 15(2) does not apply if there is evidence that the evidential device was unreliable or that a blood or urine specimen is unreliable (eg because analysis of the part given to D is below the limit). The House of Lords held that justices can assume that a device was reliable unless there is evidence that it was not, and that if there is evidence that a device was unreliable the prosecution are obliged to prove its reliability. Presumably, the same approach as taken in that case applies where the challenge is to the reliability of a blood or urine specimen. Evidence relating to unreliability may be direct evidence, but—depending on the disparity between the claimed consumption and the analysis—evidence of the unreliability may be provided by reference to the amount of alcohol allegedly consumed before the specimen was provided. Expert evidence of the reading which could be produced in the circumstances by D is not essential.

Although s 15 says that the proportion of alcohol or any drug in a specimen provided by D must in all cases be taken into account, there are exceptions. This is because s 15(4) expressly states that evidence derived from a specimen of blood must be disregarded unless:

(a) it was taken with D's consent by a 'practitioner' (ie a medical practitioner or a registered health care professional); or
(b) it was taken from D by a practitioner under RTA 1988, s 7A (without consent) and D subsequently gave permission for a laboratory test of the specimen.

In addition, as seen at **26.79**, in the case of a blood or urine specimen, the evidence derived from it is inadmissible if D's request for one of the parts into which it was divided was not properly complied with.

The Court of Appeal has held that where a specimen has been obtained in breach of the requirements of RTA 1988, s 7 its admissibility is to be decided (under PACE, s 78) by reference to the effect of non-compliance on the fairness of the proceedings. Automatic exclusion applies, however, where there has been a failure to give a warning under s 7(7) (**26.69**).

Hip-flask defence

RTOA 1988, s 15(3) allows D what is often described as the 'hip-flask' defence. It provides that, in cases under RTA 1988, s 3A, 4, 5 or 5A: **26.82**

(a) the assumption as to alcohol level must not be made if D proves that:
    (i) he consumed alcohol before he provided the specimen or had it taken from him; and
        (a) in relation to an offence under s 3A, after the time of the alleged offence, and
        (b) otherwise, after he had ceased to drive, attempt to drive or be in charge of a vehicle on a road or other public place; and
    (ii) had he not done so the proportion of alcohol in his breath, blood or urine would not have exceeded the prescribed limit and, if the proceedings are for

an offence under s 4, would not have been such as to impair his ability to drive
properly; and

(b) the assumption as to drug level must not be made if D proves:

    (i) that he took the drug before he provided the specimen or had the specimen taken from him and:

        (*a*) in relation to an offence under s 3A, after the time of the alleged offence; and

        (*b*) otherwise, after he had ceased to drive, attempt to drive or be in charge of a vehicle on a road or other public place, and

    (ii) that had he not done so the proportion of the drug in his blood or urine:

        (*a*) in the case of a specified controlled drug, would not have exceeded the specified limit for that drug; and

        (*b*) if it is alleged that he was unfit to drive through drugs, would not have been such as to impair his ability to drive properly.

It has been held by the Court of Appeal that the burden of proof placed on D is a persuasive one, ie proof on the balance of probabilities.

In such cases it will almost invariably be necessary for D to call expert medical or scientific evidence in order to discharge this burden of proof, unless in rare cases the non-expert evidence which is called is such that a layman would reliably and confidently say that the subsequent drinking was the explanation for the excess level.

Defendants who are able to prove the hip-flask defence may nevertheless be liable for the offence of wilfully obstructing a constable in the execution of his/her duty. In any event, such deliberate consumption is liable to be self-defeating since D will often still be in charge of the vehicle at the time of the additional imbibing. It is therefore good practice, in instances in which D drinks from a bottle before submitting to testing procedures, to charge D with an offence of 'being in charge, etc' should tests prove to be positive.

Use of certificates

26.83 RTOA 1988, s 16(1) provides that evidence of the proportion of alcohol in a specimen of breath may be given by the production of the print-out produced by the breath-analysis device together with the certificate signed by a constable (normally the operator) (which certification may be made on the print-out). As already explained, that certification is to the effect that the print-out relates to a specimen provided by D at the date and time shown in it. The print-out and certificate are only admissible in evidence on behalf of the prosecution if a copy of it (or both) has been handed to D when the print-out was produced, or has been served on D not later than seven days before the hearing. Moreover, the certificate is not so admissible if D, not later than three days before the hearing or within such further time as the court may allow, has served notice on the prosecution requiring the attendance at the hearing of the constable who signed the certificate. While the Act makes specific provision for print-outs to be offered in evidence, it does not require that this be done. Consequently, a constable may give oral evidence of the readings obtained on the screen of the device, provided that (s)he is able to testify that the device was working properly and was accurately self-calibrating.

Section 16(1) also provides that evidence of the proportion of alcohol or a drug in a specimen of blood or urine may be given by the production of a certificate signed by an authorised analyst as to the proportion of alcohol or any drug found in the specimen identified in the certificate. However, such a certificate is only admissible in evidence on behalf of the prosecution if a copy of it has been served on D not later than seven days before the hearing. In default of this, evidence of the analysis can be given by the authorised analyst attending as a witness. In addition, the certificate is not admissible under s 16(1) if, not later than three days before the hearing or within such further time as the court may allow, D has served notice on the prosecutor requiring the attendance at the hearing of the analyst who signed the certificate. However, a divisional court has held that, if the analyst is unable to attend, a court is entitled to admit the evidence of a certificate of analysis as hearsay under CJA 2003, s 116.

A divisional court has held that the s 16(1) requirement for an analyst to be authorised only applies to the giving of a certificate under s 16 and does not apply to, or prevent, the giving of oral evidence or a statement under CJA 1967, s 9 (**8.43**).

OFFENCE OF FAILING TO PROVIDE A SPECIMEN REQUIRED UNDER S 7

RTA 1988, s 7(6) makes it an offence for a person (D), without reasonable excuse, to fail to provide a specimen when required to do so in pursuance of s 7. Where the specimen was required to ascertain D's fitness to drive or alcohol or drug level when D was driving or attempting to drive a s 7(6) offence is punishable with up to six months' imprisonment and/or an unlimited fine, and with obligatory disqualification for driving (**24.39**). In any other case, it is punishable with up to three months' and/or a fine not exceeding level 4, and with discretionary disqualification for driving (see **24.40**).   **26.84**

Where a person is charged with failure to provide a specimen of blood for laboratory analysis in circumstances in which an approved breath-analysis device was not available, and the validity of the request for a sample of blood is challenged, the non-availability of the device must be proved in accordance with the laws of evidence. However, where D contended that there was no direct evidence to prove that there was no approved device available at the police station, as there could have been a second machine there, a divisional court ruled that evidence that the device failed to operate together with evidence that D was told that it was not therefore possible to take specimens of breath was sufficient. As in the case of the offence under s 6(6) (**26.65**), there cannot be a conviction under s 7(6) if the requirement made for a specimen is invalid for some reason; the fact that the person was brought to the police station after a wrongful arrest or after a trespass by the police does not render a requirement under s 7 invalid.

Fail

As already said 'fail' includes 'refuse'. However, where D refused to supply a specimen of breath, but within five seconds said that he wanted to change his mind, there was no refusal. A divisional court said that the justices had ignored the motorist's words five seconds after his refusal. Regard must be had to all words and conduct in reaching a conclusion. It is submitted that, in the light of this and other rulings, the right to change one's mind will exist until the constable has moved on to the next stage of the procedure. Where a person declines to provide a specimen of breath, alleging a medical reason which is discounted by a medical practitioner or registered health care practitioner, there is a failure to provide a specimen without reasonable excuse. The police officer is not obliged to start the procedure all over again. If it were otherwise, recalcitrant persons could play the system to gain some delay. Once D is required to provide a specimen an offence is committed if, without reasonable excuse, D does not provide it.   **26.85**

A person fails to provide a specimen of breath if the specimen is insufficient for analysis, or is so provided that it cannot be analysed.

Reasonable excuse

26.86   A foreigner who is unable to understand the purpose of the requirement and the penal consequences of a failure to comply has a reasonable excuse. In addition, there would be a reasonable excuse for failing to provide a specimen of blood where D refused to sign a form of consent (to providing a specimen of blood) until (s)he had read it. Apart from these instances an excuse cannot be adjudged reasonable unless D was physically or mentally unable to provide the specimen requested, or its provision would entail a substantial risk to D's health. Where D refuses to supply specimens of breath and it is subsequently found that, because of D's limited lung capacity, D would have been unable to supply specimens, this cannot amount to a reasonable excuse. There must be a causative link between the refusal and the excuse and this cannot be so where the condition was not known to exist at the time of the refusal. While an invincible repugnance, amounting to a medically recognised phobia, to blood being taken is a reasonable excuse, the fear of the sight of blood is not; D can always close his/her eyes or look away. While a fear of AIDS is not a reasonable excuse for not providing a specimen of blood, a medically recognised *phobia* in relation to contracting AIDS is.

A divisional court has held that D's non-communication of a medical condition of which D was aware at the time of the request but only belatedly proffered will normally result in the court concluding that there was no reasonable excuse for a failure to provide a specimen.

It has been held by divisional courts that stress caused by self-precipitated agitation cannot amount to a reasonable excuse for failure to provide breath, nor can mental anguish caused by a constable's conduct which D considered to be oppressive. There would have to be a causal connection between such anguish and the failure.

The fact that D was so drunk that D could not understand the procedure which was being followed does not amount to a reasonable excuse because it does not relate to D's *capacity* to supply a specimen. In a case where D was so drunk that he was physically unable to provide a specimen of breath, although he was not incapable of co-operating with the police, or comatose or unconscious, and the case was not one where it might otherwise have been unreasonable for a specimen of breath to be required, a divisional court held that in these circumstances D's physical inability could not amount to a reasonable excuse for D's failure to provide the breath specimen. This can be contrasted with another case where D failed to provide a specimen of breath and two police officers gave evidence that he appeared to be too drunk to do so, a divisional court refused to overturn a decision of the justices that he had had a reasonable excuse for not providing a specimen due to stress resulting from adverse personal and family circumstances which had led to recent breathlessness, although no medical evidence was offered. The justices had reached their decision on unimpeachable findings of fact and the divisional court would not interfere.

The wording of the relevant provisions is such that a person who fails to provide a specimen under s 7 on the grounds that (s)he has been unlawfully arrested also does not have a reasonable excuse.

Someone who has not been given time to read PACE Code C, or to refer to a law book, does not have a reasonable excuse for not providing a specimen in consequence of that lack of opportunity, nor does someone who has been advised by his solicitor not to provide a specimen. Nor does someone who refuses to supply a specimen unless he can first consult a solicitor, unless the consultation will only cause a very short delay to the taking of the specimen.

D does not have the burden of proving reasonable excuse. There must, however, be evidence in support of facts which could constitute a reasonable excuse, otherwise the prosecution does not have to prove that there was no reasonable excuse. In almost every imaginable case of alleged physical or mental incapacity, that evidence must be medical evidence; in exceptional cases, however, other evidence—even that of D—can suffice. An example of such an exceptional case was where D had provided one specimen of breath but began to lose composure, sobbing continuously, and experiencing difficulty in breathing. She was unable to provide a second specimen. A divisional court ruled that the justices were entitled to conclude that D was physically incapable of providing a specimen, whilst accepting that the case was close to the borderline. The need for medical evidence on such a point could not be accepted in absolute terms.

Other points

As already indicated, on requiring a person to provide a specimen in pursuance of RTA **26.87** 1988, s 7, a constable must warn him/her that a failure to provide it may render him/her liable to prosecution (s 7(7)). Although an omission to warn is no defence to a charge of failing to co-operate with a preliminary test, contrary to s 6(4), it is a defence to a charge under s 7(6) that such a warning has not been given or has not been understood by the person from whom the specimen is required.

Because there is only one offence under s 7(6), an information or written charge framed in terms of s 7(6) alone, without reference to the circumstances in which the requirement was made, will not fail for duplicity. It would be *good practice* to inform D of the circumstances surrounding the charge as soon as they have been established. In most cases, the issue of whether D was driving, attempting to drive, or in charge of the vehicle will be known at the time of D's arrest. It is therefore preferable that the charge should specify the offence in relation to which the investigation was being carried out.

OFFENCE OF FAILING TO PERMIT A LABORATORY TEST OF SPECIMEN TAKEN UNDER S 7A

A person who, without reasonable excuse, fails to give his/her permission for a labora- **26.88** tory test of a specimen of blood taken from him/her under RTA 1988, s 7A is guilty of an offence under s 7A(6) and punishable in the same way as a person guilty of its corresponding offence under s 7(6) (**26.84**).

The comments in **26.84–26.87** about the offence under s 7(6) are applicable to s 7A(6), although most of the instances of 'reasonable cause' cited are clearly irrelevant.

DETENTION OF PERSONS AFFECTED BY ALCOHOL OR A DRUG

RTA 1988, s 10(1) provides that a person who has been required under s 7 or 7A to **26.89** provide a specimen of breath, blood, or urine may be detained at a police station (or, if the specimen was provided otherwise than at a police station, arrested and taken to and detained at a police station) if a constable has reasonable grounds for believing that, if he was released and drove or attempted to drive, he would commit an offence against s 4, 5, or 5A. In practice this may often involve detention until a negative preliminary test. On any question under s 10 whether a person's ability to drive properly is or might be impaired through drugs, a constable must consult a medical practitioner and must act on his/her advice (s 10(3)).

Section 10(1) does not apply to a person if it ought reasonably to appear to the constable that there is no likelihood of his/her driving or attempting to drive whilst his/her ability to drive properly is impaired, or whilst the proportion of alcohol in his/her breath, blood or urine exceeds the prescribed limit, or whilst the proportion of a specified controlled drug in his/her blood or urine exceeds the specified limit for that drug (s 10(2)).

A hospital patient may not be arrested and taken to a police station if it would prejudice his/her proper care and treatment (s 10(2A)).

OTHER OFFENCES

Cycling while unfit

26.90   By RTA 1988, s 30, it is an offence to ride a cycle with two or more wheels on a road or other public place while unfit to ride through drink or drugs. This is a summary offence punishable with a fine not exceeding level 3. Being in charge of such a cycle in such circumstances is not an offence under RTA 1988, but it is an offence under the Licensing Act 1872, s 12, which is described in the next paragraph.

Drunk in charge of a carriage, horse, etc

26.91   By the Licensing Act 1872, s 12, it is a summary offence, punishable with a fine not exceeding level 1 or, at the discretion of the court, up to one month's imprisonment, for a person to be drunk while in charge on any highway or other public place of any carriage, horse, cattle, or steam engine. A mechanically propelled vehicle, trailer, bicycle, or tricycle is a 'carriage' for this purpose, but a person liable to be charged with an offence of driving or being in charge of a mechanically propelled vehicle when unfit to drive through drink or drugs should not be charged with the present offence since its maximum punishment is far less severe than that for the appropriate offence under RTA 1988.

Index